HYDRIDES OF THE ELEMENTS OF MAIN GROUPS I–IV

HYDRIDES OF THE ELEMENTS OF MAIN GROUPS I-IV

EGON WIBERG
Prof. Dr.-Ing., Dr. rer. nat. h.c., Dr. rer. techn. h.c.

and

EBERHARD AMBERGER
Prof. Dr. rer. nat.

Institute of Inorganic Chemistry,
University of Munich (Germany)

ELSEVIER PUBLISHING COMPANY
AMSTERDAM LONDON NEW YORK
1971

CHEMISTRY

ENGLISH TRANSLATION BY
EXPRESS TRANSLATION SERVICE, LONDON, S.W. 19

ELSEVIER PUBLISHING COMPANY
335 JAN VAN GALENSTRAAT
P.O. BOX 211, AMSTERDAM, THE NETHERLANDS

ELSEVIER PUBLISHING CO. LTD.
BARKING, ESSEX, ENGLAND

AMERICAN ELSEVIER PUBLISHING COMPANY, INC.
52 VANDERBILT AVENUE
NEW YORK, NEW YORK 10017

LIBRARY OF CONGRESS CARD NUMBER: 72-88081

ISBN: 0-444-40807-X

WITH 84 ILLUSTRATIONS AND 88 TABLES

COPYRIGHT © 1971 BY ELSEVIER PUBLISHING COMPANY, AMSTERDAM

ALL RIGHTS RESERVED. NO PART OF THIS PUBLICATION MAY BE REPRODUCED, STORED IN A RETRIEVAL SYSTEM, OR TRANSMITTED IN ANY FORM OR BY ANY MEANS, ELECTRONIC, MECHANICAL, PHOTOCOPYING, RECORDING, OR OTHERWISE, WITHOUT THE PRIOR WRITTEN PERMISSION OF THE PUBLISHER, ELSEVIER PUBLISHING COMPANY, JAN VAN GALENSTRAAT 335, AMSTERDAM

PRINTED IN THE NETHERLANDS

PREFACE

QD181 H1W513 Chem

The present book attempts to meet the widely expressed demand for a systematic and detailed description of the hydrides in which hydrogen is negatively polarized. Accordingly, it comprises the elements of the main groups I–IV, the hydrogen compounds of which show the typical properties of "hydrides". In some of the hydrogen compounds of these elements hydrogen is positively polarized. These are also described since, in character, they belong to the hydride-like hydrogen compounds, for instance $GeH_4{}^{\delta-} \ldots GeCl_3H^{\delta+}$, or $B_{10}H_{14}$ with its positively polarized bridge hydrogen atoms.

Most of the non-stoichiometric binary or tertiary compounds of the sub-group elements can be classified as metal compounds and/or metal alloys. These hydrogen compounds are therefore mentioned only briefly in the first introductory chapter.

The properties of hydrogen-containing π-complexes of the sub-group elements are determined almost entirely by the complex-building ligands. Systematically, they should therefore be classified with complexes and not with hydrides. Hence they are not dealt with in this book.

By presenting a readable, short survey of hydride chemistry we hope that we have provided the novice with a firm basis in the hydride field. From there, he can without difficulty progress towards his own field of interest. The book should offer the chemist working in the field of hydrides reliable and detailed information on all important publications (approximately 3800 references).

We have structured the work accordingly: an introduction to each chapter, a short, time-saving text accompanied by numerous equations and figures, which more efficiently communicate the essence than would a lengthy exposition. Because it is exceedingly time-consuming and troublesome to refer to numerical data and properties of compounds in the original literature, these have also been included. Since these data are rather extensive it was felt that they should not be included in the main text but be summarised in tables or in separate sections.

The book is divided into ten chapters. The first one (Introduction) briefly describes and compares methods of synthesis, structure and reactions of hydrides of the main-group elements and transitional elements. In the following nine chapters the main-group elements are arranged separately: Chapter 2, alkali metal hydrides; Chapter 3, alkaline earth metal hydrides; Chapter 4, boron hydrides; Chapter 5, aluminium hydrides; Chapter 6, gallium, indium and thallium hydrides; Chapter 7, silicon hydrides; Chapter 8, germanium hydrides; Chapter 9, tin hydrides; and Chapter 10, lead hydrides.

Each of the nine main chapters begins with an introduction, followed by longer sections on all known methods of synthesis of hydrides of the elements in question, accompanied by a critical evaluation of these methods, as well as a systematic description of all reactions. Further sections deal with the physical and spectroscopic properties.

It is our hope that our approach to the topic will prove to be of value to workers in the field.

Munich, Spring 1971 — Egon Wiberg, Eberhard Amberger

12836

CONTENTS

Chapter 5. Aluminium Hydrides

Chapter 8. Germanium Hydrides

Chapter 9. Tin Hydrides

Chapter 1

INTRODUCTION

1. Systematology of the hydrides

Most elements form compounds with hydrogen. Without taking the polarity of the hydrogen into account, they are called hydrides[1]. In them, the hydrogen may be involved in four types of bond which shade into one another.

(1) The covalent bond without appreciable polarisation. Hydrides of this type form discrete molecules (*e.g.*, GeH_4) or polymers [*e.g.*, $(BH)_x$].

(2) The covalent bond with highly negatively polarised hydrogen ($H^{\delta-}$, limiting case $H^{\ominus}$). Hydrides of this type form salts (*e.g.*, $Cs^{\oplus}H^{\ominus}$) or discrete molecules (*e.g.*, $SiH_4^{\delta-}$).

(3) The covalent bond with strongly positively polarised hydrogen ($H^{\delta+}$). Hydrides of this type form discrete molecules which are often associated in the condensed phase (*e.g.*, $F^{\delta-}H^{\delta+}$).

(4) A covalent bond as in (1): H°, (2): $H\delta^{-}$, or (3): $H^{\delta+}$, in which covalent delocalised bonds (metal bonds) to other metal atoms exist simultaneously. Characteristic for hydrides of this type is the low stoichiometry ($Sm^{\sim 2\oplus}$-$H^{\sim 2\ominus}_{1.93-2.55}$, $Zr^{\sim 2\oplus}D^{\sim 2\ominus}_{1.76-1.98}$). They possess a conductivity band and are accordingly solid, have a metallic appearance, and conduct electricity. The metallic state in them is certain. The state of the hydrogen is uncertain. H_1°, $H_1^{\delta-}$, $H_1^{\delta+}$, and H_2°, and in adsorptions $H_2^{\oplus}$ as well, have been discussed. As a result of this, hydrides similar to one another are given different names: interstitial hydrides, inclusion hydrides, alloy-like hydrides, solid solutions, or non-stoichiometric hydrides.

Figure 1.1 gives a review of the hydrides in the form of the periodic system of elements. In the case of the elements of the main groups (s-, p-orbital elements, Groups IA–IVA), the ionic $H^{\delta-}$ character decreases and the covalent character increases from left to right and from bottom to top. The elements of the main group on the far right (Groups VA–VIIA) form covalent hydrides with an ionic bond component but reversed polarisation ($H^{\delta+}$). This corresponds to the simultaneous increase in the ionisation energy ($E \rightarrow E^{n\oplus} + n\ e^{\ominus}$) or the increase in electronegativity (see Fig. 1.1). In the case of the elements of the subgroups (d-orbital elements, Groups IB–VIIIB), the ionic character decreases and the metallic character increases from left to right. In the Cu and Zn group (Groups IB, IIB) there is a further decrease in the metallic character and increase in the covalent character. In the lanthanide elements (f-orbital elements) the dihydrides conduct electricity, since they possess a conductivity band. This is absent in the

---

[1]The expression "hydridic hydrogen", however, denotes a negative hydrogen atom ($H^{\delta-}$ or in the limiting case $H^{\ominus}$).

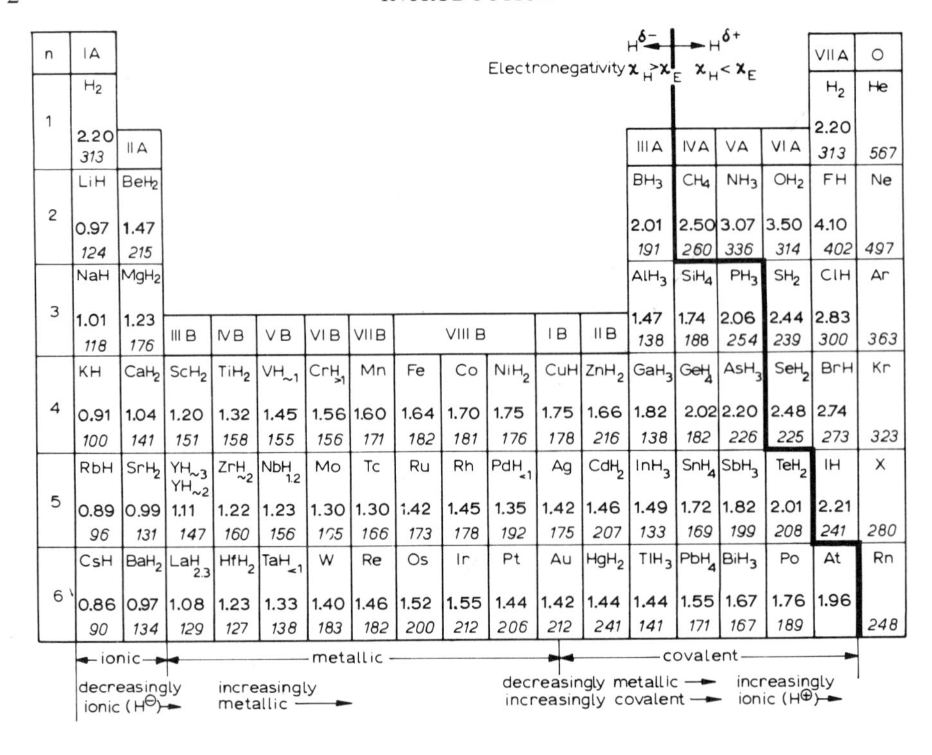

n	IA	IIA	III B	IV B	V B	VI B	VII B	VIII B			I B	II B	III A	IV A	V A	VI A	VII A	O
1	H_2 **2.20** *313*																H_2 **2.20** *313*	He *567*
2	LiH **0.97** *124*	BeH_2 **1.47** *215*											BH_3 **2.01** *191*	CH_4 **2.50** *260*	NH_3 **3.07** *336*	OH_2 **3.50** *314*	FH **4.10** *402*	Ne *497*
3	NaH **1.01** *118*	MgH_2 **1.23** *176*											AlH_3 **1.47** *138*	SiH_4 **1.74** *188*	PH_3 **2.06** *254*	SH_2 **2.44** *239*	ClH **2.83** *300*	Ar *363*
4	KH **0.91** *100*	CaH_2 **1.04** *141*	ScH_2 **1.20** *151*	TiH_2 **1.32** *158*	$VH_{\sim 1}$ **1.45** *155*	$CrH_{>1}$ **1.56** *156*	Mn **1.60** *171*	Fe **1.64** *182*	Co **1.70** *181*	NiH_2 **1.75** *176*	CuH **1.75** *178*	ZnH_2 **1.66** *216*	GaH_3 **1.82** *138*	GeH_4 **2.02** *182*	AsH_3 **2.20** *226*	SeH_2 **2.48** *225*	BrH **2.74** *273*	Kr *323*
5	RbH **0.89** *96*	SrH_2 **0.99** *131*	$YH_{\sim 3}$ $YH_{\sim 2}$ **1.11** *147*	$ZrH_{\sim 2}$ **1.22** *160*	$NbH_{1.2}$ **1.23** *156*	Mo **1.30** *165*	Tc **1.30** *166*	Ru **1.42** *173*	Rh **1.45** *178*	$PdH_{<1}$ **1.35** *192*	Ag **1.42** *175*	CdH_2 **1.46** *207*	InH_3 **1.49** *133*	SnH_4 **1.72** *169*	SbH_3 **1.82** *199*	TeH_2 **2.01** *208*	IH **2.21** *241*	X *280*
6	CsH **0.86** *90*	BaH_2 **0.97** *134*	$LaH_{2.3}$ **1.08** *129*	HfH_2 **1.23** *127*	$TaH_{<1}$ **1.33** *138*	W **1.40** *183*	Re **1.46** *182*	Os **1.52** *200*	Ir **1.55** *212*	Pt **1.44** *206*	Au **1.42** *212*	HgH_2 **1.44** *241*	TlH_3 **1.44** *141*	PbH_4 **1.55** *171*	BiH_3 **1.67** *167*	Po **1.76** *189*	At **1.96**	Rn *248*

Lanthanide elements (electronegativity: 1.01–1.14)

$LaH_{2.3}$ *129*	CeH_2 *151*	PrH_2 *134*	NdH_2 *145*	Pm	$SmH_{2.59-3}$ $SmH_{1.93-2.55}$ *129*	Eu *130*	$GdH_{2.85-3}$ $GdH_{<2.3}$ *142*
(Gd)	$TbH_{2.81-3}$ $TbH_{1.90-2.15}$ *155 ?*	$DyH_{2.68-3}$ $DyH_{1.94-2.08}$ *157*	$HoH_{2.64-3}$ $HoH_{1.95-2.24}$	$ErH_{2.82-3}$ $ErH_{1.95-2.31}$	$TmH_{2.76-3}$ $TmH_{1.99-2.41}$	YbH_2 *143*	$LuH_{2.73-3}$ $LuH_{1.35-2.23}$ *142*

Actinide elements (electronegativity: 1.00–1.22)

AcH_2	$ThH_{3.75}$ $ThH_{2.3-2.75}$	Pa	UH_3	Np	PuH_3	AmH_2	Cm
(Cm)	Bk	Cf	Es	Fm	Md	No	Lw

Fig. 1.1. Hydrides, hydride bond, electronegativity in Pauling units (heavy type) and first ionisation potential in kcal/mole (italic) in comparison with the periodic properties of the elements.

non-conducting trihydrides. Negative hydrogen atoms are generally present in the metallic hydrides of the actinide elements. Since the hydrides of elements of the subgroups and the lanthanides and actinides may possess metal bonds as well as E–H bonds, they form series of non-stoichiometric hydrides.

2. Types of hydrides

The alkali and alkaline-earth metals (Groups IA and IIA), with the exception of beryllium and magnesium[2], form crystalline, stoichiometric, predominantly ionic, salt-like hydrides. In accordance with the difference in electronegativity, their formulae resemble those of the halides:

$$E^{I\oplus}\,H^{\ominus}\;E^{II2\oplus}\,H_2^{\ominus}$$

The structures of the hydrides are determined by the covalent bond components that are undoubtedly present (and which, nevertheless, can be ignored as less important in any structural calculations) and the volumes of the large hydride ions and comparatively very small ($Li^{\oplus}$) to small ($Ca^{2\oplus}$) metal cations. The cubic body-centred alkali metals (packing density 0.68) form cubic face-centred alkali metal hydrides (packing density 0.74). The hexagonally close-packed, cubic face-centred or cubic body-centred alkaline-earth metals Ca, Sr, and Ba form orthorhombic alkaline-earth metal hydrides. Both the alkali-metal hydrides and the alkaline-earth metal hydrides are denser than the corresponding metals.

Let us proceed further in the main groups (A). The remaining IIA elements, Be and Mg, and the IIIA elements B to Tl, form the hydrides $E^{II}H_2$ and $E^{III}H_3$, corresponding to their positions in the periodic system. In these, also, the hydrogen is negative and the other element positive. However, corresponding to the increased influence of the (covalent) hybrid orbital, they do not form salt-like hydrides ($E^{II2\oplus}\,H_2^{\ominus}$ or $E^{III3\oplus}\,H_3^{\ominus}$), but hydrides with highly directed bonds:

$$^{\delta-}H{-}E^{II}{-}H^{\delta-} \qquad ^{\delta-}H{-}\underset{\displaystyle H^{\delta-}}{\underset{|}{E^{III}}}{-}H^{\delta-}$$

BeH_2 and MgH_2 have two occupied sp-hybrid orbitals and two unoccupied p-orbitals, and AlH_3, GaH_3, InH_3, and TlH_3 have three occupied sp^2-hybrid orbitals and one unoccupied p-orbital. The hydrides are therefore electron acceptors. They satisfy their electron deficiency by the formation of three-centre two-electron bonds or by the formation of dative bonds with electron donors. In the first case, this always gives oligomeric or polymeric structures, and in the second case generally monomeric but also oligomeric or polymeric structures [*e.g.*, $(BeH_2)_x$, B_2H_6, $B_{10}H_{14}$, $(AlH_3)_x$, $(InH_3)_x$, $BH_4^{\ominus}$, $THF \cdot AlH_3$, $R_3N \cdot GaH_3$]. Some structural examples are as follows:

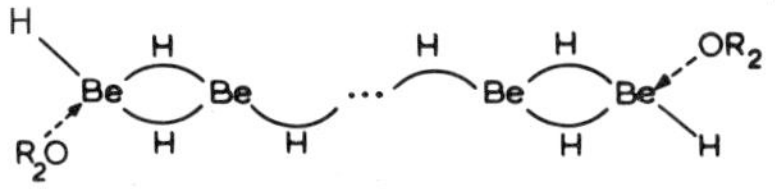

Beryllium dihydride (idealised)

[2] In addition to the crystalline $Mg^{2\oplus}H_2^{\ominus}$ related to CaH_2, magnesium also forms amorphous $MgH_2^{\delta-}$, which resembles the covalent BeH_2.

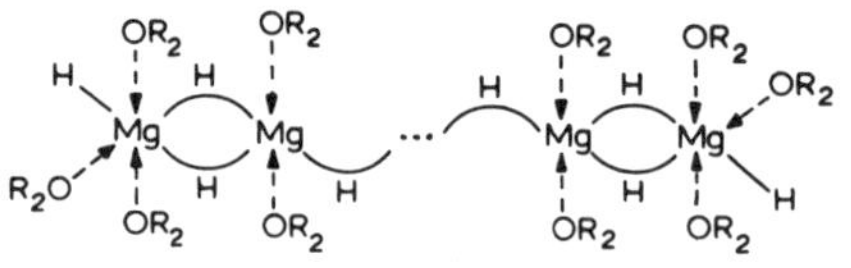

Amorphous magnesium dihydride (idealised). Crystalline MgH_2 approximates to CaH_2

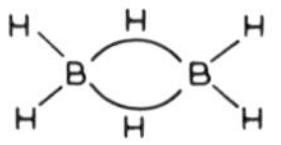

Diborane 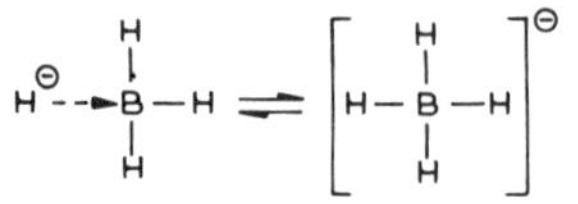   Boranate 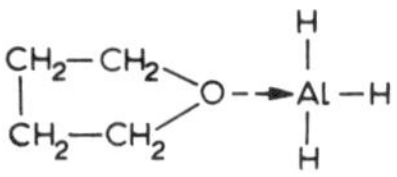   Alane tetrahydrofuranate 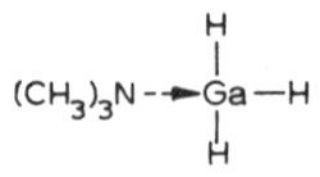   Trimethylamine-gallane

In addition, indium and thallium form the hydrides InH and TlH. Since compounds with lower oxidation stages are always more basic ("more metallic") than the higher oxidation stages, these hydrides should be related to the alkali-metal hydrides. In fact, they decompose only at similarly high temperatures. InH and TlH should therefore be compared with the likewise thermally very stable but highly covalent $(BH)_x$ only with reservations. Nevertheless, relationships must not be excluded.

The elements of the fourth main Group IVA (Si, Ge, Sn, Pb) form covalent hydrides $E^{IV}H_4$. Corresponding to their electronegativities, the hydrogen in them is negatively polarised. Since the central atoms possess four occupied sp^3-hybrid orbitals and four similar ligands, the hydrides have the tetrahedral configuration and are electronically saturated. Consequently, they form discrete monomolecular molecules held together only by Van der Waals forces (and are gaseous under normal conditions):

$$\begin{array}{c} H^{\delta-} \\ | \\ {}^{\delta-}H-E^{IV}-H^{\delta-} \\ | \\ H^{\delta-} \end{array}$$

Besides few examples (*e.g.*, $CH_2FCF_2SiH_3 \cdot NMe_3$), only in intermediate states in reactions do trigonal-bipyramidal sp^3d-configurations occur. Silicon and germanium, and also tin to a smaller extent, form alkane-like linear or branched oligomers or polymers [*e.g.*, Si_4H_{10}, Ge_5H_{12}, Sn_2H_6, $(SiH_2)_x$, $(GeH_2)_x$] as well as the two-dimensional $(SiH)_x$ and $(GeH)_x$. These are also predominantly covalent with negatively-polarised hydrogen.

The elements of the 5th to 7th main groups (Groups VA–VIIA) form covalent hydrides $E^{V}H_3$, $E^{VI}H_2$, and $E^{VII}H$. In accordance with the difference in electronegativities between E and H, the hydrogen in them is positively polarised; generally weakly in $E^{V}H_3$, more strongly in $E^{VI}H_2$, and still more strongly in $E^{VII}H$ (note also the rapid decrease in electronegativity of E and the difference of E to H from top to bottom). The hydrides of the heavy elements of the 5th main group form a very vague transition (somewhat as for AsH_3–SbH_3) from positive to negative hydrogen. They therefore follow the preceding main-group hydrides without a gap.

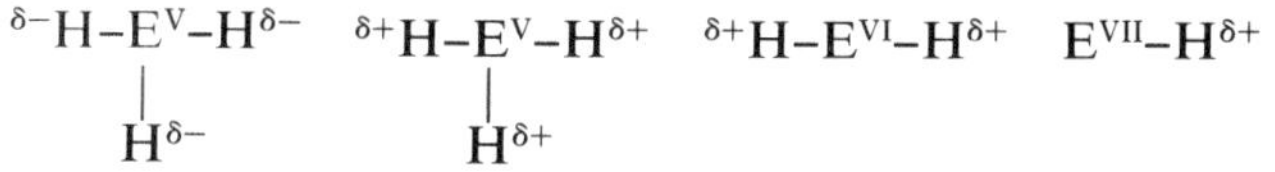

$$\begin{array}{c} ^{\delta-}H{-}E^{V}{-}H^{\delta-} \\ | \\ H^{\delta-} \end{array} \quad \begin{array}{c} ^{\delta+}H{-}E^{V}{-}H^{\delta+} \\ | \\ H^{\delta+} \end{array} \quad ^{\delta+}H{-}E^{VI}{-}H^{\delta+} \quad E^{VII}{-}H^{\delta+}$$

The central atoms of the hydrides $E^{V}H_3$, $E^{VI}H_2$, and $E^{VII}H$ have four occupied sp^3-hybrid orbitals but only three, two, or one ligand. Consequently, the central atoms are electron donors while the positively polarised hydrogen atoms are more or less strong electron acceptors. The hydrides form dative bonds with themselves (association through proton bridges) or with foreign electron acceptors. Examples:

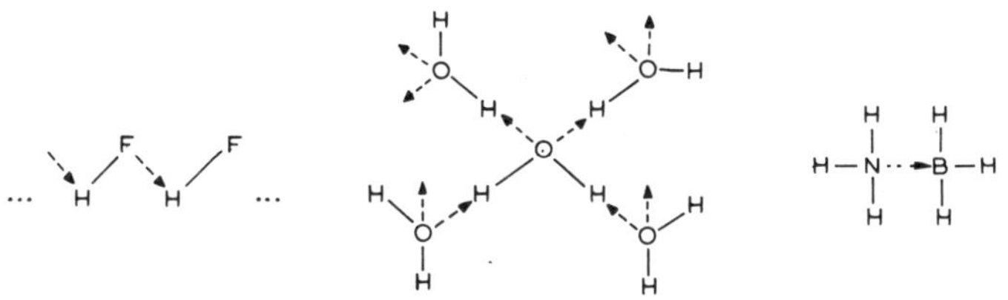

HF H_2O $H_3N \cdot BH_3$

The hydrides of the elements of the subgroups (B), beginning with the scandium group (Group IIIB), are linked with the alkaline-earth metal hydrides (Group IIA) CaH_2, SrH_2, and BaH_2. The latter form an ionic lattice. The only certain trihydride of Group IIIB, YH_3, is also ionic $Y^{3\oplus}\ H_3^{\ominus}$, although it has a metallic appearance (see Section 3.3). The electronic configuration of $Y^{3\oplus}$ is [Kr] $4d^{\circ}$ $5s^{\circ}p^{\circ}$. The elements of the subgroups form lower hydrides than corresponds to their position in the periodic system to an increasing extent on passing from left to right (from Group IIIB to VIIIB). The stoichiometry of the hydrides decreases simultaneously. The explanation of this is that the hydride-forming elements of the subgroups (IIIB → VIIIB) utilise their (hybridised) orbitals to an increasing extent not for binding hydrogen but for forming covalent but delocalised bonds (metal bonds). To an increasing extent, therefore, the properties of the hydrides formed are stamped by the properties of the corresponding metals. The hydrides of the copper group (Group IB) and zinc group (Group IIB), finally, form the transition to the unstable covalent hydrides of Group IIIA GaH_3, InH_3, and TlH_3.

Most lanthanide elements do not form strictly stoichiometric di- and trihydrides. The trihydrides are ionic ($E^{3\oplus}\ H^{\ominus}$); nevertheless, because of their non-stoichiometry, they have a metallic appearance and conduct electricity. The dihydrides appear to be more comparable with the metallic hydrides. However, ionic structures $E^{2\oplus}\ H_2^{\ominus}$ have been found for YbH_2 and EuH_2. Possibly, the dihydrides form different hydrides according to the electronic configuration of the metai.

Many transition metals form π-complex hydrides. In them, the non-hydridic ligands (the same or different) are bound to the central atom by σ- and π-bonds.

The hydrogen atoms are bound to the central atom directly. They can be subdivided into the following six classes according to the non-hydridic ligands.

1. *Mononuclear or polynuclear carbonyl hydrides.*
Examples: $Mn(CO)_5H$, $Fe(CO)_4H_2$, $Co(CO)_4H$, $Ir(CO)_4H$, $FeCo_3(CO)_{12}H$, $[Cr_2(CO)_{10}H]^{\ominus}$, $[Mo_2(CO)_{10}H]^{\ominus}$, $[W_2(CO)_6(OH)_3H]^{4\ominus}$.

2. *Complex hydrides with tertiary phosphines or similar ligands.*
Examples: $Re(PPh_3)_4H_2$, $Fe[o\text{-}C_6H_4(PEt_2)_2]H_2$, $Co(Ph_2PCH_2CH_2PPh_2)_2H$, $Pt(PPh_3)_4H_2$.

3. *Complex hydrides with nitrogen ligands.*
[Examples: $Rh(H_2NCH_2CH_2NH_2)_2(H, Cl)_2$, $[Rh(C_5H_5N)_2(Cl, H)_2]^{\oplus}$.

4. *π-Cyclopentadienylcarbonyl hydrides.*
Examples: $Cr(\pi\text{-}C_5H_5)(CO)_3H$, $Fe(\pi\text{-}C_5H_5)(CO)_2H$, $Ru(\pi\text{-}C_5H_5)(CO)_2H$.

5. *Bis-π-cyclopentadienyl hydrides.*
Examples: $Ta(\pi\text{-}C_5H_5)_2H_3$, $Mo(\pi\text{-}C_5H_5)_2H_2$, $Re(\pi\text{-}C_5H_5)_2H$, $[Fe(\pi\text{-}C_5H_5)_2H]^{\oplus}$.

6. *Cyanide hydrides.*
Examples: $[Co(CN)_5H]^{3\ominus}$, $[Rh(CN)_5H]^{3\ominus}$, $[Pt(CN)_4H]^{3\ominus}$.

3. Hydride models

When a hydrogen atom penetrates into a hydride-forming metal and occupies lattice or interstitial sites, the formation of the hydride regarded as a whole is exothermic. The fact that in by far the majority of cases this cannot be a solution of molecular hydrogen in the metal follows from the endothermic expansion of the lattice during the uptake of the hydride. The question is which of the two alternatives yields more energy and is therefore more probable: (i) the hydrogen gives the 1s-electron to the electron gas of the metal completely or partially (formation of $H^{\oplus}$), or (ii) the electron gas gives electrons to the hydrogen atom (formation of $H^{\ominus}$). Empirical rules such as electronegativity and the requirement for the screening of nuclear charges point to $H^{\ominus}$ while many PMR measurements point to $H^{\oplus}$.

Less important for the drawing up of hydride models is the question of the origin of the high heat of formation for hydrides, although the high energy of 104.2 kcal/mole must be used for the dissociation of hydrogen molecules into atoms. This requirement in fact loses its severity when we take into account the comparatively high adsorption energy of hydrogen atoms on metals.

3.1 *The interstitial atom model*

In the interstitial atom model (alloy model), the small hydrogen atoms (radius 0.3–0.7 Å) occupy octahedral or tetrahedral interstitial sites in the metal lattice. Normally, these sites are somewhat too small, so that the lattice is slightly expanded by the uptake of hydrogen. The interstitial atom model well explains the crystallographic changes in the transition from the metal to the hydride but fails to explain other physical properties, for example the change in electrical conductivity.

3.2 *The proton model*

Until recently, the proton model was generally accepted for transition-metal hydrides, particularly for $PdH_{0.7}$. The following reasons, with counterarguments in brackets, are said to demonstrate the proton model: (1) the paramagnetic susceptibility decreases with the H content (the decrease is a consequence of the expansion of the lattice); (2) H migrates to the cathode (in electron-defect conductivity mechanisms neither the sign nor the velocity of the material component determines the direction of migration); (3) the PMR spectra (today they cannot be adduced to distinguish types of solid hydride). The proton model is also opposed by the expansion of the lattice with all hydrides that such a model involves. Most hydrides can be explained without the proton model; however, many experiments can be interpreted only with the proton model.

3.3 *The hydridic model*

In the hydridic model, a helium-like configuration of the electrons ($1s^2$) surrounds the proton $H^{\oplus}$. The "$H^{\ominus}$" lattice site so obtained is embedded in an electron gas of comparatively low density. In accordance with their stoichiometry, the metal atoms possess positive integral unit charges and have noble gas configurations. If there are superfluous electrons after the formation of such a lattice, they form structure-determining directed hybride orbitals to the nearest neighbouring metal atoms. The electrons in these hybrid orbitals are sufficiently delocalised to form a conductivity band.

Although the hydridic model, like the models described above, is naive, it gives a more rational explanation of the internuclear distances and the lattice energies. It also explains, for example, the existence of $O^{2\ominus}$together with hydrogen in hydrides contaminated with oxygen: $H^{\ominus}$ and $O^{2\ominus}$, which is similar in size, repel one another. They can occupy equivalent lattice sites (sub-oxides and hydrides are crystallographically similar).

Of the many, although not very convincing facts that oppose a hydridic model we may mention two. (1) Yttrium hydride, YH_3, should be a non-conductor because of the electronic configuration of $Y^{3\oplus}$ ([Kr] $4d^\circ$ $5s^\circ p^\circ$) and $3{\times}H^{\ominus}$ ($1s^2$), since no electrons are left for the conductivity band. Why YH_3, which is bluish with a metallic lustre, nevertheless conducts like a metal is unknown. The conductivity possibly results only from a slight deviation from stoichiometry or from a slight occupancy of the conductivity band. Similar remarks apply to ytterbium dihydride, YbH_2. It conducts electricity, although the infrared spectrum resembles that of the salt-like non-metallic CaH_2. (2) The second fact opposing the hydridic model is the rapid solid-body diffusion of hydrogen. This is against the assumption of large bulky $H^{\ominus}$ ions. In spite of the assumption of $H^{\ominus}$ ions, the high mobility of the hydrogen can be explained by the migration of protons, with the $(1s^2)$ electrons remaining at their lattice sites. At the present time, the hydridic model is that which is most used.

4. Synthesis of hydrides

4.1 Adsorption and chemisorption of hydrogen on other elements

Molecular gaseous hydrogen interacts with the solid surface of a metal first by an elastic collision process. For most hydrogen molecules, the residence time in the range of action of the metal surface is insufficient for the initiation of a reaction. Many hydrogen molecules remain on the surface longer than corresponds to elastic collisions (particularly at active sites). This means that hydrogen molecules become concentrated at the surface as compared with the bulk of the gas. The concentration is called adsorption. The level of adsorption depends on the interaction between hydrogen molecules and metal.

Two groups of forces of interaction can be distinguished: (i) physical adsorption, with forces equal to those between molecules of a gas, (ii) chemisorption, with forces similar to those of the chemical bond. Transitions between the two types are possible. Hydrogen occurs in the following forms:

(1) as partially positively polarised hydrogen molecules (H_2, $H_2^{\oplus}$),

(2) as negatively polarised hydrogen atoms ($H^{\ominus}$).

(3) as hydrogen atoms which partly dissociate into protons and electrons ($H^{\oplus} + e^{\ominus}$).

In the case of physical adsorption, the (negative) energy liberated amounts to only a few kcal/mole. The energy liberated in a chemisorption may amount to –200 kcal/mole. It frequently exceeds the enthalpy of formation of the corresponding solid hydride, particularly if instead of a simple chemisorption bond (a) a multi-centre chemisorption bond is formed (b):

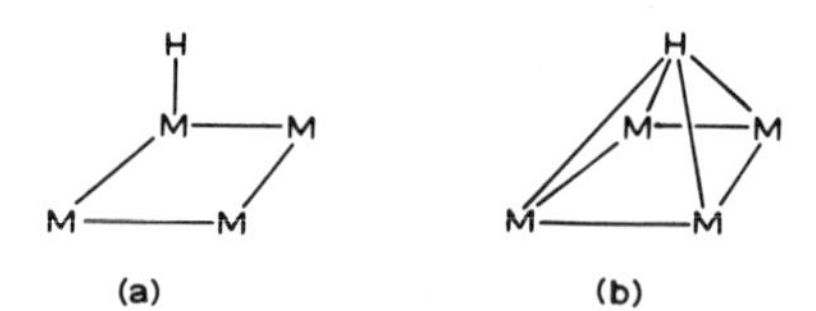

A high energy of chemisorption may lead to the detachment of individual metal atoms from the metal lattice. Consequently, the boundary between chemisorption and the formation of a compound phase with an expanded layer thickness on the surface of the metal is already reached. An example is the formation of a hydride on the solid or liquid surface of alkali metals.

For the formation of a hydride, however, metal atoms need not diffuse into the chemisorption layer. The opposite process, a diffusion of hydrogen atoms into the metal, is also possible. The simplified potential change for the uncatalysed penetration of hydrogen atoms into the metal perpendicular to the surface (Fig. 1.2, path x) is shown by Fig. 1.3a. First the heat of absorption is liberated during adsorption, generally with the formation of hydrogen atoms (Fig. 1.3a, left-hand potential minimum). In penetration into the surface of the metal, the high potential barrier (Fig. 1.3a, activation energy) must be overcome. After this, the potential energy of the hydrogen atom falls to the following energy minimum (energy of the hydrogen in the lattice). It is separated from other equal minima by the threshold of activation energy of solid-body diffusion.

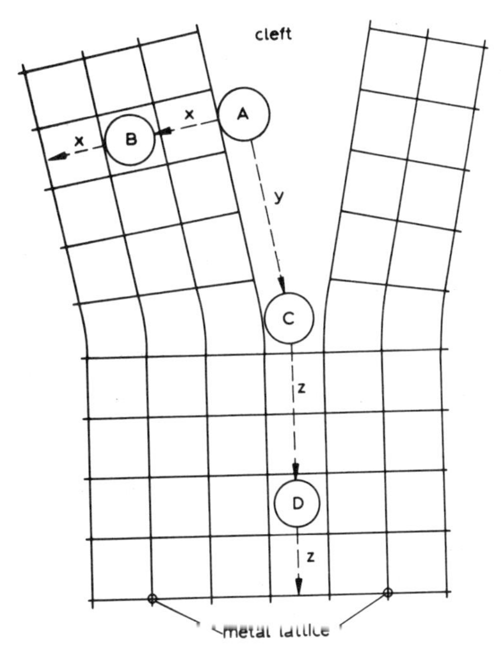

Fig. 1.2. Diffusion of a hydrogen atom perpendicular (x) and parallel (y) to the surface of a metal.

The high potential barrier of the activation energy at the phase boundary need not be overcome if the adsorption of hydrogen is catalysed. Catalysts include, inter alia, grain boundaries and clefts of molecular dimensions. In fact, atomically absorbed hydrogen – even at high heats of adsorption – can move along the metal surface (surface diffusion, Fig. 1.2, path y). Because of their mobility, hydrogen atoms can penetrate into the interior of the metal lattice at clefts in the metal surface through *surface* diffusion (Fig. 1.2, path y). Without the necessity for providing the high activation energy, according to Fig. 1.3b the potential energy of the hydrogen atom A first adsorbed on the surface falls to the potential energy of the occluded hydrogen atom D. The activation energy of surface diffusion along the wall of the cleft generally undergoes a transition into the activation energy of solid-body diffusion (Fig. 1.3b). In isotropic lattices, the thresholds of the activation energy of the movement x are equal to that of z.

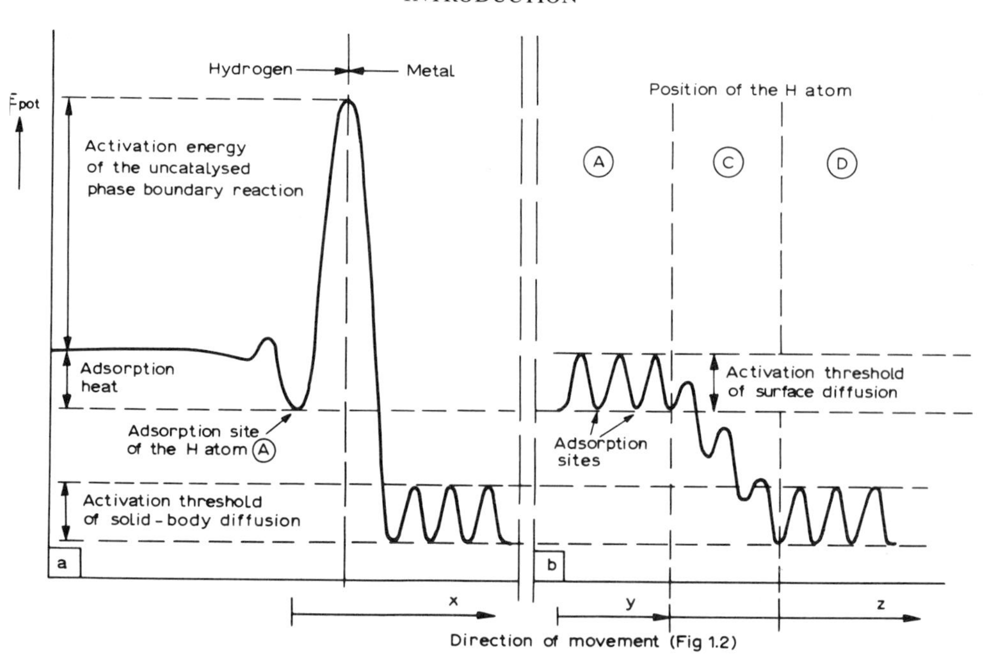

Fig. 1.3. Potential change of a hydrogen atom (a) moving perpendicular to the surface of a metal, (b) penetrating into a cleft parallel to the surface of a metal.

4.2 Reversible dissociation of binary hydrides

All hydrides with a predominantly ionic $H^{\delta-}$ bond component (salt-like hydrides) and all hydrides with a metallic bond component give off hydrogen reversibly. Figure 1.4 shows isotherms in a theoretical graph of dissociation pressure versus composition. Figures 2.2, 2.3 and 3.2 (Chapters 2 and 3) show examples for lithium, sodium and magnesium hydrides. There are three regions of interest in Fig. 1.4.

In the first region (O–a or O–a′), in the left-hand part of the graph the isotherm rises rapidly with increasing hydrogen content. Here there is a solid or liquid solution of hydrogen in the other element. The solubility rises with the temperature. In the second region (a–b or a′–b′), in the so-called plateau region in the centre of the graph, two immiscible phases exist: the saturated solid or liquid solution of EH_a or $EH_{a'}$ and the phase of a lower hydride EH_b or $EH_{b'}$. This plateau region disappears above a critical temperature T_c. The third region (b′–1.0 or b–1.0), the so-called hydride region in the right-hand part of the graph, contains one phase, the composition of which is derived from the stoichiometric hydride (in the theoretical example $EH_{1.0}$). The phase contains either interstitial metal atoms (*e.g.*, in $NaH_{0.98} = NaH + Na$) or hydrogen vacancies (*e.g.*, in $PdH_{<1} = PdH_1 - H$). If an element forms several stoichiometric hydrides, the diagram contains several sets of isotherms. As an example, Fig. 1.5 shows the $Th–H_2$ system with the hydrides ThH_2 and Th_4H_{15} (= $ThH_{3.75}$).

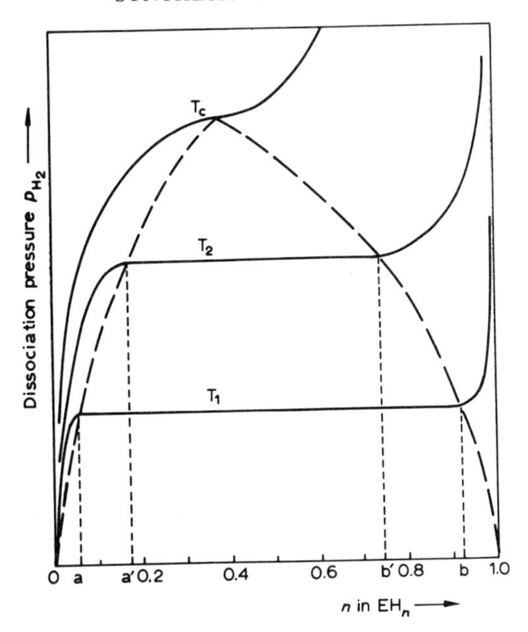

Fig. 1.4. Isotherms in the diagram of p_{H_2} versus EH_n for an element forming a stoichiometric hydride $EH_{1.0}$. $T_1 < T_2 < T_c$(each is constant).

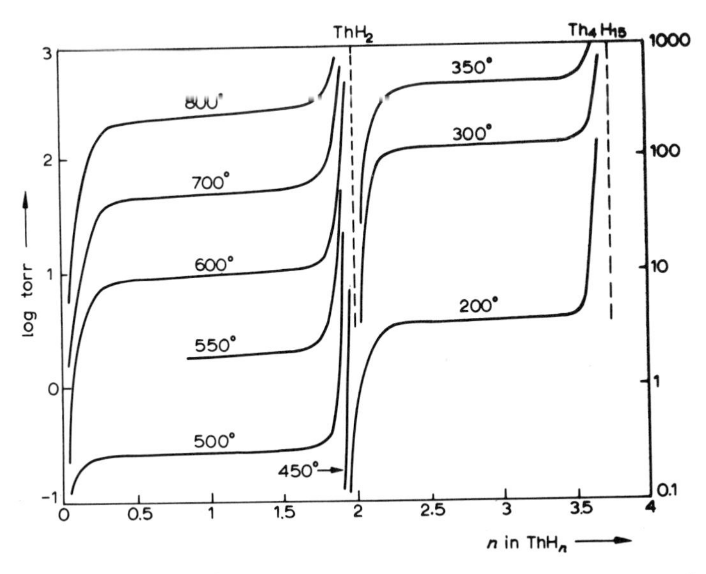

Fig. 1.5. Isotherms in the Th–H_2 system (→ ThH_n): (left) the high-temperature set leads to ThH_2, (right) the low-temperature set leads to Th_4H_{15} (at 20°).

4.3 *Synthesis of hydrides by the reaction of compounds*

In addition to the reaction of elements with hydrogen, which often does not lead to the formation of hydrides or only to non-stoichiometric hydrides, there are the following methods for synthesising hydrides. In all of them, the starting material is not the element but one of its compounds.

Method	*Prototypical reactions*
1. Partial electron transfer ($H^{\oplus}$): $H^{\oplus} + E^{\ominus} \rightarrow {}^{\delta+}E{-}H^{\delta-}$ or ${}^{\delta-}E{-}H^{\delta+}$	$3H^{\oplus} + MnB \rightarrow BH_3 + Mn^{3\oplus}$ $4H^{\oplus} + Mg_2Ge \rightarrow GeH_4 + 2Mg^{2\oplus}$ $6H^{\oplus} + Ca_3P_2 \rightarrow 2PH_3 + 3Ca^{2\oplus}$
2. Polarisation (H°): ${}^{\delta+}H{-}H^{\delta-} + E^{\oplus} \rightarrow {}^{\delta+}E{-}H^{\delta-} + H^{\oplus}$	$3H_2 + BCl_3 \rightarrow BH_3 + 3HCl$
3. Hydrogenating cleavage of the E–C bond (H°): $H_2 + E{-}\overset{\vert}{\underset{\vert}{C}}{-} \rightarrow {}^{\delta+}E{-}H^{\delta-} + H{-}\overset{\vert}{\underset{\vert}{C}}{-}$	$8H_2 + Ca(CN)_2 \rightarrow CaH_2 + 2CH_4 + 2NH_3$ $3H_2 + (C_2H_5)_3B \rightarrow BH_3 + 3C_2H_6$
4. Anionic displacement ($H^{\ominus}$): $H^{\ominus} + {}^{\delta+}E{-}X^{\delta-} \rightarrow {}^{\delta+}E{-}H^{\delta-} + X^{\ominus}$	$2(CH_3)_2AlH + (CH_3)_2Be \rightarrow BeH_2 + 2(CH_3)_3Al$ $3LiH + BCl_3 \rightarrow BH_3 + 3LiCl$ $4NaH + SiCl_4 \rightarrow SiH_4 + 4NaCl$ $LiAlH_4 + 2W(\pi\text{-}C_5H_5)_2Cl_2 \rightarrow 2W(\pi\text{-}C_5H_5)_2H_2 + LiAlCl_4$
5. Reductive cleavage of the E–C bond: $E{-}\overset{\vert}{\underset{H}{C}}{-}\overset{\vert}{\underset{\vert}{C}}{-} \rightarrow E{-}H + \rangle C{=}C\langle$	$(C_2H_5)Be \rightarrow BeH_2 + 2C_2H_4$ $[(CH_3)_2CH]_4Be_2 \rightarrow \frac{2}{n}[(CH_3)_2CHBeH]_n + 2CH_2 = CHCH_3$ $(C_2H_5)_2Mg \rightarrow MgH_2 + 2C_2H_4$
6. Degradation and building up of higher, covalent hydrides	Nucleophilic degradation: $2R_3N + B_4H_{10} \rightarrow R_3N \cdot BH_3 + R_3H_7 \cdot B_3H_7$ Thermal degradation: $Si_4H_{10} \rightarrow SiH_4 + \frac{3}{x}(SiH_2)_x$

Chapter 2

ALKALI METAL HYDRIDES

1. Historical

In 1808, one year after elementary sodium was first prepared, Davy[*21*] observed the ability of this metal to absorb hydrogen. However, well over half a century elapsed before the first hydride formula ('Na_2H') was proposed (1874): according to Troost and Hautefeuille[*151*], silver-white 'Na_2H' with a metallic lustre is formed from the elements above 200°, best at 300–350° and atmospheric pressure or at 421° under increased pressure. Only in 1902 did Moissan[*103, 104*] prepare sodium hydride NaH from the elements at 355° (at a hydrogen pressure slightly above atmospheric).

Gay-Lussac and Thenard[*42*] observed in 1809, the year of the discovery of elementary potassium by Davy, that potassium absorbs molecular hydrogen. The above authors ascribed to their reaction product the non-established formula 'K_4H', so that in 1809–1810 this hydride started off a long discussion with Davy [*22*]. Only in 1874 did Troost and Hautefeuille[*150*] recognise a more far-reaching uptake of hydrogen (at 290°) with the formation of a hydride 'K_2H'. Similarly to 'K_4H', 'K_2H' exhibited the metallic appearance of potassium. Almost a century elapsed before Moissan[*102*] discovered in 1902 that this hydrogenation product, too, was not completely hydrogenated. By the reaction of potassium with hydrogen for several hours (with an overpressure of 100 torr) at 350°, Moissan produced the end-stage of the hydrogenation process – potassium hydride KH, crystallising in colourless needles.

The absorption of hydrogen by elementary lithium was first observed by Troost and Hautefeuille[*150*] in 1874. Without going further into the formation of lithium hydride, Guntz in 1896, in connection with his investigations on the reaction of lithium with nitrogen[*51*], ethylene[*52*], and acetylene[*52*], assigned to lithium hydride the formula LiH. However, it was only in 1920 that Nernst and Moers[*111*] put the preparation and formula of lithium hydride LiH on a sound basis.

In 1903, Moissan[*105*] obtained rubidium hydride RbH and cesium hydride CsH for the first time from the elements, at 300° and under slightly increased pressure.

On the basis of copious experimental material on the dependence of the dissociation pressures of NaH, KH, RbH, and CsH on the temperature, Ephraim and Michel[*31*] in 1921 first definitely pointed out the difficulty of preparing stoichiometric hydrides from the elements.

References p. 38

2. General review

2.1 Preparation

Under suitable conditions, all alkali metals react exothermically with equimolecular amounts or with an excess of molecular hydrogen or deuterium, giving rise to stoichiometric salt-like hydrides $M^{\oplus}H^{\ominus}$ or deuterides $M^{\oplus}D^{\ominus}$ (for further details, see the individual alkali-metal hydrides). Thus

$$2M + H_2 \rightarrow 2MH + \text{energy}$$
$$2M + D_2 \rightarrow 2MD + \text{energy}$$

where M = Li, Na, K, Rb, Cs; for the energy values, see Table 2.3. Experimental investigations, which, unfortunately, have mostly aimed only at obtaining high yields of pure hydrides and less at studying the reaction kinetics of the hydride formation, show basically the existence of two mechanisms of formation: the homogeneous gas-phase reaction and the diffusion-determined uptake of hydrogen.

The homogeneous gas-phase reaction. If the reaction system contains only the alkali metal and hydrogen, hydride formation begins at temperatures which decrease in the sequence Li > Na > K > (Rb, Cs). The reason is that the temperatures corresponding with a certain vapour pressure of the alkali metals decrease in the sequence mentioned (see Table 2.1). The hydride formed by the homogeneous gas-phase reaction separates out in the coldest part of the system, and since the hydride formation is accompanied by evolution of heat, this coldest part is generally not the surface of the metal.

The diffusion-determined uptake of hydrogen. In this method, stoichiometric hydrides MH form starting from the solid or liquid surface. Homogeneous, non-stoichiometric hydrides with a lower H content are formed by the diffusion of metal and/or hydride ions when the hydrogen is in excess. The tendency to

TABLE 2.1

COMPARISON OF THE TEMPERATURES AT A NUMBER OF VAPOUR PRESSURES [*64*] AND MELTING POINTS OF THE ALKALI METALS (°C)

Vapour pressure [torr]	*Li*	*Na*	*K*	*Rb*	*Cs*
1	723	439	341	297	279
10	881	549	443	389	375
40	1003	633	524	459	449
100	1097	701	586	514	509
400	1273	823	708	620	624
760	1372	892	774	679	690
M.p.	180.54	97.90 ±0.05	63.65 ±0.05	38.8 ±0.1	28.6
Ref.	23	124	27	89	147

form hydrides primarily at the solid or liquid *surface* falls in the sequence Li ~ Na > K > Rb > Cs, corresponding to the increasing vapour pressure of the alkali metals. The diffusion-determined uptake of hydrogen is seriously inhibited by the crust of solid hydride. It is uncertain whether the hydrides are formed by the diffusion of metal cations or by the diffusion of protons. The diffusion of cations would take place in a similar manner to that in the reaction of solid silver with gaseous chlorine or bromine, which depends on the diffusion of the small metal cations through the halide layer. On the other hand, the formation of the alkali-metal hydrides from large negative $H^{\ominus}$ ions and small metal cations does not exclude diffusion of the hydrogen in the form of protons to free interstitial sites of the metal, leaving behind negative holes ($1s^2$ electrons). There the protons could form $H^{\ominus}$ ions with the ns^1 electrons of the alkali metal or with electrons of the conductivity band. In both mechanisms, the formation of the hydride and the quantitative reaction are accelerated if the hydride crust is broken by vigorous stirring with or without additives (such as mineral oil or salts of fatty acids).

2.2 *Physical properties*

Hydrogen and halogen atoms each lack one electron for a stable inert-gas configuration, and the roles of the hydrogen in the hydrides and of the halogens in the halides can therefore be compared. In this comparison, hydrogen must be arranged before fluorine in relation to its atomic weight and atomic radius, but after iodine in relation to its electronegativity (Table 2.2).
These properties of hydrogen show a certain analogy between the formal composition and the physical properties of the hydrides on the one hand and those of the halides on the other.

TABLE 2.2

PHYSICAL PROPERTIES OF HYDROGEN AND HALOGEN

	H	*F*	*Cl*	*Br*	*I*	*H*
Atomic weight	1.008	18.998	35.453	79.909	126.904	
Atomic radius	0.3707	0.717	0.994	1.142	1.334	
Electronegativity (Pauling scale)		4.10	2.83	2.78	2.21	2.20

Pure alkali-metal hydrides are colourless crystalline substances. Under normal conditions, they form cubic face-centred NaCl-type lattices, which, according to calculations, should change into a metallic state at high pressure, similarly to the halides. Like the alkali-metal halides, the hydrides have an ionic structure ($M^{\oplus}$ $H^{\ominus}$) under normal conditions. Correspondingly, in electrolysis the hydrogen separates at the anode and the metal at the cathode. In accordance with the decrease in the ionisation potential, the salt-like nature of the compounds increases from lithium to cesium. The thermal stability of the hydrides decreases in the same sequence (*cf.* the heats of formation in Table 2.3).

References p. 38

TA[BLE]

PHYSICAL PROPERTIES OF THE ALK[ALI]

	Ref.	LiH	LiD
Lattice constants (NaCl type) (Å)	[49]	4.083±0.001	4.069±0.001
	[121, 140]	4.083_5	4.068_4
Electrostatic lattice energy, U_e (kcal mole^{-1})	[49]	284.22±0.7	285.19
Lattice energy (kcal mole^{-1}) (without older data)	[49]	217.76	218.76
	[74]	219.2	220.8
	[75]	234	
Density (crystalline) (g ml^{-1})	[162]	0.769	0.879
	[121]	0.7750 (25°)	0.8826
	[62, 161]		
Compressibility (10^{-12} cm^2 dyne^{-1})	[121]	2.8	2.8
	[76]	2.36	
	[134]	2.32	
Linear coefficient of thermal expansion (10^{-5} deg^{-1})	[121]	3.6	(3.6)
	[97]		
Thermal conductivity (cal deg^{-1} cm^{-1} sec^{-1}) (100°)	[121]	0.025	(0.025)
Electrical conductivity (Ω^{-1} cm^{-1}) (600°)	[121]	0.003	0.001
Dielectric constant	[121]	12	(12)
Magentic susceptibility	[37]	-4.60×10^{-6}	
Heat of formation, $\Delta H^\circ_{cryst.}$ (kcal mole^{-1})	[49]	−21.666±0.026	−21.784±0.02
	[50]	−21.61	
	[92]	−21.34±0.15	
	[128]	−21.61	
	[101]	−21.60	
	[3]		
	[41]		
	[55]		
	[79]		
	[138, 139]		
	[137]		
	[62]		
Heat of formation, $\Delta H^\circ_{500-600^\circ}$ 100% NaH	[9]		
90% NaH	[9]		
in the Na–NaH–H_2 system 80–30% NaH	[9]		
(kcal mole^{-1} NaH) 20% NaH	[9]		
10% NaH	[9]		
Free energy of reaction, $\Delta F^\circ_{gaseous}$ (kcal mole^{-1})	[29, 30]	25.2	
Free energy of reaction, $\Delta F^\circ_{cryst.}$ (kcal mole^{-1})	[29]	−16.72	
	[30]	−16.72	
	[137]	−16.8	
Free energy of reaction, 100% NaH	[9]		
ΔF°_T (kcal mole^{-1} NaH) 30–80% NaH, 70–20% Na	[9]		
$\Delta F^\circ_T = 0$ (°K) 100% NaH	[9]		
30–80% NaH	[9]		
Entropy, $S^\circ_{298.1\ gaseous}$ (cal deg^{-1} mole^{-1})	[29, 30]	40.77	
	[110, 141]	38.77±0.3	
Entropy, $S^\circ_{298.1\ cryst.}$ (cal deg^{-1} mole^{-1})	[29, 30]	5.9±0.5	
	[128]	5.9±0.5	
	[137]	5.9±0.5	
Specific heat, $c^\circ_{p...}$ (cal deg^{-1} mole^{-1})	[29, 30]	7.06	
Specific heat, $c^\circ_{p\ cryst.}$ (cal deg^{-1} mole^{-1})	[29, 30]	8.3	
Melting point (°C) 99.8% LiH	[91]	688	
26–98% LiH+Li	[91]	685	
20% LiH+Li	[91]	666	
13% LiH+Li	[91]	624	
'LiH'	[60, 113]	680	

2.3
METAL HYDRIDES AND DEUTERIDES

NaH	*NaD*	*KH*	*KD*	*RbH*	*CsH*
4.879 ± 0.001	4.867 ± 0.001	5.708 ± 0.001	5.696 ± 0.001		
237.85 ± 0.05	238.43 ± 0.05	203.30 ± 0.04	203.73 ± 0.04		
192.74	193.47	170.12	170.42		
			—		
210		181		178	
			—		
			—		
1.36		1.43		2.59	3.41
			—		
			—		
			—		
			—		
7.2		3.6	—		
−13.487 ± 0.020	−13.339 ± 0.007	−13.819 ± 0.011	−13.238 ± 0.011		
—		—			
−13.60 ± 0.27		−15.16 ± 0.16			
−13.7		−13.6			
−13.94					
−16.6					
−12.8					
		−14.24			
		−14.15			
					−11.92 ± 0.25
					−13.48
11.610					
13.300					
15.590					
13.860					
14.250					
27.78		25.1			24.3
−9.0		−8.9		−7.3	−7.3
−8.6		−8.9			−7.0
11.610 − 17.41T					
13.590 − 19.62T					
393		47.3			51.25
419					
44.93					
7.1		10.2		17.0	20.8
11.4		14.5			19.3
7.002					

Only lithium hydride melts without decomposition, all the others decompose into metal and hydrogen below their melting points (*cf.* the hydrogen pressures in Tables 2.5, 2.6, and 2.7, and Figs. 2.2 and 2.3).

In full agreement with their salt-like nature, alkali-metal hydrides are insoluble in organic solvents. The solubility in water, which should be favoured for the same reason, cannot be observed because of hydrolysis.

The densities of the alkali-metal hydrides increase in the direction from lithium to cesium, and are all considerably greater than the densities of the corresponding free metals. The percentage increase in density on passing from the metal to the hydride is between 45 and 75%, as follows from Table 2.4.

The increase in density is explained by the change in the type of packing from alkali metal (cubic body-centred lattice, packing density 0.68) to the hydride (cubic face-centred lattice, packing density 0.74) and by the gain in volume on passing from the large uncharged metal atoms to the smaller metal cations.

TABLE 2.4
COMPARISON OF DENSITIES OF ALKALI METALS AND THEIR HYDRIDES

Density at 25° of	*Li*	*Na*	*K*	*Rb*	*Cs*
Metal	0.535	0.971	0.862	1.532	1.886
Hydride	0.769	1.36	1.43	2.59	3.41

2.3 *Chemical behaviour*

The alkali-metal hydrides differ from one another only gradually in their chemical reactions. The reactions are stamped by the nature of the negatively charged hydrogen anion, *i.e.* the donor $H^{\ominus}$. $H^{\ominus}$ reacts by electron transfer, by nucleophilic displacement, and by the formation of donor–acceptor complexes.

(1) *Electron transfer*

(a) Reduction: $E^{n\oplus} + H^{\ominus} \rightarrow E^{(n-1)\oplus} + \frac{1}{2}H_2$

Prototype examples:

$2TiO_2 + 2LiH \rightarrow Ti_2O_3 + Li_2O + H_2$

$Fe_2O_3 + 6NaH \rightarrow 2Fe + 3Na_2O + 3H_2$

$TiCl_4 + 4NaH \rightarrow Ti + 4NaCl + 2H_2$

$HOH + LiH \rightarrow LiOH + H_2$

$RSH + NaH \rightarrow NaSR + H_2$

(b) Formation of anions: $E + H^{\ominus} \rightarrow E^{\ominus} + \frac{1}{2}H_2$

Prototype examples: $O_2 + 2LiH \rightarrow Li_2O + H_2O$

(2) *Nucleophilic displacement:* $EX_n + nH^{\ominus} \rightarrow EH_n + nX^{\ominus}$

Prototype example: $SiCl_4 + 4LiH \rightarrow SiH_4 + LiCl$

(3) *Donor activity*

(a) Formation of anionic complex hydrides: Acceptor + $H^{\ominus} \rightarrow$ H Acceptor$^{\ominus}$

Prototype examples:

$AlH_3 + Li^{\oplus} H^{\ominus} \rightarrow Li^{\oplus} [AlH_4]^{\ominus}$

$B(OR)_3 + Na^{\oplus} H^{\ominus} \rightarrow Na^{\oplus} [BH(OR)_3]^{\ominus}$

(b) Donor-catalysed organic condensations
Prototype example:

$$CH_3COOC_2H_5 \xrightarrow[-H_2]{+H^\ominus} |CH_2COOC_2H_5 \xrightarrow{+CH_3COOC_2H_5}$$

$$\underset{\displaystyle OC_2H_5}{\overset{\displaystyle O^\ominus}{CH_3\underset{|}{\overset{|}{C}}CH_2COOC_2H_5}} \xrightarrow{-C_2H_5O^\ominus} CH_3\overset{\displaystyle O}{\overset{\|}{C}}CH_2COOC_2H_5$$

3. Lithium hydride

3.1 Preparation of lithium hydride and deuteride

Lithium hydride is most simply synthesised from the elements. At room temperature, hydrogen does not react with lithium, and is not absorbed even by the very finely divided metal[*65*] such as remains, for example, after the evaporation of ammonia from a lithium solution in liquid ammonia. Slight reaction begins only at 440°, far above the melting point of lithium (180.54°, [*23*]), and is vigorous between 600 and 630°[*101*]. Lithium and deuterium react similarly (preparation of lithium deuteride at 720°[*80*]). The exothermic formation of the hydride is controlled by the pressure of the hydrogen. To complete the formation of LiH, the heating is carried out finally for a short time at 700°, followed by cooling in a stream of hydrogen[*101*].

Oxidation by the reaction of Li with SiO_2 is avoided in stainless-steel apparatus, and this permits a more rapid reaction with a higher reaction temperature. At 720° and an H_2 pressure of 1 atm, well-formed white to faint grey (Li) crystals of lithium hydride 1–3 mm in size are obtained[*92*]. For further, similar preparations, see [*5, 15, 16, 54, 161*].

Very pure, clear, very faintly blue crystals of LiH some millimetres long are obtained by the reaction of lithium filtered through a steel membrane with hydrogen (purification by means of uranium at 350°) in an Armco iron crucible at 720° (1 h), then at 680° (16 h), and finally at 500° (20 h)[*49*].

Extremely pure single crystals of LiH or LiD 16 mm in diameter and 50 to 80 mm long can be obtained by the reaction of extremely pure lithium with hydrogen or deuterium in a pure iron crucible[*121*]. Single crystals are formed on cooling the crucible from 850 to 650° (at the rate of 5°/h) if a temperature gradient of 30°/cm is maintained parallel to the height of the crucible by special means. The single crystal, faintly blue owing to traces of lithium, is subsequently heated at 550° in an atmosphere of hydrogen or deuterium to eliminate lithium and stresses. The resulting crystal is clear, colourless, and free of stresses, and contains the following impurities (in molar p.p.m.): Na 20–200, Mg 0.5–6, Fe 0.5–2, Cu 0.5–2, other metals < 1, non-metallic impurities such as oxygen 10–1000[*121*].

Doped single crystals of LiH appear to be difficult to prepare, since MgH_2,

References p. 38

MgO, or Li_2O added to the LiH melt rise upwards in the melt (denser LiH is found at the top). Homogeneous Mg-doped lithium hydride is obtained when the lithium used for the hydrogenation is previously doped with magnesium[*121*].

Lithium hydride is produced in admixture with magnesium oxide by the reduction of lithium oxide with magnesium in an atmosphere of hydrogen at 500 to 900°[*45*]:

$$Li_2O \xrightarrow[-\,MgO]{+\,Mg} 2Li \xrightarrow{+\,H_2} 2LiH$$

In this form, it can be used as the starting material for lithium alanate (lithium aluminium hydride) (see Chapter 5).

Organolithium compounds LiR can readily be hydrogenated at elevated temperatures[*47*]:

$$LiR + H_2 \xrightarrow[1\text{–}7\ atm]{benzene} LiH + RH$$

With reactions in benzene solution at a pressure of 1–7 atm, the times of reaction (for 100% conversion) fall in the following sequence[*47*] of organic groups R (reaction time in hours): p-tolyl (150), n-lauryl (91), n-heptyl (66), n-butyl (61), α-naphthyl (40), methyl (38.5), phenyl (32.2).

Lithium hydride is also formed in the thermal decomposition of alkyllithiums. For example, when n-butyllithium is boiled in octane (126°) lithium hydride and, primarily, 95% of 1-butene and 5% of butane are obtained; because of subsequent reactions, however, 1,3-butadiene and resinous materials are also produced, the latter as a result of LiR-catalysed polymerisation[*160*]:

$$CH_3CH_2CH_2CH_2Li \rightarrow LiH + CH_3CH_2CH{=}CH_2$$

$$CH_3CH_2CH_2CH_2Li + CH_3CH_2CH{=}CH_2 \rightarrow CH_3CH_2CH_2CH_3 + CH_3CHLiCH{=}CH_2$$

$$CH_3CHLiCH{=}CH_2 \rightarrow LiH + CH_2{=}CHCH{=}CH_2$$

Similarly, *tert*-butyllithium decomposes in n-heptane to give lithium hydride and 2-methyl-1-propene (which is hydrogenated to a slight extent to isobutane[*156*]):

$$(CH_3)_3CLi \rightarrow LiH + (CH_3)_2C{=}CH_2$$

3.2 *Physical properties of lithium hydride and deuteride*

The most important physical properties of lithium hydride and deuteride are summarised in Table 2.3.

Commercial lithium hydride is mostly in the form of light grey pieces with dimensions of 5 to 10 mm. The hydride prepared from the elements without special precautions with respect to purity is obtained, depending on the reaction conditions, either as a white powder, as a glassy opalescent mass with a crystalline fracture, or in fine, needle-shaped crystals. Pure LiH is always pure white, becoming faintly blue on ultraviolet illumination[*46*]. Colourless single crystals of hydride become coloured on irradiation with X-rays, pile neutrons, and β-particles (from LiT)[*119*]. The elementary photochemical process (transfer of an electron from $H^{\ominus}$ to $Li^{\oplus}$ with the formation of uncharged lithium atoms or hydrogen molecules) can be reversed by irradiation with long-wave visible light or by heating, since the bulk of the hydrogen formed cannot leave the lattice[*7, 88, 119–122*]. For single-crystal lithium hydride and deuteride, see the preceding Section 3.1. Similarly to doped lithium fluoride, Mg-doped lithium hydride can be split along the crystal planes more easily than the undoped material[*121*].

Lithium hydride and lithium deuteride crystallise in the NaCl lattice. According to calculations, the hydride should be converted into a CsCl lattice at 3 to 4 kbar and into the metallic state at 35 Mbar[*13, 133*]. However, it has been impossible to confirm the appearance of the CsCl lattice experimentally in pressure experiments up to 240 kbar[*6, 48, 155*].

The following measurements have been made, particularly with respect to the colour centres, on single crystals of LiH and LiD (undoped, doped, and irradiated with X-rays or neutrons), and on mixed crystals of 40% LiT and 60% LiH: optical absorption, NMR and ESR spectra, X-ray diffraction, electrical conductivity, and diffusion[*88, 119–122*].

Lithium hydride is an ionic conductor both in the solid and in the molten states [*101, 117, 121*]. Measurements on single crystals of lithium hydride have shown, as in the case of the alkali-metal halides, a conductivity mechanism by cation-vacancy migration with an activation energy of about 0.53 eV[*121*]. On plotting the logarithm of the conductivity against the reciprocal temperature, lithium hydride fits satisfactorily into the alkali-metal halide[*71*] sequence. With respect to its electrical conductivity and its melting point, it resembles lithium chloride more than lithium fluoride, which is closer in respect of lattice constants. Lithium deuteride conducts less well, but its conductivity σ is, however, temperature-dependent in the same way (similar slope in the $\ln \sigma - 1/T$ diagram)[*121*]. In the electrolysis of molten lithium hydride, the lithium migrates to the cathode and the hydrogen to the anode. The electrolysis obeys Faraday's law almost quantitatively (on an average to 99.5%)[*101, 111, 117*].

The physical properties of lithium hydride LiH, particularly the ionic conductivity mechanism, speak convincingly in favour of a basically ionic structure for solid lithium hydride. It must not be overlooked, however, that many physical properties—particularly the UV absorption and the polarisation, as well as the low chemical stability of lithium hydride—set lithium hydride apart from typical ionic crystals (*e.g.* the alkali-metal halides)[*121*]. To what extent partial covalent character in the crystal must be assumed to be responsible for this, is still uncertain. Moreover, it must not be overlooked that the deviations from typical ionic crystals are not a sufficient criterion, although they are a necessary one, for the involvement of covalency in the LiH crystal, since the deviations in LiH can be explained by polar hypotheses without the involvement of covalent bonding[*121*]. This may be illustrated by a section through an LiH crystal[*121*] (Fig. 2.1).

References p. 38

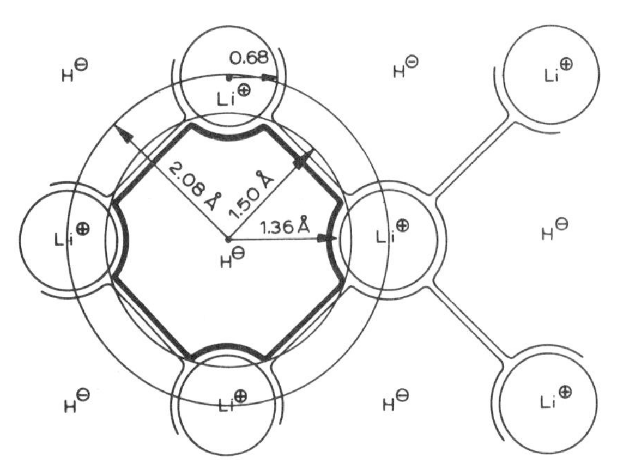

Fig. 2.1. Section through a LiH crystal[*121*]. The positive spheres of influence of the Li atoms are envisaged as 'hard' spheres. Between them are the negative highly deformed, 'soft' H atoms. Various hydrogen radii are shown for the sake of comparison: 2.08 Å, radius of a free $H^{\ominus}$ ion; 1.50 Å, effective radius of a negative H atom incorporated in a crystalline hydride of a heavy alkali metal; 1.36 Å, effective radius of a negative H atom in crystalline LiH.

The distances and positions of the lithium atoms have been established by experimental structure determinations. The following considerations are involved: the lattice sites of the $Li^{\oplus}$ ions are 'hard' spheres with $r = 0.68$ Å. The total space remaining between the $Li^{\oplus}$ ions is the sphere of influence of the $H^{\ominus}$ ions, which are compressed[1] and intensely distorted because of the extraordinarily diffuse charge distribution. The difference in charge between the Li and the H cells in LiH amounts to 0.35 electronic charges (most probable figure, calculated by the self-consistent field method[*32*] and from X-ray studies of single crystals[*4*]; a figure of 0.50 is obtained from the IR spectrum and dielectric properties[*145, 146*] and one of 0.52 from reflection measurements in the IR region[*36*]). However, a similar charge distribution is obtained on the assumption of covalent bonds[*90*]. In this respect (same charge distribution in the ionic and the covalent structure), lithium hydride forms an exception among the alkali-metal hydrides. In fact, on passing from LiH to hydrides of the higher alkali metals, the charge difference and the effective radius of the $H^{\ominus}$ ion becomes greater; as the UV absorption spectra show, in these circumstances the electronic properties approach those of the alkali-metal halides[*121*]. For further calculations on LiH, see[*12, 84, 107*]; on the theory of the electronic structure, *cf.*[*19, 25, 26, 33, 34, 68, 77, 78, 99, 112, 114, 115, 135*].

Lithium hydride dissociates at elevated temperatures (see Table 2.5). From about 700°, the degree of dissociation increases markedly (*cf.* for example the column for 90% LiH of the table). From Table 2.5 (cited in [29]) and from the p_{H_2}–% LiH diagram (Fig. 2.2) (cited in [*121*]), the dissociation pressure above

---

[1]According to Pauling, for a free $H^{\ominus}$ ion, $r = 2.08$–2.04 Å; from the anion–cation distance in LiH we obtain for the $H^{\ominus}$ ion $r = 1.36$ Å; the effective radius of the $H^{\ominus}$ ion in the higher alkali-metal hydrides is $r = 1.50 \pm 0.02$ Å.

TABLE 2.5

DISSOCIATION PRESSURES p_{H_2} [TORR] OF SOME MIXTURES OF Li AND LiH AT VARIOUS TEMPERATURES (°C)[*29*]

Mole % LiH	*700°*	*770°*	*825°*
0	0	0	0
10	25	50	50
20	40	100	150
30	40	150	280
40	40	160	360
50	40	160	410
60	40	160	430
70	40	160	430
80	40	160	430
90	40	160	440
100	260	320	720

TABLE 2.6

DISSOCIATION PRESSURE p_{D_2} [TORR] OF LiD AT VARIOUS TEMPERATURES (°K) CALCULATED FROM THERMODYNAMIC DATA FOR $LiD \rightarrow Li + \frac{1}{2}D_2$ [*149*]

Temperature (°K)	p_{D_2} *[torr]*
600	1.3×10^{-6}
700	3.6×10^{-4}
800	4.6×10^{-2}
900	9.73×10^{-1}

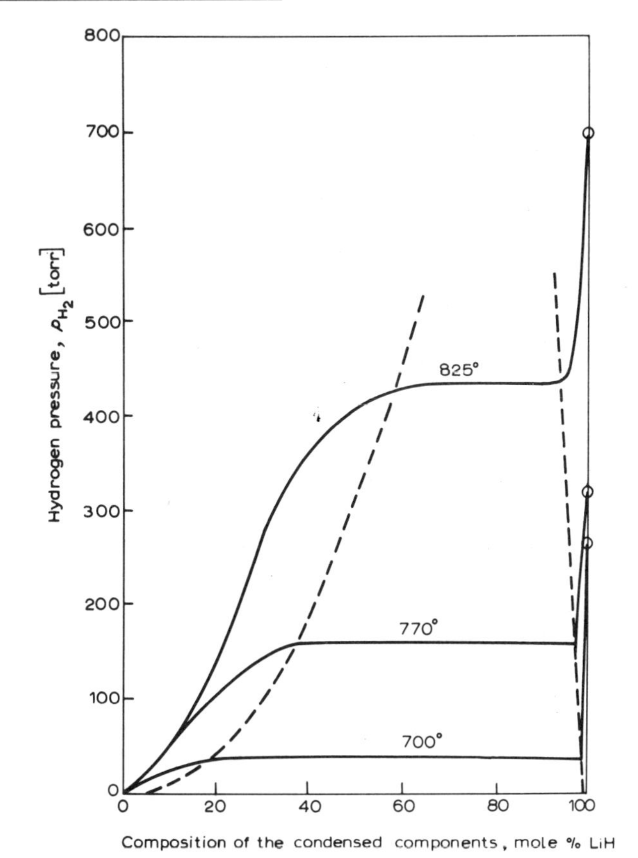

Fig. 2.2. Isotherms in the p_{H_2}–LiH_n diagram[*121*].

lithium hydride is substantially constant over a wide range of H or LiH contents in the 'hydride'.

Thus, the hydrogen pressure at 700° is 40 torr between about 20 and 95% LiH, at 770° 160 torr between 35 and 95% LiH, and at 825° about 435 torr between 60 and 90% LiH. From this, because of the univariance of the two-component system, the Gibbs phase rule shows the necessary existence of three phases. The presence of two definite condensed phases – LiH and Li – follows, since gaseous hydrogen is always present. Somewhat above the melting point of pure lithium hydride (> ~680°), the hydrogen dissociation pressure obeys the equation $\log p[\text{torr}] = -9520/T + 11.28$ (cited in [*29, 30*]). As was to be expected, the dissociation of lithium deuteride is the same as that of lithium hydride. According to Table 2.6, the D_2 dissociation pressure below 627° (900°K) is less than 1 torr [*149*].

In contrast to the other alkali-metal hydrides, lithium hydride is slightly soluble in polar organic solvents, such as diethyl ether, tetrahydrofuran, and dioxan. The systems LiH–LiCl, LiH–LiBr and LiH–NaCl form eutectics melting at 450° (68 mol.% of LiCl), ~ 486° (~ 63 mol.% of LiCl[*10*]) and 400° (71 mol.% of LiBr)[*28, 72a–c*]. Thermal analysis of systems of LiH with SrH_2, BaH_2, EuH_2, YbH_2, ScH_2, LaH_2, CeH_2 and SmH_2 shows the existence of the compounds $LiSrH_3$ and $LiEuH_3$ and eutectic compositions with mol.% MH_2: 11.2% SrH_2, 5.0% BaH_2, 6.8% YbH_2[*93a, 93b*].

3.3 Reactions of lithium hydride and deuteride

Thermal decomposition. Finely divided metals catalyse the thermal decomposition of lithium hydride. For example, LiH treated with iron powder liberates hydrogen at temperatures as low as 200 to 300°. Mercury, which does not react with LiH at 20°, forms lithium amalgam at about 360°[*101*]:

$$LiH + xHg \rightarrow LiHg_x + \tfrac{1}{2}H_2$$

Hydrogen. In an atmosphere of hydrogen with a small content of deuterium or tritium, lithium hydride exchanges its hydrogen. At a higher temperature, the isotope exchange takes place in accordance with the dissociation equilibrium. At relatively low temperatures (20–200°C), the exchange takes place through a surface reaction, with subsequent diffusion. The equilibrium constant of the process is[*158*] $K = [HT][LiH]/[H_2][LiT] = 3.66$.

Chlorine. Dry chlorine reacts vigorously with lithium hydride at elevated temperatures; depending on the experimental conditions, LiCl and HCl or LiCl and H_2 are formed[*46*]:

$$LiH + Cl_2 \rightarrow LiCl + HCl$$
$$2LiH + Cl_2 \rightarrow 2LiCl + H_2$$

Hydrogen chloride. Dry hydrogen chloride reacts with lithium hydride only at red heat[*50*]:

$$LiH + HCl \rightarrow LiCl + H_2$$

Oxygen. LiH is stable to dry air and oxygen at 20°, but it can under certain conditions ignite spontaneously in the air if it is in a finely divided state (atmospheric humidity). At red heat, it ignites in oxygen with the formation of Li_2O (subsequent reaction with H_2O!) [*50*]:

$$2LiH + O_2 \rightarrow Li_2O + H_2O$$

Water. LiH and LiD react very vigorously with water and water vapour [*49, 92, 101*]:

$$LiH_{solid} + H_2O_{liquid} \rightarrow LiOH_{\infty\ dilution} + H_{2\ gas} + Q_H$$
$$LiD_{solid} + H_2O_{liquid} \rightarrow LiOH_{\infty\ dilution} + HD_{gas} + Q_D$$
$$Q_H = 31.76 \pm 0.10\,[92];\ 31.476 \pm 0.018\,[49]\ \text{kcal/mole}\ (25°);$$
$$Q_D = 31.32 \pm 0.007\ \text{kcal/mole}\ (25°)\,[49].$$

This evolution of heat on hydrolysis could be the cause of the spontaneous ignition of lithium hydride in moist air (spontaneous ignition also occurs, for example, when a large amount of powdered LiH is treated with a little water).

Sulphur dioxide. Below 50°, sulphur dioxide is converted by lithium hydride into thiosulphate [*101, 116*]:

$$2LiH + 2SO_2 \rightarrow Li_2S_2O_3 + H_2O$$

Nitrogen. This reacts with lithium hydride at elevated temperatures to form lithium nitride [*50*]:

$$6LiH + N_2 \rightarrow 2Li_3N + 3H_2$$

Ammonia. Above 400°, gaseous ammonia enters into a rapid reaction with lithium hydride to form lithium amide and hydrogen [*46, 126*]:

$$LiH + NH_3 \rightarrow LiNH_2 + H_2$$

The same reaction takes place under certain conditions [*126*] in liquid ammonia, but the reaction cannot be predicted with certainty [*46*].

Phosphorus. At elevated temperatures, phosphorus reacts with lithium hydride to give the phosphide [*101, 116*].

Carbon dioxide. At elevated temperatures, carbon dioxide reacts with lithium hydride to give the formate [*35*]:

$$LiH + CO_2 \rightarrow HCOOLi$$

Silica. At reduced pressure, silica and lithium hydride form silicon [*101, 116*].

Boric oxide. When anhydrous boric oxide is heated with lithium hydride, there is a vigorous reaction with the production of boron hydrides [*67*].

Europium trioxide. This is converted by lithium hydride in the eutectic mixture with LiCl and KCl into its oxidation stage of +2 by heating in vacuum at 300° and then melting under helium at atmospheric pressure.

Oxides of V, Nb, Ta, Ti, Zr, Th. In a similar manner, V_2O_5, Nb_2O_5, and Ta_2O_5

References p. 38

(at 300°) and TiO_2, ZrO_2, and ThO_2 (at 500°) are reduced to lower valencies (probably VO, Nb_2O_3, TaO_2 or $TaO_2 \cdot H_2O$, Ti_2O_3, $Zr + Zr_2O_3$, and ThO)[*116*].

[β-Al(OEt)₃]₄. The hydride ions of lithium hydride add to free orbitals. Thus, on shaking in a mixture of ether and benzene (1 : 1), very finely powdered LiH and crystalline tetrameric β-triethoxidoaluminium $[Al(OEt)_3]_4$ slowly form lithium triethoxidoaluminate[*132*]:

$$4LiH + [Al(OEt)_3]_4 \rightarrow 4Li[AlH(OEt)_3]_4$$

In boiling ether, the reaction takes place with similar slowness. Lithium alanate catalyses the reaction, but the hydrogenation then proceeds further:

$$4Li[AlH_4] + [Al(OEt)_3]_4 \rightarrow 4Li[AlH(OEt)_3] + 4AlH_3$$

$$Li[AlH_4] + Li[AlH(OEt)_3] \rightarrow Li[AlH_2(OEt)_2] + Li[AlH_3(OEt)]$$

$$LiH + Li[AlH(OEt)_3] \rightarrow Li[AlH_2(OEt)_2] + LiOEt$$

$$LiH + AlH_3 \rightarrow Li[AlH_4]$$

α-Al(OEt)₃. Lithium hydride in boiling ether adds rapidly to the less highly polymeric α-triethoxidoaluminium $Al(OEt)_3$. Lithium triethoxidoalanate, which retains ether tenaciously (etherate?) is again formed, in 40% yield[*132*]:

$$LiH + Al(OEt)_3 \rightarrow Li[AlH(OEt)_3]$$

ZnR₂. With diethylzinc and diphenylzinc, lithium hydride forms etherates of the adducts $LiH \cdot ZnR_2$[*40, 159*].

Halides of 3rd and 4th main groups. The important reactions of lithium hydride with these halides, which lead to the corresponding hydrides or complex hydrides (double hydrides):

$$EX_n + nLiH \rightarrow EH_n + nLiX$$
$$EH_3 + LiH \rightarrow LiEH_4$$

(E = element of the 3rd or 4th main group; X = halogen)

are described in the chapters dealing with these elements.

Organic compounds. With organic compounds, lithium hydride has either a reducing or (because of its high donor activity) a condensing effect (*cf.* section 2.3 of this Chapter). No alkyllithiums are formed with alkyl halides[*46*]. Alcohols form lithium alkoxides. Alcoholysis takes place considerably more slowly than hydrolysis[*46*]:

$$LiH + ROH \rightarrow LiOR + H_2$$

Being a strong donor, lithium hydride catalyses the aldol condensation of aldehydes and ketones (intramolecular water subsequently splits out with the formation of a double bond) [*46*]:

$$2RCH_2CHO \xrightarrow{(LiH)} RCH_2CH(OH)CHRCHO$$
$$2CH_3COR \xrightarrow{(LiH)} CH_3CR(OH)CH_2COR$$

As was to be expected, carboxylic acids are converted into the lithium salts [*46*]. The condensation of carboxylic acid esters is catalysed by LiH only to a certain extent (in contrast to the case of NaH):

$$2RCH_2COOR \xrightarrow[-LiOR, -H_2]{+LiH} RCH_2COCHRCOOR$$

On treatment with lithium hydride, benzoyl chloride forms benzyl benzoate (condensing and reducing action) [*63*]:

$$2C_6H_5COCl + 2LiH \rightarrow C_6H_5COOCH_2C_6H_5 + 2LiCl$$

Organic compounds are better and more conveniently reduced by means of the soluble complex hydrides than by the insoluble basic hydride LiH (see, for example, Chapters 4 and 5).

Some hints may be given on the handling of lithium hydride. When the envisaged reaction permits, commercial, coarsely, powdered lithium hydride should not be milled dry in an inert gas atmosphere, but in a ball mill in a suitable well-dried solvent (suspending medium). Milling is not carried out for stock, but preferably the night before the reaction, since finely divided LiH rapidly becomes inactive through uncontrollable reactions on the surface, due possibly to moisture or to reactions with the solvent. Many ethers are decomposed relatively rapidly (*e.g.* di-n-butyl ether; diethyl ether is stable); consequently, preliminary experiments are often necessary. In the pouring of ethereal suspensions, rubber gloves, outward ventilation, and/or gas masks are necessary, since dry LiH dust (evaporating ether) intensely irritates the respiratory tracts, the mucous membrane of the eyes, and moist skin. LiH fires may be extinguished with sand and argon–carbon dioxide, nitrogen, and chlorinated hydrocarbons react with burning lithium hydride.

In working with relatively large amounts of lithium hydride, it is necessary to have protection for the eyes, the lungs, and the hands, and fire-fighting equipment.

4. Sodium hydride

4.1 Preparation of sodium hydride and deuteride

Sodium hydride is most simply obtained from the elements. In contrast to lithium, sodium forms the hydride with molecular hydrogen at temperatures as low

References p. 38

as 80°, *i.e.* well below its melting point (97.90±0.05° [*124*]), but the solid hydride crust inhibits further reaction[*31, 61*]. For this reason, when carefully purified hydrogen (H_2, D_2) is passed, with the absolute exclusion of air and moisture, over liquid sodium – generally contained in iron boats – (for NaH: 350°[*31*], 370°[*103*], 600–750°[*92*], for NaD: 320–360°[*53*]) or through liquid sodium[*3*], the hydride formed sublimes on to colder parts of the reaction tube. In addition, NaH and NaD are also formed in the gas phase.

The preparative production of sodium hydride does not take place so smoothly, since the reactions of formation and decomposition take place at almost the same temperature. As can be seen from the comparison of the vapour pressures of the alkali metals in Table 2.1 appreciable amounts of sodium already vaporise above 370°. Since the Na vapour (entrained in the current of H_2) vigorously attacks vessel walls containing SiO_2, the reaction product is more or less highly contaminated with sodium oxide and with unreacted elementary sodium (blue to brown colorations)[*103*].

Pure sodium hydride is probably obtained only when glass or porcelain components are protected by Armco iron linings[*91*]. Relatively pure hydride is obtained by the evaporation of sodium in an Armco iron apparatus at 750–790° in an atmosphere of hydrogen[*49*]. Pure sodium deuteride is obtained similarly (700°); the reaction appears to take place more slowly than with light hydrogen [*49*].

Sodium hydride contaminated with elemental sodium can be extracted with liquid ammonia, when the metal dissolves as such or as the amide. The hydride is then washed with the ether and dried[*103*]. For the preparation of pure large crystals of NaH, however, only methods similar to those used successfully in the preparation of single crystals of LiH (see section 3.1) can probably be used.

Although the absorption of H_2 begins as low as 80°, sodium hydride is best manufactured industrially at 270 to 310° in an electrically heated iron drum having a length-to-diameter ratio of 4. There the reaction mixture is kept in continuous movement by a scraper vane fixed to a shaft. The sodium is added in liquid form through a heated funnel, and, for better distribution, is diluted with an inert solid material – preferably with previously prepared sodium hydride. The hydrogen, which is fed in at the front end, must be rigorously dried and carefully freed from oxygen, since only under these conditions can a sodium hydride of 95% purity be obtained. The gas-outlet connection on the opposite side is provided with an immersion seal, so that the drum is always under a slight overpressure of hydrogen. The finished material can be removed from the drum either discontinuously, through a plate in the bottom, or continuously through a lock device. Under these conditions, sodium hydride is obtained in the form of a grey microcrystalline powder[*38, 43, 59, 136, 152*]. Because of the increasing viscosity of the Na–NaH mixture, the degree of hydrogenation depends markedly on the form of the stirrer [*24*]. The distribution of the liquid sodium on the solid inert material can be substantially improved by the addition of small amounts of surface-active substances, *e.g.* sodium or potassium salts of carboxylic acids, anthracene, phenanthrene, and fluorene[*59, 96*], or by 1–10 mol.% of potassium or potassium hydroxide[*72, 130,*

131]. Fine distribution of the sodium can also be achieved by means of suitable liquid hydrocarbons [*96*, *108*], *e.g.* kerosine, 'Bayol 85' (b.p. 185–265°C) [*95*] or vaseline oil [*96*]. In this case, sodium hydride is produced in the form of a 10 to 15% suspension. This can be concentrated to 40 to 50% by centrifuging or filtration. Sodium hydride is formed at 200° and 60 to 70 atm from sodium, hydrogen, and crystalline α-triethoxidoaluminium. When ether is added, however, Na[AlH-$(OR)_3$] is formed [*132*].

Sodium hydride can be prepared not only from the elements but also from sodium azide and hydrogen [*148*]:

$$2NaN_3 + H_2 \rightarrow 2NaH + 3N_2$$

In this reaction, the temperature (300°) must not reach the decomposition temperature of the azide (375°), since only undecomposed sodium azide forms a pure hydride.

Highly diluted sodium vapour reacts with active hydrogen at 240 to 270°. By analysis of the chemiluminescence occurring in this process it has been possible to show that the (binary) decomposition takes place not with an electronically excited hydrogen molecule H_2^* but with a vibrationally excited molecule $H_2^\dagger$ (vibrational quantum number $v = 6$) [*118*]:

$$Na + H_2^\dagger \rightarrow NaH + H$$

4.2. *Physical properties of sodium hydride and deuteride*

The most important properties of sodium hydride and deuteride are summarised in Table 2.3. About solutions of hydrogen in liquid sodium see [*3a*].

Sodium hydride is marketed as a white to dark brown or dark grey microcrystalline powder, or in pieces or grains [*94*]. Sodium hydride prepared in the laboratory (which is purer) forms a white, dense, chalk-like mass including grey whiskers of Na-containing hydride [*49*], or a white cotton-like mass with grey spots [*92*]. The grey, Na-containing parts can be separated mechanically.

Like lithium hydride, sodium hydride has an ionic structure ($Na^{\oplus} H^{\ominus}$); it crystallises in an NaCl type lattice. The melting point is masked by the lower decomposition point, but is estimated to be in the neighbourhood of the melting point of NaF, probably between NaCl (801° [*64*]) and NaF (980–997° [*64*, *11*]).

Sodium hydride dissociates at elevated temperatures (the pressure of H_2 at 422° is about one atmosphere [*62*, *91*]); under these conditions the following dissociation equilibria occur:

for 100% NaH:

$$\underset{\text{solid}}{NaH} \rightleftharpoons \underset{\substack{\text{solid}\\ \text{in infinite}\\ \text{dilution in NaH}}}{Na} + \tfrac{1}{2}H_2$$

References p. 38

for 90% NaH:

$$\underset{\substack{\text{solid}\\ +2\%\ \text{dissolved Na}}}{NaH} \rightleftharpoons \underset{\substack{\text{solid}\\ 2\%\ \text{solution in NaH}}}{Na} + \tfrac{1}{2}H_2$$

for 50% NaH (two-phase region):

$$\underset{\substack{\text{solid}\\ \text{saturated with Na}}}{NaH} \rightleftharpoons \underset{\substack{\text{liquid}\\ \text{saturated with NaH}}}{Na} + \tfrac{1}{2}H_2$$

for 10% NaH (Na-rich phase):

$$\underset{\substack{\text{liquid}\\ 10\%\ \text{solution in Na}}}{NaH} \rightleftharpoons \underset{\substack{\text{liquid}\\ +10\%\ \text{NaH}}}{Na} + \tfrac{1}{2}H_2$$

TABLE 2.7

DECOMPOSITION PRESSURE p_{H_2} OF NaH

EMPIRICAL CONSTANTS OF THE EQUATION $\log p_{H_2}$ [torr] $= -a/T + b$ FOR VARIOUS HYDRIDE CONCENTRATIONS IN MIXTURES OF NaH AND Na BETWEEN 500° AND 600°

Mole % NaH	*a*	*b*	*Ref.*
10	6222	11.70	*9*
20	6058	11.55	*9*
30–80	5958	11.47	*9*
80–90	6100	11.66	*62*
90	5806	11.32	*9*
100	5070	10.49	*9*

The dissociation pressure is independent of the composition of the solid phase over a wide range (80–30% NaH, 20–70% Na). The dissociation pressure can be described by the Clausius–Clapeyron equation $\log p_{H_2}$[torr] $= -a/T + b$. The constants a and b for hydrides of various NaH contents for the temperature range between 500 and 600° are given in Table 2.7 [*9, 62*]. In the range from 100 to 250°, the following equation holds [*3*]: $\log p_{H_2}$[torr] $= -6100/T + 11.66$. The dissociation isotherms of various mixtures of Na and NaH are plotted [*9*] in Fig. 2.3.

Sodium hydride dissolves homogeneously in sodium hydroxide up to an 18% concentration at room temperature. The thermal stability of the solutions increases with rising concentration of NaH, so that 'sodium hydride' in this form can be used up to temperatures of 500–550° [*94, 98*]. It dissolves in Na [*157*], Na–K [*94*] and salt melts [*69*]. Sodium hydride is insoluble in organic solvents and in liquid ammonia.

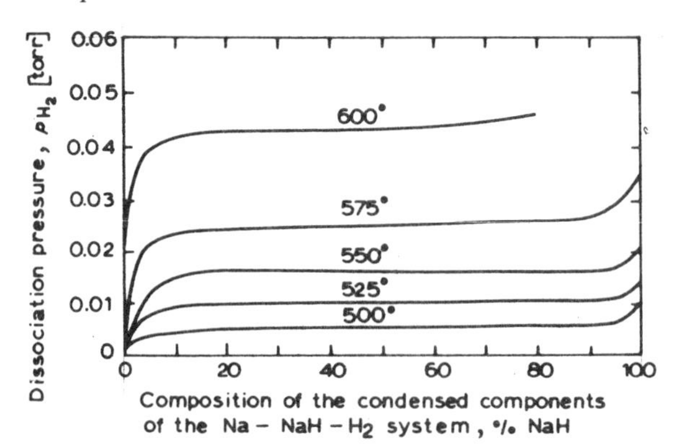

Fig. 2.3. The p_{H_2}–NaH_n isotherms [*9*].

2.3 Reactions of sodium hydride and deuteride

Halogens. Gaseous halogens (iodine vapour only above 100°) react with sodium hydride with ignition to give the corresponding sodium and hydrogen halides: $NaH + X_2 \rightarrow NaX + HX$. Liquid chlorine does not react at −35°, and neither does liquid bromine up to its boiling point [*103*].

Oxygen. Sodium hydride is stable to dry air and oxygen up to 230°, and also to liquid oxygen (?!). Above 230° it burns to Na_2O [*103*].

Water. Both sodium hydride and sodium deuteride hydrolyse extraordinarily vigorously in water [*2, 92*]:

$$NaH_{solid} + H_2O_{liquid} \rightarrow NaOH_{\infty\,dilution} + H_{2\,gas} + Q_H$$
$$NaD_{solid} + H_2O_{liquid} \rightarrow NaOH_{\infty\,dilution} + HD_{gas} + Q_D$$

$Q_H = 30.63 \pm 0.23$ [*92*]; 30.582 ± 0.019 [*49*] kcal/mole (25°);
$Q_D = 30.693 \pm 0.004$ kcal/mole (25°) [*49*].

Sodium hydride is a vigorous desiccant for inert liquids and gases, but its action is not so quantitative as that of calcium hydride. The reaction of a suspension of NaH in oil (*e.g.* 20–25% of NaH, grain size 5. . . . *25*. . . 50, in 'Bayol 85') with water is vigorous but controllable, and not so dangerous as the hydrolysis of the dry material; consequently spontaneous ignition is rare [*95*].

Sulphur. Sulphur vapour reacts vigorously with sodium hydride to form sodium sulphide [*103*]:

$$2NaH + 2S \rightarrow Na_2S + H_2S$$

Sulphur compounds. Sulphur dioxide is reduced by sodium hydride to sodium dithionite, and sulphuric acid is reduced to hydrogen sulphide or lithium sulphide [*103*].

Carbon dioxide. CO_2 gives the formate at room temperature; formate and oxalate are produced above 80° [*106*]:

$$NaH + CO_2 \rightarrow HCOON_2$$
$$2NaH + 2CO_2 \rightarrow Na_2C_2O_4 + H_2$$

Ammonia. Sodium hydride does not react with liquid ammonia at −60°. A reaction leading to sodium amide and hydrogen begins at −30°:

$$NaH + NH_3 \rightarrow NaNH_2 + H_2$$

Dry gaseous ammonia reacts similarly at 20° [*106*].

Metal oxides and halides. Many of these are reduced to the metal by sodium hydride, for example [*14*]:

$$Fe_2O_3 + 6NaH \xrightarrow{375^\circ} 2Fe + 3Na_2O + 3H_2$$
$$TiCl_4 + 4NaH \xrightarrow{400^\circ} Ti + 4NaCl + 2H_2.$$

The reaction mentioned first is of importance for the derusting of stainless steels. For this purpose, the article to be derusted is immersed in an NaOH melt containing 1.5–2% of NaH. At 375° the oxide is reduced to the metal within seconds

References p. 38

or a few minutes, depending on the thickness of the layer of rust. The new completely oxide-free iron is sprayed with water and briefly immersed in acid (this is also a pretreatment for galvanising).

NaH cleaves the Si-C-bond. Example: $(C_6H_5)_3SiCH_2C_6H_5$ and NaH in cyclohexane at 200° and 200 atm give $(C_6H_5)_3SiH$, $CH_3C_6H_5$, $(C_6H_5)_4Si$, and $(C_6H_5)_2SiH_2$ [*124a*].

$[Al(OEt)_3]_n$. The donor sodium hydride forms dative σ-bonds with acceptors. Thus, it adds to α- and β-triethoxidoaluminium $[Al(OEt)_3]_n$ ($n = 4$ or 1); the reaction takes place more rapidly than with LiH, (see section 3.3). A 90% yield of sodium triethoxidoalanate $Na[AlH(OEt)_3]$ is formed [*132*]:

$$n\text{NaH} + [\text{Al(OEt)}_3]_n \rightarrow n\text{Na}[\text{AlH(OEt)}_3]$$

$B(OMe)_3$. It also adds to trimethyl borate $B(OMe)_3$ [*18*]:

$$4\,\text{NaH} + 4\,\text{B(OMe)}_3 \xrightarrow{\text{diglyme}} 4\,\text{Na}[\text{BH(OMe)}_4]$$

dismutation, diglyme: $\rightarrow \text{Na}[\text{BH}_4]$ (soluble) $+ 3\,\text{Na}[\text{B(OMe)}_4]$ (insoluble)

dismutation, tetrahydrofuran: $\rightarrow \text{Na}[\text{BH}_4]$ (insoluble) $+ 3\text{Na}[\text{B(OMe)}_4]$ (soluble)

ZnR_2. Sodium hydride likewise adds to zinc chloride and diethylzinc in monoglyme and in diglyme, but not in diethyl ether, tetrahydrofuran, or aliphatic or aromatic hydrocarbons. Ether complexes of $NaH \cdot 2ZnCl_2$ and $NaH \cdot 2(C_2H_5)_2Zn$ are formed [*82, 40*].

Halides. For the important hydrogenations of halides (*e.g.* R_2SiCl_2) with sodium hydride, see in the sections on the corresponding elements.

Organic compounds. Because of its insolubility, the more ionic sodium hydride is less suitable for reductions of organic compounds than lithium hydride; the reactions take place with difficulty and give only poor yields. Consequently, for the reduction of organic compounds sodium hydride is better used in the form of its complexes (see, for example, Chapters 4 and 5). With alcohols sodium hydride gives alkoxides:

$$\text{NaH} + \text{HOR} \rightarrow \text{NaOR} + \text{H}_2$$

This reaction is suitable for the preparation of alcohol-free alkoxides, particularly the alkoxides of unsaturated alcohols, which are reduced when metallic sodium is used. Carboxylic acids are converted into sodium salts:

$$\text{RCOOH} + \text{NaH} \rightarrow \text{RCOONa} + \text{H}_2$$

Sodium hydride is a useful condensation catalyst. It possesses considerable advantages over the other condensing agents, such as Na, NaOH, NaOR, $NaNH_2$: (i) it acts more vigorously and more rapidly, (ii) no relatively large excess needs to be used, (iii) neither water nor alcohol is produced, (iv) there are few side reactions and reductions, (v) the hydrogen formed serves as a measure of the extent of reaction[*56, 57*]. A suspension of NaH in oil is particularly suitable, since the particles are then protected against superficial oxidation. For condensations normally requiring days, hours (and frequently only minutes) are often sufficient when they are catalysed with suspensions of NaH in oil, and generally, in addition, the reaction temperature can also be lowered. A few examples may be given.

(a) Ester condensation and metallation:

$$RCH_2COOC_2H_5 + HCH(R)COOC_2H_5 \xrightarrow[-C_2H_5ONa,\ -H_2]{+NaH}$$

$$RCH_2COCH(R)COOC_2H_5 \xrightarrow[-H_2]{+NaH} [RCH_2COCRCOOC_2H_5]Na$$

A comparison with other condensing agents is given in Table 2.8.

(b) Ethoxycarbonylation:

$$(C_2H_5O)_2CO + CH_3CH{=}CHCH_2CH_2COCH_3 \xrightarrow[-C_2H_5ONa,\ -H_2]{+NaH}$$

$$CH_3CH{=}CHCH_2CH_2COCH_2COOC_2H_5 \quad (85\%)$$

$$(C_2H_5O)_2CO + \text{cyclohexanone} \xrightarrow[-C_2H_5ONa,\ -H_2]{+NaH} \text{2-(}COOC_2H_5\text{)cyclohexanone}$$

A comparison of the NaH-catalysed reaction of cyclohexanone with diethyl carbonate (last equation) with that using other catalysts is given in Table 2.8.

(c) Stobbe reaction:

$$(C_6H_{11})_2C{=}O + \begin{array}{l}CH_2COOC_2H_5\\ |\\ CH_2COOC_2H_5\end{array} \xrightarrow[-C_2H_5ONa,\ -H_2]{+NaH}$$

$$(C_6H_{11})_2C{=}C\begin{array}{l}COOC_2H_5\\ COO^{\ominus}\end{array} \xrightarrow{+H^{\oplus}} (C_6H_{11})_2C{=}C\begin{array}{l}COOC_2H_5\\ COOH\end{array}$$

A comparison of Stobbe reactions catalysed with various catalysts is given in Table 2.8.

References p. 38

Sodium hydride is handled similarly to lithium hydride. Protection for the eyes, lungs, and hands is thus necessary, and fire-fighting equipment is required.

TABLE 2.8
COMPARISON OF NaH-CATALYSED REACTIONS WITH THOSE USING OTHER CATALYSTS

Catalyst	*Reaction temp. (°C)*	*Reaction time*	*Yield (%)*	*Ref.*
(a) *Condensation of* $CH_3COOC_2H_5$				
NaH-oil	33	7 h[a]	50[b]	*95*
NaH	50	4 h	29[b]	*58, 95, 144*
$NaNH_2$	33	2 h	4[c]	*86, 95*
Condensation of $(C_2H_5CO)_2O$				
BF_3	—	—	35	*1, 95*
(b) *Reaction of cyclohexanon with* $(C_2H_5O)_2CO$				
NaH-oil	33	1·5 h	50	*95*
NaH	20	—	37	*95, 144*
$NaNH_2$	—	2 h	18	*87, 95*
CH_3ONa	—	—	20[d]	*95, 154*
(c) *Stobbe reaction*[e]				
NaH-oil/hexane	40	1 h	95·7	*95*
$NaH/C_2H_5OC_2H_5$	20	8 h	97	*20*
C_2H_5ONa	100	—	90	*143*
C_2H_5ONa/C_2H_5OH	20	6 h	50	*143*
$C_2H_5ONa/C_2H_5OC_2H_5$	0	days	90	*8*
	20	days	60	*143*
	20	14 d	—	*142*

[a]90% conversion after 4 h.
[b]Isolation of the condensation product by distillation.
[c]Isolation of the condensation production as Cu salt.
[d]Poor product.
[e]First compound is the catalyst, second one the solvent.

5. Potassium hydride

5.1 Preparation of potassium hydride and deuteride

Similarly to LiH and NaH, potassium hydride can be obtained from the elements. At 20°, potassium (even in finely divided form) does not react with hydrogen. The reaction of hydrogen or deuterium with liquid potassium (m.p. 63.65°[*27*] begins at 260 to 280°, but is only sufficiently rapid for a preparative reaction at about 350° (gas-phase reaction) [*31, 97, 53*]. The hydride condenses in white needles on the cooler parts of the reaction tube. For the preparation of the pure hydride—as in the preparation of NaH—oxidic material of the vessels must be protected from attack by potassium vapour by Armco iron lining tubes (*e.g.* where H_2 is passed over K at 600° [*92*], or the operation is carried out in

Armco iron apparatus (640–655° on reaction with H_2, 600–605° on reaction with D_2,[*49*]). Any excess of potassium is removed by digestion with ammonia.

In the industrial preparation of potassium hydride, hydrogen reacts with potassium not in the gas phase but in finely divided form. This is achieved by adding the following materials to the potassium: alkali-metal chlorides or carbonates [*39*], tungsten disulphide [*85*] and, best, potassium hydride [*39*]; alternatively, the potassium is shaken in paraffin oil at 230 to 400° and a hydrogen pressure of 7 to 10 atm [*127, 129*], in fatty acids [*73, 109*] or in organic compounds containing triple bonds [*73, 109*]. Example: in a stirring autoclave, with the addition of 1% of machine oil or 0.5% of 'benzene aromatics', potassium is hydrogenated to 97–98% potassium hydride in 2 hours at 200 to 250° and a pressure of H_2 of 1 to 3 atm [*97*]. For further examples, the reader is referred to the literature [*97*].

5.2 *Physical properties of potassium hydride and deuteride*

The most important physical properties of potassium hydride and deuteride are summarised in Table 2.3.

Potassium hydride forms a pure white cotton-like mass frequently made grey by an excess of potassium [*92*] or a white mass interpenetrated by a few grey streaks [*49*]. Under the action of X-rays or in sunlight, with a relatively long time of irradiation, white KH becomes blue. Like all alkali-metal hydrides, KH forms an ionic lattice of the NaCl type. The melting point cannot be determined because of decomposition. On heating in vacuum, KH decomposes into potassium and hydrogen (the decomposition pressure at 420° is about 1 atm) [*29, 30*]. In the range from 315 to 390°, the pressure can be described by the equation $\log p_{H_2}$ [torr] $= -5850/T + 11.20$ [*62*].

5.3 *Reactions of potassium hydride and deuteride*

Halogens. Elementary fluorine reacts with potassium hydride at ordinary temperatures with ignition to give potassium fluoride and hydrogen fluoride: $KH + F_2 \rightarrow KF + HF$. Potassium hydride becomes red hot in a stream of chlorine, the analogous reaction taking place [*102*].

Oxygen. In dry air or dry oxygen, potassium hydride ignites spontaneously even at room temperature [*102*].

Water. Potassium hydride is vigorously hydrolysed by water:

$$KH_{solid} + H_2O_{liquid} \rightarrow KOH_{\infty\,dilution} + H_{2\,gas} + Q_H$$
$$KD_{solid} + H_2O_{liquid} \rightarrow KOH_{\infty\,dilution} + HD_{gas} + Q_D$$

$Q_H = 31.89 \pm 0.12$ [*92*], 33.081 ± 0.005 [*49*] kcal/mole (25°); $Q_D = 3.625 \pm 0.004$ kcal/mole (25°) [*49*].

Sulphur. Molten sulphur reacts with potassium hydride with ignition, forming potassium sulphide and hydrogen sulphide [*102*]:

$$2KH + 2S \rightarrow K_2S + H_2S$$

References p. 38

Hydrogen Sulphide. On gentle heating with H_2S, potassium sulphide and hydrogen are formed in a vigorous reaction [*102*]:

$$2KH + H_2S \rightarrow K_2S + 2H_2$$

Ammonia. Ammonia reacts with potassium hydride even at the melting point, forming potassium amide and hydrogen [*125*]:

$$KH + NH_3 \rightarrow KNH_2 + H_2$$

CO_2 and CO. Potassium hydride reacts with absolutely dry carbon dioxide only above 54°, with the formation of potassium formate:

$$KH + CO_2 \rightarrow HCOOK$$

In the presence of traces of water, the reaction takes place even at −85°. Above 100°, CO_2 is partially reduced to carbon; in addition, formate, carbonate, and hydrogen are formed [*106*]. Carbon monoxide adds to potassium hydride. Between 240 and 270° at a CO pressure of 30 atm, KH adds 1 mole of CO (KH·CO). Above 270°, the carbon monoxide is given off again, even at high pressures of CO [*66*].

Oxides. Like the other alkali-metal hydrides, potassium hydride is a powerful reducing agent. Thus, for example, it reduces copper oxide and lead oxide to the metals, even on gentle warming [*102*].

Organic compounds. The reducing action of potassium hydride on organic substances has so far been little investigated. It may perhaps not be particularly suitable for the purpose, because of its insolubility in organic solvents. For the reduction of organic compounds it is better used in the form of its complex compounds. With acetylene, potassium hydride reacts above 100° to form potassium acetylide and hydrogen:

$$HC{\equiv}CH + KH \rightarrow HC{\equiv}CK + H_2$$

Like the reaction with CO_2, the reaction with C_2H_2 is catalysed by water (reaction even at 20°). Methane and ethylene do not react with KH under the same conditions [*106*].

As regards handling, what has been said in the case of LiH and NaH applies to an even greater extent. KH may detonate because of its content of a small percentage of unhydrided dispersed K which then reacts with oxygen to form K_2O_2 [*6a, 6b*].

6. Rubidium and cesium hydrides

6.1 Preparation of rubidium and cesium hydrides

These compounds are formed from the elements in exothermic reactions [*93*]. Rubidium and cesium do not react with hydrogen at 20°, the hydrogenation reactions begin at 300°, and take place only very slowly between 300 and 400°.

For this reason, rubidium hydride, for example, is not prepared in a stream of hydrogen but in stationary hydrogen (vacuum-tight apparatus, Rb in Ni-boats). Both the metal and the electrolytic hydrogen must therefore be purified with particular care [*123*]. If relatively large amounts of hydride are to be prepared, the metal is diluted for better distribution, preferably with an inert material, and best with the hydride itself.

In place of the metals, the corresponding carbonates can also be used as the starting materials for the production of the hydrides. For this purpose, a 1:2 mixture of the carbonate with magnesium powder is slowly heated up to 260° (Rb_2CO_3) or 300° (Cs_2CO_3), which gives a mixture of the alkali metal with magnesium carbonate:

$$MgCO_3 + Mg \rightarrow MgCO_3 + 2M$$

This mixture can then be hydrogenated at 600 to 650° (the mixture does not take up hydrogen at 300 to 400° [*31*].

An improvement of the method consists in, for example, the reaction of a 3:1 mixture of Mg and Rb_2CO_3 in iron boats at 620° which has been well dried at 150° (reduction of the Rb_2CO_3 to the metal). Under these conditions, rubidium distils off and is condensed in a steel tube. After cooling, the metal is slowly evaporated in an atmosphere of hydrogen, at 300 to 400°. Hydrogenation takes place completely only if the metal is evaporated slowly [*161*].

6.2 *Physical properties of rubidium and cesium hydrides*

Some physical properties of rubidium and cesium hydrides are given in Table 2.3.

Rubidium hydride forms colourless prismatic needles, which appear white from a certain thickness onwards. Cesium hydride forms colourless, highly lustrous, flattened needle-shaped crystals. Any coloration of the otherwise pure white (colourless) hydrides that sometimes appears must be ascribed to the presence of small amounts of the free metals. RbH and CsH form ionic lattices of the NaCl type: $(Rb^{\oplus}, Cs^{\oplus})H^{\ominus}$. The melting points cannot be determined because of decomposition. The hydrides are insoluble in organic solvents.

6.3 *Reactions of rubidium and cesium hydrides*

Halogens. With fluorine, chlorine, and bromine, RbH and CsH burn to form the corresponding halides; iodine reacts only on heating [*106*].

Hydrogen chloride. With RbH, HCl forms rubidium chloride and hydrogen [*106*].

Oxygen. RbH and CsH are stable to dry air and oxygen at 20°; they burn to give the oxide (with pronounced emission of light in pure oxygen) [*106*].

Water. Rubidium hydride and cesium hydride react vigorously with water [*106, 137*]:

$$RbH + H_2O \rightarrow RbOH + H_2$$

$$CsH_{solid} + H_2O_{liquid} \rightarrow CsOH_{\infty dilution} + H_{2gas} + Q_H$$

$$Q_H = 48.33 \pm 0.06 \text{ kcal/mole} (25°).$$

References p. 38

S and H_2S. With sulphur, rubidium hydride burns to give the sulphide[*106*]. Hydrogen sulphide and rubidium hydride give rise to rubidium sulphide and hydrogen, in a highly exothermic reaction[*106*].

Nitrogen. With dry nitrogen, RbH reacts on gentle heating with the formation of a mixture of azide and amide[*106*], and CsH with the formation of amide and metallic Cs[*106*], while under the same conditions LiH forms lithium nitride.

Ammonia. Gaseous ammonia reacts with RbH even at 20°, forming rubidium amide and hydrogen. The reaction also takes place quantitatively with liquid ammonia[*106*].

P and As. With RbH, liquid phosphorus and liquid arsenic form the phosphide and arsenide[*106*]. On the other hand, with phosphorus CsH forms $CsPH_2$.

CO, C, Si and B. Carbon, silicon, and boron do not react with RbH below its dissociation temperature[*106*]. On gentle warming, RbH gives the formate with carbon monoxide[*106*].

Oxides. Rubidium hydride is a powerful reducing agent; it reduces the oxides of copper and lead to the corresponding metals on gentle heating, with the emission of light[*106*].

Organic compounds. The behaviour of rubidium and cesium hydrides towards organic substances has not been investigated in detail. As with the other alkali-metal hydrides, one would perhaps expect reducing and condensing properties. Alcohols and carboxylic acids form alkoxides and salts.

7. Spectra of the alkali-metal hydrides

The infrared absorption spectra and the molecular beam resonance of the alkali-metal hydrides have been measured, and the force constants, the dipole moments, the energy levels, and the ionic radii have been calculated[*17, 44, 70, 81, 83, 100, 153*].

REFERENCES

1. Adams, J. T. and C. R. Hauser, J. Am. Chem. Soc. *67*, 284 (1945).
2. Addison, C. C. and J. A. Manning, J. Chem. Soc. *1964*, 4887.
3. Addison, C. C., R. J. Pulham and R. J. Roy, J. Chem. Soc. *1964*, 4895.
3a. Addison, C. C., R. J. Pulham and R. J. Roy, J. Chem. Soc. *1965*, 116.
4. Ahmed, M. S., Nature *165*, 246 (1950).
5. Albert, P. and J. Mahé, Bull. Soc. Chim. France *1950*, 1165.
6. Alder, B. J. and R. H. Christian, Phys. Rev. *104*, 550 (1956).
6a. Ashby, E. C., Chem. Eng. News, *1969*, 9.
6b. Ashby, E. C. and B. D. James, private communication.
7. Bach, F. and K. F. Bonhoeffer, Z. Physik. Chem. *B23*, 256 (1933).
8. Bachman, G. B. and R. I. Hoaglin, J. Org. Chem. *8*, 300 (1943).
9. Banus, M. D., J. J. McSharry and E. A. Sullivan, J. Am. Chem. Soc. *77*, 2007 (1955).
10. Barber, W. A. and C. L. Sloan, J. Phys. Chem. *65*, 2026 (1961).
11. Bardwell, D. C., J. Am. Chem. Soc. *44*, 2499 (1922).
12. Baughan, E. C., Trans. Faraday Soc. *55*, 736 (1959).
13. Behringer, R. E., Phys. Rev. *113*, 787 (1959).
14. Billy, M., Ann. Chim. Phys. [9] *16*, 5 (1929).
15. Bode, H., Z. Physik. Chem. *B13*, 99 (1931).
16. Brandt, P., Acta Chem. Scand. *3*, 1050 (1949).
17. Brandt, W., Phys. Rev. *111*, 1042 (1958).

18. Brown, H. C., E. J. Mead and P. A. Tierney, J. Am. Chem. Soc. *79*, 5400 (1957).
19. Browne, J. C., J. Chem. Phys. *36*, 1814 (1962).
20. Daub, G. H. and W. S. Johnson, J. Am. Chem. Soc. *70*, 418 (1948).
21. Davy, H., Phil. Trans. *98*, 333 (1808).
22. Davy, H., Ann. Chim. *75*, 264 (1810).
23. Douglas, T. B., L. F. Epstein, J. L. Dever and W. H. Howland, J. Am. Chem. Soc. *77*, 2144 (1955).
24. Dymova, T. N. and A. A. Vysheslavtsev, Russ. J. Inorg. Chem. English Transl. *5*, 1045 (1960).
25. Ebbing, D. D. (Chem. Dept., Indiana Univ., Bloomington, Ind.), *Symposium on Molecular Structure and Spectroscopy*, Ohio State University, Columbus, 1960.
26. Ebbing, D. D., J. Chem. Phys. *36*, 1361 (1962).
27. Edmondson, W. and A. Egerton, Proc. Roy. Soc. (London) *A133*, 522 (1927).
28. Ehrlich, P. and W. Deissmann, Naturwiss. *51*, 135 (1964).
29. Elson, R. E., H. C. Horning, W. L. Jolly, J. W. Kury, W. J. Ramsey and A. Zalkin, UCRL-4519 (1955).
30. Elson, R. E., H. C. Horning, W. L. Jolly, J. W. Kury, W. J. Ramsey and A. Zalkin, UCRL-4519 revised (1956).
31. Ephraim, F. and E. Michel, Helv. Chim. Acta, *4*, 762 (1921).
32. Ewing, D. H. and F. Seitz, Phys. Rev. [*2*] *50*, 760 (1936).
33. Fallon, R. J., J. T. Vanderslice and E. A. Mason, J. Chem. Phys. *32*, 1453 (1960).
34. Fallon, R. J., J. T. Vanderslice and E. A. Mason, J. Chem. Phys. *33*, 944 (1960).
35. Ferelli, E., T. G. Pearson and P. L. Robinson, J. Chem. Soc. *1934*, 7.
36. Filler, A. S. and E. Burstein, Bull. Am. Phys. Soc. [*2*] *5*, 198 (1960).
37. Freed, S., J. Am. Chem. Soc. *52*, 2702 (1930).
38. Freudenberg, H. and H. Kloepfer, Brit. Pat. 276,313 and 283,089 (1927); C. A. *22*, 3964 (1928).
39. Freudenberg, H. and H. Kloepfer (Deutsche Gold- und Silberscheideanstalt), U.S. Pat. 1,796,265 (1927); C. A. *25*, 2528 (1931).
40. Frey, F. W., P. Kobetz, G. C. Robinson and T. O. Sistrunk, J. Org. Chem. *26*, 2950 (1961).
41. Forcrand, M. de, Compt. Rend. *140*, 990 (1905).
42. Gay-Lussac, L. J. and L. J. Thenard, Mem. Soc. Arcueil *2*. 304 (1809); cited in [*22*] and [*31*].
43. Gettert, H. and G. Hamprecht (Badische Anilin und Soda-Fabriken), Germ. Pat. 833,956 (1952); Chem. Zentr. *123*, 5153 (1952).
44. Getty, R. R. and J. C. Polanyi, Trans. Faraday Soc. *57*, 2099 (1961).
45. Gibb, T. R. P., U.S. Pat. 2,468,260 (1947).
46. Gibb, T. R. P., Trans. Electrochem. Soc. *93*, 198 (1948).
47. Gilman, H., A. L. Jacoby and H. Ludeman, J. Am. Chem. Soc. *60*, 2336 (1938).
48. Griggs, D. T., W. G. McMillan, E. D. Michael and C. P. Nash, Phys. Rev. *109*, 1858 (1958).
49. Gunn, S. R. and L. G. Green, J. Am. Chem. Soc. *80*, 4782 (1958).
50. Guntz, A., Compt. Rend. *123*, 694 (1896).
51. Guntz, A., Compt. Rend. *123*, 997 (1896).
52. Guntz, A., Compt. Rend. *123*, 1273 (1896).
53. Hackspill, L. and A. Borocco, Compt. Rend. *204*, 1475 (1937).
54. Hader, R. N., R. L. Nielsen and M. G. Herre, Ind. Eng. Chem. *43*, 2636 (1951).
55. Hagen, H. and A. Sieverts, Z. Anorg. Allgem. Chem. *135*, 254 (1929).
56. Hansley, V. L. (E. I. DuPont de Nemours & Co.). U.S. Pat. 2,158,071 (1936/39); C. A. *33*, 6342^4 (1939).
57. Hansley, V. L. (E. I. DuPont de Nemours & Co.). U.S. Pat. 2,211,419 (1939); C. A. *35*, 463 (1941). C. A. *35*, 463^8 (1941).
58. Hansley, V. L. (E. I. DuPont de Nemours & Co.), U.S. Pat. 2,218,026 (1936/40); C. A. *35*, 1066^4 (1941).
59. Hansley, V. L. (E. I. DuPont de Nemours & Co.), U.S. Pat. 2,372,670 (1945); C. A. *39*, 3129 (1945). U.S. Pat. 2,372,671 (1945); C. A. *39*, 3129 (1945); U.S. Pat. 2,504,927 (1950); C. A. *44*, 8606 (1950).
60. Hansley, V. L. and P. J. Carlisle, Chem. Eng. News *23*, 1332 (1945).
61. Herold, A., Ann. Chim. *6*, 537 (1951).
62. Herold, A., Compt. Rend. *228*, 686 (1949).
63. Hodaghian, A. and R. Levaillant, Compt. Rend. *194*, 2059 (1932).

64. Hodgman, C. D., R. C. West and S. M. Selby, Handbook of Chemistry and Physics, Chemical Rubber Publishing Co. Cleveland, Ohio, 1956.
65. Hüttig, G. F. and A. Krajewski, Z. Anorg. Allgem. Chem. *141*, 133 (1924).
66. Hugel, G., M. Boistel and M. Pane, Germ. Pat. 538,763 (1928).
67. Hurd, D. T., J. Am. Chem. Soc. *71*, 20 (1949).
68. Hurst, R. P., J. Miller and F. A. Matsen, J. Chem. Phys. *26*, 1092 (1957).
69. Jackson, H. L., F. D. Marsh and E. L. Muetterties, Inorg. Chem. *2*, 43 (1963).
70. James, T. C., W. G. Norris and W. Klemperer, J. Chem. Phys. *32*, 728 1960).
71. Jander, W., Z. Angew. Chem. *42*, 462 (1929).
72. Jenker, H. (Kali Chemie, Hannover), Germ. Pat. 1,087,118 (1959).
72a. Johnson, C. E., S. E. Wood and C. E. Crouthamel, Inorg. Chem. *3*, 1487 (1964).
72b. Johnson, C. E., S. E. Wood and C. E. Crouthamel, Chem. Phys. *44*, 880 (1966).
72c. Johnson, C. E., S. E. Wood and C. E. Crouthamel, Chem. Phys. *44*, 884 (1966).
73. Kanseley, H. I., Chem. Eng. News *23*, 1332 (1945).
74. Kapustinsky, A. F., L. M. Shamovsky and K. S. Bayushkina, Acta Physiochim. SSSR *7*, 799 (1937).
75. Kasarnowsky, J., Z. Anorg. Allgem. Chem. *170*, 311 (1928).
76. Kasarnowsky, J., Z. Physik *61*, 236 (1930).
77. Karo, A. M., J. Chem. Phys. *31*, 182 (1959).
78. Karo, A. M., J. Chem. Phys. *32*, 907 (1960).
79. Keyes, F. G., J. Am. Chem. Soc. *34*, 779 (1912).
80. Khan, I. A., Y. W. Gokhale and D. Sen, J. Sci. Ind. Res. (India) *19B*, 166 (1960); C. A. *54*, 24063i (1960).
81. Klemperer, W., J. Chem. Phys. *23*, 2452 (1955).
82. Kobetz, P. and W. E. Becker, Inorg. Chem. *2*, 859 (1963).
83. Kordes, E., Z. Physik. Chem. (Frankfurt) *33*, 1 (1962).
84. Krasnov, K. S. and V. G. Antoshkin, Zh. Neorgan. Khim. *3*, 1490 (1958); C. A. *53*, 17607 (1959).
85. Landa, C. L., F. Petru, J. Mostecky, J. Vit and V. Prochatzka, Collection Czech. Chem. Commun. *24*, 2037 (1959).
86. Levine, R., J. A. Conroy, J. T. Adams and C. R. Hauser, J. Am. Chem. Soc. *67*, 1510 (1945).
87. Levine, R. and C. R. Hauser, J. Am. Chem. Soc. *66*, 1768 (1944).
88. Lewis, W. B. and F. E. Pretzel, Phys. Chem. Solids *19*, 139 (1961).
89. Losana, L., Gazz. Chim. Ital. *65*, 855 (1935).
90. Lundquist, S. O., Arkiv Fysik *8*, 177 (1954).
91. Messer, C. E., E. B. Damon, P. C. Mayburry, J. Mellor and R. A. Seales, J. Phys. Chem. *62*, 220 (1958).
92. Messer, C. E., L. G. Fasolino and C. E. Thalmayer, J. Am. Chem. Soc. *77*, 4524 (1955).
93. Messer, C. E., H. Homonoff, R. F. Nickerson and T. R. Gibbs Jr., Tufts Coll. Issued by TIS, Oak Ridge, NYO-3955 (1953).
93a. Messer, C. E. and K. Hardcastle, Inorg. Chem. *3*, 1327 (1964).
93b. Messer, C. E. and I. S. Levy, Inorg. Chem. *4*, 543 (1964).
94. Metal Hydrides Inc., Beverly, Mass., Technical Bull. 507-C.
95. Metal Hydrides Inc., Beverly, Mass., Technical Bull. 508-A.
96. Mikheeva, V. I., T. N. Dymova and M. M. Shkrabkina, Russ. J. Inorg. Chem. English Transl. *4*, 323 (1959).
97. Mikheeva, V. I. and M. M. Shkrabkina, Russ. J. Inorg. Chem. English Transl. *7*, 238 (1962).
98. Mikheeva, V. I. and M. M. Shkrabkina, Russ. J. Inorg. Chem. English Transl. *7*, 1251 (1962).
99. Miller, J., R. H. Friedmann, R. P. Hurst and F. A. Matsen, J. Chem. Phys. *27*, 1385 (1957).
100. Mitra, S. S. and Y. P. Varshni, J. Chem. Phys. *22*, 1269 (1954).
101. Moers, K., Z. Anorg. Allgem. Chem. *113*, 173 (1920).
102. Moissan, H., Compt. Rend. *134*, 18 (1902).
103. Moissan, H., Compt. Rend. *134*, 71 (1902).
104. Moissan, H., Ann. Chim. Phys. [*7*] *27*, 349 (1902).
105. Moissan, H., Compt. Rend. *136*, 587 (1903).
106. Moissan, H., Ann. Chim. Phys. [*8*] *6*, 297 (1905).
107. Morita, A. and K. Takahashi, Progr. Theoret. Phys. (Kyoto) *19*, 257 (1958); C. A. *52*, 15997 (1958).

108. Muckenfuss, A. M. (E. I. DuPont de Nemours & Co.) U.S. Pat. 1,958,012 (1931/34); C. A. *28*, 4185 (1934).
109. Muckenfuss, A. M., U. S. Pat. 2,073,973; C. A. *31*, 3254 (1937).
110. Muslin, B., D. K. Harriss and C. W. Mitchell (Southern Illinois Univ., Carbondale, Ill.), *Symposium on Molecular Structure and Spectroscopy*, Ohio State University, Columbus, 1960.
111. Nernst, W. and K. Moers, Z. Elektrochem. *26*, 323 (1920).
112. Ormand, F. T. and F. A. Matsen, J. Chem. Phys. *29*, 100 (1958).
113. Osborg, H., Trans. Electrochem. Soc. *66*, 91 (1934).
114. Palatas, O., R. P. Hurst and F. A. Matsen, J. Chem. Phys. *31*, 501 (1959).
115. Palatas, O. and F. A. Matsen, J. Chem. Phys. *29*, 965 (1958).
116. Pearce, D. W., R. E. Burns and E. St. Clairganz, Proc. Indiana Acad. Sci. *58*, 99 (1949; C. A. *44*, 2879e (1950).
117. Peters, K., Z. Anorg. Allgem. Chem. *131*, 140 (1923).
118. Polanyi, J. and C. M. Sadowski, J. Chem. Phys. *36*, 2239 (1962).
119. Pretzel, F. E., G. V. Gritton, C. C. Rushing, R. J. Friauf, W. B. Lewis and P. Waldstein, Phys. Chem. Solids *23*, 325 (1962).
120. Pretzel, F. E., W. B. Lewis, E. G. Szklarz and D. T. Vier, J. Appl. Phys. *33*, 510 (1962).
121. Pretzel, F. E., G. N. Rupert, C. L. Mader, G. V. Gritton and C. C. Rushing, Phys. Chem. Solids *16*, 10 (1960).
122. Pretzel, F. E. and C. C. Rushing, Phys. Chem. Solids *17*, 232 (1961).
123. Proskurnin, M. and J. Kasarnowsky, Z. Anorg. Allgem. Chem. *170*, 302 (1928).
124. Rengade, E., Compt. Rend. *156*, 1897 (1913); Bull. Soc. Chim. France [*4*] *15*, 145 (1914).
124a. Rühlmann, R. and H. Heine, Z. Chem. *6*, 427 (1966).
125. Ruff, O. and E. Geisel, Ber. *39*, 842 (1906).
126. Ruff, O. and E. Geisel, Ber. *44*, 502 (1911).
127. Roessler & Hasslacher Chemical Co., Eng. Pat. 405.017 (1932); C. A. *28*, 4545 (1934).
128. Rossini, F. D. *et al.*, Natl. Bur. Std. (U.S.), Circ. No. 500 (1952).
129. Schechter, W. H., U.S. Pat. 2,929,679 (1960).
130. Schmidt, H. W. and H. Jenkner (Kali Chemie A.G.), Brit. Pat. 875,103 (1960); C.A. *55*, 22733e (1961).
131. Schmidt, H. W. and H. Jenkner (Kali Chemie A.G.), Germ. Pat. 1,078,099 (1959); C. A. *55*, 22733e (1961).
132. Schmitz-Dumont, O. and V. Habernickel, Chem. Ber. *90*, 1054 (1957).
133. Schumacher, D. P., Phys. Rev. *126*, 1679 (1962).
134. Sherman, J., Chem. Rev. *11*, 93 (1932).
135. Shull, H., J. Appl. Phys., Suppl. *33*, 290 (1962).
136. Siegmann, F., Germ. Pat. 730,329 (1939); U.S. Pat. 2,313,028 (1940).
137. Smith, M. B. and G. E. Brass, Jr., J. Chem. Eng. Data *8*, 342 (1963).
138. Sollers, E. F. and J. I. Crenshaw, J. Am. Chem. Soc. *59*, 2015 (1937).
139. Sollers, E. F. and J. I. Crenshaw, J. Am. Chem. Soc. *59*, 2724 (1937).
140. Staritzky, E. and D. Walker, Anal. Chem. *28*, 1055 (1956).
141. Stevenson, D. P., J. Chem. Phys. *8*, 898 (1940).
142. Stobbe, E., Liebigs Ann. Chem. *282*, 280 (1894).
143. Stobbe, H., Liebigs Ann. Chem. *308*, 89 (1899).
144. Swammer, F. W. and C. R. Hauser, J. Am. Chem. Soc. *72*, 1352 (1940).
145. Szigeti, B., Trans. Faraday Soc. *45*, 155 (1949).
146. Szigeti, B., Proc. Roy. Soc. (London) Ser. A *204*, 51 (1950).
147. Taylor, J. B. and I. Langmuir, Phys. Rev. [2] *51*, 756 (1937).
148. Tiede, E., Germ. Pat. 417,019 (1924/25); Chem. Zenfr. *96*, 1790 (1925).
149. Thronstadt, L., cited at 'Dissociation pressure of LiD' in Gmelin "Li", Ergänzungsbuch, Verlag Chemie, Weinheim/Bergstrasse, 1960, p. 258.
150. Troost, L. and P. Hautefeuille, Compt. Rend. *78*, 807 (1874).
151. Troost, L. and P. Hautefeuille, Ann. Chim. Phys. [5] *2*, 274 (1874).
152. Vining, W. H. (E.I. DuPont de Nemours & Co.) U.S. Pat. 2,474,021 (1949); C. A. *43*, 7202 (1949).
153. Wharton, L., L. P. Gold and W. Klemperer, J. Chem. Phys. *33*, 1255 (1960).
154. Wallingford, V. H., A. H. Homeyer and D. M. Jones, J. Am. Chem. Soc. *63*, 2252 (1941).
155. Weil, R. and A. W. Lawson, J. Chem. Phys. *37*, 2730 (1962).
156. Weiner, M., G. Vogel and R. West, Inorg. Chem. *1*, 654 (1962).
157. Williams, D. D., J. A. Grand and R. R. Miller, J. Phys. Chem. *61*, 379 (1957).

158. Wilzbach, K. E. and L. Kaplan, J. Am. Chem. Soc. *72*, 5795 (1950).
159. Wittig, G. and P. Hornberger, Ann. *577*, 11 (1952).
160. Ziegler, K. and H. G. Gellert, Ann. *567*, 179 (1950).
161. Zintl, E. and A. Harder, Z. Physik. Chem. *B14*, 265 (1931).
162. Zintl, E. and A. Harder, Z. Physik. Chem. *B28*, 478 (1935).

Chapter 3

ALKALINE-EARTH METAL HYDRIDES

1. Historical

1.1 Beryllium hydride

The first attempts to prepare a beryllium hydride go back to Winkler [*118*] who, in 1891, heated beryllium oxide together with metallic magnesium in an atmosphere of hydrogen. Under these conditions he observed a slow absorption of hydrogen and the formation of a compound with an unpleasant smell. Since with boiling water or with aqueous hydrochloric acid it evolved hydrogen, Winkler believed that he had prepared beryllium hydride. However, Lebau [*69*], who repeated this experiment eight years later (1899), was unable to confirm the formation of a hydride.

Forty-two years after Winkler's experiments – in the meantime the preparation of calcium, strontium, and barium hydrides from the elements had been fully established – Pietsch [*86*] in 1933 treated metallic beryllium with atomic hydrogen at 170 to 260° for seven hours and obtained a small amount of a white material which he took to be beryllium hydride.

Holley, Jr. and Lemons [*65*] repeated this experiment in 1954 within the framework of a wider series of investigations, planned on the large scale, on the synthesis of beryllium hydride from the elements under various experimental conditions. In a first series of experiments, they treated extremely finely subdivided beryllium for several hours with atomic hydrogen at various temperatures. In another series of experiments, they kept beryllium vapour in an atmosphere of hydrogen at 1 to 100 torr, at 1700°, for a relatively long time. In a third series of experiments, they studied the behaviour of extremely finely powdered beryllium with respect to hydrogen at 20 to 600° and up to 1000 atm. Finally, they allowed an arc between beryllium electrodes to burn for 8 hours in an atmosphere of hydrogen at a pressure of 200 atm. In the experiments at elevated H_2 pressures, pure beryllium was replaced by a beryllium-uranium alloy. The formation of a beryllium-hydrogen compound was never observed.

The white or grey substances found in minute amounts in individual cases was very probably a mixture of beryllium oxide and beryllium nitride. In fact, this substance did not appear if particularly carefully purified hydrogen was used.

On the basis of these investigations it can be said with certainty today that the synthesis of beryllium hydride from the elements has not yet been performed.

On the other hand, beryllium dihydride BeH_2 is formed in the hydrogenation of beryllium compounds. In 1951, for the first time, Wiberg and Bauer [*110*] reported the preparation of beryllium dihydride by the treatment of beryllium dichloride with lithium hydride and Barbaras, Dillard, Finholt, Wartik, Wilzbach and

References p. 77

Schlesinger[*16*] reported the preparation of beryllium hydride by the reaction of dimethylberyllium with lithium alanate, $LiAlH_4$, or dimethylaluminium hydride in ether.

In 1949 in the pyrolysis of diethylberyllium, Goubeau and Rodewald[*59*] found beryllium hydrides containing organic groups, an oily material $(HBeCH_2CH_2BeH)_x$ and a crystalline material $[BeH(C_3H_6)]_x$. A few years later (1954), Coates and Glockling[*32*] succeeded in preparing BeH_2 itself by the pyrolysis of the more unstable di-*tert*-butylberyllium.

1.2 Magnesium hydride

In 1912, Jolibois[*67*] heated ethylmagnesium iodide, and observed the evolution of a mixture of gases from 175° onwards, the main constituent of which was ethylene. On further heating, hydrogen was evolved from 280° onwards, and metallic magnesium was formed simultaneously. The non-homogeneous substance which arose from the evolution of ethylene probably contained magnesium hydride and magnesium iodide, but it was impossible to isolate magnesium hydride from the reaction products.

Clapp and Woodward[*29*], in 1938, arrived at a similar result by decomposing ethylmagnesium bromide thermally in vacuum at 220°. The grey residue formed with the evolution of ethylene reduced benzophenone to benzhydrol.

Again in admixture with other compounds, magnesium hydride was prepared by Zartmann and Adkins[*122*] in 1932, by the reduction of diphenylmagnesium with hydrogen on a Ni-catalyst. The authors did not attempt to isolate and identify the hydrogen compound.

Pure magnesium dihydride MgH_2 was prepared only in 1950. For this purpose, Wiberg and Bauer[*108*] heated diethylmagnesium. One year later (1951), Wiberg *et al.*[*116*] succeeded in synthesizing magnesium dihydride from the elements. In 1951, also, Barbaras *et al.*[*16*] reported the formation of an ether-containing magnesium hydride when, in analogy with the preparation of beryllium hydride, diethyl-magnesium was made to react with lithium alanate.

1.3 Calcium, strontium and barium hydrides

As early as 1891, Winkler[*118*] observed that, when a mixture of calcium oxide and metallic magnesium was heated in an atmosphere of hydrogen, a compound between calcium and hydrogen was formed. He ascribed to it the formula CaH. Seven years later (1898), Lengyel[*71*] reported for the first time an absorption of hydrogen by metallic calcium, beginning at 20° and vigorous at dark red heat. We owe the first preparative method for the production of CaH_2 from the elements to Moissan[*78–80*] (1899). In 1902, Gautier[*55*] succeeded in showing that CaH_2 can also be obtained from Ca–Cd alloys when these are treated with hydrogen under suitable conditions. The first patent[*42*] describing the technical production of calcium dihydride was granted as early as 1905. In the same year, Guntz[*61*] found that strontium hydride is also formed when strontium amalgam is heated with hydrogen at 500°. This process was improved by Roederer[*91*] four years later (1906); he first distilled off the bulk of the mercury at 700° and only then hydrogenated the residual porous strontium-rich residue at the same temperature.

2. General review

The hydrides of the heavy alkaline-earth metals Ca, Sr, and Ba somewhat resemble the alkali-metal hydrides. Like these, they can easily be obtained from the elements, they are also salt-like and they react in the same way as the alkali-metal hydrides (see Chapter 2, Section 2.3) via the hydride ion $H^{\ominus}$ liberated by dissociation. In addition, Ca, Sr, and Ba, and also Mg, form salt-like hydride halides EHX (X = Cl, Br, I).

Magnesium hydride MgH_2, which is more highly covalent than CaH_2, SrH_2, and BaH_2, occupies an intermediate position between the ionic salt-like Ca, Sr, and Ba hydrides and the decidedly covalent beryllium hydride BeH_2, which is closer to aluminium hydride AlH_3. Thus, MgH_2 can still be obtained from the elements, but this is not possible with beryllium. MgH_2 and its derivatives, in analogy with the preparation of BeH_2, but in contrast to CaH_2, SrH_2, and BaH_2, can also be prepared from the dialkyl compounds. Thus, they are formed from MgR_2 on pyrolysis and on hydrogenation with lithium alanate, monosilane, or diborane:

$$Mg(C_2H_5)_2 \rightarrow MgH_2 + 2C_2H_4$$

$$2Mg(C_2H_5)_2 + LiAlH_4 \rightarrow 2MgH_2 + Li[Al(C_2H_5)_4]$$

$$Mg(C_2H_5)_2 + O(C_2H_5)_2 + SiH_4 \rightarrow Mg(OC_2H_5)H + C_4H_{10} + SiH_3(C_2H_5)$$

$$6Mg(C_2H_5)Cl + (BH_3)_2 \rightarrow 6MgHCl + 2B(C_2H_5)_3$$

The reaction of dialkylmagnesiums with diborane is complex. This is again in harmony with the intermediate position of magnesium between the elements forming covalent unstable boranates, such as Al, on the one hand, and the elements forming salt-like stable boranates, such as Na, on the other hand. A summary of the reaction between $(C_2H_5)_2Mg$ and $(BH_3)_2$ is shown in Fig. 3.1 (Section 4.1.3).

The covalent beryllium dihydride BeH_2 ('beryllane') resembles the covalent hydrides of boron and aluminium. Like the latter, it cannot be obtained from the elements but only by the pyrolysis or hydrogenation of BeR_2:

$$Be(C_2H_5)_2 \rightarrow BeH_2 + 2C_2H_4$$

$$2Be(CH_3)_2 + LiAlH_4 \rightarrow 2BeH_2 + Li[Al(CH_3)_4]$$

As strong electron-acceptors, BeH_2, BeRH, and BeR_2 either form three-centre, two-electron bonds, or add electron donors. Under these conditions, the coordination number 4 always appears (*e.g.* addition of ether to terminal Be atoms)[1]:

[1]For the exact meaning of the symbolism Be–H–Be (H bridging), *cf* section 1.1 of Chapter 4.

References p. 77

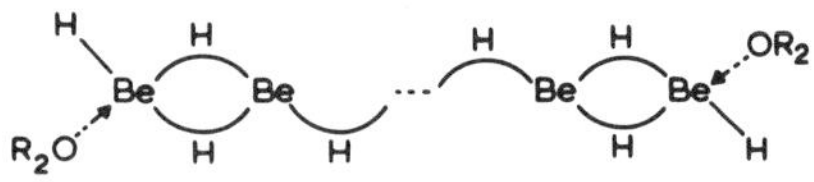

Tertiary amines do not split the Be–H^{μ}–Be bond[2], but they do split the Be–alkyl$^{\mu}$–Be bond.

According to the difference in electronegativity (H = 2.20, Be = 1.47, Mg = 1.23, Ca = 1.04, Sr = 0.99, Ba = 0.97 [Pauling units]), the hydrogen in all the alkaline-earth metal hydrides is decidedly negative. Accordingly, with proton-active compounds they evolve molecular hydrogen.

3. Beryllium hydride and deuteride

3.1 Preparation of BeH_2 and derivatives

There are two methods for the preparation of beryllium hydride:

1) The reaction of dialkylberylliums with hydrogenating agents in ether

$$BeR_2 \xrightarrow[-R^{\ominus}]{+H^{\ominus}} BeRH \xrightarrow[-R^{\ominus}]{+H^{\ominus}} BeH_2$$

2) The pyrolysis of dialkylberylliums (formation of alkenes and BeH_2). In both methods, partially alkylated beryllium hydrides may be obtained under certain conditions. The first method does not give pure beryllium hydride but always ether-containing products. The second method gives ether-free beryllium hydride only if the *whole* of the synthesis is carried out in the absence of ether and if, therefore, for example, the dialkylberyllium is prepared in dimethyl sulphide instead of diethyl ether. As with the neighbouring elements B and Al, which have a strong covalent bond component in the crystal lattice, the preparation of BeH_2 from the elements is impossible.

3.1.1 Preparation of BeH_2, $BeRH$, and BeD_2 by the hydrogenation and deuteration of BeR_2

When an ethereal solution of dimethylberyllium is added in drops to an ethereal solution of $LiAlH_4$ or $LiAlD_4$, white beryllium dihydride or dideuteride immediately separates out under evolution of a small quantity of heat in a form which settles rapidly and well [*16*].

$$2Be(CH_3)_2 + LiAlH_4 \rightarrow 2BeH_2 + Li[Al(CH_3)_4]$$
$$2Be(CH_3)_2 + LiAlD_4 \rightarrow 2BeD_2 + Li[Al(CH_3)_4]$$

If the reaction mixture is stirred vigorously for several hours, and the BeH_2 formed is then filtered off with the exclusion of moisture, washed several times

---

[2] The superscript μ indicates atoms or groups of atoms which are linked to their two neighbours through a three-centre two-electron bridge.

with ether, and dried in vacuum, the beryllium hydride contains neither methyl groups nor aluminium. Only the ether cannot be removed completely, for which reason the hydride prepared in this way has a very variable content of BeH_2 which is, however, usually about 50% by weight. BeH_2 prepared in this way is amorphous to X-rays. Since 74%-by-weight BeH_2 (which, as can be seen from the composition deduced from the analysis $BeH_2:O(C_2H_5)_2 = 19:1$, contains hydride in predominating amount) is stable, completely ether-free BeH_2 is probably stable as well[*16,65*]. The ether bound to beryllium dihydride can be removed neither by prolonged heating in vacuum or under a pressure of hydrogen nor by treatment with aluminium chloride, aluminium boranate, $Al(BH_4)_3$, or trimethylamine[*65*]. The reaction of dimethyl-beryllium with lithium alanate in solvents other than diethyl ether does not give pure beryllium dihydride either (solvents used: Me_2O, iso-Pr_2O, **n**-Bu_2O, $MeOCH_2OMe$, $MeOCH_2CH_2OMe$, $MeO(CH_2CH_2O)_3Me$, $MeO(CH_2CH_2O)_4Me$, $\overline{CH_2CH_2CH_2CH_2O}$, $\overline{CH{=}CHCH{=}CHO}$, Me_2N, Et_3N, C_5H_5N, Et_2S, and fluorinated hydrocarbons[*65*].

If, with a view to preparing ether-free beryllium hydride, dimethyl-beryllium is treated with (liquid) dimethylalane, $(CH_3)_2AlH$, in the absence of a solvent, again no pure BeH_2 is obtained[*16*]. The product in fact still contains methyl groups and aluminium:

$$Be(CH_3)_2 + AlH(CH_3)_2 \rightarrow Be(CH_3)H + Al(CH_3)_3$$
$$Be(CH_3)H + AlH(CH_3)_2 \rightarrow BeH_2 + Al(CH_3)_3$$

The content of aluminium is ascribed to the tendency of dimethylalane to polymerize or decompose with the formation of non-volatile glassy products which cannot be separated from the BeH_2.

Methylberyllium hydride, $Be(CH_3)H$, can be isolated as a solid substance when an excess of dimethylberyllium is treated with dimethylalane[*16*]. $Be(CH_3)_2$, $BeBr_2$ and LiH form also $Be(CH_3)H$[*24c*]. It reacts with water to form methane, hydrogen and beryllium hydroxide[*16*].

$$Be(CH_3)H + 2HOH \rightarrow Be(OH)_2 + CH_4 + H_2$$

The physical properties and chemical reactions of methylberyllium hydride, $Be(CH_3)H$, have not so far been investigated further.

Hydrogenating agents other than $LiAlH_4$ either do not react with dialkylberyllium or do not give pure beryllium dihydride. Thus, for example, aluminium boranate, $Al(BH_4)_3$, does not react with diethylberyllium in benzene solution with or without the addition of ether[*65*]. An ethereal solution of aluminium hydride, AlH_3, highly diluted because of its tendency to polymerization, does not give pure BeH_2[*65*]. Dimethylberyllium does not react with dimethylsilan, $SiH_2(CH_3)_2$, in dioxane or in the absence of a solvent[*65*]. With hydrogen, $Be(CH_3)_2$ at 28 kg/cm^2 and 300° and $Be(C_2H_5)_2$ at 427 kg/cm^2 and 400° give only Be_2C and no BeH_2; even in the presence of Raney nickel, no hydride is produced [*65*].

In an attempt to prepare beryllium hydride, in analogy with many other hydrides

References p. 77

(as, for example, SiH_4), by the reaction of the ether-soluble $BeCl_2$ with ether-insoluble finely ground LiH, in addition to insoluble LiCl a likewise insoluble white solid is formed [*110*]. However, this is not pure BeH_2 or $BeH_2.nLiH = Li_nBeH_{2+n}$ [*110*]. On the contrary, these hydrides are probably mixed with still unchanged particles of LiH and with LiCl–$BeCl_2$ double salts that have been formed [*65, 110*]. When the diethyl ether is replaced as solvent by triglyme[3], which dissolves LiH better, decomposition and reaction products of the solvent are also formed [*65*]. The reaction of $BeCl_2$ with $LiBH_4$ at 90–140° gives no BeH_2, but $Be(BH_4)_2$ in 80% yield [*94a*]. $Be(BH_4)_2$ forms complexes $D{\cdot}Be(BH_4)_2$ with donors D (Et_2O, Me_3P, Me_2PH, Et_3P, Ph_3P, Me_3N, Me_2NH) [*15a, 15b*].

3.1.2 Preparation of BeH_2 and BeRH by the pyrolysis of dialkylberylliums

Alkenes and beryllium dihydride are formed when dialkylberylliums are heated, and intermediate stages – alkylberyllium hydrides – can be isolated in many cases. The following is a schematic example of the reaction [15, 32–34, 59]:

$$Be(CH_2CH_3)_2 \xrightarrow{-CH_2=CH_2} BeH(CH_2CH_3) \xrightarrow{-CH_2=CH_2} BeH_2$$

The thermal stability of the dialkyl compounds $Be[CH_{3-n}(CH_3)_n]_2$ falls with increasing substitution of hydrogen in the alkyl group by methyl (increasing n [*15*]. Thus, the temperature of the beginning of decomposition falls in the sequence $Be(CH_3)_2$ (190–200°) [*34*] > $Be[CH_2(CH_3)]_2$ (85°) [*59*] > $Be[CH(CH_3)_2]_2$ (50°) [*33*] > $Be[C(CH_3)_3]_2$ (40°) [*32*]. To prepare beryllium dihydride, however, pyrolysis generally is carried out at a higher temperature, about 200°, near the decomposition point of BeH_2.

In beryllium compounds the coordination number 4 is favoured [*14*]. On heating to 190°, dimethylberyllium, which is highly associated in the solid [*70, 100*] and liquid [*15*] states and is still trimeric [*34*] and dimeric [*34*] in the gaseous state

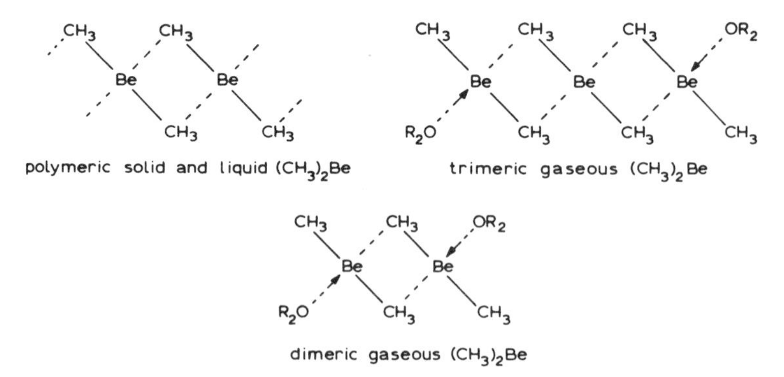

yields beryllium dihydride and ethylene [*32, 34*]:

$$Be(CH_3)_2 \rightarrow BeH_2 + CH_2{=}CH_2$$

[3]Triethyleneglycol dimethyl ether.

Calculations indicate that BeH_2 has a structure analogous to one of the polymeric forms of $Be(CH_3)_2$ (see above)[*70*]; for further details see the next section 3.2).

Between 85 and 200°, diethylberyllium, which is also associated in the liquid state and which adds ether extraordinarily tenaciously (to terminal? Be atoms) [*59*]:

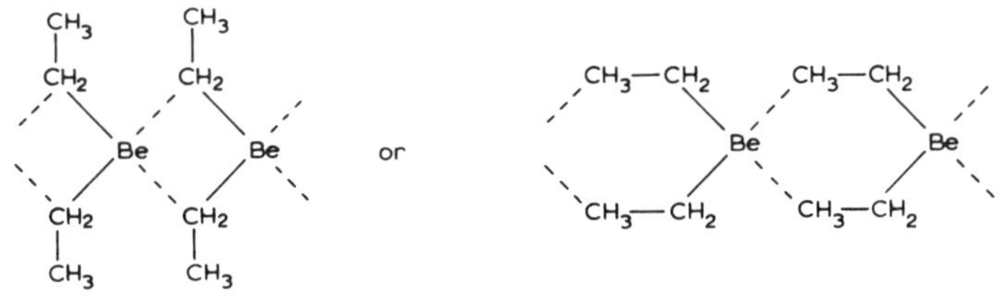

or

gives volatile products (essentially ethane and ethylene together with small amounts of methane, butane, and butene) together with a yellow viscous oil and a small amount of a yellow solid[*32, 59, 65*]. The faintly yellow oil, which can be distilled at 96–105° (0·1 torr) reacts vigorously with water to form hydrogen, ethane, ethylene, and beryllium hydroxide. This compound is a derivative of beryllium hydride which is formed in a complicated reaction and has the approximate composition $H{-}Be{-}CH_2{-}CH_2{-}Be{-}H$. The solid yellow residue, which yields hydrogen, an oily hydrocarbon, and beryllium hydroxide on hydrolysis, has the approximate composition[*59*]: $[HBe_2(CH_2)_3]_x$ or $[Be_2(CH_2)_3]_x$.

Di-isopropylberyllium, which is also highly associated and also binds ether very strongly

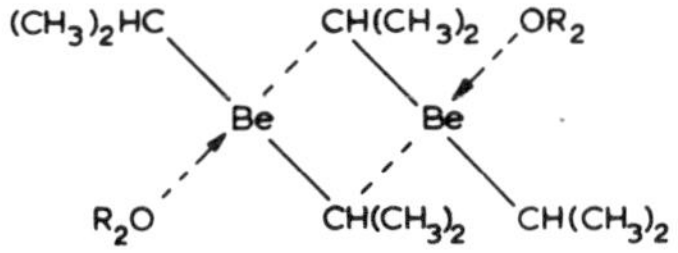

dimeric isopropylberyllium [*33*]

decomposes at 200° in a substantially simpler reaction than diethylberyllium to give isopropylberyllium hydride and propene[*33*]:

$$Be[CH(CH_3)_2]_2 \rightarrow BeH[CH(CH_3)_2] + CH(CH_3){=}CH_2$$

Isopropylberyllium hydride, $BeH[CH(CH_3)_2]$, is a non-volatile colourless viscous oil, which indicates a polymeric structure (as was to be expected). It reacts with water to form propane, hydrogen, and beryllium hydroxide. If the compound is heated to a higher temperature, no beryllium dihydride is produced, but between 220 and 250° it yields beryllium, hydrogen, propane, propylene, and an orange residue in a complicated reaction[*33*].

In contrast to the dialkyl compounds described above, the pyrolysis of the still less stable di-*tert*-butylberyllium. $Be[C(CH_3)_3]_2$, forms no pure alkylberyllium hydride[*32, 63*] but a mixture of iso- and tert-butylberyllium hydrides[*34a*]. The decomposition of a 35–45% ethereal solution, in fact, takes place smoothly with the formation of beryllium dihydride and isobutene:

$$Be[C(CH_3)_3]_2 \rightarrow BeH_2 + 2C(CH_3)_2{=}CH_2$$

References p. 77

However, the hydrolysis of the resulting coloured solution of colloidal BeH_2 still gives a little isobutane in addition to hydrogen, from which it follows that the hydride still contains t-butyl groups. Similar, pyrolysis of di-isobutylberyllium gives the glassy isobutylberyllium hydride. Its tetrahydrofuran complex in benzene is a *bis*-adduct $Bu^i_2Be_2H_2 \cdot 2THF$, but the tetramethylethylenediamine (TMED) complex is a *mono*-adduct $Bu^i_2Be_2H_2 \cdot TMED$[*34a*]. The higher the temperature of preparation (pyrolysis) is made, the smaller is the residual content of butyl groups and the purer is the beryllium dihydride formed (89% BeH_2 at 150°, and 96% BeH_2 at 210°). Pyrolysis can no longer be carried out at appreciably higher temperatures, since BeH_2 gives off hydrogen slowly at 240° and rapidly at 300°[*32*, *63*].

3.2 *Properties of* BeH_2

The BeH_2 prepared by the hydrogenation of beryllium compounds differs in its physical properties from the hydride prepared by pyrolysis. Since, up to the present time, it has not been possible to prepare either pure BeH_2 or a standard preparation of the hydride with a reproducible content of ether, caution must be used in making a comparative evaluation of the physical properties and, to a smaller extent, the chemical properties described.

The density of 97%-molar beryllium dihydride formed pyrolytically from dialkylberylliums if $0{\cdot}57 \pm 0{\cdot}02$ g/ml at −110°[*63*]. The density of the beryllium dihydride containing large amounts of ether prepared by the hydrogenation of beryllium compounds in ether averages $0{\cdot}69 \pm 0{\cdot}01$ g/ml (t°?)[*65*]. The density of the pure ether-free hydride BeH_2 estimated by comparison with $(BeCl_2)_n$, $[Be(CH_3)_2]_n$, and $[Al(CH_3)_3]_2$ is below 1·0 g/ml[*70*]. The standard enthalpy of formation, $\Delta H°$, of BeH_2 is calculated as +54 kcal/mole[*70*] (in contrast, all other hydrides of the elements of the second main group have negative enthalpies).

Similarly to beryllium dichloride[*93*] and dimethylberyllium[*94*, *100*], pure beryllium dihydride probably has a chain structure (SiS_2 type)[*24b*, *70*, *93*, *94*]. In this structure, the beryllium atoms have a tetrahedral configuration. However, they differ in that the C–Be–C angle is greater than the tetrahedral angle, while the Cl–Be–Cl and H–Be–H angles are smaller:

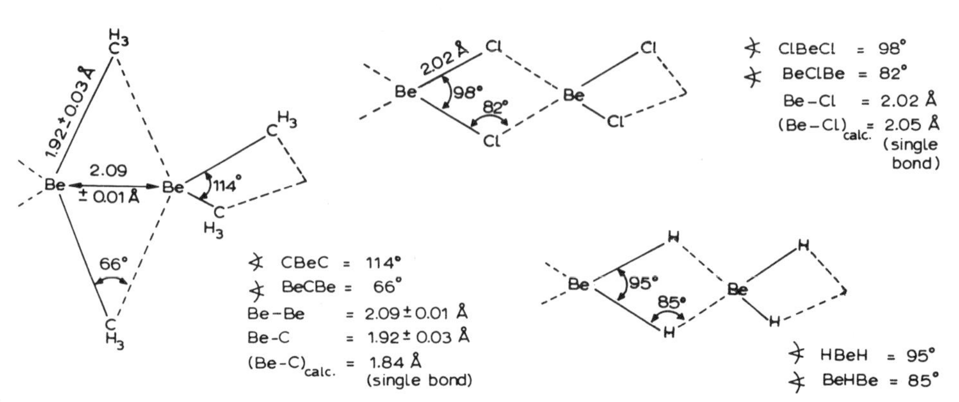

Beryllium hydride with a high ether content prepared by the hydrogenation of beryllium compounds decomposes rapidly at 125° into H_2, C_2H_6, and Be[*16*]. In contrast, hydride formed by pyrolysis and containing little ether first decomposes into the elements only at 300°[*32*]. In this respect, it resembles pyrolytically prepared magnesium dihydride. Calculations on BeH_2 and stabilities of ether-free BeH_2 and BeH see [*1b*].

BeH_2 prepared by the pyrolysis of dialkylberylliums is relatively stable to air. For example, the BeH_2 content of a 28-mg sample fell on standing for eight days only from 76 to 72% BeH_2[*63*].

Few reactions have so far been carried out with beryllium dihydride, e.g. polymerisation of ethylene[*123*]. Since BeH_2 lacks two pairs of electrons for the octet shell, its reactions are comparable with those of the 'electron-deficient compounds' BH_3 or AlH_3. In spite of pronounced covalent bond contributions in BeH_2, the hydrogen undoubtedly has a hydride nature, for which reason beryllium dihydride reacts with protons to form hydrogen.

Beryllium dihydride prepared by hydrogenation and containing 50–53·5% of BeH_2, together with ether, reacts with water vapour between −196 and 20°[*16*]. In contrast, 96% hydride prepared by pyrolysis reacts with water very slowly and incompletely even at 50°[*32*]. However, the latter hydrolyses vigorously in dilute mineral acids[*32, 63*]:

$$BeH_2 + 2H^{\oplus} \rightarrow Be^{2\oplus} + 2H_2$$

Beryllium dihydride reacts with diethylamine at 155° slowly but quantitatively, to give trimeric (or polymeric[*46a*]) bisdimethylaminoberyllium and hydrogen [*32, 33*]:

$$3BeH_2 + 6HNMe_2 \rightarrow [Be(NMe_2)_2]_3 + 6H_2$$

Similarly, beryllium dihydride reacts with trimethylethylenediamine $HMeNCH_2$-CH_2NMe_2 giving the trimeric aminoberyllium hydride $(HBeNMe{\cdot}CH_2CH_2{\cdot}NMe_2)_3$[*24e*].

Beryllium dihydride, $(BeH_2)_x$, does not react with an excess of triethylamine NMe_3 at 210°[*32*]. Obviously, the polymerisation energy of $(BeH_2)_x$ is too high to permit depolymerisation by NMe_3 somewhat as in the case of $(AlH_3)_x$. This is not surprising, since only very strong electron donors, such as trimethylamine, can depolymerise dimethylberyllium, $(BeMe_2)_x$. Since, however, metal hydrides tend to undergo even stronger polymerisation than metal alkyls [*cf.* the decreasing degree of polymerisation: $(AlH_3)_n$ (solid) > $(AlHMe_2)_n$ (viscous oil) > $(AlMe_3)_2$ (liquid), or $(GaH_3)_n$ > $GaMe_3$], the donor strength of NMe_3 is no longer adequate[*32*].

In contrast, methylberyllium hydride ether adduct $(EtBeHOEt_2)_2$ reacts with trimethylamine Me_3N or trimethylphosphane Me_3P giving the dimeric beryllium hydride $(Me_3N)PhBeH_2BePh(NMe_3)$ and the somewhat more dissociated (in benzene) analogous compound $(Me_3N)PhBeH$ respectively[*34b*], e.g.:

$$Et_2Be + BeCl_2 + 2Na[Et_3BH] + 2Et_2O \rightarrow (EtBeHOEt_2)_2 + 2Et_3B + 2NaCl$$

$$(EtBeHOEt_2)_2 + 2Me_3N \rightarrow (EtBeHNMe_3)_2 + 2Et_2O$$

References p. 77

The donor $H^{\ominus}$ adds to the acceptor BeR_2. However, this does not lead, in analogy with the formation of boranate in accordance with the equation $H^{\ominus} + BH_3 \rightarrow BH_4^{\ominus}$, to a mononuclear anion $BeR_2H_2^{2\ominus}$ ('beryllanate'): $2H^{\ominus} + BeR_2 \nrightarrow BeR_2H_2^{2\ominus}$; but to a di- or polynuclear anion with beryllium having the coordination number of 4. Thus, diethylberyllium reacts with sodium hydride in boiling diethyl ether but with lithium hydride only after the evaporation of the ether and heating to 110° to form an anion [*24, 31*]:

$$\left[Et_2Be(\mu\text{-}H)_2BeEt_2 \right]^{2\ominus}$$

$$2BeEt_2 + 2NaH \xrightarrow{\text{ether, 35°, 7 h}} Na_2[Be_2Et_4H_2]$$

$$2BeEt_2 + 2LiH \xrightarrow{\text{110°, 12 h}} Li_2[Be_2Et_4H_2]$$

$Na_2[Be_2Et_4H_2]$ crystallises in colourless needles not containing ether, and melts at 198°. It ignites in air. In 1 litre diethyl ether, 0·33 g-atoms of Be dissolve at 20°; the Li compound is more soluble.

In analogy with the formation of aluminium hydride by the reaction of alanate with aluminium trichloride, the reaction of the beryllium hydride anion $Be_2Et_4H_2^{2\ominus}$ with beryllium chloride forms the corresponding beryllium hydride $Be_3Et_4H_2 = BeH_2 \cdot 2BeEt_2$ [*31*]

$$Li_2[Be_2Et_4H_2] + BeCl_2 \xrightarrow{\text{diethyl ether}} Be_3Et_4H_2 + 2LiCl$$

and via an unstable intermediate "$Na_2Be_5Et_8H_4$" the compound Na_2BeH_4 [*24a, 24d*]:

$$2Na_2[Be_2Et_4H_2] \xrightarrow[-2NaCl]{+BeCl_2} [Na_2Be_5Et_8H_4] \xrightarrow{180°} Na_2BeH_4 + 4BeEt_2$$

Na_2BeH_4 has to be regarded as an electron-deficient polymer with strongly polarized metal–hydrogen bonds, rather than as a salt $Na_2[BeH_4]$ analogous to $Na[BH_4]$ [*1a, 24a*].

An ethereal solution of the analogous methyl compound $Be_3Me_4H_2$ contains species richer and poorer in H, since, after the more volatile $BeMe_2$ has been distilled off, the less volatile H-richer hydride remains. $Be_3Et_4H_2$ undergoes a similar dismutation. Since $Be_3Me_4H_2$ is attacked by trimethylamine, it probably contains methyl bridges rather than H-bridges [*23*]:

$$H\text{—}Be(\mu\text{-}Me)_2Be(\mu\text{-}Me)_2Be\text{—}H + \xrightarrow[-BeMe_2 \cdot NMe_3]{3\,NMe_3} (Me)(Me_3N)Be(\mu\text{-}H)_2Be(NMe_3)(Me)$$

The resulting liquid bis(trimethylamine)-dimethyldiberyllane, $(Me_3N \cdot BeMeH)_2$ (m.p. 73·0–73·2°), probably contains Be–H^{μ}–Be bridges. In fact–like the beryllium hydride described above – it is stable to decomposition by amines [*23*]. About $(Me_2N)HBeH_2BeH(NMe_2)$ (m.p. 125–126°) and similar compounds see [*97a*].

The addition of diborane, $(BH_3)_2$, to beryllium dihydride, BeH_2, at 95° with the formation of beryllium boranate, $Be(BH_4)_2 = BeH_2 \cdot 2BH_3$, takes place slowly and incompletely [*32*, *65*]. In this process, both Be–H^{μ}–Be and B–H^{μ}–B bridges must be ruptured, with the consumption of energy:

$$BeH_2 + (BH_3)_2 \rightarrow Be(BH_4)_2$$

The reaction is complicated by partial decomposition of the $(BH_3)_2$.

$Be(BH_4)_2$ has a structure like diborane with an additional BeH_2-group at the B–H^{μ}–B-double-bridge – (I) proposed structure up to 1967 which becomes obsolete; (II) structure from electron diffraction studies, 1967 and 1969 [*12a*, *35a*]; (III) structure from theoretical studies (derived from II, but with different bond length to H^a and H^b), 1968 [*13a*]; (IV) structure from i.r. and mass spectra, 1969 [*35a*].

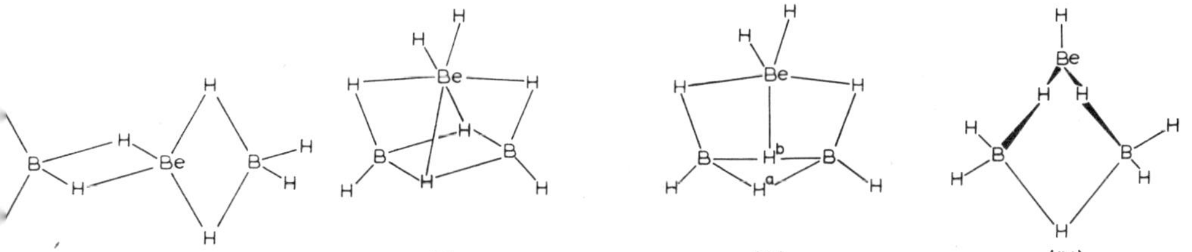

4. Magnesium hydride and deuteride

4.1 *Preparation of MgH_2, $MgHX$, and MgD_2*

4.1.1 *Preparation of MgH_2 and MgD_2 from the elements*

The reaction of high purity magnesium [*13*, *101*] in the form of turnings or powder with hydrogen or deuterium at elevated temperature and elevated pressure forms magnesium dihydride or dideuteride, respectively [*43*, *101*, *115*, *116*]:

$$Mg + H_2 \rightarrow MgH_2$$
$$Mg + D_2 \rightarrow MgD_2$$

The hydrogenation takes place via a primary dissolution of hydrogen in the magnesium since from the (almost) horizontal course of the central section of the isotherms in the Mg–H_2 and Mg–D_2 systems (Fig. 3.2, section 4.2.2) it follows that two solid phases are present – a solid solution of hydrogen or deuterium in magnesium (the atom-% hydrogen is 0–2% at 440°, 0–3% at 470°, 0–4% at 510°, 0–10% at 560°; the atom-% deuterium is 0–6% at 510°) and a solid solution of magnesium in magnesium dihydride or dideuteride. The hydrogen-rich end of the linear section of the isotherm is congruent with a compound $MgH_{1.99 \pm 0.01}$ [*101*].

At 500° and 400 atm, without a catalyst, oxide-free magnesium yields 90% of grey magnesium dihydride. When catalysts such as MgI_2 or $HgCl_2$ are added, the

References p. 77

pressure and temperature can be lowered. For example, MgH_2 is formed in 60% yield in the presence of MgI_2 at 570° and 200 atm[*116*]. For the optimum conditions of synthesis from the elements in the absence of catalysts, see[*43*]. With carbon at 1800°, magnesium oxide forms Mg which can be converted into MgH_2 with H_2 at 450° and 70 atm[*37*].

The intermetallic compound Mg_2Cu reacts with hydrogen to form MgH_2 and $MgCu_2$[*90a*]. Mg_2Ni (but not $MgNi_2$) reacts with hydrogen producing Mg_2NiH_4 [*90b*].

4.1.2 Preparation of MgH_2 by the hydrogenation of MgR_2 with $LiAlH_4$

The hydrogenation of diethylmagnesium, $MgEt_2$, with lithium alanate, $LiAlH_4$, in ethereal solution forms magnesium dihydride[*16, 65*]:

$$2MgEt_2 + LiAlH_4 \rightarrow 2MgH_2 + LiAlEt_4$$

If for this purpose the $MgEt_2$ solution is added in drops to an excess of a solution of $LiAlH_4$, the solution remains clear. White $MgH(AlH_4)$ (= $MgH_2 \cdot AlH_3$) precipitates when benzene (which precipitates neither $LiAlH_4$ nor $MgEt_2$) is added. When the order of the addition is reversed (the $LiAlH_4$ solution being added to the $MgEt_2$ solution), white solid MgH_2 precipitates, and this dissolves on further addition of $LiAlH_4$. On long standing, such a solution separates out a gelatinous solid which contains more Al than corresponds to the formula $MgH(AlH_4)$. MgH_2 prepared by hydrogenation contains a negligible content of aluminium only when the diethylmagnesium used is very pure[*16*]. (Examples: 75% MgH_2, 11–14% Et_2O, 14–11% Al compounds[*16*]; 73% MgH_2, 6% $MgEt_2$, 15% Et_2O, 0·4% AlH_3[*65*]). The reaction of $MgCl_2$ with $NaBH_4$ gives no MgH_2 but $Mg(BH_4)_2$[*77a*].

4.1.3 Preparation of MgH_2 by the hydrogenation of MgR_2 with $(BH_3)_2$

Diethylmagnesium, $MgEt_2$, can also be hydrogenated to magnesium hydride with diborane, $(BH_3)_2$. The course of this reaction (on this, *cf.* the reaction scheme of Fig. 3.1) is controlled by extremely small amounts of trialkylaluminium, AlR_3. Since normal commercial magnesium contains up to 4% of Al, all reactions with the MgR_2 prepared from it are Al-catalysed even without the addition of AlR_3.

In the reaction of (Al-containing) diethylmagnesium, $MgEt_2$, with an excess of diborane, $(BH_3)_2$, above 100° in the absence of ether, the final products are (ether-insoluble) magnesium hydride and triethylboron according to the overall equation[*19, 109*]:

$$3MgEt_2 + (BH_3)_2 \rightarrow 3MgH_2 + 2BEt_3$$

If the operation is carried out in ethereal solution (with Al-containing $MgEt_2$), magnesium boranate, $Mg(BH_4)_2$, precipitates[*21, 109*], and with an excess of diethylmagnesium, on concentrating and heating, this gives magnesium hydride and ethylborane[*19*]:

$$3MgEt_2 + 4(BH_3)_2 \xrightarrow{-2BEt_3} 3Mg(BH_4)_2 \xrightarrow[-6BH_2Et]{+3MgEt_2} 6MgH_2$$

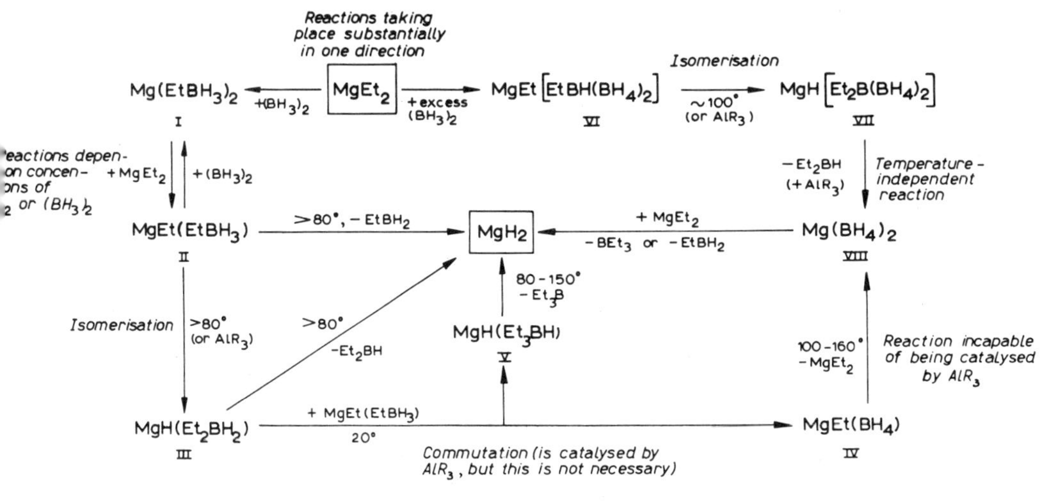

Fig. 3.1. Reactions of diethylmagnesium, $MgEt_2$, with diborane, $(BH_3)_2$; the stoichiometry must be completed suitably with the individual reactions.

If, in the reaction of $MgEt_2$ with $(BH_3)_2$ in ether, dimethylmagnesium with an aluminium content of less than 0·001% is used, three intermediate stages (I), (II), (VI) (depending on the $MgEt_2/(BH_3)_2$ molar ratio) can be found [*18*, *21*]:

$$MgEt_2 \xrightarrow{(BH_3)_2} \underset{(I)}{Mg(EtBH_3)_2} \xrightarrow{+MgEt_2} 2\ \underset{(II)}{MgEt(EtBH_3)}$$

$$MgEt_2 \xrightarrow{3/2\ (BH_3)_2} \underset{(VI)}{MgEt[EtBH(BH_4)_2]}$$

At a molar ratio of $MgEt_2$ to $(BH_3)_2$ of **1** (in practice, slightly less than 1), pure (I) is formed. (I) is a colourless ether-soluble solid stable at 20° with the exclusion of moisture. It decomposes slowly from 80° onwards and rapidly at 100 to 120° into magnesium hydride and ethylborane [*18*]:

$$\underset{(I)}{Mg(EtBH_3)_2} \xrightarrow{>80^\circ} MgH_2 + 2EtBH_2$$

At a molar ratio of $MgEt_2$ to $(BH_3)_2$ of > **1** (excess of Al-free $MgEt_2$) the solid, colourless, ether-soluble (II), also stable at 20° in the absence of moisture, is formed in addition to (I). Compound (II) which forms (I) with an excess of $(BH_3)_2$, isomerises above 80° or in the presence of AlR_3 to give (III), which decomposes into magnesium hydride and diethylborane [*18*]:

$$\underset{(II)}{MgEt(EtBH_3)} \xrightarrow{+\frac{1}{2}(BH_3)_2} \underset{(I)}{Mg(EtBH_3)_2}$$

$$\underset{(II)}{MgEt(EtBH_3)} \xrightarrow{>80^\circ} \underset{(III)}{MgH(Et_2BH_2)} \xrightarrow{>80^\circ} MgH_2 + Et_2BH$$

References p. 77

Magnesium hydride therefore remains as the residue of the thermolysis of (I) and (II), while the ethylboranes can be distilled off.

If $AlEt_3$ is added to the ethereal reaction solution containing (II) and (III), the ether-soluble (IV) and ether-insoluble (V)[4] are formed by dismutation[*19*]:

$$\underset{(II)}{MgEt(EtBH_3)} + \underset{(III)}{MgH(Et_2BH_2)} \xrightarrow{+AlEt_3} \underset{(IV)}{MgEt(BH_4)} + \underset{(V)}{MgH(Et_3BH)}$$

Under the catalytic action of $AlEt_3$, (V) decomposes further[*19*] to give magnesium hydride, MgH_2, and triethylboron, Et_3B:

$$\underset{(V)}{MgH(Et_3BH)} \xrightarrow{80-150°} MgH_2 + Et_3B$$

The rearrangement of $MgEt(BH_4)$, (IV), into $MgH(EtBH_3)$, which is conceivable in analogy with the isomerisation (II) → (III), does not occur[*19*]. This is obviously because the boranate ion, $[BH_4]^{\ominus}$ is more stable than $[EtBH_3]^{\ominus}$ or $[Et_2BH_2]^{\ominus}$.

The colourless ether-soluble compound (VI) forms at a molar ratio of $MgEt_2$ to $(BH_3)_2$ of < **1** (excess of B_2H_6, Al-free $MgEt_2$). At 100° or after the addition of a little $AlEt_3$, (VI) which is fairly stable at 20° in the absence of moisture, rapidly decomposes into magnesium diboranate, $Mg(BH_4)_2$, and diethylborane, Et_2BH. Only after the addition of diethylmagnesium, $MgEt_2$, does magnesium diboranate yield magnesium hydride[*19, 21*]:

$$\underset{(VI)}{MgEt[EtBH(BH_4)_2]} \xrightarrow[(or + AlR_3)]{\sim 100°} \underset{(VII)}{MgH[Et_2B(BH_4)_2]} \xrightarrow{-Et_2BH}$$

$$\underset{(VIII)}{Mg(BH_4)_2} \xrightarrow[-2EtBH_2]{+MgEt_2} 2MgH_2$$

Temperatures of the order of 260° must be avoided because of the decomposition of $Mg[BH_4]_2$ into Mg, B–H compounds, and hydrogen[*111*].

In the hydrogenation of diethylmagnesium with diborane, therefore, catalytic amounts of trialkylaluminium and moderately elevated temperatures favour the formation of magnesium hydride. In the final account, the molar ratio of the reactants is significant only for the isolation of the intermediate stages.

4.1.4 Preparation of MgH_2 by the hydrogenation of MgR_2 with SiH_4

Similarly to the diboranes $(BH_3)_2$, silane SiH_4, which is isosteric with the alanate ion $AlH_4^{\ominus}$, can also hydrogenate the Mg–C bond. However, in contrast to the $MgR_2/(BH_3)_2$ reaction, no magnesium dihydride, MgH_2, precipitates, but,

[4]Note that $Li(Et_3BH)$ is ether-soluble, since it is less ionic.

because of the decomposition of the ether with evolution of butane, the ether-insoluble ethoxomagnesium hydride MgH(OEt) does precipitate [*20*]:

$$n\mathrm{MgEt_2} + n\mathrm{OEt_2} + \mathrm{SiH_4} \rightarrow n\mathrm{MgH(OEt)} + n\mathrm{Et_2} + \mathrm{SiH_{4-n}Et_n}$$

With Al-free diethylmagnesium (see section 4.1.3), the reaction takes place very slowly; silane undergoes reaction with the replacement of only one hydrogen ($n = 1$) even after some weeks. Small amounts of triethylaluminium reduce the reaction time to days, with simultaneous better utilisation of the silane ($n \sim 2$). The colourless MgH(OEt) is stable at 200° in the absence of moisture [*20*]. It begins to decompose slowly at 230°, and decomposes rapidly at 300 to 350° according to the following equations [*22*]:

$$\mathrm{MgH(OC_2H_5)} \rightarrow \mathrm{MgO} + \mathrm{C_2H_6}$$
$$\mathrm{MgH(OC_2H_5)} \rightarrow \mathrm{MgO} + \mathrm{C_2H_4} + \mathrm{H_2}$$
$$\mathrm{MgH(OC_2H_5)} \rightarrow \mathrm{Mg} + \mathrm{C_2H_5OH}$$

4.1.5 Preparation of MgHX and Mg(BH$_4$)X by the reaction of MgRX with (BH$_3$)$_2$

If an excess of diborane is passed for a short time at −25° into a 4 *N* solution of ethylmagnesium chloride in tetrahydrofuran, colourless microcrystalline magnesium hydride chloride bis(tetrahydrofuranate), MgHCl·2THF, precipitates [*117*]:

$$6\mathrm{MgEtCl} + \mathrm{(BH_3)_2} \xrightarrow[-25°]{} 6\mathrm{MgHCl} + 2\mathrm{BEt_3}$$

MgHCl·2THF can be reprecipitated by dissolution in tetrahydrofuran and addition of petroleum ether.

When an excess of diborane is passed into a 4 *N* solution of ethylmagnesium bromide or iodide in ether at −25° for half an hour, the magnesium hydride halide dietherate, MgHBr·2Et$_2$O or MgHI·2Et$_2$O, also colourless and microcrystalline, precipitates out [*117*]:

$$6\mathrm{MgEtBr} + \mathrm{(BH_3)_2} \xrightarrow[-25°]{} 6\mathrm{MgHBr} + 2\mathrm{BEt_3}$$

$$6\mathrm{MgEtI} + \mathrm{(BH_3)_2} \xrightarrow[-25°]{} 6\mathrm{MgHI} + 2\mathrm{BEt_3}$$

THF solutions cannot be used for the preparation of MgHBr, since MgEtBr separates out on cooling, which would interfere with the the isolation of the hydride halid etherate.

The magnesium hydride halides, MgHX, are of interest and significance as the base substances of the Grignard compounds MgRX.

Ethylmagnesium chloride and an excess of diborane in tetrahydrofuran or (less desirably) in diethyl ether react at *higher temperatures* (+25°) exothermically (in THF quantitatively), because of addition of BH$_3$ to the MgHCl formed

References p. 77

according to the first equation of this section, yielding magnesium chloride boranate in the form of $Mg(BH_4)Cl \cdot 2.3\,THF$ or $Mg(BH_4)Cl \cdot O(C_2H_5)_2$ [*22a*]:

$$3MgEtCl + 2(BH_3)_2 \xrightarrow[-25°]{} 3Mg(BH_4)Cl + BEt_3$$

Magnesium chloride boranate, $Mg(BH_4)Cl$, can also be prepared quantitatively by conmutation [*22a*]:

$$MgCl_2 + Mg(BH_4)_2 \xrightarrow[65°]{THF} 2Mg(BH_4)Cl$$

The substitution of the chloride in magnesium chloride by boranate using alkali-metal boranates in boiling solvents, proceeds very slowly, for example [*22a*]:

$$MgCl_2 + NaBH_4 \xrightarrow[65°]{THF} Mg(BH_4)Cl + NaCl$$

Yields after one week: 50% using $LiBH_4$ in diethyl ether, 60% using $NaBH_4$ in THF, 80% using KBH_4 in THF. An excess of potassium boranate substitutes the remaining chloride even slower forming $Mg(BH_4)_2$ (yield after 10 days 25%) [*22a*].

4.1.6 *Preparation of MgH_2 by the pyrolysis of MgR_2*

When diethylmagnesium, $MgEt_2$, is heated, gases are evolved slowly from 110° onwards, and vigorously at 176° (completion of the reaction at 200°). The residue in the pyrolysis vessel consists of white, solid, non-volatile, ether-insoluble readily-oxidisable magnesium dihydride formed in the main reaction; the gases are ethylene, ethane, and distillable but rapidly polymerising $Mg(C_2H_4)$ [*51, 112, 114*]:

$$Mg(C_2H_5)_2 \begin{cases} \xrightarrow{8/10,\ -2C_2H_4} MgH_2 \\ \xrightarrow{2/10,\ -C_2H_6} \frac{1}{n}[Mg(C_2H_4)]_n \xrightarrow[n<x]{\text{polymerisation}} \frac{1}{x}[Mg(C_2H_4)]_x \end{cases}$$

volatile in high vacuum; non-volatile highly polymeric

If dibutylmagnesium, $MgBu_2$, is polymerised in a similar manner, magnesium dihydride is again formed in a main reaction, and with it, in two side-reactions, $[Mg(C_4H_8)]_x$ and Mg [*114*]:

$$Mg(C_4H_9)_2 \begin{cases} \xrightarrow{(8/10)\ -2C_4H_8} MgH_2 \\ \xrightarrow{(1/10)\ -C_4H_{10}} \frac{1}{n}[Mg(C_4H_8)]_n \xrightarrow{\text{polymerisation}} \frac{1}{x}[Mg(C_4H_8)]_x \\ \xrightarrow{(1/10)\ -C_8H_{18}} Mg \end{cases}$$

As can be seen, the magnesium hydride formed is contaminated with magnesium.

In the pyrolysis of diphenylmagnesium $MgPh_2$, the side reaction leading to the metal, which is observed in the decomposition of $MgBu_2$, becomes the main reaction. In addition, brown, powdery, non-volatile $[Mg(C_6H_4)]_x$ is formed, but no MgH_2 or other hydrides of magnesium[*114*]:

$$Mg(C_6H_5)_2 \xrightarrow{-(C_6H_5)_2} Mg$$

$$Mg(C_6H_5)_2 \xrightarrow{-C_6H_6} \frac{1}{x}[Mg(C_6H_4)]_x$$

4.1.7 Preparation of MgH_2 by the pyrolysis of MgRX

Some Grignard compounds decompose in a manner comparable with the pyrolysis of $MgPh_2$. When ethylmagnesium bromide, MgEtBr, or ethylmagnesium iodide, MgEtI, is heated at 180 to 230°, magnesium dihydride and an equimolecular mixture of magnesium halide and ethene are formed (a), together with ethane and ethyne in two minor side-reactions, (b) and (c), but no volatile Mg-containing compounds[*114*]:

$$2\,Mg(C_2H_5)X \xrightarrow{(a)\ -2\,C_2H_4} MgH_2 + MgX_2$$

$$2\,Mg(C_2H_5)X \xrightarrow{(b)\ -C_2H_4,\ -C_2H_6} Mg + MgX_2$$

$$2\,Mg(C_2H_5)X \xrightarrow{(c)\ -C_2H_2,\ -C_2H_6} MgH_2 + MgX_2$$

The extent of side-reactions (b) and (c) depends on the pyrolysis temperature. If the heating is carried out slowly to 175° they amount to 20%; conversely, they can be almost completely suppressed by rapid heating.

Butylmagnesium bromide, MgBuBr, decomposes at 200° in analogy with equation (a), to form MgH_2, $MgBr_2$, and butene[*114*]:

$$2Mg(C_4H_9)Br \xrightarrow{-2C_4H_8} MgH_2 + MgBr_2$$

Phenylmagnesium bromide, MgPhBr, decomposes at 300° in a high vacuum, in analogy with the pyrolysis of $MgPh_2$, to give biphenyl and a brown solid hydrolysing explosively with water, possibly magnesium monobromide MgBr, together with benzene and (perhaps) $C_6H_4(MgBr)_2$ or $MgC_6H_4 + MgBr_2$[*114*]:

$$2\,Mg(C_6H_5)Br \xrightarrow{(2/3)\ -(C_6H_5)_2} 2\,MgBr\ ?$$

$$2\,Mg(C_6H_5)Br \xrightarrow{(1/3)\ -C_6H_6} C_6H_4(MgBr)_2\ ?$$

4.2 Properties and reactions of MgH_2

4.2.1 Properties and reactions of MgH_2 prepared by pyrolysis

Magnesium dihydride, MgH_2, prepared by the pyrolysis of $Mg(C_2H_5)_2$ (found: $MgH_{1.97}$) is solid, white, non-volatile, and insoluble in ether[*114*]. Judging by

References p. 77

its preparation, it is probably amorphous. In a high vacuum, it is stable up to 280°; above this temperature it decomposes into the elements [*114*]:

$$MgH_2 \xrightarrow{>280^\circ} Mg + H_2$$

On contact with air, small quantities of magnesium hydride just show a rise in temperature, larger quantities ignite and burn according to the following reaction [*114*]:

$$MgH_2 + O_2 \rightarrow MgO + H_2O$$

Magnesium hydride reacts violently with water and with methyl alcohol to give, magnesium hydroxide and magnesium methoxide, respectively [*114*]:

$$MgH_2 + 2HOH \rightarrow Mg(OH)_2 + 2H_2$$
$$MgH_2 + 2HOR \rightarrow Mg(OR) + 2H_2$$

In a similar manner to LiH it reacts with diborane, $(BH_3)_2$, and alane, $(AlH_3)_3$, to produce ether-insoluble magnesium boranate, $Mg(BH_4)_2 = MgH_2{\cdot}2BH_3$, and ether-soluble magnesium alanate, $Mg(AlH_4)_2 = MgH_2{\cdot}2AlH_3$ [*113, 114*]:

$$MgH_2 + (BH_3)_2 \rightarrow Mg(BH_4)_2$$
$$MgH_2 + 2AlH_3 \rightarrow Mg(AlH_4)_2$$

4.2.2 Properties and reactions of MgH_2 synthesised from the elements

In contrast to the homologous calcium hydride, CaH_2, and in analogy with magnesium fluoride, MgF_2, MgH_2 obtained from the elements crystallises in the space-centered tetragonal (rutile) lattice with the constants $a_0 = 4.516_8 \pm 0.000_1$ Å, $c_0 = 3.020_5 \pm 0.000_2$ Å. The elementary cell contains two molecules of MgH_2. The X-ray density is 1.419 g/ml, the pycnometric density 1.45 ± 0.03 g/ml. Space group P4/mnm (D_{4h}^{14}). Two Mg atoms at (000) ($\frac{1}{2}\frac{1}{2}\frac{1}{2}$); four H atoms at ±(0.306, 0.306, 0) ($0.306+\frac{1}{2}, \frac{1}{2}-0.306, \frac{1}{2}$). In this structure, each Mg atom is surrounded by 6 H atoms at a distance of 1.95 Å, each H atom has three neighbouring Mg atoms. The H–H distances are 2.49, 2.76, and 2.76 Å. Here the distance 2.76 Å is approximately equal to the diameter of the $H^{\ominus}$ ion in LiH (see Chapter 2, section 3.2); the shorter distance (2.49 Å) is characteristic for the anion–anion distance in the rutile lattice [*43, 65*].

'Magnesium hydride' obtained from the elements contains, in addition to stoichiometric magnesium dihydride, MgH_2, a solid solution of hydrogen in magnesium, and magnesium oxide (93.1% MgH_2, 3.0% MgO, 3.9% Mg [*101*]; 60% MgH_2 [*116*]; 65% MgH_2, 30% MgO, 5% Mg [*65*]). It is very faintly grey, markedly inert to air, and reacts slowly with water, in contrast to MgH_2 prepared by pyrolysis. The difference can be ascribed satisfactorily to the different grain size and the crystalline nature [*43*]. About the mass spectrum of MgH_2 see [*37a*].

Fig. 3.2 shows the dissociation pressures of H_2 and D_2 as functions of the empirical compositions MgH_x and MgD_x, respectively, for various tempera-

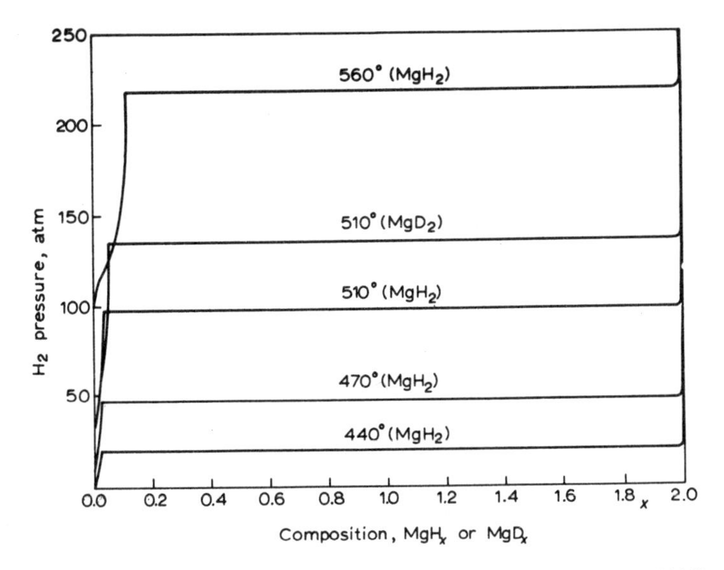

Fig. 3.2. Isotherms in the p_{H_2}–MgH_x and the p_{D_2}–MgD_x diagrams [*101*].

tures [*101*]. The dissociation pressure of H_2 by the equations $\log p_{H_2}$ [psi][5] = $-10.050/T - 20.7 \log T + 75.8$ [*65*] and $\log p_{H_2}$ [torr] $= -3790/T + 9.74$ (applicable between 337° and 417°) [cit. *44, 45*; *82*].

Standard enthalpy of formation $\Delta H° = -46000 + 41.1\,T$ [cal/mole] [*65*] (the $\Delta H°$ figures from this equation differ markedly from the figures in Table 3.1); $\Delta H°$ (450°) = 16 kcal/mole [*43*]; activation energy = 53 kcal/mole [*43*]; for other

TABLE 3.1

THERMODYNAMIC DATA FOR THE FORMATION OF MgH_2 AND MgD_2 FROM THE ELEMENTS [*101*]

		800 °K	*600 °K*	*298 °K*	*298 °K*[a]
$\Delta F°$	MgH_2	+8000 ± 120	+1580 ± 80	−8170[b]	*−7600 to −8300*
(cal mole^{-1})	MgD_2	+8620	—	−7800	*−8240*
$\Delta S°$	MgH_2	−31·47 ± 0·88	−32·21 ± 0·48	−32·3[b]	*−30·4 to −31·2*
(cal mole^{-1} deg^{-1})	MgD_2	−33·7	—	−32·5	*−32·5*
$\Delta H°$	MgH_2	−18350 ± 390	−17830 ± 80	−17790[b]	*−16900 to −17350*
(cal mole^{-1})	MgD_2	−18350	—	−17480	*−17350*
$\Delta c_p°$	MgH_2	5·8 ± 8·8	0·7 ± 1·4	0[b]	*−2*
(cal mole^{-1} deg^{-1})	MgD_2	—	—	0	*−2*

[a] The most probable figures under normal conditions.

[b] Extrapolated value.

[5] One pound per square inch = 0.070307 kp/cm² = 51.715 torr.

thermodynamic data, see Table 3.1. The measured decomposition pressures give linear curves after conversion into fugacities by the Beattie–Bridgeman equation of state [*101*]:

$$R \ln f_{H_2} = \frac{-17785 \pm 76}{T} + 32{\cdot}28 \pm 0{\cdot}45$$

$$R \ln f_{D_2} = \frac{-17480 \pm 110}{T} + 32{\cdot}48 \pm 0{\cdot}17$$

5. Calcium hydride, deuteride and tritide

5.1 *Preparation of CaH_2, CaD_2 and CaT_2*

$Ca + H_2$. Industrially, calcium hydride, C_2H_2, is obtained from the elements:

$$Ca + H_2 \rightarrow CaH_2$$

For this purpose, metallic calcium is heated in an atmosphere of hydrogen to 400° until the metal is more or less completely hydrided [*6*]. Provided that a temperature of 600° (dissociation pressure of CaH_2 about 0·1 torr) is not exceeded, unpurified hydrogen can be used, since under these conditions calcium does not react with any carbon monoxide and carbon dioxide that may be present; in fact, below this temperature the reaction can be moderated by the addition of CO or CO_2 to the hydrogen [*7, 10*]. At very high temperatures (800–900°) the reaction no longer takes place quantitatively, since the dissociation pressure of calcium hydride is already considerable in this range of temperatures (1 atm at 990°).

Finely divided calcium, such as is obtained, for example, by the evaporation of the ammonia from a solution of calcium in liquid ammonia, reacts with hydrogen to form CaH_2 even at 0°.

The formation of calcium hydride from the elements between 120 and 530° takes place without autocatalysis. The rate of formation of CaH_2 is determined by the rate of diffusion – as with the reaction of solid or liquid alkali metals with hydrogen (see Chapter 2. Section 2.1). At the same time, the temperature coefficient for the chemical process of hydride formation is substantially smaller than that for the physical process of diffusion. An activation energy of 5500 cal/mole has been found for the chemical process and one of 15000 cal/mole for the diffusion process [*98*]. According to another investigation, calcium adsorbs H_2 and D_2 in a similar manner, with almost the same activation energies (CaH_2: 13.8 kcal/mole, CaD_2: 13.9 kcal/mole, 251–325°). The specific rate constant for the formation of CaD_2 is 18% lower than that for the formation of CaH_2 [*68a*].

$Ca/Li + H_2$. In principle, calcium alloys can be hydrided with hydrogen in a similar manner to elementary calcium. The reaction of an alloy of 95% Ca and 5% Li with H_2 between 400 and 500° forms a solid solution of CaH_2 and LiH. This is said to be substantially more satisfactory than pure calcium dihydride in many cases. Thus, in the industrial production of hydrogen the hydrolysis of the mixed hydride forms a more continuous and uniform current of H_2 than the LiH-free hydride [*102*].

Ca/Al+H$_2$. Calcium–aluminium alloys with a Ca content in excess of the composition $CaAl_2$ absorb hydrogen above 200°, giving calcium hydride [*64*]:

$$CaAl_2 + H_2 \rightleftharpoons CaH_2 + 2Al$$

Above 450°, the equilibrium lies again more to the left.

Ca+CH$_4$. Above 800°, calcium reacts to a considerable extent with methane. A mixture of calcium hydride and calcium carbide is produced [*89*]:

$$5Ca + 2CH_4 \rightarrow CaC_2 + 4CaH_2$$

In addition, however, a large amount of unchanged calcium is still present. In the most favourable case, 58% of the calcium reacts in accordance with the above equation. The CaH_2 content of the mixture reaches a maximum at 900° (CaH_2:CaC_2 = 2·37). At higher temperatures, the calcium carbide predominates. Finally, CaH_2 is no longer present above 1090° which must be ascribed to the high decomposition pressure of the hydride at this temperature [*89*].

Ca+H$_2$O. In the decomposition of steam with metallic calcium between 177 and 369°, the existence of calcium hydride can be detected unambiguously [*56*]. Provided that the reaction mixture still contains an excess of calcium, the following reactions can be differentiated in the Ca–H_2O system:

$$2Ca + H_2O \rightarrow CaO + CaH_2 \qquad (a)$$
$$CaH_2 + 2H_2O \rightarrow Ca(OH)_2 + 2H_2 \qquad (b)$$
$$CaO + H_2O \rightarrow Ca(OH)_2 \qquad (c)$$

Compare this with the Na–H_2O system. Here, it is true, a hydride can form likewise, but the system of equations differs from that for the Ca–H_2O reaction:

$$Na + H_2O \rightarrow NaOH + \tfrac{1}{2}H_2 \qquad (d)$$
$$NaOH + Na \rightarrow Na_2O + \tfrac{1}{2}H_2 \qquad (e)$$
$$NaOH + 2Na \rightarrow Na_2O + NaH \qquad (f)$$

Here the NaOH produced in accordance with process (d), reacts with more Na to form the oxide and – depending on the H_2 pressure – H_2 according to (e) or NaH according to (f) [cit. *1*]. An intermediate formation of metal hydrides according to (a) or (f) of this kind in the reaction of metals with water (or mineral acids) can perhaps give an idea of the nature of the so-called 'nascent' hydrogen, which in many books and papers is still erroneously called atomic hydrogen.

CaO/Mg+H$_2$. Technical production of calcium hydride does not demand expensive calcium which has to be prepared from its compounds and isolated in a separate step before its reaction with hydrogen. It is in fact possible to start directly from calcium compounds and to hydrogenate them under suitable conditions; for example, it is possible to use a mixture of freshly calcined calcium oxide with magnesium powder at 550 to 650° [*4, 9, 13*]:

$$CaO + Mg + H_2 \rightarrow CaH_2 + MgO$$

References p. 77

When this process is carried out on industrial scale, a briquetted CaO/Mg mixture is hydrogenated at 400°. The reaction product is marketed as a mixture of CaH_2 and MgO. It can be used in this form for many technical purposes [*121*].

$CaCl_2/Zn + H_2$. Zinc can also be used as a reducing agent for calcium compounds. Calcium hydride is formed by treating calcium chloride with zinc vapour in the presence of hydrogen at 800 to 900° [*52*]:

$$CaCl_2 + Zn + H_2 \rightarrow CaH_2 + ZnCl_2$$

$CaCl_2/Na + H_2$. Further, sodium can be used as the reducing agent. According to this process, a mixture of sodium and calcium chloride is preferably deposited on a solid surface (mixture of 25% of CaH_2 and 75% of NaCl) and hydrogenated with hydrogen at 385° [*11*, *105*]:

$$CaCl_2 + 2Na + H_2 \rightarrow CaH_2 + 2NaCl$$

If necessary, the common salt present in the reaction mixture together with the CaH_2 can be removed by extraction with liquid ammonia. However, extraction is generally unnecessary since the mixture (25% CaH_2, 75% NaCl), which is marketed under the name 'Hydrimix' [*74*] is particularly reactive. The increased reactivity is probably due to increased surface of the CaH_2 [*73*].

$Ca(OR)_2 + H_2$. The calcium salts, $Ca(OR)_2$ of phenols, such as cresol, naphthol, and picric acid, are hydrogenated by hydrogen at high pressure (about 200 atm) even at 20° [*85*]:

$$Ca(OR)_2 + 2H_2 \rightarrow CaH_2 + 2ROH$$

The phenol liberated in this process can be recycled and re-used.

$CaX_2 + H_2$. On reaction with hydrogen under suitable conditions, calcium carbide, cyanide, cyanamide, phosphide, silicide, and boride give calcium hydride [*58*], for example:

$$Ca(CN)_2 + 8H_2 \rightarrow CaH_2 + 2CH_4 + 2NH_3$$

For the preparation of calcium dueteride CaD_2, see section 5.3 (*Organic compounds*, (ii), H–D exchange [*107*, *119*]). Calcium tritide, CaT_2, is formed by the pyrolysis of a mixture of CaH_2 and silica gel with subsequent absorption of tritium by the metallic calcium so produced [*102a*]:

$$CaH_2\text{-silica gel} \xrightarrow[-H_2]{\text{pyrolysis}} \text{Ca-silica gel} \xrightarrow{+T_2} CaT_2\text{-silica gel}$$

5.2 *Physical properties of calcium hydride*

Calcium dihydride is produced by synthesis from the elements in the form of greyish fine powder. The grey colour must be ascribed to the presence of unconverted metallic calcium. CaH_2 crystallises in an orthorhombic lattice constructed of positive calcium and negative hydrogen ions [*17, 87*]. The elementary cell contains four $Ca^{2\oplus}$ and eight $H^{\ominus}$ ions. The lattice constants are $a = 6{\cdot}851$, $b = 5{\cdot}948$, $c = 3{\cdot}607$ Å [*124*] or $a = 6{\cdot}838$, $b = 5{\cdot}936$, $c = 3{\cdot}600$ [*25*] respectively. In CaH_2 and CaD_2, each Ca ion is surrounded by nine H (D) ions, seven at a distance of 2·32 Å, and two at a distance of 2·85 Å [*25*]. The melting point of CaH_2 cannot be determined accurately because of decomposition; it is above 1000° [*66*].

The H_2 dissociation pressure of CaH_2 is about 0·1 torr at 600° and about 760 torr at 1000°. In mixtures of Ca and CaH_2 containing 25 to 95% of CaH_2, the H_2 dissociation pressure is independent of the composition of the base material ($Ca_{\text{sat. with } H_2} + CaH_2$). It can be represented by the following equations, which are valid between 636 and 892°: $\log p_{H_2}[\text{torr}] = -9840/T + 11{\cdot}7$ and $\log p_{H_2}[\text{torr}] = -10870/T + 11.493$. [cit. *44, 45*; *104*]. Above a concentration of 95% CaH_2, the pressure of H_2 increases considerably with increasing hydride content; the following equation applies for 100% CaH_2: $\log p_{H_2}[\text{torr}] = -7782/T + 9.070$ [cit. *44, 45*]. Below a concentration of 25% CaH_2 the decomposition pressure falls markedly with decreasing hydride content. This behaviour is ascribed to the formation of various solid solutions or eutectic phases in the Ca–CaH_2 system [*66, 104*].

Thermodynamic data of crystalline CaH_2: standard enthalpy of formation $\Delta H° = -44{\cdot}5$ [*41*], $-45{\cdot}1$ [cit. *44, 45*], $-44{\cdot}8$ [*104*], $-43{\cdot}93$ [*27*], $-46{\cdot}6$ [*62*] [kcal/mole]; free energy of reaction $\Delta F° = -35{\cdot}8$ kcal/mole; entropy $S° = 10$ cal deg^{-1} mole^{-1} [cit. *44, 45*]. About the mass spectrum of CaH_2 see [*37a*].

5.3 Reactions of CaH_2

Halogens. Elementary chlorine, bromine, and iodine do not react with calcium hydride in the cold, but even below red heat a vigorous combustion sets in [*81*]:

$$CaH_2 + 2X_2 \rightarrow CaX_2 + 2HX$$

Oxygen. Calcium hydride burns at red heat in a highly exothermic reaction (initiation of the combustion in air requires higher temperatures) [*59*]:

$$CaH_2 + O_2 \rightarrow CaO + H_2O$$

Water. With water calcium hydride reacts in a highly exothermic reaction to give calcium hydroxide:

$$CaH_2 + 2H_2O \rightarrow Ca(OH)_2 + 2H_2$$

Normally, the hydrogen evolved in this reaction does not ignite, and calcium hydride is therefore suitable for the drying of inert liquids and gases, particularly since its capacity for taking up water is greater than that of most other comparable

References p. 77

drying agents (*e.g.* P_4O_{10}). The drying time is short: thus in the case of benzene, it is decreased from six months with sodium wire to one day with CaH_2 powder [*28*]. By measuring the amount of hydrogen formed in the reaction, a water content of the order of 0·1% in organic liquids and salt hydrates can be determined with an accuracy of ± 0·001% [*92*].

Industrial drying. With calcium hydride or mixtures containing calcium hydride drying can be carried out by two methods [*8*, *74*]. Either the liquid to be dried is allowed to flow through a column containing the drying agent, or the liquid is dried batchwise in storage tanks by stirring with the drying agent. The extremely wide range of application of this process is shown by the following selection of compounds that can be dried with calcium hydride [*74*]. *Hydrocarbons*: paraffins, olefins, diolefins, aromatic hydrocarbons, petroleum distillates, coal distillates, transformer oils, vacuum-pump oils, lubricating oils, and medicinal oils. *Alcohols* and *phenols* (only at moderate temperatures in order to avoid undesirable side-reactions): isopropyl alcohol, butanols, amyl alcohols, phenols. *Ethers*: diethyl ether, dichlorodiethyl ether, diphenyl oxide, diisopropyl ether, dioxan. *Amines*: alkylamines, aniline, dimethylaniline, monomethylaniline, pyridine. *Esters* (only at moderate temperatures): ethyl acetate, methyl acetate, amyl acetate, benzyl acetate, butyl acetate, isopropyl acetate. *Halogen compounds*: methyl chloride, methylene chloride, chloroform, carbon tetrachloride, vinyl chloride, tetrachloroethylene, ethyl iodide, chlorobenzene, benzyl chloride, dichlorodiethyl ether, tetrachloroethane, trichloroethylene. *Oils*: fish oils, olive oil, linseed oil, coconut oil, insulating oils. *Gases* can be dried in a similar manner to these liquids if the hydrogen formed is not deleterious as an impurity.

Besides water, calcium hydride removes sulphur-containing compounds such as mercaptans and thioethers from organic liquids [*74*].

Hydrogen source. Particularly when it contains a few percentage parts of lithium hydride, calcium hydride is suitable as a *portable* source of hydrogen. On reaction with water, 1 kg of CaH_2 yields 1·06 Nm^3 of hydrogen (about 100 l/min). It is marketed for this purpose under various names (*e.g.* Hydrolith, Hydrogenite). Since each kg of Hydrolith yields about 1200 kcal of heat on reaction with water, the hydrogen formed contains considerable amounts of water vapour. Consequently, two generators are always connected in series, the water vapour entrained by the gas being decomposed by more Hydrolith in the second generator. Depending on their size, portable generators produce 1200 to 2000 m^3 of hydrogen per hour.

Oil industry. A new type of application of calcium hydride based on its reaction with water is shown in the oil industry. Oil wells which have been sealed because the hydrostatic pressure in the pipe is greater than the oil pressure at the lower end of the pipe can be reactivated by lowering calcium hydride to the end of the pipe. The hydrogen produced decreases the density of the oil, so that the flow of oil starts up once more [*72*].

Sulphur compounds. When calcium hydride is heated with sulphur vapour, it glows slightly and is converted into a dark-coloured CaS-containing powder [*71*]. With H_2S, it forms calcium sulphide. A simple and rapid method for determining sulphur in all organic and inorganic compounds is based on the great reactivity of calcium dihydride with respect to sulphur compounds. According to this method, the sulphur-containing compounds are mixed with CaH_2 and heated; the resulting calcium sulphide is oxidised with aqueous iodine solution to sulphur; after subsequent reaction with sulphite, the thiosulphate formed is titrated [*81*, *95*].

Nitrogen and phosphorus. N_2 and P react with calcium hydride at 500° with the formation of calcium nitride [*90*] and calcium phosphide [*81*], respectively. For example:

$$3CaH_2 + N_2 \rightarrow Ca_3N_2 + 3H_2$$

Ammonia. With ammonia, calcium amide is formed.

Carbon. Carbon reacts partially with calcium hydride between 700 and 800° with the formation of calcium carbide, while silicon and boron do not react with CaH_2 under the same conditions [*81*].

Carbon monoxide. CO does not react with calcium hydride even at 300 to 400°. Methane is formed at red heat [*76*]:

$$2CaH_2 + 2CO \rightarrow 2CaO + C + CH_4$$

The yield of methane first increases and then decreases with rising temperature. The finely divided carbon separating out at the same time also reacts with CaH_2 with the formation of carbide and hydrogen [*76*]:

$$CaH_2 + 2C \rightarrow CaC_2 + H_2$$

According to other investigations of the CaH_2–CO reaction, up to 16% of formaldehyde is formed as well as methane [*90*]:

$$CaH_2 + CO \rightarrow Ca + CH_2O$$

Carbon dioxide. CO_2 reacts similarly to CO with the formation of methane [*76*], substantially according to:

$$4CaH_2 + 2CO_2 \rightarrow 4CaO + C + CH_4 + 2H_2$$

At the same time, some formate and oxalate are always formed [*76*]. Formate is produced as the main product when a mixture of sodium bicarbonate with calcium hydride is gently heated:

$$CaH_2 + 2NaHCO_3 \rightarrow Ca(OH)_2 + 2HCOONa$$

Once started, the reaction proceeds spontaneously to completion [*90*].

Non-metal halides. E.g. boron and silicon chlorides are converted by calcium hydride at a relatively high temperature into the corresponding hydrogen compounds, for example:

$$3CaH_2 + 2BCl_3 \rightarrow 3CaCl_2 + (BH_3)_2$$

References p. 77

Here, the hydrogen compounds formed, which are not very stable to heat, must be rapidly removed from the hot reaction zone by means of a carrier gas, *e.g.* nitrogen or hydrogen. For further information on the production of hydrides, see the sections on the corresponding elements.

Metal halides. These are reduced by calcium hydride to the metal:

$$CaH_2 + 2MX \rightarrow 2M + CaX_2 + H_2$$

While silver fluoride is reduced by CaH_2 to silver even at 20°, lead fluoride and zinc fluoride must be heated to 400° for similar reactions to take place. Sodium and potassium can be distilled off from a mixture of their fluorides or chlorides with CaH_2 heated to red heat[*81*]. Molten potassium iodide is said not to react with CaH_2[*81*]. The metals U, W, Ti, Zr, V, Th, and Nb are produced in the form of powders when their chlorides are heated with calcium hydride at 100 to 140° in an inert atmosphere under slightly elevated pressure. In this process, the calcium hydride is formed *in situ* in the reaction mixture from calcium chloride and sodium hydride[*12*].

Metal oxides and sulphides. These react with calcium hydride in a similar manner to the metal halides.

(i) Thus the oxides of Ti, V, Nb, Ta, Cr, Mo, W, Fe, Mn, Cu, Sn and Pb are reduced to the elements, for example:

$$2CaH_2 + TiO_2 \rightarrow Ti + 2CaO + 2H_2$$

The process known in industry as the 'Hydrimet process' is used for the production of Ti, Zr, Nb, and Ta[*2,3,5*]. For this purpose, an intimate mixture of finely divided metal oxide and calcium hydride powder is heated in a stream of hydrogen to 600–1000°. The metal powder, produced below its melting point, can easily be freed from calcium oxide and the excess of calcium hydride by washing with water. If the metal oxide used is very pure, a very pure metal is obtained. As compared with other reductions (*e.g.* with Al, Mg, or Si) this process has the advantages of low reaction temperature and therefore better possibility of control, formation of a protective hydrogen atmosphere, very low alloying tendency of calcium, and easy separation of the metal formed from the by-products. The resulting metal powders are processed further by powder metallurgy. By using mixtures of oxides, alloys can be prepared in powder form. In vacuum, the reaction of TiO_2 and CaH_2 begins at 500°. At 900°, it is complete in only five minutes. The reaction takes place quantitatively in accordance with the above equation, without any intermediate stages. The titanium formed has a purity of 99·5%. Certain investigations suggest that it might not be the hydride as such but the metallic calcium produced by thermal decomposition which is the true reducing agent[*49*]. This assumption is confirmed by a thermodynamic analysis of the same process between 1100 and 1150°[*75*].

(ii) The titanium prepared by the reaction of TiS_2 and CaH_2 contains only 86% of Ti, since the titanium oxidises in the acid working up of the reaction product[*97*].

(iii) The reaction of alumina with calcium hydride begins at about 500° and takes place completely at 750°. It leads to the formation of aluminium, calcium oxide, and hydrogen. Under certain circumstances, aluminium reacts with calcium hydride to give $CaAl_2$. The reaction of $5CaO{\cdot}3Al_2O_3$ with CaH_2 also leads to $CaAl_2$[*30*].

(iv) Iron oxides react with calcium hydride in an atmosphere of hydrogen at 20 to 50 torr, at temperatures as low as 125°. Under these conditions, the individual oxide crystals are converted into single crystals of iron, which sinter together only above 200°[*48*].

(v) The corresponding reduction of iron sulphides can be used for the desulphurisation of crude iron. For this purpose, the calcium hydride is introduced under the surface of the iron below 1650° in a current of inert gas. In this way, the sulphur content can be reduced to 0·0009%[*83*, *99*].

(vi) A variant of the Hydrimet process is suitable for the preparation of pure tantalum and niobium sulphides. For this purpose, the pentoxides, the pentachlorides, or the pentafluorides are heated with calcium hydride and carbon disulphide in an inert atmosphere to 600 to 750°. All reaction products other than TaS_2 and NbS_2 can easily be volatilised or extracted[*53*].

(vii) The reaction of silica with calcium hydride in vacuum or in an atmosphere of a protective gas at 800 to 1300° forms 99·9% silicon[*50*].

Organic compounds. Calcium hydride reacts with organic compounds in a similar manner to other metal hydrides: with salt-forming ($H^{\oplus} + H^{\ominus} \rightarrow H_2$), reducing, and condensing actions, and as inhibitors of polymerisation. For example, in nuclear-reactor coolants such as polyphenyls and condensed ring systems, polymerisation of the liquid occurs. The radiation-induced radicals are trapped by H liberated from CaH_2, and polymerisation is thus prevented[*35*].

(i) *Salt formation.* Alcohols are converted into alkoxides:

$$CaH_2 + 2ROH \rightarrow Ca(OR)_2 + 2H_2$$

(ii) *Reduction and hydrogenation.* On treatment with CaH_2 in aqueous alcoholic or acetic acid solution in the presence of palladium chloride or platinum chloride, nitrobenzene and benzaldehyde are converted into aniline and benzyl alcohol, respectively. Diphenyl oxide can be transformed into diphenyl[*84*]. Calcium hydride is not suitable for reduction in organic solvents, because of its insolubility in these media. Double hydrides [for example, $Ca(BH_4)_2$], which are readily soluble in tetrahydrofuran, among other solvents, are recommended for this purpose. Calcium hydride exchanges no hydrogen with deuterium at 100° and hydrogenates ethylene only slowly. However, if CaH_2 or CaD_2 is activated by (partially) pumping off hydrogen in vacuum at 200°, more rapid H–D exchange with D_2 and more rapid hydrogenation with C_2H_4 takes place[*107*, *119*]. The equilibrium of the CaH_2–D_2 system is reached after only an hour above 50° (for comparison, it is reached even at −78° with the more active BaH_2)[*119*]. As a typical hydrogenation and dehydrogenation catalyst (like Pt), active CaH_2 hydrogenates ethylene at 150 to 200°[*119*], deuterates ethylene with D_2 at

References p. 77

150°[*119*], dehydrogenates isobutane to isobutene at 525°[*119*], and dehydrogenates cyclohexane to cyclohexene and benzene[*119*]; active CaD_2 deuterates n-butane, isobutane, and isobutene[*120*].

(iii) *Condensation.* As prototypes of condensation reactions, the reactions of CaH_2 with acetone and with amyl acetate may be mentioned. At the boiling points, mesityl oxide and acetoacetic ester are formed, respectively:

$$\tfrac{1}{2}CaH_2 + (CH_3)_2CO + CH_3COCH_3 \rightarrow (CH_3)_2C{=}CHCOCH_3 + \tfrac{1}{2}Ca(OH)_2 + H_2$$
$$\tfrac{1}{2}CaH_2 + CH_3COOR + CH_3COOR \rightarrow CH_3COCH_2COOR + \tfrac{1}{2}Ca(OR)_2 + H_2$$

A mixture of calcium hydride with an oxide of Cr, Mo, W, or U deposited on a difficultly reducible oxide such as Al_2O_3, TiO_2, or ZrO_2 is suitable as a catalyst for the *polymerisation* of ethylene and propene (130–260°, liquid medium such as benzene or toluene, elevated pressure)[*47*].

6. Strontium hydride

6.1 *Preparation of SrH_2*

Strontium hydride can be obtained directly from the elements[*62*]:

$$Sr + H_2 \rightarrow SrH_2$$

Extremely finely divided strontium, such as that produced in the pyrolysis of $Sr(NH_3)_6$, reacts with hydrogen even at 0°[*26*]. The absorption of hydrogen by compact strontium begins at 215° and becomes very vigorous at 260°[*36*]. However, the uptake of hydrogen never reaches the level corresponding to the formula SrH_2. The products always contain greater or lesser amounts of free metal. The preparation of the hydride from strontium amalgam or from a strontium–cadmium alloy has been mentioned previously in the historical section 1.3.

A further method of preparation is based on the reaction of strontium azide with hydrogen at relatively high temperatures[*103*]:

$$Sr(N_3)_2 + H_2 \rightarrow SrH_2 + 3N_2$$

Strontium hexafluorosilicate also reacts with hydrogen on heating to form strontium hydride[*77*]:

$$SrSiF_6 + 2H_2 \rightarrow SrH_2 + H_2SiF_6$$

6.2 *Physical and chemical properties of SrH_2*

Strontium hydride forms a white crystalline powder which is, however, generally coloured a more or less intense grey by free metal. Like calcium hydride, it crystallizes in an orthorhombic lattice, of the Pnam space group, constructed of positive strontium ions and negative hydrogen ions. The elementary cell contains four $Sr^{2\oplus}$ and eight $H^{\ominus}$ ions. The lattice constants are [*124*]: $a = 7{\cdot}358$, $b = 6{\cdot}377$, $c = 3{\cdot}882$ Å. Density 3·27 g/ml [cit. *44, 45; 65a*]. Melting point > 1000° [cit. *44, 45*].

Thermodynamic data of crystalline SrH_2: standard enthalpy of formation $\Delta H^\circ_{298\cdot1} = -42{\cdot}17$ [*62*], $-43{\cdot}0 \pm 0{\cdot}5$ [*41*], $-42{\cdot}3$ [cit. *44, 45; 68b*] [kcal/mole]; free energy of reaction $\Delta F^\circ = -33{\cdot}1$ kcal/mole [cit. *44, 45; 68b*]; entropy $S^\circ = 13$ cal deg^{-1} mole^{-1} [cit. *44, 45*].

Strontium hydride is insoluble in the usual organic and inorganic solvents. Like sodium and calcium hydrides, however, it dissolves in non-oxidizing melts of alkali-metal halides or hydroxides. Electrolysis of such melts results in evolution of hydrogen at the anode.

Strontium hydride cannot be melted without decomposition. The H_2 dissociation pressure of a 65% hydride is 153 torr at 633° and 556 torr at 759° [*46*]. The dissociation pressures are expressed by the equations $\log p_{H_2}[\text{torr}] = -6500/T + 8{\cdot}66$ (for 99% crystalline SrH_2), and $\log p_{H_2}[\text{torr}] = -10400/T + 11{\cdot}10$ (for 92·3% crystalline SrH_2) [cit. *44, 45*].

The chemical properties of strontium hydride have been studied only very incompletely up to the present time. Those that are still unknown are probably very much the same as the chemical properties of calcium hydride.

Halogens. Strontium hydride does not react with elementary chlorine in the cold. On gentle heating, a vigorous reaction takes place with the formation of strontium chloride and hydrogen chloride [*54*]. Liquid bromine does not react at its boiling point, but a vigorous reaction takes place at red heat. Elementary iodine reacts similarly [*54*]:

$$SrH_2 + 2X_2 \rightarrow SrX_2 + 2HX$$

Oxygen and sulphur. On heating in a current of oxygen, strontium hydride burns to give strontium oxide (E = O; with, of course, a subsequent reaction with H_2O). A similar reaction takes place with sulphur vapour at red heat (E = S) [*54*]:

$$SrH_2 + 2E \rightarrow SrE + H_2E$$

Water. With water, strontium hydride undergoes vigorous hydrolysis [*41, 57*]:

$$SrH_{2\,\text{cryst.}} + 2H_2O_{\text{liq.}} \rightarrow Sr(OH)_{2\,\infty\text{dil.}} + 2H_{2\,\text{gas}} + 87{\cdot}2 \text{ kcal/mole}$$

Alcohol. The analogous alcoholysis takes place less vigorously than the alcoholysis of elementary strontium [*57*].

Nitrogen. With nitrogen SrH_2 reacts from 500° onwards, to form substantially strontium nitride.

Lithium hydride. With lithium hydride SrH_2 forms the salt-like hydride $LiSrH_3$ (m.p. 745°) [*72a*].

References p. 77

Organic compounds. The behaviour of strontium hydride with respect to organic compounds has not yet been investigated in detail. The hydrogenating and the H–D exchanging action will, however, probably be between that of the relatively inactive CaH_2 and the highly active BaH_2.

7. Barium hydride

7.1 Preparation of BaH_2

Barium hydride is preferably prepared directly from the elements [62]:

$$Ba + H_2 \rightarrow BaH_2$$

Finely divided barium reacts with hydrogen from 120° onwards. At 170 to 180° the reaction takes place violently [*36*]. Metallic barium is best heated in an iron boat in vacuum to 400°, the temperature then being raised to 650° in a stream of hydrogen, with, finally, 2 hours of heating to 800° [*62, 104*]. If barium hydride is to be prepared from a Ba–Cd alloy, the alloy must be heated to 350° [*55*]. The reaction of an amalgam with hydrogen, which can be used for the preparation of strontium hydride, is not suitable for the production of BaH_2, since the very stable barium amalgam reacts with hydrogen only at such high temperatures that the hydride is largely decomposed again [*46*]. On the other hand, in analogy with strontium hydride, barium hydride can be prepared both from barium azide and hydrogen [*103*] and from barium hexafluorosilicate and hydrogen [*77*]:

$$Ba(N_3)_2 + H_2 \rightarrow BaH_2 + 3N_2$$

$$BaSiF_6 + 2H_2 \rightarrow BaH_2 + H_2SiF_6$$

7.2 Physical and chemical properties of BaH_2

Barium hydride is obtained from the elements in the form of a colourless to pale grey mass with a crystalline character. The limiting composition BaH_2 is rarely achieved. Like CaH_2 and SrH_2, BaH_2 crystallises in an orthorhombic lattice, belonging to the space group Pnam, constructed of positive barium ions and negative hydrogen ions. The elementary cell contains four $Ba^{2\oplus}$ and eight $H^{\ominus}$ ions. The lattice constants are [*124*]: $a = 6.788$, $b = 7.829$, $c = 4.167$ Å. Density 4.21 g/ml [cit. *44, 45; 60, 65a*]. The conversion of barium into barium hydride leads to a volume contraction of 13% [*88*]. Lattice energy 492 ± 7 kcal/mole [*68*]. Melting point > 1000° [cit. *44, 45*].

Thermodynamic data of crystalline BaH_2: standard enthalpy of formation $\Delta H^\circ_{298.1} = -40.9$ [*44, 45*], -42.7 ± 0.5 [*41*], -40.96 [*62*]; free energy of reaction $\Delta F^\circ = -31.6$ kcal/mole [*44, 45*]; entropy $S^\circ = 16$ cal deg^{-1} mole^{-1} [*44, 45*].

Similarly to the other saltlike metal hydrides, barium hydride is insoluble in the usual organic and inorganic solvents. Like them, it dissolves in non-oxidising salt melts of alkali-metal halides and hydroxides can be decomposed electrolytically in such melts.

Barium hydride cannot be melted without decomposition. As in the case of calcium hydride, the H_2 dissociation pressure depends on the content of free metal. Below 600°, the partial pressure of the barium is negligible. The H_2 dissociation pressure amounts to 0.01 torr at 350°, 0.03 torr at 500°, 0.16 torr at 550°, and 0.24 torr at 600°[*96*]. (Because of the high volatility of the metal and the marked dissociation, BaH_2 can be purified and crystallised by an apparent sublimation at 600 to 1000°). The H_2 dissociation pressures between 500 and 1000° are described by the equations $\log p_{H_2}$ [torr] $= -4000/T + 6.86$ (for 97% crystalline BaH_2), and $\log p_{H_2}$ [torr] $= -6450/T + 8.20$ (for 93.7% crystalline BaH_2) [cit. *44*, *45*].

The chemical properties of barium hydride, which have been investigated very incompletely up to the present time, appear to be similar to those of CaH_2 and SrH_2; because of the more highly polar nature of the bond in BaH_2, they are still more strongly expressed.

With halogens, oxygen, sulphur, nitrogen, etc., and with inorganic oxidizing agents, BaH_2 reacts analogously to the other alkaline-earth metal hydrides. The reactions set in at somewhat lower temperatures and usually take place somewhat more vigorously than with CaH_2. For example, the hydrolysis with water runs as follows[*41*]:

$$BaH_{2\ \text{cryst.}} + 2H_2O_{\text{liq.}} \rightarrow Ba(OH)_{2\ \infty\text{dil.}} + 2H_{2\ \text{gas}} + 86.0\ \text{kcal/mole}$$

Again, as was to be expected, barium hydride behaves with respect to organic compounds similarly to or somewhat more reactively than its lighter homologues. The exchange reactions with D_2 and the reaction with ethylene, described as taking place only at 150° in the case of CaH_2, takes place even at −78° with BaH_2.

8. Calcium, strontium and barium hydride halides

If one of the hydrogens in the alkaline-earth metal dihydrides (CaH_2, SrH_2, BaH_2) is replaced by a halogen atom, we obtain, in formal analogy to the case of magnesium hydride, the hydride halides MXH (M = Ca, Sr, Ba; X = Cl, Br, I). In contrast to the base substances of the Grignard compounds, which cannot be prepared free from solvent and are probably substantially covalent, the hydride halides of calcium, strontium, and barium are highly heat-stable salt-like compounds.

8.1 Preparation of the hydride halides MXH

Calcium, strontium, and barium hydride halides are produced when an equimolecular mixture of an alkaline-earth metal dihydride and the corresponding alkaline-earth metal dihalide is fused at about 900° or when equimolecular amounts

References p. 77

of the alkaline-earth metal and the halide are heated in an atmosphere of hydrogen at 900° [*38–40*]:

$$MH_2 + MX_2 \rightarrow 2MHX$$

$$M + H_2 + MX_2 \rightarrow 2MHX$$

Table 3.2 contains the enthalpies of formation of these processes.

Most conveniently, a mixture of the alkaline-earth metal with its halide is slowly heated in an iron boat in a stream of hydrogen. Vigorous absorption of hydrogen sets in between 300 and 400°, which is complete at 700°. When the mass is heated to a higher temperature, it melts. After about 0.5 to 1 hour, the reaction is completed at 900°. The yield is quantitative. On very slow cooling, the hydride halides crystallise in the form of large crystals. Since the alkaline-earth metal hydride iodides are less heat-stable, they cannot be prepared in open boats. They are obtained by allowing hydrogen to diffuse into an iron crucible closed by welding containing the M–MI_2 mixture. Naturally, this prolongs the reaction (with an iron crucible having walls 1 mm thick, for example, 12 hours/0.1 g-atom of metal [*40*].

TABLE 3.2

STANDARD ENTHALPIES OF FORMATION $\Delta H^\circ_{298.1}$ [KCAL MOLE^{-1}] FOR THE FORMATION OF ALKALINE-EARTH METAL HYDRIDE HALIDES (298.1°K, 1 ATM) [*41*]

	M	*X* = *Cl*	*X* = *Br*	*X* = *I*
$\Delta H^\circ_{298.1}$ (±0.3) for the process	Ca	−2.8	−2.6	−1.9
$\frac{1}{2}MH_2 + \frac{1}{2}MX_2 \rightarrow MHX$	Sr	−4.9	−3.7	−3.9
	Ba	−4.2	−4.5	−4.7
$\Delta H^\circ_{298.1}$ (±0.4) for the process	Ca	−25.0	−24.8	−24.2
$\frac{1}{2}M + \frac{1}{2}MX_2 + \frac{1}{2}H_2 \rightarrow MHX$	Sr	−26.4	−25.2	−25.4
	Ba	−25.5	−25.8	−26.0

8.2 *Physical and chemical properties of the hydride halides MHX*

The compounds MHX (M = Ca, Sr, Ba; X = Cl, Br, I) form large lustrous mica-like leaves. CaHI is microcrystalline (usually crystal rosettes) even when it is cooled very slowly in the process of preparation. Relatively small crystals of the hydride halides are colourless. With thicker layers, CaHCl appears colourless to faint grey at the most, SrHCl faintly brown-red, BaHCl black [*38*], CaHBr, SrHBr, and BaHBr [*39*], and CaHI, SrHI, and BaHI [*40*] 'dark'. Compare with this, the blood-red nitrogen-containing calcium subchloride Ca_3NCl [*106*]. On irradiation with X-rays, SrHCl becomes red, BaHCl green, and CaHBr, SrHBr, and BaHBr red-brown. These colours disappear very slowly in the dark [*38, 39*].

All nine hydride halides crystallise tetragonally in the PbClF type lattice (space group D^7_{4h} –P4/nmm). For the lattice constants, densities, and numbers of molecules per elementary cell, see Table 3.3. Fig. 3.3 shows the elementary cells of the hydride halides.

TABLE 3.3
LATTICE CONSTANTS, DENSITIES, AND NUMBERS OF MOLECULES PER CELL OF THE TETRAGONAL ALKALINE-EARTH METAL HYDRIDE HALIDES [*38–40*]

	a	c	d_4^{25}	*molecules/cell*
CaHCl	3.84_3	6.87_7	2.45	2
CaHBr	3.85_0	$7{\cdot}89_5$	3.38	2
CaHI	4.06_3	$8{\cdot}92_3$	3.65	2
SrHCl	4.09_2	6.94_7	3.48	2
SrHBr	4.24_6	7.27_6	4.22	2
SrHI	4.36_2	8.43_3	4.43	2
BaHCl	4.39_9	7.18_8	4.08	2
BaHBr	4.55_5	7.40_3	4.73	2
BaHI	4.81_8	7.85_1	4.80	2

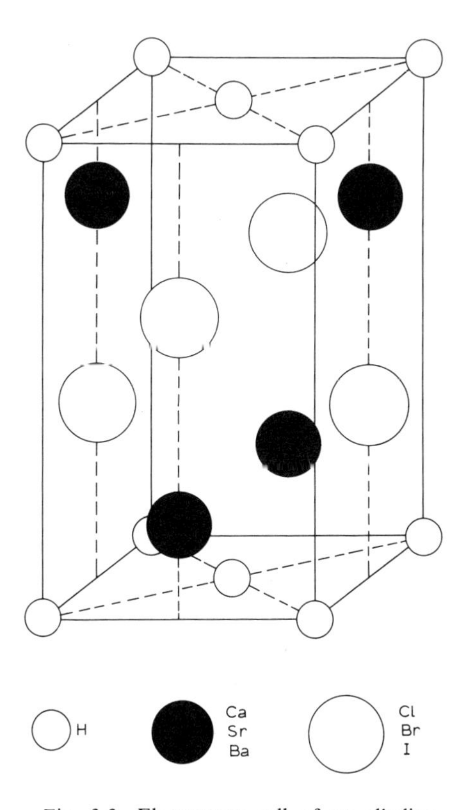

Fig. 3.3. Elementary cell of an alkaline-earth metal hydride halide [*40*].

References p. 77

According to this, the metal and halide ions each lie on the bisectors of the lateral surfaces, at different heights because of the different parameters of the alkaline-earth metal and the halide ions. The hydrogen ions are located at the corners and in the centre of the base surface. With the ionic radii inserted on the correct scale, it can be seen that the large halide ions determine the structure. The cavities which they form contain the smaller metal and hydride ions. This gives the high space occupancy of up to 70%. The lattice type present is a layer lattice, the layer sequence of which consists of five atom planes each time. As Fig. 3.4 illustrates, the (easy) cleavage probably takes place between the similarly charged halide layers abutting on one another.

Thermal analyses of the dihalide–dihydride systems of calcium, strontium, and barium show melting point maxima at the compositions MHX, and two eutectics in each case. Table 3.4 contains the melting point maxima and the lowest approximate temperatures of pyrolysis.

The alkaline-earth metal hydride halides react vigorously with water, to give hydrogen with evolution of heat [*38–41*]:

$$MHX + H_2O \xrightarrow{(+\text{ water})} M^{2\oplus}.\,aq + HO^{\ominus}.\,aq + X^{\ominus}.\,aq + H_2 + Q$$

Table 3.4 contains the enthalpies of reaction, Q, on hydrolysis in an excess of the corresponding, highly dilute, aqueous hydrohalic acids (final concentrations: 1 mole MHX/2500 moles H_2O and 0.1 mole HX/litre in the reaction above).

In air, the hydride halides decompose rapidly to a loose white powder which deliquesces in the course of time. The primary products are, possibly, definite

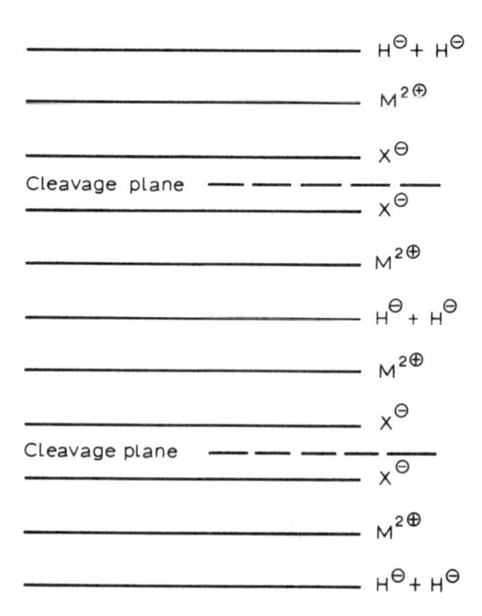

Fig. 3.4. Layer sequence in the lattice of a hydride halide (parallel to the base surface [*40*].

TABLE 3.4

MELTING POINT MAXIMA OF MHX IN A CLOSED SYSTEM AND APPROXIMATE DECOMPOSITION POINTS OF MHX IN VACUUM (10^{-3} TORR) (X = Cl[*38*], Br[*39*], I[*40*]). ENTHALPIES OF REACTION, *Q*, ON HYDROLYSIS OF MHX IN HIGHLY DILUTE, AQUEOUS HX AT 25°[*40*]

MHX	*M.p. max.* (°*C*)	*Decomp. point* (°*C*)	*Q*
CaHCl	700	630	−49.8
CaHBr	690	580	−52.6
CaHI	685	660	−55.5
SrHCl	840	740	−44.6
SrHBr	835	680	−48.2
SrHI	736	710	−51.2
VaHCl	850	760	−40.4
BaHBr	860	740	−41.2
BaHI	795	770	−43.6

alkaline-earth metal hydroxide halides. The hydride halides are insoluble in all the usual organic solvents. Their chemical behaviour, particularly their reducing action, has not yet been investigated.

REFERENCES

1. ADDISON, C. C. AND J. A. MANNING, J. Chem. Soc. *1964*, 4887.
1a. ADAMSON, G. W. AND H. M. M. SHEARER, Chem. Commun. *1965*, 240.
1b. AHLRICHS, R. AND W. KUTZELNIGG, Theor. chim. Acta. *10*, 377 (1968).
2. ALEXANDER, P. P. (General Electric Co.), U.S. Pat. 2,038,402 (1936); C. A. *30*, 3951 (1936).
3. ALEXANDER, P. P., U.S. Pat. 2,043,363 (1936); C. A. *30*, 5168 (1936).
4. ALEXANDER, P. P., U.S. Pat. 2,082,134 (1937); C. A. *31*, 5116 (1937).
5. ALEXANDER, P. P., U.S. Pat. 2,287,771 (1943); C. A. *37*, 69 (1943).
6. ALEXANDER, P. P. (Metal Hydrides, Inc.), U.S. Pat. 2,372,168 (1945); C. A. *39*, 3406 (1945).
7. ALEXANDER, P. P. (Metal Hydrides, Inc.), U.S. Pat. 2,378,368 (1945); C. A. *39*, 5056 (1945).
8. ALEXANDER, P. P. (Metal Hydrides, Inc.), U.S. Pat. 2,399,192 (1946); C. A. *40*, 4210 (1946).
9. ALEXANDER, P. P. (Metal Hydrides, Inc.), U.S. Pat. 2,401,323 (1946); C. A. *40*, 5212 (1946).
10. ALEXANDER, P. P. (Metal Hydrides, Inc.), U.S. Pat. 2,425,711/2,425,712 (1947); C. A. *41*, 7065d (1947).
11. ALEXANDER, P. P. (Metal Hydrides, Inc.), U.S. Pat. 2,702,234 (1955); C. A. *49*, 7203h (1955).
12. ALEXANDER, P. P. AND R. C. WADE (Metal Hydrides, Inc.), U.S. Pat. 2,753,255 (1956); C. A. *50*, 12798f (1956).
12a. ALMENNINGEN, A., G. GUNDERSEN AND A. HAALAND, Chem. Commun. *1967*, 557.
13. ARCHIBALD, F. R. AND P. P. ALEXANDER (Metal Hydrides, Inc.), U.S. Pat. 2,401,326 (1946); C.A. *40*, 5213 (1946).
13a. ARMSTRONG, D. R. AND P. G. PERKINS, Chem. Commun. *1968*, 352.
14. BÄHR, G. AND K. H. THIELE, Chem. Ber. *90*, 1578 (1957).
15. BALUEVA, G. A. AND S. T. IOFFE, Russ. Chem. Rev. English Transl. *31*, 439 (1962).
15a. BANFORD, L. AND G. E. COATES, J. Chem. Soc. *A1966*, 274.

15b. Banford, L. and G. E. Coates, J. Chem. Soc. *1964*, 5591.
16. Barbaras, G. D., C. Dillard, A. E. Finholt, T. Wartik, K. F. Wilzbach and H. I. Schlesinger, J. Am. Chem. Soc. *73*, 4585 (1951).
17. Bardwell, D. C., J. Am. Chem. Soc. *44*, 2499 (1923).
18. Bauer, R., Z. Naturforsch. *16b*, 557 (1961).
19. Bauer, R., Z. Naturforsch. *16b*, 839 (1961).
20. Bauer, R., Z. Naturforsch. *17b*, 201 (1962).
21. Bauer, R., Z. Naturforsch. *17b*, 277 (1962).
22. Bauer, R., Z. Naturforsch. *17b*, 626 (1962).
22a. Becker, W. C. and E. C. Ashby, Inorg. Chem. *4*, 1816 (1965).
23. Bell, N. A. and G. E. Coates, Proc. Chem. Soc. *1964*, 59.
24. Bell, N. A. and G. E. Coates, J. Chem. Soc. *1965*, 692.
24a. Bell, N. A. and G. E. Coates, Chem. Commun. *1965*, 582.
24b. Bell, N. A., G. E. Coates and F. W. Emsley, J. Chem. Soc. *A1966*, 49.
24c. Bell, N. A. and G. E. Coates, J. Chem. Soc. *A1966*, 1069.
24d. Bell, N. A. and G. E. Coates, J. Chem. Soc. *A 1968*, 628.
24e. Bell, N. A. and G. E. Coates, J. Chem. Soc. *A 1968*, 823.
25. Bergsma, J. and B. O. Loopstra, Acta Cryst. *15*, 92 (1962).
26. Biltz, W. and G. F. Hüttig, Z. Anorg. Allgem. Chem. *114*, 262 (1920).
27. Brönsted, I. N., Z Elektrochem. Angew. Physik Chem. *20*, 81 (1914).
28. Brown, A. S., P. M. Levin and E. W. Abrahamson, J. Chem. Phys. *19*, 1226 (1951).
29. Clapp, D. and R. Woodward, J. Am. Chem. Soc. *60*, 1019 (1938).
30. Chrétieu, A., W. Freundlich, M. Bichara and G. Tourné, Compt. Rend. *239*, 978 (1954).
31. Coates, G. E. and G. F. Cox, Chem. Ind. London *1962*, 269.
32. Coates, G. E. and F. Glockling, J. Chem. Soc. *1954*, 2526.
33. Coates, G. E. and F. Glockling, J. Chem. Soc. *1954*, 22.
34. Coates, G. E., F. Glockling and N. D. Huck, J. Chem. Soc. *1952*, 4496.
34a. Coates, G. E. and P. D. Roberts, J. Chem. Soc. *A 1969*, 1008.
34b. Coates, G. E. and M. Tranah, J. Chem. Soc. *A 1967*, 615.
35. Colichman, E. L., U.S. Pat. 2,909,486 (1959); C. A. *54*, 1118f (1960).
35a. Cook, T. H. and G. L. Morgan, J. Am. Chem. Soc. *91*, 774 (1969).
36. Dafert, F. W. and R. Miklauz, Monatsh. *34*, 1702 (1913).
37. Dean, L. G., C. W. McCutcheon and A. C. Doumas (Dow Chemical Co., Midland, Mich.), U.S. Pat. 3,024,091 (1962); C. A. *57*, 8198 (1962).
37a. Ehlert, T. C., R. M. Hilmer and E. A. Beauchamp, J. Inorg. Nucl. Chem. *30*, 3112 (1968).
38. Ehrlich, P., B. Alt and L. Gentsch, Z. Anorg. Allgem. Chem. *283*, 58 (1956).
39. Ehrlich, P. and H. Görtz, Z. Anorg. Allgem. Chem. *288*, 148 (1956).
40. Ehrlich, P. and H. Kulke, Z. Anorg. Allgem. Chem. *288*, 156 (1956)
41. Ehrlich, P., K. Peik and E. Koch, Z. Anorg. Allgem. Chem. *324*, 113 (1963).
42. Elektrochemische Werke, Bitterfeld, Germ. Pat. 188,570 (1905); C *1907*, II, 1283.
43. Ellinger, F. H., C. E. Holley, Jr., B. B. McInter, P. Pavone, R. M. Potter, E. Staritzky and W. H. Zachariasen, J. Am. Chem. Soc. *77*, 2647 (1955).
44. Elson, R. E., H. C. Horning, W. L. Jolly, J. W. Kury, W. J. Ramsey and A. Zalkin, UCRL-4519 (1955).
45. Elson, R. E., H. C. Horning, W. L. Jolly, J. W. Kury, W. J. Ramsey and A. Zalkin, UCRL-4519 revised (1956).
46. Ephraim, F and E. Michel, Helv. Chim. Acta *4*, 900 (1922).
46a. Fetter, N. R. and F. M. Peters, Canad. J. Chem. *43*, 1884 (1965).
47. Field, E. and M. Feller (Standard Oil Co.), U.S. Pat. 2,731,452 (1956); C. A. *50*, 6092g (1956).
48. Franklin, A. D. and R. B. Campbell, J. Phys. Chem. *59*, 65 (1955).
49. Freundlich, W. and M. Bichara, Compt. Rend. *238*, 1324 (1954).
50. Freundlich, W. and M. Bichara (Centre National de la Recherche Scientifique), Fr. Pat. 1,141,580 (1957); C. A. *54*, 6065 (1960).
51. Freundlich, W. and B. Claudel, Bull. Soc. Chim. France *1956*, 967.
52. Gardner, D., Brit. Pat. 496,294 (1938); C. A. *33*, 3542 (1939).
53. Gardner, D., U.S. Pat. 2,556,912 (1951); C. A. *45*, 7944i (1951).
54. Gautier, H., Compt. Rend. *133*, 1005 (1901).
55. Gautier, H., Compt. Rend. *134*, 1108 (1902).
56. Gibbs, D. S. and H. J. Svec, J. Am. Chem. Soc. *75*, 6052 (1953).

57. Glaseock, B. L., J. Am. Chem. Soc. *32*, 1222 (1910).
58. Goerring, D. (Farbenfabrik Bayer A. G.), Germ. Pat. 964,231 (1957); C. A. *53*, 8558 (1959).
59. Goubeau, J. and B. Rodewald, Z. Anorg. Allgem. Chem. *258*, 162 (1949).
60. Guntz, A., Compt. Rend. *132*, 965 (1901).
61. Guntz, A., Compt. Rend. *134*, 838 (1902).
62. Guntz, A. and F. Benoit, Ann. Chim. Paris [9] *20*, 1 (1923).
63. Head, E. L., C. E. Holley, Jr. and S. W. Rabideau, J. Am. Chem. Soc. *79*, 368 (1957).
64. Higuchi, I., Y. Kawana, S. Hattori and H. Chiba, J. Chem. Soc. Japan, Pure Chem. Sect. *73*, 198 (1952).
65. Holley, Jr., C. E. and J. F. Lemons (Los Alamos Scientific Laboratory), Report LA-1660 (1954, unclassified 1956).
65a. Hurd, D. T., *An Introduction to the Chemistry of the Hydrides*, Wiley, New York, 1952.
66. Johnson, W. C., M. F. Stubbs, A. E. Sidwell and A. Pechukas, J. Am. Chem. Soc. *61*, 318 (1939).
67. Jolibois, P., Compt. Rend. *155*, 355 (1912).
68. Kasarnowsky, J., Z. Physik *61*, 236 (1930).
68a. Kawana, Y., J. Chem. Soc. Japan, Pure Chem. Sect. *71*, 494 (1950); C. A. *45*, 6468h (1951).
68b. Latimer, W. M., *The Oxidation States of the Elements and their Potentials in Aqueous Solution*, 2nd Ed., Prentice–Hall, New York, 1952.
69. Lebau, P., Ann Chim. Phys. [7] *16*, 475 (1899).
70. Lemons, J. F. and W. B. Lewis, AEC Research and Development Report LA-1659 (1953, declassified 1955).
71. Lengyel, B. V., Math. Naturwiss. Ber. Ungarn *14*, 180 (1898).
72. Lissaut, K. I. (Petrolite Corp.), U.S. Pat. 2,748,867 (1956); C. A. *50*, 15057e (1956).
72a. Messer, C. E. and I. S. Levy, Inorg. Chem. *4*, 543 (1965).
73. Metal Hydrides Inc., Beverly, Mass., Technical Bulletin, ca. 1950.
74. Metal Hydrides Inc., Beverly, Mass., Technical Bulletin 'Hydrimix'.
75. Messon, G. A. and O. P. Kolchin, Sb. Nauchn. Tr. Mosk. Inst. Tsvetn. Metal. i Zolota *25*, 195 (1955).
76. Meyer, M. and V. Altmayer, Ber. *41*, 3074 (1908).
77. Meyerhofer, A. F., Brit. Pat. 250,211 (1925); C. A. *21*, 991 (1927).
77a. Mikheeva, V. I. and V. N. Konoplev, Russ. J. Inorg. Chem. *10*, 1148 (1965).
78. Moissan, H., Compt. Rend. *127*, 29 (1898).
79. Moissan, H., Ann. Chim. Phys. [7] *18*, 289 (1899).
80. Moissan, H., Bull. Soc. Chim. France [3] *21*, 876 (1899).
81. Moissan, H., Compt. Rend. *136*, 591 (1903).
82. Morrison, C. R., J. P. Pressau, P. A. Joyner and R. M. Adams (Calery Chemical Comp.), Report CCC-1024–TR-19 (1954).
83. Morrogh, H. (British Cast Iron Research Association) U.S. Pat. 2,747,990 (1956); C. A. *50*, 11926i (1956).
84. Pease, P. N. and L. Stewart, J. Am. Chem. Soc. 47, 2763 (1925).
85. Pecht, M. M. (Hardy Metallurgical Co.), U.S. Pat. 2,392,545 (1946); C. A. *40*, 1980 (1946).
86. Pietsch, E., Z. Elektrochem. *39*, 577 (1933).
87. Potter, E. C. and J. O'M. Bockris, Colloq. Intern. Centre Natl. Rech. Sci. (Paris) *39*, *Electrolyse*, C3–C6 (1952).
88. Proskurnin, M. and J. Kasarnowsky, Z. Anorg. Allgem. Chem. *170*, 308 (1928).
89. Racine, J., Bull. Soc. Chim. France *1951*, 854.
90. Reich, S. and H. O. Serpek, Helv. Chim Acta *3*, 138 (1920).
90a. Reilly, J. J. and R. H. Wiswall, Inorg. Chem. *6*, 2220 (1967).
90b. Reilly, J. J. and R. H. Wiswall, Jr., *Inorg. Chem.* **7**, 2254 (1968).
91. Roederer, G., Bull. Soc. Chim. France [3] *35*, 507 (1906).
92. Rosenbaum, C. K. and J. H. Walton, J. Am. Chem. Soc. *52*, 3568 (1930).
93. Rundle, R. E. and P. H. Lewis, J. Chem. Phys. *20*, 132 (1952).
94. Rundle, R. E. and A. I. Snow, J. Chem. Phys. *18*, 1125 (1950).
94a. Schlesinger, H. I., H. C. Brown and E. K. Hyde, J. Am. Chem. Soc. *75*, 212 (1953).
95. Schmidt, M. and G. Talsky, Chem. Ber. *90*, 1683 (1957).
96. Schumb, W. C., E. F. Sewell and A. S. Eisenstein, J. Am. Chem. Soc. *69*, 2029 (1947).
97. Schwarz, R. and A. Köster, Z. Anorg. Allgem. Chem. *285*, 1 (1956).
97a. Sheppard, Jr., G. L. Ter Haar and E. M. Marlett, *Inorg. Chem.* **8**, 976 (1969).
98. Shushunov, V. A. and A. F. Shafiev, Zh. Fiz. Khim. *26*, 672 (1952); C. A. *49*, 5939h (1955).

99. Smalley, O., Brit. Pat. 666,095 (1952); C. A. *46*, 5515f (1952).
100. Snow, A. I. and R. E. Rundle, Acta Cryst. *4*, 348 (1951).
101. Stamper, Jr., J. F., C. E. Holley, Jr. and J. F. Suttle, J. Am. Chem. Soc. *82*, 3504 (1960).
102. Steiger, L. V. (Maywood Chemical Works), U.S. Pat. 2,735,820 (1956); C. A. *50*, 815a (1956).
102a. Stöcklin, G., F. Schmidt–Bleek and W. Herr, Angew. Chem. *73*, 220 (1961).
103. Tiede, E., Germ. Pat. 417,508 (1925); Chem. Zentr. *96*, 1790 (1925).
104. Treadwell, W. D. and J. Sticher, Helv. Chim. Acta *36*, 1820 (1953).
105. Wade, R. C. and P. P. Alexander (Metal Hydrides Inc.), U.S. Pat. 2,702,740 (1955); C. A. *49*, 7204a (1955).
106. Wehner, G., Z. Anorg. Allgem. Chem. *276*, 72 (1954).
107. Weller, S. and L. Wright, J. Am. Chem. Soc. *76*, 5302 (1954).
108. Wiberg, G. and R. Bauer, Z. Naturforsch. *5b*, 396 (1950).
109. Wiberg, E. and R. Bauer, Z. Naturforsch. *5b*, 397 (1950).
110. Wiberg, E. and R. Bauer, Z. Naturforsch. *6b*, 171 (1951).
111. Wiberg, E. and R. Bauer, Z. Naturforsch. *7b*, 58 (1952).
112. Wiberg, E. and R. Bauer, Z. Naturforsch. *7b*, 129 (1952).
113. Wiberg, E. and R. Bauer, Z. Naturforsch. *7b*, 131 (1952).
114. Wiberg, E. and R. Bauer, Chem. Ber. *85*, 593 (1952).
115. Wiberg, E., R. Bauer and H. Göltzer, Germ. Pat. 862,004 (1951).
116. Wiberg, E., H. Göltzer and R. Bauer, Z. Naturforsch. *6b*, 394 (1951).
117. Wiberg, E. and P. Strebel, Ann. *607*, 9 (1957).
118. Winkler, C., Ber. *24*, 1966 (1891).
119. Wright, L and S. Weller, J. Am. Chem. Soc. *76*, 5305 (1954).
120. Wright, L. and S. Weller, J. Am. Chem. Soc. *76*, 5948 (1954).
121. Zabel, H. W., Chem. Industries *60*, 37 (1947).
122. Zartmann, W. H. and H. Adkins, J. Am. Chem. Soc. *54*, 3398 (1932).
123. Ziegler K., Brit. Pat. 713,081 (1955); U.S. Pat. 2,699,457 (1955); C. A. *49*, 3576d (1955).
124. Zintl, E. and A. Harder, Z. Elektrochem. *41*, 33 (1935).

Chapter 4

BORON HYDRIDES

1. Historical and general review

Liquid boron hydrides have been known for a long time. The existence of a volatile boron hydride was demonstrated for the first time in 1879 by Jones[*377*], who obtained it in traces when he treated magnesium boride Mg_3B_2 with hydrochloric acid. Similar results were reported in 1881 by Jones and Taylor[*380*], in 1890 by Winkler[*850*], and in 1891 by Sabatier[*631*]. However, because of the inadequate apparatus available at that time, no reactions worth mentioning of the boron hydride (which was formed only in traces in all the experiments) were carried out, nor were there any conclusive analyses. Extremely successful investigations on boron hydrides began in 1912, as a result of the techniques developed by Stock *et al.*[*731, 735–747*]. Even at that time they isolated about half of the currently known uncharged boranes containing only B and H. However, it was impossible to elucidate the bond situation with the physical means and theoretical ideas of that period. The theoretical foundations were laid only in the thirties, and in the case of the structures of boranes, they are still in a state of flux. Spectroscopy (infrared, ultraviolet, Raman, nuclear resonance, mass and electron spectroscopy), an extremely valuable tool indispensable for elucidating the bond situation, was used only in the fifties, after the development of commercial instruments. X-ray and electron diffraction studies are also used to contribute to the overall picture. For a review see[*2, 3, 295b, 354b, 467, 533b, 541c*], for the nomenclature see[*15d*].

1.1 Bond types in boranes

With its three singly-occupied orbitals, boron forms three σ-bonds, for example, in the compounds BCl_3, BMe_3, and BH_3. Monoborane(3), BH_3, is the simplest member of the boron hydride series. The fourth, unoccupied, orbital of boron is available for dative σ-bonds with electron donors (for example $H_3B \leftarrow\cdots NMe_3$). In the absence of electron-donors, the electrons and orbitals present arrange themselves into 'three-centre bonds' with a deficiency of electrons (electron-deficient molecules) (*cf.* section 2). In what follows, a $B{-}H^{\mu}{-}B$ bridge (H^{μ} denotes a bridge H-atom) produced in this way will be denoted by the bond symbol $B\overset{H}{\frown}B$, and a $B{-}B^{\mu}{-}B$ bridge will be denoted by the bond symbol $B\overset{B}{\frown}B$ (open three-centre bond) or $B\diagdown\!\diagup B$ with a vertical line to B below (closed three-centre bond). The simplest boron hydride, monoborane(3)[1] BH_3 is electron-unsaturated and therefore

[1] The names of the boron hydrides are formed as follows: the first syllable gives the number of B-atoms; the number of H-atoms is attached in brackets to the ending '. . . borane'.

References p. 355

dimerises to the relatively heat-stable diborane(6) $(BH_3)_2 = B_2H_6$, with two B–H$^{\mu}$–B bridges[2]:

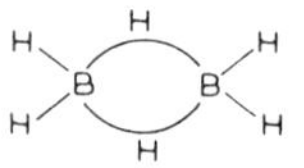

The supply of energy converts diborane(6) into higher boranes with other bond arrangements – for example, to tetraborane(10), B_4H_{10}, and decaborane(14), $B_{10}H_{14}$:

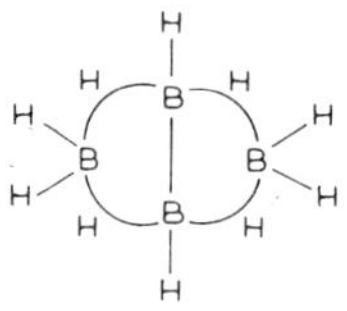

B_4H_{10}
(4 B–H$^{\mu}$–B bridges)

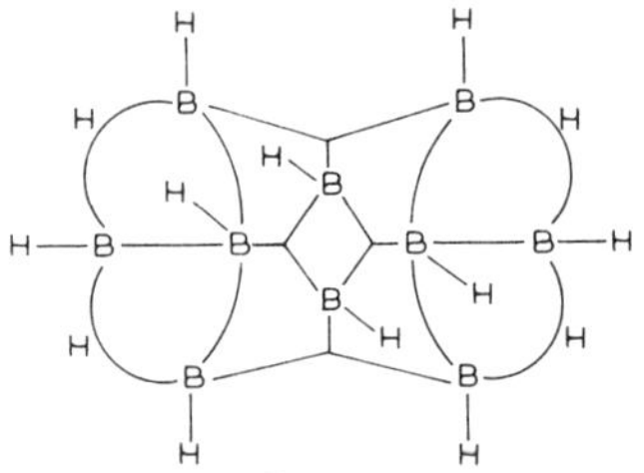

$B_{10}H_{14}$
(4 B–H$^{\mu}$–B and 6 B–B$^{\mu}$–B bridges)

1.2 Mechanisms and types of reactions

Boranes with their B–H, B–B, B–H$^{\mu}$–B, and B–B$^{\mu}$–B bonds (it must be borne in mind that such a separation into isolated bonds is only one of the possible ways of describing the structure of the skeleton) react by the following 15 mechanisms and types of reaction.

(1) Asymmetric splitting of B–H$^{\mu}$ double bridges

Positive boronium ions and negative boranate ions[3] are formed:

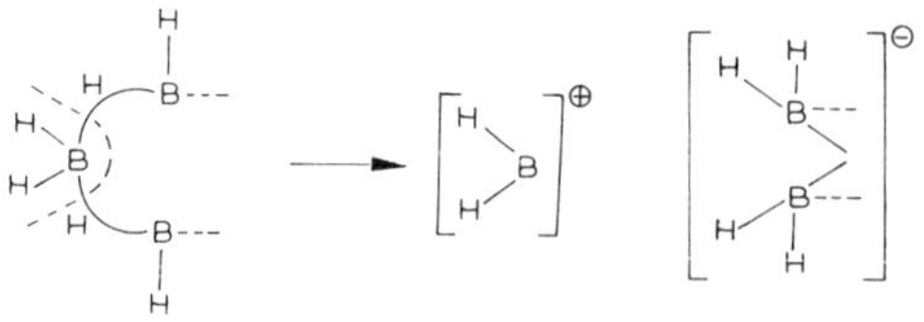

Typical examples: at low temperatures, ammonia ruptures the B–H$^{\mu}$–B double bridge in diborane(6) or tetraborane(10) with the formation of dihydrodiammineboron(III) boranate(4), $(NH_3)_2BH_2^{\oplus}BH_4^{\ominus}$, and dihydrodiammine-

[2]The number of bridge bonds in the boranes here, as in all subsequent cases, is equal to the number of boron atoms. In the case of boron anions, there is one bridge bond less for each negative charge.
[3]In the segment of a topological structure of a borane shown, each continuous straight line emerging from B indicates one electron of boron. Each broken line emerging from B signifies ½ electron of boron. Each bridge bond (curved line) emerging from B indicates ½ electron of boron.

boron(III) triboranate(8), $(NH_3)_2BH_2^{\oplus}B_3H_8^{\ominus}$:

(2) *Symmetrical rupture of B–H$^{\mu}$–B double bridges*

Donor-acceptor adducts are formed:

Typical reactions: a donor solvent—for example, tetrahydrofuran—ruptures the B–H$^{\mu}$–B double bridge in diborane with the formation of a donor–borane adduct THF · BH_3. With such solutions, ammonia (at low temperatures) forms ammonia-borane $NH_3 \cdot BH_3$ (there is no asymmetric rupture wih ammonia here, as in the previous case):

$$\xrightarrow[\text{nucleophilic displacement}]{+2NH_3,\ -2C_4H_8O} 2\ H_3N \dashrightarrow BH_3$$

The rupture requires energy. For the symmetrical rupture of diborane B_2H_6 and alkylated diboranes $B_2H_{6-n}R_n$, this energy decreases with increasing substitution (H^{μ} is substituted last)[kcal/mole]: B_2H_6 28.5, B_2H_5Me 28.0, asymm.-$B_2H_4Me_2$ 28.5, $B_2H_3Me_3$ 26.5, $B_2H_2Me_4$ 25.0, B_2HMe_5 < 5, B_2Me_6 ~ 0.

Similarly, with ether tetraborane(10) forms the borane–ether complex $R_2O \cdot BH_3$ and the triborane(7)–ether complex $R_2O_7 \cdot B_3H_7$:

References p. 355

(3) Nucleophilic attack

The non-sterically hindered B–H σ-bond (a), the B–B$^{\mu}$–B three-centre bond (b) and the B–H$^{\mu}$–B bridge bond (c) can be attacked nucleophilically. Typical examples are:

(a) The alcoholysis of $BH_4^{\ominus}$ (polar four-centre mechanism):

$$ROH + BH_4^{\ominus} \longrightarrow \begin{matrix} H_3B \cdots H^{\delta-} \\ \vdots \quad\quad \vdots \\ RO \cdots H^{\delta+} \end{matrix} \xrightarrow{-H_2} ROBH_3^{\ominus}$$

(b) Alcohol attacks the B^5–B^6–B^7 three-centre bond in $B_{10}H_{12}(N{\equiv}CMe)_2$, which is electron-poor in comparison with the donor, 'nucleophilically'. B_9H_{13} (n≡CMe)[4] is formed, with elimination of a B-atom:

$$B_{10}H_{12}(NCMe)_2 \xrightarrow[-B(OR)_3,\ -MeCN,\ -H_2]{+2\,ROH} B_9H_{12}(NCMe)$$

(c) An example of a nucleophilic attack by a donor on the B–H$^{\mu}$–B bridge is the symmetrical rupture of the B–H$^{\mu}$–B double bridge [see under (2)].

(4) Nucleophilic displacement

In donor–borane adducts, potentially stronger donor ligands (see section 1.3.2) displace more weakly bound ligands. For example:

$$(H_3C)_2O \cdots\!\rightarrow BH_3 \xrightarrow[-(CH_3)_2O]{+C_4H_8O} C_4H_8O \cdots\!\rightarrow BH_3$$

(5) Electrophilic attack

Sterically unfavourable electron-rich H atoms in higher boranes and the B–H bonds in sterically demanding donor–borane complexes are attacked electrophilically. Typical reactions are:

(a) Electrophilic attack of the electron-rich B–H bonds in the open pyridine–diphenylborane complex by the protons of water (three-centre mechanism):

$$C_5H_5N\cdot BH(C_6H_5)_2 + H_2O \longrightarrow [C_5H_5N\cdots\!\rightarrow B(C_6H_5)_2(H\cdots H\text{–}OH)] \longrightarrow (C_6H_5)_2BOH + H_2 + C_5H_5N$$

[4]The different numbering of the B-atoms in the topological structures of $B_{10}H_{12}(NCMe)_2$ and B_9H_{13} (NCMe) does not denote any isomerization of the B-skeleton. The systems of numbering are merely those accepted for the B_{10} and B_9 skeletons.

(b) Iodination of $B_{10}H_{14}$ with elementary iodine:

$$B_{10}H_{14} + I^{\oplus} \longrightarrow B_{10}H_{13}I + H^{\oplus}$$

(c) Alkylation with carbonium ions:

$$B_{10}H_{14} + R^{\oplus} \xrightarrow[-AlX_3, -HX]{+AlX_4^{\ominus}} B_{10}H_{13}R$$

(6) Elimination of protons

Lewis bases such as hydride ions, hydroxyl ions, and carbanions eliminate relatively electron-poor H-atoms in boranes as protons. Typical reactions are:

$$B_{10}H_{14} + NaH \rightarrow Na^{\oplus}B_{10}H_{13}^{\ominus} + H_2$$
$$B_{10}H_{14} + RMgI \rightarrow B_{10}H_{13}MgI + RH$$

Mechanism:

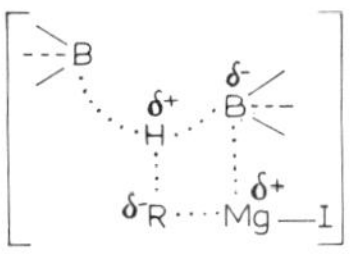

(7) Hydride elimination

Electron-rich non-sterically-hindered B–H bonds are attacked by cations (*e.g.* $R^{\oplus}$, $H^{\oplus}$, or $H^{\delta+}$) with the elimination of a hydride ion. Typical reactions:

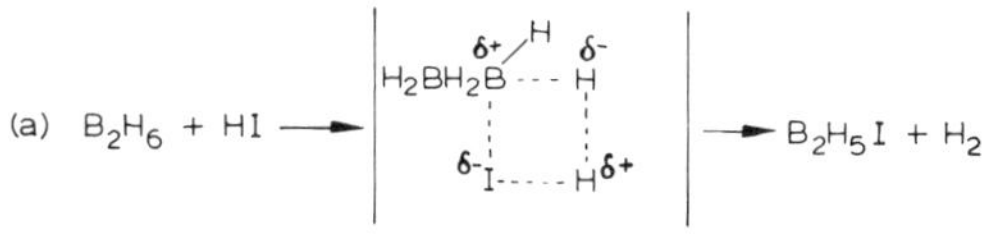

(a) $B_2H_6 + HI \longrightarrow [\ldots] \longrightarrow B_2H_5I + H_2$

(b) Another type of hydride elimination is the intramolecular abstraction of H from ammonia-borane containing N–$H^{\delta+}$ and B–$H^{\delta-}$ bonds. $NH_3 \cdot BH_3$ splits off hydrogen at elevated temperatures:

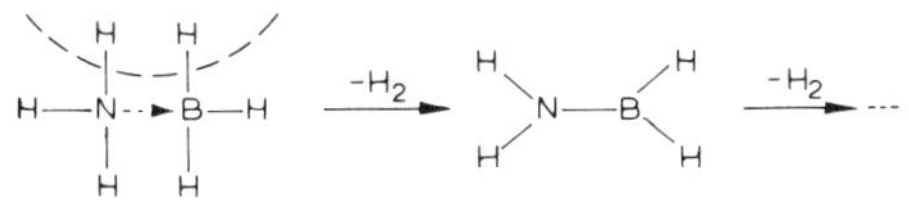

$$H_3N \rightarrow BH_3 \xrightarrow{-H_2} H_2N\text{—}BH_2 \xrightarrow{-H_2} \cdots$$

(8) Addition of protons (rare)

With HCl in ethanol, $Na_2B_{10}B_{14}$ forms $NaB_{10}H_{15}$:

$$B_{10}H_{14}^{2\ominus} + H^{\oplus} \longrightarrow B_{10}H_{15}^{\ominus}$$

(9) Elimination of electrons

$$B_{10}H_{14}^{2\ominus} + I_2 \longrightarrow B_{10}H_{14} + 2I^{\ominus}$$

References p. 355

(10) Addition of electrons

The B–H$^{\mu}$–B bridges (formed because of electron deficiency) split by taking up two electrons:

$$\text{---B}\overset{H}{\frown}\text{B---} + 2e^{\ominus} \longrightarrow \text{---B:} \quad \overset{H}{\searrow}\text{B---} \quad \text{(first stage)}$$

$$B_2H_6 + 2\,Na \longrightarrow (Na^{\oplus})_2[H_2\dot{B}\text{--}H\text{--}BH_2]^{2\ominus} \xrightarrow{\text{rearrangement}}$$

$$(Na^{\oplus})_2 [H_3B\text{—}BH_3]^{2\ominus} \quad (\xrightarrow{+\,B_2H_6} Na_2B_4H_{12} \longrightarrow NaB_3H_8 + NaBH_4)$$

$$B_4H_{10} + 2K \longrightarrow (K^{\oplus})_2 [B_4H_{10}]^{2\ominus} \quad (\xrightarrow{+B_4H_{10}} K_2B_8H_{20} \longrightarrow 2\,KB_3H_8 + BH_3 + \frac{1}{x}(BH)_x)$$

(11) Intramolecular migration of H

$$R_2C{=}\overset{H}{C}\text{—}BH_2 \rightarrow R_2\overset{H}{C}\text{—}\overset{H}{C}\text{—}BH \rightarrow \frac{1}{x}[R_2CHCHBH]_x$$

(12) Addition to multiple C—C bonds

A B–H bond adds in a four-centre *cis* addition to alkenes or alkynes with the formation of *trans* compounds (BH→BC):
δ+δ−

$$>C{=}C< + >B\text{—}H \longrightarrow \text{—}C{\cdots}C\text{—}\;(>B{\cdots}H) \longrightarrow \text{—}C(B<)\text{—}C(H)\text{—}$$

$$\text{—}C{\equiv}C\text{—} + >B\text{—}H \longrightarrow \text{—}C{\cdots}C\text{—}\;(>B{\cdots}H) \longrightarrow \text{—}C(B<){=}C(H)\text{—}$$

Corresponding to the polarisation B–H, boron is generally directed in accordance with the anti-Markovnikov rule, if there are no steric reasons against it. In the case of $BHCl_2$, on the other hand, because of the low electron density of the H-atom, the entry of boron takes place in accordance with Markovnikov's rule.

(13) Transformation, isotope exchange, isomerisation

(a) On the addition of energy, boranes undergo *transformations* according to various sequences of reactions (see section 7). Diborane forms higher boranes; higher boranes yield lower boranes. Heat, light, Hg6 ($^3P_1$) atoms[5], and electric

[5]Excited Hg-atoms.

current act as sources of energy. Typical transformation:

$$\tfrac{1}{2}(BH_3)_2 \xrightarrow{\text{heating}} \tfrac{1}{2}(BH_3)_2^* \rightleftharpoons BH_3 \xrightarrow{+B_2H_6} B_3H_9 \xrightarrow{-H_2} B_3H_7 \begin{cases} \xrightarrow{+B_2H_6,\,-BH_3} B_4H_{10} \\ \xrightarrow[+B_2H_6,\,-H_2]{} B_5H_{11} \end{cases}$$

(b) The *exchange* of H and D takes place both through H and D units and through BH_3 and BD_3 units. Typical reactions are:
exchange via BH_3 in the B_2H_6–B_2D_6 system:

$$^{11}BH_3 + (^{10}BD_3)_2 \xrightarrow[-^{10}BD_3]{} {}^{11}BH_3\,{}^{10}BD_3 \xrightarrow{\text{isomerisation}} {}^{11}B^{10}BH_3D_3$$

exchange via H in the B_5H_9–B_2D_6 system:

$$B_2^*D_6 + B_5H_9 \rightarrow B_2^*DH_5 + B_5H_4{}^{\mu}D_5$$

(c) The intramolecular *isomerisation* of higher boranes, catalysed by an external donor, takes place via borane anions:

$$1-EtB_5H_8 + Me_3N \rightarrow EtB_5H_7^{\ominus}Me_3NH^{\oplus} \rightarrow 2\text{-}EtB_5H_8 + Me_3N$$

The equilibrium of the polar $(NH_3)_2BH_2^{\oplus}BH_4^{\ominus}$ with the less polar ammonia-borane $H_3N{\cdot}BH_3$ is an internally donor-catalysed isomerization:

$$[(H_3N)_2BH_2]^{\oplus}[BH_4]^{\ominus} \rightleftharpoons 2\,H_3N{\cdot}BH_3$$

(14) Dismutation and commutation

Partially substituted boranes dismute to per-substituted and non-substituted boranes; boranes with various substituents commute. Typical reactions:

$$RBH_2{\cdot}BH_3 \text{ (H-bridged)} \rightleftharpoons \begin{cases} 2\,RBH_2 \rightleftharpoons (RBH_2)_2 \xrightarrow{\text{isomerisation}} R_2B(\mu\text{-}H)_2BH_2 \rightleftharpoons R_2BH + BH_3 \\ 2\,BH_3 \rightleftharpoons B_2H_6 \end{cases}$$

(15) Condensation and polymerisation

Ammonia-borane (borazane) and partially B-substituted borazanes containing $H^{\delta+}$ and $H^{\delta-}$ can condense with the elimination of H_2. In this way, borazane $H_3B \leftarrow NH_3$ eliminates molecular hydrogen at elevated temperatures, forming aminoborane $H_2B \rightleftharpoons NH_2$ (borazene). Further elimination of H_2 gives rise to borazine HB$=$NH, which preferentially forms a trimeric ring, $H_3B_3N_3H_3$ 'borazine'. Finally, at very high temperatures, the remaining hydrogen is eliminated with the formation of (polymeric) boron nitride BN:

MeS–BH_2 polymerises to weak coordination polymers and Me_2P–BH_2 to strong coordination polymers:

Some reactions of B_2H_6 and $B_{10}H_4$ are shown schematically in Figs. 4.1 and 4.2.

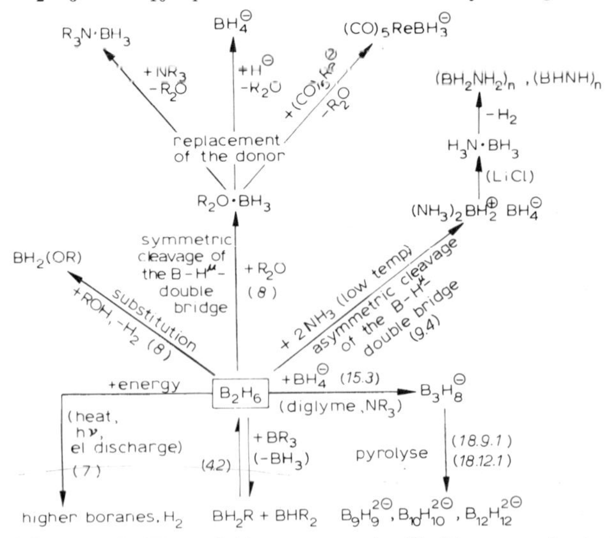

Fig. 4.1. Reactions of diborane(6). The stoichiometry must be filled in appropriately. The figures in italics refer to the corresponding sections in the text.

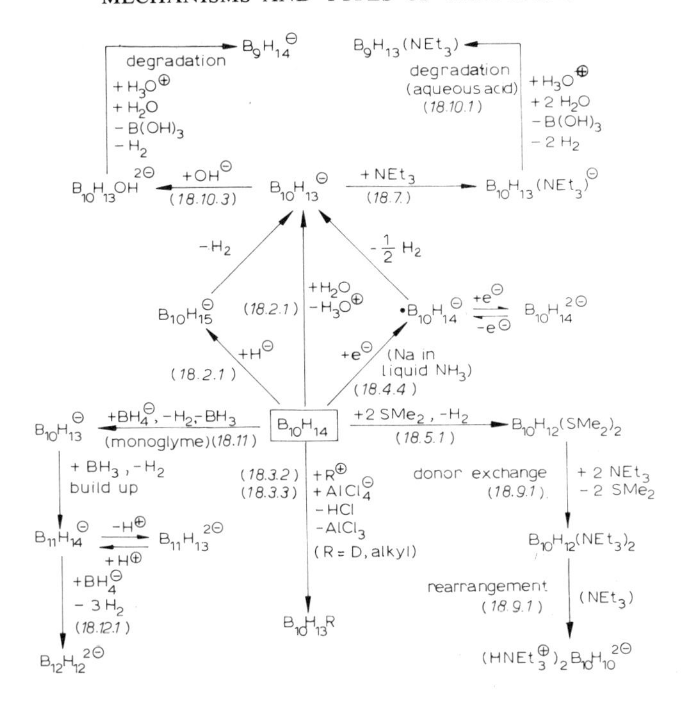

Fig. 4.2. Reactions of decaborane(14). The stoichiometry must be filled in appropriately. The figures in italics refer to the corresponding sections in the text.

1.3 Types of boranes

There are four basic types of boron hydrides: (1) boranes containing only B and H, (2) acceptor (borane)–donor compounds, (3) H-substituted boranes, (4) B-substituted boranes.

1.3.1 Boranes containing only B and H

The boranes containing only B and H are subdivided into uncharged and charged boranes. For example, with B–H, B–H$^{\mu}$–B, B–B, and B–B$^{\mu}$–B bonds[6]:

B_2H_6 (2 B–H$^{\mu}$–B bridges)

B_4H_{10} (4 B–H$^{\mu}$–B bridges)

B_6H_{10} (4 B–H$^{\mu}$–B + 2 B–B$^{\mu}$–B bridges)

[6]On this, compare footnote 2.

$B_{10}H_{14}^{2\ominus}$

(2 B–H$^{\mu}$–B and 6 B-B$^{\mu}$-B bridges)

1.3.2 Borane-donor adducts

Examples of borane–donor compounds are $R_2HB{\cdot}OR_2$, $B_3H_7{\cdot}OR_2$, $H_3B{\cdot}SMe_2$, $H_3B{\cdot}NH_3$, $H_3B{\cdot}NC_5H_5$, $H_3B{\cdot}PMe_3$, and $H_3B{\cdot}CO$. In these compounds, neither the 'donor strength' of the donor D nor the 'acceptor strength' of the acceptor A are constants. On the contrary, they depend on the combination of donor and acceptor, three factors determining the stability of the acceptor–donor adducts A–D.

(a) The nucleophilicity of the donor and the electrophilicity of the acceptor. Highly nucleophilic donors and highly electrophilic acceptors form relatively strong A–D σ-bonds, with polarisation of both atoms: $\overset{-}{A} \leftarrow\cdot\cdot \overset{+}{D}$.

(b) Occupation of free orbitals of the donor atom. If the donor atom D in the A–D adduct possesses free orbitals in a lower position, these are partially occupied by electrons from the acceptor ligand L. Partial double bonds form, with partial compensation of the polarisation $\overset{-}{A}$–$\overset{+}{D}$. The A–D σ-bond is then reinforced by a A–D π-bond: L–A⟵··D. Thus:

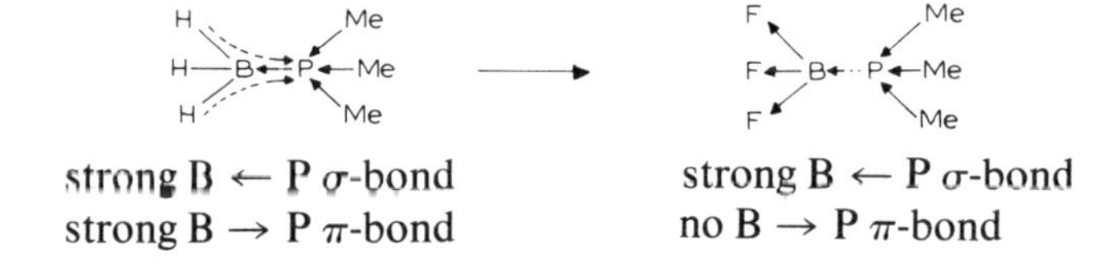

strong B ← P σ-bond
strong B → P π-bond

strong B ← P σ-bond
no B → P π-bond

(c) Occupation of free orbitals of the acceptor atom. If the acceptor atom A in the A–D adduct possesses free orbitals at a lower level, these are partially occupied by electrons from the acceptor ligand L. The electrophilicity of the acceptor A falls and the A–D σ-bond becomes weaker. For example[7]:

$Me_3B \leftarrow\cdots PMe_3$

Stability of the borane-donor adducts:

(1) $F_3B{\cdot}OR_2 > F_3B{\cdot}SR_2 > F_3B{\cdot}SeR_2$.

This shows the decrease in the strength of the A–D σ-bond in accordance with (a), because of the decreasing electronegativity of the donor (O > S > Se). The

[7]The symbol '←---' expresses the weak B ← P σ-bond, and the symbol '→' the +I effect of the methyl ligand (also a σ-bond).

A–D σ-bond is indeed strengthened by π-electron back-coordination $F \rightarrow B$ according to (b), but since all three adducts in the comparison are BF_3 adducts, this influence is eliminated when the stabilities are compared.

(2) $H_3B{\cdot}SR_2 > H_3B{\cdot}SeR_2 = H_3B{\cdot}OR_2$; $R_3B{\cdot}SR_2 > R_3B{\cdot}SeR_2 > R_3B{\cdot}OR_2$.
According to (a), the strength of the A–D bond should fall in the sequence O > S > Se. According to (b) the A–D bond is strengthened in the case of sulphur, since S possesses empty 3d orbitals in a lower position that are occupied by electrons of the acceptor ligand hydrogen in the β position ($H{-}\overset{\beta}{C}{-}\overset{\alpha}{B}{\cdot}D$).

(3) $F_3B{\cdot}NR_3 > F_3B{\cdot}PR_3$[$> F_3B{\cdot}AsR_3$ and $F_3B{\cdot}SbR_3$ (unstable)].
According to (a), the strength of the A–D bond decreases as in (1).

(4) $H_3B{\cdot}PR_3 > H_3B{\cdot}NR_3 > H_3B{\cdot}AsR_3$ [$> H_3B{\cdot}SbR_3$].
As in (2), the strength of the bond is increased predominantly owing to back-coordination according to (b).

(5) $R_3B{\cdot}NR_3 > R_3B{\cdot}PR_3 > R_3B{\cdot}AsR_3$.
Regarded from the point of view of the donor, the sequence should decrease similarly to (4). Here, however, the influence of the σ-bond masks that of the π-bond.

(6) $F_3B{\cdot}NR_3 > F_2RB{\cdot}NR_3 > FR_2B{\cdot}NR_3 > R_3B{\cdot}NR_3$.
The A–D σ-bond is strengthened by the highly electronegative acceptor ligand fluorine.

(7) $H_3B{\cdot}CO \sim H_3B{\cdot}PF_3$ [$\gg F_3B{\cdot}CO, F_3B{\cdot}PF_3$, non-existent].
Here again, the π-bond strengthens the A–D adduct according to (b).

(8) $Br_3B{\cdot}NC_5H_5 > Cl_3B{\cdot}NC_5H_5 > F_3B{\cdot}NC_5H_5$.
The reasons for the stability of the pyridine adducts are not accurately definable. The proportion of B–X π-bond ($>B \Leftarrow X$) increases in the sequence $BBr_3 < BCl_3 < BF_3$. According to (c), therefore, the boron in BF_3 is less electrophilic than the boron in BCl_3 or BBr_3. Pyridine possesses filled orbitals at a lower level, which can form B–N π-bonds with the empty boron orbitals (B–N bond strengthening).

1.3.3 H-Exchange in boranes

The H-atoms of a (charged or uncharged) borane can be replaced by elements (or groups) to the right of boron in the periodic system. Examples: $BHCl_2$, $B_{10}H_{13}I$, $BH(OR)_3^{\ominus}$, $BH_2(NR_2)$, $BH_2(AsR_2)$, BH_2R, B_5H_8R.

If a donor–borane adduct satisfies the conditions given in section 1.2 (15), it condenses with the elimination of H_2. One of the most interesting condensation products is borazine $H_3B_3N_3H_3$, which is similar in many physical properties (X-ray diffraction, electron diffraction, boiling point, melting point, etc.) and in many chemical properties to benzene. As in benzene, all the atoms lie in the same plane. The electronic structure can be described as a resonance hybrid between the form with localised π-bonds (polarisation of B and N) and the form with a delocalised π-bond system:

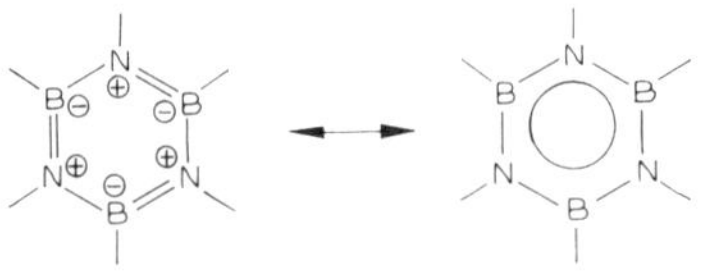

There is no phosphorus analogue of borazine ('phosphazine'). However, unusually stable six-membered and eight-membered rings of the type of $(R_2PBH_2)_3$ and $(R_2PBH_2)_4$ are formed:

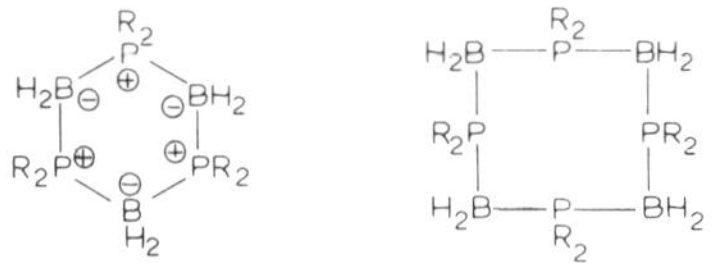

Chair structure

Corrugated 8-membered ring

In these structures, the P→B σ-bond, already strong, is strengthened further by the back-coordination of H-electrons (on the boron) to the P-atom (see section 1.3.2).

1.3.4 B-Exchange in Boranes

C-atoms can replace two B-atoms of the $B_{12}H_{12}^{2\ominus}$ icosahedron. In this way, when $B_{10}H_{12}(\text{donor})_2$ is heated with acetylene, 'carborane' $B_{10}C_2H_{12}$ is formed, with evolution of H_2 (*cf.* section 19). Its molecules form an extraordinarily thermally and chemically stable (somewhat distorted) icosahedron with neighbouring C-atoms. On prolonged heating, carborane isomerises to neocarboranes, which are also unusually stable. The neocarboranes form less distorted or undistorted icosahedra with isolated C-atoms.

2. Steric and topological structures of boranes

If in any given molecule n atomic orbitals are used for bonding purposes, it is possible, for example, to obtain from them (in addition to $n/2$ antibonding orbitals), $n/2$ bonding orbitals for two-centre bonds, which must be occupied with n electrons. If more than n electrons are available, these must occupy positions in the energetically unfavourable antibonding orbitals high in the energy level scheme.

If, again, n atomic orbitals are used in the molecule, these can form (in addition to $n/3$ antibonding orbitals) $n/3$ bonding orbitals for three-centre bonds. They must be filled with only $\frac{2}{3}n$ electrons, and not with n electrons as above. If more than $\frac{2}{3}n$ electrons are available, these must occupy the anti-bonding orbitals as before.

Superfluous electrons, *i.e.* those above the number n or $\frac{2}{3}n$ in the respective cases, may occupy instead of antibonding orbitals those orbitals that are not used for bonds. These non-bonding orbitals each offer accommodation to two electrons ('lone electron pairs').

From what has been said above it may be deduced that two-centre bonds are preferred when in a given molecule the number of orbitals occupied is equal to the number of electrons provided by the atoms (as in the hydrocarbons). If the number of orbitals that can be used for bonds is smaller than the number of electrons present, two-centre bonds are again preferred; the superfluous electrons form lone electron pairs (as in ammonia). If the number of orbitals that can be used for bonds is greater than the number of electrons present in the molecule, three-centre bonds are preferred (as in the boranes).

A few principles [*166*, *190*, *469*] must be observed for the molecular structures of the boranes to be built up from two- and three-centre bonds (in shortened form).

(1) Each H-atom provides one s-orbital and one electron. Each B-atom provides one s- and three p-orbitals, together with three electrons. The orbitals hybridise when necessary.

(2) Each external B–H bond is a localised two-centre bond. It requires the H-orbital and one hybridized B-orbital, together with one H-electron and one B-electron. The bond is non-polar because of the similar electronegatives χ of H and B($\chi_H = 2.20$, $\chi_B = 2.01$ in Pauling units). One B-atom forms only 0, 1, or 2 external B–H bonds.

(3) Each B–H^μ–B bridge bond is an occupied, localised, three-centre bond. It requires one H-orbital and one hybridised orbital from each of the two B-atoms, together with one H-electron and $\frac{1}{2}$ electron from each B. It is also, at least at first, non-polar.

(4) The orbitals and electrons of each B-atom of a borane are distributed in such a way that the requirements of the external B–H bonds and of the B–H^μ–B bridge bonds are first satisfied. The remaining orbitals and electrons are assigned to the molecular orbitals of the B-skeleton.

(5) The structure of the molecular orbitals of the B-skeleton is determined by the immediate environment of each individual B-atom.

(6) There are only four structural units in B–H boranes: BH units with tetrahedral sp^3 orbitals, BH units with trigonal sp^2 plus p orbitals, BH_2 units with trigonal sp^2 plus p orbitals, and BH_2 units with tetrahedral sp^3 orbitals.

(7) Each structure of a charged or uncharged borane molecule can be characterised by a four-figure index ('styx'). In this, the letters have the following

References p. 355

significance:

LETTER	NUMBER OF
s	$B \overset{H^\mu}{\frown} B$ bridge bonds
t	$B \overset{B^\mu}{\frown} B$ or $B \diagup\!\overset{B^\mu}{|}\!\diagdown B$ bridge bonds [8]
y	B–B simple bonds
x	$B\langle{}^{H}_{H}$ groups [9]

(8) The following equations apply to the structure of the uncharged boranes B_pH_{p+q} (*e.g.* B_2H_6 or B_4H_{10}):

	Examples	
	2002B_2H_6	4012B_4H_{10}
$q = s+x$[10]	$4 = 2+2$	$6 = 4+2$
$p = s+t$[11]	$2 = 2+0$	$4 = 4+0$
$p = t+y+q/2$	$2 = 0+0+4/2$	$4 = 0+1+6/2$

(9) For the structure of the charged borane (hydroborate anions $[B_pH_{p+q-c}]^{c\ominus}$ (e.g. $B_3H_8^{\ominus}$ or $B_{10}H_{10}^{2\ominus}$), the following equations hold[12]:

	Examples	
	2013$B_3H_8^{\ominus}$	0830$B_{10}H_{10}^{2\ominus}$
$q-c = s+x$[13]	$6-1 = 2+3$	$2-2 = 0+0$
$p-c = s+t$[14]	$3-1 = 2+0$	$10-2 = 8+0$
$p+c = t+y+q/2$	$3+1 = 0+1+6/2$	$10+2 = 8+3+2/2$

Using these rules, several valency structures, the so-called topological structures, can generally be constructed. One or more topological structures of a molecule approaches the true molecular structure. This will be shown below by comparing steric and topological structures (as far as they are known).

---

[8] *Cf.* section 1.1

[9] If a borane molecule contains, in place of x BH_2 groups, x_1 BH_3 groups (example: $[H_3B\text{–}BH_3]^{2\ominus}$: $x_1 = 1$) or x_2 BH_4 groups (example: $[BH_4]^{\ominus}$: $x_2 = 1$), in the following equations x must be replaced by the values $2x_1$ and $3x_2$, respectively.

[10] In words: the number (q) of hydrogen atoms in excess of the number of boron atoms is equal to the number of bridge hydrogen atoms plus the number of BH_2 groups.

[11] In words: the number of boron atoms (p) is equal to the number of bridge bonds (*cf.* footnote 2).

[12] c is the number of negative charges.

[13] In words: the number ($q-c$) of hydrogen atoms in excess of the number of boron atoms is equal to the number of bridge hydrogen atoms plus the number of BH_2 groups.

[14] In words: the number ($p-c$) of boron atoms less the number of negative charges is equal to the number of bridge bonds (*cf.* footnote 2).

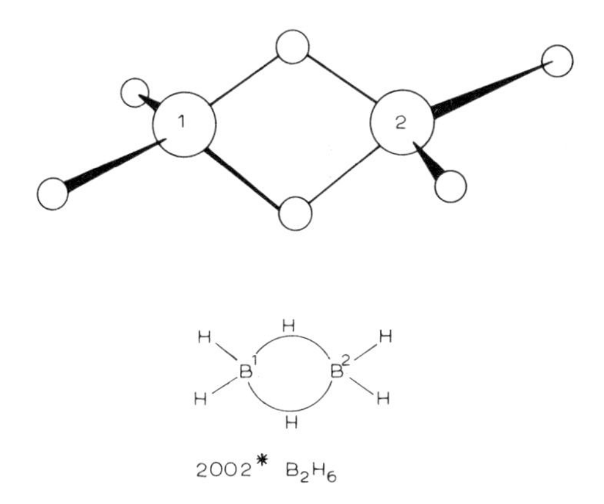

Fig. 4.3. Steric and topological structures of B_2H_6. Structure parameters [29, 327]: B–B = 1.775 Å, B–H^{μ} = 1.339 Å, B–H^{ext} = 1.196 Å; ∡ $H^{ext}BH^{ext}$ = 120.5 ± 0.9°.
*styx [*cf.* section 2 (7)]

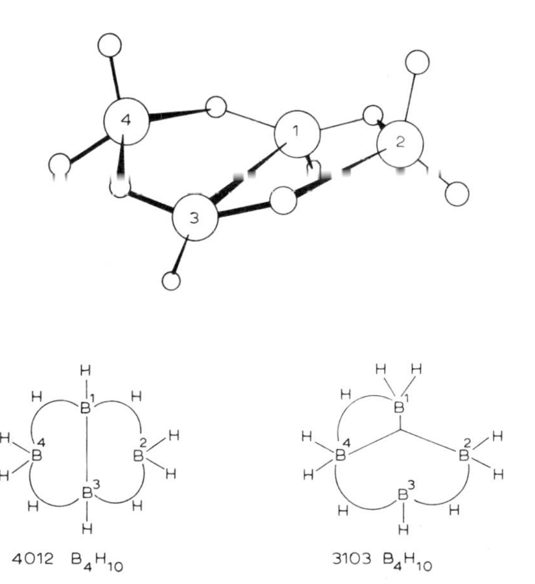

Fig. 4.4. Steric and topological structures of B_4H_{10}. Structure parameters [378, 554, 555]: B^1–B^3 = 1.71 Å, B^2–B^4 = 2.88 Å, all other B···B = 1.843–1.848 Å; B–H^{ext} = 1.19 Å, B^1–H^{μ} = H^3–H^{μ} = 1.33 Å, B^2–H^{μ} = B^4–H^{μ} = 1.43 Å; ∡ $B^2B^1B^4$ = 98°, ∡ $B^1B^3H^{ext}$ = 118°.

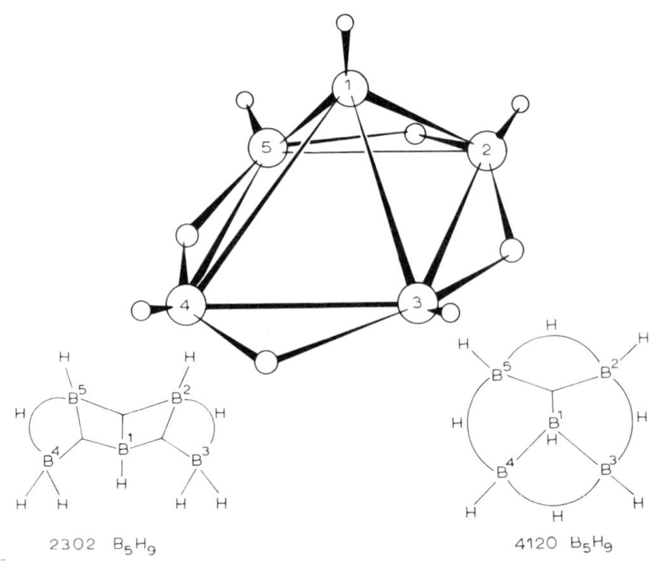

Fig. 4.5. Steric and topological structures of B_5H_9. Structure parameters[*179, 180, 325, 326, 353, 364, 365, 446, 530*]: distance of the base B-atoms B^2, B^3, B^4, B^5 from one another = 1.77–1.80 Å, distance of the base B-atoms from the apex atom B^1 = 1.66–1.70 Å, B–H^{ext} = 1.20–1.23 Å, B–H^{μ} = 1.35–1.36 Å; ∡$B^1B^2H^{ext}$ = 120 ± 20°, ∡plane $B^1B^3B^4$–plane $B^3B^4H^{\mu}$ = 187 ± 10°.

The theoretically possible plane valence structures of B_5H_9 2302, 4120, and 3211 (not shown) all exhibit the experimentally found tetragonal pyramid with C_{4v}-symmetry only approximately. The 2302 B_5H_9 with C_{2v}-symmetry approaches the correct B_5H_9 structure.

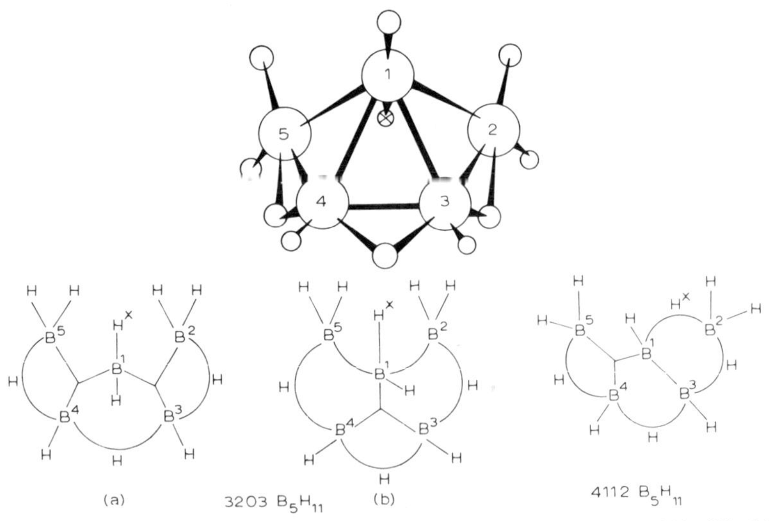

Fig. 4.6. Steric and topological structures of B_5H_{11}: Structure parameters[*353, 455, 446, 530*]: B^1–B^5 = B^1–B^2 = 1.77 Å, B^1–B^3 = 1.73 Å, B^1–B^4 = 1.70 Å, B^2–B^3 = 1.75 Å, B^3–B^4 = B^4–B^5 = 1.77 Å; B–H^{ext} = 1.07 Å, B–H^{μ} = 1.24 Å, B^1–H^x = 1.09 Å, B^2–H^x = 1.77 Å, B^5–H^x = 1.68 Å; ∡$B^2B^1B^5$ = 107°.

One of the two H-atoms of the B^1H_2 groups (H^x) possesses, in addition, very weak bonds to the B^2- and B^5-atoms. Besides 3203 B_5H_{11} and 4112 B_5H_{11}[*469*], which possibly exists only as a reactive intermediate, a 5021 B_5H_{11} is conceivable.

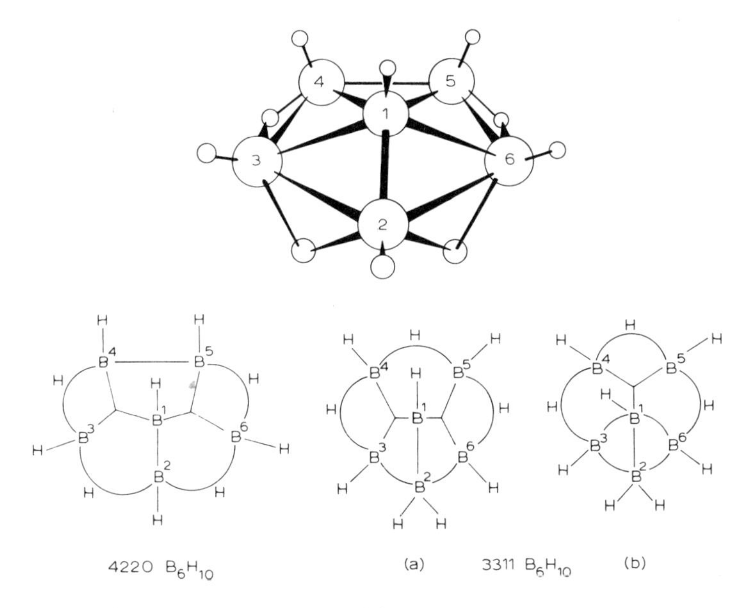

Fig. 4.7. Steric and topological structures of B_6H_{10}. Structure parameters[*168, 205, 346, 353*]: B^1–B^2 = 1.74–1.85 Å, B^1–B^3 = B^1–B^6 = 1.75–1.80 Å, B^1–B^4 = B^1–B^5 = 1.79 Å, B^4–B^5 = 1.60–1.71 Å, B^2–B^3 = B^2–B^6 = 1.79 Å, B^3–B^4 = B^5–B^6 = 1.74 Å.

The possible topological structure 2402 B_6H_{10} does not agree with the stereometry of B_6H_{10}; this form of B_6H_{10} certainly exists only as a transitional form in reactions or in the formation of B_6H_{10}.

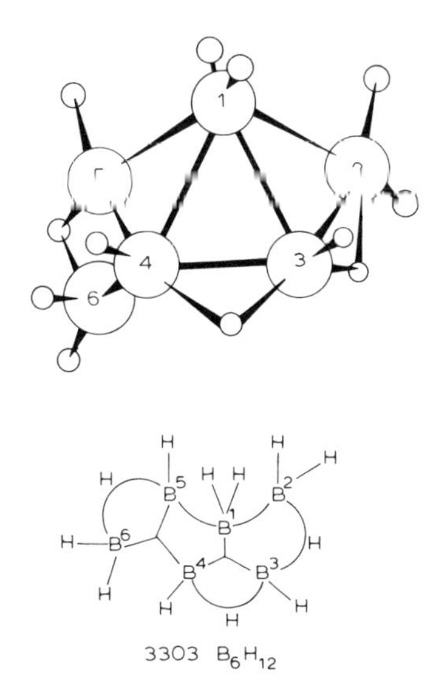

Fig. 4.8. Steric and topological structures of B_6H_{12}. Numbering of the B_5 fragment as for B_5H_{11}. The structure of a 'boranised pentaborane(11)' follows from the chemical properties of the substance and from IR and NMR spectroscopic data[*482*]. Besides 3303 B_6H_{12}, 6030, 5121, and 4212 B_6H_{12} are conceivable.

References p. 355

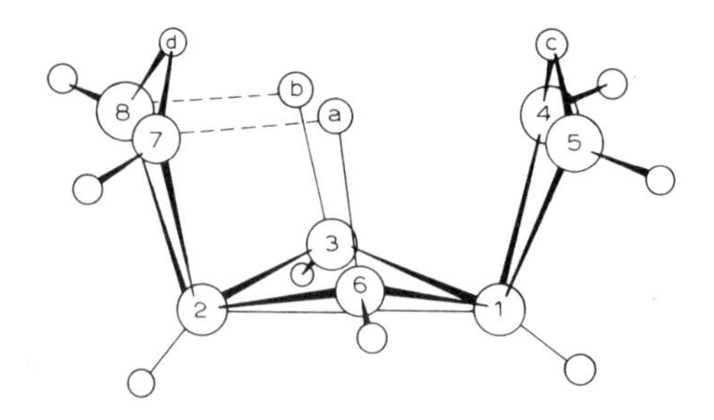

Fig. 4.9. Steric structure of B_8H_{12}. Numbering of the icosahedral fragment following the literature [*202*]. Structure parameters [*202*]: B–B = 1.672–1.803 Å (smallest separations: B^4–B^5 and B^7–B^8; greatest separation: B^1–B^2); B–H^{ext} = 1.10 Å, B^6–H^a = 1.27 Å, B^3–H^b = 1.30 Å, B^7–H^a = 1.46 Å, B^8–H^b = 1.48 Å, $B^4 \cdots H^b$ = 1.98 Å, $B^5 \cdots H^a$ = 1.99 Å.

The following topological structures are conceivable: 4420, 3510, and 2602 B_8H_{12}.

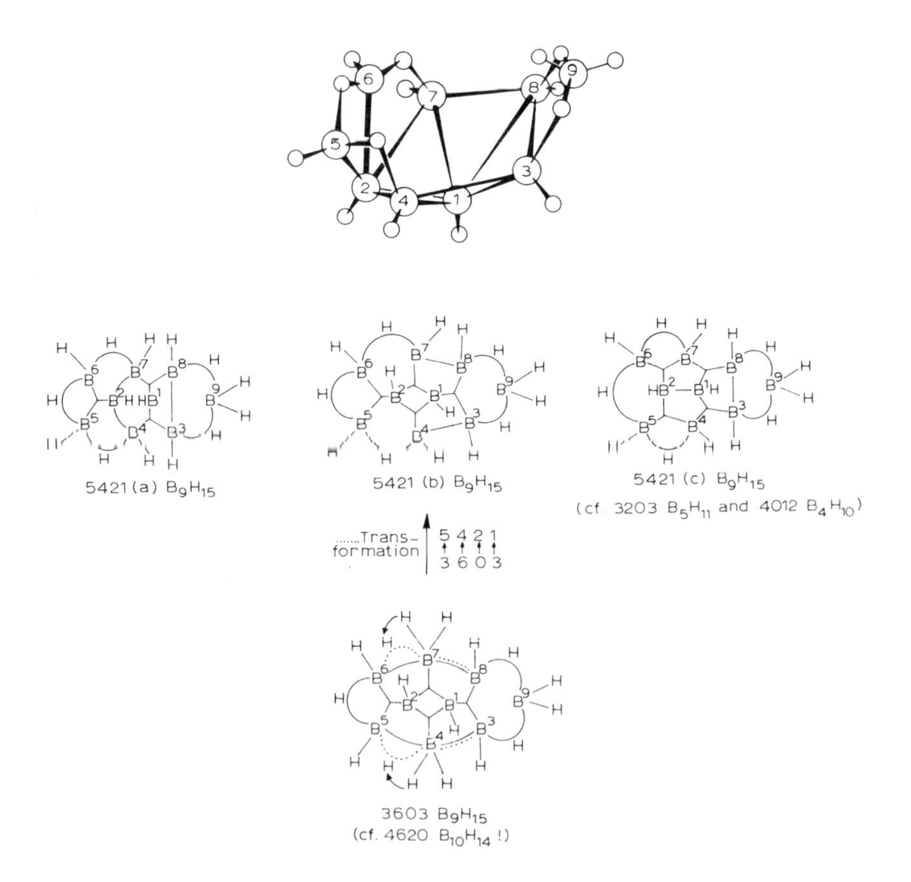

Fig. 4.10. Steric and topological structures of B_9H_{15}. 3603 B_9H_{15} can easily be transformed into 5421 (b) B_9H_{15}. In addition, 6330 and 4512 B_9H_{15} are conceivable; however, these structures do not completely satisfy the principles of the valence structure model. Structure parameters [*167*, *168*, *712*]: B–B = 1.75–1.95 Å.

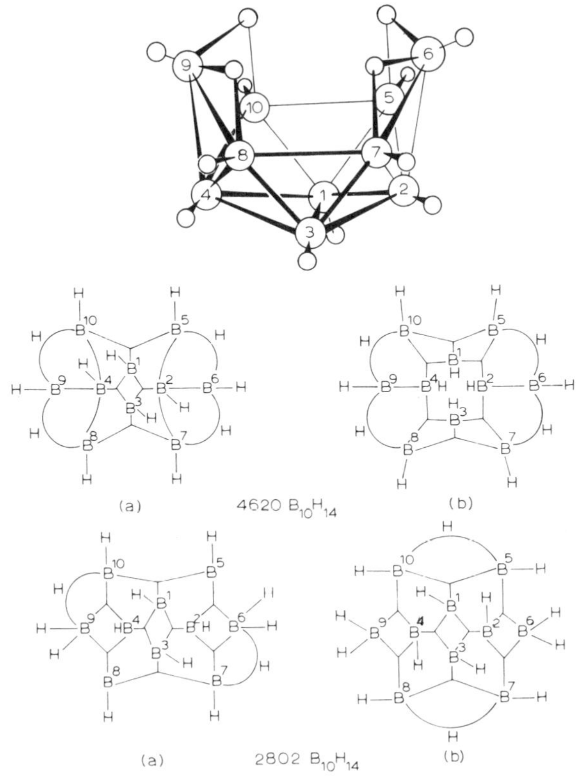

Fig. 4.11. Steric and topological structures of $B_{10}H_{14}$. Structure parameters [*353*, *387*, *480*, *530*, *632*, *709*]: B^5–B^{10}, B^7–B^8 = 1.71–2.02 Å, B–B = 1.76 Å; B–H^{ext} = 1.20–1.30 Å, B–H^{μ}–B is unsymmetrical: B–H^{μ} = 1.34 and 1.42 A. In addition, a 3711 $B_{10}H_{14}$ is conceivable.

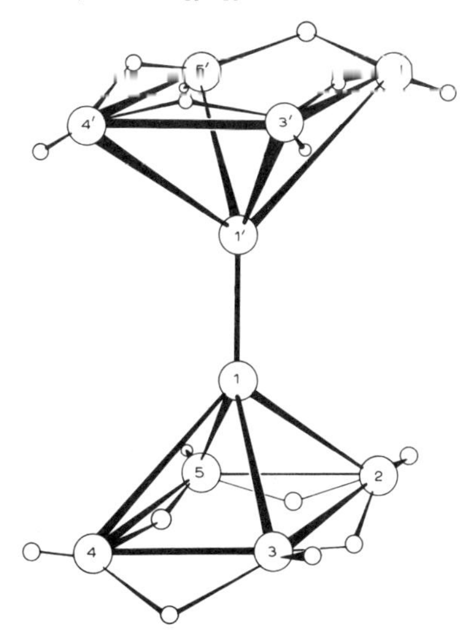

Fig. 4.12. Possible steric structure of $B_{10}H_{16}$. Structure parameters [*353*, *469*]: B^1–$B^{1'}$ = 1.71(1.66)–1.76(1.74) Å, B^2–B^3 = 1.77 Å. Possible topological forms: 6430, 5521, 4612, and 3703 $B_{10}H_{16}$.

References p. 355

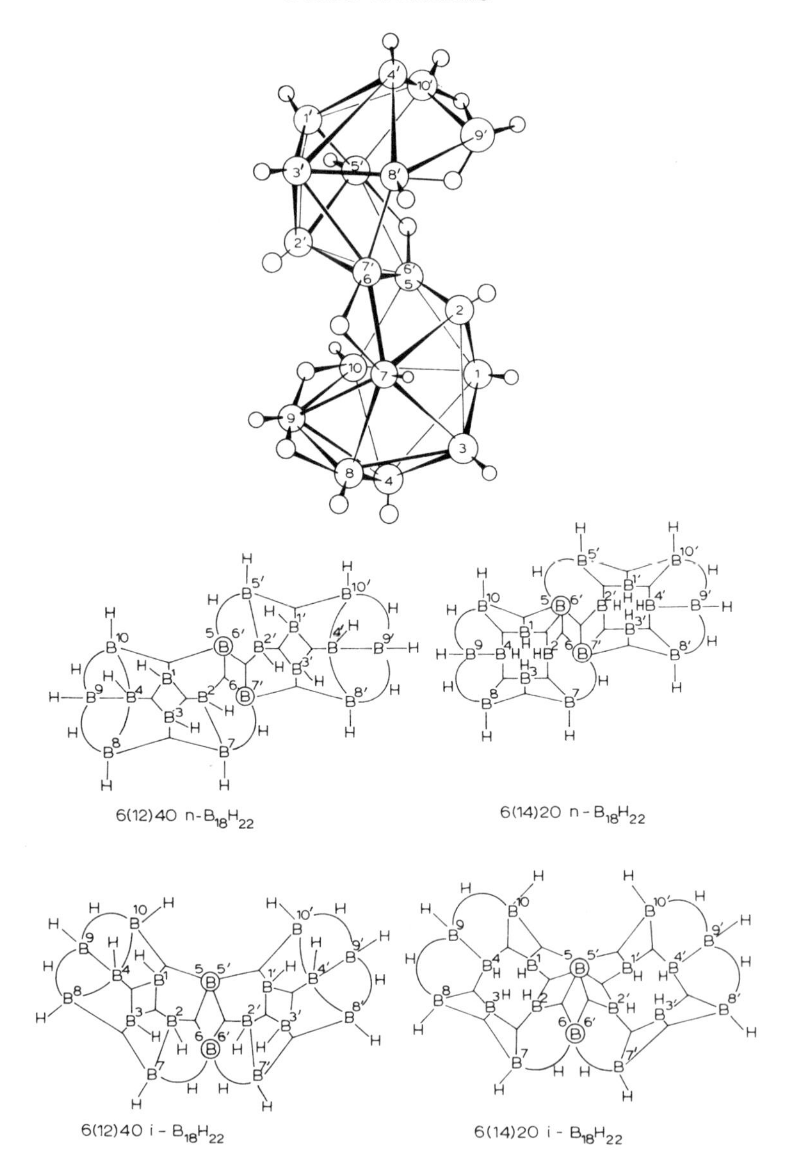

Fig. 4.13. Steric structure of n-$B_{18}H_{22}$ and the topological structures of n-$B_{18}H_{22}$ and i-$B_{18}H_{22}$. Both structures are derived from two 'basket'-shaped (see Fig. 4.11) $B_{10}H_{14}$ molecules, one fragment being given unprimed and the other primed numbers (in accordance with the numbering of $B_{10}H_{14}$). The fragments are joined to one another by common B-atoms (marked with a circle). The normal and iso forms differ only in respect of this linkage. In the iso form, the apical B-atoms B^6 and $B^{6'}$ and the neighbouring B-atoms B^5 and $B^{5'}$, are identical. (Derivation: the position of the primed $B_{10}H_{14}$ fragment is obtained by turning the unprimed $B_{10}H_{10}$ fragment about the $B^{5,5'}$–$B^{6,6'}$ axis by 180°.) In the normal form, one apical B-atom B^6 (or $B^{6'}$) is identical with the neighbouring peak atom $B^{7'}$ (or B^5). (This normal form cannot, therefore, be derived by simple rotation.) Structure parameters [*353, 711, 713*]: B–B = 1.7–2.01 Å; B–H^{ext} = 1.04–1.21 Å, B–H^{μ} = 1.20–1.37 Å.

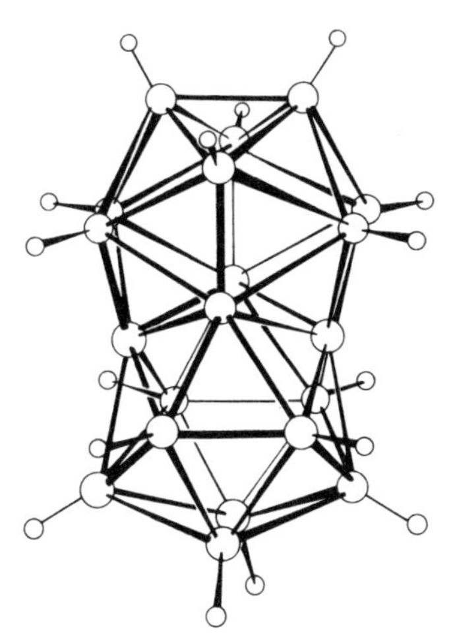

Fig. 4.14. Steric structure of $B_{20}H_{16}$. Structure parameters [*171*, *235*]: B–B = 1.77–1.88 Å.

3. Preparation of boranes

3.1 *Formation of B_2H_6 (B_2D_6)*

The following five routes are available for the formation of diborane. In the first three, B–H bonds are formed by protons, neutral hydrogen, or hydride ions. In the other two, the B–H bonds are preformed. Route(c) is the most important.

(a) Cationic displacement (by means of $H^{\oplus}$): $\rangle B^{\ominus} + H^{\oplus} \longrightarrow \rangle \overset{\delta+}{B}—\overset{\delta-}{H}$.

Typical example: $BMn + 3H^{\oplus} \rightarrow BH_3 + Mn^{3\oplus}$ (*cf.* section 3.1.4).

(b) Hydrogenating rupture (by means of H^0): $\rangle B^{\oplus} + H–H \longrightarrow \rangle \overset{\delta+}{B}—\overset{\delta-}{H} + H^{\oplus}$

Typical examples: $BCl_3 + 3H_2 \rightarrow BH_3 + 3HCl$, and $BR_3 + 3H_2 \rightarrow BH_3 + 3HR$ (*cf.* section 3.1.2).

(c) Anionic displacement (by means of $H^{\ominus}$): $\rangle B^{\oplus} + H^{\ominus} \longrightarrow \rangle \overset{\delta+}{B}—\overset{\delta-}{H}$.

Typical example: $4BCl_3 + 3LiAlH_4 \rightarrow 4BH_3 + 3LiAlCl_4$ (*cf.* section 3.1.1).

(d) Elimination of $H^{\ominus}$ from $BH_4^{\ominus}$: $BH_4^{\ominus} \rightarrow BH_3 + H^{\ominus} \xrightarrow{+H^{\oplus}} H_2$ (or $\rightarrow \frac{1}{2}H_2 + e^{\ominus}$).

Typical examples: $KBH_4 + H^{\oplus} \rightarrow BH_3 + K^{\oplus} + H_2$, and $LiBH_4$(electrolysis) $\rightarrow Li + BH_3 + \frac{1}{2}H_2$ (*cf.* section 3.1.3).

(e) Degradation of higher boranes by donors or by pyrolysis:

X = Cl, Br; M = Li, Na; n = 3, ∞

$$\text{H}_2\text{B}(\mu\text{-H})_2\text{B}_2\ldots \longrightarrow \text{Donor}\cdots\rightarrow\text{BH}_3 + \text{HB}\ldots$$

Typical examples:

$$B_4H_{10} + 2Me_2O \xrightarrow{B_3H_7\cdot OMe_2} BH_3\cdot OMe_2 \rightarrow BH_3 + OMe_2$$

$$B_4H_{10} + B_5H_{11} \xrightarrow{\text{heat}} 2(BH_3)_2 + B_5H_9$$

3.1.1 Formation of B_2H_6 by the hydrogenation of boron halides or boron alkoxides with metal hydrides

(1) $BF_3(BCl_3) + NaH(CaH_2)$, in the absence of a solvent. Diborane is formed when gaseous boron trifluoride or boron trichloride is passed with hydrogen or nitrogen as the carrier gas through a layer of powdered sodium hydride or calcium hydride (but not LiH [*658*]). Elementary boron is not formed under these conditions [*71, 264c, 369*]. Example of the reaction:

$$2BF_3 + 6NaH \xrightarrow{180°,\ \text{carrier gas}} (BH_3)_2 + 6NaF\ (\xrightarrow{+6BF_3} 6NaBF_4)$$

(2) $BF_3 + LiH$ in solution. Homogeneous or heterogeneous hydrogenations in solvents appear to be more convenient for the preparation of diborane in the laboratory than reactions without solvents. For example, diborane is formed in a smooth reaction in 87% yield from BF_3 and LiH in boiling ether [*194, 658*]:

$$2BF_3\cdot OEt_2 + 6LiH \xrightarrow[35°,\ \text{diethyl ether}]{-2OEt_2} (BH_3)_2 + 6LiF\ (\xrightarrow{+6BF_3} 6LiBF_4)$$

The reaction with sodium hydride takes place equally smoothly, but here the yield of $(BH_3)_2$ depends on the size and surface of the particles of NaH. Maximum yield 40 to 60%.

(3) $BF_3 + LiH +$ catalyst. In the reaction of BF_3 with metal hydrides described above, an unnecessary large amount of boron trifluoride is consumed by the formation of tetrafluoroborate. Furthermore, the reaction frequently sets in suddenly, with foaming, after an uncontrollable induction period. These disadvantages are avoided by the addition of catalysts forming (as intermediates) soluble boranates (hydroborates) with the sparingly soluble or insoluble metal hydride. The catalysts are methyl borate $B(OMe)_3$ and, in the reaction under pressure, (*i.e.* at high concentrations) diborane $(BH_3)_2$ itself [*194, 658*]:

$$3LiH + 3B(OMe)_3 \xrightarrow{\text{(diethyl ether)}} 3LiBH(OMe)_3 \xrightarrow[-3LiF]{+BF_3} 3B(OMe)_3 + BH_3$$

$$3LiH + 3BH_3 \xrightarrow{\text{(diethyl ether, pressure)}} 3LiBH_4 \xrightarrow[-3LiF]{+BF_3} 3BH_3 + BH_3$$

(In each scheme a dashed arrow returns the regenerated $3B(OMe)_3$ or $3BH_3$ to the start.)

(4) $BCl_3 + LiH(MgH_2, BeH_2 \cdot OEt_2, AlH_3 \cdot OEt_2)$. With boron trichloride, the hydrides mentioned form diborane in a smooth reaction (together with hydrogen) [*1, 175, 659*]:

$$2BCl_3 + 6LiH \xrightarrow[-80°]{\text{diethyl ether}} (BH_3)_2 + 6LiCl$$

$$2BCl_3 + 3MgH_2 \xrightarrow{\text{ether, aliphatic or aromatic HC}} (BH_3)_2 + 3MgCl_2$$

$$2BCl_3 + 3BeH_2 \cdot OEt_2 \xrightarrow{\text{ether}} (BH_3)_2 + 3BeCl_2 \cdot OEt_2$$

$$2BCl_3 + 2AlH_3 \cdot OEt_2 \xrightarrow{\text{ether}} (BH_3)_2 + 2AlCl_3 \cdot OEt_2$$

(5) $BF_3(BCl_3) + CaH_2$ in solution. Diborane is formed when boron trifluoride is passed with a carrier gas through a solution of CaH_2 in a $AlCl_3$–LiCl melt (23% of B_2H_6, referred to CaH_2) [*752*]:

$$2BCl_3 + 3CaH_2 \xrightarrow[\text{carrier gas}]{AlCl_3-LiCl\text{ melt}} (BH_3)_2 + 3CaCl_2$$

In monoglyme[15] at 20° BCl_3 and CaH_2 give only traces of $(BH_3)_2$ in 96 hours [*462*]. Similar low yields are obtained by the reaction of BF_3 with CaH_2 in tetrahydrofuran, monoglyme or triglyme[15] [*462*]. In diglyme[15], on the other hand, BF_3 and CaH_2 react to give a 46% yield of diborane [*462*]:

$$BF_3 + 3CaH_2 \xrightarrow[\text{diglyme}]{60-100°, 75\text{ h}} (BH_3)_2 + 3CaF_2$$

(6) $R_nBX_{3-n} + NaH(CaH_2) +$ catalyst. Industrially, the preparation of diborane (and alkylboranes) is carried out by catalysed reactions of the cheap hydrides NaH and CaH_2 with boron halides R_nBX_{3-n} ($n = 0, 1, 2$) [*336*]:

$$R_nBX_{3-n} + (3-n)NaH \longrightarrow R_nBH_{3-n} + (3-n)NaX$$

The catalysts used are substances such as trialkylborons, BR_3, which form soluble complexes with the hydride; in addition, electron donors (trialkylamines, NR_3) are added, probably to trap the boron hydride formed as an intermediate in the solution. Example: 50 parts by weight of NaH in 230 parts by weight of mineral oil (0.1 torr at 190–210°) is heated with 14 parts by weight of BEt_3 at 73°. Then, with stirring at 72 to 74°, 68 parts by weight of NEt_3 followed by 79 parts by weight of BCl_3 are added. After the temperature has been raised to 80°, a further 70 parts by weight of BCl_3 is added. Yield: 7.2 parts by weight of B_2H_6 = 88% of theory [*373*].

---

[15] Monoglyme: monoethyleneglycol dimethyl ether, $MeOC_2H_4OMe$.
Diglyme: diethyleneglycol dimethyl ether, $Me(OC_2H_4)_2OMe$.
Triglyme: triethyleneglycol dimethyl ether, $Me(OC_2H_4)_3OMe$.
Tetraglyme: tetraethyleneglycol dimethyl ether, $Me(OC_2H_4)_4OMe$.

References p. 355

The following process also uses amine adducts of borane [*424*, *429*a]:

$$\begin{array}{l} Al + 1\tfrac{1}{2}H_2 + 3C_nH_{2n} \longrightarrow Al(C_nH_{2n+1})_3 \\ BF_3 \cdot NEt_3 + Al(C_nH_{2n+1})_3 \longrightarrow B(C_nH_{2n+1})_3 \cdot NEt_3 + AlF_3 \\ B(C_nH_{2n+1})_3 \cdot NEt_3 + 3H_2 \xrightarrow{\text{pressure}} BH_3 \cdot NEt_3 + 3C_nH_{2n+2} \\ BH_3 \cdot NEt_3 + BF_3 \longrightarrow BH_3 + BF_3 \cdot NEt_3 \\ \hline BF_3 + Al + 4\tfrac{1}{2}H_2 + 3C_nH_{2n} \longrightarrow BH_3 + AlF_3 + 3C_nH_{2n+2} \end{array}$$

(7) $BCl_3 + SiH_4$. The hydrogen in silane, SiH_4, is sufficiently negative to form diborane. While silane does not attack BCl_3 without a supply of energy, in the presence of methyl radicals formed photolytically from diazomethane it hydrogenates boron trichloride with the replacement of Cl by H to give diborane (67%, calculated on BCl_3) [*649*]:

$$BCl_3 + 2SiH_4 \longrightarrow \tfrac{1}{2}(BH_3)_2 + SiH_3Cl + SiH_2Cl_2$$

On the other hand, even alone at 0°, disilane forms diborane quantitatively via a polar four-centre mechanism [*189*]:

$$\begin{array}{ccc} & \overset{\delta+}{} H_2 & \overset{\delta-}{} \\ H_3Si\text{—} & Si \ldots & H \\ & \vdots & \vdots \\ & Cl \ldots & BCl_2 \\ & \delta- & \delta+ \end{array}$$

(8) $BX_3 + M[AlH_4]_n$. The ether-soluble double hydrides $M(AlH_4)_n$ are more suitable for the preparation of diborane than the ether-insoluble simple metal hydrides MH_n. For example, with lithium alanate (lithium aluminium hydride, lithium tetrahydroaluminate), $LiAlH_4 = LiH \cdot AlH_3$, in diethyl ether, boron trifluoride [*560*, *694*], boron trifluoride-ether complex [*390*, *391*, *623*], and boron trichloride [*223*] give an almost quantitative yield of diborane. Similarly, with lithium alanate-d_4 (lithium tetradeuteroaluminate) boron trifluoride-ether gives diborane-d_6 [*390*, *391*, *451*, *690*, *694*]:

$$4BF_3 \cdot OEt_2 + 3LiAlH_4 \xrightarrow{\text{ether}} 2(BH_3)_2 + 3LiAlF_4 + 4OEt_2$$

$$4BF_3 \cdot OEt_2 + 3LiAlD_4 \xrightarrow{\text{ether}} 2(BD_3)_2 + 3LiAlF_4 + 4OEt_2$$

$$4BCl_3 + 3LiAlH_4 \xrightarrow{\text{ether}} 2(BH_3)_2 + 3LiAlCl_4$$

When BF_3–ether is added to $LiAlH_4$ solution in portions, two reactions take place successively: the formation of lithium boranate, $LiBH_4$, and the formation of diborane [*697*]:

$$3BF_3 + 3LiAlH_4 \longrightarrow 3LiBH_4 + 3AlF_3$$

$$3LiBH_4 + BF_3 \longrightarrow 2(BH_3)_2 + 3LiF(\xrightarrow{3AlF_3} 3LiAlF_4)$$

The tetrahydrofuran-soluble calcium alanate and magnesium alanate have also been used for the production of diborane [*222, 500, 800*]:

$$8BCl_3 + 3M(AlH_4)_2 \longrightarrow 4(BH_3)_2 + 3M(AlCl_4)_2$$

(g) $BX_3 + MBH_4$. Diborane is formed in very good yield by the reaction of alkali metal boranates with boron trihalides. However, the decomposition of ether by BX_3 complicates the technical application of the reaction [*2, 540, 547, 658*]:

$$BF_3 + 3LiBH_4 \xrightarrow{\text{diethyl ether}} 2(BH_3)_2 + 3LiF$$

$$4BF_3 + 3MBH_4 (\xrightarrow{\text{polyether}} 2(BH_3)_2 + 3MBF_4\ (M = Na, K)$$

$$BCl_3 + 3NaBH_4 \xrightarrow{\text{polyether}} 2(BH_3)_2 + 3NaCl$$

$$2BF_3{\cdot}OEt_2 + 6NaBH(OMe)_3 \xrightarrow[-2OEt_2]{\text{diethyl ether}} (BH_3)_2 + 6B(OMe)_3 + 6NaF$$
$$(\xrightarrow{+6BF_3} 6NaBF_4)$$

The initial step of the reaction shown last is probably, according to reactions with isotope-labelled $^{10}BF_3$, the formation of $NaBHF_3$, which forms diborane $^{10}B_2H_6$ with more $^{10}BF_3$ [*439*]:

$$^{10}BF_3 + NaBH(OMe)_3 \longrightarrow Na^{10}BHF_3 + B(OMe)_3$$
$$6Na^{10}BHF_3 + 2\,^{10}BF_3 \longrightarrow (^{10}BH_3)_2 + 6Na^{10}BF_4$$

^{10}B is replaced by ^{11}B only in a subsequent reaction [*692*].

Because of the equilibrium

$$(BH_3)_2 + 4B(OMe)_3 \rightleftharpoons 6BH(OMe)_2$$

the above reaction of sodium trimethoxoboranate with boron trifluoride–ether (6:8) gives diborane with a yield of only 50 to 60%. Almost 100% of diborane (referred to the expensive hydride) is obtained if double the amount of $BF_3{\cdot}OEt_2$ is used. Under these conditions, trimethyl borate $B(OMe)_3$, which reduces the yield in the manner shown in the above equation, is consumed by the formation of fluorine derivatives [*658*]:

$$B(OMe)_3 + BF_3{\cdot}OEt_2 \longrightarrow BF(OMe)_2 + BF_2(OMe) + OEt_2$$

(10) $B(OR)_3 + MAlH_4$. Lithium or sodium alanate reacts with methyl or phenyl borate in ethers (Et_2O, THF) or tertiary amines (Me_3N, Me_2NPh) at 20° with the formation either of diborane or of aminoborane or lithium or sodium boranate. Thus, for example, when $LiAlH_4$ is added to $B(OPh)_3$, diborane is obtained in 47% yield [*17a*]:

$$4B(OPh)_3 + 3LiAlH_4 \xrightarrow[20^\circ]{\text{diethyl ether}} 2(BH_3)_2 + 3LiAl(OPh)_4$$

On the other hand, with the (reversed) addition of $B(OPh)_3$ to $LiAlH_4$ the main product is lithium boranate with only 16% of diborane [*17a*].

3.1.2 Synthesis of B_2H_6 by the hydrogenation of boron compounds BX_3 with molecular hydrogen

(1) BX_3+H_2+M. Gaseous BF_3, BCl_3, or BBr_3 and H_2 (1:3 to 1:6) react with Al, Mg, Zn, or Na at 300 to 500° with the formation of B_2H_6, B_2H_5Cl, metal halide, HCl, and a little B and B_4H_{10}, *e.g.* in accordance with the following equation [*264b, 369*]:

$$2\,BCl_3+2Al+3H_2 \longrightarrow (BH_3)_2+2AlCl_3$$

The B_2H_5Cl produced also dismutes on fractionation, in accordance with the equation:

$$6B_2H_5Cl \rightleftharpoons 5B_2H_6+2BCl_3$$

(Best yields with $Al+BCl_3$; 30%, calculated on BCl_3) [*369*]. Metal salts forming a low-melting eutectic with the metal halide produced increase the rate of reaction [*72, 228, 264b, 370*].

(2) BX_3+H_2. The passage of an electric discharge through a mixture of BCl_3 and H_2 or BBr_3 and H_2 forms B, solid boranes, $B_2H_5Cl(Br)$, B_2H_6, and HCl (Br) [*666, 730a, 742, 796*], for example in accordance with the equation:

$$2BX_3+6H_2 \longrightarrow (BH_3)_2+6HX$$

(3) BBr_3+HCHO. Formaldehyde can also act as a source of hydrogen, giving both protons and hydride ions. If BBr_3 and HCHO are passed over copper at 400°, $(BH_3)_2$ is formed in a yield of about 25% [*261*]. A mechanism involving the giving up of a hydride ion by the 'anion' $HCO^{\ominus}$ first formed to the 'cation' $BBr_2^{\oplus}$ appears reasonable:

$$HCHO \xrightarrow[-H^{\oplus}]{} HC{=}O^{\ominus} \xrightarrow[-H^{\ominus}]{} CO$$

$$\qquad\quad + \qquad\qquad\qquad +$$

$$BBr_3 \longrightarrow Br^{\ominus} + BBr_2^{\oplus}$$

$$\qquad \downarrow \qquad\qquad \downarrow$$

$$\qquad HBr \qquad\quad BBr_2H$$

$$2BBr_3+6HCHO \xrightarrow[\text{copper}]{400^\circ} (BH_3)_2+6HBr+6CO$$

(4) BR_3+H_2. Diborane is also formed by the hydrogenation of trialkylborons with hydrogen:

$$2BR_3+6H_2 \xrightarrow[\text{Pd}]{105-140\text{ atm},\ 150^\circ} (BH_3)_2+6RH$$

BEt_3 (yield 75%) is easier to hydrogenate than BMe_3 and BBu_3. An excess of hydrogen substantially prevents the formation of higher boranes [*406*].

(5) $B_2O_3+H_2$. Boron–oxygen compounds can also be hydrogenated. Thus, at 170–180° with hydrogen at 700–800 atm sodium borate $Na_2B_4O_7$, potassium borate KB_5O_8, boric acid $B(OH)_3$, methyl borate $B(OMe)_3$, and trimethoxyboroxin $(MeOBO)_3$, in a low-melting mixture of NaCl, $AlCl_3$, and Al give a 7–30% yield of diborane (purity 98%; remainder pentaborane and hydrogen). The hydrogenation does not take place via elementary boron but with the intermediate formation of aluminium chloride hydrides (chloro alanes), $AlCl_{3-n}H_n$, for example in accordance with the following equation[*228*]:

$$2n\text{Al}+(6-2n)\text{AlCl}_3+3n\,\text{H}_2 \rightleftharpoons 6\text{AlCl}_{3-n}\text{H}_n$$
$$n\text{B}_2\text{O}_3+6\text{AlCl}_{3-n}\text{H}_n \rightarrow n(\text{BH}_3)_2+6\text{AlCl}_{3-n}\text{O}_{n/2}$$

$$2n\text{Al}+(6-2n)\,\text{AlCl}_3+n\text{B}_2\text{O}_3+3n\,\text{H}_2 \rightarrow n\,(\text{BH}_3)_2+6\text{AlCl}_{3-n}\text{O}_{n/2}$$

In the presence of secondary or tertiary amines, borazanes, aminoboranes, and borazines are formed[*228*].

(6) Boron suboxide, boron subsulphide, etc.$+H_2$. A number of patents describe a cheap method for the preparation of diborane in this way. Thus, it is formed in the reaction of boron or manganese boride with copper or iron sulphide in an atmosphere of hydrogen at 700 to 1800°[*141*], in the hydrogenation of polymeric boron suboxide BO with hydrogen at 850–*1300*–1500°[*142*], in the reaction of boron, metal boride, or boron carbide and titanium carbide with boron trioxide and an alkali-metal or an alkaline-earth metal borate in an atmosphere of hydrogen at 850 to 1500°[*142a*], and in the reaction of boron, a metal boride, or boron carbide with an oxide (steam, MgO, Al_2O_3, ThO_2, ZrO_2) in an atmosphere of hydrogen at 850 to 1500°[*142b*].

3.1.3 Synthesis of B_2H_6 by the protolysis or electrolysis of $BH_4^{\ominus}$

Protolysis of $BH_4^{\ominus}$. With concentrated sulphuric acid or phosphoric acid, sodium or potassium boranate gives a 50 to 80% yield of diborane[*178, 731*]:

$$2\text{BH}_4^{\ominus}+2\text{H}^{\oplus} \longrightarrow (\text{BH}_3)_2+2\text{H}_2$$

The gas contained 15% of SO_2 and 2% of H_2S (in the case of H_2SO_4). SO_2-free $(BH_3)_2$ is formed in the $NaBH_4$–CH_3SO_3H system.

Electrolysis of $BH_4^{\ominus}$. On electrolysis, solutions containing boranate ions evolve diborane at the anode:

$$2\text{BH}_4^{\ominus} \longrightarrow (\text{BH}_3)_2+\text{H}_2+2\text{e}^{\ominus}$$

Only a few solvents are suitable: eutectics formed by $LiBH_4$–KBH_4 (m.p. 100°), $LiBH_4$–$NaBH_4$–KBH_4 (m.p. 95°), and mono, di-, tri-, and tetraglyme. Material yields of up to 50%; current yields of 84%[*2, 83*].

References p. 355

3.1.4 Other syntheses of B_2H_6

For the synthesis of B_2H_6 by the pyrolysis of B_4H_{10} and B_5H_{11}, see section 7, Fig. 4.15, and for the formation by nucleophilic degradation of B_4H_{10}, see section 15.3. The synthesis of B_2H_6 by the decomposition of borides with acids (cationic displacement) and subsequent pyrolysis of the B_4H_{10} formed (*cf.* section 3.2.1) is obsolete. (B_4H_{10} was first obtained in a preparative amount by the protolysis of borides.)
For the separation of B_2H_6 from other boron hydrides by distillation see the literature [*438*].

3.2 Formation of B_4H_{10}

Tetraborane(10) can be prepared by degradative and synthetic methods: (a) degradation of a metal boride containing a B–B bond (potential polyborane) to give smaller molecular associations (oligoborane) under the action of acids or the degradation of a higher borane by the symmetrical rupture of a B–H double bridge under the influence of a nucleophilic agent. (b) Synthesis of tetraborane(10) (together with other higher boranes) from diborane by a radical reaction or by the Wurtz synthesis.

3.2.1 Formation of B_4H_{10} by degradation reactions

The borides of lithium [*515*], sodium [*515*], beryllium [*515, 745*], magnesium [*515, 731, 735, 744, 839*], calcium [*515*], aluminium [*515*], iron [*515*], nickel [*515*], and manganese [*515, 726*] react with hydrochloric acid [*515, 731, 735, 745*], sulphuric acid [*515*], or phosphoric acid [*515, 726, 744*] to form tetraborane and higher boranes (the maximum yield of B_4H_{10} is 15% with H_3PO_4, H_2O, and magnesium boride):

$$2Mg_3B_2 + 12H^{\oplus} \longrightarrow B_4H_{10} + H_2 + 6Mg^{2\oplus}$$

The hydrolytic degradation of pentaborane(11) gives a 98–100% yield of tetraborane(10) [*63*] (see section 17). For the degradation of higher boranes by nucleophilic agents, *cf.* section 16.

3.2.2 Formation of B_4H_{10} by Synthetic Reactions

The reaction of boron halides BX_3(X = F, Cl, Br, I) with hydrogen and hydrides containing highly hydric hydrogen (LiH, NaH, KH, BaH_2, CaH_2, SiH_4) at 150 to 400° gives rise to higher boranes, including tetraboranes(10) [*371*]. Probably diborane is first formed and this gives tetraborane and higher boranes by pyrolysis. The mixture of boron halide and hydrogen can be converted into boranes not only by the supply of thermal energy but also by means of an electric discharge [*121, 666*]. B_2Cl_4 and $LiBH_4$ yield B_4H_{10} [*787*].

The most suitable method at the present time for the synthesis of relatively large amounts of tetraborane(10) is probably the thermolysis of diborane, which is unstable even at room temperature [*121*]. In addition to tetraborane, it gives the unstable pentaborane(11).

$$\tfrac{1}{2}(BH_3)_2 \rightarrow BH_3 \xrightarrow[-H_2]{+(BH_3)_2} B_3H_7 \begin{cases} \xrightarrow{+(BH_3)_2,\,-BH_3} B_4H_{10} \\ \xrightarrow{+(BH_3)_2,\,-H_2} B_5H_{11} \end{cases}$$

which, however, in an atmosphere of hydrogen dismutes into tetraborane and diborane (see section 7).

Various types of apparatus are available for carrying out the thermolysis: in all of these the B_5H_{11} first formed from diborane (see section 7.1) is converted into B_4H_{10} in an atmosphere of H_2: U-tube 112°, 2 atm [*121*]; helical glass tube, 120°, 2 atm [*128*]; glass reactor, 180°, recirculating pump [*732*]. The best yields of tetraborane(10) are probably obtained without a pressure apparatus by a conversion in a hot-cold tube (120°/–78°). It is important in practice that under these conditions no solid reaction products are produced which could contaminate the reactor. 89 to 99% of the diborane fed in undergoes reaction. The yield of B_4H_{10} is 83% [*407*]. An increased pressure of hydrogen and a low temperature of formation favour the production of B_4H_{10} (experimental data: micro-pressure apparatus, 20°, 50 atm, 5–9 days, 20–35% B_4H_{10}; 50–75°, 50 atm, 8 hours, 20–35% of B_4H_{10}) [*169, 209*].

When B_2H_5I (lower yield with B_2H_5Br [*734*]) is shaken with sodium amalgam at −45°, tetraborane is formed by a Wurtz synthesis [*738*]:

$$B_2H_5I + 2Na \longrightarrow B_4H_{10} + 2NaI$$

This synthesis is unsuitable for the preparation of relatively large amounts of B_4H_{10}.

For the separation of B_4H_{10} from other boron hydrides by distillation see [*188*] and [*767*]; by gas chromatography, see [*64*] and [*393*]. For the mass spectrum of B_4H_{10}, see [*210*].

3.3 *Formation of B_5H_{11}*

Pentaborane(11) is formed in the short-time thermolysis of diborane at relatively low temperatures [*64, 497*]:

$$\tfrac{1}{2}(BH_3)_2 \longrightarrow BH_3 \xrightarrow[-H_2]{+(BH_3)_2} B_3H_7 \xrightarrow[-H_2]{+(BH_3)_2} B_5H_{11}$$

When a 1:5 mixture of diborane and hydrogen or nitrogen is passed through a glass or steel tube at 175 to 250°, the following boranes are produced [*497*]:

Temp. (°C)	*Carrier gas*	*% B_5H_{11}*	*% B_5H_9*	*% B_4H_{10}*	*% Solid boranes*
175	H_2	88	0	4	8
175	N_2	77	0	12	11
225	H_2	10?	78	1	11
225	N_2	7	58	5	30

A simple apparatus for the production of B_5H_{11} by the thermolysis of B_2H_6 at 120° has been described in the literature [*128*]; for an older description of the thermolysis of B_2H_6, see [*121*].

The best method for the production of B_5H_{11} specifically appears to be the commutation of tetraborane(10) and diborane in a hot–cold tube (120°/–30°). The B_5H_{11} forming in accordance with the equilibrium

$$2B_4H_{10}+B_2H_6 \rightleftharpoons 2B_5H_{11}+2H_2$$

is continuously removed from the equilibrium by condensation at $-30°$. H_2 is pumped off every 15 minutes. In addition to pentaborane(11) (yield 70%), some higher oily and solid hydrides are formed. The lower temperature limit for the commutation of B_4H_{10}–B_2H_6 into B_5H_{11} is 120° [*407*].

For the separation of B_5H_{11} from other boron hydrides, see [*407*].

3.4 Formation of B_5H_9 (B_5D_9)

In a vessel containing diborane, the temperature at one end of which is 180° and at the other end $-80°$, pentaborane(9) [*360*] and pentaborane(11) [*532*] condense out after $2\frac{1}{2}$ days:

$$\tfrac{1}{2}(BH_3)_2 \rightarrow BH_3 \xrightarrow[-H_2]{+(BH_3)_2} B_3H_7 \begin{cases} \xrightarrow{+(BH_3)_2,\ -BH_3} B_4H_{10} \\ \xrightarrow{+(BH_3)_2,\ -H_2} B_5H_{11} \end{cases} \xrightarrow{-2(BH_3)_2} B_5H_9$$

Bis(dimethylamino)borane, $BH(NMe_2)_2$ catalyses the conversion of pentaborane (11), which is easy to prepare, on slow heating from –78° to 0° into a series of boranes:

$$2B_5H_{11} \xrightarrow[-78°\rightarrow 0°]{[+BH(NMe_2)_2]} B_5H_9+B_4H_{10}+\tfrac{1}{2}(BH_3)_2$$

The main product is pentaborane(9). Example of the reaction (figures in moles):

$$\underset{1.187}{B_5H_{11}}+\underset{0.696}{BH(NMe_2)_2} \longrightarrow \underset{0.192}{B_2H_6}+\underset{0.165}{B_4H_{10}}+\underset{0.583}{B_5H_9}+\underset{0.09}{B_5H_{11}}+\underset{0.035}{B_6H_{10}}+\underset{0.074}{B_2H_5(NMe_2)},$$

together with non-volatile boranes which, on heating with dimethylamine, give 0.626 mole $BH(NMe_2)_2$ and 1.02 mole $BH_2(NMe_2)$. In the conversion, therefore, 91% of volatile boranes is formed [*63*].

Bis(dimethylamino)borane also catalyses the conversion of tetraborane(10) into a series of boranes (25% of B_5H_9) when it is allowed to warm up from $-78°$ to $-15°$ over 5 hours, for example, in accordance with the equation:

$$2B_4H_{10}+BH(NMe\ )_2 \longrightarrow B_5H_9+(BH_3)_2+2BH_2(NMe_2)+H_2$$

The empirical adduct $(MeN)_2BH{\cdot}B_4H_{10}$ forms as an intermediate [*63*]:

$$B_4H_{10} + BH(NMe_2)_2 \xrightarrow{-78°} (Me_2N)_2BH \cdot B_4H_{10} \xrightarrow{-78° \rightarrow -15°} \text{boranes and aminoboranes}$$

Example of the reaction (figures in moles):

$$\underset{0.692}{B_4H_{10}} + \underset{0.692}{BH(NMe_2)_2} \longrightarrow \underset{0.692}{(Me_2N)_2BH \cdot B_4H_{10}} \longrightarrow \underset{0.031}{B_2H_6} + \underset{0.120}{B_4H_{10}} + \underset{0.116}{B_5H_9} + \underset{0.037}{BH(NMe_2)_2} + \underset{0.092}{BH_2(NMe_2)}$$

Gaseous pentaborane(9) and H–D exchange on chromium oxide–alumina dehydrogenation catalysts. Sixfold repetition of the exchange with fresh D_2 at 20° gives B_5D_9 with 98.5% of D. No exchange takes place on molybdenum oxide–alumina catalysts, or in the absence of catalysts at 100°[*366*]. For the synthesis of partially deuterated pentaborane(9), see section 16.

For the separation of B_5H_9 from other boron hydrides, see [*64, 227, 390, 393, 854*]. For the mass spectrum of B_5H_9, see [*492, 690*].

3.5 Formation of B_6H_{10}

Under the conditions described below, up to 30% of the pentaborane(11) changes into hexaborane(10) in a complicated sequence of reactions that has not yet been elucidated. This is carried out with nucleophilic catalysts such as $BH(NMe_2)_2$, $BH_3 \cdot NMe_3$, NMe_3, Me_2O, and $Me(OC_2H_4)_2OMe$. For the mechanism of the conversion, see section 17. Examples of the conversion[*62*] are given below.

B_5H_{11}–$BH(NMe_2)_2$ conversion. A mixture of B_5H_{11} and $BH(NMe_2)_2$ (~ 2:1) is first held at –78° for 17 h and is then warmed up to 0° over 9 h. Calculated on the amount of B_5H_{11} converted (34–98%), the following products are formed (in mole%): 11% of B_2H_6, 19% of B_4H_{10}, 37% of B_5H_9, 8.4% of B_6H_{10}, and a non-volatile oil[*62*].

B_5H_{11}–NMe_3 conversion. A mixture of B_5H_{11} and NMe_3 (~ 1:1) undergoes quantitative conversion on being warmed from – 132° to – 10° over 10 h. The following products are formed (in mole%): 13% of B_2H_6, 12% of B_4H_{10}, 20% of B_5H_9, 17% of B_6H_{10}, and 13% of Me_3NBH_3. Increasing the proportion of NMe_3 decreases the yield of B_6H_{10}[*62*].

B_5H_{11}–Me_2O conversion. When the solid substance formed from pentaborane (11) and dimethyl ether at –78° (empirical formula $B_5H_{11} \cdot 1.56\ Me_2O$) is warmed to –20°, it gives, with a conversion of 77 to 90%, the following boranes (in mole%, calculated on the B_5H_{11} that has been reacted): 22–28% of B_2H_6, 21–22% of B_4H_{10}, 0–3·3% of B_5H_9, 24.5–27.3% of B_6H_{10}, 1.9–2.3% of $B_{10}H_{14}$, and about 25% of non-volatile compounds. An increase in the proportion of ether reduces

References p. 355

the rate of conversion, and a decrease in the proportion of ether lowers the yield of B_6H_{10} [62].

B_5H_{11}–diglyme conversion. At −20° is formed (in mole%): 17–18% of B_2H_6, 32–34% of B_4H_{10}, 1.1–2.8% of B_5H_9, 25–23% of B_6H_{10}, and 23% of non-volatile hydrides. Conversion of the B_5H_{11}: 40–65%. The replacement of diglyme by tetraglyme gives neither experimental advantages nor higher yields of B_6H_{10}. On the contrary, splitting of the ether takes place, leading to B–C bonds [62, 117].

B_5H_{11}–$BH_3 \cdot NMe_3$ conversion. This takes place between −20° and 0° (conversion of B_5H_{11}: 96 mole%). Conversion products in mole%: 32% of B_2H_6, 6·5% of B_4H_{10}, 13% of B_6H_{10}, 10% of B_5H_9, a little B_9H_{15}, and about 30% of non-volatile solid boron hydrides together with traces of $B_{10}H_{14}$ [117].

The decomposition of magnesium boride with acids also gives hexaborane(10) [766]. A 2.5% yield of B_6H_{10} is obtained from the reaction of $[Et_3NH]B_9H_{14}$ with polyphosphoric acid [43a].

For the separation of B_6H_{10} from other boron hydrides, see the literature [43a, 259, 766]. For the mass spectrum of B_6H_{10}, see [258].

3.6 *Formation of B_6H_{12}*

The hexaborane(12) described by Stock in 1924 [741], which was still impure, was first separated from hexaborane(10), by distillation in 1960/64 [240, 241]. It is formed (together with 40% of B_4H_{10}) in a 4% yield when tetramethylammonium triboranate(8), $[NMe_4]B_3H_8$, is stirred with polyphosphoric acid at 40° [240, 241]:

$$2B_3H_8^{\ominus} \xrightarrow[-2H_2]{+2H^{\oplus}} 2(2012\ B_3H_7) \longrightarrow 4204\ B_6H_{14} \xrightarrow{-H_2} 4212\ B_6H_{12}$$

B_6H_{12} is also formed in the preparation of pentaborane(11) from tetraborane and diborane at 110° (10 min), in the pyrolysis of tetraborane in the hot–cold tube (110°/0°) and (relatively simply) by the decomposition of liquid pentaborane(11) at 25° [482]. Sufficient amounts of B_6H_{12} for mass-spectroscopic purposes are obtained by the subjecting diborane to the electric discharge [258].

3.7 *Formation of B_8H_{12}*

When gaseous B_2H_6 and B_5H_9 (2 : 1) and hydrogen are passed at 12 torr through an electric discharge, the unstable octaborane(12) is formed in low yield. It can be isolated in pure crystalline form by fractional condensation[*202, 203*]. For preparation of B_8H_{12} and also B_8H_{14}, see section 3.9[*171b, 171c*]. B_9H_{15} decomposes by first-order kinetics to B_8H_{12}[*170a*]:

$$B_9H_{15} \underset{\text{excess diborane}}{\overset{\text{low pressure}}{\rightleftharpoons}} B_8H_{12} + BH_3$$

3.8 *Formation of B_8H_{18}*

The treatment of tetramethylammonium triboranate(8) with polyphosphoric acid gives tetraborane(10), hexaborane(10), and traces of the unstable octaborane (18)[*542a*]. The latter may have the structure of a tetraboranyl-substituted tetraborane $(B_4H_9)_2$ [compare the structure of a $B_{10}H_{16}$, a pentaboranyl-substituted pentaborane(9) $(B_5H_8)_2$].

3.9 *Formation of B_9H_{15}*

Nonaborane(15) is formed in small amount by the discharge-initiated interconversion of boranes (starting material: B_2H_6; see section 7), or the catalysed conversion of polyboranes[*117, 168, 437, 712*]. In the conversion of B_5H_{11} catalysed by donors such as $BH(NMe_2)_2$, NMe_3, or various ethers, small amounts of B_9H_{15} are formed among other products. Larger amounts of B_9H_{15} are obtained by the reaction of B_5H_{11} with hexamethylenetetramine at 0°. At a 55% conversion of the B_5H_{11}, the following products are formed (in mole%): 23.8% of B_2H_6, 10.1% of B_4H_{10}, 3.8% of B_5H_9, 0.6% of B_6H_{10}, 13.6% of B_9H_{15}, and 2.9% of $B_{10}H_{14}$. The residual percentage relates to a fraction having vapour pressures between those of B_6H_{10} and B_9H_{15}[*117*]. B_9H_{15} is also formed by reaction between liquid B_5H_{11} and gaseous B_2H_6 at 25 atm. pressure[*170a*]. At −80°, KB_9H_{14} reacts with anhydrous HCl to form i-B_9H_{15}[*171d*]. Above 35°, this decomposes with elimination of H_2 to form B_8H_{12}, $B_{10}H_{14}$, n-$B_{18}H_{22}$ and polymeric material [*171b*]. B_8H_{12} reacts with $^{10}B_2H_6$ at −30° to form $^{10}B\,^{n}B_8H_{15}$ and $^{10}B_2\,^{n}B_8H_{14}$[*493b*].

3.10 *Formation of $B_{10}H_{14}$*

Because of its laborious preparation on the laboratory scale, laboratory amounts of decaborane(14) are preferably purchased. Only where it is unobtainable is a laboratory preparation acceptable in respect to time and expense. Decaborane(14) is formed when diborane is heated to 115–120° at 1 atm[*739*] or to 180–240° at 1 to 5 atm[*545*] in 10% yield:

$$2(BH_3)_2 \xrightarrow[-4H_2]{+4(BH_3)_2} 4B_3H_7 \xrightarrow[-2BH_3,\ -4H_2]{} B_{10}H_{14}$$

References p. 355

It arises in a yield of about 30% when a sealed tube containing B_2H_6 ($p > 1$ atm) is heated at one end to 160°[*667*] or 125°[*391*]. At 90 to 95°, tetraborane(10) also gives a 10% yield of decaborane(14)[*739*]:

$$2B_4H_{10} \xrightarrow[-6H_2]{+(BH_3)_2} B_{10}H_{14}$$

For the mechanism of the thermolysis, see section 7. Solid hydrogen-poor boranes are always produced as well in such thermolyses. The formation is inhibited by the addition of not more than 5% of diethyl ether[*545*].

Boron-labelled decaborane(14) is formed by two different routes: (1) reaction of $(CH_3)_4{}^nB_9H_{12}$ (preparation[*276, 487b*]), $^{10}B_2H_6$, and HCl in monoglyme [*487b*]; (2) reaction of i-B_9H_{15} and $^{10}B_2H_6$ in a sealed vessel[*487b*]. For preparation of $^{10}B_2{}^nB_8H_{14}$, see section 3.9[*493b*].

Decaborane is preferably purified by high-vacuum sublimation[*225, 391*] or by recrystallisation from hexane[*56*].

3.11 Formation of $B_{10}H_{16}$

Decaborane(16) is formed by the slow passage of pentaborane(9) and hydrogen through an electric glow discharge between copper electrodes: $2B_5H_9 \rightarrow B_{10}H_{16} + H_2$. The yield amounts to 2 mg/h. Since $B_{10}H_{16}$ is highly volatile, it is easy to separate from the less volatile decaborane(14) formed at the same time[*280*].

3.12 Formation of $B_{18}H_{22}$

Octadecaborane(22) is formed in a 70% yield when acid aqueous solutions of $B_{20}H_{18}^{2\ominus}(EtOH_2^{\oplus})_2$ are concentrated and subsequently diluted with water, according to the overall equation[*718*]:

$$B_{20}H_{18}^{2\ominus} + 2H^{\oplus} + 6HOH \longrightarrow B_{18}H_{22} + 2B(OH)_3 + 2H_2$$

3.13 Formation of $B_{20}H_{16}$

Icosaborane(16) is formed by the slow passage of gaseous decaborane(14) through an alternating current discharge between copper electrodes[*525*]. When decaborane(14) is heated to 350° in the presence of $BMe_2(NHMe)$ as a catalyst, under pyrolysis conditions that must be accurately maintained, a 10 to 15% yield of icosaborane(16) is obtained[*521, 525*]. Fine single crystals can easily be obtained by vacuum sublimation[*235, 521, 525*].

4. Preparation of organylboranes

4.1 *Synthesis of R_nBH_{3-n} by the addition of BH_3 to alkenes*

In the gas phase, $(BH_3)_2$ reacts with ethylene above 20° to give ethylboranes Et_nBH_{3-n} ($n = 1, 2, 3$) or ethyldiboranes arising from the former by dimerization (*cf.* the scheme in section 4.2). The reaction is initiated by the addition of the B–H bond to the olefinic linkage (see section 12) and then yields, by further addition of ethylene or by commutation (see below), all the theoretically possible ethyldiboranes. At the same time, pyrolysis, particularly at higher temperatures, gives not only unsubstituted pentaborane(9) and decaborane(14) but also ethylated higher boranes, particularly derivatives of pentaborane(9) ($Et_nB_5H_{9-n}$, $n = 0$–4) and of decaborane(14)($Et_nB_{10}H_{14-n}$, $n = 0$–4)[16] [*57*]:

$$B_2H_6 \xrightarrow[+C_2H_4]{\Delta} (C_2H_5)_nB_2H_{6-n} \xrightarrow[+B_2H_6]{\Delta} (C_2H_5)_nB_5H_{9-n}, (C_2H_5)_nB_{10}H_{14-n}$$

When dissolved in donor solvents, diborane adds to alkenes with the preferential formation of the end-stage, the trialkylboron R_3B. Only with highly sterically hindered alkenes, such as 2,3-dimethylbut-2-ene, 2-methylbut-2-ene, α-pinene, etc. (or after the addition, when the compounds contain boron substituents) do the second and third addition steps take place so slowly ($k_1 \gg k_2 > k_3$) that (dimeric) monoalkylborane RBH_2 or dialkylborane R_2BH can be isolated (see section 12.5):

$$H{-}BH_2 \xrightarrow[k_1]{+\,C{=}C} H{-}C{-}C{-}BH_2 \xrightarrow[k_2]{+\,C{=}C} (H{-}C{-}C)_2BH \xrightarrow[k_3]{C{=}C} (H{-}C{-}C)_3B$$

4.2 *Synthesis of R_nBH_{3-n} by the commutation of R_3B with BH_3 or of R_2BOR' with BH_3*

Diborane and trialkylborons or triphenylboron react at 20° to 95° according to $BH_3 + BR_3 \rightleftharpoons BH_2R + BHR_2$ to give a mixture of boranes $BH_{3-n}R_n$ ($n = 0$–3) which, in turn, interact in accordance with the following scheme (individual members $BH_{3-n}R_n$ boxed) via H bonds to give alkyldiboranes or phenyldiboranes ($B_2H_{6-n}R_n$, $n = 1$–4[17] [*455–457, 669, 721, 804*] (to show the substitution stages

---

[16] At pyrolysis temperatures of 200–300°, the alkyl groups themselves may also undergo a change. Thus, for example, tripropylboron forms methylethylpropylboron and propene [*429*]:

$$(C_3H_7)_2B(C_3H_7) \xrightarrow{-C_3H_6} (C_3H_7)_2BH \longrightarrow (C_3H_7)B(C_2H_5)(CH_3)$$

better, the hydrogen atoms have been omitted from the diboranes):

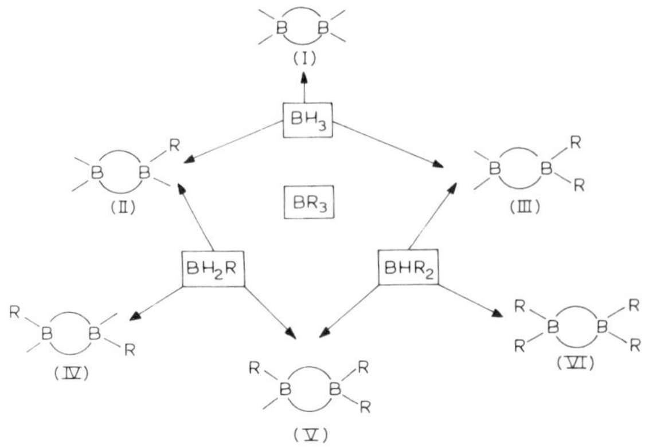

Under these conditions the main product is the asymmetric 1,1-diorganyldiborane (III), with considerably smaller amounts of the symmetrical 1,2-diorganyldiborane (IV) and the mono- (II), tri- (V), and tetraorganyldiboranes (VI)[*455–459*, *721*]. On the other hand, symmetrical 1,2-dialkyldiboranes (IV) and tetraalkyldiboranes (VI) are formed preferentially[*514*, *774*] in the commutation of $R_2B(OR')$ with BH_3.

In the gaseous or liquid phase at 25°, monoalkyldiboranes rapidly dismute without side reactions to give 1,2-dialkyldiboranes and diborane[*721*]:

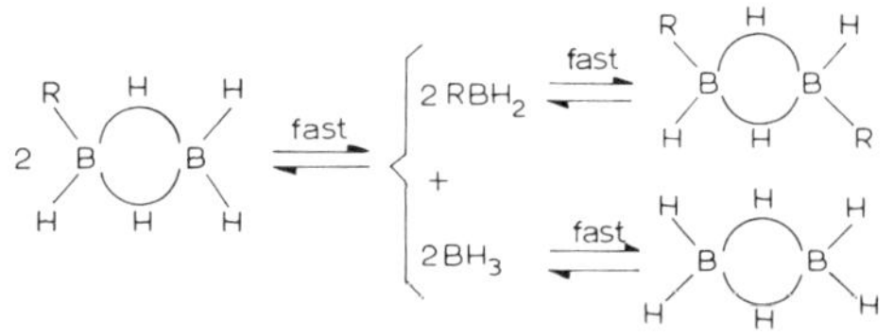

1,2-Dimethyl- and 1,2-diethyldiborane are stable[*721*] in the absence of diborane. The above equilibrium system, which rapidly becomes established in the presence of diborane, can be used for the removal of 1,2-diethyldiborane from a mixture of the 1,1- and 1,2- isomers. For this purpose, the mixture is treated several times with an excess of diborane; under these conditions, only the 1,2-diethyldiborane is converted into the monoethyldiborane, which can be separated off by distillation[*721*]. The existence of the above equilibrium system is also shown by the fact that pure $(BH_2Ph)_2$ cannot be sublimed under high vacuum,

[17]Since boron is incapable of forming stable R-bridges, the triorganylboron BR_3 does not participate in this dimerization. However, an unstable intermediate with an R bridge must be assumed in the initial step of the reaction of BH_3 with BR_3:

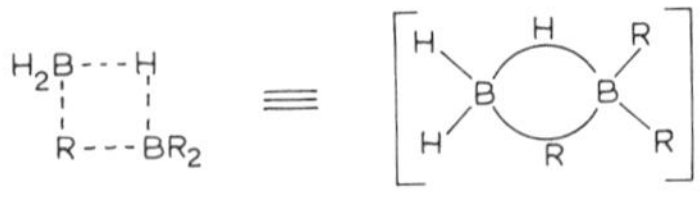

Intermediate

while it does sublime in the presence of $(BH_3)_2$ [*804*]. Diborane and the individual methyldiboranes ($B_2H_{6-n}Me_n$, $n = 0$–4, including isomers) can be separated by gas chromatography on fireclay and mineral oil [*686*].

The difficulties with the handling of gaseous diborane can be avoided when nascent diborane is permitted to undergo commutation with trialkylborons. Metal boranates, hydrogen halides, and trialkylborons react under pressure to give alkyldiboranes $B_2H_{6-n}R_n$ [*478, 478b*]:

$$(6-n)\,LiBH_4 + (6-n)\,HX \xrightarrow{-(6-n)\,LiX,\ -(6-n)\,H_2} (6-n)\,BH_3 \xrightarrow{+n\,BR_3} 3B_2H_{6-n}R_n$$

The difficult preparation and handling of the trialkylborons BR_3 can also be avoided if the latter are used in the nascent state. Under slight pressure, metal boranates, hydrogen halides, and trialkylaluminiums form alkyldiboranes B_2H_{6-n}-R_n [*477*]:

$$n\,AlR_3 + 6MBH_4 + (3n+6)\,HX \xrightarrow{150\text{–}175°} 3B_2H_{6-n}R_n + 6MX + n\,AlX_3 + (3n+6)\,H_2 \quad (M = Li, Na)$$

The reaction takes place in the presence of trialkylaluminiums at lower pressures and lower temperatures than in the absence of AlR_3, so that no alkylated higher boranes are formed in the presence of AlR_3 [*477*].

4.3 Synthesis of R_nBH_{3-n} by the hydrogenation of R_nBX_{3-n}, $(RBO)_3$ and $R_nB(OR')_{3-n}$ with metal hydrides

Alkyl or aryl boron halides (R_2BX, RBX_2) react with metal boranates or alanates (Li, Na) in ethereal solutions or without a solvent, in the presence of hydrogen halide (Cl, Br) or boron halide (Cl, Br), to form alkyldiboranes $B_2H_{6-n}R_n$ [*478, 478a 804*]. Two reactions are obviously involved: nucleophilic displacement of halogen by hydride, for example:

$$2PhBCl_2 + 4LiBH_4 \longrightarrow (PhBH_2)_2 + 2(BH_3)_2 + 4LiCl$$

and the commutation described above (see section 4.2):

$$(PhBH_2)_2 \xrightarrow[\Delta]{} (BH_3)_2, (PhBH_2)_2, (Ph_2BH)_2, Ph_3B$$

As in the preparation of alkyldiboranes $B_2H_{6-n}R_n$ by the reaction of metal boranates, hydrogen halides, and trialkylaluminiums (see section 4.2), here too, separate preparation of the alkylboron halides can be avoided. The formation of alkyldiboranes $B_2H_{6-n}R_n$ takes place at lower temperatures and lower pressures when alkylaluminium halides (R_2AlX, $RAlX_2$, and $R_3Al + AlX_3$ or $R_3Al + BX_3$) are used as the alkyl-yielding agent [*477*]; for example:

$$4n\,AlR_3 + (18-3n)\,LiBH_4 + (6+3n)\,BX_3 \xrightarrow{\sim 100°} 12B_2H_{6-n}R_n + (18-3n)\,LiX + 4n\,AlX_3$$

Temperatures above 175° favour the formation of alkylated higher boranes – for example, pentaborane(9) derivatives $B_5H_6R_3$ and $B_5H_5R_4$ [*477*].

Phenylboronic (benzeneboronic) acid, $PhB(OH)_2$, and its esters, $PhB(OR)_2$, dialkylborinic (dialkylboronous) esters, $R_2B(OR)$, tri-*tert*-butylboroxin, $(BuBO)_3$, and triphenylboroxin, $(PhBO)_3$, can be hydrogenated with lithium alanate or diborane in the presence of trimethylamine to phenylborane or *tert*-butylborane, as the case may be (as amine adducts $Me_3N \cdot RBH_2$) [*301, 302, 505–507, 804*]:

$$2PhB(OH)_2 + LiAlH_4 \longrightarrow (PhBH_2)_2 + LiAlO_2 + 2H_2O \left(\xrightarrow{+LiAlH_4} LiAlO_2 + 4H_2\right)$$

$$\left.\begin{array}{l} 2PhB(OR)_2 + LiAlH_4 \longrightarrow (PhBH_2)_2 + LiAl(OR)_4 \\ 3R_2B(OR') + 2(BH_3)_2 \longrightarrow 3(RBH_2)_2 + B(OR')_3 \end{array}\right\} \begin{array}{l} R = p-CH_3C_6H_4 \\ R' = CH_2CH(CH_3)_2 \end{array}$$

$$2\,(R'BO)_3 + 3LiAlH_4 + 6NMe_3 \longrightarrow R'BH_2 \cdot NMe_3 + 3LiAlO_2$$

$$(R' = C(CH_3)_3,\ C_6H_5)$$

5. Preparation of haloboranes

The halogen derivatives of borane BH_3 and diborane $(BH_3)_2$ can be obtained by the hydrogenation of boron trichloride ($BCl_3 + H_2$), by the halogenation of diborane ($B_2H_6 + HCl$), or by commutation ($B_2H_6 + BCl_3$). In the absence of the combined action of donors, the reactions take place only with a large supply of energy. Thus, diborane and boron trichloride on brief heating [*369*] or in an electric discharge [*666*] form dichloroborane $BHCl_2$ and monochlorodiborane B_2H_5Cl. Diborane and hydrogen chloride under elevated pressure give monochlorodiborane [*536*], and boron trichloride and hydrogen give dichloroborane [*264b, 398, 536*]:

$$B_2H_6 + BCl_3 \xrightarrow[\Delta]{\text{thermal or electrical energy}} B_2H_5Cl,\ BHCl_2$$

$$B_2H_6 + HCl \xrightarrow[\text{stainless steel}]{35^\circ,\ 11800\ \text{torr},\ 17-35\text{h}} B_2H_5Cl + H_2$$

$$BCl_3 + H_2 \xrightarrow{900^\circ,\ 1.7\ \text{sec}} BHCl_2 + HCl$$

$$2BCl_3 + H_2 + Mg \xrightarrow{400-450^\circ} 2BHCl_2 + MgCl_2$$

B_2H_5Cl and $BHCl_2$ can be separated by isobaric column distillation [*104*] or by gas chromatography [*536*]. Monochlorodiborane exists in equilibrium with diborane and boron chloride. At 35° the equilibrium is on the side of the decomposition products diborane and boron trichloride. The latter reacts with monochlorodiborane to form dichloroborane:

$$6B_2H_5Cl \rightleftharpoons 5B_2H_6 + 2BCl_3 \qquad \text{(a)}$$

$$3BCl_3 + B_2H_5Cl \rightleftharpoons 5BHCl_2 \qquad \text{(b)}$$

The dismutation (a) takes place very slowly without a catalyst (the equilibrium at 35° is achieved only after 70 h). Finely divided B_2O_3 catalyses the formation of B_2H_6 and BCl_3, but the $BHCl_2$ formed in accordance with equation (b) cannot be detected since it is absorbed on B_2O_3, possibly as $(BOCl)_x$[*536*]. In the gas phase at 25°, monochlorodiborane reacts rapidly with HCl in a wide range of molar ratios with the quantitative formation of boron trichloride[*536*]:

$$B_2H_5Cl + 5HCl \longrightarrow 2BCl_3 + 5H_2$$

$BHCl_2$ is found as an intermediate if HCl is added in small portions to B_2H_5Cl, because of equilibrium (b) which is thereby made possible[*536*].

Dichloroborane can be prepared without great expense by the commutation of borane or borane-d_3 and boron trichloride in ether at 20°. In some ethers, with an excess of diborane the monosubstituted intermediate BH_2Cl can be detected by IR spectroscopy[*568*], and in $(CH_3)_2O$, $(C_2H_5)_2O$, $(CH_2)_5O$, and $(CH_2)_4O$ it can even be isolated[*91*]:

$$(BH_3)_2 \xrightarrow{+2R_2O} 2BH_3 \cdot R_2O \xrightarrow{+BCl_3 \cdot R_2O} 3BH_2Cl \cdot R_2O \xrightarrow{+3BCl_3 \cdot R_2O} 6BHCl_2 \cdot R_2O$$

The driving force of this reaction is probably the conversion of the strong coordination bond in $R_2O \cdot BCl_3$ into three weaker coordination bonds $3R_2O \cdot BH_2Cl$, the total heat of formation of which exceeds the heat of formation of $R_2O \cdot BCl_3$. The hypothetical fluorinated boranes BH_2F and BHF_2 are less electrophilic than the corresponding chlorides, so that they do not form as ether adducts[*91*]. The simplest method of preparation of ethereal BH_2Cl solutions is probably the rapid reaction of BCl_3 with $LiBH_4$ or that of HCl with $(BH_3)_2$, in ether[*546*], or that of N-chlorsuccinimide with $BH_3 \cdot NR_3$[*25c*]:

$$BCl_3 + LiBH_4 + 2R_2O \longrightarrow 2BH_2Cl \cdot R_2O + LiCl$$

$$2BH_3 \cdot R_2O + 2HCl \longrightarrow 2BH_2Cl \cdot R_2O + 2H_2$$

$$BH_3 \cdot NR_3 + (CH_2CO)_2NCl \longrightarrow BH_2Cl \cdot NR_3 + (CH_2CO)_2NH$$

6. Preparation and solvolysis of metal boranates $M(BH_4)_n$ and their derivatives

Four routes are available for the preparation of *metal boranates* MBH_4 (M = equivalent of a metal): route (a), preparation through the synthesis of the $BH_4^{\ominus}$ ion from BH_3 and $H^{\ominus}$; route (b), preparation through the hydrogenation of substituted borates $BX_4^{\ominus}$; route (c), preparation by the dismutation of partially substituted boranates $BH_{4-n}X_n^{\ominus}$; route (d), transfer of the $BH_4^{\ominus}$ ion in tetrahydroboarates MBH_4 to other cations $M^{\oplus}$, *e.g.* by metathesis.

(a) Addition of BH_3 to a hydride ion: $BH_3 + H^{\ominus} \longrightarrow BH_4^{\ominus}$.

References p. 355

Examples:
$(BH_3)_2 + 2LiH \longrightarrow 2LiBH_4$ (*cf.* section 6.1.1)
$BCl_3 + 4NaH \longrightarrow NaBH_4 + 3NaCl$[18]
$3(BH_3)_2 + 2(Sn(OR)_2 \longrightarrow 2Sn(BH_4)_2 + 2BH(OR)_2$[19]

(b) Hydrogenation of substituted boranates $BX_4^{\ominus}$ with $(BH_3)_2$: $3B(OR)_4^{\ominus} + 2(BH_3)_2 \longrightarrow 3BH_4^{\ominus} + 4B(OR)_3$.
Example:
$3NaB(OMe)_4 + 2(BH_3)_2 \longrightarrow 3NaBH_4 + 4B(OMe)_3$ (*cf.* section 6.1.2)

(c) Dismutation of partially substituted boranate ions $BH_{4-n}X_n^{\ominus}$: $4BH(OR)_3^{\ominus} \rightarrow BH_4^{\ominus} + 3B(OR)_4^{\ominus}$.
Example:
$4NaBH(OMe)_3 \longrightarrow NaBH_4 + 3NaB(OMe)_4$ (*cf.* section 6.1.3)

(d) Metathesis: $MX + BH_4^{\ominus} \longrightarrow MBH_4 + X^{\ominus}$.
Examples:
$CsOH + NaBH_4 \longrightarrow CsBH_4 + NaOH$
$LiCl + NaBH_4 \longrightarrow LiBH_4 + NaCl$
$AgClO_4 + LiBH_4 \longrightarrow AgBH_4 + LiClO_4$
$Me_3SF + NaBH_4 \longrightarrow Me_3SBH_4 + NaF$ (*cf.* section 6.1.4).

There are two routes for the preparation of *partially substituted boranates* $BH_{4-n}X_n^{\ominus}$ (X = monovalent substituent): route (a), preparation through the additive formation of the anion $BH_{4-n}X_n^{\ominus}$; route (b), preparation through substitution in $BH_4^{\ominus}$.

(a) Addition of BX_3 or $BX_{3-n}H_n$ to hydride ions $H^{\ominus}$, or addition of BH_3 or $BH_{3-n}X_n$ to anions $X^{\ominus}$: $BX_3 + H^{\ominus} \rightarrow BX_3H^{\ominus}$ or $BH_3 + X^{\ominus} \rightarrow BH_3X$.
Examples:
$BEt_3 + NaH \longrightarrow NaBEt_3H$
$BCl_2Ph + 3LiH \longrightarrow LiBH_3Ph + 2LiCl$[20]
$B(OMe)_3 + Na + \frac{1}{2}H_2 \longrightarrow NaB(OMe)_3H$[21]
$(BH_3)_2 + 2LiPh \longrightarrow 2LiBH_3Ph$
$(BH_3)_2 + 2KF \longrightarrow 2KBH_3F$ (*cf.* Section 6.2.1)

(b) Substitution in the boranate anion $BH_4^{\ominus}$: $BH_4^{\ominus} + n/2X_2 \rightarrow BH_{4-n}X_n + n/2H_2$
Example:
$NaBH_4 + (SCN)_2 \longrightarrow NaBH_2(NCS)_2 + H_2$ (*cf.* Section 6.2.2)

---

[18]Formation of nascent BH_3.
[19]Formation of nascent SnH_2.
[20]Formation of nascent BH_2Ph.
[21]Formation of nascent NaH.

6.1 *Synthesis of unsubstituted boranates $BH_4^{\ominus}$*

6.1.1 *Synthesis of $BH_4^{\ominus}$ by the addition of BH_3 to $H^{\ominus}$*

Diborane does not react with lithium hydride in the absence of a solvent. However, in the presence of diethyl ether, which solvates Li ions, lithium boranate forms smoothly [*660–662*] at −50° with a maximum yield of 90% [*562, 663*, cit. *5*]. The more highly solvating polyethylene glycol dimethyl ethers or tetrahydrofuran must be used [*83, 137, 547, 660, 809*] to prepare sodium, potassium, calcium, strontium, and barium boranates.

$$\left.\begin{array}{l}(BH_3)_2 + 2LiH \xrightarrow{\text{diethyl ether}} 2LiBH_4 \\ (BH_3)_2 + 2MH \xrightarrow{\text{polyglycol ethers}} 2MBH_4\end{array}\right\}(M = Na, K)$$

$$(BH_3)_2 + 2MH \xrightarrow{\text{tetrahydrofuran}} 2MBH_4\ (M = Na, K, Ca/2, Sr/2, Ba/2)$$

The boron hydride may also be produced only during the reaction. Thus, boron trichloride reacts rapidly with sodium and potassium hydrides in toluene in the presence of $AlEt_3$ to give an 83 to 90% yield of the corresponding boranates [*161*]:

$$BCl_3 + 4MH \xrightarrow[\text{toluene}]{AlEt_3} MBH_4 + 3MCl\ (M = Na, K)$$

Within a few hours at 120 to 130° in the autoclave, with an excess of lithium hydride, the boron fluoride–ether adduct also gives a 90% yield of lithium boranate [*855*]:

$$BF_3 + 4LiH \xrightarrow[\text{diethyl ether}]{\text{pressure}} LiBH_4 + 3LiF$$

Here, in order to suppress substantially the formation of diborane (from $LiBH_4 + BF_3$, see section 3.1.1), other authors work at low temperatures (−5 to 34°) at atmospheric pressure [*218, 851*].

Likewise, the metal hydride MH, for example, can also be used in the nascent state:

$$2n\,(BH_3)_2 + 3M(OR)_n \longrightarrow 3M(BH_4)_n + n\,B(OR)_3$$

Thus, for example, diborane in tetrahydrofuran reacts with the methoxides of Ca, Sr, Ba [*808*], of Mg, Zn, Cd, La, Ce, Pr, Nd [*410*], of Y, Sm, Eu, Gd, Tb, Dy, Ho, Er, Tm, Yb, Lu [*863a*], of V(III), Cr(III), Mn(II), Fe(II) [*410*], of Sn(II), Pb(II), and $R_3Pb(IV)$ [*13a, 14, 15, 351*], and with the ethoxides of Zn and Ti(IV) [*410*] to give the corresponding boranates. $Ti(OBu^n)_4$ and $Ti(OPr^i)_4$ give $Ti(OBu^n)(BH_4)_2$ THF and $Ti(OPr^i)(BH_4)_2$ respectively (reduction) [*372a*].

6.1.2 *Synthesis of $BH_4^{\ominus}$ by the hydrogenation of alkoxides with $(BH_3)_2$*

An excess of diborane reacts with sodium tetramethoxyoborate, $NaB(OMe)_4$, in the absence of a solvent [*660*] or, better, in tetrahydrofuran [*84*] to give a quanti-

References p. 355

tative yield of the THF-insoluble sodium boranate, $NaBH_4$ [with a deficiency of $(BH_3)_2$, $NaBH_4$ dissolves in $NaB(OMe)_4$]. Potassium boranate is formed in the same way [*660*]:

$$3MB(OMe)_4 + 2(BH_3)_2 \longrightarrow 3MBH_4 + 4B(OMe)_3 \ (M = Na, K)$$

$(BH_3)_2$ reacts with the alkoxoborates, $M[B(OMe)_4]_2$ or $M[B(OEt)_4]_2$, which can easily be prepared from the metal, the appropriate alcohol, and trimethyl borate [*806, 807*] very sluggishly in the absence of a solvent or in diethyl ether, but it reacts quantitatively in tetrahydrofuran with the formation of the corresponding alkaline-earth metal boranates [*837*]:

$$3M[B(OR)_4]_2 + 4(BH_3)_2 \longrightarrow 3M[BH_4]_2 + 8B(OR)_3$$

Instead of expensive diborane, aluminium together with hydrogen can also be used for the hydrogenation of sodium tetramethoxoborate [*18*]:

$$3Na[B(OMe)_4] + 4Al + 6H_2 \xrightarrow{\text{diglyme}} 3Na[BH_4] + 4Al(OMe)_3$$

Often, the metathesis of metal halides with alkalimetal boranates does not lead to the metal boranates, *e.g.*:

$$SnX_2 + 2LiBH_4 \longrightarrow \underset{\text{stable}}{Li_2[SnX_2(BH_4)_2]} \xrightarrow{-LiX} \underset{\text{stable}}{Li[SnX(BH_4)_2]} \xrightarrow{-LiX} Sn(BH_4)_2$$

In these cases the reactions of B_2H_6 with the metal methoxides are useful [*13a, 14, 14a, 15*]:

$$2Sn(OCH_3)_2 + 3B_2H_6 \longrightarrow 2Sn(BH_4)_2 + 2HB(OCH_3)_2$$

$$Sn(OCH_3)_4 \begin{cases} \xrightarrow[-2HB(OCH_3)_2]{+B_2H_6} SnH_4 \\ \xrightarrow[-2HB(OCH_3)_2]{+2B_2H_6} Sn(BH_4)_2 + H_2 \end{cases}$$

$$R_3PbOCH_3 + 3B_2H_6 \longrightarrow 4R_3PbBH_4 + 2HB(OCH_3)_2$$

$$(R = CH_3, C_2H_5, n\text{-}C_3H_7, n\text{-}C_4H_9)$$

6.1.3 *Synthesis of $BH_4^{\ominus}$ by the dismutation of partially substituted boranate ions $BH_{4-n}X_n^{\ominus}$*

In the absence of a solvent, trimethyl borate, $B(OMe)_3$, adds sodium hydride [*84*] or calcium hydride [*366*] to form sodium trimethoxoborate or calcium bis-(trimethoxoborate):

$$B(OMe)_3 + NaH \longrightarrow NaBH(OMe)_3$$

$$2B(OMe)_3 + CaH_2 \longrightarrow Ca[BH(OMe)_3]_2$$

When diglyme or tetrahydrofuran is added, the $NaBH(OMe)_3$ obtained in this way rapidly dismutes into boranate and tetramethoxoborate [*84*]:

$$4NaBH(OMe)_3\,(solution) \begin{cases} \xrightarrow{\text{diglyme}} NaBH_4\,(soluble) + 3NaB(OMe)_4\,(insoluble) \\ \xrightarrow{\text{THF}} NaBH_4\,(insoluble) + 3NaB(OMe)_4\,(soluble) \end{cases}$$

However, because of the simultaneous presence of $NaB(OMe)_4$, $NaBH_4$ does not precipitate out quantitatively from tetrahydrofuran [*84*]. The preparation of lithium boranate by the reaction of $B(OMe)_3$ with LiH is not satisfactory, since the following side reactions occur:

$$B(OMe)_3 + LiH \longrightarrow BH(OMe)_2(volatile) + LiOMe$$

$$4B(OMe)_3 + 4LiH \longrightarrow 3LiB(OMe)_3 + LiBH_4$$

Consequently, $LiBH_4$ is obtained in only a 50% yield [*655*, cit. *5*].

6.1.4 *Synthesis of metal boranates by metathesis*

The metathesis of sodium boranate with potassium hydroxide, rubidium hydroxide, or cesium hydroxide in water gives a 75 to 95% yield of the corresponding boranates (*e.g.* the purity of the KBH_4 that can be obtained in this way is 99.5%) [*27, 65, 66, 219, 517, 719a, 719b*]:

$$MOH + NaBH_4 \xrightarrow{H_2O} MBH_4 + NaOH\ (M = K, Rb, Cs)$$

Lithium boranate can be prepared by the metathesis of sodium boranate with lithium chloride in dimethylformamide, $HCONMe_2$ [*153*], isopropylamine, Me_2CHNH_2 [*564*], 15% isopropylamine in diethyl ether, liquid ammonia, hydrazine, or methylamine [*655*, cit. *5*]. In the case of dimethylformamide $LiBH_4$ dissolved in $HCONMe_2$ can easily be separated from the insoluble NaCl or can easily be extracted with ether after the removal of the amine:

$$LiCl + NaBH_4 \longrightarrow LiBH_4 + NaCl$$

The metathesis of potassium boranate with lithium chloride in tetrahydrofuran or ethanol (at −10°) can be carried out similarly [*586, 587*]:

$$LiCl + KBH_4 \longrightarrow LiBH_4 + KCl$$

Lithium boranate and aluminium chloride in the solid state form aluminium boranate, $Al(BH_4)_3$, and with labelled $LiBH_4$ (^{11}B, ^{10}B, H, D) the corresponding isotopically labelled aluminium boranates [$Al(^{11}BH_4)_3$, $Al(^{10}BH_4)_3$, $Al(^{11}BD_4)_3$, and $Al(^{10}Bd_4)_3$] [*201*]. The formation of $Al(BH_4)_3$ takes place via the isolable inter-

References p. 355

mediates $Al(BH_4)Cl_2$ and $Al(BH_4)_2Cl$ (diborane and hydrogen are formed in addition) [*564*, *661*]:

$$AlCl_3 \xrightarrow[-LiCl]{+LiBH_4} AlCl_2(BH_4) \xrightarrow[-LiCl]{+LiBH_4} AlCl(BH_4)_2 \xrightarrow[-LiCl]{+LiBH_4} Al(BH_4)_3$$

In the reaction of lithium boranate with zinc chloride at 20° [*811*] or with cadmium chloride at 0° [*812*] in diethyl ether, zinc or cadmium bis(boranate) (M = Zn, Cd) containing small amounts of Cl is formed via the intermediate $ZnCl(BH_4)$, which is stable up to 120°, or via $CdCl(BH_4)$, which is stable up to 85°:

$$MCl_2 \xrightarrow[LiCl]{+LiBH_4} MCl(BH_4) \xrightarrow{+LiBH_4} Li[MCl(BH_4)_2] \xrightarrow[-LiCl]{} M(BH_4)_2$$

With the trichlorides of samarium, europium, gadolinium, terbium, dysprosium, and ytterbium, but not with lanthanum trichloride, lithium boranate in tetrahydrofuran reacts in accordance with the following equation [*102*, *624*]:

$$MCl_3 \xrightarrow[-2LiCl]{+2LiBH_4} MCl(BH_4)_2 \ (M = Sm, Eu, Gd, Tb, Dy, Yb)$$

The bis(boranate) chlorides form crystalline tetrahydrofuran adducts: $SmCl(BH_4)_2 \cdot 2.5THF$, $EuCl(BH_4)_2 \cdot xTHF$, $GdCl(BH_4)_2 \cdot 2THF$, $TbCl(BH_4)_2 \cdot 2THF$, $DyCl(BH_4)_2 \cdot xTHF$, and $YbCl(BH_4)_2 \cdot xTHF$. Above 120°, the bis(boranate) chlorides of Gd, Tb, and Yb form the orange-coloured compounds $GdCl(B_2H_6)$, $TbCl(B_2H_6)$, and $YbCl(B_2H_6)$ [*102*, *624*, *863a*]. $MgCl_2$ with $NaBH_4$ in $HCONMe_2$ give $Mg(BH_4)_2 \cdot 6HCONMe_2$ [*514d*].

While copper(I) chloride and lithium boranate in chlorobenzene evolve only diborane [*615*], copper(I) boranate is formed [*812*] in diethyl ether–tetrahydrofuran at −20°:

$$CuCl + LiBH_4 \longrightarrow CuBH_4 + LiCl$$

Silver perchlorate reacts with lithium boranate in ethereal solution at −80° with quantitative formation of silver boranate [*810*]:

$$AgClO_4 + LiBH_4 \longrightarrow AgBH_4 + LiClO_4$$

Even at very low temperatures, gold(III) chloride, $AuCl_3$, forms no gold(III) boranate, $Au(BH_4)_3$, but only its decomposition products gold, diborane, and hydrogen [*540*, *834a*]:

$$AuCl_3 + 3LiBH_4 \xrightarrow{-3LiCl} [Au(BH_4)_3] \longrightarrow Au + 3BH_3 + 1\tfrac{1}{2}H_2$$

With lithium or sodium boranate in ether below −10° iron(III) chloride forms, with evolution of diborane, stable white iron(II) boranate, which decomposes above 0° [*641*, cit. *2*]:

$$FeCl_3 \xrightarrow[-3LiCl,-BH_3,-\frac{1}{2}H_2]{+3LiBH_4,\ <-10^\circ} Fe(BH_4)_2 \xrightarrow{>0^\circ} \begin{cases} Fe + (BH_3)_2 + H_2 \\ Fe\cdot 2B + 4H_2 \end{cases}$$

With lithium boranate, dicyclopentadienyltitanium(IV) dichloride, Cp_2TiCl_2, in diethyl ether forms, with reduction, dicyclopentadienyltitanium(III) monoboranate, hydrogen, and diborane. By the addition of LiH, the latter can be used to produce more boranate [*549*]:

$$Cp_2TiCl_2 + 2LiBH_4 \longrightarrow Cp_2Ti(BH_4) + 2LiCl + \tfrac{1}{2}H_2 + \tfrac{1}{2}(BH_3)_2$$

$$Cp_2TiCl_2 + LiBH_4 + LiH \longrightarrow Cp_2Ti(BH_4) + 2LiCl + \tfrac{1}{2}H_2$$

Dicyclopentadienyltitanium(III) boranate behaves as the boranate of a simple cation, *e.g.* like $LiBH_4$ [*550*]:

$$Cp_2Ti(BH_4) + HCl \longrightarrow Cp_2TiCl + \tfrac{1}{2}(BH_3)_2 + H_2$$

$$3Cp_2Ti(BH_4) + BCl_3 \longrightarrow 3Cp_2TiCl + 2(BH_3)_2$$

With $LiBH_4$ under the same conditions, dicyclopentadienylzirconium(IV) dichloride gives $Cp_2Zr(BH_4)_2$ [*537a*].

Hexamminechromium(III) trifluoride, hexamminecobalt(III) trifluoride, and ammonium fluoride react with sodium boranate at –45° in liquid ammonia to give the corresponding boranates [*580*]:

$$MF_n + n\,NaBH_4 \xrightarrow[-45^\circ]{\text{liq. }NH_3} M(BH_4)_n + n\,NaF$$

$$[M = Cr(NH_3)_6^{3\oplus}, Co(NH_3)_6^{3\oplus}, NH_4^{\oplus}]$$

With lithium boranate in ethereal solution, mono-, di-, and trimethylammonium chlorides and ammonium chloride itself give, with evolution of hydrogen, amine-boranes, aminoboranes, and borazines [$Me_3N\cdot BH_3$, Me_2NBH_2, $(MeNBH)_3$, $(HNBH)_3$] [*26, 636*]. Examples:

$$Me_3NHCl + LiBH_4 \xrightarrow[20^\circ]{\text{diethyl ether}} Me_3N\cdot BH_3 + LiCl + H_2$$

$$Me_2NH_2Cl + LiBH_4 \xrightarrow[20^\circ]{\text{diethyl ether}} Me_2NBH_2 + LiCl + 2H_2$$

$$3MeNH_3Cl + 3LiBH_4 \xrightarrow[>20^\circ]{\text{di-n-hexyl ether}} H_3B_3N_3Me_3 + 3LiCl + 9H_2$$

In an aqueous medium (H_2O or H_2O/C_2H_5OH), however, the metathesis of the ammonium chlorides with $NaBH_4$ and KBH_4 and, with the exclusion of air, even with $LiBH_4$ takes place[22]. Thus, in aqueous systems or in diglyme, the

[22]Normally, $LiBH_4$ hydrolyses in water.

References p. 355

corresponding boranates are given by quaternary ammonium salts (yield 70–99%) [*26*], quaternary phosphonium salts [*324*], trimethylsulphonium salts [*323*], triphenylsulphonium salts [*324*], and diphenyliodonium salts [*324*]:

$$R_4NX + MBH_4 \xrightarrow{H_2O} R_4N(BH_4) + MX$$

$R = CH_3, C_2H_5, (C_6H_5CH_2)(CH_3)_3$
$X = F, Cl, Br, OH, \frac{1}{2}CO_3, CH_3COO, \frac{1}{2}(COO)_2$
$M = Li, Na$

$$Ph_4PF + KBH_4 \xrightarrow{H_2O} Ph_4P(BH_4) + KF$$

$$BzPh_3PBr + NaBH_4 \xrightarrow{\text{diglyme–chloroform}} BzPh_3P(BH_4) + NaBr$$

$$BzMe_3PBr + NaBH_4 \xrightarrow{\text{diglyme}} BzMe_3P(BH_4) + NaBr$$

$$Me_3SF + NaBH_4 \xrightarrow{H_2O} Me_3S(BH_4) + NaF$$

$$Ph_3SF + KBH_4 \xrightarrow{H_2O} Ph_3S(BH_4) + KF$$

$$Ph_2IF + KBH_4 \xrightarrow{H_2O} Ph_2IBH_4 + KF$$

The reaction mechanism of the formation of boranates in the reaction of an excess of diborane with trimethylaluminum [*675*] or dimethyl-beryllium [*123*] is complex. The overall equations are:

$$(AlMe_3)_2 + 4(BH_3)_2 \longrightarrow 2Al(BH_4)_3 + 2BMe_3$$

$$\frac{3}{x}[BeMe_2]_x + 4(BH_3)_2 \longrightarrow 3Be(BH_4)_2 + 2BMe_3$$

On high-pressure hydrogenolysis, lithium hydride and triethylboron form lithium boranate [*224*]:

$$LiH + BEt_3 + 3H_2 \xrightarrow[200\,atm]{240°} LiBH_4 + 3EtH$$

Sodium boranate can be prepared at an elevated temperature (100–300°) by the reaction of trialkylamineboranes with sodium hydride, sodium acetylide + hydrogen, or alkyl- or arylsodiums or sodium alkoxides [*428, 428a-428d, 429*], for example:

$$4H_3B{\cdot}NR_3 + 3NaOR \xrightarrow{\Delta} 3NaBH_4 + B(OR)_3 + 4NR_3$$

6.2 *Synthesis of partially substituted metal boranates $M[BH_{4-n}X_n]$*

6.2.1 *Additive formation of metal boranates $M[BH_{4-n}X_n]$ from metal and boron components*

X = *hydrocarbon substituent.* With phenylboron dichloride, $PhBCl_2$, in ether at −40°, lithium hydride forms lithium phenylboranate, $Li[BH_3Ph]$ [*805*]. With triethylboron, BEt_3, in ether or aliphatic hydrocarbons, sodium hydride forms sodium triethylboranate, $Na[BHEt_3]$ [*335*]:

$$3LiH + BCl_2Ph \longrightarrow Li[BH_3Ph] + 2LiCl$$

$$NaH + BEt_3 \longrightarrow Na[BHEt_3]$$

On boiling in diglyme, $Na[BHEt_3]$ dismutes [*335*]:

$$4Na[BHEt_3] \longrightarrow Na[BH_4] + 3Na[BEt_4]$$

Diborane reacts with phenyllithium [*805*] or with diethylmagnesium and catalytic amounts of AlR_3 [*32*] to form lithium phenylboranate or magnesium ethylboranates and magnesium boranate (see section 4.1.3 of Chapter 3):

$$2LiPh + (BH_3)_2 \longrightarrow 2Li[BH_3Ph]$$

$$2MgEt_2 + (BH_3)_2 \xrightarrow{+AlR_3} MgH(BHEt_3) + MgEt(BH_4)$$

$$2MgEt(BH_4) \longrightarrow Mg(BH_4)_2 + MgEt_2$$

X = *amino substituent.* The acceptor power of the B-atom is so weakened by the neighbourhood of the electron-rich N-atoms that aminoboranes, $B(NR_2)_3$, add no donor-active hydride ions. Thus, tris(dimethylamino)borane, $B(NMe_2)_3$, forms no tris(dimethylamino)boranate, $M[BH(NMe_2)_3]$, with lithium hydride or sodium hydride with or without a solvent, at atmospheric or elevated pressure [*350*]:

$$MH + B(NMe_2)_3 \not\longrightarrow M[BH(NMe_2)_3]$$

On the other hand, tris(dimethylamino)aluminium, $Al(NMe_2)_3$, forms tris(dimethylamino)alanate, $M[AlH(NMe_2)_3]$, with lithium hydride or sodium hydride in diethyl ether:

$$MH + Al(NMe_2)_3 \longrightarrow M[AlH(NMe_2)_3]$$

With amineborane (borazane) or phosphaneborane (borphosphane), sodium hydride forms N- or P-substituted boranates. These give rise to sparingly soluble dioxan adducts [*7*]:

$$NaH + BH_3{\cdot}NMe_2H + \tfrac{1}{2}C_4H_8O_2 \xrightarrow[35-40^\circ]{\text{monoglyme}} Na[BH_3(NMe_2)]{\cdot}\tfrac{1}{2}C_4H_8O_2 + H_2$$

$$NaH + BH_3{\cdot}PMe_2H + \tfrac{1}{2}C_4H_8O_2 \xrightarrow[35-40^\circ]{\text{monoglyme}} Na[BH_3(PMe_2)]{\cdot}\tfrac{1}{2}C_4H_8O_2 + H_2$$

X = *alkoxy substituent.* Sodium hydride reacts with trialkyl borates in tetrahydrofuran, diglyme, and triglyme, to give sodium trialkoxoboranates, $Na[BH(OR)_3]$. The reaction velocity decreases in the following sequence of R's: methyl

References p. 355

> ethyl ≫ isopropyl ≫ *tert*-butyl[*81, 89*]:

$$NaH + B(OR)_3 \longrightarrow Na[BH(OR)_3]$$

For the technical production of alkali-metal trimethoxoboranates, $M[BH(OMe)_3]$, NaH is produced in the same reaction vessel. Solvents are xylene and diglyme [*330*], for example:

$$Na + \tfrac{1}{2}H_2 + B(OMe)_3 \xrightarrow{\text{pressure}} Na[BH(OMe)_3]$$

X = *halogen or pseudohalogen*. Boron fluoride reacts with sodium hydride to give sodium trifluoroboranate[*264a*]:

$$BF_3 + NaH \xrightarrow[-70\text{ to }20^\circ]{Et_2O} Na[BHF_3]$$

Diborane in monoglyme ($MeOC_2H_4OMe{\cdot}2BH_3$) reacts with metal halides and metal pseudohalides to give halo-substituted boranates. On the addition of dioxan, they generally precipitate as dioxan adducts:

$$\tfrac{1}{2}(BH_3)_2 \xrightarrow[\text{monoglyme}]{+\ \text{O}\langle} \text{O}\langle{\cdot}BH_3 \xrightarrow[25-45^\circ]{+n\ C_4H_8O_2}$$

$$\xrightarrow{+NaSCN} Na[BH_3(SCN)]{\cdot}2C_4H_8O_2$$
$$\xrightarrow{+LiSCN} Li[BH_3(SCN)]{\cdot}1\tfrac{1}{2}C_4H_8O_2$$
$$\xrightarrow{+KF} K[BH_3F]$$
$$\xrightarrow{+NaCN} Na[BH_3{\cdot}CN{\cdot}BH_3]{\cdot}2C_4H_8O_2$$
$$\xrightarrow{+Cd(CN)_2} Cd[BH_3{\cdot}CN{\cdot}BH_3]{\cdot}3.7C_4H_8O_2$$

6.2.2 *Formation of metal boranates* $M[BH_{4-n}X_n]$ *by substitution of the basic compounds* $M[BH_4]$

In a pressure tube at 125°, carbon dioxide reacts with solid sodium boranate as follows[*789*]:

$$NaBH_4 + 2CO_2 \longrightarrow Na[BO(OCHO)(OCH_3)]$$

while in dimethyl ether at 20° it gives sodium triformatoboranate[*788*]:

$$NaBH_4 + 3CO_2 \longrightarrow Na[BH(OCHO)_3]$$

$(SCN)_2$ and $(SeCN)_2$ react with $LiBH_4$ and $NaBH_4$ in ether or diglyme with the evolution of hydrogen and the formation of pseudohaloboranates: $Li[B(NCS)_4]$, $Na[BH_{4-n}(NCS)_n]$($2 < n < 3$), and $Li[BH_{4-n}(NCSe)_n]{\cdot}$ether[*405, 405*]; for example:

$$M[BH_4] + \frac{n}{2}(SCN)_2 \longrightarrow M[BH_{4-n}(NCS)_n] + \frac{n}{2}H_2\ (M = Li, Na)$$

In the reaction of boranates with $(CN)_2$, however, the C≡N triple bond is hydrogenated with the formation of brown solids [*404*].

6.3 *Reactions of the boranate ion $BH_4^{\ominus}$*

Boranate ions, $BH_4^{\ominus}$, can react with element compounds EX in three ways.

(a) With the formation of isolable boranates EBH_4. This generally occurs when the electronegativity of the element E is less than about 2.0 [Pauling units, $\sqrt{eV}$]: $BH^{\ominus} + EX \longrightarrow E(BH_4) + X^{\ominus}$.
Example:

$$2K(BH_4) + ZnBr_2 \longrightarrow Zn(BH_4)_2 + 2KBr$$

(b) Formation of hydrides EH via boranates as intermediates[23]. This nucleophilic substitution, which leads to BH_3 and EH instead of to EBH_4 [see under (a)] generally takes place when the element E has an electronegativity above about 2.0 [Pauling units, $\sqrt{eV}$]: $H_3{}^{\ominus}B{-}H + E{-}X \rightarrow [(H_3B\cdots H\cdots E\cdots X)^{\ominus}] \rightarrow BH_3 + E{-}H + X^{\ominus}$.
Example:

$$3LiBH_4 + PCl_3 \longrightarrow \tfrac{3}{2}(BH_3)_2 + PH_3 + 3LiCl$$

(c) Substitution on the boranate ion:

$$BH_4^{\ominus} + n\,EX \longrightarrow BH_{3-n}X_n^{\ominus} + n\,HE.$$

Examples (see also section 6.2.2):

$$LiBH_4 + HN_3 \longrightarrow Li[BH_3(N_3)] + H_2$$

$$NaBH_4 + D_2 \longrightarrow Na[BH_3D] + HD$$

6.3.1 *Deuterium exchange*

A statistical *isotopic exchange* in accordance with the equation

$$M[BH_4] + n\,D_2 \longrightarrow M[BH_{4-n}D_n] + n\,HD$$

begins at 200° with lithium hydride and at 350° with sodium hydride [*101*].

6.3.2 *Main group VII and pseudohalogens*

Halogens. Halogens react with lithium boranate to form (very pure [*13*]) boron trihalides [equation (a) or (c)], together with a little diborane [equation (b)] and

[23] This reaction mechanism has not yet been studied kinetically. However, it is possible, since many (unstable) complexes of the above-formulated intermediate $(H_3B\cdots H\cdots E\cdots X)^{\ominus}$ with elements E capable of an expansion of electron shells have been isolated, *e.g.* $Li_2[(BH_4)_2(SiCl_4)]$ or $Li_2[(BH_4)_2(SiHCl_3)]$ [*347, 815*]. A predissociation of the boranate ion (which possesses no occupied non-bonding orbitals) into borane and a hydride ion (with an occupied non-bonding donor orbital) in accordance with the equation $BH_4^{\ominus} \rightarrow BH_3 + H^{\ominus}$ is less likely. In fact, the enthalpy of dissociation is 67 ± 5 kcal/mole [*11*].

References p. 355

partially halogenated monoborane (as etherate) [equation (d)][*13, 105, 543, 546, 613*]:

$$LiBH_4 + 4I_2 \xrightarrow[\text{n-hexane}]{\text{cyclohexane}} LiI + BI_3 + 4HI \qquad (a)$$

$$2LiBH_4 + Cl_2 \xrightarrow{\text{ether}} (BH_3)_2 + H_2 + 2LiCl \qquad (b)$$

$$(BH_3)_2 + 5Cl_2 + 2R_2O \xrightarrow{\text{ether}} 2BCl_3{\cdot}OR_2 + 4HCl + H_2 \qquad (c)$$

$$(BH_3)_2 + 2HCl + 2R_2O \xrightarrow{\text{ether}} 2BH_2Cl{\cdot}OR_2 + 2H_2 \qquad (d)$$

Pseudohalogens. With pseudohalogens, lithium boranate forms partially substituted boranates and partially substituted monoboranes[*399, 404, 405, 819*]:

$$MBH_4 + \frac{n}{2}(SCN)_2 \rightarrow M[BH_{4-n}(NCS)_n] + \frac{n}{2}H_2 \qquad (M = Li, Na; 2 < n < 3)$$

$$2LiBH_4 + ClCN \longrightarrow Li[H_3B \leftarrow\cdots CN \cdots\rightarrow BH_3] + LiCl + H_2$$

In contrast to $Li[BH_4]$, $Li[BH_3(CN)]$ is astonishingly stable to water and weak acids. This is explained by the fact that the CN group attracts the H–B bonding electrons, so that the negative charge of the H-atoms decreases. For further reactions with pseudohalogens and pseudohalides see sections 6.2.1 and 6.2.2.

Hydrogen halides. In the absence of ether, alkali-metal boranates react with hydrogen halides (even at −80°) with the formation of diborane[*543, 660*]:

$$2MBH_4 + 2HX \longrightarrow (BH_3)_2 + 2H_2 + 2MX \quad (M = Li, Na; X = Cl, Br)$$

Only on the addition of ether does diborane react further with HX according to equation (d) (see above), with substitution.

Hydrogen pseudohalides. The (donor) anions of the hydrogen pseudohalides substitute H-atoms of the boranate ion, for example[*821, 856*]:

$$Li[BH_4] + HCN \xrightarrow[\text{ether, 20–100°}]{-H_2} Li[BH_3(CN)]$$

$$Li[BH_4] + HN_3 \xrightarrow[\text{ether, −80°}]{-H_2} Li[BH_3(N_3)] \xrightarrow[20°]{+3HN_3} Li[B(N_3)_4] + 3H_2$$

6.3.3 *Main Group VI*

Water. With water, alkali-metal boranates, $M[BH_4]$, hydrolyse slowly at pH > 7 and rapidly at pH < 7[*502*]. The rates of hydrolysis in H_2O at 20° are as follows: $LiBH_4$ (initial velocity very fast, falls during the hydrolysis because of the increase in pH; rapid hydrolysis in acid)[*248a, 655, 657*, cit. 5]; $NaBH_4$ (fairly stable aqueous solution at pH > 9; is marketed in 40% NaOH solution[*501*]; isotope effect in hydrolysis since the velocity of hydrolysis of $NaBH_4$ is smaller than that of $NaBD_4$[*159, 161b, 248 248a, 655*]). The acid hydrolysis probably takes place as follows:

$$BH_4^{\ominus} + H^{\oplus}X^{\ominus} \underset{k_{1a}}{\rightleftharpoons} \left[\begin{array}{c} H^{\oplus}BH_4^{\ominus} \\ \vdots \\ X^{\ominus} \end{array}\right] \xrightarrow[k_{1b}]{-H_2} [BH_3X^{\ominus}] \xrightarrow{-X^{\ominus}} (BH_3)aq$$

$$\xrightarrow[\text{fast}]{+2H_2O, -H_3O^{\oplus}} BH_3(OH)^{\ominus} \xrightarrow[k_2]{+H^{\oplus}X^{\ominus}} \left[\begin{array}{c} H^{\oplus}BH_3OH^{\ominus} \\ \vdots \\ X^{\ominus} \end{array}\right] \xrightarrow[\text{fast}]{+2H_2O} B(OH)_3 + 3H_2 + X^{\ominus}$$

(X = acid radical)

The rate-determining step is either the formation of the adduct $H_5B{\cdot}X^{\ominus}$[24] or its decomposition with the formation of H_2 to give donor-stabilized monoborane $BH_3X^{\ominus}$ as the second intermediate [*162*, *163*, *247*, *248*, *499a*, *748*]. The first H-atom in $BH_4^{\ominus}$ reacts more readily than the others (IR-spectroscopic detection of $NaBH_3OH$) [*265*]. In addition, BH_3 can be detected as $BH_3{\cdot}NR_3$ [*157*, *159*]; consequently, $k_1 > k_2$. The end-product of this hydrolysis is the alkali-metal borate.

In highly alkaline solutions (pH ~ 14), when D_2O is used an isotopic exchange takes place in addition to hydrolysis [*260*, *376*]. It does not take place solely in accordance with the scheme [*376*]

$$BH_4^{\ominus} + D_2O \longrightarrow BH_3D^{\ominus} + HDO$$

since, in the presence of Pt on active carbon, $NaBH_4$ in a 1 *M* solution of Na_2CO_3 in D_2O forms exclusively D_2 [*158*]:

$$NaBH_4 + 4D_2O \xrightarrow[\text{excess of } D_2O]{\text{Pt/active carbon}} NaBO_2 + 4D_2 + 2H_2O$$

The B–H hydrolysis in D_2O/DCl also gives D_2 and not HD (see section 8).

Metal salts catalyse the hydrolysis. The catalytic activity increases in the sequence $RhCl_3 \sim RuCl_3 < H_2PtCl_6 < CoCl_2 < NiCl_2 < OsO_4 < IrCl_4 < FeCl_2 \ll PdCl_2$ [*77*, *158*].

Alcohols. Alcoholysis takes place similarly to hydrolysis:

$$BH_4^{\ominus} + PhOH \longrightarrow \left[\begin{array}{c} H_3\text{-}B\cdots H \\ \vdots \quad \vdots \\ PhO\cdots H \end{array}\right]^{\ominus} \xrightarrow{-H_2} BH_3{\cdot}OPh^{\ominus}$$

The transition states $H_3\overset{\ominus}{B}H\cdots HOPh$ (two-centre) or $H_3\overset{\ominus}{B}\begin{array}{l}\cdot\cdot H \\ \cdot\cdot H\text{–}OR\end{array}$ (three-centre) that have also been considered are less likely. The end-product of the alcoholysis

[24] The reduction of hexacyanoferrate(III) ions by $NaBH_4$ to hexacyanoferrate(II) ions takes place through a similar intermediate [*234*].

is an alkali-metal tetraalkoxoborate, $MB(OR)_4$[*161*, *164*]:

$$BH_4^{\ominus} + 4ROH \longrightarrow B(OR)_4^{\ominus} + 4H_2$$

Sulphur. The reaction of lithium boranate (but not sodium boranate) with sulphur forms unstable $LiBH_3(SH)$, $LiBH_2S$, $LiB_3S_2H_6$, Li_2S, and H_2. First the S_8 ring is degraded nucleophilically in steps to $BH_3(SH)^{\ominus}$, and this is followed by subsequent reactions[*551*, *727a*]:

$$S_8 \xrightarrow{+BH_4^{\ominus}} HS\text{-}S_6\text{-}S\overset{\ominus}{B}H_3 \xrightarrow{+BH_4^{\ominus}} HS\text{-}S_5\text{-}S\overset{\ominus}{B}H_3 + BH_3(SH)^{\ominus} \xrightarrow{+6BH_4^{\ominus}} \cdots 8BH_3(SH)^{\ominus}$$

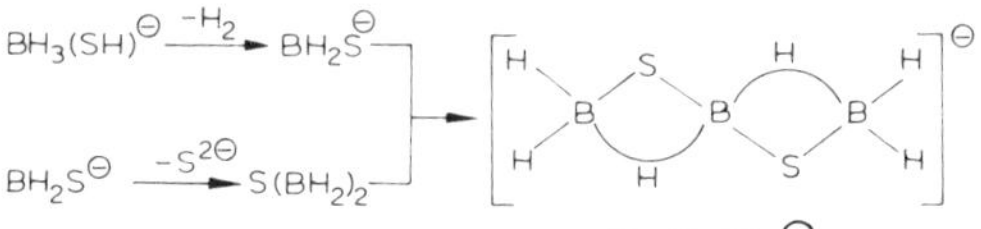

$$BH_3(SH)^{\ominus} \xrightarrow{-H_2} BH_2S^{\ominus}$$

$$BH_2S^{\ominus} \xrightarrow{-S^{2\ominus}} S(BH_2)_2$$

$$\longrightarrow B_3S_2H_6^{\ominus}$$

The reaction of sodium boranate and potassium boranate with sulphur forms $NaBS_2$ or KBS_2 respectively and hydrogen[*727a*]:

$$NaBH_4 + 2S \longrightarrow NaBS_2 + 2H_2$$

Hydrogen sulfide and thiols react with lithium boranate similarly to water or alcohols:

$$BH_4^{\ominus} + HSH \longrightarrow BH_3(SH)^{\ominus} + H_2$$
$$BH_4^{\ominus} + 4RSH \longrightarrow B(SR)_4^{\ominus} + 4H_2.$$

Thiophenol, PhSH, reacts faster than ethyl mercaptan, EtSH. In contrast to the situation in alcoholysis, the last H-atom of the boranate is more difficult to substitute, probably for steric reasons[*395a*, *544*, *835*].

6.3.4 *Main Group V*

Ammonia, amines, and hydrazine. Herewith lithium boranate forms definite adducts $LiBH_4{\cdot}NH_3$, $LiBH_4{\cdot}2NH_3$, $LiBH_4{\cdot}3NH_3$, $LiBH_4{\cdot}4NH_3$[*755*]; $LiBH_4{\cdot}NMeH_2$, $LiBH_4{\cdot}3NMeH_2$, $LiBH_4{\cdot}4NMeH_2$[*639*, cit. *5*]; $LiBH_4{\cdot}NMe_3$, $LiBH_4{\cdot}2NMe_3$[*818*]; and $LiBH_4{\cdot}N_2H_4$[*543*] (see section 20.18).

Phosphonium iodide. PH_4F and $NaBH_4$ in monoglyme as solvent at −78° give the ionic $Na^{\oplus}[BH_3PH_2BH_3]^{\ominus}$[*336a*].

[*N(Ph₂PNH₂)Cl.* When this is suspended in an aqueous solution of $NaBH_4$, the corresponding borate $[N(Ph_2PNH_2)_2]BH_4$ is formed[*58*, *677*], which can be extracted from methylene chloride:

$$[N(Ph_2PNH_2)_2]Cl + NaBH_4 \rightarrow \left[H_2N\text{-}\underset{Ph}{\overset{Ph}{P}}{=}N\text{—}{}^{\oplus}\underset{Ph}{\overset{Ph}{P}}\text{-}NH_2\right] BH_4^{\ominus} + NaCl$$

For the reactions of boranates with ammonium salts, see section 9.

Phosphorus trichloride. PCl_3 with lithium boranate in ether forms diborane and phosphane as the main products:

$$3LiBH_4 + PCl_3 \longrightarrow \tfrac{3}{2}(BH_3)_2 + PH_3 + 3LiCl$$

together with yellow to orange coloured condensation products [*543*].

Alkyl- and arylphosphorus halides. $R_{3-n}PX_n$ ($n = 1,2$) give with alkali-metal boranates the corresponding alkyl- and arylphosphanes $R_{3-n}PH_n$ and their borane adducts $R_{3-n}PH_n{\cdot}BH_3$ [*125, 233, 836*]:

$$R_2PCl + NaBH_4 \longrightarrow R_2PH{\cdot}BH_3 + NaCl \; (R = CH_3, C_2H_5)$$

$$PhPCl_2 \xrightarrow[-LiCl]{+LiBH_4} PhPClH + BH_3 \rightarrow PhPClH{\cdot}BH_3$$

$$PhPCl \xrightarrow[-2LiCl]{+2LiBH_4} PhPH_2 + 2BH_3 \rightarrow PhPH_2{\cdot}BH_3 + BH_3$$

The borane adducts can be converted at elevated temperatures (boiling in high-boiling solvents) into the corresponding organylphosphanoboranes $H_2B{-}PR_2$ [*125*]; for example:

$$3R_2PH{\cdot}BH_3 \xrightarrow[60-180°]{diglyme} (R_2PBH_2)_3 + 3H_2$$

Alkoxophosphorus chlorides and dialkylaminophosphorus chlorides. $(RO)_{3-n}$-PCl_n ($n = 1, 2$) and $(R_2N)_{3-n}PCl_n$ ($n = 1, 2$) do not (very probably because of the donor activity of the phosphorus enhanced by O or N) form the corresponding free phosphanes $(RO)_{3-n}PH_n$ or $(R_2N)_{3-n}PH_n$ but always borane adducts [*125, 533, 834*]:

$$2LiBH_4 + (RO)PCl_2 \longrightarrow \tfrac{1}{2}(BH_3)_2 + (RO)PH_2{\cdot}BH_3 + 2LiCl$$

$$LiBH_4 + (RO)_2PCl \longrightarrow (RO)_2PH{\cdot}BH_3 + LiCl$$

$$LiBH_4 + (R_2N)_2PCl \longrightarrow (R_2N)_2PH{\cdot}BH_3 + LiCl$$

Phosphorus pentachloride. The reaction of lithium boranate with PCl_5 in ether at −80° gives, instead of PH_5, only its decomposition products PH_3 and H_2 [*822*]:

$$5LiBH_4 + PCl_5 \xrightarrow[-80°]{ether} \tfrac{5}{2}(BH_3)_2 + PH_3 + H_2 + 5LiCl$$

Phosphane oxide H_3PO, or the hydroxyphosphane (phosphinous acid) PH_2-(OH) tautomeric with it, has not so far been prepared. Lithium boranate does in fact hydrogenate phosphorus oxychloride (phosphoric trichloride) (Cl_3PO) to the phosphane oxide–borane adduct $H_3PO{\cdot\cdot}{\rightarrow}BH_3$, but the latter rapidly splits

References p. 355

off hydrogen with the fomation of a highly polymeric phosphinylborane $(H_2POBH_2)_x$ [*533, 834*]:

$$3LiBH_4 + Cl_3PO \xrightarrow[-110^\circ]{\text{ether}} (BH_3)_2 + H_3PO{\cdot}BH_3 + 3LiCl$$

$$H_3PO{\cdot}BH_3 \xrightarrow{-90^\circ} \frac{1}{x}(H_2POBH_2)_x + H_2$$

Dimethylphosphorus oxide chloride (dimethylphosphinous chloride). With Me_2PClO, lithium boranate forms trimeric dimethylphosphanoborane, $(Me_2PBH_2)_3$, (for another synthesis, see section 11.1) [*125*]:

$$9LiBH_4 + 6Me_2PClO \xrightarrow[-80^\circ/+160^\circ]{\text{diglyme}} 2(Me_2PBH_2)_3 + 3LiBO_2 + 6LiCl + 12H_2$$

Halides of trivalent arsenic, antimony, and *bismuth.* The halides $R_{3-n}EX_n$ react with lithium boranate, even at low temperatures in the case of arsenic and antimony, with the formation of hydrides $R_{3-n}AsH_n$ and $R_{3-n}SbH_n$, and in the case of bismuth, with reduction to the polymeric organylbismuth free from hydride hydrogen, or to elementary bismuth [*125, 828–832*]:

$$n\,LiBH_4 + R_{3-n}EX_n \xrightarrow[-70^\circ]{\text{diethyl ether}} R_{3-n}EH_n + \frac{n}{2}(BH_3)_2 + n\,LiX$$

(E = As, Sb; R = alkyl, phenyl; $n = 1, 2, 3$)

$$2LiBH_4 + PhBiBr_2 \xrightarrow[-110^\circ]{\text{diethyl ether}} \frac{1}{x}(PhBi)_x + (BH_3)_2 + H_2 + 2LiBr$$

$$2LiBH_4 + 2Ph_2BiBr \xrightarrow[-110^\circ]{\text{diethyl ether}} (Ph_2Bi{-}BiPh_2) + (BH_3)_2 + H_2 + 2LiBr$$

$$\downarrow$$

$$2Bi + 2Ph{-}Ph$$

In reactions of alkylbismuth halides $R_{3-n}BiX_n$ with lithium alanate, which has a more powerful hydrogenating action, the corresponding hydride $R_{3-n}BiH_n$ ($n = 1, 2, 3$) can be prepared at very low temperatures (see Chap. 5, section 9.1.2).

Arsenite and antimonite. When boranate solutions are added, arsenite and antimonite solutions liberate arsane [60% yield according to eqn. (a)] and diarsane [smaller amounts, eqn. (b)] or stibane [50% yield according to eqn. (c)] [*51, 375*]:

$$3BH_4^{\ominus} + 4As(OH)_3 + 3H^{\oplus} \xrightarrow{\text{water}} 4AsH_3 + 3B(OH)_3 + 3H_2O \tag{a}$$

$$AsH_3 + AsH_2(OH) \xrightarrow{\text{water}} (AsH_2)_2 + H_2O \tag{b}$$

$$3BH_4^{\ominus} + 4Sb(OH)_3 + 3H^{\oplus} \xrightarrow{\text{water}} 4SbH_3 + 3B(OH)_3 + 3H_2O \tag{c}$$

Halides of pentavalent arsenic, antimony, and bismuth. The reactions of the halides R_nEX_{5-n} with lithium boranate – as in the case of the analogous reactions with phosphorus compounds (see section 6.3.4) – do not give hydrides of pentavalent arsenic, antimony, and bismuth. Even at low temperatures, immediate reduction to the trivalent state takes place [*823–827*]:

$$R_nEX_{5-n} + (5-n)LiBH_4 \longrightarrow R_nEH_{3-n} + H_2 + (5-n)BH_3 + (5-n)\ LiX$$

$$(E = P, As, Sb, Bi;\ R = CH_3, C_6H_5;\ X = Cl, Br, I)$$

6.3.5 Main groups IV to I

The important reactions of the boranates with multiple C–C bonds are treated in section 12. For the reactions with compounds of other elements of the IVth to Ist main groups of the periodic system, see the section on the corresponding elements, and section 6.2.

6.3.6 Sub-groups

Copper(II) chloride reacts with lithium boranate at 20° with the formation of elementary copper, and at –45° with the formation of the white, non-volatile copper monoboranate, $CuBH_4$, which decomposes above – 12° [*411*]:

$$2LiBH_4 + CuCl_2 \xrightarrow[20^\circ]{\text{diethyl ether}} Cu + (BH_3)_2 + H_2 + 2LiCl$$

$$2LiBH_4 + CuCl_2 \xrightarrow[-45^\circ]{\text{diethyl ether}} CuBH_4 + \tfrac{1}{2}(BH_3)_2 + \tfrac{1}{2}H_2 + 2LiCl$$

$$CuBH_4 \xrightarrow{>-12^\circ} Cu + \tfrac{1}{2}(BH_3)_2 + \tfrac{1}{2}H_2$$

Similarly to silver perchlorate, copper(I) chloride forms the mono-boranate, while gold(III) chloride forms only the decomposition products of the corresponding boranate: Au, $(BH_3)_2$ and H_2 (see section 6.1.4).

Zinc chloride with lithium boranate gives the complex $Li[ZnCl(BH_4)_2]$, which exists in equilibrium with LiCl and $Zn(BH_4)_2$. Consequently, the latter cannot be isolated in the pure state [*408, 543, 820*]:

$$2LiBH_4 + ZnCl_2 \xrightarrow[\text{diethyl ether}]{-LiCl} Li[ZnCl(BH_4)_2] \rightleftharpoons Zn(BH_4)_2 + LiCl$$

In contrast, reactions in the molar ratio $LiBH_4 : ZnCl_2 = 3:1$ give a chloride-free complex [*792*]:

$$3LiBH_4 + ZnCl_2 \rightarrow Li[Zn(BH_4)_3] + 2LiCl$$

Sodium boranate forms $Na[Zn(BH_4)_3]$, while with potassium boranate a compound $2KBH_4 \cdot 3Zn(BH_4)_2$ (probably the complex $K_2[(BH_4)_2Zn(BH_4)_2Zn(BH_4)_2Zn(BH_4)_2]$ [*288, 543, 552*] is produced. Potassium boranate and zinc bromide in tetrahydrofuran yield the crystalline zinc bis(boranate)-bis(tetrahydrofuran) complex $Zn(BH_4)_2 \cdot 2C_4H_8O$ [*543, 552*].

References p. 355

Cadmium chloride, in similar reactions, forms $Li[CdCl(BH_4)_2]$, $CdClBH_4$, $Cd(BH_4)_2$, and $Li_2[Cd(BH_4)_4]$ [*543, 552, 812*].

Titanium trichloride and *titanium tetrachloride* react with lithium boranate to give small yields of $Ti(BH_4)_3$, for example [*348, 543*]:

$$8LiBH_4 + 2TiCl_4 \longrightarrow 2Ti(BH_4)_3 + (BH_3)_2 + H_2 + 8LiCl$$

$$6LiBH_4 + 2TiCl_3 \longrightarrow 2Ti(BH_4)_3 + 6LiCl$$

With lithium boranate in ether, $Ti(BH_4)_3$ forms an ether complex of $Li[Ti(BH_4)_4]$:

$$LiBH_4 + Ti(BH_4)_3 \longrightarrow Li[Ti(BH_4)_4]$$

Zirconium tetrachloride, when heated with lithium boranate in vacuum, gives a 75% yield of zirconium tetrakis(boranate) [*70*]:

$$4LiBH_4 + ZrCl_4 \longrightarrow Zr(BH_4)_4 + 4LiCl$$

Manganese dichloride and *manganese diiodide* react with lithium boranate to give lithium bis(boranato)dihalomanganate(II) [*372, 817*], which can be isolated as oily ether complexes:

$$2LiBH_4 + MnX_2 \xrightarrow{\text{diethyl ether}} Li_2[Mn(BH_4)_2X_2]$$

With other molar ratios, various complexes arise; for example, $Li_3[Mn(BH_4)_3I_2]$, $Li[Mn(BH_4)I_2]$, and $LiMn[Mn(BH_4)I_4]$.

Iron(III) chloride: which is ether-soluble, reacts with lithium boranate to give iron(II) boranate (see section 6.1.4). If the ether-insoluble iron(II) chloride is treated with $LiBH_4$ at $-30°$, it is not $Fe(BH_4)_2$ itself that is formed but its lithium complex [*641*]:

$$3LiBH_4 + FeCl_2 \longrightarrow Li[Fe(BH_4)_3] + 2LiCl$$

Cobalt(II) chloride and nickel(II) chloride are reduced by lithium boranate or sodium boranate at 20° to metallic cobalt or NiB_2 [*488b, 728*]. In contrast with lithium boranate in ether at −40°, nickel(II) chloride forms $Li[Ni(BH_4)_3]$ and $Li_2[Ni(BH_4)_4]$ [*372, 488b, 817*].

7. Conversion of diborane and isotope exchange in diborane

When energy is supplied, diborane is converted into higher boranes. Possible sources of energy are heat, light, excited mercury (Hg 6^3P_1 atoms), and electric current.

Even at room temperature, in the systems D_2–$(BH_3)_2$ and $(BH_3)_2$–$(BD_3)_2$, borane

(3) is exchanged for deuterated borane(3) (not H for D) in a homogeneous reaction. In the D_2–$(BH_3)_2$ system an H–D exchange is necessary as the initial step. Here wall reactions play a part (see section 7.5).

7.1 Pyrolysis

Phenomenological description of the pyrolysis process. Diborane decomposes at about 100° into hydrogen and predominantly B_4H_{10}, with some B_5H_9 and B_5H_{11} [*204, 407, 532, 646*]. When diborane (with or without the addition of H_2) flows through a glass or steel tube heated to 175 to 250°, it decomposes into H_2, small amounts of B_4H_{10}, and B_5H_9, B_5H_{11}, and non-volatile boranes [*64, 67, 280, 468, 497*]. Under these conditions the yield of B_4H_{10} referred to the B_2H_6 is generally less than 5%, while the yields of B_5H_9 and B_5H_{11} depend markedly on the experimental conditions. When hydrogen is present, a rise in temperature leads to an increase in the amount of B_5H_9 (175°: 88% B_5H_{11}, 0% B_5H_9; 200°: 62% B_5H_{11}, 28% B_5H_9; 225°: 10% B_5H_{11}, 78% B_5H_9). With a mixture of nitrogen and $(BH_3)_2$, a rise in temperature leads to a similar increase in the yield of B_5H_9 and decrease in the yield of B_5H_{11}; in addition, particularly at 225°, large amounts of non-volatile boranes are formed. With increasing time of pyrolysis the amount of B_5H_{11} decreases, that of B_5H_9 (which is thermally more stable) increases, and the amount of B_4H_{10} remains approximately constant.

Conversion mechanism of the pyrolysis. Figure 4.15 shows the conversion

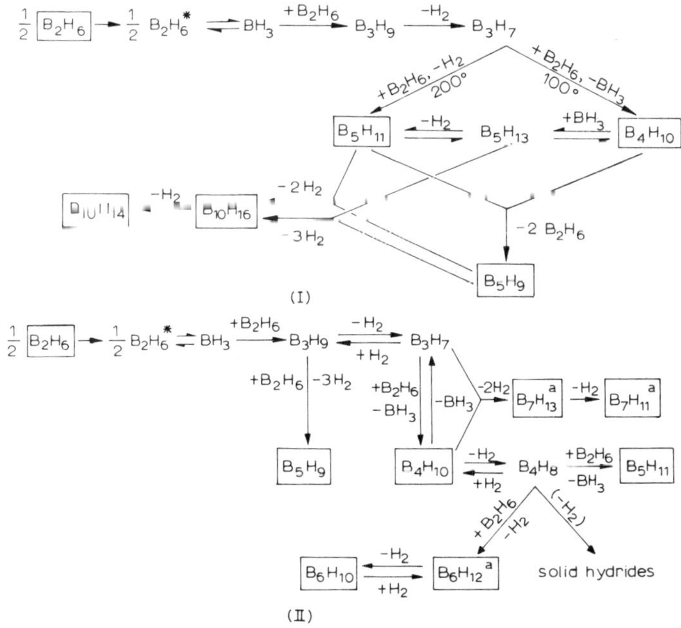

Fig. 4.15. Conversion of B_2H_5 into higher boranes during pyrolysis by heating (I) and pyrolysis in a shock tube (II). The encircled boranes were isolated. ([a]Detected by mass spectroscopy by "codistillation", a method of separation formally similar to gas chromatography [*212*, cit. *212*]).

schemes for pyrolysis by heating [*33, 34, 35b, 36, 204, 209a, 246*] and for pyrolysis in a shock tube[25] [*212*]. The initial stages of pyrolysis take place in the same way in the two methods.

The primary step is rapid dissociation of the excited $B_2H_6^*$ into BH_3 radicals (B_2H_6 (gas) $\rightleftharpoons$ $2BH_3$ (gas); dissociation energy 28.4 ± 2 kcal/mole or 37.1 ± 1 kcal/mole [*34, 214*]). This is followed not by a direct rate-determining formation of B_3H_7 ($B_2H_6 + BH_3 \nrightarrow B_3H_7 + H_2$) but by a two-stage process ($BH_3 \ldots B_3H_9 \ldots B_3H_7$). The cleavage indicated by $B_3H_9 \rightarrow B_3H_7 + H_2$ is probably the rate-determining step in pyrolysis by heating. At 100°, B_3H_7 yields mainly B_4H_{10} and a smaller amount of B_5H_{11}, but at 175° to 250° the main product is B_5H_{11}, with less B_4H_{10}. B_5H_9 does not arise by the (apparently obvious) monomolecular reaction $B_5H_{11} \rightarrow B_5H_9 + H_2$, since the presence of H_2 in the pyrolysis of B_2H_6 promotes the formation of B_5H_9 [*497*]. B_5H_9 is, rather, formed by a bimolecular reaction between boranes (probably $B_5H_{11} + B_4H_{10} \rightarrow B_5H_9 + 2B_2H_6$; however, the last step of the reaction is not absolutely certain [*647*]). Still higher boranes arise by the splitting of H_2 or B_2H_6 away from B_5 boranes. As a distinction from the pattern of formation in pyrolysis by heating, in the shock tube B_5H_9 is formed from B_3H_9; in addition, the unstable B_4H_8 has a central role for the formation of B_5H_{11}, B_6H_{10} and higher boranes.

7.2 *Non-sensitized photolysis*

The direct photolysis of diborane (without the addition of Hg vapour) at the absorbable wavelength of 1849 Å (low-pressure Hg lamp) between 0.8 and 800 torr and at 0 to 6° gives H_2, B_4H_{10}, B_5H_{11}, and higher non-volatile boranes [*438*]. Kinetic measurements indicate two independent mechanisms. The main reaction (formation of B_4H_{10}) takes place via the primary step of dissociation of H from B_2H_6 and combination of the remaining B_2H_5 radicals to form B_4H_{10}. With H radicals, however, the B_2H_5 radicals may also form the non-volatile polyborane(1) $(BH)_x$ via intermediate stages with the evolution of molecular hydrogen (1st side reaction):

$$B_2H_6 \xrightarrow[-H]{+h\nu} B_2H_5 \longrightarrow \tfrac{1}{2}B_4H_{10} \text{ (main reaction)}$$

$$B_2H_5 \xrightarrow{+H} \text{[intermediate stages]} \xrightarrow{-2H_2} [-B_2H_2-] \qquad \text{(1st side reaction)}$$

A second side reaction (formation of B_5H_{11}) takes place (like pyrolysis) through the dissociation of $B_2H_6^*$ into two BH_3 radicals (in this process the dimensions of the vessel affect the kinetics of the process, since the walls of the vessel abstract

[25] A shock tube consists of two sections separated by a thin breakable membrane: (1) a high-pressure (driver) section with a compressed, usually monatomic, gas and (2) a low-pressure section containing the gaseous compound to be investigated (here B_2H_6). After rupture of the membrane and the passage of a short band with turbulent gas, a high-pressure front propagates, leading to very high but brief temperatures in the gas to be investigated. The chemical changes taking place under these conditions are followed by mass and IR spectroscopy. For a general review of shock tube techniques, see the literature [*604a, 710a*].

energy from the excited diborane molecules $B_2H_6^*$ with the formation of unexcited diborane B_2H_6):

$$\tfrac{1}{2}B_2H_6 \xrightarrow{+h\nu} \tfrac{1}{2}B_2H_6^* \longrightarrow BH_3 \xrightarrow[-H_2]{+B_2H_6} B_3H_7 \xrightarrow[-H_2]{+B_2H_6} B_5H_{11}$$

$$\tfrac{1}{2}B_2H_6^* \xrightarrow{\text{(wall of the vessel)}} \tfrac{1}{2}B_2H_6 \qquad \text{(2nd side reaction)}$$

The H atoms not consumed by the 1st side reaction form molecular hydrogen, either by the combination of two H atoms (preferentially at low pressures of B_2H_6) or with diborane, leading to the formation of B_2H_5 radicals (preferentially at higher pressures of B_2H_6):

$$H+H \longrightarrow H_2$$
$$H+B_2H_6 \longrightarrow B_2H_5+H_2$$

7.3 *Mercury-sensitized photolysis*

Mercury-sensitized photolysis of B_2H_6 with unfiltered Hg light at a pressure of B_2H_6 of 25 to 447 torr at 30° gives rise to H_2, B_4H_{10}, and B_5H_9, this last in a yield smaller by more than a power of 10. Kinetic measurements indicate the following (provisional) reaction mechanism [*345*]:

$$B_2H_6 \xrightarrow[+Hg6^3P_1-Hg6^1S_0]{-H} B_2H_5 \longrightarrow \tfrac{1}{2}B_4H_{10} \text{ (main reaction)}$$

$$B_2H_5 \xrightarrow[\text{side reaction, dismutation}]{-\frac{1}{2}B_2H_6} \tfrac{1}{2}B_2H_4 \xrightarrow{\Delta} \text{pentaboranes, non-volatile boranes}$$

The Hg-sensitized photolysis [*538a, 624a*] of diborane therefore resembles the reaction of ethane with Hg 6^3P_1 atoms. In analogy to the formation of butane and hydrogen in that case, here tetraborane and hydrogen are the primary products. The two reactions differ mainly by the lower stability of the boranes as compared with butane.

7.4. *Electric discharge in an atmosphere of B_2H_6*

Passage of silent electric discharge through B_2H_6 in the presence of H_2 or Ar forms [*437*] higher boranes in accordance with the following scheme:

$$B_2H_6 \xrightarrow{\text{el. discharge}} B_2H_6^* \xrightarrow{\Delta} 40\%\ B_4H_{10},\ 20\%\ B_5H_9,\ 30\%\ B_5H_{11},\ \text{remainder } B_6H_{10} \text{ and } B_9H_{15}$$

For electron-impact experiments, see [*211*] and [*213*].

7.5. *Isotope exchange in diborane*

The kinetics of the exchange in the D_2–$(BH_3)_2$ system in accordance with the equation $B_2H_6+D_2 \rightarrow B_2H_5D+HD$ have been followed by means of the IR spectrum. The experimental data indicate a more or less 1.5th order for $(BH_3)_2$ and zeroth order for D_2 (at high pressure), or 1st order for D_2 (at low pressure); the activation energy is 21.8±3 kcal/mole. The exchange possibly takes place through a rapid thermal, probably homogeneous, dissociation of diborane $(BH_3)_2$ to borane BH_3. Subsequently, borane(3) reacts with deuterium at the wall of the vessel with the formation of partially deuterated borane(3), which then with borane, in a rate-determining step responsible for the fractional order of the reaction, forms deuterated diborane [*495*]:

$$\tfrac{1}{2}(BH_3)_2 \rightleftharpoons BH_3 \quad (1)$$
$$BH_3 + D_2 \rightleftharpoons BH_2D + HD \quad (2)$$
$$BH_2D + (BH_3)_2 \rightleftharpoons (BH_2D)(BH_3) + BH_3 \quad (3)$$
$$B_2H_6 + D_2 \longrightarrow B_2H_5D + HD$$

Mass-spectroscopic investigations [*692*] and ^{10}B–^{11}B exchange in diborane [*433*] support the mechanism through borane-containing units (BH_3, and possibly also B_3H_9) as compared with an H–D exchange at (terminal) H atoms; *e.g.*, $^{11}BH_3 + (^{10}BD_3)_2 \rightarrow {}^{11}B^{10}BH_3D_3 + {}^{10}BD_3$. For thermodynamic data on eq. (1), see [*34*].

The homogeneous exchange in the $(BH_3)_2$–$(BD_3)_2$ system takes place, as in the D_2–$(BH_3)_2$ system, via a rapid thermal dissociation of diborane into borane and a subsequent bimolecular rate-determining reaction of borane with diborane (activation energy approximately the same as in the D_2–$(BH_3)_2$ reaction [*495*]):

$$(BH_3)_2 \rightleftharpoons 2BH_3 \text{ (1a)} \quad \text{or} \quad (BD_3)_2 \rightleftharpoons 2BD_3 \text{ (1b)}$$
$$\left.\begin{array}{l} BH_3 + (BD_3)_2 \longrightarrow (BH_3)(BD_3) + BD_3 \\ BD_3 + BH_3)_2 \longrightarrow (BD_3)(BH_3) + BH_3 \end{array}\right\} B_2H_6 + B_2D_6 \longrightarrow 2B_2H_3D_3$$

For thermodynamic data on the BH_3–$(BD_3)_2$ exchange, see [*34*].

8. Reactions of diborane with halogens and with oxygen and sulphur compounds

Diborane reacts explosively with *chlorine* in the gas phase at 20°, and more slowly at lower temperatures [*728a, 728b*]. It reacts very slowly with *bromine* at 20° even in the light, while laboratory amounts react completely in a few hours at 100° [*728a, 728b*]. With Cl_2 and Br_2, hydrogen halides and partially halogenated substitution products are first formed, and these dismute slowly without a catalyst, and more quickly with catalysts, into products poorer and richer in halogen [*728a, 728b*] (for equilibria, see section 5). With *iodine* diborane forms boron triiodide and oily products [*738*].

For the reactions with *hydrogen chloride* and the subsequent reactions, see section 5. *Hydrogen iodide* at 50° forms B_2H_5I and more highly iodinated, unstable, boranes (diboranes?) [*738*].

For commutation and subsequent reactions in the treatment of diborane with *boron trichloride*, see [*863d*] and section 5. Similar systems of equations describe the reaction of *boron tribromide* below 0° [*666*]. For the preparation ($BF_3 + B_2H_6$) and some reactions of BHF_2 see [*149a*].

At 120 to 210°, diborane reacts with *oxygen* explosively to give boron trioxide and water:

$$B_2H_6\,(\text{gas}) + 3O_2\,(\text{gas}) \longrightarrow B_2O_3\,(\text{solid}) + 3H_2O\,(\text{gas}) + 482.9\ \text{kcal}$$

The reaction is affected by the addition of foreign gases (N_2, He, Ar, H_2 organic compounds) [*262, 625, 794*]. Oxidation experiments on B_2H_6, $Me_2B_2H_4$, $Me_4B_2H_2$, and Me_3B with oxygen at 77–170°K in solution (C_2H_6-C_3H_8, F_3CCl) show that only Me_3B which has a free sp^3 orbital available, is oxidized. In C_2H_6–C_3H_8 and F_3CCl solutions, the oxygen adds to the electron gap of the boron atom [*582*]:

$$R_3B + O_2 \longrightarrow R_3\overset{\ominus}{B}\text{–O–}\overset{\oplus}{O} \longrightarrow \begin{array}{c} R\cdots O \\ \vdots \quad \vdots \\ R_2B\cdots O \end{array} \longrightarrow R_2BOOR$$

Under suitable conditions unfavourable for a radical mechanism, boranes without free orbitals (B_2H_6, $Me_2B_2H_4$, $Me_4B_2H_2$) are not oxidized [*582*].

Diborane, alkylboranes, and donor-borane adducts *hydrolyse* with the replacement of the hydride hydrogen by hydroxide groups (evolution of H_2). The mechanism of the hydrolysis of the B–H bond [*157, 161a, 161b, 162, 376a, 393a, 394a, 626, 627*] may be illustrated on the basis of the reaction of py·$BHPh_2$ or py·$BDPh_2$ with acidified H_2O, D_2O, and T_2O [*160, 312, 395, 466, 627*]. The kinetics are of first-order on borane, first-order on water, and independent of the amount of pyridine py. The hydrolysis takes place through a 3-centre mechanism in which the water protons attack electrophilically the B–H bonding electrons:

$$\text{py}\cdot BHPh_2 + H_2O \longrightarrow \left[\begin{array}{c} \text{Ph, py, Ph}\ B\cdots H\cdots H\text{–O–H} \end{array}\right] \longrightarrow B(OH)Ph_2 + H_2 + \text{py}$$

In addition to slow hydrolysis, in D_2O/DCl a rapid H–D exchange takes place because of which D_2 and not HD is formed hydrolytically [*160*].

Water absorbed on silica gel and hydroxide groups on silica gel hydrolyse diborane in accordance with the following scheme (from IR-spectroscopic data) [*494*]:

$$\rangle\text{Si–OH}\cdots O\langle^{H}_{H} + (BH_3)_2 \longrightarrow \rangle\text{Si–OH}\cdots O\langle^{BH_2}_{BH_2} + 2H_2$$

$$\rangle\text{Si–OH} + (BH_3)_2 \longrightarrow \rangle\text{Si–OBH}_2(BH_3) + H_2$$

References p. 355

With *alcohols*, *aldehydes*, or *ketones*, diborane or borane–ether (solution of B_2H_6 in R_2O) reacts to form mono-, di- and trialkoxyboranes [*88*, *122*, *306*]:

$$\tfrac{1}{2}B_2H_6 \xrightarrow{(+R_2O)} \underset{(\text{as } R_2O\cdot BH_3)}{BH_3} \xrightarrow[-H_2]{+ROH} BH_2OR \xrightarrow[-H_2]{+ROH} BH(OR)_2 \xrightarrow[-H_2]{+ROH} B(OR)_3$$

$$(BH_3)_2 + 4OCHCH_3 \longrightarrow 2BH(OCH_2CH_3)_2$$

$$(BH_3)_2 + 4OC(CH_3)_2 \longrightarrow 2BH[OCH(CH_3)_2]_2$$

The alkoxyboranes dismute very readily, for example in the following manner [*122*]:

$$6BH(OMe)_2 \longrightarrow (BH_3)_2 + 4B(OMe)_3$$

The reaction of dimeric phenylborane with tributoxyboron leads to a similar equilibrium. Here the B–H$^{\mu}$–B double bridge in the $(PhBH_2)_2$ is partially replaced by an (electron-richer) B–O(R)–B bridge [*404a*]:

$$(PhBH_2)_2 + (n\text{-}BuO)_3B \rightleftharpoons PhBH_2 + \text{Ph(H)B}(\mu\text{-H})(\mu\text{-OR})\text{B(OBu)}_2$$

With *ethylene glycol*, diborane forms 1,3,2-dioxaborolan $BH(OCH_2)_2$ with evolution of hydrogen. At 20° this is a viscous liquid, and in the gas phase it is monomeric [*612*]:

$$(BH_3)_2 + 2\,HOCH_2CH_2OH \xrightarrow{\text{diethyl ether}} 2\,\overline{CH_2\text{–}O\text{–}CH_2\text{–}O}\text{>}BH + 4\,H_2$$

In the gas phase or when dissolved in diethyl ether or tetrahydrofuran, the borolan dismutes in the following way:

$$[\text{(CH}_2\text{O)}_2]BH \rightleftharpoons \tfrac{1}{6}(BH_3)_2 + \tfrac{1}{3}[\text{(CH}_2\text{O)}_2]BOCH_2CH_2OB[\text{(OCH}_2)_2]$$

As was to be expected, in the gas phase, $BH(OCH_2)_2$ forms (dismuting) adducts with donor compounds such as NH_3 or Me_3N, for example:

$$Me_3N + HB[\text{(OCH}_2)_2] \rightleftharpoons Me_3N\cdot BH[\text{(OCH}_2)_2]$$

With various *ethers*, diborane forms weak to medium-strong borane adducts, the stability of which decreases in the following sequence: $\overline{CH_2(CH_2)_3O}\cdot BH_3 > (CH_3)_2O\cdot BH_3 > (C_2H_5)_2O\cdot BH_3$, *i.e.* from the tetrahydrofuran adduct through the dimethyl ether adduct of the diethyl ether adduct.

At −78°, diborane forms with *dimethylsulphane*, Me_2S, the fairly stable dimethylsulphaneborane $Me_2S \cdot BH_3$. This is also produced in the pyrolysis of trimethylsulphonium boranate [*323*]:

$$Me_3S^{\oplus}BH_4^{\ominus} \longrightarrow Me_2S \cdot BH_3 + MeH$$

Alkylsulphanes RSH and diborane are in equilibrium with the very unstable alkylsulphaneboranes, $RSH \cdot BH_3$. Hydrogen sulphide itself forms no SH_2BH_3 [*131*]. When $RSH \cdot BH_3$ (or $RSH + \frac{1}{2}B_2H_6$) is heated, hydrogen is split off and polymeric $[RSBH_2]_x$ is formed [*509, 510, 513*]:

$$BH_3 \cdot HSR \xrightarrow{65-80^{\circ}} \frac{1}{x}[BH_2(SR)]_x + H_2 \quad (R = CH_3, C_2H_5, n\text{-}C_4H_9)$$

$[BH_2(SR)]_x$ is a thick syrupy liquid which on standing, or in tetrahydrofuran solution, deposits trimeric $[BH_2(SR)]_3$ [*509, 510, 513, 535*]:

$$\frac{3}{x} BH_2(SR)_x \underset{\text{rapidly in THF}}{\overset{\text{slowly}}{\rightleftharpoons}} [H_2B^{\ominus}-S^{\oplus}(R)]_3 \text{ (six-membered ring)}$$

With an excess of alkylsulphane, diborane forms the dialkylthioborane (which exists in the dimeric form as tetra(alkylthio)diborane) and tri(alkylthio)borane:

$$(BH_3)_2 \xrightarrow[-4H_2]{+4HSR} [BH(SR)_2]_2 \xrightarrow[-2H_2]{+2HSR} 2B(SR)_3$$

Depending on the ratio of the reactants, *ethane-1,2-dithiol* forms with borane the compounds $H_2BSCH_2CH_2SBH_2$, $\overline{SCH_2CH_2SB}H$, and $\overline{SCH_2CH_2SB}\overline{SCH_2CH_2SB}SCH_2CH_2S$... $\overline{SCH_2CH_2SBSCH_2CH_2S}$ [*193*]:

$$(BH_3)_2 + HSCH_2CH_2SH \xrightarrow{\text{diethyl ether}} H_2BSCH_2CH_2SBH_2 + 2H_2$$

$$(BH_3)_2 + 2HSCH_2CH_2SH \xrightarrow{\text{diethyl ether}} \overline{CH_2-S-BH-S-CH_2} + 4H_2$$

$$(BH_3)_2 + 3HSCH_2CH_2SH \xrightarrow{\text{diethyl ether}} \overline{CH_2S-B(-SCH_2)}CH_2CH_2\overline{B(-SCH_2)-SCH_2} + 6H_2$$

In the gaseous state, $H_2BSCH_2CH_2SBH_2$ and $\overline{SCH_2CH_2SB}H$ are monomeric, but in the condensed phase boron with a coordination number of 4 may possibly

References p. 355

be present, perhaps as follows [*193*]:

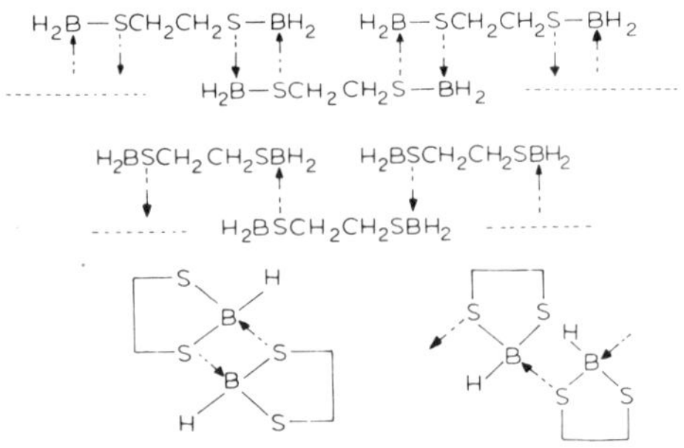

$H_2B-SCH_2CH_2S-BH_2$ $H_2B-SCH_2CH_2S-BH_2$
$H_2B-SCH_2CH_2S-BH_2$

$H_2BSCH_2CH_2SBH_2$ $H_2BSCH_2CH_2SBH_2$
$H_2BSCH_2CH_2SBH_2$

The weak poly- (donor–acceptor) coordination in $[BH_2(SMe)]_x$ is broken by strong donors such as NMe_3; $Me_3N{\cdot}BH_2(SMe)$ is formed (but compare with this the stability of $[BH_2(PMe_2)]_x$ to NMe_3, see section 11.1).

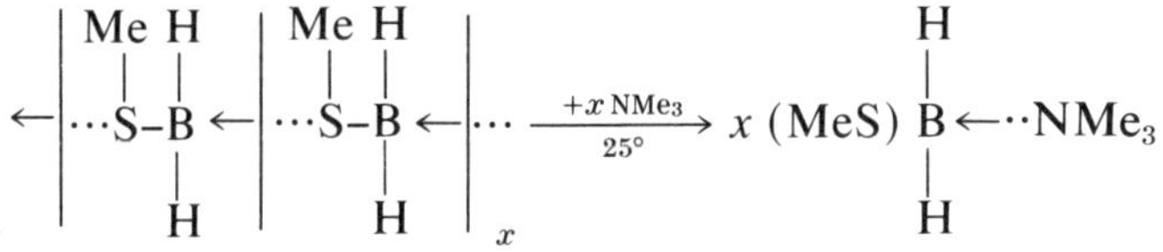

$$\leftarrow\left|\cdots \underset{\text{Me}}{S}-\underset{H}{\overset{H}{B}}\leftarrow\right|\cdots \underset{\text{Me}}{S}-\underset{H}{\overset{H}{B}}\leftarrow\left|\cdots\right._x \xrightarrow[25^\circ]{+x\,NMe_3} x\,(MeS)\,\underset{H}{\overset{H}{B}}\leftarrow\cdots NMe_3$$

Polymeric $BH_2(SMe)$ or, better, $BH_2(SMe){\cdot}NMe_3$ reacts with diborane to form $B_2H_5(SMe)$ [*131*]:

$$BH_2(SMe)+BH_3 \xrightarrow{90^\circ} BH_2(SMe){\cdot}BH_3$$

$$BH_2(SMe){\cdot}NMe_3+BH_3 \xrightarrow{90^\circ} BH_2(SMe){\cdot}BH_3+NMe_3$$

Alkylthioboron hydrides add to C=C double bonds in the same way as other boron hydrides (see section 12.1, Table 4.1) [*511*, *512*]. With primary amines ($EtNH_2$, Bu^nNH_2), $BH(SBu^n)_2$ eliminates one mole of butanethiol to give an alkylamino(butylthio)borane which then, eliminating thiol again, yields an N-alkylborazine [*511*]:

$$\tfrac{1}{2}[BH(SR)_2]_2+NR'H_2 \longrightarrow BH(SR)_2{\cdot}NR'H_2 \xrightarrow{-HSR} [BH(SR)NR'H] \xrightarrow{\mp HSR} \tfrac{1}{3}[BHNR']_3$$

$$(R = n\text{-}C_4H_9;\ R' = C_2H_5,\ n\text{-}C_4Hg)$$

Dimethyl sulphoxide cleaves the double bridge in diborane unsymmetrically [*496*]:

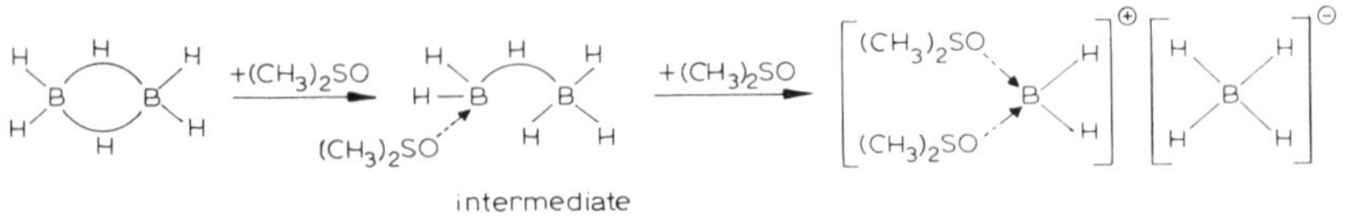

$$B_2H_6 \xrightarrow{+(CH_3)_2SO} \underset{\text{intermediate}}{(CH_3)_2SO\cdots BH_3\text{-}BH_3} \xrightarrow{+(CH_3)_2SO} [((CH_3)_2SO)_2BH_2]^{\oplus}[BH_4]^{\ominus}$$

For further information on cleavages of the double bridges in diborane and tetraborane(10), see sections 9.4.1 and 15.3, respectively.

9. Synthesis and reactions of diborane and organyldiboranes with N-compounds

9.1. *Formation of adducts of $(BH_3)_2$ and tertiary amines*

In the reaction of diborane with tertiary amines (trialkylamines and cyclic nitrogen bases containing no N–H bonds, such as pyridine and quinoline), organylamineboranes (N-organylborazanes) are formed [*76, 78, 546, 761*]:

$$\frac{1}{2}(BH_3)_2 \xrightarrow{+\text{energy}} BH_3 \begin{cases} \xrightarrow[-\text{energy}]{+NR_3} H_3B \leftarrow NR_3 \\ \xrightarrow[-\text{energy}]{+NC_5H_5} H_3B \leftarrow NC_5H_5 \end{cases}$$

For example, $(BH_3)_2$ and trimethylamine, by symmetrical cleavage of the $B{-}H^{\mu}{-}B$ double bridge, form the adduct $BH_3{\cdot}NMe_3$; the $(Me_3N)_2BH_2^{\oplus}BH_4^{\ominus}$ that would result from unsymmetrical cleavage is not formed [*701a*]. $(BH_3)_2$ and pyridine give the adduct $BH_3{\cdot}py$. The reaction is carried out either without a solvent [*124*] or, for relatively large batches, in benzene, ether, or the liquid amine itself [*76, 547*]. Unlike trimethylamine [*118*] the weak electron donor trisilylamine does not form an addition compound with diborane [*118*].

In benzene, petroleum ether, diethyl ether, or tetrahydrofuran, trialkylammonium salts and pyridinium chloride again form organylamineboranes (N-organylborazanes) with boranates. The intermediate state is the (not isolatable) ammonium boranate [*35, 207, 363, 547, 636, 761*]:

$$R_3NHX + MBH_4 \xrightarrow[-MX]{-30^\circ} \left|\underset{\text{unstable}}{[R_3NH][BH_4]}\right| \xrightarrow[-H_2]{-10\text{ to }0^\circ} R_3N{\cdot}BH_3$$

$$X = Cl, Br, \tfrac{1}{2}SO_4;\ M = Li, Na, K, \tfrac{1}{2}Ca$$

In amineboranes $\rangle N{\cdot}B\langle$, B–C bonds can be transformed into B–H. Thus, trialkylamine-trialkylboranes $R_3N{\cdot}BR'_3$ (N-trialkyl-B-trialkylborazanes) are hydrogenated under a pressure of hydrogen at 220° to trialkylamine-boranes $R_3N{\cdot}BH_3$(N-trialkylborazanes) [*428*]:

$$R_3N{\cdot}BR'_3 + 3H_2 \longrightarrow R_3N{\cdot}BH_3 + 3R'H$$

There are no intermolecular exchange reactions in the compound $BH_3{\cdot}NMe_3$ [*148a*]. Similarly, with $Me_3N{\cdot}BF_3$, lithium boranate forms the borazane Me_3N

References p. 355

$\cdot BH_3$. Here, the displacement of a whole BF_3 group by a BH_3 group (opening of the B–N bond) is more likely than a replacement of a F by H [*329*].

Alkyldiboranes or aryldiboranes react with tertiary amines (trialkylamines and triarylamines) or with pyridine to give the corresponding amineboranes (borazanes). For example, trimethylamine and 1,2-dimethyl-diborane give trimethylamine-methylborane (N-trimethyl-B-methyl-borazane) and trimethylamine and tetramethyldiborane give trimethylaminedimethylborane (N-trimethyl-B-dimethylborazane) [*671*]:

$$[BH_2Me]_2 + 2NMe_3 \longrightarrow 2MeBH_2\cdot NMe_3$$
$$[BHMe_2]_2 + 2NMe_3 \longrightarrow 2Me_2BH\cdot NMe_3$$

A mixture of mono- and trimethyldiboranes gives a mixture of the corresponding amine-alkylboranes that is very difficult to separate [*671*].

Free diborane can be avoided if alkyl- or arylboroxins $(RBO)_3$ are hydrogenated in the presence of a tertiary amine or pyridine with lithium alanate (nascent diborane), for example [*298, 299, 304, 309*]:

$$2(RBO)_3 \xrightarrow[\text{diethyl ether}]{+3LiAlH_4, -3LiAlO_2} 6\,RBH_2 \xrightarrow{+6NR'_3} 6RBH_2\cdot NR'_3$$
$$R = \text{alkyl, cycloalkyl}$$

Very useful seems to be the KBH_4 based preparation [*25a*]:

$$2KBH_4 + CO_2 + 2R_3N + H_2O \xrightarrow{0-10°} 2R_3N\cdot BH_3 + K_2CO_3 + 2H_2$$

Phenylboron dichloride, $PhBCl_2$, cannot be converted into the corresponding phenylborane with lithium alanate. On working in the presence of pyridine, hydrogenation of the B–Cl bond does in fact take place, but $PhBH_2\cdot py$ is formed, with the occurrence of side reactions. Pyridine-arylboranes $RBH_2\cdot py$ and $R_2BH \cdot py$ are formed without side reactions by the hydrogenation of di(ethoxy)arylboranes or of alkoxydiarylboranes with lithium alanate in the presence of pyridine [*298*]:

$$2RB(OEt)_2 \xrightarrow[\text{diethyl ether}]{+LiAlH_4, -LiOEt, -Al(OEt)_3} [2RBH_2] \xrightarrow{+2py} 2RBH_2\cdot py$$

$$4R_2BOR' \xrightarrow[\text{diethyl ether}]{+LiAlH_4, -LiOR', -Al(OR')_3} [4R_2BH] \xrightarrow{+4py} 4R_2BH\cdot py$$

$$R = C_6H_5, p\text{-}CH_3C_6H_4, \alpha\text{-}C_{10}H_7 \text{ etc.}; R' = C_2H_5, \text{n-}C_4H_9, NH_2CH_2CH_2$$

The hydridic hydrogen on the boron in amineboranes $R'_nBH_{3-n}\cdot NR_3$ can easily be replaced by amino groups. A prerequisite for this is that the alkyl substituents *R'* on the boron are sterically undemanding. Four examples are given below [*304, 305, 308*]:

$$RBH_2 \cdot NMe_3 + Me_2NH \xrightarrow[100-150^\circ]{(NH_4Cl)} RBH(NMe_2) + NMe_3 + H_2$$

$$RBH(NMe_2) + Me_2NH \xrightarrow[100-150^\circ]{(NH_4Cl)} RB(NMe_2)_2 + H_2$$

$R = CH_2CH_2CH_2CH_3$

$$R'BH_2 \cdot NR_3 + o\text{-}C_6H_4(NH_2)_2 \xrightarrow[80^\circ]{\text{benzene}} R'B(NH)_2C_6H_4 + NR_3 + 2\,H_2$$

$$R'BH_2 \cdot NR_3 + 3\,NH_3 \xrightarrow[-3\,R_3N]{\text{diglyme},\ 100\text{-}150^\circ\ (NH_4Cl)} [3\,R'BH_2 \cdot NH_3] \xrightarrow{-6H_2} (R'BNH)_3$$

R = sterically undemanding groups, *e.g.* n-alkyl groups

If the alkyl substituents on the boron are sterically demanding, they obstruct the attack on the primary amines. For example, the reaction of methylamine with trimethylamine-*tert*-butylborane gives not bis(methyl-amino)-*tert*-butylborane, $Bu^tB(MeNH)_2$, but only mono(methylamine)-*tert*-butylborane $Bu^tBH(MeNH)$ [*308*]

$$Bu^tBH_2 \cdot NMe_3 + MeNH_2 \xrightarrow[100-150^\circ]{(NH_4Cl)} Bu^tBH(MeNH) + NMe_3 + H_2$$

Triorganylamineboranes are obtained in very good yields by the reactions of triphenoxyboron $B(OPh)_3$, activated aluminium, a tertiary amine, and hydrogen under pressure [*18*]:

$$B(OPh)_3 + Al + \tfrac{3}{2}H_2 + NR_3 \xrightarrow[180^\circ,\ 280\ \text{atm}]{(AlCl_3)} BH_3 \cdot NR_3 + Al(OPh)_3$$

The yields are 99 mole-% with NMe_3, 92 mole-% with NEt_3, and 60 mole-% with $N(Ph)Me_2$.

In the reaction of $[(CH_3)_3N]_2BH_2^{\oplus}Cl^{\ominus}$ (see also section 9.4.1) with butyllithium BuLi, depending on the reaction conditions, $(CH_3)_3NBH_2CH_2N(CH_3)_2$ or $(CH_3)_2(CH_2Li)NBH_2CH_2N(CH_3)_2$ is formed. Both compounds are capable of the following reactions [*519a*]:

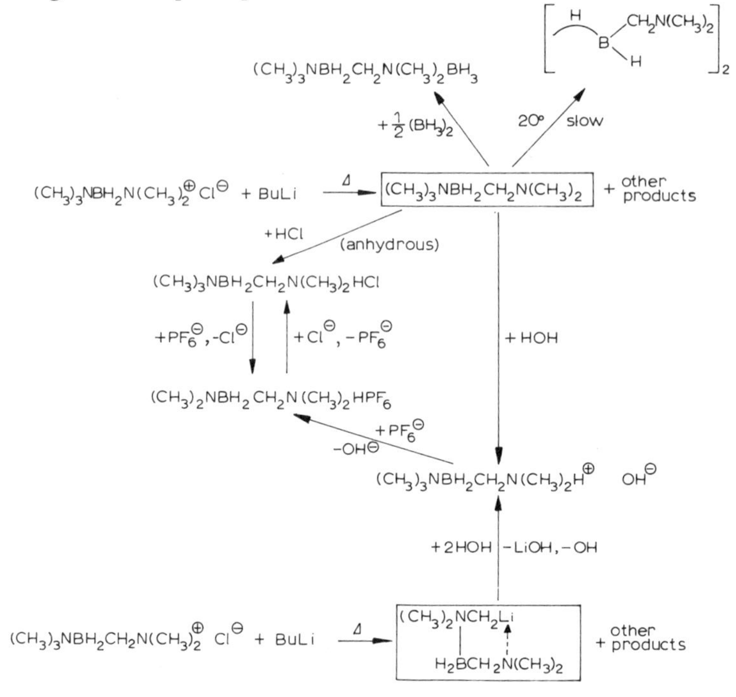

$BH_3 \cdot NR_3$ (R = Me, Et) react with $Co_2(CO)_8$ to give the dark-red crystalline solids $Co_3(CO)_{10}BH_2NR_3$ [*405c*].

9.2. *Formation of adducts of $(BH_3)_2$ with secondary amines*

Diborane reacts with dialkylamines in the absence of a solvent or in ether with the formation of dialkylamineboranes (N-dialkylborazanes)(symmetrical cleavage of the $B\text{–}H^{\mu}\text{–}B$ double bond) [*508, 802*]:

$$\tfrac{1}{2}(BH_3)_2 \xrightarrow{+\text{energy}} BH_3 \begin{cases} \xrightarrow[-\text{energy}]{+R_2NH} H_3B \leftarrow NHR_2 \\ \xrightarrow[-\text{energy}]{+NC_5H_5} H_3B \leftarrow NC_5H_5 \end{cases}$$

In the reaction with Me_2NH, a small amount of $(Me_2NH)_2BH_2^{\oplus}BH_4^{\ominus}$ is also formed through unsymmetrical cleavage of the $B\text{–}H^{\mu}\text{–}B$ double bridge [*701a*].

With boranates in benzene, petroleum ether, diethyl ether, or tetrahydrofuran, dialkylammonium salts also form dialkylamineboranes (N-dialkylborazanes) [*76, 636*]:

$$R_2NH_2X + MBH_4 \xrightarrow[-MX]{-30°} [R_2NH_2][BH_4] \xrightarrow[-H_2]{-10\text{ to }0°} BH_3 \cdot R_2NH$$

$$X = Cl, Br, \tfrac{1}{2}SO_4;\ M = Li, Na, K, \tfrac{1}{2}Ca$$

Here the reaction temperature must be kept as low as possible to avoid the splitting off of H_2 and the formation of dialkylaminoboranes (borazenes) BH_2NR_2 in place of the N-dialkylborazanes $BH_3 \cdot R_2NH$ (see sections 9.5 and 9.6). In monoglyme at 0° by condensation of the $BH_3 \cdot Me_2NH$ first formed with the elimination of hydrogen, $NaBH_4$ and Me_2NH_2Cl form 1, 1, 3, 3,-tetramethyldiborazane $H_3B \leftarrow \cdot\cdot NMe_2{-}BH_2 \leftarrow \cdot\cdot NMe_2H$ [*289*].

9.3. *Formation of adducts of $(BH_3)_2$ with primary amines*

In the reaction of diborane with an excess of primary amine (CH_3NH_2, C_2H_5-NH_2, n-$C_3H_7NH_2$) at the lowest possible temperature (–90 to –80°), by the unsymmetrical cleavage of the B–H^{μ}–B double bridge dihydrobis(methylamine)-boron(III) boranate is formed together with a considerably smaller amount of isomeric monoalkylamineborane (N-alkylborane) [*115, 672, 701a, 814*]:

$$(BH_3)_2 \xrightarrow{+2RNH_2} \begin{cases} \xrightarrow{\text{Main reaction}} (RNH_2)_2BH_2^{\oplus}BH_4^{\ominus} \\ \xrightarrow{\text{Side reaction}} 2RNH_2 \cdot\cdot\!\rightarrow BH_3 \end{cases}$$

Monoalkylamineboranes, $BH_3 \cdot RNH_2$, are also formed in the reaction of alkylamines with nascent diborane (*cf.* section 9.1) and in the reaction of monoalkylammonium chlorides, bromides, or sulphates with lithium boranate. Because of the tendency to split out H_2, the reaction temperature must be kept as low as possible and any catalytically active impurities must be avoided [*547*]:

$$RNH_3X + LiBH_4 \xrightarrow[-LiX]{-30°} \underset{\text{unstable}}{[RNH_3][BH_4]} \xrightarrow[-H_2]{-10\text{ to }0°} RNH_2 \cdot BH_3$$

$$X = Cl, Br, \tfrac{1}{2}SO_4;\ R = CH_3, C_2H_5, \text{n-}C_3H_7, \text{i-}C_3H_7, \text{n-}C_4H_9, \text{t-}C_4H_9$$

In the reaction of ethylenediamine with diborane in the absence of a solvent at –80° to 20°, $C_2H_4(NH_2)_2 \cdot 2BH_3$ is formed quantitatively in a highly exothermic reaction. Larger amounts are produced in tetrahydrofuran by the nucleophilic displacement of THF by an amine with a yield of 94% [*394*]:

$$H_2NCH_2CH_2NH_2 + (BH_3)_2 \longrightarrow BH_3 \cdot NH_2CH_2CH_2NH_2 \cdot BH_3$$

$$H_2NCH_2CH_2NH_2 + 2\,BH_3 \cdot OC_4H_8 \longrightarrow BH_3 \cdot NH_2CH_2CH_2NH_2 \cdot BH_3 + 2\,OC_4H_8$$

It is still uncertain whether this compound is an adduct, ethylenediamine-bis-borane (I) or an analogue (II) of dihydrodiammineboron(III) boranate $(NH_3)_2$-$BH_2^{\oplus}BH_4^{\ominus}$ [*394*]:

$$H_3B \cdots\!\rightarrow \overset{H}{\underset{H}{N}}CH_2CH_2\overset{H}{\underset{H}{N}} \leftarrow\!\cdots BH_3$$

I
Ethylenediaminebisborane

$$\left[\begin{matrix} CH_2{-}NH_2 \\ | \\ CH_2{-}NH_2 \end{matrix} \rightarrow BH_2 \right]^{\oplus} \left[BH_4\right]^{\ominus}$$

II
Dihydroethylenediamineboron (III) tetrahydroborate

9.4. *Formation of adducts of $(BH_3)_2$ with ammonia*

9.4.1. *Synthesis and reactions of* $[(NH_3)_2BH_2]^{\oplus}[BH_4]^{\ominus}$

Gaseous diborane and solid or liquid ammonia react under reaction conditions that must be strictly adhered to [*e.g.*, $(BH_3)_2$ in a current of nitrogen passed into liquid NH_3 at −78°][*700*] to form a "diborane-diammonia, $B_2H_6{\cdot}2NH_3$" [*635, 668, 700, 733*]:

$$B_2H_6 + 2NH_3 \xrightarrow{-128\text{ to }-78^\circ} B_2H_6{\cdot}2NH_3$$

The non-volatile compound was first regarded as the ammonium compound $(NH_4)_2B_2H_4$ [*795, 797*] and later as the ammonium compound $[NH_4]^{\oplus}[H_3BNH_2\text{-}BH_4]^{\ominus}$ [*668*]. After the discovery of the boranate ion $BH_4^{\ominus}$, it was formulated as ammonium boranate $NH_4^{\oplus}BH_4^{\ominus}$ stabilized by a separating BH_2NH_2 group $[NH_4]^{\oplus}BH_2NH_2[BH_4]^{\ominus}$ [*635*] ($[NH_4]^{\oplus}[BH_4]^{\ominus}$ is stable only below −20° [*580*]). According to the latest chemical, physical, and spectroscopic investigations diborane-diammonia has the formula $[(NH_3)_2BH_2]^{\oplus}[BH_4]^{\ominus}$ [*526*] and therefore belongs to the class of compounds $[(donor)_2BH_2]^{\oplus}X^{\ominus}$ (the donor may be NH_3, NH_2CH_3, $NH(CH_3)_2$, $(CH_3)_3N$, $(CH_3)_2(C_2H_5)N$, $\overline{CH_2(CH_2)_4N}CH_3$, $\overline{CH_2(C_4H_7CH_3)N}H$, $\frac{1}{2}(CH_3)_2NCH_2CH_2N(CH_3)_2$, $\frac{1}{2}CH_3N(CH_2CH_2)_2NCH_3$, $\frac{1}{2}(CH_3)_2NCH_2CH_2CH(CH_3)N(CH_3)_2$, $\frac{1}{2}(CH_3)_2PCH_2CH_2P(CH_3)_2$, $(CH_3)_3As$, $(CH_3)_2S$; anion X may be $Cl^{\ominus}$, $I^{\ominus}$, $ICl_2^{\ominus}$, $ClO_4^{\ominus}$, $PF_6^{\ominus}$, $AuCl_4^{\ominus}$, $\frac{1}{2}B_{12}H_{12}^{2\ominus}$, $(NC)_2CCHC(CN)_2^{\ominus}$), [*43b, 115, 174, 192, 371e, 526, 527a, 537b, 537c, 577, 578, 581, 629a, 683, 701, 701a, 702, 703, 762*]:

$$\left[(H_3N)_2BH_2\right]^{\oplus}\left[BH_4\right]^{\ominus}$$

The formation of reaction products after the unsymmetrical or symmetrical cleavage of the $B\text{–}H^{\mu}\text{–}B$ double bridge may depend on the donor activity of the donor (Me_3N, Me_2NH, $MeNH_2$, NH_3, $OSMe_2$, etc. and even $(Me_2N)_2BB(NMe_2)_2$ [*152a*]). According to NMR measurements [*189a, 701b, 701c*], the first step is the formation of a donor–acceptor bond of the donor D to the electron-poor $B\text{–}H^{\mu}\text{–}B$ double bridge. The second step is cleavage of one of the two $B\text{–}H^{\mu}\text{–}B$ bridges. The next step for weak donors (NH_3) is coordination of more donor to the same B atom (reaction path a), and for strong donors (Me_3N) to the second B atom (reaction path b):

$$B_2H_6 \xrightarrow{+D} D{\cdot}B_2H_6 \longrightarrow D{\cdot}BH_2\text{–}H\text{–}BH_3$$

$$\xrightarrow{+D_{weak}} D_2BH_2^{\oplus}\,BH_4^{\ominus}\ \text{(a)} \qquad \xrightarrow{+D_{strong}} 2\,D{\cdot}BH_3\ \text{(b)}$$

The reactions of dihydrodiammineboron(III) boranate, $[(NH_3)_2BH_2]^{\oplus}BH_4^{\ominus}$, often resemble those of the alkali metal boranates $M^{\oplus}BH_4^{\ominus}$, for example [*683*]:

$$Na^{\oplus}BH_4^{\ominus} + 2NH_4Br \xrightarrow{\text{liq. }NH_3} [(NH_3)_2BH_2]^{\oplus}Br^{\ominus} + NaBr + 2H_2$$

$$[(NH_3)_2BH_2]^{\oplus}[BH_4]^{\ominus} + 2NH_4X \xrightarrow{\text{liq. }NH_3} 2[(NH_3)_2BH_2]^{\oplus}X^{\ominus} + 2H_2$$

$X = Cl, Br$

$[(NH_3)_2BH_2][BH_4]$ reacts with ammonium chloride in diethyl ether only very slowly in the direction mentioned above. Traces of ammonia catalyse the reaction. $[(NH_3)_2BH_2]Cl$ and $[NH_4][BH_4]$ are formed metathetically; with the elimination of H_2, the latter is converted into $NH_3 \cdot BH_3$, which very slowly changes back into $[(NH_3)_2BH_2][BH_4]$ [*702*]:

$$[(NH_3)_2BH_2][BH_4] + NH_4Cl \xrightarrow[\text{catalytic amounts of }NH_3]{\text{diethyl ether}} [(NH_3)_2BH_2]Cl + [NH_4][BH_4]$$

$$[NH_4][BH_4] \xrightarrow[\text{diethyl ether, }20^{\circ}]{-H_2} NH_3 \cdot BH_3 \xrightarrow[\text{slow}]{\text{very}} \tfrac{1}{2}[(NH_3)_2BH_2][BH_4]$$

Lithium chloride, bromide, or boranate catalyses the decomposition of $[(NH_3)_2BH_2][BH_4]$ into polymeric monoaminoborane and soluble ammonia-borane [*635, 683, 703*]:

$$[(NH_3)_2BH_2][BH_4] \xrightarrow[\text{diethyl ether, }20^{\circ}]{(LiX)} \frac{1}{n}(H_2NBH_2)_n + NH_3 \cdot BH_3 + H_2$$

$$(X = Cl, Br, BH_4)$$

With sodium, $[(NH_3)_2BH_2][BH_4]$ and $[(NH_3)_2BH_2]Br$ form only (polymeric ?) monoaminoborane [*683*]:

$$[(NH_3)_2BH_2][BH_4] + Na \xrightarrow[-78^{\circ}]{\text{liq. }NH_3} H_2NBH_2 + NaBH_4 + NH_3 + \tfrac{1}{2}H_2$$

$$[(NH_3)_2BH_2]Br + Na \xrightarrow[-78^{\circ}]{\text{liq. }NH_3} H_2NBH_2 + NaBr + NH_3 + \tfrac{1}{2}H_2$$

For the reactions of $[(NH_3)_2BH_2][BH_4]$ with diborane, see section 9.5. With $NaNH_2$ or $NaC{\equiv}CH$, $[(NH_3)_2BH_2][BH_4]$ yields $(BH_2NH_2)_5$ with a decaline-like structure [*57b*].

The formation of a "diborane-diammonia II" $[(NH_3)_3BH][BH_4]_2$ (congruently $\tfrac{3}{2}[B_2H_6 \cdot 2NH_3]$) on prolonged standing of $[(NH_3)_2BH_2][BH_4]$ in liquid ammonia [*192, 701*] is less certain.

Tetramethyldiborane, $(CH_3)_2BH_2B(CH_3)_2$, reacts with ammonia in the absence of a solvent at −78° analogously to diborane; unsymmetrical cleavage of the $B{-}H^{\mu}{-}B$ double bridge leads to the formation of $[(NH_3)_2B(CH_3)_2]^{\oplus}[(CH_3)_2BH_2]^{\ominus}$ (congruently $[(NH_3)_2BH_2]^{\oplus}[BH_4]^{\ominus}$) [*527a*]:

$$2\,NH_3 + (H_3C)_2B(\mu\text{-}H)_2B(CH_3)_2 \longrightarrow [(NH_3)_2B(CH_3)_2]^{\oplus}[(CH_3)_2BH_2]^{\ominus}$$

Because of the symmetrical cleavage of tetramethyldiborane brought about by ether, the reaction of NH_3 and $(CH_3)_2BH_2(CH_3)_2$ in dimethyl ether, diethyl ether, or tetrahydrofuran takes place with the formation of amminedimethylborane [*527a*]:

$$2\,R_2O + (H_3C)_2B(\mu\text{-}H)_2B(CH_3)_2 \xrightarrow{-78°} 2\,R_2O\cdot BH(CH_3)_2$$

$$R_2O\cdot BH(CH_3)_2 + NH_3 \xrightarrow{-78°} H_3N\cdot BH(CH_3)_2 + R_2O$$

9.4.2. *Synthesis and reactions of $BH_3 \cdot NH_3$*

Ammoniaborane, $NH_3 \cdot BH_3$, is synthesized by one of the following methods.

(a) Reaction of diborane with ammonia in polar organic solvents or in water [*723*]:

$$(BH_3)_2 + 2NH_3 \xrightarrow{\text{polar solvent}} 2NH_3 \cdot BH_3$$

(b) Nucleophilic displacement of the weak donor dimethyl ether in dimethyl-ether-borane by the stronger donor ammonia (excess) (70% yield) [*702*]:

$$BH_3 \cdot OMe_2 + NH_3 \xrightarrow[-78°]{\text{dimethyl ether}} NH_3 \cdot BH_3 + Me_2O$$

Approximately equimolar amounts of $BH_3 \cdot NH_3$ and $[(NH_3)_2BH_2][BH_4]$ are formed by the displacement of the tetrahydrofuran in $BH_3 \cdot THF$ by NH_3 at $-78°$ [*700*].

(c) Metathesis of $LiBH_4$ and NH_4Cl or $(NH_4)_2SO_4$ in diethyl ether or metathesis of $NaBH_4$ and NH_4Br in diglyme with subsequent elimination of H from the unstable ammonium boranate NH_4BH_4 formed [*702, 704*]. For example, a 45% yield is obtained in the reaction:

$$LiBH_4 + NH_4Cl \xrightarrow[-LiCl]{\text{diethyl ether}} (NH_4BH_4) \xrightarrow{20°} NH_3 \cdot BH_3 + H_2$$

A 2.5% solution of ammoniaborane $NH_3 \cdot BH_3$ in water hydrolyses only very slowly (0.5–0.9%/day at 20°). As was to be expected, such a solution possesses marked reducing properties. Between 50 and 105° $NH_3 \cdot BH_3$ splits off hydrogen at increasing rate [*723*].

The H-D exchange reaction between ND_3 and $BH_3 \cdot NH_3$ reaches equilibrium very rapidly in liquid ND_3. The extent of exchange corresponds only to the N-H links in $BH_3 \cdot NH_3$ [*107*].

The insoluble isomeric dihydrodiammineboron(III) boranate precipitates slowly from solutions of dimeric ammoniaborane $(NH_3 \cdot BH_3)_2$ in diethyl ether [*702*]:

$$(NH_3 \cdot BH_3)_2 \xrightarrow{20°} [(NH_3)_2BH_2][BH_4]$$

9.5 *Synthesis and reactions of the amino derivatives of B_2H_6 and BH_3: $B_2H_5(NH_2)$ and $BH_2(NH_2)$*

In the preparation of borazine ("borazole"), $H_3N_3B_3H_3$, from diborane and ammonia, μ-aminodiborane $B_2H_5NH_2$ is formed as a by-product[*672*]. Better yields are obtained[*673*] by passing diborane over $[(NH_3)_2BH_2][BH_4]$:

$$[(NH_3)_2BH_2][BH_4] + (BH_3)_2 \xrightarrow{85-88°} 2\ H_2B(\mu\text{-}NH_2)(\mu\text{-}H)BH_2 + 2H_2$$

Electron diffraction in $B_2H_5NH_2$ shows the position of the NH_2 group in the bridge[*328*]. Aminodiborane, $B_2H_5NH_2$, adds ammonia to form $B_2H_5NH_2{\cdot}NH_3$, which gives borazine $H_3B_3N_3H_3$ in good yield at 200°[*672*].

Aminodiborane, which is fairly stable at room temperature, forms diborane and polymeric aminoborane (borazene) on being heated[*673*]:

$$2B_2H_5NH_2 \longrightarrow B_2H_6+\frac{2}{x}(H_2B\text{–}NH_2)_x$$

Polymeric aminoborane is obtained more simply by the reaction of diborane with lithium amide in ether and extraction from the insoluble aminoborane of the ether-soluble lithium boranate formed simultaneously[*637*]:

$$(BH_3)_2+2R_2O \xrightarrow{\text{diethyl ether}} 2BH_3{\cdot}OR_2 \xrightarrow[>-68°]{+LiNH_2} H_2B\text{–}NH_2+LiBH_4+2R_2O$$

Only H-free derivatives of the triborylamines $(BH_2)_3N$ [$=(BH_2)_2NBH_2$, N-diborylborazene] have been reported. This is a very interesting class of compound because of the d_π–p_π back-coordination of electrons from the nitrogen to the boron[*440*].

9.6 *Synthesis and reactions of alkylamino derivatives of B_2H_6 and BH_3*

9.6.1 *Monoamino derivatives $B_2H_5(NR_2)$ and $BH_2(NR_2)$*

When a dialkylamineborane (*e.g.*, $BH_3{\cdot}Me_2NH$, $BH_3{\cdot}HN(CH_2)_3CH_2$) is heated or when diborane is treated with an alkylamine at elevated temperatures, aminoborane (borazene) is formed[*115, 119, 508, 618, 801, 802*]:

$$\tfrac{1}{2}(BH_3)_2+R_2NH \longrightarrow H_3B\leftarrow NHR_2 \xrightarrow{120-200°} H_2B\text{–}NR_2 + H_2$$

On being heated, monoorganylamine-monoarylboranes, $R'BH_2{\cdot}RNH_2$, also split off one mole of hydrogen. The R'HBNRH so formed is in equilibrium with its

References p. 355

dismutation products $R_2'BNRH$ and H_2BNRH. In the case of the amino-boranes containing B–H and N–H bonds, because of the presence of $\delta+_H$ (on the nitrogen) and $\delta-_H$ (on the boron), further heating leads to the elimination of hydrogen and the formation of borazines $(R'BNR)_3$ or $(HBNR)_3$ [*100a, 100b, 506*]:

$$R'H_2B\cdot NRH_2 \xrightarrow{-H_2} R'HBNRH \rightleftharpoons \tfrac{1}{2}R_2'BNRH + \tfrac{1}{2}H_2BNRH$$

$$R'HBNRH \xrightarrow[40\text{–}200^\circ]{-H_2} \tfrac{1}{3}(R'BNR)_3 \qquad H_2BNRH \xrightarrow[40\text{–}200^\circ]{-\frac{1}{2}H_2} \tfrac{1}{6}(HBNR)_3$$

The reaction of alkylammonium salts with lithium boranate gives rise to alkylamineboranes $R_2NH\cdot BH_3$ [*547*] or alkylaminoboranes R_2NBH_2 [*636*] apparently in dependence on the reaction conditions:

$$R_2NH_2Cl + LiBH_4 \xrightarrow{-LiCl} [R_2NH_2][BH_4] \xrightarrow{-H_2} R_2NH\cdot BH_3 \xrightarrow{-H_2} R_2NBH_2$$

Dimethylaminoborane and diethylaminoborane are dimeric at 20° [*119, 508*]. On being heated for 1 week at 106°, dimethylaminoborane $BH_2(NMe_2)$ dismutes into mono-μ-dimethylaminodiborane, $B_2H_5(NMe_2)$, and bis(dimethylamino) borane $BH(NMe_2)_2$ [*119*]:

$$Me_2N{-}BH_2 \underset{}{\overset{\Delta}{\rightleftharpoons}} [H_2B(\mu\text{-}NMe_2)]_2 \overset{\Delta}{\rightleftharpoons} H_2B(\mu\text{-}NMe_2)(\mu\text{-}H)BH_2 + (Me_2N)_2BH$$

Electron diffraction of $B_2H_5(NMe_2)$ shows the position of the NMe_2 group in the bridge. A similar dismutation is observed in aminoboranes with unstrained cyclic amines [*e.g.*, tetramethyleneaminoborane ("pyrrolidinoborane")], while strained rings open during the formation of the aminoborane [*115, 119*]:

$$2\ (CH_2)_4NH\cdot BH_3 \xrightarrow[-H_2]{135\text{–}145^\circ} 2\ (CH_2)_4N{-}BH_2 \rightleftharpoons [H_2BN(CH_2)_4]_2 \xrightarrow{+(BH_3)_2} H_2B(\mu\text{-}N(CH_2)_4)(\mu\text{-}H)BH_2$$

$$(CH_2)_2NH\cdot BH_3 \xrightarrow{-78^\circ} CH_3CH_2N(H)BH_2 \xrightarrow{+BH_3} H_2B(\mu\text{-}N(H)C_2H_5)(\mu\text{-}H)BH_2$$

On being heated with pentaborane(9) to 103 to 110°, dimethylaminoborane $BH_2(NMe_2)$ trimerizes [*43c, 120a, 135, 770*]. (The product of heating is not $(Me_2N)_3B_3H_4$ [*109*]). The trimeric structure $[BH_2(NMe_2)]_3$ (chair form) is shown by X-ray and NMR measurements [*135, 770, 771*] (see Fig. 4.16). For $(H_2BNHMe)_3$ see [*43c*].

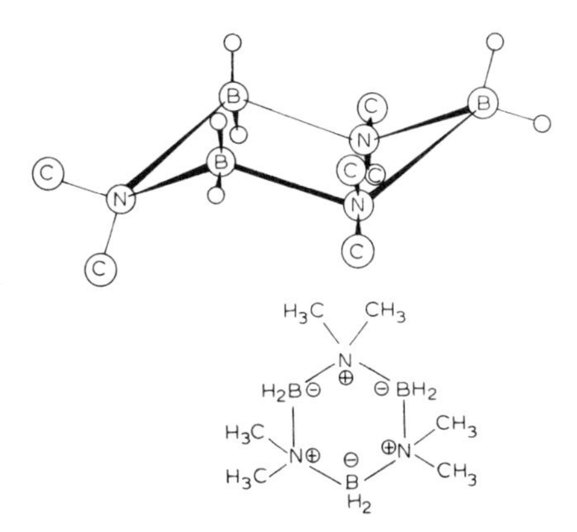

Fig. 4.16. Steric and topological structure of $[BH_2(NMe_2)]_3$. Structural parameters [*771*]:
B–N = 1.59 Å ∡BNB = 113° ∡CNC = 104°
N–C = 1.49 Å ∡NBN = 114° ∡CNB = 109.5°

9.6.2 *Synthesis and reactions of the diamino derivatives $B_2H_4(NR_2)_2$ and $BH(NR_2)_2$*

Aminoboranes can be converted with diborane or an alkylamine into amino-dimeric or amino-richer aminoboranes. The following rules apply in general: R on a nitrogen atom is not replaced, R and H on a boron atom are replaceable. H on nitrogen ($\delta+_H$) and on boron ($\delta-_H$) splits out readily (H_2). Those aminoboranes are formed which are to be expected on the basis of the molar ratio of the (formally) monomeric reactants (*e.g.*, NMe_3 and BH_3). Examples are given below.

At 205° dimethylamine, $HNMe_2$, reacts with a deficiency of diborane, $(BH_3)_2$, to form tris(dimethylamino)borane $B(NMe_2)_3$[*120*], which with an equimolecular amount of diborane $(BH_3)_2$ gives dimethylamino-borane $BH_2(NMe_2)$[*120*], and this, in turn, adds an excess of diborane to form dimethylaminodiborane[*120*]:

$$(BH_3)_2 + 6HNMe_2 \xrightarrow[-42°]{} 2B(NMe_2)_3 \xrightarrow[135°]{+2(BH_3)_2} 6BH_2(NMe_2) \xrightarrow{+3(BH_3)_2} 6B_2H_5(NMe_2)$$

The reaction of small amounts is carried out (quantitatively) in a sealed tube. For larger batches, $(BH_3)_2$ is first passed into liquid $HNMe_2$ at −42°, the solution is heated to 135°, and the $BH_2(NMe_2)$ so formed is converted in $B_2H_5(NMe_2)$ with more $(BH_3)_2$ (yield 75–85%).

With trimethylamine NMe_3 at 140°, dimethylaminoborane, $BH_2(NMe_2)$, forms bis(dimethylamino)borane $BH(NMe_2)_2$ and trimethylamineborane $Me_3N{\cdot}BH_3$ [*120*]:

$$2BH_2(NMe_2) + NMe_3 \rightleftharpoons BH(NMe_2)_2 + BH_3{\cdot}NMe_3$$

With diborane at 100°, bis(dimethylamino)borane, $BH(NMe_2)_2$, can be converted into dimethylaminoborane [*120*]:

$$2BH(NMe_2)_2 + (BH_3)_2 \longrightarrow 4BH_2(NMe_2)$$

Conversely, when dimethylaminoborane, $BH_2(NMe_2)$, is heated with dimethylamine $HNMe_2$, bis(dimethylamino)borane, $BH(NMe_2)_2$, is formed. An equimolecular ratio gives an impure product, but an excess of dimethylamine gives pure $BH(NMe_2)_2$ [*120*]:

$$(Me_2N)BH_2 \xrightarrow{+Me_2HN} (Me_2N)(Me_2HN\rightarrow)BH_2 \xrightarrow[150-200^\circ]{-H_2} (Me_2N)_2B\text{-}H$$

Bis(dimethylamino)borane, $BH(NMe_2)_2$, can also be obtained by the hydrogenation of bis(dimethylamino)chloroborane, $BCl(NMe_2)_2$, with lithium alanate [*143*]:

$$(Me_2N)_2BCl \xrightarrow[\Delta]{+LiAlH_4} (Me_2N)_2BH$$

or by heating dimethylaminoborane, $BH_2(NMe_2)$, with lithium hydride to 100°. In this reaction the LiH binds the $(BH_3)_2$ present in the equilibrium in the form of $LiBH_4$ [*120*]:

$$[H_2B(NMe_2)]_2 \rightleftharpoons 2\ BH(NMe_2)_2 + (BH_3)_2$$

$$(BH_3)_2 + 2\ NaH \longrightarrow 2\ NaBH_4$$

$$2BH_2(NMe_2) + NaH \longrightarrow BH(NMe_2)_2 + NaH\cdot BH_3$$

Bis[μ-(diethylamino)]diborane, $[BH_2(NEt_2)]_2$, which is stable to water at 20°, adds to alkenes (*e.g.*, but-1-ene, oct-1-ene) [*508*]:

$$[BH_2(NEt_2)]_2 + 4CH_2{=}CHR \xrightarrow[\text{pyridine}]{130^\circ} 2B(CH_2CH_2R)_2(NEt_2)$$

μ-(Dimethylamino)diborane, $B_2H_5(NMe_2)$, reacts with BF_3 at 100° to form $(BH_3)_2$ and $BH_2(NMe_2)\cdot BF_3$; with BCl_3, $B_2H_4Cl(NMe_2)$ is formed [*119*].

1,2-Dimethyldiborane and 1,2-diphenyldiborane, $(BH_2R)_2$, react with dimethylamine and diethylamine, $R_2'NH$, similarly to diborane [*114*]:

$$(BH_2R)_2 + 2R_2'NH \xrightarrow{-78^\circ} 2BH_2R\cdot R_2'NH \xrightarrow[-H_2]{\sim 150^\circ} 2RHBNR_2'$$

Methyl(dimethylamino)borane, $(Me_2N)BHMe$, is dimeric in the liquid state but monomeric in the gaseous state. The rate of conversion is so slow that physical data can be obtained for both forms (see section 20.9) [cit. *114*, *503*].

In aminoboranes of the form $R_2NBR'H$, R′ groups can be replaced but R_2N groups cannot. Thus, methyl(dimethylamino)borane, $(Me_2N)BHMe$, dismutes at 20° into dimethylaminoborane, $(Me_2N)BH_2$, and dimethyl(dimethylamino)borane, $(Me_2N)BMe_2$ [*114*, cit. *503*]:

$$2(Me_2N)BHMe \overset{20°}{\rightleftharpoons} (Me_2N)BH_2 + (Me_2N)BMe_2$$

Diethylamine-phenylborane, $Et_2NH{\cdot}BH_2Ph$, dismutes at 150° (probably via the intermediate stage of a phenyl(diethylamino)borane, $(Et_2N)BHPh$, into diethylaminoborane, $(Et_2N)BH_2$, and diphenyl(diethylamino)borane, $(Et_2N)BPh_2$ [*508*]:

$$2Et_2NH{\cdot}BH_2Ph \xrightarrow[-2H_2]{150°} 2(Et_2N)BHPh \rightleftharpoons (Et_2N)BH_2 + (Et_2N)BPh_2$$

9.7 *Synthesis and reactions of hydroxylamineboranes*

In the absence of a solvent, diborane and hydroxylamine, $NH_2(OH)$, react between −186 and −96° either not at all or explosively to give hydroxylamine-borane (N-hydroxyborazane), $NH_2(OH){\cdot}BH_3$. In ether the reaction takes place more gently, but not completely, below −112°. Similarly, the corresponding derivatives of hydroxylamineborane are formed by the reaction of diborane with O- or N-substituted hydroxylamines such as MeNH(OH), $Me_2N(OH)$, $H_2N(OMe)$, MeNH(OMe), and $Me_2N(OMe)$. On progressive replacement of the hydrogens by methyl groups, the hydroxylamineboranes become increasingly stable to heat. Thus the polymeric $[NH_2(OH){\cdot}BH_3]_n$ is stable only below −112°, while $NMe_2(OH){\cdot}BH_3$ and the O-methylhydroxylamineboranes are stable at room temperature [*54*, *133*]. The formation and $(BH_3)_2$-catalysed decomposition of the hydroxylamineboranes can be summarized in the following scheme:

$$\left.\begin{array}{l} n\,NH_2(OH) + \frac{n}{2}(BH_3)_2 \xrightarrow{\text{without solvent}} \\ 2n\,NH_2(OH) + 4n\,R_2O{\cdot}BH_3 \xrightarrow[-4n\,R_2O]{\text{ether}} \end{array}\right\} n\,HO{-}\underset{H}{\overset{H}{N}}{\cdots}{\rightarrow}BH_3 \xrightarrow[-n\,H_2]{>112°} \left[{\cdots}O{\cdots}\underset{H}{\overset{H}{N}}{\cdots}\underset{H}{\overset{H}{B}}{\cdots}\right]_n$$

$$NHMe(OH) + R_2O{\cdot}BH_3 \xrightarrow{-R_2O} \underset{\text{stable, monomeric liquid}}{HO{-}\underset{H}{\overset{Me}{N}}{\cdots}{\rightarrow}BH_3} \xrightarrow[>20°]{+BH_3,\,-H_2} \left[\underset{\text{rearrangement}}{H_2BO{-}\underset{H}{\overset{Me}{N}}{\cdots\cdots}BH_3}\right]$$

$$\xrightarrow{-BH_3} [H_3BONHMe] \longrightarrow \frac{1}{n}\underset{\text{yellow solid}}{[-BOH-]_n} + MeNH_2$$

9.8 *Synthesis and reactions of hydrazineboranes*

Hydrazine and diborane react to give hydrazinebis(borane), $N_2H_4 \cdot 2BH_3$ [*195*, *727*]:

$$N_2H_4 + (BH_3)_2 \longrightarrow H_3B \leftarrow \cdot\cdot NH_2\text{–}NH_2 \cdot\cdot \rightarrow BH_3$$

Hydrazinium salts N_2H_5X ($X = Cl$, HSO_4) give with boranates either hydrazinebis(borane), $H_3B \cdot N_2H_4 \cdot BH_3$, or hydrazinemonoborane, $N_2H_4 \cdot BH_3$. In diethyl ether (a relatively weak donor), $H_3B \cdot NH_2NH_2 \cdot BH_3$ is formed; in dioxan $NH_2NH_2 \cdot BH_3$ is formed [*267*]:

$$[N_2H_6]SO_4 + 2\,LiBH_4 \xrightarrow[-Li_2SO_4,\,-2\,H_2]{\text{diethyl ether}} BH_3 \cdot NH_2NH_2 \cdot BH_3$$

$$[N_2H_6]SO_4 + 2\,NaBH_4 + O(C_2H_4)_2O \xrightarrow[-Na_2SO_4,\,-2\,H_2]{\text{dioxan}} NH_2NH_2 \cdot BH_3 + O(C_2H_4)_2O \cdot BH_3$$

On being heated, $BH_3 \cdot N_2H_4 \cdot BH_3$ splits off hydrogen to form polymeric hydrazinobis(borane) [*267*]:

$$BH_3 \cdot NH_2NH_2 \cdot BH_3 \longrightarrow \frac{1}{x}(H_2B\text{–}NH\text{–}NH\text{–}BH_2)_x + 2H_2$$

There is no borane substituted fully with $NHNH_2$ groups, $B(NHNH_2)_3$. For comparison, we may mention the reaction of BCl_3 with N_2H_4 [*585*]. Here again, no $B(NHNH_2)_3$ is formed since it rapidly undergoes intermolecular condensation with the elimination of hydrazine to give $(BN_4H_5)_n$. The structure of $(BN_4H_5)_n$ may be

[Structure: chain of B atoms linked by NHN–H bridges, with NHNH₂ groups coordinated to B]

9.9 *Synthesis and reactions of aminehalogenoboranes*

Aminetrialkylboranes, $NH_3 \cdot BR_3$, are not halogenated by hydrogen chloride (no formation of $NH_{3-n}Cl_n \cdot BR_3$) but are cleaved instead [*671*]:

$$NH_3 \cdot BR_3 + HCl \longrightarrow NH_4Cl + BR_3$$

In the reaction of trialkylamineboranes, $R_3N \cdot BH_3$, with hydrogen halides with or without a solvent at −78 to 100°, all the hydrogen atoms on the boron are successively substituted by halogen by ligand exchange [*546*]:

$$R_3N \cdot BH_3 \xrightarrow{+HX} \left[R_3N \cdots \rightarrow \overset{\delta+}{B}H_2 \cdots \overset{\delta-}{H} \cdots H^{\delta+} \cdots X \right] \xrightarrow{-H_2} R_3N \cdot BH_2X \xrightarrow[-H_2]{+HX} R_3N \cdot BHX_2 \xrightarrow[-H_2]{+HX} R_3N \cdot BX_3$$

$R_3N = CH_3NH_2$, $(CH_3)_3N$, $(C_2H_5)_3N$, *tert*-$C_4H_9NH_2$, C_5H_5N; $X = Cl, Br, I$

All three alkylaminehalogenoboranes, $R_3N\cdot BH_{3-n}X_n$ ($n = 1, 2, 3$), can be isolated, since the detachment of a hydride ion from amineboranes with a small degree of halogen substitution (stronger δ^- polarization of H) takes place more rapidly than that of highly substituted amineboranes (smaller δ^- polarization of H). In the reaction with HF, only $R_3N\cdot BF_3$ can be isolated.

With boron trichloride and boron tribromide, trialkylamineboranes, $R_3N\cdot BH_3$, form, by successive ligand exchange, diborane and (isolatable) trialkylaminehalogenoboranes, $R_3N\cdot BH_{3-n}X_n$ ($n = 1, 2$, but not 3) [*546*]:

$$R_3N\cdot BH_3 \xrightarrow{+BX_3} \left[R_3N\cdots \overset{\delta+}{B}H_2(H^{\delta-})(X^{\delta-})\overset{\delta+}{B}X_2 \right] \xrightarrow{-BX_2H} R_3N\cdot BH_2X \xrightarrow[-BX_2H]{+BX_3} R_3N\cdot BHX_2$$

$$6\ BHX_2 \longrightarrow (BH_3)_2 + 4\ BX_3$$

Trialkylaminemonohalogenoboranes are formed in the reaction of trialkylamineboranes, $R_3N\cdot BH_3$, with halogen or interhalogen compounds. The hydrogen halide arising simultaneously reacts further as described above, so that again all three halogen derivatives can be isolated [*546*]:

$$R_3N\cdot BH_3 + X_2 \longrightarrow \left[R_3N\cdots \overset{\delta+}{B}H_2(H^{\delta-})(X^{\delta-})\overset{\delta+}{X} \right] \longrightarrow R_3N\cdot BH_2X + HX$$

$X_2 = Cl_2, Br_2, I_2, ICl, IBr, ICN$

Tertiary amines displace the ether in the monochloroborane–ether complex [*546*]

$$R_3N + R_2O\cdot BH_2Cl \longrightarrow R_3N\cdot BH_2Cl + R_2O$$

where R = CH_3 or C_2H_5 [C_5H_5N, α,α'-bipyridyl, NH_3, RNH_2 and R_2NH form adducts $(R_3N)_2\cdot BH_2Cl$].

9.10 Synthesis and reactions of cyanideboranes

Because of their free pair of electrons on the nitrogen atom, cyanides can form stable donor–acceptor adducts with borane at low temperatures: $R\text{–}C\equiv N| \cdots\rightarrow BH_3$. On being heated, they decompose into their components or the cyanide group is hydrogenated intramolecularly by hydride migration. Under these conditions N-substituted borazines are formed via alkylideneaminoboranes in accordance with the following scheme [*198*]:

$$R\text{—}C\equiv N + \tfrac{1}{2}(BH_3)_2 \underset{>\sim 0^\circ}{\overset{<\sim -30^\circ}{\rightleftharpoons}} R\text{—}C\equiv N\cdots\rightarrow BH_3 \xrightarrow{>\sim 0^\circ}$$

$$[R\text{—}CH\text{=}N\text{—}BH_2] \longrightarrow [R\text{—}CH_2\text{—}N\text{=}BH] \longrightarrow (HB\text{–}NCH_2R)_3$$

Borazines are formed when R = CH_3 (20°) and C_2H_5 (−10/0°). No borazines are formed when R = CH_2═CH (0°).

Dialkylboranes, BHR'_2, react similarly [*475*]. The adducts with silyl cyanide, SiH_3CN, and with trimethylsilyl cyanide, $SiMe_3CN$, are also intramolecularly hydrogenated, but here silanes and polymeric cyanoboranes are formed, the latter being converted intramolecularly into a polymer with no B–H bonds [*206*]:

$$SiH_3CN + \tfrac{1}{2}(BH_3)_2 \xrightarrow{20^\circ} SiH_3CN{\cdot}BH_3 \xrightarrow[-SiH_4]{100^\circ}$$

$$SiMe_3CN + \tfrac{1}{2}(BH_3)_2 \xrightarrow{20^\circ} SiMe_3CN{\cdot}BH_3 \xrightarrow[-SiMe_3H]{100^\circ}$$

$$\rightarrow \frac{1}{x}[(CN)BH_2]_x \xrightarrow{350^\circ} \frac{1}{x}[CHNHB]_x$$

Diborane reacts readily with NaCN to form NaH_3BCNBH_3 [*7*].

10. Synthesis and reactions of borazine and its derivatives

10.1 Introduction, structure, and physical properties

Soon after the discovery of a colourless liquid having the empirical formula BNH_2 in the reaction of gaseous diborane and gaseous ammonia by Stock and Pohland in 1926 [*737*] its structure was established: BNH_2 is a planar hexagonal ring with alternating B and N atoms and equal bond distances $(HBNH)_3$. Comparison of the physical properties already known at that time with the surprisingly similar properties of the isosteric benzene $(HCCH)_3$ led early to the German name Bor-az-ol as inorganic Benz-ol. Many investigations and a considerable volume of indicative data point convincingly to a certain aromatic nature of B–N ring (p_π–p_π bond between B and N). Some characteristics of borazine and benzene are compared in Table 4.1.

Borazine can best be formulated by means of the three Kekulé structures a–c and, in shortened form, as d:

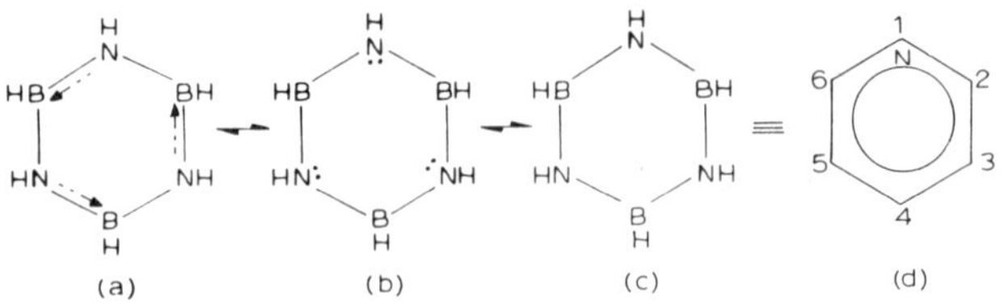

HB, BH, HN, NH, NH, B, H ... (structures)

(a) (b) (c) (d)

For additional review literature on borazine and its far more numerous partially or completely substituted derivatives, see, for example [*253, 354, 499, 541a, 699, 798, 799*].

TABLE 4.1

COMPARISON OF CHARACTERISTICS OF BORAZINE AND BENZENE

Characteristic	*Borazine*	*Benzene*
Molecular weight	80.5	78.1
Melting point, °K	216	279
Boiling point, °K	328	353
Critical temperature, °K	525	561
Density of the crystals at the m.p., g/ml	1.00	1.01
Density of the liquid at the b.p., g/ml	0.81	0.81
Trouton's constant	21.4	21.1
Parachor	208	206
Surface tension at the m.p., dyne/cm	31.1	31.1
Bond distances, Å	B⋯N 1.44	C⋯C 1.42
calculated single bond	B—N 1.54	
calculated double bond	B=N 1.36	
	N—H 1.02	
	B—H 1.20	C—H 1.08

10.2 Synthesis of borazine

There are two methods of synthesizing borazines $(HBNH)_3$ and their derivatives (alkyl, aryl, and halogen derivatives, etc.).

(1) Ring condensation of the corresponding unsubstituted or substituted amine-borane, H_3BNH_3. The syntheses differ in the method of forming the amine-borane:

$$\tfrac{1}{2}(BH_3)_2 + NH_3 \longrightarrow H_3BNH_3$$

$$\tfrac{1}{2}(BH_3)_2 + NH_4X \xrightarrow{-HX} H_3BNH_3$$

$$MBH_4 + NH_4X \xrightarrow{-MX} BH_4NH_4 \xrightarrow{-H} H_3BNH_3 \xrightarrow[\text{heat}]{-H_2} H_2BNH_2 \xrightarrow{-H_2} \tfrac{1}{n}(HBNH)_n$$

$$B_2H_5NH_2 \cdot NH_3 \xrightarrow{-BH_2NH_2} H_3BNH_3$$

$$B_4H_{10} \cdot B_5H_9 \cdot B_5H_{11} + NH_3 \xrightarrow[\text{sym. cleavage of } BH^{\mu}B \text{ double bridges}]{\text{Degradation}} H_3BNH_3$$

(2) Ring condensation of a B-trichloro- or B-trialkoxyamineboranes substituted or unsubstituted on the nitrogen, X_3BNH_3, with subsequent replacement of the substituent X by another substituent (H or R):

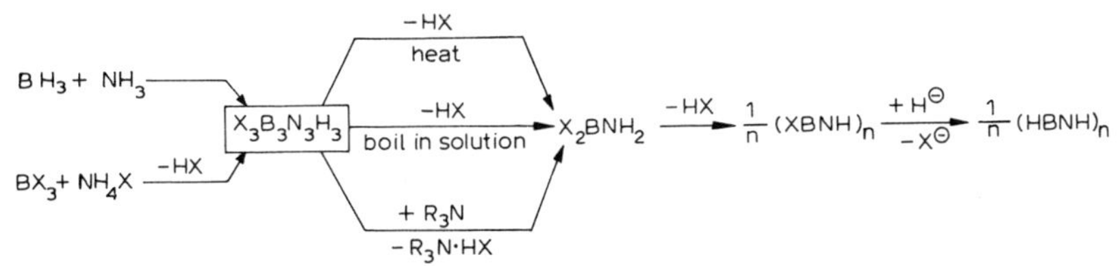

N-substitution occurs rarely; it has not yet been described for borazines with B–H bonds.

References p. 355

10.2.1 Direct synthesis by ring condensation

When diborane and ammonia in a molar ratio of 1:2 are heated, hydrogen, solid boranes, and 17 to 47% of borazine $H_3B_3N_3H_3$ are formed [*670, 672, 737, 798, 801, 814*]:

$$3(BH_3)_2 + 6NH_3 \xrightarrow[1-5\,atm]{150-300°} 2H_3B_3N_3H_3$$

The yield of borazine falls with an excess of ammonia (formation of borimide) [*737*] and with an excess of diborane (formation of aminodiborane $B_2H_5NH_2$) [*801*]. When diborane, ammonia, and methylamine are heated together for a short time, small amounts of borazine and of B-mono-, B-di-, and B-trimethylborazines are formed [*674*]:

$$(BH_3)_2 + NH_3 + MeNH_2 \xrightarrow{\Delta} H_{3-n}Me_nB_3N_3H_3$$

$(BH_3)_2$ and CH_3CN at 80–90° give $H_3B_3N_3CH_2CH_3$ (30–40% yield), μ-$(CH_3CH_2)_2N{-}B_2H_5$ and materials of higher molecular weight [*373a*]:

$$(BH_3)_2 + N{\equiv}CCH_3 \xrightarrow{\Delta} (H_2B{-}N{=}CHCH_3)_n \xrightarrow{\Delta} (HB{-}NCH_2CH_3)_m$$

When the ammonia adduct of aminodiborane, $B_2H_5NH_2 \cdot NH_3$, is heated to 200°, borazine is again formed [*673*]. It is also formed in the degradation of higher boranes (tetraborane, pentaboranes) with ammonia (primary formation of $BH_3 \cdot NH_3$) [*737, 739*].

The pyrolysis of $(NH_2CH_3)_2BH_2^{\oplus}Cl^{\ominus}$ gives $H_3B_3N_3(CH_3)_3$ [*43a*]:

$$(NH_2CH_3)_2BH_2^{\ominus}Cl^{\ominus} \xrightarrow{100-125°} \tfrac{1}{3}H_3B_3N_3(CH_3)_3 + CH_3NH_3Cl + H_2$$

10.2.2 Synthesis by ring condensation and subsequent substituent-exchange on the boron

The first stage of the synthesis is the preparation of borazines with functional groups (such as halogens or – less satisfactory – alkoxyl) on the boron.

Boron trichloride and dimethylboron fluoride react with primary amines on heating in the gas phase, or on boiling in benzene or chlorobenzene, to form B-trihalogenoborazines, with a yield of 52 to 90% [*356, 379, 625a, 816*]:

$$3BCl_3 + 3RNH_2 \longrightarrow Cl_3B_3N_3R_3 + 6HCl$$

$$3Me_2BF + 3MeNH_2 \xrightarrow{400°} F_3B_3N_3Me_3 + 6MeH$$

$$R = CH_3, C_2H_5, C_6H_5, p\text{-}CH_3C_6H_4, p\text{-}CH_3OC_6H_4$$

On being heated in the absence of a solvent or on being boiled in benzene, bromobenzene, chlorobenzene, or diglyme, boron trihalides, and ammonium halides form B-trihalogenoborazines with a yield of about 35% (almost 100% in the case of $Cl_3B_3N_3Me_3$ [*68*, *73*, *154*, *196*, *197*, *403*, *629*]:

$$3BX_3 + 3[RNH_3]X \longrightarrow X_3B_3N_3R_3 + 9HX$$

$$R = H, CH_3, CH_3OC_6H_4; X = Cl, Br$$

Iron salts, metallic iron, nickel, and best of all cobalt, lower the reaction temperature (from 165–175° [*73*] or 200° [*73*] to 80–150° and raise the yields from an average of 35 to 45% [*196*, *197*].

The formation of B-trihalogenoborazole (together with hydrogen halide and nitrogen) from boron trihalides and hydrazinium halides is catalysed in the same way (40–50% yield) [*196*, *197*].

Dehydrohalogenation is facilitated by the formation of trialkylammonium halide, for example [*773*]:

$$BCl_3{\cdot}NRH_2 + 2Et_3N \xrightarrow{\text{toluene}} \tfrac{1}{3}Cl_3B_3N_3R_3 + 2Et_3NHCl$$

$$R = CH_3, C_2H_5, n\text{-}C_4H_9$$

B-Trichloroborazines, $Cl_3B_3N_3R_3$, prepared in this way are converted by lithium boranate, sodium boranate, and lithium alanate into the corresponding B–H borazines.

Lithium alanate is suitable in only a few cases—for example, in the hydrogenation of N-triphenyl-B-trichloroborazine in the following way [*716*]:

$$4Cl_3B_3N_3Ph_3 + 3LiAlH_4 \xrightarrow{\text{diethyl ether}} 4H_3B_3N_3Ph_3 + 4LiAlCl_4$$

With N-substituents other than phenyl, apparently only a single hydride equivalent in the alanate performs hydrogenation ($AlH_4^{\ominus} \rightarrow H^{\ominus} + AlH_3$). The (non volatile) AlH_3, which prevents separation of the borazine formed [*651*] possibly forms a bridge complex:

R N ClB B Cl Al H H H RN NR B Cl

Lithium boranate in diethyl ether or sodium boranate in diglyme or triglyme, on the other hand, form the corresponding borazines in excellent yields [*651*]:

$$2Cl_3B_3N_3R_3 + 6MBH_4 \longrightarrow 2H_3B_3N_3R_3 + 3(BH_3)_2 + 6MCl$$

$$R = H, CH_3; M = Li, Na$$

In the reaction, if NMe_3 is present it binds the diborane (which is combustible and therefore not absolutely safe) in the form of $BH_3{\cdot}NMe_3$ [*154*]. Still more suitable

References p. 355

for binding $(BH_3)_2$ is $NaBH(OMe)_3$, which re-forms $NaBH_4$[*356*, *651*]:

$$(BH_3)_2 + 2NaBH(OMe)_3 \longrightarrow 2NaBH_4 + 2B(OMe)_3$$

The preparation and hydrogenation of B-trichloroborazine need not necessarily take place separately. Borazine is formed by the reaction of ammonium chlorides RNH_3Cl or hydrazinium dichloride $N_2H_6Cl_2$ with lithium or sodium boranate[*197*, *363*, *592*]:

$$MBH_4 + RNH_3Cl \xrightarrow[\text{diethylether}]{-MCl} [BH_4][RNH_3] \xrightarrow{-H_2} BH_3{\cdot}RNH_2 \xrightarrow{-2H_2} \tfrac{1}{3}H_3B_3N_3R_3$$

$$MBH_4 + N_2H_6Cl \longrightarrow \tfrac{1}{3}H_3B_3N_3H_3 + MCl + HCl + 3\tfrac{1}{2}H_2 + \tfrac{1}{2}N_2$$

$$R = H, CH_3, C_2H_5, n\text{-}C_3H_7, i\text{-}C_3H_7;\ M = Li, Na$$

10.3 Pyrolysis of borazine

The pyrolysis of borazine forms hydrogen and small amounts of the isosteres of naphthalene and biphenyl, as well as an isostere of 1,3-diamino-benzene, B-diaminoborazine[*443*]:

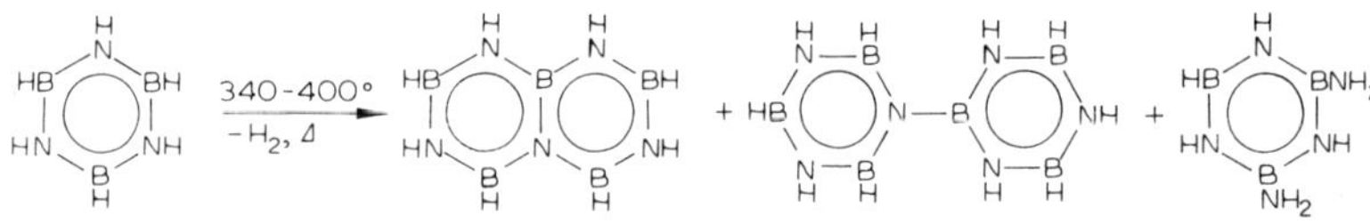

The naphthalene analogue can be prepared by passing electric discharge through an atmosphere of borazine [cit. *442*]. $[\overline{BNPhBHNPhBNPh}]_n$ is formed in the pyrolysis of $H_3B_3N_3Ph_3$[*431a*]. For the photochemical exchange reaction of borazine with deuterium see[*536b*].

10.4 Addition and elimination in borazine

Hydrogen does not hydrogenate borazine to $H_6B_3N_3H_6$ similarly to the case of benzene, but dehydrogenates it to more highly polymeric B–N–H systems with a smaller proportion of hydrogen[*801*]:

$$H_3B_3N_3H_3 + n\,H \xrightarrow[\text{Pd}]{40-50°} \frac{1}{x}(B_3N_3H_{6-n})_x + n\,H_2$$

However, hexahydroborazine, $H_6B_3N_3H_6$, is formed in the hydrogenation of $Cl_3H_3B_3N_3H_6$ ($= H_3B_3N_3H_3{\cdot}3HCl$) with $H^{\ominus}$ (see below).

With an excess of bromine at 0 to 20°, borazine forms an adduct $B_3N_3H_6{\cdot}2Br_2$. Above 60°, 2 moles of HBr are eliminated[*619*, *801*] (chlorine reacts similarly [*737*, *801*]:

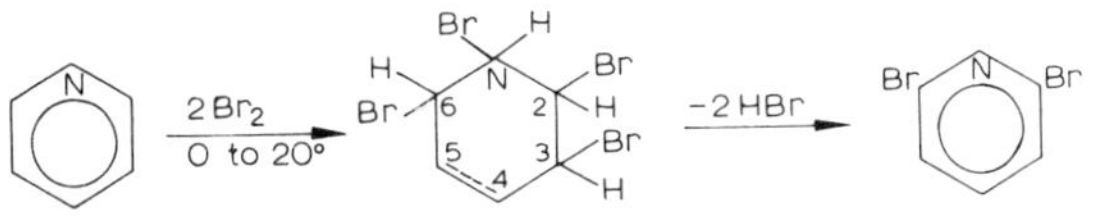

With 1 : 3 HCl or 1 : 3 HBr at 20°, borazine $H_3B_3N_3H_3$ and N-trimethyl-borazine $H_3B_3N_3Me_3$ first form the 2,4,6-trihalogenocycloborazane, $H_3X_3B_3N_3H_6$, and the 2,4,6-trihalogeno-1,3,5-trimethylcycloborazane, $H_3X_3B_3N_3Me_3H_3$. With $NaBH_4$, the latter can be converted into 1,3,5-trimethylcycloborazane, $H_6B_3N_3Me_3H_3$ [*155, 237, 701d*]. At 100°, $H_3X_3B_3N_3H_6$ and $H_3X_3B_3N_3Me_3H_3$ split off H_2 with the formation of B-trihalogenoborazine $Cl_3B_3N_3H_3$ and B-trihalogeno-N-trimethyl-borazine $Cl_3B_3N_3Me_3$, respectively, for example as follows [*237, 701d, 801, 813*]:

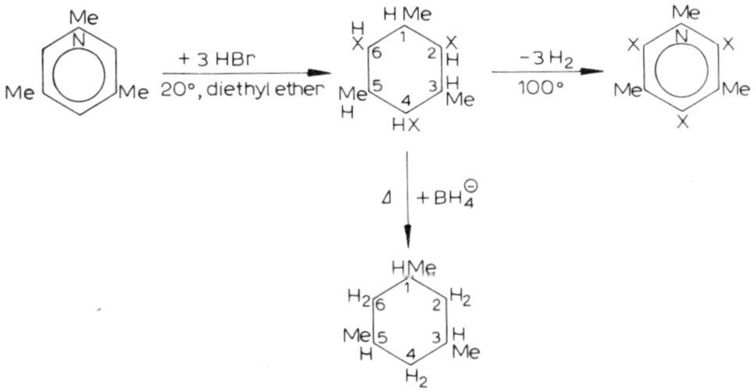

The elimination of H_2 possibly takes place at the low temperature of only 100°, with retention of the ring. The situation is different, however, in the formally comparable elimination of CH_4 from the adducts of HX and N,B-hexamethyl-borazine, $Me_3B_3N_3Me_3$ [*799, 801*]:

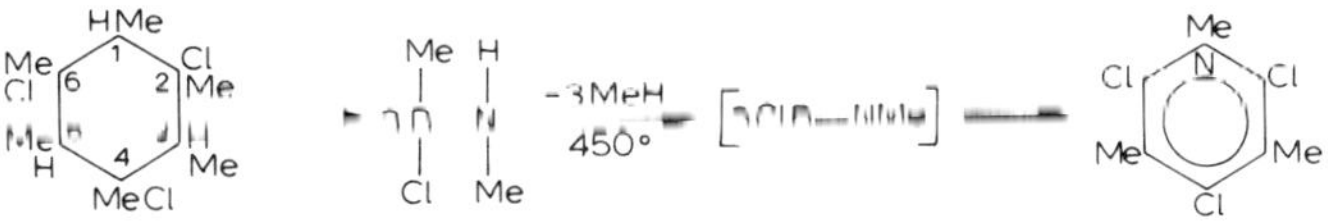

The reaction of ammonium chloride with borazine forms small amounts of B-trichloroborazine, $Cl_3B_3N_3H_3$ [*642*]. Because of the high temperatures necessary for this ($NH_4Cl \rightleftharpoons NH_3 + HCl$), a mechanism analogous to the addition of HCl is probable. Borazine with $HgCl_2$ or other halides yields $H_2ClB_3N_3H_3$ and $HCl_2B_3N_3H_3$ [*493c*].

The photochemical reaction of borazine and oxygen or steam at low pressure yields B-monohydroxyborazine, $(HO)H_2B_3N_3H_3$, and B,B′-diborazine oxide, $(H_3N_3B_3H_2)_2O$; borazine and ammonia yield B-monoaminoborazine, $(H_2N)H_2B_3N_3H_3$ [*446a*].

At 0°, borazine and N-trimethylborazine add 3 moles of water with the formation of 2,4,6-trihydroxycycloborazane, $(HO)_3H_3B_3N_3H_6$, and 2,4,6-trihydroxy-1,3,5-trimethylcycloborazane, $(HO)_3H_3B_3N_3Me_3H_3$, respectively. At 100°, 3 moles of H_2 split off with the formation of B-trihydroxyborazine, $(HO)_3B_3N_3H_3$, and B-trihydroxy-N-trimethylborazine, $(HO)_3B_3N_3Me_3$, respectively [*801, 814*]

References p. 355

(kinetics, see [*859a*]):

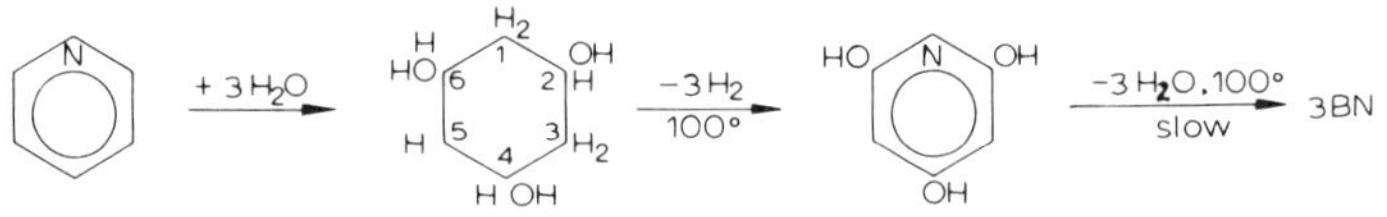

B-Trihydroxyborazine, $(HO)_3B_3N_3H_3$, which is less stable than the N-methyl derivative $(HO)_3B_3N_3Me_3$, splits off more water slowly even at 100°, forming boron nitride [*801*]. Compare the reaction of B-trimethylborazine, $Me_3B_3N_3H_3$, with water, which takes place differently. Here, trimethylcycloboroxole $(MeBO)_3$ is formed via 2,4,6-trihydroxy-2,4,6-trimethylcycloborazane, $(HO)_3Me_3B_3N_3H_6$ [*801, 814*]:

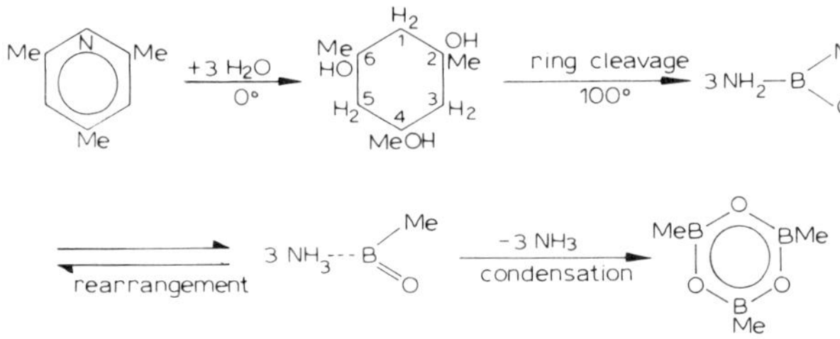

In contrast to B-trimethylborazine, B-tri(2,6-xylyl)-N-trimethylborazine and B-trimesityl-N-trimethylborazine are resistant to hydrolysis in acid and alkaline media [*536a*].

Alcohols behave with respect to borazine or N-trimethylborazine in the same way as water. At 20°, a 2,4,6-trialkoxycycloborazane $(RO)_3H_3B_3N_3H_6$ or $(RO)_3$-$H_3B_3N_3Me_3H_3$ is formed. At 100°, these compounds give off hydrogen with ring opening [*320, 801, 814*].

Ammonia and trimethylamine add to borazine primarily with the formation of a non-volatile glassy adduct $[N_3B_3H_6{\cdot}NH_3]_x$. This is probably a donor–acceptor complex to one of the boron atoms of the ring [*541a*]:

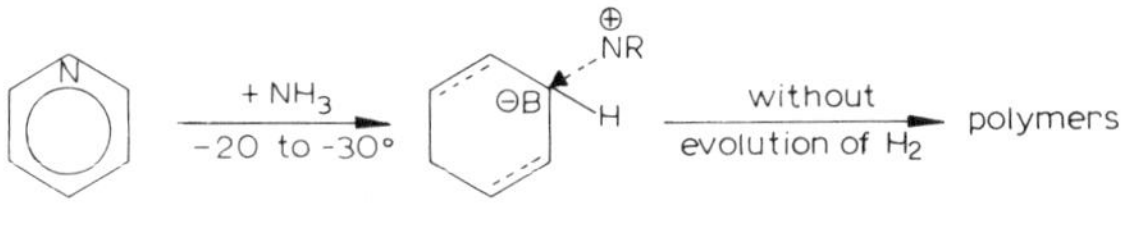

With borazine at 130°, dimethylamine yields H_2 and polymers that have not been investigated further [*798, 799, 801*].

There are two reaction mechanisms for the reaction of borazine, $H_3B_3N_3H_3$, or of N-monomethylborazine, $H_3B_3N_3MeH_2$ [*672*] with trimethylboron, BMe_3 (and similar remarks probably apply also to N-di- and N-trimethylborazines). In the first mechanism, that favoured at low temperatures, the borazine system is retained. The ligands on the boron atoms are replaced, as is always observed in mixtures of boranes with B–C and B–H bonds. In the second mechanism, that favoured at higher temperatures, the borazine ring opens. Hydrogen and methane are evolved and high-molecular products containing less hydrogen, with the approximate composition $[BNH]_x$, are formed:

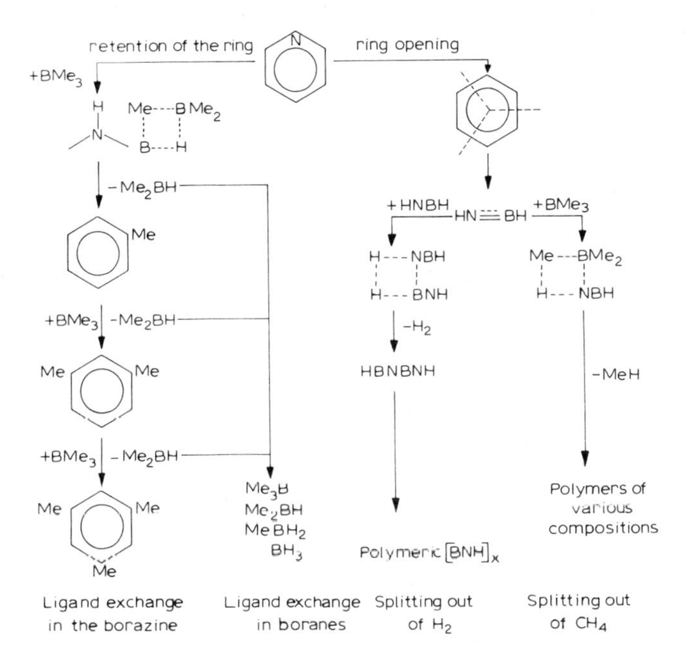

Thus, for example, with trimethylboron at 100°, borazine forms 60% of B-trimethylborazine, 5% of B-mono- and B-dimethylborazines, and 35% of "BNH". At 200 to 225°, the yield of B-methylated borazines falls in favour of "BNH", CH_4, and H_2 [*672*].

With borazine, boron trihalides give hydrogen and mono- and di-substituted borazines as well as unidentified non-volatile products [*642*]

10.5 B-Substitution in borazine (B–H → B–X)

In borazine, H can be replaced by D. Isotope exchange in $N_3B_3H_6$ takes place (without change of charge in the hydrogen isotopes) with ND_3, DCl, and DCN on the nitrogen and with D_2, $(BD_3)_2$, and $NaBD_4$ on the boron. No exchange takes place with D_2O, DC≡CD, and PD_3 [*156*].

+DCl
−HCl

+ BD_3
− BD_2H

In the reaction of borazine and N-trialkyl- or N-triarylborazines with carbanions (alkyllithiums, Grignard compounds) in diethyl ether, the hydride H atoms on the boron, like halogen atoms, are substituted successively:

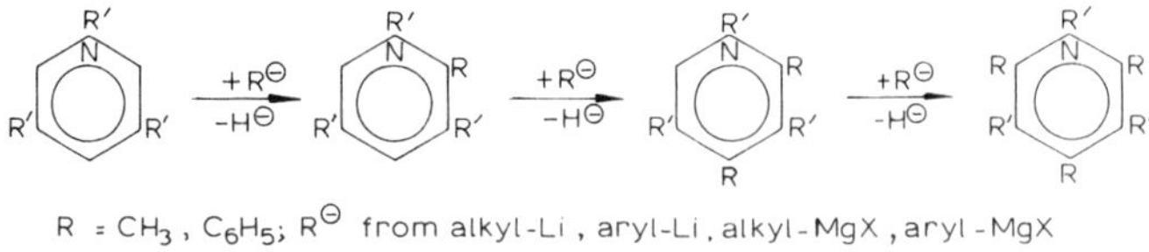

R = CH_3, C_6H_5; $R^{\ominus}$ from alkyl-Li, aryl-Li, alkyl-MgX, aryl-MgX

Some particular features of this reaction are as follows: if the N substituent R′ = CH_3, all three substitution products $H_{3-n}R_nB_3N_3Me_3$ are obtained in statistical distribution. The reaction therefore resembles the reaction of B-trihalogenoborazoles, $X_3B_3N_3R_3$, with carbanions (see section 10.6). If, however, the N substituent R′ = C_6H_5, only a single substitution product is obtained, depending on the ratio of the reactants. In the case of N-triphenylborazole, $H_3B_3N_3Ph_3$, therefore, the reactivity of the B–H bond (δ^--polarization of the H atom) with respect to aryl and alkyl anions falls off rapidly in the following sequence: B-unsubstituted > B-mono-substituted > B-disubstituted [*716*].

With N-triarylborazine, diazomethane forms successively B-mono-, B-di-, and B-trimethyl-N-triarylborazines (reactions of borazines with diazoalkanes $H_2\overset{\ominus}{C}{=}\overset{\oplus}{N}{\equiv}N|$ therefore give similar substitution products to those of B-halogenoborazines with $R^{\ominus}$) [*725*].

Borazines add to the C=O double bond in isocyanates. An elastic polymer is formed with hexamethylene-1,6-diisocyanate [*432*]:

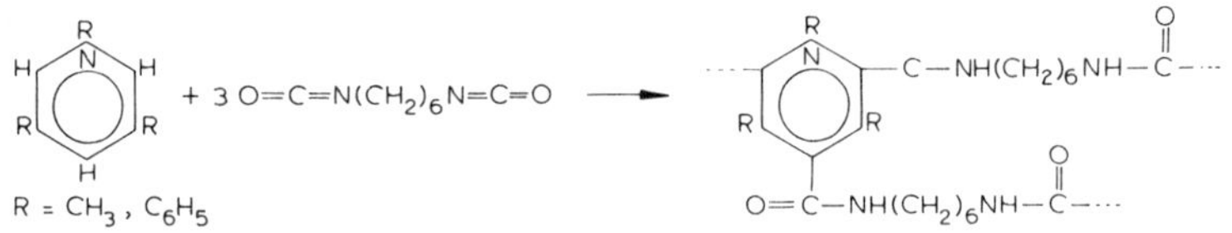

$$+\ 3\ O{=}C{=}N(CH_2)_6N{=}C{=}O \longrightarrow$$

R = CH_3, C_6H_5

10.6 B-Substitution in borazine (B–X → B–Y)

The B-trihalogenoborazines, $X_3B_3N_3R_3$ (R = H, organic substituents), which can easily be prepared in good yields, are extremely reactive and can be used for syntheses of valuable substances. For example, they react with RLi, RMgX, R_3SiK, H_2O, ROH, R_2O, KSCN, AgCN, $AgNO_2$, $AgNO_3$, $AgClO_4$, CH_2N_2, and so on, with the replacement of X by R, SiR_3, OH, OR, SCN, CN, NO_2, NO_3 ClO_4, CH_2X, etc. The reactions with $LiAlH_4$, $LiBH_4$, and $NaBH_4$ (replacement of X by H) are of great importance for the syntheses of borazines with B–H bonds (see section 10.2.2).

10.7 N-Substitution in borazoles (N–H → N–X)

Apart from the isotope-exchange reactions with DCl, DCN, and ND_3 (section 10.5), no N-substitutions have been described in borazines containing B–H bonds.

The N-metallation of B-trimethylborazine, $Me_3B_3N_3H_3$, may be mentioned because of its importance for synthesis. This synthesis could be applicable to borazine itself. For example, with methyllithium MeLi, B-trimethyl-N-dimethylborazine, $Me_3B_3N_3Me_2H$, forms almost quantitatively N-lithiopentamethylborazine, $Me_3B_3N_3Me_2Li$. B-Trimethylborazine, $Me_3B_3N_3H_3$, with methyllithium (1:1) forms mainly N-monolithio- and smaller amounts of N-dilithio- and N-trilithio-B-trimethylborazines. The N-lithioborazines can be used for coupling reactions [*777, 778*]. This will be illustrated with an example. The copyrolysis of $Me_2HB_3N_3Me_3$ and $Me_3B_3N_3Me_2H$ gives, among other products, small amounts of decamethyl-N-B′-borazabiphenyl (compare section 10.3) [*443*]:

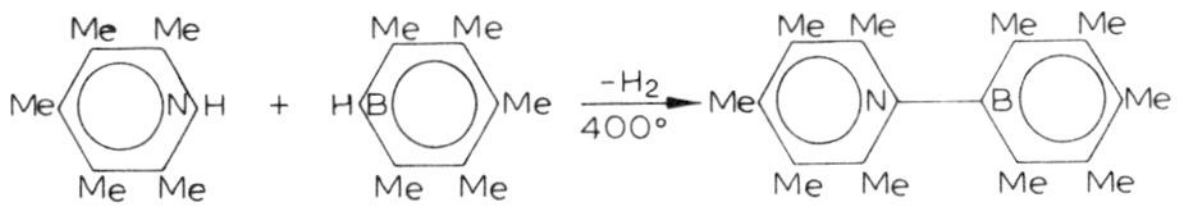

$$Me_5B_3N_3H + HB_3N_3Me_5 \xrightarrow[400°]{-H_2} Me_5B_3N_3{-}B_3N_3Me_5$$

Decamethylborazabiphenyl

Higher yields of such condensed borazine systems are obtained by coupling [*777*]:

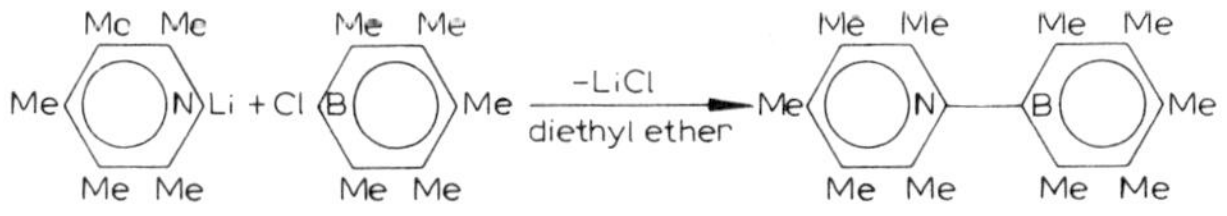

$$Me_5B_3N_3Li + ClB_3N_3Me_5 \xrightarrow[\text{diethyl ether}]{-LiCl} Me_5B_3N_3{-}B_3N_3Me_5$$

Decamethylborazabiphenyl

$$Me_5B_3N_3Cl + Li_2Me_4B_3N_3 + ClB_3N_3Me_5 \xrightarrow[\text{diethyl ether}]{-2\,LiCl} Me_5B_3N_3{-}B_3N_3Me_4{-}B_3N_3Me_5$$

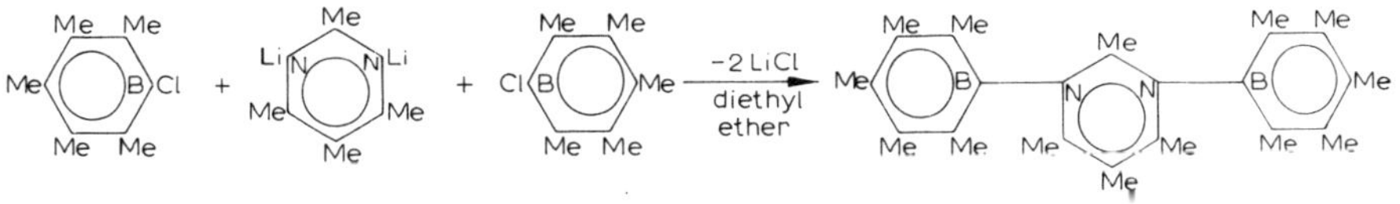

Tetradecamethylboraza-1,3-diphenylbenzene

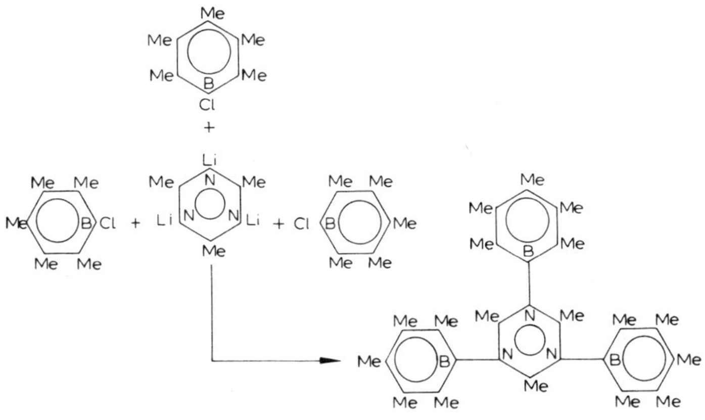

$$Me_5B_3N_3Cl + Li_3Me_3B_3N_3 + ClB_3N_3Me_5 + ClB_3N_3Me_5 \longrightarrow Me_3B_3N_3(B_3N_3Me_5)_3$$

Octadecamethylboraza-1,3,5-triphenylbenzene

References p. 355

11. Reactions of boron hydrides with P, As, and Sb compounds

11.1 Reaction of $(BH_3)_2$ with R_nPH_{3-n}

Diborane, $(BH_3)_2$, reacts with phosphanes, R_nPH_{3-n} ($n = 0, 1, 2, 3$), with the formation of phosphaneboranes (borophosphanes) $BH_3 \cdot R_nPH_{3-n}$ [*103, 129, 175f, 193a, 235d, 245*]:

$$(BH_3)_2) + 2PH_3 \xrightarrow{-110^\circ} [B_2H_6 \cdot 2PH_3?] \xrightarrow{0^\circ} 2BH_3 \cdot PH_3$$

$$\tfrac{1}{2}(BH_3)_2 + MePH_2 \xrightarrow{\text{low temperature}} BH_3 \cdot MePH_2$$

$$\tfrac{1}{2}(BH_3)_2 + (GeH_3)PH_2 \xrightarrow{-20^\circ} BH_3 \cdot (GeH_3)PH_2$$

$$\tfrac{1}{2}(BH_3)_2 + R_2PH \xrightarrow{\text{organic solvent}} BH_3 \cdot R_2PH$$

$R_2 = CH_3 + C_2H_5$, $C_2H_5 + C_2H_5$, $C_4H_9 + CH_3$, $C_3H_7 + C_2H_5$, $C_3H_7 + C_3H_7$, $C_2H_5 + CH_3CH{=}CHCH_2$, n-$C_6H_{13} + CH_3$, n-$C_8H_{17} + CH_3$, $C_2H_5 +$ n-C_7H_{15}, n-$C_5H_{11} +$ n-C_5H_{11}, n-$C_6H_{13} +$ n-C_6H_{13}, $C_{12}H_{25} + CH_3$, n-$C_8H_{17} +$ n-C_8H_{17}.

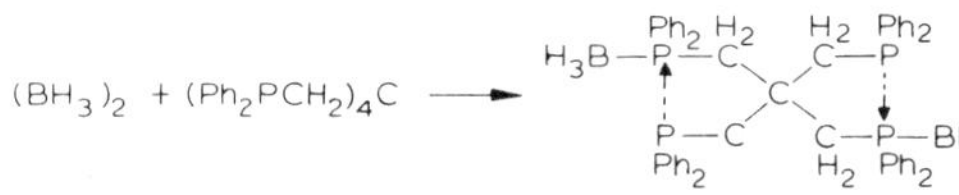

$(BH_3)_2 + (Ph_2PCH_2)_4C \longrightarrow$ [structure: $C(CH_2PPh_2)_4$ with two P atoms each coordinated to BH_3 ($H_3B \leftarrow PPh_2$ and $Ph_2P \rightarrow BH_3$)]

Trialkylphosphaneboranes can also be prepared in a simple manner by nucleophilic displacement of the amine in trimethylamine- or triethylamineboranes [*310, 311*]:

$$RBI_2 \cdot NMe_3 + PBu^n_3 \longrightarrow RBH_2 \cdot PBu^n_3 + NMe_3$$

R = n-Bu, i-Bu, *sec*-Bu, *tert*-Bu, Ph, *p*-anisyl, *p*-bromphenyl, *p*-tolyl, *o*-tolyl, mesityl

The parent substance $BH_3 \cdot PH_3$ dissociates into the components even at 0, (for comparison, $BCl_3 \cdot PH_3$ undergoes 90% decomposition only at 31°). Under carefully controlled conditions $BH_3 \cdot PH_3$ reacts with NH_3 to give $NH_4^{\oplus}[PH_2(BH_3)_2]^{\ominus}$, the BH_3 analogue of ammonium hypophosphite, $NH_4^{\oplus}[H_2PO_2]^{\ominus}$ [*259a*]. $BH_3 \cdot PH_3$ can be halogenated to give the B-halogeno derivatives as follows [*175e, 245*]:

$$BH_3 \cdot PH_3 + nHX \longrightarrow BH_{3-n}X \cdot PH_3 + nH_2$$

$$n = 1, 2, 3; X = Cl, Br$$

($BCl_3 \cdot PH_3$ is also easy to obtain from the components [*765*]). The P-alkylated phosphaneboranes, $BH_3 \cdot R_nPH_{3-n}$, split off hydrogen on being heated and form trimeric and tetrameric, and less frequently more highly polymeric or dimeric rings, as well as linear polymers of the alkylphosphanoborane, $BH_2PR_nH_{2-n}$

(for example, BH_2PMeH is a viscous oil)[*129, 130, 779*]:

$$BH_3 \cdot PMeH_2 \xrightarrow{>100^\circ} \frac{1}{n}(BH_2PMeH)_n + H_2$$

$$BH_3 \cdot PMe_2H \xrightarrow{150-200^\circ} \frac{1}{n}(BH_2PMe_2)_n + H_2$$

For another synthesis of $(BH_2PMe_2)_3$, see section 6.3.4. In the presence of donors such as $N(CH_3)_3$, $N(C_2H_5)_3$, $N(n\text{-}C_4H_9)_3$, $P(CH_3)_3$, or $PH(CH_3)_2$, high polymers (with a degree of polymerization of ~ 80) form preferentially. The strong poly (donor–acceptor) coordination in $[BH_2PMe_2]_x$ is not broken by $N(C_2H_5)_3$ (distinction from $[BH_2SMe]_x$ in section 8):

$$BH_2PMe_2 + Et_3N \longrightarrow Et_3N \cdot BH_2PMe_2 \xrightarrow{+n\,BH_2PMe_2} Et_3N \cdot [BH_2PMe_2]_{n+1}$$

```
       Et      H Me |          | H Me | H Me |
       |       |  | |          | |  | | |  | |
     Et-N ···→ B-P ··|→···  ···|→ B-P ··|→ B-P |
       |       |  | |          | |  | | |  | |
       Et      H Me |          | H Me | H Me |
```

The trimeric and tetrameric rings $(BH_2PMe_2)_n$ are unusually stable because of the high π-bond component in the P–B bond (see sections 1.3.2 and 1.3.3). They are very heat-resistant and hydrolyse slowly only at 300°[*129*]. With an excess of chlorine, N-chlorosuccinimide, N-bromosuccinimide, or iodine monochloride, the six-membered ring $(BH_2PPh_2)_3$ forms $(BX_2PPh_2)_3$ with retention of the ring structure, and with an excess of iodine, $(BHIPPh_2)_3$ is formed[*250, 250a*]. On heating at 360° $[BH_2PMe_2]_3$ does not depolymerize to the monomer, but condenses with splitting off of H_2 or MeH; at 510°, the rings are broken[*226*]. Trimeric dimethylphosphanoborane, $(BH_2PMe_2)_3$ has the chain structure[*293*], while tetrameric $(BH_2PMe_2)_4$ is a corrugated ring[*263*], as shown in Figs. 4.17 and 4.18.

At −78°, equimolar amounts of monobromodiborane BrB_2H_5 and phosphane PH_3 form a 1:1 adduct $BrB_2H_5 \cdot PH_3$ which decomposes at −45° into diborane, $(BH_3)_2$, and phosphanemonobromoborane, $BH_2Br \cdot PH_3$. The latter is stable up to 0°; above 0°, H_2 splits out with the formation of polymeric $(-BHBr \cdot PH_2-)_n$ [*175b*]:

$$BrB_2H_5 + PH_3 \xrightarrow{-78^\circ} BH_3BH_2Br \cdot PH$$

$$BH_3BH_2Br \cdot PH_3 \xrightarrow{-45^\circ} BH_3 + BH_2Br \cdot PH_3$$

$$BH_2Br \cdot PH_3 \xrightarrow{>0^\circ} \frac{1}{n}(-BHBr \cdot PH_2-)_n + H_2$$

At −63°, $BH_3 \cdot PH_3$ and $BH_2Br \cdot PH_3$ in a ratio of 1:1 are formed with a two-fold excess of PH_3[*175b*]:

$$BrB_2H_5 \xrightarrow[+PH_3]{-78^\circ} BH_3BH_2Br \cdot PH_3 \xrightarrow[+PH_3]{-63^\circ} BH_3 \cdot PH_3 + BH_2Br \cdot PH_3$$

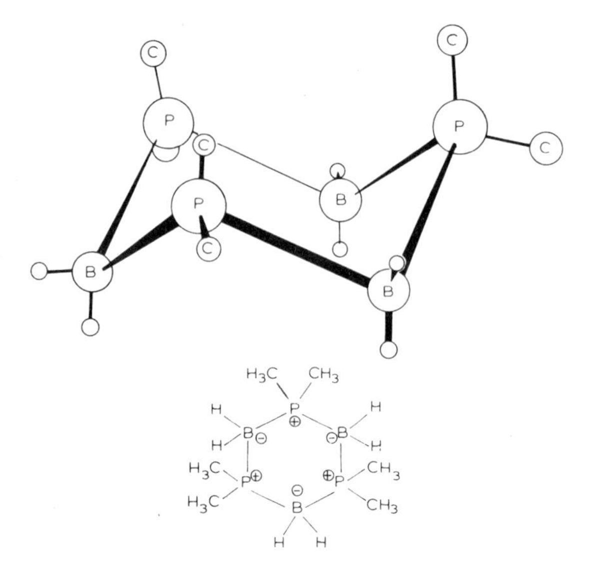

Fig. 4.17. Steric and valence structure of $[BH_2P(CH_3)_2]_3$. Structural parameters of the chair-shaped ring [*263*]: P–B = 1.935 Å; P–C = 1.837 Å.

Above 0°, the products are diborane, phosphane, $(-BHBr\cdot PH_2-)_n$ and H_2 [*175b*]

$$BH_3\cdot PH_3 \xrightarrow{0°} \tfrac{1}{2}(BH_3)_2 + PH_3$$

$$BH_2Br\cdot PH_3 \xrightarrow{0°} \frac{1}{n}(-BHBr\cdot PH_2-)_n + H_2$$

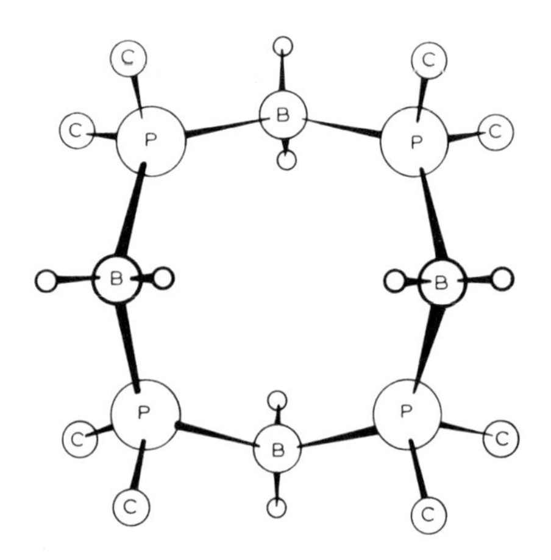

Fig. 4.18. Steric structure of $[BHP(CH_3)_2]_4$. Structural parameters of the corrugated 8-membered ring [*263*]:

P–B = 2.08 ± 0.05 Å ∡BPB = 125 ± 1°
P–C = 1.84 ± 0.04 Å ∡PBP = 104 ± 2°

11.2 Reaction of $(BH_3)_2$ with P_2H_4 and $P_2(CH_3)_4$

The reaction of diborane with diphosphane at $-78°$ forms diphosphane-bis(borane)[*47*]:

$$(BH_3)_2 + H_2PPH_2 \xrightarrow{-78°} H_3B{\cdot}H_2PPH_2{\cdot}BH_3$$

At 20°, PH_3, H_2 and a solid substance $BPH_{3.5}$ are formed[*47*]. The individual reaction steps can be followed by pyrolysis of the 1,2-adduct partially methylated at the P atoms, which is less temperature-sensitive. Thus, diborane reacts with tetramethyldiphosphane[*110*] (preparation in accordance with the equation $Me_2NPMe_2 + Me_2PH \rightarrow Me_2PPMe_2 + Me_2NH$) to form tetramethyldiphosphane-borane and tetramethyldiphosphanebis(borane)[*110*]:

$$(BH_3)_2 + 2Me_2PPMe_2 \xrightarrow{-60 \text{ to } -8°} 2BH_3{\cdot}Me_2PPMe_2$$

$$(BH_3)_2 + Me_2PPMe_2 \xrightarrow{-44 \text{ to } -5°} H_3B{\cdot}Me_2PPMe_2{\cdot}BH_3$$

On being heated, $BH_3{\cdot}Me_2PPMe_2{\cdot}BH_3$ yields, with loss of H_2, almost exclusively trimeric or tetrameric $(BH_2PMe_2)_n$; $BH_3{\cdot}Me_2PPMe_2$ yields Me_2PH and trimeric or tetrameric $(BH_2PMe_2)_n$[*110*]:

$$BH_3{\cdot}Me_2PPMe_2 \xrightarrow{174°} \frac{1}{n}[BH_2PMe_2]_n + Me_2PH \quad n = 3+4\ (47\%),\ x\ (\text{remaining } \%)$$

$$BH_3{\cdot}Me_2PPMe_2{\cdot}BH_3 \xrightarrow{170-200°} \frac{2}{n}[BH_2PMe_2]_n + H_2 \quad n = 3\ (75\%),\ 4\ (15\%)$$

$(BH_2PMe_2)_3$ reacts with HF yielding $(BF_2PMe_2)_3$[*264*].

11.3 Reaction of $(BH_3)_2$ with PF_3, $(C_6H_5)_3\overset{\oplus}{P}\overset{\ominus}{C}H_2$ or MPR_2

Diborane or diborane-d_6 reacts with phosphorus trifluoride with the primary formation of trifluorophosphaneborane $BH_3{\cdot}PF_3$ or $BD_3{\cdot}PF_3$[*576*]:

$$(BH_3)_2 + 2PF_3 \xrightarrow[4 \text{ days}]{20°,\ 8 \text{ atm}} 2BH_3{\cdot}PF_3 \xrightarrow[20°,\ \text{slow}]{\Delta} H_2,\ \text{low-volatility products}$$

With F_2PPF_2, B_2H_6 gives the relatively stable $F_2PPF_2{\cdot}BH_3$ which undergoes slow decomposition[*532d*]:

$$F_2PPF_2 + \tfrac{1}{2}(BH_3)_2 \xrightarrow{25°} F_2PPF_2{\cdot}BH_3 \xrightarrow[\text{slow}]{} PF_3{\cdot}BH_3 + \frac{1}{n}(PF)_n$$

Diborane reacts with $(CF_3)_2PH$ or $(CF_3)_2PF$ with the formation of (polymeric) $BH_2P(CF_3)_2$. The unstable intermediate stage $BH_3{\cdot}PH(CF_3)_2$ and the apparently somewhat more stable $BH_3{\cdot}PF(CF_3)_2$ cannot be isolated[*113*].

References p. 355

$$(BH_3)_2 + 2(CF_3)_2PH \xrightarrow{(CH_3)_2O\text{-catalysis}} \frac{2}{n}[BH_2P(CF_3)_2]_n + 2H_2$$

$$(BH_3)_2 + 2(CF_3)_2PF \xrightarrow[\Delta]{>0^\circ} \frac{2}{n}[BH_2P(CF_3)_2]_n + 2HF$$

and also $(CF_3)_2PH$, BF_3, H_2

At 25° gaseous PHF_2 and B_2H_6 form the unusually stable $BH_3 \cdot PHF_2$[*625c*].

With diborane the strong donor methylenetriphenylphosphorane $Ph_3\overset{\oplus\ominus}{PCH_2}$ forms methylenetriphenylphosphoraneborane[*297, 303*]:

$$(BH_3)_2 + 2Ph_3\overset{\oplus\ominus}{PCH_2} \xrightarrow[0^\circ]{\text{diethyl ether}} \overset{\ominus}{B}H_3CH_2\overset{\oplus}{P}Ph_3$$

On refluxing $BH_3CH_2PPh_3$ in chlorobenzene (b.p. 131°) or decalin (b.p. 185–190°), the following reactions occur[*428e*]:

$$BH_3\overset{\ominus}{C}H_2\overset{\oplus}{P}Ph_3 \xrightarrow{131^\circ} CH_3\overset{\ominus}{B}H_2\overset{\oplus}{P}Ph_3 \xrightarrow{190^\circ} \tfrac{2}{3}\overset{\ominus}{B}H_3\overset{\oplus}{P}Ph_3 + \tfrac{1}{3}PPh_3 + \tfrac{1}{3}B(CH_3)_3$$

The chemical properties of $BH_3 \cdot PF_3$ are characterized by the weakness of the B–P bond as compared with that in $BH_3 \cdot PH_3$. Thus, PF_3 can easily be displaced nucleophilically by the stronger donor NMe_3 with the formation of $BH_3 \cdot NMe_3$. Further action of NMe_3 on PF_3 yields "gases" and "a solid"[*576*]:

$$BH_3 \cdot PF_3 + NMe_3 \xrightarrow[20^\circ]{-BH_3 \cdot NMe_3} PF_3 \xrightarrow{+NMe_3} MeF(?) + Me_2NPF_2(?)$$

Similar to $BH_3 \cdot CO$, the adducts BH_3PF_3, $BH_3 \cdot PF_2CF_3$, and $BH_3 \cdot PF(CF_3)_2$ decompose into diborane and the corresponding phosphane[*131a*].

$BH_3 \cdot PF_3$ reacts with ammonia with the replacement of F by NH_2; $BH_3 \cdot P(NH_2)_3$ is formed[*417, 576*]:

$$BH_3 \cdot PF_3 + 6NH_3 \longrightarrow BH_3 \cdot P(NH_2)_3 + 3NH_4F$$

The reactions with partially alkylated ammonia occupy an intermediate position (with both nucleophilic displacement and replacement of F by amino groups). The reaction with methylamine resembles that with ammonia, but in this case partially substituted phosphaneboranes can be isolated, as well as the per-substituted compounds[*423*]:

$$BH_3 \cdot PF_3 + 2MeNH_2 \xrightarrow[\text{diethyl ether}]{-111^\circ} BH_3 \cdot PF_2(NHMe) + MeNH_3F$$

$$BH_3 \cdot PF_2(NHMe) + 2MeNH_2 \xrightarrow[\text{diethyl ether}]{25^\circ} BH_3 \cdot PF(NHMe)_2 + MeNH_3F$$

$$BH_3 \cdot PF(NHMe)_2 + 2MeNH_2 \xrightarrow[\text{liq. } MeNH_2]{25^\circ} BH_3 \cdot P(NHMe)_3 + MeNH_3F$$

Dimethylamine reacts with replacement of F (eqns. 1 and 2) and with displacement of the donor (eqn. 3) [*423*]:

$$BH_3 \cdot PF_3 + \tfrac{3}{2} Me_2NH \xrightarrow[\text{without a solvent}]{-78°} BH_3 \cdot PF_2(NMe_2) + \tfrac{1}{2}[Me_2NH_2]HF_2 \quad (1)$$

$$BH_3 \cdot PF_2(NMe_2) + \tfrac{3}{2} Me_2NH \xrightarrow[\text{without a solvent}]{25°} BH_3 \cdot PF(NMe_2)_2 + \tfrac{1}{2}[Me_2NH_2]HF_2 \quad (2)$$

$$BH_3 \cdot PF_3 + Me_2NH \xrightarrow[\text{diethyl ether}]{-196° \rightarrow 25°} BH_3 \cdot NHMe_2 + \text{other products} \quad (3)$$

The donor strength of the donors in borane–donor adducts increases in the sequence [*423*]: PF_3 < $(MeNH)PF_2$ ~ Me_2NPF_2 < Me_3N ~ Me_2NH < $(MeNH)_2PF$ ~ $(Me_2N)PF$. Figure 4.19 shows the structure of $BH_3 \cdot P(NH_2)_3$.

B_2H_6 and KPH_2 [*769a*] or $LiPEt_2$ react to give $KPH_2(BH_3)_2$ and $LiPEt_2(BH_3)_2$ [*235c*] respectively.

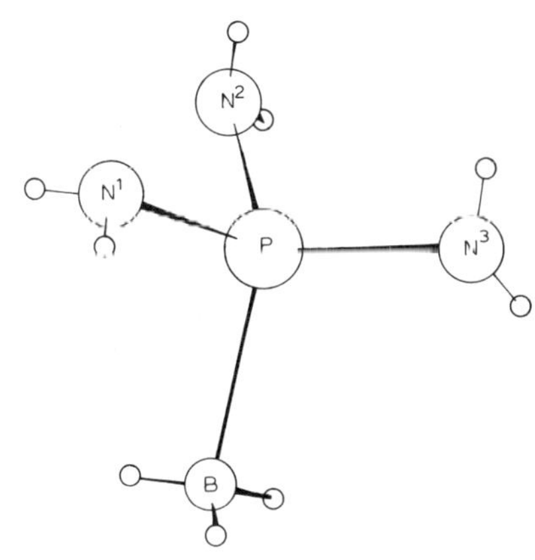

Fig. 4.19. Steric structure of $BH_3 \cdot P(NH_2)_3$. Structural parameters [*553*]:

P–N = 1.650–1.658 Å; ∡N¹PN² = [illegible]; ∡PNH = [illegible]
P–B = 1.887 Å; ∡N¹PN³ = 100.1°; ∡HNH = 99–145°
N–H = 0.8–1.0 Å; ∡N²PN³ = 116.8°; ∡PBH = 95–111°
B–H = 1.1–1.2 Å; ∡N¹PB = 123.3°; ∡HBH = 100–124°
∡N²PB = 108.4°
∡N³PB = 109.0°

11.4 Nucleophilic displacement of donors on the boron atom by P-containing donors

The very strong donors $P(NMe_2)_3$, $P(NBu^n{}_2)_3$, $P[\overline{N(CH_2)_3CH_2}]_3$, $P(OR')_3$, and $Ph_3\overset{\oplus\ominus}{PCHR''}$ displace the comparatively weaker donor NMe_3 in $BH_3 \cdot NMe_3$ or $BH_2R''' \cdot NMe_3$ [*303, 610, 611, 688*]:

$$BH_3 \cdot NMe_3 + P(NR_2)_3 \xrightarrow{20-100°} BH_3 \cdot P(NR_2)_3 + NMe_3$$

$$BH_3 \cdot NMe_3 + P(OR')_3 \longrightarrow BH_3 \cdot P(OR')_3 + NMe_3$$

$R' = CH_3, C_2H_5$, i-C_3H_7, n-C_4H_9, $ClCH_2CH_2$, $\overline{CH_2(CH_2)_4CH}$, C_6H_5

$$BH_3{\cdot}NMe_3 + PH_3\overset{\oplus\ominus}{P}CHR'' \xrightarrow[100°]{\text{diglyme or diethylether}} \overset{\ominus}{B}H_3CHR''\overset{\oplus}{P}Ph_3 + NMe_3$$

$R'' = H, CH_3$, n-C_3H_7, n-C_5H_{11}, C_6H_5

$$BH_2R'''{\cdot}NMe_3 + Ph_3\overset{\oplus\ominus}{P}CH_2 \xrightarrow[80°]{\text{benzene}} \overset{\ominus}{B}H_2R'''CH_2\overset{\oplus}{P}Ph_3 + NMe_3$$

$R''' = CH_3$, 2-C_4H_9, *tert*-C_4H_9, C_6H_5

These phosphaneboranes can also be obtained in many cases in excellent yields by the reaction of phosphane and sodium boranate with acids, for example[*610*]:

$$NaBH_4 + P(NMe_2)_3 + CO_2 \longrightarrow BH_3{\cdot}P(NMe_2)_3 + HCOONa$$

$$NaBH_4 + P(NMe_2)_3 + CH_3COOH \longrightarrow BH_3{\cdot}P(NMe_2)_3 + CH_3COONa + H_2$$

$$NaBH_4 + P(NMe_2)_3 + CH_3COCH_3 + H_2O \longrightarrow BH_3{\cdot}P(NMe_2)_3 + NaOH + C_3H_7OH$$

$$NaBH_4 + P(NMe_2)_3 + HCl \longrightarrow BH_3{\cdot}P(NMe_2)_3 + NaCl + H_2$$

To prepare phosphaneboranes $BH_3{\cdot}PR_3$ or phosphanoboranes BH_2PR_2, $NaBH_4$ can be used simultaneously as a hydrogenating agent, for example:

$$NaBH_4 + ClPPh_2 \xrightarrow{-NaCl} [BH_3{\cdot}HPPh_2] \xrightarrow{-H_2} \tfrac{1}{3}[BH_2PPh_2]_3$$

11.5 Reactions of $(BH_3)_2$ with As and Sb compounds

The reactions of diborane with arsane, AsH_3, or stibane, SbH_3, formally resemble those with phosphane, PH_3. However, arsaneboranes (and to a far greater extent stibaneboranes), are more thermally unstable. With AsH_3 or $MeAsH_2$, diborane does not form an arsaneborane $BH_3{\cdot}AsH_3$ or $BH_3{\cdot}AsH_2Me$ but (together with products having a lower hydrogen content) arsanoboranes[*751*]:

$$(BH_3)_2 + 2AsH_3 \xrightarrow{20°} 2BH_2AsH_2 + 2H_2$$

$$(BH_3)_2 + 2MeAsH_2 \xrightarrow{-78°} [2(BH_3{\cdot}AsH_2Me)] \xrightarrow{20°} 2BH_2AsHMe + 2H_2$$

With dimethyl- or trimethylarsane, diborane gives stable adducts. Dimethylarsaneborane, $BH_3{\cdot}AsHMe_2$, dissociates into its components to the extent of 96% at 24°; above 50° such a mixture of BH_3 and $AsHMe_2$ splits off H_2 and forms trimeric or tetrameric dimethylarsanoborane. Trimethylarsaneborane forms no arsanoborane[*751*]:

$$(BH_3)_2 + 2AsHMe_2 \underset{24°}{\overset{-78°}{\rightleftharpoons}} 2BH_3{\cdot}AsHMe_2 \xrightarrow{>50°} \frac{2}{n}[BH_2AsMe_2]_n + 2H_2 \ (n = 2, 3)$$

$$(BH_3)_2 + 2AsMe_3 \xrightarrow{-78^\circ} 2BH_3 \cdot AsMe_3 \xrightarrow[\Delta]{120^\circ} AsMe_3,\ H_2,\ \text{polyboranes}$$

Sodium boranate reacts with dimethylantimony bromide, Me_2SbBr, to give dimethylstibanoborane, BH_2SbMe_2. Above −78°, this decomposes with evolution of hydrogen[*116*].

12. Reactions of borane or organylboranes with C compounds

12.1 Addition of B–H to C=C

The C=C bond in alkenes adds the B–H bond of boranes [$(BH_3)_2$, nascent BH_3, donor·BH_3, $(RBH_2)_2$, $(R_2BH)_2$] with the formation of a B–C bond:

$$\geqslant C{=}C\leqslant + H{-}B\langle \longrightarrow H{-}\overset{|}{\underset{|}{C}}{-}\overset{|}{\underset{|}{C}}{-}B\langle$$

Examples of these reactions are given in Tables 4.2, 4.3, and 4.4.

12.2 Mechanism of the addition of B–H to C=C

The addition of the B–H bond to C=C is reversible[*425, 622*]. It takes place by a four-centre *cis*-addition[*93,94*]. While the rate of the diborane-alkene reaction in ether takes place much too rapidly for kinetic measurements, the reaction of dialkylborane, di(1,2-dimethylpropyl)borane, with alkenes in tetrahydrofuran has been followed reproducibly[*87*]. It is first-order on the borane and first order on the alkene. Like the monoalkylboranes, dialkylboranes are dimeric in tetrahydrofuran. Their structure probably resembles that of tetramethyldiborane with hydrogen bridges. On the other hand, diborane is present in tetrahydrofuran as the monomeric THF–borane adduct.

The following four mechanisms are conceivable, and the third and the (similar) fourth are probable [$R = (CH_3)_2CH\overset{|}{C}HCH_3$].

(1) Dissociation of the dimer into monomer, followed by a *cis*-addition to the C=C bond. This mechanism would require 3/2-order kinetics:

$$R_2B(\mu\text{-}H)_2BR_2 \underset{}{\overset{\text{fast}}{\rightleftharpoons}} 2\ R_2BH$$

$$\geqslant C{=}C\leqslant + R_2B{-}H \overset{\text{slow}}{\rightleftharpoons} \begin{matrix} -C{\cdots}C- \\ \vdots \quad \vdots \\ R_2B{\cdots}H \end{matrix} \longrightarrow R_2B{-}\overset{|}{\underset{|}{C}}{-}\overset{|}{\underset{|}{C}}{-}H$$

References p. 355

TABLE 4.2 ADDITION OF BH_3, BH_2R, BHR_2, BH_2Cl, $BHCl_2$ AND BH_2SBu^n TO ALKENES

Borane	*Alkene*	*Reaction products*[a]	*Ref.*
$(BH_3)_2$	$(CH_3)_2C{=}C(CH_3)_2$	R^1BH_2	80, 85, 865
$(BH_3)_2$	$C_6H_5CH{=}CH_2$	81% $(C_6H_5CH_2CH_2)_3B$, 19% $(C_6H_5CHCH_3)_3B$	864
$(BH_3)_2$	$CH_3CH_2CH_2CH{=}CHCH_3$	50% $(CH_3CH_2CH_2CH_2CHCH_3)_3B$, 50% $(CH_3CH_2CH_2CHCH_2CH_3)_3B$	864
$(BH_3)_2$	*cis*-$(CH_3)_2CH_2CH{=}CHCH_3$	62% $[(CH_3)_2CH_2CH_2CHCH_3]_3B$, 38% $[(CH_3)_2CH_2CHCH_2CH_3]_3B$	864
$(BH_3)_2$	*trans*-$(CH_3)_2CH_2CH{=}CHCH_3$	57% $[(CH_3)_2CH_2CH_2CHCH_3]_3B$, 43% $[(CH_3)_2CH_2CHCH_2CH_3]_3B$	864, 865
$(n\text{-}C_3H_7)_4B_2H_2$	$CH_2{=}CHOC_4H_9$-n	$(n\text{-}C_3H_7)_2BCH_2CH_2OC_4H_9$-n	508b
BH_2R^1	$CH_3(CH_2)_3CH{=}CH_2$	$R^1[CH_3(CH_2)_5]_2B$	865
BH_2R^1	$CH_3(CH_2)_2C(CH_3){=}CH_2$	92% $R^1B[CH_3(CH_2)_2CH(CH_3)CH_2]B$	865
BH_2R^1	*cis*-$CH_3(CH_2)_2CH{=}CHCH_3$	96% $R^1(C_6H_{13})_2R^1$, together with B–H/B–R exchange	865
BH_2R^1	*trans*-$CH_3(CH_2)_2CH{=}CHCH_3$	84% $R^1(C_6H_{13})_2B$, together with B–H/B–R exchange	865
BH_2R^1	*cis*-$(CH_3)_2CH_2CH{=}CHCH_3$	95% $R^1(C_6H_{13})_2B$, together with B–H/B–R exchange	865
BH_2R^1	*trans*-$(CH_3)_2CH_2CH{=}CHCH_3$	80% $R^1(C_6H_{13})_2B$	865
BH_2R^1	$CH_2(CH_2)_3CH{=}CH$	90% $R^1(C_6H_{11})_2B$	865
BH_2R^1	$C_6H_5CH{=}CH_2$	96% $R^1H(C_6H_5CH_2CH_2)B$, 4% $R^1H(C_6H_5CHCH_3)B$	865
BHR_2^2	$C_6H_5CH{=}CH_2$	98% $R_2^2(C_6H_5CH_2CH_2)B$, 2% $R_2^2(C_6H_5CHCH_3)B$	95, 864
BHR_2^3	$C_6H_5CH{=}CH_2$	99% $R_2^3(C_6H_5CH_2CH_2)B$, 1% $R_2^3(C_6H_5CHCH_3)B$	864
BHR_2^3	$CH_3CH_2CH_2CH{=}CHCH_3$	66% $R_2^3(CH_3CH_2CH_2CH_2CHCH_3)B$, 34% $R_2^3(CH_3CH_2CH_2CHCH_2CH_3)B$	864
BHR_2^4	$C_6H_5CH{=}CH_2$	99% $R_2^4(C_6H_5CH_2CH_2)B$, 1% $R_2^4(C_6H_5CHCH_3)B$	864
BHR_2^4	$CH_3CH_2CH_2CH{=}CHCH_3$	67% $R_2^4(CH_3CH_2CH_2CH_2CHCH_3)B$, 33% $R_2^4(CH_3CH_2CH_2CHCH_2CH_3)B$	864
BHR_2^5	$C_6H_5CH{=}CH_2$	99% $R_2^5(C_6H_5CH_2CH_2)B$, 1% $R_2^5(C_6H_5CHCH_3)B$	864
BHR_2^5	$CH_3CH_2CH_2CH{=}CHCH_3$	62% $R_2^5(CH_3CH_2CH_2CH_2CHCH_3)B$, 38% $R_2^5(CH_3CH_2CH_2CHCH_2CH_3)B$	864
BH_2Cl	$C_6H_5CH{=}CH_2$	78% $C_6H_5CH_2CH_2Cl$; 9% $CH_3CH(C_6H_5)Cl$	583a
$BHCl_2$	$CH_2{=}CH_2$	$(CH_3CH_2)Cl_2B$	487
$BHCl_2$	$CH_3CH{=}CH_2$	$[CH_3CHCH_3]Cl_2B$	487
$BHCl_2$	$(CH_3)_2C{=}CH_2$	$[(CH_3)_2CCH_3]B$	487
$BHCl_2$	$CH_2(CH_2)_3CH{=}CH$	$[CH_2(CH_2)_3CH_2CH]Cl_2B$	487
$(BH_2SBu^n)_x$	$CH_2{=}CH_2$	15% $(Bu^nS)(CH_3CH_2)_2B$, 35% $(CH_3CH_2)_3B$, $[(Bu^nS)_2BH]_2$	512
$(BH_2SBu^n)_x$	$CH_2{=}CHCH_2CH_2CH{=}CH_2$	76% $Bu^nSB(CH_2CH_2CH_2CH_2CH_2CH_2)$	512

[a] In the reaction products

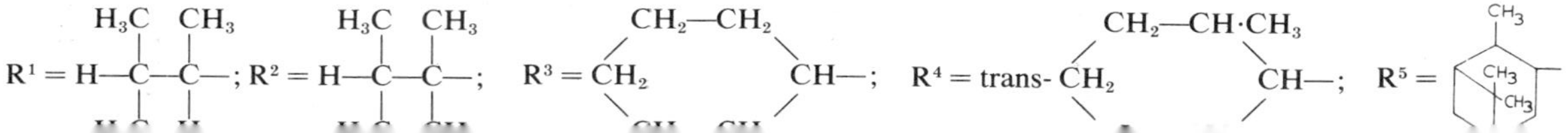

TABLE 4.3

ADDITION OF BH_3 AND BH_2R (AS AMINE ADDUCTS) TO ALKENES

Borane	*Alkene*	*Reaction products*	*Ref.*
$BH_3 \cdot NMe_3$	$CH_3(CH_2)_5CH{=}CH_2$	92.7% $[CH_3(CH_2)_7]_3B$	17
$BH_3 \cdot NEt_3$	$(CH_3)_2C{=}CH_2$	80.0% $[(CH_3)_2CHCH_2]_3B$	17
$BH_3 \cdot NEt_3$	$CH_3(CH_2)_3CH{=}CH_2$	91.3% $[CH_3(CH_2)_5]_3B$	17
$BH_3 \cdot NEt_3$	$CH_3CH_2CH_2CH{=}CHCH_3$	94.7% $[CH_3(CH_2)_5]_3B$ (n-!)	17
$BH_3 \cdot NEt_3$	$CH_2(CH_2)_3CH{=}CH$	77.6% $(C_6H_{11})_3B$	17
$BH_3 \cdot NEt_3$	$CH_3(CH_2)_5CH{=}CH_2$	85.4% $[CH_3(CH_2)_7]_3B$	17
$BH_3 \cdot NBu_3$	$CH_3(CH_2)_5CH{=}CH_2$	82.5% $[CH_3(CH_2)_7]_3B$	17
$BH_3 \cdot py$	$CH_3(CH_2)_3CH{=}CH_2$	87.3% $[CH_3(CH_2)_5]_3B$	17
$BH_2Bu^t \cdot NMe_3$	$CH_2{=}CH_2$	35% $(C_2H_5)_2Bu^tB$	302
$BH_2Bu^t \cdot NMe_3$	$CH_3CH{=}CH_2$	88% $(CH_3CH_2CH_2)_2Bu^tB$	302
$BH_2Bu^t \cdot NMe_3$	$CH_3CH_2CH{=}CH_2$	90% $(CH_3CH_2CH_2)_2Bu^tB$	302
$BH_2Bu^t \cdot NMe_3$	$(CH_3)_2C{=}CH_2$	85% $[(CH_3)_2CHCH_2]_2Bu^tB$	302
$BH_2Bu^t \cdot NMe_3$	$CH_2{=}CHCH{=}CH_2$	60% $CH_2(CH_2)_2CH_2BBu^t$	302
$BH_2Bu^t \cdot NMe_3$	$CH_2{=}C(CH_3)CH{=}CH_2$	55% $CH_2CH(CH_3)CH_2CH_2BBu^t$	302
$BH_2Bu^t \cdot NMe_3$	$(CH_2CH)_2(CH_3)_2Si$	58% $CH_2CH_2Si(CH_3)_2CH_2CH_2BBu^t$	302

TABLE 4.4

ADDITION OF NASCENT BORANE ($NaBH_4 + BF_3$) TO ALKENES

Alkene	*Reaction products*	*Ref.*
$CH_3(CH_2)_2CH{=}CH_2$	$(C_5H_{11})_3B$	86
$(CH_3)_2C{=}C(CH_3)_2$	$[(CH_3)_2CHC(CH_3)_2]BH_2$	865
$(CH_3)_3CCH{=}C(CH_3)_2$	$(C_8H_{17})_3B$	86
$CH_2(CH_2)_3CH{=}CH$	$(C_6H_{11})_3B$	86
$C_6H_5CH{=}CH_2$	$(C_6H_5C_2H_4)_3B$	86

(2) Without the dissociation stage, dimeric borane could add to two molecules of alkene in an (improbable) trimolecular reaction. The kinetics would be of third order:

$$\mathrm{R_2B(\mu\text{-}H)_2BR_2} + 2\ \mathrm{C{=}C} \underset{}{\overset{\text{slow}}{\rightleftharpoons}} [\mathrm{R_2B}\cdots\mathrm{H}\cdots\mathrm{C{=}C};\ \mathrm{BR_2}\cdots\mathrm{H}\cdots\mathrm{C{=}C}] \xrightarrow{\text{fast}} 2\ \mathrm{R_2B{-}C{-}C{-}H}$$

(3) The reaction would again take place without a dissociation step but by second-order kinetics if the B–H bridge slowly added to the first alkene molecule, which would cause rapid rupture of the bridge in R_2–BCC–H and R_2BH.

R_2BH could then quickly dimerize or add a second alkene molecule:

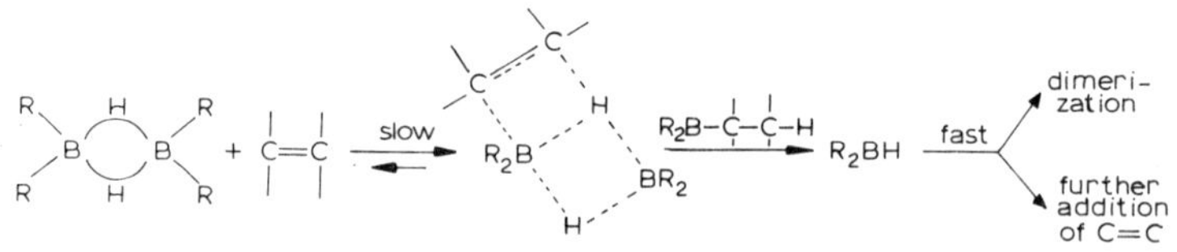

(4) Mechanism according to (3) but with the additional support of the detachment of R_2B–C–C–H and R_2BH from the disappearing bridge by the solvent (donor):

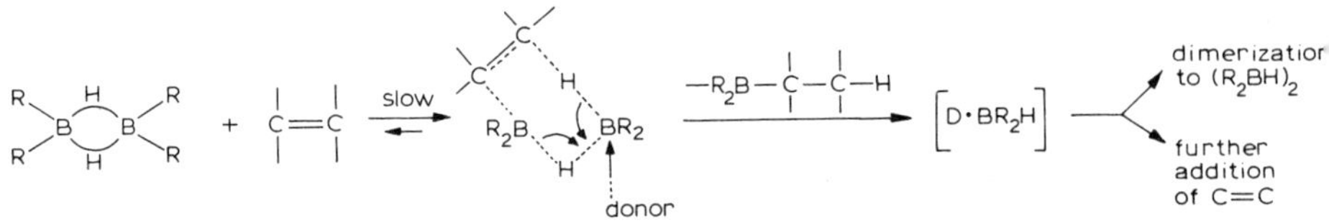

Among many other examples, the reaction of β-pinene with borane by *cis*-addition shows that the B–H bond approaches the less hindered side of the double bond so that the compound formed boron is likewise bound to the less hindered side. β-Pinene forms tri(*cis*-myrtanyl)borane, although tri(*trans*-myrtanyl)borane is favoured thermodynamically (as shown by the isomerization of the *cis*-compound by heating) [*94*, *99*]:

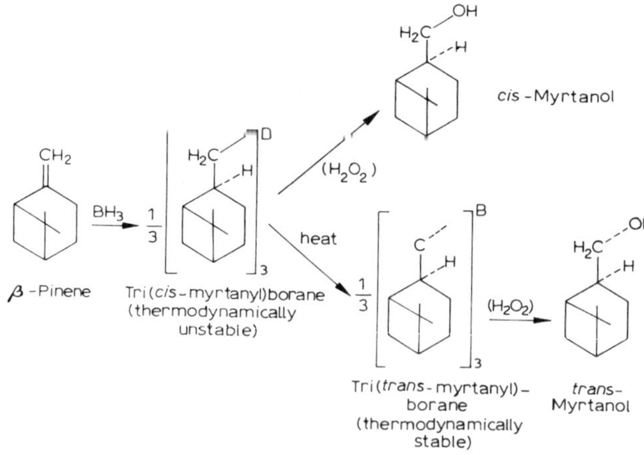

12.3 *Inductive directive effect in the addition of B–H to C=C*

The B–H bond in boranes adds to a C=C bond in accordance with its polarization (δ+ B–H δ–) by the anti-Markovnikov rule[26] (halogen-substituted boranes which contain protonic hydrogen constitute an exception):

$$R\rightarrow \overset{H}{C}{=}CH_2 \xrightarrow{+H-B<} \left| \begin{matrix} R\rightarrow \overset{\delta+}{C}\overset{\delta-}{=\!=\!=}CH_2 \\ \delta-H\text{---}B\delta+ \end{matrix} \right| \longrightarrow H{-}CHR{-}CH_2{-}B<$$

Correspondingly, in the case of terminal double bonds the boron is directed to the terminal carbon because of the inductive effect and in the case of internal double bonds it goes to the more negative carbon of the double bond. In the following examples the percentages of alkylated boranes were determined on the basis of the corresponding alcohols determined on oxidation with alkaline H_2O_2 [93]. For migration of the boron atom to the 1-position see [96].

Terminal, monosubstituted alkenes, *$RHC{=}CH_2$* (pronounced inductive directing effect):

$$CH_3CH_2CH_2CH_2CH{=}CH_2 \xrightarrow{+>BH} \begin{cases} 94\% \; CH_3CH_2CH_2CH_2CH_2CH_2B< \\ 6\% \; CH_3CH_2CH_2CH_2CH(B<)CH_3 \end{cases}$$

$$(CH_3)_3CCH{=}CH_2 \xrightarrow{+>BH} \begin{cases} 94\% \; (CH_3)_3CCH_2CH_2B< \\ 6\% \; (CH_3)_3CCH(B<)CH_3 \end{cases}$$

$$CH_3O{-}C_6H_4{-}CH{=}CH_2 \xrightarrow{+>BH} \begin{cases} 91\% \; CH_3OC_6H_4CH_2CH_2B< \\ 9\% \; CH_3OC_6H_4CH(B<)CH_3 \end{cases}$$

strong donor substituent on the phenyl ring)

$$\left[C_6H_5{-}\overset{\oplus}{C}H{=}\overset{\ominus}{C}H_2 \rightleftarrows C_6H_5{-}\overset{\ominus}{C}H{=}\overset{\oplus}{C}H_2 \right] \xrightarrow{+>BH} \begin{cases} 80\% \; C_6H_5CH_2CH_2B< \\ 20\% \; C_6H_5CH_2(B<)CH_3 \end{cases}$$

medium donor substituent on the phenyl ring)

Taken together, the Markovnikov rules comprise a single principle: in the addition of hydrogen halide to an unsymmetrical alkene the (positively polarized) hydrogen atom of the hydrogen halide adds to the negatively polarized carbon atom of the C=C bond. The (negatively polarized) halogen atom adds to the positively polarized carbon atom of the double bond. For example:

$$R\rightarrow \overset{H}{C}{=}CH_2 + \overset{\delta+}{H}{-}\overset{\delta-}{X} \longrightarrow \left| \begin{matrix} R{-}\overset{\delta+}{C}\overset{\delta-}{=\!=\!=}CH_2 \\ X\text{---}H \end{matrix} \right| \longrightarrow X{-}CHR{-}CH_2{-}H$$

The anti-Markovnikov rule results from the reversed polarization of the hydrogen atom (H δ−) in a B–H bond. Here the H atom of the boron hydride adds to the more positive carbon atom of the C=C bond. For an example, see above.

$$ClC_6H_4CH{=}CH_2 \xrightarrow{+\ >BH} \begin{cases} 65\% \rightarrow ClC_6H_4CH_2CH_2B< \\ 35\% \rightarrow ClC_6H_4CH(B<)CH_3 \end{cases}$$

(weak donor substituent on the phenyl ring)

Terminal, disubstituted alkenes, $R_2C{=}CH_2$ (highly pronounced inductive directing effect):

$$(CH_3)(CH_3CH_2)C{=}CH_2 \xrightarrow{+\ >BH} \begin{cases} 99\% \rightarrow (CH_3CH_2)(CH_3)CHCH_2B< \\ 1\% \rightarrow (CH_3CH_2)(CH_3)C(B<)(CH_3) \end{cases}$$

$$C_6H_5C(CH_3){=}CH_2 \xrightarrow{+\ >BH,\ 100\%} C_6H_5CH(CH_3)CH_2B<$$

Internal disubstituted alkenes, $RCH{=}CHR$ (no pronounced inductive directing effect):

$$(CH_3)_3C{-}CH{=}CH{-}CH_3 \xrightarrow{+\ >BH} \begin{cases} 58\% \rightarrow (CH_3)_3CCH_2CH(B<)CH_3 \\ 42\% \rightarrow (CH_3)_3CCH(B<)CH_2CH_3 \end{cases}$$

$$CH_3{-}C_6H_9 \xrightarrow{+\ >BH} 51\%\ CH_3{-}C_6H_{10}{-}B< ,\quad 49\%\ CH_3{-}C_6H_{10}{-}B<$$

Trisubstituted alkenes, $R_2C{=}CHR$ (highly pronounced inductive directing effect):

$$(CH_3)C(CH)\!=\!C(CH_3)H \xrightarrow{+\ >BH} \begin{cases} 98\% \rightarrow (CH_3)_2CHCH(B<)(CH_3) \\ 2\% \rightarrow (CH_3)_2C(B<)CH_2CH_3 \end{cases}$$

The B–H bond in dichloroborane $BHCl_2$ adds still more readily to the C=C or the C≡C bond. The reaction takes place at 20° with or without a solvent, and obeys Markovnikov's rule because of the protonic nature of the hydrogen in $BHCl_2$ [*487*]:

$$BHCl_2 + R{\rightarrow}CH{=}CH_2 \longrightarrow \left[R\overset{\sigma}{-}\overset{\delta+}{C}H\overset{\sigma}{\cdots}\overset{\delta-}{C}H_2 \;\; Cl_2\overset{\delta-}{B}\cdots H^{\delta+}\right] \longrightarrow R{-}CH(BCl_2){-}CH_2H$$

12.4 *Steric directive effect of the alkene in the addition of B–H to C=C*

The steric directive effect of the alkene is low in the first borane addition step (transition $BH_3 \rightarrow RBH_2$). The overwhelming inductive effect in comparison with the less important steric influences on direction is shown by replacement of the smaller carbon by the larger silicon which, however, only slightly inhibits *cis*-addition[*687*]:

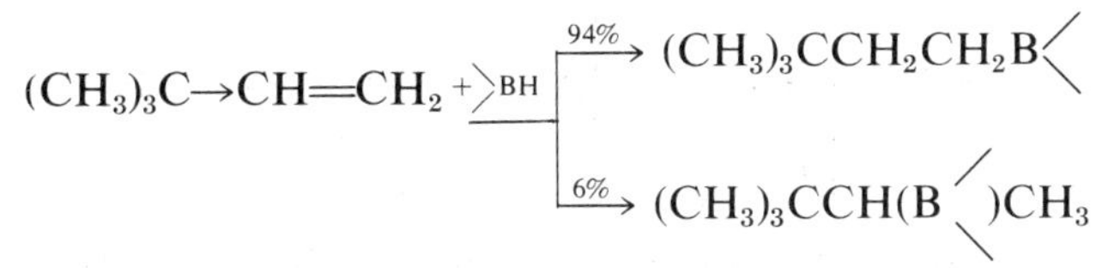

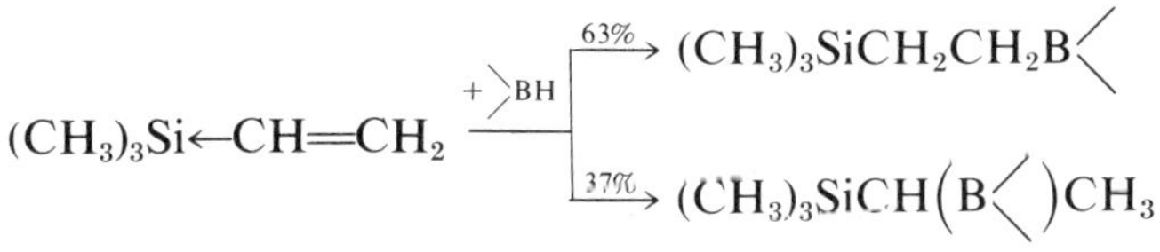

$$(CH_3)_3C{\rightarrow}CH{=}CH_2 + \rangle BH \longrightarrow \begin{cases} \xrightarrow{94\%} (CH_3)_3CCH_2CH_2B\langle \\ \xrightarrow{6\%} (CH_3)_3CCH(B\langle)CH_3 \end{cases}$$

$$(CH_3)_3Si{\leftarrow}CH{=}CH_2 \xrightarrow{+\rangle BH} \begin{cases} \xrightarrow{63\%} (CH_3)_3SiCH_2CH_2B\langle \\ \xrightarrow{37\%} (CH_3)_3SiCH(B\langle)CH_3 \end{cases}$$

Further addition of the RBH_2 formed to alkenes (conversion into R_2BH and R_3B) is however affected by the steric demands of the alkene. Most (terminal) alkenes react smoothly with diborane in diglyme at 25° within an hour to give the trialkylboron[*90*]:

$$\text{diglyme}\cdot BH_3 \xrightarrow[\text{(diglyme)}]{+RCH=CH_2} RCH_2CH_2BH_2 \xrightarrow[\text{(diglyme)}]{+RCH=CH_2} (RCH_2CH_2)_2BH$$

$$\xrightarrow[\text{(diglyme)}]{+RCH=CH_2} (RCH_2CH_2)_3B$$

Bulky alkenes, such as 2-methylbut-2-ene or 2,3-dimethylbut-2-ene, react under the same conditions, but only as far as the R_2BH or RBH_2 stage[*80, 85, 865*]:

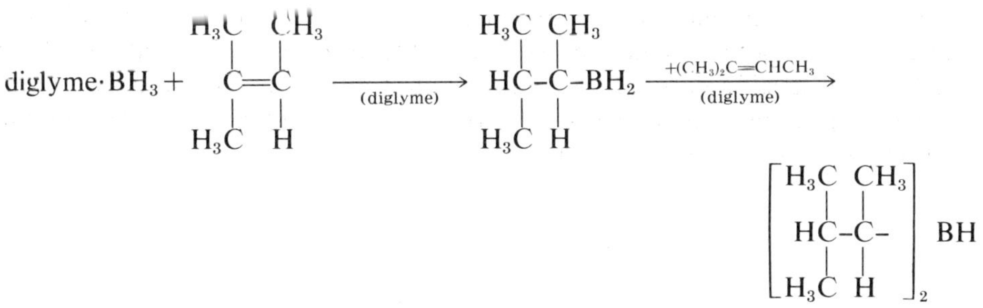

$$\text{diglyme}\cdot BH_3 + (H_3C)_2C{=}CH(CH_3) \xrightarrow{\text{(diglyme)}} HC(CH_3)(H_3C)\text{–}C(CH_3)(H)\text{–}BH_2 \xrightarrow[\text{(diglyme)}]{+(CH_3)_2C=CHCH_3} \left[HC(CH_3)(H_3C)\text{–}C(CH_3)(H)\text{–}\right]_2BH$$

bis(3-methyl-2-butyl)borane

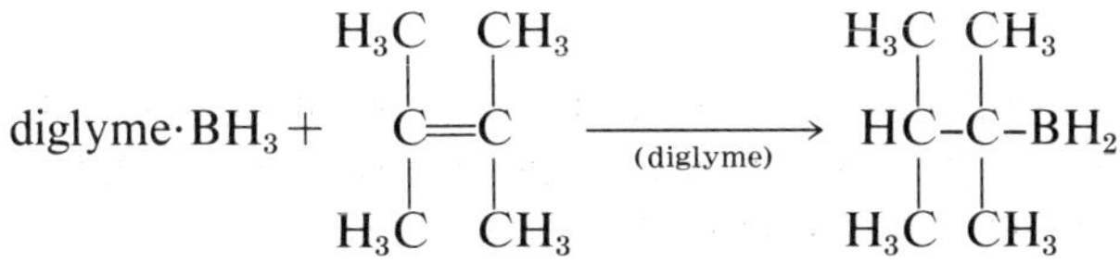

$$\text{diglyme}\cdot BH_3 + (H_3C)_2C{=}C(CH_3)_2 \xrightarrow{\text{(diglyme)}} HC(CH_3)(H_3C)\text{–}C(CH_3)_2\text{–}BH_2$$

2,3-dimethyl-2-butylborane

References p. 355

12.5 Steric directive effect of the alkylborane in the addition of B–H to C═C

The directive effect in the addition of boranes to alkenes depends on steric factors of the borane or alkylborane. With increasing substitution and bulkiness of the substituents in the borane the directive effect rises in the direction of the substitution of the B on the least substituted carbon of the alkene:

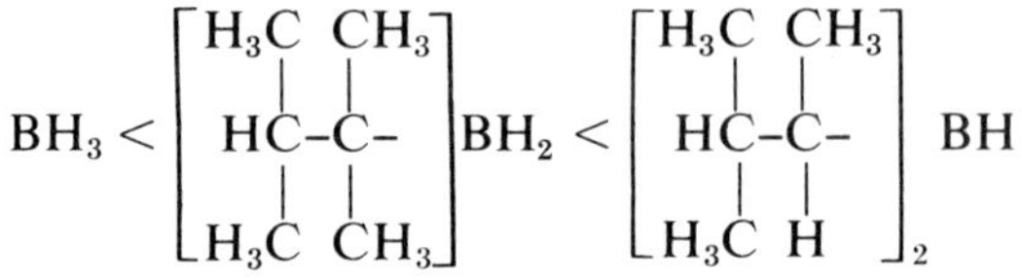

$$BH_3 < \left[\begin{array}{c} H_3C\ \ CH_3 \\ |\quad\ | \\ HC{-}C{-} \\ |\quad\ | \\ H_3C\ \ CH_3 \end{array}\right] BH_2 < \left[\begin{array}{c} H_3C\ \ CH_3 \\ |\quad\ | \\ HC{-}C{-} \\ |\quad\ | \\ H_3C\ \ H \end{array}\right]_2 BH$$

2,3-dimethyl-2-butylborane bis(3-methyl-2-butyl)borane

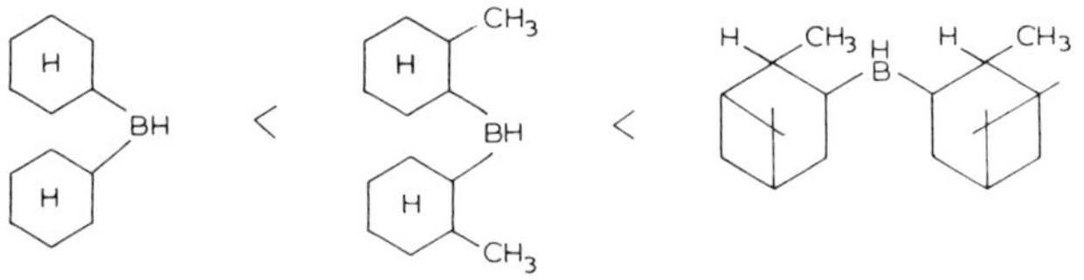

dicyclohexyl-borane < bis-(*trans*-2-methyl-cyclohexyl)-borane < diisopinocampheylborane

Thus, the addition of boranes (BH_3 or RBH_2 or R_2BH) to styrene (R′) gives phenylethylboranes, $(R'H)_3B$ or (R′H)RBH (!) or $(R'H)R_2B$, which after subsequent non-isomerizing oxidation to the alcohol yield the following proportions of 2-phenylethanol (and 1-phenylethanol), respectively. With $(BH_3)_2$: 81% (19%)[*93*]; with 2,3-dimethyl-2-butylborane [with the formation of only (R′H)-RBH]: 96% (4%)[*865*]; with bis(3-methyl-2-butyl)borane: 98% (2%)[*95*]; and with dicyclohexylborane, bis(*trans*-2-methylcyclohexyl)borane, and diisopinocampheylborane: 99% (1%)[*864*].

The reaction mechanism also influences the steric directive effect[*864*]. Because of the smooth *cis*-addition that is possible in this case, the addition of the highly stereospecific diisopinocampheylborane (in the form of the dimer: tetraisopinocampheyldiborane) to cis-*4-methyl-2-pentene* rapidly gives 98% of the corresponding 4-methyl-2-pentylborane:

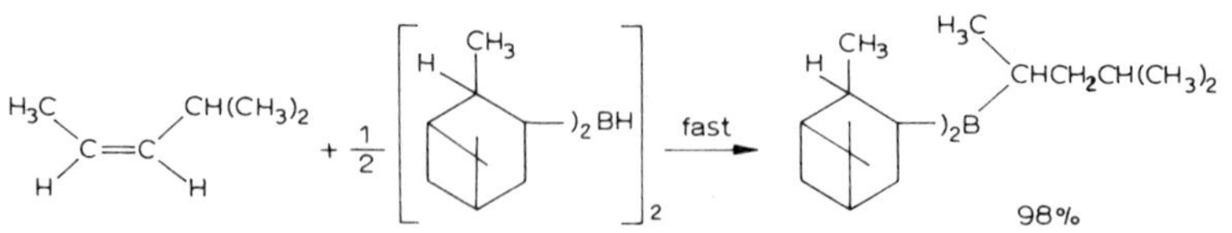

$$\text{cis-}H_3C{-}CH{=}CH{-}CH(CH_3)_2 + \tfrac{1}{2}\left[(C_{10}H_{17})_2BH\right]_2 \xrightarrow{\text{fast}} (C_{10}H_{17})_2B{-}CH(CH_3)CH_2CH(CH_3)_2 \quad 98\%$$

On the other hand, with trans-*4-methyl-2-pentene* the steospecific tetraisopinocampheyldiborane reacts slowly while the less stereospecific triisopinocampheyldiborane in equilibrium with its reacts rapidly[*864*]:

R = (pinanyl, CH₃ scheme labels)

$$R_2BH\text{ dimer} + trans\text{-}H_3C\text{-}CH{=}CH\text{-}CH(CH_3)_2 \longrightarrow R_2BCH(CH_3)CH_2CH(CH_3)_2 + R(CH_3CH_2CHCH(CH_3)_2)B\text{-}CH(CH_3)CH_2CH(CH_3)_2$$

trans

The oxidation of the reaction product therefore gives only 68% of 4-methyl-pentan-2-ol but 32% of 2-methylpentan-3-ol[*864*].

12.6 *Relative reaction rates in the addition of B–H to C═C*

Diborane $(BH_3)_2$ or the $BH_3 \cdot THF$ existing in tetrahydrofuran possesses three potential functions (B–H bonds) for the addition to C═C. The addition is not uniform, since during the successive alkylation of BH_3 to RBH_2 and R_2BH boranes with different addition characteristics (*e.g.*, stereospecificity, rate of reaction) are produced. On the other hand, the dialkylboranes R_2BH, which have only one active centre, or the monoalkylboranes RBH_2 (*e.g.*, 2,3-dimethylbut-2-ylborane) with one very active and a second much less active centre, add highly selectively.

In general, terminal alkenes react substantially faster than internal alkenes and here again *cis*-alkenes faster than *trans*-alkenes, and strained unsaturated rings faster than unstrained rings. The capacity for the addition of bis(3-methylbut-2-yl) borane to alkenes falls in the sequence: hex-1-ene ≥ 3-methylbut-1-ene > 2-methylbut-1-ene > 3,3-dimethylbut-1-ene > cyclopentene ≥ *cis*-hex-2-ene > *trans*-rex-2-ene > *trans*-4-methylpent-2-ene > cyclohexene ≥ 1-methylcyclo-pentene > 2-methylbut-2-ene > 1-methylcyclohexene ≥ 2,3-dimethylbut-2-ene [*95*]. The conversion of endocyclic to exocyclic alkenes, see [*76a*].

12.7 *Addition of B–H to C≡C*

The B–H bond adds to the C≡C triple bonds[*97, 145, 181, 307, 476*]. Mono-functional boranes R_2BH such as bis(3-methylbut-2-yl)borane add terminally in tetrahydrofuran or diglyme to terminal and non-terminal alkynes with the formation of alkenylboranes. Only a small part reacts with further R_2BH to give the trialkylboranes[*97, 145, 181, 307, 476, 487*]:

$$R'C{\equiv}CH + HBR_2 \longrightarrow R'CH{=}CHBR_2 \text{ (to a small extent } \xrightarrow{+R_2BH} R_2BR'CH\text{–}CH_2BR_2)$$

$$R'C{\equiv}CR'' + HBR_2 \longrightarrow R'CH{=}CR''BR_2 \text{ (to a small extent } \xrightarrow{+R_2BH} R_2BR'CH\text{–}CHR''BR_2)$$

$$HC{\equiv}CH + 2BHCl_2 \longrightarrow C_2H_4\,(BCl_2)_2$$

Bifunctional boranes RBH_2 (*e.g.*, in the form of $BH_2Bu^t \cdot NMe_3$) likewise add to terminal alkynes with the preferential formation of dialkenylalkyl-boranes [*307*]:

$$2R'C{\equiv}CH + H_2BBu^t{\cdot}NMe_3 \longrightarrow (R'CH{=}CH)_2BBu^t + NMe_3$$

Trifunctional borane BH_3 (*e.g.*, "nascent" in the form of $NaBH_4 + BF_3$ in THF or diglyme) adds to non-terminal alkynes with predominant formation of trialkenylboranes. On the other hand, further addition of BH_3 with polymerization, which is also possible, is diminished [*97, 98*]:

$$3R'C{\equiv}CR'' \xrightarrow[\text{fast}]{+H_3B} (R'CH{=}CR'')_3B \xrightarrow[\text{slow}]{+\,\rangle BH} BR'CH\text{-}CR''HB(R''C{=}CHR')_2$$

BH_3 adds to terminal alkynes with the formation of polymers since both a C=C bond and a B–H bond are present in the molecule for a second addition step (addition to alkenylborane). Consequently, polymerizing steps are open [*97, 98*]:

$R'C{\equiv}CH \xrightarrow{+BH_3}$ H(R')C=C(H)BH$_2$

- $\xrightarrow{+BH_3}$ R'–CH$_2$–CH(BH$_2$)$_2$ $\xrightarrow{+HC\equiv CR'}$ polymer $\xrightarrow{(H_2O_2)}$ aldehyde (26%)
- $\xrightarrow{+BH_3}$ R'–CH(BH$_2$)–CH$_2$(BH$_2$) $\xrightarrow{+HC\equiv CR'}$ polymer $\xrightarrow{(H_2O_2)}$ 1-alcohol (64%), 1,2-diol (10%)
- R'–CH$_2$–CH=BH $\rightarrow$ polymer $\xrightarrow{(H_2O_2)}$ aldehyde (26%)

(*e.g.*, R' = $CH_3CH_2CH_2CH_2$)

12.8 *Mechanism of the addition of B–H to C≡C*

The B–H bond adds to the C≡C by a four-centre mechanism (*cis*-addition):

$$RC{\equiv}CR' + \rangle B\text{–}H \longrightarrow \left| \begin{array}{cc} RC & \!\!\equiv CR' \\ \vdots & \vdots \\ B & \!\!\cdots H \end{array} \right| \longrightarrow \begin{array}{cc} RC & = CR \\ | & | \\ \rangle B & H \end{array}$$

For example, the addition of bis(3-methylbut-2-yl)borane to hex-1-yne forms trans-hexenyl-bis(3-methylbut-2-yl)borane, detected by infrared spectroscopy [*75*]:

$$CH_3CH_2CH_2CH_2C{\equiv}CH + \left[\begin{array}{c} H_3C\;\;CH_3 \\ |\quad\;\; | \\ H\text{-}C\text{-}C\text{-} \\ |\quad\;\; | \\ H_3C\;\;H \end{array}\right]_2 BH \longrightarrow \left[\begin{array}{c} H_3C\;\;CH_3 \\ |\quad\;\; | \\ H\text{-}C\text{-}C\text{-} \\ |\quad\;\; | \\ H_3C\;\;H \end{array}\right]_2 B\text{-}CH{=}HC\text{-}CH_2\text{-}CH_2CH_2CH_3$$

12.9 *Addition of B–H to fluorinated alkenes*

Fluorinated alkenes ($CF_2{=}CF_2$, $CF_2{=}CHF$, $CHF{=}CHF$, $CH_2{=}CHF$) add diborane at 20° and above. The reaction takes place via the primary exchange of F for H in the alkene. For example, the reaction of $CH_2{=}CHF$ with $(BH_3)_2$ yields 9% of BF_3, 51% of $(CH_3CH_2)BF_2$, 34% of $(CH_3CH_2)_2BF$, 6% of $(CH_3CH_2)BH_2$, and 0% of $(CH_3CH_2)_3B$ according to the (simplified) scheme[*31*]:

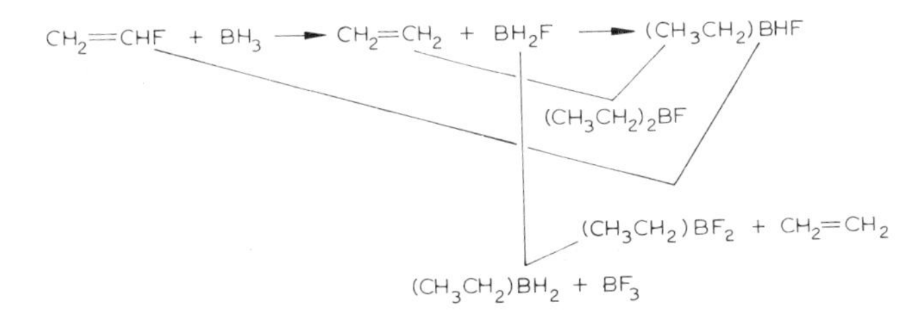

12.10 *Reaction of $(BH_3)_2$ with CO; synthesis and reactions of $BH_3{\cdot}CO$*

In the gas phase in a pressure tube at about 25 atm, diborane or diborane-d_6 and carbon monoxide form carbon monoxide-borane or carbon monoxide-borane-d_3 [*34, 52, 112, 124, 757, 860*]:

$$(BH_3)_2 + 2CO \xrightarrow{90-95°} 2BH_3{\cdot}CO$$

$BH_3{\cdot}CO$ is also formed in the reaction of boroxin $H_3B_3O_3$ [H_2O (gas) + B (solid) + B_2O_3 (liq.) $\xrightarrow[\Delta]{1100°}$ $H_3B_3O_3$] and carbon monoxide at a low pressure[*789a*]:

$$H_3B_3O_3 \longrightarrow B_2O_3 + BH_3$$
$$BH_3 + CO \longrightarrow BH_3{\cdot}CO$$

$BH_3{\cdot}CO$ is also produced in the reaction of B_5H_{11} with CO (see section 17).

The stability of $BH_3{\cdot}CO$ is probably to be ascribed to the H→C–π-bond component, $H{-}\overset{\cdot\cdot\rightarrow}{B{-}C}O$, since $BF_3{\cdot}CO$ does not exist (dissociation energy D ($BH_3{\cdot}CO$) = 23.1 ± 2 kcal/mole; for comparison, D (BH_3BH_3) = 3.71 ± 1 kcal/mole [*214*]). Carbon monoxide-borane forms a linear molecule, see Fig. 4.20.

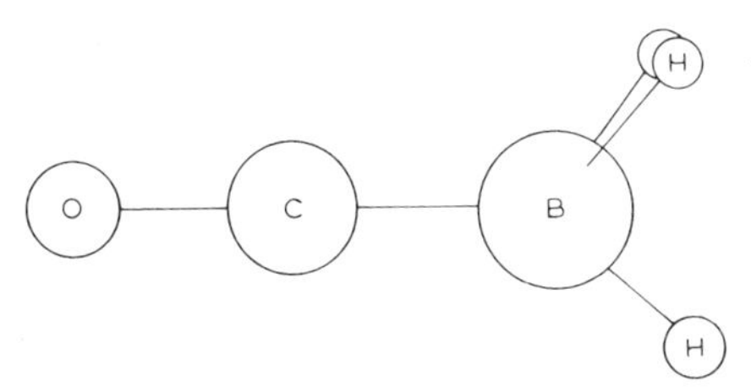

Fig. 4.20. Steric structure of $BH_3{\cdot}CO$. Structural parameters [*52*]:
B–H = 1.194 Å ∡HBH = 114°
B–C = 1.540 Å ∡CBH = 104.5°
C–O = 1.131 Å

At 50 to 60° and a low pressure, $BH_3 \cdot CO$ decomposes into diborane and carbon monoxide. Two contradictory mechanisms have been discussed for this reaction: (1) the rate-determining step is the formation of BH_3 and CO[*860*]:

$$BH_3 \cdot CO \underset{}{\overset{\text{slow}}{\rightleftharpoons}} BH_3 + CO$$

$$BH_3 + BH_3 \overset{\text{fast}}{\rightleftharpoons} (BH_3)_2$$

$$BH_3 + BH_3 \cdot CO \xrightarrow{\text{fast}} (BH_3)_2 + CO$$

(2) the rate-determining step is the replacement of the CO in $BH_3 \cdot CO$ by BH_3 [*246*]:

$$BH_3 \cdot CO \overset{\text{fast}}{\rightleftharpoons} BH_3 + CO$$

$$BH_3 + BH_3 \cdot CO \overset{\text{slow}}{\rightleftharpoons} (BH_3)_2 + CO$$

In the presence of $(CH_3)_3N$, $BH_3 \cdot CO$ decomposes with the formation of $BH_3 \cdot N(CH_3)_3$ and CO[*281b*].

In the oxidation of $BH_3 \cdot CO$ with oxygen, the following compounds are formed via a primary borane peroxide $BH_3 \cdot O_2$: boron trioxide, carbon dioxide, water, and hydrogen[*31a*]:

$$2BH_3 \cdot CO + 2O_2 \longrightarrow 2BH_3 \cdot O_2 + 2CO$$

$$2BH_3 \cdot O_2 + BH_3 \cdot CO \longrightarrow 2HOBH_2 + CO$$

$$2BH_3 \cdot O_2 + 2HOBH_2 \longrightarrow 2H_2B_2O_3 + 4H_2$$

The hydrolysis of $BH_3 \cdot CO$ in water may be represented by the following two equations[*488d*]:

$$BH_3 \cdot CO + 3H_2O \xrightarrow{100^\circ} B(OH)_3 + 3H_2 + CO$$

$$BH_3 \cdot CO + 2H_2O \xrightarrow{0^\circ} (HO)_2BCH_2OH + H_2$$

The alcoholysis of $BH_3 \cdot CO$ in alcohols [CH_3OH, C_2H_5OH, $(CH_3)_2CHOH$] may be represented by the following equations[*488a*]:

$$BH_3 \cdot CO + 3ROH \longrightarrow B(OR)_3 + 3H_2 + CO$$

$$2BH_3 \cdot CO + 4ROH \xrightarrow{-2H_2} 2(RO)_2BCH_2OH \xrightarrow{-2ROH} \text{ROB}\begin{matrix} \diagup CH_2O \diagdown \\ \diagdown OCH_2 \diagup \end{matrix}\text{BOR}$$

The BH_3 group is isoelectronic with the oxygen atom. The chemical properties of $BH_3 \cdot CO$ therefore resemble those of CO_2, and those of $BH_3CO_2^{2\ominus}$ those of $CO_3^{2\ominus}$. Examples[*136a, 488c, 488e*]:

$$BH_3\cdot CO + 2KOH \xrightarrow{\text{alcohol}} (K^{\oplus})_2\,[BH_3CO_2]^{2\ominus} + H_2O$$

$$BH_3\cdot CO + NaOH \xrightarrow[\text{adsorption at } 25°]{(NaOH + \text{asbestos})} Na^{\oplus}H^{\oplus}\,[BH_3CO_2]^{2\ominus}$$

$$BH_3\cdot CO + CaO \xrightarrow{25°} Ca^{2\oplus}\,[BH_3CO_2]^{2\ominus}$$

$$BH_3\cdot CO + NaOCH_3 \longrightarrow Na^{\oplus}\,[BH_3C(O)(OCH_3)]^{\ominus} \xrightarrow[-CH_3OH]{+NaOH} Na_2BH_3CO_2$$

$$BH_3\cdot CO + 2NH_3 \longrightarrow NH_4^{\oplus}\,[BH_3C(O)(NH_2)]^{\ominus}$$

$$BH_3\cdot CO + CH_3NH_2 \longrightarrow CH_3NH_3^{\oplus}\,[BH_3C(O)(NCH_3H)]^{\ominus}$$

$$BH_3\cdot CO + (CH_3)_2NH \longrightarrow (CH_3)_2NH_2^{\oplus}\,[BH_3C(O)(CH_3NCH_3)]^{\ominus}$$

but:

$$BH_3\cdot CO + (CH_3)PH_2 \longrightarrow BH_3\cdot PH_2(CH_3) + CO$$

12.11 Addition of B–H to C=N

Primary and secondary alkylamineboranes, $RR'NH\cdot BH_3$ ($R = H$; $R' = Me$, Et, Bu^t; $R = R' = Et$), add to the C=N bond of phenyl or ethyl isocyanate, RNCO. The BH_3 migrates to the nitrogen atom [*151a*]:

$$RR'NH\cdot BH_3 + PhNCO \longrightarrow \left[\begin{array}{c} H \\ | \\ RR'N\cdots BH_3 \\ \downarrow \quad \uparrow \\ O{=}C{=}N\text{–}Ph \end{array}\right] \longrightarrow RR'N\underset{\underset{O}{\|}}{C}\overset{\overset{BH_3}{\uparrow}}{N}HPh \longrightarrow RR'\overset{\overset{BH_3}{\uparrow}}{N}\underset{\underset{O}{\|}}{C}NHPh$$

13. Reaction of diborane with elements of the 1st, 2nd and 4th main groups and their compounds

With alkali metals (reaction mostly with sodium, more rarely with potassium) diborane forms the diamagnetic $B_2H_6^{2\ominus}$ (not 2 $BH_3^{\ominus}$!), and also the ion-pair $B_3H_8^{\ominus}$ and $BH_4^{\ominus}$. The reaction probably takes place through the following stages: (1) cleavage with a B–H$^{\mu}$–B bridge by the uptake of two sodium electrons (this takes place without a solvent and in various ethers); (2) rearrangement to the doubly negatively charged hexahydrodiborate(2−) anion $B_2H^{2\ominus}$; (3) addition of B_2H_6 with the formation of the intermediate $Na_2B_4H_{12}$ (takes place rapidly only in ether); (4) asymmetric dismutation of the intermediate:

$$B_2H_6 + 2\,Na \longrightarrow (Na^{\oplus})_2 \left[H_2B(\mu\text{-}H)\dot{B}H_3 \xrightarrow{\text{rearrangement}} H_3B\text{–}BH_3 \right]^{2\ominus} \xrightarrow{+B_2H_6} Na_2B_4H_{12} \longrightarrow NaB_3H_8 + NaBH_4$$

References p. 355

There are several methods for the practical performance of the reaction. Shaking Na amalgam[*730, 735, 740*] or K amalgam[*743*] with B_2H_6 for several days yields $M_2B_2H_6$:

$$B_2H_6 + 2Na(Hg) \longrightarrow Na_2B_2H_6 + 2(Hg)$$

If the reaction is carried out in ethyl or n-butyl ether, after two days ether-soluble sodium octahydrotriborate, NaB_3H_8, and insoluble sodium boranate, $NaBH_4$, are formed[*360*]:

$$2B_2H_6 + 2Na(Hg) \xrightarrow{R_2O} NaB_3H_8 + NaBH_4$$

The metallation of triphenylboron or naphthalene in tetrahydrofuran with alkali metal addition compounds (metal carriers) takes place substantially more rapidly (seconds) than with alkali metal amalgams[*242*]:

$$2B_2H_6 + 2M\text{-carrier} \xrightarrow{THF} MB_3H_8 + MBH_4 + 2\ \text{carrier}$$

If solvents (monoglyme, diglyme, dioxan; and in the case of Li also diethyl ether) in which alkali metal boranates are readily soluble, are used, only alkali metal octahydrotriborates MB_3H_8 are obtained, since MBH_4 reacts further with B_2H_6 to give MB_3H_8[*242, 361*]. The reaction mechanism of the last reaction is not yet elucidated. Probably it is a nucleophilic displacement of ether in the boron–ether complex $>O \cdots\rightarrow BH_3$ by $BH_4^{\ominus}$ (intermediate $B_2H_7^{\ominus}$, see [*205a*]) [*2, 24, 25, 92, 236*]:

$$(BH_3)_2 + {>}O| \longrightarrow 2\ {>}O \cdots\rightarrow BH_3$$

$$[BH_4]^{\ominus} \xrightarrow{+>O\cdots\rightarrow BH_3} [H_3B\text{–}H\text{–}BH_3]^{\ominus} \xrightarrow{+>O\cdots\rightarrow BH_3} [H_2B(H\text{–}BH_3)_2]^{\ominus} \xrightarrow{-H_2} [B_3H_8]^{\ominus}$$

Free diborane is not necessary for the preparation of $B_3H_8^{\ominus}$[*242*]:

$$4BF_3 + 5MBH_4 \longrightarrow 2MB_3H_8 + 3MBF_4 + 2H_2$$

On the reactions of $B_3H_8^{\ominus}$, see section 15.

Under controlled conditions (diglyme 100°, then boiling at 162°), diborane and sodium boranate react to give hexahydrohexaborate (2 –) $B_6H_6^{2\ominus}$. It can be precipitated as the tetramethylammonium salt (yield 5–10%)[*61*]. $(BH_3)_2$ and $(CH_3)_4$-P6 yield $(CH_3)_3PbH$ [*354a*].

14. Reactions of borane with transition metal carbonyls

Pentacarbonylrhenium anions, $Re(CO)_5^{\ominus}$, and tetracarbonyltriphenylphosphanemanganese anions, $Ph_3PMn(CO)_4^{\ominus}$ (cations: $Na^{\oplus}$, $Et_4N^{\oplus}$, $Bu_4P^{\oplus}$) displace the (donor) ether in borane–ether[*53a*, *583*]:

$$(CO)_5Re^{\ominus} \xrightarrow[-R_2O]{+R_2O \cdot BH_3} (CO)_5Re \cdots\rightarrow BH_3^{\ominus} \xrightarrow[-R_2O]{+R_2O \cdot BH_3} (CO)_5Re \cdots\rightarrow (BH_3)_2^{\ominus}$$

I II

$$(CO)_5Mn^{\ominus} \xrightleftharpoons{+THF \cdot BH_3,\ -THF} (CO)_5Mn \cdots\rightarrow BH_3^{\ominus}$$

III

$$(Ph_3P)(CO)_4Mn^{\ominus} \xrightarrow[-THF]{+THF \cdot BH_3} (Ph_3P)(CO)_4Mn \cdots\rightarrow BH_3^{\ominus}$$

According to the electronic and infrared spectra, BH_3 in the complex (I) does not add to the nucleophilic carbonyl group (structure Ia) but to the central atom (structure Ib). The structure of the bis(borane) complex (II)

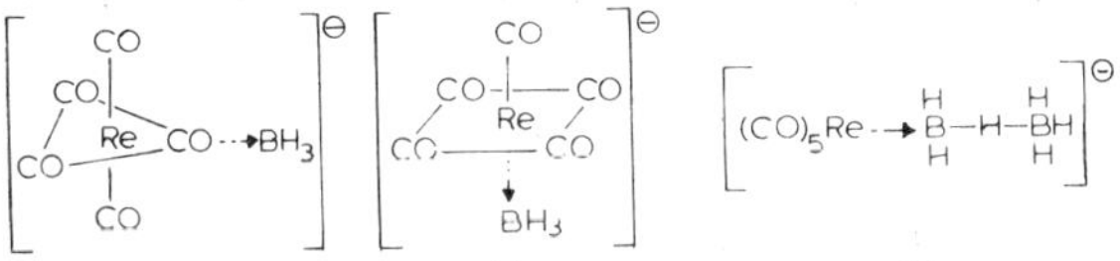

$[(CO)_4Re\text{–}CO\cdots\rightarrow BH_3]^{\ominus}$ Ia $[(CO)_5Re\cdots\rightarrow BH_3]^{\ominus}$ Ib $[(CO)_5Re\cdots\rightarrow BH_2\text{–}H\text{–}BH_3]^{\ominus}$ II

has not been elucidated; either the second BH_3 group is also coordinated with the rhenium (7-fold coordination to Re) or the adduct is derived from the hypothetical $B_2H_7^{\ominus}$ (II)[*583*], existing only in the equilibrium $2BH_4^{\ominus} + (BH_3)_2 \rightleftharpoons 2B_2H_7^{\ominus}$ [*24*, *89a*, cit. *523*].

The donor properties of the manganese in $(CO)_5Mn^{\ominus}$ are less strongly pronounced than those of the rhenium in $(CO)_5Re^{\ominus}$. Thus, when a solution of $Na^{\oplus}(CO_5)Mn \cdot BH_3^{\ominus}$ is concentrated, $Na^{\oplus}(CO)_5Mn^{\ominus}$ and BH_3 are re-formed. A tetraethylammonium salt can be precipitated at −78°, while at −25° $(BH_3)_2$ is slowly given off. Coordination of electron-donating ligands such as triphenylphosphane to the manganese stabilizes the manganese–borane adduct:

$$[(CO)_5(Ph_3P\cdots\rightarrow) Mn \cdots\rightarrow BH_3]^{\ominus}$$

is stable. As in $Ph_3PMn(CO)_4$, the P⋯→Mn bond in the borane complex possesses an almost pure σ-character without appreciable $d\pi$–$d\pi$ back-coordination [*583*].

15. Preparation and reactions of triborane(7), tetraborane(10) and triboranate(8)$^{\ominus}$ [27]

15.1. Isotope exchange in B_4H_{10}

The kinetics of the B_4H_{10}–B_2D_6 isotope exchange depends on the temperature. At relatively low temperatures (25–45°), it is stamped by the cleavage of the diborane. Here the exchange takes place by two routes (A and B). At temperatures

[27] For a comprehensive review, see [*508a*].

above 60°, the decomposition of the tetraborane (into B_2H_6, among other products) has increasing importance (route C).

In the main reaction (A) of the low-temperature exchange, boron is also exchanged. The exchange takes place through the rapid dissociation of diborane into borane which is already known from many exchange reactions with diborane. This is followed by an exchange of a BD_3 group with a BH_3 group of the B_4H_{10} [*729, 767*]:

$$(^{10}BD_3)_2 \rightleftharpoons 2\,^{10}BD_3 \qquad (A_1)$$

$$^{10}BD_3 + B_4H_{10} \longrightarrow {}^{10}BB_3H_7D_3 + BH_3 \qquad (A_2)$$

In the side-reaction (B) of the low-temperature exchange, hydrogen atoms in two positions in B_4H_{10} are excited by collision with B_4H_{10} or B_2D_6. This is followed by a H–D exchange which does not take place through BD_3 and is independent of the surface (no B exchange) [*767*]:

$$B_4H_{10} + B_4H_{10} \rightleftharpoons B_4H_{10}^{*} + B_4H_{10} \qquad (B_1)$$

$$B_4H_{10} + B_2D_6 \rightleftharpoons B_4H_{10}^{*} + B_2D_6 \qquad (B_2)$$

$$B_4H_{10}^{*} + B_2D_6 \rightleftharpoons B_4H_8D_2 + B_2D_4H_2 \qquad (B_3)$$

The high-temperature exchange takes place by routes A and B with the formation of deuterated tetraborane(10). In addition deuterated pentaborane(11) arises (reaction C). The first step (C_1) is a further development of routes B_1 and B_2. Now the two hydrogen atoms are not only excited but also split off (in a rate-determining step) (C_1). With diborane-d_6, the B_4H_8 formed gives pentaborane(11) (pentaborane-h_8–d_3) (C_2). The elimination of whole borane groups BH_3 from B_4H_{10}(C_3) taking place at the same time is, however, unimportant for the result of the reaction, since the B_3H_7 formed in it rapidly regenerates tetraborane(10) with diborane (or BH_3) [*188*, cit. *433, 767*]. The formation of B_4H_8CO in the copyrolysis of B_4H_{10} and CO at 80 to 110° (1st order for B_4H_{10}, zero order for CO) also indicates route C_1 [*69*]:

$$B_4H_{10} \longrightarrow B_4H_8 + H_2 \qquad (C_1)$$

$$B_4H_8 + (BD_3)_2 \longrightarrow B_5H_8D_3 + BD_3 \qquad (C_2)$$

$$B_4H_{10} \rightleftharpoons B_3H_7 + BH_3 \qquad (C_3)$$

15.2 Halogenation of B_4H_{10}

When bromine and an excess of tetraborane(10) are allowed to react at about −15° for 12 to 18 hours, 2-bromotetraborane(10) is formed. It can be separated from hydrogen bromide and unconverted tetraborane(10) by high-vacuum fractionation [*172*].

15.3 Nucleophilic attack on B_4H_{10}

In the reaction of tetraborane(10) and ammonia in the gaseous state or at −78° in ethereal solution, a stable diammoniate of tetraborane(10), $B_4H_{10}\cdot 2NH_3$, forms [*419*]. As in the analogous reaction with diborane, here the $B\text{–}H^{\mu}\text{–}B$ double bridge of the tetraborane splits asymmetrically with the formation of dihydridodiammineboron(III) triboranate, $(NH_3)_2BH_2^{\oplus}B_3H_8^{\ominus}$ [*420, 422*]:

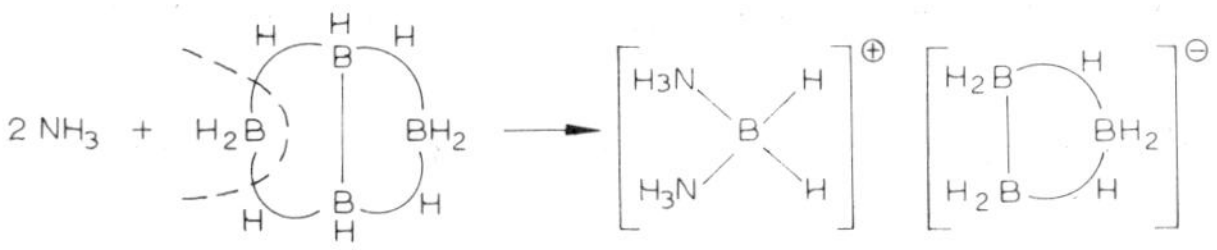

In the presence of HCl or HBr, the anion $B_3H_8^{\ominus}$ forms triborane(7) with the splitting out of hydrogen [*421*]:

$$(NH_3)_2BH_2^{\oplus}B_3H_8^{\ominus} + HX + OEt_2 \xrightarrow[-78°]{\text{diethyl ether}} (NH_3)_2BH_2^{\oplus}X^{\ominus} + B_3H_7\cdot OEt_2 + H_2$$

With dimethyl ether, diethyl ether, tetrahydropyran or tetrahydrofuran at 25° tetraborane(10) forms etherates of triborane(7) with symmetrical cleavage of the $B\text{–}H^{\mu}\text{–}B$ double bridge (see sections 1.2 and 9.4.1), and with trimethylamine, similarly, it forms trimethylamine-triborane(7) [*420, 464, 793*].
As in the cleavage of the $B\text{–}H^{\mu}\text{–}B$ double bridge in diborane (section 9.4.1), here again an unsymmetrical cleavage of the double bridge must be assumed as the initial step. The primary donor–acceptor bonding of R_2O to the electron-poor $B\text{–}H^{\mu}\text{–}B$ double bridge leads to the unstable intermediate. The further addition of R_2O then leads to symmetrical cleavage [*579, 652*]:

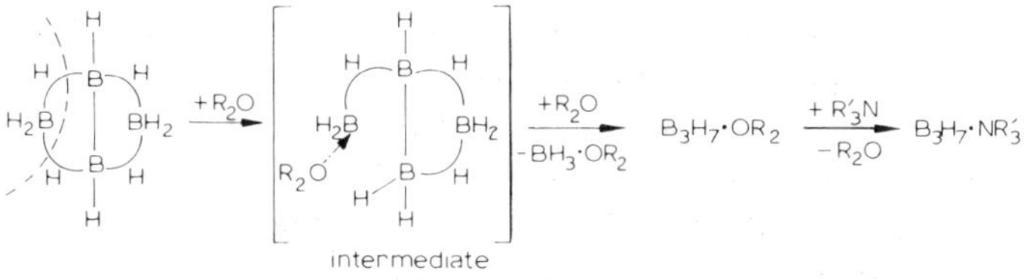

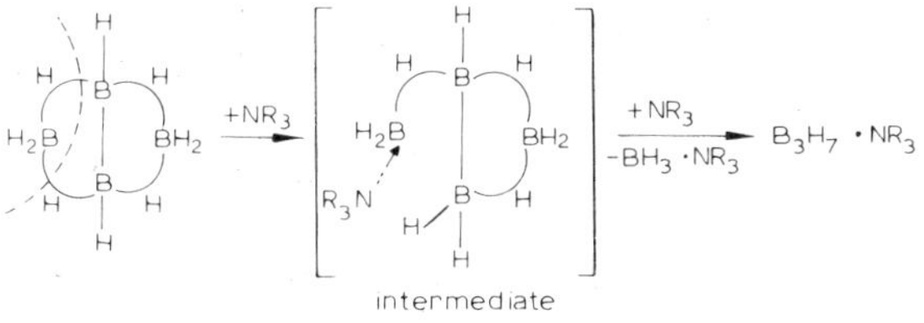

The nucleophilic displacement of the ether in $B_3H_7\cdot OR_2$ by an amine NR'_3 (R' = CH_3 or H) gives rise to an aminetriborane(7) $B_3H_7\cdot NR_3$. Here the etherate must

References p. 355

be neither too unstable (→ formation of $(NH_3)_2BH_2^{\oplus}B_3H_8^{\ominus}$, for example, see above, or decomposition of the B_3H_7) or too stable (no displacement of OR_2 by NR_3'). The best yields of $B_3H_7{\cdot}NH_3$ (81 mole-%) are accordingly obtained with tetrahydropyran, the donor activity of which is between that of dimethyl and diethyl ethers, on the one hand, and tetrahydrofuran, on the other hand [*420*].

$B_3H_7 \cdot O(CH_3)_2$ and PF_3 yield $B_2H_4(PF_3)_2$ [*164e*]. With B_2H_6, $B_2H_4(PF_3)_2$ reacts yielding B_4H_{10} and PF_3 [*164e*].

Ammoniatriborane(7), $B_3H_7{\cdot}NH_3$, forms in a yield of 20 to 30% from NaB_3H_8 (for the preparation of $B_3H_8^{\ominus}$ see section 15.5) and NH_4Cl [*420*]

$$NaB_3H_8 + NH_4Cl \xrightarrow[25^\circ]{\text{diethyl ether}} B_3H_7{\cdot}NH_3 + NaCl + H_2$$

or by a degradation reaction of decaborane(14), see section 18.10.4. $B_3H_7{\cdot}NH_3$ hydrolyses very slowly in the following way [*420*]:

$$B_3H_7{\cdot}NH_3 + 9HOH + H^{\oplus} \longrightarrow 3B(OH)_3 + 8H_2 + NH_4^{\oplus}$$

With an excess of trimethylamine in tetrahydrofuran at −78°, tetraborane(10) forms $BH_3{\cdot}NMe_3$ and a solid which has not been studied further, possibly $BH{\cdot}xNMe_3$ ($x = 0.37$–0.54) [*128*]:

$$B_4H_{10} + (3+x)\,NMe_3 \longrightarrow 3BH_3{\cdot}NMe_3 + BH{\cdot}xNMe_3$$

With the donor carbon monoxide, tetraborane(10) forms carbon monoxide-tetraborane(8) [*69*]:

$$B_4H_{10} + CO \xrightarrow{80-110^\circ} B_4H_8{\cdot}CO + H_2$$

For reactions of $B_4H_8{\cdot}CO$, see section 17.

In a hot-cold reactor B_4H_{10} and $D_2C{=}CD_2$ yield $D_4C_2B_4H_8$. This compound is more logically classified as an alkylborane than as a carborane, since its structure consists of a tetraborane skeleton bridged diagonally by a $-CD_2CD_2-$ group. $D_4C_2B_4H_8$ is not formed via a hydroboration of the C=C double bond (mechanism 1) but via an addition of the C=C double bond to the intermediate B_4H_8, formed by hydrogen release of B_4H_{10} (see also section 15.1) [*845a*]:

$$B_4H_{10} \xrightarrow{+C_2D_4} [HCD_2CD_2{-}B_4H_9] \xrightarrow[-H_2]{-HD} \begin{cases} 67\%\ D_3C_2B_4H_9 \\ 33\%\ D_4C_2B_4H_8 \end{cases} \quad (1)$$

$$B_4H_{10} \xrightarrow{-H_2} [B_4H_8] \xrightarrow{+C_2D_4} 100\%\ D_4C_2B_4H_8 \quad (2)$$

The gas-phase reaction of B_4H_{10} with $HC{\equiv}CH$, $HC{\equiv}CCH_3$ and $CH_3C{\equiv}CCH_3$ at 25–50° yields alkyl derivatives of the nido-carboranes (see section 19.1), alkylboranes, hydrocarbons and organoboron polymers [*278g*].

The cyanide ion $|C{\equiv}N|^{\ominus}$ cleaves the $B{-}H^{\mu}{-}B$ double bridge in B_4H_{10} symmetrically (see above). The BH_3 formed together with B_3H_7 (as $B_3H_7 \cdot CN^{\ominus}$) can add to both free electron pairs of the $CN^{\ominus}$ with the formation of $[H_3B \cdot CN \cdot BH_3]^{\ominus}$. The sodium salts precipitate as sparingly soluble dioxan adducts [*7*, cit. *469*]:

$$B_4H_{10} + O\langle \xrightarrow{\text{diglyme}} \begin{cases} B_3H_7 \leftarrow \cdot \cdot O\langle \xrightarrow{+NaCN,\ +C_4H_8O_2} Na^{\oplus}B_3H_7CN^{\ominus} \cdot C_4H_8O_2 \\ BH_3 \leftarrow \cdot \cdot O\langle \xrightarrow{+\frac{1}{2}NaCN,\ +C_4H_8O_2} \frac{1}{2} Na^{\oplus}BH_3CNBH_3^{\ominus} \cdot 2C_4H_8O_2 \end{cases}$$

The $B-H^{\mu}-B$ double bridge in tetraborane(10) is split symmetrically in diethyl ether, at least to a certain extent (see above: formation of $B_3H_7 \cdot OR_2 + BH_3$). The ether contained in the $BH_3 \cdot OR_2$ (as a weak donor) is displaced nucleophilically by the stronger donors PPh_3, $H^{\ominus}$, and $BH_4^{\ominus}$ [for example, in the formation of NaB_3H_8 and $(BH_3)_2$ with NaH or $NaBH_4$] [*164c*, *277*, *359*]:

$$B_4H_{10} \xrightarrow[-\frac{1}{2}(BH_3)_2]{+R_2O} B_3H_7 \cdot OR \begin{cases} \xrightarrow[-R_2O(=THF)]{+PPh_3} B_3H_7 \cdot PPh_3 \\ \xrightarrow[-R_2O]{+H^{\ominus}} B_3H_8^{\ominus} \\ \xrightarrow[-R_2O]{+BH_4^{\ominus}} B_3H_8^{\ominus} + \frac{1}{2}(BH_3)_2 \end{cases}$$

With polyphosphoric acid $Me_4N^{\oplus}B_3H_8^{\ominus}$ forms B_6H_{12} [*240*, *241*]. With triphenylphosphane PPh_3, dimethylsulphanetriborane(7), $B_3H_7 \cdot SMe_2$, forms predominantly $B_3H_7 \cdot PPh_3$. Trimethylaminetriborane(7) forms a bisphosphane adduct of the hypothetical diborane(4) [*277*]:

$$B_3H_7 \cdot NMe_3 + 3PPh_3 \xrightarrow[50^{\circ}]{\text{benzene}} B_2H_4 \cdot 2PPh_3 + BH_3 \cdot PPh_3 + NMe_3$$

15.4 *Addition of electrons to B_4H_{10}*

Tetraborane(10) reacts with alkali metal amalgams M(Hg) to give $M_2B_4H_{10}$. If ether is present, MB_3H_8, $(BH_3)_2$, and polymeric boron hydride "BH" are formed [*361*, *730*, *734*]. As in the alkali metal–diborane reactions (see section 13), a $B-H^{\mu}-B$ bridge is first opened by the addition of two electrons. This is followed by the addition of B_4H_{10} and decomposition:

$$B_4H_{10} \xrightarrow[(-Hg)]{+M(Hg)} 2M^{\oplus}[B_4H_{10}]^{2\ominus} \xrightarrow[(ether)]{+B_4H_{10}} M_2B_8H_{20} \longrightarrow 2MB_3H_8 + \frac{1}{2}(BH_3)_2 + \frac{1}{x}(BH)_x$$

References p. 355

15.5 *Preparation of $B_3H_8^{\ominus}$*

In the reaction of diborane [*360, 730, 735, 737, 850a*] (see section 13) or tetraborane [*361, 730, 734*] (see section 15.4) in ether or tetrahydrofuran with an alkali-metal amalgam, the alkali-metal triboranate(8) is formed, among other products. $Me_4NB_3H_8$ is synthesised in larger amounts and faster by the reaction of $NaBH_4$ with $(BH_3)_2$ in monoglyme at 60° in the autoclave, removal of the solvent, and precipitation of the $Me_4NB_3H_8$ with Me_4NCl in water (yield 95%) [*13b, 523*]:

$$(BH_3)_2 + NaBH_4 \xrightarrow{\text{monoglyme}} NaB_3H_8 + H_2$$

$$NaB_3H_8 + MeNB_3H_8 \xrightarrow{\text{water}} Me_4NB_3H_8 + NaCl$$

$B_3H_8^{\ominus}$ onium salts sparingly soluble in water are formed by metathesis from $Me_4NB_3H_8$ and onium compounds such as Ph_2ICl, Ph_3SCl, Ph_3SeCl, Ph_3TeBr, Ph_4PCl, Ph_4AsCl, Ph_4SbBr, and Ph_4BiNO_3 [*13b*], for example:

$$Me_4NB_3H_8 + Ph_4AsCl \xrightarrow{\text{water}} Ph_4AsB_3H_8 + Me_4NCl$$

$B_3H_8^{\ominus}$ salts readily soluble in water are formed by metathesis from the moderately sparingly soluble TlB_3H_8 and iodides such as NH_4I, RbI, CsI, MgI_2, CaI_2, SrI_2, and BaI_2 [*13b*], for example:

$$2TlB_3H_8 + BaI_2 \xrightarrow{\text{water}} Ba(B_3H_8)_2 + TlI$$

Concerning the pyrolysis of $B_3H_8^{\ominus}$ salts, see section 18.13.

$R_4N^{\oplus}B_3H_8^{\ominus}$ reacts with $Cr(CO)_6$, $Mo(CO)_6$ and $W(CO)_6$ yielding carbonyl complexes [*405d*]:

$$M(CO)_6 + B_3H_8^{\ominus} \longrightarrow M(CO)_4(B_3H_8)^{\ominus} + 2CO$$

CsB_3H_8 and $(C_5H_5)_2TiCl_2$ form $(C_5H_5)_2TiB_3H_8$ [*405d*].

15.6 *Structures of $B_3H_7{\cdot}NH_3$ and $B_3H_8^{\ominus}$*

In $B_3H_7{\cdot}NH_3$, the B atoms do not form a fully equilateral triangle (Fig. 4.21). The N atom is not located in the B_3 plane and is bound to one B atom. The molecule can therefore be regarded (a) as a highly distorted B_4H_{10} fragment (compare the topological structure of B_4H_{10}, Fig. 4.4) or (b) as a $(BH_3)_2$ substituted at the bridge H^{μ} atom by BH_2NH_3. The actual structure probably lies between the two topological structures [*557*]. Figure 4.22 shows the two possible structures for $B_3H_8^{\ominus}$.

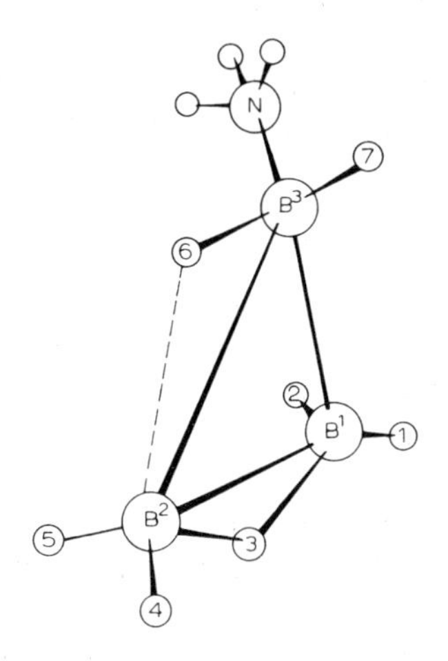

Fig. 4.21. Steric and topological structures of $B_3H_7 \cdot NH_3$. Structural parameters [557]:

$N-B^3 = 1.58$ Å	$B^1-H^1 = 1.09$ Å	$\measuredangle NB^3B^1 = 111.0°$
$B^1-B^2 = 1.74$ Å	$B^1-H^2 = 1.18$ Å	$\measuredangle NB^3B^2 = 115.3°$
$B^1-B^3 = 1.82$ Å	$B^1-H^3 = 1.23$ Å	$\measuredangle N-(B_3\text{-plane}) = 117.2°$
$B^2-B^3 = 1.80$ Å	$B^2-H^3 = 1.39$ Å	
	$B^2-H^4 = 1.12$ Å	
	$B^2-H^5 = 1.11$ Å	
	$B^2-H^6 = 1.73$ Å	
	$B^3-H^6 = 1.12$ Å	
	$B^3-H^7 = 1.14$ Å	

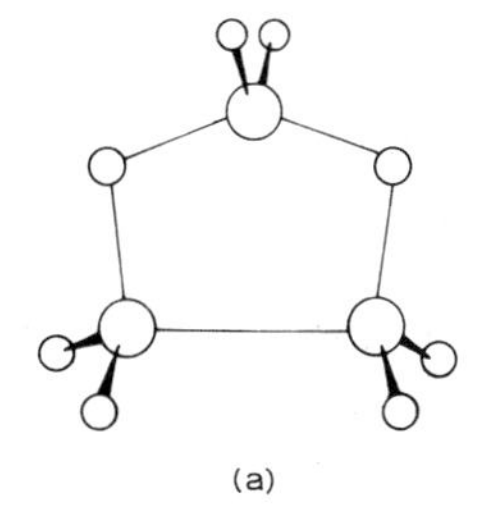

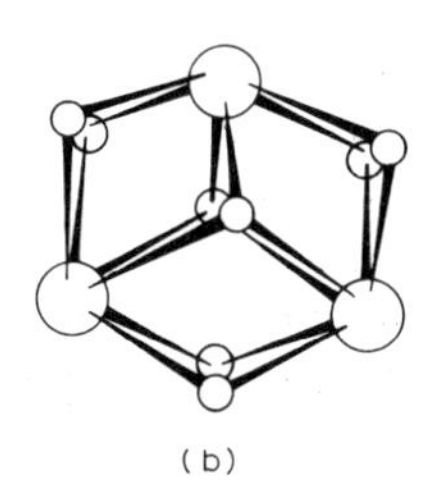

Fig. 4.22. Steric structures of $B_3H_8^{\ominus}$. (a) Structure according to the X-ray diagram. Structural parameters [590]: $B-H_{ext.} = 1.05-1.20$ Å; $B_{apex}-H^{\mu} = 1.5$ Å; $B_{base}-H^{\mu} = 1.2$ Å. (b) Possible structure according to the NMR spectrum. According to the NMR spectrum, all H atoms may be bridge atoms. Two of the eight bridge H atoms are bound to three B atoms [591].

References p. 355

16. Reactions of pentaborane(9)

In the reaction of $(BD_3)_2$ with B_5H_9 at 90°, all five terminal H atoms of the B_5H_9 are exchanged for D, while the four bridge H atoms remain unattacked [*390*]:

$$B_5H_9 \xrightarrow{(BD_3)_2,\,\Delta} B_5H_4{}^{\mu}D_5$$

The process takes place via a bimolecular reaction of BD_3 [from $(BD_3)_2$] and B_5H_9[*390*]. In addition to $B_5H_4{}^{\mu}D_5$, $B_{10}(H,D)_{14}$ is produced[*343, 344, 388*]. Here only deuterium, and not boron as well, passes from the BD_3 molecule into the B_5H_9 molecule[*434*]. The irradiation of B_5H_9 with deuterons (2MeV) gives rise to H_2, B_2H_6, $B_{10}H_{16}$, and polymers. In the presence of D_2, HD, $B_2H_{6-n}D_n$, traces of B_5H_8D, polymers, and $B_{10}H_{16}$, but no deuterated $B_{10}H_{16}$, are formed[*754*].

Concerning the formation of B_8H_{12} in the electric discharge treatment of B_5H_9 and B_2H_6, see section 3.8.

In the B_5H_9–DCl–$AlCl_3$ system, hydrogen at the apex atom (B^1) of the B_5H_9 is exchanged[*570*]:

$$B_5H_9 + DCl \xrightarrow{AlCl_3} B_5H_8\text{-1-D} + HCl$$

Concerning deuteration in the bridge position, see below.

In the halogenation of pentaborane(9) with free halogen (Cl_2, Br_2, I_2), monohalogenopentaboranes(9) halogenated in the 1- or 2-position are formed, possibly by the following reaction mechanisms. The polar reactions (reactions with Cl_2 + $AlCl_3$; Br_2; I_2) yield predominantly B_5H_8-1–X by the electrphilic replacement of a proton in position 1 by positive halogen X. If a polar chlorination with negative chlorine (ICl, ICl_3) is attempted, an at least 90% yield of B_5H_8-l-I is obtained instead of B_5H_8Cl. (Reactions with I_2[*112a, 290, 290a, 291*], with Br_2[*221, 291, 567, 650, 692b*], and with Cl_2 + $AlCl_3$[*236a, 237d*]: in a sealed flask, Cl_2, $AlCl_3$, BCl_3, and B_5H_9 are warmed from −108° to 0°; yield: 6 mmole of B_5H_8-1–Cl, about 50 mole-% calculated on the B_5H_9).

The radical reactions yield mainly B_5H_8-2–X. If, therefore, for example, a mixture of B_5H_9 and Br_2 which would normally give B_5H_8-1–Br, predominantly is irradiated with ultraviolet light, almost 100 times the amount of B_5H_8-2–Br is formed[*236a*].

Pentaborane(9) derivatives of the type B_5H_8-1–X are more stable when X is a heavier halogen, whereas B_5H_8-2–X compounds are more stable when X is a lighter halogen[*112a*]. But B_5H_8-1–I reacts with liquid $(CH_3)_2O$ at −12° yielding B_5H_8-2-(OCH_3)[*237f*].

When B_5H_8-1–Br and $AlCl_3$ are heated to 150° for 17 h, the 1-bromopentaborane is converted into 2-chloropentaborane[*567*]:

$$B_5H_8\text{-1-Br} \xrightarrow[\Delta]{+AlCl_3,\,150^\circ} B_5H_8\text{-2-Cl}$$

When B_5H_8-1–Br is heated with BCl_3 or B_5H_8-1–Br is heated with $AlBr_3$, no conversions take place[*576*]. Halogen exchanges on the B_5 skeleton are done best by gas-flow process at minimum pressure, as when $HgCl_2$ converts B_5H_8-1–I to B_5H_8-1–Cl or SbF_3 converts B_5H_8-2–I to B_5H_8-2–F[*112a*]. Hexamethylene-tetramine catalyses the isomerisation of B_5H_8-1–I to B_5H_8-2–I[*112a*]. Figure 4.23 shows the structure of B_5H_8-1–I.

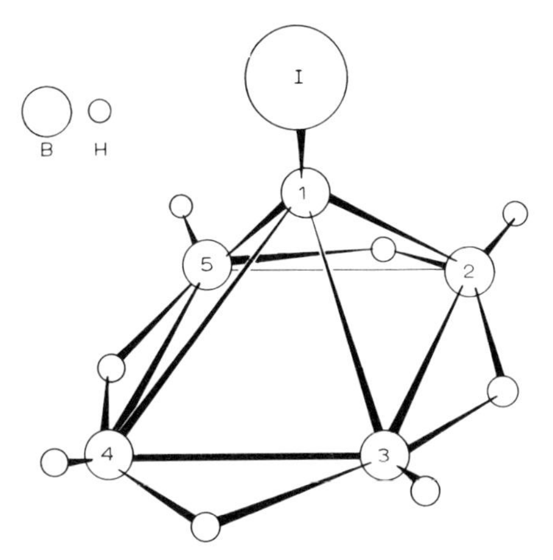

Fig. 4.23. Steric structure of B_5H_8-1–I. Structural parameters[*290*]: B^1–I = 2.20 Å (in BI_3: 2.10 Å); distance of the B atoms of the base, B–B = 1.84 Å (in B_5H_9: 1.77–1.80 Å); distance of the B atoms of the base from the apex atom B^1: B^1–B = 1.71 Å (in B_5H_9: 1.66–1.70 Å) (∡B^2–B^1–B^3 = 65°; ∡I–B^1–B^2 = 130.5°).

The donor-catalysed isomerisation of 1-deuteropentaborane(9) or 1-alkylpentaboranes(9) to position isomers at 20° (suitable donors are, for example: 2,6-dimethylpyridine and trimethylamine), takes place intramolecularly via borane anions, donor·$B_5H_7^{\ominus}$. Since pentaborane(9) has only three distinguishable positions available, a total of only three isomers can occur – the original B_5H_8-1–D, and B_5H_8-2–D and B_5H_8-μ–D[*362, 566*]:

$$B_5H_8\text{-1–D} + C_5H_3(Me_2)N \longrightarrow C_5H_3(Me_2)NH^{\oplus}B_5H_7D^{\ominus} \longrightarrow B_5H_8\text{-(2 or }\mu\text{)-D} + C_5H_3(Me_2)N$$

$$B_5H_8\text{-1–Et} + Me_3N \longrightarrow Me_3NH^{\oplus}B_5H_7Et^{\ominus} \longrightarrow B_5H_8\text{-2–Et} + Me_3N$$

Other isomerization mechanisms [B_5H_7-1,2-$(CH_3)_2$ → B_5H_7-2,3-$(CH_3)_2$] have also been discussed[*235a, 565, 568a, 844*]. Probably, however, there are at least two donor-catalyzed isomerization mechanisms, depending on the activity of the donor[*237a*]. Above 145°, 1-deuteropentaborane(9) exchanges hydrogen (H,D) intermolecularly, so that all stages of deuteration B_5H_{9-n}-(1 . . . 5,μ)-D_n can be observed (PMR spectrum) in statistical distribution[*566*]. See also [*566b*].

In the hydrolysis and alcoholysis of B_5H_9, the B–B bond is attacked more rapidly than the B–H bond[*694–696*]:

$$>B{-}B< \;\;\ominus O{-}H \longrightarrow >B{-}O^{\ominus} + >B{-}H \qquad\qquad >B{-}H \;\; \ominus O{-}H \longrightarrow >B{-}O^{\ominus} + H_2$$

The first stage of the over-all equation $B_5H_9 + 15H_2O \rightarrow 5B(OH_3) + 12H_2$ is a B_4H_6 fragment (found by mass spectroscopy). Oxygen also attacks the B–B bond more rapidly than the B–H bond. Through the intermediate stage $H_3B_3O_3$ (boroxin) diborane, tetraborane, and a compound $B_2H_2O_3$ are formed [*22, 23, 31a, 170, 280b*]. The most probable structure of the latter from its IR and mass spectra, is a trigonal pyramid III, from microwave investigations a planar molecule with C_{2v} symmetry I [*280b*]:

I II III

$H_2B_2O_3$ decomposes photochemically to yield B_2O_3 and H_2 and reacts with B_2H_6 to yield boroxine $H_3B_3O_3$ [*280b*].

Pentaborane(9) adds trimethylamine at –78°. $B_5H_9 \cdot 2NMe_3$ [$=(Me_3N)_2BH_2^{\oplus}$-$B_4H_7^{\ominus}$?] is formed. When the adduct is heated, it decomposes partly into the components and partly with the formation of trimethylamineborane and trimethylamine-containing B–H polymers [*109*]:

$$B_5H_9 + 2NMe_3 \underset{-78°}{\overset{>-78°}{\rightleftharpoons}} B_5H_9 \cdot 2NMe_3 \begin{cases} \xrightarrow{\text{asymm. cleavage}} (Me_3N)_2BH_2^{\oplus}B_4H_7^{\ominus}\,? \\ \xrightarrow[-BH_3 \cdot NMe_3]{\text{symm. cleavage}} B_4H_6 \cdot NMe_3 \longrightarrow \text{polymers} \end{cases}$$

With N,N,N′,N′-tetramethylethylenediamine, pentaborane(9) smoothly forms in an exothermic reaction the colourless crystalline adduct $B_5H_9 \cdot Me_2NCH_2CH_2$-$NMe_2$, which can be recrystallized from chlorinated hydrocarbons. In contrast to $B_5H_9 \cdot 2NMe_2$ it is stable in the air at 25°. When the adduct is heated with methanol, $B_4H_8(Me_2NCH_2CH_2NMe_2)$, trimethoxyboron, and hydrogen [*524*] are formed:

$$B_5H_9 \cdot Me_2NCH_2CH_2NMe_2 + 3CH_3OH \longrightarrow B_4H_8(Me_2NCH_2CH_2NMe_2) + B(OCH_3)_3 + 2H_2$$

When $B_4H_8(Me_2NCH_2CH_2NMe_2)$ is heated to 150°, volatile $Me_2NCH_2CH_2$-$NMe_2 \cdot (BH_3)_2$ is formed, together with the ions $B_3H_8^{\ominus}$, $B_{12}H_{12}^{2\ominus}$, and $(Me_2NCH_2$-$CH_2NMe_2)BH_2^{\oplus}$ [*524*].

$B_5H_9 \cdot Me_2NCH_2CH_2NMe_2$ and $B_4H_8(Me_2NCH_2CH_2NMe_2)$ hydrolyse in a highly acid medium to hydrogen, boric acid, and the cation [*524*]:

$$\left[\begin{array}{l} CH_2{-}N \\ | \\ CH_2{-}N \end{array} \!\!\!\rightarrow B \begin{array}{l} H \\ \\ H \end{array}\right]^{\oplus}$$

Pentaborane(9) adds to the C=C double bond in the presence of $AlCl_3$, $FeCl_3$, or $SnCl_4$. By nucleophilic attack on the apical B atom, this gives exclusively 1-alkylpentaborane in 50–85% yield, for example [*55*, *567d*, *628*, *630*]:

$$B_5H_9 + CH_2{=}CH_2 \xrightarrow{AlCl_3} B_5H_8\text{-}1\text{-}(CH_2CH_3)$$

$$B_5H_9 + CH_2{=}CHCH \xrightarrow{FeCl_3} B_5H_8\text{-}1\text{-}[CH(CH_3)_2]$$

$$B_5H_9 + CH_2{=}CHC_2H_5 \xrightarrow{FeCl_3} B_5H_8\text{-}1\text{-}[CH(CH_3)(C_2H_5)]$$

Pentaborane(9) also adds to the C≡C triple bond. Polymers with C chains to which B_5H_8 groups are appended are formed [*221*, cit. *469*]:

$$B_5H_9 + CH{\equiv}CH \longrightarrow \ldots\text{-}\underset{B_5H_8}{\overset{H}{\underset{|}{C}}}\text{———}\underset{H}{\overset{H}{\underset{|}{C}}}\text{-}\ldots$$

In the presence of Friedel-Crafts catalysts, pentaborane(9) and ethyl chloride or bromide form 1-ethylpentaborane [*625*]:

$$B_5H_9 + C_2H_5X \xrightarrow{AlCl_3} B_5H_8\text{-}1\text{-}(C_2H_5) + HX$$

However, $C_2H_5BCl_2$ and $(C_2H_5)_2BCl$, which are difficult to separate, are also formed [*567d*].

In the reaction of CH_2Cl_2, CH_2Br_2, or CH_2ClBr with B_5H_9 in the presence of $AlCl_3$, di(1-pentaboranyl)methane and dichloroboryl(1-pentaboranyl) methane are formed [*12*]:

$$B_5H_9 + CH_2Cl_2 \xrightarrow[70°,\ 5h]{AlCl_3,\ \Delta} B_5H_8\text{–}CH_2\text{–}B_5H_8,\ B_5H_8\text{–}CH_2\text{–}BCl_2,\ HCl$$

Figure 4.24 shows the structure of 2,3-dimethylpentaborane.

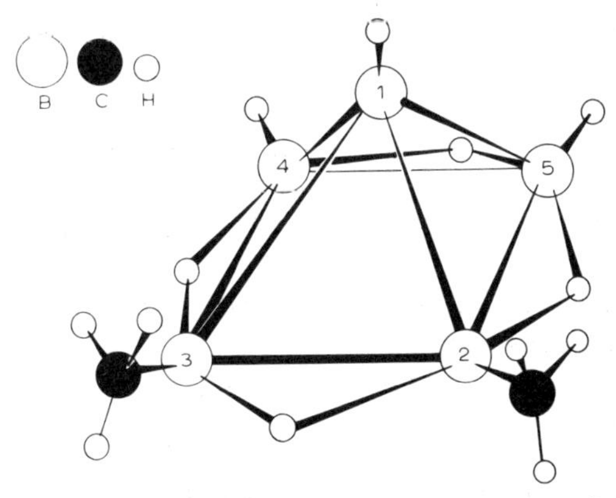

Fig. 4.24. Steric structure of B_5H_7-2,3-$(CH_3)_2$ [*235a*]. Distances: C–B = 1.55 Å,
B_{base}–B_{base} = 1.80–1.82 Å   B–H single bond = 0.98–1.11 Å
B_{apex}–B_{base} = 1.66–1.67 Å   B–H bridge = 1.32–1.50 Å

References p. 355

B_5H_8-1–Cl or B_5H_8-1–Br react with $NaMn(CO)_5$ and $NaRe(CO)_5$ to produce 2-B_5H_8-$Mn(CO)_5$ and 2-B_5H_8-$Re(CO)_5$, respectively (with σ bonds between metal and boron) [*237c*].

Alkali metals (M = Na, K) form $M_2B_5H_9$ with pentaborane(9), probably in a reaction similar to that described for the reaction of Na with B_2H_6 and B_4H_{10} (see sections 13 and 15.4) [cit. *359*]. In vacuum, the sodium derivatives releases the pentaborane(9) again:

$$Na_2B_5H_9 \xrightarrow{\text{vacuum}} 2Na + B_5H_9$$

With HCl, the potassium derivative $K_2B_5H_9$ is chlorinated; pentaborane(9) is regenerated in a side-reaction [cit. *359*]:

$$K_2B_5H_9 \begin{cases} \xrightarrow{+3HCl} K_2B_5H_6Cl_3 + 3H_2 \\ \xrightarrow{+2HCl} B_5H_9 + 2KCl + H_2 \end{cases}$$

Methyl- or n-butyllithium forms with pentaborane(9) in diethyl ether the non-volatile $Li^{\oplus}B_5H_8^{\ominus}$ as a solvated colourless viscous oil [*237a*, *249b*], *e.g.*:

$$B_5H_9 + Bu^nLi \xrightarrow{-78\text{ to }-30^\circ} LiB_5H_8 + Bu^nH$$

Lithium hydride, sodium hydride and potassium hydride form LiB_5H_8, NaB_5H_8 and KB_5H_8 with pentaborane(9) in various ethers (slowly in Et_2O) [*567d*,*249b*]:

$$MH + B_5H_9 \xrightarrow{\text{diglyme}} MB_5H_8 + H_2$$

To judge from the NMR spectrum, the structure of $B_5H_8^{\ominus}$ does not differ greatly from that of B_5H_9 [*237a*].

With an excess of HCl or DCl, LiB_5H_8 forms pentaborane(9) or μ-mono-deuteropentaborane [*237a*], and with $(CH_3)_3SiCl$ (similar with Me_3GeCl, Me_3SnCl and Me_3PbCl; H_3SiCl, H_3GeX [*237e*]) μ-trimethylsilylpentaborane with a B–Si–B three-centre two-electron bond [*237b*]:

$$LiB_5H_8 + DCl \longrightarrow B_5H_8\text{-}\mu\text{-}D + LiCl$$

$$LiB_5H_8 + (CH_3)_3SiCl \longrightarrow B_5H_8\text{-}\mu\text{-}[Si(CH_3)_3] + LiCl$$

Similarly, R_2PCl reacts with LiB_5H_8 in ether to form B_5H_8-2,3(μ)-$P(CH_3)_2$ or B_5H_8-2,3(μ)-$P(CH_3)(CF_3)$ [*116a*].

MB_5H_8 and B_2H_6 in ether or glyme react to give MBH_4 and an isomer of the normal B_6H_{10} [*249b*]:

$$MB_5H_8 + B_2H_6 \xrightarrow{-78^\circ} MBH_4 + B_6H_{10}$$

Upon standing at 20° in solution, it appears to convert, at least partially, to normal hexaborane(10). By refluxing of the B_6H_{10} product, obtained from the reaction of B_2H_6 with either NaB_5H_8 or KB_5H_8, one obtains $B_{10}H_{14}$ (yield 20–30%) [*249b*].

17. Reactions of pentaborane(11), hexaborane(10) and hexaborane(12)

The H–D exchange and the ^{10}B–^{11}B exchange in the B_5H_{11}–$(BD_3)_2$ system at 30.4° is a homogeneous first-order gas reaction [*441*].

Donors attack the B_5 skeleton of pentaborane(11) with the elimination of BH_3. Weak (potential) donors yield mainly B_4H_{10} and $(BH_3)_2$ [or, in the case where the donor is water, $B(OH)_3$]; on the other hand, strong donors also give considerable amounts of hexaborane(10) [*62, 63, 127, 558a, 558b, 724*]:

$$B_5H_{11} + H_2 \longrightarrow B_4H_{10} + \tfrac{1}{2}(BH_3)_2$$

$$B_5H_{11} + 3HOH \longrightarrow B_4H_{10} + B(OH)_3 + 2H_2$$

$$B_5H_{11} + 2D_2O \longrightarrow [BH_2(D_2O)_2^{\oplus} + B_4H_9^{\ominus}(?)] \xrightarrow{+D_2O} B_4H_9D + B(OD)_3 + 2HD$$

$$B_5H_{11} + NMe_3 \xrightarrow{-BH_3\cdot NMe_3} B_4H_8\cdot NMe_3 \begin{cases} \rightarrow \tfrac{1}{2}B_5H_9 + \tfrac{1}{2}B_3H_7\cdot NMe_3 + \tfrac{1}{2}NMe_3 \\ \rightarrow \tfrac{1}{2}B_6H_{10} + BH_3\cdot NMe_3 \end{cases}$$

$$B_5H_{11} + 2CO \xrightarrow[15-20\ atm]{22-27°} B_4H_8\cdot CO + BH_3\cdot CO$$

$$B_5H_{11} + 2PF_3 \xrightarrow{29-32°} B_4H_8\cdot PF_3 + BH_3\cdot PF_3$$

The CO in $B_4H_8\cdot CO$ can be displaced nucleophilically. However, the donor strength of PF_3 is too low for this. On the other hand, the donor strength of NMe_3 is too great; in this case the B_4H_8 skeleton is destroyed with the formation of $BH_3\cdot NMe_3$ and other products. In contrast, the weaker donor $PF_2(NMe_2)$ forms a stable adduct:

$$B_4H_8\cdot CO + PF_2(NMe_2) \xrightarrow{-20°} B_4H_8\cdot PF_2(NMe_2) + CO$$

while

$$B_4H_8\cdot PF_3 + CO \xrightarrow{18\ atm,\ 28°,\ 1\ h} B_4H_8\cdot CO + PF_3$$

$B_4H_8\cdot CO$ can add further donors, for example [*724*]:

$$B_4H_8\cdot CO + NMe_3 \longrightarrow B_4H_8CO\cdot NMe_3$$

$$B_4H_8\cdot CO + OMe_2 \longrightarrow B_4H_8CO\cdot OMe_2$$

$$B_4H_8\cdot CO + 4C_2H_4 \longrightarrow B_4H_4(C_2H_5)_4\cdot CO$$

In the gas phase, pentaborane(11) and ethene form at 25° $B_5H_{10}(C_2H_5)$ (main product), together with small amounts of $C_2H_4B_4H_8$, B_4H_{10}, $B_2H_4(C_2H_5)_2$, $B_2H_5(C_2H_5)$, B_2H_6, and H_2. The following mechanism of the formation of $B_5H_{10}(C_2H_5)$ satisfies the observations [*488*]:

$$B_5H_{11} \rightleftharpoons B_4H_8 + BH_3$$

$$BH_3 + C_2H_4 \longrightarrow BH_2(C_2H_5)$$

$$B_4H_8 + BH_2(C_2H_5) \longrightarrow B_5H_{10}(C_2H_5)$$

$$BH_3 + BH_2(C_2H_5) \rightleftharpoons B_2H_5(C_2H_5)$$

$$2BH_2(C_2H_5) \rightleftharpoons B_2H_4(C_2H_5)_2$$

The ethylation of B_5H_{11} by C_2H_4 can therefore be reduced to the replacement of the BH_3 group of the B_5H_{11} by a $BH_2(C_2H_5)$ group, with a B_4H_8 fragment occurring as an intermediate [*722*].

The gas-phase reaction of B_5H_{11} with HC≡CH, $HC{\equiv}CCH_3$ and $CH_3C{\equiv}CCH_3$ at 25–50° yields alkyl-substituted boranes (main product) and carboranes (in the B_4H_{10} reaction the carborane yields are much higher) [*278g*].

Pentaborane(11) reacts with sodium amalgam in ether. Here the rate of reaction with the donor-active ether is higher than that with sodium, so that the B_5H_{11} undergoes degradation [*359*]:

$$B_5H_{11} \xrightarrow[\Delta]{+\text{Na(Hg)} + \text{OEt}_2} NaB_3H_8\text{, polymers, other products}$$

Hexaborane(12) reacts almost quantitatively with an excess of water to give tetraborane(10) [*241*]:

$$B_6H_{12} + 6HOH \xrightarrow{0^\circ} B_4H_{10} + 2B(OH)_3 + 4H_2$$

Dimethyl ether abstracts a BH_3 group with the formation of pentaborane(9) and diborane [*241*]:

$$B_6H_{12} \xrightarrow{Me_2O} B_5H_9 + BH_3.$$

Hexaborane(10), which slowly decomposes at 20°, does not hydrolyse completely at 90° in 16 h [*258*]. Like $B_{10}H_{14}$, the deprotonation of B_6H_{10} takes place with loss of a bridge hydrogen [*374b*]:

$$LiB_5H_8 + B_6H_{10} \xrightarrow[Et_2O]{-60^\circ} B_5H_9 + LiB_6H_9$$

$$LiB_6H_9 + B_{10}H_{14} \xrightarrow[Et_2O]{-78^\circ} B_6H_{10} + LiB_{10}H_{13}$$

18. B_6 to B_{20} hydrides

For additional review literature see for example [*725g*].

18.1 Conversion in the $B_{10}H_{14}$ system

Recent investigations have shown that $B_{10}H_{14}$ is far more sensitive to heat than Stock [*739*], the discoverer of $B_{10}H_{14}$, assumed. Both in the gas phase and in the liquid phase, it undergoes 90 to 100% decomposition at 200 to 250° into hydrogen and polymeric $(BH_{0.78-1.0})_x$ [*706*]. In the gas phase, the pyrolysis takes place very slowly below 150°; at 170 to 238° it is homogeneous and first-order with respect to $B_{10}H_{14}$. Hydrogen—but not nitrogen, helium, or argon—inhibits the pyrolysis since the rate-determining H_2-forming pyrolysis step is the (reversible?) cleavage of a B–H bond. First, however, a polymeric intermediate of decaborane-like units arises without the evolution of H_2. This evolves H_2 (no volatile boranes) only on further polymerisation to $(BH)_x$ [*38*].

18.2 Elimination of protons from $B_{10}H_{14}$

18.2.1 Synthesis of $B_{10}H_{13}^{\ominus}$ and $B_{10}H_{12}^{2\ominus}$

With the monobasic acid decaborane(14), strong donors lead to the elimination of a proton and the formation of yellow salts of the singly negatively charged anion $B_{10}H_{13}^{\ominus}$. The latter slowly changes into the decaborane(14)–donor adduct $B_{10}H_{14}\cdot$ donor. In the same medium (for example, in H_2O or ROH) the B_{10} lattice is destroyed substantially more slowly:

$$B_{10}H_{14} + \text{donor} \xrightarrow{\text{fast}} [B_{10}H_{13}]^{\ominus}\,[\text{H donor}]^{\oplus} \xrightarrow{\text{slow}} B_{10}H_{14}\cdot\text{donor}$$

$$\Delta\big\downarrow \text{ very slow probably}$$

$$\text{non-ionic degradation } 10B(OH)_3 + 22H_2$$

Suitable donors are:

$OH^{\ominus}$ Reaction with H_2O, D_2O, or NaOH in ethanol, dioxan or acetonitrile [*19, 20, 285, 638, 694*], or with R_4NOH in diglyme and water [*318*]

$OR^{\ominus}$ Reaction with butanol, benzyl alcohol, cyclopentanol, sodium methoxide, lithium methoxide [*41, 42, 572*]

$CN^{\ominus}$ Reaction with NaCN in water [*411c, 411e*]

$(CH_3)_3\overset{\oplus}{N}\overset{\ominus}{C}H_2$ Reaction with $(CH_3)_4NBr + C_6H_5Li \rightarrow (CH_3)_3NCH_2 + LiBr + C_6H_6$ in diethyl ether [*318*]

$(C_6H_5)_3\overset{\oplus}{P}\overset{\ominus}{C}H_2$ In diethyl ether [*297*]

$NH_2^{\ominus}$ Reaction with $NaNH_2$ in diethyl ether [*572*]

References p. 355

$H^{\ominus}$ Reaction with NaH in diethyl ether[*15a,56,359,522,651a, 843a*]

$BH_4^{\ominus}$ Reaction with $LiBH_4$ in diethyl ether[*359, 653*]

Here, for example, oxonium, tetraorganylammonium, phosphonium, sodium, and lithium salts of decaborane(14) are formed:

$$B_{10}H_{14} + H_2O \longrightarrow H_3O^{\oplus}B_{10}H_{13}^{\ominus}$$

$$B_{10}H_{14} + (CH_3)_3\overset{\oplus}{N}CH_2^{\ominus} \longrightarrow (CH_3)_4N^{\oplus}B_{10}H_{13}^{\ominus}$$

$$B_{10}H_{14} + (C_6H_5)_3\overset{\oplus}{P}CH_2^{\ominus} \longrightarrow (C_6H_5)_3PCH_3^{\oplus}B_{10}H_{13}^{\ominus}$$

$$B_{10}H_{14} + NaOH \longrightarrow Na^{\oplus}B_{10}H_{13}^{\ominus} + H_2O$$

$$B_{10}H_{14} \xrightarrow[-H_2]{+NaH} Na^{\oplus}B_{10}H_{13}^{\ominus} \xrightarrow[-H_2]{+NaH} (Na^{\oplus})_2B_{10}H_{12}^{2\ominus}$$ [*843a*]

The formation of $B_{10}H_{13}^{\ominus}$ in the reaction of $B_{10}H_{14}$ with $LiBH_4$ or $NaBH_4$ in ether takes place via $B_{10}H_{15}^{\ominus}$ [*473, 653*]:

$$B_{10}H_{14} + M^{\oplus}BH_4^{\ominus} \xrightarrow[-BH3\cdot OR_2]{\text{(diethyl ether)}} M^{\oplus}B_{10}H_{15}^{\ominus} \xrightarrow[-H_2]{} M^{\oplus}B_{10}H_{13}^{\ominus}$$

(M = Li, Na) *hydride transfer* *elimination of hydrogen*

With $LiBH_4$ or $NaBH_4$ at an elevated temperature, however, $B_{10}H_{14}$ forms $B_{11}H_{14}^{\ominus}$ (see section 18.11):

The sodium salt $NaB_{10}H_{13}$ prepared in an aqueous medium differs from that prepared in an ethereal medium spectroscopically, in X-ray characteristics, and in its reaction products (with RX)[*56, 359*]. Likewise, samples of $LiB_{10}H_{13}$ prepared in Et_2O and in THF ($B_{10}H_{14}$ + LiH) differ[*12a*]. For the preparation of a reactive $MB_{10}H_{13}$ one must work at very low temperatures to avoid reactions with the solvents (ether, THF, diglyme)[*15a, 15b*]. Which H atom in $B_{10}H_{14}$ is split off is uncertain. Although bridge H atoms are more negative than terminal H atoms, bridge H atoms are more acidic, since in the elimination of a proton from the H bridge the B_{10} skeleton changes only slightly (B–H^{μ}–B bond → B–B bond). In the elimination of a proton in the case of a terminal H atom, on the other hand, the structure must change drastically[*470*]. There have as yet been no theoretical discussions of the structure of the $B_{10}H_{13}^{\ominus}$ ion (see Fig. 4.25)[*468*]. For degradation reactions of $B_{10}H_{14}$ by donors in an aqueous medium to give $B_9H_{13}^{\ominus}$, see section 18.10.

In the reaction of decaborane(14) with Grignard compounds, protons are eliminated (unlike the reaction with diborane)[*708*].

With alkylmagnesium halides, diborane, which contains only hydridic hydrogen, forms a magnesium halidohydride HMgX and a trialkylboron[*840*]:

$$(BH_3)_2 + 6\,RMgX \longrightarrow \left[\begin{matrix} \overset{\delta+}{B} \cdots \overset{\delta-}{H} \\ \vdots \quad\quad \vdots \\ \overset{\delta-}{R} \cdots Mg\overset{\delta+}{-}X \end{matrix}\right] \longrightarrow 2\,BR_3 + 6\,MgHX$$

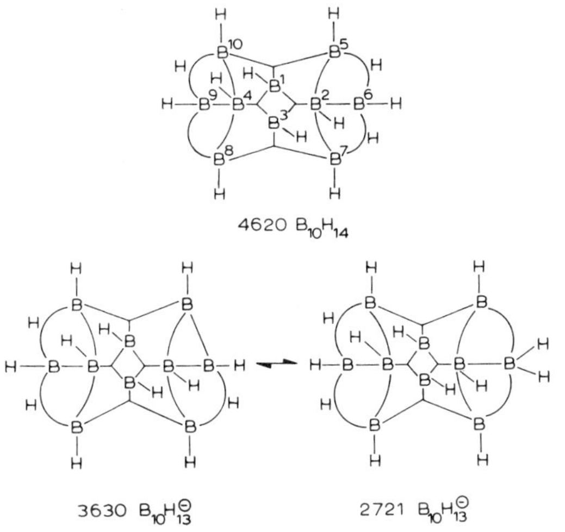

4620 $B_{10}H_{14}$

3630 $B_{10}H_{13}^{\ominus}$ 2721 $B_{10}H_{13}^{\ominus}$

Fig. 4.25. Comparison of the topological structures of $B_{10}H_{14}$ (see also section 2, Fig. 4.11) and $B_{10}H_{13}^{\ominus}$ (two tautomeric forms).

On the other hand, decaborane(14), which contains acidic hydrogen, undergoes an 80 to 90% reaction with alkylmagnesium halides[*182*, *244*, *708*] with *trans*-Grignardisation (not in the apical positions 6 and 9[*708*])[28]

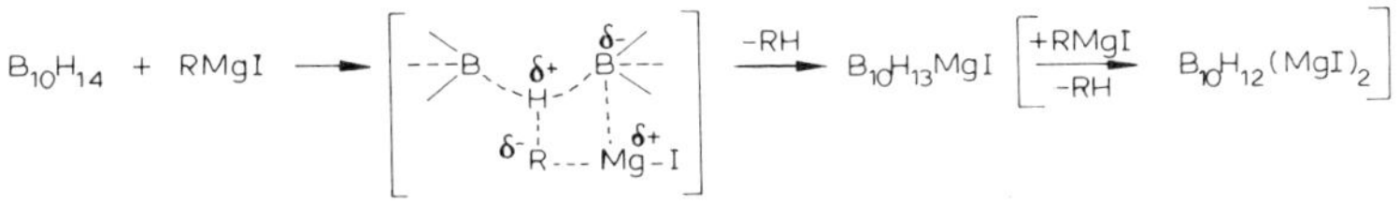

$$B_{10}H_{14} + RMgI \longrightarrow \left[\text{B}^{\delta+}\cdots\text{H}\cdots\text{B}^{\delta-};\ {}^{\delta-}\text{R}\cdots\text{Mg}^{\delta+}\text{-I} \right] \xrightarrow{-RH} B_{10}H_{13}MgI \left[\xrightarrow[-RH]{+RMgI} B_{10}H_{12}(MgI)_2 \right]$$

and only 10 to 20% reaction with the formation of a 6-alkyldecaborane:

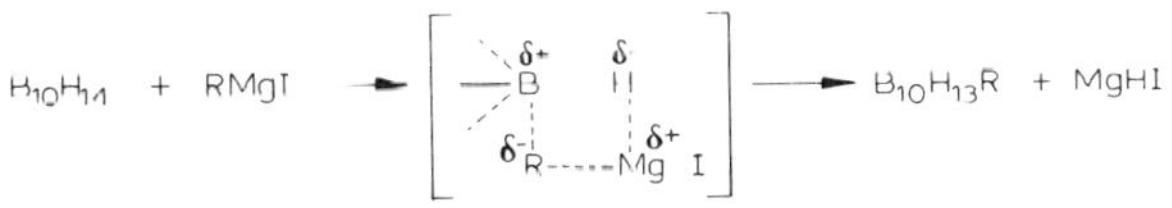

$$H_{10}H_{14} + RMgI \longrightarrow \left[\text{B}^{\delta+}\ \text{H}^{\delta-};\ {}^{\delta-}\text{R}\cdots\text{Mg}^{\delta+}\ \text{I} \right] \longrightarrow B_{10}H_{13}R + MgHI$$

Possibly, the reactions in ether take place in a similar manner to those in water via the free $B_{10}H_{13}^{\ominus}$ ions probably solvated in position 6 or 9[*56*].

Organometallic compounds, such as Et_2Mg, Me_2Zn, Et_2Zn, Ph_2Zn and Et_2Cd, react at 20° with $B_{10}H_{14}$ in diethyl ether or tetrahydrofuran to yield $MB_{10}H_{12}\cdot x$ ether ($x = 1.3$–2.0), neutral 6,9-internally bridged derivatives of $B_{10}H_{14}$[*277c*, *277d*]. Alkylmetals used are thought to be partially ionised in ether according to the equilibrium

$$2R_2M \rightleftharpoons RM^{\oplus} + R_3M^{\ominus}$$

[28]The reaction with $(CH_3)_2SO_4$, on the other hand, gives 50% of 5- and 50% of 6-$B_{10}H_{13}(CH_3)$ [*182*].

Thus, the first step in the reaction of R_2M with $B_{10}H_{14}$ will be the nucleophilic attack by $R_3M^{\ominus}$ on an acidic bridge hydrogen to form a transient $B_{10}H_{13}^{\ominus}$ anion:

$$\begin{array}{rcl} R_3M^{\ominus} + B_{10}H_{14} & \longrightarrow & B_{10}H_{13}^{\ominus} + R_2M + RH \\ RM^{\oplus} + B_{10}H_{13}^{\ominus} & \longrightarrow & RMB_{10}H_{13} \\ RMB_{10}H_{13} & \longrightarrow & MB_{10}H_{12} + RH \\ \hline R_2M + B_{10}H_{14} & \xrightarrow{\text{ether}} & 6{,}9\text{-}MB_{10}H_{12} + 2RH \end{array}$$

$MB_{10}H_{12}$ ionises in aqueous solution to yield the dianion $[(B_{10}H_{12})_2M]^{2\ominus}$ [*277c*, *277d*].

18.2.2 *Substitution in* $B_{10}H_{13}^{\ominus}$

Both $B_{10}H_{13}Na$ and $B_{10}H_{13}MgI$ resemble their organic analogues RNa and RMgI. Thus, with water, alkyl halides, benzyl chloride, dialkyl sulphates, and triethyloxonium fluoroborate they form the corresponding decarboranes(14) substituted in positions 1, 5, or 6 (unsubstituted in the case of HOH)[*56*, *244*, *572*, *708*]. Examples:

$$B_{10}H_{13}MgI + HOH \longrightarrow B_{10}H_{14} + MgIOH$$

$$B_{10}H_{13}Na + CH_3I \longrightarrow B_{10}H_{13}(CH_3) + NaI$$

Fig. 4.26 shows the structure of 1-ethyldecarborane(14).

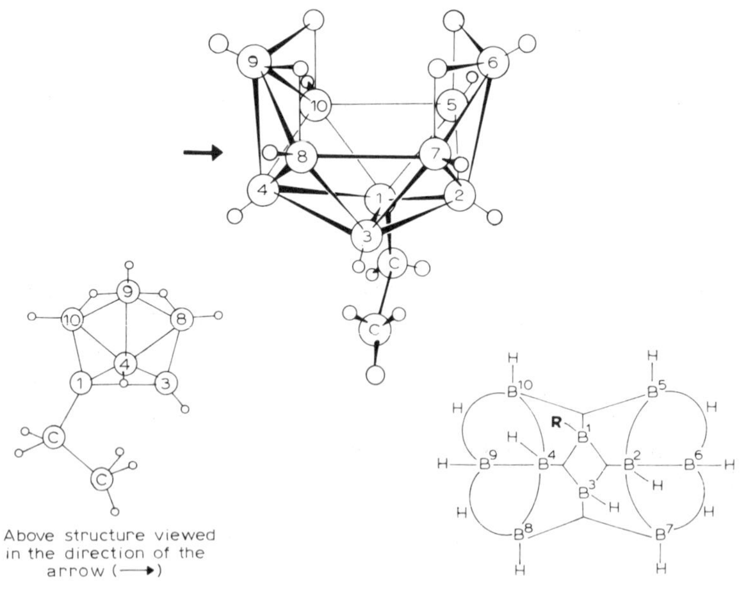

Above structure viewed in the direction of the arrow (⟶)

Fig. 4.26. Steric and topological structure of $B_{10}H_{13}$-1-C_2H_5. Structural parameters [*588*, *644*]: B–B = 1.70–1.98 Å, B–C = 1.59 Å, B–H = 0.91–1.36 Å, C–C = 1.55 Å, C–H = 0.91–1.20 Å.

18.2.3 Elimination of electrons from $B_{10}H_{13}^{\ominus}$

In ethereal solution at 20°, $B_{10}H_{13}^{\ominus}$ is oxidized by iodine but not iodinated. On the contrary, $B_{10}H_{13}Na$ rapidly forms $B_{10}H_{13}(OR)$ with a yield of 13 to 26% [*313*]:

$$B_{10}H_{13}^{\ominus}\cdot OR_2 \xrightarrow[-2I^{\ominus}]{+I_2} B_{10}H_{13}\overset{\oplus}{O}R_2 \xrightarrow[-RI]{+I^{\ominus}} B_{10}H_{13}(OR)$$

$$B_{10}H_{13}\overset{\oplus}{O}R_2 \xrightarrow[\text{side reaction}]{-H^{\oplus},\,-\text{alkene}} B_{10}H_{13}(OR)$$

$R_2O = (CH_3)_2O, (C_2H_5)_2O, (n\text{-}C_3H_7)_2O, (n\text{-}C_4H_9)_2O, (C_6H_5)(CH_3)O$
$(\rightarrow B_{10}H_{13}OC_6H_5)$.

At −80°, iodine converts $B_{10}H_{13}^{\ominus}$, which is orange-yellow in ethereal solution, into brilliant orange iodinated decarborane(14) [*572*].

18.2.4 Derivatives of the anions $B_{10}H_{13}^{\ominus}$ and $B_{10}H_{14}^{2\ominus}$: $B_{10}H_{12}X^{\ominus}$ and $B_{10}H_{13}X^{2\ominus}$

Decarborane(14) reacts with cyanide, cyanate, thiocyanate, methoxide, and cyanoformate in various ethers to form the corresponding [Na or $(CH_3)_4N$] salts of the anion $B_{10}H_{12}X^{\ominus}$. They form sparingly soluble diethyl ether or dioxan adducts [*7*]:

Examples: $B_{10}H_{14} + X^{\ominus} \xrightarrow{\text{ether}} B_{10}H_{13}^{\ominus} + HX \longrightarrow B_{10}H_{12}X^{\ominus} + H_2$

$$B_{10}H_{14} + NaCN \xrightarrow[-H_2]{\text{THF, }25°} NaB_{10}H_{12}(CN) \xrightarrow{+2OEt_2} NaB_{10}H_{12}(CN)\cdot 2O(C_2H_5)_2$$

$$B_{10}H_{14} + NaNCO \xrightarrow[-H_2]{\text{THF, }25°} NaB_{10}H_{12}(NCO) \xrightarrow{+2.5\text{ dioxan}} NaB_{10}H_{12}(NCO)\cdot 2.5\text{ dioxan}$$

In an aqueous medium, sodium cyanide forms with decarborane(14) the doubly negatively charged anion $B_{10}H_{13}(CN)^{2\ominus}$, a derivative of the anion $B_{10}H_{14}^{2\ominus}$ described in section 18.4.1 [*415*]:

$$B_{10}H_{14} + 2NaCN \xrightarrow{\text{water}} Na_2B_{10}H_{13}(CN) + HCN$$

Even Et_2NH abstracts a proton [*274*]:

$$B_{10}H_{14} + 2Et_2NH \rightarrow B_{10}H_{13}(NHEt_2)^{\ominus}NH_2Et_2^{\oplus}$$

With sodium cyanide in dimethyl sulphide solution, decarborane(14) forms the adduct $NaB_{10}H_{12}(CN)\cdot SMe_2$ [the anion of which, $B_{10}H_{12}(CN)\cdot SMe_2^{\ominus}$, is derived from the above-described anion $B_{10}H_{13}(CN)^{2\ominus}$ by the replacement of a $H^{\ominus}$ ion of the latter by the neutral donor SMe_2]:

$$B_{10}H_{14} + NaCN + SMe_2 \xrightarrow{\text{dimethyl sulphide}} NaB_{10}H_{12}(CN)\cdot SMe_2 + H_2$$

According to the ^{11}B NMR and UV spectra, $B_{10}H_{13}(CN)^{2\ominus}$ and $B_{10}H_{12}(CN)\cdot SMe_2$ appear to be derived from $B_{10}H_{14}^{2\ominus}$ substituted at positions 6 and 9, *i.e.* the apical positions.

References p. 355

18.3 *Electrophilic substitution in $B_{10}H_{14}$*

18.3.1 Halogenation with halogen cations

When decaborane(14) is heated dry with iodine above its melting point, 2-iododecaborane, 5-iododecaborane, 2,4-di-iododecaborane (for structure, see Fig. 4.27) and ?,5-di-iododecaborane are formed. They are separated by re-crystallisation from n-pentane. $B_{10}H_{13}$-2–I is the least soluble in this[*12a*]. The monoiodo derivatives, for their part, can be converted into the di-iodo derivatives [*341*, *342*, *644*, *650*, *779a*].

$$B_{10}H_{14} \xrightarrow[\Delta]{+I_2,\,-HI} 35\%\ B_{10}H_{13}\text{-}2\text{–}I + 65\%\ B_{10}H_{13}\text{-}5\text{–}I$$

$$\begin{array}{cc} +I_2,\ -HI,\ \Delta \downarrow & +I_2,\ -HI,\ \Delta \downarrow \\ \downarrow 45\% & 55\% \end{array}$$

$$B_{10}H_{14} \xrightarrow[\Delta]{+2I_2,\,-2HI} 63\%\ B_{10}H_{12}\text{-}?,5\text{–}I_2 + 37\%\ B_{10}H_{12}\text{-}2,4\text{–}I_2$$

The preparation of $B_{10}H_{13}$-2–Br takes place similarly[*350*, *650*].

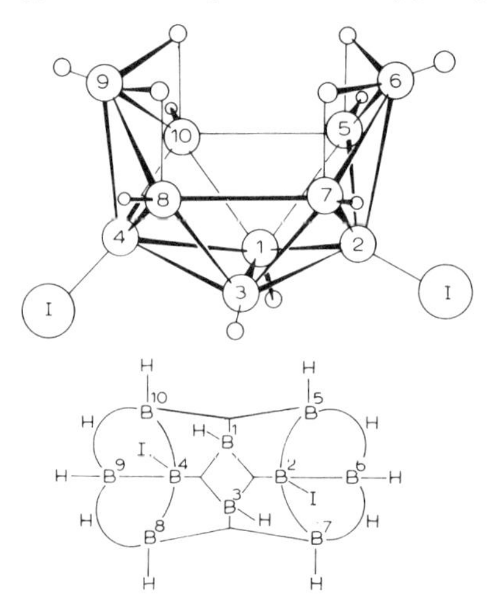

Fig. 4.27. Steric and topological structure of $B_{10}H_{12}$-2,4-I_2[*588*, *644*].

In hydrocarbons, $B_{10}H_{14}$ and I_2 do not react at 100°. In carbon disulphide in the presence of $AlCl_3$, 35% of 2- and 65% of 5-iododecaboranes are formed. The reaction takes place via an electrophilic attack of $I^{\oplus}$ on electron-rich centres of the $B_{10}H_{14}$[*342*].

The reaction of F_2CHCH_3, $AlCl_3$, and $B_{10}H_{14}$ in carbon disulphide forms isomeric chlorodecaboranes[*342a*].

18.3.2 Alkylation with carbonium ions

In analogy with a Friedel-Crafts reaction, decaborane(14) can be alkylated in carbon disulphide with an alkyl bromide in the presence of the aluminium chloride ($R^{\oplus}AlCl_3Br^{\ominus}$)[*55, 845*]:

$$B_{10}H_{14} + n\,EtBr + n\,AlCl_3 \longrightarrow [B_{10}H_{14}Et^{\oplus} + AlCl_3Br^{\ominus}] \longrightarrow B_{10}H_{14-n}Et_n + n\,HBr + n\,AlBr_3$$

(n = 1–3)

It is substituted in positions 1, 2, 3, and 4, but not in position 5 (contrast to iodination with I_2). For the kinetics of the gas phase reaction of $B_{10}H_{14}$ and EtBr, which leads to $B_{10}H_{13}$-2–Et, see [*199*].

18.3.3 Hydrogen–deuterium exchange

In the $AlCl_3$-catalysed reaction of decaborane(14) with deuterium chloride ($DCl + AlCl_3 \rightarrow D^{\oplus}AlCl_4^{\ominus}$) in carbon disulphide, hydrogen is electrophilically replaced by deuterium in positions 1, 2, 3, and 4[*184, 185*]:

$$B_{10}H_{14} \xrightarrow[-H^{\oplus}AlCl_4^{\ominus}]{+D^{\oplus}AlCl_4^{\ominus}} B_{10}H_{13}D \xrightarrow[-H^{\oplus}AlCl_4^{\ominus}]{+D^{\oplus}AlCl_4^{\ominus}} \cdots$$

On the other hand, in the base-catalysed H–D exchange in mixtures of dioxan and deuterium oxide and of dioxan and deuterium chloride, only the four acidic bridge H atoms (5/6, 6/7, 8/9, and 9/10) are rapidly substituted (proton elimination). The subsequent slow migration of deuterium to terminal positions causes the final formation of 5/6, 6/7, 8/9, 9/10, 5, 6, 7, 8, 9, 10-decaborane(14)-d_{10}, $B_{10}H_4D_{10}$[*185, 296, 522, 693*]. In this case, positions 1, 2, 3, and 4 remain unoccupied by D.

During isotope exchange in the $B_{10}H_{14}$–$(BD_3)_2$ system at 100° (as in the case of B_5H_9), terminal H atoms (and not bridge H atoms) are preferentially replaced by deuterium[*391*].

18.4 Addition of electrons to $B_{10}H_{14}$

18.4.1 $\cdot B_{10}H_{14}^{\ominus}$, $B_{10}H_{14}^{2\ominus}$, $B_{10}H_{15}^{\ominus}$

With sodium in liquid ammonia, decaborane(14) forms the white crystalline $Na_2B_{10}H_{14}$[*769*]:

$$B_{10}H_{14} + 2e^{\ominus} \xrightarrow{\text{liq. } NH_3} B_{10}H_{14}^{2\ominus}$$

In diethyl ether or tetrahydrofuran, $B_{10}H_{14}$ and Na or Na(Hg) form the red ion-radical $\cdot B_{10}H_{14}^{\ominus}$, which with further sodium forms $B_{10}H_{14}^{2\ominus}$ or, on long standing, changes into $B_{10}H_{13}^{\ominus}$ with the evolution of H_2[*769*]:

$$B_{10}H_{14} \xrightarrow{+e^{\ominus}} \cdot B_{10}H_{14}^{\ominus} \begin{cases} \overset{+e^{\ominus}}{\rightleftharpoons} B_{10}H_{14}^{2\ominus} \\ \longrightarrow B_{10}H_{13}^{\ominus} + \tfrac{1}{2}H_2 \end{cases}$$

References p. 355

$B_{10}H_{14}^{2\ominus}$ is also formed by the reaction of $B_{10}H_{14}$ with $BH_4^{\ominus}$ in an aqueous medium [*533c*, cit. *534b*].

On electrolytic reduction in glycol polyethers, 1 mole of $B_{10}H_{14}$ takes up 1.5 electrons and forms $B_{10}H_{12}^{2\ominus}$, $B_{10}H_{14}^{2\ominus}$, and $B_{10}H_{15}^{\ominus}$ [*138a, 187, 625b, 625d, 716b, 717a*]:

$$4B_{10}H_{14} + 6e^{\ominus} \rightleftharpoons B_{10}H_{12}^{2\ominus} + B_{10}H_{14}^{2\ominus} + 2B_{10}H_{15}^{\ominus}$$

With HCl, $Na_2B_{10}H_{14}$ forms decaborane(14) and halogenation products that have not been further investigated [cit. *295*], while $NaB_{10}H_{13}$ with HCl re-forms decaborane(14) quantitatively [*359*]:

$$NaB_{10}H_{13} + HCl \longrightarrow B_{10}H_{14} + NaCl$$

In 50% ethanol, on the other hand, $Na_2B_{10}H_{14}$ and HCl form $B_{10}H_{15}^{\ominus}$ by the addition of a proton [*295*]:

$$Na_2B_{10}H_{14} + HCl \xrightarrow{\text{ethanol}} NaB_{10}H_{15} + NaCl$$

$B_{10}H_{14}^{2\ominus}$ is oxidized by iodine to decaborane $B_{10}H_{14}$ [*295*]:

$$Na_2B_{10}H_{14} + I_2 \longrightarrow B_{10}H_{14} + 2NaI$$

18.4.2 Topological structures of $B_{10}H_{14}^{2\ominus}$ and $B_{10}H_{15}^{\ominus}$

It was predicted before its preparation that $B_{10}H_{14}^{2\ominus}$ would be a stable ion [*471*]. Its topological structure is comparable with that of the 4620 $B_{10}H_{14}$. The two structures differ in the apical positions 6 and 9. In 4620 $B_{10}H_{14}$, these are BH groups, and in 2632 $B_{10}H_{14}^{2\ominus}$ they are BH_2 groups (Fig. 4.28). The existence of $B_{10}H_{15}^{\ominus}$ was also predicted. Figure 4.29 shows two possible topological structures.

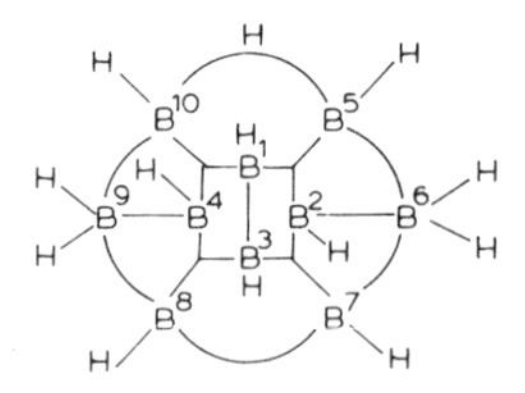

Fig. 4.28. Topological structure of $2632B_{10}H_{14}^{2\ominus}$.

18.5 Synthesis and reactions of $B_{10}H_{12}Z_2$

18.5.1 Synthesis of $B_{10}H_{12}Z_2$ by electrophilic substitution in $B_{10}H_{14}$

With decaborane(14), neutral donors Z form stable boranes of the typical formula $B_{10}H_{13}Z^{\ominus}$ or, in the overwhelming majority of cases, $B_{10}H_{12}Z_2$.

Regarded formally, the formation of $B_{10}H_{13}Z^{\ominus}$, which will be described in more detail in section 18.7, is a proton elimination with the subsequent addition

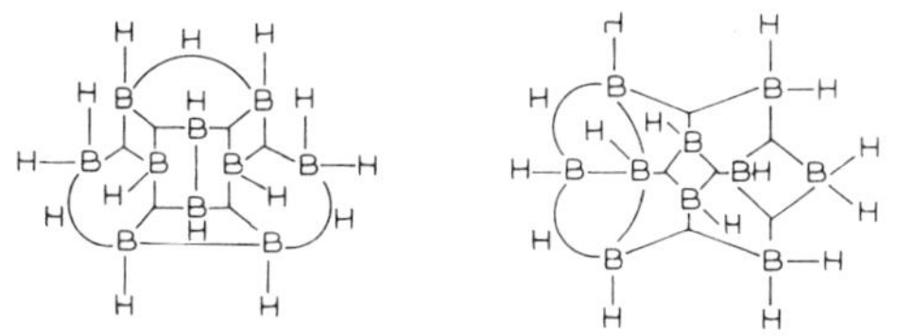

3612 $B_{10}H_{15}^{\ominus}$ 2701 $B_{10}H_{15}^{\ominus}$

Fig. 4.29. Topological structures of $B_{10}H_{15}^{\ominus}$.

of the neutral donor Z to the $B_{10}H_{13}^{\ominus}$ ion formed:

$$B_{10}H_{14} \xrightarrow[\text{cyclohexane}]{+Z,\,-ZH^{\oplus}} B_{10}H_{13}^{\ominus} \xrightarrow{+Z} B_{10}H_{13}Z^{\ominus}$$

On the other hand, the formation of $B_{10}H_{12}Z_2$ is, regarded formally, an electron addition with subsequent electrophilic displacement of two hydride ions in positions 6 and 9 (*i.e.* in the $2BH_2$ groups, *cf.* Fig. 4.28) by molecules of the donor Z[*609*]:

$$B_{10}H_{14} \xrightarrow[\text{benzene}]{+2e^{\ominus}} B_{10}H_{14}^{2\ominus} \xrightarrow{+2Z} B_{10}H_{12}\text{-6,9-}Z_2 + 2H^{\ominus} (2H^{\ominus} \longrightarrow H_2 + 2e^{\ominus})$$

$$B_{10}H_{13}\text{-2-Br} \xrightarrow{+2e^{\ominus}} B_{10}H_{13}\text{-2-Br}^{2\ominus} \xrightarrow{+2Z} B_{10}H_{11}\text{-2-Br-6,9-}Z_2 + 2H^{\ominus}$$

$$(2H^{\ominus} \longrightarrow H_2 + 2e^{\ominus})$$

A synthesis of $B_{10}H_{12}Z_2$ without using $B_{10}H_{14}$, which is difficult to prepare and expensive to buy, as starting material, consists in opening the skeleton of $B_{10}H_{10}^{2\ominus}$, which is easy to prepare[*493a*], for example:

$$3(NH_4)_2B_{10}H_{10} + 2HCl + 2Et_2S \xrightarrow[Et_2S]{20°} B_{10}H_{12}(Et_2S)_2 + 2NH_4Cl\ (NH_4)_2B_{10}H_{10}$$

Both the anions $B_{10}H_{13}Z^{\ominus}$ and $B_{10}H_{12}Z_2{}^{2\ominus}$ are derived from the anions $B_{10}H_{14}^{2\ominus}$ described previously (section 18.4) (Fig. 4.28) by the substitution of, in the first case, one, and in the second case, two negative hydride ions $:H^{\ominus}$ by neutral donor molecules :Z in the same way as, for example, the donor complex B_3H_7Z of triborane(7) is derived from $B_3H_8{}^{\ominus}$ or the donor complex $B_2H_4Z_2$ of diborane(4) B_2H_4 is derived from $B_2H_6{}^{2\ominus}$.

Suitable donors Z are: dialkyl sulphides (with SMe_2, 80 to 90% yield of $B_{10}H_{12}$-$(SMe_2)_2$[*276, 340*]), nitriles[*316, 340, 645* , dialkylcyanamides[*216, 316, 571*], alkyl isonitriles[*300*], dialkyl sulphoxides[*415*], phosphanes[*316, 340, 678, 679, 681*], phosphites[*599*], phosphane oxides[*415*], amides[*415*], thioamides[*415*], tertiary amines[*317*], aniline derivatives[*37*], heterocyclic amines[*272*], and tertrazoles[*220*]. Most derivatives are colourless, but the pyridine, tetrazole, and heterocyclic amine derivatives are coloured because of the interaction of the π-electrons of the ligands with the $B_{10}H_{12}$ skeleton[*272, 273, 319a*]. The interaction, for example in $B_{10}H_{12}(NC_5H_5)_2$, is:

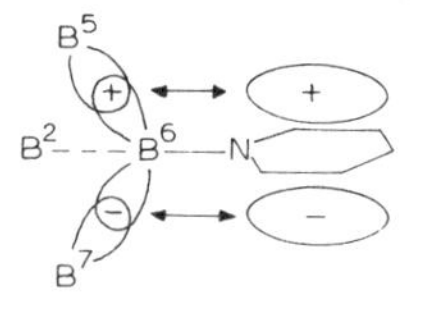

Of the two positions suitable for electrophilic attack in amides (O and N), only the oxygen is used. Thus, only 6-$R_2N{=}C(H)O$-$B_{10}H_{12}$-9-$OC(H){=}NR_2$ are formed (with the amides $HCON(CH_3)_2$, $CH_3CON(CH_3)_2$, $HCONHCH_3$, $CH_3CONHCH_3$)[*334a*].

18.5.2 Synthesis of $B_{10}H_{12}ZZ'$ by the protolysis of $B_{10}H_{12}Z^{\ominus}$ with $HZ'^{\oplus}$

$B_{10}H_{13}Z^{\ominus}$ anions (for preparation, see section 18.7) undergo protolysis with cations $HZ'^{\oplus}$ in non-aqueous solvents to give compounds of the type $B_{10}H_{12}ZZ'$. Since the ligand enters the electron-rich B^6 or B^9 position, unsymmetrical $B_{10}H_{12}Z_2$ derivatives can be obtained[*274, 534b*].

$$B_{10}H_{13}Z^{\ominus} + HZ'^{\oplus} \longrightarrow B_{10}H_{12}ZZ' + H_2$$

where $HZ'^{\oplus} = R_3NH^{\oplus}$, $CH_3CNH^{\oplus}$, $(C_2H_5)_2SH^{\oplus}$, but not $O(C_2H_4)_2OH^{\oplus}$ (anion $Cl^{\ominus}$ in each case), and $Z = N(C_2H_5)_2H$.

18.5.3 Structure of $B_{10}H_{12}Z_2$

The ^{11}B NMR spectra of $B_{10}H_{10}(NCMe)_2$, $B_{10}H_{12}(SMe_2)_2$, and other derivatives confirm the 6,9 position of the ligands Z[*571, 633*]. Figures 4.30 and 4.31 show the structure and topological structure of $B_{10}H_{12}Z_2$ (Z = SMe_2, NCMe).

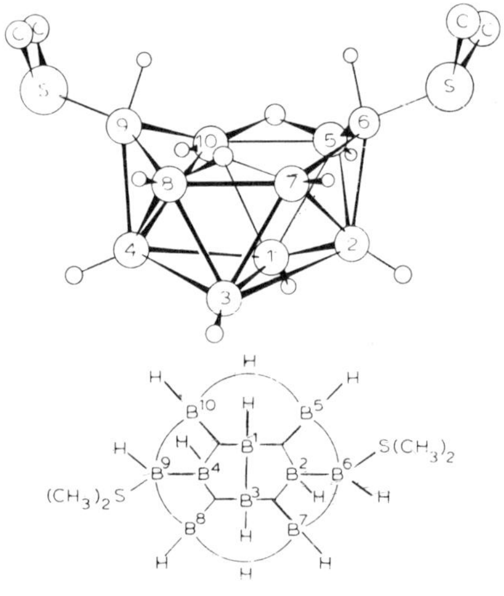

Fig. 4.30. Steric and topological structure of $B_{10}H_{12}[S(CH_3)_2]_2$. Structural parameters[*632*]:
B–B = 1.74–1.91 Å B–S = 1.92–1.93 Å
B–H = 1.02–1.35 Å S–C = 1.81–1.84 Å

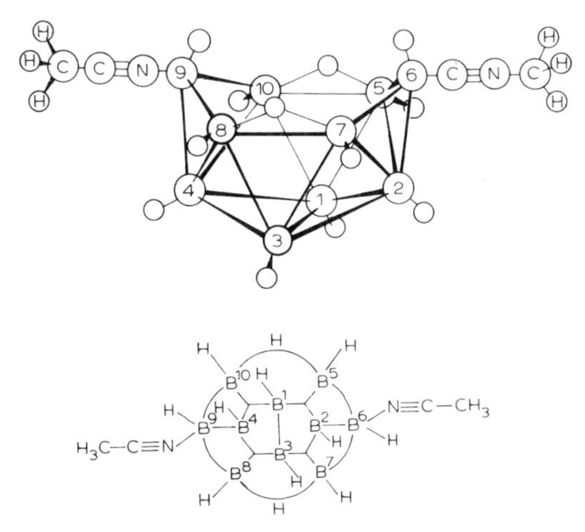

Fig. 4.31. Steric and topological structure of $B_{10}H_{12}(NCCH_3)_2$. Structural parameters [*26, 580, 609, 632, 636*]: B–B = 1.74–1.88 Å, B–H = 1.06–1.22 Å, N–C = 1.137 Å, C–C = 1.446 Å, C–H = 0.77–1.08 Å.

The arrangement of the B atoms is approximately the same as in $B_{10}H_{14}$, only in $B_{10}H_{12}Z_2$ the B–B distances differ less ($B_{10}H_{14}$: 1.71–2.01; $B_{10}H_{12}Z_2$: 1.74–1.91 Å). The arrangement of the H atoms differs as follows from that in $B_{10}H_{14}$: each B atom does in fact possess a terminal H atom, as in $B_{10}H_{14}$, but the four bridge H atoms of $B_{10}H_{14}$ (5/6, 6/7, 8/9, 9/10) are absent from $B_{10}H_{12}Z_2$. Instead of these, the molecule contains two bridge H atoms in positions 7/8 and 5/10. Both $B_{10}H_{14}^{2\ominus}$ and $B_{10}H_{12}(NCCH_3)_2$ contain in their B skeleton two electrons more and, correspondingly, a total of two bridge bonds less than $B_{10}H_{14}$.

18.5.4 Reactions of $B_{10}H_{12}Z_2$

Reactions with $B_{10}H_{12}Z_2$ take place by four basic mechanisms: (a) donor exchange ($B_{10}H_{12}Z_2 \rightarrow B_{10}H_{12}Y_2$); (b) splitting off of the donor ($B_{10}H_{12}Z_2 \rightarrow B_{10}H_{12}Z^{\ominus}$); (c) in the case of unsaturated donors, nucleophilic addition to the donor; (d) degradation of the boron skeleton ($B_{10}H_{12}Z_2 \rightarrow B_9H_{13}Z$, see section 18.10.2).

Donor exchange in $B_{10}H_{12}Z_2$. The bonds between the $B_{10}H_{12}$ skeleton and the donors Z (in the electron-rich positions 6 and 9) are strengthened by the back-coordination of electrons to the donor. Consequently, potential donors with stronger potential boron-donor bonds displace those with weaker bonds:

$$B_{10}H_{12}Z_2 + 2Y \longrightarrow B_{10}H_{12}Y_2 + 2Z$$

In the following conversion sequence, donors on the right irreversibly displace those to the left of them: $(CH_3)_2S < CH_3CN < (C_2H_5)_2NCN < HCON(CH_3)_2 = CH_3CON(CH_3)_2 < (C_2H_5)_3N = C_5H_5N = (C_2H_5)_3P$ [*316, 317, 571*]. Consequently, all the other derivatives $B_{10}H_{12}Y_2$ can be obtained from $B_{10}H_{12}(SMe_2)_2$, which is easy to prepare [*272*].

References p. 355

Splitting off of the donor in $B_{10}H_{12}Z_2$: synthesis of $B_{10}H_{12}Z$. When $B_{10}H_{12}(SMe_2)_2$ is heated in mesitylene, the stable white $B_{10}H_{12}(SMe_2)$ forms slowly [*415*]. It does not react with SMe_2 or MeCN at 20°; at higher temperatures it reacts with PPh_3 or SMe_2 to give $B_{10}H_{12}(PPh_3)_2$ or $B_{10}H_{12}(SMe_2)_2$, as the case may be. Since the UV and ^{11}B NMR spectra of $B_{10}H_{12}Z$ resemble those of the $B_{10}H_{13}^{\ominus}$ ion, $B_{10}H_{12}Z$ is to be regarded as a substituted $B_{10}H_{13}^{\ominus}$ ion (replacement of the donor $H^{\ominus}$ by the donor Z) [*415*].

Nucleophilic addition to unsaturated donors in $B_{10}H_{12}Z_2$. Proton-containing donors such as dialkylamines, hydrogen cyanide, and hydrazine, add to the C≡N triple bond in $B_{10}H_{12}(N{\equiv}CMe)_2$ or $B_{10}H_{12}(N{\equiv}CNR_2)_2$, for example [*214a, 316, 317, 571*]:

$$B_{10}H_{12}(N{\equiv}CMe)_2 + RR'NH \longrightarrow \left[B_{10}H_{12}\ \ NH{=}C\begin{matrix}\diagup Me\\ \diagdown NRR'\end{matrix}\right]_2$$

$$R = C_2H_5,\ n\text{-}C_3H_7;\ R' = H,\ C_2H_5,\ n\text{-}C_3H_7$$

$$B_{10}H_{12}(N{\equiv}CNR_2)_2 + HCN \longrightarrow \left[B_{10}H_{12}\ \ NH{=}C\begin{matrix}\diagup NR_2\\ \diagdown CN\end{matrix}\right]_2 \qquad \text{(red)}$$

18.6 *Synthesis of $B_{10}H_{11}BrZ_2$ by electrophilic substitution in $B_{10}H_{13}$-2 Br*

Like decaborane(14) itself (see section 18.5.1), 2-bromodecaborane $B_{10}H_{13}$-2-Br can be converted by neutral donors Z (diethyl sulphide, acetonitrile, triphenylphosphane) in boiling benzene into $B_{10}H_{11}BrZ_2$ [*340*]:

$$B_{10}H_{13}Br + 2Z \xrightarrow{\text{benzene}} B_{10}H_{11}BrZ_2 + H_2$$

where $Z = S(C_2H_5)_2$, CH_3CN, $P(C_6H_5)_3$.

Like $B_{10}H_{12}Z_2$ (see section 18.5.4), $B_{10}H_{11}BrZ_2$ also exchanges donor ligands, for example [*340*]:

$$B_{10}H_{11}Br[S(C_2H_5)_2]_2 + Z' \longrightarrow B_{10}H_{11}BrZ'_2 + 2S(C_2H_5)_2$$

where $Z' = CH_3CN$, or $P(C_6H_5)_3$; $N(C_2H_5)_3$ is less suitable. With trimethylamine, $B_{10}H_{11}Br[S(C_2H_5)_2]_2$ forms $[(C_2H_5)_3NH^{\oplus}]_2(B_{10}H_9Br)^{2\ominus}$ by proton elimination [*340*].

18.7 *Synthesis of $B_{10}H_{13}Z^{\ominus}$*

In the reaction of decaborane(14) with an excess of diethylamine in cyclohexane, a diethylammonium salt with the anion $B_{10}H_{13}NEt_2H^{\ominus}$ is formed [*272, 274, 318*]:

$$B_{10}H_{14} + 2Et_2NH \xrightarrow{\text{cyclohexane}} Et_2NH_2^{\oplus}B_{10}H_{13}NEt_2H^{\ominus}$$

Generally, $B_{10}H_{13}Z^{\ominus}$ can be prepared by the addition of a donor Z to $B_{10}H_{13}^{\ominus}$:

$$B_{10}H_{14} \xrightarrow[\text{ether}]{+NaH,\ -H_2} NaB_{10}H_{13} \xrightarrow[\text{ether}]{+Z} NaB_{10}H_{13}Z$$

where Z = $(C_2H_5)_3N$, $(C_2H_5)_2NH$, $C_2H_5NH_2$, C_5H_5N, $C_5H_{11}N$, $(C_2H_5)_3P$, $(CH_3)_2S$.

The structure of the $B_{10}H_{13}Z^{\ominus}$ anion is probably identical with that of $B_{10}H_{12}Z_2$ (replacement of the donor Z in the latter by the donor $H^{\ominus}$)[*274, 415*]; this is indicated by the similarity in the ^{11}B NMR spectra of $B_{10}H_{12}(SMe_2)_2$ and $B_{10}H_{13}(SMe_2)^{\ominus}$ and their chemical reactions. Reference has already been made to the analogy between $B_{10}H_{13}Z^{\ominus}$ and $B_{10}H_{14}^{2\ominus}$ (replacement of the donor $H^{\ominus}$ in the latter by the donor Z). $B_{10}H_{13}Z^{\ominus}$ readily protolyses with cations $HZ^{\oplus}$ to $B_{10}H_{12}Z_2$ (see section 18.5.2).

18.8 Amine adducts of $B_{10}H_{14}$

With gaseous ammonia at 0°, decaborane(14) forms adducts $B_{10}H_{14}\cdot nNH_3$ (n = 1–6) [*768,848a*, cit. *469*]. The salt-like $B_{10}H_{14}\cdot 3NH_3$ is particularly stable. On heating to 95°, it probably forms $B_{10}H_{18}N_2$, and at 120° probably $B_{10}H_{12}(NH_2)_2$ [*768*, cit. *469*]. With dimethylamine, decaborane(14) forms $B_{10}H_{14}\cdot nMe_2NH$ (n = 1–5)[*225*].

Predictions concerning the structures of the "amine adducts" are difficult, since NH_2 can replace a bridge H atom and NH_3 a terminal H atom. Very probably $B_{10}H_{14}\cdot NH_3 = NH_4^{\oplus}B_{10}H_{13}^{\ominus}$; $B_{10}H_{14}\cdot 2NH_3$ or $B_{10}H_{14}\cdot 2R_2NH = NH_4^{\oplus}B_{10}H_{13}(NH_3)^{\ominus}$ or $R_2NH_2^{\oplus}B_{10}H_{13}(NR_2H)^{\ominus}$; and $B_{10}H_{14}\cdot 3NH_3 = (NH_4^{\oplus})_2$-$B_{10}H_{13}(NH_2)^{2\ominus}$ or $(NH_4^{\oplus})_2B_{10}H_{12}(NH_3)^{2\ominus}$ or still more probably $(NH_3)_2$-$B_9H_{13}(NH_2)^{\ominus}$ (replacement of a bridge H atom in $B_9H_{14}^{\ominus}$ by NH_2) or $(NH_3)_2$-$BH_2^{\oplus}B_9H_{12}(NH_3)^{\ominus}$ (replacement of a terminal hydride ion $H^{\ominus}$ in $B_9H_{10}^{2\ominus}$ by NH_3)[cit. *469*].

18.9 Synthesis and reactions of $B_{10}H_{10}^{2\ominus}$

18.9.1 Synthesis of $B_{10}H_{10}^{2\ominus}$ by the elimination of protons and ligands in $B_{10}H_{12}Z_2$

When $B_{10}H_{12}Z_2$ is treated with a donor Y, donor exchange takes place with the formation of the likewise covalent $B_{10}H_{12}Y_2$ (see section 18.5.4). The further action of Y leads to the transformation of the $B_{10}H_{12}Y_2$ skeleton, with the elimination of protons and ligands ($-2H^{\oplus}$, $-2Y \rightarrow 2YH^{\oplus}$), into the ion $B_{10}H_{10}^{2\ominus}$. The same applies to $B_{10}H_{11}$-2-Br-6,9-Z_2[*315, 317, 319h, 319i, 340, 534, 595*]:

$$B_{10}H_{12}Z_2 \xrightarrow[-2Z]{+2Y} B_{10}H_{12}Y_2 \xrightarrow[\text{rearrangement}]{(+Y)} (YH^{\oplus})_2B_{10}H_{10}^{2\ominus}$$

$$BrB_{10}H_{11}Z_2 \xrightarrow[-2Y]{+2Y} BrB_{10}H_{11}Y_2 \xrightarrow[\text{rearrangement}]{(+Y)} (YH^{\oplus})_2BrB_{10}H_9{}^{2\ominus}$$

Z = $(C_2H_5)_2S$, CH_3CN; Y = NH_3, $(C_2H_5)_3N$, $C_6H_5)_3P$, $(CH_3)_4NOH(-H_2O)$

The weaker the binding of the donors Z is, the easier is the rearrangement. Naturally, because of their completely different structures, the reactions of $B_{10}H_{12}Z_2$ and the isomeric $(ZH^{\oplus})_2B_{10}H_{10}^{2\ominus}$ differ (compare Figs. 4.32a and 4.32b with Figs. 4.30 and 4.31, respectively).

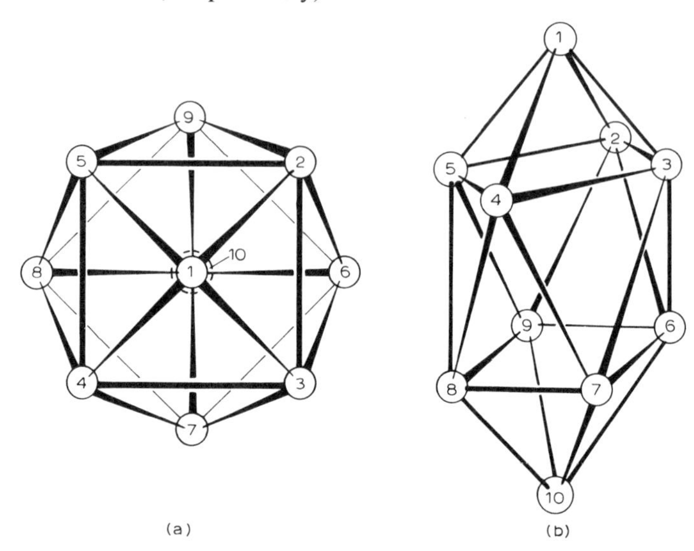

Fig. 4.32. Steric structure of the double-pyramidal tetragonal antiprism of the ion $B_{10}H_{10}^{2\ominus}$. (a) View from above, (b) view from the side.

$B_{10}H_{10}^{2\ominus}$ is formed in high yields by the controlled pyrolysis of tetraethylammonium boranates $(C_2H_5)_4NBH_4$ or $(C_2H_5)_4NB_3H_8$ at 185°[*487a*]:

$$10(C_2H_5)_4NBH_4 \xrightarrow{185°} [(C_2H_5)_4N]_2B_{10}H_{10} + 8(C_2H_5)_3N + 11H_2 + 8C_2H_6$$

$$10(C_2H_5)_4NB_3H_8 \xrightarrow{185°} 3[(C_2H_5)_4N]_2B_{10}H_{10} + 4(C_2H_5)_3N + 23H_2 + 4C_2H_6$$

Pyrolysis at 200 to 250° gives only small amounts of $B_{10}H_{10}^{2\ominus}$ together with other reaction products.

The tetramethylammonium boranates thermolyse differently [*487a*]:

$$(CH_3)_4NBH_4 \xrightarrow{185°} (CH_3)_3N \cdot BH_3 + CH_4$$

$$9(CH_3)_4NB_3H_8 \xrightarrow{185°} [(CH_3)_4N]_2B_{10}H_{10} + [(CH_3)_4N]_2B_{12}H_{12} + 5(CH_3)_3N \cdot BH_3 + 5CH_4 + 15H_2$$

The thermolysis of CsB_3H_8 forms $Cs_2B_{10}H_{10}$ (main product) and $Cs_2B_{11}H_{14}$ (by-product), together with $CsBH_4$ and hydrogen[*487a*]:

$$4CsB_3H_8 \xrightarrow{185°} Cs_2B_{10}H_{10} + 2CsBH_4 + 7H_2$$

$$5CsB_3H_8 \xrightarrow{185°} CsB_{11}H_{14} + 4CsBH_4 + 5H_2$$

In the thermolysis of KB_3H_8, approximately equal amounts of $K_2B_{10}H_{10}$ and $K_2B_{12}H_{12}$ are formed [*487a*]:

$$4KB_3H_8 \xrightarrow{185°} K_2B_{10}H_{10}+2KBH_4+7H_2$$
$$5KB_3H_8 \xrightarrow{185°} K_2B_{12}H_{12}+3KBH_4+8H_2$$

The thermolysis of $(C_2H_5)_3NHB_{10}H_{13}$, $[(C_2H_5)_3NH]_2B_{10}H_{14}$, and $[(C_6H_5)_3PCH_3]_2B_{10}H_{14}$ forms $B_{10}H_{10}^{2\ominus}$ [*705a*]:

$$2(C_2H_5)_3NHB_{10}H_{13} \xrightarrow{80°} [(C_2H_5)_3NH]_2B_{10}H_{10}+B_{10}H_{14}+H_2$$
$$[(C_2H_5)_3NH]_2B_{10}H_{14} \xrightarrow{165°} [(C_2H_5)_3NH]_2B_{10}H_{10}+2H_2$$
$$[(C_6H_5)_3PCH_3]_2B_{10}H_{14} \xrightarrow{165°} [(C_6H_5)_3PCH_3]_2B_{10}H_{10}+2H_2$$

On the separation of the borane anions, see [*487a*].

18.9.2 Reactions of $B_{10}H_{10}^{2\ominus}$ and its derivatives

The B_{10} skeleton of $B_{10}H_{10}^{2\ominus}$ and its derivatives is unusually stable. Electrophilic attacks are possible on both the electron-rich apical positions 1 and 10. In the disubstitution of $B_{10}H_{10}^{2\ominus}$, it is frequently not 1,10-derivatives that are obtained. Since skeletal isomerizations take place only at relatively high temperatures, it must be assumed that the entry of the first substituent determines the position of the following substituent. The reactions of $B_{10}H_{10}^{2\ominus}$ can frequently, but not always, be compared with those of benzene.

Acids form the stable $(H_2O^{\oplus})_2B_{10}H_{10}^{2\ominus}\cdot xH_2O$. As an acid, this is somewhat stronger than sulphuric acid. Silver salts yield insoluble covalent silver compounds without the separation of metallic silver.

The acid-catalyzed *H–D exchange* begins at the two apical positions 1 and 10. The apically substituted $B_{10}H_{10}^{2\ominus}$ isomerizes either by the migration of D (cleavage of the B–D bond) or by the transformation [*384*] of the B_{10} skeleton (without cleavage of the B–D bond) as Fig. 4.33 shows. Tropenium bromide $Tr^{\oplus}Br^{\ominus}$ forms the following salts with $(H_3O^{\oplus})_2B_{10}H_{10}^{2\ominus}$ [*294, 294b*]:

$$B_{10}H_{10}^{2\ominus}(H_3O^{\oplus})_2+Tr^{\oplus}Br^{\ominus} \xrightarrow[\text{water}]{\Delta} B_{10}H_{10}^{2\ominus}\,Tr^{\oplus}H_3O^{\oplus} \xrightarrow[-H_3O^{\oplus}]{+Cs^{\oplus}}$$
$$B_{10}H_{10}^{2\ominus}Cs^{\oplus}Tr^{\oplus} \xrightarrow[-CsBr]{+Tr^{\oplus}Br^{\ominus}} B_{10}H_{10}^{2\ominus}(Tr^{\oplus})_2$$

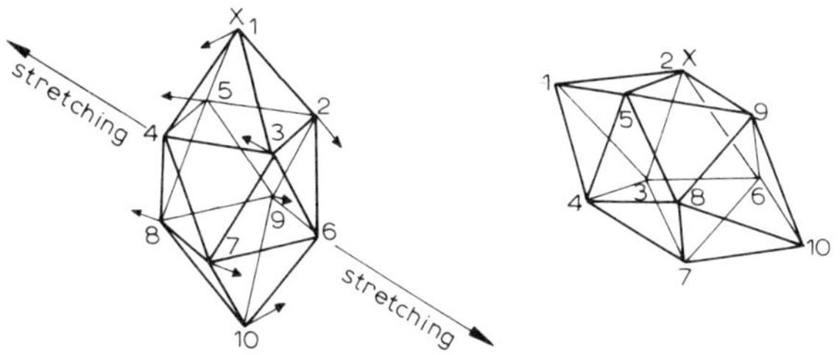

Fig. 4.33. Apex equator transformation. The B–X bond is not cleaved in this process [*384*].

References p. 355

All *halogens*, as well as N-iodosuccinimide and other similar halogenating agents, react in aqueous or alcoholic solution with the replacement of H by halogen. The rate of reaction falls with increasing substitution, so that all the halogenation products ($B_{10}H_{10-n}X_n$; $n=1$–10) can be isolated[*352,413,414*]. Their thermal stability is similar to that of $B_{10}H_{10}^{2\ominus}$. They withstand most nucleophilic attacks. Thus, for example, they are inert to MeONa + MeOH, KNH_2 + liq NH_3, NaC≡CH + liq. NH_3, and PhMgBr or MeMgBr + THF. However, with Li or Na or in liquid NH_3, $B_{10}H_7I_3^{2\ominus}$ forms $B_{10}H_{10}^{2\ominus}$. At 150° in vacuum, $(H_3O)_2$-$B_{10}Cl_{10}\cdot 5H_2O$, which recrystallises from aqueous solution, yields $(H_3O)_2B_{10}Cl_{10}\cdot 2H_2O$, and at 260° red crystals of B_9HCl_8[*232*].

With *dimethyl sulphoxide* Me_2SO, under acid conditions, $B_{10}H_{10}^{2\ominus}$ forms $B_{10}H_9SMe_2^{\ominus}$ and $B_{10}H_8(SMe_2)_2$[*412, 414, 522a*]:

$$B_{10}H_{10}^{2\ominus} + Me_2SO \xrightarrow[40-60^\circ]{(HCl)\Delta} B_{10}H_9SMe_2^{\ominus} \xrightarrow[\Delta]{+Me_2SO} B_{10}H_8(SMe_2)_2$$

With an excess of Me_2SO (as solvent) the ratio of the two compounds depends on the duration of the reaction. With MeCOOH as solvent, $B_{10}H_8(SMe_2)_2$ is formed preferentially, regardless of the stoichiometry of the reactants or the time of the reaction[*412, 414*]. The two apical positions are substituted mainly (in an excess of Me_2SO_4, for example, 77% of $B_{10}H_8$-1,10-$(SMe_2)_2$; in addition, equatorially substituted isomers are found[*412*]. With increasing occupancy of the apex positions, the reactivity with respect to electrophilic agents decreases: $B_{10}H_{10}^{2\ominus} > B_{10}H_9(SMe_2)^{\ominus} > B_{10}H_8(SMe_2)_2$. The apically substituted and equatorially substituted boranes can easily be separated by recrystallisation[*412*]. Heating leads to isomerization:

$$B_{10}H_8\text{-}1{,}10\text{-}(SMe_2)_2 \xrightarrow[45\text{ min}]{230^\circ} B_{10}H_8\text{-}2{,}7(8)\text{-}(SMe_2)_2$$

Hydroxylamine-0-sulphonic acid reacts with $B_{10}H_{10}^{2\ominus}$ in aqueous solution with the formation of $B_{10}H_9$-2-$NH_3^{\ominus}$ together with 2,4-, 2,3- and 2,7(8)-$B_{10}H_8(NH_3)_2$ [*335*]:

$$B_{10}H_{10}^{2\ominus} + NH_2OSO_3^{\ominus} \xrightarrow{-SO_4^{2\ominus}} B_{10}H_9\text{-}2\text{-}(NH_3)^{\ominus} \xrightarrow[-SO_4^{2\ominus}]{+NH_2OSO_3^{\ominus}} B_{10}H_8(NH_3)_2$$

With $B_{10}H_9$-1-$SMe_2^{\ominus}$, $B_{10}H_8$-1-(SMe_2)-6-NH_3 is formed.

The amine derivatives $B_{10}H_9$-2-$(NH_3)^{\ominus}$ and 2,4-, 2,3-, and 2,7(8)-$B_{10}H_8(NH_3)_2$ can be converted by means of *dimethyl sulphate*, Me_2SO_4, into the corresponding methylamine derivatives (formation of $B_{10}H_9$-2-$(NHMe_2)^{\ominus}$ and $B_{10}H_9$-2-$(NMe_3)^{\ominus}$ and of 2,4-, 2,3-, and 2,7(8)-$B_{10}H_8(NMe_3)_2$[*335*].

$B_{10}H_9$-2-$(NMe_3)^{\ominus}$ can be carbonylated with *oxaloyl dichloride* $(COCl)_2$ in acetonitrile[*332, 334*]:

$$B_{10}H_9(NMe_3)^{\ominus} + \begin{matrix} COCl \\ | \\ COCl \end{matrix} \xrightarrow{(CH_3CN)} B_{10}H_8(NMe_3)(CO) + CO + HCl + Cl^{\ominus}$$

The CO group can be converted with hydroxylamine-0-sulphonic acid into the NH_3 group and the latter can be methylated with dimethyl sulphate to 2,4- and 2,7(8)-$B_{10}H_8(NMe_3)_2$ [*332, 334*]:

$$B_{10}H_8\text{-2-}(NMe_3)(CO) + NH_2OSO_3^{\ominus} \xrightarrow{\Delta} B_{10}H_8\text{-2-}(NMe_3)(NH_3) \xrightarrow[\Delta(NaOH)]{+Me_2SO_4} B_{10}H_8(NMe_3)_2$$

$B_{10}H_9\text{-1-}(SMe_2)^{\ominus}$ has also been carbonylated with oxaloyl dichloride in acetonitrile, and the $B_{10}H_8\text{-1-}(SMe_2)\text{-6-}(CO)$ formed has been ammine-substituted with hydroxylamine-0-sulphonic acid to $B_{10}H_8\text{-1-}(Sme_2)\text{-6-}(NH_3)$ and has been diazotised and carboxylated with nitrous acid and subsequently reduced with zinc [formation of $B_{10}H_7\text{-1-}(SMe_2)\text{-6-}(COOH)\text{-10-}(N_2)$]. The latter can be converted with oxaloyl dichloride into the acid chloride, the acid chloride with sodium azide into the corresponding isocyanate, and the latter be hydrolyzed to the amine. Finally, the $B_{10}H_7\text{-1-}(SMe_2)\text{-6-}(NH_2)\text{-10-}N_2$ can be coupled with $H_2B_{12}Cl_{10}$ [*334*]:

$$B_{10}H_9\text{-1-}(SMe_2)^{\ominus} \xrightarrow[(CH_3CN)\ \Delta]{+(COCl)_2} B_{10}H_8\text{-1-}(SMe_2)\text{-6-}(CO) \xrightarrow[\Delta]{+NH_2OSO_3^{\ominus}} B_{10}H_8\text{-1-}(SMe_2)\text{-6-}(NH_3)$$

$$B_{10}H_8\text{-1-}(SMe_2)\text{-6-}(CO) \xrightarrow{(1)\ +HNO_2,\ +H_2O,\ \Delta;\ (2)\ +Zn,\ +HCl} B_{10}H_7\text{-1-}(SMe_2)\text{-6-}(COOH)\text{-10-}(N_2)$$

$$B_{10}H_7\text{-1-}(SMe_2)\text{-6-}(COOH)\text{-10-}(N_2) \xrightarrow[\Delta]{+(COCl)_2} B_{10}H_7(SMe_2)(COCl)(N_2) \xrightarrow[\Delta]{+NaN_3} B_{10}H_7(SMe_2)(NCO)(N_2)$$

$$B_{10}H_7(SMe_2)(NCO)(N_2) \xrightarrow[\Delta]{(1)\ +HCl,\ (2)\ +NaOH} B_{10}H_7\text{-1-}(SMe_2)\text{-6-}(NH_2)\text{-10-}(N_2) \xrightarrow[\Delta]{+H_2B_{12}Cl_{12}} [B_{10}H_7\text{-1-}(SMe_2)\text{-6-}(NH_2)\text{-10}(N_2)]_2B_{12}Cl_{10}$$

$B_{10}H_{10}^{2\ominus}$ reacts with the highly electrophilic *amide-chloride*, $(Me_2\overset{\oplus}{N}\text{–}CHCl)Cl$. However, here B–N bonds are formed and not B–C as a comparison with the "corresponding" organic reactions would lead one to expect [*331*]:

$$2\left[Me_2N\text{–}\overset{Cl}{\underset{H}{C^{\oplus}}}\right]Cl^{\ominus} + B_{10}H_{10}^{2\ominus} \longrightarrow (B_{10}\overset{\delta-}{H_8})\text{-2,7(8)-}[\overset{\delta+}{N}Me_2(CH_2Cl)]_2$$

for comparison:

$$2\left[R_2N\text{–}\overset{Cl}{\underset{R}{C^{\oplus}}}\right]Cl^{\ominus} + |\overset{\ominus}{C}\!\lt \longrightarrow R_2N\text{—}\overset{Cl}{\underset{R}{C}}\text{—}C\!\lt + Cl^{\ominus}$$

For reaction of $B_{10}H_{10}^{2\ominus}$ with aryldiazonium salts, see [*319g*].

References p. 355

The Cl atom in the neopentyl-like system $B_{10}H_8[NMe_2(CH_2Cl)]_2$ cannot be displaced nucleophilically. However, it reacts with alkali metals: sodium in alcohol reduces the Cl compound without isomerization to the $B_{10}H_{10}$ skeleton, *e.g.* [*331*]:

$$B_{10}H_8\text{-}2,7(8)\text{-}[NMe_2(CH_2Cl)]_2 \xrightarrow[(ROH)]{+Na,\,\Delta} B_{10}H_8\text{-}2,7(8)\text{-}(NMe_2)_2$$

With an excess (10–12 moles) of aqueous *nitrous acid*, HNO_2, $B_{10}H_{10}^{2\ominus}$ forms a highly explosive intermediate which can be converted with sodium boranate, $NaBH_4$, in methanol into the non-shock-sensitive colourless $B_{10}H_8\text{-}1,10\text{-}(N_2)_2$ [*411a, 418*] which has a closed polyhedral boron skeleton [*502b*]. The two N_2 groups of the latter can easily be replaced at 115–140° by amines, nitriles, ammonia, and carbon monoxide, for example [*418*]:

$$B_{10}H_8\text{-}1,10\text{-}(N_2)_2 + 2CO \xrightarrow[\text{without solvent or in } Fe(CO)_5]{120\text{–}140°,\,\Delta} B_{10}H_8\text{-}1,10\text{-}(CO)_2$$

$$B_{10}H_8\text{-}1,10\text{-}(N_2)_2 + CO \xrightarrow[\text{cyclohexane}]{\Delta} C_6H_{11}B_{10}H_7(CO)_2,\ (C_6H_{11})_2B_{10}H_6(CO)_2$$

$$B_{10}H_8\text{-}1,10\text{-}(N_2)_2 + 2NH_3 \longrightarrow B_{10}H_8(NH_3)_2 + 2N_2$$

Reactions of the carbonyl substituents. The two carbonyl substituents of $B_{10}H_8(CO)_2$ can both be converted into other substituents and be substituted, for example [*416a, 418*]:

(a) reactions of the ligand:

$$B_{10}H_8(CO)_2 + H_2O \rightleftharpoons H^{\oplus}B_{10}H_8(COOH)(CO)^{\ominus} \underset{-H_2O}{\overset{+H_2O}{\rightleftharpoons}}$$

$$(H^{\oplus})_2B_{10}H_8(COOH)_2{}^{2\ominus} \xrightarrow{(OH^{\ominus})} B_{10}H_8(COO)_2{}^{2\ominus}$$

$$B_{10}H_8(CO)_2 + LiAlH_4 \xrightarrow{\Delta} B_{10}H_8(CH_3)_2{}^{2\ominus}$$

$$B_{10}H_8(CO)_2 + 2MeOH \rightleftharpoons (H^{\oplus})_2B_{10}H_8(COOMe)_2{}^{2\ominus}$$

$$B_{10}H_8(CO)_2 + 4NH_3 \xrightarrow{\Delta} (NH_4{}^{\oplus})_2B_{10}H_8(CON_2)_2{}^{2\ominus}$$

$$B_{10}H_8(CO)_2 + NaN_3 \xrightarrow{\Delta} B_{10}H_8(NCO)_2{}^{2\ominus}$$

$$Cs_2B_{10}H_8(NCO)_2 \xrightarrow[\Delta]{300°} Cs_2B_{10}H_8(CN)_2$$

(b) ligand substitution:

$$B_{10}H_8(CO)_2 + NH_2OSO_3{}^{\ominus} + H_2O \xrightarrow{\Delta} B_{10}H_8(NH_3)_2 + CO_2$$

(c) substitution of the H atoms and reaction of the ligands:

$$B_{10}H_8(CO)_2 \xrightarrow[(water)]{+X_2,\,\Delta} (H^{\oplus})_2B_{10}X_8(COOH)_2^{2\ominus} \xrightarrow[heat]{-2H_2O} B_{10}X_8(CO)_2 \quad (X = Cl, Br)$$

The *halogen-substituted boranes*, $B_{10}H_{10-n}X_n^{2-}$, and their derivatives can react with the above-mentioned agents, for example [*411d*, *412a*, *418*]:

$$B_{10}Cl_8(CO)_2 + NaBH_4\,(LiBH_4) \xrightarrow{\Delta} B_{10}Cl_8(CH_2OH)_2^{2\ominus}$$

$$B_{10}Cl_8(CO)_2 + C_6H_5\,NMe_2 \xrightarrow{\Delta} B_{10}Cl_8(COC_6H_4NMe_2)_2^{2\ominus}$$

$$B_{10}Cl_8(CO)_2 + NH_2OSO_3^{\ominus} + H_2O \xrightarrow[\text{(water)}]{\Delta} B_{10}Cl_8(NH_3)_2 \xrightarrow[\Delta]{HNO_2}$$

$$B_{10}Cl_8(N_2)_2 \begin{cases} \xrightarrow{+NaN_3} B_{10}Cl_8(N_3)_2^{2\ominus} \\ \xrightarrow[\Delta]{+H_2S} B_{10}Cl_8(SH)_2 \end{cases}$$

With *aryldiazonium salts* ($X{-}C_6H_4N_2^{\oplus}BF_4^{\ominus}$, X = *p*-Me, *p*-Br, *p*-MeO, *p*-NO_2, *m*-NO_2, *m*-MeO, *m*-CF_3) $B_{10}H_{10}^{2\ominus}$ can easily be coupled to form coloured $X{-}C_6H_4NNB_{10}H_9^{2\ominus}$ (electrophilic substitution and proton elimination in accordance with the equation $ArNN^{\oplus} + B_{10}H_{10}^{2\ominus} \rightarrow ArNNB_{10}H_9^{2\ominus} + H^{\oplus}$) [*314*]. $B_{12}H_{12}^{2\ominus}$, which otherwise reacts similarly to $B_{10}H_{10}^{2\ominus}$ (for its reactions see section 18.12.2) cannot be coupled with aryldiazonium ions in acetonitrile [*314*].

The hydroxonium salt of $B_{10}H_{10}^{2\ominus}$ reacts with *benzoyl chloride*, C_6H_5COCl, to give $B_{10}H_9(COC_6H_5)^{2\ominus}$, which can react further in accordance with the following scheme [*416*]:

$$\begin{array}{ccccc} B_{10}H_9(COC_6H_5)^{2\ominus} & \xrightarrow[\Delta]{+H_2O_2} & B_{10}H_9(OCOC_6H_5)^{2\ominus} & \xrightarrow[\Delta]{+OH^{\ominus}} & B_{10}H_9(OH)^{2\ominus} \\ \Delta\downarrow +Cl_2 & & & & \Delta\downarrow +Cl_2 \\ B_{10}Cl_9(COC_6H_5)^{2\ominus} & \xrightarrow[\Delta]{+H_2O_2} & B_{10}Cl_9(OCOC_6H_5)^{2\ominus} & \xrightarrow[\Delta]{+HBr} & B_{10}Cl_9(OH)^{2\ominus} \end{array}$$

Potassium permanganate oxidizes $B_{10}H_{10}^{2\ominus}$ in aqueous solution at pH = 7 to boric acid [*385*]; at pH 7 there is a quantitative reaction (method of analysis) in accordance with the equation:

$$B_{10}H_{10}^{2\ominus} + 14MnO_4^{\ominus} + 18H_2O \longrightarrow 10B(OH)_4^{\ominus} + 6OH^{\ominus} + 14MnO_2$$

The oxidation of $B_{10}H_{10}^{2\ominus}$ with *copper(II) chloride* in ether gives the deep blue-violet radical $\cdot B_{10}H_9$ [*466a*]. On the opening of the skeleton and formation of $B_{10}H_{12}Z_2$, see section 18.5.1.

On being heated, $B_{10}H_8(NMe_2)_2$ *isomerises* to a mixture of various isomers. Isomerisation temperatures: 1,10-isomer: above 350°; vicinal isomers: 350° [*333*].

18.10 Degradation of $B_{10}H_{14}$

18.10.1 Synthesis of $B_9H_{13}Z$ by the protolysis of $B_{10}H_{13}^{2\ominus}$

In a non-aqueous medium, $B_{10}H_{13}Z^{\ominus}$ anions protolyse with $HZ^{\oplus}$ cations to give $B_{10}H_{12}Z_2$ (see section 18.5.2):

$$B_{10}H_{13}Z^{\ominus} + HZ^{\oplus} \longrightarrow B_{10}H_{12}Z_2 + H_2$$

In aqueous acids, on the other hand, a B atom in position 6 or 9 is split off quantitatively [*273*]:

$$B_{10}H_{13}Z^{\ominus} + H_3O^{\oplus} + 2H_2O \longrightarrow B_9H_{13}Z + B(OH)_3 + 2H_2$$

$$Z = C_2H_5NH_2, (C_2H_5)_2NH, (C_2H_5)_3N, (C_6H_5)_3P, NC_5H_5$$

18.10.2 Synthesis of $B_9H_{13}Z$ by the alcoholysis of $B_{10}H_{12}Z_2$

When $B_{10}H_{12}Z_2$ (feebly-bound ligands Z, not with NEt_3 or PPh_3) is boiled with alcohols, the B atom splits off from one of the identical positions 6 or 9. $B_9H_{13}Z$ is formed in 70% yield [*273, 276, 645*]:

$$B_{10}H_{12}Z_2 \xrightarrow[\text{relatively fast}]{+3ROH,\,-B(OR)_3,\,-H_2,\,-Z} B_9H_{13}Z\ (\xrightarrow[\text{slow}]{\Delta} B(OR)_3,\ H_2,\ Z)$$

$$Z = (C_2H_5)_2S, CH_3CN;\ R = CH_3, C_2H_5$$

In the alcoholysis of $B_{10}H_{12}(NCMe)_2$ with ethanol, an addition to the C≡N triple bond also takes place [*645*]:

$$B_{10}H_{12}(N{\equiv}CMe)_2 \xrightarrow[-B(OEt)_3,\,-MeCN,\,-H_2]{+3EtOH} B_9H_{13}(N{\equiv}CMe)$$

$$\xrightarrow{+EtOH} B_9H_{13}\left(NH{=}C\begin{matrix}\diagup Me\\ \diagdown OEt\end{matrix}\right)$$

The initial step in the splitting off of a B atom, which certainly takes place through several reaction steps, is the attack of the donor electron pair of the alcohol in a three-centre bond (see Fig. 4.34). In the same way, $B_{10}H_{11}(2\text{-}Br)[S(C_2H_5)_2]_2$ is degraded by ethanol to $B_9H_{12}(2\text{-}Br)[S(C_2H_5)_2]$ [*340*]. The structure of B_9H_{13}-(NCMe) can be seen from Fig. 4.35.

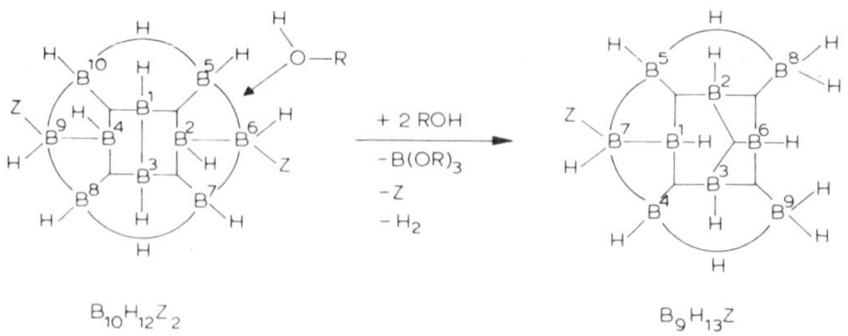

Fig. 4.34. Attack of a donor ROH on a three-centre bond in $B_{10}H_{12}Z_2$. Topological structure of $B_{10}H_{12}Z_2$ and $B_9H_{13}Z$. (Lipscomb's numbering for $B_9H_{13}Z$ [*469*] is not derived from that for $B_{10}H_{14}$ and $B_{10}H_{12}Z_2$).

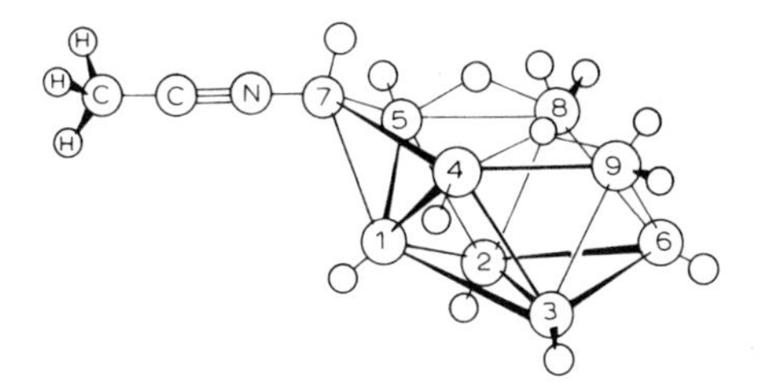

Fig. 4.35. Steric structure of $B_9H_{13}(NCCH_3)$ (numbering of the boron atoms after Lipscomb [*469*]; it has no relationship to the numbering for $B_{10}H_{14}$ and $B_{10}H_{12}Z_2$).
Structural parameters [*785, 786*]: B–B = 1.731–1.870 Å, B–H = 0.95–1.35 Å, B–N = 1.507 Å, N–C = 1.126 Å, C–C = 1.426 Å.

18.10.3 Synthesis of $B_9H_{14}^{\ominus}$ through the degradation of $B_{10}H_{14}$ by donors

With donors, decaborane forms $B_{10}H_{13}^{\ominus}$ (see section 18.2.1). Under certain other experimental conditions [aqueous solutions of NaOH, KOH, NH_4OH, $HCONHCH_3$, or $HCON(CH_3)_2$], bases degrade the B_{10} to $B_9H_{14}^{\ominus}$.

When NaOH or KOH is used, the yellow $B_{10}H_{13}^{\ominus}$ first forms without the evolution of gas, and this adds another hydroxide ion with the formation of the anion $B_{10}H_{13}OH^{2\ominus}$. After the addition of acid, degradation to $B_9H_{14}^{\ominus}$ takes place with the evolution of H_2 [*49*]:

$$B_{10}H_{14} + OH^{\ominus} \longrightarrow B_{10}H_{13}^{\ominus} + H_2O \qquad (1)$$

$$B_{10}H_{13}^{\ominus} + OH^{\ominus} \longrightarrow B_{10}H_{13}OH^{2\ominus} \qquad (2)$$

$$B_{10}H_{13}OH^{2\ominus} + H_3O^{\oplus} + H_2O \longrightarrow B_9H_{14}^{\ominus} + B(OH)_3 + H_2 \qquad (3)$$

Caesium, triphenylmethylphosphonium, and alkylammonium salts can be prepared by metathesis (*e.g.*, CsB_9H_{14} in 65% yield).

When NH_3 in H_2O is used, in analogy to eqn. (1), $B_{10}H_{13}NH_3^{\ominus}$, in which NH_3 is only feebly bound, is formed. Thus, NH_3 can easily be replaced by $CN^{\ominus}$ [formation of $B_{10}H_{13}CN^{2\ominus}$ by analogy with eqn. (2)]. Water easily hydrolyses it to $B_9H_{14}^{\ominus}$ [by analogy with eqn. (3)]. From aqueous solutions of $B_{10}H_{13}NH_3^{\ominus}$, CsF precipitates the so-called "labile $CsB_{10}H_{13}NH_3$". Because of the disturbance of the electronic structure of the skeleton due to NH_3, the IR and NMR spectra of $B_{10}H_{13}^{\ominus}$ and $B_{10}H_{13}NH_3^{\ominus}$ are different [*534a*].

Unexpectedly, the action of $OH^{\ominus}$ on "labile $B_{10}H_{13}NH_3^{\ominus}$" forms not $B_{10}H_{13}$-$OH^{2\ominus}$ [on analogy with eqn. (2)], but "stable $B_{10}H_{13}NH_3^{\ominus}$". Possibly this belongs to the structural class of $B_{10}H_{14}^{2\ominus}$: it is resistant to hydrolysis, $CN^{\ominus}$ does not replace NM_3, and acid degradation yields not $B_9H_{14}^{\ominus}$ but $B_9H_{13}NH_3$ [*534a*].

The structure of $B_9H_{14}^{\ominus}$ probably resembles that of $B_9H_{13}Z$, since when $(CH_3)_3$-NHB_9H_{14} is heated in benzene $B_9H_{13}NMe_3$ is formed with the elimination of hydrogen (yield 67%), and when $(CH_3)_4NB_9H_{14}$ is treated with HCl in acetonitrile $B_9H_{13}NCMe$ is obtained (yield 66%) [*49*].

References p. 355

18.10.4 Reactions of $B_9H_{13}Z$, $B_9H_{12}Z^{\ominus}$, and $B_9H_{12}^{\ominus}$

In $B_9H_{13}(SMe_2)$, the SMe_2 group can be replaced by other donors Z such as NEt_2H, PPh_3, NCMe, or NC_5H_4R [*273*]:

$$B_9H_{13}(SMe_2) + Z \longrightarrow B_9H_{13}Z + SMe_2$$

The $B_9H_{13}Z$ formed is identical with that prepared in accordance with sections 18.10.1 and 18.10.2 by the protolysis or alcoholysis of $B_{10}H_{12}Z_2$.

$B_9H_{13}Z$ reacts with donors such as $EtNH_2$ or NaH with the elimination of a proton and the formation of $B_9H_{12}Z^{\ominus}$ [*273, 275*]:

$$B_9H_{13}NEtH_2 + EtNH_2 \longrightarrow EtNH_3^{\oplus}B_9H_{12}NEtH_2^{\ominus}$$

$$B_9H_{13}(NC_5H_5) \underset{+HCl,\,-NaCl}{\overset{+NaH,\,-H_2}{\rightleftharpoons}} Na^{\oplus}B_9H_{12}(NC_5H_5)^{\ominus} \text{ (red)}$$

$B_9H_{13}Z$ reacts with strong Lewis bases such as Me_4NOH or Ph_3PCH_2 with the elimination of a proton and of the ligand Z. $B_9H_{12}^{\ominus}$ is formed almost quantitatively [*273, 275, 276*]:

$$B_9H_{13}(SEt_2) + Me_4N^{\oplus}OH^{\ominus} \xrightarrow{\text{ethanol}} Me_4N^{\oplus}B_9H_{12}^{\ominus} + SEt_2 + H_2O$$

$$B_9H_{13}(SEt_2) + Ph_3\overset{\oplus}{P}CH_2^{\ominus} \longrightarrow Ph_3PCH_3^{\oplus}B_9H_{12}^{\ominus} + SEt_2$$

With pyridinium chloride, $B_9H_{12}^{\ominus}$ re-forms $B_9H_{13}(NC_5H_5)$ [*273*]:

$$Me_4N^{\oplus}B_9H_{12}^{\ominus} + C_5H_5NH^{\oplus}Cl^{\ominus} \longrightarrow B_9H_{13}(NC_5H_5) + Me_4NCl$$

The methanolysis of $Me_4N^{\oplus}B_9H_{12}^{\ominus}$ leads to the precipitation of $Me_4N^{\oplus}B_3H_8^{\ominus}$ (yield referred to the $B_{10}H_{14}$ used initially: 30–35%) [*276*]:

$$Me_4N^{\oplus}B_9H_{12}^{\ominus} + 18MeOH \longrightarrow Me_4N^{\oplus}B_3H_8^{\ominus} + 6B(OMe)_3 + 11\,H_2$$

In the reaction of $B_9H_{13}(SEt_2)$ with $EtNH_2$, it is not the expected $B_9H_{12}NEtH_2^{\ominus}$ that is formed but $EtH_2N{\cdot}B_8H_{11}(NHEt)$, in an unexplained reaction [*465*].

The pyrolysis of $B_9H_{13}(SMe_2)$ yields a mixture of boranes including B_6H_{20} [*234a*].

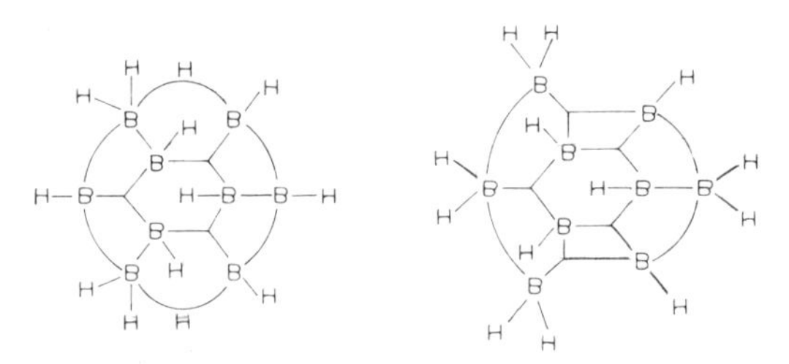

Fig. 4.36. Topological structures of $B_9H_{12}Z^{\ominus}$. In the structure shown, a H atom is replaced by the donor Z.

18.10.5 Structures of $B_9H_{12}Z^{\ominus}$, $B_9H_{12}^{\ominus}$, and $RH_2NB_8H_{11}(NHR)$

In the cases both of $B_9H_{12}Z^{\ominus}$ [*785, 786*] and of $B_9H_{12}^{\ominus}$ [*472*], there are two possible topological structures (Fig. 4.36 and Fig. 4.37). Figure 4.38 shows the structure of $EtH_2N \cdot B_8H_{11}(NHEt)$.

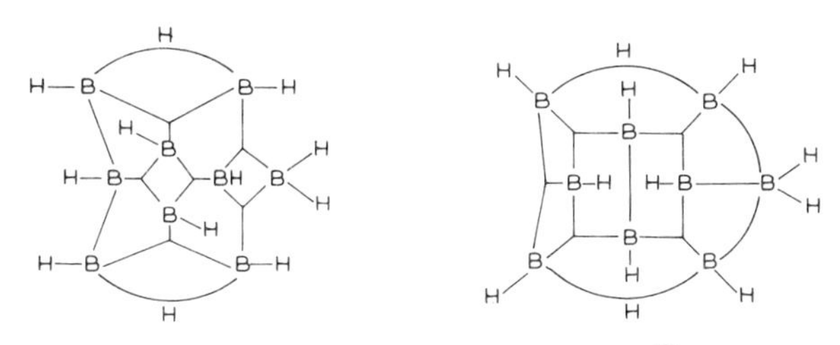

Fig. 4.37. Topological structures of $B_9H_{12}^{\ominus}$.

18.11 Synthesis and reactions of $B_{11}H_{14}^{\ominus}$, its derivatives and $B_{11}H_{13}^{2\ominus}$

In the reaction of decarborane(14) with lithium or sodium boranate in monoglyme at 90° in a pressure vessel, $B_{11}H_{14}^{\ominus}$ is formed almost quantitatively [*8, 523*]:

$$B_{10}H_{14} + MBH_4 \xrightarrow[20°,\ \text{monoglyme}]{-H_2} MB_{10}H_{13} + \text{monoglyme} \cdot BH_3 \xrightarrow{90°} MB_{11}H_{14} + H_2 + \text{monoglyme}$$

$$M = Li, Na$$

(In an aqueous medium, on the other hand, $Na_2B_{10}H_{12}$ is formed [*843a*], see section 18.2.1).

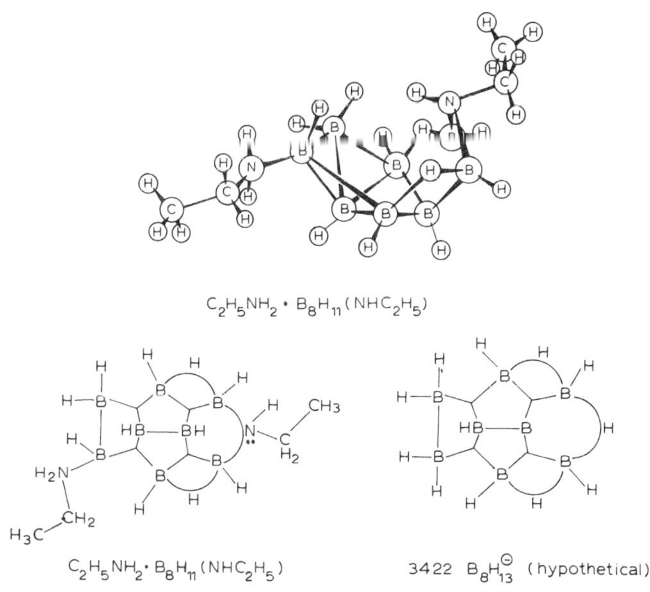

Figure 4.38. Steric and topological structure of $C_2H_5NH_2 \cdot B_8H_{11}(NHC_2H_5)$ with the topological structure of the hypothetical $B_8H_{13}^{\ominus}$ for comparison [*465, 465a*].

$B_{11}H_{14}^{\ominus}$ is also formed from $NaB_{10}H_{13}$ and $(BH_3)_2$ in ether at 45° [*8, 523*]. $B_{11}H_{14}^{\ominus}$ has been isolated as $LiB_{11}H_{14}\cdot2$ dioxan and $NaB_{11}H_{14}\cdot2.5$ dioxan. Both dioxan adducts are readily soluble in water. If RbF, CsCl, Me_4NCl, Me_3NHCl, Et_3NHCl, or Me_3SI is added to such solutions, the crystalline non-solvated $B_{11}H_{14}^{\ominus}$ salts of the corresponding cations are formed by metathesis [*8*], for example:

$$LiB_{11}H_{14}\cdot 2\,\text{dioxan} + RbF \longrightarrow RbB_{11}H_{14} + LiF + 2\,\text{dioxan}$$

The structure of $B_{11}H_{14}^{\ominus}$ is possibly derived from the icosahedral $B_{12}H_{12}^{2\ominus}$ (compare Fig. 4.39 with Figs. 4.41a and 4.41b). According to this, in $B_{11}H_{14}^{\ominus}$ an apical BH group of $B_{12}H_{12}^{2\ominus}$ would have been replaced by a $H_3^{\oplus}$ group.

In the $B_{11}H_{14}^{\ominus}$–D_2O system, H in the $H_3^{\oplus}$ position readily undergoes exchange with D [*8*].

$B_{11}H_{14}^{\ominus}$ readily eliminates protons:

$$B_{11}H_{14}^{\ominus} \rightleftharpoons B_{11}H_{13}^{2\ominus} + H^{\oplus}$$

In a little water in the presence of donors such as NaOH or NH_4OH, $B_{11}H_{12}^{2\ominus}$

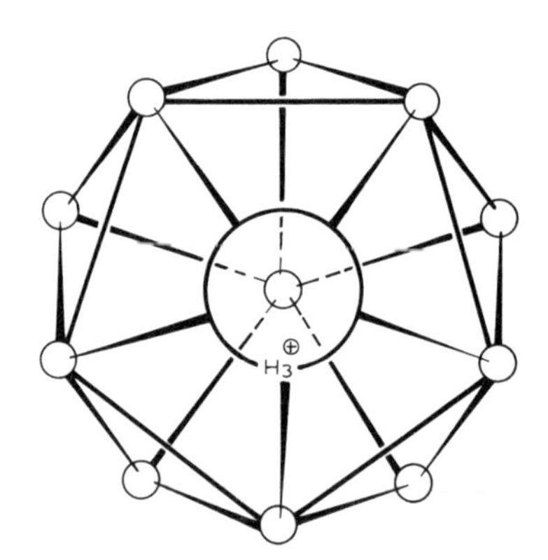

Fig. 4.39. Steric structure of $B_{11}H_{14}^{\ominus}$ (icosahedron, one apical position of which is occupied by $H_3^{\oplus}$).

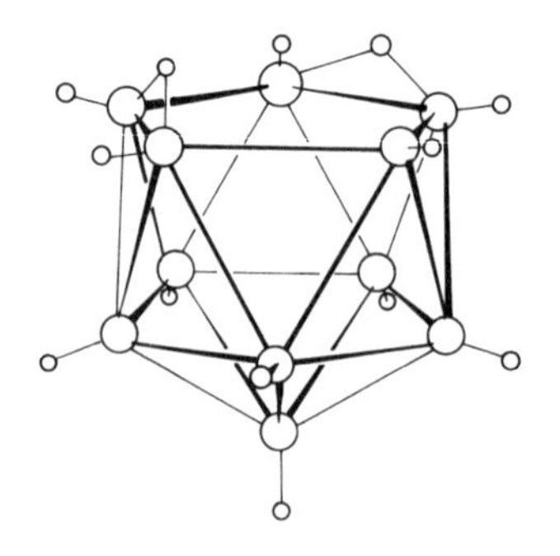

Fig. 4.40. Steric structure of $B_{11}H_{13}^{2\ominus}$ in $CsN(CH_3)_4B_{11}H_{13}$ [*235b*] (almost regular icosahedron without an apex).

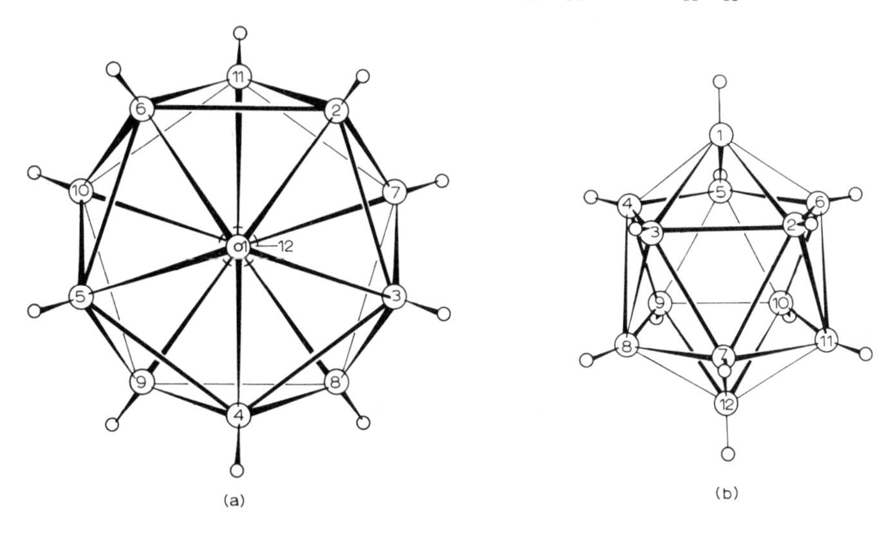

Fig. 4.41. Steric structure of the double pyramidal pentagonal antiprism (≡ icosahedron) of the boranate ion $B_{12}H_{12}^{2\ominus}$ according to [413], but the numbering of the boron atoms is according to the proposal of the ACS nomenclature committee. (a) View from above, (b) view from the side.

salts of $Rb^{\oplus}$, $Me_3S^{\oplus}$, and $(NH_3)_4Zn^{2\oplus}$ are formed [8], for example:

$$NaB_{11}H_{14} + 2\ RbF + NaOH \longrightarrow Rb_2B_{11}H_{13} + 2\ NaF + H_2O$$

Like the formation of $B_{11}H_{14}^{\ominus}$, its derivatives $B_{10}H_{12}AlH_2^{\ominus}$ and $B_{10}H_{12}GaH_2^{\ominus}$ can also be prepared [277e]:

$$Et_3N{\cdot}BH_3 + B_{10}H_{14} \rightarrow Et_3NH^{\oplus}B_{11}H_{14}^{\ominus} + H_2$$

$$Me_3N{\cdot}AlH_3 + B_{10}H_{14} \rightarrow Me_3NH^{\oplus}B_{10}H_{12}AlH_2^{\ominus} + H_2$$

$$Me_3N{\cdot}GaH_3 + B_{10}H_{14} \rightarrow Me_3NH^{\oplus}B_{10}H_{12}GaH_2^{\ominus} + H_2$$

Fig. 4.40 shows the $B_{11}H_{13}^{2\ominus}$ anion.

18.12 Synthesis and reactions of $B_{12}H_{12}^{2\ominus}$

18.12.1 Synthesis and structure of $B_{12}H_{12}^{2\ominus}$

The boiling of 2-iododecaborane in benzene and triethylamine gives $B_{10}H_{10}^{2\ominus}$ (main product) and $B_{12}H_{12}^{2\ominus}$ (4%) [317a]. However, this reaction is unsuitable for the preparation of $B_{12}H_{12}^{2\ominus}$ [596]. Polyhedral boranes including, therefore, $B_{12}H_{12}^{2\ominus}$ can be prepared quite generally by the reaction of the nucleophilic hydride ion $H^{\ominus}$ with an electrophilic borane, general formula B_xH_y [523,524a]:

$$a\,H^{\ominus} + b\,B_xH_y \longrightarrow [B_{bx}H_{by+a-z}]^{a\ominus} + \frac{z}{2}\,H_2$$

References p. 355

In the equation, $H^{\ominus}$ may be replaced as appropriate by similarly nucleophilic boronate ions, *e.g.* $BH_4^{\ominus}$, $B_3H_8^{\ominus}$, $B_{11}H_{14}^{\ominus}$, $B_{10}H_{10}^{2\ominus}$. Thus, $B_{12}H_{12}^{2\ominus}$ is formed in 60% yield by boiling $B_{10}H_{14}$ with $NaBH_4$ in diglyme [*6*]:

$$B_{10}H_{14} + BH_4^{\ominus} + OR_2 \xrightarrow{-H_2} B_{10}H_{13}^{\ominus} + BH_3{\cdot}OR_2 \xrightarrow{-H_2,\,-OR_2} B_{11}H_{14}^{\ominus} \xrightarrow[-3\,H_2]{+BH_4^{\ominus}} B_{12}H_{12}^{2\ominus}$$

Boiling $B_{10}H_{14}$ with Et_3NBH_3 yields $(Et_3NH^{\oplus})_2B_{12}H_{12}^{2\ominus}$; the similar procedure in light petroleum yields $Et_3NH^{\oplus}B_{12}H_{11}NEt_3$ and $B_{12}H_{10}(NEt_3)_2$ [*277b*].

When KB_3H_8, RbB_3H_8, or CsB_3H_8 is heated to 200–230°, hydrogen and a mixture of salts of $BH_4^{\ominus}$, $B_{10}H_{10}^{2\ominus}$, $B_{12}H_{12}^{2\ominus}$, and $B_9H_9^{2\ominus}$ are formed [*405a*]. At 165°, $Et_4NB_{10}H_{13}$ form $(Et_4N)_2B_{12}H_{12}$ [*705a*]. High yields of $B_{12}H_{12}^{2\ominus}$ are obtained by the pyrolysis of $MB_3H_8{\cdot}OR_2$ or by heating ethereal solutions of MB_3H_8 [*194a, 523*].

$B_{12}H_{12}^{2\ominus}$ has the structure of an icosahedron (double pyramidal pentagonal antiprism) (Figs. 4.41a and 4.41b) [*352, 531, 596, 858*]. In this connection compare the structure of $B_{10}H_{10}^{2\ominus}$ (Figs. 4.32a and 4.32b) as a double pyramidal tetragonal antiprism.

18.12.2 Reactions of $B_{12}H_{12}^{2\ominus}$ and its derivatives

Like $B_{10}H_{10}^{2\ominus}$, the $B_{12}H_{12}^{2\ominus}$ boranate ion is extraordinarily stable to strong bases, strong acids, and oxidising agents [*353, 413, 414*].

With *acids*, the salts $M_2^{2\oplus}B_{12}H_{12}^{2\ominus}$ form the stable hydroxonium salt $(H_3O^{\oplus})_2$-$B_{12}H_{12}^{2\ominus}{\cdot}xH_2O$. As an acid it is stronger than sulphuric acid. Silver salts give an insoluble covalent silver compound. No metallic silver is precipitated.

$B_{12}H_{12}^{2\ominus}$ reacts with the *halogens* in aqueous or alcoholic solution: fluorine (in H_2O at 0°, but in this case in addition to replacement of H by F there is also partial degradation of the B_{12} skeleton), chlorine, iodine chloride, bromine, iodine, or N-iodosuccinimide in the same way as $B_{10}H_{10}^{2\ominus}$; and, moreover, with *hydrogen halides* with the replacement of H for halogen. The reaction rates fall in the sequence Cl > Br > I. The initial reaction is very fast. The rate decreases with increasing degree of halogenation, so that all the halogenation steps can be isolated. Not all mechanisms (polar and radical mechanisms) of halogen substitution in $B_{12}H_{10}^{2\ominus}$ resemble those in $B_{10}H_{10}^{2\ominus}$. Consequently, under similar experimental conditions non-comparable halogen derivatives are often obtained [*352, 413*]. In halogenation with a halogen, for example, the following have been isolated: $B_{12}F_{12}^{2\ominus}$, $B_{12}F_{11}(OH)^{2\ominus}$, $B_{12}H_2Cl_{10}^{2\ominus}$, $B_{12}H_3Br_6Cl_3^{2\ominus}$, $B_{12}Br_{12}^{2\ominus}$, and $B_{12}I_{12}^{2\ominus}$ (tetramethylammonium salts) [*413, 414*]; on halogenation with hydrogen halides, for example, the following have been isolated: $B_{12}H_8F_4^{2\ominus}$, $B_{12}H_7F_5^{2\ominus}$, $B_{12}H_6F_6^{2\ominus}$, and $B_{12}H_{11}Cl^{2\ominus}$ [*413,414*]. Reactions of $B_{12}X_{12}^{2\ominus}$ with KCN, NaN_3 and K(OCN) see [*771c,771d*].

Under carefully controlled conditions, the action of *nitric acid* on $B_{12}H_{12}^{2\ominus}$ forms $B_{12}H_{11}(NO_2)^{2\ominus}$(Cs salt).

Organic compounds with donor 0 functions react with $B_{12}H_{12}^{2\ominus}$ (hydroxonium salt) in accordance with the following scheme:

$$B_{12}H_{11}OMe^{2\ominus} + B_{12}H_{11}(OH)^{2\ominus}$$

$$\uparrow +MeOH, +HOH, \Delta$$

$$B_{12}H_{12}^{2\ominus} \xrightarrow{+CH_3OCH_2CH_2OCH_3, HOH} B_{12}H_{10}(OCH_2CH_2OCH_3)_2^{2\ominus}$$

$$\downarrow +MeCOOH, +HOH, \Delta$$

$$B_{12}H_{11}(OH)^{2\ominus}$$

Alkenes are rapidly added by $B_{12}H_{12}^{2\ominus}$; for example, styrene, $CHPh{=}CH_2$, forms the ion $B_{12}H_{11}(CHPh{-}CH_3)^{2\ominus}$, and propene, C_3H_6, the propyl derivative $B_{12}H_{11}(C_3H_7)^{2\ominus}$ [*414*].

Carbon monoxide and $B_{12}H_{12}^{2\ominus}$ form the carbonyl derivative [*418*] $B_{12}H_{11}CO^{\ominus}$ and at least two isomers of $B_{12}H_{10}(CO)_2$. As in the case of $B_{10}H_8(CO)_2$, with $B_{12}H_{10}(CO)_2$ ligand reactions and ligand displacements are possible [*411a, 418*]:

(a) ligand reactions:

$$B_{12}H_{10}(CO)_2 \xrightleftharpoons{+H_2O} H^{\oplus}B_{12}H_{10}(COOH)(CO)^{\ominus} \xrightleftharpoons{+H_2O} (H^{\oplus})_2B_{10}H_{10}(COOH)_2^{2\ominus}$$

$$\xrightarrow[-4H_2O]{+4OH^{\ominus}} B_{12}H_{10}(COO)_2^{4\ominus}$$

$$B_{12}H_{10}(CO)_2 + 2MeOH \rightleftharpoons (H^{\oplus})_2B_{12}H_{10}(COOMe)_2^{2\ominus}$$

$$B_{12}H_{10}(CO)_2 + 4NH_3 \longrightarrow (NH_4^{\oplus})_2B_{12}H_{10}(CONH_2)_2^{2\ominus}$$

$$B_{12}H_{10}(CO)_2 + NaN_3 \longrightarrow B_{12}H_{10}(NCO)_2^{2\ominus}$$

$$B_{12}H_{10}(CO)_2 + LiAlH_4 \xrightarrow{\Delta} B_{12}H_{10}(CH_3)_2^{2\ominus}$$

(b) ligand substitutions:

$$B_{12}H_{10}(CO)_2 + 2NH_2OSO_3^{\ominus} + 2H_2O \xrightarrow{\Delta} B_{12}H_{10}(NH_3)_2 + 2CO_2 + 2HOSO_3^{\ominus}$$

(c) substitution of the H atoms and ligand reactions:

$$B_{12}H_{10}(CO)_2 \xrightarrow[(water)]{+X_2, \Delta} (H^{\oplus})_2B_{12}X_{10}(COOH)^{2\ominus} \xrightarrow[heating]{-H_2O} B_{12}X_{10}(CO)_2$$

The perhalogenated boranes $B_{12}X_{10}(CO)_2$ are still more stable than $B_{12}H_{10}(CO)_2$.

$B_{12}H_{12}^{2\ominus}$ forms with the highly electrophilic amide-chloride $(Me_2N{-}CClH)Cl$ not the B–C bonds to be expected from the organic chemistry (see section 18.9.2) but B–$\overset{\ominus}{N}$ bonds [*331*]:

$$\left[Me_2N{-}\overset{\overset{Cl}{|}}{\underset{\underset{H}{|}}{C^{\oplus}}}\right]Cl^{\ominus} + B_{12}H_{12}^{2\ominus} \longrightarrow B_{12}H_{11}[N(Me)_2(CH_2Cl)]^{\ominus} \xrightarrow{+(Me_2NCHCl)Cl}$$

$$B_{12}H_{10}[N(Me)_2(CH_2Cl)]_2$$

Hydroxylamine-O-sulphonic acid reacts with $B_{12}H_{12}^{2\ominus}$ in aqueous solution to form $B_{12}H_{11}NH_3^{\ominus}$ and 1,2-, 1,7-, and 1,12-$B_{12}H_{10}(NH_3)_2$ [*335*].

References p. 355

With *oxaloyl dichloride*, $(COCl)_2$, $B_{12}H_{11}(NMe_3)^{\ominus}$ can be carbonylated, and in the $B_{12}H_{10}(NMe_3)(CO)$ formed the CO group can be replaced by the NH_3 group using hydroxylamine-O-sulphonic acid, $NH_2OSO_3^{\ominus}$, and the latter can be methylated with dimethyl sulphate, Me_2SO_4[*334*]:

$$B_{12}H_{11}(NMe_3)^{\ominus} + \begin{array}{c} COCl \\ | \\ COCl \end{array} \xrightarrow[(MeCN)]{\Delta} B_{12}H_{10}(NMe_3)(CO) \xrightarrow[\Delta]{NH_2OSO_3^{\ominus},\ H_2O}$$

$$B_{12}H_{10}(NMe_3)(NH_3) \xrightarrow[\Delta]{+Me_2SO_4} B_{12}H_{10}(NMe_3)_2$$

Tropenium bromide, $Tr^{\oplus}Br^{\ominus}$, also reacts with $B_{12}H_{12}^{2\ominus}(H_3O^{\oplus})_2$[*293a, 293b*]:

$$B_{12}H_{12}^{2\ominus}(H_3O^{\oplus})_2 + Tr^{\oplus}Br^{\ominus} \longrightarrow B_{12}H_{11}Tr^{\ominus}H_3O^{\oplus} + HBr + H_2O$$

The electrochemical oxidation of $B_{12}H_{12}^{2\ominus}$ in acetonitrile yields $B_{24}H_{23}^{3\ominus}$ [*841a*].

18.13 Synthesis and reactions of $B_6H_6^{2\ominus}$, $B_7H_7^{2\ominus}$, $B_8H_8^{2\ominus}$, $B_9H_9^{2\ominus}$, *and* $B_{11}H_{11}^{2\ominus}$

$Na_2B_9H_9$ is formed in small amount when NaB_3H_8 is heated in diglyme to 100°[*405a*]. $Cs_2B_{11}H_{11}$ is formed in the decomposition of $Cs_2B_{11}H_{13}$ with the evolution of 1 mole of H_2 per mole of $Cs_2B_{11}H_{13}$[*405a*]. In the oxidation of a solution of $Na_2B_9H_9 \cdot xH_2O$ in tetrahydrofuran or monoglyme by atmospheric oxygen, sparingly soluble $Na_2B_8H_8$ forms via the radical anion $\cdot B_8H_8^{\ominus}$. If the oxidation is continued, other anions form as well: $B_3H_8^{\ominus}$, $B_6H_6^{2\ominus}$, $B_{10}H_{10}^{2\ominus}$, and $B_7H_7^{2\ominus}$, together with $B(OH)_3$[*405b*]. $Na_2B_6H_6$ is also formed in the reaction of $NaBH_4$ and B_2H_6 in diglyme at 160°[*61, 647a*]. Nothing has yet been reported concerning reactions of $B_6H_6^{2\ominus}$. B_6H_{10} reacts with NaH, KH or $LiCH_3$ at low temperature to liberate 1 mole of H_2 (CH_4) per mole of B_6H_{10}[*374b*]:

$$B_6H_{10} + MZ \longrightarrow MB_6H_9 + HZ$$
$$MZ = NaH,\ KH,\ LiCH_3$$

The $B_6H_9^{\ominus}$ seems to be significantly more thermally stable than its analog $B_5H_8^{\ominus}$ which is derived from B_5H_9[*374b*]. $B_5H_9^{\ominus}$ with DCl regenerates hexaborane(10) [D enters only in the bridge][*374b*].

$B_8H_8^{2\ominus}$ possesses the structure of a slightly distorted dodecahedron (like B_8Cl_8), see Fig. 4.42.

$Cs_2B_9H_9$ is stable up to 600° and $Cs_2B_{11}H_{11}$ up to 400°. Both anions hydrolyse relatively easily. In the sequence of increasing stability to oxidation they are arranged as follows: $B_9H_9^{2\ominus} < B_{11}H_{11}^{2\ominus} < B_{10}H_{10}^{2\ominus} < B_{12}H_{12}^{2\ominus}$ [*405a*].

In an alkaline medium, bromine brominates $B_9H_9^{2\ominus}$ to $B_9Br_8H^{2\ominus}$ and $B_{11}H_{11}^{2\ominus}$ to $B_{11}Br_9H_2^{2\ominus}$. In an acid medium, bromine brominates $B_{11}H_{11}^{2\ominus}$ to $B_{10}Br_{10}^{2\ominus}$ with the abstraction of a B atom, and N-chlorosuccinimide chlorinates $B_{11}H_{11}^{2\ominus}$ to $B_{10}Cl_8H_2^{2\ominus}$ [*405a*].

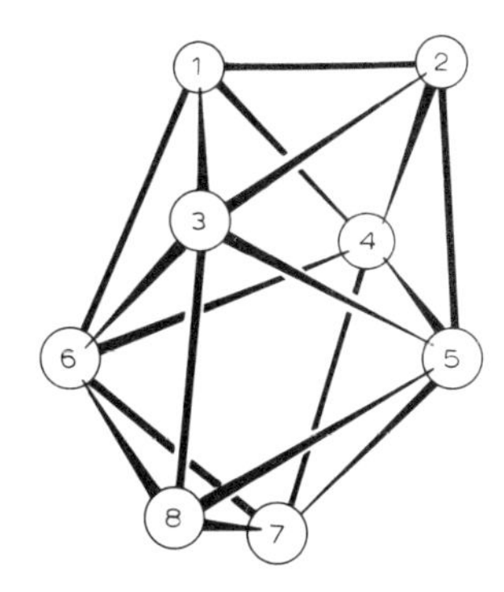

Fig. 4.42. Steric structure of the B_8 skeleton in $Zn(NH_3)_4B_8H_8$[*405b*]. Structural parameters: $B^1–B^2$ = 1.58 Å, $B^1–B^6$ = 1.74 Å, $B^2–B^3$ = 1.78 Å, $B^3–B^6$ = 1.93 Å.

18.14 Synthesis and reactions of $B_{20}H_{19}^{3\ominus}$, $B_{20}H_{18}^{2\ominus}$, $B_{20}H_{18}^{4\ominus}$, their derivatives, and $B_{18}H_{22}$, $B_{18}H_{21}^{\ominus}$, $B_{18}H_{20}^{2\ominus}$ and their derivatives

The B–H bonds in $B_{10}H_{10}^{2\ominus}$ can be converted into intermolecular $B–H^{\mu}–B$ bridges. $B_{20}H_{19}^{3\ominus}$ and $B_{20}H_{18}^{2\ominus}$ are formed[*314b, 315b*]. The electrons eliminated in this process are taken up by oxidizing agents.

In the oxidation of $B_{10}H_{10}^{2\ominus}$ by iron(III) ions, the asymmetric $B_{20}H_{18}^{2\ominus}$ (Fig. 4.44) is formed[*138, 315b, 383a, 384, 385, 563a, 597b*]. The oxidation of $[Et_3NH^{\oplus}]_2$-$B_{10}H_{10}^{2\ominus}$ with cerium(IV) ions leads to the double salt $(Et_3NH^{\oplus})_5(B_{20}H_{19}^{3\ominus}$·asym. $B_{20}H_{18}^{2\ominus})$ [*138, 315a, 351b, 472*].

The treatment of salts of asym. $B_{20}H_{18}^{2\ominus}$ in an acid ion-exchanger with ethanol EtOH leads to $(EtOH_2^{\oplus})_2B_{20}H_{18}^{2\ominus}$. After concentration of the acid solution, the addition of water gives a 70% yield of the monobasic acid $B_{18}H_{22}$[*563a, 597*]. The over-all equation of the complicated reaction is[*718*]:

$$B_{20}H_{18}^{2\ominus} + 2H^{\oplus} + 6HOH \longrightarrow B_{18}H_{22} + 2B(OH)_3 + 2H_2$$

$B_{18}H_{22}$ is also formed by the partial hydrolysis of $B_{20}H_{18}^{2\ominus}$ in diethyl ether[*597, 597a, 563a*] or by the gradual decomposition of crystalline $B_{20}H_{19}^{3\ominus}$ salts at 20°[*138*]. About the separation and the structure of n-$B_{18}H_{22}$ and i-$B_{18}H_{22}$ by crystallisation, see [*563a, 713a*].

n- or i-$B_{18}H_{22}$ react with methanol containing KOH to yield $B_{18}H_{20}^{2\ominus}$ [*563a*]. n- or i-$B_{18}H_{22}$ react with sodium amalgam in THF to produce $B_{18}H_{22}^{2\ominus}$. The transient green colour observed in this reaction may be characteristic of the $\cdot B_{18}H_{22}^{\ominus}$ radical anion (in the $B_{10}H_{14}$ reduction, the red $\cdot B_{10}H_{13}^{\ominus}$ radical anion appears) [*563a*]. n-$B_{18}H_{22}$ and I_2 yield n-$B_{18}H_{21}I$ [*563a*].

$B_{10}H_{10}^{2\ominus}$ reacts with NO_2 or $NO + Fe^{3\oplus}$ to give highly coloured NO-substituted derivatives of $B_{20}H_{19}^{3\ominus}$ or $B_{20}H_{18}^{2\ominus}$: $B_{20}H_{18}(NO)^{3\ominus}$, $B_{20}H_{14}(NO)_4^{2\ominus}$, and $B_{20}H_{12}$-$(NO)_6^{2\ominus}$ [*472, 841b, 843*]. These nitroso derivatives can be reduced to the corresponding amines with Raney nickel[*843*].

The parent substance of $B_{20}H_{18}(NO)^{3\ominus}$, $B_{20}H_{19}^{3\ominus}$, can be obtained by the reduction of $B_{20}H_{18}^{2\ominus}$ with active metals such as Mg, Al, or Zn. Ce(IV), Fe(III), and Ag(I) ions reoxidise $B_{20}H_{19}^{3\ominus}$ to $B_{20}H_{18}^{2\ominus}$. At high pH values, $B_{20}H_{18}^{4\ominus}$ is formed, the B–H^{μ}–B bridge being converted into a B–B bridge (equilibrium) [*138*]:

$$B_{20}H_{19}^{3\ominus} + H_2O \rightleftharpoons B_{20}H_{18}^{4\ominus} + H_3O^{\oplus}$$

For the electrochemical oxidation of $B_{10}H_{10}^{2\ominus}$ to $B_{20}H_{18}^{2\ominus}$, $B_{20}H_{19}^{3\ominus}$ and $B_{24}H_{23}^{3\ominus}$ see [*502a, 841a*].

The treatment of asym. $B_{20}H_{18}^{2\ominus}$ in aqueous solution with hydroxide ions does not lead to the cleavage of the molecule to give 1- or 2-substituted $B_{10}H_9(OH)^{2\ominus}$ [*139,384*], but $B_{20}H_{17}(OH)^{4\ominus}$ [*138,295,314b,315a*]:

$$B_{20}H_{18}^{2\ominus} + 2HO^{\ominus} \longrightarrow B_{20}H_{17}(OH)^{4\ominus} + H_2O$$

In this process, for kinetic reasons, an isomer is first formed, and this is converted into another, thermodynamically more stable isomer only on being heated in aqueous solution or on being treated with acids [*314b*].
Methoxide ions in methanol give an analogous anion [*138*]:

$$B_{20}H_{18}^{2\ominus} + 2MeO^{\ominus} \longrightarrow B_{20}H_{17}(OMe)^{4\ominus} + MeOH$$

Like the parent substance $B_{20}H_{18}^{4\ominus}$, $B_{20}H_{17}(OH)^{4\ominus}$ is in equilibrium with the triply charged hydrogen-richer anion [*138, 314b*]:

$$B_{20}H_{18}OH^{3\ominus} + H_2O \rightleftharpoons B_{20}H_{17}(OH)^{4\ominus} + H_3O^{\oplus}$$

Fe(III) ions oxidise $B_{20}H_{17}(OH)^{4\ominus}$ to $B_{20}H_{17}(OH)^{2\ominus}$ (in analogy with the reaction $B_{20}H_{18}^{4\ominus} \rightarrow B_{20}H_{18}^{2\ominus} + 2e^{\ominus}$) [*138*]. An excess of halogen cleaves $B_{20}H_{18}(OH)^{3\ominus}$ with the formation of $B_{10}X_{10}^{2\ominus}$ and boric acid [*138*].

If (as in the oxidation of $B_{10}H_{10}^{2\ominus}$, see above) $B_{10}H_{10}^{2\ominus}$ derivatives $B_{10}H_9Z^{\ominus}$ ($Z = SMe_2$, IC_6H_5, $O{=}C(CH_2)_3NCH_3$, $O_2S(CH_2)_3CH_2$) are oxidised with cerium(IV) ions [$Ce(NH_4)_4(SO_4)_4 \cdot 2H_2O$] in aqueous solution, the two B_{10} cages are again coupled [*138*]. The boranes $B_{20}H_{16}Z_2$ so formed, which are comparable with the $B_{20}H_{18}^{2\ominus}$ ion, are uncharged, in accordance with the donor properties of Z (at least one free electron pair per Z in comparison to one electron per substituted H). The structures of $B_{20}H_{16}(SMe_2)_2$ and $B_{20}H_{16}(IC_6H_5)_2$ (probably) resemble that of the asymmetrical $B_{20}H_{18}^{2\ominus}$, although the two boron polyhedra are not connected with one another through B–H^{μ}–B bridges. The positions of the ligands have not yet been determined [*138*].

As in the hydrolysis of $B_{20}H_{18}^{2\ominus}$, $B_{20}H_{16}Z_2$ reacts with hydroxide ions to form $B_{20}H_{15}(OH)Z_2{}^{2\ominus}$ [in analogy with the formation of $B_{20}H_{17}(OH)^{4\ominus}$]. Unlike the comparable $B_{20}H_{17}(OH)^{4\ominus}$, $B_{20}H_{15}(OH)Z_2{}^{2\ominus}$ is in equilibrium with $B_{20}H_{16}$-$(OH)Z_2{}^{\ominus}$ [*138*]:

$$B_{20}H_{16}Z_2 \xrightarrow{+OH^{\ominus}} B_{20}H_{16}(OH)Z_2^{\ominus} \xrightarrow[-H_2O]{+OH^{\ominus}} B_{20}H_{15}(OH)Z_2^{2\ominus}$$

$$B_{20}H_{16}(OH)Z_2^{\ominus} + H_2O \rightleftharpoons B_{20}H_{15}(OH)Z_2^{2\ominus} + H_3O^{\oplus}$$

Under highly alkaline conditions, the two donor ligands Z can be partially or (generally) completely displaced by hydroxide ions [*138*]:

$$B_{20}H_{15}(OH)Z_2^{2\ominus} + 2OH^{\ominus} \longrightarrow B_{20}H_{15}(OH)_3^{4\ominus} + 2Z$$

$$Z = O{=}\overline{C(CH_2)_3N}CH_3,\ O_2\overline{S(CH_2)_3C}H_2,\ SMe_2$$

Figures 4.43 to 4.45 show the close relationships between $B_{20}H_{19}^{3\ominus}$ and $B_{20}H_{18}^{2\ominus}$. Fig. 4.46 shows the structure of $B_{20}H_{16}$.

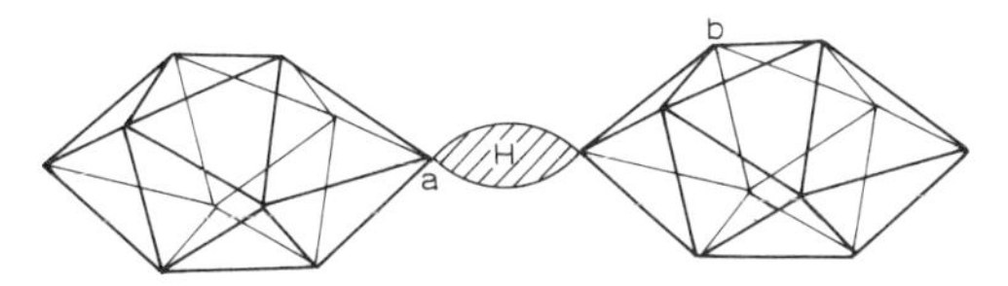

Fig. 4.43. Steric structure of the symmetrical $B_{20}H_{19}^{3\ominus}$. According to the ^{11}B NMR spectrum, this symmetrical form is less likely than that of a $B_{20}H_{19}^{3\ominus}$ ion also linked by a $B{-}H^{\mu}{-}B$ bridge but unsymmetrically (a–b) [*138, 315a, 472*].

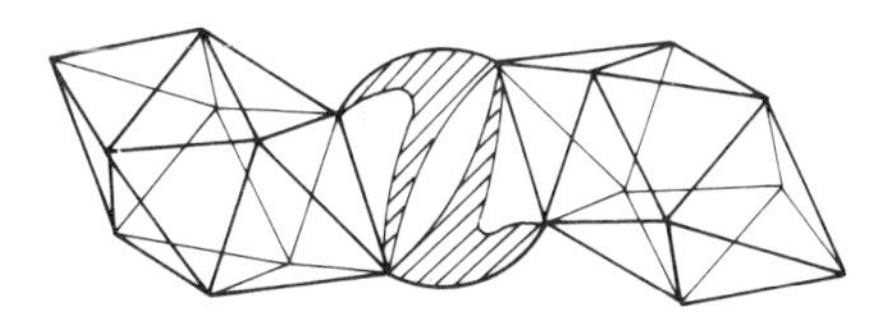

Fig. 4.44. Steric structure of the asymmetric $B_{20}H_{18}^{2\ominus}$ [*315a, 472*].

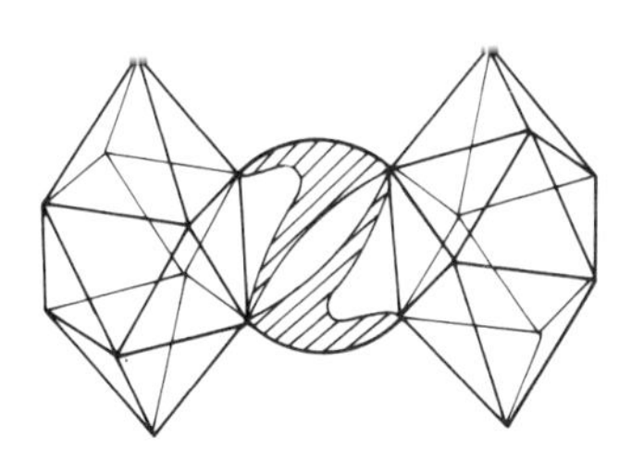

Fig. 4.45. Steric structure of photo-$B_{20}H_{18}^{2\ominus}$ [*164b, 314a*].

18.15 Reactions of $B_{10}H_{16}$ and $B_{20}H_{16}$

The decaborane(16), $B_{10}H_{16}$, that can be prepared in the manner described in section 3.11 is constructed of two B_5H_9 molecules joined at the apices by a B–B bond (see Fig. 4.12). This structure changes in reactions. All reactions so far

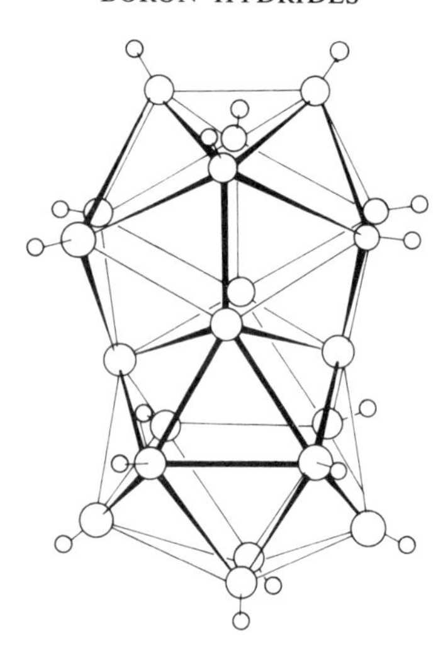

Fig. 4.46. Steric structure of $B_{20}H_{16}$ [*201b*].

studied lead either to a skeletal transformation to derivatives of decaborane(14) or to a cleavage of the peak B–B bond to give derivatives of pentaborane(9) [*279*]. Examples:

$$B_{10}H_{16} + AlCl_3 \xrightarrow[\text{carbon disulphide}]{\Delta} B_{10}H_{14} + AlCl_3$$
(no H_2 or HCl is evolved)

$$B_{10}H_{16} + 2NC_5H_5 \xrightarrow{\Delta} B_{10}H_{12}(NC_5H_5)_2$$

$$B_{10}H_{16} + I_2 \xrightarrow{150^\circ} B_{10}H_{14} + 2HI$$
(in aqueous solutions, $B_{10}H_{16}$ is inert to I_2 and Br_2)

$$B_{10}H_{16} + HI \xrightarrow[-B_5H_9]{\Delta,\ 20^\circ} B_5H_8I \xrightarrow{100^\circ} \text{polymers}$$
($B_{10}H_{16}$ reacts with HBr similarly)

The icosaborane(16), $B_{20}H_{16}$, that can be prepared as described in section 3.14 is very stable to heat, is hygroscopic, and forms highly acidic solutions with water [*235*]. After the water has been sublimed off in vacuum, $(H_3O)_2B_{20}H_{16}(OH)_2$ is left. Its aqueous solution can be titrated with tetramethylammonium hydroxide (formation of $[(CH_3)_4N]_2B_{20}H_{16}(OH)_2$); likewise, the ethanolic solution can be titrated with caesium hydroxide [formation of $Cs_2B_{20}H_{16}(OH)_2$] [*521*]. In pentane, icosaborane(16) forms with ether the adduct $B_{20}H_{16}\cdot 2OEt_2$, with trimethylamine $B_{20}H_{16}\cdot 2NMe_3$, with triphenylphosphane $B_{20}H_{16}\cdot 2PPh_3$, and with dimethyl sulphide $B_{20}H_{16}\cdot 2SMe_2$ [*521*]. $B_{20}H_{16}$ dissolves in CH_3CN without the evolution of H_2. Evaporation of the excess of CH_3CN leaves $B_{20}H_{16}(NCCH_3)_2\cdot CH_3CN$, the B skeleton of which is similar to, but not identical with the skeleton of $B_{20}H_{16}$ (see Fig. 4.14) [*201a, 201b*].

19. Carboranes

Carboranes can – in the light of present knowledge – be divided into two groups overwhelmingly on external grounds.

The *first* group contains lower carboranes with (at the present time) five- to eight-membered skeletons and one to three carbon atoms. They are less stable to heat than the second group. At the present time they can be prepared only in substantially smaller amounts than the carboranes of the second group; frequently they have been studied only by NMR and mass spectroscopy. They are prepared from lower boranes, particularly B_5H_9 (or B_4H_{10} or B_2H_6) and $HC{\equiv}CH$, (a) by the supply of high energy, *i.e.* an electrical discharge or a thermal shock (explosion-like reaction); (b) by the supply of low energy (thermolysis at 20 to about 100°). At the present time there is not yet a chemistry of derivatives of lower carboranes.

The *second* group, substantially more extensive at the present time, contains carboranes derived from $B_{12}H_{12}^{2\ominus}$: $C_2B_{10}H_{12}$ with a twelve-membered skeleton and degradation products such as $C_2B_9H_{12}^{\ominus}$. They contain two carbon atoms, possess an icosahedral structure ($C_2B_{10}H_{12}$) or a structure that can be derived from it ($C_2B_9H_{12}^{\ominus}$), and are unusually stable to heat. Their synthesis can be carried out in large amounts, *inter alia* easily by the reaction of $B_{10}H_{12}(\text{ligand})_2$ and $RC{\equiv}CR'$. There is an extensive chemistry of derivatives of these icosahedral carboranes.

19.1 Nomenclature

"Carborane" is the comprehensive term for boranes with a skeletal structure in which one or more boron atoms have been replaced by carbon atoms. The systematic nomenclature is based on the following proposals (mainly after Adams [*4*]).

1. The nature of the skeleton of a structure containing no bridge hydrogen atoms is expressed by the following prefixes: *closo* for closed skeletons (polyhedra) and *nido* for open skeletons[29].

2. The naming of this class of compounds is performed by the oxa–aza system, *i.e.*, *carba* for the carbon atom.

3. The numbering of the atoms of polyhedra (*e.g.*, 10- or 12-membered polyhedra) is performed in the following sequence: the apex of a polyhedron containing a carbon atom is given number 1. Then follow the atoms of the first belt (equator). If the carborane contains more than one carbon atom and this carbon atom is in the neighbourhood of the first carbon atom (*i.e.*, in the first belt), this is given the number 2. The sequence of numbers in this belt follows in clockwise direction when it is regarded in the direction to the (apical or polar) atom 1. The numbering of the second belt begins with the atom adjacent to atom 2 in the clockwise direction. The (apical, polar) atom opposite to atom 1 is given the last number.

[29]The prefixes *closo* and *nido* are the current terms to indicate closed and open carboranes (replacing *clovo* and *stapho*) selected by the Nomenclature Committee, Div. Inorg. Chem., June 1966; and modified by the above mentioned Committee at the 152nd National Meeting Am. Chem. Soc., New York, Sept. 1966 [cit. *686a*]. See also [*380a, 845a*].

4. The total number of atoms in the skeleton (polyhedron or open skeleton) is expressed by a Greek numeral as prefix, followed by the word *borane*.

5. In derivatives, the position is given by the prefixed number of the skeletal atom before the name of the (skeletal) substituent. Formulae are written similarly. If it is known only whether there is a B- or a C-substituent, instead of the position number a B or a C can be prefixed in formula names.

6. As in the nomenclature of the boranes, an Arabic numeral suffixed in brackets gives the number of hydrogen atoms. This number is not affected by any substitution (of hydrogen atoms).

Examples. The three distinguishable carboranes with two carbon atoms having an icosahedral structure (Fig. 4.53):

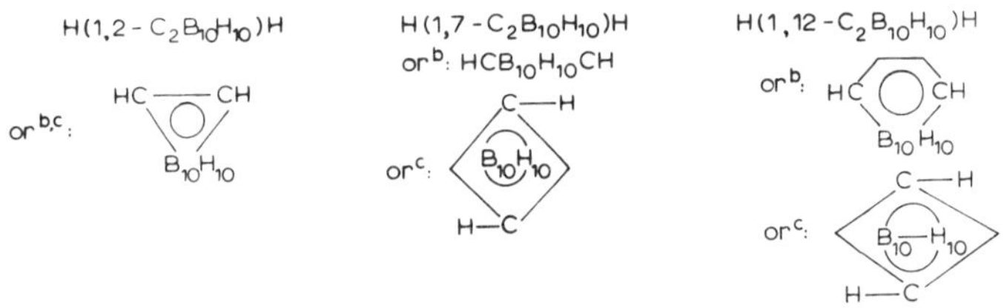

$H(1,2-C_2B_{10}H_{10})H$ — or[b,c]: HC—CH / $B_{10}H_{10}$

$H(1,7-C_2B_{10}H_{10})H$ — or[b]: $HCB_{10}H_{10}CH$ — or[c]: C—H / $B_{10}H_{10}$ / H—C

$H(1,12-C_2B_{10}H_{10})H$ — or[b]: HC / CH / $B_{10}H_{10}$ — or[c]: C—H / $B_{10}H_{10}$ / H—C

[a]All three formulae express the position of the two carbon atoms. However, the first formula has the advantage that it can be written with normal letters and in one line.
[b]Proposed by SCHROEDER [*682a*].
[c]Proposed by STANKO *et al* [*725f*].

19.2 Carboranes with 5- to 10-membered skeletons

When pentaborane(9) is shaken with an alkyne (RC≡CR′) in the presence of 2,6-dimethylpyridine for 2 to 10 hours, among other products derivatives of 2,3-dicarbahexaborane(8), 2,3-R,R′-2,3-$C_2B_4H_6$ are formed [*565a, 566a, 569*]:

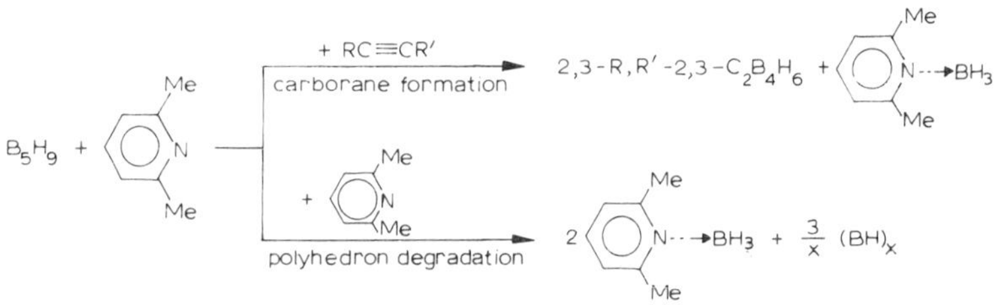

B_5H_9 + 2,6-dimethylpyridine —(+ RC≡CR′, carborane formation)→ 2,3-R,R′-2,3-$C_2B_4H_6$ + 2,6-dimethylpyridine·BH_3

B_5H_9 + 2,6-dimethylpyridine —(+ 2,6-dimethylpyridine, polyhedron degradation)→ 2 (2,6-dimethylpyridine)N→BH_3 + $\frac{3}{x}$ $(BH)_x$

RC≡CR′ = propyne, but-2-yne (40% yield) pent-1-yne (38% yield)

The competing donor-induced degradation reaction decreases the yield of carboranes such as 2-Me-2,3-$C_2B_4H_7$ (yield 8.4%) [*565a*], 2,3-Me_2-2,3-$C_2B_4H_6$ and 2-Pr^n-2,3-$C_2B_4H_7$. They do not form at room temperature without the addition of 2,6-dimethylpyridine. The dihydrocarboranes can be isolated from the reaction mixture after the addition of $Et_2O{\cdot}BF_3$ and be separated from one another by gas chromatography. The ^{11}B and ^{1}H NMR spectra indicate a pentagonal pyramid. The structures therefore resemble that of B_6H_{10} (compare Fig. 4.50 with Fig. 4.7).

On heating, the carboranes containing six skeletal atoms $Me_2C_2B_4H_6$ or $PrC_2B_4H_7$, in the undiluted state or diluted with isobutane or trimethylamine, undergo a transformation into, among other products, carboranes containing seven skeletal atoms [*565a*, *566a*]:

$$Me_2C_2B_4H_6 \xrightarrow[\text{with or without i-}C_4H_{10}]{290°,\ 20\ h,\ \Delta} \begin{cases} 25\text{–}30\%\ 2,4\text{-}Me_2\text{-}2,4\text{-}C_2B_5H_5 & \text{(Ia)} \\ 5\%\ 1,6\text{-}Me_2\text{-}1,6\text{-}C_2B_4H_4 & \text{(IIa)} \end{cases}$$

$$Pr^nC_2B_4H_7 \xrightarrow{300°,\ 20\ h,\ \Delta} \begin{cases} 30\%\ Pr^nC_2B_5H_6 & \text{(Ib)} \\ 2\%\ Pr^n\text{-}1,6\text{-}C_2B_4H_5 & \text{(IIb)} \end{cases}$$

$$\left.\begin{array}{l} Me_2C_2B_4H_6 \xrightarrow[<250°]{+Me_3N} \\ Pr^nC_2B_4H_7 \xrightarrow[<250°]{+Me_3N} \end{array}\right| \begin{array}{c} \text{no} \\ \text{abstraction} \\ \text{of } BH_3 \end{array} \left|\begin{array}{l} \xrightarrow{>250°} \\ \xrightarrow{>250°} \end{array}\right. \begin{array}{c} \text{carboranes as} \\ \text{above, but with} \\ \text{different percen-} \\ \text{tage distribution} \end{array} \left|\begin{array}{l} \longrightarrow \begin{cases} 15\% & \text{(Ia)} \\ 25\% & \text{(IIa)} \end{cases} \\ \longrightarrow \begin{cases} 15\% & \text{(Ib)} \\ 20\% & \text{(IIb)} \end{cases} \end{array}\right.$$

On pyrolysis, neither Me_2-2,4-$C_2B_4H_4$ or Pr^n-2,4-$C_2B_4H_5$ nor B-alkyl substituted boranes are formed. Silent electric discharge of 1-CH_3-B_5H_8 yields CB_5H_7[*565c*].

Other members of the series of lower carboranes can be prepared, although in low yield, by the repeated subjection to the silent electric discharge of a mixture of B_5H_{11} and ethyne. At each discharge, 0.1 to 0.2% of a carborane of the formula 1,5-$C_2B_3H_5$ is formed, together with still smaller amounts of other lower carboranes [*565a*, *690a*, *692a*]:

$$B_5H_{11} + HC{\equiv}CH \xrightarrow[\Delta]{\text{silent electric discharge}} \begin{cases} 1,5\text{-}C_2B_3H_5 \\ 1,6\text{-}C_2B_4H_6 \\ 1,2\text{-}C_2B_4H_6 \\ 2,4\text{-}C_2B_5H_7 \end{cases}$$

When gaseous B_2H_6, CH≡CH, and helium are subjected to the silent electric discharge or to an arc discharge (higher yields), the following carboranes and methylcarboranes, which can be separated by gas chromatography, are formed [*278c*]:

$$B_2H_6 + CH{\equiv}CH \xrightarrow[\text{discharge},\ \Delta]{\text{electric}} \begin{cases} 1,5\text{-}C_2B_3H_5 \\ 1,6\text{-}C_2B_4H_6 \\ 2,4\text{-}C_2B_5H_7 \\ C,3\text{-}(CH_3)_2\text{-}1,2\text{-}C_2B_3H_3 \\ 2,3\text{-}(CH_3)_2\text{-}1,5\text{-}C_2B_3H_3 \\ 2\text{-}CH_3\text{-}1,5\text{-}C_2B_3H_4 \\ 2\text{-}CH_3\text{-}1,6\text{-}C_2B_4H_5 \\ 1\text{-}CH_3\text{-}2,4\text{-}C_2B_5H_6 \\ 3\text{-}CH_3\text{-}2,4\text{-}C_2B_5H_6 \\ 5\text{-}CH_3\text{-}2,4\text{-}C_2B_5H_6 \end{cases}$$

On heating C,3-Me_2-1,2-$C_2B_3H_3$ (I) to 130° or 220° no arrangement of the skeleton occurs, but higher methylated carboranes are formed. They can be

References p. 355

separated by gas chromatography [*278f*]:

$$I_{liquid} \xrightarrow{130°} \begin{cases} 3,4,5\text{-}Me_3\text{-}1,2\text{-}C_2B_3H_2 \\ C,3,4,5\text{-}Me_3\text{-}1,2\text{-}C_2B_3H \\ C',3,4,5\text{-}Me_3\text{-}1,2\text{-}C_2B_3H \\ Me_3B \end{cases}$$

$$I_{vapour} \xrightarrow[(He)]{220°} \begin{cases} C,5\text{-}Me_2\text{-}1,2\text{-}C_2B_3H_3 \\ C,3,5\text{-}Me_3\text{-}1,2\text{-}C_2B_3H_2 \\ Me_3B \end{cases}$$

UV-irradiation of a mixture of B_5H_9 and C_2H_2 yields the carboranes $C_2B_3H_5$, 1,6-$C_2B_4H_6$ and 1,2-$C_2B_4H_6$ [*724c*]. CB_5H_7 is formed by a silent electric discharge of 1-CH_3-B_5H_8 [*565c*] or by the reaction of carbon vapour (carbon arc) with B_5H_9 [*604b*].

When lower energy than that of an electric discharge is supplied, *i.e.*, when B_5H_9 [*565a, 567c*], $B_5H_8(CH_3)$ [*568b*] or B_4H_{10} [*67a, 278d*] is heated with CH≡CH, small yields (1–2%) of carboranes with one or three carbon atoms in the skeleton are formed. They can be separated by gas chromatography and identified by IR and NMR spectroscopy:

$$B_5H_9 + CH{\equiv}CH \xrightarrow[\Delta]{215°} \begin{cases} 2\text{-}CH_3\text{-}2\text{-}CB_5H_8 \\ 3\text{-}CH_3\text{-}2\text{-}CB_5H_8 \\ 4\text{-}CH_3\text{-}2\text{-}CB_5H_8 \end{cases}$$

$$B_5H_8(CH_3) + CH{\equiv}CH \xrightarrow[\Delta]{220\text{ to }250°} \begin{cases} 1\text{-}CH_3\text{-}2,3\text{-}C_2B_4H_7 \\ 4CH_3\text{-}2,3\text{-}C_2B_4H_7 \\ 5\text{-}CH_3\text{-}2,3\text{-}C_2B_4H_7 \\ 1,2,3\text{-}(CH_3)_3\text{-}2,3\text{-}C_2B_4H_5 \\ 2,3,4\text{-}(CH_3)_3\text{-}2,3\text{-}C_2B_4H_5 \\ 2,3,5\text{-}(CH_3)_3\text{-}2,3\text{-}C_2B_4H_5 \end{cases}$$

$$B_4H_{10} + CH{\equiv}CH \xrightarrow[\Delta]{20\text{ to }50°} \begin{cases} B_5H_9 \\ 2,3,4\text{-}C_3B_3H_7 \\ 2\text{-}CH_3\text{-}2,3,4\text{-}C_3B_3H_6 \\ 2,3\text{-}CH_3\text{-}2,3,4\text{-}C_3B_3H_5 \\ 2,4\text{-}CH_3\text{-}2,3,4\text{-}C_3B_3H_5 \\ 4\text{-}CH_3\text{-}2\text{-}CB_5H_8 \end{cases}$$

In the thermolysis of B_4H_{10} and CH≡CH at 100°, which gives an explosive reaction, the lower carboranes (CH_3 derivatives of 2,3,4-$C_3B_3H_7$) are no longer formed, but a particularly large amount of 2,4-$C_2B_5H_7$ is obtained [*232b, 278d, 278h*]. The thermolysis of B_5H_9 with $CH_3C{\equiv}CCH_3$ (165°) $CH{\equiv}CCH_3$ (175°) and CH≡CH forms the corresponding carborane derivatives, *e.g.* with C_2H_2: $C_2B_3H_5$, 1,6-$C_2B_4H_6$, 2,4-$C_2B_5H_7$ (main product) and 1,6-$C_2B_8H_{10}$ [*278b*]. Prolonged heating and still higher temperatures lower the yield, since subsequent reactions take place [*565a*].

At 300°, $C_2B_4H_8$ forms $C_2B_5H_7$[*565a, 566a*]. In the presence of an excess of Me_3B, at 300° $C_2B_4H_8$ forms all possible monomethyl and most, if not actually all, dimethyl, trimethyl, and tetramethyl derivatives of 2,4-$C_2B_5H_7$. They can be separated by gas chromatography and identified by NMR, IR, and mass spectroscopy (especially 2,4-$C_2B_5H_7$ and Me_5-2,4-$C_2H_5H_2$)[*686a*].

2,3-$C_2B_4H_8$ (for structure, see Fig. 4.50) [or $(CH_3)_2C_3B_3H_5$] and NaH in diglyme form 2,3-$C_2B_4H_7^{\ominus}$[$(CH_3)_2C_3B_3H_4^{\ominus}$] with the evolution of H_2. Reaction with HCl or DCl re-forms 2,3-$C_2B_4H_7D$ [$(CH_3)_2C_3B_3H_4D$][*232b,567a*]:

$$2,3\text{-}C_2B_4H_8 + H^{\ominus} \xrightarrow{\text{diglyme}} 2,3\text{-}C_2B_4H_7^{\ominus} + H_2$$

$$2,3\text{-}C_2B_4H_7^{\ominus} + DCl \longrightarrow 2,3\text{-}C_2B_4H_7D + Cl^{\ominus}$$

$$(CH_3)_2C_3B_3H_5 + H^{\ominus} \xrightarrow{\text{diglyme}} (CH_3)_2C_3B_3H_4^{\ominus} + H_2$$

$$(CH_3)_2C_3B_3H_4^{\ominus} + DCl \longrightarrow (CH_3)_2C_3B_3H_4D + Cl^{\ominus}$$

With 2,3-$C_2B_4H_7^{\ominus}$ and DCl, 2,3-$C_2B_4H_7D$ deuterated in the B–D^{μ}–B bridge is obtained. By an exchange reaction, 2,3-$C_2B_4H_8$ and $(BD_3)_2$ form the triply B-deuterated carborane 2,3-$C_2B_4H_5$-4,5,6-D_3[*567a*]. In the presence of diglyme $(BD_3)_2$ exchanges on $C_2B_4H_8$ only at the 4 and 6 terminal positions[*724a*]. On $C_2B_4H_8$ D_2 exchanges with all boron-bounded hydrogen atoms[*724a*]. The $Me_2C_2B_4H_6$ and DCl exchange, catalysed by $AlCl_3$, occurs at both apex and base terminal positions[*724a*].

2,3-$C_2B_4H_6$-2,3-D_2 is formed by heating B_5H_9 with C_2D_2[*724a*]. In tetrahydrofuran–hexane solution $EtBH_2$ reacts with $H_2C{=}CH_2$ at 20° and 50 atm to yield C–Me–B–Et_5–CB_5H_3, C,C′–Me_2–B–Et_4–2,4-C_2B_5 (three isomers), C–Me–B–Et_4–2-CB_5H_4 (three isomers), C,C′–Me_2–B–Et_5–2,4-C_2B_5 and other carboranes (separation by gas chromatography)[*271a*]. The structure of C–Me–B–Et_5–CB_5H_3 is like that of Fig. 4.49; the terminal H atoms at the B atoms are substituted by C_2H_5, the terminal H atom at the C atom is substituted by CH_3[*271a*].

$C_2B_7H_{13}$ and $C_2B_7H_{11}Me_2$ are formed by the oxidation of $C_2B_9H_{11}$ and $C_2B_9H_9Me_2$, respectively (synthesis: polyhedral degradation of $H(C_2B_{10}H_{10})H$ and $Me(C_2B_{10}H_{10})Me$, respectively; section 19.4.2.10.) with $K_2Cr_2O_7$ in acetic acid at 0°[*763a*].

The pyrolysis of 1,3-$C_2B_7H_{13}$ at 215° in absence of B_2H_6 yields $C_2B_6H_8$, $C_2B_7H_9$ and 1,6-$C_2B_8H_{10}$, while in the presence of B_2H_6 the yield of 1,6-$C_2B_8H_{10}$ is considerably enhanced[*248b*]:

$$2(1,3\text{-}C_2B_7H_{13}) \xrightarrow[\text{diphenylether, 215°}]{\text{disproportionation}} 1,7\text{-}C_2B_6H_8 + 1,6\text{-}C_2B_8H_{10} + 4H_2$$

$$1,3\text{-}C_2B_7H_{13} \xrightarrow[\text{diphenylether, 215°}]{\text{H-elimination}} 1,7\text{-}C_2B_7H_9 + 2H_2 \text{ (low yield)}$$

Under conditions of relatively high temperatures (ca. 450°), low pressure (ca. 10 torr) and short residence time (1–3 sec.) the nido-carborane 4,5-$C_2B_4H_8$ is converted in nearly 100% yield to the three closo-carboranes 1,5-$C_2B_3H_5$, 1,6-$C_2B_4H_6$ and 2,4-$C_2B_5H_7$ (and a little 1,2-$B_2B_4H_6$) [*169b*].

The slow low-pressure pyrolysis at 48° of $C_2B_7H_{13}$ yields 2,4-$C_2B_5H_7$, 1,7-$C_2B_6H_8$ (very good yields, main product), 1,7-$C_2B_7H_9$ and 1,6-$C_2B_8H_{10}$ [*181a*] (similarly, pyrolysis of $CH_3C_2B_7H_{12}$ and $(CH_3)_2C_2B_7H_{11}$ [*181a*]). The preparation of $C_2B_6H_6(CH_3)_2$, $C_2B_7H_7(CH_3)_2$ and $C_2B_8H_8(CH_3)_2$ is given in section 19.4.2 (*10. Degradation of the polyhedron*).

The reaction of $LiC{\equiv}CCH_3$ and B_5H_9 leads to on-carbon carboranes [*568c*]:

$$LiC{\equiv}CCH_3 + B_5H_9 \xrightarrow[20^\circ]{\text{diglyme}} \begin{array}{l} 2\text{-}(C_2H_5)\text{-}2\text{-}CB_5H_8 \\ 2\text{-}(CH_3)\text{-}2,3\text{-}C_2B_4H_7 \\ B_4H_{10} \\ B_2H_6 \end{array}$$

The reaction of 1,7-$C_2B_6H_8$ and $(CH_3)_4NBH_4$ in diglyme at 100° yields CB_5H_9, 1-CH_3-2-CB_5H_8 and 3-CH_3-2-CB_5H_8 [*181b*].

$C_2B_7H_{13}$ reacts with two moles of NaH in diethyl ether to yield $C_2B_7H_{11}^{2\ominus}$. Treatment of this anion with $CoCl_2$ results in the formation of the very stable complex $Co^{III}(C_2B_7H_9)_2^{\ominus}$ [*311a*] (for thermal rearrangement, see [*252b*]):

$$2C_2B_7H_{11}^{2\ominus} + \tfrac{3}{2}Co^{2\oplus} \longrightarrow \tfrac{1}{2}Co^0 + Co(C_2B_7H_9)_2^{\ominus} + 2H_2$$

$C_2B_7H_{11}^{2\ominus}$ reacts with $BrMn(CO)_5$ to give $C_2B_6H_8Mn(CO)_3^{\ominus}$ [*318a*]. For the proposed structure, see [*318a*]. About $(\pi\text{-}C_5H_5)Ni[\pi\text{-}(3)\text{-}1,2\text{-}B_9C_2H_{11}]$ see [*849c*].

$C_2B_6H_6Me_2$ (Archimedian antiprism), $C_2B_7H_7Me_2$ (tricapped trigonal prism), $C_2B_8H_8Me_2$ (bicapped Archimedian antiprism) are formed together with hydrogen in the pyrolysis of $C_2B_7H_{11}Me_2$ at 200° in diphenyl ether [*763b*]. For the rearrangement of the nine atom family see [*535a*].

Figure 4.47 shows the trigonal bipyramidal structure of 1,5-$C_2B_3H_5$, and Fig. 4.48 the octahedral structure of 1,6-$C_2B_4H_6$. Figures 4.49, 4.50, and 4.51 show the

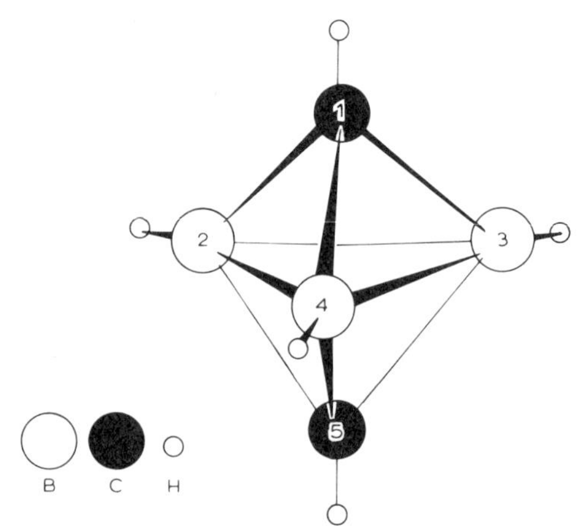

Fig. 4.47. Steric structure of the symmetrical closo-1,5-dicarbapentaborane(5), 1,5-$C_2B_3H_5$ (trigonal bipyramid) [*690a*].

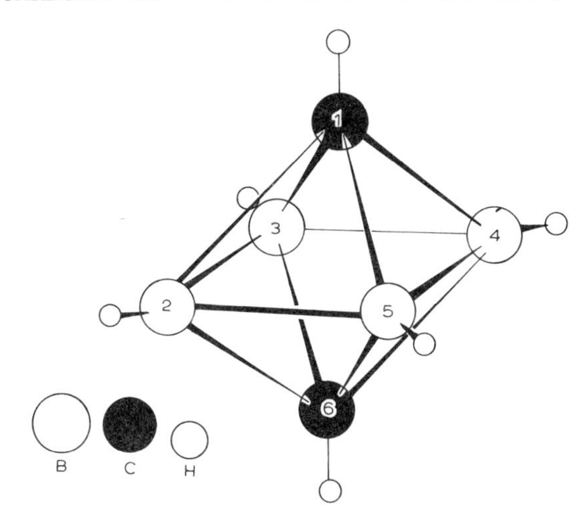

Fig. 4.48. Steric structure of the symmetrical closo-1,6-dicarbahexaborane(6), 1,6-$C_2B_4H_6$ (tetragonal bipyramid = octahedron [*692a*]). In the asymmetrical 1,2-$C_2B_4H_6$, the two carbon atoms are located at positions 1 and 2.

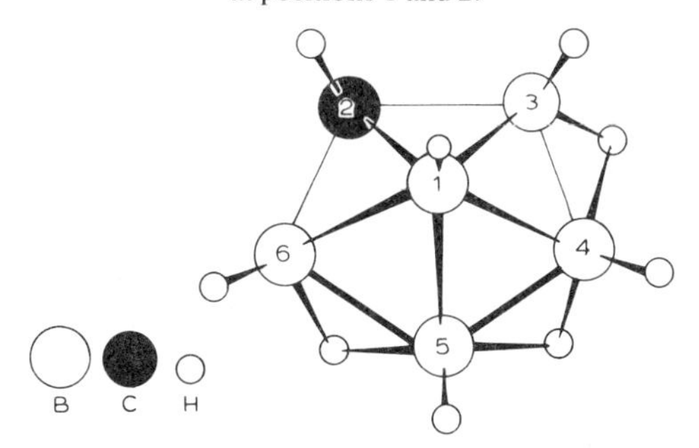

Fig. 4.49. Steric structure of nido-2-carbahexaborane(9), 2-CB_5H_9 (pentagonal pyramid) [*567c*].

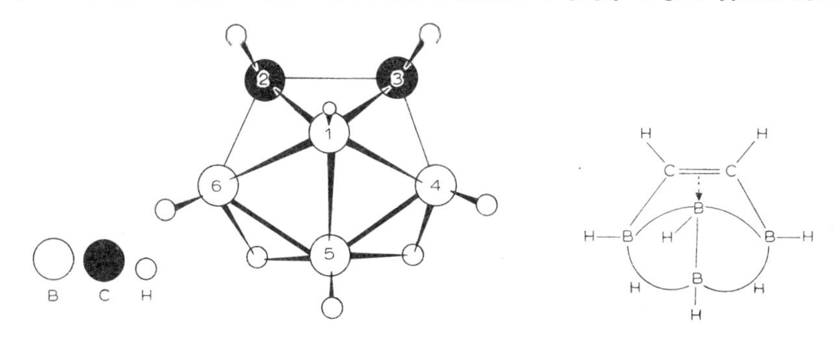

Fig. 4.50. Steric and topological structure of nido-2,3-dicarbahexaborane(8), 2,3-$C_2B_4H_8$ (pentagonal pyramid). Of the many possible topological structures, the form given here best describes the bond distances [*59, 569*]. The two, probably sp^2-hybridised, C atoms of the base ring (C^2 and C^3) form a double bond which, because of its π-bond component to the apical B atom B^1 is somewhat weakened (lengthening of the normal C=C double bond distance from 1.337 Å in C_2H_4 to 1.419 Å). Structural parameters [*59*]: B^1–B^4 = B^1–B^6 = 1.765–1.779 Å; B^4–B^5 = 1.783 Å; B^1–B^5 = 1.741 Å; B^1–C^2 = B^1–C^3 = 1.748 = 1.751 Å.

References p. 355

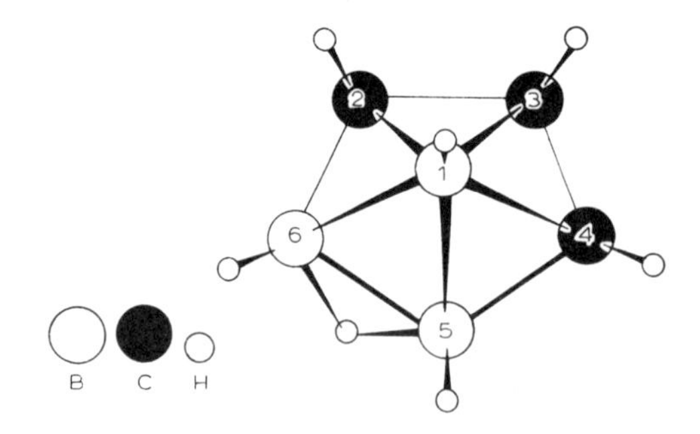

Fig. 4.51. Steric structure of nido-2,3,4-tricarbahexaborane(7), 2,3,4-$C_3B_3H_7$[*67a*] (pentagonal pyramid).

pentagonal-pyramidal structures of 2-CB_5H_9 (three B–H^μ–B bridges), 2,3-$C_2B_4H_8$ (two B–H^μ–B bridges), and 2,3,4-$C_3B_3H_7$ (one B–H^μ–B bridge); all three structures (2-CB_5H_9; 2,3-$C_2B_4H_8$; 2,3,4-$C_3B_3H_7$) are related to the likewise pentagonal-pyramidal structure of B_6H_{10} (four B–H^μ–B bridges, Fig. 4.7). Figure 4.52 shows the pentagonal-bipyramidal structure of 2,4-$C_2B_5H_7$ (for the structure of an eight-membered carborane $Me_2C_2B_6H_6$, see [*292a*] and the 9-membered carborane $Me_2C_2B_7H_9$.

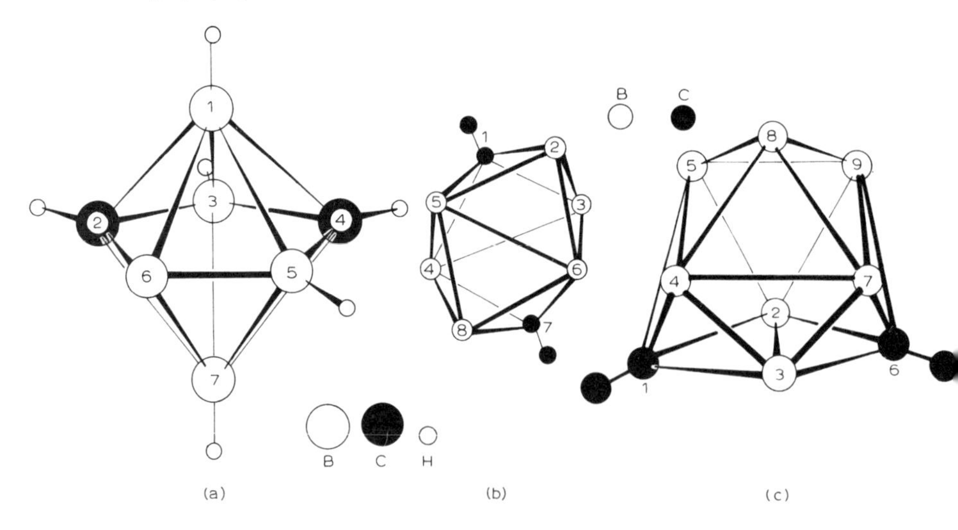

Fig. 4.52. Steric structures.

(a) Closo-2,4-dicarbaheptaborane(7), 2,4-$C_2B_5H_7$ (pentagonal bipyramid). The NMR and IR spectra of the derivative permethylated on the boron, 1,3,5,6,7-Me_5-2,4-$C_2B_5H_7$ indicate this structure[*686a*].

(b) Closo-1,7-dimethyl-1,7-dicarbaoctaborane(8), 1,7-$(CH_3)_2$-1,7-$C_2B_6H_6$ (H atoms are omitted [*293c*].

(c) Closo-1,6-dimethyl-1,6-dicarbanonaborane(9), 1,6-$(CH_3)_2$-1,6-$C_2B_7H_7$ (C_{2v} symmetry; H atoms are omitted). The C_2B_7 unit is only slightly distorted when compared with an idealized B_9 polyhedron based upon a tricapped prism with D_{3h} symmetry [*423a*].

19.3 Carboranes with 11-membered skeletons

About the degradation of the 10 boron atoms containing 12-membered skeleton $H(1,2\text{-}C_2B_{10}H_{10})H$ to 9 boron atoms containing 11-membered skeletons (*e.g.*, $C_2B_9H_{12}^{\ominus}$, $C_2B_9H_{11}^{2\ominus}$, $C_2B_9H_{13}$), see section 19.4.2 (*10. Degradation of the polyhedron*).

$B_{10}H_{12}(SEt_2)_2$ (10-membered skeleton, see Fig. 4.30) reacts with ethyl isocyanide $EtN{\equiv}C$ as described in section 18.5.4 with exchange of ligands [*319a*]:

$$B_{10}H_{12}(SEt_2)_2 + 2C{\equiv}NEt \longrightarrow B_{10}H_{12}(C{\equiv}NEt)_2 + 2SEt_2$$

On the other hand, the direct reaction of $B_{10}H_{14}$ (10-membered skeleton, see Fig. 4.11) with $CH_3CH_2N{\equiv}C$ does not yield $B_{10}H_{12}(C{\equiv}NCH_2CH_3)_2$ (likewise with a ten-membered skeleton), in analogy with the $B_{10}H_{14}$ derivative described in section 18.5.1, but a carborane with an 11-membered skeleton: $B_{10}H_{12}C$ $N(CH_3)H_2]$ (see Fig. 4.56) [*319a*]:

$$B_{10}H_{14} + CH_3CH_2N{\equiv}C \longrightarrow B_{10}H_{12}CN(CH_3)H_2 + H_2$$

$B_{10}H_{12}CNPr^nH_2$, $B_{10}H_{12}CNBu^tH_2$, $B_{10}H_{12}CNPr^nMe_2$, etc., are prepared similarly [*319a*]. On reaction with NaH and subsequently with Me_2SO_4 or MeI, $B_{10}H_{12}$-$CNMeH_2$ gives a 50% yield of $B_{10}H_{12}CNMe_3$ [*319a*].

The parent compound $B_{10}H_{12}CNH_3$ is formed by the reaction of $B_{10}H_{14}$ with NaCN [*411c, 411e*]:

$$B_{10}H_{14} + 2CN^{\ominus} \longrightarrow B_{10}H_{13}CN^{2\ominus} + HCN$$

and (i) by protonation in an acidic ion-exchange column [*411c, 411e*]:

$$B_{10}H_{13}CN^{2\ominus} + 2H^{\oplus} \longrightarrow B_{10}H_{12}CNH_3$$

or (ii) by treatment of $B_{10}H_{13}CN^{2\ominus}$ with either Me_3SiCl or Me_3SnCl, followed by base hydrolysis [*677a*], *e.g.*:

$$B_{10}H_{13}CN^{2\ominus} \xrightarrow[\Delta]{+Me_3SiCl} \begin{bmatrix}\text{hydrolysable}\\ \text{intermediate}\end{bmatrix} \xrightarrow[\Delta]{+H_2O} B_{10}H_{12}CNH_3$$

Partially methylated compounds $B_{10}H_{12}CNH_{3-n}R_n$ are obtainable from $B_{10}H_{12}CN^{2\ominus}$ by treatment with RI ($R = CH_3, C_2H_5$) in THF/H_2O [*677a*], *e.g.*:

$$B_{10}H_{13}CN^{2\ominus} + 3RI + C_4H_8O \longrightarrow B_{10}H_{12}CNR_3 + 2I^{\ominus} + C_4H_8O{\cdot}HI$$

It is possible that $B_{10}H_{13}CNR^{\ominus}$ is formed initially [*677a*]:

$$B_{10}H_{13}CN^{2\ominus} + RX \longrightarrow B_{10}H_{13}CNR^{\ominus} + X^{\ominus}$$

References p. 355

Alkylation of $B_{10}H_{12}CN^{\ominus}$·donor (donor = SMe_2, pyridine) with MeI in THF/H_2O yields $B_{10}H_{11}(OH)CNMe_3$; in THF/CH_3OH $B_{10}H_{11}(OMe)CNMe_3$ [*677a*].

With chlorine or bromine $B_{10}H_{12}CNMe_3$ gives 4(6)-$XM_{10}H_{11}CNMe_3$ or 4,6-$XB_{10}H_{10}CNMe_3$, respectively [*371d*].

The deamination and deprotonation of $B_{10}H_{12}CNMe_3$ with sodium forms the 11-membered ion $B_{10}H_{10}CH^{3\ominus}$ [*371c, 371d*]; iodine oxidation produces $B_{10}H_{10}CH^{\ominus}$ [*371d*]. Deamination of $B_{10}H_{12}CNR_3$ with NaH in refluxing THF yields $B_{10}H_{12}CH^{\ominus}$ [*371d*].

Treatment of $Na_3B_{10}H_{10}CH$ with PCl_3 leads to 1,2-$B_{10}H_{10}CHP$ with quite similar chemical and physical properties to the isoelectronic 1,2-$B_{10}H_{10}C_2H_2$ [*767b, 767c*]. $C_2B_9H_{11}$ reacts with $SnCl_2$ giving $C_2B_9SnH_{11}$ [*775b*] and $C_2B_9H_{11}$ with $Be(CH_3)_2$ leads to $C_2B_9BeH_{11}$ [*600b*]. Basic and acidic hydrolysis degrades 1,6-$C_2B_8H_{10}$ to $C_2B_7H_{12}^{\ominus}$ and boric acid respectively but similar conditions do not affect the 1,10-$C_2B_8H_{10}$ isomer [*248c*]:

$$1,6\text{-}C_2B_8H_{10} + OH^{\ominus} \xrightarrow{H_2O} C_2B_7H_{12}^{\ominus} + B(OH)_3$$

$B_{10}H_{10}CH^{3\ominus}$ reacts with PCl_3 to form 1,2-$B_{10}H_{10}CHP$ [*474b*]. Rearrangement by heating of 1,2-$B_{10}H_{10}CHP$ gives both 1,7- and 1,12-$B_{10}CHP$ [*767c*]. From the 1,7-carbaphosphaborane, while refluxing in excess piperidine, one boron atom is removed from the skeleton to form the piperidinium salt $C_5H_{10}NH_2^{\oplus}$-1,7-B_9H_{10}-$CHP^{\ominus}$ [*767a*]. Photochemical reactions of $B_9H_{10}CHP^{\ominus}$ (or similarily prepared $B_9H_{10}CHAs^{\ominus}$) with $Cr(CO)_6$, $Mo(CO)_6$ and $W(CO)_6$ lead to the formation of complexes of the general formula $[B_9H_{10}CHP \cdot M(CO)_5]^{\ominus}$ and $[B_9H_{10}CHAs \cdot M(CO)_5]^{\ominus}$ respectively (M = Cr, Mo, W) [*709a*].

19.4 Carboranes with 12-membered skeletons

For a review see [*67b*].

19.4.1 Synthesis of closo-1,2-dicarbadodecarborane(12) and derivatives

If in the anion $B_{12}H_{12}^{2\ominus}$ (section 18.12.1) a doubly negative $B_2H_2^{2\ominus}$ group is replaced by the isosteric neutral ethyne C_2H_2, we obtain a closodicarbadodecaborane(12) ("carborane") $H(C_2B_{10}H_{10})H$. This is therefore derived from $B_{12}H_{12}^{2\ominus}$ (Figs. 4.41a and 4.41b) by the replacement of the two BH groups by CH groups. Figure 4.53 shows the structure of $H(1,2\text{-}C_2B_{10}H_{10})H$ and Fig. 4.54 that of $H(1,2\text{-}C_2B_{10}H_8\text{-}9,10\text{-}Br_2)H$.

Closo-1,2-dicarbadodecaborane(12) $H(1,2\text{-}C_2B_{10}H_{10})H$ and its organyl derivatives $R(1,2\text{-}C_2B_{10}H_{10})R$ are formed when $B_{10}H_{12}Z_2$ are boiled with alkynes. It is not necessary to prepare the $B_{10}H_{12}Z_2$ separately and isolate them. It is sufficient if in the reaction of $B_{10}H_{14}$ (or its derivative) with $RC{\equiv}CR'$ 0.1 to 0.9 mole of the

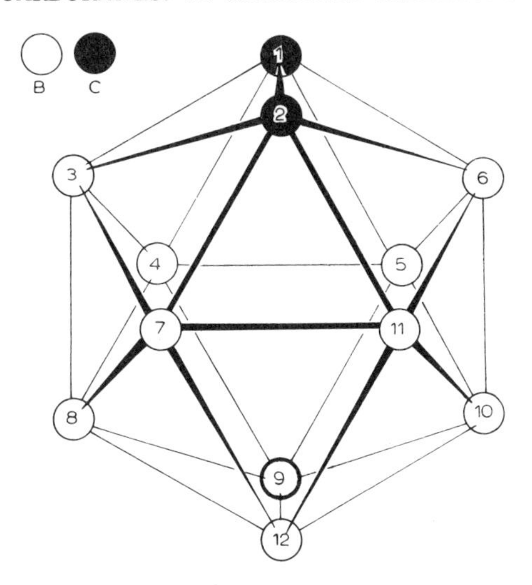

Fig. 4.53. Steric structure of closo-1,2-dicarbadodecaborane(12), $H(1,2\text{-}C_2B_{10}H_{10})H$, without the H atoms bound to the atoms of the skeleton (almost symmetrical icosahedron; numbering according to the ACS)[*4, 270*]. Closo-1,7-dicarbadodecaborane(12), $H(1,7\text{-}C_2B_{10}H_{10})H$, and closo-1,12-dicarbadocecaborane(12), $H(1,12\text{-}C_2B_{10}H_{10})H$, have similar structures.

Distances in $H(1,2\text{-}C_2B_{10}Cl_8\text{-}3,6\text{-}H_2)H$[*601, 602*]: B–B = 1.742–1.869 Å; B–C = 1.664–1.766 Å; C–C = 1.668 Å.

Distances in $BrH_2C(1,2\text{-}C_2B_{10}H_{10})CH_2Br$[*775*]: B–B = 1.69–1.86 Å; B–C = 1.68–1.77 Å; C–C = 1.64 Å.

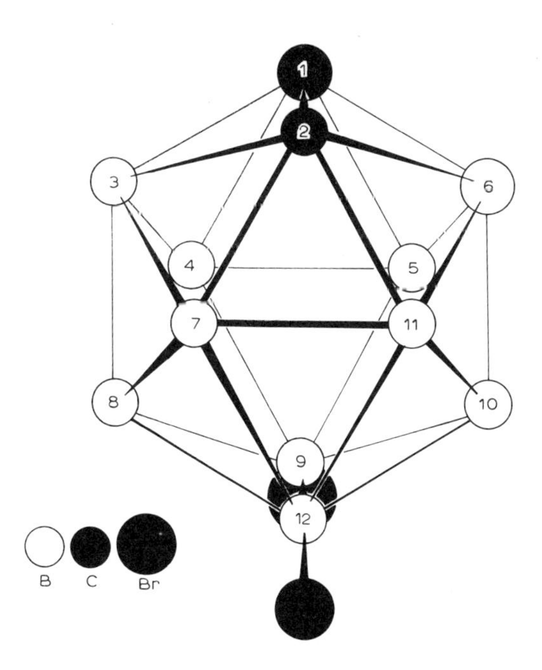

Fig. 4.54. Steric structure of $H(1,2\text{-}C_2B_{10}H_8\text{-}9,12\text{-}Br_2)H$ without the H atoms attached to the skeletal atoms[*602a*].

References p. 355

donor Z is present per mole of $B_{10}H_{14}$[*4, 9, 215, 217, 268–270, 338, 339, 573, 680, 842, 862*]:

$$B_{10}H_{14} + 2Z \xrightarrow{-H_2} B_{10}H_{12}Z_2 \xrightarrow[\text{solvent}]{+RC\equiv CR', -2Z, -H_2} R(1,2\text{-}C_2B_{10}H_{10})R'$$

$$B_{10}H_{13}R'' \xrightarrow[\text{solvent}]{+RC\equiv CR', -2H_2, (\text{donor})} R(1,2\text{-}C_2B_{10}H_9R'')R'$$

Donors Z: acetonitrile, amines, dialkyl sulphides, cyclic ethers, tetrahydrofuran, and dioxan (the latter less recommendable, since $B_{10}H_{14}$ + dioxan is sensitive to shock).

Solvents: either the donor Z itself (diethyl sulphide, ether) or, better, benzene, diethyl ether, di-n-propyl ether.

Temperature: above 70°; therefore in the case of low-boiling solvents an autoclave is necessary.

Alkyne: unsymmetrical alkynes react with high yields; many symmetrical alkynes such as 2-butyne and 2-hexyne do not react.

Substituents R,R′R″:

(a) R=R″=H; R′=H, CH_3, CH=CH_2, C(CH_3)=CH_2, C≡CH, n-C_3H_7, n-C_4H_9, $CH_2CH_2CH(CH_3)_2$, n-C_6H_{13}, C_6H_5, CH_2CH_2 ($C_2B_{10}H_{10}$)H, CH_2Br, $COOCH_3$, CH_2OOCCH_3, $CH_2CH_2OC(O)CH_3$, CH[OC(O)CH_3]CH_3, $CH_2N(C_2H_5)_2$
(b) R=R′=CH_2Cl; R″=H
(c) R=Br; R′=n-C_3H_7; R″=H
(d) R=Br; R′=n-C_4H_9; R″=H
(e) R=R′=CH_2OOCH_3; R″=H
(f) R=R′=$COOCH_3$; R″=H
(g) R=R′=H; R″=40% 1-C_2H_5, 60% 2-C_2H_5) and others

Although closo-1,2-dicarbadodecaborane(12) and decaborane(14) possess related covalent structures, they differ in their reactions. The decaborane skeleton can easily be degraded while the carborane skeleton tenaciously resists thermal and many chemical attacks. In this respect it surpasses even the chemical resistance of the $B_{12}H_{12}^{2\ominus}$ anion, which has an analogous structure. Here we give some examples of experiments on the reactions of the compound substituted on the carbon by an isopropenyl group (R=C(CH_3)=CH_2), 1-isopropenyl-closo-1,2-dicarbadodecaborane(12), a low-melting compound readily soluble in both polar and non-polar solvents: stable up to 350°, stable to boiling methanol and water with the access of air (steam distillation), stable to methanolic hydrochloric acid (but degraded by methanolic KOH), capable of crystallisation from 100% sulphuric acid at 150°. Alkaline permanganate or trifluoroperacetic acid oxidises only the propenyl ligand and not the carborane skeleton. Similarly, in the presence of Raney nickel, hydrogen hydrogenates only the propenyl ligand and not the carborane skeleton[*215*]. Thus, reactions with carboranes are limited, with a few exceptions (halogenation, degradation reactions), to reactions with the C ligands (see section 19.4.2). However, they are affected by highly electron-attracting carborane polyhedra.

19.4.2 Reactions of closodicarbadodecaborane(12) and its derivatives

The structure of closodicarbadodecaborane(12) is derived from that of decaborane(14) by the fact that it converts it into a closed polyhedron (icosahedron). The latter possesses a strong electron-attracting (—I) effect. The hydrogen atoms on the carbon are therefore positively polarised, but they can be titrated only if hydrogen atoms in the carborane polyhedron itself are overwhelmingly substituted by (—I) substituents (substitution on the boron). $H(C_2B_{10}H_2Cl_8)H$ and $H(C_2B_{10}Cl_{10})H$ are titratable dibasic acids, and $Cl(C_2B_{10}Cl_{10})H$ is a monobasic acid (for comparisons, $B_{10}H_{14}$ and $B_{10}H_{13}R$ are titratable and monobasic)[*680*]. The chlorine atoms in chlorinated carboranes $H(C_2B_{10}H_{10-n}Cl_n)H$ can be removed only in 50% aqueous potassium hydroxide in the presence of hydrogen peroxide at 100°[*680*].

The strong (—I) effect of the carborane polyhedron opposes nucleophilic displacement on the C, facilitates the formation of bis(donor–acceptor) adducts and metallation, and increases the acidity of acids, *e.g.* of $H(C_2B_{10}H_{10})(CH_2)_nCOOH$. The electron deficiency in the carborane polyhedron is so strong that electrophilic substitution (for example, Friedel-Crafts reactions) is hindered[*269*]. In general, the reactions of the carboranes resemble those of organic chemistry. Substituted closo-1,7-dicarbadodecaborane(12) derivatives react (if no steric factors oppose) like the corresponding substituted closo-1,2-dicarbadodecaborane(12) derivatives, but generally in smaller yield[*573, 574*]. Below we give some types of reactions, with examples[*4, 9, 186, 215, 217, 268–270, 278, 319, 338, 339, 389, 573, 680, 682, 718, 842, 860a, 862, 863*].

1. *Salt formation*

$H(1,2\text{-}C_2B_{10}H_{10-n}Cl_n)H + 2MOH \rightarrow$ cannot be titrated

($n = 0$–6, M = alkali metal)

$$H(1,2\text{-}C_2B_{10}H_{10-m}Cl_m)H \xrightarrow[-H_2O\,(\text{not in EtOH})]{+MOH(EtOH + H_2O)} H(\ )M \xrightarrow[-H_2O]{+MOH} M(\ )M$$

2. *Reactions in carborane ligands*

$$H(1,2\text{-}C_2B_{10}H_{10})C{=}CH_2 + H_2 \xrightarrow[\text{pressure}]{\text{Raney-nickel}} H(1,2\text{-}C_2B_{10}H_{10})CH(CH_3)$$

$$2H(1,2\text{-}C_2B_{10}H_{10})CH_2OH \xrightarrow[-2HCl]{+Me_2SiCl_2} [H(1,2\text{-}C_2B_{10}H_{10})CH_2O]_2SiMe_2$$ [*685a*]

$$3H(1,2\text{-}C_2B_{10}H_{10})CH_2OH \xrightarrow[-3HCl]{+MeSiCl_3} [H(1,2\text{-}C_2B_{10}H_{10})CH_2O]_3SiMe$$ [*685a*]

$$HOCH_2(1,2\text{-}C_2B_{10}H_{10})CH_2OH \xrightarrow[-2HCl]{+2Me_3SiCl} Me_3SiOCH_2(1,2\text{-}C_2B_{10}H_{10})CH_2OSiMe_2$$

[*685a*]

References p. 355

$$HOCH_2(1,2\text{-}C_2B_{10}H_{10})CH_2OH \xrightarrow[-2HCl]{+Me_2SiCl_2} \overline{OCH_2(1,2\text{-}C_2B_{10}H_{10})CH_2OSi}Me_2$$

$$R(1,2\text{-}C_2B_{10}H_{10})C(=O)R' \xrightarrow{+EtO^\ominus} R(1,2\text{-}C_2B_{10}H_{10})C(O^\ominus)(OEt)\text{-}R' \xrightarrow[-R'COOEt]{+EtOH} R(1,2\text{-}C_2B_{10}H_{10})H \quad [861g]$$

For comparison:

$$R(1,2\text{-}C_2B_{10}H_{10})C(=O)OR \xrightarrow[-R_2CO]{-RO^\ominus} R(1,2\text{-}C_2B_{10}H_{10})^\ominus \xrightarrow[-RO^\ominus]{+ROH} R(1,2\text{-}C_2B_{10}H_{10})H \quad [725e]$$

$$H(1,2\text{-}C_2B_{10}H_{10})C_6H_5 \xrightarrow[CCl_4,\ \Delta]{+HNO_3/H_2SO_4} H(1,2\text{-}C_2B_{10}H_{10})C_6H_4NO_2 \quad [309a, 861f]$$

$$\downarrow +Sn,\ +HCl,\ \Delta$$

(like $H(1,7\text{-}C_2B_{10}H_{10})C_6H_5$) $H(1,2\text{-}C_2B_{10}H_{10})C_6H_4NH_2$ (diazotizable!)

$$R(1,2\text{-}C_2B_{10}H_{10})CH_2Br + Mg \xrightarrow{ether} R(1,2\text{-}C_2B_{10}H_{10})CH_2MgBr \quad [725a]$$

$$H(1,2\text{-}C_2B_{10}H_{10})CH_2Br + Mg \xrightarrow{THF} CH_3(1,2\text{-}C_2B_{10}H_{10})MgBr \quad [725a, 725d]$$

3. *Metallation*

$$R(1,2\text{-}C_2B_{10}H_{10})H + RLi \xrightarrow[benzene]{ether\ or} R(1,2\text{-}C_2B_{10}H_{10})Li + RH \quad [61a, 725b]$$

$$H(1,2\text{-}C_2B_{10}H_{10})H \xrightarrow[-Bu^nH]{+Bu^nLi} H(1,2\text{-}C_2B_{10}H_{10})Li \xrightarrow[-Bu^nH]{+Bu^nLi} Li(1,2\text{-}C_2B_{10}H_{10})Li$$

$$H(1,2\text{-}C_2B_{10}H_{10})CH_3 \xrightarrow[-BuH]{+Bu^nLi} Li[1\text{-}(CH_3)\text{-}1,2\text{-}C_2B_{10}H_{10}] \quad [715a]$$

$$R(1,2\text{-}C_2B_{10}H_{10})C(=O)OR \xrightarrow[-Bu^nCOOR]{+Bu^nLi} R(1,2\text{-}C_2B_{10}H_{10})Li \quad [861g]$$

$$H(1,7\text{-}C_2B_{10}H_{10})H \xrightarrow[(liquid\ NH_3)]{+NaNH_2,\ -NH_3} H(1,7\text{-}C_2B_{10}H_{10})Na \xrightarrow{+NaNH_2,\ -NH_3} Na(1,7\text{-}C_2B_{10}H_{10})Na \quad [861f]$$

$$R(C_2B_{10}H_{10})R' + 2e^\ominus \xrightarrow[(alkali\ metal)]{(liquid\ NH_3)} R(C_2B_{10}H_{10})R'^{2\ominus} \quad [861m]$$

4. *Reactions of metallated carboranes*

$$3Bu^n(1,2\text{-}C_2B_{10}H_{10})Li + Cl_3B_3N_3Me_3 \longrightarrow [Bu^n(1,2\text{-}C_2B_{10}H_{10})]_3B_3N_3Me_3 + 3LiCl$$ [61a]

$$R(1,2\text{-}C_2B_{10}H_{10})Li + RX \longrightarrow R(1,2\text{-}C_2B_{10}H_{10})R + LiX$$ [861h]

$$H(1,7\text{-}C_2B_{10}H_{10})M + RX \longrightarrow H(1,7\text{-}C_2B_{10}H_{10})R + MX$$ [861h]

$$M(1,7\text{-}C_2B_{10}H_{10})M + 2RX \longrightarrow R(1,7\text{-}C_2B_{10}H_{10})R + 2MX$$ [861h]

$$Li(1,2\text{-}C_2B_{10}H_{10})Li + Cl_3Si(1,2\text{-}C_2B_{10}H_{10})SiCl_3 \longrightarrow 2Li(1,2\text{-}C_2B_{10}H_{10})SiCl_3$$
(commutation)

$$H(1,2\text{-}C_2B_{10}H_{10})Li + R_3SiCl \longrightarrow H(1,2\text{-}C_2B_{10}H_{10})SiR_3 + LiCl$$

$$Li(1,7\text{-}C_2B_{10}H_{10})Li + Ph_2SnCl_2 \longrightarrow \frac{1}{n}[(1,7\text{-}C_2B_{10}H_{10})\text{-}SnPh_2]_n + 2LiCl$$ [70a]

$$R(1,2\text{-}C_2B_{10}H_{10})Li + CO_2 \longrightarrow R(1,2\text{-}C_2B_{10}H_{10})COOLi$$

$$Li(1,2\text{-}C_2B_{10}H_{10})Li + 2R_2AsX \longrightarrow R_2As(1,2\text{-}C_2B_{10}H_{10})AsR_2 + 2LiX$$ [716d]

$$Li(1,2\text{-}C_2B_{10}H_{10})Li + MeAsBr_2 \longrightarrow (1,2\text{-}C_2B_{10}H_{10})\begin{matrix}\diagup AsMe \diagdown \\ \diagdown AsMe \diagup\end{matrix}(1,2\text{-}C_2B_{10}H_{10})$$ [860d]

$$2Li(1,2\text{-}C_2B_{10}H_{10})Li + 2COCl_2 \longrightarrow O{=}C\begin{matrix}\diagup (1,2\text{-}C_2B_{10}H_{10}) \diagdown \\ \diagdown (1,2\text{-}C_2B_{10}H_{10}) \diagup\end{matrix}C{=}O + 4LiCl$$ [612a]

$$Li(1,7\text{-}C_2B_{10}H_{10})Li + 2Ph_2PCl \longrightarrow Ph_2P(1,7\text{-}C_2B_{10}H_{10})PPh_2 + 2LiCl$$ [9a]

$$Li(1,2\text{-}C_2B_{10}H_{10})Li + 2PCl_3 \longrightarrow ClP\begin{matrix}\diagup (1,2\text{-}C_2B_{10}H_{10}) \diagdown \\ \diagdown (1,2\text{-}C_2B_{10}H_{10}) \diagup\end{matrix}PCl + 4LiCl$$ [9a]

$$Li(1,7\text{-}C_2B_{10}H_{10})Li + 2PCl_3 \longrightarrow Cl_2P(1,7\text{-}C_2B_{10}H_{10})PCl_2 + 2LiCl$$ [9a]

$$Li(1,2\text{-}C_2B_{10}H_{10})Li + 2RSSR \longrightarrow RS(1,2\text{-}C_2B_{10}H_{10})SR + 2RSLi$$ [716c, 718a]

$$Li(1,2\text{-}C_2B_{10}H_{10})Li \xrightarrow{+2S} LiS(1,2\text{-}C_2B_{10}H_{10})SLi \begin{cases} \xrightarrow[-2LiX]{+2RX} RS(1,2\text{-}C_2B_{10}H_{10})SR \\ \xrightarrow[-2LiOH]{+2H_2O} HS(1,2\text{-}C_2B_{10}H_{10})SH \end{cases}$$

$$R(1,2\text{-}C_2B_{10}H_{10})Li + I_2 \longrightarrow R(1,2\text{-}C_2B_{10}H_{10})I + LiI$$

$$R(1,2\text{-}C_2B_{10}H_{10})MgBr + H_2O \longrightarrow R(1,2\text{-}C_2B_{10}H_{10})H + MgBr(OH)$$

$$H(1,7\text{-}C_2B_{10}H_{10})MgBr + CO_2 \xrightarrow[H_2O]{\Delta} H(1,7\text{-}C_2B_{10}H_{10})COOH + Mg(OH)_2 \quad [725c]$$

About organomercuric derivatives of $R(1,7\text{-}C_2B_{10}H_{10})R$, see [*861n*].

$$Li(1,2\text{-}C_2B_{10}H_{10})Li + (ClCH_2)HC{=}CH(CH_2Cl) \xrightarrow{-2LiCl} \text{benzenoid ring fused to } B_{10}H_{10} \quad [358b]$$

This carborane has a benzenoid ring, structure see [*358b*].

5. *Halogenation of the polyhedron (H substitution on boron).* Under the action of light, carborane and phenylcarborane can easily be chlorinated with chlorine [*863*]:

$$R(1,2\text{-}C_2B_{10}H_{10})H + n\,Cl_2 \xrightarrow{h\nu} R(1,2\text{-}C_2B_{10}H_{10-n}Cl_n)H + n\,HCl$$

Closo-1,2-dicarbadodecaborane(12) substituted in a definite position is obtained when a substituted decaborane(14) reacts with ethyne or methylethyne; for example $B_{10}H_{13}$-1-Cl forms $H(1,2\text{-}C_2B_{10}H_9\text{-}8\text{-}Cl)H$ and $B_{10}H_{13}$-2-Br forms $1\text{-}CH_3\text{-}(1,2\text{-}C_2B_{10}H_9\text{-}8\text{-}Br)H$ [*861d*].

Under Friedel-Crafts conditions (see below, under 6), $H(1,2\text{-}C_2B_{10}H_{10})H$ is brominated first in positions B^9 and B^{12} and then at B^8, and finally at B^{10} [*602d*]. For the charge distribution in carboranes, see [*58a*].

6. *Electrophilic substitution*

$$H(1,2\text{-}C_2B_{10}H_{10})H + Cl_2 \xrightarrow[<40^\circ,\ CH_2Cl_2]{AlCl_3,\ \Delta} H(1,2\text{-}C_2B_{10}H_9Cl)H \quad [861c, 861e, 861o]$$

$$H(1,2\text{-}C_2B_{10}H_{10})H + Cl_2 \xrightarrow[40^\circ,\ CH_2Cl_2]{AlCl_3,\ \Delta} H(1,2\text{-}C_2B_{10}H_4Cl_6)H \quad [861c]$$

$$H(1,7\text{-}C_2B_{10}H_{10})H + Br_2 \xrightarrow[80^\circ]{AlCl_3,\ \Delta} H(1,7\text{-}C_2B_{10}H_4Br_6)H \quad [861c, 861e]$$

$$H(1,2\text{-}C_2B_{10}H_{10})H + Br_2 \xrightarrow[80^\circ]{AlCl_3,\ \Delta} H(1,2\text{-}C_2B_{10}H_6Br_4)H \quad [861c, 861e]$$

$$H(1,2\text{-}C_2B_{10}H_{10})H + I_2 \xrightarrow[80^\circ,\ CCl_4]{AlCl_3,\ \Delta} H(1,2\text{-}C_2B_{10}H_{10-n}I_n)H\ (n = 1,2) \quad [861c]$$

$$H(1,2\text{-}C_2B_{10}H_{10})H \xrightarrow{+SbX_3Y_2,\ 240^\circ} \text{no halogenation}$$

$$H(1,2\text{-}C_2B_{10}H_{10})H + CH_2Cl_2 \xrightarrow[135^\circ]{AlBr_3} [H(1,2\text{-}C_2B_{10}H_9)H]_2CH_2$$

(B-alkylation in the carborane polyhedron)

$$H(1,2\text{-}C_2B_{10}H_{10})H + HNO_3 \xrightarrow{\Delta} H(1,2\text{-}C_2B_{10}H_9OH)H + H(1,2\text{-}C_2B_{10}H_9\text{-}ONO_2)H \quad [861k]$$

7. *Nucleophilic substitution*

$$H(1,2\text{-}C_2B_{10}H_6Cl_4)H + R_3N \xrightarrow[\text{benzene}]{20°} \text{no reaction}$$

$$H(1,2\text{-}C_2B_{10}H_2Cl_8)H + 2Et_3N \longrightarrow (Et_3NH^{\oplus})_2(1,2\text{-}C_2B_{10}H_2Cl_8)^{2\ominus} \quad (a)$$

$$H(1,2\text{-}C_2B_{10}Cl_{10})H + 2Et_3N \longrightarrow (Et_3NH^{\oplus})_2(1,2\text{-}C_2B_{10}Cl_{10})^{2\ominus} \quad (b)$$

$$H(1,2\text{-}C_2B_{10}Cl_{10})Cl + Et_3N \longrightarrow Et_3NH^{\oplus}(1,2\text{-}C_2B_{10}Cl_{10})Cl^{\ominus} \quad (c)$$

$$\downarrow +[Ph_3PMe]I \quad -[Et_3NH]I$$

$$Ph_3PMe^{\oplus}(1,2\text{-}C_2B_{10}Cl_{10})Cl^{\ominus} \quad (d)$$

(nucleophilic displacement)

$$H(1,2\text{-}C_2B_{10}Cl_{10})Cl + Et_3N \longrightarrow H(1,2\text{-}C_2B_{10}Cl_{10})NEt_3^{\oplus}Cl^{\ominus} \quad (e)$$

$H(1,2\text{-}C_2B_{10}H_2Cl_8)H$ forms no acid salts $R_3NH(1,2\text{-}C_2B_{10}H_2Cl_8)H$ even with an excess of R_3N. A comparison of the NMR spectra of (c) and (d) with that of $H(1,2\text{-}C_2B_{10}Cl_{10})Cl$ and of the NMR spectrum of (b) with that of $H(1,2\text{-}C_2B_{10}Cl_{10})H$ shows that the negative charge in (b), (c), and (d) is not distributed uniformly over the carborane polyhedron [*680*].

8. *Uptake of electrons and isomerisation*

$$Ph(1,7\text{-}C_2B_{10}H_{10})Ph \xrightarrow[\Delta]{+2\text{Na. naphthalene}} Ph(1,7\text{-}C_2B_{10}H_{10})Ph^{2\ominus}$$

$$\downarrow \text{isomerisation}$$

$$\underset{\text{Stable radical-ion}}{Ph(1,2\text{-}C_2B_{10}H_{10})Ph\cdot^{\ominus}} \xleftarrow[\Delta]{+O_2} Ph(1,2\text{-}C_2B_{10}H_{10})Ph^{2\ominus} \xrightarrow[\Delta]{+H_2O} \underset{\text{stable}}{Ph(1,2\text{-}C_2B_{10}H_{11})Ph^{\ominus}}$$

9. *Transformation of the polyhedron.* The carboranes $H(1,2\text{-}C_2B_{10}H_{10})H$ and $H(1,2\text{-}C_2B_{10}H_{10})CH_3$ undergo transformations on heating (for structures, see Fig. 4.53) [*249, 270, 287, 385b, 574a, 575, 602, 634, 861a, 861i*]:

$$H(1,2\text{-}C_2B_{10}H_{10})H \xrightarrow[\text{inert atmosphere}]{450\text{–}500°,\ 1\text{–}2\text{ days}} \underset{\text{m.p. 263–265°}}{H(1,7\text{-}C_2B_{10}H_{10})H} \xrightarrow{514\text{–}620°} H(1,12\text{-}C_2B_{10}H_{10})H + \text{polymers}$$

$$\underset{\text{m.p. 218–219°}}{H(1,2\text{-}C_2B_{10}H_{10})Me} \xrightarrow[\text{vacuum}]{400\text{–}472°} \underset{\text{m.p. 208–209.5°}}{H(1,7\text{-}C_2B_{10}H_{10})Me}$$

If the two H atoms bound to the C atoms are replaced by bulky groups (*e.g.*,

$MePh_2Si$–), the temperature of transformation falls considerably [*634*]:

$$\underset{\text{m.p. } 241\text{–}243°}{MePh_2Si(1,2\text{-}C_2B_{10}H_{10})SiPh_2Me} \xrightarrow{260°} \underset{\text{m.p. } 138\text{–}140°}{MePh_2Si(1,7\text{-}C_2B_{10}H_{10})SiPh_2Me}$$

1,2-Carboranes containing exocyclic rings such as, for example,

$$Me_2Si\text{–O–}SiMe_2 \text{ bridging } C\text{—}C \text{ of } B_{10}H_{10} \quad \text{or} \quad Me_2Si\text{–NH–}SiMe_2 \text{ bridging } C\text{—}C \text{ of } B_{10}H_{10}$$

are naturally more resistant to transformations [*575*]. About the rearrangement of icosahedral monohalo-*m*-carboranes with cuboctahedral intermediates see [*292b*].

10. *Degradation of the polyhedron* [*308a,314c,319,319e,570a,763,763c,842, 861b,863c*]

$$C_2B_9H_{12}^{\ominus} \underset{+H_2O,\ -OH^{\ominus}}{\overset{+Na \text{ or } +NaH,\ -H^{\oplus}}{\rightleftarrows}} C_2B_9H_{11}^{2\ominus}$$

$$H(1,2\text{-}C_2B_{10}H_{10})H \xrightarrow[-B(OMe)_3,\ -H_2]{+MeO^{\ominus},+2MeOH,\ 40°} C_2B_9H_{12}^{\ominus} \xrightarrow[-HI]{+I_2} C_2B_9H_{11}I^{\ominus}$$

$$C_2B_9H_{12}^{\ominus} \xrightarrow[\text{diethyl ether}]{+H^{\oplus}} C_2B_9H_{13} \xrightarrow[-H_2]{\sim 100°} C_2B_9H_{11}$$

Treatment of 1,2-$C_2B_9H_{12}^{\ominus}$ and 1,2-$(CH_3)_2$-1,2-$C_2B_9H_{10}^{\ominus}$ with $FeCl_3$ in presence of a donor L leads to the formal substitution of the donor for a hydride ion. The resulting $C_2B_9H_{11}L$ species appears to carry the attached ligand on one of the three boron atoms associated with the open pentagonal face. Two isomers are formed [*859b*]:

$$M^{\oplus}(3)\text{-}1,2\text{-}C_2B_9H_{12}^{\ominus} + 2FeCl_3 + L \xrightarrow{\text{benzene}} (3)\text{-}1,2\text{-}C_2B_9H_{11}L + 2FeCl_2 + HCl + M^{\oplus}Cl^{\ominus}$$

M = K, Cs, Me_4N; L = THF, Pyridine, diethylsulfide, acetonitrile

About preparation of $K^{\oplus}[Co(C_2B_9H_{11})_2]^{\ominus}$ and the reaction with $AlCl_3$ in CS_2 yielding $(1,2\text{-}C_2B_9H_{10})_2CoS_2CH$ see [*140a*]. (3)-1,2-$C_2B_9H_{11}^{2\ominus}$ reacts with $Cr(CO)_6$ and $Fe(CO)_5$ giving transition metal carbonyl complexes [*164d,318b,625f*]:

$$2\ (3)\text{-}1,2\text{-}C_2B_9H_{11}^{2\ominus} + 2Fe(CO)_5 \longrightarrow [(3)\text{-}1,2\text{-}C_2B_9H_{11}]Fe(\mu\text{-}CO)_2Fe(CO)\text{-}[(3)\text{-}1,2\text{-}C_2B_9H_{11}]$$

$$(3)\text{-}1,2\text{-}C_2B_9H_{11}^{2\ominus} + Cr(CO)_6 \longrightarrow [\pi\text{-}(3)\text{-}1,2\text{-}C_2B_9H_{11}]Cr(CO)_3^{2\ominus} + 3CO$$

$(Na^{\oplus})_2$-1,2-$C_2B_9H_{11}^{2\ominus}$ reacts with $PhBCl_2$ to yield 1,2-$C_2B_{10}H_{11}$-3-Ph (boron atom insertion)[*319c*,*319f*]:

$$1,2\text{-}C_2B_9H_{11}^{2\ominus} + PhBCl_2 \xrightarrow{THF} 1,2\text{-}C_2B_{10}H_{11}\text{-}3\text{-}Ph + 2Cl^{\ominus}$$

The phenyl derivative of 1,2-$C_2B_{10}H_{12}$ reacts accordingly to the upper reaction scheme[*319c*, *763c*]:

$$C_2B_9H_{10}Ph^{2\ominus} \xrightarrow[-2Cl^{\ominus}]{+PhBCl_2} C_2B_{10}H_{10}Ph_2$$

$$1,2\text{-}C_2B_9H_{11}Ph^{\ominus} \xrightarrow{+NaH,\ -H_2} C_2B_9H_{10}Ph^{2\ominus}$$

$$1,2\text{-}C_2B_{10}H_{11}Ph \xrightarrow[-B(OEt)_3,\ -H_2]{+EtO^{\ominus},+2EtOH} 1,2\text{-}C_2B_9H_{11}Ph^{\ominus} \underset{-H^{\oplus}}{\overset{+H^{\oplus}}{\rightleftharpoons}} 1,2\text{-}C_2B_9H_{12}Ph$$

$$1,2\text{-}C_2B_9H_{12}Ph \xrightarrow{110^\circ,\ -H_2} C_2B_9H_{10}Ph$$

$$1,2\text{-}C_2B_9H_{11}Ph^{\ominus} \xrightarrow{350^\circ} 1,7\text{-}C_2B_9H_{11}Ph^{\ominus} \underset{H^{\oplus}}{\overset{+H^{\oplus}}{\rightleftharpoons}} 1,7\text{-}C_2B_9H_{12}Ph$$

$$1,7\text{-}C_2B_9H_{12}Ph \xrightarrow{75^\circ,\ -H_2} C_2B_9H_{10}Ph$$

The degradation of the polyhedron takes place without C, C′ isomerisation [*249*]:

$$Ph(1,2\text{-}C_2B_{10}H_{10})H \xrightarrow[-B(OEt)_3,\ -H_2]{+EtO^{\ominus},+2EtOH} 1,2\text{-}C_2B_9H_{11}Ph^{\ominus} \xrightarrow{+NaH,\ \Delta} 1,2\text{-}C_2B_9H_9Ph^{2\ominus}$$

$$Ph(1,2\text{-}C_2B_{10}H_{10})H \xrightarrow{410^\circ} Ph(1,7\text{-}C_2B_{10}H_{10})H \qquad 1,2\text{-}C_2B_9H_{11}Ph^{\ominus} \xrightarrow{300^\circ} 1,7\text{-}C_2B_9H_{11}Ph^{\ominus}$$

$$Ph(1,7\text{-}C_2B_{10}H_{10})H \xrightarrow[-B(OEt)_3,\ -H_2]{+EtO^{\ominus},+2EtOH} 1,7\text{-}C_2B_9H_{11}Ph^{\ominus} \xrightarrow[+H^{\oplus}\ -H_2,\ 135^\circ]{\text{polyphosphoric acid}} C_2B_9H_{10}Ph$$

Piperidine also degrades closo-1,2-dicarbadodecaborane(12) derivatives[*319b*, *861b*]:

$$H(1,2\text{-}C_2B_{10}H_{10})H + 4C_5H_{10}NH \xrightarrow[C_6H_6]{20^\circ} 1,2\text{-}C_2B_9H_{12}N\cdot C_5H_{10}NH + (C_5H_{10}N)_2BH$$

References p. 355

Sodium dichromate in 2 *N* H_2SO_4 degrades $C_2B_9H_{11}$ and $C_2B_9H_9RR'$ (R, R' = CH_3, C_6H_5, *p*-BrC_6H_4) to a 9-membered skeleton [*763c*], *e.g.*:

$$C_2B_9H_{11} \xrightarrow[\Delta]{Na_2Cr_2O_7} C_2B_7H_{13}$$

$C_2B_7H_{13}$ derivatives, when pyrolyzed at 215°, yield $C_2B_6H_8$, $C_2B_7H_9$ and $C_2B_8H_{10}$ derivatives [*763c*], *e.g.*:

$$C_2B_7H_{11}Me_2 \xrightarrow[-H_2,\,\Delta]{215°} C_2B_6H_6Me_2,\ C_2B_7H_7Me_2,\ C_2B_8H_8Me_2$$

$C_2B_9H_{11}$ and the derivatives $C_2B_9H_9RR'$ (R, R' = CH_3, C_6H_5, *p*-BrC_6H_4) add donors (CH_3OH, PEt_3, PPh_3, NEt_3, $OH^{\ominus}$) [*763c*], *e.g.*:

$$C_2B_9H_{11} + \text{donor} \rightleftharpoons C_2B_9H_{11}\cdot\text{donor}$$

19.4.3 π-Carborane complexes

The $C_2B_9H_{11}^{2\ominus}$ ion that can be prepared as described in section 19.4.2 (*10. Degradation of the polyhedron*) resembles the $C_5H_5^{\ominus}$ ion since it possesses six electrons in the five atomic orbitals of the open skeleton (see Fig. 4.55). Like the $C_5H_5^{\ominus}$ ion, $C_2B_9H_{11}^{2\ominus}$ and its C-substituted derivatives can form π-complexes with transition metals. Some examples:

$$C_2B_9H_{11}^{2\ominus} + [Cl_2Pd\text{-}\pi\text{-}CPh{=}CPhCPh{=}CPh]_2 \longrightarrow 2(\pi\text{-}C_2B_9H_{11})Pd(\pi\text{-}C_4Ph_4) + 4\,NaCl \quad [790a]$$

$$C_2B_9H_{11}^{2\ominus} + BrMn(CO)_5 \longrightarrow (\pi\text{-}C_2B_9H_{11})Mn(CO)_3^{\ominus} + 2\,CO + Br^{\ominus} \quad [308a]$$

$$C_2B_9H_{11}^{2\ominus} + BrRe(CO)_5 \longrightarrow (\pi\text{-}C_2B_9H_{11})Re(CO)_3^{\ominus} + 2\,CO + Br^{\ominus} \quad [308a]$$

$$C_2B_9H_{11}^{2\ominus} + FeCl_2 \longrightarrow (\pi\text{-}C_2B_9H_{11})_2\overset{+2}{Fe}{}^{2\ominus} + 2\,Cl^{\ominus} \quad [319, 319d, 786a]$$

$$(\pi\text{-}C_2B_9H_{11})_2\overset{+2}{Fe}{}^{2\ominus} \underset{+e^{\ominus}}{\overset{-e^{\ominus}}{\rightleftharpoons}} (\pi\text{-}C_2B_9H_{11})_2\overset{+3}{Fe}{}^{\ominus}$$

Similarly: $[(1,2\text{-}C_2B_9H_{11})_2\overset{+3}{Co}]^{\ominus}$ [*232a*]

$$C_2B_9H_{11}^{2\ominus} + C_5H_5^{\ominus} + FeCl_2 \longrightarrow (\pi\text{-}C_2B_9H_{11})\overset{+2}{Fe}(\pi\text{-}C_5H_5)^{\ominus} \quad [314c, 319d]$$

$$(\pi\text{-}C_2B_9H_{11})\overset{+2}{Fe}(\pi\text{-}C_5H_5)^{\ominus} \xrightarrow{-e^{\ominus}} (\pi\text{-}C_2B_9H_{11})\overset{+3}{Fe}(\pi\text{-}C_5H_5)$$

Like $C_2B_9H_{11}^{2\ominus}$, $PhC_2B_9H_{10}^{2\ominus}$ and $B_{10}H_{10}CH^{3\ominus}$ form π-complexes (compare Fig. 4.61 with Fig. 4.62) [*319c, 371c*], *e.g.*:

$$4\,B_{10}H_{10}CH^{3\ominus} + 3\,CoCl_2 \xrightarrow{THF} 2\,(\pi\text{-}B_{10}H_{10}CH^{3\ominus})_2Co^{3\oplus} + Co + 6\,Cl^{\ominus}$$

Figures 4.55, 4.56, and 4.57 show the 11-membered skeleton of $C_2B_9H_{11}^{2\ominus}$, $B_{10}H_{12}C(NRH_2)$, and $C_2B_9H_{10}R_2I$. Figures 4.58 to 4.62 show π-complexes with 11-membered carboranes. For $(C_2B_9H_{11})_2Cu^{2\ominus}$, see [*849b*].

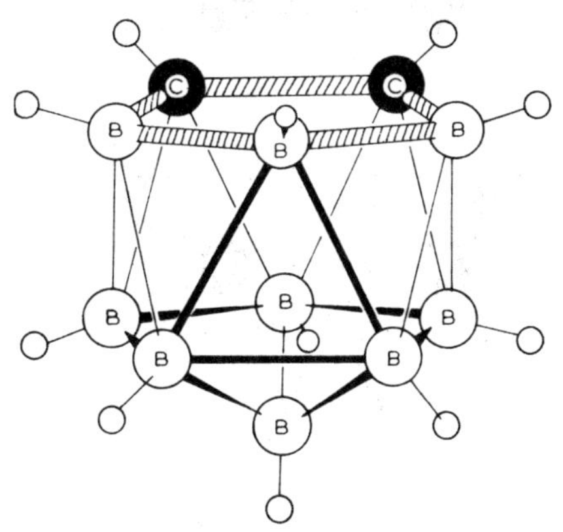

Fig. 4.55. Proposed steric structure of $C_2B_9H_{11}^{2\ominus}$ [*51a*].

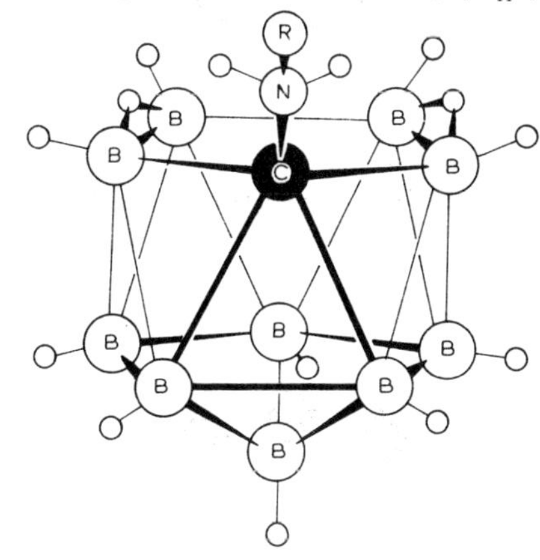

Fig. 4.56. Proposed steric structure of $B_{10}H_{12}CNRH_2$ [*319a*].

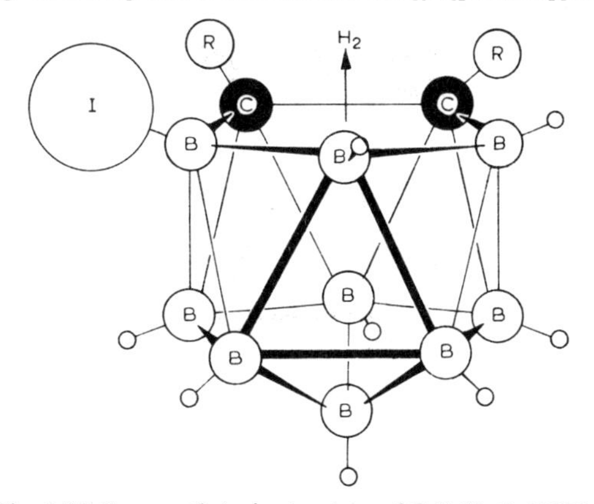

Fig. 4.57. Proposed steric structure of $C_2B_9H_{10}R_2I$ [*570a*].

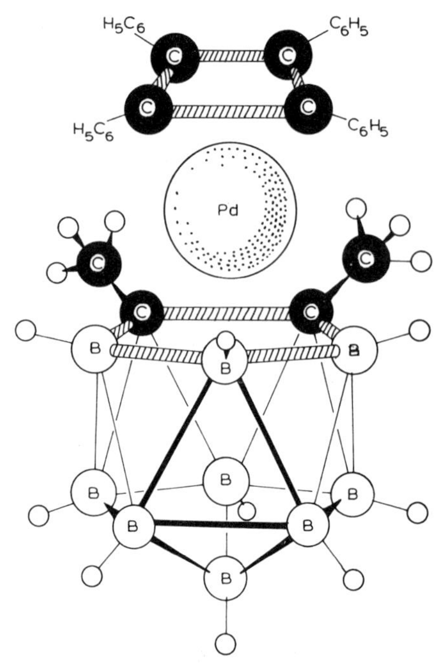

Fig. 4.58. Steric structure of $[\pi\text{-}C_2B_9H_9(CH_3)_2]Pd[\pi\text{-}C_4(C_6H_5)_4]$ [*790a*].

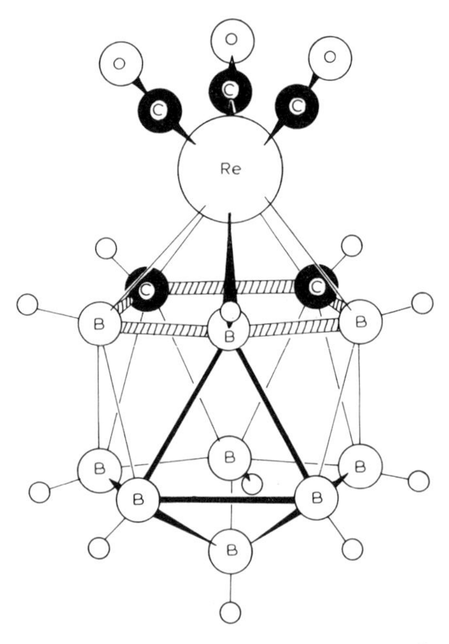

Fig. 4.59. Steric structure of $(\pi\text{-}C_2B_9H_{11})Re(CO)_3^{\ominus}$ [*860b*].

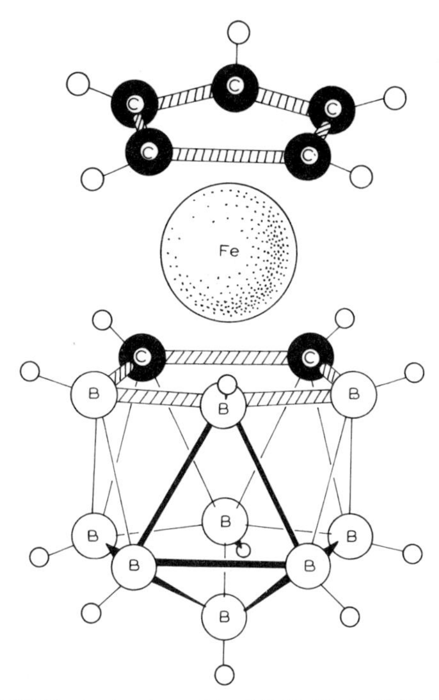

Fig. 4.60. Steric structure of (π-$C_2B_9H_{11}$)Fe(π-C_5H_5) [*863c*].

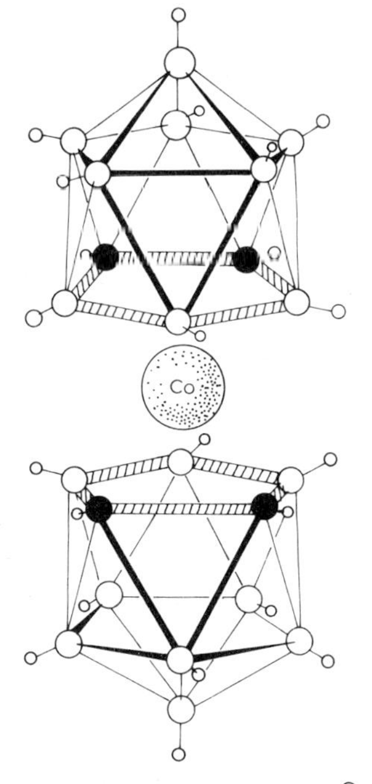

Fig. 4.61. Steric structure of $(\pi\text{-}C_2B_9H_{11})_2Co^{\ominus}$ [*860c, 863b*].

References p. 355

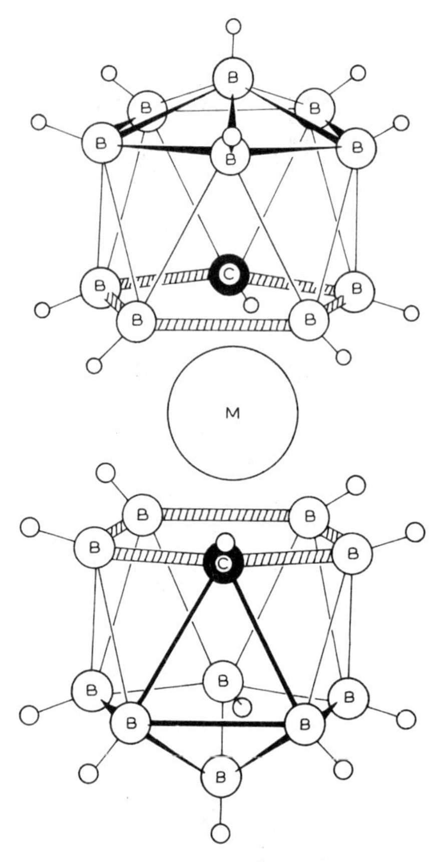

Fig. 4.62. Proposed steric structure of $(\pi\text{-}B_{10}H_{10}CH^{3\ominus})_2Co^{3\ominus}$ and $(\pi\text{-}B_{10}H_{10}CH^{3\ominus})_2Ni^{2\ominus}$ (M = Co, Ni) [*371c, 411c*].

20. Physical properties of the boranes and their derivatives

20.1 Uncharged boranes containing only B and H or D

B_2H_6. Under normal conditions, diborane is gaseous. Solid B_2H_6 forms two modifications that can be identified by their X-ray diagrams. The α-form, the low temperature modification, is produced in the condensation of B_2H_6 at 4.2°K. Above 60°K, the α-form (the so-called low-temperature modification) slowly changes into the β-form. The latter is also obtained by the condensation of B_2H_6 at 77°K followed by tempering at 90°K (to eliminate traces of α-B_2H_6). A third (low-temperature) modification that can be identified by X-ray diagrams is obtained by the condensation (4.2°K) of diborane that has been passed through a microwave discharge. It possibly contains radical hydrogen in a B_2H_6 matrix [*60, 718b*].

Liquid B_2H_6 is non-polar. Dielectric constant $\epsilon = 2.3721 - 0.002765_3\, T$ [*853*].

In the range from 113 to 181°K and 0 to 40 atm, increasing amounts of hydrogen dissolve in liquid B_2H_6 with rising temperature. Enthalpy of solution 386–600 cal/mole[*367*].

Gaseous diborane is absorbed exclusively physically on palladium black[*39*], on palladium on wood charcoal[*43*], and on hexagonal boron nitride[*43*] at 180 to 750 torr and 180 to 300°K. An activated physical adsorption takes place on wood charcoal at 115°[*39*]. On [100] surfaces of NaCl, the B_2H_6 layer forms a polymeric coating through an exothermic surface reaction (irreversible formation). The second and subsequent layers form reversibly[*623*].

M.p. −165.7°[*742*]; 108.14±0.02°K (=−165.04°) (probably a metastable phase)[*853*]; b.p. 180.63±0.02°K (=−92.52°)[*853*]. Vapour pressures: $\log p$ [torr] $=-674.82/(T-15.02)+6.968$ (valid between 108.22 and 147.00°K); $\log p$ [torr] $=-583.120/(T-24.63)+6.1885$ (valid between 150.03 and 180.66°K). Enthalpy of evaporation at the b.p. 3.413 and at ~125°K 3.420 kcal/mole[*495, 853*].

Densities: gaseous, 1.2389±0.0006 g/1 (at 275.16±0.01°K, 760 torr[*136*]; liquid, d_4 (g/ml) $= 0.4371+0.0010115\ (180.6-T)$[*746*]; d_4 (g/ml) $= 0.3140 - 0.001296\, t$[*444*]; solid: 0.577 g/ml (−183°)[*740*].

Energy of cleavage $B_2H_{6\,gas} \longrightarrow 2\ BH_{3\,gas}$, 28.5[*498, 698*]; 28.4±2[*34*] kcal/mole. For the solubility of B_2H_6 in Me_2O and $MeOC_2H_4OMe$, see[*264a*].

B_2D_6. B.p. −93.5° (extrapolated). Vapour pressure equation: $\log p$ [torr] $= -(962.983/T) - 0.00601480\,T + 9.31798$. Enthalpy of evaporation 3516 cal/mole [*169a*]. Heat of formation −585.29 kcal/mole[*283, 284*].

B_4H_{10}. M.p. −119.9°. B.p. 16.11°[*654a*]. Vapour pressure at 0°: 388±3[*407, 790*]; 387[*188*]; 386.8[cit. *188*]; 387[*724*] torr. Refractive index n_D^{25} 1.4971 to 1.4918[*790*]. Dielectric constant ϵ 2.274 to 2.286 (25°)[*790*]. Dipole moment in benzene 0.56±0.1 debye[*790*]. In order to avoid decomposition, B_4H_{10} is preferably stored at −196°[*188*]

B_4D_{10}. Vapour pressure at 0° 423 torr[*724*].

B_5H_9. M.p. −46.8±0.1°[*690*]; −46.77°[*854*]. B.p. 60.0° (calculated)[*690*]; 58.4°[*854*]. Vapour pressures: $\log p$ [torr] $= -1881.29/T - 0.0028956\,T + 9.49191$ (valid between −25° and +57°)[*690*]; $\log p$ [torr] $= -1951.14/T - 0.003688_4\,T + 9.96491$ (valid between 226 and 298°K = −47 at 25°)[*854*]. Density d_4 0.61 [cit. *243*], 0.643[*719*], 0.637±0.002[*243*] g/ml; $d = -0.000733\ T + 0.637$ (−8° to +25°)[*243*].

The dielectric constant of liquid B_5H_9 is high: $\epsilon = 21.1$ (25°) to 53.1 (at the m.p., −47°[*854*]). The dielectric constant of solid B_5H_9 depends on the setting in of the free rotatability of the B_5H_9 molecule[*854*]. Dipole moment, μ 3.37 (20°) to 5.54 (at the m.p. −46.77°) debye[*854*]; 2.13±0.04 debye (−73°)[*364, 365*].

B_5D_9. M.p. −47.0±0.1°. B.p. 59.0° (calculated)[*690*]. Vapour pressures: $\log p$ [torr] $= -1870.70/T - 0.0028783T + 9.46916$ (valid between −25° and +57°)[*690*]; vapour pressure at 0° 66.0 torr[*237a*].

References p. 355

B_5H_{11}. M.p. $-123.2 \pm 0.4°$. B.p. 63°[*121*]. Vapour pressure at 0° 52.8[*121*]; 52.5[*407*]; 52.8[*724*]. Vapour pressures: $\log p$ [torr] $= -1690.3/T + 7.901$. Enthalpy of evaporation 7734 cal. Trouton's constant 23.0 cal deg^{-1} $mole^{-1}$[*121*].

B_5D_{11}. M.p. −122.9°/122.0°. Vapour pressure at 0° 56.7 torr[*724*].

B_6H_{10}. M.p. −63 to −63.3°[*259, 736*]; −62.3°[*117*]; −62.2°[*766*]. B.p. 108° (calculated)[*117*]. Vapour pressures: $\log p$ [torr] $= -1566.3/T + 0.005841\,T + 5.2124$ (valid between −40° and +16°)[*259*] (according to [*117*], this sample of B_6H_{10} contained a more volatile compound); $\log p$ [torr] $= 6.4529 + 1.75 \log T - 0.00561\,T - 2270/T$[*117*]; Trouton's constant 21.0 cal $deg^{-1}mole^{-1}$[*117*].

B_6H_{12}. M.p. −82.2 to −82.3°; vapour pressures (selection): −30.7°, 2.6 torr; 0°, 17 torr[*482*].

B_8H_{12}. M.p. about −20°; thermally unstable[*203*].

B_9H_{15}. M.p. 2.5 to 2.7°; vapour pressure at 28.2° 0.80 torr. Can be stored below 0° for 1 h without appreciable change[*117*].

$B_{10}H_{14}$. M.p. 98.8°, b.p. 213°[*655*]; vapour pressure at 25° 0.048 torr. Vapour pressures: $\log p$ [torr] $= -4225.345/T - 0.0107975\,T + 6.63911$[*519*]; $\log p$ [torr] $= -0.013198/T + 10.577$[*374a, 519*]. Solubilities (in g/100g of solvent): ethyl acetate 50, n-propyl bromide 20, carbon disulphide 20, butyraldehyde 20, benzene 20, acetic anhydride 20, acetic acid 10–20, ethyl borate 10, carbon tetrachloride 10, cyclohexane 4–5, n-heptane 2[*252a*].

$B_{10}H_{16}$. Density 0.80 to 0.87 g/ml[*280*]; white hygroscopic solid; m.p. 196–199°; soluble in carbon tetrachloride, hydrocarbons, and alcohols; can be detached by gas chromatography.

$B_{18}H_{22}$. Strong monobasic acid[*597*].

$B_{20}H_{16}$. M.p. 196–199°[*235, 565*].

20.2 *Halogenated boranes: B_2H_5X, BHX_2, $BH_{3-n}X_n$·donor*

B_2H_5Cl	M.p. −143.1 ± 0.8°[*536*]; −143.7±0.6°[*536*]; −142.0 [*104*]. B.p. −11.0° (calculated)[*536*]. Vapour pressures: 18 (−78.5°)[*104*]; 20.3 (−78.2°)[*536*] torr. Vapour pressures: $\log p$ [torr] $= -1188/T + 7.412$ [*536*]. No decomposition at 25° (1h) or 0° (8 h)[*536*].
$BHCl_2$	Separable from B_2H_5Cl, B_2Cl_4, and B_2H_6 by low-temperature gas chromatography[*536*]. $\Delta H^0_{formation}$ (298°K) -60.37 ± 0.05 kcal[*486*].
($BHCl_2 + B_2H_5Cl$)	Vapour pressures of the reaction mixture from $BCl_3 + B_2H_5$ before the achievement of the equilibrium: −80° 8.5 torr; −23° 158–226 torr. After the achievement of

the equilibrium: −80° 53 torr; −23° 254 torr. (Selection) [*398*].

$(C_2H_5)_2BF$ (for comparison) — M.p. −121.5°. B.p. 43.6°. Vapour pressures: $\log p\,[\text{torr}] = -1479/T + 7.550$. Enthalpy of evaporation 6.77 kcal/mole. Trouton's constant 21.4 cal deg^{-1} mole^{-1} [*31*].

$(C_2H_5)BF_2$ (for comparison) — M.p. −101.0°. B.p. −25.4°. Vapour pressures: $\log p\,[\text{torr}] = -1325/T + 8.299$; enthalpy of evaporation 6.06 kcal/mole. Trouton's constant 24.5 cal deg^{-1} mole^{-1} [*31, 541b*].

$BH_2Cl \cdot O(CH_3)_2$ — M.p. −21/−20° [*91*].

$BCl_3 \cdot O(CH_3)_2$ (for comparison) — M.p. 76° [*841*].

$BH_2Cl \cdot O(C_2H_5)_2$ — M.p. −105/−85°. Vapour pressure at 0° 3 torr [*91*].

$BHCl_2 \cdot O(C_2H_5)_2$ — M.p. −25/−30°. Possibly mixture of $BH_2Cl \cdot OR_2$ and $BCl_3 \cdot OR_2$ [*91*].

$BCl_3 \cdot O(C_2H_5)_2$ (for comparison) — M.p. 56° [*539*].

$BH_2Cl \cdot \overline{O(CH_2)_3C}H_2$ — M.p. −38/−36°. Large triangular crystals [*91*].

$BCl_3 \cdot \overline{O(CH_2)_3C}H_2$ (for comparison) — M.p. 38–52° [*281*].

$BH_2Cl \cdot \overline{O(CH_2)_3C}H_2$ — M.p. 19–20°. Vapour pressure at 20° ~ "0" torr [*91*].

$BHCl_2 \cdot \overline{O(CH_2)_4C}H_2$ — M.p. 10–5°. Possibly a mixture of $BH_2Cl \cdot OR_2$ and $BCl_3 \cdot OR_2$ [*91*].

$BCl_3 \cdot \overline{O(CH_2)_4C}H_2$ (for comparison) — M.p. 49–52° [*281*].

$BH_2Cl \cdot CH_3NH_2$ — Crystals readily soluble in organic solvents. M.p. 47° [*546*].

$BH_2Cl \cdot tert\text{-}C_4H_9NH_2$ — M.p. 102° [*546*].

$BH_2Cl \cdot (CH_3)_2NH$ — M.p. 18° [*546*].

$BH_2Cl \cdot (CH_3)_3N$	M.p. 85° [*546*].
$BHCl_2 \cdot (C_2H_5)_3N$	M.p. 43° [*546*].
$BH_2Cl \cdot C_5H_5N$	M.p. 45° [*546*].
$BH_2Br \cdot CH_3NH_2$	M.p. 10° [*546*].
$BH_2Br \cdot tert\text{-}C_4H_9NH_2$	M.p. 98° [*546*].
$BH_2Br \cdot (CH_3)_2NH$	M.p. 5–6° [*546*].
$BH_2Br \cdot (CH_3)_3N$	M.p. 67° [*546*].
$BH_2I \cdot CH_3NH_2$	M.p. 8–9° [*546*].
$BH_2I \cdot tert\text{-}C_4H_9NH_2$	M.p. 99° [*546*].
$BH_2I \cdot (CH_3)_2NH$	M.p. 25° [*546*].
$BH_2I \cdot (CH_3)_3N$	M.p. 73–74° [*546*].

20.3 Boranes with B–O, B–S, and B–Se bonds: $BH_3 \cdot OR_2$, $BH_{3-n}(OR)_n$, $BH_3 \cdot SR_3$, $BH_{3-n}(SR)_n$, $BH_3 \cdot SeR_2$

$BH_3 \cdot O(CH_3)_2$	Present in relatively high concentration in ether only at low temperatures. M.p. ~−60°. Dissociation pressure at −78.5° 18 torr [*108*, *615*, *668*, 852].
$BD_3 \cdot O(CH_3)_2$	M.p. −70°. Dissociation pressure at −78.5°C 32 torr [*108*, *111*, *271*].
$2\ BH_3 \cdot O(CH_3)(C_2H_5)$	Present in relatively high concentration in ether only at low temperatures [*852*].
$3\ BH_3 \cdot O(C_2H_5)_2$	Present in relatively high concentrations in ether only at low temperatures [*852*].
$BH_3 \cdot O(C_2H_5)_2$	Present in relatively high concentration in ether only at low temperatures [*852*].
$BH_3 \cdot \overline{O(CH_2)_3CH_2}$	Dissociation pressure at −78°C 1.6 torr [*615*, *668*, *852*] $\log p\,[\text{torr}] = -1244/T + 6.592$ [*151*].

Compound	Properties
$BH_3 \cdot \overline{O(CH_2)_4CH_2}$	No details available [*852*].
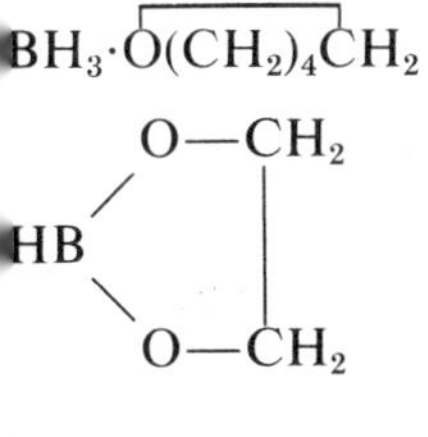	Viscous liquid which becomes vitreous on standing. Evaporates to a monomeric gas [*621*].
$BH(OCH_3)_2$	M.p. −130.6°. B.p. 26°. Vapour pressure at 0° 275 torr. Vapour pressures: $\log p$ [torr] $= -1849.0/T + 1.75 \log T - 0.008221T + 7.1895$ [*122*].
$BH_3 \cdot S(CH_3)_2$	M.p. −38 to 40° [*108, 131*]. B.p. 97° (calculated [*131*]. Dissociation pressure at 0° 4.32 torr [*108, 131*]. Vapour pressures: $\log p$ [atm$^{1/2}$] $= -1127/T + 3.190$ [*271*]; $\log p$ [torr] $= -2346/T + 9.220$. Trouton's constant 29.0 cal deg^{-1} mole^{-1} [*131*].
$BH_3 \cdot \overline{S(CH_2)_3CH}$	M.p. −45°, vapour pressure at 19.5° = 4.64 torr [*151*].
$H_2BSCH_2CH_2SBH_2$	White crystals, decomposition point 70° [*193*].
$(CH_3)_3N \cdot H_2BSCH_2$-$CH_2SBH_2 \cdot N(CH_3)_3$	White non-volatile solid stable at 20° [*193*].
$BH_3 \cdot CH_3SH$	Very unstable gas readily giving off H_2 [$\rightarrow BH_2$-(SCH_3)]. "Gaseous" at −78° [*108, 131*].
$BH_2(SCH_3) \cdot (CH_3)_3N$	Stable up to 55°. M.p. 13–15° [*131*].
B_2H_5-μ-SH	Extrapol. b.p. 27° [*395c*].
$B_2H_5(SCH_3)$	M.p. −101.5 to −100.7°. B.p. ~60°. Vapour pressures: $\log p$ [torr] $= -1666/T + 7.991$ [*131*].
$[BH_2(SCH_3)]_3$	B.p. 80 to 81° (1.5 torr); d_4^{20} 1.0121; n_D^{20} 1.5483 [*513*].
$[BH_2(SC_2H_5)]_3$	B.p. 94 to 96° (1 torr); d_4^{20} 0.9772; n_D^{20} 1.5323 [*513*].
$[BH_2(SC_3H_7)]_3$	B.p. 104 to 105° (0.009 torr); d_4^{20} 0.9508; n_D^{20} 1.5210 [*513*].
$[BH_2(SC_4H_9)]_3$	d_4^{20} 0.9376; n_D^{20} 1.5130 [*513*].
$[BH_2(SCH_3)]_x$	White non-volatile solid insoluble in ordinary organic solvents. M.p. 65–80° (decomposition) [*131*].

References p. 355

$BH[S(C_3H_7)_2]_2$	B.p. 90° (4 torr); d_4^{20} 0.9809; n_D^{20} 1.5265 [*509*].
$BH[S(C_4H_9)_2]_2$	B.p. 95° (2 torr); d_4^{20} 0.9561; n_D^{20} 1.5170 [*509*].
$BH(SCH_2)_2$	White crystalline solid. Soluble in THF and ether [*193*].
$BH(SCH_2)_2 \cdot N(CH_3)_3$	Insoluble solid. Stable and capable of sublimation at 20° [*193*].
$BH(SCH_2)_2 \cdot P(CH_3)_3$	Insoluble solid. Stable and capable of sublimation at 20° [*193*].
$[BH(SC_3H_7)_2]_2$	B.p. 93 to 95° (4 torr); d_4^{20} 0.9808; n_D^{20} 1.5265 [*510*].
$[BH(SC_4H_9)_2]_2$	B.p. 98 to 103° (2 torr); d_4^{20} 0.9561; n_D^{20} 1.5170 [*510*].
$BH(SCH_3)N(CH_3)_2$	B.p. 122 [*551a*].
$BH(SCH_3)N(C_2H_5)_2$	[*551a*].
$BH(SC_2H_5)N(C_2H_5)_2$	B.p. 65 to 67° (19 torr); d_4^{20} 0.8502; n_4^{20} 1.4616 [*505c*].
$BH(SC_3H_7)N(CH_3)_2$	B.p. 48 to 50° (13 torr); d_4^{20} 0.8706; n_D^{20} 1.4701 [*505a, 505b*].
$BH(SC_3H_7)N(C_2H_5)_2$	B.p. 79 to 81° (17 torr); d_4^{20} 0.848; n_D^{20} 1.4628 [*508*].
$BH(SC_3H_7)N(i\text{-}C_5H_{11})_2$	B.p. 110 to 114° (3.5 torr); d_4^{20} 0.8422; n_D^{20} 1.4640 [*505b*].
$BH(SC_4H_9)N(CH_3)_2$	B.p. 55 to 57° (7 torr); d_4^{20} 0.8666; n_D^{20} 1.4699 [*505b*].
$BH(SC_4H_9)N(C_2H_5)_2$	B.p. 52 to 54° (2 torr); d_4^{20} 0.849; n_D^{20} 1.4640 [*505c, 508*].
$BH(SC_4H_9)NC_5H_5$	B.p. 73 to 74° (1.5 torr); d_4^{20} 0.9170; n_D^{20} 1.4944 [*505b*].
$BH(SC_6H_5)N(C_2H_5)_2$	B.p. 82 to 84° (1.5 torr); d_4^{20} 0.9736; n_D^{20} 1.5470 [*505c*].
$BH_3 \cdot (CH_3)_2Se$	B.p. 63.2° (calculated). Vapour pressure: $\log p$ [torr] $= -1732/T + 8.030$. Trouton's constant 23.5 cal deg^{-1} mole^{-1} [*271*].

20.4 *Amineboranes ("borazanes")* $BH_3 \cdot NR_3$

$BH_3 \cdot NH_2(CH_3)$	Vapour pressures log p [torr] = $-4114/T + 11.411$ [*556*].
$BH_3 \cdot NH(CH_3)_2$	Vapour pressures log p [torr] = $-4034/T + 12.544$ [*556*].
$BH_3 \cdot N(CH_3)_3$	Monomeric in liquid ammonia[*578*]. Dipole moment in benzene 4.42 debye[*35*]. M.p. 92–93°[*35*]; 93°[*636*]; 93.5°[*547*]. B.p. 171°[cit. *503*]; 134°[*672*]. Vapour pressure at 25° 0.9 torr [cit. *337*]; at 72.0° 96 torr [*672*]. Vapour pressures: log p [torr] = $-2962/T + 9{\cdot}894$ (valid between 0° and 90°); enthalpy of evaporation 13.6 kcal/mole[*10, 556, 672*].
$BH_3 \cdot N(C_2H_5)_3$	M.p. −2°[*547*]. B.p. 42° (10^{-4} torr)[*547*]; 96–97° (12 torr)[*207*].
$BH_3 \cdot N(n\text{-}C_3H_7)_3$	M.p. 18°. B.p. 101° (12 torr)[*547*].
$BH_3 \cdot N(n\text{-}C_4H_9)_3$	M.p. −28°[*547*]; −12°[428, 547, cit. *503*]. B.p. 80° (10^{-5} torr)[*547*].
$BH_3 \cdot N(C_2H_5)_2(C_4H_9)$	M.p. −32°. B.p. 125° (14 torr)[*428*].
$BH_3 \cdot N(CH_3)_2(C_6H_5)$	M.p. 35°[*514a* cit. *503*].
BH_3·pyridine	Faintly yellow liquid stable in dry air. Almost insoluble in water. Very soluble in alcohol and ether [*761*]. M.p. 10–11°[*78,761*]. Vapour pressures: 2.5 (100°); < 0.1 (20°) torr[*78,761*]. n_D^{25} 1.5280; n_D^{20} 1.5315[*35,761*].
2 BH_3·N,N′-dimethylpiperazine	Sublimable crystals[*249a*].
2 BH_3·triethylenediamine	Crystals capable of being recrystallised from water [*249a*].
BH_3·4-picoline	M.p. 74°. Vapour pressure at 105° 1.2 torr[*78*].
BH_3·3-picoline	M.p. 10.5°. Vapour pressure at 100° 1.1 torr[*78*].
BH_3·2-picoline	M.p. 51°. Vapour pressure at 100° 1.7 torr[*78*].

References p. 355

BH_3·2-ethylpyridine — M.p. 50°. Vapour pressure at 100° 8.6 torr [*78*].

BH_3·2-isopropylpyridine — M.p. 19°. Vapour pressure at 80° 8 torr [*78*].

BH_3·2-t-butylpyridine — M.p. 25°. Vapour pressure at 50° 22 torr [*78*].

BH_3·2,6-lutidine — M.p. 111°. Vapour pressure at 100° 4.9 torr [*78*].

BH_3·quinoline — M.p. 96° [*514b*, cit. *503*].

$Me[N(BH_3)MeCH_2]_2N(BH_3)Me$ — M.p. 186° [*779b*]

20.5 sec-*Amineboranes ("borazanes")* BH_3NHR_2

$BH_3·NH(CH_3)_2$ — Monomeric in liquid ammonia [*578*]. Dipole moment in benzene 4.87 debye [*548*]. M.p. 37° [*547*]. B.p. 49° (0.01 torr) [*547*]; 108° (760 torr, calculated) [*672*]. Vapour pressures: $\log p$ [torr] $= -4034/T + 12.544$ (valid between 0 and 35°); $\log p$ [torr] $= -1832/T + 7.685$ [*672*]. Enthalpy of evaporation 18.5 kcal/mole [*10*].

$BH_3·NH(C_2H_5)_2$ — M.p. −18° [*547*]; −20° [*508*]. B.p. 84° (4 torr) [*547*]; 77° (2.5 torr) [*508*].

$BH_3·NH(n\text{-}C_3H_7)_2$ — M.p. 30° [*547*]. Dipole moment in benzene 4.55 debye [*548*].

$BH_3·NH(i\text{-}C_3H_7)_2$ — M.p. 23°, B.p. 88° (1 torr) [*547*].

$BH_3·NH(n\text{-}C_4H_9)_2$ — M.p. 15° [*547*].

$BH_3·NH(i\text{-}C_4H_9)_2$ — M.p. 19° [*547*].

$BH_3·\overline{NH(CH_2)_2CH_2}$ — Liquid at 20° [*115*].

$BH_3·\overline{NH(CH_2)_4CH_2}$ — Vapour pressure at 40° 0.6 torr [*115*].

20.6 *Primary amineboranes ("borazanes")* $BH_3·NH_2R$ *and ammonia-borane ("borazane")* $BH_3·NH_3$

Monoalkylamineboranes, $BH_3·NH_2R$, are crystalline solids or oily liquids. They are readily soluble in ether, tetrahydrofuran, glycol ethers, dioxan, chlorobenzene, nitrobenzene, benzene, chloroform, methylene chloride, esters, dimethylformamide, and dimethyl sulphoxide.

$BH_3·NH_2(CH_3)$ — Monomeric in liquid ammonia[578]. Dipole moment in benzene 5.19 debye[548]. M.p. 56°[547]. B.p. 84°. Vapour pressure at 0° 23.5 torr[672]. Vapour pressures: log p [torr] = $-4114/T + 11.411$ (valid between 0 and 45°)[10]; log p [torr] = $-1713/T + 7.669$[672]. Enthalpy of evaporation 18.8 ± 1.0 kcal/mole[10].

$BH_3·NH_2(C_2H_5)$ — M.p. 19°[547].

$BH_3·NH_2(n\text{-}C_3H_7)$ — M.p. 45° [547]. Dipole moment in benzene 4.72 debye[548].

$BH_3·NH_2(i\text{-}C_3H_7)$ — M.p. 65° [547].

$BH_3·NH_2(n\text{-}C_4H_9)$ — M.p. > −48° (uncertain)[547].

$BH_3·NH_2(tert\text{-}C_4H_9)$ — M.p. 96°[547]. Dipole moment in benzene 4.64 debye[548].

$BH_3·NH_2CH_2CH_2NH_2·BH_3$ — White crystals stable in the air. Decompose in vacuum at 90–110°[394].

$BH_3·NH_3$ — The X-ray structure is similar to that of HCN[368, 467]. The molecular weight in liquid ammonia[702], dioxan, and diethyl ether[704] corresponds to the monomeric formula. Splits off H_2 only very slowly at 20°. A completely dry ethereal solution is stable at 20°[702]. Solubilities (g/100 g of solvent): 33.6 in water, 6.5 in alcohol, 0.76 in ether, 0.5 in dioxan, 0.04 in benzene, 0.07 in toluene, 0.02 in carbon disulphide[723]. On rapid heating m.p. 104.5 ± 0.5° (decomposition)[723]. Vapour pressure at 25° ≤ 10^{-6} torr[547]. [For $(NH_3)_2BH_2^{\oplus}BH_4^{\ominus}$ see under $Z_2BH_2^{\oplus}$; for $NH_4^{\oplus}BH_4^{\ominus}$ see under $BH_4^{\ominus}$].

$(CH_3)_2BH·NH_3$ — White solid sublimable in vacuum at 20°. M.p. with decomposition ~ 25°[527a].

20.7 *Alkylamine-alkylboranes ("borazanes")* $BH_{3-n}R_n·NH_{3-n}R_n$

$BH_2(CH_3)·N(CH_3)_3$ — M.p. 0.8°[671]: −14.5°[778]. B.p. 176.4°[671]; 161.5–162.5°[778]. Vapour pressure at 76.2°, 38 torr[778].

References p. 355

$BH_2(n\text{-}C_3H_7)\cdot N(CH_3)_3$	Capable of undergoing molecular distillation [*301*, *304*].
$BH_2(i\text{-}C_3H_7)\cdot N(CH_3)_3$	Capable of undergoing molecular distillation [*301*, *304*].
$BH_2(n\text{-}C_4H_9)\cdot N(CH_3)_3$	B.p. 72° (3.0 torr) [*301*, *304*].
$BH_2(2\text{-}C_4H_9)\cdot N(CH_3)_3$	B.p. 60° (2.5 torr) [*301*, *304*].
$BH_2(i\text{-}C_4H_9)\cdot N(CH_3)_3$	Capable of undergoing molecular distillation [*304*].
$BH_2(tert\text{-}C_4H_9)\cdot N(CH_3)_3$	B.p. 60° (3.5 torr) [*304*].
$BH_2(1\text{-}C_5H_{11})\cdot N(CH_3)_3$	Capable of undergoing molecular distillation [*304*].
$BH_2(1\text{-}C_6H_{13})\cdot N(CH_3)_3$	Capable of undergoing molecular distillation [*304*].
$BH_2(\overline{CH(CH_2)_4CH_2})\cdot N(CH_3)_3$	B.p. 40–41° [*304*].
$BH_2(CH_2C_6H_5)\cdot N(CH_3)_3$	B.p. 58–60° [*304*].
$BH(CH_3)_2\cdot N(CH_3)_3$	M.p. −18° [*671*]; 32.8° [*778*]. B.p. 171.4° [*671*] 186–187° [*778*]. Vapour pressure at 101.2° 38 torr [*778*].
$BH_2(C_6H_5)\cdot N(CH_3)_3$	M.p. 69° [*298*].
$BH_2(C_6H_5)\cdot NH_2(C_2H_5)$	Viscous liquid readily miscible with organic solvents [*506*].
$BH_2(1\text{-}C_{10}H_7)\cdot NH_2(C_2H_5)$	Crystals readily soluble in ether and sparingly soluble in hexane, m.p. 71–73° [*506*].
$BH_2(CH_2C_6H_5)\cdot N(C_2H_5)_3$	M.p. 65° [*298*].
$BH_2(C_6H_5)\cdot NH_2(i\text{-}C_4H_7)$	Liquid, n_D^{20} 1.5230 [*298*].
$BH_2(p\text{-}CH_3C_6H_4)\cdot NH_2(i\text{-}C_4H_7)$	Crystals readily soluble in ether and sparingly soluble in hexane. M.p. 51–53° [*506*].
$BH_2(1\text{-}C_{10}H_7)\cdot NH_2(i\text{-}C_4H_7)$	Crystals easily soluble in ether, sparingly in benzene. M.p. 90–92° (decomp.) [*506*].

$BH_2(C_6H_5)\cdot NC_5H_5$ M.p. 80–83° [*298, 309, 804*].

$BH_2(p\text{-}ClC_6H_4)\cdot NC_5H_5$ M.p. 61–62° [*298, 309*].

$BH_2(p\text{-}CH_3OC_6H_4)\cdot NC_5H_5$ 78–79° [*298, 309*].

$BH_2(o\text{-}CH_3OC_6H_4)\cdot NC_5H_5$ M.p. 55–56° [*298*].

$BH_2(p\text{-}CH_3C_6H_4)\cdot NC_5H_5$ M.p. 63–65° [*298, 309*].

$BH_2(\alpha\text{-}C_{10}H_7)\cdot NC_5H_5$ M.p. 140–141° [*298, 309*].

$BH_2(CH_2C_6H_5)\cdot NC_5H_5$ M.p. 106° [*309*].

$BH(C_6H_5)_2\cdot NC_5H_5$ M.p. 109° [*299, 309, 446*].

$BH(p\text{-}ClC_6H_4)\cdot NC_5H_5$ M.p. 104° [*299, 309*].

$BH(p\text{-}CH_3C_6H_4)_2\cdot NC_5H_5$ M.p. 113° [*299, 309*].

$BH(p\text{-}CH_3OC_6H_4)_2\cdot NC_5H_5$ M.p. 110° [*299, 309*].

$BH(p\text{-}BrC_6H_4)_2\cdot NC_5H_5$ M.p. 124° [*299, 309*].

20.8 *Monoaminodiboranes* $B_2H_5(NH_{2-n}R_n)$

$B_2H_5(NH_2)$ μ-Aminodiborane is stable at 20°. M.p. −66.5°; b.p. 76.2° (calculated). Vapour pressure at 0° 32.3 torr. Vapour pressures: $\log p\ [\text{torr}] = -2097/T + 1.75 \log T - 0.00642\,T + 6.677$. Enthalpy of evaporation 7300 cal/mole. Trouton's constant 21.0 cal deg^{-1} $mole^{-1}$ [*328, 673*].

$B_2H_5[NH(C_2H_5)]$ μ-Monoethylaminodiborane. M.p. −96.4°; b.p. 86.6°. Vapour pressures: $\log p\ [\text{torr}] = -2096.3/T + 1.75 \log T - 0.00533\,T + 6.15245$. Vapour pressure at 0° 19.31 torr. Trouton's constant 21.0 cal deg^{-1} $mole^{-1}$ [*115*].

$B_2H_5[NH(n\text{-}C_3H_7)]$ μ-Monopropylaminodiborane. Softening point of the vitreous substance −146 to −142°. B.p. 121°.

Vapour pressure at 0° 6.14 torr. Vapour pressures: log p [torr] $= -2117/T + 8.537$. Trouton's constant 25.9 cal mole^{-1} deg^{-1}[*115*].

$B_2H_5[N(CH_3)_2]$ — M.p. $-54.4°$; b.p. 50.3°[cit. *503*]. Vapour pressure at 0° 101 torr[*395b*].

$B_2H_5[N(CH_3)(SiH_3)]$ — B.p. $-39°$. Vapour pressure 0° 81.5 torr[*79*].

$B_2H_5[N(SiH_3)_2]$ — M.p. $-68.8°$. Vapour pressure at 0° 74.5 torr[*79*].

$B_2H_5[\overline{N(CH_2)_2CH_2}]$ — μ-Trimethylaminodiborane. M.p. $-63.5°$; b.p. 121.8°. Trouton's constant 20.6 cal deg^{-1} mole^{-1}. Vapour pressure at 2.80° 5.69 torr. Vapour pressures: log p [torr] $= -2205/T + 1.75 \log T - 0.004539\ T + 5.7217$ [*115*].

$B_2H_5[\overline{N(CH_2)_3CH_2}]$ — M.p. $-45.4°$; b.p. 101°. Trouton's constant 20.2 cal deg^{-1} mole^{-1}. Vapour pressure at 0° 11.01 torr. Vapour pressures: log p [torr] $= -2248/T + 1.75 \log T - 0.00616\ T + 6.6928$[*115*].

$B_2H_5[\overline{N(CH_2)_4CH_2}]$ — μ-Pentamethyleneaminodiborane. B.p. 148°. Vapour pressure at 24.73° 7.60 torr. Vapour pressures: log p [torr] $= -2410/T + 1.75 \log T - 0.00510\ T + 6.1606$. Trouton's constant ~ 19.6 cal deg^{-1} mole^{-1}[*115*].

20.9 *Aminoboranes ("borazenes"; "cycloborazanes")* $[BH_{2-m}R_mNH_{2-n}R_n]_p$

$[BH_2NH_2]_3$ — Chair structure[*135*]. M.p. 97.0–97.8°[*135*].

$[BH_2N(CH_3)_2]_2$ — M.p. 73.5°. Vapour pressure at 20.0° 8 torr. Vapour pressures: log p [torr] $= -2721.4/T + 10.1970$. Enthalpy of sublimation 12.4 kcal/mole[*802*].

$[BH_2N(CH_3)_3]_3$ — (Erroneously given as $B_3H_4N(CH_3)_2$[*109*]). Trimeric dimethylaminoborane (trimeric N-dimethylborazene) = N-hexamethylcycloborazane. M.p. 95.0°[*109*]; 97.8°[*503*]. B.p. 220° (calculated). Vapour pressure at 80.2° 7.76 torr. Vapour pressures; log p [torr] $= -3723/T + 11.424$. Enthalpy of sublimation 6750 cal/mole[*109*].

$BH[N(CH_3)_2]_2$ — M.p. $-45°$; b.p. 109°. Vapour pressure at 0° 11 torr. Vapour pressures: log p [torr] $= -2126.3/T + 8.8204$

(valid between −25 and +15°); log p [torr] = −1684.7/T + 7.2838 (valid between 15 and 63°) [*803*].

$[BH_2N(C_2H_5)_2]_2$ — Readily sublimes in a high vacuum at 45–50°. Colourless crystals [*802*]. M.p. 44° [*508*].

$BH_2N(SiH_3)_2$ — M.p. −78°. Vapour pressure at −78° 7 torr [*79*].

$[BH_2N(SiH_3)_2$ — Vapour pressure at 25° 10 torr [*79*].

$[BH_2\overline{N(CH_2)_2C}H_2]_2$ — Bis(μ-trimethyleneaminodiborane). M.p. 51.8°. Vapour pressures: log p [torr] = −3544/T + 11.81 (solid; valid between 24 and 48°); log p [torr] = −2469/T + 8.503 (liquid; valid between 52 and 76°). Trouton's constant 25.7 cal deg^{-1} mole^{-1} [*115*].

$[BH_2\overline{N(CH_2)_3C}H_2]$ — Tetramethyleneaminoborane. M.p. 33.8–34.2°; b.p. 194°. Vapour pressure at 86.9° 14.46 torr. Vapour pressures: log p [torr] = −2705/T + 8.675. Trouton's constant 26.5 cal deg^{-1} mole^{-1} [*115*].

$[BH_2\overline{N(CH_2)_4C}H_2]$ — M.p. 111°; b.p. 202° (calculated). Vapour pressure at 80.0° 1.37 torr. Vapour pressures: log p [torr] = −4584/T + 13.131 (solid; valid between 68 and 106°); log p [torr] = −3368/T + 9.967 (liquid; valid between 114 and 142°). Trouton's constant 32.4 cal deg^{-1} mole^{-1} [*115*].

$[BH_2\overline{NCHCHCHN}]_2$ — M.p. 80–81° [*771b*].

$[BH(CH_3)NH(CH_3)]_3$ — N-Trimethyl-B-trimethylcycloborazane has a chair structure [*135, 770*].

$BH(CH_3)N(CH_3)_2$ — M.p. −136.2°; b.p. 42° (calculated) [*503*, cit. *114*].

$[BH(CH_3)N(CH_3)_2]_2$ — M.p. ~20°. Vapour pressure at 0° 1.2 torr [*503*, cit. *114*].

BH(*tert*-C_4H_9)$NH(CH_3)$ — B.p. 42° (110 torr). Dimeric in the liquid state [*308*].

BH(*tert*-C_4H_6)$N(CH_3)_2$ — B.p. 40–48° (80 torr). Monomeric in benzene [*308*].

BH(*tert*-C_4H_9)$N(C_2H_5)_2$ — B.p. 68° (46 torr). Monomeric in benzene [*308*].

BH(*tert*-C_4H_9)N(i-C_3H_7)$_2$ — B.p. 74° (31 torr). Monomeric in benzene [*308*].

$BH(\textit{tert}\text{-}C_4H_9)N(i\text{-}C_4H_9)$ B.p. 54° (3 torr). Monomeric in benzene [*308*].

$BH(\textit{tert}\text{-}C_4H_9)N(n\text{-}C_4H_9)_2$ B.p. 77° (3.3 torr) Monomeric in benzene [*308*].

$BH(i\text{-}C_4H_9)N(C_2H_5)_2$ B.p. 52° (21 torr). Monomeric in benzene [*308*].

$BH(\textit{sec}\text{-}C_4H_9)N(C_2H_5)_2$ B.p. 54° (27 torr). Monomeric in benzene [*308*].

$BH(CH(CH_2)_4CH_2)\text{-}N(CH_3)_2$ B.p. 130–132° (11 torr) [*207*].

$BH(C_6H_5)N(i\text{-}C_4H_7)_2$ B.p. 96–98° (2 torr); $d_4^{20} = 0.8892$; $n_D^{20} = 1.4930$ [*506*].

$BH(1\text{-}C_{10}H_7)N(i\text{-}C_4H_7)_2$ B.p. 156–157° (2.5 torr); $d_4^{20} = 0.9845$; $n_D^{20} = 1.5478$ [*506*].

20.10 Diborazanes $BH_3{\cdot}NR_2BH_2{\cdot}NH_{3-n}R_n$

$BH_3{\cdot}N(CH_3)_2BH_2{\cdot}NH_3$ B.p. 96.5–97.5°. Vapour pressures in torr: 1.1 (40°), 2.8 (60°), 7.6 (80°), 19.8 (102°) [*289*].

$BH_3{\cdot}N(CH_3)_3{\cdot}BH_2\text{-}NH_2CH_3$ B.p. 89–90°. Vapour pressures in torr: 1.2 (40°), 1.6 (60°), 2.9 (80°), 7.7 (102°) [*289*].

$BH_3{\cdot}N(CH_3)_2BH_2{\cdot}NH(CH_3)_2$ Vapour pressures in torr: 0.6 (40°), 1.6 (60°), 4.1 (80°), 13.3 (102°) [*289*].

$BH_3{\cdot}N(CH_3)_2BH_2{\cdot}N(CH_3)_3$ B.p. 36.8–38°. Vapour pressures in torr: 14.4 (40°), 24.7 (60°), 49.3 (80°), 64.5 (102°) [*289*].

20.11 Hydroxylamine-, hydrazine-, and cyanide-boranes

$BH_3{\cdot}NH_2OH$ Non-volatile; solid at −78°; decomposes slowly above −112° [*113*].

$BH_3{\cdot}NH(CH_3)OH$ Non-volatile; liquid at −78°; decomposes above −78° [*113*].

$BH_3{\cdot}N(CH_3)_2OH$ Monomeric liquid. 5% decomposition in 5 days in the absence of catalytic impurities. M.p. 2–4°. Vapour pressure at 25° 6 torr [*113*].

HB(NPhNPh)$_2$BH M.p. 110–112° [*551c*].

HBN(Ph)NNNPh M.p. 116–117° [*277g*].

$BH_3 \cdot NH_2OCH_3$	Volatile at 40°. M.p. 55°[*54*].
$BH_3 \cdot NH(CH_3)OCH_3$	Volatile at 25°. M.p. −23 to −21°[*54*].
$BH_3 \cdot N(CH_3)_2OCH_3$	M.p. −16.5°. Vapour pressure at 26° 3.8 torr[*54*].
$BH_3 \cdot N_2H_4$	Colourless rhombic crystals. M.p. 61°. d_{25} 0.9475. Dipole moment in dioxan 4.18 debye[*267*].
$H_2BNHNHBH_2$	Crystalline polymeric substance. From 200° it forms $(NBH)_n$[*267*].
$H_2BN(CH_3)N(CH_3)BH_2$	Vapour pressure: $\log p$ [torr] $= -2027/T + 7.8005$ [*727*].
$BH_3 \cdot NCCH_3$	Solid[*108*].
$BH_3 \cdot NCSiH_3$	M.p. 35° (decomposition)[*206*].
$BH_3 \cdot NCSi(CH_3)_3$	Vapour pressures 0.2 (26.8°); 0.9 (44.6°); 5.3 (69.3°) torr[*206*].

20.12 Borazoles (with at least one B–H bond)

$H_3B_3N_3R'_3$ $H_3B_3N_3H_3$	M.p. −56.2°[*191*]; b.p. 55.0°[*801*]. Vapour pressures: $\log p$ [torr] $= -1828.2/T + 8.6829$ (from −55 to −15°) [*801*]; $\log p$ [torr] $= -1565.5/T + 7.6616$ (from −15 to +20°)[*801*]; $\log p$ [torr] $= -1538.0/T + 7.5668$ (from 20 to 50°)[*737, 801*]. $d = 1.1551 - 0.001074T$ (T from 238.2 to 313.9°K)[*191*]; $d = 0.8613 - 0.00097\,t$ (t from −40 to +10°)[*363, 801*]. $n_D^{20} = 1.3821$[*191, 363*]. Enthalpy of evaporation 7034 cal/mole[*363, 801*]. Trouton's constant 21.2[*363*]; 21.4[*801*]. Dipole moment 0.67 debye (gas phase)[*605, 606*]; 0.50 debye (benzene solution)[*479, 781*]. With $(C_2H_5)_2O$ it forms an etherate or azeotrope[*651*]. Standard heat of formation $\Delta H_f^\circ - 204.9$ kcal/mole[*716a*].
$H_3B_3N_3H_2(CH_3)$	B.p. 84°[*672*]. Vapour pressures: $\log p$ [torr] $= -1713/T + 7.669$. Enthalpy of evaporation 7975 cal/mole. Trouton's constant 22.3 cal deg^{-1} mole^{-1}[*672*].
$H_3B_3N_3H(CH_3)_2$	M.p. 108°. Vapour pressures: $\log p$ [torr] $= -1832/T + 7.685$. Enthalpy of evaporation 8375 cal/mole. Trouton's constant 21.9 cal deg^{-1} mole^{-1}[*672*].

References p. 355

$H_3B_3N_3(CH_3)_3$	M.p. −78°[356]; b.p. 133°[363]. Vapour pressures: $\log p$ [torr] $= -2009/T + 7.812$[363, 672, 814]. Enthalpy of evaporation 9200 cal/mole [363]. Trouton's constant 22.9 cal deg^{-1} mole^{-1} [363]. Standard heat of formation ΔH_f° −231.9 kcal/mole [716a].
$H_3B_3N_3(C_2H_5)_3$	M.p. −49.1°[356]; b.p. 184°[363]. Vapour pressures: $\log p$ [torr] $= -2012/T + 7.238$. Enthalpy of evaporation 9210 cal/mole. Trouton's constant 20.2 cal deg^{-1} mole^{-1}. d 0.8604 (0°). n_D^{20} 1.4380 [363].
$H_3B_3N_3(n\text{-}C_3H_7)_3$	B.p. 225°. Vapour pressures: $\log p$ [torr] $= -2190.4/T + 7.2793$. Enthalpy of evaporation 1002 cal/mole. Trouton's constant 20.2 cal deg^{-1} mole^{-1}. d 0.8485 (0°). n_D^{20} 1.4484 [363].
$H_3B_3N_3(i\text{-}C_3H_7)_3$	M.p. −6.7°[30]; b.p. 203°. Vapour pressures: $\log p$ [torr] $= -2249.0/T + 7.6001$. Enthalpy of evaporation 1029 cal/mole. Trouton's constant 21.6 cal deg^{-1} mole^{-1}. d 0.8648 (0°). n_D^{20} 1.4434 [363].
$H_3B_3N_3(CH_2CF_3)_3$	B.p. 106–111° (92) [447].
$H_3B_3N_3(n\text{-}C_4H_9)_3$	B.p. 109–111° (3.5 torr); n^{20} 1.4524 [504].
$H_3B_3N_3(i\text{-}C_4H_9)_3$	B.p. 92° (3 torr); d 0.8245 (20°); n^{20} 1.4466 [504].
$H_3B_3N_3(tert\text{-}C_4H_9)_3$	M.p. 94° in vacuum. Vapour pressure at 70° 0.001 torr. Not attacked by H_2O at 100° [498b].
$H_3B_3N_3(\overline{CH(CH_2)_4CH_2})_3$	M.p. 98.9°[356].
$H_3B_3N_3(C_6H_5)_3$	M.p. 157–158°[356]; 154–155°[45].
$H_3B_3N_3(4\text{-}BrC_6H_4)_3$	M.p. 234–235 [725].
$H_3B_3N_3(3\text{-}BrC_6H_4)_3$	M.p. 125–125.4°[725].
$H_3B_3N_3(4\text{-}CH_3C_6H_4)_3$	M.p. 149–150°[356]; 150–152°[45].
$H_3B_3N_3(3\text{-}CH_3C_6H_4)_3$	M.p. 165.5–166.5°[725].
$H_3B_3N_3(4\text{-}CH_3OC_6H_4)_3$	M.p. 137–138°[356]; 139–140°[725].

$RH_2B_3N_3R'_3$

$(CH_3)H_2B_3N_3H_3$	M.p. −59°; b.p. 87°. Vapour pressures: $\log p$ [torr]

$=-1800/T+7.880$. Enthalpy of evaporation 8230 cal/mole. Trouton's constant 22.8 cal deg^{-1} $mole^{-1}$ [*670*].

$(CH_3)H_2B_3N_3H_2(CH_3)$ — B.p. 124°. Vapour pressures: log p [torr] $=-1732/T+7.245$. Enthalpy of evaporation 8013 cal/mole. Trouton's constant 22.2 cal deg^{-1} $mole^{-1}$ [*672*].

$(CH_3)H_2B_3N_3(CH_3)_3$ — M.p. −14.5° [*778*]; b.p. 162 [*716*]. d 0.85 (25°) [*778*].

$(CH_3)H_2B_3N_3(C_6H_5)_3$ — M.p. 140–142° [*716*]; 138–141° [*725*].

$(CH_3)H_2B_3N_3(4\text{-}CH_3C_6H_4)_3$ — M.p. 150–151° [*725*].

$(CH_3)H_2B_3N_3(3\text{-}CH_3C_6H_4)_3$ — M.p. 137–138° [*725*].

$(C_6H_5)H_2B_3N_3H_3$ — M.p. 73.5–75° [*489*].

$(C_6H_5)H_2B_3N_3(C_6H_5)_3$ — M.p. 215° [*716*].

$BrH_2B_3N_3H_3$ — M.p. 34.8°; b.p. 122.3°. Vapour pressures: log p [torr] $=-2172/T+8.373$. Enthalpy of evaporation 9939 cal/mole. Trouton's constant 25.1 cal deg^{-1} $mole^{-1}$ [*642*].

$ClH_2B_3N_3H_3$ — M.p. −34.6°; b.p. 109.5°. Vapour pressures log p [torr] $=-1846/T+7.703$. Enthalpy of evaporation 8445 cal/mole. Trouton's constant 22.1 cal deg^{-1} $mole^{-1}$ [*642*].

$R_2HB_3N_3R'_3$

$(CH_3)_2HB_3N_3H_3$ — M.p. −48°; b.p. 107°. Vapour pressure: log p [torr] $=-2019/T+8.200$. Enthalpy of evaporation 9230. Trouton's constant 24.3 cal deg^{-1} $mole^{-1}$ [*670*].

$(CH_3)_2HB_3N_3H_2(CH_3)$ — B.p. 139°. Vapour pressures: log p [torr] $=-1965/T+7.652$. Enthalpy of evaporation 8916. Trouton's constant 21.6 cal deg^{-1} $mole^{-1}$ [*672*].

$(CH_3)_2HB_3N_3(CH_3)_3$ — M.p. 32.8° [*778*]; b.p. 187° [*716*]. d 0.87 (35°) [*778*].

$(CH_3)_2HB_3N_3(C_6H_5)_3$ — M.p. 205–207° [*725*]; b.p. 101.2 (38 torr) [*778*].

$(CH_3)_2HB_3N_3(4\text{-}BrC_6H_4)_3$ — M.p. 211–213° [*725*].

References p. 355

$(C_2H_5)(CH_3)HB_3N_3$-$(C_6H_5)_3$	M.p. 128°[*716*].
$(C_6H_5)_2HB_3N_3(C_6H_5)_3$	M.p. 207°[*716*].
$Cl_2HB_3N_3H_3$	M.p. 33.0–33.5°; b.p. 151.9°. Vapour pressures: $\log p$ [torr] $= -1994/T + 7.572$. Enthalpy of evaporation 9125 cal/mole. Trouton's constant 21.5 cal deg^{-1} mole^{-1}[*642*].
$Cl_2HB_3N_3(2,6$-$(CH_3)_2C_6H_3)_3$	No data apart from spectra[*772*].
$Br_2HB_3N_3H_3$	M.p. 49.5–50.0°; b.p. 1671.1°. Vapour pressures: $\log p$ [torr] $= -2849/T + 9.352$. Enthalpy of evaporation 13037 cal/mole. Trouton's constant 29.6 cal deg^{-1} mole^{-1}[*642*].
$(NH_2)_2HB_3N_3H_3$	Only mass spectra[*489*].

$R_3B_3N_3R_3$ (selection); for other persubstituted borazoles see, for example, [*499*]

$(CH_3)_3B_3N_3(CH_3)$	M.p. 99°; b.p. 221° (760 torr; extrapolated). Vapour pressures: $\log p$ [torr] $= -2523.7/T + 7.9871$. Enthalpy of evaporation 11540 cal/mole. Trouton's constant 23.4 cal deg^{-1} mole^{-1}[*813*].
$(C_2H_5)_3B_3N_3(CH_3)_3$	M.p. 1–2°; b.p. 98° (1.8 torr). $n = 1.4791$ (22.5°)[*629*]
$(CH_3)_3B_3N_3(C_6H_5)_3$	M.p. 267–269°[*281a*].
$(C_2H_5)_3B_3N_3(C_2H_5)_3$	M.p. 88.5–89.5°[*294a*].
$(C_6H_5)_3B_3N_3(C_6H_5)_3$	M.p. 413–415°[*281a*].
$Cl_3B_3N_3(CH_3)_3$	M.p. 162–164°[*356*].
$Cl_3B_3N_3(C_2H_5)_3$	M.p. 57–59°[*356*].
$Cl_3B_3N_3(n$-$C_4H_9)_3$	M.p. 30°; b.p. 115–120° (0.5 torr)[*773*].
$Cl_3B_3N_3(C_6H_{11})_3$	M.p. 217–219°[*356*].
$Cl_3B_3N_3(C_6H_5)_3$	M.p. 273–275°[*356*].

$Cl_3B_3N_3H(C_6H_4CH_3)$ M.p. 147–149° [*402*].

$Cl_3B_3N_3(4\text{-}CH_3C_6H_4)_3$ M.p. 307–309° [*356*].

$Cl_3B_3N_3(4\text{-}CH_3OC_6H_4)_3$ M.p. 233–238° [*356*].

$F_3B_3N_3(CH_3)_3$ M.p. 85°; b.p. 224° (760 torr) [*816*]. Vapour pressures: log p [torr] $= -3165.2/T + 10.8256$. Enthalpy of evaporation 14500 cal/mole [*754a*].

20.13 Boranes with B–P bonds: $BH_3 \cdot PH_{3-n}R_n$ and BH_2PR_2

$BH_3 \cdot PH_3$ Colourless crystals. Dissociates even at 0° into the components (90% dissociated at 31°). Vapour pressures log p [torr] $= -2810/T + 11.20$. Soluble in benzene [*103, 245, 765*].

$BH_3 \cdot PH_2(CH_3)$ M.p. −50° [*108*]; −49.9 to −49.3° [*129*]; b.p. 150° (calculated) [*129*]. Dissociation pressure at 24° 3.6 torr [*108*]. Vapour pressures: log p [torr] $= -2329/T + 8.400$. Trouton's constant 25.3 cal deg^{-1} mole^{-1} [*129, 533, 834*].

$BH_3 \cdot PH_2(SiH_3)$ Stable up to −78° [*175c*].

$BH_3 \cdot PH(CH_3)_2$ M.p. −22° [*108*]; −22.6° [*129, 779*]. B.p. 85.5° [*129, 779*]; 174° (calculated) [*129*]. Dissociation pressure at 30° 2.5 torr [*108*]; at 29.5° 2.93 torr [*129*]. Vapour pressures: log p [torr] $= -2337/T + 8.100$ [*129*]. Trouton's constant 23.9 cal deg^{-1} mole^{-1} [*129*].

$BH_3 \cdot PH(OR)_2$ [*533, 834*].

$BH_3 \cdot P(CH_3)_3$ M.p. 103.0–103.5° [*129*]; 100° (decomposition) [*337*]. Vapour pressures in torr: 2.94 (50.5°); 69.2 (100.1°) [*108, 129*]. Vapour pressures: log p [torr] $= -2337/T + 8.100$ [*129*].

$BH_3 \cdot P(C_2H_5)_3$ M.p. 46°. Dissociation pressure at 25° 62 torr [*337*].

$BH_3 \cdot \overline{PHCH_2CH_2CH_2CH_2}$ M.p. −66 to −65° [*532c*].

$BH_3 \cdot P(n\text{-}C_3H_7)_3$ M.p. 11°. Dissociation pressure at 25° 10 torr [*337*].

$BH_3 \cdot P(C_6H_5)_3$ M.p. 187–188°. May be $[(C_6H_5)_3P]_2BH_2^{\oplus}BH_4^{\ominus}$ [*277*].

References p. 355

$BH_3 \cdot PF_3$

Colourless gas igniting in air. M.p. $-116.1 \pm 0.2°$; b.p. $-61.8°$ (calculated). Vapour pressures: $\log p$ [torr] $= -1038.9/T + 7.8061$. Enthalpy of evaporation 4760 cal/mole. Trouton's constant 22.5 cal deg^{-1} $mole^{-1}$ [*576*].

$BD_3 \cdot PF_3$

Colourless gas inflaming in the air. M.p. $-115.1 \pm 0.1°$; b.p. $-59.8°$ (calculated). Vapour pressures: $\log p$ [torr] $= -1010.8/T + 7.6171$. Enthalpy of evaporation 4630 cal/mole. Trouton's constant 22 cal deg^{-1} $mole^{-1}$ [*576*].

$BH_3 \cdot PHF_2$

Unusually stable. B.p. 6.2°. Vapour pressure: $\log p$ [torr] $= -1407/T + 7.917$ [*625c*].

$BH_3 \cdot P(OCH_3)_3$

B.p. 86° (23 torr). Vapour pressure at 45° 5 torr. n_D^{25} 1.4162 [*611*].

$BH_3 \cdot P(OC_2H_5)_3$

B.p. 39° (0.4 torr). Vapour pressure at 106° 22 torr. n_D^{25} 1.4195–1.4200 [*611*].

$BH_3 \cdot P(Oi\text{-}C_3H_7)_3$

B.p. 42–43° (0.1 torr). n_D^{25} 1.4171 [*611*].

$BH_3 \cdot P(On\text{-}C_3H_7)_3$

B.p. 83–85° (0.15 torr). n_D^{25} 1.4329 [*611*].

$BH_3 \cdot P(OC_4H_5)_3$

M.p. 54.0° [*611*].

$BH_3 \cdot P(OCH_2CH_2Cl)_3$

Decomposes before melting [*611*].

$BH_3 \cdot P(O\overline{CH(CH_2)_4C}H_2)_3$

M.p. 69.0° [*611*].

$BH_3 \cdot PF_2(NHCH_3)$

Clear stable liquid. Does not decompose in the air at 25°. Undergoes thermolysis in a closed tube at 50° (yellow deposit). Vapour pressures: $\log p$ [torr] $= -2237.4/T + 8.4472$. Vapour pressure at 0° 1.8 torr. Enthalpy of evaporation 10230 ± 100 cal/mole [*432*].

$BH_3 \cdot PF(NHCH_3)_2$

Clear non-volatile liquid. Vapour pressures before the decomposition point too small to be measured. Stable for 2 days at 20°. Soluble in ether and carbon tetrachloride [*432*].

$BH_3 \cdot P(NHCH_3)_3$

Crystalline. Soluble in various ethers, benzene, toluene, carbon tetrachloride, chloroform, and acetone. Soluble in water without the formation of H_2 [*423*].

$BH_3 \cdot PF_2[N(CH_3)_2]$	Volatile liquid. M.p. −56.7°; b.p. (extrapolated) 119.4°. Vapour pressures: $\log p$ [torr] = −2064.4/T + 8.1393. Enthalpy of evaporation 9430 ± 100 cal/mole [*432*].
$BH_3 \cdot PF[N(CH_3)_2]_2$	Volatile [*432*].
$BH_3 \cdot P(CH_3)_2N(CH_3)_2$	M.p. 12° [*126*].
$BH_3 \cdot P[N(CH_3)_2]_3$	Crystals soluble in dioxan. M.p. 32.5° [*610*].
$BH_3 \cdot P[N(CH_3)_2]_3 \cdot HCl$	Crystals soluble in conc. aqueous HCl [*610*].
$BH_3 \cdot P[N(C_2H_5)_2]_3$	Soluble in tetrahydrofuran and chloroform. B.p. 89° (0.15 torr) [*610*].
$BH_3 \cdot P[N(n\text{-}C_4H_9)_2]_3$	Soluble in tetrahydrofuran and chloroform [*610*].
$BH_3 \cdot P[\overline{N(CH_2)_3CH_2}]_3$	M.p. 58° [*610*].
$\overset{\ominus}{B}H_3CH_2\overset{\oplus}{P}(C_6H_5)_3$	M.p. 191–192° [*297, 303, 610, 611, 688*].
$\overset{\ominus}{B}H_3CH(CH_3)\overset{\oplus}{P}(C_6H_5)_3$	M.p. 171–172° [*297, 303, 610, 611, 688*].
$\overset{\ominus}{B}H_3CH(n\text{-}C_3H_7)\overset{\oplus}{P}(C_6H_5)_3$	M.p. 154–156° [*297, 303, 610, 611, 688*].
$\overset{\ominus}{B}H_3CH(n\text{-}C_5H_{11})\overset{\oplus}{P}(C_6H_5)_3$	M.p. 139–140° [*297, 303, 610, 611, 688*].
$\overset{\ominus}{B}H_3CH(C_6H_5)\overset{\oplus}{P}(C_6H_5)_3$	M.p. 145–147° [*297, 303, 610, 611, 688*].
$\overset{\ominus}{B}H_2(CH_3)CH_2\overset{\oplus}{P}(C_6H_5)_3$	M.p. 151–155° [*297, 303, 610, 611, 688*].
$\overset{\ominus}{B}H_2(2\text{-}C_4H_9)CH_2\overset{\oplus}{P}(C_6H_5)_3$	M.p. 100–104° [*297, 303, 610, 611, 688*].
$\overset{\ominus}{B}H_2(tert\text{-}C_4H_9)CH_2\overset{\oplus}{P}\text{-}(C_6H_5)_3$	M.p. 130–134° [*297, 303, 610, 611, 688*].
$\overset{\ominus}{B}H_2(C_6H_5)CH_2\overset{\oplus}{P}(C_6H_5)_3$	M.p. 153–156° [*297, 303, 610, 611, 688*].
$BH_3 \cdot (CH_3)_2PP(CH_3)_2 \cdot BH_3$	Stable to 154°. At 170–200° forms 75% of $[BH_2P(CH_3)_2]_3$ and 25% of $[BH_2P(CH_3)_2]_4$. Vapour

	pressures: log p [torr] $= -5200/T + 13.82$ (valid between 63 and 93°) [*110*].
$(BH_2PF_3)_2$	M.p. −114.2 to −114.5°; log p [torr] $= -1536/T + 8.20$; Enthalpy of evaporation 7.03 kcal/mole; Trouton's constant 23.5 cal deg^{-1} $mole^{-1}$ [*164e*].
$[BH_2P(CH_3)_2]_3$	M.p. 86°; b.p. 235° (calculated). Vapour pressure at 79.67° 2.00 torr. Vapour pressures: log p [torr] $= -4515/T + 13.10$ (solid; valid between 60 and 85°); log p [torr] $= -2887/T + 8.557$ (liquid; valid between 85 and 125°). Trouton's constant 26.0 cal deg^{-1} $mole^{-1}$ [*110, 126, 129*].
$[BH_2P(CF_3)_2]_3$	Stable in the air. M.p. 30.5°; b.p. 176.6° (calculated). Slow decomposition at 200°. Vapour pressure at 30.2° 1.50 torr. Vapour pressures: log p [torr] $= -4135/T + 13.806$ (solid; valid between 6 and 30°); log p [torr] $= -3330/T + 1.75 \log T - 0.00800\,T + 9.2405$ (liquid; valid between 32 and 82°). Enthalpy of evaporation 9.50 kcal/mole. Trouton's constant 21.0 cal deg^{-1} $mole^{-1}$ [*113*].
$[BH_2P(CH_3)_2]_{\sim 80} \cdot N(C_2H_5)_3$	Insoluble in most common solvents. Soluble in hot benzene. M.p. 170–172° [*779*].
$[BH_2P(CH_3)_2]_x$	White translucent brittle mass. M.p. 164–168° [*779*].
$[BH_2P(CH_3)_2]_4$	M.p. 161°; b.p. 310° (calculated). Vapour pressure at 161.1° 11.8 torr. Vapour pressures: log p [torr] $= -4900/T + 12.28$ (solid); log p [torr] $= -3196/T + 8.364$ (liquid) [*110, 129*].
$[BH_2P(CF_3)_2]_4$	M.p. 116°. Vapour pressure at 82.0° 1.10 torr. Vapour pressures: log p [torr] $= -4489.2/T + 12.683$ (solid; valid between 45 and 105°) [*113*].
$[BH_2P(C_2H_5)_2]_3$	B.p. 134° (1.5 torr) [*126*].
$[BH_2P(CH_3)(C_2H_5)]_x$	M.p. 118–126°. White translucent plastic mass [*779*].
$(BH_2)_3P_3(NH_2)(CH_3)_5$	M.p. 35°. Vapour pressure at 77.5° 2.08 torr. Vapour pressures: log p [torr] $= -5407.9/T + 1.75 \log T - 0.01949\,T + 18.121$. Preparation: $[BH_2P(CH_3)_2]_3$ + Na in liquid ammonia [*129*].

$Na^{\oplus}[BH_3PH_2BH_3]^{\ominus}$ — Air-sensitive colourless crystalline solid [*336a*].

$NH_4^{\oplus}[BH_3PH_2BH_3]^{\ominus}$ — White solid [*259a*].

$(CH_3)_2NH_2^{\oplus}[BH_3PH_2BH_3]^{\ominus}$ — M.p. 65–66° [*259a*].

$K^{\oplus}[BH_3PHBH_3]^{\ominus}$ — White solid [*259a*].

20.14 Boranes with B–As and B–Sb bonds: $BH_3 \cdot AsR_3$, $(BH_2AsR_2)_n$, and $BH_3 \cdot SbR_3$

$BH_3 \cdot As(CH_3)_3$ — White solid. M.p. 73.5–74.5° [*751*]; 72° (decomposition) [*337*]. B.p. 154° (calculated) [*751*]. Vapour pressure at 25° ~ 3 torr [*337*], at 47° 7 torr [*108*], at 52.6° 10.13 torr [*751*]. Vapour pressures: $\log p$ [torr] $= -2920/T + 9.966$ (solid); $\log p$ [torr] $= -2420/T + 8.553$ (liquid) [*751*]. Enthalpy of evaporation 2.29 kcal/mole. Trouton's constant 26.0 cal mole^{-1} deg^{-1} [*751*].

$BH_3 \cdot As(C_2H_5)_3$ — M.p. 40° (decomposition). Vapour pressure at 25° ~ 1 torr [*337*].

$BH_3 \cdot AsH(CH_3)_2$ — M.p. −22.4 to −21.5°; b.p. 85.5° (calculated). Vapour pressure at 0° 8.11 torr. Vapour pressures: $\log p$ [torr] $= -2263/T + 9.191$. Trouton's constant 28.9 cal deg^{-1} mole^{-1} [*108, 751*].

$BH_3 \cdot AsH_2(CH_3)$ — Readily decomposes into $[BH_2AsH(CH_3)]_x$ and H_2. Vapour pressure at −78.5° 10.5 torr [*108, 751*].

$[BH_2As(CH_3)_2]_3$ — M.p. 49.7–50.6°; b.p. 250° (calculated). Vapour pressure at 94.0° 2.40 torr. Vapour pressures: $\log p$ [torr] $= -3074/T + 8.752$. Trouton's constant 26.7 cal deg^{-1} mole^{-1} [*751*].

$[BH_2As(CH_3)_2]_4$ — M.p. 149.5–150.5°; b.p. 352° (calculated). Vapour pressure at 140.3° 1.06 torr. Vapour pressures: $\log p$ [torr] $= -7692/T + 18.635$ (solid); $\log p$ [torr] $= -3179/T + 7.973$ [*751*].

$[BH_2As(CF_3)_2]_3$ — M.p. 3.0–3.5°; $\log p$ [torr] $= -3754/T + 1.75 \log T - 0.0100\,T + 11.0454$ [*439b*].

$BH_3 \cdot Sb(CH_3)_3$ — M.p. −35° (decomposition) [*108, 337*].

References p. 355

20.15 Boranes with B–C bonds: $B_2H_{6-n}R_n$, $BH_{3-n}R_n$, $BH_{3-n}R_n$·*donor*, BH_3·*CO*

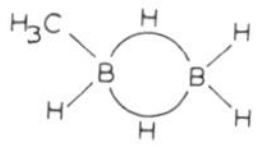

$MeHB(H_2)BH_2 \rightarrow MeBH_2 + BH_3 - 28.0$ kcal [*498*].

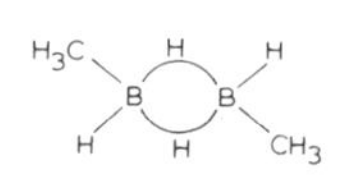

Very stable, contrary to earlier observations, even more stable than the ethyl compound. Only very slight dismutation on standing for several months at 20°. No formation of H_2 or polymers [*457*].

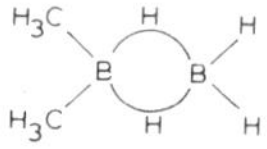

$Me_2B(H_2)BH_2 \rightarrow Me_2BH + BH_3 - 28.5$ kcal [*498*].

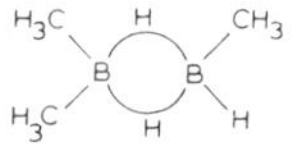

At 20°, no formation of H_2 or polymers, but dismutation (*e.g.*, formation of 1,1-dimethyldiborane). $Me_2B(H_2)BHMe \rightarrow Me_2BH + H_2BMe - 26.5$ kcal [*498*].

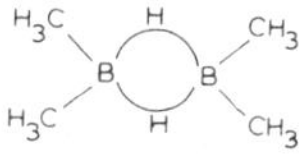

Dismutes faster than $B_2H_3Me_3$, within minutes at 20°: $3\ B_2H_2Me_4 \rightleftharpoons 2\ BMe_3 + 2\ B_2H_3Me_3$ [*459*]. $Me_2B(H_2)BMe_2 \rightarrow 2\ Me_2BH - 25.0$ kcal [*498*].

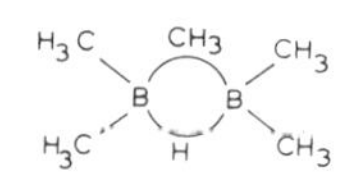

$Me_2B(MeH)BMe_2 \rightarrow Me_3B + Me_2BH - (< 5)$ kcal [*498*].

$B(CH_3)_3$

$B_2Me_6 \rightarrow 2\ BMe_3 + \sim 0$ kcal [*498*]. M.p. −159.85°; b.p. −20.1° (760 torr). Vapour pressure at −74.7° 40 torr [*747*].

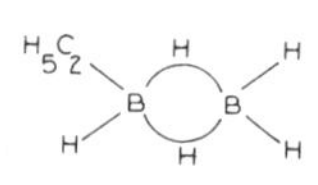

B.p. −9.53°. Vapour pressures: $\log p$ [torr] $= -883.25/T + 0.011376\,T + 3.2343$ (valid between −110 and −58°). Forms a glass at $t < -156°$ [*456, 669*].

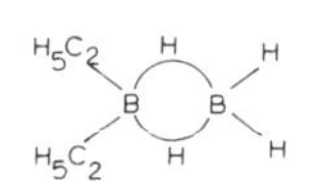

B.p. (calculated) 67.1°. Vapour pressure at 0° 46.1 torr [*458*]. Vapour pressures: $\log p$ [torr] $= -1760/T + 8.055$. Enthalpy of evaporation 8045 cal/mole. Trouton's constant 23.6 cal deg^{-1} $mole^{-1}$ (uncertain whether 1,1- or 1,2-diethyldiborane) [*669*].

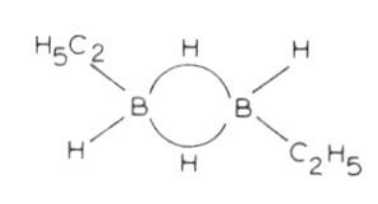

Pure B_2H_6-free 1,2-diethyldiborane is very stable, contrary to earlier observations. After standing for 2 weeks at 20°: very little dismutation, no evolution of H_2, no polymers. After several months at 20° or 30 min at 80°, formation of the thermodynamically more stable 1,1-diethyldiborane [*457*].

H_5C_2 H H B B H_5C_2 H C_2H_5

Vapour pressure at 0° 4 torr [*669*]. No formation of H_2 or polymers at 20° but dismutation into 1,1-diethyldiborane and other boranes [*459*].

H_5C_2 H C_2H_5 B B H_5C_2 H C_2H_5

M.p. −56.5 to −56.0° [*669*]. Vapour pressure at 0° 0.5 torr [*669*]. Dismutes faster than $B_2H_3Et_3$, in minutes at 20°: $3\ B_2H_2Et_4 \rightleftharpoons 2BEt_3 + 2B_2H_3Et_3$ [*459*].

$B(C_2H_5)_3$

M.p. −92.5°; b.p. 95° (760 torr) [*425*].

n-C_3H_7 H H B B H H H

Vapour pressure at −60° 6.2 torr [*669*].

n-C_3H_7 H H B B n-C_3H_7 H H

Vapour pressure at 0° 2.8 torr [*669*].

n-C_3H_7 H n-C_3H_7 B B n-C_3H_7 H n-C_3H_7

B.p. 37–42° (2–3 torr); n_D 1.4260–4285 [*514*].

CH_2—CH_2 CH_2 H CH_2 B B H H H

Vapour pressure at 0° 26.3 ± 1.0 torr [*791*].

CH_2 CH_2 $CH(CH_3)$ B B H H H

Vapour pressure at 0° 39.2 ± 1.0 torr [*791*].

p-$CH_3C_6H_4$ H H B B H H (p-$CH_3C_6H_4$)

M.p. 128–132° (decomposition). Soluble in benzene, slightly soluble in ether and hexane [*506*].

$BH_2[C(CH_3)_2CH(CH_3)_2]$

The hydride activity of a 0.8 *M* solution in THF or diglyme does not change in 18 days. In THF or diglyme there is isomerisation to $BH_2CH_2CH(CH_3)$-$CH(CH_3)_2$ within 16 days [*865*].

$B(n\text{-}C_4H_9)_3$

In 10 days at 125–130°, $B_2H_4Bu_2$ + *trans*-butene are formed [*622*].

H BH_2 H BH_2 H BH_2 H_2B H

Colourless liquid. Vapour pressure at 20° 1 torr [*867*].

$(C_2H_5)_2BC(CH_3)_3$

B.p. 60° (70 torr) [*302*].

References p. 355

Compound	Properties
$B(n\text{-}C_8H_{17})_3$	B.p. 141–146° (0.03 torr), with 50% decomposition into *trans*-2-octene and $B_2H_2(n\text{-}C_8H_{17})_4$[*17*].
H_5C_6, H, H, B, B, H, H, C_6H_5	Needles incapable of sublimation in a high vacuum. Sublimes in the presence of B_2H_6. M.p. 82–84°. Soluble in aromatic hydrocarbons, ether, and tetrahydrofuran; less soluble in dioxan and petroleum ether[*804*].
$BH_3{\cdot}CO$	M.p. −137°. Vapour pressure at −78.5° 320 torr [*108*]. Energy of dissociation 23.1 ± 2 kcal/mole[*214*].

20.16 *Hydrides with boronium ions $Z_2BH_2^{\oplus}$, $Z_2BHR^{\ominus}$*

Compound	Properties
$[(NH_3)_2BH_2]^{\oplus}BH_4^{\ominus}$	White substance sparingly soluble (insoluble?) in ether, crystalline to X-rays. Stable up to 80°. Splits off H_2 at 20° only very slowly[*700, 702*].
$[(NH_2Me)_2BH_2]^{\oplus}BH_4^{\ominus}$	Non-volatile colourless liquid slowly decomposing at 20°[*43b*].
$[(NH_2Me)_2BH_2]^{\oplus}Cl^{\ominus}$	Crystals[*43b*].
$[(NH_3)_2B(CH_3)_2]^{\oplus}$-$[(CH_3)_2BH_2]^{\ominus}$	Solid melting at 20° with decomposition[*527a*].
$[(Me_3N)_2BH_2^{\oplus}]_2B_{12}H_{12}^{2\ominus}$	M.p. 260–270° (decomposition)[*526*].
$[(Me_3N)_2BH_2]^{\oplus}PF_6^{\ominus}$	M.p. 200–205° (decomposition)[*526*].
$[(Me_3N)_2BH_2]^{\oplus}N_3^{\ominus}$	At 120–150° in vacuum forms $Me_3NBH_2N_3$ without an explosion[*520*].
$[(Me_3N)_2BH_2]^{\oplus}I^{\ominus}$	M.p. 190–195°[*174*].
$[(Me_3N)_2BH_2]^{\oplus}ICl_2^{\ominus}$	Yellow-orange crystals. B.p. 150°[*526*].
$[(Me_3N)_2BH_2]^{\oplus}ClO_4^{\ominus}$	M.p. 167–170°
$[(Me_3N)_2BH_2]^{\oplus}AuCl_4^{\ominus}$	Yellow needles[*526*].
$[(Me_2EtN)_2BH_2^{\oplus}]_2B_{12}$-$H_{12}^{2\ominus}$	M.p. 218° (decomposition)[*526*].
$[(Me_2EtN)_2BH_2]^{\oplus}PF_6^{\ominus}$	M.p. 65°[*526*].

$[(MeEt_2N)_2BH_2]^{\oplus}PF_6^{\ominus}$	M.p. 135–136°[*526*].
$[\overline{CH_2(CH_2)_4N}(CH_3)BH_2]^{\oplus}$ $PF_6^{\ominus}$	M.p. 108–109°[*526*].
$[Me_2NCH_2CH_2NMe_2$-$BH_2^{\oplus}]_2B_{12}H_{12}^{2\ominus}$	M.p. 340–350° (decomposition) [*526*].
$[Me_2NCH_2CH_2NMe_2$-$BH_2]^{\oplus}PF_6^{\ominus}$	M.p. 240–244° (decomposition) [*526*].
[MeN H NMeBH$_2$]$^{\oplus}$ $PF_6^{\ominus}$	M.p. 240–250° (decomposition) [*526*].
$[Me_2NCH_2CH_2CH$-$(CH_3)NMe_2BH_2]^{\oplus}$ $(NC)_2CCHC(CN)_2^{\ominus}$	Light-sensitive weakly yellow needles. M.p. 95–96° [*526*].
$[(C_9H_7N)_2BH_2]^{\oplus}I^{\ominus}$	M.p. 154–158°[*174*].
$[(C_9H_7N)_2BH_2]^{\oplus}ClO_4^{\oplus}$	M.p. 125–127°[*174*].
$[(Me_3P)_2BH_2]^{\oplus}PF_6^{\ominus}$	M.p. 178°[*526*].
$[(Me_3P)_2BH_2^{\oplus}]_2B_{12}H_{12}^{2\ominus}$	M.p. 169–173°[*526*].
$[Me_2PCH_2CH_2PMe_2$-$BH_2]^{\oplus}PF_6^{\ominus}$	M.p. 240–260° (decomposition) [*526*].
$(CH_3)_3NBH_2CH_2N$-$(CH_3)_2H^{\oplus}PF_6^{\ominus}$	Fairly sparingly soluble in H_2O [*519a*].
$[(C_5H_5N)_2BH(C_6H_5)]^{\oplus}I^{\ominus}$	White solid becoming yellow in the air. Can be recrystallized from $CHCl_3$–C_5H_{12}; m.p. 199–202° (decomposition). Contains protonic hydrogen [*174*].
$[(C_5H_5N)_2BH$-$(C_6H_5)]^{\oplus}ClO_4^{\ominus}$	White needles stable in the air. Crystallises from CH_3COCH_3–$CHCl_3$; m.p. 182–184° (decomposition) [*174*].
$[(C_5H_5N)_2BH$(cyclo-$C_6H_{11})]^{\oplus}I^{\ominus}$	M.p. 206–209°[*174*].
$[(C_5H_5N)_2BH$(cyclo-$C_6H_{11})]^{\oplus}ClO_4^{\ominus}$	M.p. 220–222°[*174*].

References p. 355

20.17 Boranes with transition metal–B bonds

Compound	Properties
$[(CO)_5Mn{\cdot}BH_3]N(C_2H_5)_4$	Yellow crystals decomposing at −25° [*583*].
$[(C_6H_5)_3P(CO)_4$-$Mn{\cdot}BH_3]N(C_2H_5)_4$	Yellow crystals [*583*].
$HMn_3(CO)_{10}(BH_3)_2$	Deep red needles [*385a*].
$[(CO)_5Re{\cdot}BH_3]N(C_2H_5)_4$	Faintly yellow crystals [*583*].
$[(CO)_5Re{\cdot}(BH_3)_2{\cdot}N$-$(C_2H_5)_4$	Yellow-orange crystals [*583*].
$[(CO)_5Re{\cdot}BH_3]P(C_4H_9)_4$	Red-orange crystals [*583*].
$[(CO)_5Re{\cdot}(BH_3)_2]P$-$(C_4H_9)_4$	Red-orange crystals [*583*].

20.18 Boranates $BH_4^{\ominus}$

Compound	Properties
$BH_4^{\ominus}$-ion	Radius of the free ion 2.04 Å [*1*] (for comparison, bromide ion 1.95 Å). Free energy of reaction ΔF°_{aq} −228 kcal/mole (for the dissolved $BH_4^{-}{\cdot}aq$). This gives the redox potential −1.24 V [*749*]; experimental redox potential −1.23 V [*589*]; both potentials for the process: $BH_4^{\ominus} + 8\,OH^{\ominus} \rightarrow BO_2^{\ominus} + 6\,H_2O + 8e^{\ominus}$.
$LiBH_4$	Enthalpy of formation ΔH°_f −46.36 kcal/mole (solid state) [cit. *640, 717*]. M.p. 282° [*367a*, cit. *5*]. Solubilities (in 100 g of solvent): in $(CH_3)_2O$ 1.6 g (−45.2°) [*640*]; in $(C_2H_5)_2O$ 4.28 *g* (25°), 1.28 g (0°), 0.73 g (−23°), 0.23 g (−78°) [*430*]; in THF 22.2 g (25°) [*431*]. Density 0.66 g/cm³ [*5, 448, 720, 780*].
$LiBH_4{\cdot}2H_2O$	Stable up to 36° [*660, 745*].
$LiBH_4{\cdot}2O(CH_3)_2$	Dissociation pressure in accordance with the equation $LiBH_4{\cdot}2OMe_2$ (solid) ⇌ $LiBH_4{\cdot}OMe_2$ (solid) + OMe_2 (gas): $\log p$ [torr] $= -2276.5/T + 10.6212$. ΔH_{diss} 10.42 kcal/mole [*640*].
$LiBH_4{\cdot}O(CH_3)_2$	Dissociation pressure according to the equation 2 $LiBH_4{\cdot}OMe_2$ (solid) ⇌ $(LiBH_4)_2{\cdot}OMe_2$ (solid) + OMe_2 (gas): $\log p$ [torr] $= -2431.9/T + 9.8159$. ΔH_{diss} 11.13 kcal/mole [*640*].

$LiBH_4)_2 \cdot O(CH_3)_2$	Dissociation pressure according to the equation $(LiBH_4)_2 \cdot OMe_2$ (solid) $\rightleftharpoons$ 2 $LiBH_4$ (solid) + OMe_2 (gas): $\log p$ [torr] $= -2676.1/T + 10.2378$. ΔH_{diss} 12.25 kcal/mole [*640*].
$LiBH_4 \cdot O(C_2H_5)_2$	Dissociation pressure according to the equation $LiBH_4 \cdot OEt_2$ (solid) = $(LiBH_4)_2 \cdot OEt_2$ (solid) + OEt_2 (gas): $\log p$ [torr] $= -2918.9/T + 11.7447$. ΔH_{diss} 13.36 kcal/mole [*430*].
$LiBH_4)_2 \cdot O(C_2H_5)_2$	Dissociation pressure according to the equation $(LiBH_4)_2 \cdot OEt_2$ (solid) $\rightleftharpoons$ 2 $LiBH_4$ (solid) + OEt_2 (gas): $\log p$ [torr] $= -2601.2/T + 10.3818$. ΔH_{diss} 11.90 kcal/mole [*430*].
$LiBH_4 \cdot O[CH(CH_3)_2]_2$	Dissociation pressure according to the equation $LiBH_4 \cdot OPr_2{}^i \rightleftharpoons LiBH_4$ (solid) + $OPr_2{}^i$ (gas): $\log p$ [torr] $= -2772.0/T + 11.274$. ΔH_{diss} 12.68 kcal/mole [*132*].
$LiBH_4 \cdot 2\overline{O(CH_2)_3CH_2}$	Stable up to 0°. M.p. −35°. Dissociation pressure at 0° 36 torr [*838*].
$LiBH_4 \cdot \overline{O(CH_2)_3CH_2}$	Colourless crystals giving off almost all the THF at 150°. Dissociation pressure at 18° 1 torr [*838*].
$LiBH_4 \cdot NH_3$	[*748, 755*].
$LiBH_4 \cdot 2NH_3$	[*748, 755*].
$LiBH_4 \cdot 3NH_3$	[*748, 755*].
$LiBH_4 \cdot 4NH_3$	[*748, 755*].
$LiBH_4 \cdot N_2H_4$	M.p. 160° (decomposition). Slightly soluble in diethyl ether [*534*].
$LiBH_4 \cdot N(CH_3)H_2$	[*639*, cit. *5*].
$LiBH_4 \cdot 3N(CH_3)H_2$	[*639*, cit. *5*].
$LiBH_4 \cdot 4N(CH_3)H_2$	[*639*, cit. *5*].
$LiBH_4 \cdot 2N(CH_3)_3$	Changes into $LiBH_4 \cdot N(CH_3)_3$ at −30 to −10° [*818*].

$LiBH_4 \cdot N(CH_3)_3$ — Slowly loses $N(CH_3)_3$ at 20°. Soluble in water, moderately soluble in diethyl ether, insoluble in petroleum ether[*818*].

$NaBH_4$ — Cubic body-centred lattice at 90°K; cubic face-centred lattice at 293°K. X-irradiation leads to purple colour centres which bleach at 120°[*1*]. For thermal and thermodynamic data, see[*282*, *717*]. M.p. about 400° (decomposition)[*367a*, cit. *5*] 505°[*514c*]. Decomposition point 315°[*514c*]. Solubilities in 100 g of solvent in i-PrOH 0.37 g (25°); in *tert*-BuOH 0.11 g (25°)[*11 749*]; in diglyme 0.35 *M* soln. (0°), 1.26 *M* soln (25°)[*84*]; also readily soluble in triglyme; sparingly soluble in monoglyme[*82*]. According to [*5*], however, 2.6 g dissolves in 100 g of H_2O. Maximum solubilities in polyglycol ethers are between 35 and 40° (in wt. %): diglyme 11, triglyme 13, tetraglyme 16, while the solubility maximum in monoglyme is at −8° (4 wt. %)[*5*]. Solubilities in the $NaOH–H_2O$ system (wt. %): 28.92% NaOH: 20.08%, 2.88% NaOH: 34.47%; in the 1 *N* NaOH–1.66% NaCl–H_2O system: ~ 35%[*481*]. For further data on solubilities, see[*5*].

$NaBH_4 \cdot 1$ diglyme

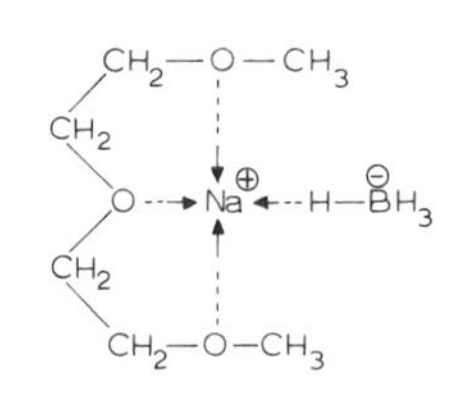

Stable at 25–40°[*84*].

KBH_4 — Enthalpy of formation ΔH_f^0 −58 kcal/mole (solid state)[*11*]. No change in the cubic face-centred lattice between 90 and 293°K. X-irradiation gives blue colour centres[*1*, *27*]. Not hygroscopic. M.p. 590°. Slow decomposition from ~ 500°[*367*, *517* cit. *5*]. d^{20} 1.11. Solubility (g/100 ml) at 25° in ROH + H_2O: 100% MeOH 0.5, 50% MeOH 6, 0% MeOH 19; in 95% EtOH 0.25. Insoluble in isopropylamine, benzene, hexane, diethyl ether, dioxan, tetrahydrofuran, acetonitrile[*27*]. Thermodynamic data[*11 717*]. Solubility in 11.89% KOH (+H_2O): 6.35% [*516*]. M.p. 585°[*514c*]. Eutectic at 347°: 5.3 wt. % KBH_4 + 94.7 wt. % KOH[*517*]. Decomposition point 584°[*514c*].

$RbBH_4$	d^{20} 1.71 [*11, 27*]. Enthalpy of formation $\Delta H_f^0 -59$ kcal/mole [*11*]. Decomposition point 600° [*514c*].
$CsBH_4$	d^{20} 2.11 [*11, 27*]. Enthalpy of formation $\Delta H_f^0 -63$ kcal/mole [*11*]. Decomposition point 600° [*514c*].
$Be(BH_4)_2$	Ignites in air. M.p. ~ 115° [*123*].
$Be(BH_4)_2 \cdot N(CH_3)_3$	Decomposes irreversibly at 140° [*123*].
$ClMgBH_4 \cdot 2THF$	White solid. In vacuum at 100° it forms $ClMgBH_4 \cdot THF$ [*46a*].
$ClMgBH_4 \cdot THF$	Rubber-like solid [*46a*].
$Mg(BH_4)_2$	Dismutes in the presence of MgR_2 into mixed boranates $Mg(R_nBH_{4-n})_2$ [*32*].
$Mg(BH_4)_2 \cdot 2THF$	White crystals which give off THF at 200°. MgH_2 forms at 280–400° [*837*].
$Mg(BH_4)_2 \cdot 6HCONMe_2$	Long colourless needles sublimable in vacuum [*514d*].
$Ca(BH_4)_2 \cdot 2THF$	Crystallises from a THF solution above 28.6°. White hydroscopic scaly crystals which give off THF at 200°. Insoluble in benzene, ether, and dioxan. Readily soluble in tetrahydrofuran (stable for some time). At 280–400°, CaH_2 forms [*462, 518, 808, 809, 837*].
$Ca(BH_4)_2 \cdot 4THF$	Crystallises from a saturated solution in THF below 29°. At 20° slowly changes into $Ca(BH_4)_2 \cdot 2THF$ [*518, 766b*].
$Sr(BH_4)_2 \cdot 2THF$	White crystals which give off THF at 200°. At 280–400° CaH_2 is formed [*808, 809, 837*].
$Ba(BH_4)_2 \cdot 2THF$	White needles which give off THF at 200°. At 280–400° BaH_2 is formed [*808, 809, 837*].
$Al(BH_4)_3$	M.p. $-64.5 \pm 0.5°$; b.p. 44.5° (extrapolated). Vapour pressure at 0° 119.5 torr. Vapour pressures: $\log p$ [torr] $= -1565/T + 7.808$. Enthalpy of evaporation 7160 cal/mole. Trouton's constant 22.5 cal deg^{-1} $mole^{-1}$ [*675*]. Vapour pressures: $\log p$ [torr] $= -1799.1/T + 9.56795$ [*411*].

References p. 355

$Al(BD_4)_3$	Vapour pressure at 0.3° 128 torr [*202*].
$Al(BH_4)_3 \cdot O(CH_3)_2$	Vapour pressure at 20° 0.5 torr. At 50° CH_4 is formed among other products [*675*].
$Al(BH_4)_3 \cdot THF$	M.p. 24–26° [*551b*].
$Al(BH_4)_3 \cdot 2THF$	[*551b*].
$Al(BH_4)_3 \cdot 3N(CH_3)_3$	Decomposes slowly at 20° [*675*].
$Al(BH_4)_3 \cdot N(CH_3)_3$	White solid rapidly subliming in vacuum or colourless crystals. M.p. 79° [*35a, 35d, 675*].
$AlCl(BH_4)_2$	Very unstable liquid. M.p. ~ −10° [*564*].
$HAl(BH_4)_2 \cdot O(C_2H_5)_2$	Compound sensitive to hydrolysis and capable of being transported in high vacuum [*18a, 551b*].
$HAl(BH_4)_2 \cdot 2O(C_2H_5)_2$	Unstable crystals. M.p. ~ +10°. Gives $HAl(BH_4)_2 \cdot OEt_2$ in vacuum [*18a, 551b*].
$HAl(BH_4)_2 \cdot THF$	M.p. −6° [*551b*].
$HAl(BH_4)_2 \cdot 2THF$	M.p. 78° [*551b*].
$AlCl_2(BH_4)$	Unstable white solid [*564*].
$H_2Al(BH_4) \cdot THF$	M.p. 55–58° [*551b*].
$H_2Al(BH_4) \cdot 2THF$	M.p. 79–80° [*551b*].
$LiBH_4 \cdot AlH_3 \cdot nO(C_2H_5)_2$	Very weakly bound crystalline adduct [*18a*].
$(LiBH_4)_2 \cdot AlH_3 \cdot nO(C_2H_5)_2$	Very weakly bound crystalline adduct [*18a*].
$(C_4H_9O)_2AlBH_4$	White solid stable up to 198° [*18a*].
$CuBH_4$	Readily soluble in diethyl ether + tetrahydrofuran (1 : 1) [*812*]. Decomposes into CuH above 0° [*812*] or above −12° [*711*]. Non-volatile white solid [*411*].
$CuBH_4 \cdot x$ pyridine	The green complex is stable up to 20° [*812*].
$AgBH_4$	White solid decomposing above −30°. Insoluble in

	diethyl ether, tetrahydrofuran, and triethylamine. Soluble in pyridine [*810*].
$ZnCl(BH_4)$	White solid soluble in diethyl ether and decomposing at 120° [*811*].
$Zn(BH_4)_2$	White solid soluble in ether. Decomposition at 85° [*811*] or 50° [*28*].
$Zn(BH_4)_2 \cdot 2\,THF$	Crystalline [*534, 552*].
$Li[ZnCl(BH_4)_2]$	[*408, 534, 820*]
$Li[Zn(BH_4)_3]$	[*534, 792*].
$Na[Zn(BH_4)_3]$	[*288, 534, 552*].
$K[(BH_4)Zn(BH_4)_2Zn$-$(BH_4)_2Zn(BH_4)]$	[*288, 534, 552*].
$K_2[Zn(BH_4)_4]$	[*514e*].
$[Zn(NH_3)_4](BH_4)_2$	[*208*].
$Cd(BH_4)_2$	White solid soluble in diethyl ether. Decomposition at 20° [*812*].
$Li[CdCl(BH_4)_2]$	White solid soluble in diethyl ether. Decomposition at 85° [*812*].
$Li_2[Cd(BH_4)_4]$	[*534, 552*].
$[Cd(NH_3)_6](BH_4)_2$	[*208*].
$Ti(BH_4)_3$	Green, sublimable [*372, 534*].
$Li[Ti(BH_4)_4] \cdot nO(C_2H_5)_2$	Blue etherate [*348, 534*].
(Cyclopentadienyl)$_2$ $TiBH_4$	Violet crystals which are soluble in dioxan and diethyl ether. Sublimes in vacuum at 130° [*549*].
(Cyclopentadienyl)$_2$-$Zr(BH_4)_2$	M.p. 155°; very faintly yellow solid [*537a*].

References p. 355

(Cyclopentadienyl)$_2$-$ZrBH_4$	Solid [*537a*].
$Zr(BH_4)_4$	M.p. 28.7°. Vapour pressures: log *p* [torr] = $-2983/T + 10.919$ [*70, 348, 349*]. d^{30} 1.01 g/cm^3 [*775a*].
$Hf(BH_4)_4$	M.p. 29°; b.p. 118°. Vapour pressures: log *p* [torr] = $-2844/T + 10.719$ [*70, 348, 349*]. d^{30} 1.60 g/cm^3 [*775a*].
$[Cr(NH_3)_6](BH_4)_3 \cdot 0.5NH_3$	Yellow needles. Decomposition in vacuum at 60° [*580*].
$Li_3[Mn(BH_4)_3I_2] \cdot nO(C_2H_5)_2$	Orange oil [*372, 817*].
$Li_2[Mn(BH_4)_2I_2] \cdot nO(C_2H_5)_2$	Orange oil [*372, 817*].
$Li_2[Mn(BH_4)_2Cl_2] \cdot nO(C_2H_5)_2$	Light yellow oil [*372, 817*].
$Li[Mn(BH_4)I_2] \cdot nO(C_2H_5)_2$	Orange yellow oil [*372, 817*].
$LiMn[Mn(BH_4)I_4]$	Orange yellow oil [*372, 817*].
$Fe(BH_4)_2$	Colourless non-volatile solid [*641*].
$Li[Fe(BH_4)_3]$	[*641*].
$[Co(NH_3)_6](BH_4)_3 \cdot NH_3$	Yellow needles. Decomposition in vacuum at 25° [*580*].
$Li[Ni(BH_4)_3]$	[*372, 817*].
$Li_2[Ni(BH_4)_4]$	[*372, 817*].
$[Ni(NH_3)_6](BH_4)_2$	[*656*].
$[Ni(en)_3](BH_4)_2$	[*208*].

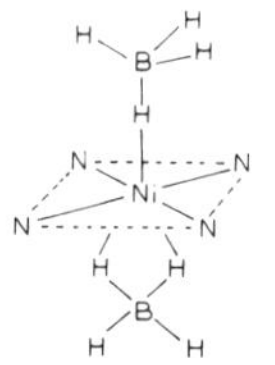

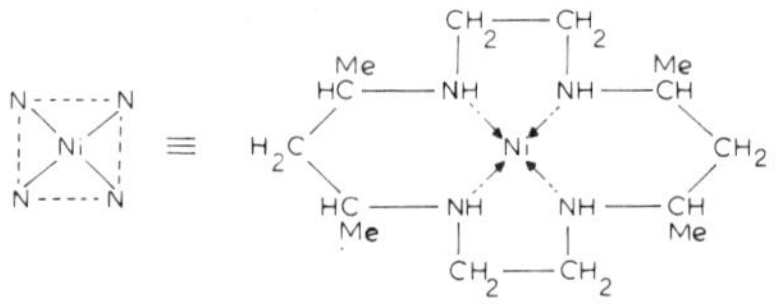

and similar complexes [*153a*].

$y(BH_4)_3 \cdot n$THF	Crystals [*863a*].

$SmCl(BH_4)_2 \cdot 2.5THF$	Crystalline; stable up to 80° under N_2; forms orange $SmCl(B_2H_6)$ above 120°[*102, 624, 863a*].
$EuCl(BH_4)_2 \cdot x THF$	Crystalline[*102, 624, 863a*].
$GdCl(BH_4)_2 \cdot 2THF$	Crystalline; forms orange $GdCl(B_2H_6)_2$ above 120° [*102, 624, 863a*].
$TbCl(BH_4)_2 \cdot 2THF$	Crystalline; forms orange $TbCl(B_2H_6)_2$ above 120° [*102, 624, 863a*].
$Dy(BH_4)_3 \cdot 3THF$	Crystalline[*863a*].
$DyCl(BH_4)_2 \cdot x THF$	Crystalline[*102, 624, 863a*].
$Ho(BH_4)_3 \cdot x THF$	Crystalline[*863a*].
$Er(BH_4)_3 \cdot 4.24THF$	Crystalline[*863a*].
$Tm(BH_4)_3 \cdot x THF$	Crystalline[*863a*].
$YbCl(BH_4)_2 \cdot x THF$	Crystalline; forms orange $YbCl(B_2H_6)_2$ above 120°[*102, 624, 863a*].
$Lu(BH_4)_3 \cdot x THF$	Crystalline[*863a*].
$Th(BH_4)_4$	Sublimable white crystalline mass. M.p. 204° (decomposition). Vapour pressures: $\log p$ [torr] $= -2844/T + 10.719$. Enthalpy of evaporation 21 kcal/mole [*348*].
$TH(BH_4)_4 \cdot 2O(C_2H_5)_2$	The etherate cannot be completely freed from ether [*348*].
$U(BH_4)_4$	Dark green crystals that can be sublimed at 20° and are fairly resistant to dry air. Only slight decomposition at 20° in years, and only 1–4% decomposition at 70° in 5–10 days. Rapid reaction at 100° according to the equation 2 $U(BH_4)_4 \rightarrow 2\ U(BH_4)_3 + (BH_3)_2 + H_2$[*664, 665*].
$U(BH_4)_3$	Red-brown pyrophoric solid[*664, 665*].
$U(BH_4)_3(BH_3CH_3)$	Green crystals[*664, 665*, cit. *5*].
$U(BH_3CH_3)_4$	Lavender-coloured crystals[*664, 665*, cit. *5*].

References p. 355

NH_4BH_4	Stable up to −20° [580].
$N(CH_3)_4BH_4$	Stable hygroscopic solid. Comparable with Na(Li)-BH_4. Insoluble in diethyl ether, isopropylamine, pyridine, chloroform, dioxan, and tetrahydrofuran. Decomposes above 150°. d^{25} 0.813 [26].
$N(C_2H_5)_4BH_4$	As NMe_4BH_4; d^{25} 0.927 [26].
$N(CH_2C_6H_5)_4BH_4$	As NMe_4BH_4; d^{25} 0.638 [26].
$[NH_2(Ph)_2P\cdots\overset{\oplus}{N}\cdots P(Ph)_2\text{-}NH_2]BH_4^{\ominus}$	M.p. 132–135° (decomposition): stable up to 100°, after which it evolves H_2; soluble in methylene chloride [58, 677].
$P(C_6H_5)_4BH_4$	Colourless, non-hygroscopic needles. Soluble in water, chloroform, methylene chloride, and ethanol. Insoluble in benzene, carbon tetrachloride, diethyl ether, and monoglyme [324].
$(C_6H_5)_3P\cdots\overset{\oplus}{C}H\cdots P(C_6\text{-}H_5)_3BH_4^{\ominus}$	Colourless needles very resistant to atmospheric moisture. Soluble in alcohols, nitromethane, acetonitrile, dimethylformamide, pyridine, benzene–ethanol, and hexane–ethanol. M.p. 211–211.5° [176].
$S(CH_3)_3BH_4$	White hygroscopic solid. M.p. 120–124° (decomposition). Readily soluble in water and ethanol [323].
$S(C_6H_5)_3BH_4$	As for $P(C_6H_5)_4BH_4$ [324].
$I(C_6H_5)_2BH_4$	Stable in vacuum at 20°; explosive decomposition at 110° [323].

20.19 *Substituted boranates $BH_{4-n}X_n^{\ominus}$*

KBH_3F	Crystals, d^{25} 1.5197 [7].
$LiBH(OCH_3)_3$	Dismutes very easily into $BH_{4-n}(OMe)_n^{\ominus}$ [81].
$LiBH(OC_2H_5)_3$	Dismutes less easily than $LiBH(OMe)_3$ [81].
$LiBH[OCH(CH_3)_2]_3$	Does not dismute [81].
$LiBH[OC(CH_3)_3]_3$	Does not dismute [81].

$NaBH(OCH_3)_3$	Readily dismutes into $BH_{4-n}(OMe)_n^{\ominus}$. Readily soluble in diglyme [*84*].
$NaBH(OCH_3)_3 \cdot THF$	To be regarded as $Na \cdot x THF^{\oplus}B(OR)_4^{\ominus} \cdot Na^{\oplus}BH_4^{\ominus}$ [*84*].
$NaBH[OC(O)H]_3$	Fairly stable. Decpmposition at m.p. (~125°) into methyl formate [*788*].
$LiBH_3(SCN) \cdot 1.5$dioxan	Crystals [*7*].
$NaBH_3(SCN) \cdot 2$dioxan	Crystals decomposing at 330° [*7*].
$NaBH(NCS)_3$	White powder. Decomposition at 125–130°. Soluble in ether, tetrahydrofuran, dioxan, and dimethyl sulphoxide. Insoluble in benzene and nitrobenzene [*404*].
$NaBH(NCS)_3 \cdot 2O(C_2H_5)_2$	Colourless oil. Liberates one Et_2O at 70–80° and two Et_2O at 100° [*404, 405*].
$NaBH(NCS)_3 \cdot 2$dioxan	White crystals. Liberates 2 dioxan at 120° [*404, 405*].
$CaBH(CH_3)_2NH_2$	[*134a*].
$LiBH_3(CN)$	Stable in water [*856, 854a*].
$Na[BH_3 \cdot CN \cdot BH_3]$	White mass; d^{25} 1.648 [*7*].
$Cd[BH_3 \cdot CN \cdot BH_3] \cdot 3.7$-dioxan	Decomposition at 35° [*7*].
$NaBH_3(NH_2)$	No details [*668*].
$Na[BH_3N(CH_3)_2] \cdot 0.5$-dioxan	Crystals [*7*].
$Na[BH_3P(CH_3)_2] \cdot 0.5$-dioxan	Crystals [*7*].
$NaBH(C_2H_5)_3$	Colourless viscous oil. Soluble in various ethers and saturated hydrocarbons. Decomposes at 135° and 1–2 torr into NaH and $B(C_2H_5)_3$. Extremely unstable in moist air (inflammation) [*81, 355*].
$LiBH_3(C_6H_5)$	White, somewhat hygroscopic powder [*805*].

References p. 355

$LiBH_3(C_6H_5)\cdot 2$dioxan	Colourless crystals. Insoluble in ether [*805*].
$LiBH_3(C_6H_5)\cdot 2O(C_2H_5)_2$	M.p. 5–9°. Soluble in diethyl ether. Loses Et_2O in vacuum at 60–70° [*805*].
$LiBH(C_6H_5)_3$	[*854b*].
$NaBH(C_6H_5)_3$	[*854b*].

20.20 B_3 to B_9 borohydrides

$B_3H_7\cdot P(C_6H_5)_3$	M.p. 161° [*277*].
$B_3H_7\cdot NH_3$	White crystals stable in air and sublimable in high vacuum. Soluble in diethyl ether, liquid ammonia, benzene, acetone, and alcohol. Sparingly soluble in petroleum ether. M.p. 73–75° (decomposition) [*420*]. Vapour pressures: log p [torr] $= -3739/T + 9.200$ (valid between 30 and 55°) [*10*].
$B_3H_7\cdot N(C_2H_5)_3$	Can be recrystallized from toluene [*420*].
$Li^{\oplus}B_3H_8^{\ominus}\cdot xO(C_2H_5)_2$	Yellow viscous oil [*850a*].
$Na^{\oplus}B_3H_8^{\ominus}$	White solid soluble in water and liquid ammonia resistant to hydrolysis and stable up to 200°. In ether, $NaB_3H_8\cdot 0.5Et_2O$ and $NaB_3H_8\cdot Et_2O$ are present at 0° [*13b, 360, 361*].
$M^{\oplus}B_3H_8^{\ominus}$ (M = NH_4, Rb, Cs)	Readily soluble in water [*13b*].
$M^{\oplus}B_3H_8^{\ominus}$ (M = Ph_2I, Ph_3S, Ph_3Se, Ph_3Te, Ph_4P, Ph_4As, Ph_4Sb, Ph_4Bi)	Very sparingly soluble in water [*13b*].
$N(CH_3)_4^{\oplus}B_3H_8^{\ominus}$	Insoluble in diethyl ether [*360, 361*].
$[Me_3NCH_2\text{-}]_2N(BH_3)Me$	M.p. > 300° [*779b*].
$Tl^{\oplus}B_3H_8^{\ominus}$	Less soluble in water than the alkali-metal triboranates. Readily soluble in mono- and diglymes and tetrahydrofuran [*13b*].
$M^{2\oplus}(B_3H_8^{\ominus})_2$ (M = Mg, Ca, Sr, Ba)	Readily soluble in water [*13b*].

$B_4H_{10}\cdot 2NH_3$ — White microcrystalline mass stable in the air. Soluble in cold water with the evolution of a small amount of H_2. Soluble in diethyl ether and liquid ammonia [*419*].

$B_4H_8\cdot PF_3$ — Vapour pressure at −47.7° 6.0–6.2 torr [*724*]. Purification by fractional condensation [*558b*].

$B_4H_8\cdot PF_2N(CH_3)_2$ — Colourless liquid becoming faintly yellow on storage at 20°. Decomposes rapidly about 55°. M.p. −18°. Vapour pressure at 27.8° 2.5 torr. Vapour pressures: $\log p$ [torr] $= -1727/T + 6.128$. Enthalpy of evaporation 8000 cal/mole [*440a, 764*].

$B_4H_{10}\cdot C_2H_4$ — Vapour pressure at 0° 14.5 torr [*488*].

$B_4H_8\cdot CO$ — M.p. −114.4 to −114.6° [*127*]; −110.3 to −113.2° [*724*]. B.p. 59.6° [*127*]. Vapour pressure at 0° 69.4 [*69*]; 71.2 [*127*]; 71.3 [*724*] torr. Vapour pressures: $\log p$ [torr] $= -1649/T + 1.75 \log T - 0.00340\,T$ [*127*]. Purification by fractional condensation [*558b*].

$B_4D_8\cdot CO$ — M.p. −112.2°. Vapour pressure at 0° 74.2 torr [*724*].

B_5H_8-2-F — M.p. −63.6°. Vapour pressures: $\log p$ [torr] $= -1981/T - 0.0055T + 1.75 \log T + 6.0979$ [*112a*].

B_5H_8-1-Cl — Material, condensable at −30° [*237d*].

B_5H_8-2-Cl — M.p. −41 to −40°. Vapour pressure at 28° 19 torr [*237d, 567*].

B_5H_8Br — M.p. 32–34° [*650*].

B_5H_8-1-Br — M.p. 38.5–39° [*567*].

B_5H_8-1-I — M.p. >56°. Vapour pressures: $\log p$ [torr] $= -3448/T + 11.079$ [*112a*].

B_5H_8-2-I — M.p. −39.3°. Vapour pressures: $\log p$ [torr] $= -2712/T - 0.052T + 1.75 \log T + 6.7038$ [*112a*].

B_5H_8I — M.p. 53° [*650*].

B_5H_7-1,2-Cl_2 — M.p. 18.2–19.5°. Vapour pressure at 20° $\leqslant$ 1 torr [*237d*].

References p. 355

$B_5H_9 \cdot 2N(CH_3)_3$	Vitreous mass decomposing at 100°[*109*].
$B_5H_{10}(C_2H_5)$	Vapour pressure at 0° 5 torr[*488*].
B_5H_8-1-(CH_3)	M.p. −56 to −55°, b.p. 75.2°. Vapour pressure at 30° 144 torr. Vapour pressures: $\log p$ [torr] $= -1710/T + 7.80$. Enthalpy of evaporation 7.82 kcal/mole[*628*].
B_5H_8-1-(C_2H_5)	M.p. −85 to −84°; b.p. 106°[*628*]. Vapour pressure at 0° 0.5 torr[*488*]. Vapour pressures: $\log p$ [torr] $= -1830/T + 7.42$ (valid between 0 and 110°). Enthalpy of evaporation 8.37 kcal/mole[*628*].
B_5H_8-1-$[CH(CH_3)_2]$	M.p. −93.5 to −92.5°; b.p. (extrapolated) 124°. Vapour pressures: $\log p$ [torr] $= -1941/T + 7.764$. Enthalpy of evaporation 8880 cal/mole[*628*].
B_5H_8-1-$[CH(CH_3)(C_2H_5)]$	B.p. (extrapolated) 148°. Vapour pressures: $\log p$ [torr] $= -2160/T + 8.006$. Enthalpy of evaporation 9.88 kcal/mole[*628, 630*].
B_5H_7-1-(CH_3)-2-$[CH(CH_3)(C_2H_5)]$	B.p. (extrapolated) 164°. Vapour pressure at 30° 4.6 torr. Vapour pressures: $\log p$ [torr] $= -2140/T + 7.75$ (valid between 28 and 150°). Enthalpy of evaporation 9.78 kcal/mole[*628*].
B_5H_8-1-(n-C_4H_9)	B.p. (extrapolated) 154.5°. Vapour pressures: $\log p$ [torr] $= -2340/T + 8.34$. Enthalpy of evaporation 10.71 kcal/mole[*630*].
B_5H_8-1-$[C(CH_3)_3]$	B.p. (extrapolated) 152°. Vapour pressures: $\log p$ [torr] $= -2250/T + 8.17$. Enthalpy of evaporation 10.29 kcal/mole[*630*].
B_5H_8-1-$P(CF_3)_2$	M.p. −42°; b.p. 184°[*116a*].
B_5H_8-2,3(μ)-$P(CH_3)_2$	M.p. 17°; b.p. 191°[*116a*].
B_5H_8-2,3(μ)-$P(CH_3)(CF_3)$	Form A: m.p. 14°; vapour pressure 1 torr at 25°[*116a*]. Form B: m.p. −28°. Vapour pressure 9 torr at 25° [*116a*].
B_5H_8-2-$Mn(CO)_5$	M.p. −11 to −10°[*237c*].
B_5H_8-2-$Re(CO)_5$	M.p. 10–11°[*237c*].

$B_6H_{10}\cdot P(C_6H_5)_3$	M.p. 168–171°[*845b*].
B_8H_{12}	Poor thermal stability [*171c*].
$\cdot B_9H_{15}$	Decomposes at −35°[*171b*].
$B_9H_{13}P(C_6H_5)_3\cdot CHCl_3$	Colourless crystals [*171b*].
B_9H_{12}-2-Br-?-$[S(C_2H_5)_2]$	Faintly orange solid. M.p. 74.0–75.5 [*340*].

20.21 Substituted decaboranes(14), decaboranate(10) salts, and derivatives

$B_{10}H_{13}Br$	M.p. 105°[*650*].
$B_{10}H_{13}$-1-I	M.p. 100°[*686b*].
$B_{10}H_{13}$-2-I	M.p. 116°[*650, 686b*].
$B_{10}H_{13}P(C_6H_5)_2$	M.p. 136–137°[*533d*].
$B_{10}H_{13}As(C_6H_5)_2$	M.p. 180° (decomp.)[*533d*].
$B_{10}H_{13}[N(C_2H_5)_2H]^{\ominus}$ $N(C_2H_5)_2H_2]^{\oplus}$	White needles; m.p. 204°[*274*].
$K[B_{10}H_{12}SiH_3]\cdot 1.1$ Monoglyme	Red Oil [*15b*].
$B_{10}H_{13}Si(CH_3)_3\cdot THF$	Red, hygroscopic oil. Stable at 20°[*15a*].
$Zn(B_{10}H_{12})_2\cdot 4Et_2O$	Yellow crystals [*277d*].
$Mg(B_{10}H_{12})_2\cdot 4Et_2O$	Cream solid [*277d*].
$Cd(B_{10}H_{12})_2\cdot 2Et_2O$	Fine, white needles, no m.p. up to 126°[*277c*].
$B_{10}H_{10}^{2\ominus}$ (aq)	Standard enthalpy of formation $\Delta H^{\circ}_{f,298}+22\pm5$ kcal/mole. Standard free energy of formation $\Delta G^{\circ}_{f,298}$ 65 ± 5 kcal/mole [*384a*].
$B_{10}H_{10}^{2\ominus}(H^{\oplus})_2\cdot 4H_2O$	Syrupy liquid soluble in water. Tends to explode [*534*].
$B_{10}H_{10}^{2\ominus}(H^{\oplus})_2\cdot 2H_2O$	Sparingly water-soluble solid. Readily soluble in $CsOH + H_2O$. Tends to explode [*534*].

References p. 355

$B_{10}H_{10}^{2\ominus}$ (heavy metal$^{2\ominus}$)	Solid [*534*].
$B_{10}H_{10}^{2\ominus}[(C_2H_5)_3NH^{\oplus}]_2$	White solid. M.p. 232–234° [*340*].
$B_{10}H_{10}^{2\ominus}(C_5H_5NH^{\oplus})_2$	Yellow plates. M.p. 221–222° [*340*].
$B_{10}H_{10}^{2\ominus}[(C_6H_5)_3PC_2H_5^{\oplus}]_2$	M.p. 233° (decomposition) [*340*].
$B_{10}H_{10}^{2\ominus}[(C_6H_5)_3$-$P(n\text{-}C_3H_7)^{\oplus}]_2$	M.p. 226° (decomposition) [*340*].
$B_{10}H_{10}^{2\ominus}[(C_6H_5)_3$-$P(n\text{-}C_4H_9)^{\oplus}]_2$	M.p. 167° (decomposition) [*340*].
$B_{10}H_{10}^{2\ominus}[(C_6H_5)_3$-$P(CH_2C_6H_5)^{\oplus}]_2$	M.p. 235° (decomposition) [*340*].
$B_{10}H_{10}^{2\ominus}[(C_6H_5)_3$-$P(CH_2C_6H_4CH_3)^{\oplus}]_2$	M.p. 210° (decomposition) [*340*].
$B_{10}Cl_8$-1,10-$(CH_2OH)_2^{2\ominus}[(CH_3)_4N^{\oplus}]_2$	No m.p. up to 360° [*411b*].
$B_{10}H_3Cl_5[S(CH_3)_2]_2$	Extractable with ethanol. M.p. 314–317° [*412*].
$B_{10}H_2Cl_6[S(CH_3)_2]_2$	Crystals. M.p. 368–369°, 360–361°, 340–341° (isomers?) [*412*].
$B_{10}H_2Cl_6(SCH_3)_2^{2\ominus}$-$[(CH_3)_4N^{\oplus}]_2$	Recrystallisable from water [*412*].
$B_{10}H_2Cl_8^{2\ominus}(NH_4^{\oplus})_2$	White solid [*413*].
$B_{10}Cl_8(OH)_2^{2\ominus}$-$[(CH_3)_4N^{\oplus}]_2$	No m.p. up to 400° [*416*].
$B_{10}Cl_8(OCHO)_2^{2\ominus}$ $[(CH_3)_4N^{\oplus}]_2$	No m.p. up to 400° [*416*].
$B_{10}Cl_8(CO)$	No decomposition at 300° [*418, 294*].
$B_{10}Cl_8$-1,10-$[COC_6H_4N(CH_3)_2]_2^{2\ominus}$-$[(C_6H_5N(CH_3)_2H^{\oplus}]_2$	Decomposition point 233° [*414a*].

$B_{10}Cl_9(OH)^{2\ominus}$-$[(CH_3)_4N^{\oplus}]_2$	No m.p. up to 400° [*416*].
$B_{10}Cl_9(COC_6H_5)$	Light yellow to brown solid [*416*].
$B_{10}Cl_9[OC(O)C_6H_5]^{2\ominus}$-$[(CH_3)_4N^{\oplus}]_2$	Can be recrystallised from H_2O [*416*].
$B_{10}Cl_{10}^{2\ominus}(Cs^{\oplus})_2$	White needles [*413*].
$B_{10}Cl_{10}^{2\ominus}[(CH_3)_4N^{\oplus}]_2$	White needles [*413*].
$B_{10}H_{11}$-2-(Br)$[S(C_2H_5)_2]$	Crystals. M.p. 100–101.5° [*340*].
$B_{10}H_9$-2-$Br^{2\ominus}$ $[(C_2H_5)_3NH^{\oplus}]_2$	Crystals that can be recrystallised from acetone-ether [*340*].
$B_{10}H_{11}$-2-(Br)$[P(C_6H_5)_3]_2$	Crystals, m.p. 243–244° [*340*].
$B_{10}H_{11}$-2-(Br)$(CNCH_3)_2$	Crystals, which do not melt below 400° but become orange at 220° [*340*].
$B_{10}H_7Br[S(CH_3)_2]_2$	Can be extracted with propanol. Isomers, m.p. 205–208°, 215–217°, 222–224°, 225–227°, 230–233° [*412*].
$B_{10}H_6Br_2[S(CH_3)_2]_2$	Can be recrystallised from methanol. M.p. 193–195° [*412*].
$B_{10}H_5Br_3[S(CH_3)_2]_2$	Can be recrystallised from acetonitrile. M.p. 281–282° [*412*].
$B_{10}H_3Br_7{}^{2\ominus}[(CH_3)_4N^{\oplus}]_2$	Solid [*413*].
$B_{10}Br_9COC_6H_5$	Crystalline solid [*416*].
$B_{10}Br_{10}^{2\ominus}(Cs^{\oplus})_2$	Solid that can be recrystallised from H_2O [*413*].
$B_{10}H_7I[S(CH_3)_2]_2$	M.p. 195–198° [*412*].
$B_{10}H_8I[S(CH_3)_2]^{\ominus}$-$[(CH_3)_4N^{\oplus}]$	Can be recrystallised from water [*412*].
$B_{10}H_8I_2{}^{2\ominus}[(CH_3)_4N^{\oplus}]_2$	Solid that can be recrystallised from H_2O–CH_3OH [*413*].

References p. 355

$B_{10}H_7I_3^{2\ominus}(Cs^{\oplus})_2$	Rubber-like or crystalline solid [*413*].
$B_{10}H_6I_4^{2\ominus}[(CH_3)_4N^{\oplus}]_2$	Solid that can be recrystallised from H_2O [*413*].
$B_{10}I_{10}^{2\ominus}(Cs^{\oplus})_2 \cdot CsI$	Crystals [*413*].
$B_{10}H_9(OH)^{2\ominus}[(CH_4)_4N^{\oplus}]_2$	Crystals which darken at 280° but do not melt below 400° [*416*].
$B_{10}H_8(OH)_2^{2\ominus}[(CH_3)_4N^{\oplus}]_2$	Crystalline mass [*416*].
$B_{10}H_9(OCH_2CH_2$-$OCH_3)^{2\ominus}[(CH_3)_4N^{\oplus}]_2$	Soluble in methanol and ether [*416*].
$B_{10}H_9[OCH{=}N(CH_3)_2]^{\ominus}$-$Cs^{\oplus}$	No m.p. up to 400° [*416*].
$B_{10}H_9[OC(O)$-$C_6H_5]^{2\ominus}[(CH_3)_4N^{\oplus}]_2$	Slightly yellow to brown solid. Decomposition point 360–365° [*416*].
$B_{10}H_9(OCHO)^{2\ominus}(Cs^{\oplus})_2$	White solid. Decomposition point 330–332° [*416*].
$B_{10}H_8$-1,6-(OH)-$[S(CH_3)_2]^{\ominus}Cs^{\oplus}$	Can be recrystallised from water [*412*].
$B_{10}H_8$-1,6-(OH)-$[S(CH_3)_2]^{\ominus}[(CH_3)_4N^{\oplus}]$	Plates [*333*, *334*].
$B_{10}H_8$-1,6-$[OS(CH_3)_2]$-$[S(CH_3)_2]$	Can be recrystallised from 50% ethanol [*412*].
$B_{10}H_9[S(CH_3)_2]^{\ominus}Cs^{\oplus}$	Can be recrystallised from a large amount of water [*412*].
$B_{10}H_9[S(CH_3)_2]^{\ominus}$-$[(CH_3)_4N^{\oplus}]$	Can be recrystallised from a large amount of water [*412*].
$B_{10}H_8$-1,6-$[S(CH_3)_2](NH_3)$	Long needles. M.p. 282–283° [*333*, *334*].
$B_{10}H_8$-1,6-$[S(CH_3)_2]_2$ and $B_{10}H_8$-1,10-$[S(CH_3)_2]_2$	Can be separated by chromatography (Al_2O_3; benzene–ethylene chloride) [*412*]. M.p. of the 1,6-isomer 235–236° [*412*].
$B_{10}H_8$-2,7(8)-$[S(CH_3)_2]_2$	Can be recrystallised from chloroform–carbon tetrachloride. M.p. 226–227.5° [*412*].

$B_{10}H_8(SCH_3)[S(CH_3)_2]^{\ominus}$-$(CH_3)_4N^{\oplus}$	Can be recrystallised from water [*412*].
$B_{10}H_8[S(H)CH_3][S(CH_3)_2]$	Can be recrystallised from benzene–petroleum ether. Decomposition point 190–195° [*412*].
$B_{10}H_7[S(CH_3)_2](NH_2)(N_2)$	No m.p. up to 400° [*333, 334*].
$B_{10}H_7[S(CH_3)_2](N_2)$-$(COOC_2H_5)$	Solid [*333, 334*].
$B_{10}H_8$-1,10-$[S(CH_3)_2](N_2)$	Can be recrystallised from ethanol. Extremely sensitive and explosive substance [*412*].
$B_{10}H_8[S(CH_3)_2](NO)$	Yellow solid [*412*].
$B_{10}H_8[S(CH_3)_2](NC_5H_5)$	M.p. 266–268° [*412*].
$B_{10}H_8$-1,10-$[S(CH_3)_2]$-(NH_3)	Crystals [*412*].
$B_{10}H_8$-1,10-$[S(CH_3)_2]$-$[N(CH_3)_3]$	White crystals; m.p. 268.5–269.5° [*412*].
$B_{10}H_8$-1,2-$[S(CH_3)_2]$-$[NH(CH_3)_2]$	M.p. 286–288° [*412*].
$B_{10}H_8$-1,6-$[S(CH_3)_2]$-$[NH(CH_3)_2]$	M.p. 221–222.5° [*412*].
$B_{10}H_8$-1,6-$[S(CH_3)_2]$-$[N(CH_3)_3]$	M.p. 203.5–204.5° [*412*].
$B_{10}H_8$-1,2-$[S(CH_3)_2]$-$[N(CH_3)_3]$	M.p. 267.5–268° [*412*].
$B_{10}H_8$-1,6-$[S(CH_3)_2]$-$(NH{=}CH$-*p*-$C_6H_4Cl)$	M.p. 282–284° [*412*].
$B_{10}H_8$-1,6-$[S(CH_3)_2]$-$[NH{=}CH$-*p*-C_6H_4N-$(CH_3)_2]$	Yellow crystals that can be recrystallised from acetonitrile [*412*].
$B_{10}H_8$-1,6-$[S(CH_3)_2]$ (CO)	White plates. M.p. 108–109° [*333, 334*].

$B_{10}H_8$-1,6-$[S(CH_3)_2]$-$(COOH)$	M.p. 264–265°[*333, 334*].
$B_{10}H_8$-1,6-(NH_3)-$[N(CH_3)_3]$	Crystals that can be recrystallised from acetonitrile–benzene. M.p. 305–307°[*333*].
$B_{10}H_8$-1,6-$[N(CH_3)_3]_2$	Crystals that can be recrystallised from acetone–ethanol. M.p. 250–252°[*333, 334*].
$B_{10}H_8$-2,3-$[N(CH_3)_3]_2$	Solid[*333*].
$B_{10}H_8$-1,6-$[N(CH_3)_3](N_2)$	Orange crystals. M.p. >400°[*333*].
$B_{10}H_8[N(CH_3)_3](CO)$	M.p. 175–178°[*334*].
$B_{10}H_7(N_2)(CN)^{\ominus}N(CH_3)_4^{\oplus}$	Faintly yellow crystals[*333*].
$B_{10}H_8$-1,10-$(N_2)_2$	White solid sensitive to shock. Sublimable in vacuum at 90–100°. Decomposes at about 125° without melting [*418*].
$B_{10}H_8(CH_2CH_2C_6H_5)_2^{2\ominus}$-$(Cs^{\oplus})_2$	Can be recrystallised from water[*416*].
$B_{10}H_8$-1,10-$(CN)_2^{2\ominus}$-$(Cs^{\oplus})_2$	Can be recrystallised from water[*333*].
$B_{10}H_8(CO)_2$	Readily sublimable in vacuum at about 100°. M.p. (in closed tube) 155–156°. Stable up to 200°; decomposes at 250°[*418*].
$B_{10}H_9$(tropenylium)$^{\ominus}Cs^{\oplus}$	Orange plates. Does not melt or decompose below 360°[*294*].
$B_{10}H_8$(tropenylium)$_2$	Chestnut-brown crystals. Remarkably stable. Insoluble in most solvents. Somewhat soluble in $(CH_3)_2SO$[*294*].
$B_{10}H_9$(tropenylium)$^{\ominus}$-$H_3O^{\oplus}$	Orange-coloured oil[*294*].
$B_{10}H_9(COC_6H_5)^{2\ominus}(Na^{\oplus})_2$	Solid[*331*].
$B_{10}H_9(COC_6H_5)^{2\ominus}$-$[(CH_3)_4N^{\oplus}]_2$	Cream-coloured, water-soluble solid. Decomposition point 285–290°[*416*].

$B_{10}H_9(COC_6H_5)^{2\ominus}$-$[(CH_3)_4N^{\oplus}]H_3O^{\oplus}$	Brick-red solid [*416*].

20.22 *Dodecaboranate(12) salts and derivatives*

$B_{12}H_{12}^{2\ominus}$ (aq)	Standard enthalpy of formation $\Delta H^\circ_{f,298}$ + 11 ± 10 kcal/mole. Standard free energie of formation $\Delta G^\circ_{f,298}$ 63 ± 10 kcal/mole [*384a*].
$B_{12}H_{12}^{2\ominus}$ (heavy metal$^{2\oplus}$)	Solid [*534*].
$B_{12}H_8F_3I^{2\ominus}(Cs^{\oplus})_2$	Recrystallisable from water [*413*].
$B_{12}H_8F_4{}^{2\ominus}(Cs^{\oplus})_2$	Recrystallisable from water [*413*].
$B_{12}F_4Br_8{}^{2\ominus}(Cs^{\oplus})_2$	Recrystallisable from water [*413*].
$B_{12}H_7F_5{}^{2\ominus}(Cs^{\oplus})_2$	Recrystallisable from water [*413*].
$B_{12}H_6F_6{}^{2\ominus}[(C_2H_5)_3$-$NH^{\oplus}]_2$	Recrystallisable from water [*413*].
$B_{12}F_{11}(OH)^{2\ominus}(Cs^{\oplus})_2$	Recrystallisable from water [*413*].
$B_{12}H_{11}Cl^{2\ominus}[(CH_3)_4N^{\oplus}]_2$	Recrystallisable from water [*413*].
$B_{12}H_{11}Cl^{2\ominus}(Cs^{\oplus})_2 \cdot CsCl$	Recrystallisable from water [*413*].
$B_{12}H_3Cl_3Br_6{}^{2\ominus}(Cs^{\oplus})_2 \cdot 2H_2O$	Recrystallisable from water [*413*].
$B_{12}H_6Cl_6{}^{2\ominus}[(CH_3)_4N^{\oplus}]_2$	Solid [*413*].
$B_{12}H_4Cl_7(C_3H_7)^{2\ominus}(Cs^{\oplus})_2$	Solid [*416*].
$B_{12}HCl_{10}(OCH_3)^{2\ominus}$-$(Cs^{\oplus})_2 \cdot H_2O$	Solid [*416*].
$B_{12}HCl_{10}(OH)^{2\ominus}(Cs^{\oplus})_2 \cdot H_2O$	Recrystallisable from water [*416*].
$B_{12}H_6Br_6{}^{2\ominus}(Na^{\oplus})_2 \cdot \frac{10}{3}$ dioxan	Precipitates from THF solutions on the addition of dioxan [*413*].
$B_{12}H_6Br_6{}^{2\ominus}[(CH_3)_3S^{\oplus}]_2$	Decomposition point 290° [*413*].

References p. 355

$B_{12}Br_6(OH_6^{2\ominus}(Cs^{\oplus})_2$	Crystals [*416*].
$B_{12}H_2Br_{10}^{2\ominus}(H_3O^{\oplus})_2 \cdot 6H_2O$	Crystals [*413*].
$B_{12}H_2Br_{10}^{2\ominus}(Cs^{\oplus})_2 \cdot 4H_2O$	Recrystallisable from water [*413*].
$B_{12}H_2Br_{10}^{2\ominus}(Ag^{\oplus})_2$	Solid [*413*].
$B_{12}H_2Br_{10}^{2\ominus}[(CH_3)_4N^{\oplus}]_2$	Recrystallisable from water [*413*].
$B_{12}Br_{10}(OCH_2CH_2\text{-}OCH_3)_2^{2\ominus}(Cs^{\oplus})_2$	Recrystallisable from water [*416*].
$B_{12}Br_{10}(OH)_2^{2\ominus}[(CH_3)_4N^{\oplus}]_2$	Recrystallisable from water [*416*].
$B_{12}Br_{11}(OH)^{2\ominus}(Cs^{\oplus})_2$	Solid [*416*].
$B_{12}Br_{11}(OH)^{2\ominus}[(CH_3)_4N^{\oplus}]_2$	Solid [*416*].
$B_{12}Br_{12}^{2\ominus}(H_3O^{\oplus})_2 \cdot 6H_2O$	White crystalline solid [*413*].
$B_{12}Br_{12}^{2\ominus}(Cs^{\oplus})_2$	Recrystallisable from water [*413*].
$B_{12}Br_{12}^{2\ominus}(Ag^{\oplus})_2$	Solid [*413*].
$B_{12}Br_{12}^{2\ominus}[(CH_3)_4N^{\oplus}]_2$	Recrystallisable from water [*413*].
$B_{12}H_{11}I^{2\ominus}[(CH_3)_4N^{\oplus}]_2$	Recrystallisable from water [*413*].
$B_{12}H_{10}I_2^{2\ominus}(Na^{\oplus})_2$	Water-soluble solid [*413*].
$B_{12}H_{10}I_2^{2\ominus}(Cs^{\oplus})_2$	Solid [*413*].
$B_{12}H_{10}I_2^{2\ominus}(Cs^{\oplus})_2 \cdot CsI$	Recrystallisable from water [*413*].
$B_{12}I_{12}^{2\ominus}(Na^{\oplus})_2 \cdot 2H_2O$	Soluble in water [*413*].
$B_{12}I_{12}^{2\ominus}(Ag^{\oplus})_2$	Solid [*413*].
$B_{12}I_{12}^{2\ominus}[(CH_3)_4N^{\oplus}]_2$	Soluble in water [*413*].
$B_{12}I_{12}^{2\ominus}[(CH_3)_4N^{\oplus}]_2 \cdot (CH_3)_4NI$	Recrystallisable from acetonitrile–water (1 : 1) [*413*].

$B_{12}H_{11}(OH)^{2\ominus}(Cs^{\oplus})_2$	No m.p. up to 400° [*416*].
$B_{12}H_{10}(OH)_2^{2\ominus}(Cs^{\oplus})_2$	Recrystallisable from water [*416*].
$B_{12}H_{11}(OCH_3)^{2\ominus}(Cs^{\oplus})_2$	Solid [*416*].
$B_{12}H_{10}(OCH_3)_2^{2\ominus}(Cs^{\oplus})_2$	Crystalline solid [*416*].
$B_{12}H_{11}(OCH_2CH_2Cl)^{2\ominus}$-$(Cs^{\oplus})_2$	Recrystallisable from water [*416*].
$B_{12}H_{11}(OCH_2CF_3)^{2\ominus}(Cs^{\oplus})_2$	Recrystallisable from water [*416*].
$B_{12}H_{11}(OCH_2C_6H_5)^{2\ominus}$-$[(CH_3)_4N^{\oplus}]_2$	Recrystallisable from water [*416*].
$B_{12}H_{11}(OCH_2CH_2$-$CH_3)^{2\ominus}(Cs^{\oplus})_2$	Crystalline solid [*416*].
$B_{12}H_{11}[OCH_2$-$(CH_3)_2]^{2\ominus}(Cs^{\oplus})_2$	Recrystallisable from water [*416*].
$B_{12}H_{11}[O(CH_2)_3$-$CH_3]^{2\ominus}(Cs^{\oplus})_2$	Crystalline solid [*416*].
$B_{12}H_{10}(OCH_2CH_2$-$OCH_3)_2^{2\ominus}(Cs^{\oplus})_2$	Recrystallisable from water [*416*].
$B_{12}H_{10}(OCHO)_2^{2\ominus}$-$[(C_3H_7)_4N^{\oplus}]_2$	Platelets, decomposition point 309–310° [*416*]
$B_{12}H_{11}(SH)^{2\ominus}(Cs^{\oplus})_2$	Recrystallisable from water [*416*].
$B_{12}H_{11}(SCH_3)^{2\ominus}(Cs^{\oplus})_2$	Recrystallisable from water [*416*].
$B_{12}H_{10}(SCH_3)_2^{2\ominus}(Cs^{\oplus})_2$	Recrystallisable from water [*416*].
$B_{12}H_9(SCH_3)_3^{2\ominus}(Cs^{\oplus})_2$	Recrystallisable from water [*416*].
$B_{12}H_{10}[N(CH_3)_3](CO)$	Recrystallisable from benzene [*333, 334*].
$B_{12}H_{11}(C_3H_7)^{2\ominus}(Cs^{\oplus})_2$	Recrystallisable from water [*416*].
$B_{12}H_9(CH_2CH_2C_6H_5)_3^{2\ominus}$-$(Cs^{\oplus})_2$	Crystals [*416*].

References p. 355

$B_{12}H_{11}$-1,12-$(COOH)^{2\ominus}$-$(Cs^{\oplus})_2$	No m.p. up to 400° [*416a*].
$B_{12}H_{10}(CO)_2$	Sublimable in vacuum at about 100° [*418*].
$B_{12}H_{10}$-1,12-$(CN)^{2\ominus}(Cs^{\oplus})_2$	Recrystallisable from water [*333*].

20.23 B_{18} to B_{24} borohydrides

n-$B_{18}H_{22}$	Pale yellow solid; m.p. 179–180° [*563a*]; 175–177° [*493b*].
i-$B_{18}H_{22}$	Pale yellow solid; m.p. 128–129° [*563a*].
n-$B_{18}H_{21}I$	Pale yellow plates; m.p. 184–185° [*563a*].
$(CH_3)_4N^{\oplus}$n-$B_{18}H_{21}^{\ominus}$	Deep yellow crystals [*563a*].
$(CH_3)_4N^{\oplus}$ i-$B_{18}H_{21}^{\ominus}$	Bright yellow crystals [*563a*].
$[(CH_3)_4N^{\oplus}]_2$n-$B_{18}H_{20}^{2\ominus}$	Pale yellow crystals [*563a*].
$[(CH_3)_4N^{\oplus}]_2$i-$B_{18}H_{20}^{2\ominus}$	Pale yellow crystals [*563a*].
$B_{20}H_{16}[S(CH_3)_2]_2$	M.p. 228–230°. Recrystallisable from methanol–acetonitrile [*138*].
$B_{20}H_{16}(IC_6H_5)_2$	Decomposition at 160°. Recrystallisable from methanol–acetonitrile [*138*].
$B_{20}H_{16}[O\overline{C(CH_2)_3N}CH_3]_2$	White substance recrystallisable from acetonitrile Decomposes at 260° [*138*].
$B_{20}H_{16}[O_2\overline{S(CH_2)_3C}H_2]_2$	Recrystallisable from methanol–acetonitrile, decomposes at 245° [*138*].
$(NH_4)_2B_{20}H_{15}(OH)$-$[S(CH_3)_2]_2 \cdot H_2O$	Colourless substance [*138*].
$[(CH_3)_4N]_2B_{20}H_{18}$	Yellow crystals recrystallisable from water–acetonitrile [*138*].
$[(CH_3)_4N]_2B_{20}H_{12}Cl_6$	M.p. 285° [*138*].
$[(CH_3)_4N]_2B_{20}H_{13}Br_5$	M.p. 260° (decomposition) [*138*].

$Cs_4B_{20}H_{18}\cdot H_2O$	White substance recrystallisable from hot water [*138*].
$[(CH_3)_4N]_3B_{20}H_{19}$	Insoluble white solid [*138, 383a*].
$[(C_2H_5)_4N]_3B_{20}H_{19}$	White solid; m.p. 163–164° [*383a*].
$[(CH_3)_4N]_4B_{20}H_{17}(OH)\cdot 7H_2O$	White solid [*138*].
$(NH_4)_4B_{20}H_{17}(OH)\cdot 0.5H_2O$	White crystals [*138*].
$[(CH_3)_4N]_3B_{20}H_{18}(OH)\cdot H_2O$	Solid [*138*].
$Cs_4B_{20}H_{17}(OCH_3)$	White solid [*138*].
$[(CH_3)_4N]_2B_{20}H_{17}(OH)$	M.p. 263° (decomposition) [*138*].
$[(CH_3)_3S]_2B_{20}H_{17}(OH)$	Yellow crystals. Decomposes at 250° [*138*].
$[(CH_3)_4N]_2B_{20}H_{17}OCOH$	M.p. 205° (decomposition) [*138*].
$[(CH_3)_4N]_2B_{20}H_{16}(OH)_2$	Long yellow needles [*138*].
$(Cs^{\oplus})_3B_{24}H_{23}^{3\ominus}$	Does not melt below 300° [*841a*].
$[(C_2H_5)_3NH^{\oplus}]_5B_{20}H_{18}\cdot B_{20}H_{19}^{5\ominus}$	Very clear large crystals; m.p. 203–204° [*383a*].

20.24 Carboranes with 5-membered skeletons

$1,5\text{-}C_2B_3H_5$	Can be stored at 20°. Does not react with air, H_2O, CO_2, $N(CH_3)_2$ and CH_3COCH_3 at 20°. M.p. $-126.40 \pm 0.05°$; b.p. $-3.7°$ (calculated). Vapour pressure at $-3.5°$ 510 torr. Vapour pressures: $\log p$ [torr] $= -1459.18/T - 0.002986T + 9.10154$. Enthalpy of evaporation 5684 cal/mole. Trouton's constant 21.2 cal deg^{-1} mole^{-1} [*690a*]. Stable to 150° [*565a*].

20.25 Carboranes with 6-membered skeletons

CB_5H_7	Vapour pressure at 26° 503 torr [*565c*].
$1,6\text{-}C_2B_4H_6$	Does not react with air, H_2O, NH_3, $N(CH_3)_3$,

References p. 355

$N(C_2H_5)_3$, and CH_3COCH_3 at 20°. M.p. −32.63 ± 0.05°; b.p. 22.7° (calculated). Vapour pressure at −32.63° 63 torr. Vapour pressures: $\log p$ [torr] = $-1896.94/T - 0.00602\,T + 11.1324$ (solid); $\log p$ [torr] = $-1676.40/T - 0.004068\,T + 9.7513$ (liquid). Enthalpy of evaporation 6042 cal/mole. Trouton's constant 20.4 cal deg^{-1} $mole^{-1}$ [*692a*].

20.26 *Carboranes with 8-membered skeletons*

$C_2B_6H_6(CH_3)_2$ — M.p. −40.4 to −39.4° [*763b, 763c*], −58 to −63° [*845b*]; b.p. 62° (134 torr) [*763c*].

20.27 *Carboranes with 9 membered skeletons*

$C_2B_7H_7(CH_3)_2$ — M.p. −22 to −21.3° [*763b, 763c*].

20.28 *Carboranes with 10-membered skeletons*

1,6-$C_2B_8H_8(CH_3)_2$ — M.p. 1.0–1.6° [*763b, 763c*]; b.p. 73° (32 torr) [*763c*].

1,10-$C_2B_8H_8(CH_3)_2$ — M.p. 26.5–27.5° [*763b, 763c*].

1,7-$CHPB_9H_{10}^{\ominus}C_6H_{10}$-$NH^{\oplus}$ — M.p. 112–113° [*767a*].

(1,7-$CHPB_9H_{10}^{\ominus})_2$-$Fe[(CH_3)_4N^{\oplus}]_2$ — M.p. 233–234° or 239.5–240.5° [*767a*].

20.29 *Carboranes with 11-membered skeletons*

$CH_3NH_2(CB_{10}H_{12})$ — M.p. 292–293° [*319a*].

$C_2H_5NH_2(CB_{10}H_{12})$ — M.p. 234–235° (decomposition) [*319a*].

n-$C_3H_7NH_2(CB_{10}H_{12})$ — M.p. 220–222° [*319a*].

tert-$C_4H_9NH_2(CB_{10}H_{12})$ — M.p. 237–238° (decomposition) [*319a*].

$(CH_3)_3N(CB_{10}H_{12})$ — M.p. 344–345° (decomposition) [*319a*].

$C_2H_5N(CH_3)_2(CB_{10}H_{12})$ — M.p. 310–311° [*319a*].

n-$C_3H_7N(CH_3)_2$-$(CB_{10}H_{12})$ — M.p. 185–187° [*319a*].

Compound	Properties
$B_{10}H_{12}CN(CH_3)_3$	M.p. 345–346° [*371d*], 345–350 [*411e*].
$Cl_2B_{10}H_{10}CH^{\ominus}$ $[(CH_3)_4N]^{\oplus}$	Solid [*371d*].
$BrB_{10}H_{11}CH^{\ominus}$ $[(CH_3)_4N]^{\oplus}$	Solid [*371d*].
$Br_2B_{10}H_{10}CH^{\ominus}$ $[(CH_3)_4N]^{\oplus}$	Solid [*371d*].
tert-$C_4H_9N(CH_3)_2$-$(B_{10}H_{12})$	M.p. 202–204° [*319a*].
$C_2B_9H_{11}$	M.p. 212–213° [*763, 763c*].
$C_2B_9H_{10}CH_3$	M.p. 84–84.5° [*763, 763c*].
$C_2B_9H_9(CH_3)_2$	M.p. 57–58° [*763, 763c*].
$C_2B_9H_9(H)(C_6H_5)$	M.p. 37–37.8° [*763, 763c*].
$C_2B_9H_{10}(p\text{-}BrC_6H_4)$	M.p. 100.5–101.5° [*763c*].
$C_2B_9H_9(CH_3)_2Br^{\ominus}$-$N(CH_3)_4^{\oplus}$	Tablets recrystallisable from ethanol [*570a*].
$C_2B_9H_{11}I^{\ominus}N(CH_3)_4^{\oplus}$	Tablets recrystallisable from ethanol [*570a*].
$C_2B_9H_9R_2I^{\ominus}N(CH_3)_4^{\oplus}$ (R = H, CH_3)	Tablets recrystallisable from ethanol [*570a*].

20.30 Carborane $H(C_2B_{10}H_{10})H$ and its B-substituted derivatives $H(C_2B_{10}H_{10-n}X_n)H$

Compound	Properties
$H(1,2\text{-}C_2B_{10}H_{10})H$	M.p. 320° [*338*]; 285–287° [*215*]. Volatile. Readily soluble in CCl_4 [*338*].
$H(1,7\text{-}C_2B_{10}H_{10})H$	M.p. 263–265° [*270*].
$H(1,2\text{-}C_2B_{10}H_9\text{-}9\text{-}Cl)H$	M.p. 237–238°. Sublimable. Recrystallisable from heptane [*861c*].
$H(1,7\text{-}C_2B_{10}H_9\text{-}9\text{-}Cl)H$	M.p. 215–216°. Sublimable. Recrystallisable from heptane [*861c*].

References p. 355

$H(1,2\text{-}C_2B_{10}H_9\text{-}9\text{-}I)H$	M.p. 117–118°. Sublimable. Recrystallisable from heptane [*861c*].
$H(1,7\text{-}C_2B_{10}H_9\text{-}9\text{-}I)H$	M.p. 109–110°. Sublimable. Recrystallisable from heptane [*861c*].
$H(1,2\text{-}C_2B_{10}H_8Cl_2)H$	M.p. 250–251° [*680*].
$H(1,7\text{-}C_2B_{10}H_8Cl_2)H$	M.p. 232° [*680*].
$H(1,7\text{-}C_2B_{10}H_8\text{-}9,10\text{-}Cl_2)H$	M.p. 217–218° [*861c*].
$H(1,2\text{-}C_2B_{10}H_8\text{-}9,12\text{-}I_2)H$	M.p. 185–186° [*861c*].
$H(1,2\text{-}C_2B_{10}H_8\text{-}9\text{-}Cl\text{-}12\text{-}Br)H$	M.p. 238–239° [*861c*].
$H(1,2\text{-}C_2B_{10}H_7Cl_3)H$	M.p. 241–243° [*680*].
$H(1,2\text{-}C_2B_{10}H_6\text{-}8,9,10,12\text{-}Cl_4)H$	M.p. 351–352°. The least soluble in CCl_4 of the chlorinated carboranes. Sublimable. Recrystallisable from heptane–benzene [*680, 861c*].
$H(1,7\text{-}C_2B_{10}H_6Cl_4)H$	M.p. 250° [*680*].
$H(1,2\text{-}C_2B_{10}H_6\text{-}8,9,10,12\text{-}Br_4)H$	M.p. 351–352°. Sublimable in vacuum [*861c*].
$H(1,7\text{-}C_2B_{10}H_6\text{-}8,9,10,12\text{-}Br_4)H$	M.p. 324–325° [*861c*].
$H(1,2\text{-}C_2B_{10}H_4Cl_6)H$	M.p. 306° [*680*].
$H(1,7\text{-}C_2B_{10}H_4Br_6)H$	M.p. 314–315° [*861c*].
$H(1,2\text{-}C_2B_{10}H_2Cl_8)H$	M.p. 272° [*680*].
$H(1,2\text{-}C_2B_{10}Cl_{10})H$	M.p. 259° [*680*].
$H(1,2\text{-}C_2B_{10}Cl_{10})H \cdot NEt_3$	M.p. 165–166° [*861o*].
$H(1,2\text{-}C_2B_{10}Cl_{10})CH_3 \cdot NEt_3$	M.p. 120–140° [*861o*].
$H(1,2\text{-}C_2B_{10}Cl_{10})Cl$	M.p. 279 [*680*].

20.31 C-Monosubstituted carboranes $H(1,2\text{-}C_2B_{10}H_{10})R$ (selection of the most important compounds)

R in $H(1,2\text{-}C_2B_{10}H_{10})R$	
CH_3	M.p. 114–115°[*338*], 214–215°[*215*], 218–219°[*270*].
neo-CH_3	M.p. 208–209.5°[*270*] [H(**1,7**-$C_2B_{10}H_{10}$)CH_3!].
C_2H_5	B.p. 75–80° (0.5 torr)[*217*].
n-C_3H_7	M.p. 62°[*338*], 68–69°[*217*].
n-C_4H_9	B.p. 40° (0.3 torr)[*339*], 75° (0.01 torr). n_D^{25} 1.5298 [*339*].
C_5H_{11}	M.p. 33°[*338*].
n-C_6H_{13}	B.p. 101–102° (0.5 torr)[*217*]. n_D^{25} 1.5211 [*338*].
C_6H_5	M.p. 66–67°[*338*], 69.5–70°[*217*].
$CH{=}CH_2$	M.p. 78–79°[*338*], 76–77°[*217, 339*].
$C(CH_3){=}CH_2$	M.p. 45–46°[*338*], 46.7–47.7°. d_4^{25} 0.942; n_D^{50} 1.5432 [*215*].
$CH_2CH{=}CH_2$	M.p. 63–65°[*339*].
CH_2Cl	M.p. 83–85°[*217*].
CH_2Br	M.p. 30°[*338*], 47–49°[*215*].
CH_2CH_2Cl	M.p. 80–82°[*217*].
CH_2CH_2Br	M.p. 114–115°[*217*].
$CH_2CH_2CH_2Cl$	M.p. 53–55°; b.p. 170° (10 torr)[*217*].
CH_2OH	M.p. 225°[*339*].
$CH_2OCH_2(1,2C_2B_{10}H_{10})H$	M.p. 342–344.5 [*268*].
$CH_2CH_2OCH_2CH_2$-$(1,2\text{-}C_2B_{10}H_{10})H$	M.p. 117–118°[*268*].

References p. 355

$CH_2CH(OH)CH_2OH$	M.p. 89–93°[*339*].
$COOCH_3$	M.p. 73°[*338*].
$CH_2(OOCCH_3)$	M.p. 42–43°; b.p. 82–84° (0.2 torr)[*338*].
$CH(OOCCH_3)CH_3$	B.p. 85–95° (0.2 torr). n_D^{21} 1.5291 [*338*].
$CH_2CH_2(OOCCH_3)$	M.p. 61–63°; b.p. 146 (1.6 torr)[*338*].
$CH(OOCCH_3)$	M.p. 206–207°[*339*].
$CH_2N(C_2H_5)_2$	M.p. 33–35°[*338*].
$Si(CH_3)_3$	M.p. 94–95°[*339*].
$Si(C_6H_5)_3$	M.p. 165–167°[*339*].
$Si(CH_3)_2[(1,2\text{-}C_2B_{10}\text{-}H_{10})H]$	M.p. 195–196.5°[*339*].
$CH_2Si(CH_3)_3$	M.p. 52° (not sharp)[*339*].

20.32 C,C′-Disubstituted carboranes $R(1,2\text{-}C_2B_{10}H_{10})R'$ and $CH_3(1,7\text{-}C_2B_{10}H_{10})CH_3$ (selection of the most important compounds)

R	*R′*	
C_6H_6	R	M.p. 148–149°[*215*].
$C(CH_3)=CH_2$	CH_3	M.p. 77–82°[*339*].
$C(CH_3)=CH_2$	R	M.p. 82–83°[*217*].
CH_2Cl	R	M.p. 119–120°[*338*], 114–115°[*217*].
CH_2Br	R	M.p. 68–69.5°[*215*].
n-C_3H_7	Br	M.p. 44–45°; b.p. 107° (1 torr)[*338*].
n-C_4H_9	Br	B.p. 85–90° (0.5 torr); n_D^{23} 1.5500[*338*].

CH_2OH	R	M.p. 303–304°[*339*].
	$H_2C{-}O{-}CH_2$ $(CB_{10}H_{10}C)$	M.p. 170°[*339*].
	$H_2C{-}O{-}CH_2$ $(CB_{10}H_9BrC)$	M.p. 127–129°[*339*].
$CH(OH)CH_3$	R	M.p. >400°[*339*].
CH_2CH_2OH	R	M.p. 124–125°[*339*].
COOH	R	M.p. 232°[*339*].
	$O{=}C{-}O{-}C{=}O$ $(CB_{10}H_{10}C)$	M.p. 180°[*339*].
CH_2COOH	R	M.p. 203–204°[*339*].
$C(CH_3){=}CH_2$	COOH	M.p. 175°[*339*].
$-\overline{C{=}CH(CH{=}CH)_2CH_2}$	CH_3	M.p. 57.5–58°[*293b*].
COCl	R	M.p. 69–70°[*339*].
$COOCH_3$	R	M.p. 66–67°[*338*].
CH_2COOCH_3	R	M.p. 43–44°[*338*], 42–43° [*215*], 47–48°[*339*].
CH_2OCOCH_3	R	M.p. 48–48.5°[*217*].
$NHCOOCH_3$	R	M.p. 257–258°[*339*].
$Si(CH_3)_2Cl$	R	M.p. 112.5–113.5°[*573*].
$Si(CH_3)Cl_2$	R	M.p. 119–120°[*573*].
$Si(C_6H_5)_2Cl$	R	M.p. 244–245°[*573*].
$SiCl_3$	R	M.p. 121–122°[*573*].
$1,2\text{-}C_2B_{10}H_{10})SiCl_3$	$SiCl_3$	M.p. 271–272°[*573*].

$1,2\text{-}C_2B_{10}H_{10})Si(CH_3)_3$	$Si(CH_3)_3$	M.p. 309–310°[*573*].
	$H_2C—Si(CH_3)_2—CH_2$ $(CB_{10}H_{10}C)$—	M.p. 149–150°[*339*].
	$(CH_3)_2Si—O—Si(CH_3)_2$ $(CB_{10}H_{10}C)$	M.p. 160–161°[*573*].
	$(CH_3)_2Si—NH—Si(CH_3)_2$ $(CB_{10}H_{10}C)$	M.p. 190–192°[*573*].
	$H_2N(CH_3)Si—NH—Si(CH_3)NH_2$ $(CB_{10}H_{10}C)$	M.p. 189–191.5°[*573*].
	$H(CH_3)N(CH_3)Si—N(CH_3)—Si(CH_3)N(CH_3)H$ $(CB_{10}H_{10}C)$—	M.p. 128–129.5°[*573*].
SH	R	M.p. 265–267°[*718a*].
SH(**neo**)	R	M.p. 164–165° [R(**1,7**-$C_2B_{10}H_{10}$)R][*718a*].
SCH_3	R	M.p. 101–102°[*718a*].
SC_2H_5(**neo**)	R	B.p. 114° (0.1 torr) [R(**1,7**-$C_2B_{10}H_{10}$)R][*718a*].
$CH_2C_6H_5$	R	M.p. 102–104°[*718a*].
C_6H_5	R	M.p. 189–191°[*718a*].
$C_6H_4CH_3$(**neo**)	R	M.p. 131–132° [R(**1,7**-$C_2B_{10}H_{10}$)R][*718a*].
$P(C_6H_5)_2$	R	M.p. 219°[*9*].
$P(C_6H_5)Cl$	R	M.p. 172–174°[*9*].
$P(C_6H_5)N_3$	R	M.p. 126–128°[*9*].
	$C_6H_5P—NH—PC_6H_5$ $(CB_{10}H_{10}C)$	M.p. 222–224°[*9*].

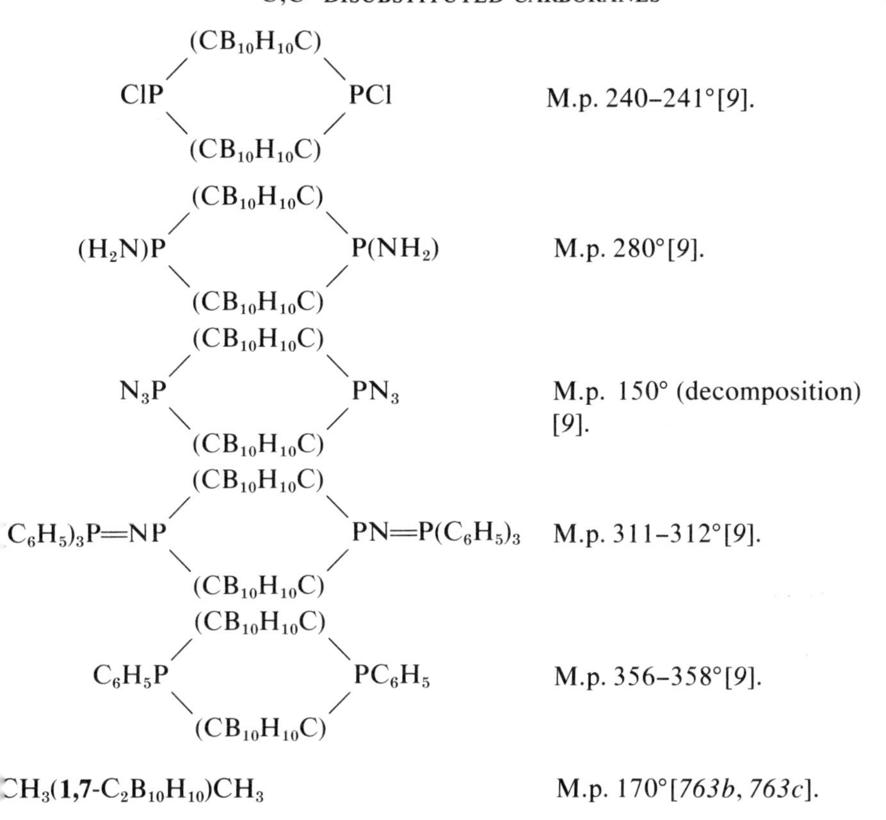

21. Spectra and structure of the boranes

The following sections give a review (numbers = literature references) of investigations of the following type: infrared spectra (*IR*), nuclear magnetic resonance (*NMR*), Raman spectra (*RA*), ultraviolet spectra (*UV*), electronic spectra (*ES*), microwave spectra (*MW*), mass spectra (*MS*), structural determinations (*STR*), molecular orbital calculations (*MO*), and electron spin resonance (*ESR*).

21.1 Mono- and diboranes with halogen and chalcogen bonds (⋆ for comparison)

H on boron surfaces — *IR* 532e

$B_2H_{6-n}D_n$ — *IR* 40, 372c, 386, 392, 450, 451, 600, 718c; *theory of the vibrations* 74; *RA* 16, 760; *NMR* 236, 396, 591, 654, 760; *STR* 292, 718b, 774a, 859; electron diffraction 709; *MS* 435, 715, dissociation[*165*].

BHF_2 — *IR* 483, 587a; *NMR* 794a

References p. 355

BDF_2	*IR* 483, 587a
B_2H_5Cl	*IR* 536
$BHCl_2$	*IR* 483a, 484, 486, 536
$(BHCl_2 + B_2H_5Cl)$	*IR* 398
$BHCl_2 \cdot O(C_2H_5)_2$	*IR* 31b, 568; *NMR* 568
$BDCl_2$	*IR* 31b, 484
$BHBr_2$	*IR* 483a, 485
$BDBr_2$	*IR* 485
$BH_3 \cdot O(CH_3)_2$	*RA* 616
$BH_3 \cdot O(C_2H_5)_2$	*RA* 150, 616; *NMR* 150
$BH_3 \cdot$THF	*RA* 616; *NMR* 591
$BH(OCH_3)_2$	*IR* 452
$BD(OCH_3)_2$	*IR* 452
$^{11}BH(OCD_3)_2$	*NMR* 57a, 208a
⋆$B(OCH_3)_3$	*IR* 452
⋆$B(OCH_2CH_2NH_2)_3$	*NMR* 498a
$BH(OC_2H_5)_2$	*IR* 453
$BD(OC_2H_5)_2$	*IR* 453
⋆$B(OC_2H_5)_3$	*IR* 453
$BH(O\text{-}i\text{-}C_3H_7)_2$	*IR* 460
⋆$B(O\text{-}i\text{-}C_3H_7)_3$	*IR* 460
CH_2—O—BH—O—CH_2 (ring: CH_2—O, CH_2—O joined to BH)	*IR* 621

$H_3B_3O_3$	*IR* 145a, 280a, 280b, 280c; *MW* 72a
$H_2B_2O_3$	*UV*, *MS* 280b; *STR* 138a
B_2H_5-μ-SH	*NMR* 395c
$BH_3 \cdot S(CH_3)_2$	*NMR* 150
$BH_3 \cdot S(C_2H_5)_2$	*NMR* 150
$H_2BSCH_2CH_2SBH_2$	*IR* 193
$(CH_3)_3N \cdot H_2BSCH_2CH_2$-$SBH_2 \cdot N(CH_3)_3$	*IR* 193
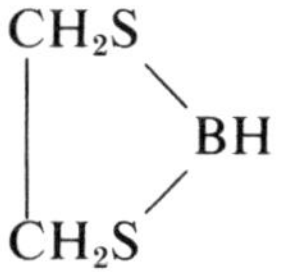 CH_2S—BH—SCH_2 (cyclic: CH_2S / BH / CH_2S)	*IR* 193
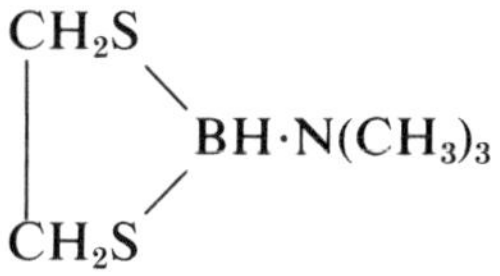 CH_2S / $BH \cdot N(CH_3)_3$ / CH_2S (cyclic)	*IR* 193
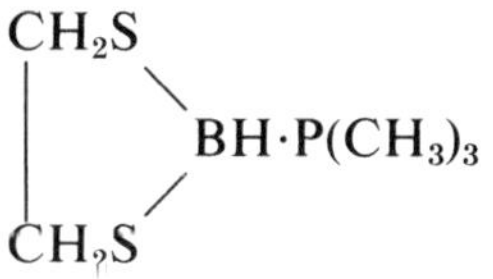 CH_2S / $BH \cdot P(CH_3)_3$ / CH_2S (cyclic)	*IR* 193

21.2 Mono- and diboranes with B–N bonds [for bis(amine-boronium) compounds, see section 21.7] (⋆ for comparison)

review of IR spectra [498c]

$B_2H_5(NH_2)$	*NMR* 239
$B_2H_5(NHCH_3)$	*NMR* 239
$B_2H_5[N(CH_3)_2]$	*IR* 490, 491; *RA* 490; *NMR* 239, 395b, 591
$B_2D_5[N(CH_3)_2]$	*IR* 490, 491; *RA* 490
$[BH_2N(CH_3)_2]_2$	*IR* 604; *NMR* 591, 604c

References p. 355

BH_2NH_2	*MS* 439a; *MO* 16c
$(BH_2NH_2)_3$	*NMR* 604c, 724b
$[BH_2N(CH_3)_2]_3$	*STR* 770, 771
$(BH_2NHR)_3$ $R = CH_3, C_2H_5,$ $C_3H_7,$ C_4H_9	*NMR* 100b, 604 c
⋆ $BF_2N(C_2H_5)_2$	*IR* 277b
⋆ $BCl_2N(C_2H_5)_2$	*IR* 277b
⋆ $BBr_2N(C_2H_5)_2$	*IR* 277b
$BH_3{\cdot}NH_3$	*NMR* 239, 329a, 604c, 604d, 724b; *MS* 439a; *STR* 16b
$BH_3{\cdot}NH_2(CH_3)$	*NMR* 239, 329a, 604c, 604d, 724b
$BH_3{\cdot}NH(CH_3)_2$	*NMR* 239, 329a, 604c, 604d, 591, 724b; *IR* 73a
$BH_3{\cdot}N(CH_3)_2BH_2{\cdot}NH_3$	*NMR* 289, 604c, 604d, 724b
$BH_3{\cdot}N(CH_3)_2BH_2{\cdot}$ $NH_2(CH_3)$	*NMR* 289, 604c, 604d, 724b
$BH_3{\cdot}N(CH_3)_2BH_2{\cdot}$ $NH(CH_3)_2$	*NMR* 289, 604c, 604d, 724b
$BH_3{\cdot}N(CH_3)_2BH_2{\cdot}$ $N(CH_3)_3$	*NMR* 289, 604c, 604d, 724b
$BH_3{\cdot}N(CH_3)_3$	*IR* 73a, 614; *NMR* 150, 239, 329a, 591, 604c, 604d, 724b; *RA* 614; *STR* 252
$BD_3{\cdot}N(CH_3)_3$	*IR* 614; *RA* 614
$BH_3{\cdot}N(CH_3)_2(C_2H_5)$	*NMR* 150
$BH_3{\cdot}N(C_2H_5)_3$	*NMR* 150, 329a, 604c, 604d, 724b
⋆ $BX_3{\cdot}NH_2Ar$	*IR* 256

$BH_2(C_6H_5)\cdot$pyridine	*IR* 173
$BH(CH_3)_2\cdot NH_3$	*IR* 527a
⋆ $B(CH_3)_3\cdot NH_3$	*NMR* 604c, 604d
⋆ $B(CH_3)_3\cdot N(CH_3)_3$	*STR* 251
$BH_3\cdot NC_5H_5$	*NMR* 591
$BH_3\cdot N(H)(C_6H_5)C(O)$-$N(C_2H_5)_2$	*IR, NMR* 151a
$BH_3\cdot N(H)(C_6H_5)C(O)N$-$(H)(CH_3)$	*IR, NMR* 151a
$BH_3\cdot HN{=}C(C_6H_5)_2$	*IR, NMR, MS* 585a
$BH_3\cdot C_6H_{12}N_4$	*STR* 292c
BH_2(*tert*-$C_4H_9)NH_2NH_2$	*NMR* 521a
⋆ $B(CH_3)_2NH_2$	*IR* 44a, 44b
⋆ $B(CH_3)_2N(CH_3)_2$	*IR* 44a, 44b
$[(CH_3)_3N(BH_2)]N_3$	*IR* 520
$BH(i\text{-}C_4H_9)N(C_2H_5)_2$	*NMR* 308
BH(*sec*-$C_4H_9N(C_2H_5)_2$	*NMR* 308
BH(*tert*-$C_4H_9)N(C_2H_5)_2$	*NMR* 308
BH(*tert*-$C_4H_9)N(CH_3)_2$	*NMR* 308
BH(*tert*-$C_4H_9)N(i\text{-}C_3H_7)_2$	*NMR* 308
BH(*tert*-$C_4H_9)N(i\text{-}C_4H_9)_2$	*NMR* 308
BH(*tert*-$C_4H_9)N(n\text{-}C_4H_9)_2$	*NMR* 308
BH(*tert*-$C_4H_9)NH(CH_3)$	*NMR* 308
$BH[N(CH_3)_2]_2$	*IR, RA* 266a

$\overline{BHN(CH_3)N{=}NN}(CH_3)$ *IR, RA* 175a; *STR* 138b

$\overline{BHN(CH_3)N{=}NN}C_6H_5$ *STR* 532a, 532b

$\overline{BHN(C_6H_5)N{=}NN}C_6H_5$ *STR* 532a, 532b; *IR, NMR, UV* 277g

$HBNCH_2C_6H_5$ *IR* 140

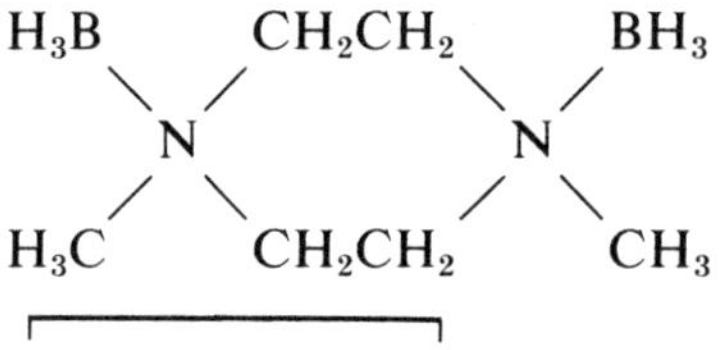

$H_3B(CH_3)N(CH_2CH_2)_2N(CH_3)BH_3$ *IR* 249a

$\overline{N(CH_3)[N(CH_3)]_3B}H$ *IR* 446b

$\overline{N(C_6H_5)[N(C_6H_5)]_3B}H$ *IR* 446b

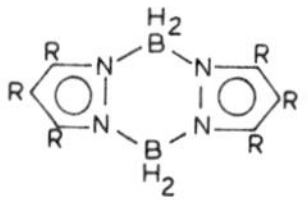

(pyrazabole structure, R-substituted, with BH_2 bridges) *IR, NMR* 771a

21.3 *Borazines* (see also the monograph [*499*])

$H_3B_3N_3H_3$ *IR* 152, 499, 381, 783; *RA* 152; *UV* 584, 598; *NMR* 382, 383, 499, 591; *STR* 294c, 499, 737

$H_3B_3N_3HD_2$ *IR* 784

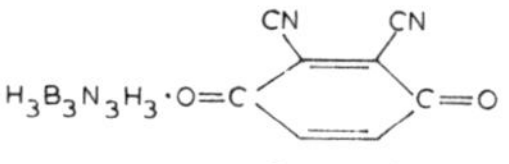

$H_3B_3N_3H_3 \cdot O{=}C_6H_2(CN)_2{=}O$ *UV* 231

$H_3B_3N_3H_3 \cdot O{=}C_6(CN)_2Cl_2{=}O$ *UV* 231

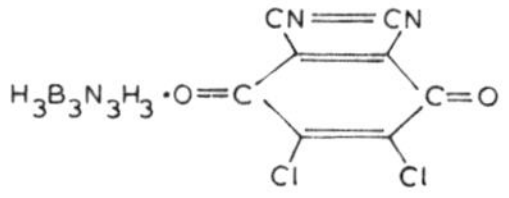

$H_3B_3N_3(CH_3)_3$ *IR* 322, 778, 782; *RA* 782; *UV* 607; *NMR* 382, 591, 620; *STR* 144

$H_3B_3N_3(C_2H_5)_3$ *IR* 50

$H_3B_3N_3[CH(C_6H_5)_2]_3$ *IR, NMR, MS* 585a

$H_3B_3N_3(C_6H_5)_3$ *IR* 45, 263a; *UV* 46, 705

$(CH_3)H_2B_3N_3(CH_3)_3$ *IR* 778

$(CH_3)_2HB_3N_3(CH_3)_3$ *IR* 778

$(NH_2)H_2B_3N_3H_3$ *IR, NMR* 600a

$(ND_2)D_2B_3N_3H_3$ *IR, NMR* 600a

$Cl_2HB_3N_3[2,6\text{-}(CH_3)_2\text{-}C_6H_3]_3$ *IR* 772; *NMR* 772

$Cl_3B_3N_3H_3$ *IR* 21, 773a, 783, 784; *UV* 357; *MW* 382, 538, 710; *STR* 146

$Cl_3B_3N_3(CH_3)_3$ *IR* 778, 783; *NMR* 238, 382

$Cl_3B_3N_3(C_2H_5)_3$ *IR* 783; *NMR* 382

$Cl_3B_3N_3(n\text{-}C_4H_9)_3$ *IR* 783

$Cl_3B_3N_3(C_6H_5)_3$ *IR* 45, 783

$(CH_3)Cl_2B_3N_3(CH_3)_3$ *IR* 778

$(CH_3)Br_2B_3N_3(CH_3)_3$ *IR* 778

$(CH_3)_2BrB_3N_3(CH_3)_3$ *IR* 778

$(n\text{-}C_4H_9)_2ClB_3N_3(CH_3)_3$ *IR* 629

$(CH_3)_3B_3N_3H_3$ *IR* 778; *RA* 266; *UV* 607; *STR* 814

$(CH_3)_3B_3N_3H_2(CH_3)$ *IR* 778

$(CH_3)_3B_3NH(CH_3)_2$ *IR* 778

$(CH_3)_3B_3NH(CH_3)_2$ *IR* 778

$(CH_3)_3B_3N_3(CH_3)_3$ *IR* 778, 783; *NMR* 382

$(CH_3)_3B_3N_3(C_2H_5)_3$ *NMR* 382

$(CH_3)_3B_3N_3(C_6H_5)_3$ *IR* 45, 103a, 783; *NMR* 382; *UV* 46

$(CH_3)_3B_3N_3(o\text{-}CH_3C_6H_4)_3$ *IR* 103a

$(CH_3)_3B_3N_3(m\text{-}CH_3C_6H_4)_3$ *IR* 103a

References p. 355

$(CH_3)_3B_3N_3$ $(p\text{-}CH_3C_6H_4)_3$	*IR* 103a
$(CH_2Cl)_3B_3N_3H_3$	*IR* 772
$(C_2H_5)_3B_3N_3H_3$	*IR* 321
$(C_2H_5)_3B_3N_3(CH_3)_3$	*IR* 629; *NMR* 382
$(C_2H_5)_3B_3N_3(C_2H_5)_3$	*IR* 783; *UV* 783; *NMR* 382
$(C_2H_5)_3B_3N_3(C_6H_5)_3$	*IR* 783; *NMR* 382
$(n\text{-}C_3H_7)_3B_3N_3H_3$	*IR* 305; *UV* 305; *NMR* 305
$(i\text{-}C_3H_7)_3B_3N_3H_3$	*IR* 305; *UV* 305; *NMR* 305
$(n\text{-}C_4H_9)_3B_3N_3H_3$	*UV* 305; *NMR* 305
$(n\text{-}C_4H_9)_3B_3N_3(CH_3)_3$	*IR* 629, 689
$(n\text{-}C_4H_9)_3B_3N_3(C_6H_5)_3$	*IR* 783
$(i\text{-}C_4H_9)_3B_3N_3H_3$	*IR* 305; *UV* 305; *NMR* 305
$(tert\text{-}C_4H_9)_3B_3N_3H_3$	*IR* 305; *UV* 305; *NMR* 305
$(n\text{-}C_6H_{13})_3B_3N_3H_3$	*IR* 305; *UV* 305; *NMR* 305
$(CH_2{=}CHCH_2)_3\text{-}B_3N_3(CH_3)_3$	*IR* 629
$(C_6H_5)_3B_3N_3H_3$	*IR* 45, 541; *UV* 46
$(C_6H_5)_3B_3N_3(CH_3)_3$	*IR* 45, 103a, 541
$(C_6H_5)_3B_3N_3(C_2H_5)_3$	*IR* 103a
$(C_6H_5)_3B_3N_3(n\text{-}C_3H_7)_3$	*IR* 103a
$(C_6H_5)_3B_3N_3(n\text{-}C_4H_9)_3$	*IR* 103a
$(C_6H_5)_3B_3N_3(C_6H_5)_3$	*IR* 783
$(o\text{-}CH_3C_6H_4)_3B_3N_3(CH_3)_3$	*IR* 103a

$(m\text{-}CH_3C_6H_4)_3B_3N_3(CH_3)_3$	*IR* 103a
$(p\text{-}CH_3C_6H_4)_3B_3N_3(CH_3)_3$	IR 103a
$(C_6H_5CH_2)_3B_3N_3H_3$	*IR* 305; *UV* 305; *NMR* 305
$(SiMe_3CH_2)_3B_3N_3(CH_3)_3$	*IR* 689
$(SiMe_3OSiMe_2CH_2)_3\text{-}B_3N_3(CH_3)_3$	*IR* 689
$(SiPh_3)_3B_3N_3(CH_3)_3$	*IR* 149, 689
$(SiPh_3)_3B_3N_3(C_6H_5)_3$	*IR* 149
$(GeR_3)_{3-n}(CH_3)_nB_3N_3\text{-}(CH_3)_3$ R = H, C_6H_5; $n = 0, 1, 2$	*IR* 15c
$(NH_2)_3B_3N_3H_3$	*IR* 255
$(NH_2)_3B_3N_3(C_2H_5)_3$	*IR* 542
$(CH_3NH)_3B_3N_3(CH_3)_3$	*IR* 21
$(C_2H_5NH)_3B_3N_3(C_2H_5)_3$	*IR* 21
$(i\text{-}C_3H_7NH)_3B_3N_3(i\text{-}C_3H_7)_3$	*IR* 21
$(n\text{-}C_4H_9NH)_3B_3N_3(n\text{-}C_4H_9)_3$	*IR* 21
$(i\text{-}C_4H_9NH)_3B_3N_3(i\text{-}C_4H_9)_3$	*IR* 21
$(tert\text{-}C_4H_9)_3B_3N_3(tert\text{-}C_4H_9)_3$	*IR* 21
$(C_6H_5NH)_3B_3N_3(C_6H_5)_3$	*IR* 21
$(C_6H_5CH_2NH)_3B_3N_3\text{-}(CH_2C_6H_5)_3$	*IR* 21

References p. 355

$(NRR')_3B_3N_3H_3$ — *IR* 254, 255, 257, 542
$[R,R' = CH_3, C_2H_5,$
$n\text{-}C_3H_7, i\text{-}C_3H_7, n\text{-}C_4H_9,$
$i\text{-}C_4H_9, \overline{CH_2(CH_2)_4CH},$
$C_6H_5, CH_2C_6H_5]$

$(N_2HMe_2)_3B_3N_3(CH_3)_3$ — *IR* 542

$(N_2HMe_2)_3B_3N_3(C_2H_5)_3$ — *IR* 542

$(R'O)_3B_3N_3R_3$ — *IR* 21, 255, 542
$[R = H, CH_3, C_2H_5;$
$R' = CH_3, C_2H_5, n\text{-}C_3H_7,$
$i\text{-}C_3H_7, n\text{-}C_4H_9, i\text{-}C_4H_9,$
$C_6H_5, (CH_3)_3Si, (C_6H_5)_3Si]$

21.4 Mono- and diboranes with B–P bonds

$BH_3{\cdot}PH_3$ — *NMR* 175e, 175f, 625e; *IR, Ra* 625e

$BH_3{\cdot}PH(CH_3)_2$ — *NMR* 591; *IR* 106

$BH_3{\cdot}P(CH_3)_3$ — *NMR* 329a

$BH_3{\cdot}P(C_6H_5)_3$ — *IR, NMR* 235d

$BH_3{\cdot}PH_2(SiH_3)$ — *NMR* 175f

$BH_3{\cdot}PH_2(GeH_3)$ — *NMR* 175f

$BH_3{\cdot}PH_2(GeD_3)$ — *NMR* 175f

$BH_3{\cdot}PD_2(GeH_3)$ — *NMR* 175f

$BH_3{\cdot}PF_3$ — *RA* 759

$BD_3{\cdot}PF_3$ — *RA* 759

$BH_3{\cdot}PHF_2$ — *NMR, MS* 625c

$BH_3{\cdot}PF_2(NHCH_3)$ — *IR* 432

$(BH_3)_x{\cdot}P_4O_6$ — *NMR* 613a
$(x = 1, 2, 3, 4)$

$BH_3 \cdot PF(NHCH_3)_2$	*IR* 432
$BH_3 \cdot P(NHCH_3)_3$	*IR* 432
$BH_3 \cdot PF_2N(CH_3)_2$	*IR* 432
$BH_3 \cdot PF[N(CH_3)_2]_2$	*IR* 432
$BH_3 \cdot P(NH_2)_3$	*STR* 553
$BH_3 \cdot P[N(CH_3)_2]_3$	*IR* 610
$BH_3 \cdot P[N(n\text{-}C_4H_9)_2]_3$	*IR* 610
$BH_3 \cdot P[\overline{N(CH_2)_3CH_2}]_3$	*IR* 610
$BH_3 \cdot P_2F_4$	*IR*, *NMR*, *MS* 532d
$B_2H_4(PF_3)_2$	*IR* 164, 175d
$\overset{\ominus}{B}H_3CH_2\overset{\oplus}{P}(C_6H_5)_3$	*NMR* 303
$\overset{\ominus}{B}H_2(CH_3)CH_2\overset{\oplus}{P}$-$(C_6H_5)_3$	*NMR* 303
$BH_2Cl \cdot PH_3$	*NMR* 175e
$BH_2Br \cdot PH_3$	*NMR* 175b, 175e
$BHCl_2 \cdot PH_3$	*NMR* 175e
$BHBr_2 \cdot PH_3$	*NMR* 175e
$BHClBr \cdot PH_3$	*NMR* 175e
$[BH_2P(CH_3)_2]_3$	*IR* 139; *RA* 139
$[BX_2P(CH_3)_2]_3$ ($X = CH_3$, F, Cl, Br)	*IR* 139; *RA* 139
$Br_nH_{6-n}B_3P_3(C_6H_5)_6$ ($n = 1$–6)	*IR* 250a
$I_nH_{6-n}B_3P_3(C_6H_5)_6$ ($n = 1$–3)	*IR* 250a

References p. 355

$Br_3IH_2B_3P_3(C_6H_5)_6$ — *IR* 250a

$[BH_2P(CH_3)_2]_4$ — *IR* 139; *RA* 139

$[BH_2P(CH_3)_2]_n \cdot N(C_2H_5)_3$ — *IR* 48, 779

$(C_6H_5)_3P \cdot B_2H_4 \cdot P(C_6H_5)_3$ — *NMR* 277

$NH_4^{\oplus}[BH_3PH_2BH_3]^{\ominus}$ — *IR*, *NMR* 259a

21.5 Mono- and diboranes with B–C bonds

$B_2H_5(CH_3)$ — *IR* 455, 479b

$B_2H_5(CD_3)$ — *IR* 455

$B_2D_5(CH_3)$ — *IR* 455

$B_2D_5(CD_3)$ — *IR* 455

$B_2H_5(C_2H_5)$ — *IR* 456

$B_2H_5(C_2D_5)$ — *IR* 455

$B_2D_5(C_2D_5)$ — *IR* 455

$B_2H_4(CH_3)_2$(sym.) — *IR* 457

$B_2H_4(CH_3)_2$(asym.) — *IR* 458

$B_2H_4(CD_3)_2$(asym.) — *IR* 458

$B_2D_4(CH_3)_2$(sym.) — *IR* 457

$B_2D_4(CH_3)_2$(asym.) — *IR* 458

$B_2D_4(CD_3)_2$(asym.) — *IR* 458

$B_2H_4(C_2H_5)_2$(sym.) — *IR* 426, 427, 457, 458, 721

$B_2H_4(C_2H_5)_2$(asym.) — *IR* 458, 721

$B_2H_4(C_2D_5)_2$(sym.) — *IR* 457

$B_2H_4(C_2D_5)_2$(asym.) — *IR* 458

$B_2D_4(C_2H_5)_2$(sym.)	*IR* 457
$B_2D_4(C_2H_5)_2$(asym.)	*IR* 458
$B_2D_4(C_2D_5)_2$(sym.)	*IR* 457
$B_2D_4(C_2D_5)_2$(asym.)	*IR* 458
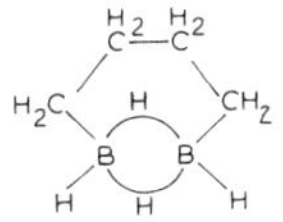	*IR* 791, 859c; *NMR* 791, 859c; *MS* 791
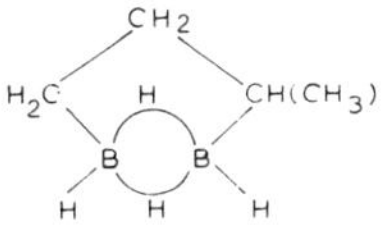	*IR* 791; *NMR* 791, 859c; *MS* 791
$B_2H_3(CH_3)_3$	*IR* 459
$B_2D_3(CH_3)_3$	*IR* 459
$B_2H_3(C_2H_5)_3$	*IR* 459
$B_2H_2(CH_3)_4$	*IR* 459; *RA* 459
$B_2H_2(C_2H_5)_4$	*IR* 459
$B_2D_2(C_2H_5)_4$	*IR* 459
$B_2D_2(C_2H_5)_4$	*IR* 459
$B(CH_3)_3$	*IR* 454
$B(CD_3)_3$	*IR* 454
$B(C_2H_5)_3$	*IR* 454
$B(C_2D_5)_3$	*IR* 454
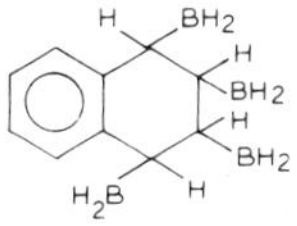	*IR* 867
$BH_3 \cdot CO$	*IR* 52, 147, 148; *RA* 757, 758; *calculation of* ν 53, 756, *STR* 16b

$BD_3 \cdot CO$	*IR* 52, 147; *RA* 757, 758; *calculation of* ν 53, 756
$BH_2(CN)$	*IR* 206

21.6 Mono- and diboranes with transition metal–B bonds

$(C_2H_5)_4N^{\oplus}[(CO)_5Mn \cdot\cdot\rightarrow BH_3]^{\ominus}$	*IR* 583 *STR* 385a
$(C_2H_5)_4N^{\oplus}$ $[(C_6H_5)_3P(CO)_4Mn \cdot\cdot\rightarrow BH_3]^{\ominus}$	*ES* 583
$Na^{\oplus}[(C_6H_5)_3P(CO)_4Mn \cdot\cdot\rightarrow BH_3]^{\ominus}$	*IR* 583
$Na^{\oplus}$ $[(CO)_5Re \cdot\cdot\rightarrow BH_3]^{\ominus}$	*IR* 583
$(C_2H_5)_4N^{\oplus}[(CO)_5Re \cdot\cdot\rightarrow BH_3]^{\ominus}$	*IR* 583; *ES* 583
$(C_2H_5)_4N^{\oplus}[(CO)_5Re \cdot\cdot\rightarrow (BH_3)_2]^{\ominus}$	*ES* 583
$(C_4H_9)_4P^{\oplus}[(CO)_5Re \cdot\cdot\rightarrow BH_3]^{\ominus}$	*IR* 583
$(CO)_{10}HMn_3(H_3B{-}BH_3)_2$	*STR* 385a
$(CO)_{10}Co_3BH_2N(C_2H_5)_3$	*STR* 405c

21.7 Boronium compounds

$(NH_3)_2BH_2^{\oplus}$	*NMR* 604c, 604d
$[N(CH_3)_3]_2BH_2^{\oplus}$	*NMR* 604c, 604d
$[N(CH_3)_2(C_2H_5)]BH_2^{\oplus}$	*NMR* 604c, 604d
$[N(CH_3)(C_2H_5)_2]BH_2^{\oplus}$	*NMR* 604c, 604d
$[(NH_3)_2BH_2]^{\oplus}BH_4^{\ominus}$	*RA* 700, 762; *NMR* 762
$[(NH_2CH_3)_2BH_2]^{\oplus}BH_4^{\ominus}$	*IR* 43b, 371e; *STR* 371e
$[NH(CH_3)_2]_2BH_2^{\oplus}BH_4^{\ominus}$	*IR* 371e; *STR* 371e
$[(NH_3)_2BH_2]^{\oplus}Cl^{\ominus}$	*STR* 251, 558
$[(NH_2CH_3)_2BH_2]^{\oplus}Cl^{\ominus}$	*IR* 43b, 371e; *STR* 371e
$[NH(CH_3)_2]_2BH_2^{\oplus}Cl^{\ominus}$	*IR* 371e; *STR* 371e

$[(NH_3)_2BH_2]^{\oplus}I^{\ominus}$	*NMR* 329a
$(CH_3C_5H_4N)_2BH_2^{\oplus}PF_6^{\ominus}$	*IR, UV, NMR* 537b
$(CH_3C_5H_4N)_2BH_2^{\oplus}$-$AsF_6^{\ominus}$	*IR, UV, NMR* 537b
$[(CH_3)_2C_5H_3N]_2$-$BH_2^{\oplus}PF_6^{\ominus}$	*IR, UV, NMR* 537b
$[(NH_3)_2BH_2]^{\ominus}B_3H_8^{\ominus}$	*STR* 422, 590
$[(CH_3)_2SO]_2BH_2^{\oplus}Cl^{\ominus}$	*RA, NMR* 496
$[(CH_3)_3NBH_2]_2SCH_3^{\ominus}$-$PF_6^{\ominus}$	*NMR* 625b
$[(CH_3)_3NBH_2SCH_3]_2$-$BH_2^{\oplus}PF_6^{\ominus}$	*NMR* 625b

21.8 Boranates: 1- and 2-membered skeletons (⋆ for comparison)

$\cdot BH_3^{\ominus}$	*MS* 358a; *ESR* 756a
$LiBH_4$	*IR* 94, 100, 200, 265, 358, 449, 603, 685, 762, 776, 857, cit. 5; *RA* 100, 685; *NMR* 164a, 762; *ESR* 136b
$NaBH_4$	*IR* 94, 230, 372c, 400, 401, 684, 762; *NMR* 236, 372c, 563, 591; *STR* 360, 750; *ESR* 136b
$BH_4^{\ominus}$	*MS* 358a
$NaBH_3(CH_3)$	*IR* 94
$NaBH_3(NMe_2)$	*NMR* 395b
$NaBH(NCS)_3$	*IR* 404, 405
⋆ $NaAlH(NCS)_3$	*IR* 404
NaB_2H_6	*STR* 360
NaB_2H_7	*NMR* 408; *STR* 360
$B_2H_7^{\ominus}$, $B_2H_5^{\ominus}$, $B_2H_3^{\ominus}$,	*MS* 358a

References p. 355

KBH_4	*RA* 245; *NMR* 230; *STR* 229; *ESR* 136b
$RbBH_4$	*NMR* 230
$BeH_2 \cdot B_2H_6$	*IR* 94; *STR* 16a, 144a, 753
$H_2Al(BH_4) \cdot O(C_2H_5)_2$	*IR* 18a
$H_2Al(BH_4) \cdot THF$	*IR* 551b
$H_2Al(BH_4) \cdot 2THF$	*IR* 551b
$HAl(BH_4)_2 \cdot nO(C_2H_5)_2$	*IR* 18a
$HAl(BH_4)_2 \cdot THF$	*IR* 551b
$HAl(BH_4)_2 \cdot 2THF$	*IR* 551b
$(C_4H_9O)_2AlBH_4$	*IR* 18a
$(LiBH_4)_2 \cdot AlH_3 \cdot nO(C_2H_5)_2$	*IR* 18a
$LiBH_4 \cdot AlH_3 \cdot nO(C_2H_5)_2$	*IR* 18a
$Al(BH_4)_3 \cdot O(CH_3)_2$	*IR*, *NMR* 53b
$Al(BH_4)_3 \cdot O(C_2H_5)_2$	*IR*, *NMR* 53b
$Al(BH_4)_3 \cdot THF$	*IR* 551b
$Al(BH_4)_3 \cdot 2THF$	*IR* 551b
$Al(BH_4)_3$	*IR* 93, 94, 156a, 372c, 494a; *RA* 201; *NMR* 156a, 372c, 444b, 494a, 591
$Al(BD_4)_3$	*RA* 201
$Al(BH_4)_3 \cdot 6NH_3$	*STR*, *NMR* 494b
$Al(BH_4)_3 \cdot N(CH_3)_3$	*STR* 35a, 35d; *NMR* 53b
$Al(BH_4)_3 \cdot P(CH_3)_3$	*IR*, *NMR* 53b
$Al(BH)_3 \cdot As(CH_3)_3$	*IR*, *NMR* 53b

$Al_3B_3(NMe_2)_7H_5$ — *IR* 290b

$(NH_3)_4AlH_2^{\oplus}$-$(NH_3)_2BH_2^{\oplus}(BH_4^{\ominus})_2$ — *NMR*, *STR* 494b

$(C_6H_5)_2I^{\oplus}BH_4^{\ominus}$ — *IR* 324

$(CH_3)_3S^{\oplus}BH_4^{\ominus}$ — *IR* 323

$(C_6H_5)_3S^{\oplus}BH_4^{\ominus}$ — *IR* 324

$[(NH_3)_2BH_2]^{\oplus}BH_4^{\ominus}$ — *RA* 700, 762; *NMR* 762

$(C_6H_5)_4P^{\oplus}BH_4^{\ominus}$ — *IR* 324

$[(C_6H_5)_3P]_2AgBH_4$ — *IR*, *NMR* 372c

$[(C_6H_5)_3P]_2CuBH_4$ — *IR* 372c, 467a, 467c; *STR* 467b; *NMR* 372c

$Zr(BH_4)_2$ — *IR*, *NMR* 372c

$(C_5H_5)_2Zr(BH_4)_2$ — *IR* 372b, 372c, 537a; *NMR* 372b, 372c

$Hf(BH_4)_2$ — *IR*, *NMR* 372c

$(C_4H_9O)Ti(BH_4)_2 \cdot THF$ — *IR* 372a

Ni(*Schiff base*)BH_4 — *IR* 153a

$Na^{\oplus}[BH_3(PH_2BH_3)]^{\ominus}$ — *IR*, *RA* 336a

21.9 Polyhedral boron hydrides: 3-membered skeletons

$B_3H_7 \cdot O(C_2H_5)_2$ — *NMR* 591

$B_3H_7 \cdot THF$ — *NMR* 620a

$B_3H_7 \cdot NH_3$ — *STR* 251, 557, 558

$B_3H_7 \cdot N(CH_3)_3$ — *STR* 559; *NMR* 620a

$B_3H_8^{\ominus}Na^{\oplus}$ — *IR*, *STR* 360; *NMR* 591

$(B_3H_8^{\ominus})_2[Me_3NCH_2\text{-}]_2^{2\ominus}$ — *IR* 779b

References p. 355

$B_3H_8^{\ominus}N(CH_3)_4^{\oplus}$	*NMR* 276
$B_3H_8^{\ominus}Cu[P(C_6H_5)_3]_2$	*IR* 467c
$B_3H_8M(CO)_4^{\ominus}$ (M = Cr, Mo, W)	*IR* 405d
$B_3H_8^{\ominus}[(NH_3)_2BH_2]^{\oplus}$	*STR* 422, 590
$B_3H_{10}^{\ominus}$, $B_3H_9^{\ominus}$, $B_3H_8^{\ominus}$, $B_3H_7^{\ominus}$, $B_3H_6^{\ominus}$	*MS* 330a, 358a

21.10 Polyhedral boron hydrides: 4-membered skeletons (derivatives of B_4H_{10}

B_4H_{10}	*IR* 438; *NMR* 356a, 591, 617, 846; *STR* 378, 554, 555; *MS* 35a
$B_4H_9D^{\mu}$	*IR* 558a; *NMR*, *MS* 558c
$B_4H_8D_2$	*IR, NMR* 558b
$B_4H_2D_8$	*IR, NMR* 558b
$^{10}B_4H_{10}$	*NMR, MS* 558c
$^{10}B_4H_9D$	*NMR, MS* 558c
$^{10}B_4D_{10}$	*NMR, MS* 558c
B_4H_8-2,4-(-CD_2CD_2-)	*IR, NMR, MS* 845a
$B_4H_8 \cdot CO$	*NMR, MS* 558b
$B_4H_8 \cdot PF_3$	*NMR, MS* 558b
$B_4H_8 \cdot PF_2N(CH_3)_2$	*STR* 440a
$B_4H_4(C_2H_5)_4 \cdot CO$	*IR, NMR* 724
$B_4H_{11}^{\ominus}$, $B_4H_{10}^{\ominus}$, $B_4H_9^{\ominus}$	*MS* 358a
$B_4H_8^{\ominus}$, $B_4H_7^{\ominus}$	*MS* 330a, 358a

21.11 Polyhedral boron hydrides: 5-membered skeletons (derivatives of B_5H_9 and B_5H_{11}

B_5H_9	*IR* 221, 366, 488; 600; *NMR* 112a, 436, 566b, 567c, 591, 643; *MW* 221, 364, 365, 848; *MS* 330a; *MO* 353, 528; *STR* 177, 180, 325, 364, 446, 474, 709
$B_5H_8D^{\mu}$	*IR*, *NMR* 237a
B_5D_9	*IR* 366; *MW* 364
B_5H_8-1-CH_3	*IR* 221, 628; *NMR* 566b, 567d
B_5H_8-2-CH_3	*NMR* 112a, 566b, 567d
B_5H_8-1-C_2H_5	*IR* 221, 628; *RA* 221; *NMR* 566b, 567d
B_5H_8-2-C_2H_5	*NMR* 566b, 567d
B_5H_8-1-$CH(CH_3)_2$	*IR* 628
B_5H_8-1-$CH(CH_3)(C_2H_5)$	*IR* 628
B_5H_7-1,2-$(CH_3)_2$	*IR*, *NMR* 567d, 568a, 771e; *STR* 235a
B_5H_7-2,3-$(CH_3)_2$	*IR*, *NMR* 567d, 568a
B_5H_7-1-CH_3-2-$CH(CH_3)$-(C_2H_5)	*IR* 628
B_5H_6-1,2,3-$(CH_3)_3$	*IR*, *NMR* 567d
B_5H_8-2-F	*IR*, *NMR* 112a
B_5H_8-1-Cl	*NMR* 237d, 567d
B_5H_8-2-Cl	*NMR* 112a, 237d, 567, 567d; *IR* 112a
B_5H_8-1-Br	*IR* 221; *NMR* 567d, 650; *MS* 291
B_5H_8-2-Br	*NMR* 112a, 567d; *IR* 112a
B_5H_8-1-I	*NMR* 567d, 650; *MS* 291; *STR* 290b
B_5H_8-2-I	*IR*, *NMR* 112a

References p. 355

B_5H_7-1-Br-2-CH_3	*NMR* 567d
B_5H_7-1,2-Cl_2	*NMR* 237d
B_5H_8-2-(OCH_3)	*IR, NMR, MS* 237f
B_5H_8-μ-EH_3 [E = Si, Ge]	*IR, NMR* 237e
B_5H_8-μ-$E(CH_3)_3$ [E = Si, Ge, Sn, Pb]	*IR, NMR* 237b, 137e
B_5H_8-2,3(μ)-$P(CH_3)_2$	*IR, NMR* 116a
B_5H_8-2,3(μ)-$P(CH_3)(CF_3)$	Form A and B *IR, NMR* 116a
B_5H_8-1-$P(CF_3)_2$	*IR, NMR* 116a
B_5H_8-2-$Mn(CO)_5$	*NMR* 237c
B_5H_8-2-$Re(CO)_5$	*NMR* 237c
$B_5H_8^{\ominus}Li^{\oplus}$	*NMR* 237a
B_5H_{11}	*IR* 438, 488; *NMR* 591, 643, 847; *MO* 353; *STR* 445, 446, 474
$B_5H_{10}(C_2H_5)$	*IR* 488
$B_5H_7^{\ominus}$, $B_5H_{11}^{\ominus}$, $B_5H_{10}^{\ominus}$	*MS* 330a, 358a

21.12 Polyhedral boron hydrides: 6-membered skeletons

B_6H_{10}	*NMR* 249b, 848; *STR* 205, 346, 353
neo-B_6H_{10}	*NMR* 249b
B_6H_{12}	*IR* 482; *NMR* 482; *STR* 482

21.13 Polyhedral boron hydrides: 7-membered skeletons

B_7H_{13}?	*MS* 648
$B_7H_7^{2\ominus}(Cs^{\oplus})_2$	*IR, UV, NMR* 405b
$B_7H_7^{2\ominus}[Zn(NH_3)_3(H_2O)]$	*IR* 405b

21.14 Polyhedral boron hydrides: 8-membered skeletons

B_8H_{12}	*MS* 202, 691; *STR* 203; *NMR* 171c
B_8H_{14}	*NMR* 171c
$B_8H_8^{2\ominus}(Rb^{\oplus})_2$	*IR* 405b
$B_8H_8^{2\ominus}(Cs^{\oplus})_2$	*NMR* 405b
$B_8H_8^{2\ominus}Zn(NH_3)_4^{2\oplus}$	*STR* 405b
$B_8Br_6H_2^{2\ominus}[N(CH_3)_4^{\oplus}]_2$	*IR* 405b
$\cdot B_8H_8^{\ominus}$	*ESR* 405b

21.15 Polyhedral boron hydrides: 9-membered skeletons

$C_2H_5NH_2B_8H_{11}NHC_2H_5$	*STR* 465, 465a
B_9H_{13}?	*MS* 561
$B_9H_{14}^{\ominus}$	*STR* 395e
B_9H_{15}	*MS* 170a; *STR* 203; *MO* 353
i-B_9H_{15}	*NMR* 171d; *STR* 395e
$(B_9H_{12}S)M[P(C_6H_5)_3]_3$ (M = Cu, Ag, Au)	*IR*, *UV*, *STR* 405d
$B_9H_9^{2\ominus}(Cs^{\oplus})_2$	*UV* 405a
$B_9H_9^{2\ominus}(Rb^{\oplus})_2$	*IR*, *Ra* 405a; *STR* 281c
$B_9H_9^{2\ominus}[S(CH_3)_3^{\ominus}]_2$	*UV* 405a; *NMR* 405a

21.16 Polyhedral boron hydrides: 10-membered skeletons

I. $B_{10}H_{14}$ *derivatives*

$B_{10}H_{14-n}D_n$	*IR* 184, 296, 461, 522, 571, 600; *UV* 286; *NMR* 56, 395d, 536c, 591, 594, 643, 693, 708, 845, 849a; *STR* 387, 709, 766a; *MO* 353, 529; *MS* 561, 693

References p. 355

$B_{10}H_4$-μ, μ'', μ'', μ''', 5, 6, 7, 8, 9, 10-D_{10}	*NMR* 779a
$^{10}B^nB_9H_{14}$	*NMR*, *MS* 487b
$B_{10}H_{13}$-1-Cl	*NMR* 849a
$B_{10}H_{13}$-2-Cl	*NMR* 849a
$B_{10}H_{13}$-2-Br	*NMR* 536c, 650
$B_{10}H_{13}$-1-I(?)	*NMR* 650; *STR* 686b
$B_{10}H_{13}$-2-I	*NMR* 650, 849; *IR* 48
$B_{10}H_{12}$-2,4-I_2	*NMR* 591, 849; *UV* 715b
$B_{10}H_{13}$-μ-OC_2H_5	*NMR* 844a
$B_{10}H_{13}(NC_5H_5)$	*ES* 319a
$B_{10}H_{12}[S(CH_3)_2]_2$	*IR* 571; *NMR* 571
$B_{10}H_{12}[S(C_2H_5)_2]_2$	*NMR* 371b, 536c
$B_{10}H_{11}$-2-Br$[S(C_2H_5)_2]_2$	*NMR* 536c
$B_{10}H_{12}(NCCH_3)_2$	*IR* 371b, 571; *NMR* 571; *STR* 608
$B_{10}H_{11}$-2-Br$(NCCH_3)_2$	*NMR* 371b
$B_{10}H_{11}$-2-Br$(NCC_2H_5)_2$	*NMR* 371b
$B_{10}H_{12}[N(C_2H_5)_2CN]_2$	*IR* 571; *NMR* 571
$B_{10}H_{12}[C(O)N(CH_3)_2]_2$	*IR* 571; *NMR* 571
$B_{10}H_{12}[N(C_2H_5)_3]_2$	IR 571; *NMR* 571
$B_{10}H_{12}\cdot 2P(C_6H_5)_3$	*IR* 571; *NMR* 571
$B_{10}H_{12}\cdot 2NC_5H_4Br$	*IR* 571; *NMR* 571
$B_{10}H_{12}(NC_5H_5)_2$	*IR* 571; *NMR* 571; *ES* 319a [see also $B_{10}H_{10}(HNC_5H_5)_2$!]

$B_{10}H_{13}$-2-CH_3	*NMR* 845
$B_{10}H_{13}$-6-CH_3	*NMR* 182, 183
$B_{10}H_{13}$-5-CH_3	*NMR* 182, 183
$B_{10}H_{13}$-5-C_2H_5	*NMR* 182; *IR* 48
$B_{10}H_{13}$-6-C_2H_5	*NMR* 183; *IR* 48
$B_{10}H_{13}(CH_2C_6H_5)$	*NMR* 708; *IR* 708
$B_{10}H_{12}(C_2H_5)_2$	*IR* 48
$B_{10}H_{12}$-1,2-$(CH_3)_2$	*NMR* 845
$B_{10}H_{12}$-2,4-$(CH_3)_2$	*NMR* 845
$B_{10}H_{12}$-6,5-(8)-$(CH_3)_2$	*NMR* 183
$B_{10}H_{12}$-6,9-$(CH_3)_2$	*NMR* 183
$B_{10}H_{11}$-1,2,3-$(CH_3)_3$	*NMR* 845
$B_{10}H_{11}$-1,2,4-$(CH_3)_3$	*NMR* 845
$B_{10}H_{10}$-1,2,3,4-$(CH_3)_4$	*NMR* 845
$B_{10}H_{13}(MgI)$	*IR* 56; *UV* 56; *NMR* 56
$B_{10}H_{12}Na_2$	*IR* 843a
$B_{10}H_{12}AlH^{\ominus}(CH_3)_3$-$NH^{\oplus}$	*IR* 278a
$(B_{10}H_{12})_2Hg^{2\ominus}Mg^{2\oplus}$	*IR* 278b
$(B_{10}H_{12})_2M[N(CH_3)_4]_2$ (M = Co, Ni, Pd, Pt, Zn)	*STR, NMR, UV* 405e
$B_{10}H_{12}Cd \cdot 2O(C_2H_5)_2$	*IR* 277a
$(B_{10}H_{12})_2Zn \cdot 4O(C_2H_5)_2$	*IR, UV, NMR* 277d
$(B_{10}H_{12})_2Mg$	*IR* 277d

References p. 355

$B_{10}H_{13}Na$	*IR* 15a, 15b, 56; *UV* 56; *NMR* 56
$B_{10}H_{13}Na \cdot x$ solvent	*IR* 15a, 277b
$(B_{10}H_{13}\text{-2-Br})Rb$	*NMR* 371b
$B_{10}H_{13}NH_3^{\ominus}Cs^{\oplus}$	*UV, NMR* 534a
$B_{10}H_{15}Na$	*NMR* 653
$B_{10}H_{15}^{\ominus}N(C_4H_9)_4^{\oplus}$	*IR* 625d
$B_{10}H_{14}^{2\ominus}[N(C_4H_9)_4^{\oplus}]_2$	*IR* 625d
$B_{10}H_{13}^{\ominus}N(CH_3)_4^{\oplus}$	*IR* 56, 277h; *UV* 56
$B_{10}H_{13}^{\ominus}NH(C_2H_5)_3^{\oplus}$	*IR* 277h
$B_{10}H_{13}^{\ominus}N(C_4H_9)_4^{\oplus}$	*IR, UV* 625d
$B_{10}H_{13}[N(C_2H_5)_2H]^{\ominus}$ $[N(C_2H_5)_2H_2]^{\oplus}$	*NMR* 274
$B_{10}H_{15}^{\ominus}$	*IR* 187
$B_{10}H_{16}$	*MO* 528

21.17 Polyhedral boron hydrides: 10-membered skeletons II. $B_{10}H_{10}^{2\ominus}$ *derivatives*

$B_{10}H_{10}^{2\ominus}(H^{\oplus})_2 \cdot 4H_2O$	*IR* 534
$B_{10}H_{10}^{2\ominus}(H^{\oplus})_2 \cdot 2H_2O$	*IR* 534
$B_{10}H_{10}^{2\ominus}(Cs^{\oplus})_2$	*NMR* 189b
$B_{10}H_{10}^{2\ominus}N(CH_3)_4^{2\oplus}$	*IR* 474a; *NMR* 189b, 474a
$B_{10}H_{10}^{2\ominus}[NH(C_2H_5)_3^{\oplus}]_2$	*IR* 571; *NMR* 571
$B_{10}H_{10}^{2\ominus}(C_5H_5NH^{\oplus})_2$	*IR* 534; *UV* 534
$B_{10}H_{10}^{2\ominus}(Cu^{\oplus})_2$	*STR* 171a
$B_{10}H_2Cl_6[S(CH_3)_2]_2$	*IR* 412

$B_{10}H_2Cl_8^{2\ominus}(NH_4^{\oplus})_2$	*UV* 413
$B_{10}Cl_8(OH)_2^{2\ominus}$-$[N(CH_3)_4^{\oplus}]_2$	*IR, UV* 413, 416
1,10-$B_{10}Cl_8[COC_6H_4N$-$(CH_3)_2]_2^{2\ominus}(H^{\oplus})_2$	*NMR, UV* 412a
1,10-$B_{10}Cl_8(CH_2NH_3)_2$	*IR, UV* 411d
$B_{10}Cl_9(OH)^{2\ominus}$-$[N(CH_3)_4^{\oplus}]_2$	*UV* 416; *IR* 416
$B_{10}Cl_9(OCOC_6H_5)^{2\ominus}$-$[N(CH_3)_4]_2$	*IR* 416
$B_{10}Cl_9(COC_6H_5)^{2\ominus}$-$[N(CH_3)_4^{\oplus}]_2$	*IR* 416
$B_{10}H_7Br[S(CH_3)_2]_2$	*IR* 412
$B_{10}H_5Br_3[S(CH_3)_2]_2$	*UV* 412
$B_{10}H_2Br_7^{2\ominus}$-$[N(CH_3)_4^{\oplus}]_2$	*UV* 413
$B_{10}Br_9(COC_6H_5)^{2\ominus}$-$[N(CH_3)_4^{\oplus}]_2$	*IR* 416
$B_{10}Br_{10}^{2\ominus}(Cs^{\oplus})_2$	*UV* 413
$B_{10}H_8I_2^{2\ominus}[N(CH_3)_4^{\oplus}]_2$	*UV* 413
$B_{10}H_7I_3^{2\ominus}(Cs^{\oplus})_2$	*UV* 413
$B_{10}H_9(OH)^{2\ominus}$-$[N(CH_3)_4^{\oplus}]_2$	*IR* 416
$B_{10}H_8(OH)_2^{2\ominus}$-$[N(CH_3)_4^{\oplus}]_2$	*IR* 416
$B_{10}H_9(OCHO)^{2\ominus}(Cs^{\oplus})_2$	*IR* 416
$B_{10}H_9(OCOC_6H_5)^{2\ominus}$-$[N(CH_3)_4^{\oplus}]_2$	*IR* 416

References p. 355

$B_{10}H_9(OCH_2CH_2$-$OCH_3)^{2\ominus}[N(CH_3)_4^{\oplus}]_2$	*IR* 416
$B_{10}H_9]OCH{=}N$-$(CH_3)_2]^{\ominus}Cs^{\oplus}$	*IR* 416; *UV* 416; *NMR* 416
$B_{10}H_9[S(CH_3)_2]^{\ominus}Na^{\oplus}$	*NMR* 412
$B_{10}H_8$-1,6-$[S(CH_3)_2]_2$	*NMR* 412
$B_{10}H_8$-1,10-$[S(CH_3)_2]_2$	*NMR* 412
$B_{10}H_8$-2,7(8)-$[S(CH_3)_2]_2$	*NMR* 412
$B_{10}H_8[S(CH_3)_2]$-$(COOH)^{\ominus}N(CH_3)_4^{\oplus}$	*IR* 333, 334
$B_{10}H_8$-1,6-$[S(CH_3)_2](CO)$	*IR* 333, 334; *NMR* 333, 334
$B_{10}H_8$-1,6-$[S(CH_3)_2]$-(NH_3)	*IR* 333, 334; *NMR* 335; *STR* 333, 334
$B_{10}H_8$-1,6-$[S(CH_3)_2]$-$(OH)^{\oplus}N(CH_3)_4^{\oplus}$	*IR* 333, 334; *STR* 333, 334
$B_{10}H_7[S(CH_3)_2]$-$(COOC_2H_5)(N_2)$	*IR* 333, 334; *UV* 333, 334
$B_{10}H_7[S(CH_3)_2](NH_2)$-$(N_2)$	*UV* 333, 334
$B_{10}H_8$-1,10-$[S(CH_3)_2](N_2)$	*IR* 412; *UV* 412; *NMR* 412
$B_{10}H_8[S(CH_3)_2](NO)$	*UV* 412
$B_{10}H_8[S(CH_3)_2](NC_5H_5)$	*UV* 412
$B_{10}H_8$-1,10-$[S(CH_3)_2]$-$[N(CH_3)_3]$	*IR*, 412; *NMR* 412
$B_{10}H_8$-1,6-$[S(CH_3)_2]$-$[NH(CH_3)_2]$	*NMR* 412
$B_{10}H_8$-1,6-$[S(CH_3)_2]$-$[N(CH_3)_3]$	*NMR* 412

$B_{10}H_9$-2-$(NH_3)^{\oplus}$	*NMR* 335
$B_{10}H_9$-2-$[(CH_3)_2NH]^{\ominus}$	*NMR* 335
$B_{10}H_9$-2-$[(CH_3)_3N^{\ominus}]$	*NMR* 335
$B_{10}H_8$-2,3-$(NH_3)_2$	*NMR* 335
$B_{10}H_8$-2,4-$(NH_3)_2$	*NMR* 335
$B_{10}H_8$-2,7(8)-$(NH_3)_2$	*NMR* 335
$B_{10}H_8$-1,6-(NH_3)-$[(CH_3)_3N]$	*IR* 333
$B_{10}H_8$-2,3-$[(CH_3)_3N]_2$	*IR* 333; *NMR* 333, 335
$B_{10}H_8$-2,4-$[(CH_3)_3N]_2$	*NMR* 331, 335
$B_{10}H_8$-1,6-$[(CH_3)_3N]_2$	*NMR* 333
$B_{10}H_8$-2,7(8)-$[(CH_3)_3N]_2$	*NMR* 335
$B_{10}H_8$-2,7(8)-$[(CH_3)_2$-$N(CH_2Cl)]_2$	*NMR* 331
$B_{10}H_8$-1,6-$[(CH_3)_3N](N_2)$	*IR* 333; *UV* 333
$B_{10}H_8[(CH_3)_3N](CO)$	*IR* 334; *NMR* 334
$B_{10}H_7(N_2)(CN)_2{}^{\ominus}$-$N(CH_3)_4{}^{\oplus}$	*UV* 333
$B_{10}H_8$-1,10-$(CN)_2{}^{2\oplus}$ $(Cs^{\oplus})_2$	*IR* 333; *NMR* 333
$B_{10}H_8(N_2)_2$	*MS* 502b
$B_{10}H_8(CO)_2$	*IR* 418
$B_{10}H_9(\text{tropenium})^{\ominus}Cs^{\oplus}$	*IR* 294; *UV* 294; *NMR* 293a
$B_{10}H_8(\text{tropenium})_2$	*IR* 294
$B_{10}H_9(COC_6H_5)^{2\ominus}(Na^{\oplus})_2$	*NMR* 331

References p. 355

$B_{10}H_9(COC_6H_5)^{2\ominus}$-$[N(CH_3)_4^{\oplus}]_2$	*IR* 416; *UV* 416
$B_{10}H_9(COC_6H_5)^{2\ominus}$-$[N(n\text{-}C_3H_7)_4][H_3O]$	*IR* 416; *UV* 416

21.18 Polyhedral boron hydrides: 11-membered skeletons

$B_{11}H_{13}^{2\ominus}Cs^{\oplus}(CH_3)_4N^{\oplus}$	*STR* 235b; *NMR* 405a
$(B_{11}H_{14})M[P(C_6H_5)_3]_3$-(M = Cu, Au)	*NMR*, *UV* 405d

21.19 Polyhedral boron hydrides: 12-membered skeletons

$B_{12}H_{12}^{2\ominus}$	*NMR* 189b, 194c, 319g, 384, 533a, 596; *STR* 479a, 858; *diamagnetic susceptibility* 384a
$B_{12}D_{12}^{2\ominus}$	*IR* 534
$B_{12}H_2Br_{10}^{2\ominus}[(CH_3)_4N^{\oplus}]_2$	*UV* 413
$B_{12}Br_{10}(OH)_2^{2\ominus}$ $[(CH_3)_4N^{\oplus}]_2$	*IR* 416; *UV* 416
$B_{12}Br_{12}^{2\ominus}(Cs^{\oplus})_2$	*UV* 413
$B_{12}H_{11}I^{2\ominus}[(CH_3)_4N^{\oplus}]_2$	*UV* 413
$B_{12}H_{10}I_2^{2\ominus}(Na^{\oplus})_2$	*NMR* 413
$B_{12}H_{10}I_2^{2\ominus}(Cs^{\oplus})_2$	*UV* 413
$B_{12}I_{12}^{2\ominus}(Na^{\oplus})_2$	*UV* 413
$B_{12}H_{11}(OH)^{2\ominus}(Cs^{\oplus})_2$	*IR* 416; *NMR* 416
$B_{12}H_{10}(OH)_2^{2\ominus}(Cs^{\oplus})_2$	*IR* 416; *NMR* 416
$B_{12}H_{11}(OCH_2C_6H_5)^{2\ominus}$-$[(CH_3)_4N^{\oplus}]_2$	*IR* 416; *NMR* 416
$B_{12}H_{10}]N(CH_3)_3](CO)$	*IR* 333, 334
$B_{12}H_{10}(CO)$	*IR* 418

$B_{12}H_{10}$-1,12-$(CO)_2^{2\ominus}$-$(Cs^{\oplus})_2$ — *NMR* 416a

$B_{12}H_{10}$-1,12-$(CN)_2^{2\ominus}$-$(Cs^{\oplus})_2$ — *IR* 333; *NMR* 333

$B_{12}H_{11}$(tropenium)$^{\ominus}$-$H_3O^{\oplus}$ — *UV*, *NMR* 293a

21.20 *Polyhedral boron hydrides: 16- and higher-membered skeletons*

$B_{16}H_{20}$ — *STR* 234a

n-$B_{18}H_{22}$ — *IR*, *NMR*, *UV* 563a

i-$B_{18}H_{22}$ — *IR*, *NMR*, *UV* 563a; *STR* 713, 714

$(CH_3)_4N^{\oplus}$n-$B_{18}H_{21}^{\ominus}$ — *IR*, *NMR*, *UV* 563a

$(CH_3)_4N^{\oplus}$ i-$B_{18}H_{21}^{\ominus}$ — *IR*, *NMR*, *UV* 563a

$[(CH_3)_4N^{\oplus}]_2$n-$B_{18}H_{20}^{2\ominus}$ — *IR*, *NMR*, *UV* 563a

$[(CH_3)_4N^{\oplus}]_2$i-$B_{18}H_{20}^{2\ominus}$ — *IR*, *NMR*, *UV* 563a

n-$B_{18}H_{21}I$ — *IR*, *NMR*, *UV* 563a

$B_{20}H_{16}$ — *IR* 521; *NMR* 235, 521, 525; *STR* 171, 525; *MS* 525

$B_{20}H_{16}\cdot 3.5C_2H_5OH$ — *IR* 521; *NMR* 521

$B_{20}H_{16}(donor)_2$ (donor: OEt_2, NMe_3, PPh_3, SMe_2) — *IR* 521; *NMR* 521

$B_{20}H_{16}(NCCH_3)_2$ CH_3CN — *STR* 201a, 201b

$B_{20}H_{16}(OH)_2^{2\ominus}(H_3O^{\oplus})_2$ — *IR* 521; *NMR* 521

$B_{20}H_{16}(OH)_2^{2\ominus}$ $[(CH_3)_4N^{\oplus}]_2$ — *IR* 138, 521; *NMR* 521

$B_{20}H_{16}(OH)_2^{2\ominus})Cs^{\oplus})_2$ — *IR* 521; *NMR* 521

$B_{20}H_{16}[S(CH_3)_2]_2$ — *IR* 138; *UV* 138

References p. 355

$B_{20}H_{19}^{2\ominus}[(CH_3)_4N^{\oplus}]_2\cdot H_2O$	*IR* 138, 383a; *UV* 383a
$B_{20}H_{19}^{2\ominus}[(C_2H_5)_4N^{\oplus}]_2$	*UV* 383a
$B_{20}H_{18}^{4\ominus}(K^{\oplus})_4\cdot 1.16H_2O$	*IR* 138; *NMR* 138
$B_{20}H_{17}(OH)^{4\ominus}(NH_4^{\oplus})_4\cdot 0.5H_2O$	*IR* 138; *NMR* 138
$B_{20}H_{18}(OH)^{3\ominus}[(CH_3)_4N^{\oplus}]_3\cdot H_2O$	*IR* 138; *NMR* 138
$B_{20}H_{17}(OCH_3)^{4\ominus}(Cs^{\oplus})_4$	*IR* 138
$B_{20}H_{17}(OCHO)_2{}^{\ominus}$-$[(CH_3)_4N^{\oplus}]_2$	*IR* 138
$B_{20}H_{18}^{2\ominus}$	*IR* 138; *UV* 138; *NMR* 314b, 593; *STR* 164b, 314a, 685b; *diamagnetic susceptibility* 384a
$[(C_2H_5)_3NH^{\oplus}]_3B_{20}$-$H_{18}NO$	*STR* 685b
$B_{20}H_{12}Cl_6{}^{2\ominus}$	*UV* 138
$(Cs^{\oplus})_3B_{24}H_{23}^{3\ominus}$	*IR*, *NMR* 841a
$[(C_2H_5)_4N^{\oplus}]_3B_{24}H_{23}^{3\ominus}$	*IR*, *NMR* 841a
$[(C_2H_5)_3NH^{\oplus}]_5B_{20}H_{18}\cdot B_{20}H_{19}^{5\ominus}$	*IR*, *UV* 383a

21.21 *Carboranes: 5-membered skeletons*

2-CH_3-1,2-$C_2B_3H_4$	*IR* 278f
C,3-$(CH_3)_2$-1,2-$C_2B_3H_3$	*IR* 278e, 278f
C,5-$(CH_3)_2$-1,2-$C_2B_3H_3$	*IR* 278f
C,3,5-$(CH_3)_3$-1,2-$C_2B_3H_2$	*IR* 278f
3,4,5-$(CH_3)_3$-1,2-$C_2B_3H_2$	*IR* 278f

C,3,4,5-$(CH_3)_4$-1,2-C_2H_3H	*IR* 278f
C′,3,4,5-$(CH_3)_4$-1,2-C_2H_3H	*IR* 278f
1,5-$C_2B_3H_5$	*IR, NMR, MS* 278c, 278f, 690a
2-CH_3-1,5-$C_2B_3H_4$	*IR, NMR, MS* 278c
C,3-$(CH_3)_2$-1,5-$C_2B_3H_3$	*IR, NMR, MS* 278c

21.22 *Carboranes: 6-membered skeletons*

CB_5H_7	*NMR* 565c, 604b
2-CH_3-2-CB_5H_8	*IR, NMR, STR* 567c, 278g
3-CH_3-2-CB_5H_8	*IR, NMR, STR* 567c
4-CH_3-2-CB_5H_8	*IR, NMR, STR* 567c
C-CH_3-B-$(C_2H_5)_5$-2-CB_5H_3	*IR, NMR* 271a
1,6-$C_2B_4H_6$	*IR, NMR, MS* 278c, 692a
2,3-$C_2B_4H_8$	*NMR, STR* 59, 565b, 567a, 569
2,3-$C_2B_4H_7D^{\mu}$	*NMR* 567a
2,3-$C_2B_4H_6$-2,3-D_2	*IR* 724a
2,3-$C_2B_4H_5$-4,5,6-D_3	*NMR* 567a
2-CH_3-1,6-$C_2B_4H_5$	*IR, NMR, MS* 278c
1-CH_3-2,3-$C_2B_4H_7$	*NMR* 568b
2-CH_3-2,3-$C_2B_4H_7$	*IR, NMR, MS* 278g
4-CH_3-2,3-$C_2B_4H_7$	*NMR* 568b
5-CH_3-2,3-$C_2B_4H_7$	*NMR* 568b

References p. 355

1,2,3-$(CH_3)_3$-2,3-$C_2B_4H_5$	*NMR* 568b
2,3,4-$(CH_3)_3$-2,3-$C_2B_4H_5$	*NMR* 568b
2,3,5-$(CH_3)_3$-2,3-$C_2B_4H_5$	*NMR* 568b
$(CH_3)_2$-2,3-$C_2B_4H_4$	*NMR* 569, 849; *STR* 59
2,3-$(CH_3)_2$-2,3-$C_2B_4H_6$	*IR, NMR, MS* 278g
2-CH_3-2,3,4-$C_3B_3H_6$	*IR, NMR, MS* 67a, 278d, 278g
2-CD_3-2,3,4-$C_3B_3H_6$	*IR, NMR* 232b
2,3-$(CH_3)_2$-2,3,4-$C_3B_3H_5$	*IR, NMR, MS* 67a, 278d, 278g
2,3-$(CD_3)_2$-2,3,4-$C_3B_3H_5$	*IR, NMR* 232b
2,4-$(CH_3)_2$-2,3,4-$C_3B_3H_5$	*IR, NMR, MS* 67a, 278d, 278g
2,4-$(CD_3)_2$-2,3,4-$C_3B_3H_5$	*IR, NMR* 232b
2-C_2H_5-3,4-$(CH_3)_2$-2,3,4-$C_3B_3H_5$	*IR, NMR, MS* 278g

21.23 Carboranes: 7-membered skeletons

2,4-$C_2B_5H_7$	*IR, NMR, MS* 278c, 567b, 567d, 686a; *STR* 44, 43e; *MW* 43e
1-CH_3-2,4-$C_2B_5H_6$	*IR, NMR, MS* 278c
3-CH_3-2,4-$C_2B_5H_6$	*IR, NMR, MS* 278c
5-CH_3-2,4-$C_2B_5H_6$	*IR, NMR, MS* 278c
$(CH_3)_5$-2,4-$C_2B_5H_2$	*IR, NMR, MS* 686a

21.24 Carboranes: 8-membered skeletons

$(CH_3)_2C_2B_6H_6$	*STR* 292a, 293c, 763b; *NMR* 763b, 763c, 845b; *IR* 763c; *MS* 845b
$[C_2B_6H_8Mn(CO)_3]^{\ominus}Cs^{\oplus}$	*NMR, STR* 318a

21.25 *Carboranes: 9-membered skeletons*

1,6-C_2B_7-1,6-$(CH_3)_2$	*NMR, STR* 423a, 763b
$C_2B_7H_{13}$	*NMR* 763a, 763c; *IR* 763c
$C_2B_7H_{12}(CH_3)$	*NMR* 763a, 763c; *IR* 763c
$C_2B_7H_{12}(C_6H_5)$	*IR, NMR* 763c
$C_2B_7H_{11}(CH_3)_2$	*NMR* 763a, 763c; *IR* 763c
$(C_2B_7H_9{}^{2\ominus})_2Co^{3\oplus}$-$[(CH_3)_4N]^{\oplus}$	*UV, NMR* 252b, 311a

21.26 *Carboranes: 10-membered skeletons*

1-$B_9H_9CH^{\ominus}Cs^{\oplus}$	*NMR* 411e
$B_9H_9CH^{\ominus}N(CH_3)_4^{\oplus}$	*NMR* 411e
$B_9H_{11}CN(CH_3)_3$	*NMR* 411e
1,6-$C_2B_8H_8(CH_3)_2$	*IR, NMR, STR* 763b, 763c
1,10-$C_2B_8H_8(CH_3)_2$	*IR, NMR, STR* 763b, 763c

21.27 *Carboranes: 11-membered skeletons*

$B_{10}H_{10}CH^{\ominus}$	*NMR* 411e
$B_{10}H_{12}CH^{\ominus}Cs^{\oplus}$	*IR, UV* 371d
$B_{10}H_{12}CH^{\ominus}[(CH_3)_4N]^{\oplus}$	*IR, UV, NMR* 371a, 371d
$Cl_2B_{10}H_{10}CH^{\ominus}$ $[(CH_3)_4N]^{\oplus}$	*IR* 371d
$BrB_{10}H_{11}CH^{\ominus}[(CH_3)_4N]^{\oplus}$	*IR* 371d
$Br_2B_{10}H_{10}CH^{\ominus}[(CH_3)_4N]^{\oplus}$	*IR* 371d
$B_{10}H_{10}CN(CH_3)_3$	*IR, NMR* 371d
$B_{10}H_{10}CN(CH_3)_2(C_3H_7)$	*IR, NMR* 371d

References p. 355

$B_{10}H_{12}CNH_2(C_2H_5)$ *IR*, *NMR*, *STR* 319a

$B_{10}H_{12}CN(CH_3)_3$ *IR*, *NMR*, *STR* 319a, 677a, 411e

4(6)-$ClB_{10}H_{11}CN(CH_3)_3$ *IR*, *MS* 371d

4,6-$Cl_2B_{10}H_{10}CN(CH_3)_3$-$(C_3H_7)$ *IR*, *MS* 371d

4(6)-$BrB_{10}H_{11}CN(CH_3)_3$ *IR*, *NMR* 371d

4(6)-$BrB_{10}H_{11}CN(CH_3)_2$-(C_3H_7) *IR*, *MS* 371d

[7,8-$B_9H_{10}CHP \cdot M(CO)_5]^{\ominus}$ $N(CH_3)_4^{\oplus}$ (M = Cr, Mo, W) *IR*, *NMR* 709a

[7,9-$B_9H_{10}CHAs \cdot Mo$-$(CO)_5]^{\ominus}$ $N(CH_3)_4^{\oplus}$ *IR*, *NMR* 709a

[7,8-$B_9H_{10}CHE \cdot M$-$(CO)_5]^{\ominus}$ $[N(CH_3)_4]^{\oplus}$ (E = P, As; M = Cr, Mo, W) *IR*, *NMR* 709a

$C_2B_9H_{11}$ *IR* 763c

$C_2B_9H_{11}^{\ominus}$ *NMR* 411e

$C_2B_9H_{10}(CH_3)$ *IR* 763c

$C_2B_9H_9(CH_3)_2$ *IR* 763c

$C_2B_9H_{10}(C_6H_5)$ *IR* 763c

$C_2B_9H_{12}I$ *NMR* 570a

$C_2B_9H_{11}I^{\ominus}$-$[(CH_3)_4N]^{\oplus}$ *NMR* 570a

$C_2B_9H_{11}Sn$ *IR*, *NMR* 775b

$(C_2B_9H_{11})Fe(\pi$-$C_5H_5)$ *Mössbauer spectr.* 329a; *NMR* 314c, 487c; *STR* 863c

$[(C_2B_9H_{11})Fe(\pi\text{-}C_5H_5)]^{\ominus}$ $(CH_3)_4N^{\oplus}$	*NMR* 314c
$[(C_2B_9H_{11})_2Fe]^{\ominus}$-$(CH_3)_4N^{\oplus}$	*Mössbauer spectr.* 329a; *NMR, ES* 319, 487c
$[(C_2B_9H_{11})_2Fe]^{2\ominus}$ $[(CH_3)_4N^{\oplus}]_2$	*NMR* 319
$[(C_2B_9H_{11})_2Co]^{\ominus}Cs^{\oplus}$	*STR* 860c; *UV* 319d; *NMR* 319d
$[(C_2B_9H_{11})Mn(CO)_3]^{\ominus}Cs^{\oplus}$	*IR, NMR, ES* 308a
$[(C_2B_9H_{11})Re(CO)_3]^{\ominus}Cs^{\oplus}$	*IR, NMR, ES* 308a

21.28 Carboranes: 12-membered skeletons

$B_{11}H_{11}CH^{\ominus}Cs^{\oplus}$	*NMR* 411e
$H(1,2\text{-}C_2B_{10}H_{10})H$	*NMR* 58b, 594, 774b; *dipole moment* 444a, 493d; *RA* 102a
$H(1,7\text{-}C_2B_{10}H_{10})H$	*NMR* 594, 774b; *dipole moment* 444a, 493d, *RA* 102a
$H(1,12\text{-}C_2B_{10}H_{10})H$	*IR* 574a; *NMR* 574a, 594, 774b; *dipole moment* 444a
$H(1,2\text{-}C_2B_{10}H_{10})$-$(1,2\text{-}C_2B_{10}H_{10})H$	*STR* 290c
$H(1,2\text{-}C_2B_{10}H_{10-n}Cl_n)H$ ($n = 0, 2, 3, 4, 6, 8, 10$)	*IR, NMR, MS* 680
$CH_3(1,7\text{-}C_2B_{10}H_{10})CH_3$	*NMR, STR* 763b
$H(1,2\text{-}C_2B_{10}H_{10})C$-$(CH_3)$=$CH_2$	*IR, UV* 215
$H(1,2\text{-}C_2B_{10}Cl_{10})Cl$	*IR, NMR* 680
$H(1,2\text{-}C_2B_{10}H_{10})Br$	*dipole moment* 493d
$H(1,2\text{-}C_2B_{10}H_8\text{-}9,12\text{-}Br_2)H$	*STR* 602a
$H(1,7\text{-}C_2B_{10}H_8\text{-}9,10Br_2)H$	*STR* 43d

References p. 355

$H(1,2\text{-}C_2B_{10}H_7\text{-}8(10),9,12\text{-}Br_3)H$	*STR* 602b
$H(1,2\text{-}C_2B_{10}H_6\text{-}8,9,10,12\text{-}Br_4)H$	*STR* 602c
$(1,2\text{-}C_2B_{10}Cl_{10})Cl^{\ominus}\text{-}(C_2H_5)_3NH^{\oplus}$	*NMR* 680
$(1,2\text{-}C_2B_{10}Cl_{10})Cl^{\ominus}\text{-}(C_6H_5)_3P(CH_3)^{\oplus}$	*NMR* 680
$[H(1,2\text{-}C_2B_{10}H_{10})CH_2]_2O$	*IR* 268
$OCN(1,2\text{-}C_2B_{10}H_{10})NCO$	*IR*, *MS* 339
$(C_6H_5)ClP(1,2\text{-}C_2B_{10}H_{10})\text{-}PCl(C_6H_5)$	*IR* 9
$N_3(C_6H_5)P(1,2\text{-}C_2B_{10}H_{10})\text{-}P(C_6H_5)N_3$	*IR* 9
$(C_6H_5)P(1,2\text{-}C_2B_{10}H_{10})P(C_6H_5)$ (P and P bridged by NH)	*IR* 9
$R_2P(1,2\text{-}C_2B_{10}H_{10})PR_2$ (P and P bridged by $Ni(CO)_2$)	*IR* 620b
$ClP[(1,2\text{-}C_2B_{10}H_{10})]_2PCl$ (ring: ClP and PCl linked by two $(1,2\text{-}C_2B_{10}H_{10})$ units)	*IR* 9
$(H_2N)P[(1,2\text{-}C_2B_{10}H_{10})]_2P(NH_2)$ (ring: two $(1,2\text{-}C_2B_{10}H_{10})$ units linking the P atoms)	*IR* 9
$[n\text{-}C_4H_9(1,2\text{-}C_2B_{10}H_{10})\text{-}BNCH_3]_3$	*IR* 61a
$B_{10}H_{10}\text{-}1\text{-}CH\text{-}2\text{-}P$	*NMR* 474b
$B_{10}H_{10}\text{-}7\text{-}CH\text{-}1\text{-}P$	*NMR* 474b, 767c; *STR* 767c

REFERENCES

1. Abrahams, S. C. and J. Kalnajs, J. Chem. Phys. *22*, 434 (1954).
2. Adams, R. M., in *Advances in Chemistry*, Ser. 32, Ed. R. F. Gould, Am. Chem. Soc., Washington D.C., 1961, p. 60.
3. Adams, R. M., in *Boron, Metallo-Boron Compounds and Boranes*, Ed. R. M. Adams, Interscience Publishers, New York, 1964, p. 507.
4. Adams, R. M., Inorg. Chem. *2*, 1087 (1963).
5. Adams, R. M., in *Boron, Metallo-Boron Compounds and Boranes*, Ed. R. M. Adams, Interscience Publishers, New York, 1964, p. 373.
6. Adams, R. M., A. R. Siedle and J. Grant, Inorg. Chem. *3*, 461 (1964).
7. Aftandilian, V. D., H. C. Miller and E. L. Muetterties, J. Am. Chem. Soc. *83*, 2471 (1961).
8. Aftandilian, V. D., H. C. Miller, G. W. Parshall and E. L. Muetterties, Inorg. Chem. *1*, 734 (1962).
9. Alexander, R. P. and H.-J. Schroeder, Inorg. Chem. *2*, 1107 (1963).
9a. Alexander, R. P. and H.-J. Schroeder, Inorg. Chem. *5*, 493 (1966).
0. Alton, E. R., R. D. Brown, J. C. Carter and R. C. Taylor, J. Am. Chem. Soc. *81*, 3550 (1959).
1. Altschuller, A. P., J. Am. Chem. Soc. *77*, 5455 (1955).
2. Altwicker, E. R., G. E. Ryschkewitsch, A. B. Garrett and H. H. Sisler, Inorg. Chem. *3*, 454 (1964).
2a. Amberger, E., unpublished results.
3. Amberger, E. and W. Dietze, Z. Anorg. Allgem. Chem. *332*, 131 (1964).
3a. Amberger, E. and R. Hönigschmid-Grossich, Chem. Ber. *99*, 1673 (1966).
3b. Amberger, E. and E. Gut, Chem. Ber. *101*, 1200 (1968).
4. Amberger, E. and M.-R. Kula, Chem. Ber. *96*, 2556 (1963).
4a. Amberger, E. and M.-R. Kula, Chem. Ber. *96*, 2560 (1963).
5. Amberger, E. and M.-R. Kula, Angew. Chem. *75*, 476 (1963).
5a. Amberger, E. and P. Leidl, J. Organometal. Chem. *18*, 345 (1969).
5b. Amberger, E. and P. Leidl, Chem. Ber. *102*, 2764 (1969).
5c. Amberger, E. and W. Stoeger, J. Organometal. Chem. *17*, 287 (1969).
5d. American Chemical Society, Inorg. Chem. *7*, 1945 (1968).
6. Anderson, T. F. and A. B. Burg, J. Chem. Phys. *6*, 586 (1939).
6a. Armstrong, D. R. and P. G. Perkins, Chem. Commun. *1968*, 382.
6b. Armstrong, D. R. and P. G. Perkins, J. Chem. Soc. *A1969*, 1044.
6c. Armstrong, D. R., B. J. Duke and P. G. Perkins, J. Chem. Soc. *A1969*, 2566.
17. Ashby, E. C., J. Am. Chem. Soc. *81*, 4791 (1959).
17a. Ashby, E. C., J. Organometal. Chem. *3*, 371 (1965).
18. Ashby, E. C. and W. E. Forster, J. Am. Chem. Soc. *84*, 3407 (1962).
18a. Ashby, E. C. and W. E. Forster, J. Am. Chem. Soc. *88*, 3248 (1966).
19. Atteberry, R. W., J. Phys. Chem. *62*, 1457 (1958).
20. Atteberry, R. W., J. Phys. Chem. *62*, 1458 (1958).
21. Auberry, W. D., M. F. Lappert and H. Pyszora, J. Chem. Soc. *1961*, 1931.
22. Baden, H. C., W. H. Bauer and S. E. Wiberley, J. Phys. Chem. *62*, 331 (1958).
23. Baden, H. C., S. E. Wiberley and W. H. Bauer, J. Phys. Chem. *59*, 287 (1955).
24. Baker, E. B., R. B. Ellis and W. S. Wilcox, J. Inorg. Nucl. Chem. *23*, 41 (1961).
25. Baker, E. B., R. B. Ellis and W. S. Wilcox (Callery Chem. Co.), U.S. Pat. 2,921,963 (1960); C.A. *54*, 11412d (1960).
25a. Baker, C. S. L., J. Organometal. Chem. *19*, 287 (1969).
26. Banus, M. D., R. W. Bragdon and T. R. P. Gibb, Jr., J. Am. Chem. Soc. *74*, 2346 (1952).
27. Banus, M. D., R. W. Bragdon and A. A. Hinckley, J. Am. Chem. Soc. *76*, 3848 (1954).
28. Barbaras, G. D., C. Dillard, A. E. Finholt, T. Wartik, K. E. Wilzbach and H. I. Schlesinger, J. Am. Chem. Soc. *73*, 4585 (1951).
29. Bartell, L. S. and B. L. Carroll, J. Chem. Phys. *42*, 1135 (1965).
30. Bartlett, R. K., H. S. Turner, R. J. Warne, M. A. Joung and W. S. McDonald, Proc. Chem. Soc. *1962*, 153.
31. Bartocha, B., W. A. G. Graham and F. G. A. Stone, J. Inorg. Nucl. Chem. *6*, 119 (1958).
31a. Barton, L., C. Perrin and R. F. Porter, Inorg. Chem. *5*, 1446 (1966).
31b. Bass, C. D., L. L. Lynds, T. Wolfram and R. E. DeWames, J. Chem. Phys. *40*, 3611 (1964).

32. BAUER, R., Z. Naturforsch. *16b*, 839 (1961).
33. BAUER, S. H., J. Am. Chem. Soc. *80*, 294 (1958).
34. BAUER, S. H., J. Am. Chem. Soc. *78*, 5775 (1956).
35. BAX, C. M., A. R. KATRITZKY AND L. E. SUTTON, J. Chem. Soc. *1958*, 1258.
35a. BAYLEY, N. A., P. H. BIRD AND M. G. H. WALLBRIDGE, Chem. Commun. *1966*, 286.
35b. BAYLIS, A. B., G. A. PRESSLEY, JR. AND F. E. STAFFORD, J. Am. Chem. Soc. *88*, 2428 (1966).
35c. BAYLIS, A. B., G. A. PRESSLEY, JR., M. E. GORDON AND F. E. STAFFORD, J. Am. Chem. Soc. *88*, 929 (1966).
35d. BAYLEY, N. A., P. H. BIRD AND M. G. H. WALLBRIDGE, Chem. Commun. *1965*, 438.
36. BAYLIS, A. B., G. A. PRESSLEY, JR., E. J. SINKE AND F. E. STAFFORD, J. Am. Chem. Soc. *8* 5358 (1964).
37. BEACHELL, H. C. AND B. F. DIETRICH, J. Am. Chem. Soc. *83*, 1347 (1961).
38. BEACHELL, H. C. AND J. F. HAUGH, J. Am. Chem. Soc. *80*, 2939 (1958).
39. BEACHELL, H. C. AND K. R. LANGE, J. Phys. Chem. *60*, 307 (1956).
40. BEACHELL, H. C. AND E. J. LEVI, J. Chem. Phys. *23*, 2168 (1955).
41. BEACHELL, H. C. AND T. R. MEEKER, J. Am. Chem. Soc. *78*, 1796 (1956).
42. BEACHELL, H. C. AND W. C. SCHAR, J. Am. Chem. Soc. *80*, 2943 (1958).
43. BEACHELL, H. C. AND H. S. VELORIC, J. Phys. Chem. *60*, 102 (1956).
43a. BEACHLEY, O. T., Inorg. Chem. *7*, 701 (1968). BEALL, H. A. AND W. N. LIPSCOMB, Ino Chem. *3*, 1783 (1964).
43b. BEACHLEY, O. T., Inorg. Chem. *4*, 1823 (1965).
43c. BEACHLEY, O. T., Inorg. Chem. *6*, 870 (1967).
43d. BEALL, H. A. AND W. N. LIPSCOMB, Inorg. Chem. *6*, 874 (1967).
43e. BEAUDET, R. A. AND R. L. POYNTER, J. Chem. Phys. *43*, 2166 (1965).
44. BEAUDET, R. A. AND R. L. POYNTER, Inorg. Chem. *3*, 1258 (1964).
44a. BECHER, J., Z. Anorg. Allg. Chem. *288*, 235 (1955).
44b. BECHER, J., Spectrochim. Acta, *19*, 575 (1963).
45. BECHER, H. J. AND S. FRICK, Z. Anorg. Allgem. Chem. *295*, 83 (1958).
46. BECHER, H. J. AND S. FRICK, Z. Physik. Chem., Frankfurt *12*, 241 (1957).
46a. BECKER, W. E. AND E. C. ASHBY, Inorg. Chem. *4*, 1816 (1965).
47. BEICHL, G. J. AND C. E. EWERS, J. Am. Chem. Soc. *80*, 5344 (1958).
48. BELLAMY, L. J., W. GERRARD, M. F. LAPPERT AND R. L. WILLIAMS, J. Chem. Soc. *1958*, 2412
49. BENJAMIN, L. E., S. F. STAFIEJ AND E. A. TAKACS, J. Am. Chem. Soc. *85*, 2674 (1963).
50. BENSON, L. A., Thesis, St. Louis University, 1959, Univ. Microfilms L.C. No. 60-321, A Arbor, Mich., U.S.A.
51. BERKA, L., T. BRIGGS, M. MILLARD AND W. L. JOLLY, J. Inorg. Nucl. Chem. *14*, 190 (1960).
51a. BERRY, T. E., F. N. TEBBE AND M. F. HAWTHORNE, Tetrahedron Letters *1965*, 715.
52. BETHKE, G. W. AND M. K. WILSON, J. Chem. Phys. *26*, 1118 (1957).
53. BETHKE, G. W. AND M. K. WILSON, J. Chem. Phys. *27*, 978 (1957).
53a. BIRD, P. H. AND M. G. H. WALLBRIDGE, Chem. Commun. *1968*, 687.
53b. BIRD, P. H. AND M. G. H. WALLBRIDGE, J. Chem. Soc. *1965*, 3923.
54. BISSOT, T. C., D. H. CAMPBELL AND R. W. PARRY, J. Am. Chem. Soc. *80*, 1868 (1958).
55. BLAY, N. J., I. DUNSTAN AND R. L. WILLIAMS, J. Chem. Soc. *1960*, 430.
56. BLAY, N. J., R. J. PACE AND R. L. WILLIAMS, J. Chem. Soc. *1962*, 3416.
57. BLAY, N. J., J. WILLIAMS AND R. L. WILLIAMS, J. Chem. Soc. *1960*, 424.
57a. BODEN, N., H. S. GUTOWSKY, J. R. HANSEN AND T. C. FARRAR, J. Chem. Phys. *46*, 28 (1967).
57b. BODDEKER, K. W., S. G. SHORE AND R. K. BUNTING, J. Am. Chem. Soc. *88*, 4396 (1966).
58. BÖHM, R., Thesis, München University, 1965.
58a. BOER, F. P., J. A. POTENZA AND W. N. LIPSCOMB, Inorg. Chem. *5*, 1301 (1966).
58b. BOER, F. P., R. A. HEGSTROM, M. D. NEWTON, J. A. POTENZA AND W. N. LIPSCOMB, J. A Chem. Soc. *88*, 5340.
59. BOER, F. P., W. E. STREIB AND W. N. LIPSCOMB, Inorg. Chem. *3*, 1666 (1964).
60. BOLZ, L. H., F. A. MAUER AND H. S. PEISER, J. Chem. Phys. *31*, 1005 (1959).
61. BOONE, J. L., J. Am. Chem. Soc. *86*, 5036 (1964).
61a. BOONE, J. L., R. J. BROTHERTON AND L. L. PETTERSON, Inorg. Chem. *4*, 910 (1965).
62. BOONE, J. L. AND A. B. BURG, J. Am. Chem. Soc. *81*, 1766 (1959).
63. BOONE, J. L. AND A. B. BURG, J. Am. Chem. Soc. *80*, 1519 (1958).
64. BORER, K., A. B. LITTLEWOOD AND C. S. G. PHILLIPS, J. Inorg. Nucl. Chem. *15*, 316 (1960).
65. BRAGDON, R. W. AND M. D. BANUS, U.S. Pat. 2,741,540 (1956); C.A. *50*, 10994b (1956).

66. Bragdon, R. W. and M. D. Banus, U.S. Pat. 2,720,444 (1955); C.A. *50*, 3718a (1956).
67. Bragg, J. K., L. V. McCarty and F. J. Norton, J. Am. Chem. Soc. *73*, 2134 (1951).
67a. Bramlett, C. L. and R. N. Grimes, J. Am. Chem. Soc. *88*, 4269 (1966).
67b. Bregadze, V. I. and O. Yu. Okhlobystin, Russ. Chem. Rev. *37*, 173 (1968).
68. Brennan, G. L., G. H. Dahl and R. Schaeffer, J. Am. Chem. Soc. *82*, 6248 (1960).
69. Brennan, G. L. and R. Schaeffer, J. Inorg. Nucl. Chem. *20*, 205 (1961).
70. Brenner, A., W. E. Reid, Jr. and J. M. Bish, J. Electrochem. Soc. *104*, 21 (1957).
70a. Bresadola, S., F. Rossetto and G. Tagliavini, Chem. Commun. *1966*, 623.
71. British Thomson-Houston Co. Ltd., Brit. Pat. 623,760 (1949); C.A. *44*, 291g (1950).
72. British Thomson-Houston Co. Ltd., Brit. Pat. 623,761 (1949); C.A. *44*, 291g (1950).
72a. Brooks, W. V. F., C. C. Costain and R. F. Porter, J. Chem. Phys. *47*, 4186 (1967).
73. Brown, C. A. and A. W. Laubengayer, J. Am. Chem. Soc. *77*, 3699 (1955).
73a. Brown, M. P. and R. W. Hesltine, Chem. Commun. *1968*, 1551.
74. Brown, D. A. and H. C. Longuett-Higgins, J. Inorg. Nucl. Chem. *1*, 352 (1955).
75. Brown, H. C., *Hydroboration*, W. A. Benjamin, Inc., New York, 1962, p. 233.
76. Brown, H. C., U.S. Pat. 2,860,167 (1958); C.A. *53*, 9060 (1959).
76a. Brown, H. C., M. V. Bhatt, T. Munekata and G. Zweifel, J. Am. Chem. Soc. *89*, 567 (1967).
77. Brown, H. C. and C. A. Brown, J. Am. Chem. Soc. *84*, 1493 (1962).
78. Brown, H. C. and L. Domash, J. Am. Chem. Soc. *78*, 5384 (1956).
79. Brown, H. C., D. Gintis and L. Domash, J. Am. Chem. Soc. *78*, 5387 (1956).
80. Brown, H. C. and G. J. Klender, Inorg. Chem. *1*, 204 (1962).
81. Brown, H. C., E. J. Mead and C. J. Shoaf, J. Am. Chem. Soc. *78*, 3616 (1956).
82. Brown, H. C., E. J. Mead and B. C. Subba Rao, J. Am. Chem. Soc. *77*, 6209 (1955).
83. Brown, H. C., E. J. Mead and B. C. Subba Rao, J. Am. Chem. Soc. *77*, 6200 (1955).
84. Brown, H. C., E. J. Mead and P. A. Tierney, J. Am. Chem. Soc. *79*, 5400 (1957).
85. Brown, H. C. and A. W. Moerikofer, J. Am. Chem. Soc. *84*, 1478 (1962).
86. Brown, H. C. and A. W. Moerikofer, J. Am. Chem. Soc. *82*, 2063 (1963).
87. Brown, H. C. and A. W. Moerikofer, J. Am. Chem. Soc. *83*, 3417 (1961).
88. Brown, H. C., H. I. Schlesinger and A. B. Burg, J. Am. Chem. Soc. *61*, 673 (1939).
89. Brown, H. C., H. I. Schlesinger, I. Sheft and D. M. Ritter, J. Am. Chem. Soc. *75*, 192 (1953).
89a. Brown, H. C., P. F. Stehle and P. A. Thierney, J. Am. Chem. Soc. *79*, 2020 (1957).
90. Brown, H. C. and B. C. Subba Rao, J. Am. Chem. Soc. *81*, 6428 (1959).
91. Brown, H. C. and P. A. Tierney, J. Inorg. Nucl. Chem. *9*, 51 (1959).
92. Brown, H. C. and P. A. Tierney, J. Am. Chem. Soc. *80*, 1552 (1958).
93. Brown, H. C. and G. Zweifel, J. Am. Chem. Soc. *82*, 4708 (1960).
94. Brown, H. C. and G. Zweifel, J. Am. Chem. Soc. *83*, 2544 (1961).
95. Brown, H. C. and G. Zweifel, J. Am. Chem. Soc. *83*, 1241 (1961).
96. Brown, H. C. and G. Zweifel, J. Am. Chem. Soc. *82*, 1504 (1960).
97. Brown, H. C. and G. Zweifel, J. Am. Chem. Soc. *83*, 3834 (1961).
98. Brown, H. C. and G. Zweifel, J. Am. Chem. Soc. *81*, 1512 (1959).
99. Brown, J. C. and G. S. Fischer, Tetrahedron Letters *21*, 9 (1960).
00. Brown, D. A., Nature *190*, 804 (1961).
00a. Brown, M. P., R. W. Heseltine and D. W. Johnson, J. Chem. Soc. *A1967*, 597.
00b. Brown, M. P., R. W. Heseltine and L. H. Sutcliffe, J. Chem. Soc. *A1968*, 612.
01. Brown, W. G., L. Kaplan and K. E. Wilzbach, J. Am. Chem. Soc. *74*, 1343 (1952).
02. Brukl, A. and K. Rossmanith, Monatsh. *90*, 481 (1959); C.A. *54*, 2068h (1960).
02a. Bukalov, S. S., L. A. Leites and V. T. Aleksanyan, Bull. Akad. Sci. USSR, Div. Chem. Sci. *1968*, 896.
03. Brumberger, H. and R. Markus, J. Chem. Phys. *24*, 741 (1956).
03a. Burch, J. E., W. Gerrard, M. Goldstein, E. Mooney, D. E. Pratt and H. A. Willis, Spectrochim. Acta *19*, 889 (1963).
04. Burg, A. B., J. Am. Chem. Soc. *56*, 499 (1934).
05. Burg, A. B., Angew. Chem. *72*, 183 (1960).
06. Burg, A. B., Inorg. Chem. *3*, 1325 (1964).
07. Burg, A. B., J. Am. Chem. Soc. *69*, 747 (1947).
08. Burg, A. B., Record Chem. Progr., Kresge-Hooker Sci. Lib. *15*, 159 (1954).
09. Burg, A. B., J. Am. Chem. Soc. *79*, 2129 (1957).
10. Burg, A. B., J. Inorg. Nucl. Chem. *11*, 258 (1959).

111. Burg, A. B., J. Am. Chem. Soc. *74*, 1340 (1952).
112. Burg, A. B., J. Am. Chem. Soc. *74*, 3482 (1952).
112a. Burg, A. B., J. Am. Chem. Soc. *90*, 1407 (1968).
113. Burg, A. B. and G. Brendel, J. Am. Chem. Soc. *80*, 3198 (1958).
114. Burg, A. B. and J. L. Boone, J. Am. Chem. Soc. *78*, 1521 (1956).
115. Burg, A. B. and C. D. Good, J. Inorg. Nucl. Chem. *2*, 237 (1956).
116. Burg, A. B. and L. R. Grant, J. Am. Chem. Soc. *81*, 1 (1959).
116a. Burg, A. B. and H. Heinen, Inorg. Chem. *7*, 1021 (1968).
117. Burg, A. B. and R. Kratzer, Inorg. Chem. *1*, 725 (1961).
118. Burg, A. B. and E. Kuljian, J. Am. Chem. Soc. *72*, 3101 (1950).
119. Burg, A. B. and C. L. Randolph, J. Am. Chem. Soc. *71*, 3451 (1949).
120. Burg, A. B. and C. L. Randolph, J. Am. Chem. Soc. *73*, 953 (1951).
120a. Burg, A. B. and J. S. Sandhu, J. Am. Chem. Soc. *89*, 1626 (1967).
121. Burg, A. B. and H. I. Schlesinger, J. Am. Chem. Soc. *55*, 4009 (1933).
122. Burg, A. B. and H. I. Schlesinger, J. Am. Chem. Soc. *55*, 4020 (1933).
123. Burg, A. B. and H. I. Schlesinger, J. Am. Chem. Soc. *62*, 3425 (1940).
124. Burg, A. B. and H. I. Schlesinger, J. Am. Chem. Soc. *59*, 780 (1937).
125. Burg, A. B. and P. J. Slota, J. Am. Chem. Soc. *82*, 2145 (1960).
126. Burg, A. B. and P. J. Slota, J. Am. Chem. Soc. *82*, 2148 (1960).
127. Burg, A. B. and J. R. Spielman, J. Am. Chem. Soc. *81*, 3479 (1959).
128. Burg, A. B. and F. G. A. Stone, J. Am. Chem. Soc. *75*, 228 (1953).
129. Burg, A. B. and R. I. Wagner, J. Am. Chem. Soc. *75*, 3872 (1953).
130. Burg, A. B. and R. I. Wagner, U.S. Pat. 2,916,518 (1959); C.A. *54*, 5464i (1960).
131. Burg, A. B. and R. I. Wagner, J. Am. Chem. Soc. *76*, 3307 (1954).
131a. Burg, A. B. and Yuan-Chin Fu, J. Am. Chem. Soc. *88*, 1147 (1966).
132. Burns, J. J. and G. W. Schaeffer, J. Phys. Chem. *62*, 380 (1958).
133. Campbell, D. H., T. C. Bissot and R. W. Parry, J. Am. Chem. Soc. *80*, 1549 (1958).
134. Campbell, G. W., The Structure of the Boron Hydrides, Progress in Boron Chemistry 167–201 (1964).
134a. Campbell, G. W. J. Am. Chem. Soc. *79*, 4023 (1957).
135. Campbell, G. W. and L. Johnson, J. Am. Chem. Soc. *81*, 3800 (1959).
136. Carr, E. M., J. T. Clarke and H. L. Johnston, J. Am. Chem. Soc. *71*, 740 (1949).
136a. Carter, J. C. and R. W. Parry, J. Am. Chem. Soc. *87*, 2354 (1965).
136b. Catton, R. C., M. C. R. Symons and H. W. Wardale, J. Chem. Soc. *A1969*, 2622.
137. Chamberlain, D. L., U.S. Pat. 3,029,128 (1962); C.A. *57*, 4317g (1962).
138. Chamberland, B. L. and E. L. Muetterties, Inorg. Chem. *3*, 1450 (1964).
138a. Chambers, J. Q., A. D. Norman, M. R. Bickell and S. H. Cadle, J. Am. Chem. Soc. *9* 6056 (1968).
138b. Chang, C. H., R. F. Porter and S. H. Bauer, Inorg. Chem. *8*, 1677 (1969).
138c. Chang, C. H., R. F. Porter and S. H. Bauer, Inorg. Chem. *8*, 1689 (1969).
139. Chapman, A. C., Trans. Faraday Soc. *59*, 806 (1963).
140. Chopard, P. A. and R. F. Hudson, J. Inorg. Nucl. Chem. *25*, 801 (1963).
140a. Churchill, M. R., K. Gold, J. N. Francis and M. F. Hawthorne, J. Am. Chem. Soc. *9* 1222 (1969).
141. Clark, C. C., K. F. A. Kanada and A. J. King (Olin Mathieson Chem. Corp.), U.S. Pa 3,017,248 (1962); C.A. *56*, 11207c (1962).
142. Clark, C. C., K. F. A. Kanada and A. J. King (Olin Mathieson Chem. Corp.), U.S. Pa 3,021,197 (1962); C.A. *56*, 13810b (1962).
142a. Clark, C. C., K. F. A. Kanada and A. J. King (Olin Mathieson Chem. Corp.) U.S. Pa 3,022,138 (1962); C.A. *56*, 15158b (1962).
142b. Clark, C. C., K. F. A. Kanada and A. J. King (Olin Mathieson Chem. Corp.), U.S. Pa 3,022,139 (1962); C.A. *56*, 15158c (1962).
143. Coates, G. E., J. Chem. Soc. *1950*, 3481.
144. Coffin, K. P. and S. H. Bauer, J. Phys. Chem. *59*, 193 (1955).
144a. Cook, Th. H. and G. L. Morgan, J. Am. Chem. Soc. *91*, 774 (1969).
145. Cope, A. C., G. A. Berchtold, P. E. Peterson and S. H. Sharman, J. Am. Chem. Soc. *8* 6370 (1960).
145a. Coulson, C. A., Acta Cryst. *B25*, 807 (1969).
146. Coursen, D. L. and J. L. Hoard, J. Am. Chem. Soc. *74*, 1742 (1952).
147. Cowan, R. D., J. Chem. Phys. *17*, 218 (1949).

148. COWAN, R. D., J. Chem. Phys. *18*, 1101 (1950).
148a. COWLEY, A. H. AND J. L. MILLS, J. Am. Chem. Soc. *91*, 2911 (1969).
149. COWLEY, A. H., H. H. SISLER AND G. E. RYSCHKEWITSCH, J. Am. Chem. Soc. *82*, 501 (1960).
149a. COYLE, T. D., J. COOPER AND J. J. RITTER, Inorg. Chem. *7*, 1014 (1968).
150. COYLE, T. D. AND F. G. A. STONE, J. Am. Chem. Soc. *83*, 4138 (1961).
151. COYLE, T. D., H. D. KAESZ AND F. G. A. STONE, J. Am. Chem. Soc. *81*, 2989 (1959).
151a. CRAGG, R. H. AND N. N. GREENWOOD, J. Chem. Soc. *A1967*, 961.
152. CRAWFORD, JR., B. AND J. T. EDSALL, J. Chem. Phys. *7*, 223 (1939).
152a. CUMMINS, J. D., M. YAMAUCHI AND B. WEST, Inorg. Chem. *6*, 2259 (1967).
153. CUNNINGHAM, G. L., J. M. BRYANT AND E. M. GAUSE (Callery Chem. Co.), U.S. Pat. 2,829,946 (1958); C.A. *52*, 13206c (1958).
153a. CURTIS, N. F., J. Chem. Soc. *1965*, 924.
154. DAHL, G. H. AND R. SCHAEFFER, J. Inorg. Nucl. Chem. *12*, 380 (1960).
155. DAHL, G. H. AND R. SCHAEFFER, J. Am. Chem. Soc. *83*, 3032 (1961).
156. DAHL, G. H. AND R. SCHAEFFER, J. Am. Chem. Soc. *83*, 3034 (1961).
156a. DAVIES, N., P. H. BIRD AND M. G. H. WALLBRIDGE, J. Chem. Soc. *A1968*, 2269.
157. DAVIS, R. E., J. Am. Chem. Soc. *84*, 892 (1962).
158. DAVIS, R. E., J. A. BLOOMER, D. R. COSPER AND A. SABA, Inorg. Chem. *3*, 460 (1964).
159. DAVIS, R. E., E. BROMELS AND C. L. KIBBY, J. Am. Chem. Soc. *84*, 885 (1962).
160. DAVIS, R. E., A. E. BROWN, R. HOPMANN AND C. L. KIBBY, J. Am. Chem. Soc. *85*, 487 (1963).
161. DAVIS, R. E. AND J. A. GOTTBRATH, J. Am. Chem. Soc. *84*, 895 (1962).
161a. DAVIS, R. E. AND J. A. GOTTBRATH, Inorg. Chem. *4*, 1512 (1965).
161b. DAVIS, R. E., R. E. KENSON, C. L. KIBBY AND H. H. LLOYD, Chem. Commun. *1965*, 593.
161c. DAVIS, R. E. AND R. E. KENSON, J. Am. Chem. Soc. *89*, 1384 (1967).
162. DAVIS, R. E., C. L. KIBBY AND C. G. SWAIN, J. Am. Chem. Soc. *82*, 5950 (1960).
163. DAVIS, R. E. AND C. G. SWAIN, J. Am. Chem. Soc. *82*, 5949 (1960).
164. DESSY, R. E. AND E. GRANNEN, JR., J. Am. Chem. Soc. *83*, 3953 (1961).
164a. DE MOOR, J. E. AND G. P. VAN DER KELEN, J. Organometal. Chem. *6*, 235 (1966).
164b. DE BOER, B. G., A. ZALKIN AND D. H. TEMPLETON, Inorg. Chem. *7*, 1085 (1968).
164c. DEEVER, W. R. AND D. M. RITTER, Inorg. Chem. *7*, 1036 (1968).
164d. DE BOER, B. G., A. ZALKIN AND D. H. TEMPLETON, Inorg. Chem. *7*, 2288 (1968).
164e. DEEVER, W. R., E. R. LORY AND D. M. RITTER, Inorg. Chem. *8*, 1263 (1969).
165. DIBLER, V. H. AND F. L. MOHLER, J. Am. Chem. Soc. *70*, 987 (1948).
166. DICKERSON, R. E. AND W. N. LIPSCOMB, J. Chem. Phys. *27*, 212 (1957).
167. DICKERSON, R. E., P. J. WHEATLEY, P. A. HOWELL, N. W. LIPSCOMB AND R. SCHAEFFER, J. Chem. Phys. *25*, 606 (1956).
168. DICKERSON, R. E., P. J. WHEATLEY, P. A. HOWELL AND N, W. LIPSCOMB, J. Chem. Phys. *27*, 200 (1957).
169. DILLARD, C. R., Thesis, University of Chicago, 1949; cit. J. P. FAUST, Adv. Chem. Ser. *32*, 73 (1961).
169a. DITTER, J. F., J. C. PERRINE AND J. SHAPIRO, J. Chem. Eng. Data *6*, 271 (1961).
169b. DITTER, J. F., Inorg. Chem. *7*, 1748 (1968).
170. DITTER, J. F. AND I. SHAPIRO, J. Am. Chem. Soc. *81*, 1022 (1959).
170a. DITTER, J. F., J. R. SPIELMAN AND R. E. WILLIAMS, Inorg. Chem. *5*, 118 (1966).
171. DOBROTT, R. D., L. B. FRIEDMAN AND W. N. LIPSCOMB, J. Chem. Phys. *40*, 866 (1964).
171a. DOBROTT, R. D. AND W. N. LIPSCOMB, J. Chem. Phys. *37*, 1779 (1962).
171b. DOBSON, J., P. C. KELLER AND R. SCHAEFFER, Inorg. Chem. *7*, 399 (1968).
171c. DOBSON, J. AND R. SCHAEFFER, Inorg. Chem. *7*, 402 (1968).
171d. DOBSON, J., P. C. KELLER AND R. SCHAEFFER, J. Am. Chem. Soc. *87*, 3522 (1965).
172. DOBSON, J. AND R. SCHAEFFER, Inorg. Chem. *4*, 593 (1965).
173. DOUGLASS, J. E., J. Am. Chem. Soc. *84*, 121 (1962).
174. DOUGLASS, J. E., J. Am. Chem. Soc. *86*, 5431 (1964).
175. DOW CHEMICAL CO., Midland, Mich., U.S. Pat. 3,024,091 (1958); C.A. *57*, 8198b (1962).
175a. DOWNS, A. J. AND J. H. MORRIS, Spectrochim. Acta *22*, 957 (1966).
175b. DRAKE, J. E. AND J. SIMPSON, Inorg. Nucl. Chem. Letters *3*, 87 (1967).
175c. DRAKE, J. E. AND J. SIMPSON, Inorg. Chem. *6*, 1984 (1967).
175d. DREEVER, W. R. AND D. M. RITTER, J. Am. Chem. Soc. *89*, 5073 (1967).
175e. DRAKE, J. E. AND J. SIMPSON, J. Chem. Soc. *A1968*, 974.
175f. DRAKE, J. E. AND C. RIDDLE, J. Chem. Soc. *A1968*, 1675.
176. DRISCOLL, J. S. AND C. N. MATTHEWS, Chem. Ind. (London) *1963*, 1282.

177. DUFFEY, G. H., J. Chem. Phys. *20*, 194 (1952).
178. DUKE, B. J., J. R. GILBERT AND I. A. READ, J. Chem. Soc. *1964*, 540.
179. DULMAGE, W. J. AND W. N. LIPSCOMB, J. Am. Chem. Soc. *73*, 3539 (1951).
180. DULMAGE, W. J. AND W. N. LIPSCOMB, Acta Cryst. *5*, 260 (1952).
181. DULOU, R. AND Y. CHRÉTIEN-BESSIÈRE, Bull. Soc. Chim. France *1959*, 1362.
181a. DUNKS, G. B. AND M. F. HAWTHORNE, Inorg. Chem. *7*, 1038 (1968).
181b. DUNKS, G. B. AND M. F. HAWTHORNE, J. Am. Chem. Soc. *90*, 7355 (1968).
182. DUNSTAN, I., N. J. BLAY AND R. L. WILLIAMS, J. Chem. Soc. *1960*, 5016.
183. DUNSTAN, I., R. L. WILLIAMS AND N. J. BLAY, J. Chem. Soc. *1960*, 5012.
184. DUPONT, J. A. AND M. F. HAWTHORNE, J. Am. Chem. Soc. *81*, 4998 (1959).
185. DUPONT, J. A. AND M. F. HAWTHORNE, J. Am. Chem. Soc. *84*, 1804 (1962).
186. DUPONT, J. A. AND M. F. HAWTHORNE, J. Am. Chem. Soc. *86*, 1643 (1964).
187. DUPONT, J. A. AND M. F. HAWTHORNE, Chem. Ind. London *1962*, 405.
188. DUPONT, J. A. AND R. SCHAEFFER, J. Inorg. Nucl. Chem. *15*, 310 (1960).
189. DYKE, C. H. VAN AND A. G. MACDIARMID, J. Inorg. Nucl. Chem. *25*, 1503 (1963).
189a. EASTHAM, J. F., J. Am. Chem. Soc. *89*, 2237 (1967).
189b. EATON, D. R., Inorg. Chem. *4*, 1520 (1965).
190. EBERHARDT, W. H., B. CRAWFORD, JR. AND W. N. LIPSCOMB, J. Chem. Phys. *22*, 989 (1954).
191. EDDY, L. B., S. H. SMITH, JR. AND R. R. MILLER, J. Am. Chem. Soc. *77*, 2105 (1955).
192. EGAN, B. Z. AND S. G. SHORE, J. Am. Chem. Soc. *83*, 4717 (1961).
193. EGAN, B. Z., S. G. SHORE AND J. E. BONNELL, Inorg. Chem. *3*, 1024 (1964).
193a. ELLERMANN, J. AND W. H. GRUBER, Chem. Ber. *102*, 1 (1969).
194. ELLIOT, J. R., W. L. ROTH, G. F. ROEDEL AND E. M. BOLDENBUCK, J. Am. Chem. Soc. *74*, 5211 (1952).
194a. ELLIS, I. A., D. F. GAINES AND R. SCHAEFFER, J. Am. Chem. Soc. *85*, 3885 (1963).
195. EMELÉUS, H. J. AND F. G. A. STONE, J. Chem. Soc. *1951*, 840.
196. EMELÉUS, H. J. AND G. J. VIDELA, Proc. Chem. Soc. *1957*, 288.
197. EMELÉUS, H. J. AND G. J. VIDELA, J. Chem. Soc. *1959*, 1306.
198. EMELÉUS, H. J. AND K. WADE, J. Chem. Soc. *1960*, 2614.
199. EMERY, F. W., P. L. HAROLD AND A. J. OWEN, J. Chem. Soc. *1964*, 4931.
200. EMERY, A. R. AND R. C. TAYLOR, J. Chem. Phys. *28*, 1029 (1958).
201. EMERY, A. R. AND R. C. TAYLOR, Spectrochim. Acta *16*, 1455 (1960).
201a. ENEMARK, J. H., L. B. FRIEDMAN, J. A. HARTSUCK AND W. N. LIPSCOMB, J. Am. Chem. Soc. *88*, 3659 (1966).
201b. ENEMARK, J. H., L. B. FRIEDMAN AND W. N. LIPSCOMB, Inorg. Chem. *5*, 2165 (1966).
202. ENRIONE, R. E., F. P. BOER AND W. N. LIPSCOMB, J. Am. Chem. Soc. *86*, 1415 (1964).
203. ENRIONE, R. E., F. P. BOER AND W. N. LIPSCOMB, Inorg. Chem. *3*, 1659 (1964).
204. ENRIONE, R. E. AND R. SCHAEFFER, J. Inorg. Nucl. Chem. *18*, 103 (1961).
205. ERINKS, K., W. N. LIPSCOMB AND R. SCHAEFFER, J. Chem. Phys. *22*, 754 (1954).
205a. EVANS, W. G., C. E. HOLLOWAY, K. SUKUMARABANDHU AND D. H. MCDANIEL, Inorg. Chem. *7*, 1446 (1968).
206. EVERS, E. C., W. O. FREITAG, J. N. KEITH, W. A. KRINER, A. G. MACDIARMID AND S. SUJISHI, J. Am. Chem. Soc. *81*, 4493 (1959).
207. FARBENFABRIKEN BAYER A.G., Brit. Pat. 822,229 (1959); C.A. *54*, 8634d (1960).
208. FARBENFABRIKEN BAYER A.G., Germ. Pat. 1,070,148 (1959); C.A. *55*, 18039e (1961).
208a. FARRAR, T. C., J. COOPER AND T. D. COYLE, Chem. Commun. *1966*, 610.
209. FAUST, J. P., Advan. Chem. Ser. *32*, 74 (1961).
209a. FEHLNER, T. P., J. Am. Chem. Soc. *87*, 4200 (1965).
210. FEHLNER, T. P. AND W. S. KOSKI, J. Am. Chem. Soc. *85*, 1905 (1963).
211. FEHLNER, T. P. AND W. S. KOSKI, J. Am. Chem. Soc. *86*, 581 (1964).
212. FEHLNER, T. P. AND W. S. KOSKI, J. Am. Chem. Soc. *86*, 1012 (1964).
213. FEHLNER, T. P. AND W. S. KOSKI, J. Am. Chem. Soc. *86*, 2733 (1964).
214. FEHLNER, T. P. AND W. S. KOSKI, J. Am. Chem. Soc. *87*, 409 (1965).
214a. FEIN, M. M., J. BOBINSKI, J. E. PAUSTIAN, D. GRAFSTEIN AND M. S. COHEN, Inorg. Chem. *4*, 422 (1965).
215. FEIN, M. M., J. BOBINSKI, N. MAYERS, N. SCHWARTZ AND M. S. COHEN, Inorg. Chem. *2*, 1111 (1963).
216. FEIN, M. M., J. GREEN, J. BOBINSKI AND M. S. COHEN, Inorg. Chem. *4*, 583 (1965).
217. FEIN, M. M., D. GRAFSTEIN, J. E. PAUSTIAN, J. BOBINSKI, B. M. LICHSTEIN, N. MAYES, N. N. SCHWARTZ AND M. S. COHEN, Inorg. Chem. *2*, 1115 (1963).

218. Fedneva, E. M., Zh. Neorgan. Khim. *4*, 286 (1959); C.A. *53*, 12911 (1959).
219. Feodor, W. S., M. D. Banus and D. P. Ingalls, Ind. Eng. Chem. *49*, 1664 (1957).
220. Fetter, N. R., Chem. Ind. London *1959*, 1548.
221. Figgis, B. and R. L. Williams, Spectrochim. Acta *15*, 331 (1959).
222. Finholt, A. E. (Metal Hydrides, Inc.), U.S. Pat. 2,550,985 (1951); C.A. *45*, 7757e (1951).
223. Finholt, A. E., A. C. Bond, Jr. and H. I. Schlesinger, J. Am. Chem. Soc. *69*, 1199 (1947).
224. Fisher, N. G., U.S. Pat. 2,729,540 (1956); C.A. *50*, 6758b (1956).
225. Fitch, S. J. and A. W. Laubengayer, J. Am. Chem. Soc. *80*, 5911 (1958).
226. Flodin, R. E., L. A. Wall, F. L. Mohler and E. Quinn, J. Am. Chem. Soc. *76*, 3344 (1954).
227. Ford, M. D. (Callery Chem. Co., Pittsburgh, Pa.) U.S. Pat. 3,032,480 (1962); C.A. *57*, 5764e (1962).
228. Ford, T. A., G. H. Kalb, A. L. McClelland and E. L. Muetterties, Inorg. Chem. *3*, 1032 (1964).
229. Ford, P. T. and H. M. Powell, Acta Cryst. *7*, 604 (1954).
230. Ford, P. T. and R. E. Riohards, Discussions Faraday Soc. *19*, 230 (1955).
231. Forster, R., Nature *195*, 490 (1962).
232. Forstner, J. A., T. E. Haas and E. L. Muetterties, Inorg. Chem. *3*, 155 (1964).
232a. Francis, J. N. and M. F. Hawthorne, J. Am. Chem. Soc. *90*, 1663 (1968).
232b. Franz, D. A., J. W. Howard and R. N. Grimes, J. Am. Chem. Soc. *91*, 4010 (1969).
233. Freedman, L. D. and G. O. Doak, J. Am. Chem. Soc. *74*, 3414 (1952).
234. Freund, T., J. Inorg. Nucl. Chem. *9*, 246 (1959).
234a. L. B. Friedman, R. E. Cook and M. D. Glick, J. Am. Chem. Soc. *90*, 6862 (1968).
235. Friedman, L. B., R. D. Dobrotb and W. N. Lipscomb, J. Am. Chem. Soc. *85*, 3505 (1963).
235a. Friedman, L. B. and W. L. Lipscomb, Inorg. Chem. *5*, 1752 (1966).
235b. Fritchie, Jr., C. J., Inorg. Chem. *6*, 1199 (1967).
235c. Fritz, G. and F. Pfannerer, Z. Anorg. Allg. Chem. in press.
235d. Frisch, M. A., H. G. Heal, H. Mackle and I. O. Madden, J. Chem. Soc. *1965*, 899.
236. Gaines, D. F., Inorg. Chem. *2*, 523 (1963).
236a. Gaines, D. F., J. Am. Chem. Soc. *88*, 4528 (1966).
237. Gaines, D. F. and R. Schaeffer, J. Am. Chem. Soc. *85*, 395 (1963).
237a. Gaines, D. F. and T. V. Iorns, J. Am. Chem. Soc. *89*, 3375 (1967).
237b. Gaines, D. F. and T. V. Iorns, J. Am. Chem. Soc. *89*, 4249 (1967).
237c. Gaines, D. F. and T. V. Iorns, Inorg. Chem. *7*, 1041 (1968).
237d. Gaines, D. F. and J. A. Martens, Inorg. Chem. *7*, 704 (1968).
237e. Gaines, D. F. and T. V. Iorns, J. Am. Chem. Soc. *90*, 6617 (1968).
237f. Gaines, D. F., J. Am. Chem. Soc. *91*, 1230 (1969).
238. Gaines, D. F. and R. Schaeffer, J. Am. Chem. Soc. *85*, 3592 (1963).
239. Gaines, D. F. and R. Schaeffer, J. Am. Chem. Soc. *86*, 1505 (1964).
240. Gaines, D. F. and R. Schaeffer, Proc. Chem. Soc. *1963*, 267.
241. Gaines, D. F. and R. Schaeffer, Inorg. Chem. *3*, 438 (1964).
242. Gaines, D. F., R. Schaeffer and F. Tebbe, Inorg. Chem. *2*, 526 (1963).
243. Gakle, P. S. and S. Tannenbaum, J. Am. Chem. Soc. *77*, 5289 (1955).
244. Gallaghan, J. and B. Siegel, J. Am. Chem. Soc. *81*, 504 (1959).
245. Gamble, E. L. and P. Gilmont, J. Am. Chem. Soc. *62*, 717 (1940).
246. Garabedian, M. E. and S. W. Benson, J. Am. Chem. Soc. *86*, 176 (1964).
247. Gardiner, J. A. and J. W. Collat, J. Am. Chem. Soc. *86*, 3165 (1964).
248. Gardiner, J. A. and J. W. Collat, J. Am. Chem. Soc. *87*, 1692 (1965).
248a. Gardiner, J. A. and J. W. Collat, Inorg. Chem. *4*, 1208 (1965).
248b. Garrett, P. M., J. C. Smart, G. S. Ditta and M. F. Hawthorne, Inorg. Chem. *8*, 1907 (1969).
248c. Garrett, P. M., J. C. Smart and M. F. Hawthorne, J. Am. Chem. Soc. *91*, 4707 (1969).
249. Garrett, P. M., F. N. Tebbe and M. F. Hawthorne, J. Am. Chem. Soc. *86*, 5016 (1964).
249a. Gatti, A. R. and T. Wartik, Inorg. Chem. *5*, 2075 (1966).
249b. Geanangel, R. A. and S. G. Shore, J. Am. Chem. Soc. *89*, 6771 (1967).
250. Gee, W., J. B. Holden, R. A. Shaw and B. C. Smith, J. Chem. Soc. *1964*, 3171.
250a. Gee, W., J. B. Holden, R. A. Shaw and B. C. Smith, J. Chem. Soc. *A1967*, 1545.
251. Geller, S., J. Chem. Phys. *32*, 1569 (1960).
252. Geller, S., R. E. Hughes and J. L. Hoard, Acta Cryst. *4*, 380 (1951).
252a. General Electric, Rept. No. 55248 (March 20, 1950).
252b. George, T. A. and M. F. Hawthorne, J. Am. Chem. Soc. *90*, 1661 (1968).

253. GERRARD, W., *The Organic Chemistry of Boron*, Academic Press, London, 1961, p. 174.
254. GERRARD, W., H. R. HUDSON, E. F. MOONEY, I. M. STRIPP AND H. A. WILLIS, Spectrochim. Acta *18*, 149 (1962).
255. GERRARD, W., H. R. HUDSON AND E. F. MOONEY, J. Chem. Soc. *1962*, 113.
256. GERRARD, W. AND E. F. MOONEY, J. Chem. Soc. *1960*, 4028.
257. GERRARD, W., E. F. MOONEY AND H. A. WILLIS, Spectrochim. Acta *18*, 155 (1962).
258. GIBBINS, S. G. AND I. SHAPIRO, J. Am. Chem. Soc. *82*, 2968 (1960).
259. GIBBINS, S. G. AND I. SHAPIRO, J. Chem. Phys. *30*, 1483 (1959).
259a. GILJE, J. W., K. W. MORSE AND R. W. PARRY, Inorg. Chem. *6*, 1761 (1967).
260. GIRARDOT, P. AND R. W. PARRY, J. Am. Chem. Soc. *73*, 2368 (1951).
261. GLEMSER, O., Germ. Pat. 949,943 (1956); C.A. *51*, 14785h (1957).
262. GOBBETT, E. AND J. W. LINNETT, J. Chem. Soc. *1962*, 2893.
263. GOLDSTEIN, P. AND R. A. JACOBSON, J. Am. Chem. Soc. *84*, 2457 (1962).
263a. GOLDSTEIN, M. AND E. F. MOONEY, Chem. Commun. *1966*, 104.
264. GOODROW, M. H., R. I. WAGNER AND R. D. STEWART, Inorg. Chem. *3*, 1212 (1964).
264a. GORBUNOV, A. I., G. S. SOLOV'EVA, I. S. ANTONOV AND M. S. KHARSON, Russ. J. Inorg. Chem. (English Transl.) *10*, 1074 (1965).
264b. GORBUNOV, A. I. AND G. S. SOLOV'EVA, Russ. J. Inorg. Chem., (English Transl.) *12*, 1 (1967).
264c. GOUBEAU, J. AND R. BERGMANN, Z. Anorg. Allgem. Chem. *263*, 68 (1950).
265. GOUBEAU, J. AND K. KALLFASS, Z. Anorg. Allgem. Chem. *299*, 160 (1959).
266. GOUBEAU, J. AND H. KELLER, Z. Anorg. Allgem. Chem. *272*, 303 (1953).
266a. GOUBEAU, J., E. BESSLER AND D. WOLF, Z. Anorg. Allgem. Chem. *352*, 285 (1967).
267. GOUBEAU, J. AND E. RICKER, Z. Anorg. Allgem. Chem. *310*, 123 (1961).
268. GRAFSTEIN, D., J. BOBINSKI, J. DVORAK, J. E. PAUSTIAN, H. F. SMITH, S. KARLAN, C. VOGEL AND M. M. FEIN, Inorg. Chem. *2*, 1125 (1963).
269. GRAFSTEIN, D., J. BOBINSKI, J. DVORAK, H. SMITH, N. SCHWARTZ, M. S. COHEN AND M. M. FEIN, Inorg. Chem. *2*, 1120 (1963).
270. GRAFSTEIN, D. AND J. DVORAK, Inorg. Chem. *2*, 1128 (1963).
271. GRAHAM, W. A. G. AND F. G. A. STONE, J. Inorg. Nucl. Chem. *3*, 164 (1956).
271a. GRASSBERGER, M. A., E. G. HOFFMANN, G. SCHOMBURG AND R. KÖSTER, J. Am. Chem. Soc. *90*, 56 (1968).
272. GRAYBILL, B. M. AND M. F. HAWTHORNE, J. Am. Chem. Soc. *83*, 2573 (1961).
273. GRAYBILL, B. M., A. R. PITOCHELLI AND M. F. HAWTHORNE, Inorg. Chem. *1*, 626 (1962).
274. GRAYBILL, B. M., A. R. PITOCHELLI AND M. F. HAWTHORNE, Inorg. Chem. *1*, 622 (1962).
275. GRAYBILL, B. M., A. R. PITOCHELLI AND M. F. HAWTHORNE, Abstr. Papers 138th Meeting Am. Chem. Soc., New York, 1960.
276. GRAYBILL, B. M., J. K. RUFF AND M. F. HAWTHORNE, J. Am. Chem. Soc. *83*, 2669 (1961).
277. GRAYBILL, B. M. AND J. K. RUFF, J. Am. Chem. Soc. *84*, 1062 (1962).
277a. GREENWOOD, N. N. AND N. F. TRAVERS, J. Chem. Soc. *A1967*, 880.
277b. GREENWOOD, N. N. AND J. WALKER, J. Chem. Soc. *A1967*, 959.
277c. GREENWOOD, N. N. AND N. F. TRAVERS, J. Chem. Soc. *A1967*, 880.
277d. GREENWOOD, N. N. AND N. F. TRAVERS, J. Chem. Soc. *A1968*, 15.
277e. GREENWOOD, N. N. AND J. A. MCGINNETY, J. Chem. Soc. *1965*, 331.
277f. GREENWOOD, N. N. AND J. H. MORRIS, Proc. Chem. Soc. *1963*, 338.
277g. GREENWOOD, N. N. AND J. H. MORRIS, J. Chem. Soc. *1965*, 6205.
277h. GREENWOOD, N. N. AND D. N. SHARROCKS, J. Chem. Soc. *1969*, 2334.
278. GREEN, J. AND A. P. KOTLOBY, Inorg. Chem. *4*, 599 (1965).
278a. GREENWOOD, N. N. AND J. A. MCGINNETY, J. Chem. Soc. *A 1966*, 1090.
278b. GREENWOOD, N. N. AND N. F. TRAVERS, Chem. Commun. *1967*, 216.
278c. GRIMES, R. N., J. Am. Chem. Soc. *88*, 1895 (1966).
278d. GRIMES, R. N. AND C. L. BRAMLETT, J. Am. Chem. Soc. *89*, 2557 (1967).
278e. GRIMES, R. N., J. Am. Chem. Soc. *88*, 1070 (1966).
278f. GRIMES, R. N., J. Organometal. Chem. *8*, 45 (1967).
278g. GRIMES, R. N., C. L. BRAMLETT AND R. L. VANCE, Inorg. Chem. *7*, 1066 (1968).
278h. GRIMES, R. N., CH. L. BRAMLETT AND R. L. VANCE, Inorg. Chem. *8*, 55 (1969).
279. GRIMES, R. N. AND W. N. LIPSCOMB, Proc. Natl. Acad. Sci. U.S. *48*, 496 (1962).
280. GRIMES, R., F. E. WANG, R. LEWIN AND W. N. LIPSCOMB, Proc. Natl. Acad. Sci. U.S. *47*, 996 (1961).
280a. GRIMM, F. A., L. BARTON AND R. F. PORTER, Inorg. Chem. *7*, 1309 (1968).
280b. GRIMM, F. A. AND R. F. PORTER, Inorg. Chem. *7*, 706 (1968).

280c. GRIMM, F. A. AND R. F. PORTER, Inorg. Chem. *8*, 731 (1969).
281. GRIMLEY, J. AND A. K. HOLLIDAY, J. Chem. Soc. *1954*, 1212.
281a. GROSZOS, S. J. AND S. F. STAFIEJ, J. Am. Chem. Soc. *80*, 1357 (1958).
281b. GROTEWOLD, J., E. A. LISSI AND A. E. VILLA, J. Chem. Soc. *A 1966*, 1038.
281c. GUGGENBERGER, L. J., Inorg. Chem. *7*, 2260 (1968).
282. GUNN, S. R. AND LEROY G. GREEN, J. Am. Chem. Soc. *77*, 6197 (1955).
283. GUNN, S. R. AND L. G. GREEN, J. Chem. Phys. *36*, 1118 (1962).
284. GUNN, S. R. AND L. G. GREEN, J. Chem. Phys. *37*, 2724 (1962).
285. GUTER, G. A. AND G. W. SCHAEFFER, J. Am. Chem. Soc. *78*, 3546 (1956).
286. HAALAND, A. AND W. H. EBERHARD, J. Chem. Phys. *36*, 2386 (1962).
287. HAAS, T. E., Inorg. Chem. *3*, 1053 (1964).
288. HAGENMULLER, P. AND M. RAULT, Compt. Rend., *248*, 2758 (1959).
289. HAHN, G. A. AND R. SCHAEFFER, J. Am. Chem. Soc. *86*, 1503 (1964).
290. HALL, L. H., J. Am. Chem. Soc. *86*, 4729 (1964).
290a. HALL, L. H., W. S. KOSKI AND V. V. SUBBANNA, J. Am. Chem. Soc. *86*, 1304 (1964).
290b. HALL, R. E. AND E. P. SCHRAM, Inorg. Chem. *8*, 270 (1969).
290b. HALL, L. H., S. BLOCK AND A. PERLOFF, Acta. Cryst. *19*, 658 (1965).
290c. HALL, L. H., A. PERLOFF, F. A. MAUER AND S. BLOCK, J. Chem. Phys. *43*, 3911 (1965).
291. HALL, L. H., V. V. SUBBANNA AND W. S. KOSKI, J. Am. Chem. Soc. *86*, 3969 (1964).
292. HAMILTON, W. C., J. Chem. Phys. *29*, 460 (1958).
292a. HART, H. V. AND W. N. LIPSCOMB, J. Am. Chem. Soc. *89*, 4220 (1967).
292b. HART, H. V. AND W. N. LIPSCOMB, J. Am. Chem. Soc. *91*, 771 (1969).
292c. HANIC, F. AND V. SUBRTOVÁ, Acta Cryst. *B25*, 405 (1969).
293. HAMILTON, W. C., Acta Cryst. *8*, 199 (1955).
293a. HARMON, A. B. AND K. M. HARMON, J. Am. Chem. Soc. *88*, 4093 (1966).
293b. HARMON, K. M., A. B. HARMON AND B. C. THOMPSON, J. Am. Chem. Soc. *89*, 5309 (1967).
293c. HART, H. AND W. N. LIPSCOMB, Inorg. Chem. *7*, 1070 (1968).
294. HARMON, K. M., A. B. HARMON AND A. A. MACDONALD, J. Am. Chem. Soc. *86*, 5036 (1964).
294a. HAWORTH, D. T. AND L. F. HOHNSTEDT, J. Am. Chem. Soc. *82*, 3860 (1960).
294b. HARMON, K. M., A. B. HARMON AND A. A. MACDONALD, J. Am. Chem. Soc. *91*, 323 (1969).
294c. HARSHBARGER, W., G. LEE, R. F. PORTER AND S. H. BAUER, Inorg. Chem. *8*, 1683 (1969).
295. HAWTHORNE, M. F., Advan. Inorg. Chem. Radiochem. *5*, 307 (1963).
295b. HAWTHORNE, M. F., in *The Chemistry of Boron and its Compounds*, Ed. E. L. MUETTERTIES, John Wiley & Sons, Inc., New York/London/Sydney, 1967.
296. HAWTHORNE, M. F. AND J. J. MILLER, J. Am. Chem. Soc. *80*, 754 (1958).
297. HAWTHORNE, M. F., J. Am. Chem. Soc. *80*, 3480 (1958).
298. HAWTHORNE, M. F., J. Am. Chem. Soc. *80*, 4291 (1958).
299. HAWTHORNE, M. F., J. Am. Chem. Soc. *80*, 4293 (1958).
300. HAWTHORNE, M. F., Abstr. Papers 135th Meeting Am. Chem. Soc., Boston, 1959.
301. HAWTHORNE, M. F., J. Am. Chem. Soc. *81*, 5836 (1959).
302. HAWTHORNE, M. F., J. Am. Chem. Soc. *82*, 748 (1960).
303. HAWTHORNE, M. F., J. Am. Chem. Soc. *83*, 367 (1961).
304. HAWTHORNE, M. F., J. Am. Chem. Soc. *83*, 831 (1961).
305. HAWTHORNE, M. F., J. Am. Chem. Soc. *83*, 833 (1961).
306. HAWTHORNE, M. F., J. Am. Chem. Soc. *83*, 1345 (1961).
307. HAWTHORNE, M. F., J. Am. Chem. Soc. *83*, 2541 (1961).
308. HAWTHORNE, M. F., J. Am. Chem. Soc. *83*, 2671 (1961).
308a. HAWTHORNE, M. F. AND T. D. ANDREWS, J. Am. Chem. Soc. *87*, 2496 (1965).
309. HAWTHORNE, M. F., Chem. Ind. (London) *1957*, 1242.
309a. HAWTHORNE, M. F., T. E. BERRY AND P. A. WEGNER, J. Am. Chem. Soc. *87*, 4746 (1965).
310. HAWTHORNE, M. F. AND W. L. BUDDE, J. Am. Chem. Soc. *86*, 5337 (1964).
311. HAWTHORNE, M. F., W. L. BUDDE AND D. WALMSLEY, J. Am. Chem. Soc. *86*, 5337 (1964).
311a. HAWTHORNE, M. F. AND T. A. GEORGE, J. Am. Chem. Soc. *89*, 7114 (1967).
312. HAWTHORNE, M. F. AND E. S. LEWIS, J. Am. Chem. Soc. *80*, 4296 (1958).
313. HAWTHORNE, M. F. AND J. J. MILLER, J. Am. Chem. Soc. *82*, 500 (1960).
314. HAWTHORNE, M. F. AND F. P. OLSEN, J. Am. Chem. Soc. *86*, 4219 (1964).
314a. HAWTHORNE, M. F. AND R. L. PILLING, J. Am. Chem. Soc. *88*, 3873 (1966).
314b. HAWTHORNE, M. F., R. L. PILLING AND P. M. GARRETT, J. Am. Chem. Soc. *87*, 4740 (1965).
314c. HAWTHORNE, M. F. AND R. L. PILLING, J. Am. Chem. Soc. *87*, 3987 (1965).
315. HAWTHORNE, M. F., R. I. PILLING AND R. N. GRIMES, J. Am. Chem. Soc. *86*, 5338 (1964).

315a. HAWTHORNE, M. F., R. I. PILLING, P. F. STOKELY AND P. M. GARRETT, J. Am. Chem. Soc. *85*, 3704 (1963).
315b. HAWTHORNE, M. F., R. L. PILLING, AND P. F. STOKELY, J. Am. Chem. Soc. *87*, 1893 (1965).
316. HAWTHORNE, M. F. AND A. R. PITOCHELLI, J. Am. Chem. Soc. *80*, 6685 (1958).
317. HAWTHORNE, M. F. AND A. R. PITOCHELLI, J. Am. Chem. Soc. *81*, 5519 (1959).
317a. HAWTHORNE, M. F. AND A. R. PITOCHELLI, J. Am. Chem. Soc. *82*, 3228 (1960).
318. HAWTHORNE, M. F., A. R. PITOCHELLI, R. D. STRAHM AND J. J. MILLER, J. Am. Chem. Soc. *82*, 1825 (1960).
318a. HAWTHORNE, M. F. AND A. D. PITTS, J. Am. Chem. Soc. *89*, 7115 (1967).
318b. HAWTHORNE, M. F. AND H. W. RUHLE, Inorg. Chem. *8*, 176 (1969).
319. HAWTHORNE, M. F., D. C. YOUNG AND P. A. WEGNER, J. Am. Chem. Soc. *87*, 1818 (1965).
319a. HAWTHORNE, M. F. AND P. A. WEGENER, Inorg. Chem. *3*, 774 (1964).
319b. HAWTHORNE, M. F., P. A. WEGENER AND R. C. STAFFORD, Inorg. Chem. *4*, 1675 (1965).
319c. HAWTHORN, M. F. AND P. A. WEGENER J. Am. Chem. Soc. *90*, 896 (1968).
319d. HAWTHORN, M.F., D. C. YOUNG, T. D. ANDREWS, D. V. HOWE, R. L. PILLING, A. D. PITTS, M. REINTJES, L. F. WARREN, JR. AND P. A. WEGNER, J. Am. Chem. Soc. *90*, 879 (1968).
319e. HAWTHORNE, M. F., D. C. YOUNG, P. M. GARRETT, D. A. OWEN, S. G. SCHWERIN, F. N. TEBBE AND P. A. WEGNER J. Am. Chem. Soc. *90*, 862 (1968).
319f. HAWTHORNE, M. F.AND P. A. WEGNER, J. Am. Chem. Soc. *87*, 4392 (1965).
319g. HAWTHORNE, M. F. AND F. P. OLSEN, J. Am. Chem. Soc. *87*, 2366 (1965).
319h. HAWTHORNE, M. F., R. L. PILLING AND R. N. GRIMES, J. Am. Chem. Soc. *89*, 1067 (1967).
319i. HAWTHORNE, M. F., R. L. PILLING AND R. C. VASAVADA, J. Am. Chem. Soc. *89*, 1075 (1967).
320. HAWORTH, D. L. AND L. F. HOHNSTEDT, J. Am. Chem. Soc. *81*, 842 (1959).
321. HAWORTH, D. L. AND L. F. HOHNSTEDT, J. Am. Chem. Soc. *82*, 3860 (1960).
322. HAWORTH, D. L. AND L. F. HOHNSTEDT, Chem. Ind. London *1960*, 559.
323. HEAL, H. G., J. Inorg. Nucl. Chem. *12*, 255 (1960).
324. HEAL, H. G., J. Inorg. Nucl. Chem. *16*, 208 (1961).
325. HEDBERG, K., M. E. JONES AND V. SCHOMAKER, Proc. Natl. Acad. Sci. U.S. *38*, 679 (1952).
326. HEDBERG, K., M. E. JONES AND V. SCHOMAKER, J. Am. Chem. Soc. *73*, 3538 (1951).
327. HEDBERG, K. AND V. SCHOMAKER, J. Am. Chem. Soc. *73*, 1482 (1951).
328. HEDBERG, K. AND A. STOSIK, J. Am. Chem. Soc. *74*, 954 (1952).
329. HEITSCH, C. W., Inorg. Chem. *3*, 767 (1964).
329a. HEITSCH, C. W., Inorg. Chem. *4*, 1019 (1965). HERBER, R. H., Inorg. Chem. *8*, 174 (1969).
330. HENLE, W. (Deutsche Gold- und Silberscheideanstalt), Germ. Pat. 1, 025, 854 (1958).
330a. HERTEL, G. R. AND W. S. KOSKI, J. Am. Chem. Soc. *87*, 404 (1965).
331. HERTLER, W. R., Inorg. Chem. *3*, 1195 (1964).
332. HERTLER, W. R., J. Am. Chem. Soc. *86*, 2949 (1964).
333. HERTLER, W. R., W. H. KNOTH AND E. L. MUETTERTIES, J. Am. Chem. Soc. *86*, 5434 (1964).
334. HERTLER, W. R., W. H. KNOTH AND E. L. MUETTERTIES, Inorg. Chem. *4*, 288 (1965).
334a. HERTLER, W. R. AND E. L. MUETTERTIES, Inorg. Chem. *5*, 160 (1966).
335. HERTLER, W. R. AND M. S. RAASCH, J. Am. Chem. Soc. *86*, 3661 (1964).
336. HESSE, G. AND H. JÄGER, Chem. Ber. *92*, 2022 (1959).
336a. HESTER, R. E. AND E. MAYER, Spectrochim. Acta *23A*, 2218 (1967).
337. HEWITT, F. AND A. K. KOLLIDAY, J. Chem. Soc. *1953*, 530.
338. HEYING, T. L., J. W. AGER, JR., S. L. CLARK, D. J. MANGOLD, H. L. GOLDSTEIN, M. HILLMAN, R. J. POLAK AND J. W. SZYMANSKI, Inorg. Chem. *2*, 1089 (1963).
339. HEYING, T. L., J. W. AGER, JR., S. L. CLARK, R. P. ALEXANDER, S. PAPETTI, J. A. REID AND S. I. TROTZ, Inorg. Chem. *2*, 1097 (1963).
340. HEYING, T. L. AND C. NAAR-COLIN, Inorg. Chem. *3*, 282 (1964).
341. HILLMAN, M., J. Am. Chem. Soc. *82*, 1096 (1960).
342. HILLMAN, M., J. Inorg. Nucl. Chem. *12*, 383 (1960).
342a. HILLMAN, M. AND D. J. MANGOLD, Inorg. Chem. *4*, 1356 (1965).
343. HILLMAN. M. AND D. J. MANGOLD AND J. H. NORMAN, J. Inorg. Nucl. Chem. *24*, 1565 (1962).
344. HILLMAN, M., D. J. MANGOLD AND J. H. NORMAN, Advan. Chem. Ser. *32*, 157 (1961).
345. HIRATA, T. AND H. E. GUNNING, J. Chem. Phys. *27*, 477 (1957).
346. HIRSHFELD, F. L., K. ERIKS, R. E. DICKERSON, E. L. LIPPERT, JR. AND W. N. LIPSCOMB, J. Chem. Phys. *28*, 56 (1958).
347. HÖCKELE, G., Thesis München University, 1960.
348. HOEKSTRA, H. R. AND J. J. KATZ, J. Am. Chem. Soc. *71*, 2488 (1949).

349. HOEKSTRA, H. R. AND J. J. KATZ, U.S. Pat. 2, 575, 760 (1951); C.A. *46* 2248f (1952).
350. HÖNIGSCHMID-GROSSICH, R., personal communication.
351. HÖNIGSCHMID-GROSSICH, R., Thesis, München University, 1964.
352. HOFFMANN, R. H. AND W. N. LIPSCOMB, J. Chem. Phys. *37*, 520 (1962).
353. HOFFMANN, R. H. AND W. N. LIPSCOMB, J. Chem. Phys. *37*, 2872 (1962).
354. HOHNSTEDT, L. F. AND G. W. SCHAEFFER, Advan. Chem. Ser. *32*, 232 (1961).
354a. HOLLIDAY, A. K. AND G. N. JESSOP, J. Organometal. Chem. *10*, 291 (1967).
354b. HOLZMANN, R. T. (Editor), R. L. HUGHES, I. C. SMITH AND E. W. LAWLESS, *Production of the Boranes and Related Research*, Academic Press, New York/London 1967.
355. HONEYCUTT, J. B. AND J. M. RIDDLE, J. Am. Chem. Soc. *83*, 369 (1961).
356. HOHNSTEDT, L. F. AND D. T. HAWORTH, J. Am. Chem. Soc. *82*, 89 (1960).
356a. HOPKINS, R. C., J. D. BALDESCHWIELER, R. SCHAEFFER, F. N. TEBBE AND A. NORMAN, J. Chem. Phys. **43**, 973 (1965).
357. HORAU, J., N. LUMBROSO AND A. PACAULT, Compt. Rend. *242*, 1702 (1956).
358. HORNING, D. F., Discussions Faraday Soc. *9*, 115 (1950).
358a. HORTIG, G., O. MÜLLER, K. R. SCHUBERT AND E. FLUCK, Z. Naturforsch. *21b*, 609 (1966).
358b. HOTA, N. K. AND D. S. MATTESON, J. Am. Chem. Soc. *90*, 3570 (1968).
359. HOUGH, W. V. AND L. J. EDWARDS, Advan. Chem. Ser. *32*, 184 (1961).
360. HOUGH, W. V., L. J. EDWARDS AND A. D. MCELROY, J. Am. Chem. Soc. *80*, 1828 (1958).
361. HOUGH, W. V., L. J. EDWARDS AND A. D. MCELROY, J. Am. Chem. Soc. *78*, 689 (1956).
362. HOUGH, W. V., L. J. EDWARDS AND A. F. STANG, J. Am. Chem. Soc. *85*, 831 (1963).
363. HOUGH, W. V., G. W. SCHAEFFER, M. DZURUS AND A. C. STEWART, J. Am. Chem. Soc. *77*, 864 (1955).
364. HROSTOWSKI, H. J. AND R. J. MYERS, J. Chem. Phys. *22*, 262 (1954).
365. HROSTOWSKI, H. J., R. J. MYERS AND G. C. PIMENTEL, J. Chem. Phys. *20*, 518 (1952).
366. HROSTOWSKI, H. J. AND G. C. PIMENTEL, J. Am. Chem. Soc. *76*, 998 (1954).
367. HU, J.-H. AND G. E. MACWOOD, J. Phys. Chem. *60*, 1483 (1956).
367a. HUFF, G. F. AND R. H. SHAKELY (Callery. Chem. Co.), Rept. No. CCC-1024–TR-2 (1953).
368. HUGHES, E. W., J. Am. Chem. Soc. *78*, 502 (1956).
369. HURD, D. T., J. Am. Chem. Soc. *71*, 20 (1949).
370. HURD, D. T. (General Electric Co.), U.S.Pat. 2,469,879; C.A. *43*, 6795e (1949).
371. HURD, D. T., U.S. Pat. 2,596,690 (1952); C.A. *46*, 11602c (1952).
371a. HYATT, D. E., D. A. OWEN AND L. J. TODD, Inorg. Chem. *5*, 1749 (1966).
371b. HYATT, D. E., F. R. SCHOLER AND L. J. TODD, Inorg. Chem. *6*, 630 (1967).
371c. HYATT, D. E. J. L. LITTLE, J. T. MORAN, F. R. SCHOLER AND L. J. TODD, J. Am. Chem. Soc. *89*, 3342 (1967).
371d. HYATT, D. E., F. R. SCHOLER, L. J. TODD AND J. L. WARNER, Inorg. Chem. *6*, 2229 (1967).
371e. INOUE, M. AND G. KODAMA, Inorg. Chem. *7*, 430 (1968).
372. JAHN, A., Thesis, München University 1954.
372a. JAMES, B. D. AND M. G. H. WALLBRIDGE, J. Inorg. Nucl. Chem. 28, 2456 (1966).
372b. JAMES, B. D., R. K. NANDA AND M. G. H. WALLBRIDGE, Inorg. Chem. *6*, 1979 (1967).
372c. JAMES, B. D., R. K. NANDA AND M. G. H. WALLBRIDGE, J. Chem. Soc. *A1966*, 182.
373. JENKNER, H. (Kali Chemie, Hannover), Germ. Pat. 1,038,019 (1958); C.A. *54*, 18908d (1960).
373a. JENNINGS, J. R. AND K. WADE, J. Chem. Soc. A1968, 1946.
374. JENSEN, E., *A Study of $NaBH_4$*, Nyt Nordish Forley, Copenhagen, 1954.
374a. JOHN, JR., A., Nucl. Sci. Abstr. *10*, 1295 (1956).
374b. JOHNSON II, H. D., S. G. SHORE, N. L. MOCK AND J. C. CARTER, J. Am. Chem. Soc. *91*, 2131 (1969).
375. JOLLY, W. L., J. Am. Chem. Soc. *83*, 335 (1961).
376. JOLLY, W. L. AND R. E. MESMER, J. Am. Chem. Soc. 83, 4470 (1961).
376a. JOLLY, W. L. AND T. SCHMITT, J. Am. Chem. Soc. *88*, 4282 (1966).
377. JONES, F., J. Chem. Soc. *35*, 41 (1879).
378. JONES, M. E., K. HEDBERG AND V. SCHOMAKER, J. Am. Chem. Soc. *75*, 4116 (1953).
379. JONES, R. G. AND C. R. KINNEY, J. Am. Chem. Soc. *61*, 1378 (1939).
380. JONES, F. AND R. L. TAYLOR, J. Chem. Soc. *39*, 213 (1881).
380a. ISSLEIB, K., R. LINDNER AND A. TZSCHACH, Z. Chem. *6*, 1 (1966).
381. ITO, K., H. WANATABE AND M. KUBO, J. Chem. Phys. *32*, 947 (1960).
382. ITO, K., H. WANATABE AND M. KUBO, J. Chem. Phys. *34*, 1043 (1961).
383. ITO, K., H. WANATABE AND M. KUBO, Bull. Chem. Soc. Japan *33*, 1588 (1960).
383a. KACZMARCZYK, A., Inorg. Chem. *7*, 164 (1968).

384. KACZMARCZYK, A., R. DOBROTT AND W. N. LIPSCOMB, Proc. Natl. Acad. Sci. U.S. *48*, 729 (1962).
384a. KACZMARCZYK, A., W. C. NICHOLS, W. H. STOCKMAYER AND T. B. EAMES, Inorg. Chem. *7*, 1057 (1968). KACZMARCZYK, A. AND G. B. KOLSKI, Inorg. Chem. *4*, 665 (1965).
385. KACZMARCZYK, A., G. B. KOLSKI AND W. P. TOWNSEND, J. Am. Chem. Soc. *87*, 1413 (1965).
385a. KAESZ, H. D., W. FELLMANN, G. R. WILKENS AND L. F. DAHL, J. Am. Chem. Soc. *87*, 2753, 2755 (1965).
385b. KAESZ, H. D., R. BAU, H. A. BEALL AND W. N. LIPSCOMB, J. Am. Chem. Soc. *89*, 4218 (1967).
386. KAPSHTAL', V. N. AND L. M. SVERDLOV, Russ. J. Phys. Chem., (English transl.), *37*, 680 (1963).
387. KASPER, J. S., C. M. LUCHT AND D. HARKER, Acta Cryst. *3*, 436 (1950).
388. KAUFMAN, J. J., Inorg. Chem. *1*, 973 (1962).
389. KAUFMAN, J. M., J. GREEN, M. S. COHEN, M. M. FEIN AND E. L. COTTRILL, J. Am. Chem. Soc. Soc. *86*, 4210 (1964).
390. KAUFMAN, J. J. AND W. S. KOSKI, J. Chem. Phys. *24*, 403 (1956).
391. KAUFMAN, J. J. AND W. S. KOSKI, J. Am. Chem. Soc. *78*, 5774 (1956).
392. KAUFMAN, J. J., W. S. KOSKI AND R. ANACREON, J. Mol. Spectry. *11*, 1 (1963).
393. KAUFMAN, J. J., J. E. TODD AND W. S. KOSKI, Anal. Chem. *29*, 1033 (1957).
393a. KELLY, H. C., Inorg. Chem. *5*, 2173 (1966).
394. KELLY H. C. AND J. O. EDWARDS, J. Am. Chem. Soc. *82*, 4842 (1960).
394a. KELLY, H. C., M. B. GIUSTO AND F. R. MARCHELLI, J. Am. Chem. Soc. *86*, 3882 (1964).
395. KELLY, H. C., F. R. MARCHELLI AND M. B. GIUSTO, Inorg. Chem. *3*, 431 (1964).
395a. KELLER, P. C., Inorg. Chem. *8*, 1695 (1969).
395b. KELLER, P. C., J. Am. Chem. Soc. *91*, 1230 (1969).
395c. KELLER, P. C., Chem. Commun. *1969*, 209.
395d. KELLER P. C., D. MACLEAN AND R. SCHAEFFER, Chem. Commun. *1965*, 204.
395e. KELLER, P. C., Inorg. Chem. *9*, 75 (1970).
396. KERN, W. C. AND W. N. LIPSCOMB, J. Chem. Phys. *37*, 275 (1962).
397. KERR, E. C., N. C. HALLETT AND H. L. JOHNSTON (Ohio State University), Tech. Rept. No. 10, Contract N 6 ONR-225, T. O. IX22, Sept. 1952.
398. KERRIGAN, J. V., Inorg. Chem. *3*, 908 (1964).
399. KERSCHER, U., Thesis, München University, 1960.
400. KETELAAR, J. A. A. AND C. J. H. SCHUTTE, Spectrochim. Acta *17*, 815 (1961).
401. KETELAAR, J. A. A. AND C. J. H. SCHUTTE, Spectrochim. Acta *17*, 1240 (1961).
402. KINNEY, C. R. AND M. J. KOLBEZEN, J. Am. Chem. Soc. *64*, 1585 (1942).
403. KINNEY, C. R. AND C. L. MAHONEY, J. Org. Chem. *8*, 526 (1943).
404. KLANBERG, F., Z. Anorg. Allgem. Chem. *316*, 197 (1961).
405. KLANBERG, F., Proc. Chem. Soc. *1961*, 203.
405a. KLANBERG, F. AND E. L. MUETTERTIES, Inorg. Chem. *5*, 1955 (1966).
405b. KLANBERG, F. D. R. EATON, L. J. GUGGENBERGER AND E. L. MUETTERTIES, Inorg. Chem. *6*, 1271(1967).
405c. KLANBERG, F., W. B. ASKEW AND L. J. GUGGENBERGER, Inorg. Chem. *7*, 2265 (1968).
405d. KLANBERG, F., E. L. MUETTERTIES AND L. J. GUGGENBERGER, Inorg. Chem. *7*, 2272 (1968).
405e. KLANBERG, F., P. A. WEGNER, G. W. PARSHALL AND E. L. MUETTERTIES, Inorg. Chem. *7*, 2072 (1968).
406. KLEIN, R., A. BLISS, L. SCHOEN AND H. NADEAU, J. Am. Chem. Soc. *83*, 4131 (1961).
407. KLEIN, M. J., B. C. HARRISON AND I. J. SOLOMON, J. Am. Chem. Soc. *80*, 4149 (1958).
408. KLEJNOT, O., Thesis ("Diplomarbeit"), München University, 1953.
409. KLEJNOT, O., Thesis, München University, 1955.
410. KLEMM, L. I. AND W. KLEMM, Z. Anorg. Allgem. Chem. *225*, 258 (1935).
411. KLINGEN, T. J., Inorg. Chem. *3*, 1058 (1964).
411a. KNOTH, W. H., J. Am. Chem. Soc. *88*, 935 (1966).
411b. KNOTH, W., J. Am. Chem. Soc. *89*, 4850 (1967).
411c. KNOTH, W., J. Am. Chem. Soc. *89*, 3344 (1967).
411d. KNOTH, W. H., J. Am. Chem. Soc. *89*, 4850 (1967).
411e. KNOTH, W. H. J. Am. Chem. Soc. *89*, 1274 (1967).
412. KNOTH, W. H., W. R. HERTLER AND E. L. MUETTERTIES, Inorg. Chem. *4*, 280 (1965).
412a. KNOTH, W. H., N. E. MILLER AND W. R. HERTLER, Inorg. Chem. *6*, 1977 (1967).
413. KNOTH, W. H., H. C. MILLER, J. C. SAUER, J. H. BALTHIS, Y. T. CHIA AND E. L. MUETTERTIES, Inorg. Chem. *3*, 159 (1964).

414. Knoth, W. H., H. C. Miller, D. C. England, G. W. Parshall and E. L. Muetterties, J. Am. Chem. Soc. *84*, 1056 (1962).
414a. Knoth, W. H., N. E. Miller and W. R. Hertler, Inorg. Chem. *6*, 1977 (1967).
415. Knoth, W. H. and E. L. Muetterties, J. Inorg. Nucl. Chem. *20*, 66 (1961).
416. Knoth, W. H., J. C. Sauer, D. C. England, W. R. Hertler and E. L. Muetterties, J. Am. Chem. Soc. *86*, 3973 (1964).
416a. Knoth, W. H., J. S. Sauer, J. H. Balthis, H. C. Miller and E. L. Muetterties, J. Am. Chem. Soc. *89*, 4842 (1967).
417. Kodama, G., Abstr. Papers 135th Meeting Am. Chem. Soc., Boston, 1959, p. 33 M.
418. Knoth, W. H., J. C. Sauer, H. C. Miller and E. L. Muetterties, J. Am. Chem. Soc. *86*, 115 (1964).
419. Kodama, G. and R. W. Parry, J. Am. Chem. Soc. *79*, 1007 (1957).
420. Kodama, G., R. W. Parry and J. C. Carter, J. Am. Chem. Soc. *81*, 3534 (1959).
421. Kodama, G. and R. W. Parry, Congr. Intern. Chim. Pure Appl., 16^{e}, Paris, 1957, Mem. Sect. Chim. Minerale (1958), p. 483.
422. Kodama, G. and R. W. Parry, J. Am. Chem. Soc. *81*, 6250 (1960).
423. Kodama, G. and R. W. Parry, Inorg. Chem. *4*, 410 (1965).
423a. Koetzle, T. F., F. E. Scarbrough and W. N. Lipscomb, Inorg. Chem. *7*, 1076 (1968).
424. Köster, R. (Studiengesellschaft Kohle mbH, Mühlheim), Germ. Pat. 1,028, 100 (1958); C.A. *54*, 15872c (1960).
425. Köster, R., Ann. *618*, 31 (1958).
426. Köster, R., Angew. Chem. *71*, 520 (1959).
427. Köster, R., Angew. Chem. *72*, 626 (1960).
428. Köster, R., Angew. Chem. *69*, 94 (1957).
428a. Köster, R., Germ. Pat. 1,018, 397 (1957); C.A. *54*, 13580a (1960).
428b. Köster, R., Germ. Pat. 1, 034, 596 (1958); C.A. *54*, 18907g (1960).
428c. Köster, R., Germ. Pat. 1, 035, 109 (1958); C.A. *54*, 18907g (1960).
428d. Köster, R., Germ. Pat. 1, 080, 983 (1960); C.A. *55*, 22731d (1961).
428 e. Köster, R. and B. Rickborn, J. Am. Chem. Soc. *89*, 2782 (1967).
429. Köster, R., W. Larburg and G. W. Rotermund, Ann. *682*, 21 (1965).
429a. Köster, R., and K. Ziegler, Angew. Chem. *69*, 94 (1957).
430. Kolski, T. L., H. B. Moore, L. E. Roth, K. J. Martin and G. W. Schaeffer, J. Am. Chem. Soc. *80*, 549 (1958).
431. Kolski, T. L. and G. W. Schaeffer, J. Phys. Chem. *64*, 1696 (1960).
431a. Korshak, V. V., V. A. Zamyatina, N. I. Bekasova and L. G. Komarova, Bull. Acad. Sci. USSR, Div. Chem. Sci., English Transl. *1964*, 2123.
432. Korshak, V. V., V. A. Zamyatina and R. M. Oganesyan, Bull. Acad. Sci. USSR, Div. Chem. Sci., English Transl. *1962*, 1580.
433. Koski, W. S., Advan. Chem. Ser. *32*, *78*, (1961).
434. Koski, W. S. and J. J. Kaufman, J. Chem. Phys. *24*, 221 (1956).
435. Koski, W. S., J. J. Kaufman, C. F. Pachucki and F. J. Shipko, J. Am. Chem. Soc. *80*, 3202 (1958).
436. Koski, W. S., J. J. Kaufman and P. C. Lauterbur, J. Am. Chem. Soc. *79*, 2382 (1957).
437. Kotlenski, W. V. and R. Schaeffer, J. Am. Chem. Soc. *80*, 4517 (1958).
438. Kreye, W. C. and R. A. Marcus, J. Chem. Phys. *37*, 419 (1962).
439. Kury, J. W. and W. L. Jolly, UCRL-4432, Dec. 27, 1954.
439a. Kuznetsof, P. M., D. F. Shriver and F. E. Stafford, J. Am. Chem. Soc. *90*, 2557 (1968).
439b. Lane, A. P., and A. B. Burg, J. Am. Chem. Soc. *89*, 1040 (1967).
440. Lappert, M. F. and G. Srivastava, Proc. Chem. Soc. *1964*, 120.
440a. LaPrade, M. D. and C. E. Nordman, Inorg. Chem. *8*, 1669 (1969).
441. Larson, T. E. and W. S. Koski, Congr. Intern. Chim. Pure Appl., 16^{e}, Paris, 1957, Mem. Sect. Chim. Minerale (1958), p. 453.
442. Laubengayer, A. W., O. T. Beachley and R. F. Porter, Inorg. Chem. *4*, 578 (1965).
443. Laubengayer, A. W., P. C. Moews and R. F. Porter, J. Am. Chem. Soc. *83*, 1340 (1961).
444. Laubengayer, A. W., R. P. Ferguson and A. E. Newkirk, J. Am. Chem. Soc. *63*, 559 (1941).
444a. Laubengayer, A. W. and W. R. Rysz, Inorg. Chem. *4*, 1513 (1965).
444b. Lauterbur, P. C., R. C. Hopkins, R. W. King, O. V. Ziebarth and C. W. Heitsch, Inorg. Chem. *7*, 1026 (1968).
445. Lavine, L. R. and W. N. Lipscomb, J. Chem. Phys. *21*, 2087 (1953).

446. LAVINE, L. R. AND W. N. LIPSCOMB, J. Chem. Phys. *22*, 614 (1954).
446a. LEE, G. H. AND R. F. PORTER, Inorg. Chem. *6*, 648 (1967).
446b. LEACH, J. B. AND J. H. MORRIS, J. Chem. Soc. *A1967*, 1590.
447. LEFFER, A. J., Inorg. Chem. *3*, 145 (1964).
448. LEHMANN, W. J., M.S. Thesis, St. Louis University, 1954.
449. LEHMANN, W. J., Ph.D. Thesis, St. Louis University, 1954.
450. LEHMANN, W. J. AND J. F. DITTER, J. Chem. Phys. *31*, 549 (1959).
451. LEHMANN, W. J., J. F. DITTER AND I. S. SHAPIRO, J. Chem. Phys. *29*, 1248 (1958).
452. LEHMANN, W. J., T. P. ONAK AND I. S. SHAPIRO, J. Chem. Phys. *30*, 1215 (1959).
453. LEHMANN, W. J., H. G. WEISS AND I. S. SHAPIRO, J. Chem. Phys. *30*, 1222 (1959).
454. LEHMANN, W. J., C. O. WILSON AND I. S. SHAPIRO, J. Chem. Phys. *31*, 1071 (1959).
455. LEHMANN, W. J., C. O. WILSON, JR. AND I. S. SHAPIRO, J. Chem. Phys. *32*, 1088 (1960).
456. LEHMANN, W. J., C. O. WILSON, JR. AND I. S. SHAPIRO, J. Chem. Phys. *32*, 1786 (1960).
457. LEHMANN, W. J., C. O. WILSON, JR. AND I. S. SHAPIRO, J. Chem. Phys. *33*, 590 (1960).
458. LEHMANN, W. J., C. O. WILSON, JR. AND I. S. SHAPIRO, J. Chem. Phys. *34*, 476 (1961).
459. LEHMANN, W. J., C. O. WILSON, JR. AND I. S. SHAPIRO, J. Chem. Phys. *34*, 783 (1961).
460. LEHMANN, W. J., H. G. WEISS AND I. S. SHAPIRO, J. Chem. Phys. *30*, 1226 (1959).
461. LE SECH, M. AND M. P. GARRIGUES, Chim. Anal. (Paris) *43*, 174 (1961).
462. LEVI, A., J. B. VETRANO, D. E. TRENT AND J. F. FORSTER, J. Inorg. Nucl. Chem. *13*, 32[illegible] (1960).
463. LEVIN, I. R., J. Chem. Phys. *42*, 1244 (1965).
464. LEVITIN, N. E., E. F. WESTRUM, JR. AND J. C. CARTER, J. Am. Chem. Soc. *81*, 3547 (1959).
465. LEWIN, R., P. G. SIMPSON AND W. N. LIPSCOMB, J. Chem. Phys. *39*, 1532 (1963).
465a. LEWIN, R., P. G. SIMPSON AND W. N. LIPSCOMB, J. Am. Chem. Soc. *85*, 478 (1963).
466. LEWIS, E. S. AND R. H. GRINSTEIN, J. Am. Chem. Soc. *84*, 1158 (1962).
466a. LEWIS, J. S. AND A. KACZMARCZYK, J. Am. Chem. Soc. *88*, 1068 (1966).
467. LIPPERT, E. L. AND W. N. LIPSCOMB, J. Am. Chem. Soc. *78*, 503 (1956).
467a. LIPPARD, S. J. AND K. M. MELMED, J. Am. Chem. Soc. *89*, 3929 (1967).
467b. LIPPARD, S. J. AND K. M. MELMED, Inorg. Chem. *6*, 2223 (1967).
467c. LIPPARD, S. J. AND D. A. UCKO, Inorg. Chem. *7*, 1051 (1968).
468. LIPSCOMB, W. N., Advan. Inorg. Chem. Radiochem. *1*, 117 (1959).
469. LIPSCOMB, W. N., *Boron Hydrides*, Benjamin, New York/Amsterdam, 1963.
470. LIPSCOMB, W. N., J. Inorg. Nucl. Chem. *11*, 1 (1959).
471. LIPSCOMB, W. N., J. Phys. Chem. *62*, 381 (1958).
472. LIPSCOMB, W. N., Proc. Natl. Acad. Sci. U.S. *47*, 1791 (1961).
473. LIPSCOMB, W. N., Inorg. Chem. *3*, 1638 (1964).
474. LIPSCOMB, W. N., J. Chem. Phys. *25*, 38 (1956).
474a. LIPSCOMB, W. N., A. R. PITOCHELLI AND M. F. HAWTHORNE, J. Am. Chem. Soc. *81*, 583[illegible] (1959).
474b. LITTLE, J. L., J. T. MORAN AND L. J. TODD, J. Am. Chem. Soc. *89*, 5495 (1967).
475. LLOYD, J. E. AND K. WADE, J. Chem. Soc. *1964*, 1649.
476. LOGAN, T. J. AND T. J. FLAUT, J. Am. Chem. Soc. *82*, 3446 (1960).
477. LONG, L. H. AND A. C. SANHUEZA, Chem. Ind. London *1961*, 588.
478. LONG, L. H. AND M. G. H. WALLBRIDGE, Chem. Ind (London) *1959*, 295.
478a. LONG, L. H. AND M. G. H. WALLBRIDGE, J. Chem. Soc. *1963*, 2181.
478b. LONG, L. H. AND M. G. H. WALLBRIDGE, J. Chem. Soc. *1965*, 3513.
479. LONGSDALE, K. AND E. W. TOOR, Acta Cryst. *12*, 1048 (1959).
479a. LONGUET-HIGGINS AND M. DE V. ROBERTS, Proc. Roy Soc. (London) *A230*, 110 (1955).
479b. LOW, M. J. D., R. EPSTEIN AND A. C. BOND, Chem. Commun. *1967*, 226.
480. LUCHT, C. M., J. Am. Chem. Soc. *73*, 2372 (1951).
481. LUK'YANOVA, E. I. AND V. F. KOKHOVA, Russ. J. Inorg. Chem., English Transl. *8*, 111 (1963).
482. LUTZ, C. A., D. A. PHILLIPS AND D. M. RITTER, Inorg. Chem. *3*, 1191 (1964).
483. LYNDS, L., J. Chem. Phys. *42*, 1124 (1965).
483a. LYNDS, L., Spectrochim. Acta *22*, 2123 (1966).
484. LYNDS, L. AND C. D. BASS, J. Chem. Phys. *40*, 1590 (1964).
485. LYNDS, L. AND C. D. BASS, J. Chem. Phys. *41*, 3165 (1964).
486. LYNDS, L. AND C. D. BASS, Inorg. Chem. *3*, 1147 (1964).
487. LYNDS, L. AND D. R. STERN, J. Am. Chem. Soc. *81*, 5006 (1959).
487a. MAKHLOUF, J. M., W. V. HOUGH AND G. T. HEFFERAN, Inorg. Chem. *6*, 1196 (1967).
487b. MACLEAN, D. B., J. D. ODOM AND R. SCHAEFFER, Inorg.Chem. *7*, 408 (1968).

487c. Maki, A. H. and T. E. Berry, J. Am. Chem. Soc. *87*, 4437 (1965).
488. Maguire, R. G., I. J. Solomon and M. J. Klein, Inorg. Chem. *2*, 1133 (1963).
488a. Malone, L. J., Inorg. Chem. *7*, 1039 (1968).
488b. Mal'tseva, N. N., Z. K. Sterlyadkina and V. I. Mikheeva, Russ. J. Inorg. Chem. English Transl. *11*, 392 (1966).
488c. Malone, L. J. and R. W. Parry, Inorg. Chem. *6*, 817 (1967).
488d. Malone, L. J. and M. R. Manley, Inorg. Chem. *6*, 2260 (1967).
488e. Malone, L. J. and R. W. Parry, Inorg. Chem. *6*, 176 (1967).
489. Mamantov, G. and J. L. Margrave, J. Inorg. Nucl. Chem. *20*, 348 (1961).
490. Mann, D. E., J. Chem. Phys. *22*, 70 (1954).
491. Mann, D. E., J. Chem. Phys. *22*, 762 (1954).
492. Margrave, J. L., J. Chem. Phys. *32*, 1889 (1960).
493. Margrave, J. L., J. Phys. Chem. *61*, 38 (1957).
493a. Marshall, M. D., R. M. Hunt, G. T. Hefferan, R. M. Adams and J. M. Makhlouf, J. Am. Chem. Soc. *89*, 3361 (1967).
493b. Maruca, R., J. D. Odom and R. Schaeffer, Inorg. Chem. *7*, 412 (1968).
493c. Maruca, R., O. T. Beachley, Jr. and A. W. Laubengayer, Inorg. Chem. *6*, 573 (1967).
493d. Maruca, R., H. Schroeder and A. W. Laubengayer, Inorg. Chem. *6*, 572 (1967).
494. Mathieu, M. -V. and B. Imelik, J. Chim. Phys. *59*, 1189 (1962).
494a. Maybury, P. C. and J. E. Ahnell, Inorg. Chem. *6*, 1286 (1967).
494b. Maybury, P. C., J. C. Davis, Jr. and R. A. Patz, Inorg. Chem. *8*, 160 (1969).
495. Maybury, P. C. and W. S. Koski, J. Chem. Phys. *21*, 742 (1953).
496. McAchran, G. E. and S. G. Shore, Inorg. Chem. *4*, 125 (1965).
497. McCarthy, L. V. and P. A. DiGiorgio, J. Am. Chem. Soc. *73*, 3138 (1951).
498. McCoy, R. E. and S. H. Bauer, J. Am. Chem. Soc. *78*, 2061 (1956).
498a. Meek, D. W. and C. S. Springer, Jr., Inorg. Chem. *5*, 445 (1966).
498b. Meller, A. and E. Schaschel, Inorg. Nucl. Chem. Letters *2*, 41 (1966).
498c. Meller, A., Organomet. Chem. Rev. *2*, 1 (1967).
499. Mellon, Jr., E. K. and J. J. Lagowski, Advan. Inorg. Chem. Radiochem. *5*, 259 (1963).
499a. Mesmer, R. E. and W. L. Jolly, Inorg. Chem. *1*, 608 (1962).
500. Metal Hydrides Inc., Beverley, Mass., Final Development Report, Dec. *23*, 1949.
501. Metal Hydrides Inc., Beverly, Mass., Tech. Bull. No. 550 (1958).
502. Mesmer, R. E. and W. L. Jolly, Inorg. Chem. *1*, 608 (1962).
502a. Middaugh, R. L. and F. Farha, Jr., J. Am. Chem. Soc. *88*, 4147 (1966).
502b. Middaugh, R. L., Inorg. Chem. *7*, 1011 (1968).
503. Mikhailov, B. M., Russ. Chem. Rev., English Transl. *31*, 207 (1962).
504. Mikhailov, B. M. and V. A. Dorokhov, Izv. Akad. Nauk SSSR, Otd. Khim. Nauk *1961*, 1346; C.A. *56*, 2463i (1962).
504a. Mikhailov, B. M. and V. A. Dorokhov, Zh. Obshch. Khim. *31*, 4020 (1961).
505. Mikhailov, B. M. and V. A. Dorokhov, Bull. Acad. Sci. USSR, Div. Chem. Sci., English Transl. *1962*, 576.
505a. Mikhailov, B. M. and V. A. Dorokhov, Izv. Akad. Nauk SSSR, Otd. Khim. Nauk *1961*, 1163.
505b. Mikhailov, B. M. and V. A. Dorokhov, Izv. Adak. Nauk SSSR, Otd. Khim. Nauk *1961*, 2084.
505c. Mikhailov, B. M. and V. A. Dorokhov, Zh. Obshch. Khim. *31*, 3750 (1969).
505c. Mikhailov, B. M. and V. A. Dorokhov, Zh. Obshch. Khim. *31*, 3750 (1961).
506. Mikhailov, B. M. and V. A. Dorokhov, Dokl. Akad. Nauk SSSR *130*, 782 (1960).
507. Mikhailov, B. M. and V. A. Dorokhov, Dokl. Akad. Nauk SSSR *133*, 119 (1960).
508. Mikhailov, B. M. and V. A. Dorokhov, Dokl. Akad. Nauk SSSR *136*, 356 (1961).
508a. Mikhailov, B. M. and M. E. Kuimova, Russ. Chem. Rev., English Transl. *35*, 569 (1966).
508b. Mikhailov, B. M. and E. N. Safonova, Bull. Acad. Sci. USSR, Div. Chem. Sci., English Transl. *1965*, 1452.
509. Mikhailov, B. M. and T. A. Shchegoleva, Izv. Akad. Nauk SSSR, Otd. Khim. Nauk *1959*, 1868.
510. Mikhailov, B. M. and T. A. Shchegoleva, Dokl. Akad. Nauk SSSR *131*, 843 (1960).
511. Mikhailov, B. M., T. A. Shchegoleva and A. N. Blokhina', Izv. Akad. Nauk SSSR, Otd. Khim. Nauk *1960*, 1307.
512. Mikhailov, B. M., T. A. Shchegoleva, V. D. Sheludyakov and A. N. Blokhina', Bull. Acad. Sci. USSR, Div. Chem. Sci., English Transl. *1963*, 579.

513. MIKHAILOV, B. M., T. A. SHCHEGOLEVA, E. M. SHASHKOVA AND V. D. SHELUDYAKOV, Izv Akad. Nauk SSSR, Otd. Khim. Nauk *1961*, 1163.
514. MIKHAILOV, B. M. AND L. S. VASIL'EV, Bull. Acad. Sci. USSR, Div. Chem. Sci., English Transl. *1962*, 580.
514a. MIKHEEVA, V. I. AND E. M. FEDNEVA, Zh. Neorgan. Khim. *2*, 604 (1957).
514b. MIKHEEVA, V. I. AND E. M. FEDNEVA, Zh. Neorgan. Khim. *1*, 894 (1956).
514c. MIKHEEVA, V. I. AND S. M. ARKHIPEV, Russ. J. Inorg. Chem., English Transl. *11*, 805 (1966).
514d. MIKHEEVA, V. I. AND V. N. KONOPLEV, Russ. J. Inorg. Chem., English Transl. *11*, 923 (1966).
514e. MIKHEEVA, V. I., N. N. MAL'TSEVA AND L. S. ALEKSEEVA, Russ. J. Inorg. Chem. (Engl. Transl.) *13*, 682 (1968).
515. MIKHEEVA, V. I. AND V. YU. MARKINA, Zh. Neorgan. Khim. *1*, 619 (1956); Advan. Chem. Ser. *32*, 77 (1961).
516. MIKHEEVA, V. I. AND M. S. SELIVOKHINA, Russ. J. Inorg. Chem., English Transl. *8*, 227 (1963).
517. MIKHEEVA, V. I., M. S. SELIVOKHINA AND O. N. KRYUKOVA, Russ. J. Inorg. Chem., English Transl. *7*, 838 (1962).
518. MIKHEEVA, V. I. AND L. V. TITOV, Russ. J. Inorg. Chem., English Transl. *9*, 440 (1964).
519. MILLER, G. A., J. Phys. Chem. *67*, 1363 (1963).
519a. MILLER, N. E., J. Am. Chem. Soc. *88*, 4284 (1966).
520. MILLER, N. E., B. L. CHAMBERLAND AND E. L. MUETTERTIES, Inorg. Chem. *3*, 1064 (1964).
521. MILLER, N. E., J. A. FORSTNER AND E. L. MUETTERTIES, Inorg. Chem. *3*, 1690 (1964).
521a. MILLER, J. J. AND F. A. JOHNSON, J. Am. Chem. Soc. *90*, 218 (1968).
522. MILLER, J. J. AND M. F. HAWTHORNE, J. Am. Chem. Soc. *81*, 4501 (1959).
522a. MILLER, H. C., W. R. HERTLER, E. L. MUETTERTIES, W. H. KNOTH AND N. E. MILLER, Inorg. Chem. *4*, 1216 (1965).
523. MILLER, H. C., N. E. MILLER AND E. L. MUETTERTIES, Inorg. Chem. *3*, 1456 (1964).
524. MILLER, N. E., H. C. MILLER AND E. L. MUETTERTIES, Inorg. Chem. *3*, 866 (1964).
524a. MILLER, H. C., N. E. MILLER AND E. L. MUETTERTIES, J. Am. Chem. Soc. *85*, 3885 (1963).
525. MILLER, N. E. AND E. L. MUETTERTIES, J. Am. Chem. Soc. *85*, 3506 (1963).
526. MILLER, J. E. AND E. L. MUETTERTIES, J. Am. Chem. Soc. *86*, 1033 (1964).
527. MOEWS, JR., P. C., Thesis, Cornell University, 1960; Univ. Microfilms, L.C.No. 60-4877, Ann Arbor, Mich., U.S.A.
527a. MOEWS, JR., P. C. AND R. W. PARRY, Inorg. Chem. *5*, 1552 (1966).
528. MOORE, JR., E. B., J. Am. Chem. Soc. *85*, 676 (1963).
529. MOORE, JR., E. B., J. Chem. Phys. *37*, 675 (1962).
530. MOORE, JR., E. B., R. E. DICKERSON AND W. N. LIPSCOMB, J. Chem. Phys. *27*, 209 (1957).
531. MOORE, JR., E. B., L. L. LOHR AND W. N. LIPSCOMB, J. Chem. Phys. *34*, 1329 (1961).
532. MORREY, J. R., A. B. JOHNSON, YUAN-CHIN FU AND G. R. HILL, Advan. Chem. Ser. *32*, 157 (1961).
532a. MORRIS, J. H. AND P. G. PERKINS, J. Chem. Soc. *A 1966*, 576.
532b. MORRIS, J. H. AND P. G. PERKINS, J. Chem. Soc. *A 1966*, 580.
532c. MORRIS, J. H., M. TAMRES AND S. SEARLES, Inorg. Chem. *5*, 2156 (1966).
532d. MORSE, K. W. AND R. W. PARRY, J. Am. Chem. Soc. *89*, 172 (1967).
532e. MORTERRA, C. AND M. J. D. LOW, Chem. Commun. *1969*, 862.
533. MÜLLER, G., Thesis, München University, 1959.
533a. MUETTERTIES, E. L., Inorg. Chem. *4*, 769 (1965).
533b. MUETTERTIES, E. L., in *The Chemistry of Boron and its Compounds*, Ed. E. L. MUETTERTIES, John Wiley & Sons Inc. New York/London/Sydney, 1967.
533c. MUETTERTIES, E. L., Inorg. Chem. *2*, 647 (1963).
533d. MUETTERTIES, E. L. AND V. D. AFTANDILAN, Inorg. Chem. *1*, 731 (1962).
534. MUETTERTIES, E. L., J. H. BALTHIS, Y. T. CHIA, W. H. KNOTH AND H. C. MILLER, Inorg. Chem. *3*, 444 (1964).
534a. MUETTERTIES. E. L. AND F. KLANBERG, Inorg. Chem. *5*, 315 (1966).
534b. MUETTERTIES, E. L. AND W. H. KNOTH, Inorg. Chem. *4*, 1498 (1965).
535. MUETTERTIES, E. L., N. E. MILLER, K. J. PACKER AND H. C. MILLER, Inorg. Chem. *3*, 870 (1964).
535a. MUETTERTIES, E. L. AND A. T. STORR, J. Am. Chem. Soc. *91*, 3098 (1969).
536. MYERS, H. W. AND R. F. PUTNAM, Inorg. Chem. *2*, 655 (1963).
536a. NAGASAWA, K., Inorg. Chem. *5*, 442 (1966).
536b. NADLER, M. P. AND R. F. PORTER, Inorg. Chem. *8*, 599 (1969).
536c. NAAR-COLIN, C. AND T. L. HEYING, Inorg. Chem. *2*, 659 (1963).

537. Nagasawa, K., T. Yoshizaki and H. Wanatabe, Inorg. Chem. *4*, 275 (1965).
537a. Nanda, R. K. and M. G. H. Wallbridge, Inorg. Chem. *3*, 1798 (1964).
537b. Nainan, K. C. and G. E. Ryschkewitsch, Inorg. Chem. *7*, 1316 (1968).
537c. Nainan, K. C. and G. E. Ryschkewitsch, J. Am. Chem. Soc. *91*, 330 (1969).
538. Nakamura, D., H. Wanatabe and M. Kubo, Bull. Chem. Soc. Japan *34*, 142 (1961).
539. Nespital, W., Z. Physik. Chem. *B16*, 153 (1932).
540. Neu, J. T. and K. S. Pitzer, J. Chem. Phys. *17*, 1007 (1949).
540a. Neumaier, H., Thesis, München University, 1961.
541. Newsom, H. C., W. D. English, A. L. McCloskey and W. G. Woods, J. Am. Chem. Soc. *83*, 4134 (1961).
541a. Niedenzu, K. and J. W. Dawson, *Borone–Nitrogen Compounds*, Springer, Berlin, 1965, p. 85.
541b. Niedenzu, K., Organometa. Chem. Rev. *1*, 305 (1966).
541c. Niedenzu, K. and J. W. Dawson, in *The Chemistry of Boron and its Compounds*, Ed. E. L. Muetterties, John Wiley & Sons, Inc., New York/London/Sydney, 1967.
542. Niedenzu, K., D. W. Harrelson and J. W. Dawson, Chem. Ber. *94*, 671 (1961).
542a. Note on 149th Meeting Am. Chem. Soc. (J. Dobson, R. Schaeffer and D. F. Gaines), April *1965*, 46.
543. Nöth, H., Angew. Chem. *73*, 371 (1961).
544. Nöth, H., München University 1954.
545. Nöth, H. (Imp. Chem. Ind. Ltd.), Brit. Pat. 818,426 (1959); C.A. *54*, 12520e (1960).
546. Nöth, H. and H. Beyer, Chem. Ber. *93*, 2251 (1960).
547. Nöth, H. and H. Beyer, Chem. Ber. *93*, 928 (1960).
548. Nöth, H. and H. Beyer, Chem. Ber. *93*, 939 (1960).
549. Nöth, H. and R. Hartwimmer, Chem. Ber. *93*, 2238 (1960).
550. Nöth, H. and R. Hartwimmer, Chem. Ber. *93*, 2246 (1960).
551. Nöth, H. and G. Mikulaschek, Z. Anorg. Allgem. Chem. *311*, 241 (1961).
551a. Nöth, H. and G. Mikulashek, Chem. Ber. *94*, 634 (1961).
551b. Nöth, H. and H. Suchy, J. Organometal. Chem. *5*, 197 (1966).
551c. Nöth, H. and W. Regnet, Chem. Ber. *102*, 167 (1969).
552. Nöth, H. and P. Winter, unpublished results.
553. Nordman, C. E., Acta Cryst. *13*, 535 (1960).
554. Nordman, C. E. and W. N. Lipscomb, J. Chem. Phys. *21*, 1856 (1953).
555. Nordman, C. E. and W. N. Lipscomb, J. Am. Chem. Soc. *75*, 4116 (1953).
556. Nordman, C. E. and C. R. Peters, J. Am. Chem. Soc. *81*, 3551 (1959).
557. Nordman, C. E. and C. Reimann, J. Am. Chem. Soc. *81*, 3538 (1959).
558. Nordman, C. E., C. Reimann and C. R. Peters, Advan. Chem. Ser. *32*, 204 (1961).
558a. Norman, A. D. and R. Schaeffer, Inorg. Chem. *4*, 1225 (1965).
558b. Norman, A. D. and R. Schaeffer, J. Am. Chem. Soc. *88*, 1143 (1966).
558c. Norman, A. D., R. Schaeffer, A. B. Bayliss, G. A. Pressley, Jr. and F. E. Stafford, J. Am. Chem. Soc. *88*, 2151 (1966).
559. Norment, H. G., Acta Cryst. *14*, 1216 (1961).
560. Norton, F. J., J. Am. Chem. Soc. *71*, 3488 (1949).
561. Norton, F. J., J. Am. Chem. Soc. *72*, 1849 (1950).
562. *Office of Naval Research*, London, Tech. Rept. ONRL-52-52 (1952).
563. Ogg, Jr., R. A., J. Chem. Phys. *22*, 1933 (1954).
563a. Olsen, F. P., R. C. Vasavada and M. F. Hawthorne, J. Am. Chem. Soc. *90*, 3946 (1968).
564. Olson, W. M. and R. T. Sanderson, J. Inorg. Nucl. Chem. *7*, 228 (1958).
565. Onak, T. P., J. Am. Chem. Soc. *83*, 2584 (1961).
565a. Onak, T. P., R. P. Drake and G. B. Dunks, Inorg. Chem. *3*, 1686 (1964).
565b. Onak, T. P., Inorg. Chem. *7*, 1043 (1968).
565c. Onak, T. P., R. Drake and G. Dunks, J. Am. Chem. Soc. *87*, 2505 (1965).
566. Onak, T. P., F. J. Gerhart and R. E. Williams, J. Am. Chem. Soc. *85*, 1754 (1963).
566a. Onak, T. P., F. J. Gerhart and R. E. Williams, J. Am. Chem. Soc. *85*, 3378 (1963).
566b. Onak, T. P. and F. J. Gerhard, Inorg. Chem. *1*, 742 (1962).
567. Onak, T. P. and G. B. Dunks, Inorg. Chem. *3*, 1060 (1964).
567a. Onak, T. P. and G. B. Dunks, Inorg. Chem. *5*, 439 (1966).
567b. Onak, T. P., G. B. Dunks, R. A. Beaudet and R. L. Poynter, J. Am. Chem. Soc. *88*, 4622 (1966).
567c. Onak, T. P., G. B. Dunks, J. R. Spielmann, F. J. Gerhard and R. E. Williams, J. Am. Chem. Soc. *88*, 2061 (1966).

567d. Onak, T. P., G. B. Dunks, I. W. Searcy and J. Spielman, Inorg. Chem. *6*, 1465 (1967).
568. Onak, T. P., H. Landesman and I. Shapiro, J. Phys. Chem. *62*, 1605 (1958).
568a. Onak, T. P., L. B. Friedman, J. A. Hartsuck and W. N. Lipscomb, J. Am. Chem. Soc. *88*, 3439 (1966).
568b. Onak, T., D. Marynick, P. Mattschei and E. Groszek, Inorg. Chem. *7*, 1754 (1968).
568c. Onak, T., P. Mattschei and E. Groszek, J. Chem. Soc. *A1969*, 1990.
569. Onak, T. P., R. E. Williams and H. G. Weiss, J. Am. Chem. Soc. *84*, 2830 (1962).
570. Onak, T. P. and R. E. Williams, Inorg. Chem. *1*, 106 (1962).
570a. Olsen, F. P. and M. F. Hawthorne, Inorg. Chem. *4*, 1839 (1965).
571. Pace, R. J., J. Williams, and R. L. Williams, J. Chem. Soc. *1961*, 2196.
572. Palchak, R. J. F., J. H. Norman and R. E. Williams, J. Am. Chem. Soc. *83*, 3380 (1961).
573. Papetti, S. and T. L. Heying, Inorg. Chem. *2*, 1105 (1963).
574. Papetti, S. and T. L. Heying, Inorg. Chem. *3*, 1448 (1964).
574a. Papetti, S. and T. L. Heying, J. Am. Chem. Soc. *86*, 2295 (1964).
575. Papetti, S., B. B. Schaeffer, H. J. Troscianiec and T. L. Heying, Inorg. Chem. *3*, 1444 (1964).
576. Parry, R. W. and T. C. Bissot, J. Am. Chem. Soc. *78*, 1524 (1956).
577. Parry, R. W., P. R. Girardot, D. R. Schulz and S. G. Shore, J. Am. Chem. Soc. *80*, 1 (1958).
578. Parry, R. W., G. Kodama and D. R. Schulz, J. Am. Chem. Soc. *80*, 24 (1958).
579. Parry, R. W., R. W. Rudolph and D. F. Shriver, Inorg. Chem. *3*, 1479 (1964).
580. Parry, R. W., D. R. Schulz and P. R. Girardot, J. Am. Chem. Soc. *80*, 1 (1958).
581. Parry, R. W. and S. G. Shore, J. Am. Chem. Soc. *80*, 15 (1958).
582. Parts, L. and J. T. Miller, Jr., Inorg. Chem. *3*, 1483 (1964).
583. Parshall, G. W., J. Am. Chem. Soc. *86*, 361 (1964).
583a. Pasto, D. J. and P. Balasubramanijan, J. Am. Chem. Soc. *89*, 295 (1967).
584. Patel, J. C. and S. Basu, Naturwiss, *47*, 302 (1960).
585. Paterson, W. G. and M. Onyszchuk, Can. J. Chem. *41*, 1872 (1963).
585a. Pattison, I. and K. Wade, J. Chem. Soc. *A1968*, 842.
586. Paul, R. and N. M. Joseph, Bull. Soc. Chim. France *1953*, 758.
587. Paul, R. and N. M. Joseph, U.S.Pat. 2,726, 926 (1955); C.A. *50*, 8979h (1956).
587a. Perec, M. and L. N. Becka, J. Chem. Phys. *43*, 721 (1965).
588. Perloff, A., Acta Cryst. *17*, 332 (1964).
589. Pescock, R. C., J. Am. Chem. Soc. *75*, 2862 (1953).
590. Peters, C. R. and C. E. Nordman, J. Am. Chem. Soc. *82*, 5758 (1960).
591. Phillips, W. D., H. C. Miller and E. L. Muetterties, J. Am. Chem. Soc. *81*, 4496 (1959).
592. Phillips, C. S. G., P. Powell and J. A. Semlyen, J. Chem. Soc. *1963*, 1202.
593. Pilling, R. L., M. F. Hawthorne and E. A. Pier, J. Am. Chem. Soc, *86*, 3568 (1964).
594. Pilling, R. L., F. N. Tebbe, M. F. Hawthorne and E. A. Pier, Proc. Chem. Soc. *1964*, 402.
595. Pitochelli, A. R., R. Ettinger, J. A. Dupont and M. F. Hawthorne, J. Am. Chem. Soc. *84*, 1057 (1962).
596. Pitochelli, A. R. and M. F. Hawthorne, J. Am. Chem. Soc. *82*, 3228 (1960).
597. Pitochelli, A. R. and M. F. Hawthorne, J. Am. Chem. Soc. *84*, 3218 (1962).
597a. Pitochelli, A. R. and M. F. Hawthorne, J. Am. Chem. Soc. *84*, 2318 (1962).
597b. Pitochelli, A. R., W. N. Lipscomb and M. F. Hawthorne, J. Am. Chem. Soc. *84*, 3026 (1962).
598. Platt, J. R., H. B. Klevens and G. W. Schaeffer, J. Chem. Phys. *15*, 598 (1947).
599. Polak, R. J. and T. L. Heying, J. Org. Chem. *27*, 1483 (1962).
600. Pondy, P. R. and H. C. Beachell, J. Chem. Phys. *25*, 238 (1956).
600a. Porter, R. F. and E. S. Yeung, Inorg. Chem. *7*, 1306 (1968).
600b. Popp, G. and M. F. Hawthorne, J. Am. Chem. Soc. *90*, 6553 (1968).
601. Potenza, J. A. and W. N. Lipscomb, J. Am. Chem. Soc. *86*, 1874 (1964).
602. Potenza, J. A. and W. N. Lipscomb, Inorg. Chem. *3*, 1673 (1964).
602a. Potenza, J. A. and W. N. Lipscomb, Inorg. Chem. *5*, 1471 (1966).
602b. Potenza, J. A. and W. N. Lipscomb, Inorg. Chem. *5*, 1478 (1966).
602c. Potenza, J. A. and W. N. Lipscomb, Inorg. Chem. *5*, 1483 (1966).
602d. Potenza, J. A., W. N. Lipscomb, G. D. Vickers and H. Schroeder, J. Am. Chem. Soc. *88*, 628 (1966).
603. Price, W. C., H. C. Longuet-Higgins, B. Rice and T. F. Young, J. Chem. Phys. *17*, 217 (1949).

604. Price, W. C., A. Fraser, B. Robinson and H. C. Longuet-Higgins, Discussions Faraday Soc. *9*, 131 (1950).
604a. Pritchard, H. O., Quart. Rev. (London) *14*, 46 (1960).
604b. Prince, S. R. and R. Schaeffer, Chem. Commun. *1968*, 451.
604c. Purser, J. M. and B. F. Spielvogel, Chem. Commun. *1968*, 386.
604d. Purser, J. M. and B. F. Spielvogel, Inorg. Chem. *7*, 2156 (1968).
605. Ramaswamy, K. L., Proc. Indian Acad. Sci. *A2*, 364 (1935).
606. Ramaswamy, K. L., Proc. Indian Acad. Sci. *A2*, 630 (1935).
607. Rector, C. W., G. W. Schaeffer and J. R. Platt, J. Chem. Phys. *17*, 460 (1949).
608. Reddy, J. and N. W. Lipscomb, J. Am. Chem. Soc. *81*, 754 (1959).
609. Reddy, J. and N. W. Lipscomb, J. Chem. Phys. *31*, 610 (1959).
610. Reetz, T. and B. Katlafsky, J. Am. Chem. Soc. *82*, 5036 (1960).
611. Reetz, T., J. Am. Chem. Soc. *82*, 5039 (1960).
612. Reeves, R. B. and D. W. Robinson, J. Chem. Phys. *41*, 1699 (1964).
612a. Reiner, J. R. and R. P. Alexander, Inorg. Chem. *5*, 1460 (1966).
613. Renner, T., Angew. Chem. *69*, 478 (1957).
613a. Riess, J. G. and J. R. Van Wazer, *89*, 851 (1967).
614. Rice, B., R. Galiano and W. J. Lehmann, J. Phys. Chem. *61*, 1222 (1957).
615. Rice, B., J. A. Livasy and G. W. Schaeffer, J. Am. Chem. Soc. *77*, 2750 (1955).
616. Rice, B. and H. S. Uchida, J. Phys. Chem. *59*, 650 (1955).
617. Rigden, J. S., R. C. Hopkins and J. D. Baldeschwieler, J. Chem. Phys. *35*, 1532 (1961).
618. Rigden, J. S. and W. S. Koski, J. Am. Chem. Soc. *83*, 552 (1961).
619. Riley, R. F. and C. J. Schack, Inorg. Chem. *3*, 1651 (1964).
620. Ring, M. A. and W. S. Koski, J. Chem. Phys. *35*, 381 (1961).
620a. Ring, M. A., E. F. Witucki and R. C. Greenough, Inorg. Chem. *6*, 395 (1967).
620b. Rohrscheid, F. and R. H. Holm, J. Organometal. Chem. *4*, 335 (1965).
621. Rose, S. H. and S. G. Shore, Inorg. Chem. *1*, 744 (1962).
622. Rosenblum, L., J. Am. Chem. Soc. *77*, 5016 (1955).
623. Ross, S. and H. Clark, J. Am. Chem. Soc. *76*, 4297 (1954).
624. Rossmanith, K., Monatsh. *92*, 768 (1961).
624a. Rousseau, Y., O. P. Strausz and H. E. Gunning, J. Chem. Phys. *39*, 962 (1963).
625. Roth, W. and W. H. Bauer, J. Phys. Chem. *60*, 639 (1956).
625a. Rothgery, E. F. and L. F. Hohnstedt, Inorg. Chem. *6*, 1065 (1967).
625b. Rowatt, R. J. and N. E. Miller, J. Am. Chem. Soc. *89*, 5509 (1967).
625c. Rudolph, R. W. and R. W. Parry, J. Am. Chem. Soc, 89, 1621 (1967).
625d. Rupp, E. B., D. E. Smith and D. F. Shriver, J. Am. Chem. Soc. *89*, 5562 (1967).
625e. Rudolph, R. W., R. W. Parry and C. F. Farran, Inorg. Chem. *5*, 723 (1966).
625f. Ruhle, H. W. and M. F. Hawthorne, Inorg. Chem. *7*, 2279 (1968).
626. Ryschkewitsch, G. E., J. Am. Chem. Soc. *82*, 3290 (1960).
627. Ryschkewitsch, G. E. and E. R. Birnbaum, Inorg. Chem. *4*, 575 (1956).
628. Ryschkewitsch, G. E., S. W. Harris, E. J. Mezey, H. H. Sisler, E. W. Weilmuenster and A. B. Garrett, Inorg. Chem. *2*, 890 (1963).
629. Ryschkewitsch, G. E., J. J. Harris and H. H. Sisler, J. Am. Chem. Soc. *80*, 4515 (1958).
629a. Ryschkewitsch, G. E. and W. W. Lochmaier, J. Am. Chem. Soc. *90*, 6260 (1968).
630. Ryschkewitsch, G. E., E. J. Mezey, E. R. Altwicker, H. H. Sisler and A. B. Garrett, Inorg. Chem. *2*, 893 (1963).
631. Sabatier, P., Compt. Rend. *112*, 865 (1891).
632. Sands, D. E. and A. Zalkin, Acta Cryst. *15*, 410 (1962).
633. Sands, D. E. and A. Zalkin, Acta Cryst. *13*, 1030 (1960).
634. Salinger, R. M. and C. L. Frye, Inorg. Chem. *4*, 1815 (1965).
635. Schaeffer, G. W., M. D. Adams and F. J. Koenig, J. Am. Chem. Soc. *78*, 725 (1956).
636. Schaeffer, G. W. and E. R. Anderson, J. Am. Chem. Soc. *71*, 2143 (1949).
637. Schaeffer, G. W. and L. J. Basile, J. Am. Chem. Soc. *77*, 331 (1955).
638. Schaeffer, G. W., J. J. Burns, T. J. Klingen, L. A. Martincheck and R. W. Rosett, Abstr. Papers 135th Meeting Am. Chem. Soc., Boston, 1959.
639. Schaeffer, G. W. and D. J. Hunt, Abstr. Papers 135th Meeting Am. Chem. Soc., Boston, April 1959, p. 28M.
640. Schaeffer, G. W., T. L. Kolski and D. L. Ekstedt, J. Am. Chem. Soc. *79*, 5912 (1957).
641. Schaeffer, G. W., J. S. Roscoe and A. C. Stewart, J. Am. Chem. Soc. *78*, 729 (1956).
642. Schaeffer, G. W., R. Schaeffer and H. J. Schlesinger, J. Am. Chem. Soc. *73*, 1612 (1951).

643. SCHAEFFER, G. W., J. N. SHOOLERY AND R. JONES, J. Am. Chem. Soc. *79*, 4606 (1957).
644. SCHAEFFER, R., J. Am. Chem. Soc. *79*, 2726 (1957).
645. SCHAEFFER, R., J. Am. Chem. Soc. *79*, 1006 (1957).
646. SCHAEFFER, R., J. Inorg. Nucl. Chem. *15*, 190 (1960).
647. SCHAEFFER, R., J. Chem. Phys. *26*, 1349 (1957).
647a. SCHAEFFER, R., Q. JOHNSON AND G. S. SMITH, Inorg. Chem. *4*, 917 (1965).
648. SCHAEFFER, R., K. H. LUDLUM AND S. E. WIBBERLEY, J. Am. Chem. Soc. *81*, 3157 (1959).
649. SCHAEFFER, R. AND L. ROSS, J. Am. Chem. Soc. *81*, 3486 (1959).
650. SCHAEFFER, R., J. N. SHOOLERY AND R. JONES, J. Am. Chem. Soc. *80*, 2670 (1958).
651. SCHAEFFER, R., M. STEINDLER, L. HOHNSTEDT, H. S. SMITH, JR., L. B. EDDY AND H. I. SCHLESINGER, J. Am. Chem. Soc. **76**, 3303 (1954).
651a. SCHAEFFER, R. AND F. TEBBE, J. Am. Chem. Soc. *85*, 2020 (1963).
652. SCHAEFFER, R., F. TEBBE AND C. PHILLIPS, Inorg. Chem. *3*, 1475 (1964).
653. SCHAEFFER, R. AND F. TEBBE, Inorg. Chem. *3*, 1638 (1964).
654. SCHAEFFER, R. AND F. TEBBE, Inorg. Chem. *3*, 904 (1964).
654a. SCHECHTER, W. H., U.S. Pat. 3,033,766 (1962); C.A. *57*, 12257e (1962).
655. SCHECHTER, W. H., G. B. JACKSON AND R. M. ADAMS, *Boron Hydrides and Related Compounds*, Callery Chemical Co., May 1954.
656. SCHENK, P. W. AND W. MÜLLER, Angew. Chem. *71*, 457 (1959).
657. SCHLESINGER, H. I. (University of Chicago), Navy Contract N 173s–9820, Final Report (1944–45).
658. SCHLESINGER, H. I., H. C. BROWN, J. R. GOLDBREATH AND J. J. KATZ, J. Am. Chem. Soc. *75*, 195 (1953).
659. SCHLESINGER, H. I., H. C. BROWN, A. B. BURG AND R. T. SANDERSON, J. Am. Chem. Soc. *62*, 3421 (1940).
660. SCHLESINGER, H. I., H. C. BROWN, H. R. HOEKSTRA AND L. R. RAPP, J. Am. Chem. Soc. *75*, 199 (1953).
661. SCHLESINGER, H. I., H. C. BROWN AND E. K. HYDE, J. Am. Chem. Soc. *75*, 209 (1953).
662. SCHLESINGER, H. I. AND H. C. BROWN, J. Am. Chem. Soc. *62*, 3429 (1940).
663. SCHLESINGER, H. I. AND H. C. BROWN, U.S. Pat. 2,545,633 (1951); C.A. *45*, 6811b (1951).
664. SCHLESINGER, H. I. AND H. C. BROWN, U.S. Pat. 2,600,370 (1952); C.A. *46*, 10561i (1952).
665. SCHLESINGER, H. I. AND H. C. BROWN, J. Am. Chem. Soc. *75*, 219 (1953).
666. SCHLESINGER, H. I. AND A. B. BURG, J. Am. Chem. Soc. *53*, 4321 (1931).
667. SCHLESINGER, H. I. AND A. B. BURG, Chem. Rev. *31*, 1 (1942).
668. SCHLESINGER, H. I. AND A. B. BURG, J. Am. Chem. Soc. *60*, 290 (1938).
669. SCHLESINGER, H. I., L. HORWITZ AND A. B. BURG, J. Am. Chem. Soc. *58*, 407 (1936).
670. SCHLESINGER, H. I., L. HORWITZ AND A. B. BURG, J. Am. Chem. Soc. *58*, 409 (1936).
671. SCHLESINGER, H. I., N. W. FLODIN AND A. B. BURG, J. Am. Chem. Soc. *61*, 1078 (1939).
672. SCHLESINGER, H. I., D. M. RITTER AND A. B. BURG, J. Am. Chem. Soc. *60* 1296 (1938).
673. SCHLESINGER, H. I., D. M. RITTER AND A. B. BURG, J. Am. Chem. Soc. *60*, 2297 (1938).
674. SCHLESINGER, H. I., D. M. RITTER AND A. B. BURG, J. Am. Chem. Soc. *60*, 1300 (1938).
675. SCHLESINGER, H. I., R. T. SANDERSON AND A. B. BURG, J. Am. Chem. Soc. *62*, 3421 (1940).
676. SCHLESINGER, H. I. AND G. W. SCHAEFFER (U.S. At. Energy Comm.), U.S. Pat. 2,494, 267 (1950); C.A. *44*, 1882g (1950).
677. SCHMIDPETER, A. AND R. BÖHM, unpublished results.
677a. SCHOLER, F. R. AND L. J. TODD, J. Organometal. Chem. *14*, 261 (1968).
678. SCHROEDER, H. -J., J. R. REINER AND T. L. HEYING, Inorg. Chem. *1*, 618 (1962).
679. SCHROEDER, H. -J., Inorg. Chem. *2*, 390 (1963).
680. SCHROEDER, H. -J., T. L. HEYING AND J. R. REINER, Inorg. Chem. *2*, 1092 (1963).
681. SCHROEDER, H. -J., J. R. REINER AND T. A. KNOWLES, Inorg. Chem. *2*, 393 (1963).
682. SCHROEDER, H. -J., J. R. REINER, R. P. ALEXANDER AND T. L. HEYING, Inorg. Chem. *3*, 1464 (1964).
682a. SCHROEDER, H. -J., 3rd International Symposium on Organometallic Chemistry, Munich September 1967.
683. SCHULTZ, D. R. AND R. W. PARRY, J. Am. Chem. Soc. *80*, 4 (1958).
684. SCHUTTE, C. J. H., Spectrochim. Acta *16*, 1054 (1959).
685. SCHUTTE, C. J. H., Nature *190*, 805 (1961).
685a. SCHWARTZ, N. N., E. O'BRIEN, S. KARLAN AND M. M. FEIN, Inorg. Chem. *4*, 661 (1965).
685b. SCHWALBE, C. H. AND W. N. LIPSCOMB, J. Am. Chem. Soc. *91*, 194 (1969).
686. SEELY, G. R., J. P. OLIVER AND D. M. RITTER, Anal. Chem. *31*, 1993 (1959).

686a. SEKLEMIAN, H. V. AND R. E. WILLIAMS, Inorg. Nucl. Chem. Letters *3*, 289 (1967).
686b. SEKQUEIRA, A. AND W. C. HAMILTON, Inorg. Chem. *6*, 1281 (1967).
687. SEYFERT, D., J. Inorg. Chem. *7*, 152 (1958).
688. SEYFERT, D., Angew. Chem. *72*, 36 (1960).
689. SEYFERT, D. AND H. P. KÖGLER, J. Inorg. Nucl. Chem. *15*, 99 (1960).
690. SHAPIRO, I. AND J. F. DITTER, J. Chem. Phys. *26*, 798 (1957).
690a. SHAPIRO, I., C. D. GOOD AND R. E. WILLIAMS, J. Am. Chem. Soc. *84*, 3837 (1962).
691. SHAPIRO, I. AND B. KEILIN, J. Am. Chem. Soc. *76*, 3864 (1954).
692. SHAPIRO, I. AND B. KEILIN, J. Am. Chem. Soc. *77*, 2663 (1955).
692a. SHAPIRO, I., B. KEILIN, R. E. WILLIAMS AND C. D. GOOD, J. Am. Chem. Soc. *85*, 3167 (1963).
692b. SHAPIRO, I. AND H. LANDESMAN, J. Chem. Phys. *33*, 1590 (1960).
693. SHAPIRO, I., M. LUSTIG AND R. E. WILLIAMS, J. Am. Chem. Soc. *81*, 838 (1959).
694. SHAPIRO, I. AND H. G. WEISS, J. Phys. Chem. *63*, 1319 (1959).
695. SHAPIRO, I. AND H. G. WEISS, J. Am. Chem. Soc. *76*, 1205 (1954).
696. SHAPIRO, I. AND H. G. WEISS, J. Am. Chem. Soc. *76*, 6020 (1954).
697. SHAPIRO, I., H. G. WEISS, M. SCHMICH, S. SKOLNIK AND G. B. L. SMITH J. Am. Chem. Soc. *74*, 901 (1952).
698. SHEPP, A. AND S. H. BAUER, J. Am. Chem. Soc. *76*, 265 (1954).
699. SHELDON, J. C. AND B. C. SMITH, Quart, Rev. (London) *14*, 200 (1960).
700. SHORE, S. G. AND K. W. BÖDDEKER, Inorg. Chem. *3*, 914 (1964).
701. SHORE, S. G., P. R. GIRARDOT AND R. W. PARRY, J. Am. Chem. Soc. *80*, 20 (1958).
701a. SHORE, S. G., C. W. HICKAM, JR. AND D. COWLES, J. Am. Chem. Soc. *87*, 2755 (1965).
701b. SHORE, S. G. AND C. L. HALL, J. Am. Chem. Soc. *89*, 3947 (1967).
701c. SHORE, S. G. AND C. L. HALL, J. Am. Chem. Soc. *88*, 5346 (1966).
701d. SHORE, S. G. AND C. W. HICKAM, Inorg. Chem. *2*, 838 (1963).
702. SHORE, S. G. AND R. W. PARRY, J. Am. Chem. Soc. *80*, 8 (1958).
703. SHORE, S. G. AND R. W. PARRY, J. Am. Chem. Soc. *80*, 12 (1958).
704. SHORE, S. G. AND R. W. PARRY, J. Am. Chem. Soc. *77*, 6084 (1955).
705. SHRIVER, D. F. AND P. M. KUZNETSOF, Inorg. Chem. *4*, 434 (1965).
705a. SIEDLE, A. R., J. GRANT AND M. D. TREBLOW, Inorg. Chem. *6*, 1602 (1967).
706. SIEGEL, B. AND J. L. MACK, J. Phys. Chem. *62*, 373 (1958).
707. SIEGEL, B. AND J. L. MACK, J. Phys. Chem. *63*, 1212 (1959).
708. SIEGEL, B., J. L. MACK, J. U. LOWE, JR. AND J. GALLAGHAN, J. Am. Chem. Soc. *80*, 4523 (1958).
709. SILBIGER, G. AND S. H. BAUER, J. Am. Chem. Soc. *70*, 115 (1948).
709a. SILVERSTEIN, H. T., D. C. BEER AND L. J. TODD. J. Organometal. Chem. 21, 139 (1970).
710. SILVER, A. H. AND P. J. BRAY, J. Chem. Phys. *32*, 288 (1960).
710a. SIMPSON, C., Ann. Rept. Progr. Chem. (Chem. Soc. London) *58*, 46 (1962).
711. SIMPSON, P. G. AND W. N. LIPSCOMB, Proc. Natl. Acad. Sci. U.S. *48*, 1490 (1962).
712. SIMPSON, P. G. AND W. N. LIPSCOMB, J. Chem. Phys. *35*, 1340 (1961).
713. SIMPSON, P. G. AND W. N. LIPSCOMB, J. Chem. Phys. *39*, 26 (1963).
713a. SIMPSON, P. G., K. FOLTING, R. D. DOBROTT AND W. N. LIPSCOMB, J. Chem. Phys. *39*, 2339 (1963).
714. SIMPSON, P. G., K. FOLTING AND W. N. LIPSCOMB, J. Am. Chem. Soc. *85*, 1879 (1963).
715. SINKE, E. J., G. A. PRESSLEY, JR., A. B. BAYLIS AND F. E. STAFFORD, J. Chem. Phys. *41*, 2207 (1964).
715a. SMART, J. C., P. M. GARRETT AND M. F. HAWTHORNE, J. Am. Chem. Soc. *91*, 1031 (1969).
715b. SMALLWOOD, E. AND W. H. EBERHARD, J. Chem. Phys. *43*, 1259 (1965).
716. SMALLEY, J. H. AND S. F. STAFIEJ, J. Am. Chem. Soc. *81*, 582 (1959).
716a. SMITH, B. C., L. THAKUR AND M. A. WASSEF, J. Chem. Soc. *A1967*, 1616.
716b. SMITH, D. E., E. B. RUPP AND D. F. SHRIVER, J. Am. Chem. Soc. *89*, 5568 (1967).
716c. SMITH, H. D., JR. AND L. F. HOHNSTEIN, Inorg. Chem. *7*, 1061 (1968).
716d. SMITH, H. D., Inorg. Chem. *8*, 676 (1969).
717. SMITH, M. B. AND G. E. BRASS, JR., J. Chem. Eng. Data *8*, 342 (1963).
717a. SMITH, D. E., E. B. RUPP AND D. F. SHRIVER, J. Am. Chem. Soc. *89*, 5568 (1967).
718. SMITH, JR., H. D., J. Am. Chem. Soc. *87*, 1817 (1965).
718a. SMITH, JR., H. D., C. D. OBENLAND AND S. PAPETTI, Inorg. Chem. *5*, 1013 (1966).
718b. SMITH, H. W. AND W. N. LIPSCOMB, J. Chem. Phys. *43*, 1060 (1965).
718c. SMITH, W. L. AND I. M. MILLS, J. Chem. Phys. *41*, 1479 (1964).
719. SMITH, JR., H. D. AND R. R. MILLER, J. Am. Chem. Soc. *72*, 1452 (1950).

719a. SOCIETÉ DES USINES CHIMIQUES RHÔNE-POULENCE, Brit. Pat. 717, 541 (1954); C.A. *49*, 3488h (1955).
719b. SOCIETÉ DES USINES CHIMIQUES RHÔNE-POULENCE, Brit. Pat. 746, 612 (1960); C.A. *50*, 13384e (1956).
720. SOLDATE, A.M., J. Am. Chem. Soc. *69*, 987 (1947).
721. SOLOMON, I. J., M. J. KLEIN AND K. HATTORI, J. Am. Chem. Soc. *80*, 4520 (1958).
722. SOLOMON, I. J., M. J. KLEIN, R. G. MAGUIRE AND K. HATTORI, Inorg. Chem. *2*, 1136 (1963).
723. SOROKIN, V. P., B. I. VESNINA AND N. S. KLIMOVA, Russ. J. Inorg. Chem., English Transl. *8*, 32 (1963).
724. SPIELMAN, J. R. AND A. B. BURG, Inorg. Chem. *2*, 1139 (1963).
724a. SPIELMAN, F. R., R. WARREN, G. B. DUNKS. J. E. SCOTT, JR. AND T. ONAK, Inorg. Chem. *7*, 216 (1968).
724b. SPIELVOGEL, B. F. AND J. M. PURSER, J. Am. Chem. Soc. *89*, 5294 (1967).
724c. SPIELMAN, J. R. AND J. E. SCOTT, J. Am. Chem. Soc. *87*, 3513 (1965).
725. STAFIEJ, S. F., U.S.Pat. 2, 998, 449 (1959); C.A. *56*, 1479e (1962).
725a. STANKO. V. I., J. Gen. Chem. USSR, English Transl. *35*, 1140 (1965).
725b. STANKO, V. I. AND A. I. KLIMOVA, J. Gen. Chem. USSR, English Transl. *35*, 1142 (1965).
725c. STANKO, V. I. AND A. I. KLIMOVA, J. Gen. Chem. USSR (English Transl.) *36*, 165 (1966).
725d. STANKO, V. I., G. A. ANOROVA AND T. P. KLIMOVA, Zh. Obshch. Khim. *36*, 1774 (1966).
725e. STANKO, V. I., A. I. KLIMOVA, YU. A. CHAPOVSKII AND T. P. KLIMOVA, Zh. Obshch. Khim. *36*, 1779 (1966).
725f. STANKO, V. I., V. A. BRATTSEV, G. A. ANOROVA AND A. M. TSUKERMAN, Zh. Obshch. Khim. *36*, 1865 (1966).
725g. STANKO, V. I., YU. A. CHAPOVSKII, V. A. BRATTSEV AND L. I. ZAKHARKIN, Russ. Chem. Rev. (English Transl.) *34*, 424 (1965).
726. STEELE, B. D. AND J. C. MILLS, J. Chem. Soc. *1930*, 74.
727. STEINDLER, N. J. AND H. I. SCHLESINGER, J. Am. Chem. Soc. *75*, 756 (1953).
727a. STERLYADKINA, Z. K., O. N. KRYUKOVA AND V. I. MIKHEEVA, Russ. J. Inorg. Chem., English Transl. *10*, 316 (1965).
728. STEWART, A. C. AND G. W. SCHAEFFER, J. Inorg. Nucl. Chem. *3*, 194 (1956).
728a. STOCK, A., Ber. *53*, 827 (1920).
728b. STOCK, A., E. KUSS AND A. PRIESS, Ber. *47*, 3115 (1914).
729. STOCK, A. AND K. FRIEDERICI, Ber. *46*, 1959 (1913).
730. STOCK, A. AND H. LAUDENKLOS, Z. Anorg. Allgem. Chem. *228*, 178 (1936).
730a. STOCK, A., H. MARTIN AND W. SÜTTERLIN, BER. *67*, 396 (1934).
731. STOCK, A. AND C. MASSENEZ, Ber. *45*, 3539 (1912).
732. STOCK, A. AND W. MATHING, Ber. *69*, 1456 (1936).
733. STOCK, A. AND E. KUSS, Ber. *53B*, 789 (1923).
734. STOCK, A., F. KURZEN AND H. LAUDENKLOS, Z. Anorg. Allgem. Chem. *225*, 243 (1935).
735. STOCK, A. AND E. KUSS, Ber. *56*, 789 (1923).
736. STOCK, A. AND E. KUSS, Ber. *56*, 803 (1923).
737. STOCK, A. AND E. POHLAND, Ber. *59*, 2215 (1926).
738. STOCK, A. AND E. POHLAND, Ber. *59B*, 2223 (1926).
739. STOCK, A. AND E. POHLAND, Ber. *62*, 90 (1922).
740. STOCK, A. AND E. POHLAND, Ber. *59*, 2210 (1926).
741. STOCK, A. AND W. SIECKE, Ber. *57*, 562 (1924).
742. STOCK, A. AND W. SÜTTERLIN, Ber. *67*, 407 (1934).
743. STOCK, A., W. SÜTTERLIN AND F. KURZEN, Z. Anorg. Allgem. Chem. *225*, 225 (1935).
744. STOCK, A., E. WIBERG AND H. MARTINI, Ber. *63*, 2927 (1930).
745. STOCK, A., E. WIBERG AND H. MARTINI, Z. Anorg. Allgem. Chem. *188*, 32 (1930).
746. STOCK, A., E. WIBERG AND W. MATHING, Ber. *69*, 2811 (1936).
747. STOCK, A. AND F. ZEIDLER, Ber. *54*, 531 (1921).
748. STOCKMAYER, W. H., R. R. MILLER AND R. J. ZETO, J. Phys. Chem. *65*, 1076 (1961).
749. STOCKMAYER, W. H., D. W. RICE AND C. C. STEPHENSON, J. Am. Chem. Soc. *77*, 1980 (1955).
750. STOCKMAYER, W. H. AND C. C. STEPHENSON, J. Chem. Phys. *21*, 1311 (1953).
751. STONE, F. G. A. AND A. B. BURG, J. Am. Chem. Soc. *76*, 386 (1954).
752. STONE, H. W. AND B. GRIBBS, Congr. Intern. Chim. Pure Appl., 16e, Paris, 1957, Mem. Sect. Chim. Minerale (1958), p. 503.
753. STOSICK, A. J., Acta Cryst. *5*, 151 (1952).
754. SUBBANNA, V. V., L. H. HALL AND W. S. KOSKI, J. Am. Chem. Soc. *86*, 1304 (1964).

754a. SUJISHI, S. AND S. WITZ, J. Am. Chem. Soc. *79*, 2447 (1957).
755. SULLIVAN, E. A. AND S. JOHNSON, J. Phys. Chem. *93*, 233 (1959).
756. SUNDRAM, S. AND F. F. CLEVELAND, J. Chem. Phys. *32*, 166 (1960).
756a. SYMONS, M. C. R. AND H. W. WARDALE, Chem. Commun. *1967*, 758.
757. TAYLOR, R. C., J. Chem. Phys. *26*, 1131 (1957).
758. TAYLOR, R. C., J. Chem. Phys. *27*, 979 (1957).
759. TAYLOR, R. C. AND T. C. BISSOT, J. Chem. Phys. *25*, 780 (1956).
760. TAYLOR, R. C. AND A. R. EMERY, Spectrochim. Acta *10*, 419 (1958).
761. TAYLOR, M. D., L. R. GRANT AND C. A. SANDS, J. Am. Chem. Soc. *77*, 1506 (1955).
762. TAYLOR, R. C., D. R. SCHULTZ AND A. R. EMERY, J. Am. Chem. Soc. *80*, 27 (1958).
763. TEBBE, F., P. M. GARRETT AND M. F. HAWTHORNE, J. Am. Chem. Soc. *88*, 4222 (1966).
763a. TEBBE, F., P. M. GARRETT, AND M. F. HAWTHORNE, J. Am. Chem. Soc. *88*, 607 (1966).
763b. TEBBE, F., P. M. GARRETT, D. C. YOUNG AND M. F. HAWTHORNE, J. Am. Chem. Soc. *88*, 609 (1966).
763c. TEBBE, F., P. M. GARRETT AND M. F. HAWTHORNE, J. Am. Chem. Soc. *90*, 869 (1968).
764. TERHAAR, G., M. A. FLEMING AND R. W. PARRY, J. Am. Chem. Soc. *83*, 1767 (1962).
765. TIERNEY, P. A., D. W. LEWIS AND D. BERG, J. Inorg. Nucl. Chem. *24*, 1163 (1962).
766. TIMMS, P. L. AND C. S. G. PHILLIPS, Inorg. Chem. *3*, 297 (1964).
766a. TIPPE, A. AND W. C. HAMILTON, Inorg. Chem. *8*, 464 (1969).
766b. TITOV, L. V., Russ. J. Inorg. Chem. (Engl. Transl.) *13*, 936 (1968).
767. TODD, J. E. AND W. S. KOSKI, J. Am. Chem. Soc. *81*, 2319 (1959).
767a. TODD, L. J., I. C. PAUL, J. L. LITTLE, P. S. WELCKER AND C. R. PETERSON, J. Am. Chem. Soc. *90*, 4489 (1968).
767b. TODD, L. J., A. R. BURKE, H. T. SILVERSTEIN, J. L. LITTLE AND G. S. WIKHOLM, J. Am. Chem. Soc. *91*, 3376 (1969).
767c. TODD, L. J., J. L. LITTLE AND H. T. SILVERSTEIN, Inorg. Chem. *8*, 1698 (1969).
768. TOENISKOETTER, R. H., Thesis, St. Louis University, 1956.
769. TOENISKOETTER, R. H., G. W. SCHAEFFER, E. C. EWERS, R. E. HUGHES AND G. E. BAGLEY, Abstr. Papers 134th Meeting Am. Chem. Soc., Chicago, 1958.
769a. THOMPSON, N. R., J. Chem. Soc, *1963*, 6290.
770. TREFONAS, L. M. AND W. N. LIPSCOMB, J. Am. Chem. Soc. *81*, 4435 (1959).
771. TREFONAS, L. M., F. S. MATHEWS AND W. N. LIPSCOMB, Acta Cryst. *14*, 273 (1961).
771a. TROFIMENKO, S., J. Am. Chem. Soc. *89*, 3165 (1967).
771b. TROFIMENKO, S., J. Am. Chem. Soc. *88*, 1842 (1966).
771c. TROFIMENKO, S., J. Am. Chem. Soc. *88*, 1899 (1966).
771d. TROFIMENKO, S. AND H. N. CRIPPS, J. Am. Chem. Soc. *87*, 653 (1965).
771e. TUCKER, P. M. AND T. ONAK, J. Am. Chem. Soc. *91*, 6869 (1969).
772. TURNER, H. S., Chem. Ind. (London) *1958*, 1405.
773. TURNER, H. S. AND R. J. WARNE, Chem. Ind. (London) *1958*, 526.
773a. TURNER, H. S. AND R. J. WARNE, J. Chem. Soc. *1965*, 6421.
774. UCHIDA, H. S., H. B. KREIDER, A. MURCHISON AND J. F. MASI, J. Phys. Chem. *63*, 1414 (1959).
774a. VENKATESWARU, K. AND P. THIRUGNANASAMBANDAM, Proc. Indian Acad. Sci. *48*, 344 (1958).
774b. VICKERS, G. D., H. AGAHIGIAN, E. A. PIERS AND H. SCHROEDER, Inorg. Chem. *5*, 693 (1966).
775. VOET, D. AND W. N. LIPSCOMB, Inorg. Chem. *3*, 1679 (1964).
775a. VOLKOV, V. V. AND O. I. SHAPIRO, Russ. J. Inorg. Chem., English Transl. *11*, 789 (1966).
775b. VOORHEES, R. L. AND R. W. RUDOLPH, J. Am. Chem. Soc. *91*, 2174 (1969).
776. WADDINGTON, T. C., J. Chem. Soc. *1958*, 4783.
777. WAGNER, R. I. AND J. L. BRADFORD, Inorg. Chem. *1*, 99 (1962).
778. WAGNER, R. I. AND J. L. BRADFORD, Inorg. Chem. *1*, 93 (1962).
779. WAGNER, R. I. AND F. F. CASERIO, JR., J. Inorg. Nucl. Chem. *11*, 259 (1959).
779a. WALLBRIDGE, M. H. G., J. WILLIAMS AND R. L. WILLIAMS, J. Chem. Soc. *A1967*, 132.
779b. WALKER, F. E. AND R. K. PEARSON, J. Inorg. Nucl. Chem. *27*, 1981 (1965).
780. WALLER, M. C., *Inorganic Chemistry of Borohydrides,* Metal Hydrides Inc., Beverly, Mass., U.S.A.
781. WANATABE, H. AND M. KUBO, J. Am. Chem. Soc. *82*, 2428 (1960).
782. WANATABE, H., Y. KURODA AND M. KUBO, Spectrochim. Acta *17*, 454 (1961).
783. WANATABE, H., M. NARISADA, T. NAKAGAWA AND M. KUBO, Spectrochim. Acta *16*, 78 (1960).
784. WANATABE, H., T. TOTANI, T. NAKAGAWA AND M. KUBO, Spectrochim. Acta *16*, 1076 (1960).
785. WANG, F. E., P. G. SIMPSON AND W. N. LIPSCOMB, J. Am. Chem. Soc. *83*, 491 (1961).
786. WANG, F. E., P. G. SIMPSON AND W. N. LIPSCOMB, J. Chem. Phys. *35*, 1335 (1961).

786a. WARREN, JR., L. F., AND M. F. HAWTHORNE, J. Am. Chem. Soc. *89*, 470 (1967).
787. WARTIK, T., R. MOORE AND H. I. SCHLESINGER, J. Am. Chem. Soc. *71*, 3265 (1949).
788. WARTIK, T. AND R. K. PEARSON, J. Am. Chem. Soc. *77*, 1075 (1955).
789. WARTIK, T. AND R. K. PEARSON, J. Inorg. Nucl. Chem. *7*, 404 (1958).
789a. WASON, S. K. AND R. F. PORTER, Inorg. Chem. *5*, 161 (1966).
790. WEAVER, J. R., C. W. HEITSCH AND R. W. PARRY, J. Chem. Phys. *30*, 1075 (1959).
790a. WEGNER, P. A. AND M. F. HAWTHORNE, Chem. Commun. *1966*, 861.
791. WEISS, H. G., W. J. LEHMANN AND I. SHAPIRO, J. Am. Chem. Soc. *84*, 3840 (1962).
792. WEISS, H. G. AND I. SHAPIRO, J. Am. Chem. Soc. *81*, 6167 (1959).
793. WESTRUM, JR., E. F. AND N. E. LEVITIN, J. Am. Chem. Soc. *81*, 3544 (1959).
794. WHATLEY, A. T. AND R. N. PEASE, J. Am. Chem. Soc. *76*, 1997 (1954).
794a. WHIPPLE, E. B., T. H. BROWN, T. C. FARRAR AND T. C. DOYLE, J. Chem. Phys. *43*, 1841 (1965).
795. WIBERG, E., Ber. *69*, 2816 (1936).
796. WIBERG, E., Ber. *77*, 75 (1944).
797. WIBERG, E., Z. Anorg. Allgem. Chem. *173*, 199 (1928).
798. WIBERG, E., Naturwiss. *35*, 182 (1948).
799. WIBERG, E., Naturwiss. *35*, 212 (1948).
800. WIBERG, E. AND R. BAUER, Z. Naturforsch. *7b*, 58 (1952).
801. WIBERG, E. AND A. BOLZ, Ber. *73*, 209 (1940).
802. WIBERG, E., A. BOLZ AND P. BUCHHEIT, Z. Anorg. Allgem. Chem. *256*, 285 (1948).
803. WIBERG, E. AND A. BOLZ, Z. Anorg. Allgem. Chem. *257*, 131 (1948).
804. WIBERG, E., J. E. F. EVANS AND H. NÖTH, Z. Naturforsch. *13b*, 263 (1958).
805. WIBERG, E., J. E. F. EVANS AND H. NÖTH, Z. Naturforsch. *13b*, 265 (1958).
806. WIBERG, E. AND R. HARTWIMMER, Z. Naturforsch. *10b*, 290 (1955).
807. WIBERG, E. AND R. HARTWIMMER, Z. Naturforsch. *10b*, 291 (1955).
808. WIBERG, E. AND R. HARTWIMMER, Z. Naturforsch. *10b*, 294 (1955).
809. WIBERG, E. AND R. HARTWIMMER, Z. Naturforsch. *10b*, 295 (1955).
810. WIBERG, E. AND W. HENLE, Z. Naturforsch. *7b*, 575 (1952).
811. WIBERG, E. AND W. HENLE, Z. Naturforsch. *7b*, 579 (1952).
812. WIBERG, E. AND W. HENLE, Z. Naturforsch. *7b*, 582 (1952).
813. WIBERG, E. AND K. HERTWIG, Z. Anorg. Allgem. Chem. *255*, 141 (1947).
814. WIBERG, E., K. HERTWIG AND A. BOLZ, Z. Anorg. Allgem. Chem. *256*, 177 (1938).
815. WIBERG, E. AND G. HÖCKELE, Unpublished results.
816. WIBERG, E. AND G. HORELD, Z. Naturforsch, *6b*, 338 (1951).
817. WIBERG, E. AND A. JAHN, unpublished results.
818. WIBERG, E. AND A. JAHN, Z. Naturforsch. *11b*, 489 (1956).
819. WIBERG, E. AND U. KERSCHER, unpublished results.
820. WIBERG, E. AND O. KLEJNOT, unpublished results.
821. WIBERG, E. AND H. MICHAUD, Z. Naturforsch. *9b*, 499 (1954).
822. WIBERG, E. AND K. MÖDRITZER, Z. Naturforsch. *11b*, 747 (1956).
823. WIBERG, E. AND K. MÖDRITZER, Z. Naturforsch. *11b*, 748 (1956).
824. WIBERG, E. AND K. MÖDRITZER, Z. Naturforsch. *11b*, 750 (1956).
825. WIBERG, E. AND K. MÖDRITZER, Z. Naturforsch. *11b*, 751 (1956).
826. WIBERG, E. AND K. MÖDRITZER, Z. Naturforsch. *11b*, 753 (1956).
827. WIBERG, E. AND K. MÖDRITZER, Z. NATURFORSCH. *11b*, 755 (1956).
828. WIBERG, E. AND K. MÖDRITZER, Z. Naturforsch. *12b*, 123 (1957).
829. WIBERG, E. AND K. MÖDRITZER, Z. Naturforsch. *12b*, 127 (1957).
830. WIBERG, E. AND K. MÖDRITZER, Z. Naturforsch. *12b*, 128 (1957).
831. WIBERG, E. AND K. MÖDRITZER, Z. Naturforsch. *12b*, 131 (1957).
832. WIBERG, E. AND K. MÖDRITZER, Z. Naturforsch. *12b*, 132 (1957).
833. WIBERG, E. AND K. MÖDRITZER, Z. Naturforsch. *12b*, 135 (1957).
834. WIBERG, E. AND G. MÜLLER, unpublished results.
834a. WIBERG, E. AND H. NEUMAIER, Inorg. Nucl. Chem. Letters *1*, 35 (1965).
835. WIBERG, E. AND H. NÖTH, unpublished results.
836. WIBERG, E. AND H. NÖTH, Z. Naturforsch. *12b*, 125 (1957).
837. WIBERG, E., H. NOTH AND R. HARTWIMMER, Z. Naturforsch. *10b*, 292 (1955).
838. WIBERG. E., H. NÖTH AND R. USON, Z. Naturforsch. *11b*, 490 (1956).
839. WIBERG, E. AND K. SCHUSTER, Ber. *67B*, 1807 (1934).
840. WIBERG, E. AND P. STREBEL, Ann. *607*, 9 (1957).

841. Wiberg, E. and W. Sütterlin, Z. Anorg. Allgem. Chem. *202*, 22 (1931).
841a. Wiersma, R. J. and R. L. Middaugh, J. Am. Chem. Soc. *89*, 5078 (1967).
841b. Wiesboeck, R. A., J. Am. Chem. Soc. *85*, 2725 (1963).
842. Wiesboeck, R. A. and M. F. Hawthorne, J. Am. Chem. Soc. *86*, 1642 (1964).
843. Wiesboeck, R. A., A. R. Pitochelli and M. F. Hawthorne, J. Am. Chem. Soc. *83*, 4109 (1961).
843a. Wilks, P. H. and J. C. Carter, J. Am. Chem. Soc. *88*, 3441 (1966).
844. Williams, R. E., Inorg. Chem. *1*, 971 (1962).
844a. Williams, R. E., Inorg. Chem. *4*, 1504 (1965).
845. Williams, R. L., I. Dunstan and N. J. Blay, J. Chem. Soc. *1960*, 5006.
845a. Williams, R. E. and F. J. Gerhart, J. Organometal. Chem. *10*, 168 (1967).
845b. Williams, R. E. and F. J. Gerhart, J. Am. Chem. Soc. *87*, 3513 (1965).
846. Williams, R. E., S. G. Gibbins and I. Shapiro, J. Am. Chem. Soc. *81*, 6164 (1959).
847. Williams, R. E., S. G. Gibbins and I. Shapiro, J. Chem. Phys. *30*, 320 (1959).
848. Williams, R. E., S. G. Gibbins and I. Shapiro, J. Chem. Phys. *30*, 333 (1959).
848a. Williams, J., R. L. Williams and J. C. Wright, J. Chem. Soc. *1963*, 5816.
849. Williams, R. E. and T. P. Onak, J. Am. Chem. Soc. *86*, 3159 (1964).
849a. Williams, R. E. and E. Pier, Inorg. Chem. *4*, 1357 (1965).
849b. Wing, R. M., J. Am. Chem. Soc. *89*, 5599 (1967).
849c. Wilson, R. J., L. F. Warren, and M. F. Hawthorne, J. Am. Chem. Soc. *91*, 758 (1969).
850. Winkler, C., Ber. *23*, 772 (1890).
850a. Winter, L. P., Thesis, München University, 1962.
851. Winternitz, P. F., U.S.Pat. 2, 532, 217 (1950); C.A. *45*, 2162 (1951).
852. Wirth, H. E., F. E. Massoth and D. X. Gilbert, J. Phys. Chem. *62*, 870 (1958).
853. Wirth, H. E. and E. D. Palmer, J. Phys. Chem. *60*, 911 (1956).
854. Wirth, H. E. and E. D. Palmer, J. Phys. Chem. *60*, 914 (1956).
854a. Wittig, G., Ann. *573*, 200 (1961).
854b. Wittig, G., Ann. *563*, 110 (1949).
855. Wittig, G. and P. Hornberger, Z. Naturforsch. *6b*, 225 (1951).
856. Wittig, G. and P. Raff, Z. Naturforsch. *6b*, 225 (1951).
857. Woodword, L. A. and H. L. Roberts, J. Chem. Soc. *1956*, 1170.
858. Wunderlich, J. A. and W. N. Lipscomb, J. Am. Chem. Soc. *82*, 4427 (1960).
859. Yamazaki, M., J. Chem. Phys. *27*, 1401 (1957).
859a. Yoshizaki, T., H. Watanabe and T. Nakagawa, Inorg. Chem. *7*, 422 (1968).
859b. Young, D. C., D. V. Howe and M. F. Hawthorne, J. Am. Chem. Soc. *91*, 859 (1969).
859c. Young, D. E. and S. G. Shore, J. Am. Chem. Soc. *91*, 3497 (1969).
860. Yuan-Chin Fu and G. R. Hill, J. Am. Chem. Soc. *84*, 353 (1962).
860a. Zakharkin, L. I. and Yu. A. Chapovskii, Bull. Acad. Sci. USSR, Div. Chem. Sci., English Transl. *1964*, 723.
860b. Zalkin, A., T. E. Hopkins and D. H. Templeton, Inorg. Chem. *5*, 1189 (1966).
860c. Zalkin, A., T. E. Hopkins and D. H. Templeton, Inorg. Chem. *6*, 1911 (1967).
860d. Zaborski, R. and K. Cohn, Inorg. Chem. *8*, 678 (1969).
861. Zakharkin, L. I. and V. V. Gavrilenko, Bull. Acad. Sci. USSR, Div. Chem. Sci., English Transl. *1962*, 157.
861a. Zakharkin, L. I. and V. N. Kalinin, J. Gen. Chem. USSR, English Transl. *36*, 376 (1966).
861b. Zakharkin, L. I. and V. N. Kalinin, Bull. Acad. Sci. USSR, Div. Chem. Sci., English Transl. *1965*, 567.
861c. Zakharkin, L. I. and V. N. Kalinin, Bull. Acad. Sci. USSR, Div. Chem. Sci., English Transl. *1965*, 549.
861d. Zakharkin, L. I. and V. N. Kalinin, Bull. Acad. Sci. USSR, Div. Chem. Sci., English Transl. *1965*, 566.
861e. Zakharkin, L. I. and V. N. Kalinin, Bull. Acad. Sci. USSR, Div. Chem. Sci., English Transl. *1965*, 1287.
861f. Zakharkin, L. I. and V. N. Kalinin, Bull. Acad. Sci. USSR, Div. Chem. Sci., English Transl. *1965*, 2173.
861g. Zakharkin, L. I. and A. I. Lvov, J. Organometal. Chem. *5*, 313 (1966).
861h. Zakharkin, L. I. and A. V. Kazantsev, J. Gen. Chem. USSR, English Transl. *35*, 1128 (1965).
861i. Zakharkin, L. I., V. N. Kalinin and L. S. Podvisotskaya, Bull. Acad. Sci. USSR, Div. Chem. Sci., English Transl. *1966*, 1444.

861k. Zakharkin, L. I., V. N. Kalinin and L. S. Podvisotskaya, Zh. Obshch. Khim. *36*, 178 (1966).
861l. Zakharkin, L. I., E. I. Kukulina and L. S. Podvisotskaya., Izv. Akad. Nauk SSSR Otd. Khim. Nauk *1966*, 1866.
861m. Zakharkin, L. I. and V. N. Kalinin, Bull. Acad. Sci. USSR, Div. Chem. Sci. (Engl. Transl.) *1967*, 2471.
861n. Zakharkin, L. I. and L. S. Podvisozkaya, J. Organometal. Chem. *7*, 385 (1967).
861o. Zakharkin, L. I. and A. Ogorodnikova, J. Organometal. Chem. *12*, 13 (1968).
862. Zakharkin, L. I., V. I. Stanko and Yu. A. Chapovskii, Bull. Acad. Sci. USSR, Div. Chem. Sci., English Transl. *1964*, 547.
863. Zakharkin, L. I., V. I. Stanko and A. I. Klimova, Bull. Acad. Sci. USSR, Div. Chem. Sci., English Transl. *1964*, 722.
863a. Zange, E., Chem. Ber. *93*, 652 (1960).
863b. Zalkin, A., T. E. Hopkins and D. H. Templeton, Inorg. Chem. *6*, 1911 (1967).
863c. Zalkin, A., D. H. Templeton and T. E. Hopkins, J. Am. Chem. Soc. *87*, 3988 (1965).
863d. Zweifel, G. J. Organometal. Chem. *9*, 215 (1967).
864. Zweifel, G., N. R. Ayyangar and H. C. Brown, J. Am. Chem. Soc. *85*, 2072 (1963).
865. Zweifel, G. and H. C. Brown, J. Am. Chem. Soc. *85*, 2066 (1963).
866. Zweifel, G., K. Nagase and H. C. Brown, J. Am. Chem. Soc. *84*, 183 (1962).
867. Zeldin, M. and T. Wartik, Inorg. Chem. *4*, 1372 (1965).

Chapter 5

ALUMINIUM HYDRIDES

1. Historical

In 1939 Wiberg and Stecher[*166*] discovered stoichiometric hydrides of aluminium by subjecting a mixture of trimethylaluminium and hydrogen to an electric discharge; the methyl groups were successively replaced by hydrogen: $R_3Al + nH_2 \rightarrow R_{3-n}AlH_n + nRH$ ($n = 1, 2, 3$). The oily discharge product contained the compounds $AlR_3 \cdot AlHR_2$, $AlHR_2 \cdot AlHR_2$, and $AlHR_2 \cdot AlH_2R$. In 1942, the solid polymeric aluminium trihydride (polyalane) $(AlH_3)_x$ was obtained from the oil via the amine adducts $AlH_3 \cdot NR_3$ and $AlH_3 \cdot 2NR_3$[*121*].

At about the same time (1940), Schlesinger, Sanderson and Burg[*112*] succeeded in preparing a double hydride, aluminium boranate $AlH_3 \cdot 3BH_3 = Al(BH_3 = Al(BH_4)_3$, by heating trimethylaluminium with diborane.

In 1947, Finholt, Bond, jr. and Schlesinger[*36*] described for the first time a simple preparation of large amounts of low-molecular ether-soluble $(AlH_3)_n$ from high-molecular ether-insoluble $(AlH_3)_x$ and from lithium alanate (lithium aluminium hydride) $LiH \cdot AlH_3 = LiAlH_4$. The same paper contains references to its importance for organic reductions. The chemistry of the aluminium hydrides began on a broad basis one year later.

2. General review

The outer electron configuration of aluminium is similar to that of boron. It forms three normal σ-bonds with its three singly-occupied 3p orbitals. However, aluminium compounds containing only such σ-bonds are even rarer than the corresponding boron compounds. Two examples are triisopropylboron/triisopropylaluminium[1] and triisobutylboron/triisobutylaluminium[1], both with sterically demanding ligands. Aluminium trihydride, alane AlH_3, contains Al–H σ-bonds and in addition, in the absence of donors (as in the case of BH_3), two-electron three-centre bonds (Al–H^{μ}–Al bridges, $Al \overset{H}{\frown} Al$). Because of the great difference between the electronegativities of the atoms (and unlike the B–H bond), the Al–H bond is highly polar, with negative hydrogen.

Similarly to borane BH_3, alane is an electron-acceptor because of its unoccupied fourth 3p orbital. In contrast to boron, however, relatively low unoccupied 3d orbitals are also available to aluminium, so that the latter's coordination number can be not only 4 (like boron) but also 5 or 6.

---

[1]While the methyl, ethyl, and n-propyl compounds of aluminium dimerize to form Al_2R_6, Pr_3^iAl and Bu_3^iAl are scarcely associated. In contrast to this, all trialkyl borons R_3B, even the methyl compound, are monomeric[*21a*].

References p. 438

This occupancy of the orbitals determines the bonds in the aluminium hydrides, the structure of the aluminium hydrides and the types and mechanisms of their reactions.

2.1 *Alane structures*

(a) With donors. With donors, D, alane forms dative σ-bonds ($D \cdot\cdot \rightarrow AlH_3$). In the case of the hydrides $R_{3-n}AlH_n$ ($n = 1, 2, 3$) containing only Al–H σ-bonds this leads to coordination numbers of 4 and 5.

Prototypes:

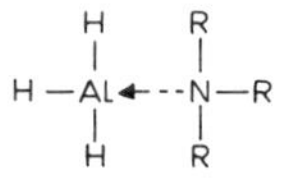

$AlH_3 \cdot NR_3$
C.N. = 4
[133]

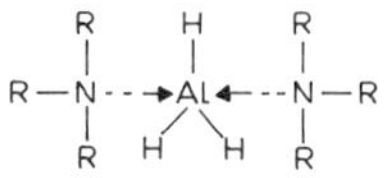

$AlH_3 \cdot 2NR_3$
C.N. = 5
[52, 57, 105]

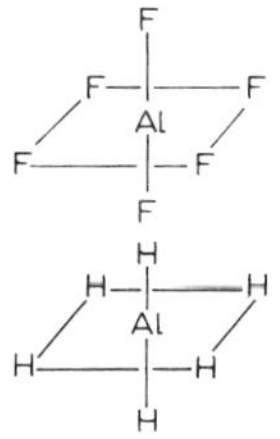

$AlF_6^{3\ominus}$
and the related
$AlH_6^{3\ominus}$
C.N. = 6
[6a, 26a, 127a]

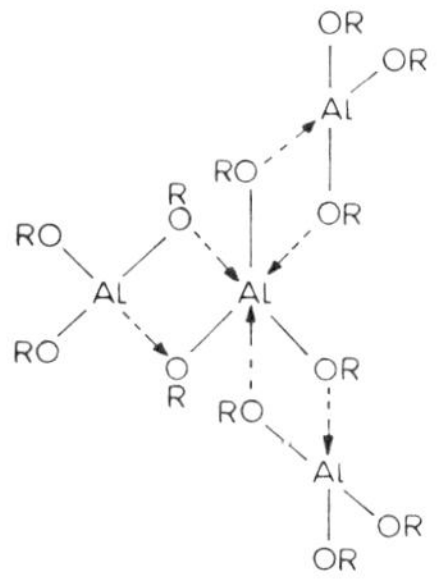

$[(Pr^iO)_3Al]_4$
1 x octahedron, C.N. = 6
3 x tetrahedron, C.N. = 4 [117]

The acceptor strength of AlH_3 and R_nAlH_{3-n} is greater than that of the corresponding boron compounds. For example, hydrogenation of R_nBX_{3-n} with $H^{\ominus}$ yields R_nBH_{3-n}, while hydrogenation of R_nAlX_{3-n} leads exclusively to the coordinatively saturated alanate $R_nAlH_{4-n}^{\ominus}$.

(b) Without donors. In the absence of donors alane forms two-electron three-centre bonds (like borane) because of its electron deficiency and for the reasons given in the boron section 2 of Chapter 4. Boron can form a maximum of two B–H^{μ}–B bridges per atom [stable $(BH_3)_2$]. On the other hand, aluminium reaches its coordinative limit only with six Al–H^{μ}–Al bridges [stable polymeric $(AlH_3)_x$].

Prototypes:

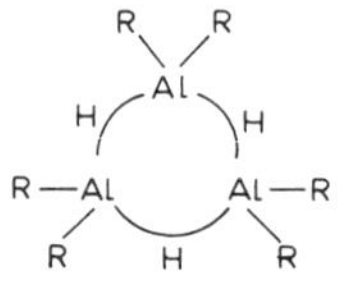

$(R_2AlH)_3$
C.N. = 4
[58]

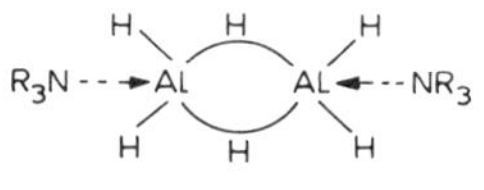

$(AlH_3 \cdot NR_3)_2$
C.N. = 5
Structure uncertain
[95, 104]

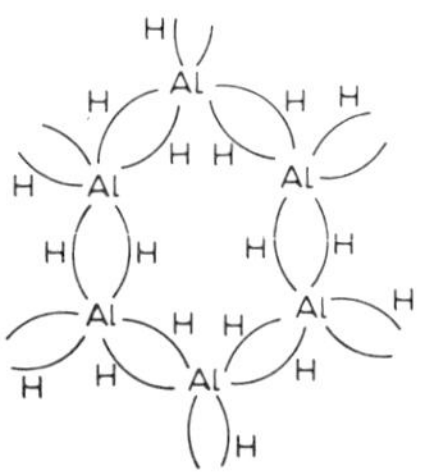

Idealized $(AlH_3)_x$
C.N. = 6
[54, 100, 126a, cit. 5]

(c) Bridge substitution. The hydrogens in Al–H^{μ}–Al bridges are substituted by donors more frequently than in the case of the boron hydrides. In other

words, in donor-substituted alanes the donor substituents frequently possess bridge functions (see section 10.4).

Prototype:

$[AlH_2(NR_2)]_2$
C.N. = 4

Alkylalanes, R_nAlH_{3-n} ($n = 1, 2, 3$), exchange alkyl groups extremely rapidly even at 20°. The exchange probably takes place via non-hydride bridges.

Prototype:

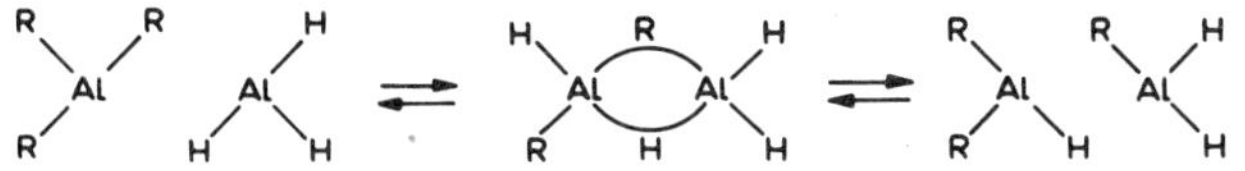

2.2 *Types of reactions*

(a) Symmetrical cleavage of the Al–H$^{\mu}$–Al double bridge. This is caused by donors *D* with formation of donor-alanes (for further information on the addition of donors see section 5):

Typical reactions:

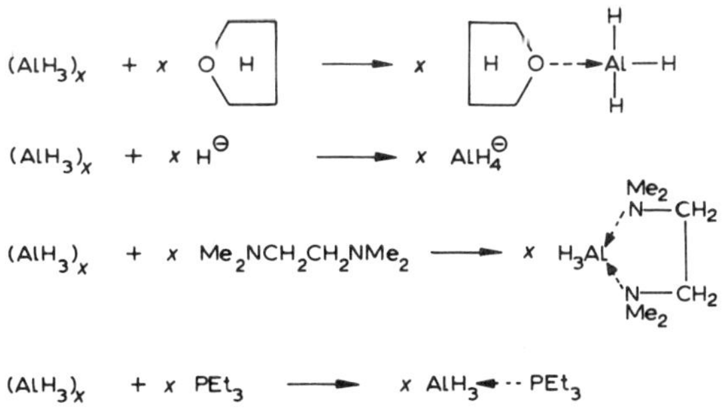

$$(AlH_3)_x + x\ H^{\ominus} \longrightarrow x\ AlH_4^{\ominus}$$

$$(AlH_3)_x + x\ Me_2NCH_2CH_2NMe_2 \longrightarrow x\ H_3Al(Me_2NCH_2CH_2NMe_2)$$

$$(AlH_3)_x + x\ PEt_3 \longrightarrow x\ AlH_3 \leftarrow PEt_3$$

(b) Nucleophilic displacement. In donor–alane adducts, potentially stronger donors displace weaker ones. Typical reactions (for further details see section 5):

$$AlH_3 \cdot NR_3 + H^{\ominus} \rightarrow AlH_4^{\ominus} + NR_3$$
$$AH_3 \cdot NR_3 + RNH_2 \rightarrow AlH_3 \cdot RNH_2 + NR_3$$
$$AlH_3 \cdot NR_3 + RCN \rightarrow AlH_3 \cdot RCN + NR_3$$

References p. 438

(c) Nucleophilic attack. Al is attacked by donors D with the abstraction of $H^{\ominus}$ or Al-substituents more negative than H (substitutions):

$$D + \rangle Al-X \rightarrow D \cdots\rightarrow -\overset{|}{\underset{|}{Al}}-X \rightarrow D-Al\langle + X$$

Typical reactions: in chloroalane AlH_2Cl, in accordance with the relatively large difference in electronegativity between Al and Cl, the Al–Cl bond is more polar than the Al–H bond. Consequently, anions first replace Cl and only then H (for further details see section 3.3):

$$AlH_2Cl \xrightarrow[\text{nucleophilic substitution}]{+H^{\ominus},\ -Cl^{\ominus}} AlH_3 \xrightarrow[\text{donor addition}]{+H^{\ominus}} AlH_4^{\ominus}$$

$$AlH_2Cl \xrightarrow[-Li^{\oplus}Cl^{\ominus}]{+Li^{\oplus}PEt_2^{\ominus}} AlH_2(PEt_2) \xrightarrow{\text{cyclization}} \tfrac{1}{3}\,[H_2Al(PEt_2)]_3$$

$$\tfrac{1}{3}\,[H_2Al(PEt_2)]_3 \xrightarrow[\text{donor addition}]{+PEt_2^{\ominus}} AlH_2(PEt_2)_2^{\ominus}$$

In the presence of protonic hydrogen, on the other hand, H is replaced with the formation of molecular hydrogen (high heat of formation of H_2), as shown in the following two equations (for further details see section 3.3):

$$AlH_2Cl + 2\overset{\delta+\delta-}{HPMe_2} \rightarrow Al(PMe_2)_2Cl + 2H_2$$

$$AlHCl_2 + \overset{\delta+\delta-}{HPMe_2} \rightarrow Al(PMe_2)Cl_2 + H_2$$

$$AlH_4^{\ominus} + 3\overset{\delta+\delta-}{HPMe_2} \rightarrow AlH(PMe_2)_3^{\ominus} + 3H_2$$

The substitution stops at the stage of $AlH(PMe_2)_3^{\ominus}$ probably because of steric hindrance

$$AlH_4^{\ominus} + 4H^{\oplus}Cl^{\ominus} \rightarrow AlCl_4^{\ominus} + 2H_2$$

$$AlH_4^{\ominus} + 4H^{\oplus}OH^{\ominus} \rightarrow Al(OH)_4^{\ominus} + 2H_2$$

$$AlH_4^{\ominus} + R \rightarrow AlH_3Cl^{\ominus} + RH$$

The last reaction is difficult to initiate and is followed by other reactions.

(d) Cyclization and polymerization. Donor–alane complexes in which the donors contain protonic hydrogen or hydrogenatable groups abstract one or more H-atoms from the aluminium as hydride. This leads either to the elimination of H_2 ($H^{\delta+}$ of the donor + $H^{\delta-}$ of the alane = H_2), hydrogenating migration of H to positively polarized C-atoms of the donor molecule, or hydrogenating ring opening. In all these cases the dative σ-bond (donor $\cdots\rightarrow$ Al, *e.g.* in amine–alane

adducts $AlH_3 \cdot NHR_2$) is converted into a dative σ-bond strengthened by π-bonds [donor-Al, *e.g.* in aminoalanes $AlH_2(NR_2)$]. Further elimination of hydride from the aluminium gives rise not to donor–aluminium double bonds as in the case of the corresponding reactions of the boranes, but to cyclic or more or less highly polymeric materials (for further details, see section 5).

Typical reactions:

$$AlH_3 \xrightarrow{+2NH_3} (H_3N)_2AlH_3 \xrightarrow{-H_2} (H_2N)(H_3N)AlH_2 \xrightarrow{-H_2} (H_2N)_2AlH \xrightarrow{+NH_3} (H_2N)_2HAl \leftarrow NH_3$$

$$\xrightarrow{-H_2} (H_2N)_2Al{-}NH_2 \quad \text{(with an excess of } NH_3\text{)}$$

$$AlH_3 \xrightarrow{+NH_3} AlH_3 \leftarrow NH_3 \xrightarrow{-H_2} AlH_2{-}NH_2 \xrightarrow{-H_2} {-}AlHNH{-}$$

$$\xrightarrow{-H_2} >AlN< \quad \text{(with a deficiency of } NH_3\text{)}$$

$$AlH_4^{\ominus} + 4PH_3 \rightarrow Al(PH_2)_4^{\ominus} + 4H_2$$

$$AlH_4^{\ominus} + 4AsH_3 \rightarrow Al(AsH_2)_4^{\ominus} \text{ (unstable)} + 2H_2$$

$$AlH_3 \leftarrow N{\equiv}CR \xrightarrow{\text{hydride migration}} AlH_2{-}N{=}CRH \xrightarrow[\text{and polymerization}]{\text{hydride migration}} \frac{1}{x}[{-}AlH{-}N(CRH_2){-}]_x$$

$$AlH_3 \leftarrow N(C_5H_5) \xrightarrow{\text{hydride migration}} H_2Al{-}N(C_5H_6)$$

$$AlH_3 \leftarrow HN(CH_2)_2 \xrightarrow{\text{elimination of } H_2} H_2Al{-}N(CH_2)_2 \xrightarrow[\text{and polymerization}]{\text{hydrogenating ring opening}} \frac{1}{x}[{-}AlH{-}N(C_2H_5){-}]_x$$

$$AlH_3 \leftarrow O(CH_2)_4 \xrightarrow{\text{hydrogenating ring opening}} H_2Al{-}O{-}CH_2CH_2CH_2CH_3$$

(*e*) *Addition to multiple bonds.* Al–H bonds of the acceptor AlH_3 (but not of the donor $AlH_4^{\ominus}$) add to (π-dative) C=C double bonds. Because of the negative nature of the hydrogen the reaction is anti-Markovnikov, the Al adding to the more negative C-atom:

$$>\overset{\delta-}{C}{=}\overset{\delta+}{C}< + >\overset{\delta+}{Al}{-}\overset{\delta-}{H} \rightarrow \left[\begin{matrix} -\overset{\delta-}{C} \cdots \overset{\delta+}{C}- \\ \vdots \quad \vdots \\ \underset{\delta+}{Al} \cdots \underset{\delta-}{H} \end{matrix}\right] \rightarrow >Al{-}\overset{|}{\underset{|}{C}}{-}\overset{|}{\underset{|}{C}}H$$

References p. 438

The additions of the Al–H bonds of AlH_3 (acceptor) to C≡C triple bonds (donor are stereospecific. In a four-centre *cis* addition, *trans* compounds are formed:

$$\overset{\delta-}{H-C}\equiv\overset{\delta+}{C}-R + \overset{\delta+}{Al}-\overset{\delta-}{H} \rightarrow \left| \begin{matrix} \overset{\delta-}{H-C}\equiv\overset{\delta+}{C}-R \\ \overset{\delta+}{Al}\cdots\overset{\delta-}{H} \end{matrix} \right| \rightarrow \begin{matrix} H \quad\quad R \\ C=C \\ Al \quad\quad H \end{matrix}$$

Even $AlH_4^{\ominus}$ (which of course is not an acceptor) converts C≡C into C=C unde certain conditions. However, as can be seen from the first two of the followin reactions, and in accordance with what was said above, hydrogenation canno be carried out directly but via the alkynyl derivatives of aluminium (blocking o the protonic hydrogen):

$$AlH_4^{\ominus} + 4\overset{\delta+}{H}-C\equiv CR \rightarrow Al(C\equiv CR)_4^{\ominus} + 4H_2$$

$$2(i\text{-}C_4H_9)_3AlH^{\ominus} + HC\equiv CH \rightarrow (i\text{-}C_4H_9)_3Al-C\equiv C-Al(i\text{-}C_4H_9)_3 + 2H_2$$

$$AlH_4^{\ominus} + 3HC\equiv CH \rightarrow AlH(CH=CH_2)_3^{\ominus}$$

The C=C bond (donor) that can be attacked by the acceptor AlH_3 generall resists hydrogenation by the donor alanate $AlH_4^{\ominus}$. For example, after hydrolysi the reaction product of lithium alanate and but-2-en-1-al gives but-2-en-1-ol with retention of the C=C bond, which corresponds to an addition of $AlH_4^{\ominus}$ t the C=O bond:

$$CH_3CH=CH-C(H)=O \xrightarrow{+AlH_4^{\ominus}} CH_3CH=CH-CH_2-\overset{\ominus}{O}\cdots\rightarrow AlH_3 \xrightarrow[\Delta]{+H_2O} CH_3CH=CH-CH_2-OH$$

(f) Elimination of electrons.

$$AlH_3 + X^{\ominus} \rightarrow AlH_2X + \tfrac{1}{2}H_2 + e^{\ominus}$$

Typical reactions:

$$AlH_3 \cdot NR_3 + \frac{n}{2}HgR_2 \rightarrow AlH_{3-n}R_n \cdot NR_3 + \frac{n}{2}H_2 + \frac{n}{2}Hg$$

$$AlH_3 \cdot NR_3 + \frac{n}{2}HgCl \rightarrow AlH_{3-n}Cl_n \cdot NR_3 + \frac{n}{2}H_2 + \frac{n}{2}Hg$$

For a review see [*89a*].

3. Preparation of aluminium hydrides

3.1 Elementary syntheses of aluminium hydrides

Molecular hydrogen does not diffuse through aluminium [9]. A thin layer abou 570 atoms thick on a glass substrate (danger of diffusion of O!) at −196° and a H_2 pressure of 1.10 to 1.26 torr takes up only a little of the atomic hydrogen pro duced by a heated tungsten wire. The amount of H_2 desorbed after the layer ha been heated to $-78°$ or 43° corresponds to a formula $AlH_{0.094}$[*118*]. We are probably dealing here only with absorbed molecular hydrogen, particularly since the total pressure during the preparation was about 4 orders of magnitude above the mean free path length.

In an atmosphere of hydrogen with a pressure of about 1 torr, evaporating aluminium (0.1–19 mg/min.) forms aluminium hydride with thermally produced (tungsten wire) atomic hydrogen. It separates out on the wall of the vessel cooled to $-196°$ together with unchanged aluminium as $AlH_{0.077}$ to $AlH_{1.02}$, depending on the experimental conditions (particularly the dimensions of the apparatus). The process appears to be a homogeneous gas reaction. The H content of the true hydride is probably higher. The highest empirical H content is in fact obtained at low rates of evaporation of the aluminium (rate of hydrogenation ~ rate of evaporation). $AlH_{0.9-1.0}$ is stable at $-78°$. Hydrogen is evolved on thermolysis at 0, 13, and 43° (activation energy 14.4 kcal/mole), and on hydrolysis[*118*].

Relatively stable gaseous AlH_3 and $(AlH_3)_2$ are formed in a continuous stream of partially dissociated hydrogen with aluminium vapour (which is few directly into a mass spectrometer). In similar experiments with gallium and indium, only GaH_3 and InH_3, which are more unstable, are formed[*11*, *11a*].

Hexagonal aluminium deuteride is formed[*5*] when ultra-pure 99.999% aluminium is irradiated with deuterons accelerated in an 88-inch cyclotron (12.5 MeV) below 140°.

3.2 *Preparation of alanes, AlH_3 and AlH_nX_{3-n}*

3.2.1 *From $LiAlH_4$ and AlX_3*

The reaction of aluminium halides AlX_3 (X = Cl, Br, I) with lithium alanate in diethyl ether below 4° or above 12° forms aluminium trihydride $(AlH_3)_x$ (polyalane) in accordance with the overall equation[*36*, *134*]:

$$AlX_3 + 3LiAlH_4 \rightarrow 4AlH_3 + 3LiX$$

Here the dihaloalanes $AlHX_2$ (aluminium dihalide hydrides)[*49*] and monohaloalanes AlH_2X (aluminium halide dihydrides)[*32*, *33*] are formed as intermediates. According to conductometric measurements, the reaction of the chloride in ether takes place as follows[*33*]:

$$Al_2Cl_6 \xrightarrow[-LiCl]{+LiAlH_4} Al_2Cl_5^{\oplus} + AlH_4^{\ominus} \xrightarrow[-LiCl]{+LiAlH_4} 4AlH_2Cl \xrightarrow[-4LiCl]{+4LiAlH_4} 8AlH_3$$

and the reaction of the iodide in ether as:

$$Al_2I_6 \xrightarrow{LiI} Al_2I_7^{\ominus} + Li^{\oplus} \xrightarrow[-3LiI]{+2LiAlH_4} 4AlH_2I \xrightarrow[-4LiI]{+4LiAlH_4} 8AlH_3$$

Conductometric monitoring of the course of the reaction is complicated in the case of the chloride and bromide by the formation of $AlHX_2$[*49*]:

$$Al_2X_6 + AlH_3 \rightarrow 3AlHX_2$$

Under suitable conditions (4–12°), $AlHX_2$ and AlH_2X can be isolated from the reaction of AlX_3 with $LiAlH_4$[*83*]. However, it is better to prepare the monohydride and the dihydride in two steps[*160–162*]:

(1) preparation of a solution in ether, freed from ether-insoluble LiCl by centrifuging, of low-molecular $(AlH_3)_n$; the latter cannot be kept for long because of its rapid polymerization;

(2) commutation of AlH_3 with AlX_3 to give AlH_nX_{3-n}; ether complexes of the monomeric halide hydrides are formed ($AlHX_2$ or AlH_2X or mixtures, depending on the AlH_3–AlX_3 ratio):

$$AlX_3 + AlH_3 \rightarrow AlHX_2 + AlH_2X \quad (\text{when } AlX_3 : AlH_3 = 1:1)$$
$$AlHX_2 + AlH_3 \rightarrow 2AlH_2X \quad (\text{when } AlX_3 : AlH_3 = 1:2)$$
$$AlH_2X + AlX_3 \rightarrow 2AlHX_2 \quad (\text{when } AlX_3 : AlH_3 = 2:1)$$

3.2.2 *From $LiAlH_4$ and HCl*

Although the preparation of AlH_3 by the reaction of lithium alanate with hydrogen chloride

$$LiAlH_4 + HCl \xrightarrow{\text{diethyl ether}} AlH_3 + LiCl + H_2$$

involves a loss of hydride hydrogen from the starting material, the practical yield of AlH_3 (referred to $LiAlH_4$) is higher (85–93%) than in the reaction of $LiAlH_4$ with AlX_3 (which is of the order of 30%)[*89, 125*]. For this purpose, an ethereal solution of HCl is added slowly in drops, with stirring to a 1–2 *M* solution of $LiAlH_4$ in diethyl ether at −20°. Since a certain amount of HCl always escapes with the hydrogen formed, towards the end of the reaction a supplementary amount of HCl (generally only small) must be added to make up the stoichiometric ratio. This is determined acidimetrically by the hydrolysis (formation of LiOH) of the ethereal solution separated from precipitated LiCl. If an insufficient amount of HCl is used in this reaction, the AlH_3 formed and the unconsumed $LiAlH_4$ give unstable greasy adducts $LiAlH_4 \cdot nAlH_3$. Removal of LiCl by filtration through sintered glass, addition of about 2 moles of tetrahydrofuran per mole of the alane and evaporation to dryness under vacuum result in an 85–93% yield of coarsely crystalline $AlH_3 \cdot x$THF (Cl content about 1%)[*89, 125*].

3.3 *Preparation of substituted alanes AlH_nR_{3-n}*

Partially alkylated alanes AlH_nR_{3-n} can be obtained by hydrogenation of the corresponding partially alkylated aluminium halides AlX_nR_{3-n}. Their preparation and isolation is not simple, since they undergo dismutation with extraordinary ease and easily form double hydrides. Thus, the reaction of ethylaluminium dichloride or dibromide AlX_2Et with an excess of sodium hydride in benzene at 70 to 80° gives not only the desired ethylalane AlH_2Et but also diethylalane $AlHEt_2$ and alane AlH_3 ($AlH_3 + NaH \rightarrow NaAlH_4$)[*172*]:

$$AlX_2Et + NaH \xrightarrow{\Delta} AlH_2Et,\ AlHEt_2,\ NaAlH_4 \quad (X = Cl, Br)$$

To prepare alkylaluminium hydrides, AlH_nR_{3-n}, it is necessary to avoid an excess of the hydrogenating agent (LiH, NaH) since otherwise only alanates are formed, for example [*172*]:

$$2AlCl_2Et + 6NaH \rightarrow Na[AlH_2Et_2] + Na[AlH_4] + 4NaCl$$
$$2AlBr_2Et + 6LiH \rightarrow Li[AlH_2Et_2] + Li[AlH_4] + 4LiBr$$

The dismutation product $Na[AlEt_4]$, which is also theoretically possible undergoes commutation with $Na[AlH_4]$, again giving diethylalanate [*172*]:

$$Na[AlEt_4] + Na[AlH_4] \rightarrow 2Na[AlH_2Et_2]$$

The hydrogenation of aluminium chloride, diisopropoxyaluminium chloride, and trimethylaluminium with sodium boranate or diborane [*41*, *92*, *111*, *112*] does not lead to the corresponding hydride but to boranates:

$$AlCl_3 \xrightarrow[-LiCl]{+LiBH_4} AlCl_2(BH_4) \xrightarrow[-LiCl]{+LiBH_4} AlCl(BH_4)_2 \xrightarrow[-LiCl]{+LiBH_4} Al(BH_4)_3$$

$$(PrO)_2AlCl + NaBH_4 \rightarrow (PrO)_2Al(BH_4) + NaCl$$

$$AlMe_3 + 2(BH_3)_2 \rightarrow Al(BH_4)_3 + BMe_3$$

Reactions of $R_2AlCH_2CH(CH_3)_2$ with olefins point to an equilibrium with dialkylalane and isobutane [*85*]:

$$R_2AlCH_2CH(CH_3)_2 \rightleftharpoons R_2AlH + CH_2 = C(CH_3)_2$$

The dimeric trimethylsiloxoaluminium dichloride, $[(Me_3SiO)AlCl_2]_2$, can be hydrogenated with lithium alanate to form crystalline distillable μ,μ'-bis(trimethylsiloxo)dichlorodialane $[(Me_3SiO)AlClH]_2$ [*113*]:

$$[(Me_3SiO)AlCl_2]_2 + 2LiAlH_4 \longrightarrow [(Me_3SiO)AlClH]_2 + \frac{2}{x}(AlH_3)_x + 2LiCl$$

With this hydride, too, NMR and IR spectra [*113*] confirm the rule that no $Al–H^{\mu}–Al$ bridges occur in N- or O-substituted alanes (see section 10.4).

Aluminium chloride, $AlCl_3$, and dialkylaluminium hydrides, $AlHR_2$, readily dismute to give various hydrides containing 60 to 70% of AlH_3 [*71*]:

$$AlCl_3 + 3AlHR_2 \rightarrow AlH_3 + AlClR_2$$

Molecular hydrogen hydrogenates triethylaluminium at 60° and a pressure of about 10 atm to diethylalane. Ziegler catalysts ($CrCl_3$, α-$TiCl_3$, β-$TiCl_3$) accelerate the reaction and increase the yield [84].

References p. 438

With lithium diethylphosphide, monochloroalane forms trimeric phosphorus-substituted alane which reacts with more phosphide to give a phosphorus-substituted alanate [*40*]:

$$3\,AlH_2Cl + 3\,LiPEt_2 \xrightarrow{-3\,LiCl} [H_2Al(PEt_2)]_3 \text{ (cyclic trimer: } H_2Al\text{–}PEt_2\text{ ring)}$$

$$[AlH_2(PEt_2)]_3 + 3\,LiPEt_2 \longrightarrow 3\,Li[AlH_2(PEt_2)_2]$$

Phosphorous-substituted alanes are also formed by the reaction of AlH_3[2] and AlH_nCl_{3-n} with dimethylphosphane $PHMe_2$[2]. Ether is unsuitable as a solvent for this reaction, since it competes with the phosphane in the primary formation of a complex. Phosphorus-substituted alanates are formed similarly [*18*]:

$$AlH_3 + PHMe_2 \xrightarrow{\text{(liquid } PHMe_2)} [AlH_3 \cdot PHMe_2] \xrightarrow[-H_2]{20°, \Delta} AlH_2(PMe_2),\ AlH(PMe_2)_2,\ Al(PMe_2)_3\ .$$

$$AlH_2Cl + 2PHMe_2 \xrightarrow[-2H_2]{\text{(liquid } PHMe_2)} Al(PMe_2)_2Cl$$

$$AlHCl_2 + PHMe_2 \xrightarrow[-H_2]{\text{(liquid } PHMe_2)} Al(PMe_2)Cl_2$$

$$LiAlH_4 + 3PHMe_2 \xrightarrow[-3H_2]{\text{(liquid } PHMe_2)} Li[AlH(PMe_2)_3]$$

4. Physical properties

Table 5.1 contains, inter alia, the physical properties of alane adducts and derivatives. The reversible elimination of trimethylamine from $AlH_3 \cdot 2NMe_3$ [*50*] in accordance with $AlH_3 \cdot 2NMe_3 \rightleftharpoons AlH_3 \cdot NMe_3 + NMe_3$, $\log p$ [torr] $= -3937.0/T + 11.971$; $H = -18.0 \pm 0.1$ kcal/mole, shows that the acceptor activity of $(AlH_3 \cdot NMe_3)$ is comparable with that of (BMe_3) $(BMe_3 \cdot NMe_3 \rightleftharpoons BMe_3 + NMe_3$; $H = -17.6$ kcal/mole). At the same time, irreversible processes take place with the elimination of H_2 [*50*]. For the polymerization of various trialkylaluminiums, see [*98*].

5. Reactions of alanes AlH_nX_{3-n} and AlH_nR_{3-n}

Alanes AlH_nX_{3-n} and $R_{3-n}AlH_n$ ($n = 1, 2, 3$) possess negative hydrogen and free orbitals (3p, 3d). Consequently, their reactions are governed by three factors. (1) Because of the free 3p and 3d orbitals, aluminium (in contrast to boron) can occupy the coordination numbers of 4, 5, and 6. Therefore in absence of a donor alanes form hydrogen bridge bonds. Each bridge formed raises the co-

[*text continued p. 402*]

[2] Possibly formed in the nascent state by the reaction

$$LiAlH_4 + Me_2PH_2Cl \rightarrow AlH_3 + PHMe_2 + H_2 + LiCl$$

TABLE 5.1
PREPARATION AND PROPERTIES OF ALANE ADDUCTS AND DERIVATIVES

Compound	*Preparation*	*Properties*	*References*
$AlH_3 \cdot 4O(C_2H_5)_2$	$3LiAlH_4 + AlCl_3 + 16OEt_2 \xrightarrow{\text{benzene}} 4(AlH_3 \cdot 4OEt_2) + 3LiCl$	Degree of association in benzene 1.21	26
$(AlH_3)_{2.05...2.30} \cdot O(C_2H_5)_2$	Ethereal solution of AlH_3 + pentane	White solid	78
$AlH_3 \cdot O(C_2H_5)_2 \cdot OC_4H_8$	$3LiAlH_4 + AlCl_3 + 4OEt_2 + 4THF \xrightarrow{\text{benzene}} 4(AlH_3 \cdot OEt_2 \cdot THF) + 3LiCl$	Degree of assocation in benzene 0.99	26
$AlH_3 \cdot 2OC_4H_8$	$(AlH_3)_x + 2x THF \xrightarrow{-5°} x(AlH_3 \cdot 2THF)$	Fine felted colorless needles; Stability: $AlH_3 \cdot 2OEt_2 \ll AlH_3 \cdot 2THF < AlH_3 \cdot 2NMe_3$; THF pressure at −5°: 6 torr; no ring opening at −45°, since hydrolysis gives 3 moles of H_2: $AlH_3 \cdot 2THF + 3H^{\oplus} \rightarrow Al^{3\oplus} + 2THF + 3H_2$ (for ring opening, see $AlH_3 \cdot OC_4H_8$); solubility of $AlH_3 \cdot 2THF$ in benzene at 18°: 5.5 mole per 1000 g of benzene	132 125 89
$AlH_3 \cdot OC_4H_8$	(a) $AlH_3 \cdot 2THF \xrightarrow[\text{pressure of THF 6 torr}]{-5°} AlH_3 \cdot THF + THF$	Solubility of AlH_3 in THF at 25°: 1.90 mole of AlH_3 = 64 g per litre	26 132
	(b) $3LiAlH_4 + AlCl_3 + 4THF \xrightarrow{\text{benzene}} 4(AlH_3 \cdot THF) + 3LiCl$	At 20°, such solutions are unstable. Even at −25° the hydride activity falls to about a half in 37 days (ether cleavage). Degree of association in benzene 1.55	89 125
$AlH_3 \cdot 0.25OC_4H_8$	$AlH_3 \cdot THF \xrightarrow[\text{pressure of THF 15 torr}]{20°} AlH_3 \cdot 0.25THF + 0.75THF$		132
$AlH_3 \cdot NH_3$	(a) NH_3(soln. in THF) + AlH_3(soln. in ether) $\xrightarrow{-80°} AlH_3 \cdot NH_3$	Separates from the ethereal solution in the form of a voluminous white precipitate; slow evolution of H_2 sets in above −45°	143

References p. 438

TABLE 5.1 (*cont'd*)

Compound	*Preparation*	*Properties*	*References*
	(b) NH_3(gas) + AlH_3(soln. in ether) $\xrightarrow{-196\ \text{to}\ -80°} AlH_3 \cdot NH_3$		
$AlH_3 \cdot 2NH_3$	Intermediate in the preparation of $AlH_2(NH_2) \cdot NH_3$	Slowly gives off H_2 even at −80°; see $AlH_2(NH_2) \cdot NH_3$	144
AlH_3	$AlH_3 \cdot xOR_2 \xrightarrow{-xOR_2} AlH_3$ (87–96% hydride)	octahedral coordination symmetry of Al	54, 100 127a, cit. 5
$AlH_2(NH_2)$	$AlH_3 \cdot NH_3 \xrightarrow{-35°,\ 5-6h} AlH_2(NH_2) + H_2$	White solid; structure I, section 6	143
$AlH(NH)$	$AlH_2(NH_2) \xrightarrow[\text{several hours}]{20°} AlH(NH) + H_2$	Structure II, section 6	143
AlN	$AlH(NH) \xrightarrow{100°,\ 1h,\ -H_2} AlH_{0.27}NH_{0.27} \xrightarrow[\Delta]{-150°,\ 1h,\ -H_2} AlH_{0.23}NH_{0.23}$	Structure III, section 6	143
$AlH_2(NH_2) \cdot NH_3$	AlH_3(ether) + $2NH_3$(excess) $\xrightarrow{<-80°}$ $[AlH_3 \cdot 2NH_3] \xrightarrow{-80°,-H_2} AlH_2(NH_2) \cdot NH$	White solid evolving H_2 at −50°	144
$AlH(NH_2)_2$	$AlH_2(NH_2) \cdot NH_3 \xrightarrow{-50°,\ -H_2} AlH(NH_2)_2$	White solid; forms no ammoniate under the experimental conditions	144
$Al(NH)NH_2$	$AlH(NH_2)_2 \xrightarrow{20°,\ -H_2} Al(NH)NH_2$	White solid; structure VI, section 6	144
$Al(NH_2)_3$	Difficult to prepare: (a) add ethereal AlH_3 soln. in drops to liquid NH_3 (≤ −40°) (b) $AlH(NH_2)_2$ (without solvent) + liquid NH_3 (≤ −78°)	Fine white powder free from hydride hydrogen; evolves NH_3 in vacuum even at 20°, but the last traces of NH_3 are not liberated even at 430°[→ $(AlN)_x$]	144
$AlH_3 \cdot NH_2(CH_3)$	$AlH_3 + NH_2Me \xrightarrow[\text{ether}]{-55°} AlH_3 \cdot NH_2Me$	Stable below −40°	145
$AlH_2(NHCH_3)$	$AlH_3 + NH_2Me \xrightarrow[-H_2]{>-40°} AlH_2NHMe$	Polymeric; structure VII, section 6	145

$AlH(NCH_3)$	(a) $AlH_2(NHMe) \xrightarrow[-H_2]{>20^\circ} AlH(NMe)$ (b) $2AlH_3 \cdot NMe_3 + Al(NHR)_3 \rightarrow 3AlH(NMe) + 3H_2 + 2NMe_3$	Polymeric; structure VIII, section 6	145 124a
$AlH_3 \cdot 2NH(CH_3)_2$	$AlH_3 + 2NHMe_2 \xrightarrow[ether]{-50^\circ} AlH_3 \cdot 2NHMe_2$	Freed from ether and excess of $NHME_2$ in vacuum at −40°; fairly unstable, decomposes in ethereal solution about −20° with evolution of H_2; the intermediate $AlH_3 \cdot NH(CH_3)_2$ (not isolated) is said to be produced on heating	146
$AlH_2[N(CH_3)_2]$	(a) $AlH_3(ether) + NHMe_2 \xrightarrow{-40^\circ}$ no reaction $\xrightarrow[-H_2]{-20/0^\circ} AlH_2(NMe_2)$ (b) $LiAlH_4 + [NH_2Me_2]Cl \xrightarrow{ether, -10^\circ} AlH_2(NMe_2) + LiCl + 2H_2$ (yield 63%)	Colourless lustrous, highly refractive prisms; can be sublimed in vacuum at 40–60°; m.p. 89–90°; soluble in benzene and ether; decomposition ~ 130°; degree of association in benzene 2.99, in ether 2.16; structures IV + V, section 6	104 146
$AlH[N(CH_3)_2]_2$	$AlH_3(ether) - 2NHMe_2 \xrightarrow{20^\circ} AlH(NMe_2)_2 + 2H_2$ equilibrium in ether: $[(Me_2N)HAl(\mu\text{-}NMe_2)_2AlH(NMe_2)] + 2OEt_2 \rightleftharpoons 2\,(H)(Me_2N)Al(NMe_2)\cdots OEt_2$	Colourless lustrous crystals; can be sublimed in vacuum at 35–50°; soluble in benzene and ether	146
$Al[N(CH_3)_2]_3$	$AlH_3 + 3NHMe_2 \xrightarrow{20^\circ} Al(NMe_2)_3 + 3H_2$	Colourless crystals soluble in ether and benzene; can be sublimed in vacuum at 60–100°; m.p. 87–88°; monomeric in ether	146
$AlH_3 \cdot 2N(CH_3)_3$	(a) $AlH_3(monomeric) + 2NMe_3 \xrightarrow[20^\circ]{ether} AlH_3 \cdot 2NMe_3$ (b) $3LiAlH_4 + AlCl_3 + 8NMe_3 \rightarrow 4(AlH_3 \cdot 2NMe_3) + 3LiCl$ (c) $AlH_3 \cdot NMe_3 + NMe_3 \rightarrow AlH_3 \cdot 2NMe_3$	White powder; can be sublimed at 40° (1 torr) to give colourless refractive crystals; m.p. 95° decomposes > 100°, first reversibly into $AlH_3 \cdot NMe_3$ and NMe_3 and then irreversibly into NMe_3 and $(AlH_3)_x$; at ≫ 100° it decomposes into H_2, NMe_3, and Al; soluble in ether, THF, and benzene; relatively stable solutions; degree of association in ether 1.03, in THF 1.03, in benzene 0.98	121 134 26 50 133

TABLE 5.1 (cont'd)

Compound	Preparation	Properties	References
	$+3H_2O \rightarrow Al(OH)_3 + 2NMe_3 + 3H_2$ $+5H^{\oplus} \xrightarrow{\text{ether}} Al^{3\oplus} + 2NMe_3H^{\oplus} + 3H_2$	Hydrolysis in accordance with the adjacent equations	
$AlD_3 \cdot 2N(CH_3)_3$	$AlD_3 \cdot NMe_3 + NMe_3 \rightarrow AlD_3 \cdot 2NMe_3$	White crystals	39
$AlH_3 \cdot N(CH_3)_3$	(a) $AlH_3(\text{monomeric}) + NMe_3 \xrightarrow[20^\circ]{\text{ether}} AlH_3 \cdot NMe_3$ (b) $3LiAlH_4 + AlCl_3 + 4NMe_3 \rightarrow 4(AlH_3 \cdot NMe_3) + 3LiCl$ (c) $LiAlH_4 + NMe_3 \cdot HCl \xrightarrow[-60^\circ]{\text{ether}} AlH_3 \cdot NMe_3 + LiCl + H_2(88\%)$	Colourless highly refractive crystals; can be sublimed at 50–60° (1 torr); m.p. 76°; decomposition > 100° into NMe_3 and $(AlH_3)_x$, ≫ 100° into Al, H_2, and NMe_3; soluble in ether as the monomer and in benzene as the monomer to dimer (degree of association 1.44	26 39 104 121 133 134
	$+3H_2O \xrightarrow{\text{explosively}} Al(OH)_3 + NMe_3 + 3H_2$ $+4H^{\oplus} \xrightarrow{\text{ether}} Al^{3\oplus} + NMe_3H^{\oplus} + 3H_2$	Hydrolysis in accordance with the adjacent equations	
$AlD_3 \cdot N(CH_3)_3$	$LiAlD_4 + NMe_3 \cdot DCl \xrightarrow[-60^\circ]{\text{ether}} AlD_3 \cdot NMe_3 - LiCl + D_2\ (93\%)$	M.p. 77/78°; soluble in benzene as the monomer to dimer (degree of association 1.43)	39 104
$AlH_3 \cdot N(CH_3)(C_2H_5)_2$	$LiAlH_4 + NMeEt_2 \cdot HCl \xrightarrow{\text{ether}} AlH_3 \cdot NMeEt_2 + LiCl + H_2(78\%)$	Soluble in benzene as the monomer to dimer (degree of association 1.33)	104
$AlH_3 \cdot N(C_2H_5)_3$	(a) $AlH_3\ (\text{monomeric}) + NEt_3 \xrightarrow[20^\circ]{\text{ether}} AlH_3 \cdot NEt_3$ (b) $LiAlH_4 + NEt_3 \cdot HCl \xrightarrow{\text{ether}} AlH_3 \cdot NEt_3 + LiCl + H_2(82\%)$	Can be sublimed in vacuum at 0°; distils at 30° (4.5 torr); m.p. 19.2°; soluble in ether as the monomer and in benzene as the monomer to dimer (degree of association 1.57); decomposes at 40° into $(AlH_3)_x$ and NEt_3, and at 180–200° into Al, H_2, and NEt_3; forms no diamine	104 157
	$+3H_2O \xrightarrow{\text{explosively}} Al(OH)_3 + NEt_3 + 3H_2$ $+HCl\ (\text{gas}) \rightarrow AlCl_3 + NEt_3 + 3H_2$	Hydrolysis in accordance with the adjacent equations	
$AlH_2[N(C_2H_5)_2]$	$LiAlH_4 + NHEt_2 \cdot HCl$ $\xrightarrow{\text{ether}}$ [illegible]	M.p. 42°; monomeric to trimeric in benzene (degree of association 2.16)	104

$AlH_3 \cdot N(C_3H_7)_3$	(a) $AlH_3(\text{monomeric}) + NPr_3 \xrightarrow[20°]{\text{ether}} AlH_3 \cdot NPr_3$ (b) $LiAlH_4 + NPr_3 \cdot HCl \xrightarrow{\text{ether}} AlH_3 \cdot NPr_3 + LiCl + H_2$ (56%)	Can be sublimed in vacuum at 30°; colourless crystals sensitive to moisture; m.p. 79.6°, 80/81°; decomposes > 40° into the components; Soluble in ether as the monomer and in benzene as the monomer to dimer (degree of association 1.12)	104
$AlH_3 \cdot N(C_4H_9)_3$	$AlH_3(\text{monomeric}) + NBu_3 \xrightarrow[20°]{\text{ether}} AlH_3 \cdot NBu_3$	Cannot be sublimed without decomposition high vacuum; can be recrystallized from benzene; m.p. 41.5°; soluble in ether as the monomer, slight dimerization in benzene	157
$AlH_3 \cdot N(CH_3)_2(CH_2CH{=}CH_2)$	$LiAlH_4 + N(CH_3)_2(CH_2CH{=}CH_2) \cdot HCl \xrightarrow{\text{ether}} AlH_3 \cdot N(CH_3)_2(CH_2CH{=}CH_2) + LiCl + H_2$ (89%) $H_3Al{-}N(CH_3)_2{-}CH_2CH{=}CH_2 \xrightarrow[48h]{85°}$ $[{-}AlH_2{-}N(CH_3)_2{-}CH_2CH_2CH_2{-}]_x$ and $[{-}CHCH_2{-}(CH_2{-}N(CH_3)_2{-}AlH_3)]_x$	Soluble in benzene as the monomer to dimer (degree of association 1.37); polymerizes in both the directions shown in column 2; the polymer contains 2.1 active H-atoms per Al-atom; this indicates predominant hydride addition to the double bond (upper polymer)	104
$[AlH_3 \cdot (CH_3)_2NCH_2CH_2N(CH_3)_2]_2$	$2LiAlH_4 + Me_2NCH_2CH_2NMe_2 \cdot 2HCl \xrightarrow[-2LiCl\ -2H_2]{Me_2NCH_2CH_2NMe_2} [AlH_3 \cdot Me_2NCH_2CH_2NMe_2]_2$	White solid; vapour pressures; 1.5 torr (99.3°), 10.6 torr (119.2°); no decomposition after 24 h at 133°; structure IX, section 6	23
$[AlH_3 \cdot (CH_3)_2\text{-}NCH_2CH_2CH_2N(CH_3)_2]_2$	$2AlH_3 \cdot N(C_2H_5)_3 + (CH_3)_2NCH_2CH_2CH_2N(CH_3)_2 \xrightarrow{\text{benzene}} [AlH_3 \cdot (CH_3)_2NCH_2CH_2CH_2N(CH_3)_2]_2 + N(C_2H_5)_3$	M.p. 135°	169a
$N(C_2H_4)_3N \cdot AlH_3$	Al (activated powder) + $\frac{3}{2}H_2$ + $N(C_2H_4)_3N$ $\xrightarrow[70°,\ 340\ atm,\ 6\ h]{\text{tetrahydrofuran}}$ $N(C_2H_4)_3N \cdot AlH_3$	Stable up to 200°; insoluble in the usual organic solvents	6
$AlH_3 \cdot 2NC_5H_5$	$AlH_3 + 2Py \xrightarrow[-30°/-40°]{\text{ether}} AlH_3 \cdot 2Py$	Any excess of pyridine and ether can be pumped off at −5°; white crystals	131

References p. 438

TABLE 5.1 (cont'd)

Compound	*Preparation*	*Properties*	*References*
	$+3HOR \rightarrow Al(OR)_3 + 2Py + 3H_2$	On alcoholysis forms 3 moles of H_2 in accordance with the adjacent equation	
$AlH_3 \cdot NC_5H_5$	$AlH_3 + Py \xrightarrow[-30°]{ether} AlH_3 \cdot Py$	White crystals	131
$AlH_2(NC_5H_6)$	$H_3Al \leftarrow \cdots NC_5H_5 \xrightarrow[migration]{20°, hydride} H_2Al{-}NC_5H_5H$ (quinonoid system)	Faintly yellow	
	$+3HOR \rightarrow Al(OR)_3 + HNC_5H_6 + 2H_2$	Forms 2 moles of H_2 on alcoholysis in accordance with the adjacent equation	131
$AlH(NC_5H_6)_2$	$H_3Al(\leftarrow \cdots NC_5H_5)_2 \xrightarrow[ether]{0°}$ yellow $\xrightarrow[hydride\ migration]{34°, exothermic} HAl(NC_5H_5H)_2$	Red solid insoluble in benzene, petroleum ether, and THF; soluble in pyridine	131
	$+3HOR \rightarrow Al(OR)_3 + 2HNC_5H_6 + H_2$	Forms 1 mole of H_2 on alcoholysis in accordance with the adjacent equation	131
$Al(NC_5H_6)_3$	$AlH(NC_5H_6)_2 + NC_5H_5 \xrightarrow[slow]{20-50°} Al(-NC_5H_5H)_3$	Red-brown solid mass no longer containing active hydrogen	131
$AlH_3 \cdot P(C_2H_5)_3$	$AlH_3(monomeric) + PEt_3 \xrightarrow[-60°/-70°]{ether} AlH_3 \cdot PEt_3$	Stable up to $-20°$; rapid decomposition into the components at 30°	50 37
$[AlH_2P(C_2H_5)_2]_3$	$3AlH_2Cl + 3LiPEt_2 \rightarrow [AlH_2PEt_2]_3 + 3LiCl$	Colourless liquid. M.P. 1–2°	39b
$[AlH(CH_3)_2]_2$	$AlMe_3 + H_2 \xrightarrow[\Delta]{slow\ discharge}$ $(AlHMe_2)_2$	Colourless highly viscous liquid; vapour pressure at 100°: 96 torr	121
$Al(CH_3)_3 \cdot AlH(CH_3)_2$	$AlMe_3 \cdot AlHMe_2$	Probable by-product in the glow-discharge process	121
$AlH_2(CH_3) \cdot AlH(CH_3)_2$	$AlH_2Me \cdot AlHMe_2$	Non-volatile discharge product	121
$AlH_2(CH_3) \cdot N(CH_3)_3$ by b, c, d	(a) $AlCl_nR_{3-n} \cdot NMe_3 + nLiH \rightarrow AlH_nR_{3-n} \cdot NMe_3 + nLiCl$	M.p $-35°$; b.p. 25–26°/1 torr; deg. of assn. in cyclohexane 1.95	95

$AlH(CH_3)_2 \cdot N(CH_3)_3$ by b, c	(b) $AlH_3 \cdot NMe_3 + n/2HgR_2 \rightarrow AlH_{3-n}R_n \cdot NMe_3 + n/2Hg + n/2H_2$	M.p. 33/35°; b.p. 42–43°/1 torr; deg. of assn. in cyclohexane 1.34	95
$AlH_2(C_2H_5) \cdot N(CH_3)_3$ by a, b, c, d	(c) $n/3AlH_3 \cdot NMe_3 + (3-n)/3AlR_3 \cdot NMe_3 \rightarrow AlH_nR_{3-n} \cdot NMe_3$	B.p. 39–40°/1 torr; deg. of assn. in cyclohexane 1.60	95
$AlH(C_2H_5)_2 \cdot N(CH_3)_3$ by a, b, c	(d) $AlClR_2 + LiAlH_4 + 2NMe_3 \rightarrow 2AlH_2R \cdot NMe_3 + LiCl$	M.p. −28°; b.p. 63–65°/1 torr; deg. of assn. in cyclohexane 1.18	95
$AlH_2(C_6H_5) \cdot N(CH_3)_3$	$AlH_2Cl \cdot N(CH_3)_3 + C_6H_5Li \rightarrow AlH_2(C_6H_5) \cdot N(CH_3)_3 + LiCl$	Yellow, viscous liquid which decomposes slowly at 20°, and rapidly above 50°.	27b
$AlH_2(C_6H_5) \cdot N(C_2H_5)_3$	$AlH_2Cl \cdot N(C_2H_5)_3 + C_6H_5Li \rightarrow AlH_2(C_6H_5) \cdot N(C_2H_5)_3 + LiCl$	Pale yellow, slightly viscous liquid. Decomposition without boiling below 80° at 0.01 torr	27b
$AlH(C_6H_5)_2 \cdot N(C_2H_5)_3$	$AlHCl_2 \cdot N(C_2H_5)_3 + 2C_6H_5Li \rightarrow AlH(C_6H_5)_2 \cdot N(C_2H_5)_3 + 2LiCl$	Pale yellow, slightly viscous liquid. Decomposition without boiling below 80° at 0.0001 torr.	27b
$AlH_2Cl \cdot nOC_2H_5$	$LiAlH_4 + AlCl_3 \xrightarrow{ether} 2AlH_2Cl + LiCl$		39c
$AlH_2Cl \cdot 2OC_4H_8$	(a) $AlX_3 + 2AlH_3 + 6THF \rightarrow 3AlH_2X \cdot 2THF$	M.p. 91–93°	113a
$AlD_2Cl \cdot 2OC_4H_8$	(b) $2AlX_3 + AlH_3 + 6THF \rightarrow 3AlH_2 \cdot 2THF$	M.p. 93–95°	
$AlH_2Br \cdot 2OC_4H_8$	(c) $AlX_3 + LiAlH_4 + 2THF \rightarrow 2AlH_2X \cdot 2THF$	M.p. 83–85°	
$AlH_2I \cdot 2OC_4H_8$	(d) $3AlX_3 + LiAlH_4 + 8THF \rightarrow 4AlHX_2 \cdot 2THF$	M.p. 165–166°	
$AlD_2I \cdot 3OC_4H_8$	(e) $3HCl + LiAlH_4 + 2THF \rightarrow AlH_2Cl \cdot 2THF + LiCl + 3H_2$	M.p. 161–163°	
$AlHCl_2 \cdot 2OC_4H_8$	(f) $2HCl + LiAlH_4 + 2THF \rightarrow AlHCl_2 \cdot 2THF + LiCl + 2H_2$	M.p. 90–95°	
$AlDCl_2 \cdot 2OC_4H_8$	(g) $2AlH_3 + I_2 + 4THF \rightarrow 2AlH_2I \cdot 2THF + H_2$	M.p. 86–89°	
$AlHBr_2 \cdot 2OC_4H_8$		M.p. 64–67°	
$AlHI_2 \cdot 2OC_4H_8$		M.p. 133–134°	
$AlHClBr \cdot 2OC_4H_8$		M.p. 80–82°	
$AlHClI \cdot 2OC_4H_8$		M.p. 169–170°	
$AlH_2Cl \cdot N(CH_3)_3$	$AlH_3 \cdot NMe_3 + \frac{n}{2}HgX_2 \rightarrow AlH_{3-n}X_n \cdot NMe + \frac{n}{2}Hg + \frac{n}{2}H_2$ (X = Cl, Br; n = 1, 2, 3) (92%)	M.p. 51–53°; deg. of assn. in C_6H_6 1.32	101
$AlHCl_2 \cdot N(CH_3)_3$	(30%)	M.p. 100–103°; deg. of assn. in C_6H_6 1.46	101
$AlCl_3 \cdot N(CH_3)_3$	(79%)	M.p. 156–157°; deg. of assn. in C_6H_6 1.36	101
$AlH_2Br \cdot N(CH_3)_3$	(74%)	M.p. 33–34°	101
$AlHBr_2 \cdot N(CH_3)_3$	(40%)	M.p. 104–106°	101

TABLE 5.1 (cont'd)

Compound	Preparation	Properties	References
$AlHCl[N(CH_3)_2]$	$AlH_2(NMe_2) + \frac{n}{2}HgCl_2 \rightarrow$ (59%)	M.p. 83°	101
$AlCl_2[N(CH_3)_2]$	$AlH_{2-n}Cl_n(NMe_2) + \frac{n}{2}Hg + \frac{n}{2}H_2$ (42%) $(n = 1.2)$	M.p. 151°; deg. of assn. in C_6H_6 2.12	101
$Al(BH_4)_3 \cdot O(C_2H_5)_2$	(a) AlH_3 (fresh Et_2O soln.) $+ \frac{3}{2}(BH_3)_2$ (gas) $\xrightarrow[0°]{Et_2O} Al(BH_4)_3 \cdot OEt_2$ (b) $Al(BH_4)_3 + OEt_2 \rightarrow Al(BH_4)_3 \cdot OEt_2$	An oil extremely sensitive to hydrolysis which, in contrast to $Al(BH_4)_3$, is not spontaneously inflammable in air; m.p. −17°; d^{25} 0.750 g/ml; Et_2O soln. at −25° stable for at least 2.5 months; decomposes at < 80°; volatile at 50° in high vacuum; vapour pressure at 0° < 1 torr; soluble in benzene, pentane, and diethyl ether; monomeric in benzene	89 125
$Al(BH_4)_3 \cdot OC_4H_8$	$Al(BH_4)_3 \cdot OEt_2 + THF \xrightarrow[20°]{} Al(BH_4)_3 \cdot THF + Et_2O$	M.p. 24–26°; monomeric in benzene	89 125
$Al(BH_4)_3 \cdot 2OC_4H_8$	$Al(BH_4)_3 \cdot OEt_2 + 2THF \xrightarrow[0°]{pentane} Al(BH_4)_3 \cdot 2THF$ (main reaction) $\xrightarrow[0°]{-BH_3 \cdot THF} AlH(BH_4)_2 \cdot THF$ $\rightarrow Al(OBu^n)(BH_4)_2$	Decomposes even at < 0°. (hydrogenating ring cleavage)	89 125
$Al(BH_4)_3 \cdot S(CH_3)_2$	$Al(BH_4)_3 + S(CH_3)_2 \rightarrow Al(BH_4)_3 \cdot S(CH_3)_2$		9c
$Al(BH_4)_3 \cdot N(CH_3)_3$	$Al(BH_4)_3 + E^v(CH_3)_3 \xrightarrow{n\text{-heptane}} Al(BH_4)_3 \cdot E^v(CH_3)_3$	M.p. 78–80°	9c
$Al(BH_4)_3 \cdot P(CH_3)_3$		M.p. 78°	9c
$Al(BH_4)_3 \cdot As(CH_3)_3$		M.p. 57°	9c
$AlH(BH_4)_2 \cdot 2O(C_2H_5)_2$	(a) AlH_3(soln.) $+ (BH_3)_2$ (gas) $\xrightarrow{Et_2O} AlH(BH_4)_2$ (poor method, since owing to the instability of the AlH_3 soln., it is impossible to determine its concentration) (b) Mixing of two (stable) AlH_3–BH_3–Et_2O solutions: 1. $AlH_3/BH_3 > 2$	On concentrating the ethereal solution at −25°, the two etherates are obtained successively; dietherate: m.p. 10–13°, decomposition pressure at 0° about 150 torr; monoetherate: vapour pressure at 52/53°: 0.1–0.2 torr, stable at −25°	89 125

$AlH(BH_4)_2 \cdot O(C_2H_5)_2 \cdot OC_4H_8$	$AlH(BH_4)_2 \cdot OEt_2 + 2THF \rightarrow AlH(BH_4)_2 \cdot OEt_2 \cdot THF$	Solid; easily looses Et_2O	89 125
$AlH(BH_4)_2 \cdot OC_4H_8$	$AlH(BH_4)_2 \cdot OEt_2 \cdot THF \xrightarrow{\text{vacuum}} AlH(BH_4)_2 \cdot THF + Et_2O$	Liquid	89 125
$AlH_2(BH_4) \cdot x\, O(C_2H_5)_2$	Preparation analogous to $AlH(BH_4)_2 \cdot OEt_2$ (b)	White solid when $x = 0.85$; Vapour pressure at 0°: 2 torr	89 125
$AlH_2(BH_4) \cdot OC_4H_8$	(a) $AlH_2(BH_4)(\text{in } Et_2O) + THF(\text{in } Et_2O) \xrightarrow{0°} AlH_2(BH_4) \cdot THF$ (b) $AlH_2(BH_4) \cdot 2THF \xrightarrow[\text{high vacuum}]{\text{warm slowly to 80°}} AlH_2(BH_4) \cdot THF + THF$ (ring opening of the THF frequently takes place during this process)	White solid, m.p. 55–58° [by method (a)]; m.p. 56–71° [by method (b)]; bimolecular in benzene	89 125
$AlH_2(BH_4) \cdot 2OC_4H_8$	$AlH_3(\text{in THF}) + \frac{1}{2}(BH_3)_2\ (\text{in THF}) + 2THF \xrightarrow{-20°} AlH_2(BH_4) \cdot 2THF$	Colourless crystals; m.p. 79–80°; stable at 20°, in contrast to $AlH_3 \cdot 2THF$; readily soluble in THF, Et_2O, and C_6H_6; less readily in petroleum ether, CCl_4, and $ChCl_3$; monomeric in benzene; vapour pressure at 20° about 1 torr	89 125
$Al(OBu^n)(BH_4)_2$	(a) $AlH(BH_4)_2 \cdot 2THF + Bu^nOH \rightarrow Al(OBu^n)(BH_4)_2 + H_2$ (b) $AlH(BH_4)_2 \cdot 2THF \xrightarrow[\text{ring opening}]{\text{vacuum, >50°}} Al(OBu^n)(BH_4)_2 + THF$	Colourless liquid; m.p. 13–14°, d^{27} 0.780 g/ml; vapour pressures (torr); 0.05 (65°), 0.1 (81°), 0.2 (87°), 0.5 (96°); dimeric in benzene; soluble in benzene, petroleum ether, and diethyl ether; not soluble in THF without decomposition	89 125
$Al(OBu^n)_2(BH_4)$	$Al(OBu^n)(BH_4)_2 \cdot THF \xrightarrow{120°} Al(OBu^n)_2(BH_4) + \frac{1}{2}(BH_3)_2$	Liquid	89 125
$AlH_2(BH_4) \cdot N(CH_3)_3$	$AlH_2Cl \cdot NMe_3 + LiBH_4 \xrightarrow{\text{benzene}} AlH_2(BH_4) \cdot NMe_3 + LiCl(77\%)$	M.p. 19°; can be recrystallized from hexane at −78°	103

References p. 438

TABLE 5.1 (cont'd)

Compound	*Preparation*	*Properties*	*References*
$AlH(BH_4)_2 \cdot N(CH_3)_3$	$AlHCl_2 \cdot NMe_3 + 2LiBH_4 \xrightarrow{benzene} AlH(BH_4)_2 \cdot NMe_3 + 2LiCl$ (60%)	B.p. 59–60°/0.03 torr; high-vacuum distillation	103
$Al(BH_4)_3 \cdot N(CH_3)_3$	(a) $AlCl_3 \cdot NMe_3 + 3LiBH_4 \xrightarrow{benzene} Al(BH_4)_3 \cdot NMe_3 + 3LiCl$ (51%)	M.p. 78°; vacuum sublimation at 35°	103
	(b) $AlCl_2(BH_4) \cdot NMe_3 + 2LiBH_4 \xrightarrow{benzene} Al(BH_4)_3 \cdot NMe_3 + 2LiCl$		
$AlH(BH_4)[N(CH_3)_2]$	$AlHCl(NMe_2) + LiBH_4 \xrightarrow{benzene} AlH(BH_4)(NMe_2) + LiCl$ (73%)	M.p. 69°; vacuum sublimation at 30°	103
$Al(BH_4)_2[N(CH_3)_2]$	$AlCl(NMe_2) + 2LiBH_4 \xrightarrow{benzene} Al(BH_4)_2(NMe_2) + 2LiCl$ (53%)	M.p. 156°; vacuum sublimation at 60°	103
$Al(BH_4)[N(CH_3)_2]_2$ (unstable)	$AlCl(NMe_2)_2 + LiBH_4 \xrightarrow{ether} Al(BH_4)(NMe_2)_2$(unstable) + LiCl	Transformation into $AlH(NMe_2)_2 \cdot 2BH_2(NMe_2)$ + similar compounds takes place during purification operations	103
$AlCl_2(BH_4) \cdot N(CH_3)_3$ (unstable)	$AlH_2(BH_4) \cdot NMe_3 + HgCl_2 \xrightarrow[0°]{benzene} AlCl_2(BH_4) \cdot NMe_3$(unstable) + Hg + H_2	Rearrangement takes place during purification operations; detection by conversion into the boranate [see the preparation (b) of $Al(BH_4)_3 \cdot NMe_3$]	103
$AlH[N(CH_3)_2]_2 \cdot 2BH_2[N(CH_3)_3]$	(a) $AlH(NMe_2)_2 + 2BH_2(NMe_2) \xrightarrow{hexane} AlH(NMe_2)_2 \cdot 2BH_2(NMe_2)$ (I)	Can be recrystallized from hexane; m.p. 115–157°	102
	(b) $2Al(NMe_2)_3 + 2BH_3 \cdot NMe_3 \xrightarrow{hexane}$ I + $2NMe_3 + AlH(NMe_2)_2$	M.p. 120°	102 102
	(c) $2AlH_3NMe_3 + 2B(NMe_2)_3 \xrightarrow{hexane}$ I + $AlH(NMe_2)_2 + 2NMe_3$	M.p. 121°	102
	(d) $Al(NMe_2)_3 + 3BH_2(NMe_2) \xrightarrow{hexane}$ I + $BH(NMe_2)_2$		
	(e) $AlH(NMe_2)_2 + 2BH_2(NMe_2) \xrightarrow[\Delta]{hexane}$ I + $2AlH_2(NMe_2)$ [side reaction to (a)]		102
	(f) $3AlH_2(NMe_2) + 2B(NMe_2)_3 \xrightarrow{hexane}$	M.p. 121°	102

$AlD[N(CH_3)_2]_2 \cdot 2BD_2[N(CH_3)_2]$	$AlD_3 \cdot NMe_3 - 2B(NMe_2)_3 \xrightarrow{\text{hexane}} (I-d_5) + AlD(NMe_2)_2 + 2NMe_3$	M.p. 122°	102
$AlH_2(SC_3H_7) \cdot N(CH_3)_3$	$AlH_3 \cdot NMe_3 - HSPr \rightarrow AlH_2(SPr) \cdot NMe_3 + H_2$	Clear liquid	80
$AlH(SC_3H_7)_2 \cdot N(CH_3)_3$	$AlH_3 \cdot NMe_3 - 2HSPr \rightarrow AlH(SPr)_2 + 2H_2$	Colourless oil	80
$AlHCl(SC_3H_7) \cdot N(CH_3)_3$	$AlH_2Cl \cdot NMe_3 + HSPr \rightarrow AlHCl(SPr) \cdot NMe + H_2$	White solid; m.p. 49°; sublimes at 100° (0.1 torr)	80
$AlH(NC_5H_{10})(SC_3H_7)$	$AlH_3 \cdot NMe_3 - H\overline{N(CH_2)_4C}H_2 + HSC_3H_7 \rightarrow AlH(NC_5H_{10})(SC_3H_7) + NMe_3 + 2H_2$	Oil	80
$AlH_2(SC_6H_5) \cdot N(CH_3)_3$	$AlH_3 \cdot NMe_3 - HSC_6H_5 \rightarrow AlH_2(SC_6H_5) \cdot NMe_3 + H_2$	Oil	80
$AlH(SC_6H_5) \cdot N(CH_3)_3$	$AlH_3 \cdot NMe_3 - 2HSC_6H_5 \rightarrow AlH(SC_6H_5) \cdot NMe_3 + 2H_2$	M.p. 107–110°; slightly soluble in ether and benzene	80
$\{AlClH[OSi(CH_3)_3]\}_2$		Large colourless distillable crystals; readily soluble in proton-active solvents	113

References p. 438

ordination number by one unit. (2) Electron donors add to alanes (like AlR_3[60]) through their occupied non-bonding orbitals in accordance with the general scheme $AlH_3 + nD \rightarrow AlH_3 \cdot nD$. (3) If such donors possess protonic hydrogen, molecular hydrogen splits off — as with the corresponding boron compounds — giving nucleophilic substitution.

Factor (1) AlH_nX_{3-n} and AlH_nR_{3-n} ($n = 1, 2, 3$) polymerize with the formation of $Al-H^{\mu}-Al$ and other bridges (increase in the coordination number), for example [77, 126a]:

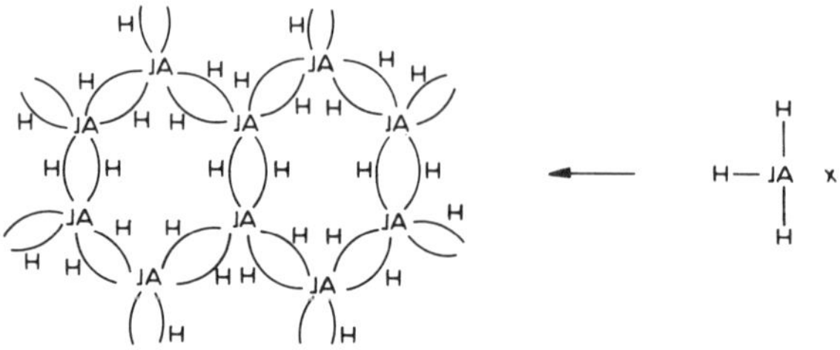

In the reaction of diborane with alane in diethyl ether or tetrahydrofuran boranates $AlH_2(BH_4)$, $AlH(BH_4)_2$, and $Al(BH_4)_3$ are obtained as adducts to the solvent D with the formation of $Al-H^{\mu}-Al$ bridges (sixfold coordination of the Al) [89, 125]:

$$AlH_3 + \tfrac{1}{2}(BH_3)_2 + 2THF \rightarrow AlH_2(BH_4)\cdot 2THF \rightarrow \tfrac{1}{2}[AlH_2(BH_4)\cdot THF]_2 + THF$$
$$AlH_3 + (BH_3)_2 + 2D \rightarrow AlH(BH_4)_2\cdot 2D \rightarrow AlH(BH_4)_2\cdot D + D$$
$$AlH_3 + \tfrac{3}{2}(BH_3)_2 + 2THF \rightarrow Al(BH_4)_3\cdot 2THF \rightarrow Al(BH_4)_3\cdot THF + THF$$

For further details on the reactions of AlH_3 and BH_3, see Table 5.1, and for the structures, section 7.

Trimethylaluminium, $AlMe_3$, and methylaluminium dichloride, $AlCl_2Me$, also form bridges with an increase of the coordination number from 3 to 4:

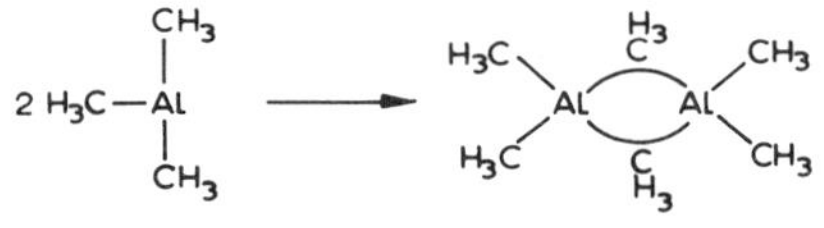

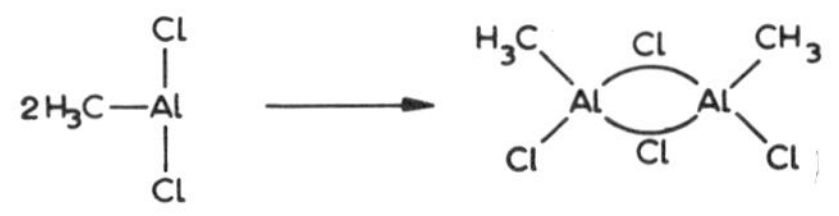

(as one of the three possible structures) [17]

Because of such bridge formations, alkylalanes (*e.g.* $R_2AlH + R'_2AlH$) readily undergo mutual exchange of hydrogen or alkyl groups [*cf.* section 2.1(c)] [17, 55, 63, 65].

Like its lower isologue BH_3 and the higher isologue GaH_3, alane AlH_3 also adds hydrogen compounds and alkyl compounds of elements belonging to main groups V and VI. The stability of the donor–acceptor adducts decreases from top to bottom of the periodic system. For example, the decomposition points fall in the following sequence[*147*]:

$$BH_3 \cdot NH_3(>90°) > AlH_3 \cdot NH_3(-45°) > GaH_3 \cdot NH_3 \text{ (not capable of isolation)}$$
$$BH_3 \cdot NH_3(>90°) > BH_3 \cdot PH_3(+65°) > BH_3 \cdot AsH_3 \text{ (not capable of isolation)}$$

Factor (2) The Al–H^{μ}–Al bridges in polymeric alane are broken only with stronger *electron donors* than the negatively polarized bridge H^{μ}-atom (*e.g.* with tetrahydrofuran, trialkylamines, trialkylphosphines, C=C and C≡C bonds, C=O and C=N bonds, and free hydride ions, but not with diethyl ether). In these reactions, donor–acceptor complexes form with both monomeric and polymeric alane. In the absence of protonic hydrogen in the donor, the complexes are fairly stable, (for details, see Table 5.1). In the following examples, AlH_3 stands for alanes $(AlH_3)_n$ with various degrees of polymerization:

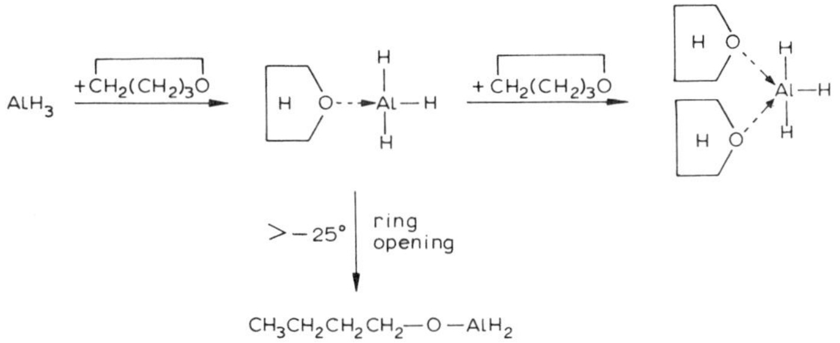

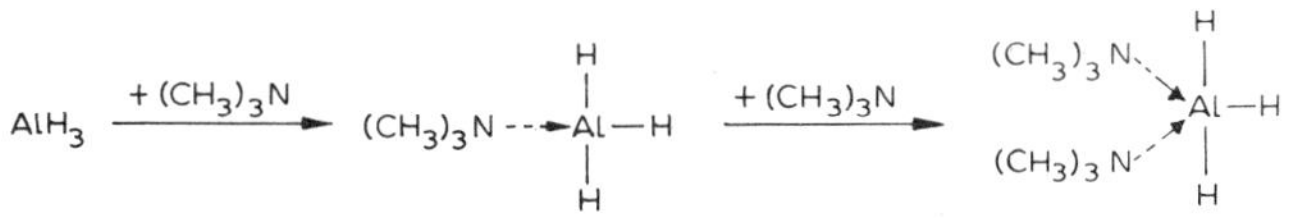

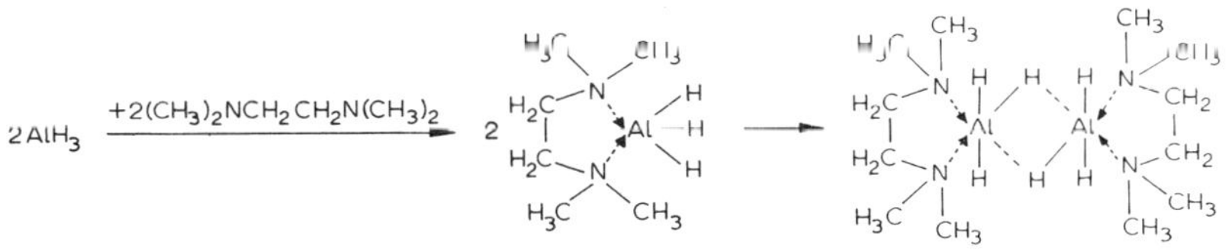

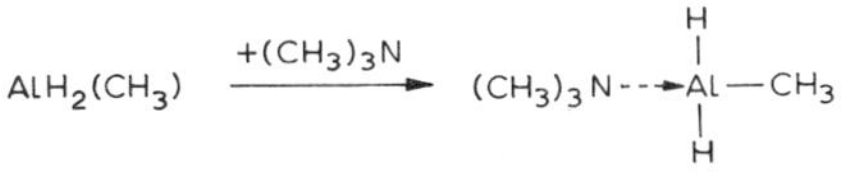

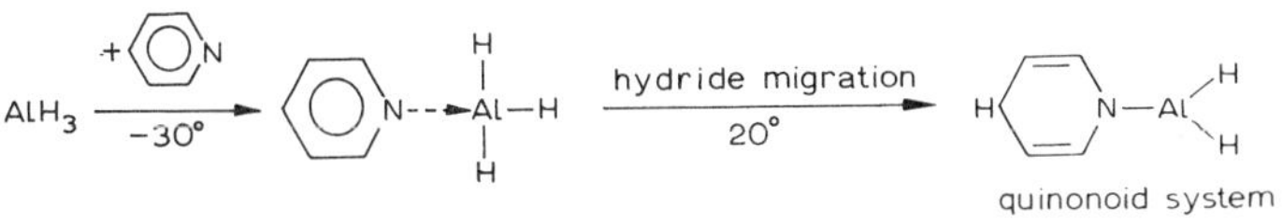

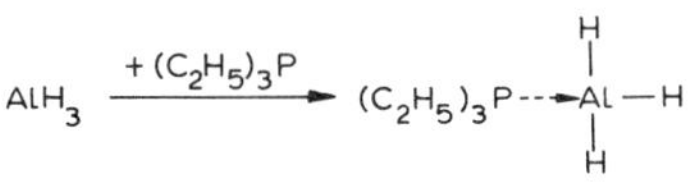

$$AlH_3 \xrightarrow{+CH_2(CH_2)_3O} (CH_2)_4O \cdots\!\to AlH_3 \xrightarrow{+CH_2(CH_2)_3O} [(CH_2)_4O]_2 \cdots\!\to AlH_3$$

$$(CH_2)_4O \cdots\!\to AlH_3 \xrightarrow[\text{ring opening}]{>-25°} CH_3CH_2CH_2CH_2{-}O{-}AlH_2$$

$$AlH_3 \xrightarrow{+(CH_3)_3N} (CH_3)_3N \cdots\!\to AlH_3 \xrightarrow{+(CH_3)_3N} [(CH_3)_3N]_2 \cdots\!\to AlH_3$$

$$2\,AlH_3 \xrightarrow{+2(CH_3)_2NCH_2CH_2N(CH_3)_2} 2\,[(CH_3)_2NCH_2CH_2N(CH_3)_2]AlH_3 \longrightarrow [(CH_3)_2NCH_2CH_2N(CH_3)_2]AlH_2(\mu\text{-}H)_2AlH_2[(CH_3)_2NCH_2CH_2N(CH_3)_2]$$

$$AlH_2(CH_3) \xrightarrow{+(CH_3)_3N} (CH_3)_3N \cdots\!\to AlH_2{-}CH_3$$

$$AlH_3 \xrightarrow[-30°]{+C_5H_5N} C_5H_5N \cdots\!\to AlH_3 \xrightarrow[20°]{\text{hydride migration}} H{-}C_5H_5N{-}AlH_2$$

quinonoid system

$$AlH_3 \xrightarrow{+(C_2H_5)_3P} (C_2H_5)_3P \cdots\!\to AlH_3$$

References p. 438

The stability of the amine adducts $AlH_3 \cdot 2NR_3$ decreases in the sequence $N(CH_3)_3 > N(CH_3)_2(CH_2CH{=}CH_2) > N(CH_3)(C_2H_5)_2 > N(C_2H_5)_3 > N(C_3H_7)_3$. An amine on the left in the series displaces one further to the right from the amine–alane adduct [*105*] (see also section 6).

Similarly to the B–H, the Al–H bond adds to *multiple bonds*. Thus, *C=C bonds* slowly add AlH_3 (at 70–80°) or $AlHEt_2$ (at about 100°) [*104, 174, 176*]. Terminal double bonds react somewhat more rapidly than internal bonds. True addition to internal C=C double bonds is generally observed only with cyclic alkenes (cyclopentene, cycloheptene) unless the double bonds migrate into the terminal position where they can add Al-H somewhat more rapidly. The C=C double bond probably adds by a polar mechanism, and the anti-Markovnikov rule applies:

$$\begin{array}{ccc} \overset{\delta-}{-C} & \cdots & \overset{\delta+}{C-} \\ \vdots & & \vdots \\ \underset{\delta+}{-Al} & \cdots & H^{\delta-} \end{array}$$

Examples [*174, 176*]:

$$CH_3(CH_2)_3-\overset{\delta+}{CH}=\overset{\delta-}{CH_2} + \overset{\delta-}{H}-\overset{\delta+}{AlEt_2} \xrightarrow{100\%} CH_3(CH_2)_4CH_2AlEt_2$$

$$\left[C_6H_5-\overset{\oplus}{CH}-\overset{\ominus}{CH_2} \rightleftharpoons C_6H_5-\overset{\ominus}{CH}-\overset{\oplus}{CH_2}\right] + \overset{\delta-}{H}-\overset{\delta+}{AlEt_2} \xrightarrow[17\,h]{100°} \begin{cases} \xrightarrow{76\text{-}78\%} C_6H_5-CH_2-CH_2AlEt_2 \\ \xrightarrow{24\text{-}22\%} C_6H_5-CH(CH_3)-AlEt_2 \end{cases}$$

For the stereospecific polymerisation of isoprene with catalytic systems including $TiCl_4$ and $AlH_3 \cdot N(CH_3)_3$, $AlH_2Z \cdot N(CH_3)_3$, $AlHZ_2 \cdot N(CH_3)_3$, AlH_2NR_2 and $AlHX \cdot N(CH_3)_3$ ($Z = Cl, OC_8H_{17}$; $R = C_6H_5, C_8H_{17}$; $X = Cl, Br$) see [*79b*].

The Al–H bond of AlH_3 (but not of $AlH_4^{\ominus}$) adds to the *C≡C bond* in a stereospecific *cis* addition with the formation of *trans* adducts (referred to R_2Al and R′). On hydrolysis, they give pure *cis*-alkenes [*167, 168*]:

$$R-C\equiv C-R' + R_2Al-H \longrightarrow \left|\begin{array}{c} R-C\equiv C-R' \\ \vdots \quad \vdots \\ R_2Al\cdots H \end{array}\right| \longrightarrow \begin{array}{c} R \quad\quad R' \\ C=C \\ R_2Al \quad\quad H \end{array} \xrightarrow{+H_2O} \begin{array}{c} R \quad\quad R' \\ C=C \\ H \quad\quad H \end{array} + R_2AlOH$$

Alane AlH_3, dialkylalanes $AlHR_2$, and haloalanes $AlH_{3-n}X_n$ add to *C=O and C=N bonds* [*86, 87, 141, 142*]. Examples of the addition of AlH_3 and $AlHR_2$ are:

$$RCH{=}O + AlH_3 \longrightarrow \left|\begin{array}{c} R \quad\quad H \\ \overset{\oplus}{C}-\overset{\ominus}{\bar{O}}|\cdots\rightarrow Al-H \\ H \quad\quad H \end{array}\right| \longrightarrow [(RCH_2O)AlH_2] \xrightarrow{+2\,RCHO} (RCH_2O)_3Al \xrightarrow[-Al(OH)_3]{+3\,HOH} 3\,RCH_2OH$$

$$R_2C{=}O + AlH_3 \longrightarrow \left|\begin{array}{c} R \quad\quad H \\ \overset{\oplus}{C}-\overset{\ominus}{\bar{O}}|\cdots\rightarrow Al-H \\ R \quad\quad H \end{array}\right| \longrightarrow [(R_2CHO)AlH_2] \xrightarrow{2\,R_2C=O} (R_2CHO)_3Al \xrightarrow[-Al(OH)_3]{+3\,HOH} 3\,R_2HCOH$$

$$RCH{=}NR + AlHR_2 \rightarrow \left| \overset{\oplus}{R}CH{-}\overset{\ominus}{\underset{R}{N}}\cdots\rightarrow \underset{R}{\overset{R}{Al}}{-}H \right| \rightarrow RCH_2N(R)AlR_2 \xrightarrow{+RCH=NR} RCH_2N(R)AlR_2 \leftarrow \overset{\oplus}{R}CH{-}\overset{\ominus}{N}R$$

The last-mentioned azomethine complexes are highly coloured and are used for spectroscopic determination of small amounts of $AlHR_2$ (linear azomethine [*86*], cyclic azomethine [*87*]; for other methods of determination see [*65, 66*]).

Haloalanes $AlH_{3-n}X_n$ add to the *C≡N triple bond* in accordance with the following mechanism [*141*]:

$$-C{\equiv}N \xrightarrow{+\,\rangle Al-H} -CH{=}N{-}Al\langle \xrightarrow{+\,\rangle Al-H} -CH_2{-}N\langle^{Al\langle}_{Al\langle} \xrightarrow[-2\,\rangle AlOH]{+2\,HOH} -CH_2{-}NH_2$$

Trimeric dialkylanes add to $Me_2N{-}C{\equiv}N$ giving the four-membered Al–N ring system [*127b*]:

$$2(R_2AlH)_3 + Me_2N{-}C{\equiv}N \rightarrow 3Me_2N{-}\overset{H}{C}{=}N{-}AlR_2 \quad \begin{matrix} \downarrow \quad \uparrow \\ R_2Al{-}N{=}\underset{H}{C}{-}NMe_2 \end{matrix}$$

Even the *hydride ion* is sufficiently nucleophilic to form the stable alanate ion $AlH_4^{\ominus}$ from AlH_3. However, this is slightly dissociated:

$$AlH_3 + H^{\ominus} \rightleftharpoons AlH_4^{\ominus}$$

$$AlH_3 \xrightarrow{+Et_2O} Et_2O\cdots\rightarrow AlH_3 \xrightarrow{+Et_2O} (Et_2O)_2 \rightarrow AlH_3 \rightleftharpoons \frac{1}{x}(AlH_3)_x + 2\,Et_2O$$

Factor (3) As indicated above (factor 2) a donor–alane complex relatively stable to heat is first formed. The thermal stability of the adduct falls considerably if the donor contains *protonic hydrogen* or double bonds hydrogenatable by hydride migration. Only in a few cases can such aminealane adducts be isolated (*e.g.* $AlH_3 \cdot NH_2R$ [*145*] or $AlH_3 \cdot NHR_2$ [*104*]; for further details see Table 5.1). At higher temperatures they eliminate hydrogen or hydrogenate the multiple bond by hydride migration from the Al atom (see section 6). Table 5.1 contains, inter alia, details of the following examples:

$$AlH_3 \xrightarrow[<-80°]{+2\,NH_3} (NH_3)_2 \rightarrow AlH_3 \xrightarrow[-80°]{-H_2} (H_2N)(H_3N)AlH_2 \xrightarrow[-50°]{-H_2} (H_2N)_2Al{-}H \xrightarrow{+NH_3}$$

References p. 438

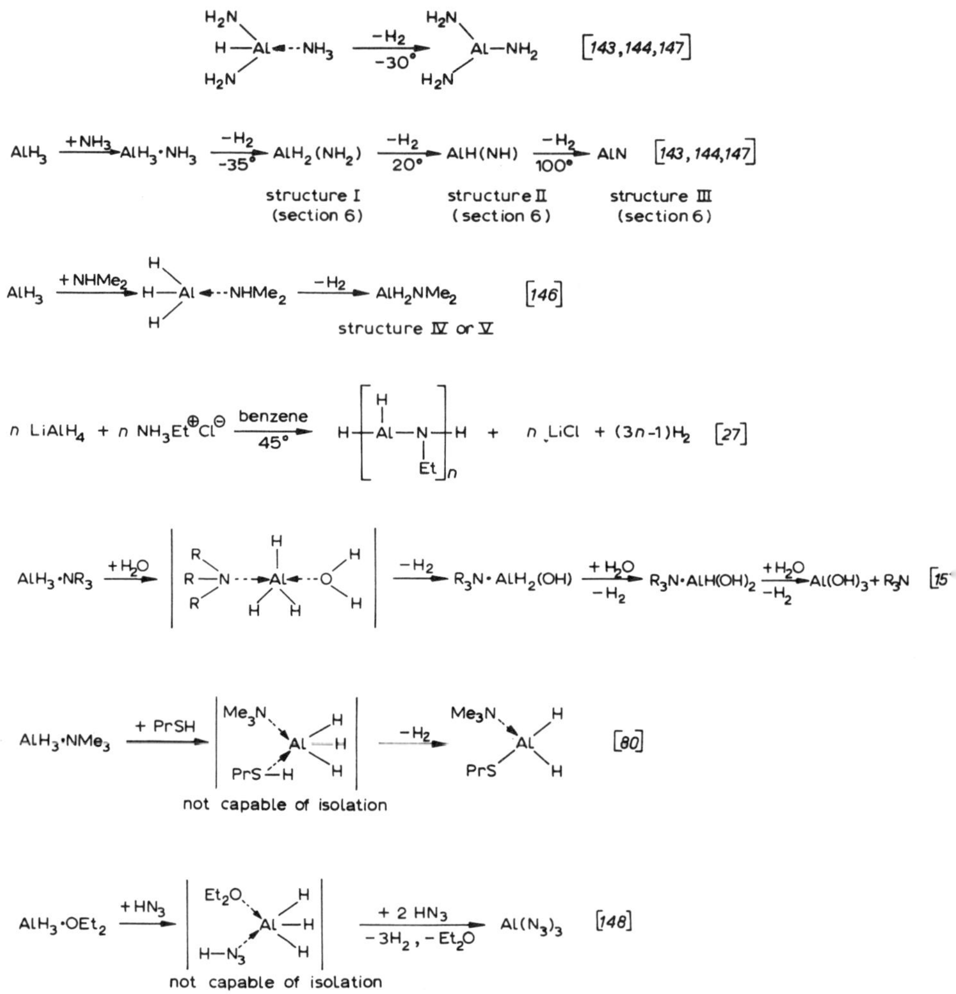

$$(H_2N)_2HAl \leftarrow NH_3 \xrightarrow[-30^\circ]{-H_2} (H_2N)_2Al{-}NH_2 \quad [143,144,147]$$

$$AlH_3 \xrightarrow{+NH_3} AlH_3{\cdot}NH_3 \xrightarrow[-35^\circ]{-H_2} AlH_2(NH_2) \xrightarrow[20^\circ]{-H_2} AlH(NH) \xrightarrow[100^\circ]{-H_2} AlN \quad [143,144,147]$$

structure I (section 6) — structure II (section 6) — structure III (section 6)

$$AlH_3 \xrightarrow{+NHMe_2} H_3Al \leftarrow NHMe_2 \xrightarrow{-H_2} AlH_2NMe_2 \quad [146]$$

structure IV or V

$$n\,LiAlH_4 + n\,NH_3Et^{\oplus}Cl^{\ominus} \xrightarrow[45^\circ]{benzene} H{+}[AlH{-}N(Et)]_n{+}H + n\,LiCl + (3n-1)H_2 \quad [27]$$

$$AlH_3{\cdot}NR_3 \xrightarrow{+H_2O} [R_3N \rightarrow AlH_3 \leftarrow OH_2] \xrightarrow{-H_2} R_3N{\cdot}AlH_2(OH) \xrightarrow[-H_2]{+H_2O} R_3N{\cdot}AlH(OH)_2 \xrightarrow[-H_2]{+H_2O} Al(OH)_3 + R_3N \quad [15$$

$$AlH_3{\cdot}NMe_3 \xrightarrow{+PrSH} [Me_3N \rightarrow AlH_3 \leftarrow PrS{-}H] \xrightarrow{-H_2} Me_3N \rightarrow AlH_2(PrS) \quad [80]$$

not capable of isolation

$$AlH_3{\cdot}OEt_2 \xrightarrow{+HN_3} [Et_2O \rightarrow AlH_3 \leftarrow N_3{-}H] \xrightarrow[-3H_2,\,-Et_2O]{+2\,HN_3} Al(N_3)_3 \quad [148]$$

not capable of isolation

6. Reactions of the aluminium hydride adducts $AlH_3{\cdot}D$

Since solutions of "free" monomeric alane are fairly unstable, the reactions of AlH_3 are preferably carried out with stable donor–acceptor adducts $AlH_3{\cdot}D$, generally with trimethylamine-alane, $AlH_3{\cdot}NMe_3$ [*68*]. Because of the importance of such adducts, their methods of preparation and properties will be outlined below with the NMe_3 adduct as the prototype compound. Experimental details and other information can be obtained from Table 5.1. The preparation of amine-alanes can take place:

(1) by adduct formation from alane and amine in a two-stage reaction:

$$3LiAlH_4 + AlCl_3 \xrightarrow{ether} 4AlH_3 + 3LiCl \text{ (filter off)}$$

$$AlH_3 \text{ (monomeric; polymeric forms less satisfactory)} + NMe_3 \xrightarrow{\text{ether, THF, etc.}} AlH_3{\cdot}NMe_3$$

(2) by adduct formation in a one-pot process:

$$LiAlH_4 \xrightarrow[\text{Me}_2\text{O or Et}_2\text{O}]{+\text{HCl (gas)}, -\text{LiCl}} AlH_3{\cdot}\text{etherate} \xrightarrow[\text{ether}]{+\text{NMe}_3} AlH_3{\cdot}NMe_3$$

(3) by a reaction between alanate and ammonium ion in a one-pot process (this can be used for deuterated compounds):

$$LiAlH_4 + NMe_3{\cdot}HCl \xrightarrow{\text{ether}} AlH_3{\cdot}NMe_3 + LiCl + H_2$$
$$LiAlD_4 + NMe_3{\cdot}DCl \xrightarrow{\text{ether}} AlD_3{\cdot}NMe_3 + LiCl + D_2$$

(4) by synthesis from the elements (aluminium powder, hydrogen) and triethylenediamine or tetramethylethylenediamine, for example:

$$Al + \frac{3}{2}H_2 + N(C_2H_4)_3N \xrightarrow[70°]{350\text{ atm}} N(C_2H_4)_3N \dashrightarrow AlH_3$$

Triethylenediamine-alane, which is stable up to 200° (!?) is insoluble in common organic solvents; it is probably not monomeric [*6*].

In amine-alanes and aminoalanes, the hydride ion $H^{\ominus}$ is capable of replacing the amine nitrogen. On the other hand, carbanions $R^{\ominus}$ and halogen ions $X^{\ominus}$ replace the hydrogen atoms of the alane.

For example, *LiH*, *NaH*, or CaH_2, form alanates with trimethylamine-alane (nucleophilic displacement) [*105*]:

$$AlH_3{\cdot}NMe_3 + LiH \rightarrow LiAlH_4 + NMe_3$$
$$AlH_3{\cdot}NMe_3 + NaH \rightarrow NaAlH_4 + NMe_3$$
$$2AlH_3{\cdot}NMe_3 + CaH_2 \rightarrow Ca(AlH_4)_2 + 2NMe_3$$

Even dimethylaminoalane, $AlH_2(NMe_2)$, which can form no adduct with additional trimethylamine [no formation of a $AlH_2(NMe_3){\cdot}NMe_3$] reacts with sodium hydride to form the alanate (dismutation) [*105*]:

$$2AlH_2(NMe_2) + NaH \rightarrow NaAlH_4 + AlH(NMe_2)_2$$

Carbanions from dialkylmercuries and alkyllithiums react with trimethylamine-alane to replace hydrogen by alkyl, for example [*95*, *101*]:

$$AlH_3{\cdot}NMe_3 + \frac{n}{2}HgR_2 \rightarrow AlH_{3-n}R_n{\cdot}NMe_3 + \frac{n}{2}Hg + \frac{n}{2}H_2$$
$$(n = 1, 2, 3;\ R = CH_3, C_4H_9, C_4H_9, CH_2{=}CH, C_3H_7C{\equiv}C, C_6H_5)$$
$$AlH_3{\cdot}NMe_3 + LiBu^n \rightarrow AlH_2Bu^n{\cdot}NMe_3 + LiH$$

References p. 438

The LiH so formed hydrogenates the $AlH_2R \cdot NMe_3$ produced to $LiAlH_4$ (see above). Aminoalanes are also alkylated by $R^{\ominus}$, for example [*101*]:

$$AlH_2(NMe_2) + 2\,LiR \rightarrow AlR_2(NMe_2) + 2\,LiH\ (R = n\text{-}C_4H_9,\ C_6H_5)$$

Halide ions from mercury(II) halides replace hydrogen in trimethylamine-alane or dimethylaminoalane [*101*]:

$$AlH_3 \cdot NMe_3 + \frac{n}{2}\,HgX_2 \rightarrow AlH_{3-n}X_n \cdot NMe_3 + \frac{n}{2}\,Hg + \frac{n}{2}\,H_2 \begin{cases} n = 1, 2, 3 \\ X = Cl, Br \end{cases}$$

$$AlH_2(NMe_2) + \frac{n}{2}\,HgX_2 \rightarrow AlH_{2-n}X_n(NMe_2) + \frac{n}{2}\,Hg + \frac{n}{2}\,H_2 \begin{cases} n = 1, 2 \\ X = Cl, Br \end{cases}$$

The reactivity falls off as follows; alanes: $LiAlH_4 > AlH_3 \cdot NMe_3 > AlH_2(NMe_2)$; mercury compounds: $Cl > C_3H_7C{\equiv}C > CH_2{=}CH \sim C_6H_5 > n{-}C_4H_9$ [*101*].

The *donors* methylamine, ethylamine, acetonitrile, and ethyleneimine displace the trialkylamine in $AlH_3 \cdot NR_3$ nucleophilically [*27*]. In the following examples, because of the high reaction temperature (40°, benzene solution), the donor alane adducts first produced immediately form polymers:

$$n\,AlH_3 \cdot NEt_3 \xrightarrow[-n\,NEt_3]{+n\,NH_2R} [n\,AlH_3 \cdot NH_2R] \xrightarrow{-(2n-1)H_2} H_2N{-}[AlH{-}N(R)]_{n-1}{-}AlH_2$$

$$n\,AlH_3 \cdot NEt_3 \xrightarrow[-n\,NEt_3]{+n\,NCMe} [n\,AlH_3 \cdot N{\equiv}CMe] \xrightarrow{\text{hydride migration}} [n\,AlH_2N{=}CHMe] \xrightarrow{\text{hydride migration}}$$

$$MeCH{=}N{-}[AlH{-}N(CH_2Me)]_{n-1}{-}AlH_2$$

$$n\,AlH_3 \cdot NEt_3 \xrightarrow[-n\,NEt_3]{+n\,NH(CH_2)_2} n\,AlH_3 \cdot NH(CH_2)_2 \xrightarrow{-n\,H_2} n\,AlH_2{-}N(CH_2)_2 \xrightarrow[\text{with ring opening}]{\text{hydride migration}}$$

$$(CH_2)_2N{-}[AlH{-}N(C_2H_5)]_{n-1}{-}AlH_2$$

In the reaction of trimethylaluminium $AlMe_3$ (with no Al-H bond) with bis-(dimethylamino)beryllium, $Be(NMe_2)_2$, adducts are formed with Al:Be ratios of 1:1, 2:1, and 4:1 [*96*]:

$$AlMe_3 + Be(NMe_2)_2 \xrightarrow{-196° \rightarrow +50°} AlMe_3 \cdot Be(NMe_2)_2 \overset{+100°}{\nrightarrow} Me_2AlBe(NMe_2) + NMe_3$$

$$2AlMe_3 + Be(NMe_2)_2 \xrightarrow{-196° \rightarrow +50°} (AlMe_3)_2 \cdot Be(NMe_2)_2$$

$$4AlMe_3 + Be(NMe_2)_2 \xrightarrow{-196° \rightarrow +70°} (AlMe_3)_4 \cdot Be(NMe_2)_2$$

If the alanes contain Al–H bonds ($AlHMe_2 \cdot NMe_3$, $AlH_2Me \cdot NMe_3$, and $AlH_3 \cdot NMe_3$), in their reaction with $Be(NMe_2)_2$ at about 100°, regardless of the ratio of Al compound to Be compound, adducts with an Al: Be ratio of 1:2 are formed in accordance with the overall equation [96]:

$$AlH_nMe_{3-n} \cdot NMe_3 + Be(NMe_2)_2 \xrightarrow{\Delta} AlH_{n-1}Me_{3-n}(NMe_2) \cdot (HBeNMe_2)_2 + \text{Al-rich residue} + NMe_3$$

Of the many possible structures of the latter adducts (Al: Be = 1:2), it has not yet been possible to pick out the correct ones by means of the PMR spectra [*96*].

7. Comparative review of structures of aluminium hydrides and their derivatives

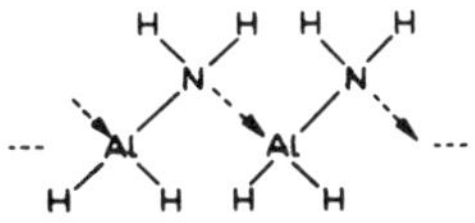

(I) ideal $[AlH_2(NH_2)]_x$ [*143*]

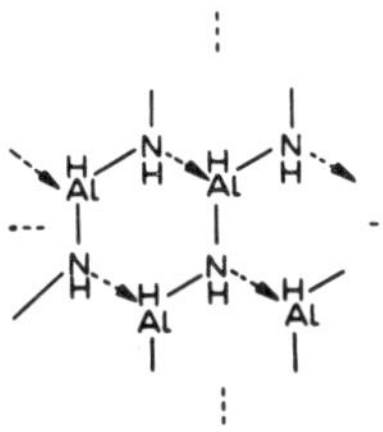

(II) ideal $[AlH(NH)]_x$ [*143*]

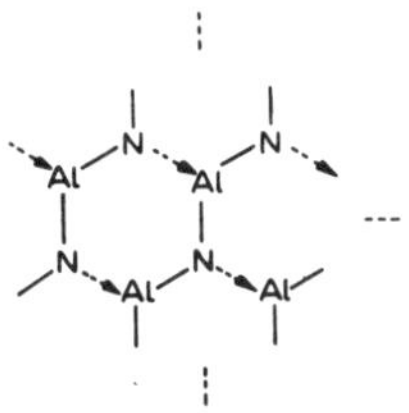

(III) three-dimensional wurtzite structure $[AlN]_x$

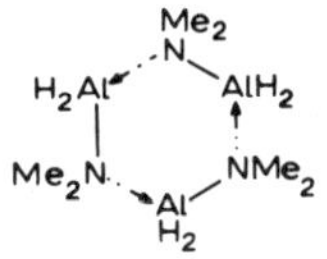

(IV) $[AlH_2(NMe_2)]_3$ in benzene [*104*] (see however [*27*])

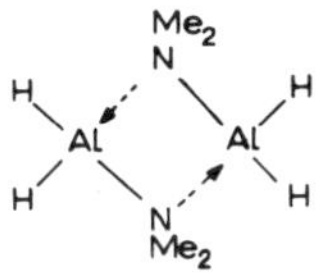

(V) $[AlH_2(NMe_2)]_2$ in ether

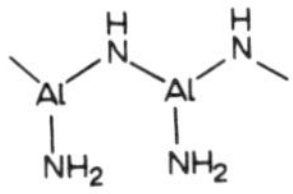

(VI) $[Al(NH)(NH_2)]_x$ [*144*]

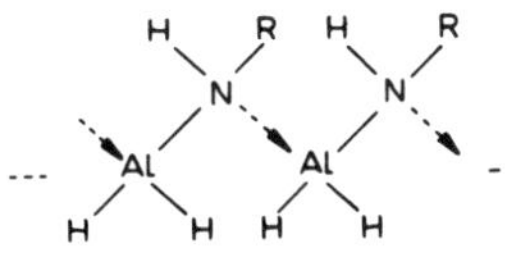

(VII) $[AlH_2(NHR)]_x$ [*144*]

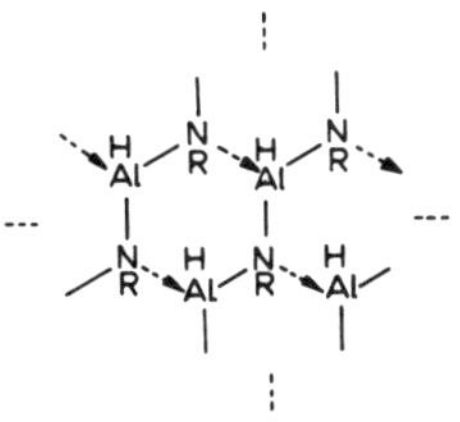

(VIII) $[AlH(NR)]_x$ [*144*]

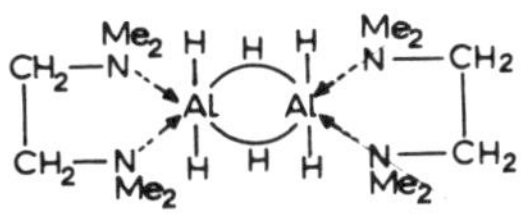

(IX) $[AlH_3 \cdot NMe_2CH_2CH_2NMe_2]_2$ (structure uncertain because of the Al–H frequency of 1695 cm^{-1} [*23*]; contains N-bridges acc. to [57]) [*23*]

$[AlH_3{\cdot}NMe_3]_2$

degree of association in benzene 1.44; however, the vibrational spectrum makes this structure questionable [*95*, *104*]

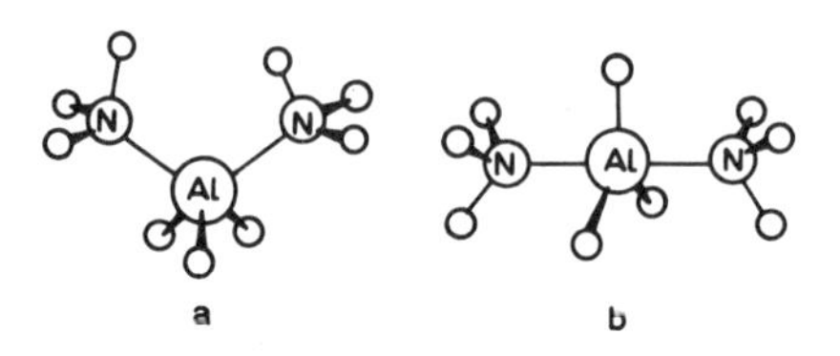

$AlH_3{\cdot}2NR_3$

(a) questionable according to dipole moment and spectrum [*57*]
(b) probable structure; ∢ N–Al–N = 180 ± 4° [*52*, *57*, *105*]

Al_2X_6
$AlMe_6$, $Al_2Me_4Cl_6$, Al_2Et_6
$Al_2Et_4Cl_2$, $Al_2Pr_6{}^n$
$Al_2Bu_4{}^nCl_2$, $Al_2\,Bu_2{}^i\,Cl_2$
(but not $AlBu_3{}^i$!) [*45*, *53*, *76*]

topological structure of $Al_2(C_2H_5)_6$

$[AlHR_2]_3$
R = any group [*58*]

$[AlMe_2(OMe)]_3$
[*58*]

$[AlMe_2(OR)]_2$
R = any group [*58*]
than CH_3 [*58*]

$Al(OPr^i)_3$
contains a central octahedral Al-atom and three (distorted) tetrahedral Al-atoms [*117*]

$[Al(Ph)N(Ar)]_4$
(a) chair form
(b) cubane structure

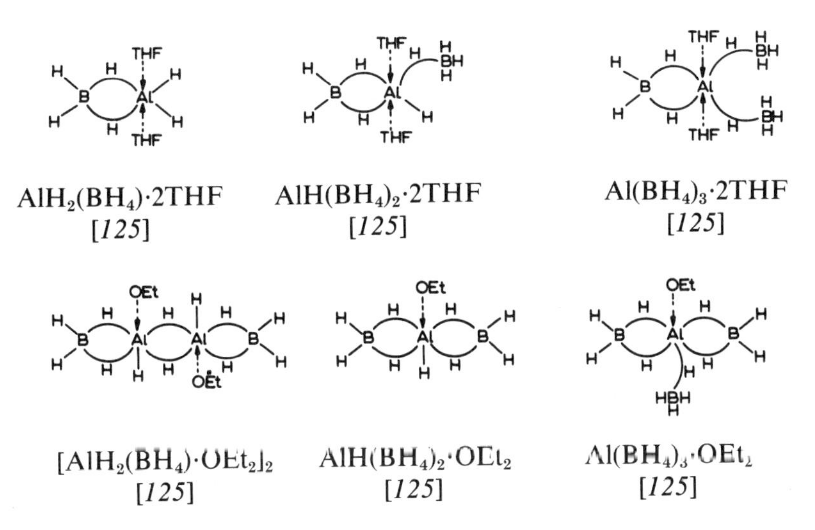

$AlH_2(BH_4)\cdot 2THF$ [*125*] $AlH(BH_4)_2\cdot 2THF$ [*125*] $Al(BH_4)_3\cdot 2THF$ [*125*]

$[AlH_2(BH_4)\cdot OEt_2]_2$ [*125*] $AlH(BH_4)_2\cdot OEt_2$ [*125*] $Al(BH_4)_3\cdot OEt_2$ [*125*]

8. Preparation and physical properties of alanates $AlH_{4-n}R_n^{\ominus}$

8.1 Preparation of unsubstituted alanates $AlH_4^{\ominus}$ and $AlH_6^{3\ominus}$

Alkali metal alanates $E^I(AlH_4)$ and $E^I(AlH_2R_2)$ [*36, 38, 109, 110, 130, 165, 171*] and alkaline earth metal alanates $E^{II}(AlH_4)_2$ [*35, 116, 128, 129*] are formed by the reaction of finely divided metal hydrides with aluminium halides in polar organic solvents (= Me_2O, Et_2O, MeOEt, Pr^iOMe, the dimethyl ethers of the ethylene glycols) in accordance with the overall equations:

$$4\,E^IH + AlX_3 \xrightarrow{\text{solvent}} E^I(AlH_4) + 3\,E^IX \begin{cases} E^I = \text{Li, Na, K, Cs} \\ X = \text{Cl, Br} \end{cases}$$

References p. 438

$$2\ E^{I}H + AlClR_2 \xrightarrow{\text{solvent}} E^{I}(AlH_2R_2) + E^{I}Cl \quad \begin{cases} E^{I} = \text{Li, Na, K} \\ R = \text{Me, Et, i-Pr} \end{cases}$$

$$4\ E^{II}H_2 + 2\ AlX_3 \xrightarrow{\text{solvent}} E^{II}(AlH_4)_2 + 3\ E^{II}X_2 \quad \begin{cases} E^{II} = \text{Ca} \\ X = \text{Cl, Br} \end{cases}$$

The addition of a little LiBr accelerates the reaction [*35*]. However, the yield is satisfactory only in the case of lithium hydride. In the reaction between LiH and $AlCl_3$ in diethyl ether, the hydride must be finely ground since otherwise the insoluble LiCl may coat the particles of LiH. The reaction is carried out both in the laboratory and industrially. Since LiBr is soluble in ether, nut-sized lumps of LiH can be used in the reaction of LiH with AlBr [*168*].

While the reaction of LiH with AlX_3 takes place smoothly and generally with high yields, the reaction of NaH with AlX_3 yields only low yields of impure $NaAlH_4$. However, if the alane is first made from $NaAlH_4$ or $LiAlH_4$ and this is treated with NaH, pure sodium alanate can be obtained with a growth factor of $\frac{4}{3}$ [*35,122*]:

$$LiAlH_4 + AlCl_3 \rightarrow 4AlH_3 + 3LiCl \quad \text{(starting reaction; unnecessary if } NaAlH_4 \text{ is available)}$$

$$4\ AlH_3 + 4\ NaH \rightarrow 4\ NaAlH_4$$

$$3\ NaAlH_4 + AlCl_3 \rightarrow 4\ AlH_3 + 3\ NaCl$$

$$\overline{4\ NaH + AlCl_3 \rightarrow NaAlH_4 + 3\ NaCl}$$

The growth factor is, however, somewhat reduced by side reactions and losses during isolation (yield 98% of theory, purity 95% [*35*]).

The metathesis of an excess of hydride with alanate can lead to the trans-metallation of alanates and alkylalanates (conversion of the less ionic alanate into the more ionic alanate) [*170, 170a*]:

$$LiAlH_4 + NaH \xrightarrow{\text{tetrahydrofuran}} NaAlH_4 + LiH \quad (90{\cdot}7\%)$$

$$LiAlH_4 + KH \xrightarrow{\text{diglyme}} KAlH_4 + LiH \quad (90\%)$$

$$NaAlH_4 + KH \xrightarrow{\text{diglyme}} KAlH_4 + NaH \quad (84{\cdot}5\%)$$

$$EAlH_{4-n}R_n + E'H \xrightarrow{\text{benzene}} E'AlH_{4-n}R_n + EH \ (80\text{–}90\%)$$

[electronegativity of the alkali metals E (Li, Na) > E′ (Na, K); R = alkyl; $n = 0, 1, 2$]

Halides may be used in place of the hydrides for the transmetallation of aluminates. Since sodium and potassium halides are less soluble in ether than lithium halides, in the metathesis of lithium chloride or bromide with sodium or potassium alanates, for example, lithium alanate is formed (conversion of the ionic alanate into a less ionic alanate) [*170*]:

$$NaAlH_4 + LiX \xrightarrow{\text{ether}} LiAlH_4 + NaX \quad \text{(with LiCl: 93.5\%)}$$

$$KAlH + LiX \xrightarrow{\text{tetrahydrofuran}} LiAlH_4 + KX \quad \text{(with LiCl: 91\%)}$$

Here the degree of grinding is very important. Thus, for example, coarsely powdered $NaAlH_4$ and LiCl can be boiled in ether for hours without exchange taking place; on fine grinding, however, they react suprisingly fast[*21*]. The reaction of $KAlH_4$ with NaBr in tetrahydrofuran leads only to an equilibrium [*170*]:

$$KAlH_4 + NaBr \rightleftharpoons NaAlH_4 + KBr$$

Such reactions are not limited to alkali metal alanates. For example, lithium or sodium alanate is converted into magnesium alanate with magnesium bromide [*21, 128, 129*]:

$$2\,LiAlH_4 + MgBr_2 \rightarrow Mg(AlH_4)_2 + 2\,LiBr$$
$$2\,NaAlH_4 + MgBr_2 \rightarrow Mg(AlH_4)_2 + 2\,NaBr$$

Alkali metal or alkaline earth metal hydrides, aluminium and hydrogen form completely halogen-free alanates in tetrahydrofuran and diethers of the polyethyleneglycols, but not in diethyl ether[*21*]:

$$E^{I}H + Al + \tfrac{3}{2}H_2 \xrightarrow{\text{tetrahydrofuran}} E^{I}(AlH_4)$$
$$E^{II}H_2 + 2Al + 3H_2 \xrightarrow{\text{tetrahydrofuran}} E^{II}(AlH_4)_2$$
$$(E^{I} = Li, Na, K;\ E^{II} = Ca)$$

The hydrogenation probably takes place in two stages[*21*]. The alkali metal hydride catalyses the uptake of H on aluminium, just as, for example, UH_3 appreciably decreases the activation energy of the absorption of H on platinum. The AlH_3 formed in traces then reacts with the alkali metal hydride to give the alanate:

$$Al + \tfrac{3}{2}H_2 \xrightarrow{E^{I}H\text{-catalysed}} AlH_3$$
$$AlH_3 + E^{I}H \rightarrow E^{I}AlH_4$$

In practice, in the reaction of NaH, Al, and H_2, commercial 85% NaH and commercial granulated aluminium are milled with steel balls in peroxide-free tetrahydrofuran. The resulting reaction mixture, which is spontaneously inflammable in air, is then hydrogenated at 150 atm and 150° (it takes up hydrogen from 125°). Lithium alanate can be obtained in the same way; in this case the uptake of hydrogen begins at 35° and an H_2 pressure of 30 atm. In general, the temperature of hydrogenation is the lower the more active is the suspension to be hydrogenated. The reaction of a mixture of 1 mole of NaH and 1 mole of LiH with 1 to 2 moles of Al forms 1 mole of $NaAlH_4$; only the residual amount of Al forms $LiAlH_4$. Diethyl ether is unsuitable as a reaction medium. Thus, for example, the non-hydrogenated ground suspension of LiH and Al in THF spontaneously decomposes an added ethereal solution of $LiAlH_4$. In contrast, a solution of $LiAlH_4$ in THF remains unchanged[*21*].

References p. 438

Sodium and aluminium can be co-hydrogenated with hydrogen under pressure in tetrahydrofuran to give completely halogen-free sodium alanate [*21*]:

$$Na + Al + 2H_2 \xrightarrow{\text{tetrahydrofuran}} NaAlH_4$$

For a simple preparation of solutions of AlH_3 in THF by the electrolysis of $LiAlH_4$ solutions in this solvent, see section 9.1.

Sodium alanate Na_3AlH_6 is formed by the reaction of NaH and $NaAlH_4$ at 160° in heptane [*171a*] and by direct synthesis [*6a*]:

$$2\,NaH + NaAlH_4 \rightarrow Na_3AlH_6,$$
$$3\,Na + Al + 3H_2 \xrightarrow[\text{165° pressure}]{\text{toluene}} Na_3AlH_6$$

Lithium alanate Li_3AlH_6 is formed by the reaction of n-C_4H_9Li and $LiAlH_4$. The over-all reaction may be [*26a*]:

$$2\,n\text{-}C_4H_9Li + LiAlH_4 \xrightarrow{\text{hexane+diethylethar}} Li_3AlH_6 + 2C_4H_8$$

Tri-n-octyl-n-propylammonium alanate $(n\text{-}C_8H_{17})_3(n\text{-}C_3H_7)NAlH_4$ and tetramethyl ammonium alanate $(CH_3)_4\,NAlH_4$ are formed by the reaction of the corresponding ammonium compound and lithium alanate, e.g.: [*27a*]

$$(C_8H_{17})_3(C_3H_7)NBr + LiAlH_4 \xrightarrow{\text{benzene}} (C_8H_{17})_3(C_3H_7)NAlH_4 + LiBr$$

8.2 *Preparation of partially substituted alanates* $AlH_{4-n}R_n^{\ominus}$

Metal hydrides (donor: $H^{\ominus}$) add to substituted alanes (acceptor: $AlH_{3-n}R_n$) with formation of metal alanates, for example [*171*, *176*]:

$$E^IH + AlR_3 \rightarrow Na(AlHR_3) \quad \begin{cases} E^I = Li, Na, K \\ R = Me, Et, n\text{-}Pr, t\text{-}Bu \end{cases}$$

$$E^IH + AlHBu^t_2 \rightarrow E^I(AlH_2Bu^t_2)\ (E^I = Li, Na, K)$$

The chlorides $AlClR_2$ may be used instead of the hydrides $AlHR_2$ (*cf.* section 8.1).

Ligands are exchanged in mixtures of alanates with different alanate ions ($AlH_{4-n}R_n^{\ominus}$; $n = 0, 1, 2, 4$). However, such metathesis do not lead to the monoalkylalanate [*70*, *172*]:

$$Na(AlEt_4) + Na(AlH_4) \xrightarrow{180\text{–}190°} 2Na(AlH_2Et_2)$$
$$Na(AlH_2Et_2) + Na(AlH_4) \xrightarrow{180\text{–}190°} \text{no } Na(AlH_3Et)$$

When lithium alanate is added to pyridine at 0°, an exothermic reaction gives rise to lithium tetrakis(*N*-dihydropyridyl)alanate, which has a mild hydrogenating action [*73–75*]:

$$LiAlH_4 + 4\,C_5H_5N \longrightarrow Li[Al(NC_5H_6)_4]$$

Lithium hydride also adds to aluminium alkoxides $Al(OR)_3$. Both the crystalline tetrameric β-$Al(OEt)_3$, which is held together by OR bridges and reacts only after previous ring opening, and the liquid α-$Al(OEt)_3$, which is soluble in benzene as the monomer and is more reactive, are suitable for this purpose.

Very finely powdered crystalline β-triethoxyaluminium reacts with lithium hydride to form lithium triethoxyalanate, $Li[AlH(OEt)_3]$, only on prolonged shaking in a 1 : 1 mixture of ether and benzene [*114*]:

$$LiH + [Al(OEt)_3]_4 \rightarrow 4Li[AlH(OEt)_3]$$

In boiling ether, the reaction takes place similarly slowly. Lithium alanate catalyses the reaction, but the hydrogenation goes further [*114*]:

$$4Li[AlH_4] + [Al(OEt)_3]_4 \rightarrow 4Li[AlH(OEt)_3] + 4AlH_3$$
$$Li[AlH_4] + Li[AlH(OEt)_3] \rightarrow Li[AlH_2(OEt)_2] + Li(OEt) + AlH_3$$
$$LiH + Li[AlH(OEt)_3] \rightarrow Li[AlH_2(OEt)_2] + Li(OEt)$$
$$LiH + AlH_3 \rightarrow LiAlH_4$$

In boiling ether, lithium hydride adds rapidly to liquid α-triethoxyaluminium. In the same way, $Li[AlH(OEt)_3]$, which tenaciously retains ether (ether complex?) is formed in 40% yield [*114*]:

$$LiH + Al(OEt)_3 \rightarrow Li[AlH(OEt)_3]$$

Sodium hydride adds to α- and β-triethoxyaluminiums $[AlOEt)_3]_n$ ($n = 4$ or 1) more rapidly than lithium hydride (see section 4.3 of Chapter 2). Sodium triethoxyalanate is formed in 90% yield [*114*]:

$$n\,NaH + [Al(OEt)_3]_n \rightarrow n\,Na[AlH(OEt)_3]$$

In ether, $AlHCl_2{\cdot}OEt_2$ reacts with KCl to form a halo-substituted alane $KAlHCl_3$ [*19a*].

8.3 *Physical properties of alanates $AlH_{4-n}R_n^{\ominus}$, $AlH_{4-n}X_n^{\ominus}$ and $AlH_6^{3\ominus}$*

8.3.1 *Lithium alanates $LiAlH_4$ and Li_3AlH_6*

The commercial lithium alanate obtained by the reaction of LiH with $AlCl_3$ is white to grey, spongy or in lumps. It contains [*21*] 97% by weight of $LiAlH_4$, up to 1.5% by weight of Cl, and 0.5% by weight of C. $LiAlH_4$ precipitated several times from concentrated ethereal solution by the addition of benzene is pure white, crystalline and non-volatile [*24*]. Crystal structure [118a]. In solutions in ether or non-highly-solvating media, it may sometimes decompose spontaneously if catalytic impurities are present [*130*]. It appears to be more stable in tetrahydrofuran, so that THF solutions are to be preferred to ethereal solutions, although the solubility in THF is lower. Care is necessary during working with solutions in dimethyl ether, since under certain circumstances they may explode with carbon dioxide condensed in the solution [*8*]. Suspensions in paraffin oil may also be used. Good solvents (see Table 5.2) take up relatively large amounts of benzene.

References p. 438

TABLE 5.2

SOLUBILITIES OF $LiAlH_4$, $NaAlH_4$ AND $LiAlH(OBu^t)_3$ [*4*, *13*, *21*, *36*]

Solvent	*B.p. of the solvent (°C)*	*Solubility (moles/l) of $LiAlH_4$ at* 0°	25°	50°	75°	100°	*$NaAlH_4$ at 20°*	LiAlH(OBu 20°
Diethyl ether	36		5.92				insoluble	0.08
Tetrahydrofuran	64		2.96				3	1.43
Ethyleneglycol dimethyl ether (monoglyme)	85	1.29	1.80	2.57	3.09	3.34	—	0.16
Diethyleneglycol dimethyl ether (diglyme)	—	0.26	1.29	1.54	2.06	2.06	2.2	1.61
Triethyleneglycol dimethyl ether (triglyme)	216	0.56	0.77	1.29	1.80	2.06	2.8	
Tetraethyleneglycol dimethyl ether (tetraglyme)	275	0.77	1.54	2.06	2.06	1.54		
Diethyleneglycol diethyl ether	181	0.77	1.03	1.03	1.29	1.54		
Dibutyl ether	142		0.56					0.02
Dioxan	101		0.03					—
Acetonitrile	82							0.01
tert-Butanol	82.8							0
Dimethoxymethane	44							0

In such mixed solutions compounds that are only soluble in benzene can be reduced in the homogeneous phase [*21*]. Mixtures of ether, tetrahydrofuran, and light hydrocarbons contain the bis-tetrahydrofuranate of $LiAlH_4$ [*48*].

Lithium alanate decomposes from 120° (145° [*82*]) onwards, the decomposition being rapid and complete at 150 to 220° [*36*] (according to [*82*] only at 580°) to give H_2 and Al–Li alloy (15–16% Li [*82*]) and melts at 150° [*21*]. For the mechanism of the decomposition see [*42*]. Up to a pressure of 250,000 atm it exhibits no metallic conductivity [*1*]. The thermodynamic data for crystalline $LiAlH_4$ are: heat of formation $\Delta H^\circ_{298} = -24.08$ [*24*], -28.4 ± 1.5 [*120*] kcal/mole; free energy $\Delta F^\circ_{298} = -12.9$ kcal/mole [*120*]; entropy $S^\circ_{298} = 23.5$ kcal/deg mole [*120*]; enthalpy and free energy for decomposition according to the equation $LiAlH_4$(crystalline) $\rightarrow$ LiH (crystalline) + Al (crystalline) + $\frac{3}{2}$ H_2 (gaseous): $\Delta H^\circ_{298} = +6.8$, $\Delta F^\circ_{298} = -3.9$ kcal/mole [*120*].

Lithium hexahydroalanate Li_3AlH_6 decomposes without melting or subliming above 210°, and, so far no solvent has been found for it [26a]. For handling with alanates see section 9.1.1.

8.3.2 *Lithium alkylalanates $LiAlH_{4-n}R_n$*

Melting points of the lithium alkylalanates [*171*]: $Li[AlMe_4]$ 260° with decomposition; $Li[HMe_3] \cdot OEt_2$, liquid; $Li[AlHMe_3]$, 200–205°; $Li[AlEt_4]$, 163–165°; $Li[AlPr_4{}^n]$, 180–182°; $Li[AlBu_4{}^i]$, 170–175°; $Li[AlHBu_3{}^i]$, 162–164°; $Li[AlH_2\text{-}Bu_2{}^i]$, 180° with decomposition. Lithium alkylalanates are partially soluble in benzene, hexane, and alkylaluminium hydrides [*177*, *178*].

8.3.3 Lithium tri-tert-butoxyalanate

$Li[AlH(OBu^t)_3]$ precipitates from ethereal solution as a white finely divided solid[*12–14*]. For solubility, see Table 5.2. $Li[Al(OMe)_4]$ and $Li[AlOEt)_4]$ are insoluble in benzene, xylene, dioxan, tetrahydrofuran, diethyl ether, acetone, acetonitrile, dimethylformamide, monoglyme and diglyme[*13*].

8.3.4 Lithium tetrakis(N-dihydropyridyl)alanate

$Li[Al(N\ \ H_2)_2(N\ \ H_2)_2]$ (see section 8.2) forms faintly yellow solutions in pyridine. It precipitates from such 0.5 *M* solutions in pyridine on standing at 20° in the form of thick, faintly yellow crystals. They are hygroscopic, smell of pyridine, and decompose without melting[*75*].

8.3.5 Sodium alanates $NaAlH_4$ and Na_3AlH_6

Sodium alanate obtained from the elements is a fine white crystalline powder. It contains > 98% by weight of $NaAlH_4$ and 0.2% by weight of C. Density 1.27 g/ml. It does not become grey under the action of light and air[*21*]. On slow heating it decomposes only very slowly between 145 and 183°. At 230 to 240°, decomposition into NaH, Al, and H_2 is rapid. By the controlled thermal decomposition of $NaAlH_4$ Na_3AlH_6 is produced[*6a*]. Melting point 183°[*35*]. For solubility, see Table 5.2. The thermodynamic data of crystalline $NaAlH_4$ are[*120*]: normal heat of formation $\Delta H^0_{298} = -27.0 \pm 1.0$ kcal/mole; free energy $\Delta F^0_{298} = -11.6$ kcal/mole; entropy $S^0_{298} = 29.6$ cal/deg mole. Enthalpy and free energy for decomposition in accordance with the equation $NaAlH_4$ (crystalline) → NaH (crystalline) + $\frac{3}{2}H_2$ (gaseous): $\Delta H^0_{298} = +13.5$ kcal/mole; $\Delta F^0_{298} = +3.0$ kcal/mole[*120*].

Na_3AlH_6 is insoluble in all solvents tested[*6a*]. The X-ray powder data are similar to those of Na_3AlF_6[*6a*]. For handling with alanates see section 9.1.1.

8.3.6 Sodium alkylalanates $NaAlH_{4-n}R_n$

The melting points of sodium alkylalanates are as follows[*171*]: $Na[AlMe_4]\cdot OEt_2$ 124° – 126°; $Na[AlMe_4]$ 221–225° with decomposition; $Na[AlHMe_3]$ 144–145°; $Na[AlH_2Me_2]$ 87–88°; $Na[AlEt_4]$ 122–124°; $Na[AlHEt_3]$ 64°; $Na[AlH_2Et_2]$ 85–87° (87–88°[*172*]); $Na[AlPr^n_4]$ 54–56°; $Na[AlHPr^n_3]$ 68–70°; $Na[AlH_2Pr^n_2]$ 95–99°; $Na[AlBu^i_4]$ 79–82°; $Na[AlHBu^i_3]$ 203–205°; $Na[AlH_2Bu^i_2]$ 180–184°. Sodium alkylalanates are partially soluble in benzene, hexane, and alkylaluminium hydrides[*177, 178*].

8.3.7 Potassium alanate $KAlH_4$

Crystalline potassium alanate is apparently only slightly soluble in diethyl ether and tetrahydrofuran[*170*]. The thermodynamic data of crystalline $KAlH_4$ are[*120*]: normal heat of formation $\Delta H^0_{298} = -39.8 \pm 0.9$ kcal/mole; free energy $\Delta F^0_{298} = -23.8$ kcal/mole; entropy $S^0_{298} = 30.8$ cal/deg mole. Enthalpy and free energy for decomposition according to the equation $KAlH_4$ (crystalline) → KH

References p. 438

(crystalline) + Al (crystalline) + $\frac{3}{2}H_2$ (gaseous): $\Delta H^0_{298} = +26.0$ kcal/mole; $\Delta F^0_{298} = +14.9$ kcal/mole [*120*]. For handling with alanates see section 9.1.1.

8.3.8 Potassium alkylalanates $KAlH_{4-n}R_n$

Melting points of the potassium alkylalanates [*171*]: $K[AlMe_4]$ 264° with decomposition; $K[AlHMe_3]$ 202–205°; $K[AlH_2Me_2]$ 135–142°; $K[AlEt_4]$ 80–82°; $K[AlHEt_3]$ 46–50°; $K[AlH_2Et_2]$ liquid; $K[AlPr_4^n]$ liquid; $K[AlHPr_3^n]$ 26–28°; $K[AlH_2Pr_2^n]$ 113–115°; $K[AlBu_4^i]$ 47–48°; $K[AlHBu_3^i]$ 60–61°; $K[AlH_2Bu_2^i]$ 163–164°.

8.3.9 Alkaline-earth metal alanates $E^{II}(AlH_4)_2$

$Mg[AlH_4]_2$ is soluble in ether. It decomposes from 140°, more rapidly at 200°, into MgH_2, Al, and H_2 [*128, 129*]. The solubility of $Ca[AlH_4]_2$ in tetrahydrofuran is 0.3 mole/l. On concentration or on the addition of light hydrocarbon it separates out in the form of rod-shaped monoclinic crystals of $Ca[AlH_4]_2 \cdot 3\,THF$. It is soluble in monoglyme and very slightly in ether, and is insoluble in dioxan and hydrocarbons. It decomposes thermally above 230° [*35, 116*].

8.3.10 Quaternary ammonium alanates R_4NAlH_4

Tri-n-octyl-n-propylammonium alanate $(n\text{-}C_8H_{17})_3(n\text{-}C_3H_7)NAlH_4$ is a white, crystalline solid, with a m.p. 61–63° (decomp.) after purification by precipitation from benzene solution with pentane [*27a*].

Tetramethylammonium alanate $(CH_3)_4NAlH_4$, a white solid, decomposes above 173° without melting; density 0.990 g/cm³. [*27a*].

9. Reactions of the alanates $AlH_{4-n}R_n^{\ominus}$

9.1 Inorganic reactions

The alanate ion $AlH_4^{\ominus}$ contains negative hydrogen by means of which hydrogenations can be effected. However, in the absence of polar solvents, alanates have a hydrogenating effect only under particularly severe conditions (dry heating, etc.). Thus, for example, under normal ("Grignard") conditions, but without a solvent, only traces of the corresponding hydrides are formed with BCl_3, $SiCl_4$, and PCl_3. In the presence of ether, however, high yields of $(BH_3)_2$, SiH_4, and PH_3 are obtained [*94*].

The role of the solvent has not yet been elucidated. Does it merely make possible the predissociation of the hydride ion $H^{\ominus}$, which then reacts with the substrate? [*94*]:

$$AlH_4^{\ominus} \rightleftharpoons AlH_3 + H^{\ominus}$$

or does $AlH_4^{\ominus}$ react primarily with the substrate, with the (donor) solvent merely facilitating the secondary detachment of AlH_3, for example:

$$R_2O + AlH_4^{\ominus} + SiCl_4 \longrightarrow \left[R_2O\rightarrow \underset{H\ \ H}{\overset{H}{Al}}-H\cdots\underset{Cl\ \ Cl}{\overset{Cl}{Si}}-Cl \right]^{\ominus} \longrightarrow R_2O\cdots\rightarrow AlH_3 + HSiCl_3 + Cl^{\ominus}$$

In any case, the yields of GeH_4 and SnH_4 in the reactions of $GeCl_4$ and $SnCl_4$, respectively, with $LiAlH_4$ or $Li[AlH(OBu^t)_3]$ increase with increasing donor character of the solvent [*3*]. The possibility of a liquid compound $LiAlH_4 \cdot nO(C_2H_5)_2$ was discussed [*19*]. About the conductometric titrations of inorganic compounds in the $LiAlH_4$–$O(C_2H_5)_2$ system, see [*67*].

Because of the equilibria mentioned above, Cl-free alane can be obtained in good yield by the electrolysis of sodium alanate in tetrahydrofuran (the specific resistance of the saturated solution is only 50 Ω cm at 30°) [*20*]:

$$NaAlH_4 \rightarrow AlH_3 + Na^{\ominus} + H^{\ominus}$$

9.1.1 *Main groups VII and VI*

Alanates react with hydrogen compounds of main groups VII and VI with the evolution of hydrogen:

$$ALH_4^{\ominus} + 4HE^{VII} \xrightarrow{\text{ether}} AlE_4^{VII\ominus} + 4H_2 \quad (E^{VII} = Cl, Br)$$

(On the existence of $AlCl_4^{\ominus}$ and $AlBr_4^{\ominus}$ see for example ref. [*164*]).

$$AlH_4^{\ominus} + 4HOH \xrightarrow{\text{ether}} Al(OH)_4^{\ominus} + 4H_2$$

Pseudohalogens such as cyanogen $(CN)_2$, selenocyanogen $(SeCN)_2$, and thiocyanogen $(SCN)_2$ form very unstable lithium tetrapseudohaloalanates with lithium alanate (the boron isologues are more stable). For example [*69, 164*]:

$$LiAlH_4 + 2(SCN)_2 \rightarrow Li[Al(NCS)_4] + 2H_2$$

A solution of lithium alanate in tetrahydro(tetrahydrofurfuryloxy)pyran absorbs carbon oxide to form the methylalanate ion $[AlH_3CH_3]^{\ominus}$. On alcoholysis, this yields methane [*81*].

With ethers (tetrahydrofuran, diethylether) lithium alanate gives adducts: $LiAlH_4 \cdot nOC_4H_8$ ($n = 1, 2, 3$) and $LiAlH_4 \cdot 2OEt_2$ [*159*].

K_3AlH_6 when kept in air for some time is very dangerous to handle. A 20 gram sample detonates with a force equivalent to 500 grams of TNT, initiated, for example, by scraping with a metal spatula. The reason is probably that unhydrided metallic potassium in the hydride reacts with oxygen (air) giving peroxide, and that this reacts with K_3AlH_6. Whenever K_3AlH_6 is employed, it should be synthesized afresh, and stored for as short a time as possible under a dry, inert diluent in a flask fitted with a well-greased ground-glass stopper and inside a dry-box in an inert atmosphere [*5a, 5b*].

9.1.2 *Main group V*

In general, ammonia, isopropylamine, phosphane, and arsane react with alkali metal alanates with replacement of H by E^VH_2 [*35, 37*]:

$$E^IAlH_4 + 4E^VH_3 \rightarrow E^I[Al(E^VH_2)_4] + 4H_2 \qquad \begin{cases} E^I = Li, Na \\ E^V = N, P, As \end{cases}$$

$$NaAlH_4 + 4NHPr_2^i \rightarrow Na[Al(NPr_2^i)_4] + 4H_2$$

(Liquid) ammonia and isopropylamine react rapidly in the absence of solvent in accordance with the above equation and without side-reactions to form the alkali metal alanate [35, 37]. In diethyl ether or diglyme the product depends on the ratio of the reactants. At a low ratio (which is generally the case because of the low concentration of NH_3 in the solvent), up to 3 moles of H_2 are evolved per mole of NH_3. The solid formed (soluble only in liquid NH_3) has not yet been investigated. The evolution of H_2 in its formation indicates a potential substitution of all the H atoms in the ammonia, possibly in the sense of the reaction:

$$3LiAlH_4 + NH_3 \rightarrow (LiAlH_3)_3N + 3H_2$$

The evolution of a smaller amount of H_2 has also been observed [36]:

$$2LiAlH_4 + 5NH_3 \rightarrow [LiAlH(NH_2)_2]_2NH + 6H_2$$

Under the same conditions, solutions of AlH_3 in diethyl ether or diglyme form various insoluble products which again depend on the NH_3/AlH_3 ratio (see also Table 5.1 under $AlH_3 \cdot NH_3$).

In contrast to ammonia, phosphane reacts with lithium alanate in digylme and tetrahydrofuran independently of the ratio of the reactants. In analogy with the above equation, the solution lithium tetrakis(dihydrogenphosphido)alanate, $Li[Al(PH_2)_4]$, is formed. In diethyl ether, however, (as in the case of a low NH_3/E^IAlH_4 ratio) almost 3 moles of H_2 per mole of PH_3 are evolved [37]. For the preparation of $Li[AlH_3P(C_2H_5)_2]$, $Li[AlH_2\{P(C_2H_5)_2\}_2]$ $Li[AlH\{P(C_2H_5)\}_3]$ see [39a, 39b].

With lithium alanate in diglyme, arsane liberates hydrogen in a molar ratio $H_2 : AsH_3 = 0.97$. The compound formed has not been isolated. Because of the H_2/AsH_3 ratio and the behaviour of the compound on hydrolysis (see below), the formula $Li[Al(AsH_2)_4]$ seems to be fairly certain [37]. The compounds $E^I[Al(E^VH_2)_4]$ hydrolyse in accordance with the following equation [37]:

$$E^I[Al(E^VH_2)_4] + 4HOH \rightarrow E^I[Al(OH)_4] + 4E^VH_3$$

With ethyl iodide, $Li[Al(PH_2)_4]$ forms ethylphosphane and phosphane. Another reaction product also evolves ethylphosphane and phosphane on hydrolysis [37].

With trimethylamine at 20° lithium alanate gives the monoaminate $LiAlH_4 \cdot NMe_3$, at >0° the diaminate $LiAlH_4 \cdot 2NMe_3$ [158].

The halides of the elements belonging to main group V (PX_3, $R_{3-n}AsX_n$, $R_{3-n}SbX_n$, $R_{3-n}BiX_n$; $n = 1, 2, 3$) react with lithium alanate with the replacement of the X by H (primary formation of the corresponding hydrides).

Examples:

$$PX_3 \xrightarrow[-\frac{3}{4}LiAlX_4]{+\frac{3}{4}LiAlH_4,\ ether} \begin{cases} PH_3 & \text{(a)} \\ \frac{1}{x}(PH)_x + H_2 & \text{(b)} \end{cases}$$

he formation of phosphane by reaction (a) increases with falling reaction temerature (80% at −115°, 30% at −30°, 20% at 20°) [*154*].

$$AsCl_3 + 3LiAlH_4 \xrightarrow[\text{diethyl ether}]{-90°} AsH_3 + 3AlH_3 + 3LiCl$$

(84% [*149*]; in addition a little As_2H_4 is formed [*34*])

$$SbCl_3 + 3LiAlH_4 \xrightarrow[\text{diethyl ether}]{-95°} SbH_3 + 3AlH_3 + 3LiCl \qquad (82\% [149])$$

$$BiCl_3 + 3LiAlH_4 \xrightarrow[\text{dimethyl ether}]{-110°} 4BiH_3 + 3LiAlCl_4 \qquad (<1\% [2])$$

$$(C_6H_5)AsCl_2 \xrightarrow[\text{diethyl ether, }-75°]{+2LiAlH_4,\ -2AlH_3,\ -2LiCl} (C_6H_5)AsH_2 \xrightarrow[-H_2]{+AlH_3} (C_6H_5)AsH\text{-}AlH_2 \qquad [150]$$

$$CH_2(CH_2)_4AsCl \xrightarrow[\text{diethyl ether, }-60°]{+LiAlH_4,\ -LiAlCl_4} 4CH_2(CH_2)_4AsH \qquad (77\% [153])$$

$$(C_6H_5)SbI_2 \xrightarrow[\text{diethyl ether, }-60°]{+2LiAlH_4,\ -2AlH_3,\ -2LiI} (C_6H_5)SbH_2 \qquad (24\% [151])$$

$$(C_6H_5)_2SbCl \xrightarrow[\text{diethyl ether, }-75°]{+LiAlH_4,\ -LiAlCl_4} 4(C_6H_5)_2SbH \qquad (22\% [152])$$

$$(CH_3)_{3-n}BiX_n \xrightarrow[\text{dimethyl ether, }-110°]{+n/4\,LiAlH_4,\ -n/4\,LiAlX_4} (CH_3)_{3-n}BiH_n \xrightarrow[-45/-15°]{-(3-n)/3\,(CH_3)_3Bi} \frac{n}{3}BiH_3$$

$$(X = Cl, Br;\ n = 1, 2) \qquad (35\% [2])$$

.1.3 *Main groups IV, III, and II*

The halides of the main groups IV and III and the halides and alkyl compounds f Be and Mg (for example, $R_{4-n}SiX_n$, GeX_4, $R_{4-n}SnX_n$, BX_n, BeR_2, etc.) are onverted by lithium alanate or lithium tri-tert-butoxyalanate into the correponding hydrides or alanates. Details can be obtained from the sections on the orresponding elements [*126, 163*].

.1.4 *Transition elements*

Lithium alanate reduces transition metal halides with the intermediate formation f transition metal alanates. Only some of the latter can be isolated, although their ormation can be postulated from the course of the reaction.

Gold. Gold(III) alanate, $Au(AlH_4)_3$, cannot be prepared by the reaction of $AuCl_3$ with $LiAlH_4$, nor can gold(III) boronate, $Au(BH_4)_3$, and gold hydride, AuH_3, be prepared by the reaction of $AuCl_3$ with $LiBH_4$ and LiH respectively, even at very low temperatures. In all cases, metallic gold precipitates with the ormation of hydrogen [*155*].

Silver. Silver perchlorate forms with $LiAlH_4$ at −80° solid yellow silver alanate, $Ag[AlH_4]$, which decomposes into the elements above −50° [*138*]:

$$AgClO_4 \xrightarrow[\text{ether, }-80°]{+LiAlH_4,\ -LiClO_4} AgAlH_4 \xrightarrow{-50°} Ag + Al + 2H_2$$

References p. 438

Cu, Zn, Cd, Hg. Copper(I) iodide [*139*], zinc iodide [*140*], cadmium iodide [*136*] and mercury(II) iodide [*135*] form the corresponding hydrides at low temperatures:

$$4CuI \xrightarrow[\text{pyridine-THF-ether, } 20°]{+LiAlH_4,\ -LiI,\ -AlI_3} 4CuH \xrightarrow{60°} 4Cu + 2H_2$$

$$ZnI_2 \xrightarrow[\text{ether, } -40°]{+2LiAlH_4,\ -2LiI,\ -2AlH_3} ZnH_2 \xrightarrow{90°} Zu + H_2$$

(on the stoichiometric formation of $ZnH_2 \cdot ZnI_2 = ZnHI$, see [*137*])

$$CdI_2 \xrightarrow[\text{ether-THF, } -70°]{+2LiAlH_4,\ -2LiI} [Cd(AlH_4)_2] \xrightarrow{-2AlH_3} CdH_2 \xrightarrow{+20°} Cd + H_2$$

$$HgI_2 \xrightarrow[\text{ether-THF-petroleum ether, } -135°]{+2LiAlH_4,\ -2LiI,\ -2AlH_3} HgH_2 \xrightarrow{-125°} Hg + H_2$$

Lithium alanate forms no hydride with dialkylmercuries or diarylmercuries, metallic mercury being produced by the reaction $LiAlH_4 + 2R_2Hg \rightarrow LiAlR_4 + 2Hg + 2H_2$ and, to a smaller extent, by the reaction $LiAlH_4 + R_2Hg \rightarrow LiAlH_2R_2 + Hg + H_2$ (decreasing reactivity with R = Ph ⋙ s-Bu > n-Bu) [*112*].

Iron. Iron(III) chloride is reduced to a Fe–Al alloy in an ethereal solution of lithium alanate at −45°, probably via ferrous alanate $Fe(AlH_4)_2$ [*108*]:

$$FeCl_3 + LiAlH_4 \xrightarrow{\text{ether}} FeCl_2 + LiCl + AlH_3 + \tfrac{1}{2}H_2$$

$$FeCl_2 + 2LiAlH_4 \xrightarrow[-2LiCl]{-45°} [Fe(AlH_4)_2] \rightarrow Fe \cdot 2Al + 4H_2$$

If, however, solid ethereal solutions of $LiAlH_4$ and $FeCl_3$ that are separated by a layer of solid ether are warmed from −196 to −116° (m.p.), black-brown ether-insoluble $Fe(AlH_4)_2$ is formed [via the unstable intermediate $Fe(AlH_4)_3$, not via $FeCl_2$] [*90, 91*]:

$$FeCl_3 + 3LiAlH_4 \xrightarrow{-3\,LiCl} [Fe(AlH_4)_3] \rightarrow Fe(AlH_4)_2 + AlH_3 + \tfrac{1}{2}H_2$$

$Fe(AlH_4)_2$ decomposes into the elements more readily than the also solid and non-volatile (but white) $Fe(BH_4)_2$ prepared from $FeCl_3$ and $LiBH_4$. According to thermolysis experiments, the evolution of H_2 begins very slowly at about −40° and becomes violent at 90 to 100° [*90, 91*].

Cobalt. The isologous cobalt compounds are apparently more unstable, since they are not produced either with $LiAlH_4$ or with $LiBH_4$. However, the reactions were carried out at much too high a temperature, only slightly below 20° [*122a, 124*].

Niobium. Niobium pentachloride reacts with lithium alanate in a molar ratio of 1 : 5, with reduction of the niobium, to give niobium alanate $Nb(AlH_4)_n$ ($n < 5$) [*90, 156*]:

$$NbCl_5 + 5LiAlH_4 \xrightarrow{\text{diethyl ether}} Nb(AlH_4)_n + (5-n)AlH_3 + \frac{5-n}{2}H_2 + 5LiCl$$

n decreases with increasing temperature, being 3.5 at −70°, 3.0 at −40°, and 2.5 at +20°; thus $Nb_2(AlH_4)_7$ is formed at −70°, $Nb_2(AlH_4)_6$ at −40°, and $Nb(AlH_4)_5$ at +20°.

Triple hydrides $Li_mNb(AlH_4)_{n+m}$ are formed [*90*, *156*] with an excess of lithium alanate (molar ratio of $NbCl_5$ to $LiAlH_4 < 1:5$):

$$NbCl_5 + (5+m)LiAlH_4 \rightarrow Li_mNb(AlH_4)_{n+m} + (5-n)AlH_3 + \frac{5-n}{2}H_2 + 5LiCl$$

Here again n decreases with increasing temperature: $n = 3.0$ at $-70°$ and 2.0 at $+25°$. The amount of $LiAlH_4$ added, m, is 0.5 when the heat of the reaction is removed rapidly and 1.0 under less mild conditions. Compounds that have been isolated are $LiNb_2(AlH_4)_7$ ($-70°$) and $LiNb_2(AlH_4)_5$ ($+25°$) or $LiNb(AlH_4)_3$ ($+25°$).

Tantalum. Tantalum pentachloride decomposes with lithium alanate in diethyl ether at 20° to give the red tantalum(IV) dihydride dialanate [*90*].

$$TaCl_5 + 5\ LiAlH_4 \rightarrow TaH_2(AlH_4)_2 + 3\ AlH_3 + 5\ LiCl + \tfrac{1}{2}H_2$$

Similar alanates with a mean degree of oxidation of the Ta between +5 and +4 are formed at lower temperatures [*90*].

9.2 *Organic reactions*

Alanate ions are suitable for the mild reductions of organic compounds in an anhydrous medium. For reductions in ethereal solution, it is preferred to use the ether-soluble $LiAlH_4$ and $Mg(AlH_4)_2$. $NaAlH_4$ is only readily soluble in diglyme and tetrahydrofuran, and in these media it too is suitable. Substrates insoluble in ether but soluble in benzene, light hydrocarbons, or pyridine are dissolved in the latter solvents and the solutions are added to an ethereal solution of $LiAlH_4$. With very adverse solution properties, the extraction technique is often helpful.

The reactions of alanates form Al complexes which must subsequently be hydrolysed with water, acids, bases, ethanol, ethyl acetate, etc. Functional groups generally react specifically with alanate. A review of reductions with $LiAlH_4$ is given in Table 5.3. As far as is known, the other alanates, which have been investigated less thoroughly, react in an analogous way.

Reductions potentially taking place in two stages can be turned in two directions by carrying out the reaction either with a deficiency of the alanate (method A, addition of alanate to the substrate) or an excess of the alanate (method B, addition of the substrate to the alanate). We may briefly anticipate such reactions with four examples:

(1) The reduction of a *nitrile* with lithium alanate takes place via the stage of an N-substituted imide. After hydrolysis, method A gives an aldehyde and method B a primary amine [*12a*]:

$$RC{\equiv}N \longleftrightarrow R\overset{\oplus}{C}{=}\overset{\ominus}{\underline{N}} \xrightarrow[\text{ether}]{+AlH_4^{\ominus}} RCH{=}N(\overset{\ominus}{A}lH_3) \xrightarrow[\text{ether}]{+AlH_4^{\ominus}} RCH_2{-}N(\overset{\ominus}{A}lH_3)_2$$

$$\downarrow \Delta,\ +H_2O \qquad\qquad \downarrow \Delta,\ +H_2O$$

$$RCH{=}O \qquad\qquad RCH_2{-}NH_2$$

Method A (aldehyde) Method B (primary amine)

References p. 438

TABLE 5.3

REDUCTION OF FUNCTIONAL ORGANIC GROUPS WITH $LiAlH_4$

Initial compound	*Reaction product*
Hydrocarbons	
Alkanes C–C	very rare bond cleavage
Alkenes C=C	formation of alkanes generally only in the neighbourhood of polar groups
Alkynes –C≡CR	alkenes –CH=CHR or $>Al^{\ominus}$–C≡CR + H_2
Halogen compounds	
Alkyl–X	alkane
Aryl–X	generally no reaction
Oxygen compounds	
Alcohols ROH	alkoxide-substituted alanates $[Al(OR)_4]^{\ominus}$ or $[AlH(OR)_3]^{\ominus}$ + H_2
Aldehydes RCHO	primary alcohols RCH_2OH
Acetals $RCH(OR')_2$	no reaction
$R_2C(OR')_2$	no reaction
$R_2C(OR')_2$	no reaction
Ketones R_2CO	secondary alcohols R_2CHOH
Quinones O=C₆H₄=O	hydroquinones HO–C₆H₄–OH
Carboxylic acids RCOOH	primary alcohols RCH_2OH
Acid anhydrides $(RCO)_2O$	primary alcohols RCH_2OH
Acid halides RCOX	primary alcohols RCH_2OH
RCOX + $Li[AlH(OBu^t)_3]$	primary alcohols RCH_2OH or aldehydes RCHO
Carboxylic acid esters RCOOR′	primary alcohols RCH_2OH + R′OH
Lactones $(CH_2)_n$–C = O (ring closed via O)	diols $HO(CH_2)_nCH_2OH$
Amides $RCONH_2$	amines RCH_2NH_2
Amides RCONHR′	amines RCH_2NHR'
Amides $RCONR'_2$	amines $RCH_2NR'_2$, or aldehydes RCHO
Epoxides R_2C—CHR (bridged by O)	alcohols $R_2(OH)C$–CH_2R or (with $LiAlH_4$ + $AlCl_3$) R_2CH–CHR(OH)
Epoxides $R(CH_3O)C$—CR'_2 (bridged by O)	$R(CH_3O)CH$–$C(OH)R'_2$
Ozonides R_2C—O—CR_2 (bridged by O—O)	alcohols R_2CHOH
Ethers ROR	no reaction
Sulphur compounds	
C–S compounds	
$R_2C(SH)_2$	R_2CHSH
R–S–S–R	R-SH
R–S_3–R	R-SH
R–S_4–R	R-SH
R_2C–CR_2 (bridged by S)	$R_2C(SH)$–CHR_2

TABLE 5.3 (cont'd)

Initial compound	*Reaction product*
R–CO–SR	$R-CH_2OH$
$R-CS-NH_2$	$R-CH_2-NH_2 + RCN$
RSCN	R–SH
S–O compounds	
R–S(→O)–R	R–S–R
$R-SO_2-R$	R–S–R
$(R-SO_2)_2O$	$R-SO_2H$
$R-SO_2X$	R–SH
$Alkyl-O-SO_2R$	alkane
$Aryl-O-SO_2R$	aryl–OH
RSO_2H	RSSR + RSH
S–Halogen compounds	
RSX	RSSR
Nitrogen compounds	
CO–NH └R┘	CH_2-NH └R┘ or $CHO-NH_2R$
CO–NH–CO └—R—┘	$CH_2-NH-CH_2$ └—R—┘
R–NH–COOR′	$R-NH-CH_3 + R'OH$
RNC	$R-NH-CH_3$
RCN	$R-CH_2-NH_2$ or R–CHO
$R_2C{=}NOH$	R_2CH-NH_2
RNCO	$R-NH-CH_3$
RNCS	$R-NH-CH_3$
⟩N→O	⟩N
R_nC-NO	R_nC-NH_n
R_2N-NO	R_2N-NH_2
$R-NO_2$	$R-NH_2$
$Aryl-NO_2$	aryl–N=N–aryl or $aryl-NH_2$
R–NHOH	$R-NH_2$
Aryl–N(→O)=N–aryl	aryl–N=N–aryl
RN_3	$R-NH_2$
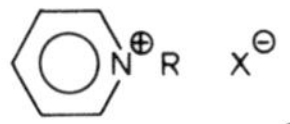 $N^{\oplus}R$ $X^{\ominus}$	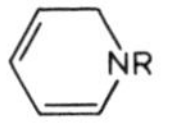 NR
$[R-CH{=}NR_2]^{\oplus} X^{\ominus}$	R–CH$_2$–NH–R

References p. 438

(2) The reaction of a *lactone* with lithium alanate gives, after the hydrolysis either the hydroxyaldehyde (method A) or the alkanediol (method B):

R₂C—(CH₂)ₙ / O—C=O ⟷ R₂C—(CH₂)ₙ / O(⊕)—C—Ō|(⊖) —(+AlH₄⊖, ether)→ R₂C—(CH₂)ₙ / O—CH—OAlH₃⊖ —(+AlH₄⊖, ether)→ R₂C—(CH₂)ₙ / H₃AlO⊖ CH₂—OAlH₃⊖

Δ + H₂O ↓ (both intermediates)

R₂C—(CH₂)ₙ / HO CHO ⟷ R₂C—(CH₂)ₙ / O—CH—OH; R₂C—(CH₂)ₙ / HO CH₂—OH

Method A hydroxyaldehyde; Method B (alkanediol)

(3) *Lactams* react with lithium alanate by method B to give the cyclic amine; using method A, hydrolysis gives an aminoaldehyde:

R₂C—(CH₂)ₙ / HN—C=O ⟷ R₂C—(CH₂)ₙ / HN(⊕)—C—Ō|(⊖) —(+AlH₄⊖)→ R₂C—(CH₂)ₙ / HN—CH—OAlH₃⊖ —(+AlH₄⊖, Δ)→ R₂C—(CH₂)ₙ / HN—CH₂

Method B (cyclic amine)

Δ + H₂O ↓

R₂C—(CH₂)ₙ / H₂N CHO

Method A (aminoaldehyde)

(4) The *C=C double bond* is generally not converted into a single bond by lithium alanate for example:

$$(CH_3)CH{=}CHCH{=}O \xrightarrow[\Delta]{+LiAlH_4} (CH_3)CH{=}CHCH_2\text{-}OH$$

If the C=C double bond is conjugated with both an aryl group and a polar group (Ar–C=C–C=O; Ar–C=C–N<), it undergoes reaction [*172b*]. Method A then leads (at low temperatures) to the alkenol, and method B to the alkanol, for example:

$(C_6H_5)CH{=}CHC(H){=}O$ ⟷ $(C_6H_5)CH{=}CH\overset{\oplus}{C}(H)$–Ō|⊖ —(+AlH₄⊖)→ $(C_6H_5)CH{=}CHCH_2{-}\overset{\ominus}{O}AlH_3$ —(+AlH₄⊖, Δ)→ $(C_6H_5)CH_2CH_2CH_2OH$

Method B (alkanol)

Δ + H₂O ↓

$(C_6H_5)CH{=}CHCH_2{-}OH$

Method A (alkenol)

In addition, a few examples may be taken from Table 5.3. Lithium alanate adds to *acetylene* to form lithium triethylenealanate [*175*]:

$$Li[AlH_4] + 3HC{\equiv}CH \rightarrow Li[AlH(CH{=}CH_2)_3]$$

On the other hand, sodium triisobutylalanate substitutes both hydrogens of acetylene [*70*]:

$$2Na[AlHBu^i_3] + HC{\equiv}CH \rightarrow Na_2[Bu^i_3AlC{\equiv}CAlBu^i_3] + 2H_2$$

With *higher alkynes* again the acidic hydrogen is substituted [*7, 67a, 80a 119*]:

$$Li[AlH_4] + 4HC{\equiv}CBu \rightarrow Li[Al(C{\equiv}CBu)_4] + 4H_2$$

Heterocyclic ring systems give the corresponding dihydro compounds [*10*], e.g.:

Alkyl halides can be converted into alkanes (with some difficulties). The first hydride ion of the alanate reacts by second-order kinetics, possibly in a S_N2 reaction [*6b, 6c, 67a*]:

$$AlH_4^{\ominus} \underset{\text{fast}}{\overset{-AlH_3(+R_2O)}{\rightleftharpoons}} H^{\ominus} \underset{\text{slow}}{\xrightarrow{+RX, -X^{\ominus}}} RH$$

The kinetics of the second, very slowly hydrogenating, hydride ion (and still more that of the third and fourth) are uncertain [79]. Aryl halides are generally resistant to lithium alanate.

Alcohols form lithium tetraalkoxyalanates with lithium alanate. Bulky alcohols such as isobutanol and isopentanol form lithium trialkoxyalanates even with an excess of the alcohol [*13, 14a*]:

$$Li[AlH_4] + 4HOMe \rightarrow Li[Al(OMe)_4] + 4H_2$$
$$Li[AlH_4] + n\,HOBu^t \rightarrow Li[AlH(OBu^t)_3] + (n-3)\,HOBu^t + 3H_2$$

In the reaction of *epoxides* with lithium alanate the hydride ion attacks the less substituted position, while in the reaction with a mixture of lithium alanate and aluminium chloride it is the more highly substituted that is attacked [*28, 29*]:

References p. 438

Similarly, the reactions of *ketones* with lithium alanate alone on the one hand, and with a mixture of $LiAlH_4$ and $AlCl_3$ on the other, form different products [*30*, *172a*]. For example, the reaction of 4-*tert*-butylcyclohexanone with $LiAlH_4$ and with $LiAlH_4 + AlCl_3$ forms *trans*- and *cis*-4-*tert*-butylcyclohexanols in various ratios [*30*]. One mole of lithium alanate converts about two moles of ketone into the alcohol. Two reduction mechanisms have been discussed for the reaction in the presence of pyridine [*72*, *74*, *75*]:

$$>C{=}O \xrightarrow{+AlH_4^{\ominus}} [HC{-}O{-}\overset{\ominus}{A}lH_3] \xrightarrow[-HCO^{\ominus}]{+\ C_5H_5N} C_5H_5\overset{\oplus}{N}{-}\overset{\ominus}{A}lH_3$$

$$AlH_4^{\ominus} \xrightarrow{C_5H_5N} (H_2)C_5H_5N{-}\overset{\ominus}{A}lH_3 \xrightarrow[-HCO^{\ominus}]{+>C{=}O} C_5H_5\overset{\oplus}{N}{-}\overset{\ominus}{A}lH_3$$

Epoxyethers form with lithium alanate 1,2-alkoxyalcohols. The hydride ion $H^{\ominus}$ therefore attacks the carbon carrying the alkoxy group [*123*]:

$$R{-}C(OCH_3){-}C(R')(R') \text{ (epoxide, } O \text{ bridging)} \xrightarrow[\Delta]{+\ LiAlH_4} R{-}C(OCH_3)(H){-}C(R')(OH){-}R'$$

Acid chlorides react with lithium *tert*-butoxyalanate in tetrahydrofuran or diglyme at −78° to form aldehydes [*12*, *13*]. Alcohols are also formed at higher temperatures:

$$RCClO + Li[AlH(Bu^t)_3] \xrightarrow[-78°]{fast} RCHO + LiCl + Al[OBu^t]_3$$

$$RCHO + Li[AlH(OBu^t)_3] \xrightarrow{slow} Li[(RCH_2O)Al(OBu^t)_3] \xrightarrow[\Delta]{+H_2O} RCH_2OH$$

Amides react with lithium alanate via an intermediate nitrile stage (!) to form primary amines [*88*]:

$$RCONH_2 \xrightarrow[-\frac{1}{4}Li_2O,-\frac{1}{4}Al_2O_3]{+\frac{1}{2}LiAlH_4,-2H_2} RCN \xrightarrow[\Delta]{+LiAlH_4} \text{amide salt} \xrightarrow[\Delta]{+H_2O} RCH_2NH_2$$

Tertiary amides generally react with lithium alanate or lithium diethoxyalanate to give aldehydes [*15*, *16*]:

$$R{-}C({=}O){-}NR'_2 \xrightarrow[\Delta]{+LiAlH_4 \text{ or } Li[AlH_2(OEt)_2]} R{-}C({=}O){-}H$$

[R = n − C_3H_5, n − C_4H_9, i − C_4H_9, n − C_6H_{13}, cyclo − C_6H_{11}, $C_{10}H_9$, *o* − ClC_6H_4; R′ = CH_3, C_2H_5, $CH_3 + C_6H_5$, $(CH_2)_5$, $(CH_2)_4$, $(CH_2)_2$]

10. Spectra of alanes and their derivatives

Because of its three valency electrons, aluminium is an element characterised by pronounced electron deficiency. Its valency shell can take up a large number of additional electrons: either eight electrons (sp^3, *e.g.* $AlH_4^{\ominus}$), or, with favourable energetic and geometrical factors, even ten (sp^3d, *e.g.* $AlH_3 \cdot 2NR_3$) or twelve (sp^3d^2, *e.g.* $AlF_6^{3\ominus}$). The tendency to occupy vacant orbitals is so great that aluminium compounds very rarely occur in the trivalent state in the condensed phase. This is the case, for example, in sterically hindered trialkylaluminiums such as $AlBu^i_3$. The electron deficiency can be eliminated in three different ways:

(1) the free Al orbital interacts with an occupied non-bonding orbital of a foreign molecule; the foreign donor may be a neutral molecule or an anion (examples: $AlH_3 + NMe_3 \rightarrow AlH_3 \leftarrow \cdot \cdot NMe_3$; $AlH_3 + R^{\ominus} \rightarrow [AlH_3 \leftarrow R]^{\ominus}$);

(2) if the Al compound itself contains an occupied non-bonding orbital, this can join with the free Al orbital to give an autocomplex:

$$\begin{array}{ccc} & \ominus & \oplus \\ H_2 & Al — & N — R_2 \\ & \uparrow & \downarrow \\ R_2 — & N — & AlH_2 \\ & \oplus & \ominus \end{array}$$

(3) the bonding orbital of an Al substituent is distributed over the orbitals of two Al atoms, *i.e.* the substituent orbital interacts with orbitals belonging to two different Al atoms (Rundle's "half bond", "three-centre bond", "bridge bond", or, better, 4-centre–4-electron bond) [*76, 77, 106, 107*]: in this situation hydrogen, halogen atoms, or C-atoms of alkyl groups (see Fig. 5.1) may function as bridge atoms.

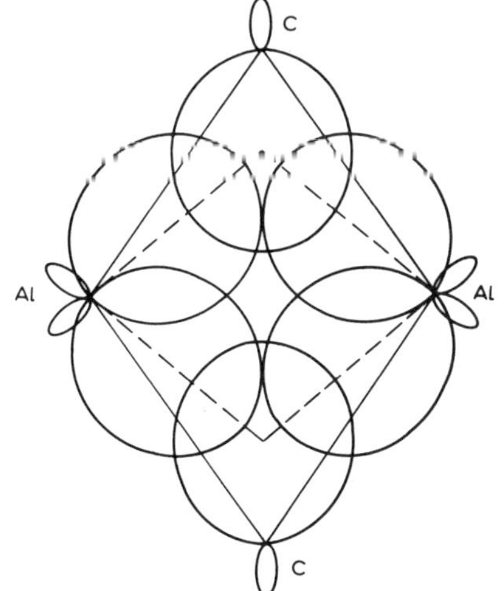

Fig. 5.1. Schematic drawing of the bonding atomic orbitals of a 4-centre–4-electron bond (Al–C^{μ}–Al double bridge) in $Al_2(CH_3)_6$ (acc. to [*76*]). The full line joins the atom centres of Al and C (bond direction). The broken line shows the direction of the bonding atomic orbitals of the Al-atoms which gives the greatest possible overlap with the orbitals of the C-atom (direction of the orbital lobes of the C-atom).

References p. 438

Both infrared absorption and nuclear resonance provide information on the nature of the association (bridge atoms or dipole action) and on which atoms form the bridges in the case of mixed substituents, as well as on the influence of the substituents.

10.1 *Proton bridges and hydride bridges*

Compounds with acidic hydrogen such as water or methanol associate through hydrogen atoms. Acceptor hydrides such as beryllium hydride and aluminium hydride also associate through hydrogen atoms. It is necessary and sufficient that the two bridge-forming partners should possess both an acceptor and a donor function. The bridges are, among other things, sensitive to donor solvents. In both cases the donor solvent competes with the donor site of the bridge-forming molecule for the acceptor site of the associated molecule. With proton bridge-forming agents the latter site is the hydrogen atom, and with hydride bridge-forming agents it is the metal atom [*56, 64*]:

$$\text{proton bridges: } \underset{\underset{\displaystyle R}{|}}{O}-H\leftarrow\cdot\cdot\underset{\underset{\displaystyle R}{|}}{O}-H + 2\underset{\underset{\displaystyle R}{|}}{O}-R \rightleftharpoons 2\underset{\underset{\displaystyle R}{|}}{O}-H\leftarrow\cdot\cdot\underset{\underset{\displaystyle R}{|}}{O}-R$$

$$\text{hydride bridges: } H-\overset{\overset{\displaystyle R}{|}}{\underset{\underset{\displaystyle R}{|}}{Al}}\leftarrow\cdot\cdot H-\overset{\overset{\displaystyle R}{|}}{\underset{\underset{\displaystyle R}{|}}{Al}} + 2\underset{\underset{\displaystyle R}{|}}{O}-R \rightleftharpoons 2H-\overset{\overset{\displaystyle R}{|}}{\underset{\underset{\displaystyle R}{|}}{Al}}\leftarrow\cdot\cdot\underset{\underset{\displaystyle R}{|}}{O}-R$$

Both bridges can be detected spectroscopically. In the case of proton bridges in water [*46*] and in methanol [*56*], the broad self-association bands change into the likewise broad mixed association bands. These are the broader the greater is the interaction with the donor solvent. In parallel with this there is a red shift of the association bands (Figs. 5.2 and 5.3). On the other hand, the hydride bridges in aluminium hydrides disappear when donor solvents are added. To the extent that the solvent complex is formed, the H-atom of the Al–H bond vibrates more freely. The broad self-association band changes into the sharp Al–H band (Fig. 5.4).

10.2 *Influence of donors on the half-width of the Al–H band*

Ether can be distilled off almost completely from mixtures of alanes ($AlHR_2$, AlH_2R, AlH_3) and ether. Ether complexes can clearly be detected spectroscopically on the basis of sharp Al–H bands. As the IR absorption of a solution of $AlHBu^i_2$ in Bu_2O/n-C_6H_{14} shows (Fig. 5.5), the hydride–autocomplex/hydride–donor solvent complex equilibrium is on the side of the ether complex only with a large excess of ether, since sharp Al–H bands can be observed only at a molar ratio of hydride to ether equal to 1 : 9.0 [*56*].

With increasing strength of the donor, the autocomplex/donor complex shifts in favour of the donor complex. According to Fig. 5.4, $AlHBu^i_2$ in dibutyl ether (*c*) is still present overwhelmingly in the form of the auto-complex (large half-width

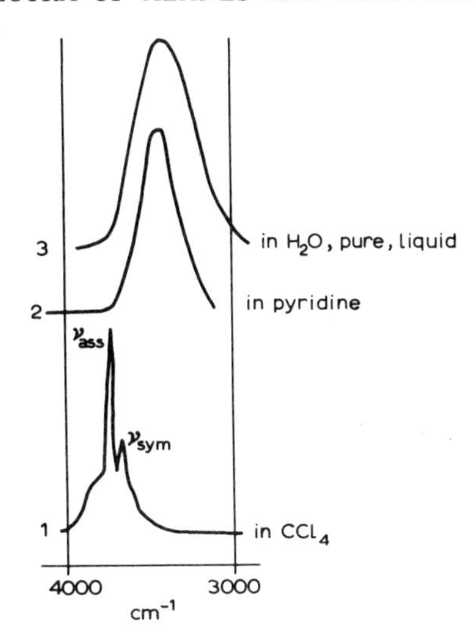

Fig. 5.2. OH bands of water[*46*]. (1) The bands of the ν_{ass} and ν_{sym} stretching vibrations of the OH bond, which still appear separately in CCl_4, are shifted towards the red when donor solvents (*e.g.* pyridine) are added. (2) Association band in pyridine (donor-acceptor complex). (3) Association band without solvent (autocomplex).

of the Al–H band). In the presence of the strong donor triethylamine (*a*), on the other hand, the equilibrium is right over on the side of the amine complex (small half-width of the Al–H band). This equilibrium position is reached even at a stoichiometric molar ratio of hydride and amine since higher concentrations of amine do not appreciably change the half-width[*56*].

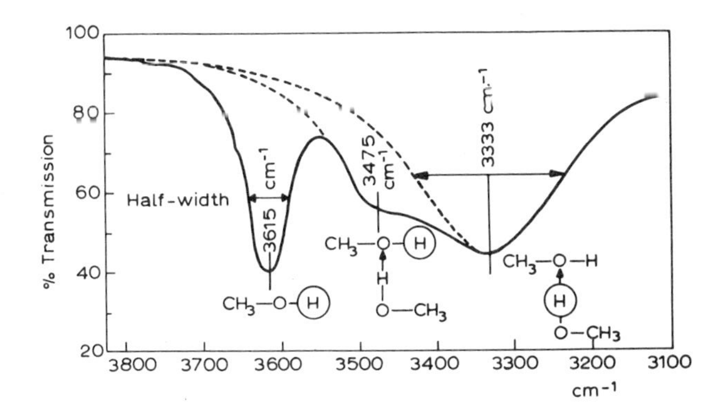

Fig. 5.3. OH-stretching vibration bands of undisturbed methanol and of methanol associates (in CCl_4)[*56*]. The sharp OH band of non-associated methanol appears at 3615 cm^{-1}. The very broad band (shoulder) of the OH-stretching vibration of the terminal H-atom in associated methanol appears at 3475 cm^{-1}. The very broad band of the OH-stretching vibration of a bridge H-atom appears at 3333 cm^{-1}. The dotted curves show the course of the bands unaffected by neighbouring bands. Half-width: width of the band, expressed *e.g.* in wave numbers, at the height of half the maximum extinction.

References p. 438

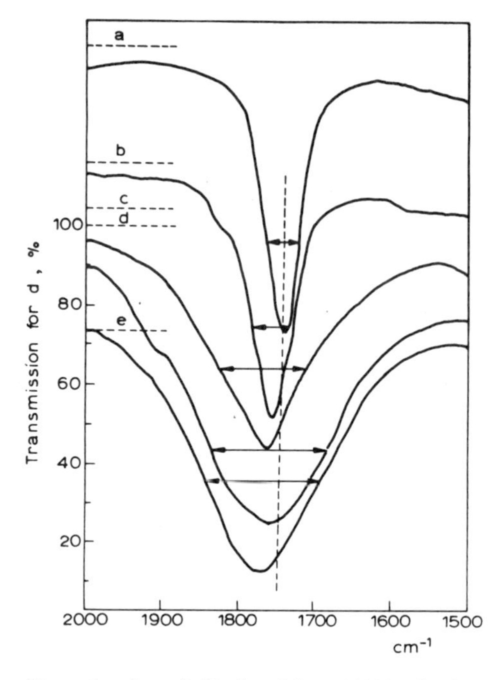

Fig. 5.4. Al–H-stretching vibration of diisobutylalalane (10% soln. by wt. in cyclohexane) without a solvent and in various donor solvents [*56*]: (a) $N(C_2H_5)_3$, (b) tetrahydrofuran, (c) dibutyl ether, (d) anisole, (e) without a donor. The broad band of the not freely-vibrating Al–H-stretching vibration of liquid Bu_2^i AlH, associated via Al–H^μ–Al bridges (band *e*) changes with a displacement to the red into the sharp bands of the freely-vibrating Al–H-stretching vibration when the autocomplex is converted by a donor solvent into a donor–aluminium hydride complex. The stronger the donor ($d < c < b < a$), the stronger is the *D*–A complex, the lower the strength of the Al–H^μ–Al bridge bond, the freer the Al–H vibration, and the sharper the Al–H band. (Sliding ordinate.)

10.3 Influence of donors on the frequency of the Al–H band

The strength of a donor is characterised not only by the half-width but also by the position of the Al–H band. Fig. 5.6 shows the position and width of some Al–H bands [*56, 57*]. The occupation of free orbitals of the aluminium by foreign (donor) electrons increases the screening of the positive aluminium nucleus and thus decreases the effective nuclear charge for the electrons of the Al–H bond. The more or less complete charge transfer from the donor to the aluminium decreases its ionization potential or electronegativity. Consequently, according to Gordy [*44*], the force constant f decreases: $f_{AB} = a(x_A x_B / r_{AB}^2)^{3/4} + b$ (in which f_{AB} = force constant, x_A and x_B = electronegativities of A and B, r_{AB} = bond distance of A–B, a and b = constants) and therefore also the Al–H frequency [*56*]. Unfortunately, the absolute magnitude of the Al–H valency vibrating completely freely and unaffected by Al ligands cannot be measured. It can be estimated (1) from the diatomic radicals AlH and BH, (2) from the force constant interpolated for AlH_3, (3) from the lowering of the Al–H frequency with the addition

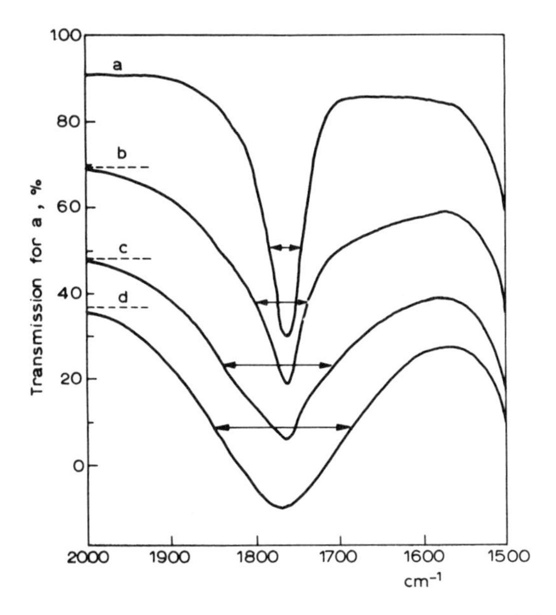

Fig. 5.5. Al–H-stretching vibrations of 10.5% of diisobutylalane by weight in mixtures of dibutyl ether and n-hexane[56]. Moles of alane/ether: (a) 1 : 9.0, (b) 1 : 2.4, (c) 1 : 0.9, (d) pure n-hexane. $Al-H^{\mu}-Al$ bridges are present in pure hexane (without the additions of the donor Bu_2O), as can be seen from the broad bands of the Al–H-stretching vibrations (band *d*). The addition of increasing amounts of diisobutyl ether leads to increasing weakening of $Al-H^{\mu}-Al$ bridges and the formation of ether–alane complexes, but the sharp final form of the Al–H band (band *a*) is reached only with an almost tenfold excess of ether. (Sliding ordinate.)

of increasing amounts of amine[57] ($AlH_3 \cdot NMe_3$, 1788 cm^{-1}; $AlH_3 \cdot 2NMe_3$, 1706 cm^{-1}; $\Delta = 82$ cm^{-1}; $1788 + 82 = 1870$ cm^{-1}). The free Al–H-stretching frequency is therefore in the region of 1850 ...*1870*... 1900 cm^{-1}.

10.4 *Influence of substituents on the frequency of the Al–H band*

The substitution of H in alane by substituents with a negative inductive effect such as OR, NR_2, Cl, Br, I, makes the aluminium positive. The Al–H bond therefore becomes stronger and the Al–H-stretching frequency higher. The $d_\pi - p_\pi$ back-coordination possible with all the negative substituents mentioned opposes this effect. In general, substituents with a positive inductive effect lower the Al–H stretching frequency. However, the latter band-displacement is generally not characteristic. Some values of $\nu_{\text{Al–H stretching}}$ [cm^{-1}] are given below for comparison: Cl substitution (in THF): $AlH_3 \cdot NEt_3$, 1740; $AlH_2Cl \cdot NEt_3$, 1793; $AlHCl_2 \cdot NEt_3$, 1848. Br substitution (solvent ?): $AlH_3 \cdot NEt_3$, 1779; $AlH_2Br \cdot NEt_3$, 1845; $AlHBr_2 \cdot NEt_3$, 1900[57]. OEt substitution: $AlHBu_2OEt_2$, 1772; AlHEt (OEt), 1824 (the different alkyl groups attached to the aluminium have little spectroscopic effect[56]: ν-$AlH_3 \cdot NEt_3 \sim \nu$-$AlHR_2 \cdot NEt_3$[57]).

In the case of alanes with oxygen and nitrogen substituents, hydrogen bridges are never observed spectroscopically (sharp Al–H bands). According to this, the

References p. 438

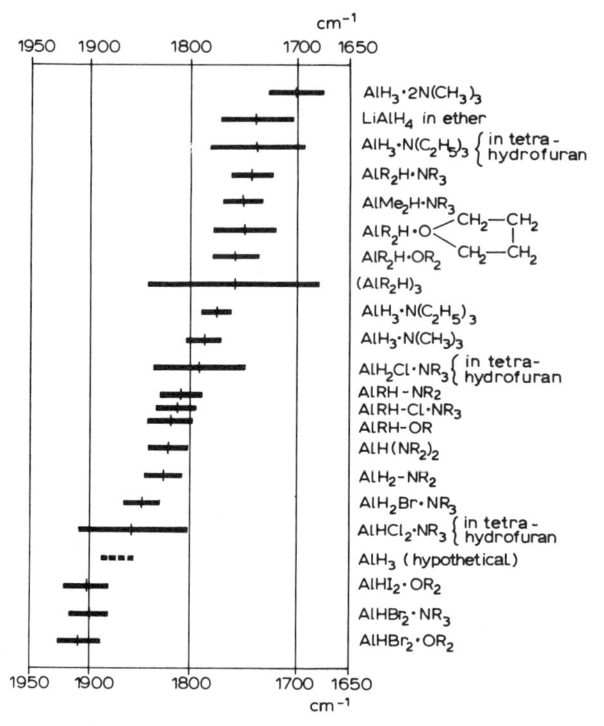

Fig. 5.6. Al–H-stretching frequency. Position and width of some Al–H bands [57]. The centres of the lines give the positions of the band maxima, and their lengths give the half-widths.

dimeric $[AlH_2(NR_2)]_2$ contains only N-bridges; on the other hand, chloroalanes can form H-bridges (broad Al–H bands) [57]:

10.5 Literature on spectra of alanes and related compounds

Abbreviations: *IR* = infrared spectrum, *NMR* = ^{1}H nuclear-resonance spectrum, *RA* = Raman spectrum.

Unsubstituted alane adducts	*Ref.*	
	IR	*NMR*
$AlH_3 \cdot \infty O(C_2H_5)_2$	26	—
$AlH_3 \cdot 4O(C_2H_5)_2$	26	—
$AlH_3 \cdot 2$dioxan	—	22
$AlH_3 \cdot \infty$THF	26, 89, 125	—

Unsubstituted alane adducts	*Ref.*	
	IR	*NMR*
$AlH_3 \cdot 2THF$	89, 125	22
$AlH_3 \cdot THF$	26, 173	—
$AlH_3 \cdot THF \cdot$dioxan	—	22
$AlH_3 \cdot THF \cdot OEt_2$	26	—
$AlH_3 \cdot 2NMe_3$	9a, 26, 39, 45c, 51, 173	45c
$AlD_3 \cdot 2NMe_3$	39	—
$AlH_3 \cdot NMe_3 \cdot THF$	—	22
$AlH_3 \cdot NMe_3$	26, 39, 45c, 173	45c, 104
$AlD_3 \cdot NMe_3$	39	104
$AlH_3 \cdot Me_2NCH_2NMe_2$	45c	45c
$AlD_3 \cdot Me_2NCH_2NMe_2$	45c	45c
$AlH_3 \cdot Me_2NCH_2CH_2NMe_3$	23	—
$AlH_3 \cdot Me_2N(CH_2)_3NMe_2$	45c	45c
$AlH_3 \cdot NMe_2(CH_2CH_2{=}CH_2)$	—	104

C-substituted alanes and alane adducts	*Ref.*		
	IR	*NMR*	*RA*
$AlH_2Me \cdot NMe_2$	—	95	—
$[AlHMe_2]_2$	—	61, 104	—
$AlHMe_2 \cdot NMe_3$	—	95	—
$AlMe_3$	45, 47, 53, 59, 61–63	17, 62, 104	62, 63
$[Al(CD_3)_3]_2$	45	—	—
$AlMe_3 \cdot NEt_3$	—	61	—
$AlH_2Et \cdot NMe_3$	—	95	—
$[AlHEt_2]_3$	—	63	—
$AlHEt_2 \cdot THF$	—	63	—
$AlHEt_2 \cdot NMe_3$	—	95	—
$AlHEt_2 \cdot NEt_3$	—	63	—
$[AlEt_3]_2$	53, 63	55, 61–6, 93, 169	—
$AlEt_3 \cdot OEt_2$	53	63	—
$AlEt_3 \cdot THF$	—	63	—
$2AlEt_3 \cdot NaF$	—	63	—

References p. 438

C-substituted alanes and alane adducts	*Ref.*		
	IR	*NMR*	*RA*
$[AlPr^n_3]_2$	53, 63	—	63
$AlPr^i_3$	—	117	—
$[AlBu^n_3]_2$	53, 63	—	—
$AlBu^i_3$	53, 62, 63	55, 62, 117	62
$Al(n\text{-}hex)_3$	63	—	—
$Al(i\text{-}oct)_3$	63	—	—
$Al(n\text{-}C_{10}H_{21})_3$	63	—	—
$Al(n\text{-}C_{12}H_{25})_3$	63	—	—

Halogen-substituted alanes and alane adducts	*Ref.*	
	IR	*NMR*
$[AlClMe_2]_2$	45, 53	17, 47, 59, 61
$[AlCl(CD_3)_2]_2$	45, 53	—
$[AlCl_2Me]_2$	45	47
$[AlCl_2(CD_3)]_2$	45	—
$[AlClEt_2]_2$	53	63, 169
$AlClEt_2 \cdot OR_2$	63	—
$AlHCl_2 \cdot$ether	79a	79a
ether:		
OMe_2, OEt_2		
OPr^i, OAm^i		
THF, $O(CH_2)_5$		
Dioxan		
Monoglyme		
$AlH_2Cl \cdot 2THF$	113a	113a
$AlH_2Br \cdot 2THF$	113a	113a
$AlH_2I \cdot 2THF$	113a	
$AlD_2Cl \cdot 2THF$	113a	
$AlD_2Br \cdot 2THF$	113a	
$AlD_2I \cdot 2THF$	113a	
$AlHCl_2 \cdot 2THF$	79a, 113a	79a, 113a
$AlHBr_2 \cdot 2THF$	113a	113a
$AlHI_2 \cdot 2THF$	113a	
$AlHClBr \cdot 2THF$	113a	

Halogen-substituted alanes and alane adducts	*Ref.*	
	IR	*NMR*
$[AlBrEt_2]_2$	—	169
$AlHBr_2 \cdot OR_2$	63	—
$[AlIEt_2]_2$	—	169
$AlHI_2 \cdot OR_2$	63	—
$[AlClBu^n_2]_2$	53	—
$[AlClBu^i_2]_2$	53	—

Chalcogen-substituted alanes and alane adducts	*Ref.*	
	IR	*NMR*
$AlMe_2(OMe)$	—	61
$AlMe_2(OEt)$	—	61
$AlMe_2(OBu^t)$	—	61
$[AlEt_2(OEt)]_2$	—	63
$[AlEt_2(OBu)]_2$	—	63
$AlH_2(SPr) \cdot NMe_3$	80	—
$AlH(SPr)_2 \cdot NMe_3$	80	—
$AlHCl(SPr) \cdot NMe_3$	80	—
$AlH(SPr)(NC_5H_{10})$	80	—
$AlH(SPh)_2 \cdot NMe_3$	80	—
$AlH_2(SPh) \cdot NMe_3$	80	—
$[AlHCl(OSiMe_3)]_2$	113	113

N- and P-substituted alanes and alane adducts	*Ref.*	
	IR	*NMR*
$[AlH_2(NMe_2)]_3$	—	104
$AlH(NMe_2)_2 \cdot 2BH_2(NMe_2)$	—	102
$[AlH_2PMe_2]_3$	39b	—

References p. 438

Boranate-substituted alanes and lithium alanate	*Ref.*		
	IR	*NMR*	*RA*
$Al(BH_4)_2(NMe_2)$	103	103	—
$AlH(BH_4)(NMe_2)$	103	103	—
$AlH_2(BH_4){\cdot}NMe_3$	103	103	—
$AlH(BH_4)_2{\cdot}NMe_3$	103	103	—
$Al(BH_4)_3{\cdot}NMe_3$	9c, 103	9c, 103	—
$Al(BH_4)_3{\cdot}PMe_3$	9c	9c	—
$Al(BH_4)_3{\cdot}AsMe_3$	9c	9c	—
$Al(BH_4)_3$	9c, 97	9c, 91 a,b; cit. 103	31
$Al(^{10}BD_4)_3$	—	—	31
$LiAlH_4$	25	42a, 93	—
$LiAlD_4$(quadrupole coupling [99a])			
$AlH_2(BH_4){\cdot}2THF$	89, 125	—	—
$AlH(BH_4)_2{\cdot}2THF$	89, 125	—	—
$AlH_2(BH_4){\cdot}THF$	89, 125	—	—
$AlH(BH_4)_2{\cdot}THF$	89, 125	—	—
$AlH(BH_4)_2{\cdot}OEt_2$	89, 125	—	—
$Al(BH_4)_3{\cdot}THF$	89. 125	—	—
$Al(BH_4)_3{\cdot}OMe_2$	9c	9c	—
$Al(BH_4)_3{\cdot}OEt_2$	9c, 89, 125	9c	—
$Al(BH_4)_3{\cdot}SMe_2$	9c	9c	—

REFERENCES

1. ALDER, B. J. AND R. H. CHRISTIAN, Phys. Rev. *104*, 550 (1956).
2. AMBERGER, E., Chem. Ber. *94*, 1447 (1961).
3. AMBERGER, E., Hydrides of Ge, Sn, Pb and Si, Thesis (in German), University of München 1961.
4. ANSUL CHEMICAL CORP., Marinette, Wisconsin, Bulletin, Febr. 1957.
5. APPEL, M. AND J. P. FRANKEL, J. Chem. Phys. *42*, 3984 (1965).

5a. ASHBY, E. C., Chem. Engineering News *1969*, 9.

5b. ASHBY, E. C. AND B. D. JAMES, private commun 1968.

6. ASHBY, E. C., J. Am. Chem. Soc. *86*, 1883 (1964).

6a. ASHBY, E. C. AND P. KOBETZ, Inorg. Chem. *5*, 1615 (1966)

6b. ASHCROFT, S. F., A. S. CARSON, W. CARTER AND P. G. LAYE, Trans. Faraday Soc. *61*, 22[illegible] (1965).

6c. ASHCROFT, S. F., A. S. CARSON AND F. B. PEDLEY, Trans. Faraday Soc. *59*, 2713 (1963).

7. BAILEY, W. J. AND C. R. PFEIFFER, J. Org. Chem. *20*, 1337 (1955).
8. BARBARAS, G., G. D. BARBARAS, A. E. FINHOLT AND H. I. SCHLESINGER, J. Am. Chem. Soc *70*, 877 (1948).
9. BAUKLOH, W. AND W. WENZEL, Arch. Eisenhuettenw. *11*, 278 (1937); C.A. *32*, 2066 (1938).

9a. BEATTIE, I. R. AND T. GILSON, J. Chem. Soc. *1964*, 3528.

9b. Block, F. and A. P. Gray, Inorg. Chem. *4*, 304 (1965).
9c. Bird, P. H. and M. G. H. Wallbridge, J. Chem. Soc. *1965*, 3923.
10. Bohlmann, F., Chem. Ber. *85*, 390 (1952).
11. Breisacher, P. and B. Siegel, J. Am. Chem. Soc. *87*, 4255 (1965).
11a. Breisacher, P. and B. Siegel, J. Am. Chem. Soc. *86*, 5053 (1964).
12. Brown, H. C. and R. F. McFarlin, J. Am. Chem. Soc. *78*, 252 (1956).
12a. Brown, H. C. and C. P. Garg, J. Am. Chem. Soc. *86*, 1085 (1964).
13. Brown, H. C. and R. F. McFarlin, J. Am. Chem. Soc. *80*, 5372 (1958).
14. Brown, H. C. and B. C. Subba Rao, J. Am. Chem. Soc. *80*, 5377 (1958).
14a. Brown, H. C. and C. F. Shoaf, J. Am. Chem. Soc. *86*, 1079 (1964).
15. Brown, H. C. and A. Tsukamoto, J. Am. Chem. Soc. *81*, 502 (1959).
16. Brown, H. C. and A. Tsukamoto, J. Am. Chem. Soc. *83*, 2016 (1961).
17. Brownstein, S., B. C. Smith, G. Erlich and A. W. Laubengayer, J. Am. Chem. Soc. *82*, 1000 (1960).
18. Burg, A. B. and K. Mödritzer, J. Inorg. Nucl. Chem. *13*, 318 (1960).
18a. Carson, A. S., W. Carter and F. B. Pedley, Proc. Roy. Soc. *A260*, 550 (1961).
19. Chattoraj, S. C., C. A. Hollingsworth, D. H. McDaniel and G. B. Smith, J. Inorg. Nucl. Chem. *24*, 101 (1962).
19a. Chini, P., L. Gatto, A. Baradel and C. Vacca, Chimica l'industria *47*, 642 (1965).
20. Clasen, H. (Metallgesellschaft A. G., Frankfurt), Germ. Pat. Appl. M 46047 IVa/12i (1960).
21. Clasen, H., Angew. Chem. *73*, 322 (1961).
21a. Coates, G. E., *Organo-Metallic Compounds*, 2nd Ed., Methuen and Co. Ltd. London, John Wiley and Sons Inc. New York, 1960, pp. 94, 132. Or 3rd Ed. Vol. I, 1967, pp. 193, 300.
21b. Cuicinella, S., Euchem Conference on the Chemistry of Metal Hydrides, Bristol 1966.
22. Dautel, R. and W. Zeil, Z. Elektrochem. *62*, 1132 (1958).
23. Davidson, J. M. and T. Wartik, J. Am. Chem. Soc. *82*, 5506 (1960).
24. Davis, W. D., L. S. Mason and G. Stegman, J. Am. Chem. Soc. *71*, 2775 (1949).
25. D'Or, L. and J. Fuger, Bull. Soc. Roy. Sci. Liege *25*, 14 (1956); C. A. *50*, 11114a (1956).
26. Ehrlich, R., A. R. Young II, B. M. Liechstein and D. D. Perry, Inorg. Chem. *2*, 650 (1963).
26a. Ehrlich, R., A. R. Young II, G. Rice, F. Dvorak, P. Shapiro and H. F. Smith, J. Am. Chem. Soc. *88*, 858 (1966).
27. Ehrlich, R., A. R. Young II, B. M. Liechstein and D. D. Perry, Inorg. Chem. *3*, 628 (1964).
27a. Ehrlich, R., A. R. Young II and D. D. Perry, Inorg. Chem. *4*, 758 (1965).
27b. Ehrlich, R., C. B. Parisek and G. Rice, Inorg. Chem. *4*, 1075 (1965).
28. Eliel, E. L. and D. W. Delmonte, J. Am. Chem. Soc. *80*, 1744 (1958).
29. Eliel, E. L. and M. N. Rerick, J. Am. Chem. Soc. *82*, 1362 (1960).
30. Eliel, E. L. and M. N. Rerick, J. Am. Chem. Soc. *82*, 1367 (1960).
31. Emery, A. R. and R. C. Taylor, Spectrochim. Acta *16*, 1455 (1960).
32. Evans, G. G. and F. P. Delgreco, J. Inorg. Nucl. Chem. *4*, 48 (1957).
33. Evans, G. G., J. K. Kennedy, Jr. and F. P. Delgreco, J. Inorg. Nucl. Chem. *4*, 40 (1957).
34. Fensham, P. J., J. Inorg. Nucl. Chem. *23*, 139 (1961).
35. Finholt, A. E., G. D. Barbaras, G. K. Barbaras, G. Urry, T. Wartik and H. I. Schlesinger, J. Inorg. Nucl. Chem. *1*, 317 (1955).
36. Finholt, A. E., A. C. Bond, Jr. and H. I. Schlesinger, J. Am. Chem. Soc. *69*, 1199 (1947).
37. Finholt, A. E., C. Helling, V. Imhof, L. Nielsen and E. Jacobson, Inorg. Chem. *2*, 504 (1963).
38. Finholt, A. E., R. Nystrom, W. G. Brown and H. I. Schlesinger, 110th Meeting Am. Chem. Soc. 10-9-1946.
39. Fraser, G. W., N. N. Greenwood and B. P. Straughan, J. Chem. Soc. *1963*, 3742.
39a. Fritz. G., Angew. Chem. *78*, 80 (1966).
39b. Fritz. G. and G. Trenczek, Z. anorg. allg. Chem. *331*, 206 (1964).
39c. Fritz. G. and G. Trenczek, Z. anorg. allg. Chem. *313*, 236 (1961).
40. Fritz, G. and G. Trenczek, Angew. Chem. *75*, 723 (1963).
41. Gál, G., I. Foldesi and I. Krasznai, Magyar Kém. Folyóirat, *63*, 5 (1958); C. A. *52*, 18308f (1958).
42. Garner, W. E. and E. W. Haycock, Proc. Roy. Soc. (London) *A211*, 335 (1952).
42a. Garsten, A., Phys. Rev. *79*, 397 (1950).
43. Gillespie, R. J., J. Chem. Soc. *1952*, 1002.

44. Gordy, W., J. Chem. Phys. *14*, 305 (1946).
45. Gray, A. P., Can. J. Chem. *41*, 1511 (1963).
45a. Greenwood, N. N., B. P. Straughan and B. S. Thomas, J. Chem. Soc. *A1968*, 1248.
46. Greinacher, E., W. Lüttke and R. Mecke, Z. Elektrochem. *59*, 23 (1955).
47. Gronewege, M. P., J. Smidt and H. de Vries, J. Am. Chem. Soc. *82*, 4425 (1960).
48. Guild, L. V., C. A. Hollingsworth, D. H. McDaniel and S. K. Podder, Inorg. Chem. *2*, 921 (1963).
49. Hayashi, T. and T. Ishida, Bull. Univ. Osaka Prefect., Ser. A *7*, 55 (1959); C. A. *54*, 7398a (1960).
50. Heitsch, C. W., Nature *195*, 995 (1962).
51. Heitsch, C. W. and R. N. Kniseley, Spectrochim. Acta *19*, 1385 (1963).
52. Heitsch, C. W., C. E. Nordman and R. W. Parry, Inorg. Chem. *2*, 508 (1963).
53. Hoffmann, E. G., Z. Elektrochem. *64*, 616 (1960).
54. Hoffman, C. J., *Aluminium Hydride*, LMSD-703150, Lockheed Missiles and Space Division, Sunnyvale, Calif., ASTIA No. AD 244583, Aug. 1960.
55. Hoffmann, E. G., Z. Elektrochem. *64*, 144 (1960).
56. Hoffmann, E. G., Z. Elektrochem. *61*, 1101 (1957).
57. Hoffmann, E. G., Z. Elektrochem. *61*, 1110 (1957).
58. Hoffmann, E. G., Ann. *629*, 104 (1960).
59. Hoffmann, E. G., Angew. Chem. *73*, 444 (1961).
60. Hoffmann, E. G., Angew. Chem. *73*, 578 (1961).
61. Hoffmann, E. G., Trans. Faraday Soc. *58*, 1 (1962).
62. Hoffmann, E. G., Bull. Soc. Chim. France *1963*, 1467.
63. Hoffmann, E. G. and G. Schomburg, *Advances in Molecular Spectroscopy*, Pergamon, Oxford, 1962, p. 804.
64. Hoffmann, E. G. and G. Schomburg, *Hydrogen Bonding*, Pergamon, Oxford, 1959, p. 509.
65. Hoffmann, E. G. and W. Tornau, Z. Anal. Chem. *186*, 231 (1962).
66. Hoffmann, E. G. and W. Tornau, Z. Anal. Chem. *188*, 321 (1962).
67. Jander, G. and K. Kraffczyk, Z. Anorg. Allgem. Chem. *283*, 217 (1956).
67a. Jacobs, T. L. and R. D. Wilcox, J. Am. Chem. Soc. *86*, 2240 (1964).
68. Jones, J. I. and W. S. McDonald, Proc. Chem. Soc. *1962*, 366.
69. Klanberg, F., Proc. Chem. Soc. *1961*, 203.
70. Kobetz, P., W. E. Becker, R. C. Pinkerton and J. B. Honeycutt, Jr., Inorg. Chem. *2*, 859 (1963).
71. Köster, R., Germ. Pat. 1,024,062 (1958); C. A. *54*, 7087c (1960).
72. Lansbury, P. T., J. Am. Chem. Soc. *83*, 429 (1961).
73. Lansbury, P. T. and J. O. Peterson, J. Am. Chem. Soc. *84*, 1756 (1962).
74. Lansbury, P. T. and J. O. Peterson, J. Am. Chem. Soc. *83*, 3537 (1961).
75. Lansbury, P. T. and J. O. Peterson, J. Am. Chem. Soc. *85*, 2236 (1963).
76. Lewis, P. H. and R. E. Rundle, J. Chem. Phys. *21*, 986 (1953).
77. Longuet-Higgins, H. C., J. Chem. Soc. *1946*, 139.
78. McLure, I. and T. D. Smith, J. Inorg. Nucl. Chem. *19*, 170 (1961).
79. Malter, D. J., J. H. Wotiz and C. A. Hollingsworth, J. Am. Chem. Soc. *78*, 1311 (1956).
79a. Marconi, W., A. Mazzei, S. Cucinella and M. Greco, Annali di Chimica *55*, 897 (1965).
79b. Marconi, W., A. Mazzei, S. Cucinella and M. de Maladé, Makromolekulare Chem. *71* 134 (1964).
80. Marconi, W., A. Mazzei, F. Bonati and M. de Maldè, Z. Naturforsch. *18b*, 3 (1963).
80a. Markova, V. V., V. A. Kormer and A. A. Petrov, J. Gen.. Chem. USSR (Engl. Transl.) *35*, 445 (1965).
81. Martin, J. F., A. J. Neale and H. S. Turner, J. Chem. Soc. *1956*, 4428.
82. Mikheeva, V. I., M. S. Selivokhina and O. N. Kryukova, Dokl. Akad. Nauk SSSR, *109*, 541 (1956); C. A. *51*, 9273e (1957).
83. Mikheeva, V. I., M. S. Selivokhina and V. V. Leonova, Zh. Neorgan. Khim. *4*, 2436 (1959); C. A. *54*, 15043e (1960).
84. Minsker, K. S., V. I. Biryukov, A. I. Graevskii and G. A. Razuvaev, Bull. Acad. Sci. USSR, Div. Chem. Sci. (English Transl.) *1963*, 572.
85. Natta, G., P. Pino, G. Mazzanti, P. Long and F. Bernardini, J. Am. Chem. Soc. *81*, 2561 (1959).
86. Neumann, W. P., Ann. *629*, 23 (1960).
87. Neumann, W. P., Ann. *667*, 1 (1963).

88. Newman, M. S. and T. Fukunaga, J. Am. Chem. Soc. *82*, 693 (1960).
89. Nöth, H. and H. Suchy, unpublished.
89a. Nöth, H. and E. Wiberg, Fortschr. chem. Forschung, Vol. **8**, 3, p. 327–436, Springer Verlag, Berlin/Heidelberg/New York, 1967.
90. Neumaier, H., Thesis, University of München, 1961.
91. Neumaier, H., D. Büchel and G. Ziegelmaier, private commun.
91a. Ogg, R. A. and J. D. Ray, Discussions Faraday Soc. *19*, 215 (1955).
91b. Ogg, R. A. and J. D. Ray, Discussions Faraday Soc. *19*, 234 (1955).
92. Olson, W. M. and R. T. Sanderson, J. Inorg. Nucl. Chem. *7*, 228 (1958).
93. O'Reilly, D. E., J. Chem. Phys. *32*, 1007 (1960).
94. Paddock, N. L., Nature *167*, 1070 (1951).
95. Peters, F. M., B. Bartocha and A. J. Bilbo, Can. J. Chem. *41*, 105 (1963).
96. Peters, F. M. and N. R. Fetter, J. Organometal. Chem. *4*, 18 (1965).
97. Price, W. C., J. Chem. Phys. *17*, 1044 (1949).
98. Pitzer, K. S. and H. S. Gutowski, J. Am. Chem. Soc. *68*, 2204 (1946).
99. Pitzer, K. S. and R. K. Sheline, J. Chem. Phys. *16*, 552 (1948).
99a. Pyykkö, P. and B. Pedersen, Chem. Phys. Letters *2*, 297 (1968).
100. Rice, Jr., M. J., *Non-Solvated Aluminium Hydride*, U.S. Office Naval Res., Contr. ONR-494(04), ASTIA No. 106967, Aug. 1956.
101. Ruff, J. K., J. Am. Chem. Soc. *83*, 1798 (1961).
102. Ruff, J. K., Inorg. Chem. *1*, 612 (1962).
103. Ruff, J. K., Inorg. Chem. *2*, 515 (1963).
104. Ruff, J. K. and M. F. Hawthorne, J. Am. Chem. Soc. *82*, 2141 (1960).
105. Ruff, J. K. and M. F. Hawthorne, J. Am. Chem. Soc. *83*, 535 (1961).
106. Rundle, R. E., J. Am. Chem. Soc. *69*, 1327 (1947).
107. Rundle, R. E., J. Chem. Phys. *17*, 671 (1949).
108. Schaeffer, G. W., J. S. Roscoe and A. C. Stewart, J. Am. Chem. Soc. *78*, 729 (1956).
109. Schlesinger, H. I. and A. E. Finholt, U.S. Pat. 2,576,311 (1947); C. A. *46*, 2761i (1952).
110. Schlesinger, H. I. and A. E. Finholt, U.S. Pat. 2,567,972 (1946); C. A. *46*, 2762i (1952).
111. Schlesinger, H. I., R. T. Sanderson and A. B. Burg, J. Am. Chem. Soc. *61*, 536 (1939).
112. Schlesinger, H. I., R. T. Sanderson and A. B. Burg, J. Am. Chem. Soc. *62*, 3421 (1940).
113. Schmidbauer, H. and F. Schindler, Chem. Ber. *97*, 952 (1964).
113a. Schmidt, D. L. and E. E. Flagg, Inorg. Chem. *6*, 1262 (1967).
114. Schmitz-Dumont, O. and V. Habernickel, Chem. Ber. *90*, 1054 (1957).
115. Schomburg, G., Thesis, University of Aachen, 1956.
116. Schwab, W. and K. Wintersberger, Z. Naturforsch. *8b*, 690 (1953).
117. Shiner, Jr., V. J., D. Whittaker and V. P. Fernandez, J. Am. Chem. Soc. *85*, 2318 (1963).
118. Siegel, B., J. Am. chem. Soc. *82*, 1535 (1960).
118a. Skalar, N. and B. Post, Inorg. Chem. *6*, 669 (1967).
119. Smith, G. B., D. H. McDaniel, E. Biehl and C. A. Hollingsworth, J. Am. Chem. Soc. *82*, 3560 (1960).
120. Smith, M. B. and G. E. Bass, Jr., J. Chem. Eng. Data *8*, 342 (1963).
121. Stecher, O. and E. Wiberg, Ber. *75*, 2003 (1942).
122. Stein, V. and R. Joppen (Metallgesellschaft A. G.), Germ. Pat. 947,702 (1954); C.A. *53*, 2551 (1959).
122a. Sterlyadkina, Z. K., N. N. Mal'tseva, O. N. Kryukova and V. I. Mikheeva, Russ. J. Inorg. Chem. (English Transl.) *11*, 531 (1966).
123. Stevens, C. L. and T. H. Coffield, J. Am. Chem. Soc. *80*, 1919 (1958).
124. Stewart, A. C. and G. W. Schaeffer, J. Inorg. Nucl. Chem. *3*, 194 (1956).
124a. Storr, A., J. Chem. Soc. *A1969*, 2605.
125. Suchy, H., Thesis, University of München, 1965.
126. Sujishi, S. and J. N. Keith, J. Am. Chem. Soc. *80*, 4138 (1958).
126a. Turley, J. W. and H. W. Rinn, Inorg. Chem. *8*, 18 (1969).
127. Traylor, T. G., Chem. Ind. London *1959*, 1223.
127a. Wells, A. F., *Structural Inorganic Chemistry*, 3rd Ed., Clarendon Press, Oxford, 1962, p. 382.
127b. Wade, K. and B. K. Wyatt, J. Chem. Soc. *A1969*, 1121.
128. Wiberg, E. and R. Bauer, Z. Naturforsch. *5b*, 397 (1950).
129. Wiberg, E. and R. Bauer, Z. Naturforsch. *7b*, 131 (1952).
130. Wiberg, E., R. Bauer, M. Schmidt and R. Usón, Z. Naturforsch. *6b*, 393 (1951).

131. Wiberg, E. and W. Gösele, Z. Naturforsch. *10b*, 236 (1955).
132. Wiberg, E. and W. Gösele, Z. Naturforsch. *11b*, 485 (1956).
133. Wiberg, E., H. Graf, M. Schmidt and R. Usón, Z. Naturforsch. *7b*, 578 (1952).
134. Wiberg, E., H. Graf and R. Usón, Z. Anorg. Allgem. Chem. *272*, 221 (1953).
135. Wiberg, E. and W. Henle, Z. Naturforsch. *6b*, 461 (1951).
136. Wiberg, E. and W. Henle, Z. Naturforsch. *6b*, 461 (1951).
137. Wiberg, E. and W. Henle, Z. Naturforsch. *7b*, 249 (1952).
138. Wiberg, E. and W. Henle, Z. Naturforsch. *7b*, 250 (1952).
139. Wiberg, E. and W. Henle, Z. Naturforsch. *7b*, 250 (1952).
140. Wiberg, E., W. Henle and R. Bauer, Z. Naturforsch. *6b*, 393 (1951).
141. Wiberg, E. and A. Jahn, Z. Naturforsch. *7b*, 580 (1952).
142. Wiberg, E. and A. Jahn, Z. Naturforsch. *7b*, 581 (1952).
143. Wiberg, E. and A. May, Z. Naturforsch. *10b*, 229 (1955).
144. Wiberg, E. and A. May, Z. Naturforsch. *10b*, 230 (1955).
145. Wiberg, E. and A. May, Z. Naturforsch. *10b*, 232 (1955).
146. Wiberg, E. and A. May, Z. Naturforsch. *10b*, 234 (1955).
147. Wiberg, E., A. May and H. Nöth, Z. Naturforsch. *10b*, 239 (1955).
148. Wiberg, E. and H. Michaud, Z. Naturforsch. *9b*, 495 (1954).
149. Wiberg, E. and K. Mödritzer, Z. Naturforsch. *12b*, 123 (1957).
150. Wiberg, E. and K. Mödritzer, Z. Naturforsch. *12b*, 127 (1957).
151. Wiberg, E. and K. Mödritzer, Z. Naturforsch. *12b*, 128 (1957).
152. Wiberg, E. and K. Mödritzer, Z. Naturforsch. *12b*, 131 (1957).
153. Wiberg, E. and K. Mödritzer, Z. Naturforsch. *12b*, 135 (1957).
154. Wiberg, E. and G. Müller-Schiedmayer, Chem. Ber. *92*, 2372 (1959).
155. Wiberg, E. and H. Neumaier, Inorg. Nucl. Chem. Letters *1*, 35 (1965).
156. Wiberg, E. and H. Neumaier, Z. Anorg. Allgem. Chem. *340*, 189 (1965).
157. Wiberg, E. and H. Nöth, Z. Naturforsch. *10b*, 237 (1955).
158. Wiberg, E., H. Nöth and R. Usón, Z. Naturforsch. *11b*, 486 (1956).
159. Wiberg, E., H. Nöth and R. Usón, Z. Naturforsch. *11b*, 487 (1956).
160. Wiberg, E. and M. Schmidt, Z. Naturforsch. *6b*, 458 (1951).
161. Wiberg, E. and M. Schmidt, Z. Naturforsch. *6b*, 459 (1951).
162. Wiberg, E. and M. Schmidt, Z. Naturforsch. *6b*, 460 (1951).
163. Wiberg, E. and M. Schmidt, Z. Naturforsch. *6b*, 333 (1951).
164. Wiberg, E., M. Schmidt and A. G. Galinos, Angew. Chem. *66*, 443 (1954).
165. Wiberg, E. and M. Schmidt, Z. Naturforsch. *7b*, 59 (1952).
166. Wiberg, E. and O. Stecher, Angew. Chem. *52*, 372 (1939).
167. Wilke, G. and H. Müller, Chem. Ber. *89*, 444 (1956).
168. Wilke, G. and H. Müller, Ann. *618*, 267 (1958).
169. Yamamoto, O., Bull. Chem. Soc. Japan *36*, 1463 (1963).
169a. Young II, A. R. and R. Ehrlich, Inorg. Chem. *4*, 1358 (1965).
170. Zakharkin, L. I. and V. V. Gavrilenko, Bull. Acad. Sci. USSR, Div. Chem. Sci. (English Transl.) *1962*, 1076.
170a. Zakharkin, L. I. and V. V. Gavrilenko, Russ. J. Inorg. Chem. (English Transl.) *11*, 529 (1966).
171. Zakharkin, L. I. and V. V. Gavrilenko, J. Gen. Chem. USSR (English Transl.) *32*, 688 (1962).
171a. Zakharkin, L. I. and V. V. Gavrilenko, Dokl. Akad. Nauk. SSSR *145*, 145,793 (1962).
172. Zakharkin, L. V., V. V. Gavrilenko and I. M. Khorlina, Bull. Acad. Sci. USSR, Div. Chem. Sci. (English Transl.) *1962*, 402.
172a. Zakharkin, L. V., D. N. Maslin and V. V. Gavrilenko, Bull. Acad. Sci. USSR, Div. Chem. Sci. (English Transl.) *1964*, 521.
172b. Zakharkin, L. V. and L. A. Savina, Bull. Acad. Sci. USSR, Div. Chem. Sci. (English Transl.) *1964*, 1600.
173. Zeil, W., R. Dautel and W. Honsberg, Z. Elektrochem. *60*, 1131 (1956).
174. Ziegler, K., Angew. Chem. *64*, 323 (1952).
175. Ziegler, K., European Scientific Notes 6 (13), 178 (1952).
176. Ziegler, K., H.-G. Gellert, H. Martin, K. Nagel and J. Schneider Ann. *589*, 91 (1954).
177. Ziegler, K., H. Lehmkuhl and E. Lindner, Chem. Ber. *92*, 2320 (1959).
178. Ziegler, K., R. Köster, H. Lehmkuhl and R. Reinert, Ann. *629*, 33 (1960).

CHAPTER 6

GALLIUM, INDIUM AND THALLIUM HYDRIDES

1. Gallium hydrides

1.1 Historical and general review

The first indication of the formation of a volatile gallium hydride was given in 1881 by experiments of Lecoq de Boisbaudran[*16*]. When gallium(II) chloride was introduced into water, a gas with a repulsive odour was evolved with loss of gallium from the solution.

Like its lighter homologues aluminium (sections 3.1 and 3.2 of Chapter 5) and boron (section 3.1.1 of chapter 4) and its heavier homologues indium (sections 2.2.1 and 2.2.2) and thallium (section 3.2.1), gallium forms a trihydride according to the general equation:

$$3LiEH_4 + EX_3 \xrightarrow{\text{diethyl ether}} 4EH_3 + 3LiX \quad (E = B, Al, Ga, In, Tl)$$

As in the case of boron and aluminium, alkyl derivatives of gallium trihydride are known (section 1.2.1).

In accordance with the increasing tendency to polymerization from borane, BH_3, to thallane, TlH_3, BH_3 is still in the monomeric, non-polymerizing form in solutions in diethyl ether. With diethyl ether at low temperatures (GaH_3, 0°; InH_3, $-30°$), gallane, GaH_3, and indane, InH_3, form solutions which are stable for some time but from which polygallane, $(GaH_3)_x$, and polyindane, $(InH_3)_x$, precipitate slowly at higher temperatures. Even at low temperature thallane, TlH_3, precipitates from ethereal solution in the form of polythallane, $(TlH_3)_x$, soon after its formation.

No gallium monohydride, $(GaH)_x$, is yet known, although polymeric hydrides of the heavy elements of the IIIrd main group with the oxidation stage 1, $(EH)_x$, are generally more thermally stable than those of the oxidation stage 3, $(EH_3)_x$. In this respect, therefore, gallium resembles aluminium, which forms a hydride $AlH_{\sim 1}$ from the elements only with difficulty (*cf.* section 3.1 of Chapter 5). The hydrides of oxidation stage 1 are formed from the trihydrides by the splitting off of hydrogen:

$$(EH_3)_x \rightarrow (EH)_x + xH_2 \quad (E = B, In, Tl) \qquad \text{(a)}$$

or from the elements (so far known only in the case of aluminium):

$$Al + H \rightarrow AlH \qquad \text{(b)}$$

References p. 460

$(BH)_x$ is formed according to eq. (a) only when electrical or thermal energy is supplied, $(InH)_x$ forms even at 20°, and $(TlH)_x$ at temperatures as low as 0°. The stability of $(EH_3)_x$ and also of $(EH)_x$ depends markedly on the degree of polymerization (magnitude of x). Thus, the oligomeric, oily $(GaH_3)_n$ [section 1.2.2(3)] decomposes slowly even at −15°, while the highly polymeric $(GaH_3)_x$ [section 1.2.2(4)] is stable up to 130°. In the form of the trimethylamine adduct $GaH_3 \cdot NMe_3$ or the trimethylphosphane adduct $GaH_3 \cdot PMe_3$ [section 1.2.2(1)], even the monomeric gallane GaH_3 is still quite stable up to the melting point of the adduct (70 or 71°, respectively).

Corresponding to the formation of the ions $BH_4^{\ominus}$ and $AlH_4^{\ominus}$, the acceptors GaH_3, InH_3, and TlH_3 also add the donor $H^{\ominus}$. The reaction of lithium hydride with the trihalides forms, via the intermediate trihydride, the anions GaH_4 (section 1.2.4), $InH_4^{\ominus}$ (section 2.2.3) and $TlH_4^{\ominus}$ (section 3.2.2):

$$EX_3 \xrightarrow[-3LiX]{+3LiH} EH_3 \xrightarrow{+H^{\ominus}} EH_4^{\ominus}$$

The stability of the compound $LiEH_4$ decreases from $LiBH_4$ to $LiTlH_4$. While $LiBH_4$ and $LiAlH_4$ still have little tendency to decompose as follows

$$EH_4^{\ominus} \rightarrow \frac{1}{x}(EH_3)_x + H^{\ominus}$$

from solutions of $LiGaH_4$, $LiInH_4$, and $LiTlH_4$ in ether, polymeric $(EH_3)_x$ precipitates with increasing readiness.

Gallium, indium, and thallium, like aluminium, form thermally unstable boranates: $Ga(BH_4)_3$ (section 1.2.5), $In(BH_4)_3$ (section 2.2.4) and $Tl(BH_4)_3$ (section 3.2.3). The predominantly ionic thallium(I) boronate, $TlBH_4$, (section 3.2.3) is, however, more stable. The alanates $E(AlH_4)_3$ are still more unstable than the boronates (sections 1.2.6, 2.2.5, and 3.2.4).

As far as is known, the chemical reactions of the gallium hydrides GaR_2H and $GaCl_2H$ (section 1.3) roughly resemble those of AlH_3.

Like AlH_3, GaH_3 undergoes typical donor–acceptor reactions. Thus, the GaH_3 can be liberated from the fairly stable trimethylamine adduct $GaH_3 \cdot NMe_3$ by BF_3 (see section 1.3). In the presence of Me_2S, $GaH_3 \cdot SMe_2$, which is somewhat less stable than the amine adduct, forms (*cf.* section 1.3):

$$GaH_3 \cdot NMe_3 \xrightarrow[-BF_3 \cdot NMe_3]{+BF_3} GaH_3 \xrightarrow{+Me_2S} GaH_3 \cdot SMe_2$$

The Ga–H bond adds to the nucleophilic C=C double bond (*cf.* section 1.3):

$$>C=C< \; + \; >Ga-H \longrightarrow \underset{\downarrow \; >Ga-H}{>C=C<} \longrightarrow \underset{>Ga \cdots H}{>C \cdots C<} \longrightarrow \underset{>Ga \quad H}{-\overset{|}{\underset{|}{C}}-\overset{|}{\underset{|}{C}}-}$$

Because of the pronounced hydride nature of the H-atom [electro-negativities: Al = 1.47, Ga = 1.82, In = 1.49, Tl = 1.44, H = 2.20 (Pauling units)], molecular hydrogen is formed with protons. Thus, for example, the hydrides of Ga, In, and Tl hydrolyse vigorously with water in the following way (intermolecular elimination of hydrogen):

$$>E^{\delta+}\text{–}H^{\delta-} + H^{\delta+}\text{–}OH^{\delta-} \rightarrow\ >E\text{–}OH + H_2$$

If the gallane contains positive and negative hydrogen, as in $GaH_3{\cdot}NMe_2H$ and $GaH_3{\cdot}PMe_2H$, hydrogen is eliminated intramolecularly:

$$GaH_3{\cdot}EMe_2H \rightarrow GaH_2EMe_2 + H_2$$

For a review of gallium hydride and its derivatives see [*7a*].

1.2 Preparation of gallium hydrides

1.2.1 Synthesis of alkylgallanes $(GaH_{3-n}R_n)_m$

All the synthesis of alkylated gallium hydrides start from trialkylgalliums GaR_3 or $(GaR_3)_2$. Consequently, the preparation of these will be briefly described. They arise (1) by the oxidative alkylation of gallium with dialkylmercury compounds [eq. (1)] [*4, 31*]; (2) by the alkylation of gallium trichloride with alkyl compounds, *e.g.* those of zinc, magnesium, and aluminium [eqs. (2a)–(2d)] [*4, 5, 14, 15, 17, 18, 31*]; (3) by the transalkylation of triisobutylgallium with higher alkenes [eq. (3)] [*5*]:

$$2Ga + 3HgR_2 \rightarrow 2GaR_3 + 3Hg \qquad (1)$$
$$2GaCl_3 + 3ZnR_2 \rightarrow 2GaR_3 + 3ZnCl_2 \qquad (2a)$$
$$GaCl_3 + 3RMgCl \rightarrow GaR_3 + 3MgCl_2 \qquad (2b)$$
$$2GaCl_3 + 3AlR_3 \rightarrow GaR_3 + Ga(AlR_2Cl_2)_3 \qquad (2c)$$
$$GaCl_3 + 3AlR_3 + 3KCl \rightarrow GaR_3 + 3K(AlR_2Cl_2) \qquad (2d)$$
$$Ga(CH_2\text{–}CHMe_2)_3 + 3CH_2{=}CHR \rightarrow Ga(CH_2CH_2R)_3 + 3CH_2{=}CMe_2 \qquad (3)$$

(1) *Synthesis of methyldigallanes, $Ga_2H(CH_3)_5$ and $Ga_2H_2(CH_3)_4$, by the hydrogenation of $Ga(CH_3)_3$.* When a mixture of trimethylgallium, $GaMe_3$, and hydrogen is subjected to the glow discharge [*27, 29, 30*], pentamethyldigallane $Ga_2HMe_5 = GaMe_3{\cdot}GaMe_2H$ [*27*] and tetramethyldigallane $Ga_2H_2Me_4 = GaMe_2H{\cdot}GaMe_2H$ [*29, 30*] are formed, depending of course, on the reaction conditions

$$GaMe_3 + H_2 \xrightarrow[\text{discharge}]{\text{electric}} GaMe_2H + MeH$$

In addition, there is a solid brown product igniting in air, which has not been investigated further, and a mixture of hydrocarbons [*30*].

(2) *Synthesis of diethylgallane, $GaH(C_2H_5)_2$, by the hydrogenation of $GaCl(C_2H_5)_2$.* Triethylgallium, $GaEt_3$, and gallium trihalides, GaX_3 (X = Cl, Br), commute to give diethylgallium halide $GaXEt_2$ [*6*]:

$$2GaEt_3 + GaX_3 \rightarrow 3GaEt_2X$$

The latter can be hydrogenated with diethylalane to diethylgallane in 25 to 30% yield[6]:

$$GaEt_2Cl + AlEt_2H + KCl \rightarrow GaEt_2H + K(AlEt_2Cl_2)$$

1.2.2 Synthesis of gallanes $(GaH_3)_n$

(1) *Synthesis of the donor adducts of the monomeric hydride GaH_3 and the deuteride GaD_3 with $N(CH_3)_3$, $N(CH_3)_2H$, $P(CH_3)_3$, $P(CH_3)_2H$, and $P(C_6H_5)_3$.* With trimethylammonium chloride $NMe_3 \cdot HCl$ or $NMe_3 \cdot DCl$ in diethyl ether, lithium gallanate, $LiGaH_4$ ($LiGaD_4$) (section 1.2.4), forms the gallanetrimethylamine adduct $GaH_3 \cdot NMe_3$ ($GaD_3 \cdot NMe_3$), which decomposes only slowly at 20°[11, 26]:

$$LiGaH_4 + NMe_3 \cdot HCl \xrightarrow[-78°/0°]{\text{diethyl ether}} GaH_3 \cdot NMe_3 + LiCl + H_2$$

$$LiGaD_4 + NMe_3 \cdot DCl \xrightarrow[-78°/0°]{\text{diethyl ether}} GaD_3 \cdot NMe_3 + LiCl + D_2$$

According to tensimetric measurements, $GaH_3 \cdot NMe_3$ adds a second molecule of NMe_3 between −45 and 23°[26]:

$$GaH_3 \cdot NMe_3 + NMe_3 \rightarrow GaH_3 \cdot 2NMe_3$$

$GaH_3 \cdot 2\ NMe_3$ is less stable than the corresponding aluminium compound $AlH_3 \cdot 2NMe_3$[11, 26]. Donor displacement by the stronger donor dimethylamine, Me_2NH, from the gallane-trimethylamine adduct $GaH_3 \cdot NMe_3$ ($GaD_3 \cdot NMe_3$) gives rise to the colourless oily gallane dimethylamine adduct, which is not stable for long at 20°[9]:

$$GaH_3 \cdot NMe_3 + NMe_2H \xrightarrow[-196°/-10°]{} GaH_3 \cdot NMe_2H + NMe_3$$

$$GaD_3 \cdot NMe_3 + NMe_2H \xrightarrow[-196°/-10°]{} GaD_3 \cdot NMe_2H + NMe_3$$

Preparative amounts of $GaH_3 \cdot NMe_2H$ are obtained in 56% yield by the reaction of lithium gallanate with dimethylammonium chloride[9]:

$$LiGaH_4 + NMe_2H \cdot HCl \xrightarrow[-78°/0°]{\text{diethyl ether}} GaH_3 \cdot NMe_2H + LiCl + H_2$$

Equimolecular amounts of lithium gallanate, $LiGaH_4$ ($LiGaD_4$), form with trimethylphosphane, PMe_3, or triphenylphosphane, PPh_3, and hydrogen chloride, the stable gallane-trimethylphosphane adduct $GaH_3 \cdot PMe_3$ (m.p. 70.5–71°) ($GaD_3 \cdot PMe_3$)[8] and the gallane-triphenylphosphane adduct, which is unstable at room temperature even in ethereal solution[8]:

$$LiGaH_4 + PMe_3 + HCl \xrightarrow[0°]{\text{diethyl ether}} GaH_3 \cdot PMe_3 + LiCl + H_2$$

$$LiGaD_4 + PMe_3 + HCl \xrightarrow[0°]{\text{diethyl ether}} GaD_3 \cdot PMe_3 + LiCl + HD$$

$$LiGaH_4 + PPh_3 + HCl \xrightarrow[-5°]{\text{diethyl ether}} GaH_3 \cdot PPh_3 + LiCl + H_2$$

The colourless oily gallane-dimethylphosphane adduct $GaH_3 \cdot PMe_2H$ ($GaD_3 \cdot PMe_2H$), which is not stable for long at 20°, is produced similarly [9]:

$$LiGaH_4 + PMe_2H + HCl \xrightarrow[<20^\circ]{\text{diethyl ether}} GaH_3 \cdot PMe_2H + LiCl + H_2$$

$$LiGaD_4 + PMe_2H + HCl \xrightarrow[<20^\circ]{\text{diethyl ether}} GaD_3 \cdot PMe_2H + LiCl + HD$$

However, no stable gallane-phosphane adducts were formed [9] in analogous experiments with PEt_2H or PPh_2H [9].

(2) *Synthesis of gaseous gallane GaH_3.* When a mixture of molecular and atomic hydrogen at 0.06 to 0.35 torr is passed over molten gallium, gaseous gallane, GaH_3, which is stable for 2.3 to 175 msec, is formed and its corresponding ion $GaH_3^{\oplus}$ can be measured in the mass spectrometer [3]. In contrast to the situation in comparable experiments with aluminium [2], in which $(AlH_3)_2$ is formed in addition to AlH_3, no gaseous digallane, $(GaH_3)_2$, is formed [3] with gallium.

(3) *Synthesis of oligogallanes, $(GaH_3)_n$, by the electrophilic displacement of NMe_3 from $GaH_3 \cdot NMe_3$.* The solid trimethylamine-gallane, $GaH_3 \cdot NMe_3$ [section 1.2.2(1)], reacts with gaseous boron fluoride at −20 to −15° to give a viscous, colourless oil [13] $(GaH_3)_n$:

$$GaH_3 \cdot NMe_3 + BF_3 \rightarrow \frac{1}{n}(GaH_3)_n + BF_3 \cdot NMe_3$$

At 20° in the gaseous phase, this $(GaH_3)_n$ forms short-lived gaseous $(GaH_3)_{n'}$ ($n' < n$; $n' = 1$ or 2?) [13].

(4) *Synthesis of polygallanes, $(GaH_3)_x$, by the hydrogenation of $GaCl_3$.* At 0° in diethyl ether, lithium gallanate (section 1.2.4) and gallium trichloride in a molar ratio of 3 : 1 form the soluble gallane ether adduct $GaH_3 \cdot nOEt_2$. At 20°, solid white polygallane, $(GaH_3)_x$, which is stable up to 130°, precipitates slowly from the solution [35]:

$$3LiGaH_4 + GaCl_3 \xrightarrow[0^\circ]{4nOEt_2,\ -3LiCl} 4GaH_3 \cdot nOEt_2 \xrightarrow[20^\circ]{4nOEt_2} \frac{4}{x}(GaH_3)_x$$

No GaH_3 is formed by the reaction of $LiGaH_4$ and $GaCl_3$ in a molar ratio of 2 : 1 at 20° [27].

1.2.3 Synthesis of partially substituted gallanes $GaH_{3-n}X_n$ or $GaH_{3-n}X_n$–$N(CH_3)_3$

(1) *Synthesis of Halogallanes $GaH_{3-n}X_n$.* There are three methods for the preparation of halogallanes: (a) partial hydrogenation of gallium trihalides GaX_3; (b) partial halogenation of gallane (in the form of the trimethylamine adduct $GaH_3 \cdot NMe_3$) with HX or Me_3NHX; (c) commutation of gallane and gallium trichloride (as amine adducts).

References p. 460

(a) At $-20°$, equimolecular amounts of gallium trichloride and trimethylsilane, Me_3SiH, give a 95% yield of dichlorogallane in accordance with the equation:

$$GaCl_3 + Me_3SiH \rightarrow GaHCl_2 + Me_3SiCl$$

after the Me_3SiCl has been pumped off at $-30°$, this remains in the form of colourless crystals stable for only a limited time at 20° [*21*, *22a*, *23*]. It first decomposes rapidly at the melting point (29°) to gallium tetrachlorogallanate $Ga(GaCl_4)$ ("gallium dichloride" $GaCl_2$), and hydrogen, the decomposition being quantitative at 150° [*21*]:

$$2GaHCl_2 \rightarrow Ga(GaCl_4) + H_2$$

It dissolves in benzene in the form of the dimer $(GaHCl_2)_2$. With trimethylamine such solutions form the dichlorogallane-trimethylamine adduct $GaHCl_2{\cdot}NMe_3$ (*cf.* section 1.3), which is more stable than the free $GaHCl_2$ [*21*].

(b) In the absence of a solvent, or in benzene or diethyl ether, the gaseous hydrogen halides HCl, HBr, and HI form with the gallane-trimethylamine adducts $GaH_3{\cdot}NMe_3$ or $GaD_3{\cdot}NMe_3$ (depending on the molar ratios used) the halogallane-trimethylamine adducts $GaH_{3-n}X_n{\cdot}NMe_3$ [*10*]:

$$GaH_3{\cdot}NMe_3 + nHX \rightarrow GaH_{3-n}X_n{\cdot}NMe_3 + nH_2 \quad (n = 1, 2;\ X = Cl, Br)$$

$$GaD_3{\cdot}NMe_3 + nDCl \rightarrow GaD_{3-n}Cl_n{\cdot}NMe_3 + nD_2 \quad (n = 1, 2)$$

$$GaD_3{\cdot}NMe_3 + nHBr \rightarrow GaD_{3-n}Br_n{\cdot}NMe_3 + nHD$$

(without H–D exchange, $n = 1, 2$)

$GaX_3{\cdot}NMe_3$ may also be prepared with 3 moles of HX. However, this persubstituted gallane-amine adduct is obtained more simply by the reaction of a solid gallium trihalide with an excess of liquid trimethylamine [*10*]:

$$GaX_3 + NMe_3 \xrightarrow{0°} GaX_3 \cdot NMe_3 \quad (X = Cl, Br, I)$$

Mono- and dihalogenation, but not perhalogenation, take place in the reaction of $GaH_3{\cdot}NMe_3$ with an excess of trimethylammonium halide, Me_3NHX. When Me_3NHCl is added, 2 moles of H_2 are formed, as follows [*10*]:

$$GaH_3{\cdot}NMe_3 + 2Me_3NHX \xrightarrow{0°} GaHX_2{\cdot}NMe_3 + 2NMe_3 + 2H_2$$

When Me_3NHBr is added, however, not more than 1.5 mole of H_2 is evolved. Equimolecular amounts of $GaD_3{\cdot}NMe_3$ and Me_3NDCl form $GaD_2Cl{\cdot}NMe_3$, NMe_3, and D_2 [*10*].

(c) In benzene, stoichiometric amounts of gallane and gallium trihalide (as the NMe_3 adducts) give monohalogallanes [*10*]:

$$2GaH_3 \cdot NMe_3 + GaX_3 \cdot NMe_3 \rightarrow 3GaH_2X \cdot NMe_3 \quad (X = Cl, Br, I)$$

Disubstitution requires an excess of the trihalide [*10*].

(2) Synthesis of aminogallanes GaH_2NR_2 and phosphanogallanes GaH_2PR_2. Like $BH_3 \cdot NMe_2H$, $AlH_3 \cdot NMe_2H$, and $BH_3 \cdot PMe_2H$, the $GaH_3 \cdot NMe_2H$, $GaD_3 \cdot NMe_2H$, and $GaH_3 \cdot PMe_2H$ split off hydrogen. On standing at 20° in an atmosphere of hydrogen, the oily gallane-amine adducts become more viscous; finally, white crystals separate out [*9*]:

$$GaH_3 \cdot NMe_2H \underset{20°}{\rightarrow} GaH_2NMe_2 + H_2$$

$$GaD_3 \cdot NMe_2H \underset{20°}{\rightarrow} GaD_2NMe_2 + HD$$

Solid trimethylamine-gallane reacts with gaseous methylamine and yields trimethylamine and methylamine-gallane (displacement reaction). Even below 20°, hydrogen elimination occurs to give the final product, the trimeric methylaminogallane [*27a*]:

$$GaH_3 \cdot NMe_3 + NMeH_2 \xrightarrow{-NMe_3} GaH_3 \cdot NMeH_2 \rightarrow GaH_2NMeH + H_2$$

Solid trimethylamine-gallane reacts with gaseous ammonia and yields aminogallane with elimination of one mole of hydrogen. In contrast, the alane system (see Table 5.1 in section 4 of Chapter 5) eliminates two moles of hydrogen [*27a*]:

$$GaH_3 \cdot NMe_3 + NH_3 \rightarrow GaH_2NH_2 + NMe_3 + H_2$$

$$AlH_3 \cdot NMe_3 + NMeH_2 \rightarrow AlHNMe + NMe_3 + 2H_2$$

The gallane-dimethylphosphane adduct also evolves hydrogen slowly at room temperature, first with the formation of (unstable) dimethylphosphanogallane, GaH_2PMe_2, [*9*]:

$$GaH_3 \cdot PMe_2H \rightarrow GaH_2PMe_2 + H_2$$

The latter decomposes as follows:

$$GaH_2PMe_2 \rightarrow Ga + PMe_2H + \tfrac{1}{2}H_2$$

the gallium formed probably also catalyses the decomposition of the $GaH_3 \cdot PMe_2H$ [*9*]:

$$GaH_3 \cdot PMe_2H \rightarrow Ga + PMe_2H + \tfrac{3}{2}H_2$$

1.2.4 Synthesis of metal gallanates $MGaH_4$

In a similar manner to the formation of boranates $BH_4^{\ominus}$ or alanates $AlH_4^{\ominus}$, gallium trichloride and finely divided lithium hydride (lithium deuteride) give a 95% yield of lithium gallanate, $LiGaH_4$ ($LiGaD_4$), [*7, 11, 33*]:

$$GaCl_3 + 4LiH \xrightarrow[0° \text{ or } 35°]{\text{diethyl ether}} LiGaH_4 + 3LiCl$$

References p. 460

After evaporation of the ether at 0°, this compound remains as the ether adduct. It decomposes slowly at 20° and rapidly at 50°, as follows:

$$LiGaH_4 \rightarrow LiH + Ga + \tfrac{3}{2}H_2$$

The metathesis of silver perchlorate and lithium gallanate in diethyl ether gives the orange ether-insoluble silver gallanate, which is stable up to −75° [*28*]:

$$AgClO_4 + LiGaH_4 \xrightarrow[-LiClO_4]{\text{diethyl ether, } -100°} AgGaH_4 \xrightarrow{-75°} Ag + Ga + 2H_2$$

1.2.5 Synthesis of gallium boranates $Ga(BH_4)_3$ and $Me_2Ga(BH_4)$

Trimethylgallium, $GaMe_3$, reacts with diborane, $(BH_3)_2$, at 20° to give *metastable* gallium tris(boranate), $Ga(BH_4)_3$. The latter decomposes rapidly and autocatalytically (metallic Ga) to gallium, diborane, and hydrogen [*20*]:

$$2GaMe_3 \xrightarrow[-6B_2H_5Me]{+6B_2H_6} 2GaH_3 \xrightarrow{+3B_2H_6} 2Ga(BH_4)_3 \rightarrow 2Ga + 3(BH_3)_2 + 3H_2$$

At −45°, trimethylgallium and diborane give dimethylgallium boranate, $Me_2Ga(BH_4)$. This is stable at −80° and decomposes only slowly at 20° [*20*]:

$$2GaMe_3 \xrightarrow[-2B_2H_5Me]{+2B_2H_6} 2GaMe_2H \xrightarrow{+B_2H_6} 2GaMe_2(BH_4)$$

1.2.6 Synthesis of gallium alanates $Ga(AlH_4)_3$ and $GaCl_{3-n}(AlH_4)_n$

With lithium alanate, $LiAlH_4$, gallium trichloride, $GaCl_3$, (molar ratio 3:1) forms ether-soluble gallium tris(alanate), $Ga(AlH_4)_3$. This is fairly unstable and dismutes in ether into insoluble polyalane, $(AlH_3)_x$, and the soluble gallane-ether adduct $GaH_3{\cdot}OEt_2$. At 35°, both gallium tris(alanate) and the gallane-ether adduct decompose [*34*]:

$$GaCl_3 + 3LiAlH_4 + nOEt_2 \xrightarrow[-30°/0°]{-3\,LiCl} Ga(AlH_4)_3{\cdot}nOEt_2$$

$$\xrightarrow[-\frac{3}{x}(AlH_3)_x,\,-(n-1)\,OEt_2]{0°} GaH_3{\cdot}OEt_2 \xrightarrow{-30°} Ga + \tfrac{3}{2}H_2 + OEt_2$$

Chlorine-containing oils [possibly $GaCl_{3-n}(AlH_4)_n{\cdot}mOEt_2$] which decompose even at 0°, form under other experimental conditions [*27*].

1.3 Reactions of gallium hydrides

The electron-acceptors GaR_2H and $GaCl_2H$ react with donors such as trimethylamine, C=C double bonds, water, or hydrazoic acid.

Thus, in benzene solution $(GaCl_2H)_2$ forms the donor–acceptor adduct $GaCl_2H{\cdot}NMe_3$ (*cf.* section 1.2.3) with trimethylamine [*21*]. Possibly the reaction takes place in a manner similar to the symmetrical cleavage of the $B{-}H^{\mu}{-}B$ double bridge in diborane by trimethylamine (see section 1.2 of Chapter 4):

$$Cl_2Ga(\mu\text{-}H)_2GaCl_2 + 2NMe_3 \longrightarrow 2GaCl_2H{\cdot}NMe_3$$

The Ga–H bond of diethylgallane Et_2GaH [section 1.2.1(2)][6] or that of dichlorogallane (section 1.2.3)[23] adds to the C=C double bond of ethylene[22], 1-decene[6], 1-hexene, cyclohexene[22, 23] or styrene[23] (anti-Markovnikov orientation[22]). The addition probably takes place through an intermediate state analogous to the protonated double bond:

$$\rangle C{=}C\langle \;\; \downarrow H^{\oplus} \qquad \rangle C{=}C\langle \;\; \downarrow GaH$$

At 65°, mixed trialkylgalliums are formed in this way. Example[6]:

$$Et_2GaH + CH_2{=}CH(n\text{-}C_8H_{17}) \xrightarrow{65°} Et_2Ga(C_{10}H_{21})$$

At 100°, a subsequent dismutation becomes more pronounced[6]:

$$3Et_2Ga(C_{10}H_{21}) \xrightarrow{100°} 2Et_3Ga + Ga(C_{10}H_{21})_3$$

Diethylgallane hydrolyses vigorously with water; complete hydrolysis to $Ga^{3\oplus}$ requires dilute acids[6]:

$$Et_2GaH \xrightarrow[-H_2]{+HOH} Et_2GaOH \xrightarrow{-EtH} \frac{1}{x}\left[\begin{array}{c}-Ga-O- \\ | \\ Et\end{array}\right]_x$$

With the gallane-ether adduct $GaH_3{\cdot}OEt_2$, hydrazoic acid forms the colourless, crystalline, ether-insoluble, tetrahydrofuran-soluble gallium triazide $Ga(N_3)_3$[32]:

$$GaH_3 + 3HN_3 \rightarrow Ga(N_3)_3 + 3H_2$$

In $GaH_3{\cdot}NMe_3$ [section 1.2.2(1)], the GaH_3 can be displaced electrophilically by BF_3[13]:

$$GaH_3{\cdot}NMe_3 + BF_3 \rightarrow GaH_3 + BF_3{\cdot}NMe_3$$

In the presence of dimethyl sulphide, SMe_2, a gallane-dimethyl sulphide adduct $GaH_3{\cdot}SMe_2$ is formed[11, 13], which decomposes at 20°:

$$GaH_3{\cdot}NMe_3 + BF_3 + SMe_2 \xrightarrow[-15°]{\text{dimethyl sulphide}} GaH_3{\cdot}SMe_2 + BF_3{\cdot}NMe_3$$

BH_3, which is a weaker acceptor than BF_3, does not liberate GaH_3[13]. (Stabilities: $GaH_3{\cdot}SMe_2 < GaMe_3{\cdot}SMe_2$; $GaH_3{\cdot}OMe_2 < GaH_3{\cdot}SMe_2$[11].)

Similarly, with dimethyl sulphide the viscous oligogallane, $(GaH_3)_n$ [section 1.2.2(3)], forms the gallane-dimethyl sulphide adduct $GaH_3{\cdot}SMe_2$[13]:

$$(GaH_3)_n + n\,SMe_2 \rightarrow n\,GaH_3{\cdot}SMe_2$$

The donor strength with respect to GaH_3 in gallane-donor adducts falls in the sequence $Me_2NH > Me_3N \sim Me_3P > Me_2PH$. Thus, for example, dimethylamine displaces the trimethylamine in $GaH_3{\cdot}NMe_3$, while the reaction of trimethylamine with $GaH_3 \cdot PMe_3$ leads to a (spectroscopically observable) equilibrium [*8, 9*]:

$$GaH_3 \cdot NMe_3 + Me_2NH \rightarrow GaH_3 \cdot NMe_2H + NMe_3$$
$$GaH_3 \cdot PMe_3 + NMe_3 \rightleftharpoons GaH_3 \cdot NMe_3 + PMe_3$$

(The equilibrium constant at room temperature for $[GaH_3 \cdot NMe_3][PMe_3]/[GaH_3 \cdot PMe_3][NMe_3] = 2.94$ [*8*].)

As with the corresponding boron compounds (see section 9.6 of Chapter 4) and aluminium compounds (see section 5 of Chapter 5), gallane adducts $GaH_3 \cdot NR_2H$ or $GaH_3 \cdot PR_2H$ containing negative hydrogen on the gallium and positive hydrogen on the donor atom, eliminate hydrogen intramolecularly with the formation of compounds isologous with aminoborane:

$$GaH_3 \cdot NMe_2H \rightarrow GaH_2NMe_2 + H_2$$
$$GaH_3 \cdot PMe_2H \rightarrow GaH_2PMe_2 + H_2$$

The evolution of H_2 begins in the decreasing sequence, with respect to temperature, B > Ga > Al ($BH_3 \cdot NMe_2H$ at 100° [*9*], $GaH_3 \cdot NMe_2H$ at 20° [*9*], $AlH_3 \cdot NMe_2H$ at −10° [*9, 19*]).

1.4 Physical properties of gallium hydrides

GaH_3	Detected by means of its ion $GaH_3^{\oplus}$ in the mass spectrometer. Gas stable for 2.3 to 175 msec at low pressures [*3*].
$(GaH_3)_n$	Viscous oils (n = large) or very unstable gases (n = small). The oily $(GaH_3)_n$, stable below −15°, do not dissolve in benzene, toluene, petroleum ether or carbon tetrachloride. At 20°, both oily and gaseous $(GaH_3)_n$ rapidly decompose into Ga and H_2 [*13*].
$(GaH_3)_x$	White, voluminous solids insoluble in diethyl ether. More stable than the oligomers $(GaH_3)_n$ [*35*].
$Ga_2H_2(CH_3)_4$	Colourless, highly viscous liquid solidifying to a glass. B.p. 172° (calculated). Vapour pressure: 0.5 torr at 0°, 64 torr at 95°. Vapour pressure equation: $\log p$ [torr] $= -2057.4/T + 1.75 \log T - 0.00044193\,T + 3.0661$. Heat of evaporation: 10.3 kcal/mole. At 84 and 103°, gaseous tetramethyldigallane has the monomeric formula $Ga_2H_2Me_4$. Between 130 and 180° decomposition takes place as follows: $\frac{3}{2}(GaHMe_2)_2 \rightarrow 2GaMe_3 + Ga + \frac{3}{2}H_2$ [*29, 30*].

$Ga_2H(CH_3)_5$ — Viscous oil that can be condensed from the gas phase at $-15°$ [*27*].

$GaH(C_2H_5)_2$ — Colourless mobile liquid. In contrast to $GaCl(C_2H_5)_2$, it ignites in air and reacts vigorously with water. B.p. 40–42° (10^{-4} torr) [*6*].

$GaHCl_2$ — Colourless crystals. M.p. 29° (decomp.). Stable for limited time at 20°. Soluble in benzene, cyclohexane, and diethyl ether at low temperatures. Dimeric in benzene [*21, 23*]. IR spectrum [*21*].

GaH_2NMe_2 — White crystals. Monomeric in the gas phase, dimeric in benzene solution. IR and PMR spectra [*9*].

GaD_2NMe_2 — IR and PMR spectra [*9*].

GaH_2PMe_2 — Soluble in diethyl ether. Decomposes when the ether is evaporated in vacuum. IR and PMR spectra [*9*].

GaD_2PMe_2 — As GaH_2PMe_2 [*9*].

$GaH_3\cdot N(CH_3)_2H$ — Colourless oil. Slowly splits off H_2 at 20°. IR and PMR spectra [*9*].

$GaD_3\cdot N(CH_3)_2H$ — Colourless mobile oil. Slowly splits off HD at 20°. IR and PMR spectra [*9*].

$GaH_3\cdot 2N(CH_3)_3$ — Exists only between -45.4 and $-22.8°$ in the equilibrium $GaH_3\cdot NMe_3 + NMe_3 \rightleftharpoons GaH_3\cdot 2NMe_3$. Dissociation pressure at $-45.5°$: 6.1 torr; $\log p\,[\text{torr}] = -2265/T + 10.867$ (valid between -28 and 0°) [*11*].

$GaH_3\cdot NMe_3$ — White sublimable solid. In the solid state forms a molecular lattice without H-bridge bonds. In the linear arrangement according to ...>N–Ga<... ...>N–Ga... with Ga–Ga = 5.91 Å and N–Ga = 1.97 Å, the tetrahedral configuration of N and Ga is retained. M.p. 70.5° [*11*]; 68.9–69.8° (slight decomposition [*26*]. Soluble in diethyl ether, dimethyl ether, nujol, and trimethylamine. Monomeric in diethyl ether. Vapour pressure at 25° about 2 torr [*11, 25, 26*]. IR spectra [*12, 24*].

$GaD_3\cdot N(CH_3)_3$ — M.p. 68.1° [*11*]. IR spectra [*12, 24*].

$GaH_2Cl\cdot N(CH_3)_3$ — White solid sublimable in vacuum. M.p. 65–66°. IR spectrum [*10*].

$GaD_2Cl\cdot N(CH_3)_3$ — Sublimable solid. IR spectrum [*10*].

$GaH_2Br\cdot N(CH_3)_3$ — Sublimable solid. IR spectrum [*10*].

$GaD_2Br\cdot N(CH_3)_3$ — Sublimable solid. IR spectrum [*10*].

$GaH_2I\cdot N(CH_3)_3$ — Sublimable solid. IR spectrum [*10*].

$GaD_2I\cdot N(CH_3)_3$ — Sublimable solid. IR spectrum [*10*].

References p. 460

$GaHCl_2 \cdot N(CH_3)_3$	Colourless crystals. M.p. 70° (decomp.). More stable than free $GaHCl_2$ [*21, 23*]. Soluble in benzene and diethyl ether. IR spectrum [*10*].
$GaDCl_2 \cdot N(CH_3)_3$	Soluble in benzene. IR spectrum [*10*].
$GaHBr_2 \cdot N(CH_3)_3$	Soluble in benzene and diethyl ether. IR spectrum [*10*].
$GaDBr_2 \cdot N(CH_3)_3$	Soluble in benzene and diethyl ether. IR spectrum [*10*].
$GaHI_2 \cdot N(CH_3)_3$	Soluble in benzene. IR spectrum [*10*].
$GaDI_2 \cdot N(CH_3)_3$	Soluble in benzene, IR spectrum [*10*].
$GaH_3 \cdot P(CH_3)_2H$	Colourless unstable (H_2, Ga) liquid. IR and PMR spectra [*9*].
$GaD_3 \cdot P(CH_3)_2H$	Stable for only a short time in diethyl ether [*9*].
$GaH_3 \cdot P(CH_3)_3$	White crystals. M.p. 70.5–71°. Vapour pressure at 22° about 1 torr. IR, PMR, and ^{71}Ga-NMR spectra [*8*].
$GaD_3 \cdot P(CH_3)_3$	White crystals. IR, PMR, and ^{71}Ga-NMR spectra [*8*].
$GaH_3 \cdot P(C_6H_5)_3$	Probably colourless crystals. Sparingly soluble in ether. In contrast to $GaH_3 \cdot PMe_3$, unstable even in ethereal solution at 20° (→ Ga) [*8*].
$GaH_3 \cdot S(CH_3)_2$	No sharp melting point because of instability. Decomposes at 20° into Ga, H_2, and $S(CH_3)_2$ [*11*].
$LiGaH_4$	The solid ether adduct decomposes slowly at 20° and rapidly at 150° into LiH, Ga, and H_2 [*33*]. ^{71}Ga-NMR spectrum [*1*].
$Ga(BH_4)_3$	Metastable at 20°. Decomposes, with catalysis by Ga, to Ga, $(BH_3)_2$, and H_2 [*20*].
$(CH_3)_2Ga(BH_4)$	Volatile crystals stable at −80°. M.p. −1.5°. Vapour pressure 14 torr at 0°, 51 torr at 24°; Trouton's constant: 23.5 kcal deg^{-1} mole^{-1}. Monomeric in the gaseous state [*20*].
$Ga(AlH_4)_3$	The ethereal solution decomposes rapidly at 0° [*34*].

2. Indium hydrides

2.1 Historical

Observations on the dissolution of indium(I) oxide in dilute hydrochloric acid (Klemm and Dierks, 1934, [*38*] first indicated the formation of a volatile indium hydride. More than 30 years elapsed before Breisacher and Siegel [*36*] succeeded in 1965 in detecting the extraordinarily unstable electron-unsaturated gaseous indane InH_3. In contrast, the involatile polyindanes, $(InH_3)_x$ and $(InH)_x$, are more stable and have been known for a longer time, having been first prepared in 1957 by Wiberg, Dittmann and Schmidt [40].

2.2 *Preparation of indium hydrides*

2.2.1 *Synthesis of gaseous monoindane, InH_3*

When a mixture of molecular and atomic hydrogen at 0.07 to 0.43 torr is passed over molten indium, gaseous indane stable for 2.8 to 215 msec is formed, and the corresponding ion $InH_3^{\oplus}$ can be measured in the mass spectrometer [36]. In contrast to similar experiments with aluminium [37], in which $(AlH_3)_2$ is formed in addition to AlH_3, no dimeric hydride $(InH_3)_2$ arises [36].

2.2.2 *Synthesis of polyindane, $(InH_3)_x$, and polyindium(I) hydride, $(InH)_x$*

The reaction of lithium indanate, $LiInH_4$ (section 2.2.3), with indium trichloride (3:1) in diethyl ether forms a solution of the indane ether adduct $InH_3 \cdot nOEt_2$. From this, after some days at $-30°$ or a few hours at 20°, solid white polyindane, $InH_3)_x$, precipitates [40]:

$$3LiInH_4 + InCl_3 \xrightarrow[-30°]{+4n\,OEt_2,\,-3LiCl} 4InH_4 \cdot nOEt_2 \xrightarrow[-30°/+20°]{-4n\,OEt_2} \frac{4}{x}(InH_3)_x$$

Polyindane, $(InH_3)_x$, moistened with ether forms, slowly at 20° and rapidly at 200°, white solid indium(I) hydride, $(InH)_x$, which is stable up to 340° [40]:

$$(InH_3)_x \xrightarrow{20°} (InH)_x + x\,H_2$$

2.2.3 *Synthesis of lithium indanates $LiInX_3H$ and $LiInH_4$*

Indium trihalides InX_3 (X = Br, I) and lithium hydride in a molar ratio of 1:1 react in ethereal solution to give lithium trihaloindanates, $LiInX_3H$. These also arise, with the formation of borane, in the reaction of indium trihalides (Cl, Br) with lithium boranate in a molar ratio of 1:1. In the form of colourless oily ether adducts ($LiInCl_3H \cdot 4OR_2$, $LiInBr_3H \cdot 6OR_2$, and $LiInI_3H \cdot 6OR_2$), they are stable at 0°. They decompose when the ether is removed at $-10°$ to 0° [39]:

$$InI_3 + LiH + 6R_2O \xrightarrow{5°/10°} LiInI_3H \cdot 6R_2O \xrightarrow[-6R_2O]{-10°} LiH + InI_3 (\rightarrow InI + I_2)$$

$$InBr_3 + LiH + 6R_2O \xrightarrow{0°} LiInBr_3H \cdot 6R_2O \; \begin{cases} \xrightarrow{-6R_2O\text{(pumping off)}} LiH + InBr_3 \\ \xrightarrow{0°} \tfrac{2}{3}LiInBr_4 + \tfrac{1}{3}LiBr + \tfrac{1}{3}In + \tfrac{1}{2}H_2 + 6R_2O \end{cases}$$

$$\uparrow\; -20°,\ -BH_3$$

$$InBr_3 + LiBH_4 + 6R_2O \xrightarrow{-20°} LiInBr_3(BH)_4 + 6R_2O$$

References p. 460

$$InCl_3 + LiBH_4 + 4R_2O \xrightarrow{-50°} LiInCl_3(BH)_4$$

$$+4R_2O \xrightarrow[-BH_3]{-20°} LiInCl_3H \cdot 4R_2O \xrightarrow[-4R_2O]{-15°} \tfrac{2}{3}LiInCl_4 + \tfrac{1}{3}LiCl + \tfrac{1}{3}In + \tfrac{1}{2}H_2$$

The reaction of indium halides (Cl, Br) with lithium hydride in a molar ratio of 1:4 gives a 60% to 65% ($InCl_3$) or 80% ($InBr_3$) yield of lithium indanate $LiInH_4$ [*43*]:

$$InX_3 + 4LiX \xrightarrow[-25°]{\text{diethyl ether}} LiInH_4 + 3LiX$$

The ethereal solution of lithium indanate is comparatively stable at −20°. However, ether-insoluble LiH and ether-insoluble polyindane, $(InH_3)_x$, gradually separate out (even at −25°) [*43*]:

$$LiInH_4 \rightarrow LiH + \frac{1}{x}(InH_3)_x$$

The white solid $LiInH_4 \cdot 3OEt_2$ freed from ether at −30° decomposes instantaneously at 0° as follows [*43*]:

$$LiInH_4 \cdot 3OEt_2 \rightarrow LiH + In + \tfrac{3}{2}H_2 + 3OEt_2$$

2.2.4 Synthesis of indium boranate $In(BH_4)_3$

Trimethylindium, $InMe_3$, reacts with diborane, $(BH_3)_2$, in diethyl ether–tetrahydrofuran to give indium tris(boranate), $In(BH_4)_3$. In solution, it is present in the form of the tetrahydrofuran adduct. At −10° it decomposes when the tetrahydrofuran is drawn off [*41*]:

$$InMe_3 + \left(3 + \frac{3}{n}\right)BH_3 + \left(n + \frac{3}{n}\right)THF \xrightarrow[\text{tetrahydrofuran/diethyl ether}]{-45°,\ -\frac{3}{n}BH_{3-n}Me_n \cdot THF} In(BH_4)_3 \cdot nTHF$$

$$\xrightarrow[-(n-3)THF]{-40°} In(BH_4)_3 \cdot 3THF \xrightarrow[-THF]{-30°} In(BH_4)_3 \cdot 2THF \xrightarrow[-2THF]{-10°} In + 3BH_3 + \tfrac{3}{2}H_2$$

Less satisfactory is the preparation of $In(BH_4)_3$ from $InCl_3$ and $LiBH_4$, since here $LiInCl_3H$ is formed preferentially (see section 2.2.3).

2.2.5 Synthesis of indium alanate $In(AlH_4)_3$

Indium trichloride, $InCl_3$, and lithium alanate, $LiAlH_4$, react in ethereal solution to give ether-insoluble indium tris(alanate), $In(AlH_4)_3$, which decomposes slowly even at −40° and rapidly at 20° [*42*]:

$$InCl_3 + 3LiAlH_4 \xrightarrow[-3LiCl]{-70°} In(AlH_4)_3 \xrightarrow{-40°} In + \tfrac{3}{2}H_2 + 3AlH_3$$

2.3 *Physical properties of indium hydrides*

InH_3 — Detected by means of its ion $InH_3^{\oplus}$ in the mass spectrum. A gas stable for 2.8 to 215 msec at low pressures [*36*].

$(InH)_x$ — White insoluble solid. Stable up to 340°. Rapid decomposition at 400° [*40*].

$InH_3 \cdot nO(C_2H_5)_2$ — An ethereal solution of indium hydride InH_3 deposits polyindane $(InH_3)_x \cdot nO(C_2H_5)_2$ in 2–3 days at −30° and in 5–7 hours at 20° [*40*].

$(InH_3)_x \cdot nO(C_2H_5)_2$ — White powder. Insoluble in ether, benzene, chloroform, acetone, tetrahydrofuran, and dioxan [*40*].

$LiInH_4 \cdot 3O(C_2H_5)_2$ — White solid stable at −30°. Decomposes instantaneously at 0°. At −20° the ethereal solution is fairly stable, but LiH and $(InH_3)_x$ form gradually [*43*].

$LiInCl_3H \cdot 4O(C_2H_5)_2$ — Colourless oil stable below −15° [*39*].

$LiInBr_3H \cdot 6O(C_2H_5)_2$ — Colourless clear oil stable at 0°. Solidifies between −86° and −94°. Unstable in the absence of ether [*39*].

$LiInI_3H \cdot 6O(C_2H_5)_2$ — Colourless oil stable at −20° to −10°. Solidifies between −85° and −95°. Unstable in the absence of ether. Insoluble in benzene, toluene, and ether at 0° to 5° [*39*].

$In(BH_4)_3 \cdot 3THF$ — White crystals which give off 1 THF in vacuum at −30° [*41*].

$In(BH_4)_3 \cdot 2THF$ — Decomposes in vacuum at −10° into In, H_2, $(BH_3)_2$, and THF [*41*].

$In(AlH_4)_3$ — The white solid, insoluble in diethyl ether, decomposes at −40° [*42*].

$InCl_2(AlH_4)$ — More stable than $In(AlH_4)_3$ [*42*].

3. Thallium hydrides

3.1 *Historical*

According to experiments due to Pietsch and Seuferling in 1931 [*44*, *45*], when a mixture of molecular and atomic hydrogen is passed over finely divided solid thallium an extremely unstable volatile thallium hydride is formed. It decomposes only a few centimetres from the position of its formation (metal surface) with the deposition of a Tl mirror. At that time (1931) it was impossible to detect the hydride (possibly TlH_3) by mass spectroscopy [as in the case of AlH_3, $(AlH_3)_2$, GaH_3, and InH_3; see section 3.1 of Chapter 5 and sections 1.2.2(2) and 2.2.1].

References p. 460

As in the case of aluminium, gallium, and indium, the non-volatile polymeric thallium hydrides are more stable. In 1957 Wiberg, Dittmann, Nöth and Schmidt [48] succeeded for the first time in preparing a polythallane $(TlH_3)_x$ and a thallium(I) hydride $(TlH)_x$.

3.2 Preparation of thallium hydrides

3.2.1 Synthesis of polythallane, $(TlH_3)_x$, and polythallium(I) hydride, $(TlH)_x$

The reaction of lithium thallanate, $LiTlH_4$ (section 3.2.2), with thallium trichloride, $TlCl_3$, in diethyl ether initially forms thallium trihydride dissolved in ether (thallane ether adduct) $TlH_3 \cdot nOEt_2$. The solution rapidly deposits white solid polythallane, $(TlH_3)_x$, [48]:

$$3LiTlH_4 + TlCl_3 \xrightarrow[20^\circ]{+4n\,OEt_2,\,-3LiCl} 4\,TlH_3 \cdot nOEt_2 \xrightarrow[0^\circ/10^\circ]{-4n\,OEt_2} \frac{4}{x}(TlH_3)_x$$

The brown solid polymeric thallium(I) hydride, $(TlH)_x$, forms soon after the polythallane has flocculated out from the ethereal solution [48]:

$$(TlH_3)_x \xrightarrow{0^\circ} (TlH)_x + x\,H_2$$

3.2.2 Synthesis of lithium thallanate, $LiTlH_4$

An excess of finely divided lithium hydride reacts with thallium trichloride in ethereal solution to give a 27% yield of lithium thallanate, $LiTlH_4$ [49]:

$$TlH_3 + 4LiH \xrightarrow[-15^\circ]{\text{deithyl ether}} LiTlH_4 + 3LiCl$$

At 0°, the solution deposits solid white polythallane, $(TlH_3)_x$. (Compare this with the substantially lower tendency of, for instance, the alanate ion to decompose by the reaction $AlH_4^{\ominus} \rightarrow AlH_3 + H^{\ominus}$). In turn, polythallane rapidly splits off hydrogen with the formation of polymeric thallium(I) hydride, $(TlH)_x$ [49]:

$$LiTlH_4 \xrightarrow[+n\,R_2O,\,-LiH]{0^\circ} TlH_3 \cdot nR_2O \xrightarrow[-n\,R_2O]{0^\circ/10^\circ} \frac{1}{x}(TlH_3)_x \xrightarrow{0^\circ/10^\circ} \frac{1}{x}(TlH)_x + H_2$$

3.2.3 Synthesis of thallium boranates $TlCl(BH_4)_2$ and $TlBH_4$

At −110°, thallium trichloride, $TlCl_3$, and an excess of lithium boranate give not $Tl(BH_4)_3$, but only thallium chloride bis(boranate), $TlCl(BH_4)_2$. This decomposes rapidly even at −95° into thallium(I) chloride, borane, and hydrogen [50]:

$$TlCl_3 \xrightarrow[+2LiBH_4,\,-2LiCl]{\text{diethyl ether},\,-110^\circ} TlCl(BH_4)_2 \xrightarrow{-95^\circ} TlCl + 2BH_3 + H_2$$

As was to be expected, the boranate of univalent thallium, $TlBH_4$, resembles the alkali-metal boranates. It can be prepared by the metathesis of thallium(I) nitrate and potassium boranate in an aqueous medium [*46*]:

$$TlNO_3 + KBH_4 \xrightarrow{\text{water}} TlBH_4 + KNO_3$$

The reaction of thallium monoethoxide, TlOEt, with lithium boranate in diethyl ether gives the ether-insoluble adduct $TlBH_4 \cdot LiOEt$ [*47*].

3.2.4 *Synthesis of thallium alanates $TlCl(AlH_4)_2$ and $TlAlH_4$*

In ethereal solution at −115°, thallium trichloride, $TlCl_3$, and an excess of alane, AlH_3, form thallium chloride bis(alanate), $TlCl(AlH_4)_2$. This decomposes rapidly into TlCl, AlH_3, and H_2 even at −95°. Lithium alanate, $LiAlH_4$, and thallium trichloride yield only decomposition products of the hypothetical $Tl(AlH_4)_3$ even at −116°; Tl, AlH_3, and H_2 are produced [*50*]. Thallium(I) alanate, $Tl(AlH_4)$, is only slightly more stable. This is formed by the metathesis of thallium(I) perchlorate and lithium alanate in diethyl ether at −100° [*47*]:

$$TlClO_4 + LiAlH_4 \rightarrow TlAlH_4 + LiClO_4$$

At −80°, it decomposes into Tl, AlH_3, and H_2 [*47*].

3.3 *Physical properties of thallium hydrides*

$(TlH_3)_x$	White solid, very unstable in suspension in diethyl ether. At 0° rapidly forms $(TlH)_x$ and H_2 [*48*].
$(TlH)_x$	Brown powder, stable at 20° in the absence of air and moisture. Decomposes into the elements at 270°. Insoluble in ether, benzene, toluene, dioxan, tetrahydrofuran [*48*].
$LiTlH_4$	White hygroscopic (ether-containing?) powder. Decomposes rapidly at 30° into LiH, Tl, and H_2. The solution in diethyl ether is stable for only a few hours at 0° [*49*].
$TlBH_4$	White crystals insoluble in water. Metallic Tl catalyses the decomposition. Infrared spectrum [*46*].
$Tl(AlH_4)_3$	Decomposes even at −100° [*47*].
$TlAlH_4$	Decomposes even at −100° [*47*].

References p. 460

REFERENCES

Gallium

1. AKITT, J. W., N. N. GREENWOOD AND A. STORR, J. Chem. Soc. *1965*, 4410.
2. BREISACHER, P. AND B. SIEGEL, J. Am. Chem. Soc. *86*, 5053 (1964).
3. BREISACHER, P. AND B. SIEGEL, J. Am. Chem. Soc. *87*, 4255 (1965).
4. DENNIS, W. M. AND W. PATNODE, J. Am. Chem. Soc. *54*, 182 (1932).
5. EISCH, J. J., J. Am. Chem. Soc. *84*, 3605 (1962).
6. EISCH, J. J., J. Am. Chem. Soc. *84*, 3830 (1962).
7. FINHOLT, A. E., A. C. BOND, JR. AND H. I. SCHLESINGER, J. Am. Chem. Soc. *69*, 1199 (1947).

7a. GREENWOOD, N. N., *Gallium Hydride and its Derivatives*, in *New Pathways in Inorganic Chemistry* (E. A. V. Ebsworth, A. G. Maddock and A. G. Sharpe, eds.), Cambridge University Press, 1968, pp. 37–62.

8. GREENWOOD, N. N., E. J. F. ROSS AND A. STORR, J. Chem. Soc. *1965*, 1400.
9. GREENWOOD, N. N., E. J. F. ROSS AND A. STORR, J. Chem. Soc. *A1966*, 706.
10. GREENWOOD, N. N. AND A. STORR, J. Chem. Soc. *1965*, 3426.
11. GREENWOOD, N. N., A. STORR AND M. G. H. WALLBRIDGE, Inorg. Chem. *2*, 1036 (1963).
12. GREENWOOD, N. N., A. STORR AND M. G. A. WALLBRIDGE, Proc. Chem. Soc. *1962*, 249.
13. GREENWOOD, N. N. AND M. G. H. WALLBRIDGE, J. Chem. Soc. *1963*, 3912.
14. KRAUS, C. A. AND F. E. TOONDER, Proc. Natl. Acad. Sci. US *19*, 292, 298 (1933).
15. KRAUS, C. A. AND F. E. TOONDER, J. Am. Chem. Soc. *55*, 3547 (1933).
16. LECOQ DE BOISBAUDRAN, P. E., Compt. Rend. *93*, 294 (1881).
17. OLIVER, J. P. AND L. G. STEVENS, J. Inorg. Nucl. Chem. *24*, 953 (1962).
18. RENEWANZ, G., Ber. *65B*, 1208 (1932).
19. RUFF, J. K. AND M. F. HAWTHORNE, J. Am. Chem. Soc. *82*, 2141 (1960).
20. SCHLESINGER, H. I., H. C. BROWN AND G. W. SCHAEFFER, J. Am. Chem. Soc. *65*, 1786 (1943).
21. SCHMIDBAUR, H., W. FINDEISS AND E. GAST, Angew. Chem. *77*, 170 (1965).
22. SCHMIDBAUER, H. AND H. F. KLEIN, Angew. Chem. *78*, 306 (1966).

22a. SCHMIDBAUER, H. AND H. F. KLEIN, Chem. Ber. *100*, 1129 (1967).

23. SCHMIDBAUR, H., private communication.
24. SHRIVER, D. F., R. L. AMSTER AND R. C. TAYLOR, J. Am. Chem. Soc. *84*, 1321 (1962).
25. SHRIVER, D. F. AND C. E. NORDMAN, Inorg. Chem. *2*, 1298 (1963).
26. SHRIVER, D. F. AND R. W. PARRY, Inorg. Chem. *2*, 1039 (1963).
27. SHRIVER, D. F., R. W. PARRY, N. N. GREENWOOD, A. STORR AND M. G. H. WALLBRIDGE, Inorg. Chem. *2*, 867 (1963).

27a. STORR, A., J. Chem. Soc. *A1969*, 2605.

28. WIBERG, E. AND W. HENLE, Z. Naturforsch. *7b*, 576 (1952).
29. WIBERG, E. AND T. JOHANNSEN, Naturwiss. *29*, 320 (1942).
30. WIBERG, E. AND T. JOHANNSEN, Angew. Chem. *55*, 38 (1942).
31. WIBERG, E., T. JOHANNSEN AND O. STECHER, Z. Anorg. Allgem. Chem. *251*, 114 (1943).
32. WIBERG, E. AND H. MICHAUD, Z. Naturforsch. *9b*, 502 (1954).
33. WIBERG, E. AND M. SCHMIDT, Z. Naturforsch. *6b*, 171 (1951).
34. WIBERG, E. AND M. SCHMIDT, Z. Naturforsch. *6b*, 172 (1951).
35. WIBERG, E. AND M. SCHMIDT, Z. Naturforsch. *7b*, 577 (1952).

Indium

36. BREISACHER, P. AND B. SIEGEL, J. Am. Chem. Soc. *87*, 4255 (1965).
37. BREISACHER, P. AND B. SIEGEL, J. Am. Chem. Soc. *86*, 5053 (1964).
38. KLEMM, W. AND F. DIERKS, Z. Anorg. Allgem. Chem. *219*, 45 (1934).
39. WIBERG, E., O. DITTMANN, H. NÖTH AND M. SCHMIDT, Z. Naturforsch. *12b*, 56 (1957).
40. WIBERG, E., O. DITTMANN AND M. SCHMIDT, Z. Naturforsch. *12b*, 57 (1957).
41. WIBERG, E. AND H. NÖTH, Z. Naturforsch. *12b*, 59 (1957).
42. WIBERG, E. AND M. SCHMIDT, Z. Naturforsch. *6b*, 172 (1951).
43. WIBERG, E. AND M. SCHMIDT, Z. Naturforsch. *12b*, 54 (1957).

Thallium

44. PIETSCH, E., Z. Elektrochem. *39*, 577 (1933).
45. PIETSCH, E. AND F. SEUFERLING, Naturwiss. *19*, 574 (1931).

46. WADDINGTON, T. C., J. Chem. Soc. *1958*, 4783.
47. WIBERG, E., O. DITTMANN, H. NÖTH AND M. SCHMIDT, Z. Naturforsch. *12b*, 62 (1957).
48. WIBERG, E., O. DITTMANN, H. NÖTH AND M. SCHMIDT, Z. Naturforsch. *12b*, 61 (1957).
49. WIBERG, E., O. DITTMANN AND M. SCHMIDT, Z. Naturforsch. *12b*, 60 (1957).
50. WIBERG, E. AND H. NÖTH, Z. Naturforsch. *12b*, 63 (1956).

Chapter 7

SILICON HYDRIDES

1. Historical

Wöhler and Buff discovered the first silicon hydride in 1857. In a paper "Über eine Verbindung von Silicium mit Wasserstoff" ("On a compound of silicon with hydrogen")[*689*] they gave an account of a gas formed when silicon-containing aluminium is immersed in an aqueous salt solution at the positive pole of a galvanic circuit. The gas ignited in air with the separation of brown flocs. According to further investigations by Wöhler[*690*], on pyrolysis it decomposed into silicon and hydrogen, and it precipitated the metals from solutions of heavy metal salts (*e.g.*, $Cu^{2\oplus}$, $Pt^{4\oplus}$). Thus, even at that time the essential differences between a silane and an alkane were recognized: the reducing properties of the (highly negative) silane hydrogen and the sensitivity of silane to oxidation. The same period saw the discovery of trichlorosilane $SiHCl_3$ by Buff and Wöhler[*80*]. However, they described the compound as $Si_2Cl_3 \cdot 2\,HCl$ ("$SiHCl_{2.5}$").

Only 10 years later (1867) did Friedel and Ladenburg[*189*] establish the formulae of the two compounds: SiH_4 and $SiHCl_3$. In the same paper the two authors reported another property of the silane derivatives. Triethoxysilane $SiH(OEt)_3$, obtained by Buff and Wöhler[*80*] by the alcoholysis of trichlorosilane, dismutes under the catalytic influence of sodium to give silane and tetraethoxysilane:

$$4\,SiH(OEt)_3 \xrightarrow{(Na)} SiH_4 + 3\,Si(OEt)_4$$

The dismutation of silicon hydrides and their derivatives described here for the first time was investigated further only considerably later, in the fifties of the present century.

Almost 50 years passed after the discovery of monosilane, SiH_4, before Moissan and Smiles[*402*] succeeded (in 1902) in isolating the first homologue disilane, Si_2H_6, from the gas formed by the protolysis of magnesium silicide. At this time, however, the experimental means for intensive investigation of the air-sensitive silanes had not yet been created. Only in 1916 did Stock open up the route to the study of silanes. Using a completely closed grease-free air-vacuum apparatus, Stock and his group, particularly in the twenties, succeeded in preparing many silanes and their derivatives (*e.g.*, halides, pseudohalides, ether adducts, and amine adducts). Investigation of many polymeric silanes, *e.g.* polysilane and siloxene, took place in the same decade. The synthesis of numerous, but only peralkylated, metal silyls, $M(SiR_3)_m$, was performed around 1930. Hydridic metal silyls, $M(SiH_3)_n$, have been dealth with to a relatively large extent only since about 1960. The most important series of conversions of the silyl compounds (see section 6.1.5) were drawn up in the fifties. They show, *inter alia* which silane derivatives can react with metal salts. The main emphasis of the current theoretical investigations and their experimental confirmation is being placed on studies of the Si bond (p_π–p_π, d_π–p_π, d_π–d_π bonds) (section 9.3).

2. General review

2.1 Delimination

As an element of the fourth main group of the periodic system, silicon forms "saturated" hydrides, "silanes", with the general formula Si_nH_{2n+2} (SiH_4, Si_2H_6, . . . , Si_8H_{18}, and $Si_{10}H_{22}$), which correspond superficially to the alkanes. As in the case of the alkanes, in the silanes one or more hydrogen atoms can be replaced by other elements or groups, such as halogens, nitrogen groups, oxygen groups, alkali metals, or organic radicals. The number of silicon compounds in which all hydrogen atoms have been replaced by organic ligands is extremely large. Such perorganosilanes are derived mainly from monosilane (SiR_4) and, to a far smaller extent, from disilane (Si_2R_6) and higher silanes.

The number of organosilicon compounds increases severalfold if another element—for example, oxygen, sulphur, or nitrogen—is introduced. We then obtain the groupings:

$$\begin{array}{c} R' \\ | \\ -Si-O- \\ | \\ R \end{array} \qquad \begin{array}{c} R' \\ | \\ -Si-S- \\ | \\ R \end{array} \qquad \begin{array}{c} R' \quad R'' \\ | \quad\;\; | \\ -Si-N- \\ | \quad\quad \\ R \quad\quad \end{array}$$

Compounds with Si–O bonds are called siloxanes. With increasing number of alternating Si–O bonds, the series of siloxanes converges to the highly polymeric "silicones" $(R_2SiO)_x$. Similarly, compounds with Si–S bonds, the silsulphanes (silthianes) and those with Si–N bonds, the silazanes, can also polymerize, although to a smaller extent.

If all the hydrogen atoms of silane are replaced simultaneously by oxygen $\left(\gt Si\text{–}O\right)$ and oxygen–metal groups $\left(\gt Si\text{–}OM\right)$ polymeric (vitreous or crystalline) metal silicates are obtained. With increasing proportion of oxygen in the O–OM combination, the series of compounds converges to silicon dioxide (silica). The organosilicon compounds, the siloxanes, and their transitions to the silicates form the subject matter of numerous books and comprehensive reviews [*135, 211, 220, 230, 249, 345, 427, 482, 501, 503*]. In the present *Silicon* section only compounds with at least one Si–H bond will be considered (comprehensive reviews, see [*25a, 352a, 355*]). In general, Si–H bonds in siloxanes, silsulphanes and silazanes change the physical properties only slightly, since the degree of polymerization or the crystal bond is generally retained. The chemical properties change only to the extent that the (reactive) Si–H bond is capable of undergoing reactions, while the

References p. 621

somewhat less reactive Si–O, Si–S, or Si–N skeleton generally remains unchanged

2.2 *Electronegativity, ionic character, bond energy*

Silicon is less electronegative than carbon, and consequently the overwhelming number of the bonds that silicon can form with electronegative elements E are more polar than the bonds in the corresponding carbon compounds. This is shown by a comparison of corresponding silicon and carbon compounds (columns 4 and 6 of Table 7.1).

TABLE 7.1

ELECTRONEGATIVITIES OF SOME ELEMENTS (in Pauling units) AND CALCULATED FORMAL IONIC NATURE OF THE Si–E AND C–E BONDS (calculated from the expression $16(\chi_A-\chi_B)+3.5\,(\chi_A-\chi_B)^2$ [*249*])

Element, E 1	*Electro-negativity,* χ 2	$\chi_E-\chi_{Si}$ 3	*% Ionic nature of the Si–E bond* 4	$\chi_E-\chi_C$ 5	*% Ionic nature of the C–E bond* 6
Si	1.74	—	—	0.76	12
C	2.50	0.76	14	—	—
H	2.20	0.46	8	0.30	35
F	4.10	2.36	57	1.60	5
Cl	2.83	1.09	22	0.33	6
Br	2.74	1.00	19	0.24	4
I	2.21	0.47	8	0.29	5
O	3.50	1.70	39	1.00	19
S	2.44	0.70	13	0.06	1
N	3.07	1.33	27	0.50	9

The ionic nature of the bond decreases, however, because of the d_π–p_π back-coordination of electrons to the silicon (see d_π–p_π bonding *e.g.* ref. [*107*]) that is possible in the case of silicon (see section 9.3). Apart from bonds to elements of pronounced metallic nature, silicon is always positive (*e.g.* $\overset{+}{Si}-\overset{-}{C}$, $\overset{+}{Si}-\overset{-}{H}$, $\overset{+}{Si}-\overset{-}{F}$; but $\overset{-}{C}-\overset{+}{Si}$, $\overset{\pm}{C}-\overset{\mp}{H}$, and $\overset{+}{C}-\overset{-}{F}$). As was to be expected, the highly polar silicon bonds are more readily cleaved by electrophilic attack than the corresponding carbon bonds:

$$CH_3I + HOH \xrightarrow{\text{slow}} CH_3OH + HI$$
$$SiH_3I + HOH \xrightarrow{\text{fast}} SiH_3OH + HI$$

The rapid hydrolysis of silyl iodide is due not only to the polarity of the Si–I bond but also to the involvement of the silicon d-orbitals, discussed further below, in the attack of the water. Another factor is the small screening effect exerted by the hydrogens on the large silicon atom. In accordance with the different electro-negativities of silicon and carbon, the reaction products of comparable silicon and carbon hydrides are frequently different:

$$Ph_3\overset{+}{Si}\overset{-}{H} + \overset{-}{Me}\overset{+}{Li} \rightarrow Ph_3SiMe + LiH$$
$$Ph_3\overset{-}{C}\overset{+}{H} + \overset{-}{Me}\overset{+}{Li} \rightarrow Ph_3CLi + MeH$$

The higher reactivity of the silicon bond as compared with the carbon bond is not merely a consequence of lower bond energy. This is shown by comparison of the bond energies in Table 7.2.

TABLE 7.2

BOND ENERGIES

VALUES IN KCAL/MOLE AFTER Cotterell [*100*] AND (IN BRACKETS) AFTER Pitzer [*474*]

Standard: $C_{solid} \rightarrow C_{atomic} = 170$ kcal/mole

Silicon bond	*Bond energy*	*Carbon bond*	*Bond energy*
Si–Si	53 (45)	C–C	82.6
Si–C	76	C–Si	76
Si–H	76	C–H	98.7
Si–F	135	C–F	116
Si–Cl	91 (87)	C–Cl	81
Si–Br	74	C–Br	68
Si–I	56	C–I	51
Si–O	108	C–O	85.5
Si–N	77	C–N	72.8

According to this, only the energies of the Si–Si and Si–H bonds are lower than those of the corresponding C–C and C–H bonds. The lower stability of the higher silanes as compared with alkanes, again, cannot be ascribed directly to the smaller bond strength of Si–Si as compared with C–C. Peralkylated or perarylated disilane derivatives in fact possess a high thermal stability, similar to that of the corresponding ethane derivatives. Consequently, thermodynamic factors are less responsible for the greater or smaller ease of pyrolysis (apart from the enthalpies of the pyrolysis products) than kinetic factors.

2.3 Participation of d-orbitals of silicon in bonds

The outer electron shell of silicon ($3s^2 3p^2 3d^0$) resembles that of carbon ($2s^2 2p^2$). The normal covalencies of silicon and carbon are therefore the same, namely four. As in the case of carbon, with silicon the electron functions generally combine to give sp^3 hybrids. However, unlike carbon, silicon still possesses relatively low unoccupied d-orbitals, which makes possible (i) an increase in the covalency to more than four, (ii) $d_\pi{}^{Si}$–$p_\pi{}^{E}$ or $d_\pi{}^{Si}$–$d_\pi{}^{E}$ double bond components in addition to the silicon–element σ-bonds, (iii) sp^3d or sp^3d^2 hybridization in intermediate stages of reactions, and (iv) different reaction mechanisms for silicon than for carbon.

Of its total of five d-orbitals, silicon uses a maximum of only two, like the other elements of the second short period. The maximum covalency is therefore six. Here the electron functions of the silicon form sp^3d^2 hybrid functions. The most typical example of this is the octahedral ion $SiF_6{}^{2\ominus}$. As an example of this configuration in a hydride, a similar configuration is possessed by the compound $SiH_3I \cdot 2NMe_3$, which is non-conducting in acetonitrile [*32*]. Compounds with

References p. 621

pentacovalent silicon have not yet been isolated, apart from the $CHF_2CF_2SiH_3 \cdot NMe_3$ described below. The 1:1 adducts of monohalogenosilanes with amines or phosphanes such as $SiH_3I \cdot NMe_3$[*32*], $SiH_3I \cdot PEt_3$[*32*], $SiH_3Cl \cdot NMe_3$[*168*] and $MeSiH_2X \cdot NMe_3$ (X = F, Cl, Br, I)[*142*] do not appear to be interco-ordinated sp^3d complexes of silicon. Because of their conductivity in non-aqueous solvents, it must rather be assumed that they are ammonium or phosphonium salts, at least in the dissolved state (*e.g.* $[SiH_3NMe_3]^{\oplus}I^{\ominus}$). On the other hand according to NMR and IR measurements $CHF_2CF_2SiH_3 \cdot NMe_3$ appears to contain pentacovalent Si[*104*]:

```
      NMe3
       |  H
       | /
    H-Si
       | \
       |  H
      CF2CHF2
```

It is very probable that sp^3d states of silicon play a part in substitution reactions by lowering the activation energy (see below).

The participation of d-orbitals of silicon in bonds in silanes will be illustrated and compared with the bonds in the corresponding carbon compounds (d_π–p_π bond; on the p_π–p_π and d_π–d_π bonds, see section 2.5):

(a) trisilylamine, $(SiH_3)_3N$, is planar[*147*], trimethylamine, $(CH_3)_3N$, is tetrahedral[*5*];

(b) silyl isothiocyanate, SiH_3NCS[*284, 353*], and silyl isocyanate, SiH_3NCO [*182*], are linear, while methyl isocyanate, CH_3NCO[*178*], and methyl azide, CH_3N_3[*178*], are angular;

(c) the Si–O–Si angle in disiloxane, $H_3Si–O–Si_3H$, is larger (140–155°)[*111, 369*] than the C–O–C angle in dimethyl ether, $H_3C–O–CH_3$ (111.5°) [*110, 111, 303*];

(d) the dipole moments of the chlorosilanes, $SiH_{4-n}Cl_n$, are lower than those of the corresponding methane derivatives $CH_{4-n}Cl_n$;

(e) In the neighbourhood of the π-electron system of benzene or of a C≡C triple bond, a silyl group is electron-attracting as compared with the methyl isologue; this would not be expected from consideration of electronegativity (Si = 1.74, C = 2.50 Pauling units).

Re(*a*). Since silicon is less electronegative than carbon, we should assume that trisilylamine, $(SiH_3)_3N$, will be a stronger donor than trimethylamine, $(CH_3)_3N$. However, the opposite is the case. Unlike trimethylamine, trisilylamine forms only very unstable complexes with BH_3 or BF_3[*81, 624*]. The electron pair of the nitrogen not used for σ-bonds or one of the orbital lobes of the sp^3-nitrogen orbitals occupied by two electrons, is non-bonding only in trimethylamine. In trisilylamine, the planar sp^2 configuration has lower energy, since here the doubly occupied p_z-orbital of the nitrogen can, if it is assumed that the Si–N bonds lie in the xy-plane, overlap with the free d_{xz}-orbitals of the silicon. This leads to the planar $(SiH_3)_3N$[*259, 353*]. Compare, however, the literature[*486a*].

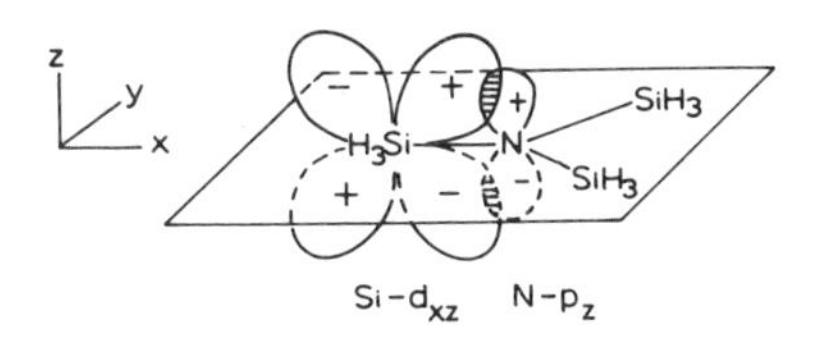

Re (*b*). For analogous reasons, silyl isothiocyanate, SiH_3NCS[*284*, *353*], and silyl isocyanate, SiH_3NCO[*182*], in contrast to the angular methyl isocyanate, CH_3NCO[*178*], and methyl azide, CH_3N_3[*445*], are linear, (sp configuration on the nitrogen of the Si compounds).

Re (*c*). The Si–O–Si bond angle in disiloxane, $H_3Si–O–SiH_3$[*111*, *369*], is larger than the C–O–C angle in dimethyl ether, $H_3C–O–CH_3$[*110*, *111*, *303*]. It is just through the d_π–p_π bond components that the Si–O σ-bond acquires some double-bond nature: $H_3Si–O–SiH_3$. Analogous considerations apply to the Si–N σ-bond: $(SiH_3)_2N–SiH_3$, S=C=N–SiH_3. However, care should be taken in predicting structures on the basis of a co-involvement of the silicon d-orbitals in bonds. According to what has been said above, silyl azide, SiH_3N_3, for example, should be linear. However, according to its microwave spectrum it possesses an angular structure, like the isologous methyl compound CH_3N_3[*148*].

Re (*d*). Since the electronegativity of silicon is lower than that of carbon, the dipole moment of the silicon–halogen bond should be substantially higher than that of the carbon–halogen bond. Comparison of the measured figures shows the opposite to be the case (Table 7.3).

TABLE 7.3
DIPOLE MOMENTS OF CHLORINE DERIVATIVES OF SILANE AND METHANE[*75*, *113*]

	[*Debye*]		[*Debye*]
SiH_3Cl	1.28	CH_3Cl	1.87
SiH_2Cl_2	1.17	CH_2Cl_2	1.56
$SiHCl_3$	0.85	$CHCl_3$	1.00

Here again, therefore, the electron density of the atom (Cl) bound to the silicon is lowered by back-coordination to the silicon[*113*, *490*]: $H_3Si–Cl$. The capacity for back-coordination (*i.e.* for the overlapping of p- and d-orbitals) decreases with increasing bond length, *i.e.* in the sequence SiF > SiCl > SiBr[*113*, *490*]. However, according to the Raman spectrum of $(CH_3)_3SiI$, the Si–I bond also possesses a considerable doublebond component[*243*].

Re (*e*). ESR measurements on $(CH_3)_3C(C_6H_5)$, $(CH_3)_3Si(C_6H_5)$, and $(CH_3)_3$-$Ge(C_6H_5)$ show that, as compared with the $(CH_3)_3C$ group, the $(CH_3)_3Si$ group, and to a smaller extent the $(CH_3)_3Ge$ group, have an electron-attracting effect (+I effect < −M effect). If the silyl or germyl group is separated from the π-electron system of the benzene ring by a methylene bridge, both groups show – as

References p. 621

was to be expected from the electronegativities – a fairly similar electron-releasing effect (+I effect > – M effect) [*56a*] (about hyperconjugation effect in $(CH_3)SiCH_2C_6H_5$ see ref. [*409a*]):

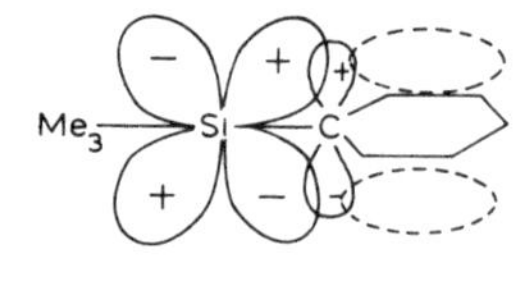

d_π–p_π double bond component of the Si–C bond in phenyltrimethylsilane, Me_3SiPh

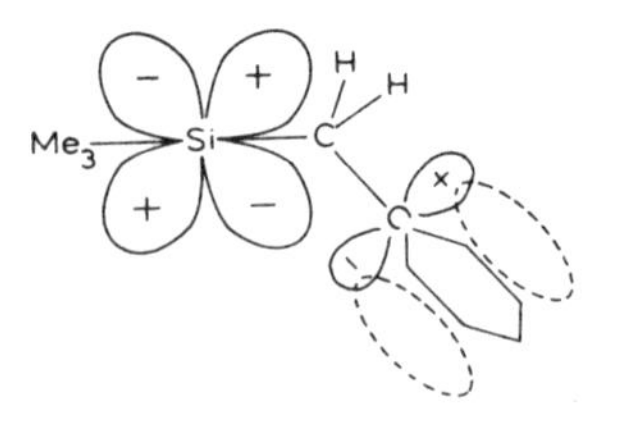

No double bond component of the Si–C bond in benzyltrimethylsilane, Me_3SiCH_2Ph

ESR measurements on

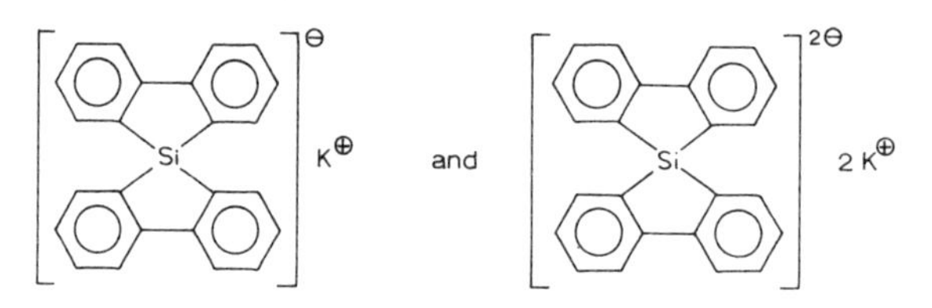

clearly show the involvement of the 3d-orbitals of the silicon in the Si–C bonds and conjugation of the two biphenyl groups through the silicon atom (coplanarity is not necessary here) [*106*].

On the basis of systematic hydrolysis experiments on p-substituted triarylsilanes, $(ZC_6H_4)_3SiH$, some σ_{Si}-constants, analogous to Hammett's or Taft's σ_C-constants, have been determined empirically. The following equations apply [*538*]:

$$\sigma_{si} = \sigma^{i} + \sigma^{m}_{si}; \quad \sigma^{m}_{si} \sim 0.75\,\sigma^{m}_{c}$$

Interactions may also take place between C≡C triple bonds and the d-orbitals of silicon and, to a smaller extent the d-orbitals of tin. Comparative measurements of ν_{CH} and ν_{CC} in $(CH_3)_3CC{\equiv}CH$, $(CH_3)_3SiC{\equiv}CH$; and $(CH_3)_3SnC{\equiv}CH$ show the situation in silicon: +I effect < –M effect (electron attraction); and in tin: +I effect > –M effect (reduced electron liberation) [*668a*].

In general, the tendency of a silyl group to liberate electrons corresponding to the electronegativitiy of silicon is shown only when no occupied non-bonding orbitals of suitable symmetry are in the neighbourhood. On the other hand, the silyl group has an electron-attracting effect when occupied (atomic or molecular) orbitals can overlap with the d-orbitals of silicon.

2.4 Mechanism of substitution on silicon

There are no true S_N1 reactions in silicon chemistry since the primary step $X_3SiY \rightarrow X_3Si^{\oplus} + Y^{\ominus}$ would require a considerably higher activation energy han the formation of a pentacovalent intermediate stage of lower energy $\overset{*}{Y}SiX_3Y$) using d-orbitals of silicon. Since the separation of the reactants generally akes place slowly, a pseudo-S_N1 mechanism usually predominates over a S_N2 nechanism in which addition of the reactant $\overset{*}{Y}$ would be the rate-determining tep. This is shown by a comparison of the S_N1 and the S_N2 mechanisms for carbon nd the pseudo-S_N1 mechanism for silicon:

$$S_N1 \quad CH_3Y \xrightarrow{slow} CH_3^{\oplus} + Y_K \xrightarrow{+\overset{*}{Y}} \qquad \rightarrow \overset{*}{Y}CH_3 + Y^{\ominus}$$

$$S_N2 \quad \overset{*}{Y}^{\ominus} + CH_3Y \rightarrow [\overset{*}{Y} \ldots CH_3\text{–}Y]^{\ominus} \rightarrow \qquad \rightarrow [\overset{*}{Y}\text{–}CH_3 \ldots Y]^{\ominus} \rightarrow \overset{*}{Y}CH_3 + Y^{\ominus}$$

$$\text{pseudo-}S_N1 \quad \overset{*}{Y}^{\ominus} + SiH_3Y \rightarrow \underset{\text{transition state I}}{[\overset{*}{Y} \ldots SiH_3\text{–}Y]^{\ominus}} \rightarrow \underset{\text{low-energy intermediate stage (sp}^3\text{d)}}{\overset{*}{Y}\text{–}SiH_3\text{–}Y} \xrightarrow{slow} \underset{\text{transition state II}}{[\overset{*}{Y}\text{–}SiH_3 \ldots Y]^{\ominus}} \rightarrow \overset{*}{Y}SiH_3 + Y^{\ominus}$$

The (idealized) transition states I and II, the sp^3d complex, and the energy poten- ial in the S_N2 and pseudo-S_N1 mechanisms, can be represented by the following liagrams:

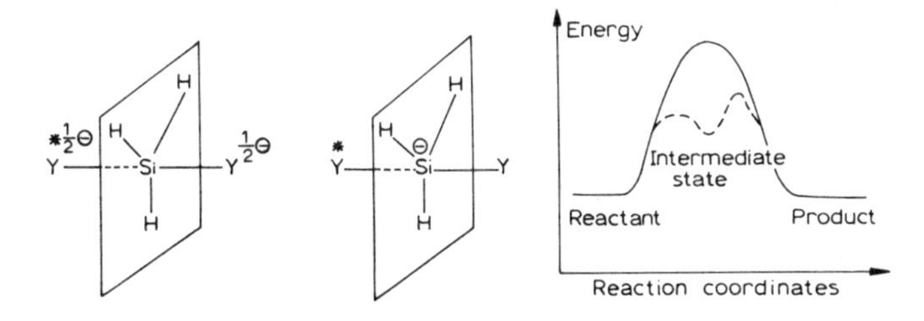

ransition state I or II in the tate of maximum energy (sp^2 onfiguration on the Si or C)

Stabilized transition state (sp^3d configuration on the Si)

—— Energy profile for an S_N2 mechanism
--- Energy profile for a pseudo-S_N1 mechanism
The broken curve would be symmetrical only if the two transition states I and II had the same life times

The attack of a nucleophilic agent must not necessarily take place "from the ack" (the Y–Si–Y angle in the sp^3d complex is 180°). In contrast to most organic eactions, in organosilicon hydrides with large organic substituents, the silicon can lso be attacked "from the side" (Y–Si–Y angle 90°). This is illustrated by the ollowing comparison: in 1-chlorobicyclo [2, 2, 1] heptane (I) the chlorine is extra- rdinarily unreactive [*52*].

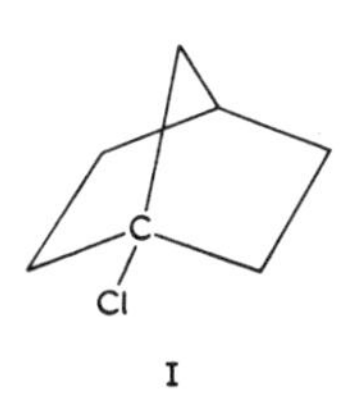

I

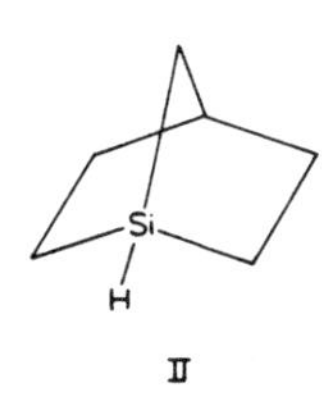

II

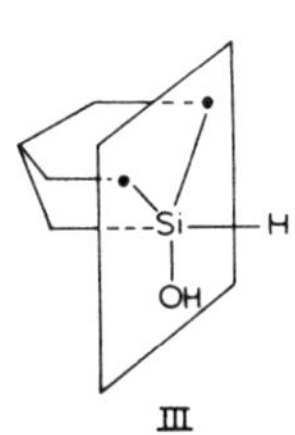

III

References p. 621

The reason for this is that because of the rigid ring angle the carbon atom at the bridgehead cannot assume the planar sp^2 configuration energetically favourable for a carbonium ion (*i.e.* for an S_N1 mechanism). Reaction by the S_N2 mechanism is inhibited by the impossibility of an attack from the back with subsequent inversion through the carbon ring. The analogously constructed 1-silabicyclo[2, 2, 1]heptane (II) behaves quite differently [*584*]. It hydrolyses rapidly by a pseudo-first-order reaction: $-d[\text{silane}]/dt = k_2[\text{silane}][OH^{\ominus}]$, and this substantially more rapidly than monocyclic five- or six-membered rings with one ring-Si atom. This is because, as can be seen in the case of III, a nucleophilic agent attacking from the side can form the low-energy intermediate stage III (with the sp^3d configuration). Similar considerations apply to monocyclic Si–C rings trimethylenemethylsilane, $\overline{CH_2CH_2CH_2Si}(CH_3)H$, hydrolyses to bis(trimethylenemethylsilyl) oxide through a pentacovalent intermediate stage IV with retention of the ring, or else the highly strained four-membered ring of the intermediate stage IV opens with retention of the Si–H bond (which is very stable here) [*583a*]

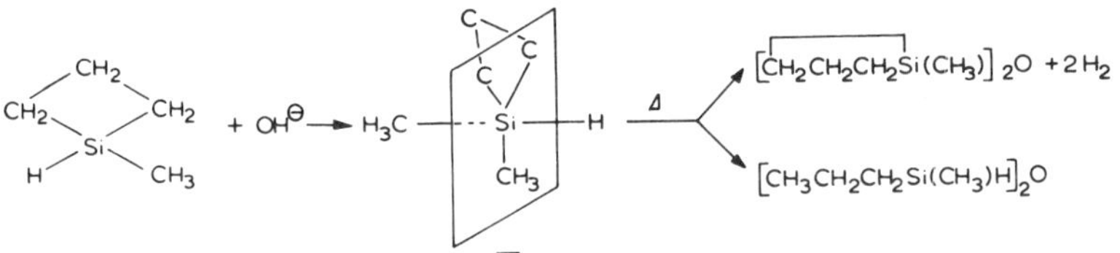

$$\text{(CH}_2)_3\text{Si(CH}_3)\text{H} + OH^{\ominus} \rightarrow \text{IV} \xrightarrow{\Delta} [\overline{CH_2CH_2CH_2Si}(CH_3)]_2O + 2H_2 \quad \text{or} \quad [CH_3CH_2CH_2Si(CH_3)H]_2O$$

IV

2.5 *Silicon double bonds*

As can be seen from section 2.4, silicon bonds with a higher order than 1 presuppose non-bonding p- or d-orbitals of the bond partner. In spite of many attempts to prepare a prototype of the class of compounds with p_π–p_π double bonds of silicon (such as $SiH_2{=}SiH_2$ or aromatic systems with silicon atoms incorporated), which would be very interesting from the theoretical point of view, no success has so far been achieved. The cyclic double-bond-containing products of the reaction of $(CH_3)_2Si$ (from the reaction $x\ Me_2SiCl_2 + 2x\ Na \rightarrow (Me_2Si)_x + 2x\ NaCl$) or of GeI_2 with $C_6H_5C{\equiv}CC_6H_5$ [*339, 655–657, 668*] (first described as the three-membered rings V and VI [*339, 655–657*] and later as the six-membered ring VII [*668*]) contain no benzenoid π-electron system. However, it is likely that the olefinic π-electrons are partially delocalized on d-orbitals of silicon (d_π–p_π bond). In fact, the C=C double bonds in VII are extraordinarily resistant to double bond agents (Br_2, $KMnO_4$, H_2–Pt). Possibly the attacks are also sterically

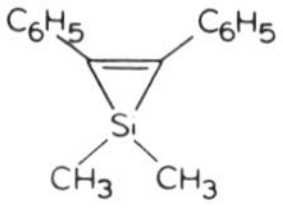

V 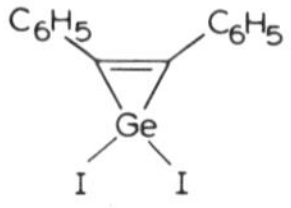 VI 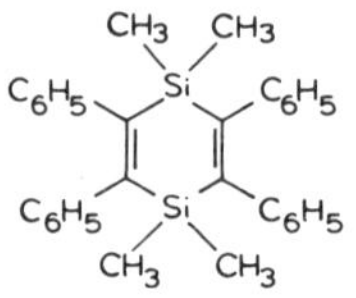 VII

hindered. So far there is no satisfactory explanation for the non-existence of a benzenoid system with the silicon incorporated. The greater covalent radius of the elements of high period numbers in comparison with the elements of the first short period has been discussed in this connection. This hinders an adequate overlapping of the 3p-orbitals with other 3p-orbitals or 2p-orbitals [*406, 474*]. Furthermore, a system would gain no energy by the formation of p_π–p_π double bonds precisely because partial d_π–p_π double bonds on silicon are possible [*406, 407*]. Many chemical, and particularly optical, properties of polysilanes and siloxene derivatives indicate a pseudo-benzenoid system of electrons through the interaction of the d-orbitals of neighbouring silicon atoms (d_π–d_π bond) [*533, 534*].

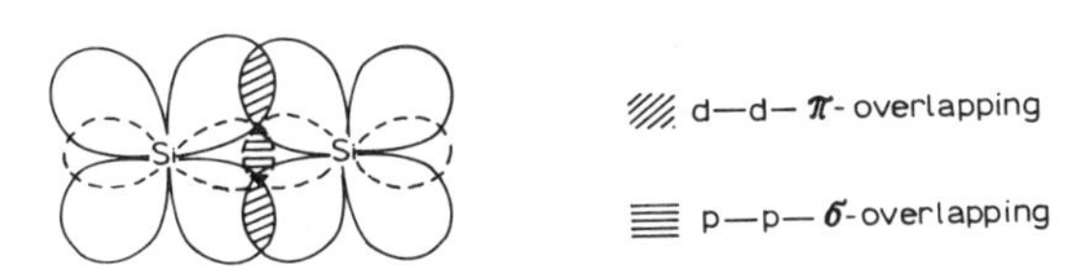

2.6 *Silyl anions, cations, and radicals*

With silanes, alkali metals form silyl anions (*e.g.*, R_3SiK, R = H, alkyl, aryl [*635*]). The stability of the silyl anions (with alkali metals as cations) increases in the sequence $H_3Si:^{\ominus}$ < $(alkyl)_3Si:^{\ominus}$ < $(aryl)_3Si:^{\ominus}$. Thus, it increases with increasing delocalization of the free electrons.

Because of the lower electronegativity of silicon, relatively stable silonium ions $SiH_3^{\oplus}$ or $(CH_3)_3Si^{\oplus}$, corresponding to the carbonium ions, are expected. However, such silyl cations have been observed only as particles in mass spectra.

The decomposition of silanes or partially alkylated silanes by pyrolysis or electric discharge, as well as many additions and substitutions in C–C multiple bond compounds, takes place through silyl radicals of the type $\cdot SiH_3$. Monosilanes, $R_{4-n}SiH_n$ (n = 1–4), form radicals by abstraction of $\cdot$H and silanes with Si–Si bonds by homolytic cleavage of these bonds.

2.7 *Summary*

The following factors are responsible for the higher reactivity of silicon compounds as compared with the corresponding carbon compounds.

From the thermodynamic point of view:

polarity: Si–X > C–X

bond energy: Si–Si < C–C; Si–H < C–H; Si–O ≫ C–O

From the kinetic point of view:

covalent radius: Si > C (lower screening of the silicon atom against nucleophilic attacks)

d-orbitals: they can be used to increase the coordination number in the intermediate stages only in the case of silicon, leading to a lowering of the activation energy

3. Silanes Si_nH_{2n+2} and SiH_n

There are two main groups of silanes containing only hydrogen and differing mainly by their physical properties: the low-molecular mainly volatile silanes ("*oligosilanes*" Si_nH_{2n+2}; $n = 1–8$; unbranched and branched chains; no cyclic silanes), and the high-molecular non-volatile silanes ("*polysilanes*" SiH_n; $n = 0–2$). The polysilanes form an almost continuous series from SiH_2 with infinite chains through the lepidoidal SiH with isolated Si layers and $SiH_{<1}$ with three-dimensional Si skeletons—with linkage of the $—SiH_2—$, $>SiH—$ and $>Si<$ groups in statistical sequence—to SiH_0 = silicon; in addition, there are all the amorphous forms of the same empirical formula and *adsorbates of hydrogen on silicon*. The following five routes can be used preparatively.

1. In compounds of the type $>Si–Z$ (Z = Hal, –OR, –O–, –SR), the ligand Z is replaced by H through the action of hydrogenating agents. The method is used mainly for the synthesis of oligosilanes (see section 3.1.).

2. In a metal silicide or in a metal–silicon alloy, one equivalent of metal is replaced by hydrogen through the action of protons, with uptake of (alloy) electrons and preservation of the Si skeleton. Oligosilanes are produced if the alloys contain short Si chains, while if they contain long chains or extended Si networks, the products are polysilanes (see section 3.2).

3. No new Si–Si bonds arise in either of the methods of preparation of silicon hydrides mentioned above. On the contrary, in hydrogenation operations care must be taken that they are retained. New Si–Si bonds can be synthesized by the reaction of halogenated monosilanes with metals in analogy with the Wurtz synthesis in organic chemistry (see section 3.3) and by the coupling of metallated hydrides with halogenated hydrides (see section 10.2.).

4. Dismutation of silanes. (a) Polysilanes dismute into oligosilanes and polysilanes with a lower hydrogen content ($SiH_x \rightarrow Si_nH_{2n+2} + SiH_y$; $x > y \geqslant 0$) (see section 3.4). (b) Oligosilanes dismute into monosilane and polysilane (*e.g.*, $Si_2H_6 \rightarrow SiH_4 + SiH_2$ (see section 3.3).

5. Higher silanes are produced from lower silanes, particularly monosilanes, by the supply of energy (spark discharge, photo-excited mercury atoms, and in particular silent electric discharge [*232*, *233*]). For further details see section 10. Figure 7.1 shows schematically the synthesis and degradation of the basic silanes.

3.1 *Special syntheses for monosilane and disilane*

3.1.1 *Monosilane*

Monosilane can be obtained, generally in very good yields, by the reaction of halogenosilanes, alkoxysilanes, or silica with metal hydrides (*e.g.*, LiH) or double hydrides (*e.g.*, $NaBH_4$, $LiAlH_4$, $LiAlD_4$) in anhydrous organic solvents (*e.g.*,

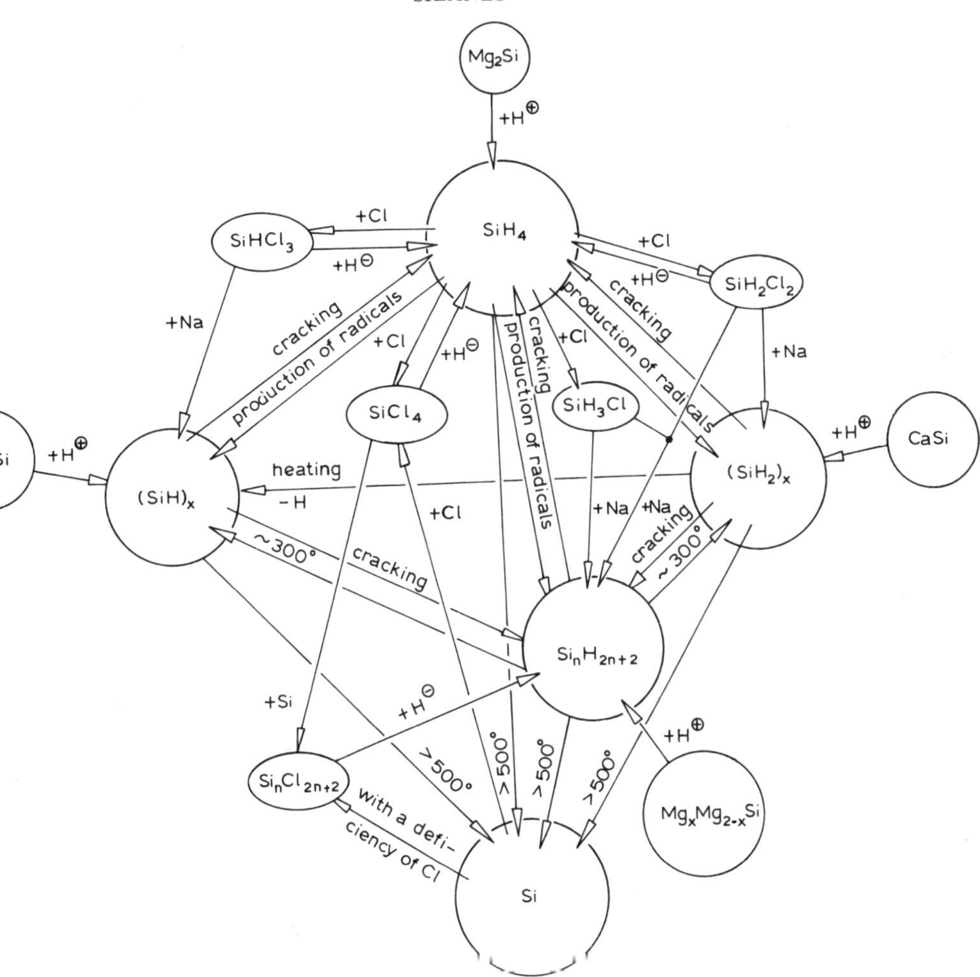

Fig. 7.1. Scheme of the synthesis, formation, and degradation of the basic silanes (in this scheme, *cracking* generally denotes cleavage of an Si–Si bond and *production of radicals* generally denotes cleavage of an Si–H bond)

diethyl ether, tetrahydrofuran, diethylene glycol dimethyl ether[*443*]), for example:

$$SiCl_4 + 4\ LiH \rightarrow SiH_4 + 4\ LiCl$$
$$SiHCl_3 + \tfrac{3}{4}\ Li\ AlD_4 \rightarrow SiHD_3 + \tfrac{3}{4}\ Li\ AlCl_4$$

Higher-boiling solvents are advantageous, as they can more easily be separated from the silane by distillation. The reactions of silicon chlorides ($SiCl_4$, $SiHCl_3$) with simple ether-insoluble hydrides take place with lower yields than those with ether-soluble double hydrides, since the solid hydride particles become covered with insoluble solid metal chloride. If $SiCl_4$ is passed over heated NaH or CaH_2, *i.e.* working completely without a solvent, elementary silicon and only a very

References p. 621

TABLE 7.4

PREPARATION OF MONOSILANE BY THE HYDROGENATION OF MONOSILANE DERIVATIVES

Abbreviations: Et = diethyl ether; Buet = dibutyl ether, Diox = 1,4-dioxan, THF = tetrahydrofuran, Teglyme = tetraethyleneglycol dimethyl ether, Diglyme = diethyleneglycol dimethyl ether, and Eut = eutectic

Compound to be hydrogenated A	*Hydrogenating agent B*	*Solvent*	*Reaction temperature (°C)*	*Yield of silane (mole-% referred to A)*	*Notes*	*Refs.*
$SiCl_4$	$LiAlH_4$	Et	−196...0	99		186
$SiCl_4$	$LiAlH_4$	Buet	—	—		123
$SiCl_4$	$LiAlH_4$	Diox	20...105	90–97		
$SiCl_4$	$LiAlH_4$	Et	20	99	High loss of Et with SiH_4 evolved	447
$SiCl_4$	$LiAlH_4$	Diox	20	< 99	Small loss of Diox, but smaller yield	447
$SiCl_4$	$LiAlH_4$	Teglyme	0	—	Simple separation of Teglyme and SiH_4	342
$SiCl_4$	$NaAlH_4$	Diglyme	< 0	100		703
$SiHCl_3$	$NaAlH_4$	Diglyme	−5	100	Exothermic	703
$SiCl_4$	LiH	Et	−196...0	≤ 89		186
$SiCl_4$	LiH	LiCl–KCl–Eut	359	73–100	Semicontinuous } Regeneration through electrolysis + H_2	629
$SiCl_4$	LiH	LiCl–KCl–Eut	359	88–97	Continuous } Regeneration through electrolysis + H_2	629
$SiCl_4$	$Al–AlCl_3–H_2$	$AlCl_3–NaCl$	170	27	Eut, 800 atm	280
$SiCl_4$	$Al–AlCl_3–H_2$	$AlCl_3–NaCl$	170	84	Eut, 750 atm	280
$SiCl_4$	$Al–LiAlH_4–H_2$	$AlCl_3–NaCl$	175	100	Eut, 950 atm	280
$SiCl_4$	$Al–LiAlH_4–H_2$	$AlCl_3–NaCl$	150	95	Eut, 850 atm	280
$SiCl_4$	$Al–NaH–H_2$	$AlCl_3–NaCl$	175	64	Eut, 800 atm	280
$SiCl_4$	$Al–BaH_2–H_2$	$AlCl_3–NaCl$	175	75	Eut, 820 atm	280
$SiCl_4$	$Al–SrH_2–H_2$	$AlCl_3–NaCl$	175	64	Eut, 850 atm	280
$SiCl_4$	$Al–i–Bu_2AlH–H_2$	$AlCl_3–NaCl$	175	75	Eut, 800 atm	280
$SiCl_4$	$Al–i–Bu_3Al–H_2$	$AlCl_3–NaCl$	150	9	Eut, 990 atm	280
$SiCl_4$	$Al–C_2H_5I–H_2$	$AlCl_3–NaCl$	150	10	Eut, 800 atm	280
K_2SiF_6	$Al–AlCl_3–H_2$	$AlCl_3–NaCl$	200	92	Eut, 800 atm	280
$BaSiF_6$	$Al–AlCl_3–H_2$	$AlCl_3–NaCl$	200	<92	Eut, 800 atm	280
$ZnSiF_6$	$Al–AlCl_3–H_2$	$AlCl_3–NaCl$	200	<92	Eut, 800 atm	280
Silicagel	$AlCl_3–NaCl–H_2$	—	200	62	400 atm	280
Silicagel	$AlCl_3–NaCl–H_2$	—	200	76	800 atm	280

SiO_2	$AlCl_3$–NaCl–H_2	—	250	20	850 atm	280
$NaSiO_3$	$AlCl_3$–NaCl–H_2	—	250	16	900 atm	280
$Si(OC_2H_5)_4$	$AlCl_3$–NaCl–H_2	—	200	12	750 atm	280
$Si(SC_2H_5)_4$	$AlCl_3$–NaCl–H_2	—	200	35	800 atm, together with SiH_4 and $(C_2H_5)_2SiH_2$	280
$Si(C_2H_5)_4$	$AlCl_3$–H_2	—	300	53	2500 atm, together with SiH_4 and ethylsilanes	280
$(SiCl_3)_2O$	$LiAlH_4$	Et	−196 . . . 20	100		194, 546
$(SiCl_3)_2O$	$LiAlH_4$	THF	0; −40	—		546
$(SiCl_3)_2O$	$LiAlH_4$	none	120	0	No formation of SiH_4	546
$(SiCl_3)_2O$	$LiAlH_4$	CCl_4	76.8	0	No formation of SiH_4	546
$(SiCl_3)_2O$	$LiAlH_4$	CCl_4	76.8	>0	Silane formation after addition of Diox	546
SiO_2	$LiAlH_4$	none	200	10	1 torr	665
SiO_2	$LiAlH_4$	none	200	10	1 atm in a current of helium	665
SiO_2	$LiAlH_4$	none	148–170	7		57a
$Si(OC_2H_5)_4$	$NaAlH_4$	Diglyme	−5 . . . 0	100		703
$SiH(OC_2H_5)_3$	$NaAlH_4$	Diglyme	−10 . . . −5	100		703
$SiH(OC_2H_5)_3$	$KAlH_4$	Diglyme	20 . . . 25	93		703
$(SiCl_3)_2O$	$LiBH_4$	Et	—	—	More B_2H_6 than SiH_4 is formed (3 : 1)	546
$(SiCl_3)_2O$	$NaBH_4$	Diox	105	0	No reaction	546
Na_2SiO_3	$K(Na)BH_4$	H_2O	0	0	No formation of SiH_4	294
$(SiCl_3)_2O$	NaH	Diox	105	0	No reaction	546
$CH_2:CH_2SiCl_3$	$NaBH_4$	Diglyme?	—	>0	(but $CH_2:CH_2SiMe_3 \rightarrow SiH_4$ [*551*])	664
$SiCl_4$	$LiAlD_4$	Teglyme	0	—	Simple separation of Teglyme and SiD_4	342
$SiHCl_3$	$LiAlD_4$	Et	−196 . . . 20	—		69
$SiHCl_3$	$LiAlD_4$	Et?	—	—		158
$SiHCl_3$	$LiAlD_4$	Teglyme	0	—	Simple separation of Teglyme and $SiHD_3$	342
$SiHBr_3$	$LiAlD_4$	Et	−196 . . . 20	—		447
$SiHI_3$	$LiAlD_4$	Et?	—	—		158
$Si(OR)_4$	Et_2AlH	—	20	80		290
$SiCl_4$	NaH–Et_3B	THF	20	94		286
SiF_4	NaH–Et_3B	THF	20	90		286
$SiCl_4$	NaH–Et_3B	Mineral oil	78	95		530
$Si(OEt)_2Cl_2$	NaH	Octane	250–320	80		288
$Si(OEt)_2F_2$	NaH	Octane	250–320	>80		288

small amount of silane are formed [*304, 625*]. In the absence of a donor solvent, lithium alanate has no hydrogenating effect. Even with a large excess of alanate, no lithium silanate is formed ($6\ LiAlH_4 + SiCl_4 \rightarrow Li_2SiH_6 + 4\ LiCl + 6\ AlH_3$) [*54a*]. Silicate anions do not react with the relatively water-resistant sodium or potassium boranate in aqueous systems. (In contrast, germanium compounds are hydrogenated to germane in such systems with high yields). Alkali metal or alkaline-earth metal hydrides dissolved in salt melts form silane in high yields with silicon halides. In this process, the metal hydride can be regenerated after (semi-continuous process) or simultaneously with (continuous process) the formation of silane by electrolysis and the introduction of hydrogen [*629*]:

$$SiCl_4 + 4\ LiH \rightarrow SiH_4 + 4\ LiCl$$
$$4\ LiCl \rightarrow 4\ Li + 2\ Cl_2$$
$$4\ Li + 2\ H_2 \rightarrow 4\ LiH$$
$$\overline{SiCl_4 + 2\ H_2 \rightarrow SiH_4 + 2\ Cl_2}$$

Silicon tetrachloride, fluorosilicates, silicates, and silicic acid esters and thioesters can be hydrogenated to monosilane with hydrogen in aluminium-containing salt melts (*e.g.*, $Al–AlCl_3–NaCl–H_2$) with often quantitative yields. Since aluminium chloride plays an essential role as activator (without $AlCl_3$ no SiH_4 forms), it must be assumed that chloroalanes (AlH_nCl_{3-n}) formed as intermediates are the true hydrating agent. The reaction mechanisms are not completely elucidated; very probably the hydrogenations do not take place via elementary silicon [*280*]. For examples of the reaction and references, see Table 7.4.

3.1.2 Disilane

Disilane and trisilane can be obtained similarly to monosilane:

$$2Si_2Cl_6 + 3LiAlH_4 \rightarrow 2Si_2H_6 + 3LiAlCl_4$$
$$Si_3Cl_8 + 2LiAlH_4 \rightarrow Si_3H_8 + 2LiAlCl_4$$

The Si–Si bond of the easily obtained hexachlorodisilane is preserved only if the operation is carried out at low temperatures, if the residence time of Si_2H_6 in the reaction solution is short (Si_2H_6 pumped off), and if an excess of lithium alanate is scrupulously avoided. This can be achieved in a simple manner either by spraying an $LiAlH_4$ solution or scattering dry powdered $LiAlH_4$ into a solution of Si_2Cl_6. Similar considerations apply to the hydrogenation of Si_3Cl_8. The technically important hydrogenation of hexachlorodisilane with hydrogen in salt melts has not yet been brought to a state of maturity. For examples of the reaction and references, see Table 7.5.

Disilane can also be synthesized in a Wurtz reaction. While only 1.7% of disilane is produced when silyl chloride is shaken with sodium amalgam at 20°, and little or none with liquid sodium–potassium alloy [*611*], liquid and gaseous silyl iodide react with sodium amalgam at 20° with yields of 40% of Si_2H_6 and

TABLE 7.5

PREPARATION OF DISILANE AND TRISILANE BY THE HYDROGENATION OF SUITABLE SILANE DERIVATIVES

Compound to be hydrogenated A	*Hydrogenating agent B*	*Solvent*	*Reaction temperature (°C)*	*Yield of Si_2H_6 or Si_3H_8 (mole-% referred to A)*	*Notes*	*Refs.*
Si_2Cl_6	$LiAlH_4$	EtOEt	0	87		186, 247
Si_2Cl_6	$LiAlH_4$	EtOEt	0	40–45	5–10% of SiH_4	64
Si_2Cl_6	$LiAlH_4$	EtOEt	$-120 \rightarrow -65$	98	Addition of B to A; pumping off of Si_2H_6	680a
Si_2Cl_6	$LiAlH_4$	EtOEt	−50	88	Addition of B to A; 12% of $SiH_4+(SiH_2)_x$	680a
Si_2Cl_6	$LiAlH_4$	EtOEt	20	35	Addition of B to A; 65% of $SiH_4+(SiH_2)_x$	680a
Si_2Cl_6	$LiAlH_4$	EtOEt	20	0	Addition of A to B; 100% of $SiH_4+(SiH_2)_x$	680a
Si_2Cl_6	$LiAlH_4$	EtOBu	0	40–45	5–10% of SiH_4	64
Si_2Cl_6	$LiAlH_4$	$(n\text{-}Bu)_2O$	0	40–45	5–10% of SiH_4	64
Si_2Cl_6	$LiAlH_4$	$(n\text{-}Bu)_2O$	0	80	7% of SiH_4; dry $LiAlH_4$ to Si_2Cl_6-soln.	1, 64
Si_2Cl_6	LiH	THF	20	0		
Si_2Cl_6	LiH	LiCl–KCl–LiH	359	?	Passage of $Si_2Cl_6+H_2$ into the melt	629
Si_2Cl_6	AlH_3	EtOEt	0	90	Addition of B to A; 10% of $SiH_4+(SiH_2)_x$	680a
Si_2Cl_6	Alanes	$AlCl_3$–NaCl–Al–H_2	?	"low"	Pressure	99
Si_2Cl_6	NaH–Et_3B	Mineral oil	−10	75	5% of SiH_4	286
$Si_2(OEt)_6$	$LiAlH_4$	EtOEt	20	"low"	Much SiH_4, $(SiH_2)_x$	683
Si_3Cl_8	$LiAlH_4$	EtOEt	$-120 \rightarrow -65$	97	Addition of B to A; pumping off of Si_3H_8	680a
Si_3Cl_8	$LiAlH_4$	EtOEt	20	28	72% of $SiH_4+2(SiH_2)_x$	680a

33% of SiH_4 and of 67% of Si_2H_6 and 13% of SiH_4, respectively. A reaction mechanism involving free silyl radicals is unlikely. Intermediate formation of silyl metal compounds [reactions (1) and (2)] appears more probable:

$$SiH_3I + Na/Hg\,(excess) \xrightarrow{\text{fast}} SiH_3HgI + Na \quad (1)$$

$$2SiH_3I + 2Na/Hg \xrightarrow{\text{fast}} (SiH_3)_2Hg + 2NaI + Hg \quad (2)$$

$$SiH_3I + (SiH_3)_2Hg \xrightarrow{\text{slow}} Si_2H_6 + SiH_3HgI \quad (3)$$

$$(SiH_3)_2Hg \xrightarrow{\text{slow}} SiH_4 + (SiH_2)_x + Hg \quad (4)$$

In the reaction of liquid SiH_3I the silylmetal concentration is high, so that reaction (4) comes to the fore, while gaseous SiH_3I (at low concentrations of silylmetal) forms disilane preferentially according to reaction (3).

3.2 *Synthesis of oligosilanes from silicides*

The action of acids (HCl or H_2SO_4 in H_2O; NH_4Cl or NH_4Br in liquid NH_3; N_2H_5Cl in N_2H_4) on magnesium–silicon alloys (2 moles of Mg + 1 mole of Si $\triangleq$ $[Mg_x \cdot Mg_{2-x}Si]_n$) forms homologous oligosilanes Si_nH_{2n+2} ($n = 1$–8) [*70, 99, 165, 171, 182, 183, 291, 292, 608*]. Examples of such reactions are given in Table 7.6. In aqueous systems the total yields of hydrides are low, and the products contain relatively large amounts of higher silanes, while in the ammonia and hydrazine systems the total yields are high and monosilane is formed almost exclusively. Both the total yields of hydrides and also the proportions of the homologous silanes depend markedly on the preparation of the alloys or silicides (particularly in the case of aqueous systems). The main role here is played by the temperature and time of tempering. The acid reactions probably take place by a mechanism similar to that in the decompositions of Mg–Ge alloys [*12*] (see Chapter 8, section 3.4). According to this, the Si–Si bonds are not formed in the course of the reaction with acid, as was previously assumed, but are preformed in the alloy or the silicide:

$$Mg_2Si \xrightarrow[-2\,Mg^{2\oplus}]{+4\,H^{\oplus}} SiH_4$$

$$[Mg_x \cdot Mg_{2-x}Si]_n \xrightarrow[-2n\,Mg^{2\oplus}]{+4n\,H^{\oplus}} Si_nH_{2n+2} + (n-1)\,H_2$$

where $x = 1 - 1/n = 0$ to 1, and correspondingly $n = 1$ to ∞.
This mechanism described in Chapter 8, section 3.4, transferred to silicon, probably applies both to aqueous and non-aqueous systems. The low yields of higher silanes in the NH_3 or N_2H_4 system are a consequence of the donor-catalysed cleavage of the Si–Si bond (see section 5.17).

The individual silanes SiH_4 to Si_6H_{14} can be separated by vacuum distillation or gas chromatography [*165, 608, 706*]. The isolation of isomeric silanes (*e.g.*, n-Si_4H_{10}, i-Si_4H_{10}) is achieved by gas chromatography [*70, 180, 181, 250*] (see section 3.7.10). About the separation of the silicon isotopes in SiH_4 by thermal diffusion see [*121a, 121b*].

TABLE 7.6

PREPARATION OF LIQUID SILANES BY THE REACTION OF Mg–Si ALLOYS OR Mg_2Si WITH ACIDS

Alloy or silicide			*Acid*			*Yield (mole-%) of volatile hydrides %Si of hydrides/ % Si of alloy*	*Notes*	*Refs.*
Moles Mg/ mole Si	*Preparation temp. (°C)*	*Tempering time (h)*	*Compound*	*System*	*Temp. (°C)*			
2.1	650	22	25% H_2SO_4	H_2O	100	Si_nH_{2n+2} ($n \geq 2$) 50, + SiH_4[a]	Apparatus for large amounts of higher silanes	182, 182a
2	620–650	24	HCl	H_2O	—	Si_nH_{2n+2} 35		171
2.2	400	24	15 wt.% HCl[b]	H_2O	<65	Si_nH_{2n+2} very low		165
2.2	500	24	15 wt.% HCl[b]	H_2O	<65	Si_nH_{2n+2} 16	Fractionation scheme given	165
2.2	600	24	15 wt.% HCl[b]	H_2O	<65	Si_nH_{2n+2} 27		165
2.2	650	24	15 wt.% HCl[b]	H_2O	<65	Si_nH_{2n+2} 38		165
2.2	700	24	15 wt.% HCl[b]	H_2O	<65	Si_nH_{2n+2} 30		165
2.2	800	24	15 wt.% HCl[b]	H_2O	<65	Si_nH_{2n+2} 17		165
2.2	—	—	—	—	—	SiH_4 4, Si_2H_6 3, Si_3H_8 1, Si_4H_{10} 0.5	Together with small amounts of Si_5H_{12}, Si_6H_{14}	608
2	620–650	24	15 wt.% HCl	H_2O	—	SiH_4 10, Si_2H_6 7, Si_3H_8 4	Together with higher silanes	183
2	620–650	24	N_2H_5Cl	N_2H_4	50–60	SiH_4 70, Si_2H_6 1–2,		183
2	1050	24?	N_2H_5Cl	N_2H_4	50–60	SiH_4 41, Si_2H_6?		183
2	—	—	20% H_3PO_4	H_2O	—	Si_nH_{2n+2} ($n = 1–8$)	Isomeric silanes	70
2	300–500	6	NH_4Cl	lig. NH_3	20	SiH_4	Preparation of SiH_4 on the semi-technical scale	99

[a]Composition of the crude gas (mole-%): SiH_4 40, Si_2H_6 25, Si_3H_8 16, Si_4H_{10}10, Si_5H_{12} 6, higher silanes (all the isomers of Si_6H_{14} and Si_7H_{16} and some isomers of Si_8H_{18}) 2.

[b]15 wt.-% HCl gives the highest yields in each case.

References p. 621

3.3 *Synthesis of polysilanes*

The in the ideal case linear polymeric silicon dihydride, polysilane(2), $(SiH_2)_x$, and the in the ideal case two-dimensional polymeric silicon monohydride, polysilane(1), $(SiH)_x$, are two special cases in the series of solid polymeric silanes of the general formula SiH_n ($2 \geq n > 0$).

3.3.1. *Polysilanes with an indefinite content of hydrogen ($SiH_{0.4-2.0}$)*

These can be prepared by the dismutation of higher volatile silanes by the supply of energy. Where *thermal energy* is supplied, even 20° is adequate, e.g. for pentasilane, Si_5H_{12}. Disilane, Si_2H_6, requires temperatures of up to 400°. Hydrogen, monosilane, and polysilanes are formed [*171, 276, 609*], for example:

$$Si_2H_6 \xrightarrow{\Delta} SiH_4, SiH_2, SiH, H_2$$

Since monosilane is the most thermally stable silicon hydride, it forms (alkyl) polysilanes to a relatively large extent only on pyrolysis in the presence of alkenes, probably through the intermediate addition of Si–H to C=C; for example, at 450° SiH_4 and $CH_2{=}CH_2$ give polymeric, probably two-dimensional $[SiCH_3]_x$ [*198*].

Other than by the supply of thermal energy, the dismutation can also be brought about by *electrical energy*; again polysilanes of variable composition ($SiH_{1.2-1.7}$) are formed [*498a, 547, 593*]. Their degree of polymerization appears to be lower than with polysilane of the same empirical formula formed in other ways. At 20° they slowly dismute into volatile silanes and polysilanes with a lower content of hydrogen [*547*]. In (sun)*light*, in addition to hydrogen and monosilane, the solid hydrides $SiH_{1.22}$, $SiH_{1.13}$, and $SiH_{1.12}$ are formed [*610*]. When a mixture of silane and mercury is irradiated with a mercury lamp (*reaction* of SiH_4 *with* $Hg6(^3P_1)$ *atoms*), polymeric $SiH_{0.4-0.9}$ is formed [*174, 498a*].

Polysilanes of variable composition are also obtained in the reaction of sodium or sodium amalgam with monochlorosilane or dichlorosilane:

$$2SiH_3Cl \xrightarrow[-2NaCl]{+2Na} SiH_4 + SiH_2$$

$$3SiH_2Cl_2 \xrightarrow[-6NaCl]{+6Na} SiH_4 + 2SiH$$

Amorphous polysilane(1) is also formed from tribromosilane and magnesium in ether, under conditions that must be accurately maintained [*540, 541*]:

$$2SiHBr_3 \xrightarrow[-3MgBr_2]{+3Mg} \frac{2}{x}(SiH)_x$$

It always contains bromine (although only in a very small amount). Since the polymerization takes place through the groups $-SiHBr_2$, $>SiHBr$, and $\rightarrow SiH$, not only bromine-free but also bromine-containing partners can occur. If the optimum reaction conditions are not scrupulously maintained, polysilanes with a higher content of hydrogen (*e.g.*, $SiH_{1.47}$) are obtained, since then, for example, $SiHBr_2$ groups can undergo dismutation:

$$2\ \text{-SiHBr}_2 \rightarrow \text{-SiH}_2\text{Br} + \text{-SiBr}_3$$

The completely amorphous habit, the lack of swellability, and the absence of a discontinuity in the oxidation reaction show the amorphous structure of the polysilane(1) prepared from $SiHBr_3$[*267, 539*].

3.3.2 *Lepidoidal polysilane(1)*

This is formed when calcium disilicide $CaSi_2$ (preformed layers of Si) is treated with acids:

$$CaSi_2 \xrightarrow[-Ca^{2\oplus}]{+2H^{\oplus}} \frac{2}{n}(SiH)_n$$

Suitable acids are HCl in absolute C_2H_5OH, $t < 15°$[*533, 543*], or $t = -50°$ to $200°$[*268*], and HBr in $AlBr_3$ melts, $t = 100$ to $200°$[*533, 544*]. The reaction in ethanol leads to sheet-like polysilane(1) still containing small amounts of chlorine and alkoxy groups. It is very likely that at reaction temperatures below 25° the silicon sheets remain unaffected (no Si–O–Si bonds). The Si sheets are only oxidized above 25°[*533*]. In the case of the reaction in aluminium bromide, the reaction product includes increasing amounts of bromine when the temperature is raised, as well as an increasing number of Si bonds of the three-dimensional type. When Ca_2Si is decomposed in $HCl + C_2H_5OH$, amorphous $SiH_{0.7}$ to $SiH_{0.9}$ is found[*506*].

3.3.3 *Polysilane(2)*

This is obtained by the reaction of calcium monosilicide CaSi with glacial acetic acid or hydrogen chloride in absolute ethanol[*547*]:

$$CaSi \xrightarrow[-Ca^{2\oplus}]{+2H^{\oplus}} SiH_2$$

According to the literature[*506*], the limiting formula SiH_2 is not achieved. In $HCl + C_2H_5OH$, amorphous $SiH_{0.7}$ or $SiH_{0.9}$ is found[*506*]. The decomposition of $Ca(Si_{0.5}Ge_{0.5})$ forms a large amount of SiH_4 together with the volatile mixed hydrides $SiGeH_6$, Si_2GeH_8, and $SiGe_2H_8$[*507*]. The decomposition of calcium silicides with acid probably resembles the decomposition of the calcium germanides (see Chapter 8, section 3.4).

In the hydrogenation of the benzene-soluble polymeric silicon dibromide $(SiBr_2)_x$, prepared by the reaction of $SiBr_4$ with Si at 1200°, with $LiAlH_4$ in ethereal solution, a 30% yield of (amorphous?) polysilane(2), $(SiH_2)_x$, is formed in an exothermic reaction. The small amount of LiBr which it contains cannot be decreased even by day-long extraction with ether. In addition, other, unidentified, silanes are formed[*529*].

References p. 621

3.4 *Synthesis of oligosilanes from polysilanes*

Like the isologous germanium compounds, both the polysilane(2) SiH_2 formed by the acid treatment of CaSi and the polysilane(1.68) $SiH_{1.68}$ formed from monosilane by electric discharge yield volatile silanes and hydrogen on being heated to 380°. For example, SiH_2 gives a 15% total yield of the following oligo-silanes: 25–30% of SiH_4, 10–16% of Si_2H_6, and 64–45% of the higher silanes [*547*]

$$SiH_2 \xrightarrow[\Delta]{380^\circ} SiH_4, Si_nH_{2n+2}, H_2$$

In the treatment of polysilane with acid, only hydrogen and no volatile silane is formed. In contrast, similar reactions of polygermane(2) yield volatile germanes (see Chapter 8, section 3.2).

3.5 *Synthesis of higher oligosilanes from monosilane*

Higher volatile silanes are produced in 63% yield together with polysilane(1.2)-(1.7) and hydrogen by the passage of monosilane through a discharge tube (ozonizer type)[*547, 593*]. The volatile component contains 66% of Si_2H_6, 23% of Si_3H_8, and 11% of higher silanes [*593*]. Disilane is formed when mono-silane is passed through a tube heated to 470–480° and the reaction gases are quenched [*195*]. Higher hydrides do not arise on heating in a static system (see section 3.7.1).

3.6 *Adsorption of hydrogen on silicon*

Although the adsorption of hydrogen on silicon has been little studied, the experimental material so far available shows with great probability that different types of adsorption may occur. The scale of the Si–H bond ranges from van der Waal's forces to something like a chemical bond. However, comparative measure-ments of the contact potential, photothreshold, and the changes in the resistance of thin silicon layers coated with hydrogen, and in particular uncoated, are lacking; they could show the influence of the adsorbed hydrogen on the silicon electrons and give further information on the Si–H bond. Three mutually overlapping but distinguishable Si–H bonds occur in the adsorption.

(a) With molecular hydrogen, saturation is reached in adsorption before the surface is covered with a monomolecular layer (in adsorption on Si wires at 300°K the Si : H_2 ratio is ~ 10^5 [*337*]). The conductivity and the photothreshold of a thin layer of silicon do not change, in contrast to adsorption with atomic hydro-gen [*160, 335, 336*] (probably van der Waals' bonding of molecular hydrogen to the silicon surface).

(b) With atomic hydrogen, a monomolecular coverage is achieved [*56, 335*] and the photothreshold rises [*160*]. The dipole moment of a hydrogen atom bound to the surface of the silicon is 0.15×10^{-18} esu [*160*]. Spectroscopic studies of the adsorption of H atoms in the range from 4.3 to 5.2 μ (Si–H stretching vibrations in $SiH_4 = 4.57\ \mu$) on highest-purity silicon surfaces showed the presence of Si–H bonds between the adsorbed H atoms and the atoms of the Si surface that were of different strengths but were always substantially weaker than the Si–H bond in silanes. In fact, a broad band with a maximum at 4.85 μ was observed. The H atoms bound only to the surface can easily be removed by being sputtered away with Ar atoms[*56*]. (Binding of atomic hydrogen to the silicon surface).

(c) Hydrogen ions penetrate the silicon layer to a depth of up to 300 Å with decreasing concentration. Starting from hydrogen relatively weakly bound to silicon in the neighbourhood of the surface (characterized by ν(Si–H) = 4.75 μ), with increasing depth of penetration the Si–H bond strength approaches that of an Si–H bond in silanes (characterized by ν(Si–H) = 4.68 μ).

Hydrogen bound in the interior of the layer of silicon can be removed only with great difficulty by sputtering with Ar ions[*56*] (almost true Si–H bonding).

Hydrogen passed over silicon at 1100 to 1200° transports silicon but the transport effect is lower than in similar experiments with germanium, although it is a hundred times greater than would have been expected for a pure vaporization process[*528*]. In the case of germanium the species undergoing transportation is the metastable GeH. According to thermodynamic calculations, supported by spectroscopic observations, it is unlikely that the isologous (metastable) gaseous SiH is the species undergoing transport[*50*].

3.7 *Physical properties of the silanes*

Thermodynamic bond energy E (Si–Si) = 46.4 kcal/mole (in Si_2H_6)[*245*]; 46.3 kcal/mole (in Si_2H_6)[*244*]; 45 kcal/mole[*474*]. The bond energy E(Si–H) = 76 (?) kcal/mole[*474*]. Activation energy for the dissociation of the Si–Si bond D (H_3Si–SiH_3) ~ 50 kcal/mole (in pyrolysis); D (H_3Si–SiH_3) = 81.3 kcal/mole (from the critical potentials of mass spectroscopy[*600*]; D (Cl_3Si–SiCl_3) = 85 kcal/mole[*600*]. For comparison: D (H_3C–CH_3) = 83 kcal/mole[*cit. 600*]. For further thermodynamic data, see the literature[*437*].

3.7.1 *Monosilane*

At room temperature SiH_4 is a colourless gas resistant to mercury, igniting in air in the pure state but not in the presence of a little nitrogen, carbon dioxide, or ethylene[*99*]. M.p. −185°[*355,608*]. B.p.: −111.5° (calculated)[*608*]: −111.86 ± 0.002°[*707*]; −94.16° (calculated from a modified boiling point equation by Egloff [*175*]: T_{boil} [°K] = $a \ln(n+b)+k$, where n is the number of central atoms of a volatile hydride, and a, b, and k are constants from the Table 7.7). Apart from the values for $n=1$, the equation gives figures agreeing very well with experiment. It is very suitable for the pre-calculation of boiling points of similar compounds. Density 0.68 g/ml (−186°, liquid)[*608*]. At −179.3° the vapour pressure is 1 torr [*248a*]. Vapour pressures: $\log p$ [torr] = $-662.6/T+6.9961$[*197*]; $\log p$ [torr] = $-929.05/T+1.75 \log T-0.017084\,T+7.5350$[*687*]; $\log p$ [torr] = $1054.380/T+6.1839 \log T-0.0346\,T+1.3470$[*707*]. For mixtures of CH_4, PH_3 and AsH_3 in SiH_4 see [*122a*]. Enthalpy of evaporation 3029[*687*]; 2687±2[*707*] cal/mole. Trouton's constant 17.3 cal mole^{-1}deg^{-1}[*687*]. Molar enthalpy of formation

References p. 621

TABLE 7.7
THE CONSTANTS a, b, AND k OF THE MODIFIED Egloff BOILING POINT EQUATION T_b [°K] $= a \ln(n+b)+k$[175]

	a	b	k
n-Alkanes	323.73	4.4	−416.31
Iso-2-methylalkanes	323.73	4.4	−424.51
Silanes	395.8	3.5	−416.31
Germanes	446.1	3.0	−416.31
Mono-n-alkylsilanes	321.1	5.2	−416.3
Poly-n-alkylsilanes	322.0	5.0	−416.3
Dimethyl-n-silanes	322.0	5.0	−424.5

(296.16°K, 1 atm): +7.8 ± 3.5 kcal (direct calorimetric method[*74*]; +7.3 kcal [*244*]. The equilibrium constant of the equilibrium $Si_{solid}+2H_{2\,gas} \rightleftharpoons SiH_{4\,gas}$: $K_p = 1.9\times10^{-10}$ (25°); 1.4×10^{-7} (450°) (calculated for $\Delta H_0 = 7.8$ kcal/mole). Consequently, at 450° only 1.4×10^{-5} (1 atm) or 1.4×10^{-3} (100 atm) mole-% of SiH_4 is present in the gas phase[*74*]. SiH_4 should decompose completely into the elements between 300 and 1300°K[*122b*]. For mass spectroscopic data and critical potentials, see [*517*], section 4.7 and Chapter 8, section 3.8, and also Table 7.18 of this chapter and Table 8.4 of chapter 8. For comparison with the critical potentials of the alkanes, see the literature[*319*]. SiH_4 dissolves only slightly in ether, benzene, carbon tetrachloride, chloroform, and carbon disulphide[*608*]. For retention times in gas chromatography, see under disilane (section 3.7.3). Heat of formation (gaseous) −11.3 kcal/mole (293.16°K)[*179a*]. For the viscosity of SiH_4, see the literature[*654a*].

Pyrolysis. Together with small amounts of hydrides of B, P, As, or Sb, silane can be decomposed thermally. In this process semiconductor layers of doped silicon are formed on surfaces of a semiconducting material[*550*]. At red heat, SiH_4 decomposes into silicon and hydrogen[*608*]. When silane is passed rapidly at a low pressure through a tube heated to 470−480°, yellow to brown and black solids soluble in alkali and disilane are formed in addition to silicon[*195*]. The pyrolysis of silane at 288° and 461° is a homogeneous first-order reaction[*122*].

3.7.2 Deuteromonosilane, SiD_4

M.p. −186.4°. B.p. −111.35° (extrapolated). At −159.95° the vapour pressure is 11.0 torr. Vapour pressures: $\log p$ [torr] $= -773.0/T + 1.75 \log T - 9.4026\times10^{-3}\,T + 5.31421$. Enthalpy of vaporization 2962 cal/mole. Trouton's constant 18.31 cal mole^{-1} deg^{-1}[*617a*]. SiD_4 and B_5H_9 undergo hydrogen transfer at 125°[*636a*].

3.7.3 Disilane

At room temperature Si_2H_6 is a colourless gas which decomposes very slowly, is resistant to mercury, and ignites in air. M.p. −132.5°[*355*]. B.p.: −15°[*608*]; −14.30[*242*]. At −118° the vapour pressure is 1 torr[*608*]. Vapour pressures: $\log p$ [torr] $= -1133/T + 7.2578$[*687*]; $\log p$ [torr] $= -1190.8/T + 1.75 \log T - 0.0040475\,T + 4.3099$[*687*]. Enthalpy of vaporization 5182 cal/mole. Trouton's constant 19.8 cal $mole^{-1}$ deg^{-1}[*687*]. Density 0.68 g/ml (−25°, liquid)[*608*]. Si_2H_6 dissolves in benzene and carbon disulphide. Mixtures with carbon tetrachloride or chloroform explode. Heat of formation: +17.1 kcal/mole (298.16° K, 1 atm)[*244*]; −35.8 kcal/mole (293.16°K)[*179a*].

Pyrolysis. On silicon-coated glass surfaces, gaseous disilane decomposes into silicon, a little polysilane, monosilane and hydrogen[*171,632a*]. The pyrolysis probably takes place through a closed sequence of reactions with atomic hydrogen as the chain carrier:

$$Si_2H_6 \rightarrow 2\cdot SiH_3$$
$$\cdot SiH_3 + Si_2H_6 \rightarrow SiH_4 + \cdot Si_2H_5$$
$$\cdot Si_2H_5 \rightarrow SiH_4 + Si + \cdot H$$
$$\cdot H + Si_2H_6 \rightarrow \cdot Si_2H_5 + H_2$$

The considerable induction period in the pyrolysis is reduced by the addition of hydrogen (→ formation of hydrogen atoms); this leads to a larger amount of silane:

$$\cdot SiH_3 + H_2 \rightarrow SiH_4 + \cdot H$$

During the pyrolysis of Si_2H_6, ethylene present is polymerized, 1 mole of C_2H_4 being polymerized by 1 mole of Si_2H_6; it is more likely that the polymerization-inducing agents are Si-containing radicals than H atoms[*171*].

Heat stability. Disilane purified by gas chromatography is more thermally stable than that purified by high-vacuum distillation. Thus, the concentration of Si_2H_6 in mole-% on storage in stainless steel cylinders decreases as follows (Si_2H_6 purified by gas chromatography) [Si_2H_6 purified by distillation]. At 30°: start (100) [97.7], after 49 days (100) [93.9], after 154 days (97.0) [84.2], after 244 days (97.5) [75.0]; at −78°: start (100), after 49 days (100), after 154 days (98.5) and after 244 days (98.6)[*250*]. About the mechanism of the thermolysis and photolysis see [*433a,522a*].

Gas chromatography. Disilane prepared by the hydrogenation of Si_2Cl_6 with $LiAlH_4$ and purified by fractionation in a low-temperature column still contains chlorosilanes and monosilane, according to both mass-spectroscopic and gas-chromatographic analyses [*e.g.* (mole-%), 87–88 Si_2H_6, 6–7 SiH_4, 0.1 SiH_2Cl_2, 0.1 $SiHCl_3$, 0.1 $SiCl_4$, 2–3 $(C_2H_5)_2O$, 1–2 H_2O, and 1–2 N_2 or air]. Pure Si_2H_6 can be isolated on copper columns (3660 × 19 mm) packed with silica gel. At 1500 ml of helium/min and a temperature of 75°, the retention time for Si_2H_6 is 35.5 min[*250*].

Mass-spectroscopic data[*518*]. The critical potential for the formation of $SiH_3^{\oplus}$ is 11.85 ± 0.05 eV and for that of $SiH_2^{\oplus}$ 11.94 ± 0.04 eV from Si_2H_6. For other critical potentials, species frequencies, etc., see Chapter 8, section 3.8, and for the critical potentials of the alkanes for comparison, see the literature[*319*].

3.7.4 Deuterodisilane, Si_2D_6

M.p. −130.15°. B.p. −15.35°. At −84.95° the vapour pressure is 14.74 torr. Vapour pressures: $\log p$ [torr] $= -1394.3/T - 7.1510 \times 10^{-3}T + 5.91428$. Enthalpy of vaporization 5081 cal/mole. Trouton's constant 19.71 cal $mole^{-1}$ deg^{-1}[*617a*].

3.7.5 *Trisilane.*

At room temperature, Si_3H_8 is a colourless mobile liquid which decomposes slowly but is stable to mercury[*608*]. M.p.: −117.2°[*248a, 355*]; −117.4°[*610*]. B.p. 53.05°[*617a*]. At −70°, the vapour pressure is 1 torr[*608, 610*]. Vapour pressures: $\log p$ [torr] $= -1559.2/T + 7.6764$[*687*]; $\log p$ [torr] $= -1713.6/T + 1.75 \log T - 0.0047506\, T + 5.2282$[*610, 687*]. Enthalpy of evaporation 7127 cal/mole. Trouton's constant 20.5 cal $mole^{-1}$ deg^{-1}[*610, 687*]. Heat of formation: +25.9 kcal/mole (298.16°K, 1 atm)[*244*]. Bond energy (Si–Si) 46.8 kcal/mole [*244*]. Density: 0.725 g/ml (0°, liquid)[*608*]; 0.743 g/ml (0°, liquid)[*610*]; 4.15 g/l (18°, gaseous)[*610*]. Si_3H_8 explodes on addition of carbon tetrachloride.

Pyrolysis. The thermal decomposition of Si_3H_8 takes place, like that of Si_2H_6, through a chain of reactions the chain carrier of which is atomic hydrogen. As in the case of disilane, silyl radicals play an important role. The primary step appears to be decomposition of Si_3H_8 into SiH_3 and SiH_2 radicals:

$$Si_3H_8 \rightarrow 2 \cdot SiH_3 + :SiH_2$$

Subsequently, $\cdot SiH_3$ is hydrogenated and $:SiH_2$ dehydrogenated[*171, 632a*].

3.7.6 *Deuterotrisilane, Si_3D_8*

Melting point −116.75°. Boiling point 51.85°[*617a*].

3.7.7 *Tetrasilane*

At room temperature, Si_4H_{10} (mixture of isomers) is a colourless liquid which decomposes fairly rapidly. Its vapour changes the form of a mercury meniscus (attack?)[*608*]. M.p.: −93.5°[*608*]; −84.3°[*165*]; ~−90°[*610*]; −88.2°[*633*]. B.p.: 100.0° (calculated)[*608*]; 107.4° (calculated)[*165*]; 109° (calculated)[*610*]. At 0° the vapour pressure is 7.8 or 9.1 torr[*165, 608*]. Vapour pressures: $\log p$ [torr] $= -2008.2/T + 8.2479$[*687*]; $\log p$ [torr] $= -2247.3/T + 1.75 \log T - 0.057617\, T + 6.4472$[*610*]. At 0° the vapour pressure of n-Si_4H_{10} is 10.0 ± 1 torr[*233*] and that of i-Si_4H_{10} is >5 torr[*233*]. Heat of evaporation: 8500 cal/mole (at the b.p.) [*165*]; 9179 cal/mole[*687*]. Trouton's constant 22.2 cal $mole^{-1}$ deg^{-1}[*165*]. Density: 0.79 g/ml (0°, liquid)[*608*]; 0.825 g/ml[*610*] (0°, liquid). Energy of formation for liquid n-Si_4H_{10}: −70.4 kcal/mole (293.16°K)[*179a*].

3.7.8 *Deuterotetrasilane, Si_4D_{10}*

Melting point −89.35°. Boiling point 106.85°[*617a*].

3.7.9 *Pentasilane to octasilane*

Si_5H_{12} (mixture of isomers): vapour pressure at 0° 1.5 torr[*608*]. Three isomers detected: n-Si_5H_{12}, m.p. −74.5°[*633*]. Si_6H_{14} (mixture of isomers): vapour pressure at 20° about 1 torr[*608*]; four isomers separated and detected[*182a, 633*]. Si_7H_{16} has been separated into two isomers (n- and iso-) and a mixture of isomers; Si_8H_{18} has been separated into n-Si_8H_{18} and a mixture of isomers[*182a, 633*].

3.7.10 *Isomeric silanes*

Isomeric silanes can be separated by gas chromatography[*70, 182a, 233, 250*]. A mixture of isomers prepared from $Mg_2Si + H_3PO_4$ can be separated on Celite-silicon oil 702 in a current of hydrogen at 110° into the following fractions (analytical amounts) 1 SiH_4, 1 Si_2H_6, 1 Si_3H_8, 2 Si_4H_{10}, 3 Si_5H_{12}, 4 Si_6H_{14}, 5 Si_7H_{16}, 3 Si_8H_{18}. As for the n-alkanes, plots of the logarithms of the specific retention volumes of these n-silanes against the number of silicon atoms are linear ($\log V_g = kn_{Si}$). Comparison of the sizes of the elution zones shows that linear n-silanes are formed preferentially in the acid decomposition of alloys[*70*]. Table 7.8 contains the retention times on helium-bathed squalene-kieselguhr columns [*182a*]. Large amounts of higher silanes that are liquid at room temperature and inflame in the air can be handled without danger in injection syringes, cannulas, and piercing caps[*179–181*].

TABLE 7.8

GAS CHROMATOGRAPHY OF SILANES[*182a*]; RETENTION TIMES IN MINUTES

Column: 15% squalane on kieselguhr
Carrier gas: helium
Column temperature: 110°

	A		*B*
Rate of flow (ml/min)	85		119
Pressure at the head of the column (kp/cm^2)	1.33		1.6
SiH_4	1.1	i-Si_7H_{16}	24
Si_2H_6	1.4		31
Si_3H_8	2.0		37
i-Si_4H_{10}	3.4		43
n-Si_4H_{10}	3.9		48
i-Si_5H_{12}	5.4		57
	7.6		62
n-Si_5H_{12}	9.5		76
i-Si_6H_{14}	13.5	n-Si_7H_{16}	95
	14.4	i-Si_8H_{18}	117
	19.5		132
	22.5		159
n-Si_6H_{14}	24.5	n-Si_8H_{18}	200

3.7.11 *Polysilanes*

Silicon dihydride, $(SiH_2)_x$, prepared by the hydrogenation of $(SiBr_2)_x$ with $LiAlH_4$ and stable at room temperature, forms ivory-coloured ether-insoluble flocs which ignite spontaneously in air[*529*]. The polysilanes $SiH_{1.2}$ and $SiH_{1.69}$, prepared from SiH_4 and Si_3H_8 respectively by the supply of electrical energy, are gold-coloured deposits with a metallic lustre, which also inflame in air. The polysilanes $SiH_{1.12-1.22}$ formed by the action of light on higher liquid silanes are solid, yellow, and insoluble in organic solvents[*547*]. The polysilane $(SiH_2)_x$ obtained by the acid treatment of CaSi is a light brown solid inflaming in air[*547*]. Amorphous polysilane(1) prepared from $SiHBr_3$ and Mg is a compact, hard, brittle,

References p. 621

lemon-yellow solid. It can be subdivided without a preferred direction of cleavage, inflames in air at 98 to 102°, and is insoluble in $(C_2H_5)_2O$, C_6H_6, CCl_4, $CHCl_3$, C_2H_5OH, and H_2O. It fluoresces weakly in UV light[*541*].

Depth of colour of the polysilanes. The depth of colour increases from light brown, almost white polysilane(2), $(SiH_2)_x$, to the dark-coloured "polysilane(O)" ("$(SiH_0)_x$", silicon or lepidoidal silicon $[Si_6]_x$[*304*]). The decisive factor for the depth of colour is the accumulation of Si–Si bonds (weak chromophore). For the Si–Si bond the involvement of one bond due to d-orbitals in addition to σ-bonding is postulated (d–d overlapping)[*534*]. Polysilanes absorb strongly in the visible region if one or more of the following conditions is satisfied[*534*]:

(1) very long Si–Si chains;
(2) $[Si_6]$ rings;
(3) interaction of (1) and/or (2) in the crystal association;
(4) as many as possible Si–Si bonds starting from *one* silicon atom; the absorption of light increases in the sequence from primary to quaternary silicon atoms:

$$\text{Si}-\overset{|}{\underset{|}{\text{Si}}}- \;<\; \text{Si}-\overset{|}{\underset{|}{\text{Si}}}-\text{Si} \;<\; \text{Si}-\text{Si}-\overset{|}{\underset{|}{\text{Si}}} \;<\; \text{Si}-\overset{\overset{\text{Si}}{|}}{\underset{\underset{\text{Si}}{|}}{\text{Si}}}-\text{Si}$$

(5) electron-donating ligands on the Si–Si bond system.

Structures of the polysilanes. Three structures have been proposed and discussed for polysilane(1)[*541*].

(a) A hexagonal layer lattice of trivalent SiH groups $([Si_6]H_6)_n$. This structure is probable for the lepidoidal polysilane(1), since the starting material $(CaSi_2)$ has preformed layers.

(b) Small three-dimensional sections of a silicon lattice. In these, silicon atoms on the surface of the section are saturated externally by hydrogen. The silicon atoms in the interior of the section form only Si–Si bonds. An Si:H ratio of 1:1 can be realized theoretically by a molecule of the size $Si_{48}H_{48}$ and larger. Such a structure should lead to a certain solubility of the polysilane, which, however, is not the case.

(c) A completely irregular polymerizate, preferably of trivalent SiH groups, which form rings with different numbers of members in statistical distribution. This structure is likely for the amorphous polysilane(1). X-ray interferences should be found only with the lepidoidal polysilane(1) (a). Because of the irregular corrugation of the layers, however, this structure is also amorphous to X-rays. In this it resembles the siloxenes prepared from crystalline calcium silicide, which are also amorphous to X-rays. Here, also, an irregular corrugation of the layers is assumed[*304*].

In the polysilane(2) obtained from calcium silicide CaSi, in the ideal case the zigzag chain of silicon atoms preformed in the CaSi should be present. However, this has not as yet been confirmed experimentally.

Production of Si–H groups on silica surfaces. Silica surfaces can be methyated by reaction with methanol. Degassing at 750°, the infrared bands due to the stretching vibrations of the methyl groups declined rapidly; simultaneously a strong Si–H stretching frequency appears [*401a*].

4. Organosilicon hydrides

The following five preparative routes can be used to obtain organosilicon hydrides.

1. Nucleophilic displacement of a negative ligand in organosilanes by a hydride ion $H^{\ominus}$ (silyl cation + $H^{\ominus}$ anion):

$$R_{3-n}X_n\overset{\delta+}{Si}-\overset{\delta-}{X} + H^{\ominus} \rightarrow R_{3-n}X_nSi-H + X^{\ominus}$$

Prototype reactions:

$$EtSiCl_3 + 3NaH \rightarrow EtSiH_3 + 3NaCl$$

$$EtSi(OEt)_2 + 2LiH \rightarrow EtSiH + 2LiOEt$$

2. Nucleophilic displacement of a negative ligand in substituted silicon hydrides by a carbanion $R^{\ominus}$ (H-containing silyl cation + carbanion):

$$R'_{3-n}H_n\overset{\delta+}{Si}-\overset{\delta-}{X} + R^{\ominus} \rightarrow R'_{3-n}H_nSi-R + X^{\ominus}$$

$$R'_{3-n}H_nSiX + RX + 2M \rightarrow [R'_{3-n}H_nSiX + RM + MX] \rightarrow R'_{3-n}H_nSiR + 2MX$$

$$(X = H, Hal;\ R = \text{alkyl, aryl, silyl};\ M = Li, Na)$$

Prototype reactions:

$$PhSiH_2X + PhMgBr \rightarrow Ph_2SiH_2 + BrMgX$$

$$SiHCl_3 + 3MeC_6H_4Li \rightarrow (MeC_6H_4)_3SiH + 3LiCl$$

$$SiHCl_3 + N_2CH_2 \rightarrow (CH_2Cl)SiHCl_2 + N_2$$

$$SiHCl_3 + 3BuCl + 6Na \rightarrow Bu_3SiH + 6NaCl$$

3. Protonation of an organosilylmetal compound:

$$>\overset{\delta-}{Si}-\overset{\delta+}{M} + H^{\oplus} \rightarrow >Si-H + M^{\oplus} \quad (M = Li, Na, K)$$

Prototype reaction:

$$Li_2(SiPh_2)_4 + 2HCl \rightarrow H_2(SiPh_2)_4 + 2LiCl$$

4. Abnormal Grignard reaction:

(a) $>Si-X + RCH_2CH_2MgX \rightarrow >Si-H + RCH{=}CH_2 + MgX_2$ (R = H, alkyl)

(b) $RH_2SiOSiH_2R + R'MgX$ → ($\rightarrow RH_2SiR' + RH_2SiOMgX$)
→ $RR'HSiOSiH_2R + HMgX \rightarrow RSiH_3 + RR'HSiOMgX$

References p. 621

Prototype reactions:

(a) $PhSiCl_3 + 3C_3H_7MgBr \rightarrow Ph(C_3H_7)_2SiH + C_3H_6 + 3MgClBr$

(b) $2[PhH_2Si]_2O + 3MeMgI \xrightarrow{\Delta} PhMeSiH_2, PhSiH_3, Ph_2SiOMgI, PhMeHSiOMgI$

5. Hydrosilation of alkenes or alkynes:

$$>Si{-}H + {-}C{\equiv}C{-} \rightarrow >Si{-}\overset{|}{C}{=}\overset{|}{C}{-}H$$

$$>Si{-}H + >C{=}C< \rightarrow >Si{-}\underset{|}{\overset{|}{C}}{-}\underset{|}{\overset{|}{C}}{-}H$$

Prototype reactions:

$$SiH_4 + C_2H_2 \rightarrow CH_2{=}CHSiH_3$$
$$Et_2SiH_2 + C_9H_{18} \rightarrow Et_2(C_9H_{19})SiH$$

4.1 *Synthesis of organosilicon hydrides by hydrogenation of organosilyl cations*

In the reaction of organohalogenosilanes $R_{4-n}SiX_n$ with metal hydrides (*e.g.*, LiH, R_2AlH) or double hydrides (*e.g.*, $NaBH_4$, $LiAlH_4$, $Al(BH_4)_3$, $LiAlD_4$) in anhydrous solvents (*e.g.*, diethyl ether, tetrahydrofuran, diglyme, dioxan, LiCl–KCl melts) the corresponding organosilicon hydride is generally formed in excellent yield, for example:

$$R_{4-n}SiX_n + n\,LiH \rightarrow R_{4-n}SiH_n + nLiX$$
$$R_{4-n}SiX_n + n\,R'_2AlH \rightarrow R_{4-n}SiH_n + n\,R'_2AlX$$
$$R_{4-n}SiX_n + \frac{n}{4}\,LiAlD_4 \rightarrow R_{4-n}SiD_n + \frac{n}{4}\,LiAlX_4$$

Table 7.9 contains examples of these reactions and the corresponding references.

In order to obtain good yields the (very dry) solvent and the reaction temperature must be adapted to the hydride to be prepared. Highly volatile air-sensitive organosilanes require strict exclusion of air, similarly to monosilane. Higher-boiling organosilanes insensitive to air can be obtained by mixing the hydride and the halide under nitrogen between 0° and the boiling point of the solvent. The beginning of the reaction between alanate and chloride, which usually takes place rapidly and smoothly, can be recognized from the formation of voluminous white complexes which do not settle. Only towards the end of the reaction does lithium chloride with good settling properties separate out. Direct distillation is not recommended for the separation of the silicon hydride and the reaction solution [*414*]. For example, in the presence of $AlCl_3$, excess $LiAlH_4$ cleaves the Si–C bonds of many organosilanes with the formation of SiH bonds. In addition, $AlCl_3$ catalyses the dismutation $R_3SiH \rightarrow R_2SiH_2 + R_4Si$ [*241*]. Refluxing to complete

[*Text continued p. 496*]

TABLE 7.9

SYNTHESIS OF ORGANOSILANES BY THE HYDROGENATION OF ORGANOSILICON HALIDES (Si–X → Si–H; Si–X → Si–D)

Halide	Hydrogenating agent, solvent, reaction temperature (°C)	Hydride	Yield (mole-%)	Ref.
		$RSiH_3$		
SiF_4	$(C_2H_5)_2AlH$, without solvent, pressure, 260°	$(C_2H_5)_{4-n}SiH_n$ (n=1,2,3)	?	290
CH_3SiCl_3	$LiAlH_4$, dioxan, 105°	CH_3SiH_3	80–90	632
CH_3SiCl_3	$LiAlH_4$, di-n-butyl ether	CH_3SiH_3	—	560
CH_3SiCl_3	$LiBH_4$, without solvent, 25°	—	0	315
$C_2H_5SiCl_3$	$LiAlH_4$, dioxan, 105°	$C_2H_5SiH_3$	80–90	632
$C_2H_5SiCl_3$	$NaH–(C_2H_5)_3B$, THF, 20°	$C_2H_5SiH_3$	96	286
n-$C_4H_9SiCl_3$	$LiAlH_4$, dioxan, 105°	n-$C_4H_9SiH_3$	80–90	632
i-$C_4H_9SiCl_3$	$LiAlH_4$, dioxan, 105°	i-$C_4H_9SiH_3$	80–90	632
n-$C_5H_{11}SiCl_3$	$LiAlH_4$, dibutyl ether	n-$C_5H_{11}SiH_3$	80	598
cyclo-$C_6H_{11}SiCl_3$	$LiAlH_4$, diethyl ether, 35°	cyclo-$C_6H_{11}SiH_3$	63	15, 136, 379
$CH_2{=}CHSiCl_3$	$LiAlH_4$, diethyl ether	$CH_2{=}CHSiH_3$	—	441
$CH_2{=}CHSiCl_3$	$LiAlH_4$, dioxan, 105°	$CH_2{=}CHSiH_3$	80–90	632
$C_6H_5SiCl_3$	$LiAlH_4$, diethyl ether, 20–35°	$C_6H_5SiH_3$	70	63, 186, 379, 598, 206
$ClCH_2SiCl_3$	$LiAlH_4$, di-n-butyl ether, 0°	$ClCH_2SiH_3$ (+SiH_4)	80–90	298
Cl_3CSiCl_3	$LiAlH_4$, di-n-butyl ether, 0° (explosion!)	SiH_4 (!)	0	298
$Cl_3Si(CH_2)_4SiCl_3$	$LiAlH_4$	$SiH_3(CH_2)_4SiH_3$	—	451
$CH_3CHClSiCl_3$	LiH, dioxan	$CH_3CHClSiH_3$	—	453
		$RSiH_{3-n}D_n$		
CH_3SiH_2Cl	$LiAlD_4$	CH_3SiH_2D	—	313
CH_2DSiH_2Cl	CH_2DSiH_2D	CH_2DSiH_2D	—	313
$CH_2DSiHCl_2$	$LiAlD_4$	CH_2DSiHD_2	—	313
CH_2DSiCl_3	$LiAlD_4$	CH_2DSiD_3	—	313
$CH_2{=}CHSiCl_3$	$LiAlD_4$, diethyl ether	$CH_2{=}CHSiD_3$	—	441
		R_2SiH_2		
$(CH_3)_2SiCl_2$	$LiAlH_4$ diethyl ether, 0°	$(CH_3)_2SiH_2$	—	462
$(CH_3)_2SiCl_2$	$LiAlH_4$, dioxan, 105°	$(CH_3)_2SiH_2$	80–90	632
$(CH_3)_2SiCl_2$	$(C_2H_5)_2AlH$, 20,	$(CH_3)_2SiH_2$, $(C_2H_5)_2SiH_2$	~100	290
$(CH_3)_2SiCl_2$	$(i\text{-}C_4H_9)_2AlH$	$(CH_3)_2SiH_2$, $(i\text{-}C_4H_9)_2SiH_2$	~100	290

References p. 621

TABLE 7.9 (cont'd)

Halide	*Hydrogenating agent, solvent, reaction temperature (°C)*	*Hydride*	*Yield (mole-%)*	*Ref.*
$(C_2H_5)_2SiCl_2$	$LiAlH_4$, diethyl ether	$(C_2H_5)_2SiH_2$	—	186
$(C_2H_5)_2SiCl_2$	$LiAlH_4$, dioxan, 105°	$(C_2H_5)_2SiH_2$	80–90	632
$(C_2H_5)_2SiCl_2$	LiH, dioxan, 105°	$(C_2H_5)_2SiH_2$	66	186
$(C_2H_5)_2SiCl_2$	NaH, dioxan, 105°	—	0	186
$(C_2H_5)_2SiCl_2$	$NaH + AlCl_3$, dioxan, 105°	$(C_2H_5)_2SiH_2$	22	186
$(C_2H_5)_2SiCl_2$	LiH, LiCl–KCl melt, 359° (regeneration by electrolysis + H_2)	$(C_2H_5)_2SiH_2$	89	629
$(C_2H_5)_2SiF_2(Cl_2)$	NaH–$(C_2H_5)_3B$, THF, 20°	$(C_2H_5)_2SiH_2$	95	286
$(C_2H_5)_2SiF_2$	NaH, without solvent, 280–320°	$(C_2H_5)_2SiH_2$	88	288
$(C_2H_5)_2SiF_2$	$Na + H_2$, without solvent or in octane 135 atm, 290°	$(C_2H_5)_2SiH_2$	93	288
$(C_3H_7)_2SiCl_2$	$LiAlH_4$, diethyl ether, 35°	$(C_3H_7)_2SiH_2$	80	186
$(C_6H_5)_2SiCl_2$	$LiAlH_4$, diethyl ether, 35°	$(C_6H_5)_2SiH_2$	76	63, 598
$(C_6H_5)_2SiCl_2$	$LiAlH_4$, THF	$(C_6H_5)_2SiH_2$	> 76	229
$\overline{CH_2(CH_2)_3Si}Cl_2$	$LiAlH_4$, di-n-propyl ether	$\overline{CH_2(CH_2)_3Si}Cl_2$	—	666
$\overline{CH_2(CH_2)_3Si}Cl_2$	LiH, diisopentyl ether	$\overline{CH_2(CH_2)_3Si}Cl_2$	—	475
$\overline{CH_2(CH_2)_4Si}Cl_2$	$LiAlH_4$, diethyl ether, 35°	$\overline{CH_2(CH_2)_4Si}H_2$	31	666
$\overline{CH_2(CH_2)_5Si}Cl_2$	$LiAlH_4$, diethyl ether, 35°	$\overline{CH_2(CH_2)_5Si}H_2$	27	666
		R_2SiD_2		
$(CH_3)_2SiCl_2$	$LiAlD_4$, diethyl ether, 0°	$(CH_3)_2SiD_2$	—	462
		R_3SiH		
$(CH_3)_3SiCl$	$LiAlH_4$, dioxan, 105°	$(CH_3)_3SiH$	80–90	632
$(CH_3)_3SiCl$	$LiAlH_4$, di-n-butyl ether, 0°	$(CH_3)_3SiH$	—	465
$(CH_3)_3SiCl$	$NaBH_4$, KBH_4, $\leq$ 130°	—	0	241
$(CH_3)_3SiCl$	$Al(BH_4)_3$	Me_2SiH_2, Me_3SiH, Me_4Si	—	241
$(CH_3)_3SiCl$	$Na + H_2$, without solvent or in octane, 130 atm, 280°	$(CH_3)_3SiH$	$\leq$ 90	288
$(C_2H_5)_3SiF$	$LiAlH_4$	$(C_2H_5)_3SiH$	—	38, 137
$(C_2H_5)_3SiCl$	$LiAlH_4$, dioxan, 105°	$(C_2H_5)_3SiH$	80–90	632
$(C_2H_5)_3SiI$	LiH, dioxan	$(C_2H_5)_3SiH$	—	138
$(C_2H_5)_3SiF$	$(C_2H_5)_2AlH$ or $(i\text{-}C_4H_9)_2AlH$, 20°	$(C_2H_5)_3SiH$	100	290
$(C_2H_5)_3SiF$	$Na + H_2$, without solvent, 130 atm, 300°	$(C_2H_5)_3SiH$	90	288

$(C_2H_5)_3SiCl$	$Na + H_2$, THF, 60–80°	$(C_2H_5)_3SiH$	100	286
$\overline{CH_2(CH_2)_4}Si(CH_3)Cl$	$LiAlH_4$, diethyl ether, 35°	$\overline{CH_2(CH_2)_4}Si(CH_3)H$	—	666
CH_2—CH_2—$SiCl$—CH_2—CH_2—CH—CH_2 (ring)	$LiAlH_4$, diethyl ether, 0°	CH_2—CH_2—SiH—CH_2—CH_2—CH—CH_2 (ring)	60	585
$[\overline{CH_2(CH_2)_4CH}]_3SiCl$	$LiAlH_4$	$(C_6H_{11})_3SiH$	—	415
$(CH_2{=}CH)(CH_3)_2SiCl$	$LiAlH_4$, diethyl ether	$(CH_2{=}CH)(CH_3)_2SiH$	—	114
$(CH_3)_2(C_6H_5)SiF$	$LiAlH_4$	$(CH_3)_2(C_6H_5)SiH$	—	38, 137
$(C_6H_5)_3SiCl$	NaH, octane, 150 atm, 180–200°	$(C_6H_5)_3SiH$	81	288
$(ZC_6H_4)(C_6H_5)_2SiCl$	$LiAlH_4$, diethyl ether, 20–35°	$(ZC_6H_4)(C_6H_5)_2SiH$	55–83	219
[Z = *p*-Cl, *m*- or *p*-CH_3, *m*- or *p*-$(CH_3)_2N$, *p*-CH_3O]				
		R_3SiD		
$(CH_3)_3SiCl$	$LiAlD_4$, di-n-butyl ether, 0°	$(CH_3)_3SiD$	—	465
$(CH_2Cl)(CH_3)_2SiCl$	$LiAlD_4$, di-n-butyl ether, 45° (compare CH_2ClSiH_3!)	$(CH_2D)(CH_3)_2SiD$	—	465
$(CHCl_2)(CH_3)_2SiCl$	$LiAlD_4$, di-n-butyl ether, 45° (compare CH_2ClSiH_3!)	$(CHD_2)(CH_3)_2SiD$	—	465
		Silanes with several Si atoms	—	
$[(C_6H_5)_3Si](C_6H_5)_2SiCl$	$LiAlH_4$, diethyl ether	$[(C_6H_5)Si](C_6H_5)_2SiH$	—	227, 298
$(CH_3)_3SiCH_2SiCl_3$	$LiAlH_4$	$(CH_3)_3SiCH_2SiH_3$	—	200
$(CH_3)_2ClSiCH_2SiCl_3$	$LiAlH_4$	$(CH_3)_2HSiCH_2SiH_3$	—	200
$Cl_2HSiCH_2SiCl_3$	LiH, isopentanol	$H_3SiCH_2SiH_3$	—	53, 452
$CH_3Cl_2SiCH_2SiCl_3$	$LiAlH_4$	$CH_3H_2SiCH_2SiH_3$	—	200
$Cl_3SiCH_2SiCl_3$	$LiAlH_4$, di-n-butyl ether	$H_3SiCH_2SiH_3$	—	200, 451, 452
$(CH_3)_3SiCH_2Si(CH_3)_2Cl$	$LiAlH_4$	$(CH_3)_3SiCH_2Si(CH_3)_2H$		200
$(CH_3)_2ClSiCH_2Si(CH_3)_2Cl$	$LiAlH_4$	$(CH_3)_2HSiCH_2Si(CH_3)_2H$	—	200
$(CH_3)_3SiCH_2SiCH_3Cl_2$	$LiAlH_4$	$(CH_3)_3SiCH_2SiCH_3H_2$	—	200
$(CH_3)_2ClSiCH_2SiCH_3Cl_2$	$LiAlH_4$	$(CH_3)_2HSiCH_2SiCH_3H_2$	—	200
$(CH_3)_3SiCH_2SiCl_2CH_2SiCl_3$	$LiAlH_4$	$(CH_3)_3SiCH_2SiH_2CH_2SiH_3$	—	200
$Cl_3SiCH_2SiCl_2CH_2SiCl_3$	$LiAlH_4$	$H_3SiCH_2SiH_2CH_2SiH_3$	—	200
$Cl_3SiCH_2SiCl_2CH_2SiCH_3Cl_2$	$LiAlH_4$	$H_3SiCH_2SiH_2CH_2SiCH_3H_2$	—	200
$[Cl_2Si\text{-}CCl_2]_3$	—	$[H_2Si\text{-}CCl_2]_3$	—	200a

References p. 621

TABLE 7.9 (cont'd)

Halide	*Hydrogenating agent, solvent, reaction temperature (°C)*	*Hydride*	*Yield (mole-%)*	*Ref.*
Cl_2Si–CH_2–$SiCl_2$–CH_2–$SiCl_2$–H_2C	$LiAlH_4$	H_2Si–CH_2–SiH_2–CH_2–SiH_2–H_2C	—	200
Cl_2Si–CH_2–$SiCl_2$–CH_2–$Si(Cl)(CH_3)$–H_2C	$LiAlH_4$	H_2Si–CH_2–SiH_2–CH_2–$Si(H)(CH_3)$–H_2C	—	200
Cl_2Si–CH_2–$Si(CH_3)(Cl)$–CH_2–$Si(Cl)(CH_3)$–H_2C	$LiAlH_4$	H_2Si–CH_2–$Si(CH_3)(H)$–CH_2–$Si(H)(CH_3)$–H_2C	—	200
$(Cl)(H_3C)Si$–CH_2–$Si(CH_3)(Cl)$–CH_2–$Si(Cl)(CH_3)$–H_2C	$LiAlH_4$	$(H)(H_3C)Si$–CH_2–$Si(CH_3)(H)$–CH_2–$Si(H)(CH_3)$–H_2C	—	200
Cl_2Si–CH_2–$Si(CH_3)_2$–CH_2–…–H_2C	$LiAlH_4$	H_2Si–CH_2–$Si(CH_3)_2$–CH_2–Si…–H_2C	—	200

$\overline{Si(Cl)(CH_3)-CH_2-Si(CH_3)(Cl)-CH_2-Si(CH_3)_2-CH_2}$	$LiAlH_4$	$\overline{Si(H)(CH_3)-CH_2-Si(CH_3)(H)-CH_2-Si(CH_3)_2-CH_2}$	—	200
$\overline{(H_3C)_2Si-CH_2-Si(CH_3)(Cl)-CH_2-Si(CH_3)_2-CH_2}$	$LiAlH_4$	$\overline{(H_3C)_2Si-CH_2-Si(CH_3)(H)-CH_2-Si(CH_3)_2-CH_2}$	—	200
$C_6H_4(SiCl_2)_2CH_2$ (benzo-fused ring)	$LiAlH_4$	$C_6H_4(SiH_2)_2CH_2$ (benzo-fused ring)	—	200
$C_6H_4[Si(CH_3)Cl][SiCl_2]CH_2$ (benzo-fused ring)	$LiAlH_4$	$C_6H_4[Si(CH_3)H][SiH_2]CH_2$ (benzo-fused ring)	—	200
$C_6H_4[Si(CH_3)Cl]_2CH_2$ (benzo-fused ring)	$LiAlH_4$	$C_6H_4[Si(CH_3)H]_2CH_2$ (benzo-fused ring)	—	200

References p. 621

the reaction is therefore also problematic. Consequently, before fractionation the reaction mixture should be hydrolysed or (milder treatment) the hydride be extracted from the reaction solution (ether, LiCl, $AlCl_3$, $LiAlH_4$) with petroleum ether.

In the hydrogenation of organosilicon halides containing unsaturated ligands with boronates, under certain circumstances additions occur as well as substitution (Si–X → Si–H; Si–C → Si–H). For example, with $NaBH_4$ in tetraethylene glycol dimethyl ether $CH_2{=}CHSiCl_3$ does not form $CH_2{=}CHSiH_3$ but SiH_4[*664*]. On the other hand, in the reaction with $NaBH_4+AlCl_3$, $CH_2{=}CHSiMe_3$ adds to the B–H bonds of BH_3 with the formation of $[Me_3SiCH_2CH_2]_2B[CH(CH_3)SiMe_3]$ [*551*].

Generally, organosilicon *chlorides* are hydrogenated. Triorganosilicon *fluorides*, which can be prepared rapidly and easily by passing gaseous SiF_4 into ethereal Grignard solutions, can also be hydrogenated to triorganosilanes. However, in many cases the similar volatilities of fluoride and the hydride complicate the latter's isolation[*38*, *137*, *290*]. In general, because of the solubility of all the reaction products in ether, organosilicon *bromides* and *iodides* react with lithium alanate even more readily than the chlorides. However, such reactions have rarely been described (for example see [*477*]).

Although the ether-insoluble lithium hydride and deuteride and sodium hydride react with greater difficulty than the ether-soluble double hydrides (*e.g.* lithium alanate) (incubation period), in many cases they offer advantages since in hydrogenation with $LiAlH_4$ the aluminium chloride formed often interferes with the process: as mentioned above, it catalyses the dismutation of the hydrides formed (Si–C cleavage). Furthermore, under certain circumstances it acts on the organic groups themselves, for example by catalysing the polymerization of C=C double bonds. Hydrogenation with LiH, LiD, or $NaH+AlCl_3$ is always carried out in high-boiling ethers or in the absence of a solvent at high temperatures. On prolonged boiling, the ethers are attacked. Previously prepared sodium hydride is unnecessary, since for the (technical) synthesis of silanes the hydrogenating agent is prepared in the reaction vessel itself from sodium and hydrogen. The incubation period is avoided by the addition of triethylboron (intermediate formation of boranate). Alkylalkoxysilanes can be converted into the corresponding alkylsilicon hydrides by means of dialkylalanes, lithium alanate, or lithium hydride [*293*, *479*, *671*]:

$$R_{4-n}Si(OR')_n + n\,R''_2AlH \rightarrow R_{4-n}SiH_n + n\,R''_2Al(OR')$$

$$R_{4-n}Si(OR')_n + \frac{n}{4}\,LiAH_4 \rightarrow R_{4-n}SiH_n + \frac{n}{4}\,LiOR' + \frac{n}{4}\,Al(OR')_3$$

$$R_{4-n}Si(OR')_n + n\,LiH \rightarrow R_{4-n}SiH_n + n\,LiOR'$$

The silicon–alkoxide bond is resistant to lithium boranate so that alkylalkoxyhalogenosilanes, $R_aSiX_b(OR')_c$ ($a+b+c = 4$), can be hydrogenated to alkylalkoxysilanes, $R_aSiH_b(OR')_c$[*315*]. Peralkylated siloxanes are converted into trialkylsilanes by reaction with dialkylalanes, with sodium and hydrogen under pressure (→ NaH), and with triethylaluminium at elevated temperatures (→ $(C_2H_5)_2AlH + C_2H_4$)[287–289]:

$$R_3SiOSiR_3 + R'_2AlH \rightarrow R_3SiH + R'_2AlOSiR_3$$
$$R_3SiOSiR_3 + NaH \rightarrow R_3SiH + NaOSiR_3$$

Examples of the hydrogenation and references are contained in Table 7.10.

TABLE 7.10

SYNTHESIS OF SILANES AND ORGANOSILANES BY THE HYDROGENATION OF ORGANOALKOXYSILANES AND SILOXANES WITH $LiAlH_4$ AND LiH (Si–OR → Si–H) AND WITH $LiBH_4$ (Si–OR ↛ Si–H)

Alkoxide, siloxane	*Hydrogenating agent, solvent, reaction temperature* (°C)	*Hydride*	*Yield (mole-%)*	*Ref.*
$SiH(OC_2H_5)_3$	$KAlH_4$	SiH_4	—	485
$(C_3H_7)_3SiOC_2H_5$	$LiAlH_4$, di-n-butyl ether	$(C_3H_7)_3SiH$	—	671
$(C_2H_5)_2Si(OC_2H_5)_2$	$LiAlH_4$	—	0	671
$(C_2H_5)_2Si(OC_2H_5)_2$	LiH, diisopentyl ether	$(C_2H_5)_2SiH_2$	—	671
n-$C_6H_{13}Si(OC_2H_5)_3$	$LiAlH_4 + AlCl_3$	n-$C_6H_{13}SiH_3$	66	671
$(C_6H_5)_2Si(OC_2H_5)_2$	$LiAlH_4$, di-n-butyl ether	$(C_6H_5)_2SiH_2$	—	671
$[(C_6H_5)H_2Si]_2O$	$LiAlH_4$, diethyl ether	$C_6H_5SiH_3$	59	252
$SiHCl_2(OC_2H_5)$	$LiBH_4$, without solvent, 20°	$SiH_3(OC_2H_5)$	40	315
$SiHCl(OCH_3)_2$	$LiBH_4$, without solvent, 25°	$SiH_2(OCH_3)_2$	63	315
$SiHCl(OC_2H_5)_2$	$LiBH_4$, without solvent, 20°	$SiH_2(OC_2H_5)_2$	22	315
$CH_3SiCl_2OC_2H_5$	$LiBH_4$, ether, 20°	$CH_3SiH_2OC_2H_5$	99	315
$CH_3SiCl_2(OC_2H_5)$	$LiBH_4$, without solvent, 25°	$CH_3SiH_2(OC_2H_5)$	high	315
$(CH_3)_3SiOSi(CH_3)_3$	$(i\text{-}C_4H_9)_2AlH$, without solvent	$(CH_3)_3SiH$	81	287
$(CH_3)_3SiOSi(CH_3)_3$	Na + H_2, without solvent, 195 atm, 255°	$(CH_3)_3SiH$	—	288
$(CH_3)_3SiOSi(CH_3)_3$	$(C_2H_5)_3Al$, without solvent	$(CH_3)_3SiH(+ C_2H_4)$	5	289
$(CH_3)_3SiOSi(CH_3)_3$	$(C_2H_5)_3Al$, without solvent, pressure, 200–240°	$(CH_3)_3SiH(+ C_2H_4)$	72–93	289
$(C_2H_5)_3SiOSi(C_2H_5)_3$	$(C_2H_5)_3Al$, without solvent, 190–200°	$(C_2H_5)_3SiH(+ C_2H_4)$	90	289
$(CH_3)_2Si(OC_2H_5)_2$	$(C_2H_5)_2AlH$, without solvent, 20°	$(CH_3)_2SiH_2$, $(C_2H_5)_2SiH_2$	~ 100	290
$(CH_3)_2Si(OC_2H_5)_2$	$(i\text{-}C_4H_9)_2AlH$, without solvent,	$(CH_3)_2SiH_2$, $(i\text{-}C_4H_9)_2SiH_2$	~ 100	290

4.2 *Synthesis of organosilicon hydrides by alkylation (arylation) of silyl cations*

In the reaction of (organo)silanes $R_{3-n}H_n\overset{\delta+\delta-}{Si\text{–}H}$, (organo)halogenosilanes $R_aH_b\overset{\delta+}{Si}\text{–}\overset{\delta-}{X_c}$ $(a+b+c=4)$ or alkoxysilanes $SiH(OR)_3$ with organylmetals (R′Li, R′MgX) or N_2CH_2 in organic solvents, organosilicon hydrides are formed by replacement of hydrogen, halogen, or alkoxyl by the organic substituent. Examples of such reactions and the references can be found in the tables mentioned in connection with the equations below.

$R_{4-n}SiH_n + m\,R'Li \rightarrow R_{4-n}R'\,SiH_{n-m} + m\,LiH$ (Table 7.11)
$(n = 3,4;\ m = 2,3)$

$SiH_4 + n\,(C_2H_5)_2Mg + n\,(C_2H_5)_2O \xrightarrow{(C_2H_5)_3Al} (C_2H_5)_nSiH_{4-n} + n\,C_4H_{10} + n\,C_2H_5OMgH$ (Table 7.11)

$R_{4-n}SiH_n + m\,R'MgX \rightarrow R_{4-n}\,R'_mSiH_n + m\,HMgX$ (Table 7.12)
$(n = 1,2,3;\ m = 1,2)$

$R_aSiH_bX_c + c\,R'Li \rightarrow R_aSiH_bR'_c + c\,LiX$ (Table 7.13)
$(a = 0,2;\ c = 1,2;\ a+b+c = 4)$

$R_aSiH_bX_c + c\,R'MgY \rightarrow R_aSiH_bR'_c + c\,XMgY$ (Table 7.14)
$(a = 0,1,2;\ b = 1,2,3;\ c = 1,2,3;\ a+b+c = 4)$

$SiH(OR')_3 + 3\,RMgX \rightarrow SiHR_3 + 3\,(R'O)MgX$ (Table 7.14)

Alkyl halides and sodium can be used ("Wurtz reaction") in place of the alkyl-metals (see Table 7.15):

$SiHX_3 + 3RX + 6Na \rightarrow SiHR_3 + 6NaX$

TABLE 7.11

SYNTHESIS OF ORGANOSILANES BY THE ALKYLATION OF SILANES WITH ALKYLMETALS (Si–H → Si–R)

Silane	*Alkylmetal, solvent reaction temperature* (°C)	*Organosilane*	*Yield (mole-%)*	*Ref.*
SiH_4	CH_3Li, OPr_2, 20°	$(CH_3)_{4-n}SiH_n$ $(n = 0–3)$	—	243a
SiH_4	C_2H_5Li, petroleum ether, 20–40°	$(C_2H_5)_2SiH_2$	27	415, 447
SiH_4	C_2H_5Li, petroleum ether, 20–40°	$(C_2H_5)_3SiH$	36	447
SiH_4	i-C_3H_7Li, petroleum ether, 20–40°	(i-$C_3H_7)_3SiH$	38	447
SiH_4	1-$C_{10}H_7Li$, petroleum ether, 20–40°	(1-$C_{10}H_7)_3SiH$	39	447
$C_6H_5SiH_3$	C_2H_5Li, petroleum ether	$(C_6H_5)(C_2H_5)_2SiH + (C_6H_5)(C_2H_5)_3Si + (C_6H_5)_4Si$	—	407, 414
SiH_4	$(C_2H_5)_2Mg + (C_2H_5)_3Al$, diethyl ether	$(C_2H_5)_nSiH_{4-n} + nC_4H_{10} + C_2H_5OMgH$	—	54

The size of Tables 7.12 and 7.14 shows clearly that the reaction of a hydride or halide with the Grignard reagent is the most important method of preparation for organosilicon hydrides by alkylation. Alkylsilicon hydrides have also been obtained from the reaction of disilane with alkyllithium (Si–Si bond cleavage) [*243a*].

4.3 *Synthesis of organosilicon hydrides by protonation of organosilyl anions*

The action of acids (HCl in water; $CH_3COOH + C_2H_5OH$ in petroleum ether; cyclohexyl bromide $C_6H_{11}Br$ = "$C_6H_{10} \cdot HBr$") on peralkylsilylmetals, which are

TABLE 7.12

SYNTHESIS OF ORGANOSILANES BY THE ALKYLATION OF ORGANOSILICON HYDRIDES WITH GRIGNARD REAGENTS (Si–H → Si–R)

Silane	Grignard compound, solvent, reaction temperature (°C)	Organosilane	Yield (mole-%)	Ref.
		R_2SiH_2		
$C_6H_5SiH_3$	C_6H_5MgBr, THF, 20°	$(C_6H_5)_2SiH_2$	66	228, 229
$C_6H_5SiH_3$	C_6H_5MgBr, diethyl ether, 20°	$(C_6H_5)_2SiH_2$	52	228
$C_6H_5SiH_3$	n-$C_{12}H_{25}MgBr$, THF, 20–65°	$(C_6H_5)(C_{12}H_{25})SiH_2$	78	228
$C_6H_5SiH_3$	$C_{10}H_{21}MgBr$, THF, 65°	$(C_6H_5)(C_{10}H_{21})SiH_2$	62	229
$C_6H_5SiH_3$	$C_6H_5CH_2MgCl$, THF, 65°	$(C_6H_5)(C_6H_5CH_2)SiH_2$	67	229
$C_6H_5SiH_3$	4-$CH_3OC_6H_4MgBr$, THF, 65°	$(C_6H_5)(4\text{-}CH_3OC_6H_4)SiH_2$	75	229
$C_6H_5SiH_3$	2-$CH_3C_6H_4MgBr$, THF, 65°	$(C_6H_5)(2\text{-}CH_3C_6H_4)SiH_2$	87	229
$C_6H_5SiH_3$	4-$C_6H_5OC_6H_4MgBr$, THF, 65°	$(C_6H_5)(4\text{-}C_6H_5OC_6H_4)SiH_2$	74	229
		R_3SiH		
$\overline{CH_2(CH_2)_4Si}H_2$	CH_3MgBr	$\overline{CH_2(CH_2)_4Si}(CH_3)H$	< 47	666
$(C_6H_5)(n\text{-}C_{12}H_{25})SiH_2$	$C_6H_5CH_2MgCl$, THF, 65°	$(C_6H_5)(n\text{-}C_{12}H_{25})(C_6H_5CH_2)SiH$	63	228
$(C_6H_5)_2SiH_2$	n-C_4H_9MgCl, THF, 65°	$(C_6H_5)_2(C_4H_9)SiH_2$	72	228, 229
$(C_6H_5)_2SiH_2$	$C_{10}H_{21}MgBr$, THF, 65°	$(C_6H_5)_2(C_{10}H_{21})SiH$	70	229
$(C_6H_5)_2SiH_2$	$BrMg(CH_2)_5MgBr$, THF, 65°	$(C_6H_5)_2HSi(CH_2)_5SiH(C_6H_5)_2$	43	229
$(C_6H_5)_2SiH_2$	$CH_2{=}CHCH_2MgCl$, THF, 65°	$(C_6H_5)_2(CH_2{=}CHCH_2)SiH$	77	229
$(C_6H_5)_2SiH_2$	$C_6H_5CH_2MgCl$, THF, 65°	$(C_6H_5)_2(C_6H_5CH_2)SiH$	70	229
$(C_6H_5)(C_{12}H_{25})SiH_2$	$C_6H_5CH_2MgBr$, THF, 65°	$(C_6H_5)(C_{12}H_{25})(C_6H_5CH_2)SiH$	63	229
$C_6H_5SiH_3$	C_6H_5MgBr, THF, 65°	$(C_6H_5)_3SiH$	67	229
$(C_6H_5)_2SiH_2$	C_6H_5MgBr, THF, 65°	$(C_6H_5)_3SiH$	79	228, 229
$(C_6H_5)_2SiH_2$	C_6H_5MgBr, diethyl ether, 35°	$(C_6H_5)_3SiH$	31	228, 229
$(C_6H_5)(2\text{-}CH_3C_6H_4)SiH_2$	3-$CH_3C_6H_4MgBr$, THF, 65°	$(C_6H_5)(2\text{-}CH_3C_6H_4)(3\text{-}CH_3C_6H_4)SiH$	73	229
$(C_6H_5)(4\text{-}CH_3OC_6H_4)SiH_2$	2-$CH_3C_6H_4MgBr$, THF, 65°	$(C_6H_5)(4\text{-}CH_3OC_6H_4)(2\text{-}CH_3C_6H_4)SiH$	32	229
$(C_6H_5)(2\text{-}CH_3C_6H_4)SiH_2$	4-$C_6H_5OC_6H_4MgBr$, THF, 65°	$(C_6H_5)(2\text{-}CH_3C_6H_4)(4\text{-}C_6H_5OC_6H_4)SiH$	56	229
$(C_6H_5)(4\text{-}CH_3OC_6H_4)SiH_2$	4-$C_6H_5OC_6H_4MgBr$, THF, 65°	$(C_6H_5)(4\text{-}CH_3OC_6H_4)(4\text{-}C_6H_5OC_6H_4)SiH$	60	229
$(C_6H_5)(4\text{-}C_6H_5OC_6H_4)SiH_2$	2-$C_6H_5OC_6H_4MgBr$, THF, 65°	$(C_6H_5)(4\text{-}C_6H_5OC_6H_4)(2\text{-}C_6H_5OC_6H_4)SiH$	43	229

TABLE 7.13

SYNTHESIS OF ORGANOSILANES BY THE ALKYLATION OR SILYLATION OF HALOGENOSILANES WITH ALKYLMETALS, SILYLMETALS, OR DIAZOMETHANE

[Si–X → Si–R; Si–X → Si–(SiR$_2$H)]

Halogenosilane	*Alkylmetal or silylmetal or CH_2N_2*	*Solvent reaction temperature (°C)*	*Organosilane*	*Ref.*
$SiHCl_3$	o-$CH_3C_6H_4Li$	diethyl ether, 35°	$(o\text{-}CH_3C_6H_4)_3SiH$	226
$(C_6H_5)_2SiHCl$	$(C_6H_5)_3SiLi$	diethyl ether, 20°	$(C_6H_5)_3SiSi(C_6H_5)_2H$	686
$(C_6H_5)_2SiCl_2$	$(C_6H_5)_2HSiLi$	diethyl ether, 20°	$H[Si(C_6H_5)_2]_3H$	686
$Cl[Si(C_6H_5)_2]_3Cl$	$(C_6H_5)_2HSiLi$	diethyl ether, 20°	$H[Si(C_6H_5)_2]_5H$	686
$Cl[Si(C_6H_5)_2]_{2n+1}Cl$	$(C_6H_5)_2HSiLi$	diethyl ether, 20°	$H[Si(C_6H_5)_2]_{2n+3}H$	686
$(C_6H_5)_2ClSiSiCl(C_6H_5)_2$	$(C_6H_5)_2HSiLi$	diethyl ether, 20°	$H[Si(C_6H_5)_2]_{2n}H$	686
$(C_6H_5)_2SiHCl$	$Li[Si(C_6H_5)_2]_4Li$	diethyl ether, 20°	$H[Si(C_6H_5)_2]_6H$	686
$SiHCl_3$	$N{\equiv}NCH_2$	diethyl ether −60 to −70° Cu-powder	$(CH_2Cl)SiHCl_2$	555

easy to prepare, forms alkylsilanes through the replacement of the metal by hydrogen[*686*, *77*]:

$$PhSiK + C_6H_{11}Br \longrightarrow [PhSi \cdots K, Br, H\text{–}C_6H_{10}] \longrightarrow (C_6H_5)_3SiH + C_6H_{10} + KBr$$

$$LiSiPh_2SiPh_2SiPh_2SiPh_2Li + 2H^{\oplus} \rightarrow HSiPh_2SiPh_2SiPh_2SiPh_2H + 2Li^{\oplus}$$

4.4 *Synthesis of organosilicon hydrides by abnormal Grignard reactions*

In the successive replacement of the chlorine in phenyltrichlorosilane, certain sterically hindered Grignard compounds RMgBr (R = cyclohexyl, cyclopentyl, isopropyl, isobutyl) replace the last chlorine atom not by the alkyl but by hydrogen. With the reduction of alkyl to alkene, phenyldialkylsilane[1] is formed[*253*, *375*, *542*, *647*] for example:

$$PhSiX_3 + 2\ \overline{CH_2(CH_2)_4CH}MgBr \xrightarrow{-2\,MgBrX} Ph(C_6H_{11})_2SiX$$

$$Ph(C_6H_{11})_2SiX \xrightarrow{+C_6H_{11}MgBr} [Ph(C_6H_{11})_2Si \cdots X, MgBr, H\text{–}C_6H_{10}] \longrightarrow Ph(C_6H_{11})_2SiH + C_6H_{10} + MgBrX$$

[1] Compare the reduction that frequently takes place in the reaction of ketones with Grignard compounds:

$$R_2C{=}O + H\text{–}\overset{|}{\underset{|}{C}}\text{–}\overset{|}{\underset{|}{C}}\text{–}MgBr \longrightarrow [R_2C^{\oplus}\cdots H\cdots C\text{–}C\text{–}MgBr\cdots O] \longrightarrow R_2C(H)\text{–}O\text{–}MgBr + {>}C{=}C{<}$$

TABLE 7.14

SYNTHESIS OF ORGANOSILANES BY THE ALKYLATION OF HALOGENOSILANES OR ALKOXYSILANES WITH GRIGNARD REAGENTS (Si–X → Si–R or Si–OR → Si–R)

Halogenosilane	*Grignard compound, solvent, reaction temperature* (°C)	*Organosilane*	*Yield (mole-%)*	*Ref.*
		$RSiH_3$		
SiH_3Br	CH_2DMgBr	CH_2DSiH_3	—	313
SiH_3Br	CD_3MgBr	CD_3SiH_3	—	313
SiH_3Br	$^{13}CH_3MgBr$	$^{13}CH_3SiH_3$	—	313
SiH_3Br	C_3H_7MgBr, di-n-butyl ether	$C_3H_7SiH_3$	62	439
SiH_3Br	C_4H_9MgBr, di-n-butyl ether	$C_4H_9SiH_3$	53	439
SiH_3Br	*tert*-C_4H_9MgX, di-n-butyl ether	*tert*-$C_4H_9SiH_3$	low	439
SiH_3Br	$C_5H_{11}MgX$, di-n-butyl ether	$C_5H_{11}SiH_3$	58	439
SiH_3Br	$\overline{CH_2(CH_2)_4CH}MgBr$, di-n-butyl ether	$\overline{CH_2(CH_2)_4CH}SiH_3$	38	439, 15
SiH_3I	$CH{\equiv}CMgBr$, THF, 20°	$CH{\equiv}CSiH_3$	15	146
SiH_3I	$CH_3C{\equiv}CMgBr$, di-n-butyl ether, 20°	$CH_3C{\equiv}CSiH_3$	80	456, 500
SiH_3Br	$CH{\equiv}CMgBr$, di-n-butyl ether	$SiH_3C{\equiv}CSiH_3$	35	348
SiH_3Br	$BrMgC{\equiv}CMgBr$, di-n-butyl ether	$SiH_3C{\equiv}CSiH_3$	—	348
SiH_3Br	*p*-$CH_3OC_6H_4MgBr$, di-n-butyl ether	*p*-$CH_3C_6H_4SiH_3$	low	439
SiH_3Br	1-$C_{10}H_7MgBr$, di-n-butyl ether	1-$C_{10}H_7SiH_3$	25	439
		R_2SiH_2		
SiH_2Cl_2	$BrMg(CH_2)_5MgBr$, diethyl ether, 35°	$\overline{CH_2(CH_2)_4Si}H_2$	27	666
SiH_2Br_2	$C_5H_{11}MgX$, di-n-butyl ether	$(C_5H_{11})_2SiH_2$	87	439
$C_6H_5SiH_2Br$	CH_3MgBr, diethyl ether, 20°	$(C_6H_5)(CH_3)SiH_2$	57	251
$C_6H_5SiH_2Br$	C_2H_5MgBr, diethyl ether, 20°	$(C_6H_5)(C_2H_5)SiH_2$	26	251
$C_6H_5SiH_2Br$	n-C_3H_7MgBr, diethyl ether, 20°	(C_6H_5)(n-C_3H_7)SiH_2	28	251
$C_6H_5SiH_2Br$	i-C_3H_7MgBr, diethyl ether, 20°	(C_6H_5)(i-C_3H_7)SiH_2	41	251
$C_6H_5SiH_2Br$	n-C_4H_9MgBr, diethyl ether, 20°	(C_6H_5)(n-C_4H_9)SiH_2	58	251
$C_6H_5SiH_2Br$	n-$C_5H_{11}MgBr$, diethyl ether, 20°	(C_6H_5)(n-C_5H_{11})SiH_2	50	251
$C_6H_5SiH_2Br$	n-$C_6H_{13}MgBr$, diethyl ether, 20°	(C_6H_5)(n-C_6H_{13})SiH_2	36	251
$C_6H_5SiH_2Br$	$\overline{CH_2(CH_2)_4CH}MgBr$, diethyl ether, 20°	$(C_6H_5)[\overline{CH_2(CH_2)_4CH}]SiH_2$	25	251
		R_3SiH		
$SiHCl_3$	CH_3MgCl, diethyl ether, 0-35°	CH_3SiHCl_2	—	172
CH_3SiHCl_2	CD_3MgI, di-n-butyl ether, 0°	$CH_3(CD_3)_2SiH$	—	54
$SiHCl_3$	$C_6H_5CH_2MgBr$, diethyl ether, 0-35°	$(C_6H_5CH_2)_2SiHCl$	—	49, 172

TABLE 7.14 (cont'd)

Halogenosilane	Grignard compound, solvent, reaction temperature (°C)	Organosilane	Yield (mole-%)	Ref.
$SiHCl_3$	C_2H_5MgBr, diethyl ether, 0°	$C_2H_5SiHCl_2$	—	172, 512
CH_3SiHCl_2	C_2H_5MgBr, diethyl ether	$(C_2H_5)_2(CH_3)SiH$	20	343, 483
$SiHCl_3$	n-$C_3H_7MgBr + CH_3MgBr$, diethyl ether	(n-C_3H_7)$(CH_3)_2SiH$	12	483
CH_3SiHCl_2	n-C_3H_7MgBr, diethyl ether	(n-C_3H_7)$_2(CH_3)SiH$	49	483
CH_3SiHCl_2	$BrMg(CH_2)_4MgBr$, diethyl ether, 35°	$\overline{CH_2(CH_2)_3Si}(CH_3)H$	40	666
CH_3SiHCl_2	$BrMg(CH_2)_5MgBr$, diethyl ether, 35°	$\overline{CH_2(CH_2)_4Si}(CH_3)H$	47	666
CH_3SiHCl_2	$BrMg(CH_2)_6Mg Br$, diethyl ether, 35°	$\overline{CH_2(CH_2)_5Si}(CH_3)H$	4	666
$Cl_3SiCH_2SiHCl_2$	$C_2H_5Br + Mg$	$(C_2H_5)_3SiCH_2SiH(C_2H_5)_2$	—	413
$SiHCl_3$	$(CH_3)_3SiCH_2MgCl$	$[(CH_3)_3SiCH_2]_3SiH$	—	171
$SiHCl_3$	C_2H_5MgBr, diethyl ether, 0–35°	$(C_2H_5)_2SiHCl$	—	172
$SiHCl_3$	C_2H_5MgBr, diethyl ether, 20–35°	$(C_2H_5)_3SiH$	70–78	674
$SiH(OC_2H_5)_3$	C_2H_5MgBr	$(C_2H_5)_3SiH$	—	328
$SiHCl(OC_2H_5)_2$	C_2H_5MgBr	$C_2H_5SiH(OC_2H_5)_2$	—	328
$SiHCl_2(OC_2H_5)$	C_2H_5MgBr	$(C_2H_5)_2SiH(OC_2H_5)$	—	328
$SiHCl_3$	n-C_3H_7MgBr, diethyl ether	(n-C_3H_7)$_3SiH$	43	483
$SiHCl_3$	i-C_3H_7MgCl	(i-C_3H_7)$_2SiHCl$	—	218
$SiHCl_3$	C_4H_9MgBr, diethyl ether, 0°	$C_4H_9SiHCl_2$	—	512
$SiHCl_3$	$C_5H_{11}MgBr$, diethyl ether, 0°	$C_5H_{11}SiHCl_2$	—	512
$SiHCl_3$	$C_6H_{13}MgBr$, diethyl ether, 0°	$C_6H_{13}SiHCl_2$	—	512
$C_2H_5SiHCl_2$	$CH_2{=}CHCH_2I + Mg$	$(CH_2{=}CHCH_2)_2(C_2H_5)SiH$	—	449
$SiHCl_3$	$CH_2{=}CHCH_2MgBr$	$(CH_2{=}CHCH_2)_3SiH$	—	282
$SiHCl_3$	$CH{\equiv}CC(CH_3)_2OH + C_2H_5MgBr + Cu_2Cl_2 + HgCl_2$	$[Me_2(OH)CC{\equiv}C]_3SiH$	—	566
$SiHCl_3$	C_6H_5MgBr, diethyl ether, 0–35°	$(C_6H_5)SiHCl_2$	26	49, 343, 512
$SiHCl_3$	C_6H_5MgBr	$(C_6H_5)_2SiHCl$	—	49, 173
$SiHCl_3$	p-$CH_3C_6H_4MgBr$	(p-$CH_3C_6H_4$)$_2SiHCl$	41	49
$SiHCl_3$	C_6H_5MgBr, diethyl ether, 35°	$(C_6H_5)_3SiH$	73–88	223, 495

TABLE 7.15

SYNTHESIS OF ORGANOSILANES BY THE ALKYLATION OF HALOGENOSILANES WITH ALKYL HALIDES AND SODIUM (Si–X → Si–H)

Halogenosilane	*Alkylating agent, solvent*	*Organosilane*	*Ref.*
$SiHCl_3$	i-C_4H_9Cl + Na, diethyl ether	(i-$C_4H_9)_3SiH$	510
$SiHCl_3$	i-$C_5H_{11}Cl$ + Na, diethyl ether	(i-$C_5H_{11})_3SiH$	510
$SiHCl_3$	C_6H_5Cl + Na	$(C_6H_5)_3SiH$; $(C_6H_5)_4Si$	360
$SiHCl_3$	*p*-$N(CH_3)_2C_6H_4Br$ + Na	[*p*-$N(CH_3)_2C_6H_4]_3SiH$	118, 511

Similarly, with $\overline{CH_2(C_2H_4)C}HMgBr$, $SiCl_4$ forms $(C_6H_{11})_2SiHCl$[*647*]:

$$SiCl_4 + C_6H_{11}MgBr \xrightarrow{100^\circ} (C_6H_{11})_2SiHCl + C_6H_{11}SiCl_3 + C_6H_{10}$$

Two routes are available in the reaction of Grignard compounds with bis(phenylsilyl) oxide: asymmetrical cleavage of the Si–O–Si bond and the abnormal Grignard reaction [*252*]:

$$PhH_2SiOSiH_2Ph + RMgX \rightarrow PhRSiH_2 + PhH_2SiOMgX$$

$$PhH_2SiOSiH_2Ph + RMgX \rightarrow PhRHSiOSiH_2Ph + HMgX \rightarrow$$

$$\text{[transition state: } PhRHSi{-}O\cdots SiH_2Ph \text{ with } X{-}Mg\cdots H\text{, } sp^3d] \rightarrow PhRHSiOMgX + PhSiH_3$$

$$PhRHSiOMgX \xrightarrow{\Delta,\ \text{hydrolysis}} (PhRHSi)_2O + MgXOH$$

For examples of the reaction see Table 7.16.

Alkyllithiums react with bis(phenylsilyl) oxide with asymmetric cleavage of the Si–O–Si bond in all cases. Alkylsilanes and lithium silanolates are formed. For example, bis(phenylsilyl) oxide and alkyllithiums react to give phenyltrialkylsilanes, lithium phenyldialkylsilanolates and lithium hydride:

$$PhH_2SiOSiH_2Ph + 5RLi \rightarrow PhR_3Si + PhR_2SiOLi + 4LiH$$

4.5 *Synthesis of organosilicon hydrides by hydrosilation of multiple C–C bonds*

An excellent synthetic route for organosilicon hydrides is the addition of silicon hydride to multiple C–C, C–N, and C–O bonds, particularly to alkenes and substituted alkenes:

$$\geqslant Si{-}H + CH_2{=}CHR \rightarrow \geqslant SiCH_2CH_2R$$

This is especially recommended when the substituents to be introduced contain functional groups. Further details can be found in sections 5.11–5.13.

TABLE 7.16

PREPARATION OF ORGANOSILANES BY THE ANOMALOUS GRIGNARD REACTION WITH STERICALLY DEMANDING GRIGNARD COMPOUNDS

Halıde or oxide	*Grignard compound, solvent, reaction temperature* (°C)	*Organosilicon hydride (reaction product)*	*Yield (mole-%)*	*Ref.*
$C_6H_5SiCl_3$	i-C_3H_7MgBr, diethyl ether, 20°, then 170°	(C_6H_5)(i-$C_3H_7)_2SiH(+C_3H_6)$	50	253
$C_6H_5SiCl_3$	*tert*-C_4H_9MgBr, diethyl ether, 20°, then 170°	(C_6H_5)(*tert*-$C_4H_9)_2SiH(+C_4H_8)$	10	253
$C_6H_5SiCl_3$	$CH_2(CH_2)_3CHMgBr$, diethyl ether, 20°, then 170°	$(C_6H_5)[CH_2(CH_2)_3CH]_2SiH$ $(+CH_2(CH_2)_2CH{=}CH)$	49	253
$C_6H_5SiCl_3$	$CH_2(CH_2)_4CHMgBr$, diethyl ether, 20°, then 170°	$(C_6H_5)[CH_2(CH_2)_4CH]_2SiH$ $(+CH_2(CH_2)_3CH{=}CH)$	40–66	116, 253
$[(C_6H_5)H_2Si]_2O$	CH_3MgI	$C_6H_5SiH_3$, $(C_6H_5)(CH_3)SiH_2$, $[(C_6H_5)(CH_3)HSi]_2O$	—	252
$[(C_6H_5)]H_2Si]_2O$	C_2H_5MgBr	37% $C_6H_5SiH_3$, 12% $(C_6H_5)(C_2H_5)SiH_2$, $[(C_6H_5)(C_2H_5)HSi]_2O$		252
$[(C_6H_5)H_2Si]_2O$	C_6H_5MgBr	30% $C_6H_5SiH_3$, 64% $(C_6H_5)_2SiH_2$, $[(C_6H_5)HSiO]_3$		252
$[(C_6H_5)H_2Si]_2O$	$C_6H_5CH_2MgCl$	30% $C_6H_5SiH_3$, 40%$(C_6H_5)(C_6H_5CH_2)SiH_2$		252
$SiCl_4$	$CH_2(CH_2)_3CHMgCl$	$[CH_2(CH_2)_3CH]_2SiHCl$	30	375
$SiCl_4$	$CH_2(CH_2)_4CHMgCl$	$[CH_2(CH_2)_4CH]_2SiHCl$	40	375
$SiCl_4$	i-C_3H_7MgCl	(i-$C_3H_7)_2SiHCl$	45	375
$SiCl_4$	C_6H_5MgBr	$(C_6H_5)_2SiHCl$	12	375

4.6 *Reactions of organic substituents of silicon*

There are only a few possibilities of introducing functional groups into organic substituents of silicon hydrides or of carrying out reactions of silicon hydrides with preservation of the Si–H bond. Thus, for example, alkyl or aryl substituents on silicon hydrides cannot be halogenated by direct photohalogenation, since the Si–H bond is also halogenated (and substantially more readily than the C–H bond). The preparation of $>Si(H)(CH_2)_nX$, for example, must therefore always be carried out in the following sequence of reactions:

$$>Si(Cl)(CH_2)_nH \xrightarrow[\Delta]{X_2,\,h\nu} >Si(Cl)(CH_2)_nCl \xrightarrow[\Delta]{LiAlH_4} >SiH(CH_2)_nCl$$

(for photochlorination, see [*323*], and for a simple chlorination apparatus see [*298*]). The halogen in the α-position of halogenated alkylsilanes is fairly labile (activation energy for $(CH_3)_3SiCH_2I + {}^{131}I^{\ominus} \rightleftharpoons (CH_3)_3SiCH_2{}^{131}I + I^{\ominus}$ is only 19.0 kcal/mole, while for $(CH_3)_3CCH_2I$ it is 26 kcal/mole and for $(CH_3)_3SnCH_2I$ it is 10.8 kcal/mole [*275*]). Because of the nearness of two positive charges, in contrast to all other alkyl halides, $(CH_3)_3SiCH_2Cl$ rearranges readily, particularly in the presence of traces of metal salts or the metals themselves [*675*]:

$$(CH_3)_3\overset{\delta+}{Si}\overset{\oplus}{C}H_2 + E^{III}Cl_4^{\ominus} \rightarrow (CH_3)_2Si(CH_2CH_3)Cl + E^{III}Cl_3$$

The following synthetic steps are available for the synthesis of complicated chain-like compounds consisting of C members and peralkylated or perarylated Si members [*203*]:

$$RSiCl_2(CH_2Cl) \xrightarrow[\Delta]{+R'MgBr} RR'_2SiCH_2Cl \xrightarrow[\Delta]{+Mg} RR'_2SiCH_2MgCl$$

$$RR'_2SiCH_2MgCl \xrightarrow[\Delta]{+HgCl_2} (RR'_2SiCH_2)_2Hg$$

$$\xleftarrow[\text{so on}]{\text{and}} RR'_2SiCH_2\underset{|}{\overset{|}{Si}}CH_2Cl \xleftarrow[\Delta]{>Si(CH_2Cl)Cl} RR'_2SiCH_2Li \xleftarrow[\Delta]{+Li} (RR'_2SiCH_2)_2Hg$$

The transfer of such a synthetic route to similar Si–C chains containing Si–H bonds (R, R′ = H) is substantially more difficult because of the increased reactivity of the Si–H bond as compared with the Si–C bond. The reaction with $HgCl_2$ must probably be omitted because of the danger of halogenation of the Si–H bond. However, the crystallizable Hg compound would be important for an intermediate purification.

4.7 *Physical properties of the organosilanes*

The physical properties of an organosilicon hydride R_nSiH_{4-n} lie between those of the corresponding tetralkylsilane R_4Si and silane SiH_4. Depending on the ligands, these compounds can be separated suitably either by high vacuum or normal column distillation, by gas chromatography (*e.g.*, for R_4Si Apiezon L on

 [*Text continued p. 512*]

TABLE 7.17

PHYSICAL PROPERTIES OF ORGANO SILICON HYDRIDES $RSiH_3$ AND R_2SiH_2

([a]Vapour pressure equation: $\log p\,[\text{torr}] = -A/T + B$ or $\log p\,[\text{torr}] - A/T - B \log T + C$)

Organosilane	*M.p.* (°C)	*B.p.* [°C (torr)]	*Constant of vapour pressure equation*[a]				*Enthalpy of evaporation* (*cal/mole*)	*Trouton's constant T, density d (g/ml), refractive index n, enthalpy of formation ΔH°* (*kcal/mole*)	*Ref.*
			A	*B*	*C*	*validity limits* (°C)			
CH_3SiH_3	−156.81	−57.5 (760)	1329.5	4.1741	10.7071	−117/−58	4390		142
		−47.8 (16.8)							661a
$CH_3OCH_2SiH_3$	—	—	919.05	7.1454	—	−29/47	4205		560
			948.4	7.2789	—				687
$C_2H_5SiH_3$	−179.7	13.7 (760)	1784.3	6.0372	24.3222	−75/−16	5330	ΔH° −15; ΔH° −21 (gas)	142, 334
		55.5 (739.5)							
$C_3H_7SiH_3$	—	21.3 (760)	1363	7.510	—	—	—		439
n-$C_4H_9SiH_3$	−138.24	56.42; 56.7 (760)	2896.8	10.5930	38.3121	−33/11	7300	n_D^{25} 1.3895; n_D^{20} 1.3927;	142, 439
		55.5 (739.5)						d^{25} 0.6699; ΔH° ⊖18;	6
i-$C_4H_9SiH_3$	vitreous	48.60; 48.2 (760)	2398.1	7.0214	27.9253	−40/20	7050	ΔH°, −18; n_D^{20} 1.3876	142, 6
tert-$C_4H_9SiH_3$	—	34.4 (760)	1136	6.575	—	—	—		439
$C_5H_{11}SiH_3$	—	86.8 (740.5)						d^{25} 0.7043; n_D^{25} 1.4012	439
$C_6H_{13}SiH_3$	—	114.5 (751)						d^{20} 0.7182; n_D^{20} 1.4129	53
n-$C_7H_{15}SiH_3$	—	140.7 (760)						d_4^{20} 0.737; n_{white}^{20} 1.4199	17
($\overline{CH_2(CH_2)_4}CHCH_2SiH_3$	—	143–5 (760)						n_D^{25} 1.4540	559
$\overline{CH_2(CH_2)_4}CHSiH_3$	—	119.5 (739.5)						d_4^{20} 0.805; 0.7958;	439
		119.0 (760)						n_{white}^{20} 1.4495; n_D^{25} 1.4464	15
$CH_3\overline{CH(CH_2)_4}CHSiH_3$	—	144–145 (760)						n_D^{25} 1.4540	558
CH_2ClSiH_3	—	31.5–32	1436	7.593	—	—	6572	T 21.6; n_D^{25} 1.4149	298
$CHCl_2SiH_3$	—	67.3–68	1698	7.869	—	—	7771	T 22.8 n_D^{25} 1.4478	298
$CH_2{=}CHSiH_3$	−171.61	−22.8 (760)	1893.9	7.7159	28.9480	−87/−23	5120	ΔH° +12; ΔH° +6 (gas)	632, 334
	−179.08								
$CH_2{=}CHCH(CH_3)SiH_3$	—	43 (750)						d^{20} 0.6846; n_D^{20} 1.4050	452
$CH(CH_3){=}CHCH_2SiH_3$	—	56–76 (752)						d^{20} 0.7042; n_D^{20} 1.4178	452
$CH{\equiv}CSiH_3$	−90.7 ± 0.5	−22.4 ± 0.5	1150	7.465	—	—	5240 ± 50	T 20.9	146
$SiH_3C{\equiv}CSiH_3$	~ −59	43 (760)	1434	7.417	—	—	—	d_4^{25} 0.743; n_D^{25} 1.4234	348
$C_6H_5SiH_3$	—	120							206
p-$CH_3C_6H_4SiH_3$	—	180 (739.5)						d^{25} 0.9797; n_D^{25} 1.5251	439
		75–76 (21)							
1-$C_{10}H_7SiH_3$	—	49.5 (0.025)						d^{25} 1.0054; n_D^{25} 1.6030	439

$(CH_3)_2SiH_2$	150.22	19.6 (760)	1889.2	7.5510	28.4812	−86/−22	5100	T 19.1; $\Delta H°$ −36; $\Delta H°$ −42 (gas)	142, 334
			1186	7.5844	—	—	5421		687
$(C_2H_5)_2SiH_2$	−134.39	55.99; 53–57 (76[illegible]	2564.5	7.9119	30.5841	−32/22	7180	$\Delta H°$ −30; $\Delta H°$ −36 (gas);	149, 186, 447, 669
		55 (763)						n_D^{20} 1.3917	6, 334
		53–56 (739)							
$(C_3H_7)_2SiH_2$	—	111 (758); 110.5 ([illegible]69)						d^{20} 0.7183; n_D^{20} 1.4110	53, 669
$(C_5H_{11})_2SiH_2$	—	204.5 (739.5)						d^{25} 0.7610; n_D^{25} 1.4311	439
$CH_2{=}CHCH(CH_3)Si(CH_3)H_2$	—	70–71 (749)						d^{20} 0.7165; n_D^{20} 1.4162	452
$\overline{CH_2(CH_2)_3Si}H_2$	—	71						n_D^{20} 1.444; d_4^{25} 0.780	466
$\overline{CH_2(CH_2)_4Si}H_2$	—	102						n_D^{20} 1.4533; d_4^{25} 0.818	666
$\overline{CH_2(CH_2)_5Si}H_2$	—	135						n_D^{20} 1.4533; d_4^{25} 0.818	666
$(C_6H_5)(CH_3)SiH_2$	—	53 (30)						n_D^{25} 1.505	251
$(C_6H_5)(C_2H_5)SiH_2$	—	60–61 (18)						n_D^{25} 1.504	251
$(C_6H_5)(n\text{-}C_3H_7)SiH_2$	—	48–49 (4); 44–4[illegible] (5)						d_4^{25} 0.9430; n_D^{25} 1.4838; 1.500	251, 379
$(C_6H_5)(i\text{-}C_3H_7)SiH_2$	—	48 (6)						n_D^{25} 1.498	251
$(C_6H_5)(n\text{-}C_4H_9)SiH_2$	—	50 (1.5)						n_D^{25} 1.499	251
$(C_6H_5)n\text{-}C_5H_{11})SiH_2$	—	46 (0.29)						n_D^{25} 1.496	251
$(C_6H_5)(n\text{-}C_6H_{13})SiH_2$	—	56 (0.29)						n_D^{25} 1.486	251
$(C_6H_5)[\overline{CH_2(CH_2)_4CH}]SiH_2$	—	56–58 (0.29)						n_D^{25} 1.509	251
$(C_6H_5)(C_{10}H_{21})SiH_2$	—	108–111 (0.6)						n_D^{20} 1.4906 d_{20}^{20} 0.8648	229
$(C_6H_5)(C_{12}H_{25})SiH_2$	—	130–131 (0.6)						n_D^{20} 1.4881; d_{20}^{20} 0.8629	229
$(C_6H_5)(C_6H_5CH_2)SiH_2$	—	97–98 (0.9)						n_D^{20} 1.5738	229
$(C_6H_5)_2SiH_2$		58 (0.6), 69–73 (0.5)						n_D^{25} 1.581; n_D^{20} 1.5797; 1.5792	229, 251
$(C_6H_5)(4\text{-}CH_3OC_6H_4)SiH_2$	45–47	118–121 (0.15)							229
$(C_6H_5)(2\text{-}CH_3C_6H_4)SiH_2$		85–86 (0.1)						n_D^{20} 1.5808; d_{20}^{20} 1.0015	229
$(C_6H_5)(4\text{-}C_6H_5OC_6H_4)SiH_2$	—	145–147 (0.02)						n_D^{20} 1.6093	229

References p. 621

TABLE 7.18

PHYSICAL PROPERTIES OF TRIORGANOSILICON HYDRIDES R_3SiH

Organosilane	*M.p. (°C)*	*B.p. [°C (torr)]*	*Density d (g/ml); refractive index n; enthalpy of formation ΔH° (kcal/mole)*	*Ref.*
$(CH_3)_3SiH$	−135.86	6.7 (760)	$\Delta H°$ −54; −55; log p [torr] = −1628.3/T −2.9574 log T + 15.9404 (valid between −68 and 0°); enthalpy of evaporation 5820 cal/mole	142, 243b
$(C_2H_5)_3SiH$	−156.90	108.77 (760)	n_D^{20} 1.4120; 14117; $\Delta H°$ −39 (gas)	142, 186
		105–106 (739)		6, 674
		107 (733)		334
$(i\text{-}C_3H_7)_3SiH$	—	60 (10); 60–61 (3.5)		218, 447
$\overline{CH_2(CH_2)_3}Si(CH_3)H$	—	91.5	n_D^{20} 1.4390; d_4^{25} 0.798	666
$\overline{CH_2(CH_2)_4}Si(CH_3)H$	—	118	n_D^{20} 1.4462; d_4^{25} 0.809	666
$\overline{CH_2(CH_2)_5}Si(CH_3)H$	—	114	n_D^{20} 1.423	666
$(\overline{CH_2(CH_2)_4CH})_3SiH$	< −78	183–185 (9)	n_D^{25} 1.5132	415
CH_2CH_2 CH_2—CH_2—SiH CH_2CH_2	63	131 (732)		585
$(CH_3)_2(CH_2Cl)SiH$	—	80–81 (763)	n_D^{25} 1.4150	298, 555
$(C_2H_5)_2(CH_2Cl)SiH$	—	135–138 (760)	d_4^{25} 0.888; n 1.4357	255
$(C_6H_5)(i-C_3H_7)_2SiH$		56–60 (1.2)	n_D^{25} 1.495	253
$(C_6H_5)(tert\text{-}C_4H_9)_2SiH$	—	46 (0.4)	n_D^{25} 1.489	253
$(C_6H_5)[\overline{CH_2(CH_2)_3CH}]_2SiH$	—	100 (0.03)	n_D^{25} 1.529	253
$(C_6H_5)[\overline{CH_2(CH_2)_4CH}]_2SiH$	—	124 (0.03)	n_D^{25} 1.538	253
$(C_6H_5)(n\text{-}C_{12}H_{25})(C_6H_5CH_2)SiH$	—	180–183 (0.12)	n_D^{20} 1.5233; d_4^{20} 0.9209	228
$(C_6H_5)_2(n\text{-}C_4H_9)SiH$	—	110–112 (1)	n_D^{20} 1.5541; d_4^{20} 0.9604	228, 229
$(C_6H_5)_2(C_{10}H_{21})SiH$	—	164–168 (1.5)	n_D^{20} 1.5253; d_{20}^{20} 0.9262	229
$(C_6H_5)_2(CH_2{=}CHCH_2)SiH$	—	112–116 (1.5)	n_D^{20} 1.5743; d_{20}^{20} 0.9940	229
$(C_6H_5)_2(C_6H_5CH_2)SiH$	—	140–143 (0.)	n_D^{20} 1.6073; d_{20}^{20} 1.0494	229
$(C_6H_5)_3SiH$	—	44–45		223, 229
		36–37		495
$(1\text{-}C_{10}H_7)_3SiH$	—	236–237		447

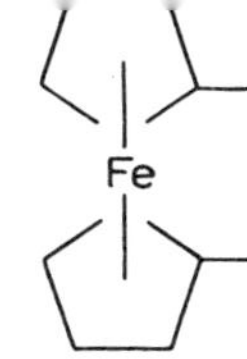

R	*R′*				
$Si(CH_3)(C_2H_5)H$	H	—	92–94 (0.06)	d_4^{30} 1.1043; n_D^{20} 1.5015	410
$Si(CH_3)(C_2H_5)H$	(=R)	—	138–140 (0.04)	d_4^{30} 1.0880; n_D^{20} 1.4790	410
$Si(C_2H_5)_2H$	H	—	112–114 (0.06)	d_4^{30} 1.1061; n_D^{20} 1.5520	410
$Si(C_2H_5)_2H$	$Si(C_2H_5)_2H$	—	150–152 (0.04)	d_4^{30} 1.0811; n_D^{20} 1.5326	410
$Si(C_2H_5)_3$	H	—	102–103 (0.02)	d_4^{30} 1.1418; n_D^{20} 1.5661	410
$Si(C_2H_5)_3$	$Si(C_2H_5)_3$	—	148–150 (0.03)	d_4^{30} 1.0670; n_D^{20} 1.5480	410

TABLE 7.19
PHYSICAL PROPERTIES OF SILANES CONTAINING SEVERAL SILICON ATOMS

Organosilane	*B.p.* [°*C (torr)*]	*Density d (g/ml), refractive index n*	*Ref.*
$CH_3SiH_2SiH_3$	16.6 (760)	m.p. −134.9°	2a
$(CH_3H_2H_2SiSiH_2(CH_3)$	−46.2 (11.7); 0.0 (106.8)	$\log p$ [torr] = −1288.10/T + 6.74345	53, 193, 108, 200
$SiH_3CH_2SiH_3$	17 (757); 15.2–15.4 (770)	d^4 0.6929; n_D^4 1.4115	
$SiH_3CH_2CH_2SiH_3$	46 (746.6); 45 (760)	m.p. - 102°; $\log p$ [torr] = −1563/T + 7.793; d^{20} 0.6987; n_D^{20} 1.4140	53, 673
$(CH_3)_2SiHCH_2SiH_3$	70.5–71 (768)		193, 200
$(CH_3)_3SiCH_2SiH_3$	91–92 (768)		193, 200
$[(CH_3)_3SiCH_2]_3SiH$	103–108 (8–9)		117
$(CH_3)SiH_2CH_2Si(CH_3)H_2$	71 (768)		193, 200
$(CH_3)_2SiHCH_2Si(CH_3)H_2$	88.5 (768)		193, 200
$(CH_3)_3SiCH_2Si(CH_3)H_2$	103 (768)		193, 200
$(CH_3)_2SiHCH_2Si(CH_3)_2H$	107 (768)		193, 200
$(CH_3)_3SiCH_2Si(CH_3)_2H$	120 (768)		193, 200
$SiH_3CH(SiH_3)_2$	57.1 (760)	$\log p$ [torr] = −1644.3/T + 7.860 valid between −61 and 10°; Q 7514 cal/mole	68
$SiH_3CH_2SiH_2CH_2SiH_3$	100 (760)		193, 200
$(CH_3)SiH_2CH_2SiH_2CH_2SiH_3$	123 (758)		193, 200
$(CH_3)_2SiHCH_2SiH_2CH_2SiH_3$	133 (762)		193, 200
$(CH_3)_3SiCH_2SiH_2CH_2SiH_3$	154 (756)		193, 200
$(C_6H_5)_2SiH(CH_2)_5Si(C_6H_5)_2H$	217–222 (0.008)	n_D^{20} 1.5946; d_{20}^{20} 1.0390	229
$CH_2SiH_2CH_2SiH_2CH_2SiH_2$	142 (760)		193, 200
$CH_2SiH_2CH_2Si(CH_3)HCH_2SiH_2$	159 (766)		193, 200
$CH_2Si(CH_3)HCH_2Si(CH_3)HCH_2SiH_2$	166 (764)		193, 200
$CH_2Si(CH_3)HCH_2Si(CH_3)HCH_2Si(CH_3)H$	180 (764)		193, 200
$CH_2Si(CH_3)HCH_2Si(CH_3)_2CH_2Si(CH_3)H$	190 (764)		193, 200
$CH_2Si(CH_3)HCH_2Si(CH_3)_2CH_2Si(CH_3)_2$	201 (767)		193, 200

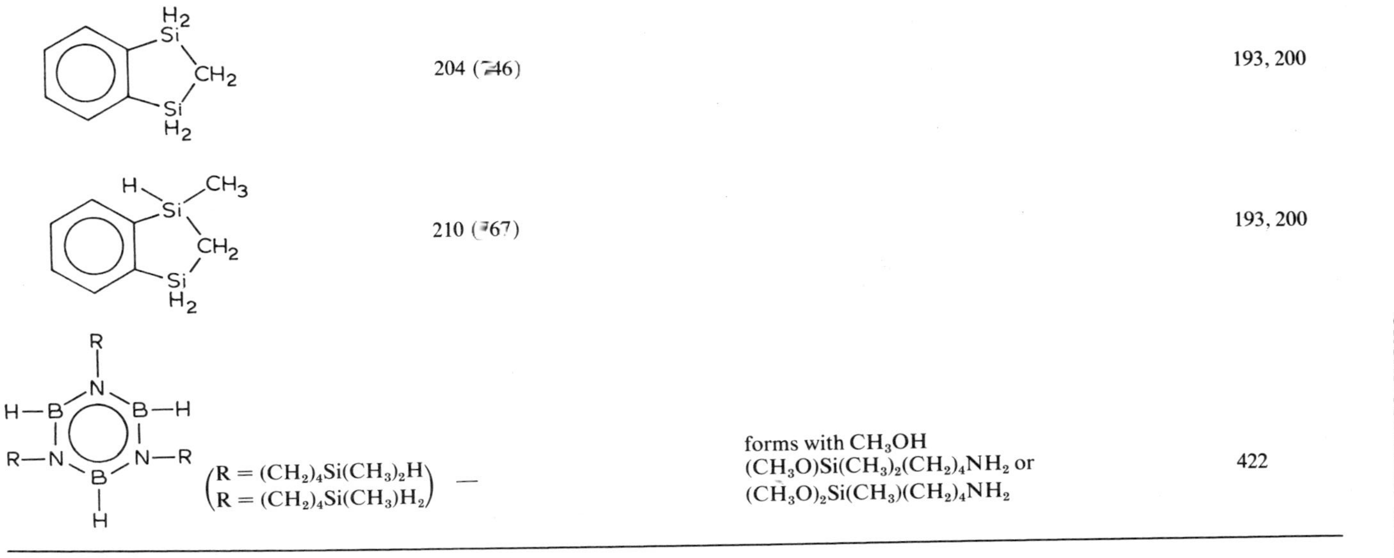

		204 (746)		193, 200
		210 (767)		193, 200
	$\left(\begin{array}{l} R = (CH_2)_4Si(CH_3)_2H \\ R = (CH_2)_4Si(CH_3)H_2 \end{array}\right)$	—	forms with CH_3OH $(CH_3O)Si(CH_3)_2(CH_2)_4NH_2$ or $(CH_3O)_2Si(CH_3)(CH_2)_4NH_2$	422

References p. 621

Silocel C 22[*3*])[*203b*], or by thin-layer chromatography[*644a*]. Tables 7.17–7.19 contain the physical properties of the hydrides R_nSiH_{4-n} and of hydrides containing several silicon atoms.

The alkylsilanes $(C_2H_5)SiH_3$, $(C_2H_5)_2SiH_2$, and $(C_2H_5)_3SiH$ are only slightly more thermally stable than monosilane SiH_4. They decompose at 440–460° into hydrogen, hydrocarbons, silane, methylsilane, higher silanes, and methylpolysilane(1). The pyrolysis probably takes place through a radical mechanism (Si and C radicals)[*196*].

The mass-spectroscopic data of silane, organosilanes, and halogenosilanes can be found in Table 7.20 and in the literature[*270, 271, 419a, 486*].

TABLE 7.20
CRITICAL POTENTIALS OF MONOSILANE, MONOALKYLMONOSILANES, AND HALOGENOSILANES[*601*]

Silane	*Ion*	*Crit. pot. (eV)*	*Silane*	*Ion*	*Crit. pot. (eV)*
SiH_4	SiH_3^+	$12.4_0 \pm 0.02$	$SiCl_4$	$SiCl_3^+$	$12.4_8 \pm 0.02$
	SiH_2^+	$11.9_1 \pm 0.02$	$HSiCl_3$	$SiCl_3^+$	$11.9_1 \pm 0.03$
$(CH_3)SiH_3$	SiH_3^+	$12.8_0 \pm 0.1$	$(CH_3)SiCl_3$	$CH_3SiCl_3^+$	$11.3_6 \pm 0.03$
	SiH_2^+	$11.6_2 \pm 0.1$		$SiCl_3^+$	$11.9_0 \pm 0.08$
	CH_3^+	15.1 ± 0.3		CH_3^+	15.0 ± 0.2
$(C_2H_5)SiH_3$	$C_2H_5SiH_3^+$	$10.1_8 \pm 0.05$	$(C_2H_5)SiCl_3$	$C_2H_5SiCl_3^+$	$10.7_4 \pm 0.04$
	SiH_3^+	12.8 ± 0.2		$SiCl_3^+$	$12.1_0 \pm 0.03$
	SiH_2^+	12.0 ± 0.1		$C_2H_5^+$	$12.7_7 \pm 0.05$
	$C_2H_5^+$	12.6 ± 0.2		$C_2H_4^+$	$12.4_8 \pm 0.05$
$(CH_3)_2HCSiH_3$	$C_3H_7SiH_3^+$	$9.8_5 \pm 0.1$	$(CH_3)_2HCSiCl_3$	$C_3H_7SiCl_3^+$	$10.2_8 \pm 0.1$
	$C_3H_7^+$	$11.3_3 \pm 0.03$		$SiCl_3^+$	13.1 ± 0.2
	$C_3H_6^+$	$10.8_1 \pm 0.04$		$C_3H_7^+$	$11.3_6 \pm 0.1$
	SiH_3^+	13.1 ± 0.2		$C_3H_6^+$	$10.9_2 \pm 0.1$
$(CH_3)_3CSiH_3$	$C_4H_9SiH_3^+$	9.5 ± 0.2	$(CH_3)_3CSiCl_3$	$C_4H_9SiCl_3^+$	—
	$C_4H_9^+$	$10.2_5 \pm 0.02$		$SiCl_3^+$	13.0 ± 0.1
	$C_4H_8^+$	$9.8_9 \pm 0.05$		$C_4H_9^+$	$10.7_2 \pm 0.1$
	SiH_3^+	13.7 ± 0.2		$C_4H_8^+$	$10.2_6 \pm 0.2$

5. Reactions of Si–H and Si–Si bonds

Figure 7.2 shows schematically the reactions of the Si–H bond.

5.1 *Reaction of the Si–H bond with halogens*

Silicon hydrides (R_nSiH_{4-n}, $n = 0$–3) react with free halogens (Cl_2, Br, I_2) with the replacement of hydrogen on the silicon by halogen and the formation of the corresponding hydrogen halide:

$$\geq Si{-}H + X_2 \rightarrow \geq Si{-}X + HX$$

If the silane contains several Si–H bonds, it can be halogenated in steps, via

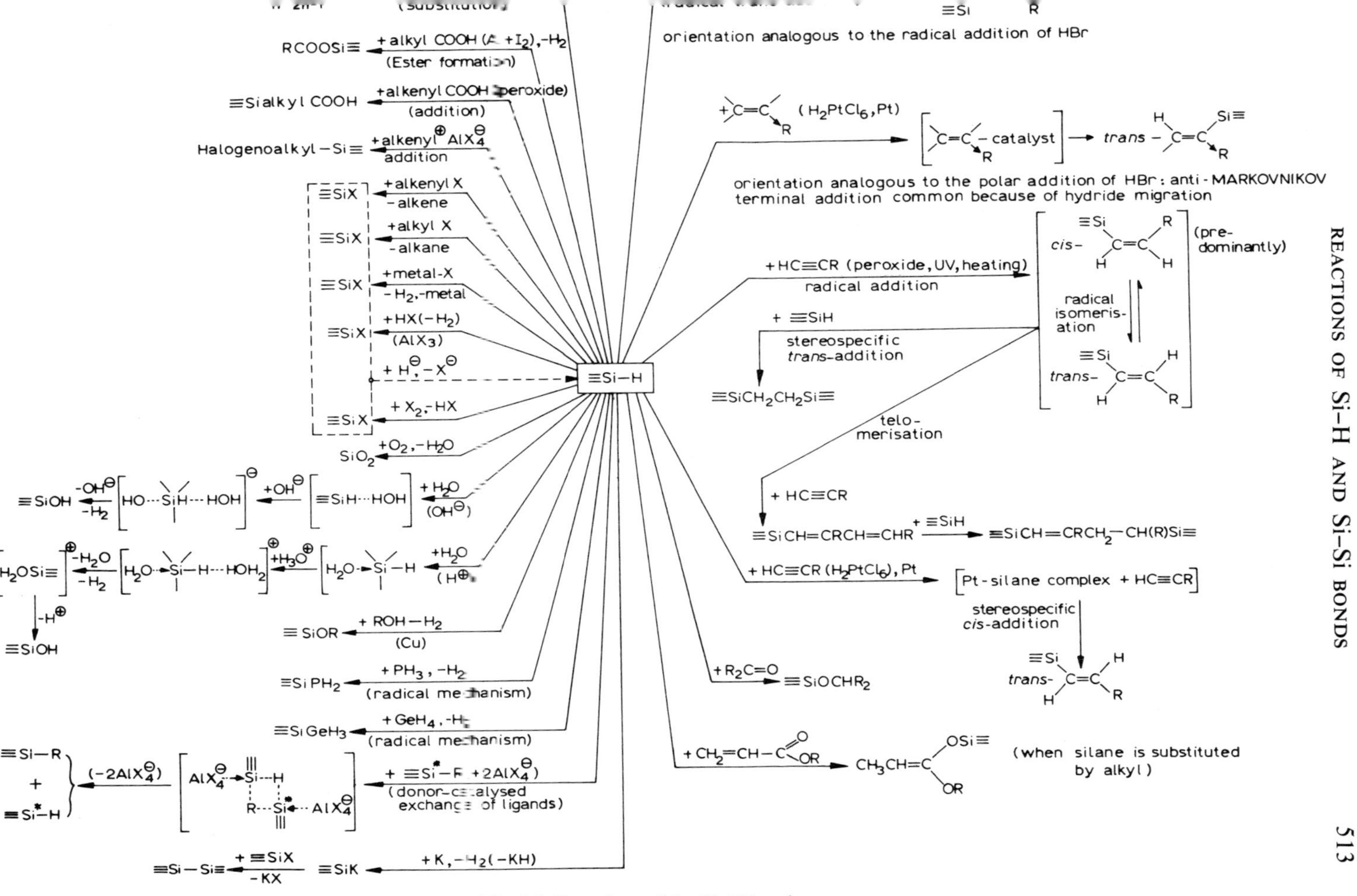

Fig. 7.2. Reactions of the Si–H bond.

References p. 621

isolatable intermediate compounds, to perhalogenated derivatives. Section 6.1.1 contains details and Table 7.21 references and examples of such reactions.

5.1.1 Reaction in the gas phase

In the reaction of silanes with halogens at 20° in the gas phase, the rate of the reaction increases with an increasing number of Si–H bonds. For example, a mixture of SiH_4 and Br_2 explodes [*612*]. When the partial pressure of at least one of the reactants is lowered (most simply by cooling), the reaction can be modified. Mono- to perhalogenated substitution products are always formed, and under certain circumstances even Si–C bonds are attacked. The gas-phase reaction very probably takes place through a (weakly polar) four-centre reaction:

$$SiH_4 + X_2 \rightarrow \begin{array}{ccc} \overset{\delta+}{H_3Si} & \cdots & \overset{\delta-}{H} \\ \vdots & & \vdots \\ {}_{\delta-}X & \cdots & X_{\delta+} \end{array} \rightarrow H_3SiX + HX$$

The silyl halide formed is more stable than the corresponding germyl halide. It has a substantially smaller tendency to split off hydrogen halide and form a dihydride than the isologous germanium compound, since the Si–X bond is less strongly polarized than the Ge–X bond because of the p_π–d_π back-coordination of electrons $\rangle Si \leqslant X$ that is possible in the case of the silicon:

$$SiH_3X \not\rightarrow SiH_2 + HX$$
$$GeH_3X \rightarrow GeH_2 + HX$$

5.1.2 Reaction in solution

The substitution reaction with free halogens can also be substantially moderated by means of inert solvents (CCl_4, C_6H_6, $CHCl_3$, dioxan, or the liquid organosilane itself). If explosions nevertheless occur (*e.g.*, $C_2H_5SiH_3 + Br_2$ in C_2H_5Br at $-30°$) [*368*], reactions in the overlying gas space are responsible.

The halogenations of the organosilicon hydrides, R_nSiH_{4-n}, in solution depend on the organic substituent and on the solvent. For example, at room temperature liquid *alkyl*silanes form monoiodoalkylsilanes with an equimolar amount of iodine, and with an excess of iodine diiodoalkylsilanes as well as the monoiodoalkylsilanes (replacement of H by I) [*13*]:

$$BuSiH_3 + I_2 \rightarrow BuSiH_2I + HI$$
$$BuSiH_3 + 2I_2 \rightarrow BuSiHI_2 + 2HI$$

In the treatment of *aryl*silanes with iodine, two reactions take place. Replacement of H of the Si–H bond (by I_2) competes with the substitution of C of the Si–C bond (by the HI formed). In ethyl halide solution, the Si–C cleavage reaction is slight. Thus, phenylsilane in ethyl iodide or ethyl bromide reacts with iodine

206, 207] or bromine[*368*] without appreciable Si–C cleavage, to give monoiodophenylsilane or monobromo- and dibromo-phenylsilanes respectively:

$$PhSiH_3 + I_2 \xrightarrow{EtI} PhSiH_2I + HI$$

$$PhSiH_3 + Br_2 \xrightarrow{EtBr} PhSiH_2Br + HBr$$

$$PhSiH_2Br + Br_2 \xrightarrow{EtBr} PhSiHBr_2 + HBr$$

In the absence of a solvent, the Si–C cleavage reaction comes to the fore:

$$PhSiH_3 + I_2 \rightarrow PhSiH_2I + HI \quad (1)$$

$$PhSiH_2I + I_2 \rightarrow PhSiHI_2 + HI \quad (2)$$

$$PhSiH_3 + HI \rightarrow SiH_3I + PhH \quad (3)$$

$$PhSiH_2I + HI \rightarrow SiH_2I_2 + PhH \quad (4)$$

$$SiH_2I_2 + 2I_2 \rightarrow SiI_4 + 2HI \quad (5)$$

Here further iodine substitution (reaction 2) of the $PhSiH_2I$ formed according to (1) is overshadowed by the Si–C cleavage reactions (3, 4) and by the HI-yielding reaction (5)[*206, 207*]. In liquid hydrogen iodide as solvent, as was to be expected from the equations given above, the Si–C cleavage reaction predominates (preparation of SiH_3I)[*206, 207*]:

$$PhSiH_3 + HI \xrightarrow[\text{fast}]{\text{liquid HI}} SiH_3I + PhI$$

$$PhSiH_3 + I_2 \xrightarrow[\text{slow}]{\text{liquid HI}} PhSiH_2I + HI$$

$$PhSiH_2I + HI \xrightarrow[\text{slow}]{\text{liquid HI}} SiH_2I_2 + PhI$$

Silanes substituted with various halogens can be obtained by the reaction of phenylhalogenosilanes with hydrogen halides[*208*], for example:

$$PhSiH_2Cl + HBr \rightarrow SiH_2ClBr + PhH$$

Phenyltrichlorosilane does not react with iodine[*206*]. Phenyltrimethylsilane undergoes cleavage in the following way[*485*]:

$$PhSiMe_3 + I_2 \rightarrow PhI + ISiMe_3$$

In solution, the substitutions take place by a polar reaction mechanism. Kinetic measurements of the reaction of triorganosilanes with iodine in solution gave the following results:

(a) The reaction of Et_3SiH with I_2 is kinetically of fourth order in the con-

References p. 621

centration range of 0.033 to 0.005 M in CCl_4 (first order with respect to [silane] and third order with respect to $[I_2]$:

$$-\,d[I_2]/dt = k_4[R_3SiH][I_2]^3$$

The order decreases at higher temperature (third to fourth order at 55°).

(b) The reaction rate increases with increasing capacity of the aryl substituent to give up electrons (decreasing value of Hammett's σ-constant $p\text{-}ClC_6H_4 < C_6H_5 < p\text{-}CH_3C_6H_4$):

$$(p\text{-}ClC_6H_4)_3SiH < (C_6H_5)_3SiH < (p\text{-}CH_3C_6H_4)_3SiH$$

$$(C_6H_5)(CH_3)_2SiH < (p\text{-}CH_3C_6H_4)(CH_3)_2SiH$$

Cl_3SiH does not react with iodine. The rate-determining step can therefore hardly pass through an intermediate containing positive silicon.

(c) The steric effect is low, but may nevertheless frequently mask the inductive effect. Thus, in benzene solution the reaction rates fall in the following sequence of alkyl substituents: $C_2H_5 > n\text{-}C_4H_9 > n\text{-}C_3H_7 > i\text{-}C_3H_7 > i\text{-}C_4H_9$. Consequently, in the halogenation of the Si–H bond with halogen, as for example in the hydrolysis of the Si–H bond (see section 7.5.6), the electrophilic attack of the halogen on the hydrogen atom is not the rate-determining step.

(d) Oxygen, light, and surface changes do not affect the reaction (homogeneous radical-free reaction).

(e) The rate of the reaction increases with increasing polarity of the solvent: $CCl_4 < C_6H_6 < CHCl_3 < C_6H_5NO_2 \sim$ dioxan.

(f) Hydrogen iodide does not affect the rate of the reaction in the absence of catalysts (AlI_3).

Consequently, the halogenation of an Si–H bond by elementary iodine very probably takes place through an electrophilic attack of the halogen on the hydrogen atom and a subsequent, rate-determining, nucleophilic attack of the halogen on the silicon [*118*]:

$$I_2 + I_2 \rightleftharpoons I_4\,(= \overset{\delta+}{I}\text{–}\overset{\delta-}{I_3})$$

$$\gt\overset{\delta+}{Si}\text{–}\overset{\delta-}{H} + \overset{\delta+}{I}\text{–}\overset{\delta-}{I_3} \overset{\text{fast}}{\rightleftharpoons} [\gt Si\text{–}H \cdots I\text{–}I_3]$$

$$I_2 + [\gt Si\text{–}H \cdots I\text{–}I_3] \xrightarrow{\text{slow}} [I\text{–}I \cdots Si\text{–}H \cdots I\text{–}I_3] \rightarrow I^{\oplus} + \gt SiI + HI + I_3^{\ominus}$$

In non-polar solvents a cyclic process is not excluded either, although it is less likely:

[cyclic scheme: Si, H, I, I, I, I, I with curved arrows]

The postulated transition state $R_3SiH \cdot I_4$ is formally comparable with the complex $R_3SiI \cdot I_2$ [*140*], but the Si–H bond, less polar than the Si–I bond, requires the more highly polarizable I_4 for "complex formation".

Bromine and iodine chloride react with triorganosilanes substantially faster than iodine. In the reaction of ICl with $(C_2H_5)_3SiH$ in CCl_4, the following two reactions compete (because of the polarization $\overset{\delta+}{Si}$–$\overset{\delta-}{H}$ and $\overset{\delta+}{I}$–$\overset{\delta-}{Cl}$, the sequence of reactions (b) is more probable):

(a) $R_3SiH + ICl \rightarrow R_3SiI + HCl$
$R_3SiI + ICl \rightarrow R_3SiCl + I_2$
$R_3SiH + 2ICl \rightarrow R_3SiCl + HCl + I_2$

(b) $R_3SiH + ICl \rightarrow R_3SiCl + HI$
$HI + ICl \rightarrow HCl + I_2$
$R_3SiH + 2ICl \rightarrow R_3SiCl + HCl + I_2$

5.2 *Reaction of the Si–H bond with hydrogen halides*

Silicon hydrides (R_nSiH_{4-n}, $n = 0$–3) react with hydrogen halides (HCl, HBr, HI) with the replacement of the hydrogen on the silicon by the halogen and with formation of molecular hydrogen:

$$\equiv Si\text{–}H + HX \rightarrow \equiv Si\text{–}X + H_2$$

Examples of the reaction and references are given in Tables 7.21 and 7.22. The presence of aluminium halides increases the rate of the reaction. A polar mechanism is fairly certain:

$$\geq SiH + AlX_3 \rightarrow [\geq \overset{\oplus}{Si}\text{–}H \ldots \overset{\ominus}{A}lX_3]$$

$$HX + AlX_3 \rightarrow H^{\oplus} + AlX_4^{\ominus}$$

$$AlX_4^{\ominus} + [\geq \overset{\oplus}{Si}\text{–}H \ldots \overset{\ominus}{A}lX_3] \rightarrow [X_3Al\text{–}X \ldots Si\text{–}H \ldots AlX_3]^{\ominus} \xrightarrow{H^{\oplus}} AlX_3 + \geq SiX + H_2 + AlX_3$$

It is difficult to decide with certainty whether the electrophilic attack of AlX_3 on the hydrogen atom (a) or the nucleophilic attack of $AlX_4^{\ominus}$ ions on the silicon atom (b) is the rate-determining step, but it is very likely that the last reaction step is rate-determining (detachment of anionic hydrogen from the sp^3d complex of silicon). If this were not so, the reaction steps on the silonium ion would be rate-determining, and these must be excluded, *e.g.*, by kinetic measurements in the hydrolysis of the Si–X bond.

5.3 *Reaction of the Si–H bond with metal and non-metal halides and with metal pseudohalides*

Silicon hydrides (R_nSiH_{4-n}, $n = 0$–3) react with certain metal halides (MX_n) and non-metal halides (*e.g.*, HSO_3Cl) and with metal pseudohalides (*e.g.*, AgNCO,

References p. 621

TABLE 7.21

REACTION OF SILICON HYDRIDES (SiH_4, $RSiH_3$) AND SILOXENES WITH HALOGENS OR HYDROGEN HALIDES PREPARATION OF HALOGENOSILANES ($Si{-}H \rightarrow Si{-}X$)

(Moles of) silane used	(Moles of) halogenating agent	Phase or solvent	Catalyst	Reaction temp. (°C)	Reaction time (h)	Yield (mole-% referred to silane) SiR_3X^a	$SiR_2X_2^a$	$SiRX_3^a$	SiX_4	Conversion of the hydride (mole-%)	Ref.
(1)n-$C_4H_9SiH_3$	(1)I_2	liquid	none	10	3	82	—			—	13, 658
(1)n-$C_4H_9SiH_3$	(2)I_2	liquid	none	10	3	48	47			—	13
(1)SiH_4	(0.6)Br_2	solid/gas	none	−78	2.5	25	13	4	3	—	612
(1)SiH_4	(1)HBr	gas	$AlBr_3$	100	—	25	33	~0	~0	—	420, 613
(1.1)SiH_4	(1)HCl	gas	$AlCl_3$	100	10	49	3			53	420, 614
(1.1)SiH_4	(1)HCl	gas	$AlCl_3$	100	30	58	14			72	614
(4)CH_3SiH_3	(1)HI	gas	AlI_3	100	12	—	—			—	169
(1)SiH_4	(1)HI	gas	AlI_3	80	24	~40	~20			75	153, 167, 347
n-$C_7H_{15}SiH_3$	I_2	liquid	—	~140	3	49	36	—	—	—	17
n-$C_7H_{15}SiH_3$	I_2 excess	liquid	—	~140	2	—	—	89	—	—	17
$(C_2H_5)_2SiH_2$	I_2	liquid	—	0–20	1	X					14
SiD_4	I_2	gas	AlI_3	50–100	—	X					153
SiH_3SiH_3	HI	gas	AlI_3	20	2.5	76					2, 662
cyclo-$C_6H_{11}SiH_3$	I_2	liquid	—	119	4	X	X	X	—	—	15
SiH_3SiH_3	HCl	—	$AlCl_3$	≤ 20	—	X					1, 352
SiH_3SiH_3	HBr	—	$AlBr_3$	≤ 20	—	X					1, 2, 352
$C_6H_5SiH_3$	Br_2	C_2H_5Br	—	—	2	33	25	—	—	—	368
$C_6H_5SiH_3$	I_2	C_2H_5I	—	—	—	X					206
$C_6H_5SiH_3$	I_2	without solvent	—	—	—	X^b					206
$C_6H_5SiH_3$	I_2	liq. HI	—	—	—	low[b]					206
(2)SiD_4	(1)DCl	gas	$AlCl_3$	200	3	X	X	—	—	—	421
SiH_4	Br_2	solid/gas	none	−40/−120	3–5	80–92					624
SiD_4	DI	gas	AlI_3	—	—	—					33
(1)$[Si_6O_3]H_6$	(2)Br_2	CS_2	—	—	—	$[\,]H_4Br_2$					264
(1)$[Si_6O_3]H_4Br_2$	(1)Br_2	CS_2	—	—	—	$[\,]H_3Br_3$					264
(1)$[Si_6O_3]H_3Br_3$	(2)Br_2	CS_2	—	—	—	$[\,]HBr_5$					264
(1)$[Si_6O_3]HBr_5$	(1)Br_2	CS_2	—	—	—	$[\,]Br_6$					264
$[Si_6O_3]H_6$	HBr	—	—	—	—	$[\,]H_5Br$					264, 308
$[Si_6O_3]H_6$	I_2	CS_2	—	—	—	$[\,]H_5I$					264, 305

[a] R = H, alkyl, aryl, silyl

TABLE 7.22

REACTION OF SILICON HYDRIDES R_3SiH WITH HALOGENS AND HYDROGEN HALIDES (Si–H → Si–X)

Hydride	*X_2 or HX*	*Solvent or phase*	*Reaction product (by-product)*	*Yield (mole-%)*	*Ref.*
$(C_2H_5)_3SiH$	Cl_2	CCl_4	$(C_2H_5)_3SiCl$	60	285
$(C_2H_5)_3SiH$	I_2	Petroleum ether	$(C_2H_5)_3SiI(H_2)$	91	658
$(C_2H_5)_3SiH$	$I_2 + AlI_3$	Petroleum ether	$(C_2H_5)_3SiI(H_2)$; AlI_3 accelerates the reaction	91	658
$(C_2H_5)_3SiH$	Br_2	CCl_4	$(C_2H_5)_3SiBr$	30; 90	285, 658
$(C_3H_7)_3SiH$	I_2	CCl_4	$(C_3H_7)_3SiI(H_2)$	87	658
$(C_3H_7)_3SiH$	$I_2 + AlI_3$	CCl_4	$(C_3H_7)_3SiI(H_2$, some $R_3SiCl)$	87	658
$(C_4H_9)_3SiH$	I_2	Petroleum ether	$(C_4H_9)_3SiI(H_2)$	91	658
$(C_6H_{13})_3SiH$	I_2	Petroleum ether	$(C_6H_{13})_3SiI(H_2)$	89	658
$(Cyclohexyl)_3SiH$	Br_2	CCl_4	$(Cyclohexyl)_3SiBr$	< 30	285
$(C_6H_5CH_2)_3SiH$	Cl_2	CCl_4	$(C_6H_5CH_2)_3SiCl$	< 60	285
$(C_6H_5CH_2)_3SiH$	Br_2	CCl_4	$(C_6H_5CH_2)_3SiBr$	—	285
$(C_2H_5)_3SiH$	HI	—	$(C_2H_5)_3SiI$	—	658
$(i\text{-}C_5H_{11})_3SiH$	HI	—	$(i\text{-}C_5H_{11})_3SiI$	91	658

AgNCS) with hydrogen being replaced by halogen. In this process the metal halide is reduced to metal or a metal halide of a lower oxidation stage, and hydride hydrogen is oxidized to molecular hydrogen or a proton [*13–15, 17, 99, 102, 165*]:

$$\gt Si\text{–}\overset{-1}{H} + \overset{+n}{E}X_n \rightarrow \gt Si\text{–}X + \overset{+(n-1)}{E}X_{n-1} + \tfrac{1}{2}\overset{0}{H_2}$$

$$\gt Si\text{–}\overset{-1}{H} + \overset{+n}{E}X_n \rightarrow \gt Si\text{–}X + \overset{+(n-2)}{E}X_{n-2} + \overset{+1}{H}X$$

The yields of the halogenosilanes are between 70 and 100%. Table 7.23 contains examples of the reaction.

5.4 *Reaction of the Si–H bond with organic halides*

Alkyl halides, RX, alkenyl halides, $CH_2{=}CHCH_2X$, and acid halides, RCOX, react with triarylsilanes on boiling under reflux, often even without the addition of aluminium halides (which favour heterolysis); the reaction with trialkylsilanes requires the presence of aluminium halides:

$$R_3SiH + RX \rightarrow R_3Si + RH$$

The halogenation catalysed with Friedel-Crafts catalysts very probably take place through carbonium and not through silonium ions, since carbonium ions are formed in reactions of $AlCl_3$ with compounds in which both mechanisms are possible (*e.g.*, $(CH_3)_2ClSiCH_2Cl$) [*84*]:

$$(CH_3)_2ClSiCH_2Cl \xrightarrow[-AlCl_4^{\ominus}]{+AlCl_3} [(CH_3)_2ClSi\overset{\oplus}{C}H_2] \xrightarrow{+PCl_3} \text{P-complex} \xrightarrow{\text{alcoholysis}} (CH_3)_2(RO)SiCH_2OP(OR)_2$$

TABLE 7.23

REACTION OF SILICON HYDRIDES WITH METAL AND NON-METAL HALIDES AND WITH PSEUDOHALIDES ($Si\text{-}H \rightarrow Si\text{-}X$)

Hydride	*Halide*	*Reaction product (by-product)*	*Yield (mole-%)*	*Ref.*
n-$C_4H_9SiH_3$	$HgCl_2$	n-$C_4H_9SiH_2Cl$	88	13
n-$C_4H_9SiH_3$	$HgBr_2$	n-$C_4H_9SiH_2Br$	80	13
n-$C_7H_{15}SiH_3$	$HgCl_2$	n-$C_7H_{15}SiH_2Cl$ (Hg, HCl)	89	17
n-$C_7H_{15}SiH_3$	$HgBr_2$	n-$C_7H_{15}SiH_2Br$ (Hg, HBr)	97	17
$(C_2H_5)_3SiH$	$CuCl_2$	$(C_2H_5)_3SiCl$ (CuCl, H_2)	91	14
$(C_2H_5)_3SiH$	$CuBr_2$	$(C_2H_5)_3SiBr$ (CuBr, H_2)	88	14
$(C_2H_5)_3SiH$	$KAuCl_4$	$(C_2H_5)_3SiCl$ (KCl, Au, H_2)	92	14
$(C_2H_5)_3SiH$	$HgBr_2$	$(C_2H_5)_3SiBr$ (Hg_2Br_2, H_2)	78	14
$(C_2H_5)_3SiH$	VCl_4	$(C_2H_5)_3SiCl$ (VCl_3, VCl_2, H_2)	87	14
$(C_2H_5)_3SiH$	$VOCl_3$	$(C_2H_5)_3SiCl$ ($VOCl_2$, VOCl, H_2)	87	14
$(C_2H_5)_3SiH$	CrO_2Cl_2	$(C_2H_5)_3SiCl$ (Cr_2O_3, H_2O)	49	14
$(C_2H_5)_3SiH$	$MoCl_5$	$(C_2H_5)_3SiCl$ ($MoCl_4$, $MoCl_3$, $MoCl_2$, H_2)	79	14
$(C_2H_5)_3SiH$	$RuCl_2$	$(C_2H_5)_3SiCl$ (Ru, H_2)	74	14
$(C_2H_5)_3SiH$	$PdCl_2$	$(C_2H_5)_3SiCl$ (Pd, H_2)	90	14
$(C_2H_5)_3SiH$	$PtCl_4$	$(C_2H_5)_3SiCl$ (Pt, H_2)	68	14
$(C_2H_5)_3SiH$	$SnCl_4$	$(C_2H_5)_3SiCl$ ($SnCl_2$, H_2)	72	14
$(C_2H_5)_3SiH$	$SbCl_3$	$(C_2H_5)_3SiCl$ (Sb, H_2)	85	14
$(C_2H_5)_3SiH$	$BiCl_3$	$(C_2H_5)_3SiCl$ (Bi, H_2)	79	14
$(C_2H_5)_3SiH$	$SeCl_4$	$(C_2H_5)_3SiCl$ (Se, H_2, HCl)	86	14
$(C_2H_5)_3SiH$	$TeCl_4$	$(C_2H_5)_3SiCl$ (Te, H_2, HCl)	76	14
$(C_2H_5)_3SiH$	HSO_3Cl	$(C_2H_5)_3SiCl$ (SO_2, HCl, H_2)	77	14
$(C_2H_5)_2SiH_2$	$CuCl_2$	$(C_2H_5)_2SiHCl$ (CuCl, Cu)	—	14
$(C_2H_5)_2SiH_2$	AgCl	$(C_2H_5)_2SiHCl$	100	14
$(C_2H_5)_2SiH_2$	$HgCl_2$	$(C_2H_5)_2SiHCl$ (Hg_2Cl_2, $(C_2H_5)_2SiCl_2$)	100?	14
$(C_2H_5)_2SiH_2$	$HgBr_2$	$(C_2H_5)_2SiHBr$	—	14
$(C_2H_5)_2SiH_2$	$PdCl_2$	$(C_2H_5)_2SiCl_2$	91	14
$(C_2H_5)_2SiH_2$	$PdCl_2$	(deficiency of $PdCl_2$; $(C_2H_5)_2SiHCl$, $(C_2H_5)_2SiCl_2$	10	14
Si_4H_{10}	PI_3	Iodotetrasilanes (H_2)		165
$Z_{4-n}SiH_n$	$HgCl_2 + H_2O$	$Z_{4-n}Si(OH)_n + Hg_2Cl_2 + 2HCl$ (for determining Si–H)	100.0	99
$CH_2(CH_2)_4CHSiH_3$	$HgCl_2$	$CH_2(CH_2)_4CHSiH_2Cl$	79	15
$CH_2(CH_2)_4CHSiH_3$	$HgBr_2$	$CH_2(CH_2)_4CHSiH_2Br$	79	15
$CH_2(CH_2)_4CHSiH_3$	AgCl	$CH_2(CH_2)_4CHSiH_2Cl$	80	15
$CH_2(CH_2)_4CHSiH_3$	$PdCl_2$	$CH_2(CH_2)_4CHSiH_2Cl$	low	15

SiD_4	PI_3 (P_{red} + I_2)	SiD_3I		346
$CH_2Si(CH_3)_2CH_2Si(CH_3)_2CH_2Si(CH_3)H$	$HgCl_2$	$CH_2Si(CH_3)CH_2Si(CH_3)_2CH_2Si(CH_3)Cl$	—	202
$CH_2(CH_2)_4CHSiH_3$	AgNCO	$CH_2(CH_2)_4CHSiH_{3-n}(NCO)_n$; $n = 1(44\%), 3$	—	15
$CH_2(CH_2)_4CHSiH_3$	AgNCS	$CH_2(CH_2)_4CHSiH_{3-n}(NCS)_n$; $n = 1(41\%), 3$	—	15
$(C_6H_5)_3SiH$	$PhHgCCl_2Br$	$(C_6H_5)_3SiCCl_2H$	90	553
$(C_6H_5)_3SiH$	$PhHgCBr_3$	$(C_6H_5)_3SiCBr_2H$	89	553
$(C_2H_5)_3SiH$	$PhHgCCl_2Br$	$(C_2H_5)_3SiCCl_2H$	79	553
$(C_6H_5)_2SiH_2$	1 $PhHgCCl_2Br$	$(C_6H_5)_2Si(CCl_2H)H$	77	553
$(C_6H_5)_2SiH_2$	3 $PhHgCCl_2Br$	$(C_6H_5)_2Si(CCl_2H)_2$	83	553
SiH_4	BBr_3	60% SiH_3Br, 40% SiH_2Br_2	50	130
Si_2H_6	BBr_3	40% SiH_3SiH_2Br, 48% SiH_2BrSiH_2Br, 2% SiH_3SiHBr_2, 9% $SiH_2BrSiHBr_2$, 1% $SiHBr_2SiHBr_2$	60	130

Another point in favour of the carbonium ion mechanism is the formation of $PhC^{\oplus}BBr_4^{\ominus}$ in the reaction of BBr_3 with PhCBr, while PhSiBr forms no silonium analogue with BBr_3 [*105*].

The reaction of the Si–H bond with organic halides appears to take place through a nucleophilic attack of halide on the silicon and an electrophilic attack of the carbonium ions on the hydride ion in a polar four-centre reaction [*105*]:

$$Ph_3SiH + Ph_3\overset{\oplus}{C}\overset{\ominus}{Cl} \xrightarrow{\text{benzene}} \begin{matrix} Ph_3\overset{\delta+}{Si} \ldots \overset{\delta-}{H} \\ \vdots \qquad \vdots \\ \underset{\delta-}{Cl} \cdots \underset{\delta+}{C}Ph_3 \end{matrix} \longrightarrow Ph_3SiCl + Ph_3CH$$

The halogen atoms in alkenyl halides, which are more mobile than those in alkyl halides, generally react with the Si–H bond even without a catalyst:

$$R_3SiH + CH_2{=}CHCH_2Br$$
$$\downarrow$$
$$R_3SiH + [CH_2{=}CH{-}\overset{\oplus}{C}H\overset{\ominus}{Br} \rightleftharpoons \overset{\oplus}{C}H_2{-}CH{=}CH_2\overset{\oplus}{Br}]$$
$$\downarrow$$
$$R_3SiBr + CH_2{=}CHCH_3$$

However, the addition of an aluminium halide promotes the reaction. Since the rate of the process in the reaction of allyl bromide, $AlCl_3$, and a triorganosilane ($R = C_2H_5$, C_3H_7, C_6H_5) does not depend on the organic substituents on the Si, the assumption is obvious that the formation of the carbonium ion is the rate-determining step [*535a*]:

$$RX + AlCl_3 \rightarrow R^{\oplus}AlCl_3X^{\ominus}$$

In the presence of platinum salts or peroxides, or on UV irradiation, the Si–H bond is not halogenated but adds to the C=C bond (see section 5.11).

The reaction of $C_6H_5HgCX_2Br$ ($X = Cl$, Br) with $(C_6H_5)_3SiH$ [or $(C_6H_5)_3GeH$] in benzene at 80° results in insertion of CCl_2 and CBr_2, respectively, into the Si–H (Ge–H) bonds to give Si–CX_2H (Ge–CX_2H) compounds [*553a*]:

$$(C_6H_5)_3SiH + C_6H_5HgCCl_2Br \rightarrow (C_6H_5)_3SiCCl_2H + C_6H_5HgBr$$

Acid halides have no acylating effect, in contrast to the aromatic acylation observed in organic chemistry (for example, by the acylium ion mechanism):

$$RCOCl \underset{-AlCl_4^{\ominus}}{\overset{+AlCl_3}{\rightleftharpoons}} RCO^{\oplus} \xrightarrow{ArH} RCOArH^{\oplus} \xrightarrow{+AlCl_4^{\ominus}} HCl + ArCRO{\cdot}AlCl_3$$

Rather, in accordance with the highly hydridic character of the hydrogen, the silane is halogenated:

$$\mathrm{RCOCl} \underset{-\mathrm{AlCl}_4^{\ominus}}{\overset{+\mathrm{AlCl}_3}{\rightleftharpoons}} \mathrm{RCO}^{\oplus}$$

$$\mathrm{AlCl}_4^{\ominus} + {>}\mathrm{Si}\text{–}\overset{\delta-}{\mathrm{H}} + \mathrm{RCO}^{\oplus} \rightarrow [\mathrm{Cl_3Al\text{–}Cl} \ldots {>}\mathrm{Si\text{–}H} \ldots \mathrm{OCR}] \rightarrow \mathrm{AlCl_3} + {>}\mathrm{Si\text{–}Cl} + \mathrm{RCHO}$$

Di-, tri-, and tetrasilanes do not react with carbon tetrachloride or chloroform in the absence of $AlCl_3$. The addition of traces of oxygen leads to explosive formation of chlorinated silanes [*615, 617*]. The halogenation of the Si–H bond with halogenoalkanes therefore does not take place ionically but radically in the absence of a donor ($AlX_4^{\ominus}$). Since monosilane does not react with chloroalkanes even in the presence of traces of oxygen, it must be assumed that the primary step of the radical chain is the formation of silyl radicals by homolytic cleavage of the Si–Si bond [*617*].

The reaction of $SiHCl_3$ and C_6H_5Cl in the gas phase at 380–430° is initiated by dissociation of the Si–H bond at the surface. Radical chains are then propagated in the gas phase and terminated at the surface [*117e*].
Table 7.24 contains examples of the reactions.

5.5 *Reaction of the Si–H bond with oxygen*

Pure monosilane does not explode when it is mixed with oxygen at atmospheric pressure [*173*]. When the pressure is lowered, it ignites at a well-defined pressure, which is higher at higher temperatures [*173*], silica being formed explosively:

$$SiH_4 + 2O_2 \rightarrow SiO_2 + 2H_2O$$

Monosilane containing small amounts of CO_2, N_2, C_2H_4(!), or H_2(!) does not ignite in air [*99*]. Furthermore, in the undiluted state it can be condensed in the free atmosphere with liquid nitrogen, this giving liquid silane [aerial oxygen condenses in the trap only at the freezing point of SiH_4 (– 184.7°)] [*99*]. For the thermodynamics of the combustion of SiH_4, see [*434*].

The air-stable methylsilane explodes in air in the presence of mercury [*173, 614*]. Monoorganosilanes with relatively large organic substituents [*e.g.*, $C_2H_5SiH_3$, cyclo-$C_6H_{11}SiH_3$, $C_6H_5SiH_3$, $CH_2(SiH_3)_2$] can generally be distilled satisfactorily in the free atmosphere. Metals (Hg manometer) or metal halides that catalyse a dismutation, for example $2\,RSiH_3 \rightleftharpoons R_2SiH_2 + SiH_4$, should be scrupulously excluded. Di- and tri-organosilanes are still more stable than the monosubstitution derivatives. Thus, diethylsilane is resistant to air in the pure state. It is hardly attacked by PbO_2 and MnO_2. Ozone oxidizes it relatively slowly, the ethyl groups also being attacked. Ag_2O and HgO oxidize it surprisingly rapidly (cool to –80° or work in petroleum ether). Tributylsilane can be distilled without oxidation at 220°, and tri(*m*-trifluoromethylphenyl)silane at 322–325°.

TABLE 7.24

REACTION OF SILICON HYDRIDES WITH ORGANIC HALIDES (Si–H → Si–X)

Hydride	*Halogenating agent*	*Reaction products*	*Yield (mole %)*	*Ref.*
$(C_2H_5)_3SiH$	C_6H_5COCl	—	0	285
$(C_2H_5)_3SiH$	$C_6H_5COCl + AlCl_3$	$(C_2H_5)_3SiCl$	30	285
$(C_2H_5)_3SiH$	C_6H_5COBr	$(C_2H_5)_3SiBr$	15	285
$(C_2H_5)_3SiH$	o-$ClC_6H_4COCl + AlCl_3$	$(C_2H_5)_3SiCl$	—	285
$(C_2H_5)_3SiH$	p-$C_2H_5OC_6H_4COCl$	$(C_2H_5)_3SiCl$	—	285
$(C_2H_5)_3SiH$	p-$C_2H_5OC_6H_4COBr$	—	0	285
$(C_6H_5CH_2)_3SiH$	C_6H_5COCl	$(C_6H_5CH_2)_3SiCl$	—	285
$(C_6H_5CH_2)_3SiH$	C_6H_5COBr	$(C_6H_5CH_2)_3SiBr$	—	285
$(C_6H_5CH_2)_3SiH$	o-ClC_6H_4COCl	—	0	285
$(C_6H_5CH_2)_3SiH$	p-ClC_6H_4COCl	—	0	285
$(C_6H_5CH_2)_3SiH$	p-$C_2H_5OC_6H_4COCl$	$(C_6H_5CH_2)_3SiCl$	—	285
$(C_6H_5CH_2)_3SiH$	p-$C_2H_5OC_6H_4COBr$	$(C_6H_5CH_2)_3SiBr$	—	285
$(C_2H_5)_3SiH$	n-$C_6H_{13}Cl + AlCl_3$	$(C_2H_5)_3SiCl + C_6H_{14}$	—	590, 674
$(C_2H_5)_3SiH$	i-$C_3H_7Cl + AlCl_3$	$(C_2H_5)_3SiCl +$ i-C_3H_8	—	126
$(C_2H_5)_3SiH$	$CH_2{=}CHCH_2Cl + AlCl_3$	$(C_2H_5)_3SiCl + CH_2{=}CHCH_2CH_3$	—	126
$(C_2H_5)_3SiH$	$CH_2{=}CHCH_2Br$	$(C_2H_5)_3SiBr + CH_2{=}CHCH_2CH_3$	—	672
$(C_6H_5)_3SiH$	$CH_2{=}CHCH_2Br$	—	—	535a
$(C_2H_5)_3SiH$	$CH_3COOCH_2CH_2Br$	$(C_2H_5)_3SiBr + CH_3COOCH_2CH_3$	—	672
$Si_2H_6, Si_3H_8, Si_4H_{10}$	CCl_4 or $CHCl_3$	—	0	617
$Si_2H_6, Si_3H_8, Si_4H_{10}$	CCl_4 or $CHCl_3 (+ O_2)$	explosion, partially chlorinated di-, tri-, and tetrasilanes	100	617
Si_3H_8	CCl_4 or $CHCl_3 + AlCl_3$	partially chlorinated trisilane, H_2, CH_4, CH_nCl_{4-n}	—	617
$(C_2H_5)_3SiH$	$(CH_3)_3CCH_2Cl + AlCl_3$	$(C_2H_5)_3SiCl + (CH_3)_4C$	—	674
	$(CH_3)_3CCH_2CH_2Cl + AlCl_3$	$(C_2H_5)_3SiCl$ (+ i-C_6H_{14}, $(CH_3)_3CC_2H_5$)	—	674
$(C_2H_5)_3SiH$	$CH_2{=}CHCH_2Br + AlBr_3$	$(C_2H_5)_3SiBr$	$\leqslant 94$	126
$(C_2H_5)_3SiH$	$CH_2{=}CHCH_2Br$	—	—	535a
$(C_3H_7)_3SiH$	$CH_2{=}CHCH_2Br$	—	—	535a
$(C_2H_5)_3SiH$	$C_4H_9Br + AlBr_3$	$(C_2H_5)_3SiBr$	$\leqslant 94$	126
$(C_2H_5)_3SiH$	$C_3H_7Cl + AlCl_3$	$(C_2H_5)_3SiCl$	$\leqslant 94$	126
$(C_2H_5)_3SiH$	i-$C_3H_7Br + AlBr_3$	$(C_2H_5)_3SiBr$	$\leqslant 94$	126
$(C_2H_5)_3SiH$	$C_3H_7Br + AlBr_3$	$(C_2H_5)_3SiBr$	$\leqslant 94$	126
$(C_2H_5)_3SiH$	$BrCH_2CH_2Br + AlBr_3$	$(C_2H_5)_3SiBr$	$\leqslant 94$	126

$(C_2H_5)_3SiH$	$C_2H_5I + AlI_3$	$(C_2H_5)_3SiI$	⩽ 94	126
$(C_2H_5)_3SiH$	C_6H_5Cl or $C_6H_5Br + AlX_3$	no reaction	0	126
$(CH_3)(C_3H_7)_2SiH$	$C_6H_{11}Cl + AlCl_3$	$(CH_3)(C_3H_7)_2SiCl$	69	126
$(C_6H_5)_3SiH$	$(C_6H_5)_3CBr$, cyclohexane	—	0	105
$(C_6H_5)_3SiH$	$(C_6H_5)_3CBr$, dioxan	$(C_6H_5)_3SiBr$ (at 101°); at 25°: no reaction	100	105
$(C_6H_5)_3SiH$	$(C_6H_5)_3CBr$, CCl_4	$(C_6H_5)_3SiBr$ (at 76°); at 25°: no reaction	100	105
$(C_6H_5)_3SiH$	$(C_6H_5)_3CBr$, C_6H_6	$(C_6H_5)_3SiBr$, slow reaction 12–48 h	23–60	105
$(C_6H_5)_3SiH$	$(C_6H_5)_3CBr$, CCl_2CCl_2	—	0	105
$(C_6H_5)_3SiH$	$(C_6H_5)_3CBr$, CS_2	—	0	105
$(C_6H_5)_3SiH$	$(C_6H_5)_3CBr$, $(C_2H_5)_2O$	—	0	105
$(C_6H_5)_3SiH$	$(C_6H_5)_3CBr$, C_2H_5I	$(C_6H_5)_3SiBr$	100	105
$(C_6H_5)_3SiH$	$(C_6H_5)_3CBr$, THF	—	0	105
$(C_6H_5)_3SiH$	$(C_6H_5)_3CBr$, CH_2Cl_2	$(C_6H_5)_3SiBr$, fast reaction (15 min)	100	105
$(C_6H_5)_3SiH$	$(C_6H_5)_3CBr$, $C_2H_4Cl_2$	$(C_6H_5)_3SiBr$, fast reaction (15 min)	100	105
$(C_6H_5)_3SiH$	$(C_6H_5)_3CBr$, $C_6H_5NO_2$	$(C_6H_5)_3SiBr$, fast reaction (15 min)	100	105
$(C_6H_5)_3SiH$	$(C_6H_5)_3CBr$, CH_3NO_2	$(C_6H_5)_3SiBr$, fast reaction (15 min)	100	105
$(C_2H_5)_3SiH$	$(C_6H_5)_3CBr$, CH_2Cl_2	$(C_2H_5)_3SiBr$, fast reaction (25°, 10 min)	100	105
$(C_2H_5)_3SiH$	$(C_6H_5)_3CBr$, C_6H_6	$(C_2H_5)_3SiBr$ (25°, 24 h)	80	105
$(C_2H_5)_3SiH$	$(C_6H_5)_3CBr$, CH_3CN	$(C_2H_5)_3SiBr$ (25°, 18 h)	100	105
$(C_2H_5)_3SiH$	$(C_6H_5)_3CBr$, CCl_4	$(C_2H_5)_3SiBr$ (76°, 18 h)	100	105
$(C_2H_5)_3SiH$	$(C_6H_5)_3CBr$, THF	— (65°, 18 h)	0	105
$(C_2H_5)_3SiH$	$(C_6H_5)_3CCl$, CH_2Cl_2	$(C_2H_5)_3SiCl$, fast reaction (25°, 10 min)	100	105
$(C_2H_5)_3SiH$	$(C_6H_5)_3CCl$, C_6H_6	$(C_2H_5)_3SiCl$ (25°, 24 h)	70	105
$(C_2H_5)_3SiH$	$(C_6H_5)_3CCl$, CH_3CN	$(C_2H_5)_3SiCl$ (25°, 18 h)	45	105
$(C_2H_5)_3SiH$	$(C_6H_5)_3CCl$, CCl_4	$(C_2H_5)_3SiCl$ (76°, 18 h)	100	105
$(C_2H_5)_3SiH$	$(C_6H_5)_3CCl$, THF	— (65°, 18 h)	0	105
$(1)(C_6H_5)_2SiH_2$	$(1)(C_6H_5)_3CBr$	$(C_6H_5)_2SiHBr$	—	105
$(1)(C_6H_5)_2SiH_2$	$(1)(C_6H_5)_3CCl$	$(C_6H_5)_2SiHCl$	—	105
$(1)(C_6H_5)_2SiH_2$	$(2)(C_6H_5)_3CBr$	$(C_6H_5)_2SiBr_2$	—	105
$(1)(C_6H_5)_2SiH_2$	$(2)(C_6H_5)_3CCl$	$(C_6H_5)_2SiCl_2$	—	105
$SiHCl_3$	C_2H_5C	$C_2H_5SiCl_3$	—	121
$SiHCl_3$	$(C_6H_5)_3CCl$	$SiCl_4$	—	105

5.6 *Base-catalysed solvolysis of the Si–H bond*

In alkali-free water, silane or organosilanes either do not undergo solvolysis or do so only very slowly[*447, 604, 608*]. In the presence of hydroxide or alkoxide ions, the Si–H bond hydrolyses substantially faster[*127*]. Its rate of solvolysis in compounds which contain only one Si–H bond and organic ligands which are not too large is low enough at room temperature to enable the course of the reaction to be followed conveniently by evolution of the hydrogen (alkane). In solvolysis only the reaction according to equation (1) is important:

$$\geq SiH + R'O^{\ominus} + ROH \rightarrow \geq SiOR' + RO^{\ominus} + H_2 \quad (1)$$

$$\geq SiH + R'O^{\ominus} + ROH \rightarrow \geq SiOH + RO^{\ominus} + R'H \quad (2)$$

This is shown by reactions of SiH_4 in EtOH + EtOLi[*447*]; SiH_4 in n-PrOH + PrOLi[*447*]; SiH_4 in n-BuOH + BuOLi[*447*]; Et_3SiH, n-Pr_3SiH, *sec*-Pr_3SiH, n-Bu_3SiH, *sec*-Bu_3SiH, i-Bu_3SiH, *tert*-Bu_3SiH, n-Am_3SiH, $(p\text{-}YC_6H_4)_3SiH$ (Y = H, CH_3, CH_3O, Cl, F), $(C_6H_5CH_2)_3SiH$, and $(1\text{-}C_{10}H_7)_3SiH$ in EtOH + H_2O + KOH[*538*]; Ph_3SiH in piperidine + H_2O[*302*]; Ph_3SiD in piperidine + H_2O [*221, 302*]; Ph_3SiH in EtOH + H_2O + KOH[*302*]; Ph_3SiD in EtOH + H_2O + KOH[*302*]; $(YC_6H_4)_3SiH$ [Y = *p*-Cl, *m*-CH_3, *p*-CH_3, *m*-$N(CH_3)_2$, *p*-$N(CH_3)_2$, *p*-OCH_3] in piperidine + H_2O[*219*]; Ph_3SiT in piperidine + H_2O[*309*]; and n-Pr_3Si in EtOH + H_2O + KOH[*302*]. The alkane component of the gas formed in the solvolysis of $(C_2H_5)_3SiH$, for example, is in fact vanishingly small[*483*].

The rate of hydrolysis of the Si–H bond increases in the following sequence from left to right[*584, 667*]:

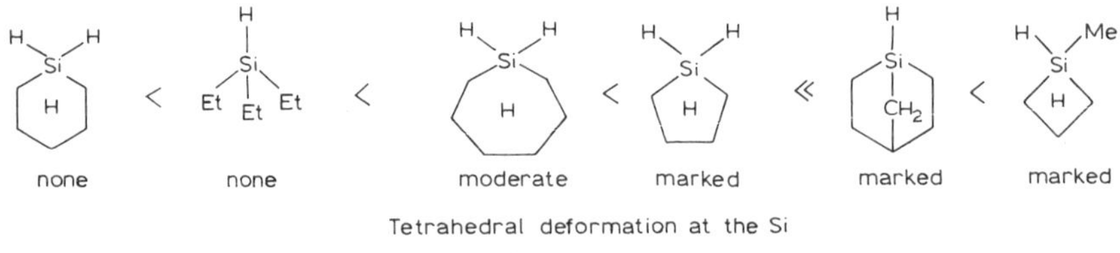

The gradation in the rates of solvolysis is a consequence of the more or less pronounced deformation of the normal tetrahedral angle at the silicon in the intermediate stage to the angles of a trigonal bipyramid (90° or 120°). The five-coordinated transition state that must be assumed in solvolysis (see below) is substantially preformed in the compounds on the right of the solvolysis sequence, for example:

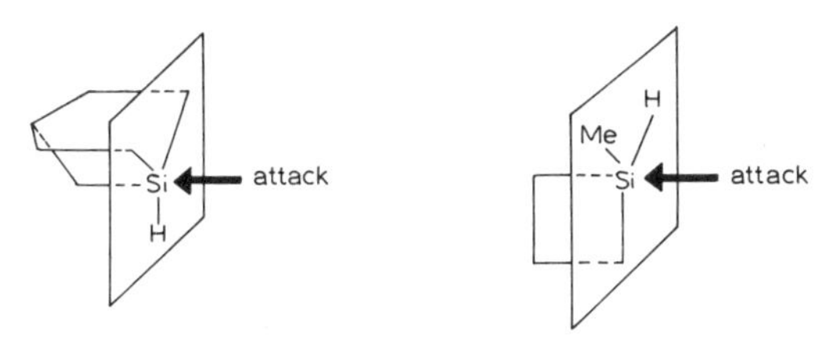

The rate of hydrolysis of tri-organosilanes in a solution of potassium hydroxide in aqueous ethanol or in aqueous piperidine rises with increasing electron-releasing action of the ligands[*37, 219, 221, 272–274, 301, 302, 483, 535–537*]. These hydrolysis experiments give two results: (i) the charge of the silicon atom in the transition state, which has higher energy, must be more negative than in the ground state, since the donation of electrons to the silicon opposes hydrolysis; (ii) steric factors affect the hydrolysis less strongly than the inductive effect. When all the experimental results are taken into account, the following three reaction mechanisms for the base-catalysed solvolysis of the Si–H bond are conceivable.

(1) Nucleophilic attack of the hydroxide ion or alkoxide ion on silicon and electrophilic attack of water or alcohol on the hydride hydrogen, electron re-arrangement, and Si–H and H–O cleavage (the dotted bonds have variable strengths):

$$OH^{\ominus} + SiR_3H \longrightarrow [HO \ldots \overset{\ominus}{Si}(R_3){-}H]$$

$$[HO \ldots \overset{\ominus}{Si}(R_3){-}H] + HOH \longrightarrow [HO \ldots \overset{\ominus}{Si}(R_3){-}H \ldots H{-}OH] \longrightarrow HO{-}SiR_3 + H_2 + OH^{\ominus}$$

$$R_3SiH + H_2O \xrightarrow{OH^{\ominus}} R_3SiOH + H_2$$

(2) Triple collision of hydroxide ion or alkoxide ion, silane and water or alcohol, electron rearrangement, and decomposition of the intermediate with the formation of hydrogen:

$$HO^{\ominus} + SiR_3H + H_2O \longrightarrow [HO \ldots \overset{\ominus}{Si}(R_3){-}H \ldots H{-}OH]$$

$$[HO \ldots \overset{\ominus}{Si}(R_3){-}H \ldots H{-}OH] \longrightarrow [HO \ldots Si(R_3){-}H \ldots H{-}\overset{\ominus}{O}H] \longrightarrow HO{-}SiR_3 + H_2 + OH^{\ominus}$$

$$R_3SiH + H_2O \xrightarrow{OH^{\ominus}} R_3SiOH + H_2$$

(3) Division of the (rare) triple collision into two steps: continuous association of water with the hydride hydrogen, nucleophilic attack of hydroxide ions on silicon, electron rearrangement, cleavage of the Si–H and O–H bonds:

$$R_3SiH + H_2O \rightleftharpoons R_3Si{-}H \ldots H{-}OH$$

$$HO^{\ominus} + SiR_3H \ldots H{-}OH \longrightarrow [HO \ldots \overset{\ominus}{Si}(R_3){-}H \ldots H{-}OH] \longrightarrow HO{-}SiR_3 + H_2 + OH^{\ominus}$$

$$R_3SiH + H_2O \xrightarrow{OH^{\ominus}} R_3SiOH + H_2$$

References p. 621

According to the experimental material so far available, the reaction sequenc according to (3) is the most likely. For a similar mechanism, see [*492*]. On th hydrolysis of optically active silanes, see [*588, 586*].

5.7 *Acid-catalysed hydrolysis of the Si–H bond*

Organosilanes undergo solvolysis in an acid medium with the evolution o hydrogen [*38*]:

$$\rangle\text{Si–H} + H_2O \xrightarrow{H^{\oplus}} \rangle\text{Si–OH} + H_2$$

The rate of hydrolysis of the first Si–H bond in $(CH_3CH_2CH_2)_2SiH_2$ is 18 time higher than that of the second [*38*]. With low concentrations of acid (HCl), th Si–H bond hydrolyses by first-order kinetics with respect to $[H_3O^{\oplus}]$. As in base catalysed hydrolysis, the rate falls with increasing donation of electrons by th ligands to the silicon [*38*], a proof that here again silicon in the transition state i more negative than in the ground state; in the acid-catalysed hydrolysis, as well the rate falls with increasing steric demands of the ligands.

After an unlikely silonium ion has been excluded for the transition state, ther are two possible sequences of reactions (dotted bonds denote variable bon strengths).

(1) Nucleophilic attack of water on silicon, electrophilic attack of oxoniun ions on the hydride, electron rearrangement:

$$\begin{aligned}
H_2O + SiR_3H &\rightleftharpoons [H_2O\ldots Si(R_3)\text{–}H]\\
[H_2O\ldots Si(R_3)\text{–}H] + H\text{–}\overset{\oplus}{O}H_2 &\xrightarrow{a} [H_2O\ldots Si(R_3)\text{–}H\ldots H\text{–}OH_2]^{\oplus}\\
&\longrightarrow H_2OSiR_3^{\oplus} + H_2 + H_2O\\
H_2OSiR_3^{\oplus} &\rightarrow HOSiR_3 + H^{\oplus}\\
\hline
R_3SiH + H_2O &\xrightarrow{H^{\oplus}} R_3SiOH + H_2
\end{aligned}$$

(2) Electrophilic attack of the oxonium ion on the hydride, nucleophilic attack of water on silicon, electron rearrangement:

$$\begin{aligned}
SiR_3H + H_2O^{\oplus} &\xrightarrow{b} [Si(R_3)\text{–}H\ldots H\text{–}\overset{\oplus}{O}H_2]\\
H_2O + [Si(R_3)\text{–}H\ldots H\text{–}\overset{\oplus}{O}H_2] &\longrightarrow [H_2O\ldots Si(R_3)\text{–}H\ldots OH_2]^{\oplus}\\
&\longrightarrow H_2OSiR_3^{\oplus} + H_2 + H_2O\\
H_2OSiR_3^{\oplus} &\longrightarrow HOSiR_3 + H^{\oplus}\\
\hline
R_3SiH + H_2O &\longrightarrow R_3SiOH + H_2
\end{aligned}$$

A distinction between the two sequences of reaction cannot be made with the available experimental material. It is merely likely to some extent that the attack of the oxonium ion on the hydride (a, b) is the rate-determining step.

5.8 *Reaction of the Si–H bond with anhydrous proton-active compounds*

Monosilane does not react with methanol [*269, 447*] or ethanol [*608*] without a catalyst or in the presence of methoxylithium. Copper catalyses the alcoholysis (Table 7.25 contains examples of the reaction) [*447, 603, 604*]:

$$\rangle Si{-}H + ROH \xrightarrow{Cu} \rangle Si{-}OR + H_2$$

In this reaction, monosilane forms di-, tri-, and tetramethoxysilanes but no monomethoxysilane [*604*]. On being boiled with alcohols in the absence of a catalyst, phenylsilane reacts only very sluggishly [*379*]. In the presence of metallic copper, by the dropwise addition of alcohols (CH_3OH, C_2H_5OH, n-C_3H_7OH, n-C_4H_9OH) to phenylsilane at 20°, all the hydrogen atoms in the silane are substituted by alkoxy groups successively in a controllable fast reaction:

$$PhSiH_3 \xrightarrow[Cu]{+ROH,\,-H_2} PhSiH_2(OR) \xrightarrow[Cu]{+ROH,\,-H_2} PhSiH(OR)_2 \xrightarrow[Cu]{+ROH,\,-H_2} PhSi(OR)_3$$

The rate of solvolysis and the tendency to the formation of monoalkoxy derivatives increase in the sequence $CH_3OH < C_2H_5OH <$ n-$C_3H_7OH <$ n-C_4H_9OH [*379*]. Thus, Cu-catalysed alcoholysis, in contrast to base-catalysed alcoholysis [*447*] (see section 5.6) is also suitable for the partial alkoxylation of silicon hydrides. But there are other methods for the preparation of Si–H-containing silicon alkoxides, for example [*651a*]:

$$[(CH_3)_2SiH]_2S + 2CH_3OH \xrightarrow[5\,min]{20°} 2(CH_3)_2(CH_3O)SiH + H_2S$$

The partially alkoxylated silanes $RSiH_2(OR')$ and $RSiH(OR')_2$ hydrolyse still more readily than the Si–H-free compounds $R_{4-n}Si(OR')_n$.

It follows from the experiments that the rate of substitution in uncatalysed alcoholysis becomes greater with increasing number of alkoxy groups on the silicon. The action of metallic copper has not been studied further.

Under the catalytic influence of $Al + I_2$ [*cit. 440*] or colloidal Ni [*440*] or in pyridine, diethylaniline, or dimethylformamide as a solvent without a catalyst [*493*], with a silicon hydride a saturated carboxylic acid forms hydrogen and a silyl acetate (examples of the reaction are given in Table 7.25):

$$\rangle \overset{\delta+}{Si}{-}\overset{\delta-}{H} + {}^{\ominus}O{-}C(=O){-} + H^{\oplus} \xrightarrow{Al, I_2; Ni} \rangle SiO{-}C(=O){-} + H_2$$

An unsaturated carboxylic acid offers the additional possibility of the addition of the Si–H bond to the double C=C bond (see section 5.11.5). In spite of the small amount of experimental material, it may be assumed that in this case the carboxyl

References p. 621

group remains unaffected only if metal catalysts (formation of H_2) are absent and catalysts favouring addition to the double C=C bond are present. Example:

$$Ph_3SiH + CH_2{=}CH(CH_2)_8COOH \xrightarrow{(PhCOO)_2} Ph_3Si(CH_2)_{10}COOH$$

TABLE 7.25

REACTION OF SILICON HYDRIDES WITH PROTON-ACTIVE NON-AQUEOUS COMPOUNDS (Si–H → Si–OX)

(Moles of) hydride	*(Moles of) acid*	*Catalyst*	*Reaction temp. (°C)*	*Reaction product*	*Yield (mole-%)*	*Ref.*
$(1)C_6H_5SiH_3$	$(1)CH_3OH$	Cu	20	$C_6H_5SiH_2OCH_3$ $C_6H_5SiH(OCH_3)_2$	26 12	379
$(1)C_6H_5SiH_3$	$(1)C_2H_5OH$	Cu	20	$C_6H_5SiH_2OC_2H_5$ $C_6H_5SiH(OC_2H_5)_2$	11 5	379
$(1)C_6H_5SiH_3$	$(1)n\text{-}C_3H_7OH$	Cu	20	$C_6H_5SiH_2OC_3H_7$ $C_6H_5SiH(OC_3H_7)_2$	34 8	379
$(1)C_6H_5SiH_3$	$(1)n\text{-}C_4H_9OH$	Cu	20	$C_6H_5SiH_2OC_4H_9$ $C_6H_5SiH(OC_4H_9)_2$	47 14	379
$(1)SiH_4$	$(1)CH_3OH$	—	20	$Si(OCH_3)_4$	20	604
$(1)SiH_4$	$(0.3)CH_3OH$	Cu	20	$SiH_2(OCH_3)_2$ $SiH(OCH_3)_3$ $(Si(OCH_3)_4$	~ 2 ~ 11 ~ 2	604
$(1)SiH_4$	$(1)CH_3OH$	Cu	20	$SiH(OCH_3)_3$ $Si(OCH_3)_4$	~ 34 ~ 1	604
SiH_4	CH_3OH	C_2H_5OLi	—	—	—	447
SiH_4	C_2H_5OH	CH_3OLi	—	—	—	447
$(1)SiH_4$	$(37)C_2H_5OH$	C_2H_5OLi	—	$Si(OC_2H_5)_4$	67	447
$(1)SiH_4$	$(28)n\text{-}C_3H_7OH$	C_3H_7OLi	—	$Si(OC_3H_7)_4$	78	447
$(1)SiH_4$	$(26)n\text{-}C_4H_9OH$	C_4H_9OLi	—	$Si(OC_4H_9)_4$	41	447
R_3SiH	H_3PO_4	$Al + I_2$	—	$(R_3SiO)_3PO$	—	cit. 440
R_3SiH	$CH_3PO(OH)_2$	$Al + I_2$	—	$(R_3SiO)_2(CH_3)PO$	—	cit. 440
R_3SiH	$B(OH)_3$	$Al + I_2$	—	$B(OSiR_3)_3$	—	cit. 440
R_3SiH	C_6H_5OH	Ni	—	$R_3SiOC_6H_5$	—	cit. 440
R_3SiH	R'_3SiOH	Ni	—	$R_3SiOSiR'_3$	—	cit. 440
$(C_2H_5)_3SiH$	HCOOH	Ni	> 20	$(C_2H_5)_3Si(OCOH)$	50	440
$(C_2H_5)_3SiH$	CH_3COOH	Ni	> 20	$(C_2H_5)_3Si(OCOCH_3)$	84	440
$(C_2H_5)_3SiH$	$n\text{-}C_3H_7COOH$	Ni	> 20	$(C_2H_5)_3Si(OCOC_3H_7)$	79	440
$(C_2H_5)_3SiH$	$n\text{-}C_5H_{11}COOH$	Ni	> 20	$(C_2H_5)_3Si(OCOC_5H_{11})$	78	440
$(CH_3)(C_6H_5)_2SiH$	CH_3COOH	Ni		$(CH_3)(C_6H_5)_2Si(OCOCH_3)$	59	440
$C_6H_5SiH_3$	CH_3COOH	Pyridine	40–50	$C_6H_5SiH_{3-n}[OC(CH_3)O]_n$	—	493

5.9 *Reaction of the Si–H bond with R_2S, RSH, N_2^*, RN_2, RNH_2, R_2NH, PH_3, AsH_3, and GeH_4*

Diphenylsilane reacts with *thioheterocycles* with the formation of the corresponding Si-heterocycles (in low yields) [*688*]:

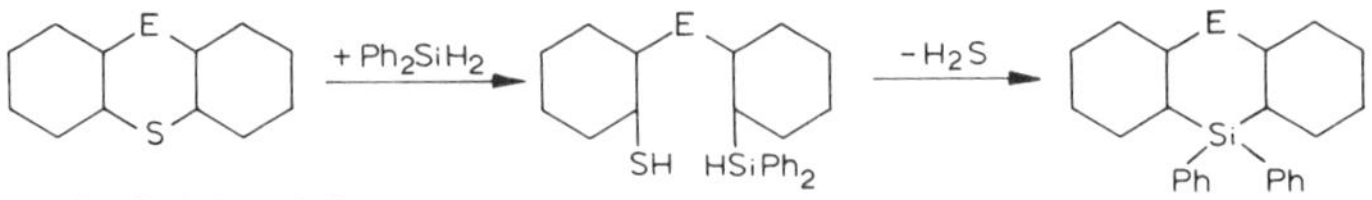

E = O, S, NH, NC_2H_5

The reaction of *thioalcohols* with silicon hydride (second reaction step) does not, however, generally lead to C–Si linkage with the elimination of H_2S. The reaction of triphenylsilane with *p*-thiocresol does not give triphenyl-*p*-tolylsilane but an approximately 69% yield of triphenyl(*p*-tolylthio)silane [*197*]:

$$Ph_3SiH + HSC_6H_4CH_3 \nrightarrow Ph_3Si(C_6H_4CH_3) + H_2S$$

$$Ph_3SiH + HSC_6H_4CH_3 \rightarrow Ph_3Si(SC_6H_4CH_3) + H_2$$

In the reaction of monosilane with active *nitrogen* at 28° and at 250°, hydrogen is produced but no ammonia. Methylsilanes, CH_3SiH_3, $(CH_3)_2SiH_2$, $(CH_3)_3SiH$ and $(CH_3)_4Si$ form hydrogen cyanide, ammonia, and hydrogen. The ratio of the amounts of the reaction products to that of the silanes that have reacted excludes the obvious mechanism via imide radicals ($SiH_4 + N^* \xrightarrow{-12\,cal} SiH_3 + NH$; $NH + N \xrightarrow{-cal} N_2 + H$). Ammonia is probably formed via amide radicals [*123*].

Diazoethane, CH_3CHN_2, ethylates the Si–H bond via ethylidene as an intermediate according to the following scheme [*322*]:

$$CH_3CHN_2 \rightarrow CH_3CH\!: + N_2$$
$$CH_3CH\!: + C_6H_5SiH_3 \rightarrow (C_6H_5)(CH_3CH_2)SiH_2$$

The silanes $C_6H_5SiH_3$, $ClC_6H_4SiH_3$, and $C_4H_9SiH_3$ react with *amines* [aliphatic amines: $(C_2H_5)_2NH$, $NH_2C_2H_4NH_2$, $HOC_2H_4NH_2$, $(HOC_2H_4)_2NH$; aromatic amines: *p*-$NO_2C_6H_4NH_2$, $C_6H_5NH_2$, *p*-$C_2H_5OC_6H_4NH_2$] in the following way [*492, 651*]:

$$\equiv\text{Si–H} + \text{HN}\langle \rightarrow \equiv\text{Si–N}\langle + H_2$$

The reactions with aliphatic amines take place by a molecular mechanism and those with aromatic amines by an ionic mechanism [*492*]. About the reaction of silane with amide ions see ref. [*513a*].

At 450°, monosilane reacts with *phosphane*, PH_3, to give silylphosphane, SiH_3PH_2, non-volatile Si–P hydrides, disilicon phosphide, and hydrogen [*197, 199*]. The supply of electrical energy to a gaseous mixture of silane and phosphane gives, together with higher silanes (and phosphanes?) the mixed hydrides SiH_3PH_2 and Si_2PH_7 (SiH_3PHSiH_3 and $Si_3SiH_2PH_2$) [*665*] (see section 10.1).

Likewise, the silent electric discharge of silane, SiH_4, and *arsane*, AsH_3, gives, in addition to higher silanes, the compounds SiH_3AsH_2, which decomposes explosively in the light [*130, 515*].

The supply of electrical energy to a mixture of silane and *germane* forms, in addition to higher silanes and germanes mixed hydrides from which it has been possible to isolate silylgermane SiH_3GeH_3 [*582*] (see section 10.1).

5.10 Reaction of the Si–H bond with alkanes

At relatively high temperatures and high pressures with or without a catalyst silicon hydrides react with alkanes or aromatics. In a reaction which is complicated and more suitable for industrial use than for the laboratory, the Si–H bond is converted into an Si–C bond (substitution):

$$HSiCl_3 + CH_4 \rightarrow CH_3SiCl_3 + H_2$$

$$CH_3HSiCl_2 + C_6H_6 \rightarrow (CH_3)(C_6H_5)SiCl_2 + H_2$$

The reaction conditions are: pressure 1–150 atm, temperature without a catalyst 400 to 850°, with a catalyst generally somewhat lower. Table 7.26 contains references and examples of the reaction.

TABLE 7.26
REACTION OF TRICHLOROSILANE WITH ALKANES (Si–H → Si–R)

Silane	*Alkane*	*Reaction conditions*	*Reaction product*	*Ref.*
$HSiCl_3$	CH_4	500–600°, 1 atm	22% $(CH_3)SiCl_3$	385
$HSiCl_3$	CH_4	475°, 165 atm, BCl_3	$(CH_3)SiCl$	561
$HSiCl_3$	C_2H_6	375°, 120 atm, BCl_3	$(CH_3)SiCl_3$, $(CH_3)_2SiCl_2$, $(C_2H_5)SiCl_3$	561
$HSiCl_3$	C_3H_8	500–600°, 1 atm	$(CH_3)SiCl_3$, $(CH_2{=}CH)SiCl_3$	385
$HSiCl_3$	C_4H_{10}	500–600°, 1 atm	$(CH_3)SiCl_3$, $(CH_2{=}CH)SiCl_3$	385
$HSiCl_3$	C_6H_{14}	500–600°, 1 atm	$(CH_3)SiCl_3$, $(CH_2{=}CH)SiCl_3$, $(CH_2{=}CHCH_2)SiCl_3$, $(C_6H_5)SiCl_3$	385
$HSiCl_3$	C_9H_{20}	500–600°, 1 atm	$(CH_3)SiCl_3$, $(CH_2{=}CH)SiCl_3$, $(CH_2{=}CHCH_2)SiCl_3$, $(C_6H_5)SiCl_3$	385
$HSiCl_3$	$(CH_3)_3CCH_2CH(CH_3)_2$	500–600°, 1 atm	$(CH_3)SiCl_3$, $(CH_2{=}CH)SiCl_3$, $(CH_2{=}CHCH_2)SiCl_3$	385
$HSiCl_3$	$\overline{CH_2(CH_2)_4CH_2}$	500–600°, 1 atm	$(CH_3)SiCl_3$, $(CH_2{=}CH)SiCl_3$, $(CH_2{=}CHCH_2)SiCl_3$, $(C_6H_5)SiCl_3$	385
$HSiCl_3$	C_2H_5Cl	316°, 120 atm	$(C_2H_5)SiCl_3$	28

5.11 *Reaction of the Si–H bond with alkenes*

Silicon hydrides react with alkenes to give substitution (at relatively high temperatures):

$$\geq Si\text{–}H + XC(|){=}C(|)\text{—} \rightarrow \geq Si\text{–}C(|){=}C(|)\text{—} + HX$$

or with radical or polar addition ("hydrosilation"):

$$\geq Si\text{–}H + \overset{|}{\underset{|}{C}}{=}\overset{|}{\underset{|}{C}} \rightarrow \geq Si\text{–}\overset{|}{\underset{|}{C}}\text{—}\overset{|}{\underset{|}{C}}\text{—}H$$

5.11.1 *Substitution in the alkene*

While at relatively low temperatures (200–400°), a silicon hydride adds to an alkene without a catalyst (on this, compare the addition of halogen to the double

C=C bond at low temperatures), at relatively higher temperatures (400–650°) substitution takes place (on this, compare the substitution of alkenes with halogen with retention of the double bond at high temperatures):

$$\rangle Si\text{-}H + CH_2{=}CH(CH_2)_nX \nearrow \rangle Si(CH_2)_nCH{=}CH_2 + HX \quad (1)$$

$$\rangle Si\text{-}H + CH_2{=}CH(CH_2)_nX \searrow \rangle SiX + CH_2{=}CH(CH_2)_nH \quad (2)$$

X = F, Cl, Br; F or Cl may be present in the alkene in place of H; $n = 0, 1\ldots$)

Table 7.27 contains examples of the reaction and references. The stronger the electron attraction of the silicon substituents is, the more alkenylsilane (eq. 1) and the less halogenosilane (eqn. 2) are formed. Example [*448*]: the yields of alkenylsilane according to eqn. (1) or of halogenosilane according to eqn. (2) in the reaction of $(p\text{-}YC_6H_4)_3SiH$ with $CH_2{=}CHCH_2Cl$ with various *para* substituents [Y] are: 4% [CH_3], 8% [H], 42% [Cl] (eqn. 1); 32% [CH_3], 20% [H], 6–10% [Cl] (eqn. 2); the tendency to give up electrons is in the following sequence according to the σ constants: $p\text{-}CH_3C_6H_4 > C_6H_5 > p\text{-}ClC_6H_4$. Steric factors are also said to play a part in the selection of the route according to eqn. (1) or (2). Thus, the yields of the addition reaction (eqn. 1) in RR'_2SiH fall in the sequence Cl > CH_3 > $ClCH_2$ > $C_6H_5 \geqslant ClC_6H_4$ (at a given R′ = Cl, CH_3, C_6H_5) [*59*].

5.11.2. Radical addition of silicon hydride to the double C=C bond

The radical addition is stereospecific, like the addition of HBr [*231*]. The bridge radical formed as an intermediate adds to the hydrogen of a second molecule of silane in the *trans*-position so that *cis*-adducts are formed:

$$\rangle Si\cdot + \rangle C{-}C\langle_R \rightarrow \rangle C{-\!-\!-}C\langle_R \text{ (bridged by } \dot{Si}\text{)} \xrightarrow{+\rangle Si-H} {-}\underset{\underset{|}{\rangle Si}}{C}{-}\underset{R}{C}{-}H + \rangle Si\cdot$$

Tables 7.28–7.32 contain references and examples of the reaction. The stereospecificity depends largely on the way in which the silyl radicals are produced. It has been studied in connection with the addition of trichlorosilane to 1-methylcyclohex-1-ene [*558*]. UV irradiation and peroxide catalysis give only a small amount of (forbidden) *trans*-1-methyl-2-(trichlorosilyl)-cyclohexane (*cis*:*trans* ~ 6:1). In the thermally catalysed addition, the amount of *trans*-adduct is still higher (*cis*:*trans* ~ 3:1). In fact, here at the higher temperature the bridge radical changes to an increasing extent into the non-stereospecific open radical $\rangle\dot{C}{-}\underset{|}{\overset{|}{C}}{-}Si\langle$. The formation of isomeric cyclohexylmethyltrichlorosilane, $C_6H_{11}CH_2SiCl_3$, cannot be explained by a radical mechanism. Possibly a polar

References p. 621

mechanism competes with an addition not radically catalysed to an outstanding extent (for further details see section 5.11.3).

The telomerisation of the alkenes (particularly in the case of the thermally induced and the photosensitised additions) and the formation of higher silanes compete with the addition [*196*]:

$$\geq\text{Si–H} \longrightarrow \geq\text{Si}\cdot \xrightarrow{+\,>\text{C=C}<} \geq\text{Si–}\overset{|}{\underset{|}{\text{C}}}\text{–}\overset{|}{\underset{|}{\text{C}}}\cdot$$

$$\xrightarrow{+n\,>\text{C=C}<} \geq\text{Si(}\overset{|}{\underset{|}{\text{C}}}\text{–}\overset{|}{\underset{|}{\text{C}}}\text{)}_n\cdot \xrightarrow{+\text{H}\cdot} \geq\text{Si(}\overset{|}{\underset{|}{\text{C}}}\text{–}\overset{|}{\underset{|}{\text{C}}}\text{)}_n\text{H}$$

$$\xrightarrow{+\,\geq\text{Si–H}} \geq\text{Si(}\overset{|}{\underset{|}{\text{C}}}\text{–}\overset{|}{\underset{|}{\text{C}}}\text{)}_n\text{Si}\leq + \text{H}\cdot$$

$$\geq\text{Si}\cdot + \geq\text{Si–H} \rightarrow \geq\text{Si–Si}\leq + \text{H}\cdot$$

The radical addition is initiated by the production of chain-propagating silyl radicals. They are formed by irradiation, photosensitised irradiation, radical-forming agents, and a rise in temperature.

(1) *Irradiation*

$$\geq\text{Si–H} \xrightarrow{h\nu} \geq\text{Si}\cdot + \text{H}\cdot$$

(2) *Photosensitised irradiation* (if the reactants possess no inherent adsorption in the necessary energy range, *e.g.* SiH_4):

$$\geq\text{Si–H} + \text{Hg}(^3\text{P}_1) \rightarrow \geq\text{Si}\cdot + \text{H}\cdot + \text{Hg}(^1\text{S}_0)$$

(3) *Radical-forming agents*

$$(\text{RCOO})_2 \rightarrow 2\text{RCOO}\cdot$$

$$\geq\text{Si–H} + \text{RCOO}\cdot \rightarrow \geq\text{Si}\cdot + \text{RCOOH}$$

or [*310, 311*]:

$$\text{CH}_3\text{N=NCH}_3 \xrightarrow{h\nu} 2\text{CH}_3\cdot + \text{N}_2$$

$$\geq\text{Si–H} + \text{CH}_3\cdot \rightarrow \geq\text{Si}\cdot + \text{CH}_4$$

Acetyl peroxide is used most frequently. Benzoyl peroxide, which is a still better catalyst, leads to explosions with many alkenes. *tert*-butyl perbenzoate has proved very satisfactory. Apart from peroxides, other radical-forming agents such as ozone or azobis(isobutyronitrile) have been used. Iron (from stainless steel in the case of halogenoolefins) and iron salts inhibit catalysis by peroxide. Nickel, lead, zinc, carbon dioxide, alcohol, water, silicon grease, and air do not affect the addition. The reactions of oxidation-sensitive silanes are, however, preferably carried out under nitrogen. Tin promotes the catalysis by peroxides.

(4) *Rise in temperature*

$$\geq Si{-}H + \text{thermal energy} \rightarrow \geq Si\cdot + H\cdot$$

The orientation in the radical addition of silicon hydride resembles that in the addition of hydrogen halide to the double C=C bond. The unpaired electron most probably remains at the position of greatest possible electron delocalisation (stability of a radical: primary < secondary < tertiary). Consequently, the silyl residue becomes attached to the carbon atom adjacent to the radical carbon atom:

$$\geq Si\cdot + RCH{=}CHR' \rightarrow \geq SiRCH{-}\dot{C}HR'$$

$$\geq SiRCH{-}\dot{C}HR' + \geq Si{-}H \rightarrow \geq SiRCH{-}CH_2R' + \geq Si\cdot$$

in which R, R′ = alkyl, H; R < R′. Examples: [*57*, *83*, *216*, *255*, *597*].

$$SiH_4 \xrightarrow{UV} H_3Si\cdot + H\cdot$$

$$H_3Si\cdot + CH_2{=}CHCF_3 \rightarrow H_3SiCH_2\dot{C}HCF_3 \xrightarrow[-H_3Si]{+SiH_4} H_3SiCH_2CH_2CF_3 \rightarrow$$

$$\xrightarrow[-H\cdot]{UV} H_2\dot{Si}CH_2CH_2CF_3 \xrightarrow{+CH_2=CHCF_3} CF_3CH_2CH_2SiH_2CH_2\dot{C}HCF_3 \rightarrow$$

$$\xrightarrow[-H_3Si\cdot \text{ or } H_2\dot{Si}CH_2CH_2CF_3]{+SiH_4 \text{ or } H_3SiCH_2CH_2CF_3} H_2Si(CH_2CH_2CF_3)_2 \xrightarrow{\text{similarly}} Si(CH_2CH_2CF_3)_4$$

$$Cl_2SiH_2 + CH_2{=}CHCF_3 \xrightarrow{UV} Cl_2SiHCH_2CH_2CF_3$$

$$Cl_3SiH + CF_2{=}CF_2 \xrightarrow{UV} Cl_3SiCF_2CHF_2$$

$$Cl_3SiH + CH_2{=}CF_2 \xrightarrow{UV} Cl_3SiCH_2CHF_2$$

$$Cl_3SiH + CH_2{=}CHCF_3 \xrightarrow{UV} Cl_3SiCH_2CH_2CF_3$$

$$Cl_3SiH + CH_2CH{=}CHCH_2CH_3 \xrightarrow{(CH_3)_3COOCOC_6H_5} Cl_3Si[CH(CH_3)CH_2CH_2CH_3] + Cl_3Si[CH(CH_2CH_3)_2] \, (70:30)$$

$$Cl_2Si(CH_3)H + CH_3CH{=}CHCH_2CH_2 \xrightarrow{(CH_3COO)_2} Cl_2Si(CH_3)[CH(CH_3)CH_2CH_2CH_3]$$

If a silane contains several Si–H bonds, all add to double C=C bonds (often successively with rising temperature), for example [598]:

$$PhSiH_3 + nCH_2{=}CH(CH_2)_5CH_3 \xrightarrow{(CH_3)_3COOCOC_6H_5} PhSiH_{3-n}[(CH_2)_7CH_3]_n \quad (n = 1, 2, 3)$$

5.11.3 Polar addition of silicon hydride to the double C=C bond

The polar addition is catalysed by *metal salts*, $H_2PtCl_6{\cdot}6H_2O$ (most active catalyst), K_2PtCl_6, $NiCl_2 \cdot C_5H_5N$, $NiCl_2 \cdot 4C_5H_5N$, $RuCl_3 \cdot OsO_4$ and $PdCl_2$ are inactive. The mechanism by which metal-catalysed addition takes place is uncertain. Active *metals* are: Pt black, Pt/active carbon, Pt/γ-Al_2O_3, Pt/asbestos, Pd/γ-Al_2O_3, and Pd/active carbon; Pd/asbestos is inactive.

In the polar addition of the Si–H bond to the double C=C bond, the hydrogen atom goes – in the opposite way to the polar addition of hydrogen halide – to the carbon atom with the lowest electron density (*anti*-Markovnikov rule). Alkene substituents with a negative effect direct the silyl group to the carbon atom adjacent to the substituent, and substituents with a positive effect to the carbon atom of the double bond not adjacent to the substituent

$$\equiv\overset{\delta+}{Si}-\overset{\delta-}{H} + \overset{\delta+}{CH_2}{=}\overset{\delta-}{CH}\rightarrow R \rightarrow \equiv Si-CHR(CH_3)$$

where R = substituent with a negative (over-all) substituent effect, *e.g.*, an acyl or an ester group

$$\equiv\overset{\delta+}{Si}-\overset{\delta-}{H} + \overset{\delta-}{CH_2}{=}\overset{\delta+}{CH}\leftarrow R' \rightarrow \equiv Si-CH_2-CH_2R'$$

where R′ = substituent with a positive (over-all) substituent effect, *e.g.*, an alkyl group or chlorine.

Tables 7.28–7.32 contain references and examples of the reactions.

In the addition of silicon hydrides to alkenes with non-terminal double bonds, the silyl group becomes attached to a terminal carbon atom even in a reaction obviously taking place by a polar mechanism. It is unlikely that the addition takes place through a heterolytic cleavage of the Si–H bond into a free silonium ion and a hydride ion, followed by a nucleophilic attack of $H^{\ominus}$ on the alkene and hydride migration with the formation of the most stable tautomeric carbanion with a terminal negative charge:

$$\equiv Si-H + CH_3CH{=}CHR \nrightarrow \equiv\overset{\oplus}{Si} + (CH_3\overset{\ominus}{C}H-CH_2R \leftrightarrow \overset{\ominus}{C}H_2CH_2-CH_2R) \nrightarrow \equiv SiCH_2CH_2CH_2R$$

This is because all accurately investigated hydrolyses of silane, for example, do *not* take place via a silonium ion; furthermore, in the addition of oct-1-ene to an optically active silane R_3Si^*H, the configuration at the Si^* is retained [589]. Such polar additions more probably take place by the following chain of reactions: The catalyst forms the active catalyst [2]A with the alkene or with the silane or with the silane + alkene. It yields the catalyst-free anionic unstable chain propagating agent the carbanion $C_1^{\ominus}$. This changes into the tautomeric stable carbanion $C_n^{\ominus}$ with a terminal charge. The stable carbanion $C_n^{\ominus}$ reacts preferentially with silane to form the anionic complex $B^{\ominus}$ ($RHSi\lessgtr^{\ominus}$). By hydride transfer, with more alkene it forms the chain-propagating agent $C_1^{\ominus}$ and the alkylated silane. At the same time, an isomerisation of the alkene takes place [*521*]:

$$\rangle SiH + pt + alkene \rightarrow A \xrightarrow{-\rangle Si.pt} C_1^{\ominus} \quad \text{(chain initiation)}$$

$$C_1^{\ominus} \rightarrow C_2^{\ominus} \rightarrow \cdots C_n^{\ominus} \quad \text{(tautomerism)}$$

$$\rangle SiH + C_n^{\ominus} \rightarrow B^{\ominus} \xrightarrow{+alkene_1} \rangle SiR_n + C_1^{\ominus}$$

$$C_n^{\ominus} + alkene_1 \rightleftharpoons C_1^{\ominus} + alkene_n \quad \text{(isomerism of the alkene)}$$

where $\rangle SiH$ = silicon hydride; pt = platinum catalyst (*e.g.*, $H_2PtCl_6 \cdot 6H_2O$); A = active catalyst of unknown composition formed from SiH + pt + alkene or SiH + pt or alkene + pt; $C_1^{\ominus}$ = carbanion formed from the alkene, the negative charge of which is located on one of the carbon atoms which originally was doubly bound (*e.g.*, $CH_3CH{=}CHR + H^{\ominus} \rightarrow CH_3\overset{\ominus}{C}HCH_2R$); $C_n^{\ominus}$ = most stable carbanion with terminal charge (*e.g.*, $\overset{\ominus}{C}H_2CH_2CH_2R$).

A similar chain of reactions which also explains the H–D exchange when the deuterated silane Cl_3SiD is used is given in the literature [*505*].

The polar addition of silane to alkenes also appears to be stereo-specific. In distinction from the polar *trans* addition of the hydrogen from HBr to the three-membered bromonium ion $\rangle C\!\!-\!\!\!\underset{Br^{\oplus}}{}\!\!\!-\!\!C\langle$ which leads to *cis* adducts, the addition of silane (via carbanions) gives *trans* adducts, and isomers. Example [*559*]:

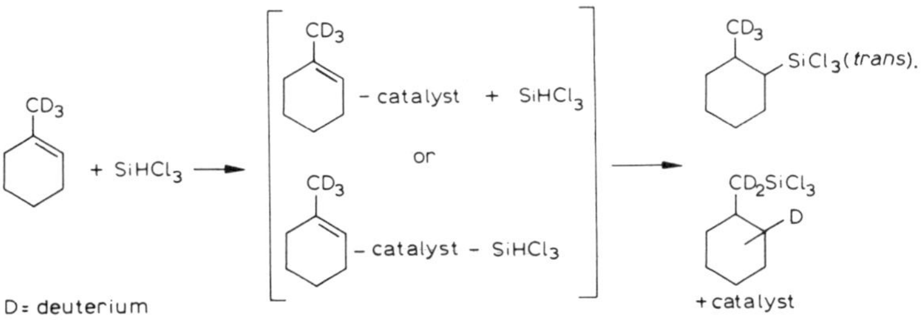

CD3-cyclohexene + SiHCl3 → [CD3-cyclohexene – catalyst + SiHCl3 or CD3-cyclohexene – catalyst – SiHCl3] → CD3-cyclohexyl-SiCl3 (*trans*). and CD2SiCl3-cyclohexane-D + catalyst

D = deuterium

[2] A could perhaps be the equilibrium [*96*]:

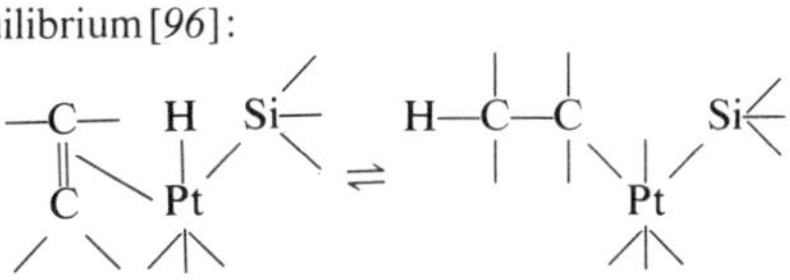

$$\underset{\substack{\| \\ C}}{-C-}\;\; \underset{Pt}{H}\;\; Si\lessgtr \;\rightleftharpoons\; H-\overset{|}{\underset{|}{C}}-\overset{|}{\underset{|}{C}}\;\; \underset{Pt}{}\;\; Si\lessgtr$$

Since 1-methyl-2-(trichlorosilyl)cyclohexane does not isomerise under the experimental conditions and since in the isomerisation of the methyl-d_3 compound the whole of the deuterium content is preserved, the above-mentioned hydride migration is indicated:

CD_3-cyclohexane $\xrightarrow{+H^{\ominus}}$ [chair cyclohexane with CD_2 bridged by D, or CD_2/D/H bridged form]$^{\ominus}$ $\xrightarrow[-H^{\ominus}]{+SiHCl_3}$ CD_2SiCl_3-cyclohexane–D

It is difficult to explain the fact that trichlorosilane does add to 1-methyl- or 1-ethylcyclohex-1-ene (95–98% terminal addition) but not to 1-n-propyl- or 1-n-octylcyclohex-1-ene [*559*].

5.11.4 Donor-catalysed addition of a silicon hydride to the double C–C bond

Donors such as dimethylamine, pyridine, piperidine, quinoline, and mixtures of tri-n-butylamine, N,N,N′,N′-tetramethylethylenediamine, and copper(I) chloride catalyse the addition of silicon hydride to the C=C double bond in the absence of a solvent and also in benzene or acetonitrile solution [*67*, *431–433*, *461*]. In favour of a polar mechanism is the dependence of the rate of the reaction on the above-mentioned catalysts and on the solvent, and also the absence of telomerisation even with alkenes which readily undergo polymerisation. The mechanism itself is, however, even less clear than those of all additions catalysed in other ways. Of the two mechanisms given, which are not the only possible ones, the second appears to be the more probable, since it avoids an abnormal polarity in heterolysis $\left(\rangle\overset{\delta-}{Si}-\overset{\delta+}{H}\right)$ and postulates the sp^3d or sp^3d^2 state of the silicon atom:

1. $Cl_3SiH + \text{donor} \rightarrow Cl_3Si^{\ominus} + \text{H-donor}^{\oplus}$

$$CH_2{=}CH{-}C{\equiv}N \rightarrow CH_2{-}\overset{\oplus}{C}H{=}C{=}\underline{N}^{\ominus}$$

$$Cl_3Si^{\ominus} + \overset{\oplus}{C}H_2CH{=}C{=}\underline{N}^{\ominus} \rightarrow Cl_3SiCH_2CHC\underline{N}^{\ominus}$$

$$Cl_3SiCH_2CHCN^{\ominus} + \text{H-donor}^{\oplus} \rightarrow Cl_3SiCH_2CH_2CN + \text{donor}$$

2. $R_3N + SiHCl_3 \rightarrow \left[R_3\overset{\oplus}{N}{-}SiHCl_3\right]$ (Si bearing H and three Cl)

$$R_3N\cdots\rightarrow SiHCl_3 + \overset{\oplus}{C}H_2CH{=}C{=}\underline{N}^{\ominus} \rightarrow R_3N\cdots\rightarrow Si(H)(Cl_3)\cdots\rightarrow CH_2CH{=}C{=}\underline{N}$$

$$\rightarrow Cl_3SiCH_2CH_2CN + R_3N$$

5.11.5 Examples of the substitution and addition reactions between silicon hydrides and alkenes

Silicon hydrides, $Z_{4-n}SiH_n$ ($n = 1$–4; Z = alkyl, aryl, halogen, $\frac{1}{2}$O), substitute or add alkenes by the following prototypical reactions.

Unsubstituted alkenes (for literature and other examples of the substitution and addition, see Tables 7.27 and 7.28).

$$SiH_4 + CH_2{=}CHMe \xrightarrow[\text{radical}]{120°} H_3SiCH_2CH_2Me$$

$$Cl_3SiH + CH_2{=}CHPh \xrightarrow[\text{polar?}]{\text{Pt–C} + (CH_3)_3CC_6H_3(OH)_2} Cl_3SiCH_2CH_2Ph$$

$$Cl_3SiH + CH_2{=}CHPh \xrightarrow[\text{(also telomerisation)}]{(C_6H_5COO)_2,\ \text{radical}} Cl_3Si(CH_2CHPh)_{59}H$$

$$Cl_3SiH + CH_2{=}CHPh \xrightarrow[\text{substitution instead of addition}]{625°} Cl_3SiCH{=}CHPh + H_2$$

$$Cl_3SiH + PhH \xrightarrow[\text{substitution}]{350°} \begin{cases} PhCl_2SiH + HCl \\ Cl_3SiPh + H_2 \end{cases}$$

$$SiHCl_3 + CH_2{=}CHCH{=}CH_2 \xrightarrow[\text{polar 1,4-addition}]{\text{Pt–C}} Cl_3SiCH_2CH{=}CHCH_3$$

Alkenyl halides. Halogenoalkenes can add by a radical mechanism to all Si–H bonds of a silane. In addition, telomerisation takes place; the values of *l*, *m*, and *n* depend on the ratio of the reactants[*215*]. References and further examples are given in Table 7.29.

$$CH_3SiH_3 \xrightarrow{+n\,CF_2{=}CF_2} CH_3SiH_2(CF_2CF_2)_nH \xrightarrow{+m\,CF_2{=}CF_2} CH_3SiH(CF_2CF_2)_mH(CF_2CF_2)_nH \xrightarrow{+l\,CF_2{=}CF_2} CH_3Si(CF_2CF_2)_lH(CF_2CF_2)_mH(CF_2CF_2)_nH$$

$$CH_3Cl_2SiH + n\,CF_2{=}CFCl \xrightarrow[\text{polar}]{H_2PtCl_6 \cdot 6H_2O} CH_3Cl_2Si(C_2F_3Cl)_nH$$

$$SiHCl_3 + CF_2{=}CFCF{=}CF_2 \xrightarrow[\text{radical(?) 1,2-addition}]{200–250°} Cl_3SiCF_2CHFCF{=}CF_2$$

Alkenyl cyanides. References and further examples are given in Table 7.30.

$$Cl_3SiH + CH_2{=}CHCH_2CN \xrightarrow[\text{radical}]{\gamma\text{-rays}} Cl_3Si(CH_2)_3CN$$

Terminally directed polar addition:

$$CH_3Cl_2SiH + CH_2{=}CHCH_2CN \xrightarrow{H_2PtCl_6 \cdot 6H_2O} CH_3Cl_2Si(CH_2)_3CN$$

References p. 621

Non-terminally directed polar addition:

$$Cl_3SiH + CH_2{=}CHCN \xrightarrow{NiCl_2 \cdot 4C_5H_5N} (Cl_3Si)(CH_3)CHCN$$

Alkenylcarboxylic acids, alkenylcarboxylic esters, alkenylcarboxylic acid chlorides, alkenylcarboxylic acid anhydrides, alkenyl acetates. Table 7.30 contains references and further examples. (For the 1,4-addition of R_3SiH and 1,2-addition of Cl_3SiH to the C=C–C=O system, see section 5.13).

$$Ph_3SiH + CH_2{=}CH(CH_2)_8COOH \xrightarrow[\text{radical}]{(C_6H_5COO)_2} Ph_3Si(CH_2)_{10}COOH$$

$$Et_3SiH + CH_2{=}CH(CH_2)_8COOCH_3 \xrightarrow[\text{radical}]{\text{UV}} Et_3Si(CH_2)_{10}COOCH_3$$

By a polar mechanism, the silyl group adds to esters of unsaturated acids non-terminally, as was to be expected from the substituent influence [*587*]:

$$MeCl_2SiH + \overset{\delta+}{C}H_2{=}\overset{\delta-}{C}H{\rightarrow}COO(CH_2)_3CH_3 \rightarrow MeCl_2Si[CH(CH_3)COO(CH_2)_3CH_3]$$

If, however, the negative substituent influence of the CO group in esters of unsaturated acids is diminished by the propinquity of a methyl group or if the double C=C bond is no longer conjugated with the C=O double bond, terminal addition takes place [*587*]:

$$MeCl_2SiH + CH_2{=}C({\leftarrow}CH_3)COO(CH_2)_3CH_3 \xrightarrow[\text{polar}]{\text{Pt–C}} MeCl_2SiCH_2CH(CH_3)COO(CH_2)_3CH_3$$

$$MeCl_2SiH + CH_2{=}CHCH_2COOEt \xrightarrow[\text{polar}]{\text{Pt–C}} MeCl_2Si(CH_2)_3COOEt$$

In both radical and polar addition to alkenyl acetate the silyl group always becomes attached to the terminal carbon atom [*596, 598*]:

$$PhSiH_3 + CH_2{=}CHCH_2OCOCH_3 \xrightarrow[\text{radical}]{CH_3COOCOC_6H_5} PhH_2Si(CH_2CH_2CH_2OCOCH_3)$$

$$MeCl_2SiH + CH_2{=}CHOCOCH_3 \xrightarrow[\text{polar}]{H_2PtCl_6 \cdot 6H_2O} MeCl_2Si(CH_2CH_2OCOCH_3)$$

Allyl and isopropenyl compounds [$CH_2{=}CHCH_2Br(Cl)$, $CH_2{=}CHCH_2OCOCH_3$, $CH_2{=}CHCH_2OCOC_6H_5$, $(CH_2{=}CHCH_2OCO)_2C_6H_4$, $CH_2{=}CH(CH_3)OCOCH_3$] react with many silicon hydrides [CH_3SiHCl_2, $\{(CH_3)_2SiH\}_2O$, $(CH_3HSiO)_{4-5}$] with the formation of 25–64% propene [*513, 596*]:

$$\rangle Si{-}H + C_3H_5Z \rightarrow \rangle Si{-}Z + C_3H_6$$

Alkenols. Because of the acidic hydrogen in alcohols, hydrogen and alkenyloxysilanes are formed [*22*]:

$$(EtO)_3SiH + CH_2{=}CHCH_2C_6H_4OH \xrightarrow{KOH} (EtO)_3Si(OC_6H_4CH_2CH{=}CH_2) + H_2$$

If the acidic hydrogen is replaced by alkyl (see below under *Alkenyl ethers*) or by trisilyl, addition takes place [*22*]:

$$Me(EtO)_2SiH + CH_2{=}CHCH_2C_6H_4OSiMe_3 \rightarrow Me(EtO)_2Si(CH_2)_3C_6H_4OSiMe_3$$

Alkenyl ethers. References and further examples are given in Table 7.30.

$$PhSiH_3 + (CH_2{=}CHCH_2)_2O \xrightarrow[\text{radical}]{(n\text{-}C_8H_{17})_2SnH_3} PhH_2Si(CH_2)_3OCH_2CH{=}CH_2$$

$$Cl_3SiH + CH_2{=}CHOMe \xrightarrow[\text{polar}]{\text{Pt–C + Ionol}} Cl_3SiCH_2CH_2OMe$$

Alkenylamines. Because of the acidic hydrogen atom of the NH_2 group, no addition takes place to the double C=C bond but H_2 is evolved. After the H has been replaced by trimethylsilyl [$NH_2 \rightarrow N(SiMe_3)_2$], alkenylamines add silicon hydrides. References and further examples are given in Table 7.31.

$$PhSiH_3 + CH_2{=}CHCH_2N(SiMe_3)_2 \xrightarrow[\text{radical}]{CH_3COOCOC_6H_5} PhH_2Si(CH_2)_3N(SiMe_3)_2$$

B-Alkenylborazines. Polar addition gives terminally silylated compounds. Table 7.31 contains references and further examples.

$$3(CH_3)_2ClSiH + [CH_2{=}CH{\rightarrow}\overset{\delta-}{B}{\leftarrow}\overset{\delta+}{N}Ph]_3 \xrightarrow[\text{polar}]{H_2PtCl_6 \cdot 6H_2O} [(CH_3)_2ClSiCH_2CH_2BNPh]_3$$

Nitroalkenes. Further examples and references are given in Table 7.31.

$$Cl_3SiH + CH_2{=}CHCH_2NO_2 \xrightarrow{H_2PtCl_6 \cdot 6H_2O} Cl_3Si(CH_2)_3NO_2$$

Alkenylsilanes. Other examples and references are given in Table 7.32

$$Ph_3SiH + CH_2{=}CHCH_2Si(Ph_3)_3 \xrightarrow[\text{radical}]{(C_6H_5COO)_2} Ph_3Si(CH_2)_3SiPh_3$$

The polar addition of silicon hydrides to alkenylsilanes has not been elucidated. Thus, alkenylsilanes polymerise in the presence of platinum. Under these conditions the silyl group enters the β-position with respect to the double C=C bond [*114*]:

$$Me_2(CH_2{=}CH)SiH + CH_2{=}CHSiHMe_2 \xrightarrow{Pt\text{–}C} Me_2\overline{SiCH_2CH_2SiMe_2CH_2CH_2}$$

According to the polar mechanism, $Me_2\overline{SiCH(CH_3)SiMe_2CHCH_3}$ should be formed, since the addition of hydrogen chloride, diborane [*554*], or dialkylalanes

References p. 621 [*Text continued p. 557*]

TABLE 7.27

REACTION OF SILICON HYDRIDES WITH ALKENES OR AROMATICS (SUBSTITUTION)

Silicon hydride	*Substituent*	*Temperature (°C) catalyst*	*Reaction product*	*Ref.*
Cl_3SiH	$\overline{CH{=}CH(CH_2)_3CH_2}$	625°	$Cl_3SiCH{=}CH_2, Cl_3SiCH{=}CHCH_3, Cl_3SiC_6H_5, Cl_3SiCH_3$	385
Cl_3SiH	$CH_2{=}CH_2$	625°	$Cl_3SiCH{=}CH_2$ (10–30%)	383, 385, 391
Cl_3SiH	$CH_2{=}CHCH_3$	625°	$Cl_3SiCH{=}CHCH_3$ $(+H_2)$(10–30%)	385, 391
Cl_3SiH	$CH_2{=}CHC_6H_5$	625°	$Cl_3SiCH{=}CHC_6H_5$ $(+H_2)$(10–30%)	385
Cl_3SiH	$CH_2{=}CHSiCl_3$	625°	$Cl_3SiCH{=}CHSiCl_3$ $(+H_2)$(10–30%)	385
Cl_3SiH	$CH_2{=}C(CH_3)_2$	600°	$Cl_3SiCH_2C(CH_3){=}CH_2$ (10%)	385
Cl_3SiH	$CH_2{=}CH{-}CH{=}CH_2$	600°	$Cl_3SiCH{=}CH_2, Cl_3SiCH_2CH{=}CH_2, Cl_3SiCH_3, Cl_3SiCH_2CH{=}CHCH_2$	385
Cl_3SiH	C_6H_6	600°	$Cl_3SiC_6H_5$	385
Cl_3SiH	$CCl_2{=}CClSi(CH_3)_3$	$[(CH_3)_3CO]_2$	$Cl_3SiCCl{=}CClSi(CH_3)_3$	366
Cl_3SiH	$CCl_2{=}CClSiCl_3$	$[(CH_3)_3CO]_2$	$Cl_3SiCCl{=}CClSiCl_3$	366
Cl_3SiH	$CCl_2{=}CHCl$	Pt–C	$Cl_3SiCH{=}CCl_2$	660
Cl_3SiH	$CH_2{=}CHCl$	550–650°	$Cl_3SiCH{=}CH_2$	228
RCl_2SiH	$CH_2{=}CHCl$	550–650°	$RCl_2SiCH{=}CH_2$	71d
$C_6H_5Cl_2SiH$	$CH_2{=}CHCl$	550–650°	$(C_6H_5)Cl_2SiCH{=}CH_2$	71d
Cl_3SiH	$CHCl{=}CCl_2$	500	$Cl_3SiCH{=}CCl_2$	4, 384
Cl_3SiH	$CCl_2{=}CCl_2$	300	$Cl_3SiCCl{=}CCl_2$	71e
Cl_2SiH_2	$CCl_2{=}CCl_2$	300	$Cl_2Si(CCl{=}CCl_2)_2$	71e
Cl_3SiH	$CCl_2{=}CCl_2$	600° or UV or $[(CH_3)_3CO]_2$	$Cl_3SiCCl{=}CCl_2$	366
Cl_3SiH	$CCl_2{=}CClCF_3$	UV or $[(CH_3)_3CO]_2$	$Cl_3SiC(CF_3){=}CCl_2$	365
CH_3Cl_2SiH	$CCl_2{=}CClCF_3$	$[(CH_3)_3CO]_2$	$(CH_3)Cl_2SiC(CF_3){=}CCl_2$	365
Cl_3SiH		300–380° + pressure	$SiCl_3$ $SiCl_3$ substitution and addition	44

Cl_3SiH	C_6H_6	350°	$Cl_3SiC_6H_5$ (+ H_2), $C_6H_5Cl_2SiH$ (+ HCl)	71i
(p-$CH_3C_6H_4)_3SiH$	$CH_2{=}CHCH_2Cl$	580°?	(p-$CH_3C_6H_4)_3SiCH_2CH{=}CH_2$ (4%); (p-$CH_3C_6H_4)_3SiCl$ (32%)	448
(m-$CH_3C_6H_4)_3SiH$	$CH_2{=}CHCH_2Cl$	580°	(m-$CH_3C_6H_4)_3SiCH_2CH{=}CH_2$ (5%); (m-$CH_3C_6H_4)_3SiCl$ (32%)	448
$(C_6H_5)_3SiH$	$CH_2{=}CHCH_2Cl$	600°	$(C_6H_5)_3SiCH_2CH{=}CH_2$ (8%); $(C_6H_5)_3SiCl$ (20%)	448
(p-$ClC_6H_4)_3SiH$	$CH_2{=}CHCH_2Cl$	580°	(p-$ClC_6H_4)_3SiCH_2CH{=}CH_2$ (42%); $(C_6H_5)_3SiCl$ (6–10%)	448
(p-$CH_3C_6H_4)_3SiH$	$CH_2{=}CHCH_2Br$	570°	(p-$CH_3C_6H_4)_3SiCH_2CH{=}CH_2$ (3%); (p-$CH_3C_6H_4)_3SiBr$ (58–61%)	448
$(C_6H_5)_3SiH$	$CH_2{=}CHCH_2Br$	570°	$(C_6H_5)_3SiCH_2CH{=}CH_2$ (5–7%); $(C_6H_5)_3SiBr$ (24%)	448
(p-$ClC_6H_4)_3SiH$	$CH_2{=}CHCH_2Br$	570°	(p-$ClC_6H_4)_3SiCHCH_2{=}CH_2$ (16%); (p-$ClC_6H_4)_3SiBr$ (12%)	448
(p-$CH_3C_6H_4)_3SiH$	$CH_2{=}CHCl$	600°	(p-$CH_3C_6H_4)_3SiCH{=}CH_2$ (< 1%); (p-$CH_3C_6H_4)_3SiCl$ (32%)	448
$(C_6H_5)_3SiH$	$CH_2{=}CHCl$	600°	$(C_6H_5)_3SiCH{=}CH_2$ (8%); $(C_6H_5)_3SiCl$ (18%)	448
(p-$ClC_6H_4)_3SiH$	$CH_2{=}CHCl$	600°	(p-$ClC_6H_4)_3SiCH{=}CH_2$ (36%); (p-$ClC_6H_4)_3SiCl$ (8%)	448
$HSiCl_3$	$\overline{CH{=}CH(CH_2)_3CH_2}$	600°	no substitution product: $\overline{CH{=}CH(CH_2)CHSiCl_3}$; CH_3SiCl_3; $CH_2{=}CHSiCl_3$	385
$HSiCl_3$	$CH_2{=}CH{-}CH{=}CH_2$	600°	CH_3SiCl_3; $CH_2{=}CHSiCl_3$; $CH_2{=}CHCH_2SiCl_3$; $CH_3CH{=}CHCH_2SiCl_3$	385
$HSiCl_3$	$CH_2{=}C(CH_3)_2$	600°	CH_3SiCl_3; $CH_2{=}C(CH_3)CH_2SiCl_3$	385
$HSiCl_3$	C_6H_6	600°	$C_6H_5SiCl_3$	385
$HSiCl_3$	C_6H_6	500–850°, 1 atm	$C_6H_5SiCl_3$	380, 385
$HSiCl_3$	C_6H_6	300°, pressure	$C_6H_5SiCl_3$	98
$HSiCl_3$	C_6H_6	353°, 80 atm or 275–400°, 60 atm, BCl_3	$C_6H_5SiCl_3$, $HSiCl_2C_6H_5$	128, 376
$HSiCl_3$	C_6H_5Cl	379°, 73 atm	$C_6H_5SiCl_3$	43
$HSiCl_3$	C_6H_5Cl	290, 55 atm, BCl_3	$C_6H_5SiCl_3$, $ClC_6H_4SiCl_3$	42
$HSiCl_3$	$C_6H_5C_6H_4Cl$	255°, 20 atm, BCl_3	$(C_6H_4SiCl_3)_2$	42
CH_3HSiCl_2	C_6H_6	[illegible]00°, pressure, Ni; [illegible]00°, pressure, $AlCl_3$	$(CH_3)(C_6H_5)SiCl_2$	98
CH_3HSiCl_2	C_6H_6	[illegible]50°, pressure, H_3BO_3	$(CH_3)(C_6H_5)SiCl_2$	358
CH_3HSiCl_2	$CH_3C_6H_5$	136°, 30 atm, BCl_3	$(CH_3)(CH_3C_6H_4)SiCl_2$, $(CH_3)(CH_3C_6H_4)_2SiCl$	71
$C_2H_5HSiCl_2$	C_6H_6	[illegible]50°, pressure, H_3BO_3	$(C_6H_5)(C_2H_5)SiCl_2$	358
CH_3HSiCl_2	$CH_2{=}CH_2$	6[illegible]0°	$CH_3Cl_2SiCH{=}CH_2$ (+ H_2)	383
Cl_3SiH	$(CH_3)_2C{=}CH_2$	6[illegible]0°	10% $Cl_3SiCH{=}C(CH_3)_2$, 10% $Cl_3SiCH_2C(CH_3){=}CH_2$	388
CH_3HSiCl_2	$CHCl{=}CCl_2$	5[illegible]0°	$CH_3Cl_2SiCH{=}CCl_2$ + HCl	397
$C_2H_5HSiCl_2$	$CHCl{=}CCl_2$	5[illegible]0°	$C_2H_5Cl_2SiCH{=}CCl_2$ + HCl	397

TABLE 7.27 (cont'd)

Silicon hydride	*Substituent*	*Temperature (°C) catalyst*	*Reaction product*	*Ref.*
Cl_3SiH	$ClCH{=}CHCl$	600°	$Cl_3SiCH{=}CHCl$, $Cl_3SiCH{=}CHSiCl_3$	392
Cl_3SiH	2-chlorothiophene (ring structure with Cl)	570	67% 2-thienyl-$SiCl_3$ + HCl ; 2% thiophene + $SiCl_4$; 6% Cl_3Si-2,5-thienylene-$SiCl_3$ + HCl + H_2	97

TABLE 7.28

REACTION OF SILICON HYDRIDES WITH UNSUBSTITUTED ALKENES (ADDITION TO THE DOUBLE C═C BOND)

Silicon hydride	Alkene	Catalyst, temperature (°C)	Reaction product	Ref.
$R_{4-n}SiH_n$, $(RO)_3SiH$, $(R_{3-n}SiH_n)_2O$				
SiH_4	$CH_2{=}CH_2$	120°	$H_3SiCH_2CH_3$	278, 694
SiH_4	$CH_2{=}CH_2$	370°	$H_2Si(CH_2CH_3)_2$	278, 694
SiH_4	$CH_2{=}CH_2$	450°	$H_3SiCH_2CH_3$, $H_2Si(CH_2CH_3)_2$	637
SiH_4	$CH_2{=}CH_2$	440–460°	$H_3SiCH_2CH_3$, $H_2Si(CH_2CH_3)_2$, $HSi(CH_2CH_3)_3$ (+ H_2, hydrocarbons, CH_3SiH_3, $(SiCH_3)_x$, etc.)	196, 198
SiH_4	$CH_2{=}CH_2$	$Hg\,^3P_0 \rightarrow Hg\,^1S_0$	$H_3SiCH_2CH_2$, $H_3Si\text{-n-}C_4H_9$, $H_3Si(CH_2)_4SiH_3$	673
SiH_4	$CH_2{=}CHCH_3$	120°	$H_3SiCH_2CH_2CH_3$	694
$(C_2H_5)_2SiH_2$	$CH_2{=}CH(CH_2)_6CH_3$	Pt–C	$(C_2H_5)_2HSi\text{-n-}C_9H_{19}$	637
$(C_6H_5)_2SiH_2$	$CH_2{=}CH(CH_2)_7CH_3$	Pt–C	$(C_6H_5)_2HSi\text{-n-}C_{10}H_{21}$, $(C_6H_5)_2Si(n\text{-}C_{10}H_{21})_2$	637
$(n\text{-}C_4H_9)_3SiH$	$CH_2{=}CH_2$	340°	$(n\text{-}C_4H_9)_3Si(CH_2CH_2)_nH$ ($n = 3–5$)	357
$(C_6H_5)SiH_3$	$CH_2{=}CH(CH_2)_5CH_3$	$C_6H_5COOOC(CH_3)_3$	$(C_6H_5)H_2Si(CH_2)_7CH_3$; $(C_6H_5)HSi[(CH_2)_7CH_3]_2$; $(C_6H_5)Si[(CH_2)_7CH_3]_2$	598
$(C_6H_5)SiH_3$	cyclo-C_6H_{10}	$C_6H_5COOOC(CH_3)_3$	$(C_6H_5)H_2SiC_6H_{11}$	598
$[O(CH_3)SiH]_4$	$CH_2{=}CH$–(cyclohexenyl ring)	$H_2PtCl_6 \cdot 6H_2O$	$[O(CH_3)SiCH_2CH_2$–(cyclohexenyl ring)$]_4$	596
$(C_2H_5O)_3SiH$	$CH_2{=}CH_2$	Pt–C	$(C_2H_5O)_3SiC_2H_5$	359
$[(CH_3)_2SiH]_2O$	$CH_3CH_2CH{=}CHCH_2CH_2CH_3$	$H_2PtCl_6 \cdot 6H_2O$	$[(CH_3)_2Si(n\text{-}C_7H_{15})]_2O$ (terminal addition of silyl)	520
$(C_6H_5)_2HSiCH_2CH{=}CH_2$	$(C_6H_5)_2HSiCH_2CH{=}CH_2$	Pt–C	$[\text{–}Si(C_6H_5)_2CH_2CH_2CH_2\text{–}]_n$	396
$(C_2H_5)_2SiH_2$	$\overline{CH{=}CH(CH_2)_5CH_2}$	Pt–γ-Al_2O_3	$(C_2H_5)_2HSi\overline{CH(CH_2)_6CH_2}$; no $(C_2H_5)_2Si[\overline{CH_2(CH_2)_6CH_2}]_2$	467
$(C_2H_5)_3SiH$	$\overline{CH{=}CHCH_2CH_2CH{=}CHCH_2CH_2}$	Pt–γ-Al_2O_3	$(C_2H_5)_3Si\overline{CH(CH_2)_3CH{=}CHCH_2CH_2}$	467
$(C_2H_5)_3SiH$	$(CH{=}CH)_4$	Pt–γ-Al_2O_3	$(C_2H_5)_3Si\overline{CHCH{=}CHCH_2CH{=}CHCH{=}CH}$; $(C_2H_5)_3Si\overline{CHCH_2(CH)_5CH}$	468
$(C_2H_5)_3SiH$	deficiency of $(CH{=}CH)_4$	Pt–γ-Al_2O_3	Bis(triethylsilyl)cyclooctadiene	468

TABLE 7.28 (cont'd)

Silicon hydride	Alkene	Catalyst, temperature (°C)	Reaction product	Ref.
$R_{3-n}SiH_nX$				
$(CH_3)(C_6H_5)ClSiH$	$CH_2{=}CHCH_3$	160–400°	$(CH_3)(C_6H_5)ClSiCH_2CH_2CH_3$	45, 46
$(C_6H_5)_2ClSiH$	$CH_2{=}CHCH_3$	160–400°	$(C_6H_5)_2ClSiCH_2CH_2CH_3$	45, 46
H_3SiCl	$CFCl{=}CF_2$	O_3	$(C_2F_3ClH)_3SiCl$	71f
$(CH_3)_2ClSiH$	$CH_3CH_2CH{=}CHCH_2CH_2CH_3$	$H_2PtCl_6 \cdot 6H_2O$	$(CH_3)_2ClSi(CH_2)_6CH_3$ (terminal addition of silyl)	520
$(C_3H_7)(CH_3)FSiH$	$CH_2{=}CHC_6H_5$	$H_2PtCl_6 \cdot 6H_2O$	$(C_3H_7)(CH_3)FSiCH(CH_3)C_6H_5$; $(C_3H_7)(CH_3)FSiCH_2CH_2C_6H_5$ (small amount)	573
$(i\text{-}C_3H_7)(CH_3)FSiH$	$CH_2{=}CHC_6H_5$	$H_2PtCl_6 \cdot 6H_2O$	$(i\text{-}C_3H_7)(CH_3)FSiCH(CH_3)C_6H_5$; $(i\text{-}C_3H_7)(CH_3)FSiCH_2CH_2C_6H_5$ (small amount)	573
$(C_4H_9)(CH_3)FSiH$	$CH_2{=}CHC_6H_5$	$H_2PtCl_6 \cdot 6H_2O$	$(C_4H_9)(CH_3)FSiCH(CH_3)C_6H_5$; $(C_4H_9)(CH_3)FSiCH_2CH_2C_6H_5$ (small amount)	573
$(i\text{-}C_4H_9)(CH_3)FSiH$	$CH_2{=}CHC_6H_5$	$H_2PtCl_6 \cdot 6H_2O$	$(i\text{-}C_4H_9)(CH_3)FSiCH(CH_3)C_6H_5$; $(i\text{-}C_4H_9)(CH_3)FSiCH_2CH_2C_6H_5$ (small amount)	573
$R_{2-n}SiH_nX_2$				
$(CH_3)Cl_2SiH$	$CH_2{=}CH_2$	300°	$CH_3Cl_2SiCH_2CH_3$	45, 46, 418
$(CH_3)Cl_2SiH$	$CH_2{=}CH_2$	260–275°, 560 atm	$(CH_3)Cl_2Si(CH_2CH_2)_nH$ ($n = 1$–6)	416, 417
$(CH_3)Cl_2SiH$	$CH_2{=}CHCH_3$	160–400°	$(CH_3)Cl_2SiCH_2CH_2CH_3$	45, 46
$(CH_3)Cl_2SiH$	$CH_2{=}CHCH_2CH_3$	160–400°	$(CH_3)Cl_2SiCH_2CH_2CH_2CH_3$	45
$(CH_3)Cl_2SiH$	$CH_2{=}CH(CH_2)_3CH_3$	160–400°	$(CH_3)Cl_2Si(CH_2)_5CH_3$	45
$(CH_3)Cl_2SiH$	$\overline{CH{=}CH(CH_2)_3CH_2}$	160–400°	$(CH_3)Cl_2SiC_6H_{11}$	45
$(CH_3)Cl_2SiH$	$\overline{CH{=}CH(CH_2)_3CH_2}$	γ-rays	$(CH_3)Cl_2SiC_6H_{11}$	162
$(C_6H_5)Cl_2SiH$	$CH_2{=}CH_2$	280–300°, 60 atm	$(C_6H_5)Cl_2Si(CH_2CH_2)_nH$ ($n = 1$–3)	419
$(C_6H_5)Cl_2SiH$	$CH_2{=}CHCH_3$	160–400°	$(C_6H_5)Cl_2SiCH_2CH_2CH_3$	45, 46
$(CH_3)Cl_2SiH$	$CH_3CH{=}CHCH_2CH_3$	$(CH_3COO)_2$	$(CH_3)Cl_2Si[CH(CH_3)CH_2CH_2CH_3]$	83
$CH_2(SiCl_2H)_2$	$CH_2{=}CH(CH_2)_3CH_3$	$(C_6H_5COO)_2$	$CH_2[SiCl_2(CH_2)_5CH_3]_2$	640
CH_3SiCl_2H	$CH_2{=}CHCH_3$	$H_2PtCl_6 \cdot 6H_2O$	$(CH_3)Cl_2SiCH_2CH_2CH_3$	478
$(CH_3)Cl_2SiH$	cyclo-C_6H_{10}	$H_2PtCl_6 \cdot 6H_2O$	$(CH_3)Cl_2SiC_6H_{11}$	596
$(CH_3)Cl_2SiH$	$CH_3CH_2CH_2CH{=}CHCH_2CH_3$	$H_2PtCl_6 \cdot 6H_2O$	$(CH_3)Cl_2Si(CH_2)_6CH_3$ (terminal addition of silyl)	520
$(CH_3)Cl_2SiH$	$CH_2{=}CH\overline{CHCH_2CH{=}CHCH_2CH_2}$	$H_2PtCl_6 \cdot 6H_2O$	$(CH_3)Cl_2SiCH_2CH_2\overline{CHCH_2CH{=}CHCH_2CH_2}$	596
Cl_3SiH	$CH_2{=}CH_2$	Pt–C	$Cl_3SiCH_2CH_3$	359
$(CH_3)Cl_2SiH$	$CH_2{=}CH_2$	Pt–C	$(CH_3)Cl_2SiCH_2CH_3$	190
$(CH_3)Cl_2SiH$	$CH_3CH{=}CHCH_2CH_3$	$H_2PtCl_6 \cdot 6H_2O$; Pt–C; $RuCl_3$	$(CH_3)Cl_2Si(CH_2)_4CH_3$ (terminal addition of silyl)	596
$(CH_3)Cl_2SiH$	$CH_2{=}CHCH_2CH_2CH_3$	$H_2PtCl_6 \cdot 6H_2O$	$(CH_3)Cl_2Si(CH_2)_4CH_3$	596

$(CH_3)Cl_2SiH$	$CH_2{=}C(CH_3)CH_2CH_3$	$H_2PtCl_6 \cdot 6H_2O$	$(CH_3)Cl_2SiCH_2CH(CH_3)CH_2CH_3$	521
$(CH_3)Cl_2SiH$	$CH_2{=}CHCH(CH_3)CH_3$	$H_2PtCl_6 \cdot 6H_2O$	$(CH_3)Cl_2SiCH_2CH_2CH(CH_3)CH_3$	521
$(CH_3)Cl_2SiH$	$CH_3CH{=}C(CH_3)CH_3$	$H_2PtCl_6 \cdot 6H_2O$	$(CH_3)Cl_2SiCH_2CH(CH_3)CH_2CH_3$(30%), $CH_3Cl_2SiCH_2CH_2CH(CH_3)CH_3$(70%)	521
$(CH_3)Cl_2SiH$	$CH_2{=}\overline{C(CH_2)_4C}H_2$ (deficiency)	$H_2PtCl_6 \cdot 6H_2O$	$(CH_3)Cl_2SiCH_2\text{-}\overline{CH(CH_2)_4C}H_2$	521
$(CH_3)Cl_2SiH$	$CH_2{=}\overline{C(CH_2)_4C}H_2$ (excess)	$H_2PtCl_6 \cdot 6H_2O$	(=CH_2 → —CH_3) $(CH_3)Cl_2SiCH_2\overline{CH(CH_2)_4C}H_2$; $(CH_3)Cl_2Si\overline{CHCH_2CH(CH_3)(CH_2)_2C}H_2$	521
$(CH_3)Cl_2SiH$	$\overline{CH_2(CH_2)_3CH{=}C}CH_3$	$H_2PtCl_6 \cdot 6H_2O$	$(CH_3)Cl_2SiCH_2\overline{CH(CH_2)_4C}H_2$; $(CH_3)Cl_2Si\overline{CHCH_2CH(CH_3)(CH_2)_2C}H_2$	521
$(CH_3)SiHCl_2$	$\overline{CH_2CH{=}CHCH_2CH_2CHC}H_3$	$H_2PtCl_6 \cdot 6H_2O$	$(CH_3)Cl_2SiCH_2\overline{CH(CH_2)_4C}H_2$; $(CH_3)Cl_2Si\overline{CHCH_2CH(CH_3)(CH_2)_2C}H_2$	521
$(CH_3)F_2SiH$	$CH_2{=}CHC_6H_5$	$H_2PtCl_6 \cdot 6H_2O$	$(CH_3)F_2SiCH(CH_3)C_6H_5$ (small amount); $(CH_3)F_2SiCH_2CH_2C_6H_5$	573
$(CH_3)Cl_2SiH$	$CH_2{=}CHC_6H_5$	Pt–C	$(CH_3)Cl_2SiCH_2CH_2C_6H_5$	514
$(C_3H_7)F_2SiH$	$CH_2{=}CHC_6H_5$	$H_2PtCl_6 \cdot 6H_2O$	$(C_3H_7)F_2SiCH_2CH_2C_6H_5$	573
$(CH_3)SiCl_2H$	$CH_2{=}CH(CH_2)_5CH_3$	^{60}Co-γ-radiation	$(CH_3)Cl_2Si(CH_2)_7CH_3$	161
$(CH_3)Cl_2SiH$	$\overline{CH{=}CH(CH_2)_3C}H_2$	^{60}Co-γ-radiation	$(CH_3)Cl_2SiC_6H_{11}$	161
$(CH_3)Cl_2SiH$	$CH_3CH{=}CHCH_3$	^{60}Co-γ-radiation	$(CH_3)Cl_2SiCH(CH_3)CH_2CH_3$	161
$(CH_3)Cl_2SiH$	$CH_3CH{=}C(CH_3)_2$	^{60}Co-γ-radiation	$(CH_3)Cl_2SiC_5H_{11}$	161
$(CH_3)Cl_2SiH$	$\overline{CH{=}CH(CH_2)_2C}H_2$	^{60}Co-γ-radiation	$(CH_3)Cl_2SiC_5H_9$	161
$(CH_2{=}CHCH_2)F_2SiH$	$(CH_2{=}CHCH_2)F_2SiH$	$H_2PtCl_6 \cdot 6H_2O$	$[-CH_2CH_2CH_2Si(F)_2-]_x$	382
X_3SiH				
Cl_3SiH	$CH_2{=}CH_2$	160–400°	$Cl_3SiCH_2CH_3$	45
Cl_3SiH	$CH_2{=}CH_2$	270–388°	$Cl_3Si(CH_2CH_2)_nH$ (n = 1–5)	356
Cl_3SiH	$CH_2{=}CH_2$	285°, 200 atm	$Cl_3Si(CH_2CH_2)_nH$ (n = 1–5)	416, 417
Cl_3SiH	$CH_2{=}CH_2$	Pt–C	$Cl_3SiCH_2CH_3$	359
Cl_3SiH	$CH_2{=}C(CH_3)_2$	160–400°	$Cl_3SiCH_2CH(CH_3)_2$	45, 46
Cl_3SiH	$CH_3CH{=}CHCH_3$	160–400°	$Cl_3SiCH(CH_3)CH_2CH_3$	45, 46
Cl_3SiH	$CH_2{=}CHCH{=}CH_2$	160–400°	$Cl_3SiCH_2CH_2CH{=}CH_2$	45, 394
Cl_3SiH	$CH_2{=}CH(CH_2)_2CH{=}CH_2$	160–400°	$Cl_3Si(CH_2)_6SiCl_3$	45
Cl_3SiH	$\overline{CH{=}CH(CH_2)_3C}H_2$	160–400°	$Cl_3SiC_6H_{11}$	45
Cl_3SiH	$CH_2{=}C(CH_3)_2$	UV	$Cl_3SiCH_2CH(CH_3)_2$	466
Cl_3SiH	$(CH_3)_2C{=}C(CH_3)_2$	$(CH_3COO)_2$	$Cl_3Si[C(CH_3)_2CH(CH_3)_2]$	466
Cl_3SiH	$CH_2{=}C(CH_3)(CH_2)_5CH_3$	$(CH_3COO)_2$	$Cl_3SiCH_2CH(CH_3)(CH_2)_5CH_3$	466

References p. 621

TABLE 7.28 (cont'd)

Silicon hydride	*Alkene*	*Catalyst, temperature (°C)*	*Reaction product*	*Ref.*
Cl_3SiH	$CH_2{=}CH(CH_2)_4CH_3$	$(CH_3COO)_2$	$Cl_3Si(CH_2)_6CH_3$	466
Cl_3SiH	$CH_2{=}CH(CH_2)_5CH_3$	UV; $(CH_3COO)_2$; γ-rays	$Cl_3Si(CH_2)_7CH_3$	162, 466
Cl_3SiH	$CH_2{=}CH(CH_2)_2CH_3$	$(CH_3COO)_2$	$Cl_3Si(CH_2)_4CH_3$	83
Cl_3SiH	$CH{=}CH(CH_2)_3CH_2$	$(CH_3COO)_2$	$Cl_3SiC_6H_{11}$	83
Cl_3SiH	$CH{=}CH(CH_2)_3CH_2$	Pt–C	$Cl_3SiC_6H_{11}$	359
$SiHCl_3$	$CH_2{=}CHCH_2CH_2CH_3$	Pt–C;	$Cl_3Si(CH_2)_4CH_3$	359
Cl_3SiH	$CH_2{=}CHC_6H_5$	Pt–C + *tert*-$BuC_6H_4(OH)_2$	$Cl_3SiCH_2CH_2C_6H_5$	359, 661
Cl_3SiH	$CH_2{=}CHC_6H_5$	$(C_6H_5COO)_2$	$Cl_3Si(C_6H_5CHCH_2)_nH$ (n = about 59)	297
Cl_3SiH	$CH_2{=}CH\text{-}CH{=}CH_2$	Pt–C	$Cl_3SiCH_2CH{=}CHCH_3$; $(Cl_3SiCH_2CH_2)_2$	36, 660
Cl_3SiH	$CH_2{=}CH(CH_2)_2CH_3$	Pt–C; K_2PtCl_6	$Cl_3Si(CH_2)_4CH_3$	596
Cl_3SiH	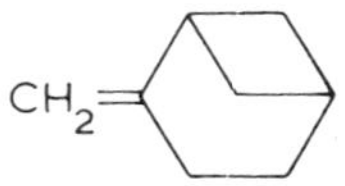 $CH_2{=}$	UV or peroxides	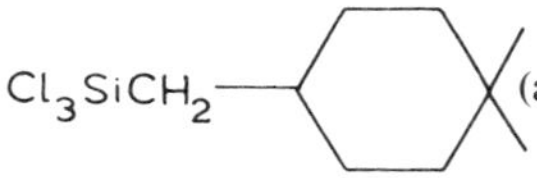 Cl_3SiCH_2- (addition and ring opening)	93, 192, 513
Br_3SiH	$CH_2{=}CH(CH_2)_3CH_3$	UV	$Br_3Si(CH_2)_5CH_3$	640
Br_3SiH	$CH_2{=}C(CH_3)_2$	$(C_6H_5COO)_2$	$Br_3SiCH_2CH(CH_3)_2$	638
Br_3SiH	$CH{=}CH(CH_2)_3CH_2$	UV	$Br_3SiC_6H_{11}$	640
Cl_3SiH	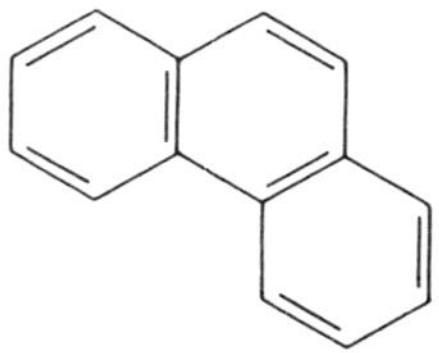	300–380° + pressure	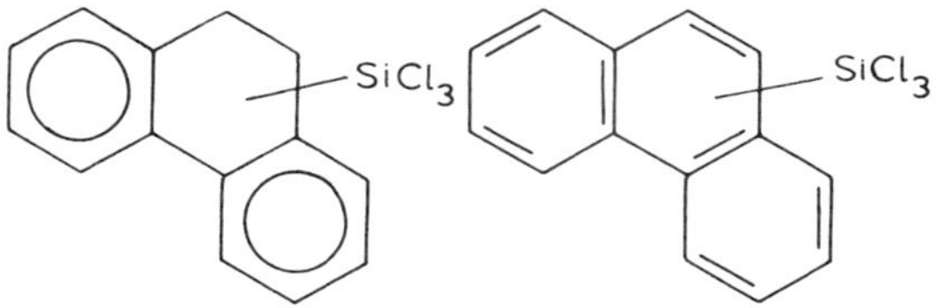 $SiCl_3$; $SiCl_3$ Addition and substitution	44
Cl_3SiH	$CH_2{=}CH(CH_2)_5CH_3$	$(CH_3COO)_2$ or γ-rays	$Cl_3Si(CH_2)_7CH_3$	162, 495, 591
Cl_3SiH	$CH_3CH{=}CHCH_2CH_3$	*tert*-$BuOOCOC_6H_5$	$Cl_3Si[CH(CH_3)CH_2CH_2CH_3]$; $Cl_3Si[CH(CH_2CH_3)_2]$ (70:30)	596, 597

$SiHCl_3$	$CH_3CH_2CH{=}CHCH_2CH_2CH_3$	$(CH_3COO)_2$	$Cl_3Si[CH(CH_2CH_3)CH_2CH_2CH_2CH_3]$; $Cl_3Si[CH(CH_2CH_2CH_3)_2]$	520
Cl_3SiH	$CH_2{=}CH(CH_2)_5CH_3$	$NiCl_2 \cdot 4C_5H_5N$	$Cl_3Si(CH_2)_7CH_3$; $Cl_3Si[CH(CH_3)(CH_2)_5CH_3]$	433
Cl_3SiH	$CH_2{=}CHC_6H_5$	$NiCl_2 \cdot 4C_5H_5N$	$Cl_3SiCH_2CH_2C_6H_5$; $Cl_3Si[CH(CH_3)C_6H_5]$	433
Cl_3SiH	$CH_3CH{=}CHCH_2CH_3$	$H_2PtCl_6 \cdot 6H_2O$	$Cl_3Si(CH_2)_4CH_3$ (terminal addition of silyl)	596
Cl_3SiH	$CH_3CH_2CH{=}CHCH_2CH_2CH_3$	$H_2PtCl_6 \cdot 6H_2O$	$Cl_3Si(CH_2)_6CH_3$ (terminal addition of silyl)	520
Cl_3SiH	$CH{=}CH(CH_2)_5CH_2$	$Pt{-}\gamma{-}Al_2O_3$	$Cl_3SiCH(CH_2)_6CH_2$	467
Cl_3SiH	$CH{=}CHCH_2CH_2CH{=}CHCH_2CH_2$	$Pt{-}\gamma{-}Al_2O_3$	$Cl_3SiCH(CH_2)_3CH{=}CHCH_2CH_2$	467
Cl_3SiH	$(CH{=}CH)_4$	$Pt{-}\gamma{-}Al_2O_3$	$Cl_3SiCHCH_2(CH{=}CH)_2CH{=}CH$	468
Cl_3SiH	$CH_3{-}C{=}CH(CH_2)_3CH_2$	$(CH_3COO)_2$	$Cl_3SiCH(CH_2)_4CHCH_3$ (85% *cis*, 15% *trans*; total yield 100%)	558
Cl_3SiH	$CH_3C{=}CH(CH_2)_3CH_2$	UV	$CH_3CH(CH_2)_4CHSiCl_3$ (89% *cis*, 11% *trans*; total yield 49%)	558
Cl_3SiH	$CH_3C{=}CH(CH_2)_3CH_2$	$H_2PtCl_6 \cdot 6H_2O$	$Cl_3SiCH_2CH(CH_2)_4CH_2$; 98% terminal (total yield 78%)	558
Cl_3SiH	$CH_3C{=}CH(CH_2)_3CH_2$	300°	$Cl_3SiCH_2CH(CH_2)_4CH_2$ (19%); $Cl_3SiCH(CH_2)_4CHCH_3$ (58% *cis*, 23% *trans*)	558
Cl_3SiH	$CH_3C{=}CH(CH_2)_3CH_2$	300° + chloranil	$Cl_3SiCH_2CH(CH_2)_4CH_2$ (18%); $Cl_3SiCH(CH_2)_4CHCH_3$ (57% *cis*, 25% *trans*)	558
Cl_3SiH	$CH_3C{=}CH(CH_2)_3CH_2$	300° + Fe or $FeCl_3$	$Cl_3SiCH_2CH(CH_2)_4CH_2$ (21%); $Cl_3SiCH(CH_2)_4CHCH_3$ (44 *cis*; 35% *trans*)	558
Cl_3SiH	$CH_2{=}C(CH_3)_2$	^{60}Co-γ-radiation	$Cl_3SiCH_2CH(CH_3)_2$	161
Cl_3SiH	$CH_3CH{=}CHCH_3$	^{60}Co-γ-radiation	$Cl_3SiCH(CH_3)CH_2CH_3$	161
Cl_3SiH	$CH_3CH{=}C(CH_3)_2$	^{60}Co-γ-radiation	$Cl_3SiC_5H_{11}$	161
Cl_3SiH	$CH_2{=}CH(CH_2)_5CH_3$	^{60}Co-γ-radiation	$Cl_3Si(CH_2)_7CH_3$	161
Cl_3SiH	$CH{=}CH(CH_2)_3CH_2$	^{60}Co-γ-radiation	$Cl_3SiC_6H_{11}$	161
Cl_3SiH	$CH_2{=}CHC_6H_5$	^{60}Co-γ-radiation	Polymerisation	161
Cl_3SiH	$CH{=}CH(CH_2)_2CH_2$	^{60}Co-γ-radiation	$Cl_3SiC_5H_9$	161
Cl_3SiH	$CH_3C{=}CH(CH_2)_3CH_2$	^{60}Co-γ-radiation	$Cl_3Si(C_6H_{10}CH_3)$	161
Cl_3SiH	$CH_2{=}C(CH_3)(C_6H_5)$	^{60}Co-γ-radiation	Polymerisation	161

References p. 621

TABLE 7.29

REACTION OF SILICON HYDRIDES WITH HALOGENOALKENES (ADDITION TO THE DOUBLE C=C BOND)

Silicon hydride	*Alkene*	*Catalyst*	*Reaction product*	*Ref.*
$R_{3-n}SiH_nX (X = H, F, Cl, OR)$				
$ClSiH_3$	$CF_2{=}CFCl$	O_3	$ClSi(C_2F_3ClH)_3$	71f
$(CH_3)(OC_2H_5)_2SiH$	$CH_2{=}CHCF_2CF_2CF_3$	UV	$(CH_3)(C_2H_5O)_2Si(CH_2CH_2C_3F_7)$	216
$(CH_3)_2SiH_2$	$CF_2{=}CF_2$	UV	$(CH_3)_2SiH(CF_2CF_2H)$, $(CH_3)_2SiH(CF_2CF_2)_nH$	214
$(C_2H_5)_3SiH$	$CH_2{=}CHCH_2Br$	UV	$(C_2H_5)_3Si(CH_2)_2CH_2Br$	535a
$(C_3H_7)_3SiH$	$CH_2{=}CHCH_2Br$	UV	$(C_3H_7)_3Si(CH_2)_2CH_2Br$	535a
$(C_6H_5)_3SiH$	$CH_2{=}CHCH_2Br$	UV	$(C_6H_5)_3Si(CH_2)_2CH_2Br$	535a
$(C_2H_5)_2SiHF$	$CH_2{=}CHCH_2Cl$	$H_2PtCl_6 \cdot 6H_2O$	$(C_2H_5)_2Si(CH_2CH_2CH_2Cl)F$	213, 573
$R_{2-n}SiH_nCl_2$				
$(CH_3)Cl_2SiH$	$CH_2{=}CHCF_3$	UV or $[(CH_3)_3CO]_2$	$(CH_3)Cl_2Si(CH_2CH_2CF_3)$	365
$(CH_3)Cl_2SiH$	$CH_2{=}CHCH_2CF_3$	UV, $[(CH_3)_3CO]_2$, H_2PtCl_6	$(CH_3)Cl_2Si[(CH_2)_3CF_3]$	365, 458
Cl_2SiH_2	$CF_2{=}CF_2$	200°	$Cl_2Si(CF_2CF_2H)_2$	71g
$(CH_3)Cl_2SiH$	$CF_2{=}CF_2$	UV	$(CH_3)Cl_2Si(CF_2CF_2H)$, $(CH_3)Cl_2Si(CF_2CF_2)_nH$	273
Cl_2SiH_2	$CH_2{=}CHCF_3$	UV, $H_2PtCl_6 \cdot 6H_2O$	$Cl_2SiH(CH_2CH_2CF_3)$, $Cl_2Si(CH_2CH_2CF_3)_2$	216, 458
$(CH_3)Cl_2SiH$	$CH_2{=}CHCF_2CF_2CF_3$	UV	$(CH_3)Cl_2Si(CH_2CH_2C_3F_7)$	216
$(CH_3)Cl_2SiH$	$CH_2{=}CHCl$	Pt–C	$(CH_3)Cl_2Si(CHClCH_3)$, $(CH_3)Cl_2Si(CH_2CH_3)$	480
$(C_2H_5)Cl_2SiH$	$CH_2{=}CHCl$	Pt–C	$(C_2H_5)Cl_2Si(CHClCH_3)$, $(C_2H_5)Cl_2Si(CH_2CH_3)$	237, 480
$(CH_3)Cl_2SiH$	$CH_2{=}CHCH_2Cl$	Pt–C	$(CH_3)Cl_2Si(CH_2)_3Cl$, $(CH_3)SiCl_3$, $(CH_3)Cl_2Si(CH_2CH_2CH_3)$	481
$(C_2H_5)Cl_2SiH$	$CH_2{=}CHCH_2Cl$	Pt–C	$(C_2H_5)Cl_2Si(CH_2)_3Cl$, $(C_2H_5)SiCl_3$, $(C_2H_5)Cl_2Si(CH_2CH_2CH_3)$	481, 660
$(CH_3)Cl_2BiH$	$CH_2{=}C(CH_3)CH_2Cl$	Pt–C	$(CH_3)Cl_2Si[CH_2CH(CH_3)CH_2Cl]$	481
$(C_2H_5Cl_2SiH$	$CH_2{=}C(CH_3)CH_2Cl$	Pt–C	$(C_2H_5)Cl_2Si[CH_2CH(CH_3)CH_2Cl]$	481
$(CH_3)Cl_2SiH$	$CF_2{=}CFCl$	Pt–C	$(CH_3)Cl_2Si(CFClCF_2)$	480
$(C_2H_5)Cl_2SiH$	$CF_2{=}CFCl$	Pt–C	$(C_2H_5)Cl_2Si(CFClCF_3)$	480
$(CH_3)Cl_2SiH$	$CF_2{=}CF_2$	Pt–C	$(CH_3)Cl_2Si(CF_2CF_2H)$	480
$(C_2H_5)Cl_2SiH$	$CF_2{=}CF_2$	Pt–C	$(C_2H_5)Cl_2Si(CF_2CF_2H)$	480
$(CH_3)Cl_2SiH$	CF_2CFCl	$H_2PtCl_6 \cdot 6H_2O$	$(CH_3)Cl_2Si[(C_2F_3Cl)_nH]$	478
$(CH_3)Cl_2SiH$	$CH_2{=}CHCH_2Cl$	150–200°, $H_2PtCl_6 \cdot 6H_2O$	$(CH_3)Cl_2SiCH_2CH_2CH_2Cl$	393
$(CH_3)Cl_2SiH$	$CH_2{=}CHCH_2Cl$	^{60}Co-γ-radiation	$(CH_3)Cl_2Si[(CH_2)_3Cl]$	161
$(CH_3)Cl_2SiH$	$CH_2{=}C(CH_3)CF_2CF_3$	^{60}Co-γ-radiation	$(CH_3)Cl_2Si[CH_2CH(CH_3)CF_2CF_3]$	161
$(CH_3)Cl_2SiH$	$CH_2{=}(CH_3)CF_2CF_2CF_3$	^{60}Co-γ-radiation	$(CH_3)Cl_2Si[(CH_2CH(CH_3)CF_2CF_2CF_3]$	161
$(CH_3)Cl_2SiH$	Allyl derivatives of siloxanes	$H_2PtCl_6 \cdot 6H_2O$	Addition	23

Cl_3SiH	$CH_2{=}CHCH_2Cl$	$(CH_3COO)_2$	$Cl_3Si(C_3H_6Cl)$	466
Cl_3SiH	$CH_2{=}CHCF_3$	UV, $[(CH_3)_3CO]_2$, H_2PtCl_6	$Cl_3Si(CH_2CH_2CF_3)$	365, 458
Cl_3SiH	$CH_2{=}CHCH_2CF_3$	UV or $[(CH_3)_3CO]_2$	$Cl_3Si[(CH_2)_3CF_3]$	365
Cl_3SiH	$CFCl{=}CF_2$	UV or $[(CH_3)_3CO]_2$	$Cl_3Si(C_2F_3ClH)$	248, 365
Cl_3SiH	$CFCl{=}CF_2$	O_3	$Cl_3Si[(C_2F_3Cl)_nH]$	71f
Cl_3SiH	$CF_2{=}CF_2$	200°	$Cl_3Si(CF_2CF_2H)$	71g
Cl_3SiH	$CF_2{=}CF_2$	UV	$Cl_3Si(CF_2CF_2H)$, $Cl_3Si(CF_2CF_2)_nH$	254
Cl_3SiH	$CH_2{=}CHCF_2CF_3$	γ-rays	$Cl_3Si(CH_2CH_2C_2F_5)$	162
Cl_3SiH	$CF_2{=}CFCF{=}CF_2$	250°	$Cl_3Si(CF_2CHFCF{=}CF_2)$	71g, 248
Cl_3SiH	$CF_2{=}CFCF_2CF_3$	250°	$Cl_3Si(CF_2CFHCF_2CF_3)$	71g
Cl_3SiH	$CH_2{=}CHCl$	Pt–C	$Cl_3Si(CH_2CH_2Cl)$	359
Cl_3SiH	$CH_2{=}CHCH_2Cl$	Pt–C	$Cl\ Si(CH_2CH_2CH_3)$, $[Cl_3Si(CH_2)_3Cl]$	660
Cl_3SiH	$CF_2{=}CH_2$	Pt–C	$Cl_3Si(CH_2CF_2H)$	660
Cl_3SiH	$CH_2{=}CHCH_2Cl$	^{60}Co-γ-radiation	$Cl_3Si(CH_2CH_2CH_2Cl)$	161
Cl_3SiH	$CHCl{=}CHCl$	^{60}Co-γ-radiation	$Cl_3Si(CHClCH_2Cl)$	161
Cl_3SiH	$CH_2{=}CHCF_2CF_3$	^{60}Co-γ-radiation	$Cl_3Si(CH_2CH_2CF_2CF_3)$	161
Cl_3SiH	$CH_2{=}C(CH_3)CF_2CF_3$	^{60}Co-γ-radiation	$Cl_3Si[CH_2C(CH_3)CF_2CF_3]$	161
Cl_3SiH	$CH_2{=}CHCF_2CF_2CF_3$	^{60}Co-γ-radiation	$Cl_3Si(CH_2CH_2CF_2CF_2CF_3)l$	161
Cl_3SiH	$CH_2{=}C(CH_3)CF_2$	^{60}Co-γ-radiation	$Cl_3Si[CH_2C(CH_3)CF_2CF_2CF_3]$	161
Cl_3SiH	$Cl_3SiCH{=}CH_2$ + $Cl_3SiCH_2CH{=}CH_2$	$H_2PtCl_6\cdot 6H_2O$	$Cl_3SiCH_2CH_2SiCl_3$, $Cl_3SiCH_2CH_2CH_2SiCl_3$	398
Cl_3SiH	$CH_2{=}CHCHClCH_2Cl$	$H_2PtCl\cdot 6H_2O$	$Cl_3SiCH_2CH_2CHClCH_2Cl$	359

TABLE 7.30

REACTION OF SILICON HYDRIDES WITH ALKENYL CYANIDES, UNSATURATED ACIDS, ESTERS, CHLORIDES, OR ANHYDRIDES OF UNSATURATED ACIDS, ALKENYL ACETATES, UNSATURATED ETHERS, ACETATES, AND ALKENYLOXYSILANES (ADDITION TO THE DOUBLE C=C BOND)

Silicon hydride	*Alkene*	*Catalyst (inhibitor)*	*Reaction product*	*Ref.*
	Alkenyl cyanides			
Cl_3SiH	$CH_2{=}CHCN$	Raney-Ni	$Cl_3Si(CH_2CH_2CN)$	450
$(CH_3)Cl_2SiH$	$CH_2{=}CHCN$	Pt–C (Ionol*)	$(CH_3)Cl_2Si[CH(CH_3)CN]$	239
Cl_3SiH	$CH_2{=}CHCN$	Pt-C (Ionol*)	$Cl_3Si[CH(CH_3)CN]$	239
$(CH_3)Cl_2SiH$	CH_2CHCH_2CN	$H_2PtCl_6 \cdot 6H_2O$	$(CH_3)Cl_2Si[(CH_2)_3CN]$	596
$(CH_3)Cl_2SiH$	$CH_2{=}CHCH_2CN$	Pt-C	$(CH_3)Cl_2Si[(CH_2)_3CN]$	587
$(CH_3)Cl_2SiH$	$CH_2{=}CH(CH_2)_2CN$	Pt–C	$(CH_3)Cl_2Si[(CH_2)_4CN]$	587
Cl_3SiH	$CH_2{=}CHCN$	$Ni + C_5H_5N$; $NiCl_2 \cdot 4C_5H_5N$	$Cl_3Si[CH(CH_3)CN]$	432, 433
Cl_3SiH	$CH_2{=}HCCH_2CN$	$NiCl_2 \cdot 4C_5H_5N$	$Cl_3Si(CH_2CH_2CH_2CN)$; $(Cl_3Si)[CH(CH_3)CH_2CN]$	433
Cl_3SiH	$CH_2{=}CHCH_2CN$	^{60}Co-γ-radiation	$Cl_3Si[(CH_2)_3CN]$	161, 162
Cl_3SiH	$CH_2{=}CHCN$	$H_2PtCl_6 \cdot 6H_2O$	0%	58
Cl_3SiH	$CH_2{=}CHCH_2CN$	$H_2PtCl_6 \cdot 6H_2O$	99% $Cl_3SiCH_2CH_2CH_2CN$	58
Cl_3SiH	$CH_2{=}CHCN$	$HCON(CH_3)_2$	65% $Cl_3SiCH_2CH_2CN$	58
Cl_3SiH	$CH_2{=}CHCH_2CN$	$HCON(CH_3)_2$	7% $Cl_3SiCH_2CH_2CH_2CN$	58
Cl_3SiH	$CH_2{=}CHCN$	$B(C_4H_9)_3$	80% $Cl_3SiCH_2CH_2CN$	58
Cl_3SiH	$CH_2{=}CHCH_2CN$	$B(C_4H_9)_3$	0% $Cl_3SiCH_2CH_2CH_2CN$	58
Cl_3SiH	$CH_2{=}CHCN$	BCl_3	17% $Cl_3SiCH_2CH_2CN$	58
Cl_3SiH	$CH_2{=}CHCH_2CN$	BCl_3	17% $Cl_3SiCH_2CH_2CH_2CN$	58
	Unsaturated acids and esters of unsaturated acids			
$(C_6H_5)_3SiH$	$CH_2{=}CH(CH_2)_8COOH$	$(C_6H_5COO)_2$	$(C_6H_5)_3Si[(CH_2)_{10}COOH]$	209
$(C_2H_5)Cl_2SiH$	$CH_2{=}CH(CH_2)_8COOCH_3$	UV	$(C_2H_5)Cl_2Si[(CH_2)_{10}COOCH_3]$	89, 90, 91
$(C_2H_5)_3SiH$	$CH_2{=}CH(CH_2)_8COOCH_3$	UV	$(C_2H_5)_3Si[\ CH_2)_{10}COOCH_3]$	89, 90, 91
$(C_6H_5)SiH_3$	$CH_2{=}CH(CH_2)_{15}COOCH_3$	$C_6H_5COOOC(CH_3)_3$	$C_6H_5)SiH_2[(CH_2)_{17}COOCH_3]$	594, 598
$(ClC_6H_4)SiH_3$	$CH_2{=}CH(CH_2)_8COOCH_3$	$C_6H_5COOOC(CH_3)_3$	$(ClC_6H_4)SiH_2[(CH_2)_{10}COOCH_3]$	594
Br_3SiH	CH_2 $COOCH_3$	$C_6H_5COOOC(CH_3)_3$	Br_3Si CH_2 $COOCH_3$	598

$(CH_3)Cl_2SiH$	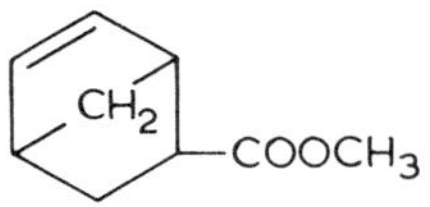 CH$_2$, COOCH$_3$	$H_2PtCl_6 \cdot 6H_2O$	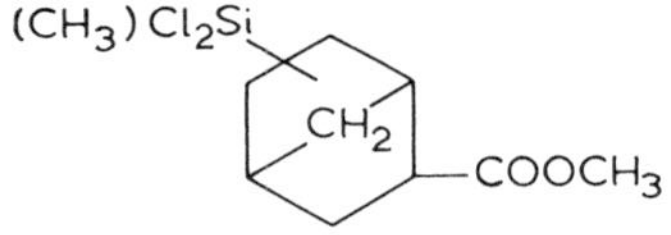 $(CH_3)Cl_2Si$, CH$_2$, COOCH$_3$	596
Cl_3SiH	$CH_2{=}CHCOOCH_3$	Pt–C (Ionol*)	$Cl_3Si[CH(CH_3)COOCH_3]$	239
$(CH_3)Cl_2SiH$	$CH_2{=}CHCOOCH_3$	Pt–C (Ionol*)	$(CH_3)ClSi[CH(CH_3)COOCH_3]$	239
Cl_3SiH	$CH_2{=}CHCOOCH_2CH_2OCH_3$	Pt–C (Ionol*)	$Cl_3Si[CH(CH_3)COOCH_2CH_2OCH_3]$	239
$(CH_3)Cl_2SiH$	$CH_2{=}C(CH_3)COOCH_3$	Pt–C	$(CH_3)Cl_2Si[CH(CH_3)COOCH_3]$	596
$(C_2H_5)_3SiH$	$CH_2{=}CHCOOCH_3$	Pt–C	$(C_2H_5)_3Si[CH(CH_3)COOCH_3]$	115
$(CH_3)Cl_2SiH$	$CH_2{=}C(CH_3)COOCH_3$	Pt–C	$(CH_3)Cl_2Si[CH_2CH(CH_3)COOCH_3]$	587
$(CH_3)Cl_2SiH$	$CH_2{=}C(CH_3)COO(CH_2)_4H$	Pt–C	$(CH_3)Cl_2Si[CH_2CH(CH_3)COO(CH_2)_4H]$	587
$(CH_3)Cl_2SiH$	$CH_2{=}CHCH_2CH(COOC_2H_5)_2$	Pt–C	$(CH_3)Cl_2Si[(CH_2)_3CH(COOC_2H_5)_2]$	587
	Chlorides of unsaturated acids			
Cl_3SiH	$CH_2{=}CH(CH_2)_8COCl$	UV	$Cl_3Si[(CH_2)_{10}COCl]$	89, 90, 91
	Anhydrides of unsaturated acids			
$(C_6H_5)_2SiH_2$	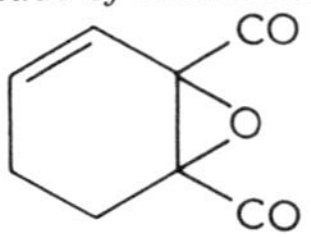 CO, O, CO	$(CH_3)_3COOCOC_6H_5$	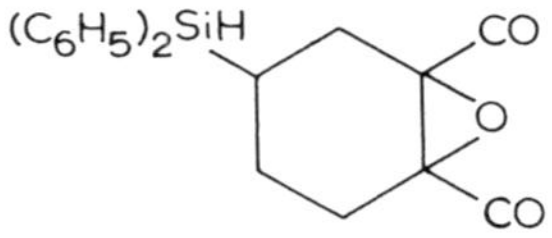 $(C_6H_5)_2SiH$, CO, O, CO	598
	Alkenyl acetates			
$(C_6H_5)SiH_3$	$CH_2{=}CHCH_2OCOCH_3$	$C_6H_5COOOC(CH_3)_3$	$(C_6H_5)SiH_2[(CH_2)_3OCOCH_3]$	598
$(C_6H_5)SiH_3$	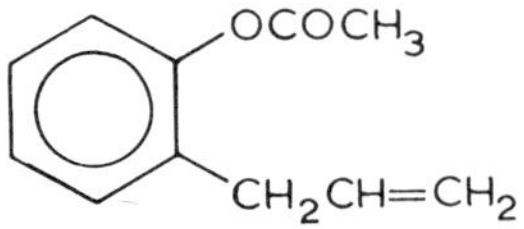 OCOCH$_3$, CH$_2$CH=CH$_2$	$C_6H_5COOOC(CH_3)_3$	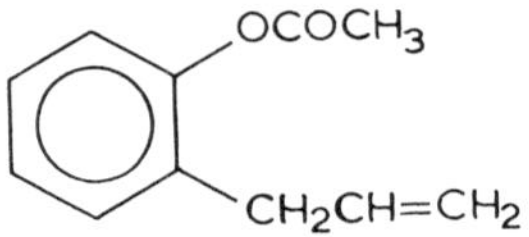 OCOCH$_3$, CH$_2$CH=CH$_2$	598
$(CH_3)Cl_2SiH$	$CH_2{=}CHCH_2OCOCH_3$	Pt–C (Ionol*)	$(CH_3)Cl_2Si[(CH_2)_3OCOCH_3]$	239
$(C_6H_5)Cl_2SiH$	$CH_2{=}CHCH_2OCOCH_3$	Pt–C	$(C_6H_5)Cl_2Si[(CH_2)_3OCOCH_3]$	596
$(CH_3)Cl_2SiH$	$CH_2{=}CHOCOCH_3$	$H_2PtCl_6 \cdot 6H_2O$	$(CH_3)Cl_2Si(CH_2CH_2OCOCH_3)$	596
$(C_2H_5O)_2(CH_3)SiH$	$CH_2{=}CHCH(OCOCH_2)_2$	Pt–C (Ionol*)	$(C_2H_5O)_2(CH_3)Si[CH_2CH_2CH(OCOCH_3)_2]$	239
$[(CH_3)_2SiH]_2O$	$CH_2{=}CHCH_2OCOCH_3$	Pt–C	$[(CH_3)_2Si(CH_2)_3OCOCH_3]_2O$	596
$(CH_3)Cl_2SiH$	$CH_2{=}CHCH_2OCOCH_3$	^{60}Co-γ-radiation	$(CH_3)Cl_2Si[(CH_2)_3OCOCH_3]$	161, 162
Cl_3SiH	$CH_2{=}CHCH_2OCOCH_3$	^{60}Co-γ-radiation	$Cl_3Si[(CH_2)_3OCOCH_3]$	161

*2,6-Di-*tert*-butyl-4-methylphenol (polymerisation inhibitor)

References p. 621

TABLE 7.30 (cont'd)

Silicon hydride	*Alkene*	*Catalyst (inhibitor)*	*Reaction product*	*Ref.*
	Unsaturated ethers and acetals			
Cl_3SiH	$CH_2{=}CHOC_2H_5$	UV	$Cl_3Si(CH_2CH_2OC_2H_5)$	88
Cl_3SiH	$CH_2{=}CHCH_2OCH_3$	UV or peroxides	$Cl_3Si[(CH_2)_3OCH_3]$	646
$(C_6H_5)SiH_3$	$(CH_2{=}CHCH_2)_2O$	$C_6H_5COOOC(CH_3)_3$	$(C_6H_5)SiH_2[(CH_2)_2CH(OC_2H_5)_2]$	598
$(CH_3)Cl_2SiH$	$CH_2{=}CHOC_2H_5$	Pt–C (Ionol*)	$(CH_3)Cl_2Si(CH_2CH_2OC_2H_5)$	239
$(C_2H_5O)_2(CH_3)SiH$	$CH_2{=}CHOC_2H_5$	Pt–C (Ionol*)	$(C_2H_5O)_2(CH_3)Si(CH_2CH_2OC_2H_5)$	239
$(CH_3)Cl_2SiH$	$CH_2{=}CHCH_2O(CH_2)_2CN$	$H_2PtCl_6 \cdot 6H_2O$	$(CH_3)Cl_2Si[(CH_2)_3O(CH_2)\ CN]$	596
$(CH_3)Cl_2SiH$	$CH_2{=}C\langle{}^{O-CH_2}_{O-CHCH_3}\rangle$ (ring: $O-CH_2$ and $O-CHCH_3$ joined)	$H_2PtCl_6 \cdot 6H_2O$	$(CH_3)Cl_2Si\left(CH_2CH\langle{}^{O-CH_2}_{O-CHCH_3}\rangle\right)$ (ring: $O-CH_2$ and $O-CHCH_3$ joined)	596
	Alkenyloxysilanes			
$(C_6H_5)SiH_3$	$CH_2{=}CHCH_2OSi(CH_3)_3$	$C_6H_5COOOC(CH_3)_3$	$(C_6H_5)SiH_2)_3OSi(CH_3)_3]$	71c, 598
$(C_2H_5O)_2(CH_3)SiH$	o-$(CH_2{=}CHCH_2)C_6H_4OSi(CH_3)_3$	$H_2PtCl_6 \cdot 6H_2O$	$(C_2H_5O)_2(CH_3)Si[(CH_2)_3C_6H_4OSi(CH_3)_3]$	22
$(C_4H_9O)_3SiH$	o-$(CH_2{=}CHCH_2)C_6H_4OSi(CH_3)_3$	$H_2PtCl_6 \cdot 6H_2O$	$(C_4H_9O)_3Si[(CH_2)_3C_6H_4OSi(CH_3)_3]$	22

*2,6-Di-*tert*-butyl-4-methylphenol (polymerisation inhibitor)

TABLE 7.31

REACTION OF SILICON HYDRIDES WITH AMINOALKENES, NITROALKENES, OR B-ALKENYLBORAZINES

Silicon hydride	*Alkene*	*Catalyst*	*Reaction product*	*Ref.*
	Aminoalkenes			
$(C_6H_5)SiH_3$	$CH_2{=}CHCH_2N[Si(CH_3)_3]_2$	$C_6H_5COOOC(CH_3)_3$	$(C_6H_5)H_2Si(CH_2)_3N[Si(CH_3)_3]_2$	519, 595, 598
$(C_6H_5)SiH_3$	$CH_2{=}CH(CH_2)_9NHSi(CH_3)_3$	$C_6H_5COOOC(CH_3)_3$	$(C_6H_5)H_2Si(CH_2)_{11}NHSi(CH_3)_3$	519, 595
$(CH_3)Br_2SiH$	(cyclohexenyl)–$CH_2N[Si(CH_3)_3]_2$	$C_6H_5COOOC(CH_3)_3$	$(CH_3)Br_2Si$–(cyclohexyl, H)–$CH_2N[Si(CH_3)_3]_2$	519, 595
Br_3SiH	$CH_2{=}CHCH_2N[Si(CH_3)_3]_2$	$C_6H_5COOOC(CH_3)_3$	$Br_3Si(CH_2)_3N[Si(CH_3)_3]_2$	640
	Nitroalkenes			
Cl_3SiH	$CH_2{=}CHCH_2NO_2$	$H_2PtCl_6 \cdot 6H_2O$	$Cl_3Si(CH_2)_3NO_2$	430
$(CH_3)Cl_2SiH$	$CH_2{=}CHCH_2NO_2$	$H_2PtCl_6 \cdot 6H_2O$	$(CH_3)Cl_2Si(CH_2)_3NO_2$	430
	B-*Alkenylborazines*			
$(CH_3)Cl_2SiH$	$[CH_2{=}CHBNC_6H_5]_3$	$H_2PtCl_6 \cdot 6H_2O$	$[(CH_3)Cl_2SiCH_2CH_2BNC_6H_5]_3$	557
$(CH_3)_3SiOSi(CH_3)_2H$	$[CH_2{=}CHBNC_6H_5]_3$	$H_2PtCl_6 \cdot 6H_2O$	$[(CH_3)_3SiOSi(CH_3)_2CH_2CH_2BNC_6H_5]_3$	557
$[(CH_3)_3SiO]_2(CH_3)SiH$	$[CH_2{=}CHBNC_6H_5]_3$	$H_2PtCl_6 \cdot 6H_2O$	$[\{(CH_3)_3SiO\}_2(CH_3)Si(CH_2CH_2BNC_6H_5)]_3$	557

References p. 621

TABLE 7.32

REACTION OF SILICON HYDRIDES WITH ALKENYLSILANES (ADDITION TO THE DOUBLE C=C BOND)

Silicon hydride	*Alkenylsilane*	*Catalyst*	*Reaction product*	*Ref.*
R_3SiH				
$(C_6H_5)_3SiH$	$(C_6H_5)_3SiCH_2CH{=}CH_2$	$(C_6H_5COO)_2$	$(C_2H_5)_3Si[(CH_2)_3Si(C_6H_5)_3]$	639
$(C_6H_5)_3SiH$	$(CH_3)_3SiCH_2CH{=}CH_2$	$(C_6H_5COO)_2$	$(C_6H_5)_3Si[(CH_2)_3Si(CH_3)_3]$	639
$(CH_2{=}CH)(CH_3)_2SiH$	$(CH_2{=}CH)(CH_3)_2SiH$	Pt–C	$(CH_3)_2\overline{SiCH_2CH_2Si(CH_3)_2CH_2C}H_2$, polymers	114
$(CH_2{=}CH)(C_2H_5)_2SiH$	$(CH_2{=}CH)(C_2H_5)_2SiH$	Pt–C	$(C_2H_5)_2\overline{SiCH_2CH_2Si(C_2H_5)_2CH_2C}H_2$, polymers	114
$RSiHCl_2$				
$(CH_3)Cl_2SiH$	$(CH_3)_3SiCH{=}CH_2$	$(C_6H_5COO)_2$	$(CH_3)Cl_2Si[CH_2CH_2Si(CH_3)_3]$	556
$(CH_3)Cl_2SiH$	$(C_3H_7)Cl_2SiCH{=}CH_2$	300°	$(CH_3)Cl_2Si[CH_2CH_2Si(C_3H_7)Cl_2]$	71b
$(CH_3CH_2CH_2)Cl_2SiH$	$(CH_3CH_2CH_2)Cl_2SiCH{=}CH_2$	300°	$(CH_3CH_2CH_2)Cl_2Si[CH_2CH_2Si(CH_2CH_2CH_3)Cl_2]$	71b
$SiHCl_3$				
Cl_3SiH	$CH_2{=}CHSiCl_3$	$(CH_3COO)_2$	$Cl_3Si(CH_2CH_2SiCl_3)$	71a, 83
Cl_3SiH	$CH_2{=}CHSiCl_3$	300°	$Cl_3Si(CH_2CH_2SiCl_3)$	71b
Cl_3SiH	$CH_2{=}CHSi(CH_3)_3$	$(C_6H_5COO)_2$	$Cl_3Si[CH_2CH_2Si(CH_3)_3]$	299, 556
Cl_3SiH	$(CH_2{=}CH)Si(C_2H_5)_3$	$(CH_3COO)_2$	$Cl_3Si[CH_2CH_2Si(C_2H_5)_3]$	299
Cl_3SiH	$(CH_2{=}CH)_2Si(CH_3)_2$	$(CH_3COO)_2$	$(Cl_3SiCH_2CH_2)_2Si(CH_3)_2$	299
Cl_3SiH	$(CH_2{=}CHCH_2)SiCl_3$	$(CH_3COO)_2$, $(C_6H_5COO)_2$	$Cl_3Si[(CH_2)_3SiCl_3]$	71a, 83, 639
Cl_3SiH	$(CH_2{=}CHCH_2)Si(CH_3)_3$	$(C_6H_5COO)_2$	$Cl_3Si[(CH_2)_3Si(CH_3)_3]$	639

[*704*] to trialkylvinylsilanes forms, for example, $Me_3SiCH_2CH_2Cl$ or a mixture of $(Me_3SiCH_2CH_2)_3B$ and $[Me_3SiCH(CH_3)]_3B$ or $Me_3SiCH_3CH_2AlR_2$. This is explained by means of a four-centre mechanism (an ionic mechanism would lead to the same results): although secondary carbonium ions (I, III, and V) should be more stable than primary carbonium ions (II, IV, and VI), because of the propinquity of the positive charges (electronegativities: Si = 1.74, C = 2.50) they are less stable.

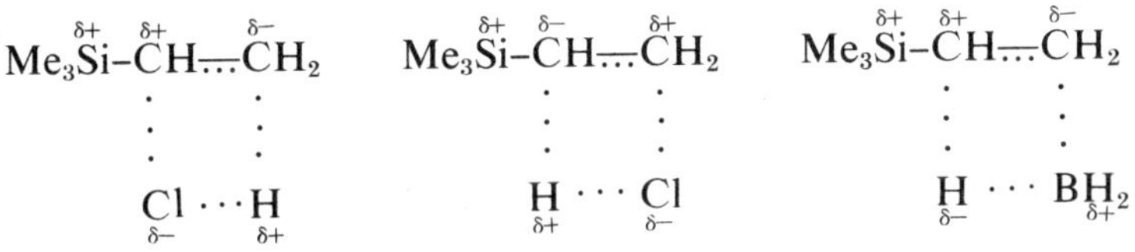

$$\begin{array}{ccc}
\overset{\delta+}{Me_3Si}\text{–}\overset{\delta+}{C}H\cdots\overset{\delta-}{C}H_2 & Me_3\overset{\delta+}{Si}\text{–}\overset{\delta-}{C}H\cdots\overset{\delta+}{C}H_2 & Me_3\overset{\delta+}{Si}\text{–}\overset{\delta+}{C}H\cdots\overset{\delta-}{C}H_2 \\
\vdots \quad \vdots & \vdots \quad \vdots & \vdots \quad \vdots \\
\underset{\delta-}{Cl}\cdots\underset{\delta+}{H} & \underset{\delta+}{H}\cdots\underset{\delta-}{Cl} & \underset{\delta-}{H}\cdots\underset{\delta+}{BH_2} \\
\text{I} & \text{II} & \text{III}
\end{array}$$

$$\begin{array}{ccc}
Me_3\overset{\delta+}{Si}\text{–}\overset{\delta-}{C}H\cdots\overset{\delta+}{C}H_2 & Me_3\overset{\delta+}{Si}\text{–}\overset{\delta+}{C}H\cdots\overset{\delta-}{C}H_2 & Me_3\overset{\delta+}{Si}\text{–}\overset{\delta-}{C}H\cdots\overset{\delta+}{C}H_2 \\
\vdots \quad \vdots & \vdots \quad \vdots & \vdots \quad \vdots \\
\underset{\delta+}{H_2B}\cdots\underset{\delta-}{H} & \underset{\delta-}{H}\cdots\underset{\delta+}{AlR'_2} & \underset{\delta+}{R_2Al}\cdots\underset{\delta-}{H} \\
\text{IV} & \text{V} & \text{VI}
\end{array}$$

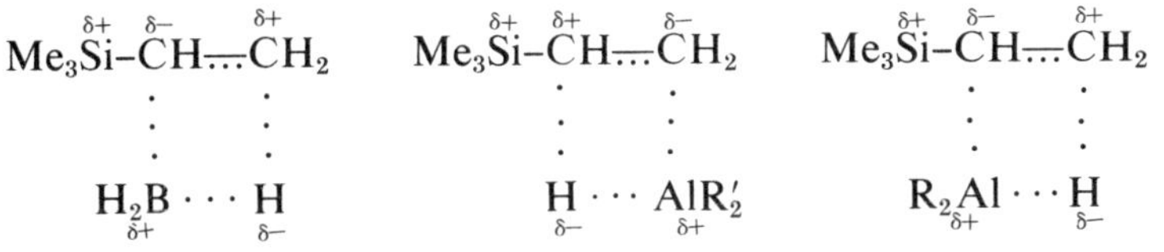

The stable secondary carbonium ion completely determines the addition of diborane to a double C═C bond not adjacent to silicon. Allylsilane forms tris(γ-trimethylsilylpropyl)borane exclusively [*554*]:

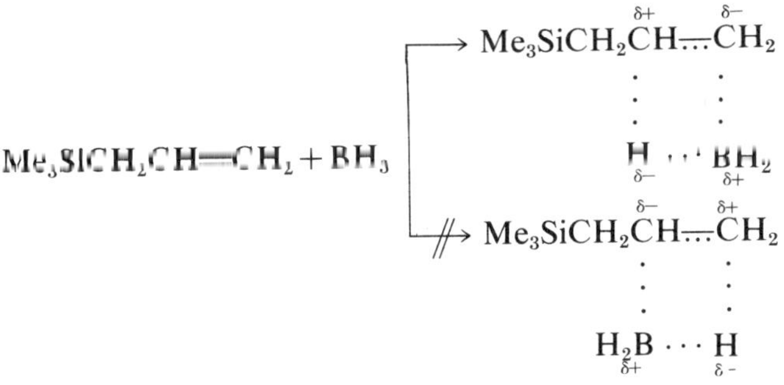

$$Me_3SiCH_2CH{=}CH_2 + BH_3 \left\{\begin{array}{l}
\longrightarrow Me_3SiCH_2\overset{\delta+}{C}H\cdots\overset{\delta-}{C}H_2 \\
\qquad\qquad\qquad \vdots \qquad \vdots \\
\qquad\qquad\qquad \underset{\delta-}{H}\cdots\underset{\delta+}{BH_2} \\
\not\longrightarrow Me_3SiCH_2\overset{\delta-}{C}H\cdots\overset{\delta+}{C}H_2 \\
\qquad\qquad\qquad \vdots \qquad \vdots \\
\qquad\qquad\qquad \underset{\delta+}{H_2B}\cdots\underset{\delta-}{H}
\end{array}\right.$$

5.12 Reaction of the Si–H bond with alkynes

Silicon hydrides R_3SiH add in two stages to triple C≡C bonds by a radical or polar mechanism. Telomerisation frequently occurs, as well. Table 7.33 contains examples and references.

$$HC{\equiv}CH \xrightarrow{+\,\rangle SiH} \rangle SiCH{=}CH_2 \xrightarrow{+\,\rangle SiH} \rangle SiCH_2{-}CH_2Si\langle \quad \text{addition}$$

$$\rangle SiCH{=}CH_2 \xrightarrow{+HC\equiv CH} \rangle SiCH{=}CHCH{=}CH_2 \xrightarrow{+\,\rangle SiH} \rangle SiCH{=}CHCH_2{-}CH_2Si\langle \quad (R = H, \text{alkyl}, C_6H_5)$$

telomerisation

The *radical addition* is initiated by the production of chain-propagating silyl radicals. They are formed: (1) by photosensitised irradiation [*673*], (2) by radical-forming agents [*62, 72, 82, 83*], or (3) by raising the temperature [*357, 673*]:

1. $\rangle Si{-}H + Hg(^3P_1) \rightarrow \rangle Si\cdot + H\cdot + Hg(^1S_0)$
2. $(RCOO)_2 \rightarrow 2RCOO\cdot$

   $\rangle Si{-}H + RCOO\cdot \rightarrow \rangle Si\cdot + RCOOH$
3. $\rangle Si{-}H \xrightarrow{\text{thermal energy}} \rangle Si\cdot + H\cdot$

In the radical reaction, at least in the first addition stage of the benzoyl-peroxide-catalysed reaction, which gives alkenylsilanes with a yield of 40 to 50%, *cis* and *trans* isomers are formed in a ratio of 3:1. Either the reaction takes place through a strongly sterospecific *trans* addition [*60, 62*], which, however, because of a partial radical isomerisation does not give *cis* adducts exclusively

$$\rangle Si\cdot + H{-}C{\equiv}C{-}H \rightarrow (\rangle Si)(H)C{=}\dot{C}(R) \xrightarrow{+\,\rangle Si{-}H} (\rangle Si)(H)C{=}C(R)(H) + \rangle Si\cdot \quad \textit{cis}\text{ adduct}$$

$$\downarrow\uparrow$$

$$(\rangle Si)(H)C{=}\dot{C}(R) \xrightarrow{+\,\rangle Si{-}H} (\rangle Si)(H)C{=}C(H)(R) + \rangle Si\cdot \quad \textit{trans}\text{ adduct}$$

or *cis* and *trans* additions compete, with reaction rates in a ratio of 3:1. The second addition stage, an alkene addition, also leads to a *cis* adduct (see section 5.11.2).

The *polar addition*, at least the first step of the Pt–C- and H_2PtCl_6-catalysed reaction, which gives alkenylsilanes with a yield of 70%, is a sterospecific *cis* addition [*60, 62*]. Via a platinum-silane complex A which has not been studied in detail (chemisorption?), to which ethyne can add from only one side (*cis* addition), it gives *trans* adducts exclusively:

$$\geq Si{-}H + Pt \rightarrow A \xrightarrow{+HC\equiv CR} (\geq Si)(H)C{=}C(H)(R)$$

Since in the H_2PtCl_6-catalysed addition only the black solids that form appear to be active, here again it is perhaps finely divided platinum and not the acid itself which is the active catalyst. However, further investigation is needed to elucidate this point.

Substitution takes place more rarely and only at relatively high temperatures [*386*]:

$$\geq Si{-}H + HC\equiv CH \rightarrow \geq SiC\equiv CH + H_2$$

It is very difficult to understand by a radical mechanism. A primary substitution (formation of alkenylsilane) and subsequent synchronous elimination (formation of alkynylsilane and hydrogen or re-formation of the starting materials):

$$\geq Si{-}H + HC\equiv CH \xrightarrow{\text{addition}} (\geq Si)HC{=}CH_2 \rightarrow$$

$$\begin{array}{ccc} \begin{array}{c} H\cdots H \\ \vdots \quad\;\; \vdots \\ C \overset{\cdots}{—} C \\ | \qquad | \\ \geq Si \quad H \end{array} & \text{or} & \begin{array}{c} H \quad H \\ | \qquad | \\ C \overset{\cdots}{—} C \\ \vdots \qquad \vdots \\ Si\cdots H \end{array} \\ \downarrow & & \downarrow \\ \geq Si{-}C\equiv CH + H_2 & & \geq SiH + HC\equiv CH \end{array}$$

is unlikely. The reaction of phenylethyne and trichlorosilane certainly takes place differently. Trichlorosilylstyrene which must be formed here as an intermediate, is not in fact dehydrogenated under the same conditions [*386*]. Because of the

TABLE 7.33

REACTION OF SILICON HYDRIDES WITH ALKYNES (ADDITION AND SUBSTITUTION)

Silicon hydride	*Alkyne*	*Catalyst*	*Reaction product*	*Ref.*
SiH_4	$CH{\equiv}CH$	460–510°	$H_3Si(CH{=}CH_2)$, $HSi(CH{=}CH_2)_2(C{\equiv}CH)$	673
SiH_4	$CH{\equiv}CH$	$Hg(^3P_1) \rightarrow Hg(^1S_0)$	$H_3Si(CH{=}CH_2)$	673
$(C_2H_5)_2CH_3SiH$	$CH{\equiv}C(CH_2)_2OH$		$(C_2H_5)_2CH_3SiC[C(CH_2)_2OH]{=}CH_2$	389
$(CH_3)Cl_2SiH$	$CH{\equiv}CH$	$Pd–Al_2O_3$ (Pt, Ni), 300°	$(CH_3)Cl_2Si(CH{=}CH_2)$, $CH_3Cl_2Si(CH_2CH_2SiCl_2CH_3)$	570, 571
$(C_2H_5)Cl_2SiH$	$CH{\equiv}CH$	$Pd–Al_2O_3$ (Pt, Ni), 170°	$(C_2H_5)Cl_2Si(CH{=}CH_2)$, $(C_2H_5)Cl_2Si(CH_2–CH_2)SiCl_2(C_2H_5)$ $(C_2H_5)Cl_2Si(CH{=}CHCH{=}CH_2)$, $(C_2H_5)Cl_2Si(CH_2–CH_2CH{=}CH)SiCl_2(C_2H_5)$	571, 572
$(C_2H_5O)_3SiH$	$CH{\equiv}CH$	Pt	$(C_2H_5O)_3Si(CH{=}CH_2)$	359, 660
Cl_3SiH	$CH{\equiv}CH$	$(CH_3)_3COOC(CH_3)_3$	$Cl_3Si(CH{=}CH_2)$	83
Cl_3SiH	$CH{\equiv}CH$	$(CH_3)_3COOC(CH_3)_3$	$Cl_3Si(CH_2CH_2)SiCl_3$	83
Cl_3SiH	$CH{\equiv}CH$	$(C_6H_5COO)_2$	$Cl_3Si(CH_2CH_2)SiCl_3$	556
Cl_3SiH	$CH{\equiv}CCH_2CH_2CH_3$	350°, pressure	$Cl_3Si(CH{=}CHCH_2CH_2CH_3)$	357
Cl_3SiH	$CH{\equiv}CCH_2CH_2CH_3$	Pt–C	$Cl_3Si(CH{=}CHCH_2CH_2CH_3)$	62
Cl_3SiH	$CH{\equiv}C(CH_2)_3CH_3$	Pt–C	$Cl_3Si[CH{=}CH(CH_2)_3CH_3]$	62
Cl_3SiH	$CH{\equiv}C(CH_2)_4CH_3$	Pt–C	$Cl_3Si[CH{=}CH(CH_2)_4CH_3]$	62
Cl_3SiH	$CH{\equiv}CCH_2CH_2CH_3$	$(C_6H_5COO)_2$	$Cl_3Si(CH{=}CHCH_2CH_2CH_3)$	62
Cl_3SiH	$CH{\equiv}C(CH_2)_3CH_3$	$(C_6H_5COO)_2$	$Cl_3Si[CH{=}CH(CH_2)_3CH_3]$	62, 82
Cl_3SiH	$CH{\equiv}C(CH_2)_4CH_3$	$(C_6H_5COO)_2$	$Cl_3Si[CH{=}CH(CH_2)_4CH_3]$	62
Cl_3SiH	$CH{=}C(CH)_3CH_3$	350°, pressure	$Cl_3Si[CH{=}CH(CH_2)_3CH_3]$	357
Cl_3SiH	$CH{\equiv}CH$	$Pd–Al_2O_3$ (Pt, Ni), 300°	$Cl_3Si(CH{=}CH_2)$, $Cl_3Si(CH_2CH_2)SiCl_3$	570
Cl_3SiH	$CH{\equiv}CH$	Pt	$Cl_3Si(CH{=}CH_2)$, $Cl_3Si(CH_2CH_2)SiCl_3$	359, 660
Cl_3SiH	$CH{\equiv}CC_6H_5$	Pt–C	$Cl_3Si(CH{=}CHC_6H_5)$	62
Cl_3SiH	$CH{\equiv}CC_6H_5$	$(C_6H_5COO)_2$		62
Cl_3SiH	$CH{\equiv}CC_6H_5$	$(C_4H_9)_3N$	$Cl_3SiCH(C_6H_5)CH_2SiCl_3$	61
$(C_2H_5)_3SiH$	$\overline{CH_2(CH_2)_4C}(OH)C{\equiv}CH$	Pt, 200°	$(C_2H_5)_3Si[(CH{=}CH_2)\overline{C(OH)(CH_2)_4CH_2}]$	457
Cl_3SiH	$CH{\equiv}CC(CH_3)_3$	$(C_6H_5COO)_2$	$Cl_3Si[CH{=}CHC(CH_3)_3]$ (83% *trans*, 17% *cis*)	675
Cl_3SiH	$CH{\equiv}CCH(CH_3)_2$	$(C_6H_5COO)_2$	$Cl_3Si[CH{=}CHCH(CH_3)_3]$ (72% *trans*, 28% *cis*)	675
Cl_3SiH	$CH{\equiv}CCH(CH_3)_2$	$H_2PtCl_6 \cdot 6H_2O$	$Cl_3Si[CH{=}CHCH(CH_3)_3]$ (~ 100% *trans*)	675

Cl_3SiH	$CH{\equiv}CC(CH_3)_3$	$H_2PtCl_6 \cdot 6H_2O$	$Cl_3Si[CH{=}CHC(CH_3)_3]$ (50% *trans*)	675
Cl_3SiH	$CH{\equiv}C(CH_2)_2CH_3$	$(C_6H_5COO)_2$	$Cl_3Si[CH{=}CH(CH_2)_2CH_3]$ (21% *trans*, 79% *cis*)	675
Cl_3SiH	$CH{\equiv}C(CH_2)_2CH_3$	Pt–C	$Cl_3Si[CH{=}CH(CH_2)_2CH_3]$ (~ 100% *trans*)	675
Cl_3SiH	$CH{\equiv}C(CH_2)_2CH_3$	$H_2PtCl_6 \cdot 6H_2O$	$Cl_3Si[CH{=}CH(CH_2)_2CH_3]$ (~ 100% *trans*)	675
Cl_3SiH	$CH{\equiv}C(CH_2)_3CH_3$	$(C_6H_5COO)_2$	$Cl_3Si[CH{=}CH(CH_2)_3CH_3]$ (23% *trans*, 77% *cis*)	675
Cl_3SiH	$CH{\equiv}C(CH_2)_3CH_3$	Pt–C	$Cl_3Si[CH{=}CH(CH_2)_3CH_3]$ (~ 100% *trans*)	675
Cl_3SiH	$CH{\equiv}C(CH_2)_3CH_3$	$H_2PtCl_6 \cdot 6H_2O$	$Cl_3Si[CH{=}CH(CH_2)_3CH_3]$ (~ 100% *trans*)	675
Cl_3SiH	$CH{\equiv}C(CH_2)_4CH_3$	$(C_6H_5COO)_2$	$Cl_3Si[CH{=}CH(CH_2)_4CH_3]$ (25% *trans*, 75% *cis*)	67:
Cl_3SiH	$CH{\equiv}C(CH_2)_4CH_3$	Pt–C	$Cl_3Si[CH{=}CH(CH_2)_4CH_3]$ (~ 100% *trans*)	675
Cl_3SiH	$CH{\equiv}C(CH_2)_4CH_3$	$H_2PtCl_6 \cdot 6h_2O$	$Cl_3Si[CH{=}CH(CH_2)_4CH_3]$ (~ 100% *trans*)	675
Cl_3SiH	$CH{\equiv}CCH_2Cl$	$H_2PtCl_6 \cdot 6H_2O$	$Cl_3SiCH{=}CHCH_2Cl$, $Cl_3SiC(CH_2Cl){=}CH_2$	390
Cl_3SiH	$CH{\equiv}C(C_6H_5)$	500°	$Cl_3SiCC(C_6H_5)$ (substitution)	386, 387

References p. 621

polarity of $\rangle C{-}H^{\delta+}$ and $\rangle Si{-}H^{\delta-}$, the synchronous elimination of hydrogen is indicated:

$$\rangle \overset{\delta+}{Si}{-}\overset{\delta-}{H} + HC{\equiv}\overset{\delta-}{C}{-}\overset{\delta+}{H} \rightarrow \begin{matrix} HC{\equiv}C \cdots H^{\delta+} \\ \vdots \qquad \vdots \\ \rangle Si \cdots H^{\delta-} \end{matrix} \rightarrow HC{\equiv}CSi\langle + H_2$$

5.13. *Reaction of the Si–H bond with the double C═O and C═N bonds*

Silicon hydrides add to saturated ketones on irradiation in UV light with the formation of the corresponding silyl ethers [*92*, cit. *454*], possibly by a radical mechanism:

$$\rangle Si{-}H \xrightarrow{h\nu} \rangle Si\cdot + H\cdot$$

$$\rangle Si\cdot + R{-}\underset{\underset{O}{\|}}{C}{-}R' \rightarrow R{-}\underset{\underset{OSi\langle}{|}}{\dot{C}}{-}R' \xrightarrow{\rangle Si{-}H} R{-}\underset{\underset{OSi\langle}{|}}{\overset{H}{C}}{-}R' + \rangle Si\cdot$$

In the presence of $ZnCl_2$, the reaction takes place without irradiation by a polar mechanism to give the same result [*94*]:

$$\rangle \overset{\delta+}{Si}{-}\overset{\delta-}{H} + |\overset{\ominus}{\underline{O}}{-}\overset{\oplus}{C}\langle \rightarrow \rangle Si{-}O{-}\underset{|}{\overset{|}{C}}H$$

References and examples are given in Table 7.34.

Unsaturated acids offer silicon hydrides two possibilities of addition. Tri*alkyl*silanes add by a polar mechanism (catalyst $H_2PtCl_6 \cdot 6H_2O$) to the C═C–C═O conjugated system of unsaturated acids and their esters in the 1,4-position to form saturated esters or unsaturated "acetals" [*454*]:

$$R_3Si{-}H + CH_2{=}CH{-}C\begin{matrix}{/\!\!/O} \\ {\backslash OH}\end{matrix} \rightarrow \left[CH_3{-}CH{=}C\begin{matrix}{/OSiR_3} \\ {\backslash OH}\end{matrix} \right] \rightarrow CH_3{-}CH_2{-}C\begin{matrix}{/OSiR_3} \\ {\backslash\!\!\backslash O}\end{matrix}$$

$$R_3Si{-}H + CH_2{=}CH{-}C\begin{matrix}{/\!\!/O} \\ {\backslash OR'}\end{matrix} \rightarrow CH_3{-}CH{=}C\begin{matrix}{/OSiR_3} \\ {\backslash OR'}\end{matrix}$$

Tri*chloro*silane, on the other hand, adds to the C═C–C═O conjugated system in the 1,2-position to the double C═C bond (for further details see section 5.11.5 and Table 7.30).

$$\overset{\delta+}{Cl_3Si}\text{–}\overset{\delta-}{H} + \overset{\delta+}{C}H_2{=}\overset{\delta-}{C}H \rightarrow C(=O)(OR) \rightarrow Cl_3Si[CH(CH_3)COOR]$$

Partially chlorinated alkylsilanes occupy an intermediate position between R_3SiH and Cl_3SiH. Correspondingly, they add in both the 1,4- and the 1,2-positions [*454*]:

$$RCl_2Si\text{–}H + CH_2{=}CH\text{–}C(=O)(OR') \rightarrow \begin{cases} RCl_2Si\text{–}OC(OR'){=}CH_2\text{–}CH_3 \\ RCl_2Si\text{–}CH\{(CH)_3\}COOR' \\ RCl_2Si\text{–}CH_2CH_2COOR' \end{cases}$$

The C=O group in urea does not add the Si–H bond in trichlorosilane. The reactants form the adduct $SiHCl_3 \cdot 6(NH_2)_2CO$ [*535b*]. Possibly the two neighbouring electron-attracting amide groups prevent the negative charging of the oxygen that is necessary for an addition

$$\overset{\delta+}{O}{=}C(\rightarrow NH_2^{\delta-})(\rightarrow NH_2^{\delta-})$$

but the hydroxyl hydrogen of the iso-form $HO\text{–}C(NH_2)(NH)$ should be capable of replacement (see below).

In amides $R'C(O)NHR$, the negative substituent effect of the amide group is lower. Silicon hydride therefore reacts with amides [*191*] with replacement of the hydroxyl hydrogen (with a N-phenylated amide, a conjugated system):

$$\geq Si\text{–}H + \left[C_6H_5\text{–}C(=O)\text{–}N(H)\text{–}C_6H_5 \leftrightarrow C_6H_5\text{–}C(\text{–}\bar{O}|^{\ominus})\text{=}N(H^{\oplus})\text{–}C_6H_5 \right] \xrightarrow{-H_2} C_6H_5\text{–}C(\text{–}O\rightarrow Si\leq)\text{=}N\text{–}C_6H_5$$

or with replacement of the amide hydrogen (in N-alkylated amides):

$$\geq Si\text{–}H + alkyl\text{–}C(=O) \rightarrow \underset{\delta-}{N}(H^{\delta+})\text{–}C_6H_5 \xrightarrow{-H_2} alkyl\text{–}C(=O)\text{–}N(\uparrow Si\leq)\text{–}C_6H_5$$

(and possibly $\text{–}C(OSi\leq)\text{–}NH\text{–}$ as well)

References p. 621

where R = phenyl (conjugated system possible): more O-silyl compound, or R = alkyl (conjugated system impossible): more N-silyl compound.

The double C=N bond of the silyl ether imide, -C(OSi≡)=N-, formed adds more silane substantially more slowly than in the preceding substitution (formation of a N-silyl compound):

$$-C(OSi\equiv)=NC_6H_5 \xrightarrow{+\equiv SiH} -CH(OSi\equiv)-N(C_6H_5)(Si\equiv) \xrightarrow[-\,O(Si\equiv)_2]{+\equiv SiH} -CH_2-N(C_6H_5)(Si\equiv)$$

As can be seen, in the addition process, the positive silyl group becomes attached to the nitrogen of the double C=N bond; correspondingly, the neighbouring π-electron system of benzene appears to have only a slight influence on the charge of the nitrogen (no mesomerism in the sense of the formula $\overset{\delta+}{=N}$–C_6H_5 ($\delta-$) . Table 7.34 contains examples of the reaction.

TABLE 7.34

REACTIONS OF SILICON HYDRIDES WITH COMPOUNDS CONTAINING DOUBLE C=O OR C=N BONDS (ADDITION TO THE DOUBLE C=O OR C=N BOND)

Silicon hydride	*C=O or C=N compound*	*Catalyst*	*Reaction product*	*Re*
$(C_2H_5)_3SiH$	$C_6H_5CONHC_6H_5$	$ZnCl_2$, 140°, 8 h	$(C_2H_5)_3Si[OC(C_6H_5)=NC_6H_5] + H_2$	19
$(C_2H_5)_3SiH$	$CH_3CONHC_6H_5$	$ZnCl_2$, 140°, 2 h	$(C_2H_5)_3Si[OC(CH_3)=NC_6H_5] + H_2$; a little $(C_2H_5)_3Si[N(C_6H_5)COCH_3]$	19
$(C_2H_5)_3SiH$	$CH_3CONH(CH_2CH_2CH_3)$	$ZnCl_2$, 150°, 4 h	$(C_2H_5)_3Si[N(C_3H_7)COCH_3]$; $(C_2H_5)_3Si[OC(CH_3)=NC_3H_7]$	19
$(C_2H_5)_3SiH$	$C_6H_5C(OCH_3)=NC_6H_5$	$ZnCl_2$, 140°, 72 h	$(C_2H_5)_3Si[N(C_6H_5)CH_2H_5]$	19
$(C_2H_5)_3SiH$	$C_6H_5C[OSi(C_2H_5)_3]=NC_6H_5$	$ZnCl_2$, 150°, 60 h	$(C_2H_5)_3Si[N(C_6H_5)CH_2C_6H_5]$, $[(C_2H_5)_3Si]_2O$	19
$(C_2H_5)_3SiH$	$C_6H_5CH=NC_6H_5$	$ZnCl_2$, 150°	$(C_2H_5)_3Si[N(C_6H_5)CH_2C_6H_5]$	19
$(C_2H_5)_3SiH$	CH_3COCH_3	$ZnCl_2$	$(C_2H_5)_3Si[OCH(CH_3)_2]$	9
$(C_2H_5)_3SiH$	cyclohexanone (C_6H_{10}=O)	$ZnCl_2$, 95°, 48 h	$(C_2H_5)_3Si(O-C_6H_{11})$	9
$(C_2H_5)_3SiH$	2-methylcyclohexanone	$ZnCl_2$, 24 h	$(C_2H_5)_3Si[O\overline{CHCH(CH_3)(CH_2)_3CH_2}]$	94
$(C_2H_5)_3SiH$	$C_6H_5COCH_3$	95°, 150 h	$(C_2H_5)_3Si[OCH(CH_3)(C_6H_5)]$	94
$(C_2H_5)_3SiH$	$CH_3COCH_2CH_2CH=C(CH_3)_2$	105°, 16 h	$(C_2H_5)_3Si[OCH(CH_3)CH_2CH_2CH=C(CH_3)_2]$	94
$(C_2H_5)_3SiH$	$CH_3COCH_2CH_2CH=CH_2$	$H_2PtCl_6 \cdot 6H_2O$	$(C_2H_5)_3Si(CH_2)_4COC\ H_3$; $(C_2H_5)_3Si[OCH(CH_3)CH_2CH_2CH=CH_2]$	32

5.14 *Exchange of ligands in silanes*

With silane, energy-rich "hot" tritium atoms form the monosubstitution product SiH_3T and HT, together with still more highly tritiated silanes [*95a*]. Organosilicon hydrides or organosilicon hydride–halides exchange ligands by a similar mechanism to that of peralkylated, perarylated, or perhalogenated silanes. Since the latter conversions have been investigated in substantially more detail, the summarised results may be premised.

5.14.1 *The metal-halide-catalysed exchange of alkyl groups*

Alkylhalogenosilanes do not dismute in the absence of traces of metals, metal halides, or nucleophilic solvents[*184*]. Dismutation takes place only in the presence of catalysts. Four different sequences of reactions are conceivable for the metal-halide-catalysed dismutation (*e.g.*, by AlX_3) of tetraalkylsilanes, $R_nR'_{4-n}Si$ ($R \neq R'$), into a number of other tetraalkylsilanes, $R_mR'_{4-m}Si$ ($m \neq n$):

1. Reaction via carbonium ions, $R^{\oplus}$, with the primary step:

$$R'_3RSi + AlX_3 \rightleftharpoons R^{\oplus} + R'_3SiAlX_3^{\ominus}$$

2. Reaction via silonium ions, $R'_3Si^{\oplus}$, with the primary step:

$$R'_3RSi + AlX_3 \rightleftharpoons R'_3Si^{\oplus} + RAlX_3^{\ominus}$$

3. Reaction via alkylaluminium halide, $RAlX_2$, with the primary step:

$$R'_3RSi + AlX_3 \rightleftharpoons R'_3SiX + RAlX_2$$

4. Reaction via a polar 4-centre reaction induced by $AlX_4^{\ominus}$:

$$2R'_3RSi + 2AlX_4^{\ominus} \rightarrow \begin{array}{ccc} & R'_3 & \\ & | & \\ AlX_4^{\ominus}\cdots\rightarrow & Si\cdots & R \\ & \vdots & \vdots \\ & R'\cdots & Si \leftarrow\cdots AlX_4^{\ominus} \\ & R & R'_2 \end{array} \rightarrow R'_4Si + R'_2R_2Si + 2AlX_4^{\ominus}$$

Catalysts, cocatalysts, inhibitors, and solvents affect the rate of exchange and the kinetics as follows[*212, 360–362, 469, 470, 508–511, 698, 699, 705*]. The catalytic action of metal halides on the exchange of ligands falls in parallel with their catalytic action in the isomerisation of hydrocarbons or the alkylation of benzene and with their Lewis-acid strengths in the following sequence: $(AlBr_3)_2 > (AlCl_3)_2 > (AlI_3)_2 > EtAlCl_2 . AlCl_3 > (GaBr_3)_2 \gg (GaCl_3)_2 > BCl_3 > FeCl_3$ [*511*]. The $AlBr_3$-catalysed dismutation is greatly affected by a number of co-catalysts (relative rate of the reaction without a catalyst = 1): $(C_2H_5)_2O$ (> 5.5), CH_3Br (5.6), NH_3 (> 7.5), H_2S (> 86), H_2O (> 116), AlOBr (> 118), O_2 (> 137. The dismutation takes place twice as fast in cyclohexane and 100 times as fast in benzene as without a solvent.

The experimental results show that the metal-catalysed disproportionation of $(CH_3)_3(C_2H_5)Si$ takes place by mechanism 4. Mechanism 1 (carbonium ion) is opposed by, in particular, four facts. (a) In the dismutation of substituted silanes

References p. 621

an isopropyl group is not rearranged. The dismutation therefore does not take place via carbonium ions in the following way: $R^{\oplus} + R_3R'Si \rightarrow R_4Si + R'^{\oplus}$ [*212*]. (b) Benzene accelerates the dismutation [*508*]. However, benzene should trap carbonium ions [*469, 470, 511*]. (c) In cyclohexane solution no methylcyclopentane is formed [*508*], although cyclohexane should be isomerised by carbonium ions [*511*]. (d) The catalytic effects of the cocatalysts isopropyl bromide and methyl bromide are equal [*508*]. Since i-C_3H_7Br forms carbonium ions in the presence of Lewis acids more rapidly than CH_3Br [*510*], they should have different effects.

Mechanism 2 (silonium ion) is opposed by, in particular, two facts. (a) In the dismutation of $R_3R'Si$ in benzene, no silylbenzene is produced, although benzene is alkylated by carbonium ions. (b) The Lewis bases NH_3, $(CH_3)_3N$, and $(CH_3)_3$-NHBr used as cocatalysts accelerate the reaction although they should trap both silonium and carbonium ions [*508*]:

$$R_3Si^{\oplus} + R'_3N \rightarrow R'_3NSiR_3^{\oplus}$$

$$Me_3NHBr + Al_2Br_6 \rightarrow Me_3NH^{\oplus}Al_2Br_7^{\ominus}$$

$$R_3Si^{\oplus} + Al_2Br_7^{\ominus} \rightarrow R_3SiBr + Al_2Br_6$$

Mechanism 3 ($RAlX_2$ intermediate stage) is opposed by the following facts. The presence of a "cocatalyst" $(CH_3)_6Al_2$ diminishes the dismutation catalysed by Al_2Br_6, although the methylaluminium bromides formed from Al_2Br_6 and $(CH_3)_6Al_2$ should have an accelerating effect. Oxygen [*508*] and water [*25, 508*] raise the rate of the reaction, although they should destroy methylaluminium chlorides.

The following facts are in favour of reaction mechanism (4) (polar 4-centre reaction). (a) The activity of the catalysts rises with their donor activity. (b) The dismutation reaction possesses an extremely low steric factor [*508*]. Consequently R'_3RSi molecules exchange alkyl groups on collision only when they have a quite definite narrowly delimited mutual geometrical position. This points to a polar mechanism according to (I) since reactions which take place by polar mechanisms in non-polar media quite generally possess a small steric factor. (c) The marked acceleration of the reaction by the cocatalysts H_2O, H_2S, and O_2 is not surprising, since, for example, the polymerisation of alkenes in hydrocarbons as solvents polymerised by Friedel-Crafts catalysts is also accelerated by substances such as water. This can be explained by the formation of a strong donor $H^{\oplus}AlCl_4^{\ominus}$:

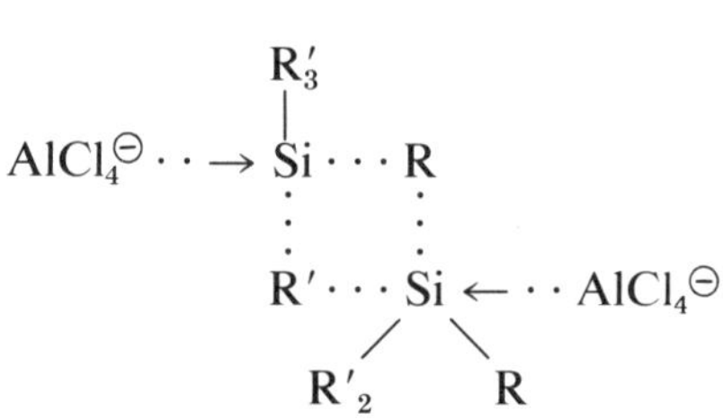

```
                 R'3
                 |
AlCl4⊖ · · → Si · · · R
                 :       :
                 :       :
                 R' · · · Si ← · · AlCl4⊖
                        /   \
                     R'2     R
```

5.14.2 *The metal-halide-catalysed exchange of hydrogen*

The rate of the dismutation of trimethylsilicon hydride [into $(CH_3)_4Si$, $(CH_3)_2SiH_2$, CH_3SiH_3, and SiH_4], like that of $(CH_3)_3(C_2H_5)Si$, decreases in the sequence of catalysts $(AlBr_3)_2 > (GaBr_3)_2 > (GaCl_3)_2 > BCl_3$. Likewise, the cocatalyst H_2O accelerates the $(AlBr_3)_2$-catalysed dismutation reaction. Similarly, cyclohexane as a solvent has a slight accelerating effect and benzene a pronounced one. The rate of exchange of the silicon ligands in $(CH_3)_3SiH$, $(CH_3)_3(C_6H_5)Si$, $(CH_3)_3(C_2H_5)Si$, and $(CH_3)_3SiBr$ decreases in benzene and cyclohexane in the sequence CH_3 for H > CH_3 for C_6H_5 > CH_3 for C_2H_5, CH_3 for Br [*509*].

Consequently, dismutation takes place in the gas phase and in solution through a more or less polar 4-centre reaction. Because of the hydridic hydrogen atom, an electrophilic catalysis (a) by AlX_3 is possible. In the presence of water, however, the nucleophilic catalysis (b) by $AlX_4^{\ominus}$ will probably control the dismutation process:

$$2R_3SiH + AlX_3 \rightarrow \left[\begin{array}{ccc} R_3Si \cdots & H & \cdots \rightarrow AlX_3 \\ \vdots & \vdots & \\ R \cdots & SiR_2H & \end{array}\right] \rightarrow R_4Si + R_2SiH_2 + AlX_3 \qquad \text{(a)}$$

$$Al_2X_6 + 3H_2O \rightarrow 6HX + Al_2O_3 \qquad \text{(b)}$$

$$Al_2X_6 + 2HX \rightarrow 2H^{\oplus}AlX_4^{\ominus}$$

$$2R_3SiH + 2AlX_4^{\ominus} \rightarrow \left[\begin{array}{ccc} & R_3 & \\ & | & \\ AlX_4^{\ominus} \cdots \rightarrow & Si \cdots & H \\ & \vdots & \vdots \\ & R \cdots & Si \leftarrow \cdots AlX_4^{\ominus} \\ & & / \quad \backslash \\ & & R_2 \quad H \end{array}\right]$$

$$\rightarrow R_4Si + R_2SiH_2 + 2AlX_4^{\ominus}$$

In the presence of $H_2PtCl_6 \cdot 6H_2O$, PtO_2, or Pt, $(C_6H_5)_2SiH_2$ dismutes into $(C_6H_5)_nSiH_{4-n}$ ($n = 0$–4). The equilibrium is achieved in a few hours [*224*]. In ethereal solution, $LiAlD_4$ deuterates phenylsilanes with the rate constants 5.7×10^{-5} ($PhSiH_3$), ~ 6×10^{-3} (Ph_2SiH_2), and 2.6×10^{-6} (Ph_3SiH) sec^{-1}. The increase in the rate constants from PhSiH to Ph_2SiH_2 by two powers of 10 is ascribed to the electron-attracting effect of the phenyl groups, so that the nucleophilic attack of $AlD_4^{\ominus}$ is facilitated. The marked decrease to Ph_3SiH is explained by steric hindrance [*330*].

5.15 *Reaction of the Si–H bond with metal salts and alkylmetals*

Triorganosilanes R_3SiH [R = C_2H_5, n-C_3H_7, n-C_4H_9, i-C_3H_7, i-C_4H_9, $(C_6H_5)+(CH_3)_2$, $(p\text{-}CH_3C_6H_4)+(CH_3)_2$, $p\text{-}ClC_6H_4$, C_6H_5, $p\text{-}CH_3C_6H_5$, $p\text{-}ClC_6H_4CH_2$, $C_6H_5CH_2$, $p\text{-}CH_3C_6H_4CH_2$] react with silver perchlorate in toluene or benzene

References p. 621

with the primary formation of a golden yellow solid which becomes black with the evolution of hydrogen [*139*]. The reaction is represented by the approximate empirical equation

$$R_3SiH + AgClO_4 \rightarrow R_3SiClO_4 + Ag + \tfrac{1}{2}H_2$$

although the yields of silver and hydrogen differ by up to 20%. The experimental results permit no clear conclusion concerning the mechanism. The intermediate formation of AgH (yellow solids) is likely. Consequently, the following reaction sequence is reasonable [*139*]:

$$R_3SiH + AgClO_4 \rightarrow R_3SiClO_4 + AgH$$
$$AgH \rightarrow Ag + H\cdot$$
$$H\cdot + H\cdot \rightarrow H_2$$

$$R_3SiH + AgClO_4 \rightarrow R_3SiClO_4 + Ag + \tfrac{1}{2}H_2$$

Some of the hydrogen radicals can also form protons with silver ions with the precipitation of silver, and this would explain the deficiency of hydrogen according to the empirical equation: $Ag^{\oplus} + H\cdot \rightarrow Ag + H^{\oplus}$. On the other hand, the formation of silyl radicals according to equation $R_3SiH + H\cdot \rightarrow R_3Si\cdot + H_2$ is less likely. In the above sequence of reactions, no account was taken of the undoubtedly strong influence of the catalytically active silver precipitate, the amount of which increases during the reaction:

$$R_3SiH + Ag^{\oplus} \rightarrow [R_3Si\text{–}H\cdot\cdot\!\rightarrow Ag]^{\oplus}$$
$$ClO_4^{\ominus} + [R_3Si\text{–}H\cdot\cdot\!\rightarrow Ag]^{\oplus} \rightarrow [ClO_4\cdot\cdot\!\rightarrow Si(R_3)\text{–}H\cdot\cdot\!\rightarrow Ag] \rightarrow R_3SiClO_4 + AgH$$
$$AgH + Ag_n \rightarrow Ag_{n+1} + H_{ads}$$
$$H_{ads} + H_{ads} \rightarrow H_2$$

$$R_3SiH + AgClO_4 \rightarrow R_3SiClO_4 + Ag + \tfrac{1}{2}H_2$$

This mechanism could explain the indefinite order of the reaction that has been found.

Diethylcadmium and diethylmercury, like triethylantimony and triethylbismuth, react with triethylsilane with the formation of the corresponding silylcadmiums or silylmercuries [*487*]:

$$(C_2H_5)_3SiH + (C_2H_5)_2M \rightarrow [(C_2H_5)_3Si]_2M + 2\,C_2H_6 \quad (M = Hg, Cd)$$
$$(C_2H_5)_3SiH + (C_2H_5)_3M \rightarrow [(C_2H_5)_3Si]_3M + 3\,C_2H_6 \quad (M = Sb, Bi)$$

About the insertion of carbenes ($PhHgCBr_3 \rightarrow Br_2C\!: + PhHgBr$ and $PhHgCBrH \rightarrow Br_2C\!: + PhHgH$) in the Si–H bonds see [*499, 552a*].

5.16 *Reaction of the Si–H bond with alkali metals*

When germane or stannane is passed into the blue solution of an alkali metal in liquid ammonia it becomes decolorised. First the dimetallic hydride and then on further reaction the monometallic hydride, for example, Na_2GeH_2 and $NaGeH_3$, are said to be formed. The formation of the monometallic hydride is not affected by the ammonolysis which takes place slowly since $NaNH_2$ also gives the monometallic hydride with GeH_4:

$$2\,Na_{(NH_3)} \xrightarrow[\text{fast}]{+GeH_4,\ -H_2} Na_2GeH_2 \begin{cases} \xrightarrow[\text{fast}]{+GeH_4} 2\,NaGeH_3 \\ \xrightarrow[\text{slow}]{+NH_3,\ -NaGeH_3} NaNH_2 \xrightarrow[-NH_3]{+GeH_4} NaGeH_3 \end{cases}$$

On the other hand, silane, when it is passed into a solution of an alkali metal in liquid ammonia, decolorises it substantially more slowly. Because of ammonolysis, which takes place faster, no metallated silane but only alkali-metal amide can be isolated[*164, 262, 512a*]:

$$SiH_4 + E^I_{(NH_3)} \xrightarrow[\text{slow}]{-\frac{1}{2}H_2} E^ISiH_3 \xrightarrow[\text{fast}]{+NH_3,\,-SiH_4} E^INH_2 \quad (E^I = Li, Na, K)$$

On the other hand, silanes partially substituted with sterically demanding ligands can be metallated in ethylamine[*381*]:

$$(Ph_3Ge)_3SiH + Li \xrightarrow{(C_2H_5NH_2)} (Ph_3Ge)_3SiLi + \tfrac{1}{2}\,H_2$$

However, the synthesis of metallated silanes requires the exclusion of ammonia or amine. Ether-soluble, colourless to faintly yellow, silylpotassium is formed by the reaction of monosilane and disilane with potassium in monoglyme[*11, 498*], diglyme, or triglyme[*11*], with or without the addition of catalytic amounts of tetraphenylethylene[*11*]. In addition, hydrogen and potassium hydride are formed in various amounts (solv = solvent):

$$K \xrightarrow{\text{solv}} K^{\oplus} + e^{\ominus}_{\text{solv}}$$

$$2e^{\ominus}_{\text{solv}} \rightleftharpoons e^{2\ominus}_{2,\,\text{solv}}$$

$$SiH_4 + e^{\ominus}_{\text{solv}} \rightarrow H_3\dot{S}iH^{\oplus} \rightarrow SiH_3^{\ominus} + H\cdot$$

$$H\cdot + H\cdot \rightarrow H_2$$

$$SiH_4 + e^{2\ominus}_{2,\,\text{solv}} \rightarrow H_3\ddot{S}iH^{2\ominus}. \rightarrow SiH_3^{\oplus} + H^{\ominus}$$

The reaction of 50 mmole of SiH_4 requires several months[*498*]. With liquid K–Na alloy (only the K reacts), and with the scrupulous exclusion of air and moisture, the reaction time can be reduced to a few hours[*11a, 11*]. Although the metallation takes place faster by some orders of magnitude in triglyme, this solvent cannot be recommended since the cleavage of the ether with the formation of coloured polymers frequently occurs for no apparent reason[*11*]. $RbSiH_3$ and $CsSiH_3$ are formed similarly[*11*].

References p. 621

With potassium[*498*] in monoglyme (reaction of 3 mmole takes 2 months at −78°) or with potassium hydride[*401, 497, 647a*], disilane forms SiH_3K, Si_2H_5K, Si_3H_7K, and SiH_4 in a complex reaction. Whether the Si–H or the Si–Si bond is attacked first in the reaction of Si_2H_6 with K is uncertain. In the reaction of Si_2H_6 with KH, it is probably not the Si–H but the Si–Si bond which is attacked first. Under the action of donors D such as KH (and to a smaller extent with LiD or LiCl), in fact, disilane dismutes into monosilane and silene: SiH_2[*70a, 401, 497*]. With more KH the latter forms silylpotassium, $KSiH_3$, or with more Si_2H_6 trisilane, Si_3H_8, which, in turn, can undergo further dismutation:

$$Si_2H_6 \xrightarrow{D} SiH_4 + :SiH_2$$
$$KH + :SiH_2 \rightarrow \underline{KSiH_3}$$
$$Si_2H_6 + :SiH_2 \rightarrow Si_3H_8$$

$$Si_3H_8 \xrightarrow{D} SiH_4 + :SiHSiH_3$$
$$KH + :SiHSiH_3 \rightarrow \underline{KSi_2H_5}$$
$$Si_2H_6 + :SiHSiH_3 \rightarrow Si_4H_{10}$$

$$Si_4H_{10} \xrightarrow{D} SiH_4 + :SiHSiH_2SiH_3$$
$$KH + SiHSiH_2SiH_3 \rightarrow \underline{KSi_3H_7}$$

With sodium amalgam, the main products are silane and polysilane(2) possibly together with silylsodium[*611*]:

$$Si_2H_6 + Na(Hg) \xrightarrow{\Delta} SiH_4, SiH_2, SiH_3Na$$

Here the mechanism is quite obscure. Because of the different reactivities of the Si–H and Si–C bonds, only with reservations should the reaction of Si_2H_6 with E^I be compared with the reactions of permethylated higher silanes (Me_8Si_3, $Me_{10}Si_4$) with liquid sodium-potassium[*222, 618*]:

$$Me_{10}Si_4 \xrightarrow[\Delta]{+Na/K} \begin{cases} 2\,Me_5Si_2M \xrightarrow{+2\,Me_{10}Si_4} \begin{cases} 2\,Me_7Si_3M + 2\,Me_8Si_3 \\ 2\,Me_5Si_2M + 2\,Me_{10}Si_4 \\ 2\,Me_3SiM + 2\,Me_{12}Si_5 \end{cases} \\ Me_3SiM + Me_7Si_3M \end{cases}$$

$$Me_{15}Si_7M \begin{cases} \xrightarrow{+Me_3SiM} M[Me_2Si]_6M + Me_6Si_2 \\ \xrightarrow{+Na/K} M[Me_2Si]_6M + Me_3SiM \end{cases} \quad [M = K(\text{or Na?}), Me = CH_3]$$

5.17 *Reaction of the Si–Si bond*

The Si–Si bond in higher silanes is attacked by many donors (alanate ions, pyridine, hydroxide ions). If an ethereal solution of Si_2H_6 is allowed to stand with small amounts of *lithium alanate*, SiH_4 is formed slowly at $-60°$ and rapidly at 20°. After all volatile materials have been distilled off, in addition to the unchanged solid catalyst $LiAlH_4$, a colourless, non-volatile, ether-soluble, oily, low-molecular-weight ether adduct of polysilane(2), $(SiH_2)_n$, which explodes in the air, is left [*680a*]:

$$Si_2H_6 \xrightarrow{LiAlH_4} SiH_4 + SiH_2$$

At 20°, with an excess of pyridine, Si_2H_6 separates out solid, white, polymeric SiH_2 within a few hours, and within a week half of this dismutes further to yellow (polymeric) SiH and SiH_4 [*680a*]:

$$Si_2H_6 \xrightarrow[-SiH_4]{\text{pyridine, 20°, hours}} SiH_2 \xrightarrow{\text{pyridine, 20°, weeks}} \tfrac{1}{3}SiH_4 + \tfrac{2}{3}SiH.$$

Water and trimethylamine (formation of *hydroxide ions*) decompose Si_2H_6 quantitatively in half an hour at $-78°$ to form SiH_4 and $(SiH)_x$. The SiH_4 is hydrolysed partially at $-78°$ and quantitatively at 20° to silicic acid [*680a*].

LiH, AlH_3, $LiBH_4$, and $LiAl_2H_7$ do not catalyse the dismutation of disilane [*680a*]. About the cleavage of $Ph_3SiSiPh_3$ with Li see [*222*].

6. Halogenosilanes

6.1 *Synthesis of halogenosilanes*

Five methods are available for the preparation of silicon hydride–halides or organosilicon hydride–halides.

(1) Formation of the Si–X bond in a silicon hydride:

$$SiH_4 + X_2, + HX, + RX, + MX_n$$

(2) Formation of the Si–H bond in a silicon halide:

$$SiCl_4 + H_2, HCHO, M + H_2$$

(3) Formation of the Si–X bond in an arylsilicon hydride:

$$C_6H_5SiH_3 + HX \rightarrow SiH_3X + C_6H_6$$

(4) Formation of Si–X *and* Si–H bonds; radical gas-phase hydrogenation:

$$Si + HX, RX$$

(5) Transhalogenation of silicon hydride halides:

$$\rangle Si\text{–}X + MX'_n, HX'$$

6.1.1. *Partial halogenation of silicon hydrides*

The hydrogen atoms of a silicon hydride can be replaced by halogen. For this, more or less polarised halogen molecules or halogen anions are required:

References p. 621

(1) $\overset{\delta+}{X}-\overset{\delta-}{X}$ (reaction of a silicon hydride with halogen in the gas phase);

(2) $\overset{\delta+}{H}-\overset{\delta-}{X}$ or $H^{\oplus}X^{\ominus}$, or $H^{\oplus}$[X·Friedel-Crafts catalyst]$^{\ominus}$ (reaction of a silicon hydride with a hydrogen halide in the gas phase or in solution with or without catalysts);

(3) $R^{\oplus}X^{\ominus}$ or $R^{\oplus}$[X·Friedel-Crafts catalyst]$^{\ominus}$ (reaction of a silicon hydride with an organic halide in solution);

(4) $M^{\oplus}X^{\ominus}$ (reaction of a silicon hydride with a metal halide).

Re (1), $\geqslant$*SiH* + X_2. The reaction of a silicon hydride R_nSiH_{4-n} with free halogen (Br_2, I_2) without the addition of catalysts leads (explosively in the case of brominations above −30° [*624*]) to the series of halogenosilanes $R_nSiH_{4-n-m}X_m$ [*13–15, 17, 153, 206, 368, 612, 624*]. Examples of the reaction are given in Table 7.21; for the reaction mechanism, see section 5.1. Partially halogenated silanes are obtained in good yields only when, in the sequence of reactions

$$SiH_4 \xrightarrow[-HX]{+X_2} SiH_3X \xrightarrow[-HX]{+X_2} SiH_2X_2 \xrightarrow[-HX]{+X_2} SiHX_3 \xrightarrow[-HX]{+X_2} SiX_4$$

$$RSiH_3 \xrightarrow[-HX]{+X_2} RSiH_2X \xrightarrow[-HX]{+X_2} RSiHX_2 \xrightarrow[-HX]{+X_2} RSiX_3$$

the partially halogenated silanes are removed from the further action of halogen. This is conveniently achieved by the directed condensation of the halogenosilane, which is less volatile than the silane, and higher yields of halogenosilanes can be achieved in spite of higher reaction temperatures (high reaction rates), *e.g.*, in the reaction of SiH_4 with Br_2: whole reaction vessel at −78°, 3 h, 25% of SiH_3Br, 13% of SiH_2Br_2 [*612*]; on the other hand: reaction vessel at 50°, one part of it at −120°, 3 h, 80–92% of SiH_3Br [*624*]. In higher-boiling silanes that are liquid at 20°, bromine or iodine halogenates smoothly to give partially halogenated silanes. References and examples of the reaction are given in Table 7.21.

Re (2), $\geqslant$*SiH* + *HX*. The aluminium-halide-catalysed reaction of silane or an alkylsilane with a hydrogen halide (HCl, HBr, HI) in the gas phase is less vigorous than halogenation with free halogen. At 100° it leads to partially halogenated silanes [*153, 167, 169, 346, 420, 421, 613, 614*]:

$$SiH_4 \xrightarrow[-H_2]{+HX} SiH_3X \xrightarrow[-H_2]{+HX} SiH_2X_2 \xrightarrow[-H_2]{+HX} SiHX_3 \xrightarrow[-H_2]{+HX} SiX_4$$

Examples of the reaction are given in Table 7.21. For the reaction mechanism, see section 5.2. The individual halogenosilanes can be isolated from the reaction product by fractional high-vacuum distillation. With HBr or HI, phenylsilane gives not $PhSiH_{3-n}X_n$ but SiH_3X (see below).

Re (3), $\geqslant$*SiH* + *RX*. Silanes can be halogenated with organic halogen compounds which are capable of undergoing heterolytic cleavage with or without the aid of catalysts:

$$\geqslant Si{-}H + RX \rightarrow \geqslant Si{-}X + RH$$

Suitable halogenation systems are (R = alkyl): RX + AlX_3; R_3CX; $CCl_4 + O_2$, or $AlCl_3$; $CHCl_3 + O_2$, or $AlCl_3$; alkenyl-X + AlX_3; YC_6H_4COX; $RCOOCH_2$-CH_2X. Table 7.24 gives examples of the reactions and references. For reaction mechanisms, see section 5.4.

Re (4), $>$*SiH* + MX_n. Many metal and non-metal halides capable of giving up halogen replace hydrogen in silicon hydrides by halogen (see section 5.3). Some of them – $HgCl_2$, $HgBr_2$, $CuCl_2$, AgCl, $PdCl_2$, BBr_3, and PI_3 (see Table 7.23) – are suitable for partial substitution, for example:

$$>\text{Si–H} + \text{HgX}_2 \rightarrow\ >\text{Si–X} + \text{Hg} + \text{HX}$$

When BBr_3 is used, the Si–Si bond in $Si_2H_{6-n}Br_n$ is retained [*130*].

6.1.2. *Partial hydrogenation of silicon halides*

Silicon halides cannot be partially hydrogenated with the usual hydrogenating agents, (*e.g.*, LiH, $LiBH_4$, $LiAlH_4$) [*251, 368, 438*]. With a deficiency of hydrogenating agent, perhydrogenated silanes are produced together with unconverted halide. However, partial hydrogenation can be achieved by means of atomic hydrogen as well as by the hydrogenation of the unstable $SiCl_2$ formed by the radical decomposition of $SiCl_4$.

Thus, silicon hydride–halides arise in the reaction of silicon tetrahalides, particularly $SiCl_4$, with hydrogen in a direct-current arc (> 4000°) with quenching of the reaction product to −120° (yield of $SiHCl_3$ about 60 wt.%) [*249a*]:

$$\text{SiX}_4 + \tfrac{1}{2}\text{H}_2 \xrightarrow[>4000^\circ]{\text{arc}} \text{SiHX}_3 + \text{HX}$$

When $SiCl_4$ and HCHO are passed over α-Al_2O_3 at 350 to 500°, no trichlorosilane or silane, but dichlorosilane and, surprisingly, a high yield of monochlorosilane are formed [*229a*]:

$$\text{SiCl}_4 + 3\text{HCHO} \rightarrow \text{SiH}_3\text{Cl} + 3\text{HCl} + 3\text{CO}$$
$$\text{SiCl}_4 + 2\text{HCHO} \rightarrow \text{SiH}_2\text{Cl}_2 + 2\text{HCl} + 2\text{CO}$$
$$2\text{CO} \rightarrow \text{CO}_2 + \text{C (poisons the Al}_2\text{O}_3\text{)}$$

Possibly, the equilibrium

$$\text{SiCl}_4 \rightleftharpoons \text{SiCl}_2 + 2\text{Cl}$$

which is far over on the left-hand side at 350–500°, is shifted to the right by the trapping of chlorine (→ HCl) by H radicals

$$\text{HCHO} \xrightarrow{350-500^\circ} 2\text{H} + \text{CO}$$

$SiCl_2$ adds hydrogen:

$$\text{SiCl}_2 + 2\text{ H} \rightarrow \text{SiH}_2\text{Cl}_2$$

The formation of SiH_3Cl (86% of SiH_3Cl at an $SiCl_4$:HCHO ratio of 1:3) is, however, difficult to explain by an $SiCl_2$ mechanism. Possibly in this process, H(H_2) and HCl add to active silicon [*693*].

If silicon tetrachloride and hydrogen are passed over metals with a lower electronegativity than silicon (for example, Al, Zn, and, less satisfactory, Mg and Fe), in addition to metal halides the hydrides SiH_4, SiH_2Cl_2, and $SiHCl_3$ are formed. The question of whether the hydrogenation takes place (1) through an $SiCl_2$ radical mechanism or (2) through the intermediate formation of a metal hydride (MH) is still open:

(1) *Formation of* $SiCl_2$

$$SiCl_4 \rightleftharpoons SiCl_2 + 2\,Cl$$
$$3\,Cl + Al \rightarrow AlCl_3$$
$$Cl + \tfrac{1}{2}\,H_2 \rightarrow HCl$$
$$SiHCl_3 \rightarrow SiCl_2 + HCl$$

Addition to $SiCl_2$:

$$SiCl_2 + H_2 \rightarrow SiH_2Cl_2$$
$$SiCl_2 + HCl \rightarrow SiHCl_3$$

$AlCl_3$-catalysed dismutation, *e.g.*:

$$2\,SiH_2Cl_2 \rightarrow SiH_4 + SiCl_4$$

(2) *Formation of a metal hydride*

$$M + \tfrac{1}{2}\,H_2 \rightarrow [MH]$$

$$[MH] + \rangle Si\text{–}Cl \rightarrow \rangle Si\text{–}H + [MCl]$$

In contrast to partial hydrogenation, in the silicon hydride–halide, halogen can be partially replaced by alkyl [*172*]:

$$SiHCl_3 + nRMgBr \rightarrow R_nSiHCl_{3-n} + nMgBrCl$$
$$(R = CH_3, C_2H_5, C_6H_5, C_6H_5CH_2)$$

6.1.3 *Halogenation of the Si–C bond in arylsilicon hydrides*

The tendency of the Si–C bond in alkylsilanes to fissure under the influence of HX increases in the following sequence of ligands: $n\text{-}C_3H_7 \sim i\text{-}C_3H_7 < C_2H_5 < CH_3 \ll C_6H_5$ [*137*]. While, in fact, phenylsilane is halogenated by *halogen* (formation of $C_6H_5SiH_2X$), the Si–C bond is cleaved by *hydrogen halide* without the Si–H bond being attacked. Consequently, preparative amounts of halogenosilanes can be made conveniently by the reaction of phenylsilane with hydrogen halide [*27, 204–208, 654*] and (chlorophenyl) silane with hydrogen halides [*31b*]:

$$C_6H_5SiH_3 + HX \rightarrow C_6H_6 + SiH_3X$$
$$(X = [F], Cl, Br, I)$$

$$(C_6H_5)_2SiH_2 + HX \rightarrow C_6H_6 + C_6H_5SiH_2X$$
$$(X = Br, I)$$

$$C_6H_5SiH_2X + HX \rightarrow C_6H_6 + SiH_2X_2$$
$$(X = Br, I)$$

$$ClC_6H_4SiH_3 + HI \rightarrow ClC_6H_5 + SiH_3I$$

The rate of the reaction falls with decreasing electronegativity of X. $C_6H_5SiCl_3$ can no longer be cleaved, and HCl cleaves only $C_6H_5SiH_3$. The reaction of $C_6H_5SiH_2X$ with HY forms only primary halogenosilanes with different halogen atoms SiH_2XY; they readily dismute to SiH_2X_2 and SiH_2Y_2[*208*]. Examples of the reaction are given in Table 7.35.

TABLE 7.35

PREPARATION OF HALOGENOSILANES BY CLEAVING THE PHENYL–SILICON BOND IN PHENYLSILANES WITH HX

Phenylsilane	*HX*	*Temp. (°C)*	*Time*	*Halogenosilane*	*Ref.*
SiH_2Ph_2	HI	−40°	6 days	88% SiH_2I_2	204
SiH_3Ph	HBr	−78°	3 h	100% SiH_3Br	204
SiH_2BrPh	HBr	−78°	6 days	63% SiH_2Br_2	204
SiH_2Ph_2	HBr	−78°	24 h	85% $C_6H_5SiH_2Br$	204
SiH_3Ph	HI	120°	16 h	SiH_3I, together with $C_6H_5SiH_2I$	206
SiH_3Ph	HI	−40°	1 h	SiH_3I	206
SiH_3Ph	HCl	−78°	7 days	72% SiH_3Cl (after 2 days 8%)	205
SiH_3Ph	HF	0°		Mainly Si–C cleavage together with side reaction	
SiH_3Ph	HF	20°, CuO		Benzene, Si fluorides vigorous reaction	205
SiH_3Ph	HCl	20–50°	a few hours	75% SiH_3Cl	654
SiH_3Ph	HBr	20–50°	a few hours	68% SiH_3Br	654
SiD_3Ph	DCl	20–50°	a few hours	~75% SiD_3Cl	654
SiD_3Ph	DBr	20–50°	a few hours	68% SiD_3Br	654

6.1.4 *Addition of hydrogen and halogen to elementary silicon*

The combined formation of Si–H and Si–X bonds is achieved by the reaction of elementary pure silicon or, with high yields, silicon alloyed with Cu, Ag, Ca, or Fe, with hydrogen chloride or an alkyl halide at an elevated temperature. HX gives SiH_nX_{4-n} ($n = 0, 1, 2$), and RX gives SiH_nX_{4-n} ($n = 0, 1$), and $R_aSiH_bX_c$ ($a + b + c = 4$; $b = 0, 1$). The reaction of HCl with silicon first gives an active intermediate $H \cdots Si \cdots Cl$[*24, 48, 238, 277, 295, 502*].

In the reaction of RCl with silicon, RCl decomposes primarily into HCl and alkene:

$$HCl + Si \rightarrow [H \cdots Si \cdots Cl]$$
$$[H \cdots Si \cdots Cl] + 2HCl \rightarrow SiHCl_3 + H_2$$
$$[H \cdots Si \cdots Cl] + 3HCl \rightarrow SiCl_4 + 2H_2$$
$$[H \cdots Si \cdots Cl] + C_2H_5Cl \rightarrow C_2H_5SiHCl_2$$
$$[H \cdots Si \cdots Cl] + 2C_2H_5Cl \rightarrow C_2H_5SiCl_3 + C_2H_4 + H_2$$

References p. 621

The reactivity of the silicon alloy depends largely on the ratio of the foreign atoms to the silicon atoms [*582*]. The Cu–Si alloy is a layer lattice. At a ratio of 4 Cu to 11 Si, the layer separation is 55 Å and the alloy is active. If more than four copper atoms are present, the layer distance separates to 13.5 Å and the alloy is inactive. With a decreasing number of foreign atoms in *interlattice* sites, the activity increases [*582*]. In order to activate HCl, copper need not necessarily be in contact with silicon. Remarkably, it is sufficient to pass HCl over a copper spiral before it is brought into contact with the silicon [*134*]. Examples of the reactions are given in Table 7.36.

TABLE 7.36

DIRECT SYNTHESIS OF HALOGENOSILANES BY THE REACTION OF HALIDES WITH SILICON (+ CATALYST)

Halide	*Catalyst*	*Temp. (°C)*	*Reaction product (% referred to the halides used)*	*Ref.*
HCl	—	—	Copper not in contact with Si, high-purity $SiHCl_3$	134
HCl	—	280–300°	$SiHCl_3$	404
C_6H_5Br	Ag	136–190°	$C_6H_5SiHCl_2$ (a little)	343
$CH_3Cl + HCl$	10% Cu	300°	CH_3SiCl_3 (main product), $SiCl_4$, $(CH_3)_2SiCl_2$, CH_3SiHCl_2	47, 425
CH_3Cl	10% Cu	300°	9% $SiCl_4$, 37% CH_3SiCl_3, 42% $(CH_3)_2SiCl_2$, 12% CH_3SiHCl_2	425
C_2H_5Cl	5% Cu	300°	$C_2H_5SiCl_3$, $C_2H_5SiHCl_2$, $(C_2H_5)_2SiHCl$, $(C_2H_5)_2SiCl_2$	20, 21
$C_2H_5Cl + HCl$	10% Cu	275°	$C_2H_5SiCl_3$ (main product), $(C_2H_5)_2SiCl_2$, $(C_2H_5)_3SiCl$, $C_2H_5SiHCl_2$	47
$C_2H_5Br + H_2$	20% Cu	290–300°	$C_2H_5SiBr_3$, $C_2H_5SiHBr_2$	327
n-C_3H_7Cl	20% Cu	350–360°	19% n-$C_3H_7SiCl_3$, 10% (n-$C_3H_7)_2SiCl_2$, 32% (n-C_3H_7)$SiHCl_2$	460
n-C_3H_7Br	20% Cu	340°	$SiBr_4$, n-$C_3H_7SiBr_3$, $SiHBr_3$, 80% C_3H_6	411, 412
n-C_4H_9Cl	20% Cu	350–360°	18% n-$C_4H_9SiCl_3$, 30% n-$C_4H_9SiHCl_2$	460
n-C_4H_9Br	20% Cu	340°	$SiBr_4$, n-$C_4H_9SiBr_3$, $SiHBr_3$, 80% C_4H_8	411
i-C_3H_7Cl	20% Cu	350–360°	21% i-$C_3H_7SiCl_3$, 11% (i-$C_3H_7)_2SiCl_2$, 38% i-$C_3H_7SiHCl_2$	460
i-C_4H_9Cl	20% Cu	350–360°	15% i-$C_4H_9SiCl_3$, 29% i-$C_4H_9SiHCl_2$	460
CH_2Cl_2	Cu	320–350°	$CH_2(SiCl_3)_2$, $CH_2(SiHCl_2)_2$, $Cl_3SiCH_2SiHCl_2$	556, 641
CH_2Cl_2	10% Cu	300°	$CH_2(SiCl_3)_2$, $Cl_3SiCH_2SiHCl_2$, $(CH_2SiCl_2)_3$	446
$CHCl_3$	10% Cu	300°	$CH_2(SiCl_3)_2$, $Cl_3SiCH_2SiHCl_2$, $Cl_3SiCH(SiHCl_2)_2$	403
$(CH_3)_2CCl_2$	20% Cu	360–380°	10% $Cl_3SiC(CH_3){=}CH$, 8% $Cl_2HSiC(CH_3){=}CH_2$, 14% $(Cl_2HSi)_2CCH_3$, 12% $Cl_2HSi(CH_3)_2SiCl_3$	455
$(CH_3)HCCl_2$	20% Cu	360–380°	5% $SiCl_4$, 6% $Cl_2HSiCH{=}CH_2$, 16% $Cl_3SiCH{=}CH_2$, 7% $(Cl_2HSi)_2CHCH_3$, 19% $(Cl_3Si)_2CH(CH_3)$, 19% $(CH_3)HC(SiCl_3)(SiHCl_2)$	455
$Cl_2CH_3Si(CH_2)_3Cl$	20% Cu	370–380°	14% CH_3SiCl_3, 24% $CH_3Cl_2SiCH{=}CHCH_3$ + $CH_3Cl_2SiCH_2CH{=}CH_2$, 14% $CH_3Cl_2Si(CH_2)_3SiCl_3$, 7% $CH_3Cl_2Si(CH_2)_3SiHCl_2$, 4% $Cl_2Si[(CH_2)_3SiCl_2CH_3]$	640
$HBr + H_2$	Cu	380°	50–60% $SiBr_4$, 40% $SiHBr_3$, 5% SiH_2Br_2	692

6.1.5 *Transhalogenation of the Si–X bond*

By means of the corresponding heavy-metal salts, particularly antimony and silver salts, silanes on the left of an arrow in the following conversion series are, apart from a few exceptions (Si–H) converted into the silanes on the right of the arrow [*18, 138, 150, 152*]:

$$\rangle SiI \rightarrow (\rangle Si)_2S \rightarrow \rangle SiBr (\rightarrow \rangle SiH \rightarrow \rangle SiNCSe) \rightarrow \rangle SiNC \text{ and}$$

$$\rangle SiCl \rightarrow \rangle SiNCS \rightarrow \rangle SiNCO \rightarrow (\rangle Si)_2O \rightarrow \rangle SiF$$

Such reactions (together with the conversion of silicon hydride halides into silicon hydride pseudohalides) are important mainly for the preparation of silicon hydride–fluorides, since this avoids anhydrous HF and F_2 which are relatively difficult to handle:

$$3 \rangle SiX + SbF_3 \rightarrow 3 \rangle SiF + SbX_3 \quad (X = Cl, Br, I)$$

Chlorosilanes are converted into fluoro derivatives by aqueous hydrofluoric acid with retention of the Si–H bond [*573*]. Possibly silanol is formed as an intermediate. ($\rangle Si_2O$ is to the left of $\rangle SiF$):

$$\rangle SiCl \xrightarrow[-HCl]{+H_2O} [\rangle SiOH] \xrightarrow[-H_2O]{+HF} \rangle SiF$$

Table 7.37 contains examples of the reaction.

TABLE 7.37
TRANSHALOGENATION OF HALOGENOSILANES

Halogenosilane	*Reactant*	*Reaction product*	*Ref.*
SiH_3Br	SbF_3	SiH_3F	420
CH_3SiH_2Cl	SbF_3	CH_3SiH_2F	463
CH_3SiD_2Cl	SbF_3	CH_3SiD_2F	463
CD_3SiH_2Cl	SbF_3	CD_3SiH_2F	463
$^{13}CH_3SiH_2Cl$	SbF_3	$^{13}CH_3SiH_2F$	463
CH_2DSiH_2Cl	SbF_3	CH_2DSiH_2F	463
$SiDCl_3$	SbF_3	$SiDF_3$	421
SiD_2Cl_2	SbF_3	SiD_2F_2	421
$(CH_3)(C_2H_5)SiHCl$	40% HF aq.	$(CH_3)(C_2H_5)SiHF$ (60%)	573
$(CH_3)(C_3H_7)SiHCl$	40% HF aq.	$(CH_3)(C_3H_7)SiHF$ (66%)	573
$(CH_3)(i\text{-}C_3H_7)SiHCl$	40% HF aq.	$(CH_3)(i\text{-}C_3H_7)SiHF$ (73%)	573
$(CH_3)(C_4H_9)SiHCl$	40% HF aq.	$(CH_3)(C_4H_9)SiHF$ (80%)	573
$(CH_3)(i\text{-}C_4H_9)SiHCl$	40% HF aq.	$(CH_3)(i\text{-}C_4H_9)SiHF$ (74%)	573
CH_3SiH_2I	$HgCl_2$	CH_3SiH_2Cl (90%)	142
CH_3SiH_2I	AgBr	CH_3SiH_2Br (93%)	142

6.2 *Physical properties of the halogenosilanes*

The physical properties of the silicon hydride halides are given in Table 7.38. The boiling points of SiH_3X and Si_2H_5X follow the relations:

$$\text{B.p.}(SiH_3X) = [1.110 \times \text{b.p.}(CH_3X)] - 2.10$$
$$\text{B.p.}(Si_2H_5X) = [1.079 \times \text{b.p.}(C_2H_5X)] + 26.10$$

Examples:

$$\text{B.p.}(SiH_3Br) = 1.110 \times 3.56^\circ (CH_3Br) - 2.10 = 1.85^\circ \text{ (found } 1.90^\circ)$$
$$\text{B.p.}(Si_2H_6Cl) = 1.079 \times 13.1^\circ (C_2H_5Cl) + 26.10 = 40.23^\circ \text{ (found } 40.1^\circ)$$

On irradiation by vacuum ultra violet $SiHCl_3$ decomposes by detachment of the H atoms [*280a*] and SiH_2Cl_2 into $SiCl_2 + H_2$ [*381a*].

7. Reactions of the Si–X bond

7.1. *Reaction of the Si–X bond with halides and pseudohalides*

If in the conversion series (see section 6.1.5) X is to the left of Y or near Y, the halogenosilanes $R_{4-m-n}SiH_mX_n$ (n, m = 1, 2, 3; R = alkyl, aryl; X = halogen) are converted by hydrogen halides, hydrogen pseudohalides, metal halides, or metal pseudohalides into the corresponding halogeno- or pseudohalogenosilanes

$$\rangle Si\text{–}X + Y^{\ominus} \rightarrow \rangle Si\text{–}Y + X^{\ominus} \quad (X = \text{halogen; } Y = \text{halogen, pseudohalogen})$$

For example, SiH_3I and AgCN give SiH_3CN, or SiH_3I and $Pb(NCN)_2$ give SiH_3NCN; however, SiH_3Cl and AgCN give no SiH_3CN:

$$SiH_3I + AgCN \rightarrow SiH_3CN + AgI$$
$$2SiH_3I + Pb(NCN)_2 \rightarrow 2SiH_3NCN + PbI_2$$

Further examples of the reactions and references are given in Table 7.39.

7.2 *Reaction of the Si–X bond with oxides, sulphides, and selenides*

According to the conversion series, halogenosilanes form with water, alcohols, sodium alkoxides, mercury sulphides (HgX, $HgSC_nF_{2n+1}$) [*143, 163*], mercury selenides ($HgSeC_nF_{2n+1}$) [*143*], and silver selenide [*163*], generally in very good yields, alkoxysilanes $SiH_{4-n}(OR)_n$, disilyl oxides $(R_{3-n}SiH_n)_20$, or bis(disilanyl)-oxide $(Si_2H_5)_2O$, disilyl sulphides $(R_{3-n}SiH_n)_2S$, and disilyl selenide $(SiH_3)_2Se$.

The compounds mentioned cannot be prepared by any other method than that given above, except in a few special cases. Thus, bis(trichlorosilyl) oxide, $Cl_3SiOSiCl_3$, cannot be hydrogenated to disilyl oxide, $H_3SiOSiH_3$ (siloxane), since both the Si–Cl and the Si–O bonds are hydrogenated. Alkoxysilanes, SiH_3OR, are formed from the corresponding halogen compounds $SiCl_3OR$ only by hydrogenation with lithium boranate, while other hydrogenating agents (*e.g.*, LiH or

TABLE 7.38

PHYSICAL PROPERTIES OF THE HALOGENOSILANES

([a]Vapour pressure equation: log p[torr] $= -A/T + B$ or log p[torr] $= -A/T + 1.75 \log T - BT + C$)
Halogenosilanes with Si–O bonds as well are given in Table 7.42

Halogenosilane	*M.p. (°C)*	*B.p. [°C (torr)]*	*Constants of vapour pressure equation[a]* *A*	*B*	*C*	*Enthalpy of evaporation Q (cal/mole), Trouton's constant T (cal mole⁻¹ deg⁻¹), Density d (g/ml), refractive index n*	*Ref.*
$SiH_{4-n}X_n$							
SiH_3F	—	−88.1; −98.6 (760)	984.86	8.2001	—	Q 4500; T 25.8; vapour pressure at −122.0° = 100 torr	166, 623
SiH_2F_2	−122.0	−77.8 (760)				Q 4751; T = 25.7; vapour pressure at −93.0° = 296 torr	166
SiH_3Cl	−118	−30.3 (763)	1435.3	0.0088305	6.5558	d_{liq}^{-113} 1.145	614, 687
SiH_2Cl_2	−122.0	8.2 (756)	1297.2	0.0024827	3.9022	d_{liq}^{-122} 1.42; UV photolysis: $SiCl_2 + H_2$	381a, 614, 687
$SiHCl_3$	—	—				Enthalpy of formation ΔH°_{298} −123.9 (liq.)(kcal/mole); $d^{19.5}$ 1.3453	424, 691
$SiHCl_3 \cdot 2(C_2H_5)_2PH$	−58						208a
SiH_3Br	—	1.9 (760)	1178.3	0.001435	3.2914	d_{liq}^{-80} 1.72; d^0_{gas} 1.533 (g/litre)	612, 687
$SiHBr_3$	−73.5	111.8	1819.5	7.6079	—		545, 692
SiH_3I	−57.0	42.8 (668.6); 45.[illegible] (760)				Q 7130; d 2.0718 (1 − 0.00204 t)	26, 167
SiH_2I_2	—	138.0 (567.5)				Q 8050; d 2.7943 (1 − 0.00320 t)	169
SiH_3SiH_2X							
SiH_3SiH_2F	−100.7	−10.0 (760)	1376.015	8.110310	—	Q 6300; T 23.9; no decomp. in 1 day, 0°	2, 352
SiH_3SiH_2Cl	—	40.1 (760)	1529.80	7.76416	—	Q 7000; T 22.4	2, 134
SiH_3SiH_2Br	—	—	—	—	—		2
SiH_3SiH_2I	—	102.8 (760)	1767.9	7.5843	—	Q 8087; T 21.5	662
$RSiH_2X$							
CH_3SiH_2F	—	—	—	—	—	Vapour pressure at −78° = 104 torr	142
$\overline{CH_2(CH_2)_4CH}SiH_2F$	—	118.1 (760)				d_4^{20} 0.925; n^{20}_{white} 1.420_7	15
n-$C_7H_{15}SiH_2F$	—	138.1 (760)				d_4^{20} 0.828; n^{20}_{white} 1.399_7	17
CH_3SiH_2Cl	−135	—	1354.6	7.7263	—	Q 6192	142, 687
n-$C_4H_9SiH_2Cl$	—	102.8				d_4^{20} 0.888; n^{20}_{white} 1.419_2	13

References p. 621

TABLE 7.38 (cont'd)

Halogenosilane	*M.p. (°C)*	*B.p. [°C (torr)]*	*Constants of vapour pressure equation*[a] *A*	*B*	*C*	*Enthalpy of evaporation Q (cal/mole), Trouton's constant T (cal mole^{-1} deg^{-1}), Density d (g/ml), refractive index n*	*Ref.*
$\overline{CH_2(CH_2)_4C}HSiH_2Cl$	—	158.0 (760)				d_4^{20} 0.980; n_{white}^{20} 1.465$_5$	15
n-$C_7H_{15}SiH_2Cl$	—	176.8 (760)				d_4^{20} 0.884; n_{white}^{20} 1.435$_4$	17
CH_3SiH_2Br	−119	34 ± 0.5 (760)	1490	7.732	—	Q 6800; T 21.8	142
n-$C_4H_9SiH_2Br$	—	124.6				d_4^{20} 1.183; n_{white}^{20} 1.451$_4$	13
$\overline{CH_2(CH_2)_4C}HSiH_2Br$	—	182.0 (760)				d_4^{20} 1.243; n_{white}^{20} 1.493$_2$	15
n-$C_7H_{15}SiH_2Br$	—	194.5 (760)				d_4^{20} 1.097; n_{white}^{20} 1.457$_1$	17
$C_6H_5SiH_2Br$	−30.0	118.0–118.2 (760)				d^{25} 1.3632; n_D^{25} 1.5555	368
CH_3SiH_2I	−109.5	71.8 (760)	1663	7.730	—	Q 4514; T 22.1; d_{liq}^{20} 1.768	2, 169
n-$C_4H_9SiH_2I$	—	155.4				d_4^{20} 1.454; n_{white}^{20} 1.502$_8$	13
$\overline{CH_2(CH_2)_4C}HSiH_2I$	—	210 (760)				d_4^{20} 1.494; n_{white}^{20} 1.537$_0$	15
n-$C_7H_{15}SiH_2I$	—	219.3 (760)				d_4^{20} 1.303; n_{white}^{20} 1.494$_5$	17
$RSiHX_2$							
n-$C_7H_{15}SiHF_2$	—	132.6 (760)				d_4^{20} 0.916; n_{white}^{20} 1.375$_9$	17
$GeH_3SiF_2SiHF_2$						orange-yellow solid	582a
$GeH_3SiF_2SiF_2SiHF_2$						orange-yellow solid	582a
CH_3SiHCl_2	~ −93 −92.5	15 (250) 41.1 (760)	1506.1	7.6784	—	d^0 ~ 0.93; d^0 1.62; Q 6891; T 21.92; vapour pressure at 0° = 146.5 torr	172, 315, 614
$ClCH_2SiHCl_2$	—	97.0–97.5 (773)					555
$C_2H_5SiHCl_2$	—	74; 74.2	1648.2	7.6271	—	d_4^{25} 1.095	172, 512
$C_4H_9SiHCl_2$	—	127–128					512
$C_5H_{11}SiHCl_2$	—	151–152					512
$C_6H_{13}SiHCl_2$	—	172–175					512
$\overline{CH_2(CH_2)_4C}HSiHCl_2$	—	270 (760)				d_4^{20} 1.979; n_{white}^{20} 1.604$_7$	15
$C_6H_5SiHCl_2$	—	180–184 (760) 100–102 (45)				d_4^{25} 1.225	172, 343, 512 375
$Cl_3SiCH_2SiHCl_2$	—	161 (755)				d^{20} 1.458	53
$(CH_3)_3SiCH_2SiHCl_2$	—	52.0–53.5 (24)					117
$C_2H_5SiHBr_2$	—	119.5 (760)				d^{20} 1.728	327

$C_6H_5SiHBr_2$	−51.0	146.7–147.0 (58)				d^{25} 1.7293; n_D^{25} 1.5778	368
CH_3SiHI_2	−80.5	159 (760)	2045	7.625	—	Q 9358; T 21.7	169
n-$C_4H_9SiHI_2$	—	219				d_4^{20} 2.007; n_{white}^{20} 1.579_1	13
n-$C_7H_{15}SiHI_2$	—	269 (760)				d_4^{20} 1.730; n_{white}^{20} 1.550_3	17
R_2SiHX							
$(CH_3)(C_2H_5)SiHF$	—	36 (728)				d_4^{20} 0.8161; n_D^{20} 1.3630	573
$(CH_3)(C_3H_7)SiHF$	—	57 (730)				d_4^{20} 0.8036; n_D^{20} 1.3617	573
$(CH_3)(i\text{-}C_3H_7)SiHF$	—	47 (730)				d_4^{20} 0.8025; n_0^{20} 1.3621	
$(CH_3)(C_4H_9)SiHF$	—	83 (727)				d_4^{20} 0.8039; n_D^{20} 1.3770	573
$(CH_3)(i\text{-}C_4H_9)SiHF$	—	78 (729)				d_4^{20} 0.8052; n_D^{20} 1.3749	573
$(CH_3)_2SiHCl$	−103.0	34.9 (760)	1458.0	7.6161	—	d^0 0.910; Q 6672; T 21.66; vapour pressure at 0° = 189.6 torr	315
$(C_2H_5)_2SiHCl$	—	99; 99.2	1689.5	7.4195	—		14, 172
$(CH_3)(i\text{-}C_3H_7)SiHCl$	—	89–92 (760)				d_4^{20} 0.8732; n_D^{20} 1.4104	375
$(i\text{-}C_3H_7)_2SiHCl$	—	54–55 (45)				d_4^{20} 0.8720; n_D^{20} 1.427	375
$[\overline{CH_2(CH_2)_3CH}]_2SiHCl$	—	93–94 (2)				d_4^{20} 1.0073; n_D^{20} 1.4919	375
$[\overline{CH_2(CH_2)_4CH_2}]SiHCl$	—	105 (1.5)				d_4^{20} 1.0070; n_D^{20} 1.4988	375
$(CH_3)(C_6H_5)SiHCl$	—	113 (100)				d_4^{20} 1.043; n_D^{20} 1.5157	49
$(C_2H_5)(C_6H_5)SiHCl$	—	111 (50)				n_D^{20} 1.5103	49
$(C_6H_5CH_2)_2SiHCl$	—	102–110 (0.4); 146–148 (∼1)				n_D^{20} 1.5721	49, 173
$(C_6H_5CH_2)(C_2H_5)SiHCl$	—	114–116.5 (30)				d_4^{20} 1.019; n_D^{20} 1.5130	49
$(p\text{-}ClC_6H_4)(Me_2CH)SiHCl$	—	134.0–134.75 (30)				d_4^{20} 1.115; n_D^{20} 1.5239	49
$(C_6H_5)_2SiHCl$	—	143 (10); 140 (7)				d_4^{20} 1.118; n_D^{20} 1.581	49, 172, 375
$(p\text{-}CH_3C_6H_4)_2SiHCl$	—	112–117 (0.5)				d_4^{20} 1.096; n_D^{20} 1.5701	49
$(\overline{CH_2(CH_2)_4CH})_2SiHCl$	—	100–105 (2)				d_4^{20} 1.0094; n_D^{20} 1.4942	647
$[(CH_3)_3SiCH_2]_2SiHCl$	—	79–80 (5)					117
$(C_2H_5)_2SiHBr$	—	121.3 (760)				d_4^{20} 1.193; n_{white}^{20} 1.449	14
$(C_2H_5)_2SiHI$	—	149.2 (760)				d_4^{20} 1.458; n_{white}^{20} 1.500_2	14

References p. 621

TABLE 7.39

REACTION OF SILYL HALIDES WITH PSEUDOHALIDES (CONVERSION OF SILYL HALIDES INTO SILYL PSEUDOHALIDES)

Halogenosilane	*Pseudohalide*	*Reaction product*	*Yield (mole-%)*	*Ref.*
SiH_3I	AgNC	SiH_3NC	90	167, 350
SiH_3I	Hg(CN), >20°	SiH_3NC	–	350
SiH_3I	AgNCS, 20°	SiH_3NCS	66	153, 350
SiH_3I	AgCNO, 20°	no SiH_3CNO! $Si(NCO)_4$, HOCN, H_2, Ag, AgI	–	350
SiH_3I	KCNO, 20°	–	–	350
SiH_3NCS	AgCNO, 20°	vigorous reaction: $Si(NCO)_4$, HOCN	–	350
$SiH_3I + N_2$	AgCNO	SiH_3NCO	25	152
$Si(CH_3)_3Si(CH_3)_2Cl$	AgCNO	$Si(CH_3)_3Si(CH_3)_2NCO$	48	645
$(CH_3)_2SiHI$	AgCN	$(CH_3)_2SiH{\cdot}NC$	77	327
CH_3SiH_2I	AgCNS	CH_3SiH_2NCS	–	327
$SiH_3I + N_2$	AgSeCN	$SiH_3 \cdot NCSe$	70	150
SiH_3I	$Ag_2N \cdot CN$, ≥20°	$(SiH_3)_2NCN$	25	151
SiH_3I	PbNCN	$(SiH_3)_2NCN$	10–20	151
CH_3SiH_2I	AgCN	$CH_3SiH_2(CN)$	65	169
$(CH_3)_3SiI$	AgCN	$(CH_3)_3SiCN$	–	367
$(CH_3)_3SiBr$	AgCN	$(CH_3)_3SiCN$	–	138, 367
$(CH_3)_3SiCl$	AgCN	–	–	138
$(CH_3)_3SiCl$	AgCN	$(CH_3)_3SiNC$	14	367
$(CH_3)_3SiCl$	AgCN, 125°, pressure	$(CH_3)_3SiCN$	–	66, 177, 260, 674
$(CH_3)_3SiCl$	HCN; Li	$(CH_3)_3SiCN$	–	177
$(CH_3)_3SiCl$	NaN_3 ($ZnCl_2$–KCl)	$(CH_3)_3SiN_3$	–	626, 628
$(CH_3)_2SiCl_2$	NaN_3 ($ZnCl_2$–KCl)	$(CH_3)_2Si(N_3)_2$	60	626, 628
CH_3SiCl_3	NaN_3 ($ZnCl_2$–KCl)	$CH_3Si(N_3)_3$; $CH_3Si(N_3)_2Cl$; $CH_3SiN_3Cl_2$	–	627, 628
$(C_6H_5)_3SiCl$	NaN_3 ($AlCl_3$–THF)	$(C_6H_5)_3SiN_3$	90	670
$(C_6H_5)(CH_3)_2SiCl$	NaN_3 ($AlCl_3$–THF)	$(C_6H_5)(CH_3)_2SiN_3$	–	670
$(CH_3)_3SiCl_3$	NaN_3 ($AlCl_3$–THF)	$(CH_3)_3SiN_3$	–	670
$(CH_3)_3SiBr$	AgCN	$(CH_3)_3SiNC$	79.5	367
$(CH_3)_3SiI$	AgCN	$(CH_3)_3SiNC$	80	367
$(CH_3)_2SiBr_2$	AgCN	$(CH_3)_2Si(NC)_2$	55	367
SiH_3I	AgCN	SiH_3CN	–	347
SiD_3I	AgCN	SiD_3CN	–	347
Si_2H_5I	AgCN	Si_2H_5CN	–	109

$LiAlH_4$) attack both the Si–Cl and the Si–O bonds (see section 4.1 and Table 7.10).

7.2.1 Reaction of the Si–X bond with water

Halogenosilanes $R_{4-m-n}SiH_mX_n$ ($m = 0$–3; $n = 1$–4; R = alkyl, aryl; X = halogen) hydrolyse in water, aqueous hydrochloric acid, and aqueous alcohol in accordance with eqn. (a)[*169, 279, 349, 484, 647 681, 682*] and in anhydrous alcohol according to eqn. (b)[*172, 603*] to give the corresponding disilyl oxide (disilyl ether, siloxane):

$$2 \gt SiX + H_2O \rightleftharpoons \gt Si_2O + 2\,HX \qquad (a)$$

$$\gt SiX + ROH \rightleftharpoons \gt SiOR + HX$$
$$ROH + HX \rightleftharpoons RX + H_2O$$
$$\gt SiOR + H_2O \rightarrow \gt SiOH + ROH \qquad (b)$$
$$2 \gt SiOH \rightarrow \gt Si_2O + H_2O$$

Table 7.40 contains examples of the solvolysis of silicon hydride–halides and references; on the mechanism of the solvolysis of Si–X, see the literature[*279, 484, 681, 682*].

The hydrolysis of $SiHCl_3$ with water-containing benzene forms polymeric amorphous $SiHO_{1.5}$ and hydrolysis with water-containing ether gives polymeric mica-like $SiHO_{1.5}$[*682*]. On hydrolysis, CH_3SiHCl_2 telomerises to more or less viscous oils depending on the amount and nature of the telogen [ROH, $(CH_3)_3$-SiCl][*426*]. If the H in $SiHCl_3$ is replaced by bulky alkyl groups (i-C_3H_7, *tert*-C_4H_9), the hydrolysis does not give polymers but low-molecular-weight silimantanes: 2, 4, 6, 8, 9, 10-hexaoxa-1, 3, 5, 7-tetraalkylsilamantane[*681*]:

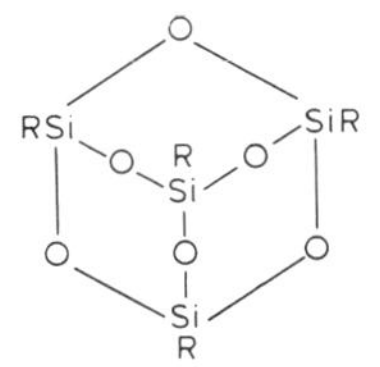

In the solvolysis of siloxene bromides, the corresponding siloxene hydroxides are formed[*264*], while methanol in pyridine gives siloxene methoxides[*264, 307*]:

$$[Si_6O_3]H_{6-n}Br_n + n\,H_2O \rightarrow [Si_6O_3]H_{6-n}(OH)_n + n\,HBr$$
$$[Si_6O_3]H_{6-n}Br_n + n\,CH_3OH + n\,C_5H_5N \rightarrow [Si_6O_3]H_{6-n}(OCH_3)_n + n\,C_5H_5NHBr$$

7.2.2 Reaction of the Si–X bond with alcohols and ethers

In the conversion of the Si–X bond into the Si–OR bond by means of alcohols, side reactions take place in accordance with Section 7.2.1. They are avoided by

References p. 621

TABLE 7.40

REACTION OF SILYL HALIDES WITH CHALCOGENIDES (CONVERSION OF THE Si–X BOND TO THE $Si–E^{VI}$ BOND)

Halogenosilane	*Reactant*	*Reaction product*	*Ref.*
SiH_3I	HgS_{red}, 20°	$(SiH_3)_2S$ (92%)	163
SiH_3I	H_2O, 0°	$(SiH_3)_2O$ (67%)	163
SiH_3I	Ag_2Se	$(SiH_3)_2Se$	163
SiH_3I	$HgSCF_3$	SiH_3SCF_3	129
SiH_3I	$HgSeCF_3$, 20°	SiH_3SeCF_3 (70%)	143
SiH_3I	$HgSeC_3F_7$, 20°	$SiH_3SeC_3F_7$ (87%)	143
SiH_3Cl	H_2O, 30°	$(SiH_3)_2O$	349
SiD_3Cl	D_2O, 30°	$(SiD_3)_2O$ (94%)	349
CH_3SiH_2I	H_2O	$(CH_3SiH_2)_2O$	169
SiH_3I	CH_3OH	$(SiH_3)_2O$	172
$SiHCl_3$	H_2O in benzene	amorphous $HSiO_{1.5}$	680
$SiHCl_3$	H_2O in ether	lepidoidal $HSiO_{1.5}$	680
$[Si_6O_3]H_{6-n}Br_n$	H_2O	$[Si_6O_3]H_{6-n}(OH)_n$	264
$[Si_6O_3]H_{6-n}Br_n$	CH_3OH + pyridine	$[Si_6O_3]H_{6-n}(OCH_3)_n$	264
SiH_3I	$CH_2(CH_2)_3O$	$I(CH_2)_4OSiH_2I$	27
Si_2H_5I	HgS_{red}, 20°	$(Si_2H_5)_2S$ (87%)	663
$(CH_2(CH_2)_4CH)HSiCl$	H_2O	$[(CH_2(CH_2)_4CH)_2SiH]O$	647

treating the halogenosilanes with alcohol-free sodium alkoxide:

$$\gt Si–X + RONa \rightarrow \gt Si–OR + NaI$$

or with trialkoxymethanes [*569*]:

$$\gt Si–X + HC(OR)_3 \rightarrow \gt Si–OR + RX + HCOOR$$

$$(R = CH_3, C_2H_5; X = F \text{ or better Cl})$$

The reaction with ethylene oxide gives alkoxysilane [*525*] (the ionic equations certainly do not give the true reaction mechanism):

$$R_{4-n}SiCl_n \rightleftharpoons R_{4-n}SiCl_{n-1}^{\oplus} + Cl^{\ominus}$$

$$\underset{\diagdown O \diagup}{CH_2\text{-}CH_2} + Cl^{\ominus} \rightarrow ClCH_2CH_2O^{\ominus}$$

$$R_{4-n}SiCl_{n-1}^{\oplus} + ClCH_2CH_2O^{\ominus} \rightarrow R_{4-n}SiCl_{n-1}(OCH_2CH_2Cl)$$

The shape of the vapour pressure curves of mixtures of SiH_3I and $(CH_3)_2O$, $(CH_3)_2S$, $CH{=}CHCH{=}CHO$, $CH{=}CHCH{=}CHS$, $CH_2(CH_2)_3O$(THF), and $CH_2(CH_2)_3S$ shows that only tetrahydrofuran forms an adduct (at −46°: $SiH_3I \cdot$ THF); no other ethers or sulphides react [*26, 27*]. $SiH_3I \cdot$ THF decomposes above −30° with the liberation of half a molecular proportion of monosilane:

$$SiH_3I \rightarrow \tfrac{1}{2}SiH_2I_2 + \tfrac{1}{2}SiH_4$$

Since SiH_2I_2 should dismute further with the formation of SiH_4, it must be assumed that tetrahydrofuran has a stabilising effect, possibly with the opening of the ring ($I(CH_2)_4OSiH_2I$).

7.2.3 *Reaction of the Si–X bond with sulphides and selenides*

The reaction of silyl halides with metal sulphides or selenides leads to the replacement of halogen by S or Se:

$$2 \gt Si{-}X + M{-}S{-}M \rightarrow \gt Si{-}S{-}Si \lt + 2MX$$

$$2 \gt Si{-}X + M{-}Se{-}M \rightarrow \gt Si{-}Se{-}Si \lt + 2MX$$

Table 7.40 contains references and examples of the reaction.

7.3 *Reaction of the Si–X bond with N-bases, phosphorus, arsenic, phosphanes, and arsanes*

The reaction of silyl halides with N-bases takes place via compounds with quaternary nitrogen. If the bases contain no N–H bond (R_3N, R_2NNR_2), they are relatively stable. If they do contain N–H bonds, they very readily split off hydrogen halide, but partial replacement of H by R on the silicon has a stabilising effect [*375, 647*]:

$$SiH_3X + NR_3 \rightarrow [(SiH_3)NR_3]X$$
$$SiH_3X + R_2NH \rightarrow [(SiH_3)NR_2 \cdot HX] \rightarrow SiH_3(NR_2) + HX(\xrightarrow{NR_2H} R_2NH \cdot HX)$$

$$2\,[\overline{CH_2(CH_2)_4CH}](CH_3)SiHCl \xrightarrow[-2NH_4Cl]{+4NH_3} 2\,[\overline{CH_2(CH_2)_4CH}](CH_3)SiHNH_2 \text{ (non-isolatable)}$$

$$\xrightarrow{-NH_3} \{[\overline{CH_2(CH_2)_4CH}](CH_3)SiH\}_2NH \text{ (isolatable)}$$

Examples of the reaction with the yields given in brackets [*2, 27–29, 32, 32a–c, 32e, 35, 55, 81, 142, 147, 168, 352, 559a, 624, 647, 648, 663*]:

$$3SiH_3Cl + 4NH_3 \rightarrow N(SiH_3)_3 + 3NH_3 \cdot HCl \quad (80\%)$$
$$2SiH_3Cl + 3NH_2Me \rightarrow N(SiH_3)_2Me + 2NH_2Me \cdot HCl$$
$$SiH_3Cl + NHMe_2 \xrightarrow{-NH_2Me_2Cl} N(SiH_3)Me_2 \xrightarrow{+HCl} N(SiH_3)Me_2 \cdot HCl \text{ (unstable)}$$
$$SiH_3Cl + NMe_3 \rightarrow [NMe_3(SiH_3)]Cl$$
$$SiH_3I + NMe_3 \rightarrow [NMe_3(SiH_3)]I \xrightarrow[\Delta]{+NMe_3} (Me_3N)_2SiH_3I$$
$$SiH_3I + NEt_3 \rightarrow [NEt_3(SiH_3)]I$$
$$2MeSiH_2Cl + 3NH_3 \xrightarrow{-2NH_4Cl} NH(SiH_2Me)_2 \text{ (very unstable)}$$
$$2[\overline{CH_2(CH_2)_4CH}]SiCl + 3NH_3 \xrightarrow{-2NH_4Cl} NH[SiH(CH(CH_2)_4CH_2)_2]_2 \text{ (stable)}$$
$$3MeSiH_2Cl + 4NH_3 \rightarrow N(SiH_2Me)_3 + 3NH_4Cl \quad (65–70\%)$$

References p. 621

$2MeSiH_2Cl + 3NH_2Me \rightarrow N(SiH_2Me)_2Me + 2NH_2Me \cdot HCl$ (65%)

$3\,RHSiCl_2 + 10\,MeNH_2 \rightarrow [(HMeN)RHSi]_2NMe + \frac{1}{n}[RHSiNMe]_n + 6\,MeNH_2 \cdot HCl$

$MeSiH_2I + 2NHMe_2 \rightarrow N(SiH_2Me)Me_2 + NHMe_2 \cdot HI$ (90%)

$SiH_2I_2 + 2NH_2R \xrightarrow{-30°,\ pentane} \frac{1}{x}(SiH_2NR)_x + 2NH_2R \cdot HI$

$SiH_2I_2 + 4(CH_3)_2NH \rightarrow SiH_2[N(CH_3)_2]_2 + 2(CH_3)_2N \cdot HI$

$SiHCl_3 + 6(CH_3)_2NH \rightarrow SiH[N(CH_3)_2]_3 + 3(CH_3)_2N \cdot HCl$

$3Si_2H_5I + 4NH_3 \rightarrow N(Si_2H_5)_3 + 3NH_3 \cdot HI$ (58–64%)

$Si_2H_5Br + 2NHMe_2 \rightarrow N(Si_2H_5)Me_2 + NHMe_2 \cdot HBr$

$Si_2H_5I + NMe_3 \rightarrow [NMe_3(Si_2H_5)]I$

$SiH_3I + 2H\overline{N(CH_2)_nC}H_2 \rightarrow SiH_3 \cdot \overline{N(CH_2)_nC}H_2 + [H_2\overline{N(CH_2)_nC}H_2]I$ ($n = 3$ or 4)

$4SiH_3I + 5H_2NNH_2 \rightarrow (SiH_3)_2NN(SiH_3)_2 + 4N_2H_4 \cdot HI$ (40%)

$SiH_3I + Me_2NNMe_2 \xrightarrow{gas\ phase} SiH_3I \cdot N_2Me_4$

$SiH_3I + Me_2NNMe_2 \xrightarrow[\Delta]{ether\ or\ hexane} SiH_3I \cdot N_2Me_4, SiH_3I \cdot N_2Me_4 \cdot SiH_3I$

The SiH_3I–pyridine adducts that form at low temperatures (1:2, increasing stability in the sequence 2-n-hexylpyridine < pyridine < trimethylpyridine) are good silylating agents, since they decompose at higher temperatures. At 50°, $SiH_3I \cdot Py$ forms a polysilane containing pyridine ($SiH_2 \cdot 0.6Py_x$) that has not been fully elucidated [*27*]. Me_4NCl catalyses the redistribution of H and Cl on the silicon; for example [*672a*]:

$$Me_2SiHCl + MeSiCl_3 \xrightarrow{Me_4NCl} Me_2SiCl_2 + MeSiHCl_2$$

Silyl iodide reacts with phosphorus and arsenic with the formation of silylated phosphorus and arsenic iodides [*32*]:

$$SiH_3I + P \xrightarrow[\Delta]{20-100°} SiH_3PI_2$$

$$SiH_3I + As \xrightarrow[\Delta]{} SiH_3AsI_2$$

With trialkylphosphanes, PR_3, and trialkylarsanes, AsR_3, unstable quaternary compounds are formed [*32*]:

$$SiH_3I + PMe_3 \rightarrow [PMe_3(SiH_3)]I$$

$$SiH_3I + PEt_3 \rightarrow [PEt_3(SiH_3)]I$$

$$SiH_3I + AsMe_3 \rightarrow [AsMe_3(SiH_3)]I$$

The phosphorus derivatives split off SiH_4 at 20° with the formation of non-volatile products. The arsenic derivatives dissociate rapidly and reversibly.

7.4 *Reaction of the Si–X bond with diazomethane*

The Si–Cl bonds in $SiCl_4$[*555, 563, 697*], $SiHCl_3$[*13, 555*], and $C_6H_5SiCl_3$[*321*], but not in $(C_6H_5)_2SiCl_2$ and $(C_6H_5)_3SiCl$[*321*] react with diazomethane to give the corresponding chloromethylsilanes. The reaction takes place (a) by a radical mechanism (carbene addition) when catalysed by UV light[*697, 700*] or (b) by an ionic mechanism when catalysed by copper (rapid reaction in ether, slow reaction in petroleum ether)[*321, 563*]:

$$CH_2N_2 \xrightarrow[-N_2]{\text{UV light}} \cdot CH_2 \cdot \xrightarrow{\geq Si-Cl} \geq SiCH_2Cl \qquad (a)$$

$$\overset{\delta-}{Cl}-\overset{\delta+}{Si}\leq + |\overset{\delta-}{C}H_2-\overset{\delta+}{N}\equiv N| \rightarrow Cl-\overset{\delta-}{\underset{/|\backslash}{Si}} \leftarrow \cdot\cdot |CH_2-N\equiv\overset{\delta+}{N}| \xrightarrow{-N_2}$$

$$Cl-\underset{/|\backslash}{\overset{|}{Si}}-CH_2 \rightarrow \geq SiCH_2Cl \qquad (b)$$

7.5 *Reaction of the Si–X bond with organometals*

Organometals (RLi, R_2Zn, R_2Hg, RMgX, R_3Al) replaces halogen in halogensilanes by the anionic substituent R:

$$\geq \overset{\delta+}{Si}-\overset{\delta-}{X} + \overset{\delta-}{R}\overset{\delta+}{M} \rightarrow \geq Si-R + MX$$

Examples [*55, 226, 348, 410*]:

$$SiHCl_3 + 3LiC_6H_4Me \rightarrow SiH(C_6H_4Me)_3$$

$$2Et_3SiI + Hg(CH_2COOMe)_2 \rightarrow 2Et_3Si(CH_2COOMe) + HgI_2$$

$$SiH_3Br + Mg(C{\equiv}CH)Br \rightarrow SiH_3(C{\equiv}CH) + MgBr_2$$

$$Fe(C_5H_4Li)_2 \xrightarrow[-LiCl]{+R_2SiHCl} Fe(C_5H_4SiR_2H)(C_5H_4Li) \xrightarrow[-LiCl]{+R_2SiHCl} Fe(C_5H_4SiR_2H)_2$$

$$3\,Et_3GeLi + SiHCl_3 \longrightarrow (Et_3Ge)_3SiH + 3\,LiCl$$

Further examples of these reactions are given in Tables 7.13 and 7.14. The Grignardation of the Si–Cl bond takes place through a polar four-centre mechanism [*489*], and alkylation with trialkylaluminiums takes place through an electrophilic attack of aluminium on the halogen atom on the silicon[*340, 341*].

References p. 621

7.6 *Reaction of the Si–X bond with metals of the 1st and 2nd main groups*

The reaction of trialkyl- or triarylchlorosilanes with Li, Na, K, Rb, Cs, or Mg in tetrahydrofuran, 2-methyltetrahydrofuran, and tetrahydropyran forms, via silylmetals or via non-isolatable Grignard compounds, depending on the reactant, silylmetals, disilane, or products of ether cleavage [*219, 225, 605, 630*]. Thus, a silylmetal is produced with Li, K, Rb, and Cs via the disilane, which can be isolated in 63% yield when the reaction is broken off prematurely:

$$Ph_3SiCl \xrightarrow[-E^ICl]{+2E^I} Ph_3SiE^I \xrightarrow[-E^ICl]{+Ph_3SiCl} Ph_3SiSiPh_3 \xrightarrow{+2E^I} 2Ph_3SiE^I$$

On the other hand, the Si–Si bond is not cleaved on treatment with sodium; the same applies if magnesium is treated with phenyl-substituted disilanes (synthesis from chlorosilane). In the reaction of alkyl-substituted silanes with magnesium, the ether is split:

$$Ph_2SiHCl \xrightarrow[-NaCl]{+2Na} Ph_2SiHNa \xrightarrow[-NaCl]{+Ph_2SiHCl} Ph_2HSiSiHPh_2$$

$$PhRSiHCl \xrightarrow[(THF)]{+Mg} [PhRHSiMgCl] \xrightarrow[-MgCl_2]{+PhRSiHCl} PhRHSiSiHRPh$$

$$Me_3SiCl \xrightarrow[(THF)]{+Mg} [Me_3SiMgCl] \xrightarrow[+Me_3SiCl,\ -MgCl_2]{+\overline{CH_2CH_2CH_2CH_2O}} MeSi(CH_2)_4OSiMe_3$$

With potassium, liquid potassium–sodium alloy, or liquid sodium amalgam [*108*], silyl halides (SiH_3Cl, SiH_3Br) form monosilane and polysilane(1–2). The intermediate occurrence of silylmetals and disilane according to the reactions mentioned above must be assumed:

$$SiH_3Cl \xrightarrow[-KCl]{+2K} [SiH_3K] \xrightarrow[-KCl]{+SiH_3Cl} [H_3SiSiH_3] \xrightarrow{+nK} SiH_4 + \frac{1}{x}(SiH_{2-n})_x + n\,KH$$

$$(n = 0\text{–}1)$$

On the reaction of halogenated hydrides with metallated hydrides, see section 10.2. On the reaction of sodium amalgam with SiH_3Cl and Me_2SiHBr, see [*683a*].

8. Pseudohalogenosilanes and carbonylsilanes

8.1 *Synthesis of pseudohalogenosilanes and carbonylsilanes*

Pseudohalogenosilanes can be prepared by the reaction of silicon hydrides with hydrogen pseudohalides: Si–H → Si-pseudohalogen, or by the reaction of halogenosilanes with pseudohalide ions on the right of them in the conversion series: Si–X → Si-pseudohalogen (see sections 6.1.5 and 7.1). Silylamines are used as starting materials for special synthesis; reactions occur only when the acid to be treated is not too weak. Examples [*65, 138, 148, 149, 156, 177, 626, 628*]:

$$(SiH_3)_3N + 3HN_3 \xrightarrow{(n\text{-}C_4H_9)_2O} 3SiH_3N_3 + NH_3$$

TABLE 7.41

PHYSICAL PROPERTIES OF THE PSEUDOHALOGENOSILANES AND CARBONYLSILANES

Pseudohalogenosilane	*M.p.* (°C)	*B.p.* [°C (torr)]	*Vapour pressure equation* log p [torr] = $-A/T+B$: *A*	*B*	*valid between* (°C)	*Enthalpy of evaporation, Q (cal/mole), Trouton's constant, T (cal mole⁻¹deg⁻¹), density, d (g/ml), refractive index, n*	*Ref.*
SiH_3CN	32.4	—	2550	10.951	−20/31	*Q* 11670, for solid SiH_3CN	350
		49.6 ± 0.3 (760)	1567	7.735	26/46	*Q* 7169, for liquid SiH_3CN	
CH_3SiH_2CN	−23.1	—	—	—	—	Vapour pressure at 0° 26.5 torr; polymer at temp. > 40°	169
$(CH_3)_2SiHNC$	−60.9	—	—	—	—	< 20° fairly stable	327
SiH_3NCO	−88.6 ± 0.5	18.1 ± 0.2 (760)	1428	7.783	−10/18	*Q* 6540 ± 50; *T* 22.5	152
$\overline{CH_2(CH_2)_4CH}SiH_2NCO$	—	163.7 (760)	—	—	—	d_4^{20} 0.990; n_{white}^{20} 1.466$_8$	15
SiH_3NCS	−51.8 ± 0.2	84.0 ± 4 (760)	1950	8.340	−33/10	*Q* 8923; *T* 25.0; d^{20} 1.05; 74% decomp./12 days	350
CH_3SiH_2NCS	—	—	—	—	—	Vapour pressure at 0° 8 torr; slow decomp. at 0°	327
$\overline{CH_2(CH_2)_4CH}SiH_2NCS$	—	231 (760)	—	—	—	d_4^{20} 1.018; n_{white}^{20} 1.533$_6$	15
SiH_3NCSe	−15.1 ± 0.5	111 ± (760)	2345	8.968	0/60	*Q* 10800; *T* 28; polymer at 20°	150
SiH_3N_3	−81.8 ± 0.5	~ 28 (760) 25.8 ± 1 (760)	1459	7.745	0/26	Vapour pressure at 0° 253 torr; *Q* 6680; *T* 22.3	148
SiH_3OCHO	−60.4 ± 5	31.4 (760)	1590	8.101	0/20	*Q* 7280; *T* 23.9	156
$SiH_3OC(CH_3)O$	−62.4 ± 5	56.2	1685	7.997	0/20	*Q* 7710; *T* 23.4	156
$SiH_3OC(CF_3)O$	−114.8 ± 5	30.2	1604	8.168	0/20	*Q* 7340; *T* 24.1	156
$SiH_3Co(CO)_4$	−53·5	112 (760)	1754	7.432	22/90	Decomposition above 85°	30, 31
$SiH_3Mn(CO)_4$	25.5	134 (760)	—	—	—	Decomposition above 100°	30
$(SiH_3)_2Fe(CO)_4$	52	145 (760)	—	—	—		31a
$(SiH_3)Fe(CO)_2(C_5H_5)$	no	—	—	—	—	Subl. at −40°, high vacuum	11b

References p. 621

$$Et(SiH_3)_2N + 2HN_3 \xrightarrow{(n\text{-}C_4H_9)_2O} 2SiH_3N_3 + NH_2Et$$

$$(SiH_3)_3N + 3HNCO \rightarrow 3SiH_3NCO + NH_3 \quad \text{(yield 95\% [156])}$$

$$(SiH_3)_3N + H_2S \rightarrow \text{no reaction [156]}$$

$$(SiH_3)_3N + HCN \rightarrow \text{no reaction [156]}$$

$$2\,(SiH_3)_3N + 3\,H_2S + 3\,Hg(NCS)_2 \rightarrow 6\,SiH_3NCS + 3\,HgS + 2\,NH_3 \quad \text{(yield 58\% [156])}$$

$$(SiH_3)_3N + 3CF_3COOH \rightarrow 3SiH_3OCOCF_3 + NH_3 \quad \text{(yield 97\% [156])}$$

$$(SiH_3)_3N + 3HCOOH \rightarrow 3SiH_3OCHO + NH_3 \quad \text{(yield 52\%, together with } SiH_4 \text{ [156])}$$

$$(SiH_3)_3N + 3CH_3COOH \rightarrow 3SiH_3OCOCH_3 + NH_3 \quad \text{(yield 20\%, together with } SiH_4 \text{ [156])}$$

$NaCo(CO)_4$ and $NaMn(CO)_4$ react rapidly with SiH_3I at low temperatures to form $SiH_3Co(CO)_4$ or $SiH_3Mn(CO)_4$ [*25c–e, 30, 31, 31a*]:

$$NaCo(CO)_4 + SiH_3I \xrightarrow[-40°]{Et_2O \text{ or } Me_2O} SiH_3Co(CO)_4 + NaI$$

$$NaMn(CO)_4 + SiH_3I \xrightarrow[-30°]{Et_2O} SiH_3Mn(CO)_4 + NaI$$

Table 7.41 contains the physical properties of the products.

8.2 *Reactions of the pseudohalogenosilanes and carbonylsilanes*

The reactions of the pseudohalogenosilanes are governed by the free d-orbitals of the silicon. They permit the addition of donors or the formation of π-bonds with neighbouring elements having occupied non-bonding orbitals (*e.g.*, N or O). The reactions therefore take place differently from those of the corresponding alkyl pseudohalides. (For a summary of the reactions of peralkylated silyl cyanides R_3SiCN, see the literature [*650*]).

With diborane and boron halides, methyl cyanide forms fairly stable 1 : 1 adducts. When the borane adduct is heated, hydrogen migrates from boron to the cyanide carbon with the formation of polymers, while the boron fluoride adduct, for example, decomposes into the initial compounds [*cit. 176, cit. 177*]:

$$MeCN + \tfrac{1}{2}(BH_3)_2 \rightarrow MeCN \cdot BH_3 \xrightarrow{heat} \frac{1}{x}(MeCH_2NBH)_x$$

$$MeCN + BX_3 \rightarrow MeCN \cdot BX_3 \xrightarrow{heat} MeCN + BX_3 \quad (X = F, Cl)$$

The analogous 1 : 1 adducts of the *silicon* are unstable; they decompose even at -126 to $-78°$ [*152, 176, 177, 350*]:

$$2R_3SiCN + (BH_3)_2 \xrightarrow{-126/-78^\circ} \left[\begin{array}{ccc} R_3SiCN \cdots \rightarrow & BH_2 \\ \uparrow \vdots & | \\ H & H \\ | & \vdots \downarrow \\ H_2B \leftarrow \cdots & NCSiR_3 \end{array}\right] \begin{array}{l} \xrightarrow{\text{warm}} 2\,R_3SiH + 2\,BH_2CN \\ \\ \xrightarrow{+NMe_3} 2\,R_3SiCN + 2\,BH_3 \cdot NMe_3 \end{array}$$

$(R = H, CH_3)$

$$2H_3SiCN + BX_3 \xrightarrow{-126/-78^\circ} \left[\begin{array}{cc} H_3SiCN & BX_2 \\ \uparrow \vdots & | \\ X & X \\ | & \vdots \downarrow \\ X_2BNC & SiH_3 \end{array}\right] \xrightarrow{\text{warm}} 2\,H_3SiX + 2\,BX_2NC$$

$(X = F, Cl, Br)$

$SiH_3CN \cdot BF_3$ is extraordinarily unstable, while $CH_3CN \cdot BF_3$ and $GeH_3CN \cdot BF_3$ are stable (see Section 7.3 of Chapter 8).

$$SiH_3CN + H_2O \rightarrow \left[\begin{array}{cc} H_3Si \cdots & CN \\ \vdots & \vdots \\ HO \cdots & H \end{array}\right] \xrightarrow[-2HCN]{+SiH_3CN} (SiH_3)_2O$$

$$2xSiH_3CN \xrightarrow{20^\circ,\ \text{slow}} x\,SiH_4 + (SiHCN)_x + x\,HCN$$

$$SiH_3NCS + H_2O \rightarrow \left[\begin{array}{cc} H_3Si \cdots & NCS \\ \vdots & \vdots \\ HO \cdots & H \end{array}\right] \xrightarrow[-2HCNS]{+SiH_3NCS} (SiH_3)_2O$$

$$SiH_3NCO + (BH_3)_2 \xrightarrow{20^\circ} \text{polymers}$$

$$SiH_3NCO + BF_3 \xrightarrow{-78^\circ} \left[\begin{array}{cc} H_3Si \cdots & NCO \\ \vdots & \vdots \\ F \cdots & BF_2 \end{array}\right] \xrightarrow{-SiH_3F} BF_2(NCO) \xrightarrow{20^\circ} \tfrac{2}{3}BF_3 + \tfrac{1}{3}B(NCO)_3$$

$$SiH_3NCO + HOCN \xrightarrow{20^\circ} \text{polymers}$$

$$SiH_3NCO + AgNCO \xrightarrow{20^\circ} \text{rapid decomposition}$$

$SiH_3NCO + PH_3 \rightarrow$ no reaction
$SiH_3NCO + PEt_3 \rightarrow$ decomposition

References p. 621

$$SiH_3NCO + NH_3 \xrightarrow{-78°} SiH_4 + \text{non-volatile products}$$

$SiH_3NCO + H_2O \rightarrow CO_2 + $ non-volatile products

$SiH_3NCO + MeOH \rightarrow$ complex reaction because of attack on the Si–H and Si–N bonds

$$SiH_3NCSe \xrightarrow[\text{BF}_3\text{-catalyst}]{\text{warming, UV light}} SiH_4, SiH_3CN, (SiH_3)_2Se$$

Cobaltcarbonylsilane and manganesecarbonylsilane react in a similar manner to the pseudohalogenosilanes [*25c–e, 30, 31*]:

$$SiH_3Co(CO)_4 \xrightarrow[\text{hours}]{20°, \Delta} SiH_4, (SiH_2)_x, [Co(CO)_4]_2$$

$SiH_3Co(CO)_4 \xrightarrow[15\ \text{min}]{120°, \Delta} SiH_3Co(CO)_4$ (30%), H_2, CO, traces of SiH_4, $HCo(CO)_4$, dark brown residue ~ $[SiHCo(CO)_2]_x$ containing no Si–Si bond

$SiH_3Co(CO)_4 + HCl \rightarrow SiH_3Cl + HCo(CO)_4$

$2SiH_3Co(CO)_4 + H_2O \rightarrow (SiH_3)_2O + 2HCo(CO)_4$

$SiH_3Co(CO)_4 + 5\,Br_2 \rightarrow CoBr_2 + SiBr_4 + 3\,CO + COBr + 3\,HBr$

$2SiH_3Co(CO)_4 + HgI_2 \rightleftharpoons 2SiH_3I + Hg[Co(CO)_4]_2$

$SiH_3Co(CO)_4 + PPh_3 \rightarrow SiH_3Co(CO)_3(PPh_3) + CO$

$SiH_3Mn(CO)_4 + HCl \rightarrow$ no reaction

$SiH_3Mn(CO)_4 + HgX_2 \xrightarrow{20°}$ no reaction

9. Silicon hydrides with E^{VI} and E^{V} bonds

9.1 Synthesis of the silyl compounds of O, S, Se, N, P, As, and Sb

The following synthetic routes lead to silyl derivatives of the elements of the 6th and 5th main groups.

(1) Nucleophilic displacement of negative silyl substituents as follows:

$$H_3\overset{\delta+}{Si}\text{–}\overset{\delta-}{X} + Z^{\ominus} \rightarrow H_3\overset{\delta+}{Si}\text{–}\overset{\delta-}{Z} + X^{\ominus}$$

where $Z = OH^{\ominus}(H_2O, NaOH)$, $OR^{\ominus}$ (NaOR), $\frac{1}{2}S^{2\ominus}[HgS, CF_3S^{\ominus}(HgSCF_3)]$, $\frac{1}{2}Se^{2\ominus}(Ag_2Se)$, $CF_3Se^{\ominus}(HgSeCF_3)$ (see sections 7.1 and 7.2), $NH_2^{\ominus}$ (NH_3), $PH_2^{\ominus}$ (KPH_2), $AsH_2^{\ominus}$ ($KAsH_2$), $SbH_2^{\ominus}$ ($KSbH_2$) (see sections 7.3 and 10.2.2). About the reaction of $(CH_3)_3SiPH_2$ with $PH_2^{\ominus}$ see [*338*].

(2) Radical [*131, 197, 199*] and polar syntheses (see section 5.8). Replacement

of H in the silane with the formation of H_2 (not generally applicable):

$$SiH_4 + PH_3 \xrightarrow{\text{energy}} SiH_3PH_2 + H_2$$
$$PhSiH_3 + MeOH \xrightarrow{(Cu)} PhSiH_2(OMe) + H_2$$

(3) Coupling [*9*, *68*, *198a*, *429a*]:

$$SiH_3Br + Li_3Sb \rightarrow Sb(SiH_3)_3 + 3LiBr$$
$$SiH_3Br + LiPEt_2 \rightarrow SiH_3PEt_2 + LiBr$$

$$4SiH_3X + LiAl(PH_2)_4 \rightarrow 4SiH_3PH_2 + LiAlX_4$$
$$4Si_2H_5X + LiAl(PH_2)_4 \rightarrow 4Si_2H_5PH_2 + LiAlX_4$$

(4) Cleavage of the silicon–phenyl bond [*27*] (analogy with cleavage with HX, see section 6.1.3):

$$SiH_3Ph + HX \rightarrow SiH_3X + PhH$$
$$SiH_3(C_6H_4Cl) + NH_2Me \rightarrow SiH_3NHMe + C_6H_5Cl$$
$$SiH_3(C_6H_4Cl) + NH_2Me \cdot HI \xrightarrow[\Delta]{200^\circ} (SiH_2NMe)_x$$

(5) Hydrolysis of $CaSi_2$. Polymeric silicon hydride with Si–O bonds can be obtained by the reaction of suitable metal silicides with acids. The acid hydrolysis (propanol, dil. HCl, 0°) of a calcium disilicide prepared from the elements forms siloxene [*263*, *264*, *306*]:

$$3\,CaSi_2 + 6HCl + 3H_2O \rightarrow [Si_6O_3]H_6 + 3CaCl_2 + 3\,H_2$$

Siloxene is a highly polymeric insoluble acid. In it, Si_6 rings are bound two-dimensionally through oxygen bridges. Each silicon atom also bears an hydrogen atom (not shown) which can be substituted in accordance with section 5.1 or Table 7.21. These hydrogen atoms project on either side of the $[Si_6O_3]$ plane alternately.

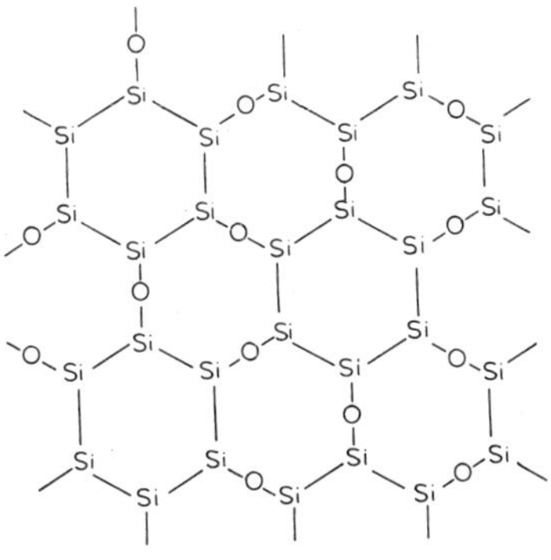

$Si_{28}O_{14}H_{28}$

References p. 621

TABLE 7.42
PHYSICAL PROPERTIES OF SILANES WITH Si–CHALCOGEN BONDS (O, S, Se)

Silane	*M.p.* (°C)	*B.p.* [°C(*torr*)]	*log* p [*torr*] = −*A*/T + *B*		*Enthalpy of evaporation, Q (cal/mole); Trouton's constant, T (cal mole⁻¹ deg⁻¹); density, d (g/ml); refractive index, n*	*Ref.*
			A	*B*		
$SiH_3(OCH_3)$	−98.5 ± 0.1	−21.1 ± 0.2(760)	1320	8.1196	*Q* 6040; *T* 23.9	603
$SiH_2(CH_3)(OCH_3)$	—	6.8(760)	1487.7	8.1956	*Q* 6807.6; *T* 24.3	661a
$SiH_2(OCH_3)_2$	−98.0	38.2(400)	1558.6	7.8897	*Q* 7133; *T* 2290; d_0 0.857; n_D^{25} 1.3515	315
	−99.8 ± 0.2	33.5 ± 0.2(760)	1645	8.243	*Q* 7528	604
$SiH_2(OC_2H_5)_2$	−123.0	81.0	1810.0	8.0006	*Q* 8282; *T* 23.40; d_0 0.864	315
$(C_6H_5)SiH_2(OCH_3)$	—	60–65(10)	—	—	d_4^{25} 1.0051; n_D^{25} 1.4845	379
$(C_6H_5)SiH_2(OC_2H_5)$	—	60–67(10)	—	—	d_4^{25} 0.9350; n_D^{25} 1.4771	379
$(C_6H_5)SiH_2(n\text{-}C_4H_9O)$	—	52–57(0.7)	—	—	d_4^{25} 0.9408; n_D^{25} 1.4820	379
$(CH_3)_2SiH(OCH_3)$	< −134	33(760)	1558	7.966	*Q* 7130; *T* 22	651a
$(CH_3)SiH(OCH_3)_2$	~ −136	61.1	1712.5	8.0139	*Q* 7836; *T* 23.44; d_0 0.896; n_D^{25} 1.3574	315
$(CH_3)SiH(OC_2H_5)_2$	vitreous	99.4	1922.5	8.0628	*Q* 8799; *T* 23.62; d_0 0.875	315
$(C_6H_5)SiH(OCH_3)_2$	—	78–84(10)	—	—	d_4^{25} 1.0298; n_D^{25} 1.5400	379
$(C_6H_5)SiH(OC_2H_5)_2$	—	80–84(10)	—	—	d_4^{25} 0.9533; n_D^{25} 1.4814	379
$(C_6H_5)SiH(n\text{-}C_3H_7O)_2$	—	90–96(1.4)	—	—	d_4^{25} 0.9545; n_D^{25} 1.4781	379
$(C_6H_5)SiH(n\text{-}C_4H_9O)_2$	—	90–96(0.7)	—	—	d_4^{25} 0.9662; n_D^{25} 1.4785	379
$SiH(OCH_3)_3$	−114.8 ± 0.2	81.1 ± 0.3(760)	1919	8.297	*Q* 8782; *T* 24.8	604
	−113.5	83.9	1929.0	8.3100	*Q* 8827; *T* 24.73; d_0 0.932	315
SiH_3OCHO	−60.4 ± 5	31.4(760)	1590	8.101	*Q* 7280; *T* 23.9	156
$SiH_3OC(CH_3)O$	−62.4 ± 5	56.2	1685	7.997	*Q* 7710; *T* 23.4	156
$SiH_3OC(CF_3)O$	−114.8 ± 5	30.2	1604	8.168	*Q* 7340; *T* 24.1	156
$SiH_3OP(CF_3)_2$	−116	40(760)	—	—		81a
$SiH[OP(CF_3)_2]_3$	—	161(760)	—	—		81a
$(CH_3)SiHCl(OC_2H_5)$	−108.0	73.6(760)	1737.8	7.9007	Vapour pressure at 0° 33.4 torr; *Q* 7951; *T* 22.93; d^0 1.016	315
$SiHCl(OC_2H_5)_2$	—	107–110(743)	—	—	d_4^{25} 0.9806	491
$SiHCl(n\text{-}C_3H_7O)_2$	—	26–28(7); 48–49(8)	—	—	d_4^{25} 0.9587; n_D^{25} 1.3978–1.4004	491
$(SiH_3)_2O$	−144	−15(760–770)	1232.2	7.6864	d^{-80} 0.881; at 20°, 4.75 atm in ampoule: 3% decomp./4 months	349, 616
$(CH_3)_3SiOSiH_3$	—	46.5	—	—		352
$(n\text{-}C_7H_{15}SiH_2)_2O$	—	282(760)	—	—	d_4^{20} 0.827; n_{white}^{20} 1.4374	17
$[(C_2H_5)_2SiH]_2O$	—	170.6(760)	—	—	d_4^{20} 0.820; n_{white}^{20} 1.4162	14
$[(CH_2(CH_2)_4CH_2SiH]_2O$	—	195–200(1.75)	—	—	d_4^{20} 0.9728; n_D^{20} 1.5052	647

$[(C_6H_5)(CH_3)SiH]_2O$	—	80(0.12)	—	—	n_D^{25} 1.519	252
$[(C_6H_5)(C_2H_5)SiH]_2O$	—	107(0.03)	—	—	n_D^{25} 1.523	252
$(C_6H_5SiHO)_3$	—	168(0.02)	—	—	n_D^{25} 1.566	252
$(Si_2H_5O)SiH_3$	—	43	—	—		352
SiH_3SH	~ −134	14.2(760)	1355	7.596	*Q* 6201; *T* 21.9; (decomp.: slow at −78°, fast at −30°	163
$(CH_3S)SiH_3$	−116.7 ± 0.4	46.8 ± 0.2(760)	1604	7.893	*Q* 7340; *T* 22.9; 6% decomp. at 20° + light in 4 days	602
$(CF_3S)SiH_3$	−127 ± 0.5	13.6 ± 0.2(760)	1339	6.5482	*Q* 6150 ± 50; *T* 21.4	602
$(SiH_3)_2S$	−70.0 ± 0.2	58.8 ± 0.7	1692	7.977	*Q* 7743; *T* 23.3; *d* 1.21211(1 − 0.00082350 t)	163
$(Si_2H_5)_2S$	—	144(760)	2487.6	8.8494	*Q* 11380; *T* 27.3; stable at −78°, stable for 1 week at 20°	663
$(CF_3Se)SiH_3$	−125.8	35(760)	1460	7.615	*Q* 6700; *T* 22; vapour stable for 36 h at 20°	143
$(C_3F_7Se)SiH_3$	−98	76(760)	1729	7.833	*Q* 7900; *T* 23; liquid stable for 72 h at 20°	143
$(SiH_3)_2Se$	−68.0 ± 0.2	85.2 ± 1(760)	1796	7.894	*Q* 8219; *T* 22.9; d^{20} 1.36	143

The physical properties of the silicon hydrides with Si–E^{VI} and Si–E^{V} bonds are given in Tables 7.42 and 7.43.

The mica-like silicon oxide-hydride $(HSiO_{1.5})_x$ also possesses a sheet structure. It is formed by the hydrolysis of $SiHCl_3$ in moist ether [*682*] (hydrogen atoms on silicon again not shown):

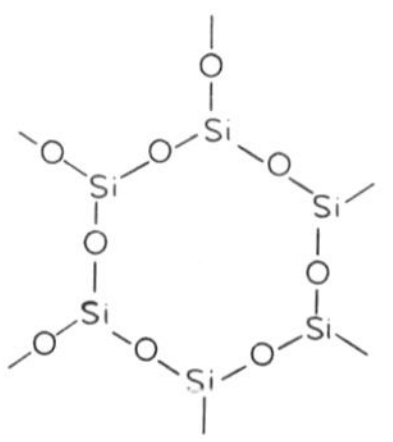

$(HSiO_{1,5})_x$

9.2 *Reactions of the Si–O, Si–S, and Si–Se bonds*

The reactions of the Si–O, Si–S, and Si–Se bonds are governed by the electrophilic attack on the more or less occupied non-bonding orbitals on the O, S, or Se atom. In distinction from the C–O–C bond, here the non-bonding pair of electrons is used more or less intensively for the $p_\pi \rightarrow d_\pi$ back-coordination of electrons to the silicon $\left(>C{-}\overline{\underline{O}}{-}C<,\ >Si{\cdots}O{\cdots}Si< \right)$. Consequently, such silicon compounds are less nucleophilic than isologous carbon compounds, as is shown by reactions of the silyl compounds $(SiH_3)_2O$, $(SiR_3)_2O$, SiR_3OH, and $(SiH_3)_2S$ with BH_3, BF_3, BBr_3, and $Al(CH_3)_3$ [*485, 603, 620, 621, 677–679*]. While $(CH_3)_2O$ and $(CH_3)_2S$ form adducts with $(BH_3)_2$, BF_3, and $Al(CH_3)_3$ that are easy to isolate, the silicon isologues either do not react or form only intermediate addition compounds [*620*].

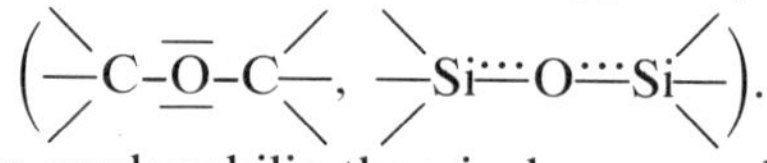

The occurrence or non-occurrence of donor-acceptor adducts described below shows the following gradations in the nucleophilicity of the silyl oxides and sulphides: $(CH_3)_2O > (SiH_3)_2O$; $(CH_3)_2S > (SiH_3)_2S$; and $(SiH_3)_2O > (SiH_3)_2S$. A comparison with peralkylated or perarylated silyl oxides is useful. The nucleophilicity of the oxygen atom in them falls in the sequence [*695*]: C–O–C > C_{alk}–O–Si > C_{ar}–O–Si ≫ Si–O–Si ≫ (O)Si–O–Si, and the electrophilicity of the silicon atom increases in the sequence: C_{ar}–O–Si < C_{alk}–O–Si ≪ Si–O–Si (about the nucleophilicity see also [*661a*]).

The subsequent reactions of the silyl–acceptor adducts very probably take place in accordance with the same mechanism:

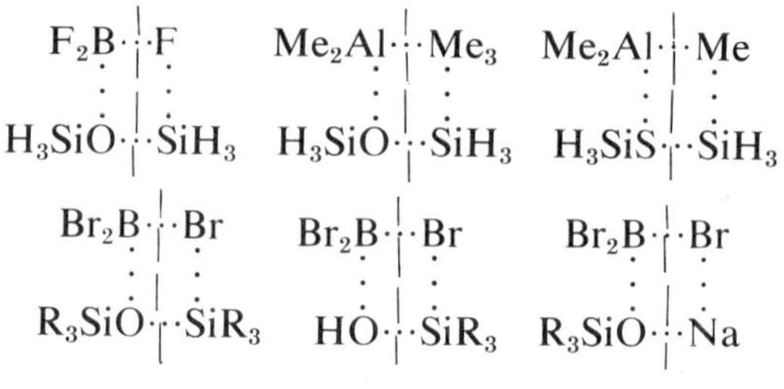

$$\begin{matrix} F_2B\cdots F & Me_2Al\cdots Me_3 & Me_2Al\cdots Me \\ \vdots \quad \vdots & \vdots \quad \vdots & \vdots \quad \vdots \\ H_3SiO\cdots SiH_3 & H_3SiO\cdots SiH_3 & H_3SiS\cdots SiH_3 \end{matrix}$$

$$\begin{matrix} Br_2B\cdots Br & Br_2B\cdots Br & Br_2B\cdots Br \\ \vdots \quad \vdots & \vdots \quad \vdots & \vdots \quad \vdots \\ R_3SiO\cdots SiR_3 & HO\cdots SiR_3 & R_3SiO\cdots Na \end{matrix}$$

TABLE 7.43

PHYSICAL PROPERTIES OF SILANES WITH Si–E^{V} BONDS (E^{V} = N, P, As, Sb)

Silane	M.p. (°C)	B.p. [°C(torr)]	log p[torr] = −A/T + B: A	B	Enthalpy of evaporation, Q (cal/mole); Trouton's constant, T (cal mole^{-1} deg^{-1}); density, d (g/ml); refractive index, n	Ref.
$(i\text{-}C_3H_7)_2SiHNH_2$	—	45–46 (40)			d_4^{20} 0.7890; n_D^{20} 1.4260	375
$[\overline{CH_2(CH_2)_3CH}]_2SiHNH_2$	—	90–94 (2.5)			d_4^{20} 0.9293; n_D^{20} 1.4888	375
$[\overline{CH_2(CH_2)_4CH}]_2SiHNH_2$	—	97 (1)			d_4^{20} 0.9324; n_D^{20} 1.4968	375
$[\overline{CH_2(CH_2)_4CH}](CH_3)SiHNH_2$	—	70–71 (25)			d_4^{20} 0.8759; n_D^{20} 1.4640	375
$(C_6H_5)_2SiHNH_2$	—	125 (0.6)			d_4^{20} 1.0711; n_D^{20} 1.6010	375
$[(i\text{-}C_3H_7)(CH_3)SiH]_2NH$	—	79 (30)			d_4^{20} 0.8048; n_D^{20} 1.4392	375
$[(i\text{-}C_3H_7)_2SiH]_2NH$	—	121–122 (25)			d_4^{20} 0.8303; n_D^{20} 1.4450	375
$[\overline{CH_2(CH_2)_4CH}SiH(CH_3)]_2NH$	—	118–119 (1.5)			d_4^{20} 0.9136; n_D^{20} 1.4861	375
$\{[\overline{CH_2(CH_2)_3CH}]_2SiH\}_2NH$	—	174 (2)			d_4^{20} 0.9761; n_D^{20} 1.4991	375
$\{[\overline{CH_2(CH_2)_4CH}]_2SiH\}_2NH$	—	190–195 (2)			d_4^{20} 0.9709; n_D^{20} 1.5077	647
$[(HCH_3N)(CH_3)HSi]_2NCH_3$	—	48.5 (5)	—	—	d_4^{20} 0.8871; n_D^{20} 1.4425	559a
$[(C_6H_5)_2SiH]_2NH$	—	210 (1)			d_4^{20} 1.0963; n_D^{20} 1.6123	375
$(SiH_3)_2NH$	−132	36 (760)	1220	6.832	Q 5580; T 18.0; the liquid decomp. slowly at 0°	32d, 34
$(CH_3)_2(SiH_3)N$	3.3–3.4	—	3070	13.726	Q 14000; very unstable	624
	2.2	19.2 (760)	2756	12.36 (sol.)	Q_{subl} 6340; Q_{vap} 12600; T 21.8	32a
			1385	7.622 (liq.)		
$(C_2H_5)_2(SiH_3)N$	−149	78 (760)	1620	7.495	Q 7410; T 21.2; d_4^{20} 0.751	32a
$(CH_3)(SiH_3)_2N$	−124.1 to −124.6	32.3 (760)	—	—	Q 7410; vapour pressures (torr): 192 (0°)	168
					2 (−72.1°)	624
$(C_2H_5)_2(SiH_3)N \cdot SiH_3I$	27–38.5	—	—	—		174a
$(C_4H_8NSiH_3) \cdot SiH_3I$	47–49	—	—	—		174a
$(C_5H_{10}NSiH_3) \cdot SiH_3I$	53.5–56	—	—	—		174a
$\overline{CH_2(CH_2)_3N}SiH_3$	−46	81.5 (760)	1935	8.344	Q 8860; T 25.0; d_4^{20} 0.810	32b
$\overline{CH_2(CH_2)_4N}SiH_3$	−109	104 (760)	1700	7.385	Q 7780; T 20.6; d_4^{20} 0.775	32b
$(C_2H_5)_2NSiH_3$	−149	78 (760)	1620	7.495	Q 7410; T 21.2; d_4^{20} 0.751	32b
$(C_2H_5)(SiH_3)_2N$	−127	65.9 (760)	—	—	Q 6760; vapour pressures (torr): 68 (1.5°), 2 (−45.4°)	168
$[(CH_3)_2SiH]N[Si(CH_3)_3]_2$					vapour pressure at 84°: 26 torr	83a
$[(CH_3)_2SiH]_2N[Si(CH_3)_3]$					vapour pressure at 60°: 20 torr	63a
$[(CH_3)_2SiH]_3N$					vapour pressure at 153°: 760 torr	83a

TABLE 7.43 (cont'd)

Silane	*M.p.* (°C)	*B.p.* [°C(torr)]	*log p*[torr] = −*A*/T + *B*		*Enthalpy of evaporation, Q* (cal/mole); *Trouton's constant, T* (cal mole^{-1} deg^{-1}); *density, d* (g/ml); *refractive index, n*	*Ref.*
			A	*B*		
$(SiH_3)_3N$	—	—	—	—	*Q* 7000; vapour pressure at 0°: 108.8, 110.6 torr; dipole moment = 0	624, 648
$(CH_3)_2(CH_3SiH_2)N$	−150 · · 160	45.3 ± 0.2(760)	1474	7.505	*Q* 6750; *T* 21.3	142
$(CH_3)(CH_3SiH_2)_2N$	−115	80.1 ± 0.2(760)	1683	7.646	*Q* 7700; *T* 21.7	142
$(CH_3SiH_2)_3N$	−107	108.6 ± 0.2(760)	1760	7.491	*Q* 8100; *T* 20.7	142
$(CH_3)_2(Si_2H_5)N$	—	65.8	—	—		352
$(Si_2H_5)_3N$	—	176(760)	2328.9	8.0645	*Q* 10660; *T* 23.7; stable at 20° for 17 days	663
$SiH_3SiH_2N(CH_3)_2$	—	65(760)	1851.04	8.34160	*Q* 8470; *T* 25.0	2
$(CH_3)_2NSiH_2SiH_2N(CH_3)_2$	—	129.3	—	—		352
$SiH_3SiH[N(CH_3)_2]_2$	—	129.3(760)	2054.63	7.98561	*Q* 9400; *T* 23.4	2
$(SiH_3)_2NN(SiH_3)_2$	−24.0 to −24.1	109.5 ± 0.5(760)	—	—	*Q* 8190; *T* 21.4; d^{20} 0.83; vapour pressure at 0°: 6.2 torr	28
$(SiH_3NSiH_2)_3$	—	133.0(760)	2057.3	7.9456	*Q* 9414; *T* 23.2	665a
$(CH_3HSiNCH_3)_3$	—	56 (5)	—	—	d_4^{20} 0.9297; n_D^{20} 1.4580	559a
$(CH_3HSiHCH_3)_4$	—	88 (2)	—	—	d_4^{20} 0·9776; n_D^{20} 1.4810	559a
$(CH_3)_2(CH_3SiH_2)N \cdot CH_3SiH_2I$	—	—	—	—	Non-volatile at −64°	142
$(CH_3)_2[(CH_3)_3Si]N \cdot CH_3SiH_2I$	—	—	—	—	Vapour pressure at 0°; 43 torr; decomposition at 20°	142
$(CH_3)_3(SiH_3)NCl$	—	—	—	—	Dissociation pressure (torr): 531 (86.5°), 27 (18.6°)	168
$(CH_3)_2(SiH_3)NHCl$	—	—	—	—	Unstable	168
$(C_2H_5)_3(SiH_3)NI$	—	—	—	—	Dissociation pressure at 20° < 1 torr	32
$(CH_3)_3(SiH_3)NI$	—	—	—	—	Dissociation pressure at 20°: 1 torr; vapour pressure at 0°: 83 torr	32
$(SiH_3)_2NCN$	−74.8 ± 0.5	84.7 ± 0.5(760)	1761	7.800	*Q* 8100 ± 50; *T* 22.6; stable for a few days at 20°	151
$(SiH_3)_3N \cdot BF_3$	⩽ −80	—	—	—	Equilibrium pressures (torr): 10.0 (−41°), 1.9 (−80°)	81, 623
$(CH_3)_2(CH_3SiH_2)N \cdot B(CH_3)_3$	~ −35	—	—	—	Dissociation pressures (torr): 180 (0°), 6.5 (−64°)	142
$(SiH_3)_2NBH_2$	—	—	—	—	Vapour pressure at −78°: 7 torr	81

$[(SiH_3)_2NBH_2]_2$	—	—	—	—	Vapour pressure at 25°: 10 torr	81
$(SiH_3)_2NB_2H_5$	−69.4 to −68.8	≦ 54 (760)	1669	7.974	*Q* 7640; *T* 23.3	81
$(CH_3)(SiH_3)NB_2H_5$	−39.0	—	1910	9.01	Solid; decomposes at 20°;	81
	—	51 (760)	1686	8.081	*Q* 7716; *T* 23.8; liquid	
$(SiH_3)_2NBF_2$	—	—	1563	8.102	*Q* 7200	623
$(CH_3SiH_2)_2NBF_2$	—	—	—	—	Stable at 20°; fat catalyses decomposition; non-volatile at −96°	142
$(CHF_2CF_2)SiH_3N(CH_3)_3$ ($Sisp^3d$)	—	2.8 (37.0)	—	—		104
$[(CH_3)_2N]_2SiH_2$	−105/−103	93 (760)	1691	7.499	d^{20} 0.788; *Q* 7710; *T* 21.1	35
$[(CH_3)_2N]_3SiH$	−91/−89	142 (760)	2147	8.052	d^{20} 0.850; *Q* 9830; *T* 23.7	35
$(SiH_3)_2NBCl_2$	< −78?	—	—	—	Vapour pressure at 25°: 22 torr	81, 623
SiH_3PH_2	< −135	12.7 (760)	—	—	Vapour pressures (torr): 487 (0°), 12 (−77°)	197
$(SiH_3)_3P$	—	114 (760)	1901.8	7.792	*Q* 8697	10, 68
$(CH_3)_3(SiH_3)PI$	—	—	—	—	Very slightly volatile in vacuum	32
$(C_2H_5)_3(SiH_3)PI$	—	—	—	—	Very slightly volatile in vacuum	32
SiH_3PI_2	—	90 ± 0.5 (760)	—	—	*Q* 9300; *T* 20.5; vapour pressures (torr): 299 (109°), 1.2 (0°)	32
$(SiH_3)_3As$	−1.8	20 (760)	2142.6	8.333	*Q* 9798; d_4^{20} 1.201	8, 68
$(CH_3)_3(SiH_3)AsI$	8.1–9.6	—	—	—	Vapour pressure at −0.7°: 105 torr	32
SiH_3AsI_2	−4.0 ± 0.5	210 ± 0.5 (760)	—	—	*Q* 9200; *T* 19.3; vapour pressures (torr): 238 (107.9°), 50 (59°)	32
$(SiH_3)_3Sb$	—	255 (760)	1670.3	6.041	*Q* 7638	9, 68

References p. 621

In subsequent reactions of the adduct of silyl disilyl oxide and boron trichloride, there are two possibilities of cleavage [*351*]:

$$\begin{array}{ccccc} H_3Si & \vdots & O & | & Si_2H_5 \\ \uparrow & \vdots & \downarrow & | & \uparrow \\ Cl & \vdots & B & | & Cl \\ & \vdots & | & | & \\ & & Cl & & \end{array}$$

In the case of the cleavage shown by the dotted line, SiH_3Cl and unstable $Si_2H_5OBCl_2$ are produced, and in the case of the cleavage shown by the broken line the products are Si_2H_5Cl and unstable SiH_3OBCl_2. The cleavage actually takes place as shown by the broken line, since here the electrons of the chlorine can be distributed over two silicon d-orbitals. Examples of the reaction are given below.

$(SiH_3)_2O$ and $(SiH_3)_2S$ do not react with $(BH_3)_2$ [*620*]. $(SiH_3)_2S$ does not react with BF_3, while $(SiH_3)_2O$ does so with the intermediate formation of $(SiH_3)_2O \cdot BF_3$ to give SiH_3F and a non-volatile compound of the approximate formula BOF [*620*]; SiH_3OCH_3 also gives SiH_3F:

$$(SiH_3)_2O + BF_3 \xrightarrow[\text{[m.p. (SiH}_3)_2\text{O]}]{<-144^\circ} (SiH_3)_2O \cdot BF_3 \xrightarrow[-SiH_3F]{-127^\circ} SiH_3 \cdot OBF_2 \xrightarrow{>-127^\circ} SiH_3F + \frac{1}{x}(BOF)_x$$

$$CH_3OSiH_3 + BF_3 \rightarrow \text{adduct is not observable} \xrightarrow{>-127^\circ} SiH_3F + CH_3OBF_2$$

Both $(SiH_3)_2O$ and $(SiH_3)_2S$ react with $Al(CH_3)_3$ [*620*]. $(SiH_3)_2O$ forms CH_3SiH_3, SiH_4, and a solid that has not been investigated further via an adduct which cannot be isolated and which decomposes slowly at −80° and rapidly at 25°:

$$(SiH_3)_2O + AlMe_3 \xrightarrow{<-80^\circ} (SiH_3)_2O \cdot AlMe_3 \xrightarrow[-MeSiH_3]{>-80^\circ} SiH_3OAlMe_2 \xrightarrow[-SiH_4]{\sim-80^\circ} \text{solid product}$$

With $Al(CH_3)_3$, $(SiH_3)_2S$ forms almost exactly one equivalent of CH_3SiH_3 and a solid I which can be detected by vapour pressure measurements. Solid I polymerises at 0° to a second solid II, which decomposes at 25° into more CH_3SiH_3 and CH_3AlS:

$$(SiH_3)_2S + AlMe_3 \xrightarrow[-MeSiH_3]{-80/-23^\circ} \underset{\text{I}}{SiH_3SAlMe_2} \xrightarrow{0^\circ} \frac{1}{x}\underset{\text{II}}{[SiH_3SAlMe_2]_x} \xrightarrow[-MeSiH_3]{25^\circ} \frac{1}{x}MeAlS_x$$

BCl_3, BBr_3, and $R_{3-n}BBr_n$ react with $[(CH_3)_3Si]_2O$, $Si(OCH_3)_4$, $Si(OC_2H_5)_4$, $(CH_3)_3SiOC_2H_5$, and $(C_2H_5)_3SiOH$ with the replacement of oxygen by bromine. On the other hand, $(C_2H_5)_3SiONa$, with highly nucleophilic oxygen ($R_3SiO^{\ominus}Na^{\oplus}$) forms silyl borates:

$$(R_3Si)_2O + BBr_3 \longrightarrow (R_3Si)_2O \cdot BBr_3 \xrightarrow[-R_3SiBr]{20^\circ} R_3SiOBBr_2 \rightarrow B_2O_3, R_3SiBr$$

$$\left.\begin{array}{l} Si(OR)_4 + R_{3-n}BBr_n \rightarrow \\ 3R_3SiOH + BBr_3 \rightarrow \\ 3R_3SiONa + BBr_3 \rightarrow \end{array}\left\{\begin{array}{l}\text{adducts}\\ \text{are not}\\ \text{observable}\end{array}\right.\right\}\begin{array}{l} \rightarrow SiBr_n(OR)_{4-n} + R_{3-n}B(OR)_n \\ \rightarrow 3R_3SiBr + B(OH)_3 \\ \rightarrow B(OSiR_3)_3 + 3NaBr \end{array}$$

Donors such as NH_3, Me_2NH, PH_3, Me_3N and $H^{\ominus}$ catalyses the condensation of $(SiH_3)_2O$ and CH_3OSiH_3 with elimination of SiH_4[*700a*].

9.3 *Reactions of the Si–N, Si–P, and Si–As bonds*

9.3.1 *General*

Acceptors add to the free pair of electrons of the nitrogen in ammonia, amines, and hydrazine, to form four-coordinated nitrogen:

$$R_3N + H^{\oplus} \rightarrow R_3NH^{\oplus}$$
$$R_3N + BH_3 \rightarrow R_3NBH_3$$
$$(R = H, \text{organyl}, R_2N)$$

If such nitrogen bases are silylated to an increasing extent, the electron density on the nitrogen decreases. Measurements of dipole moments show that the nucleophilicity of nitrogen decreases with the degree of electron impoverishment on the nitrogen caused by electronegative silicon substituents[*526*]. The non-σ-bonding nitrogen electrons are included in a $p_\pi \rightarrow d_\pi$ bond system[*141, 649*] and are therefore no longer available on the nitrogen without the consumption of energy. Simultaneously, with increasing silylation the nitrogen–ligand angle of 106.8° (I) widens to 120° in the planar[*147*] $(SiH_3)_3N$ (II)[3].

I II III IV

Only with strong electron acceptors is it possible to outweigh the resonance stabilisation in III, for example. Thus, the negative heat of formation of such adducts decreases with increasing number of silyl substituents [by 9 kcal/mole per silyl substitution: $(CH_3)_3N \cdot BF_3$ −28, $(CH_3)_2(SiH_3)N \cdot BF_3$ −19, $(CH_3)(SiH_3)_2N \cdot BF_3$ −10, $(SiH_3)_3N \cdot BF_3$ −1 kcal/mole]. Consequently, in the co-condensation of $(SiH_3)_3N$ and BF_3 no adduct can be observed but only products following cleavage according to IV. In the case of $(CH_3)(SiH_3)_2N$ and $(CH_3)_2(SiH_3)N$, adducts can be detected but here again cleavage similar to IV also takes place[*623*]. The nucleophilicity of the nitrogen is affected by steric factors as well as by the induction and resonance effects[*281*].

---

[3]In contrast, the heavy atom skeletons of $(SiH_3)_3P$ and $(SiH_3)_3As$ are not planar (∡ Si–P–Si: 96.45°, ∡ Si–As–Si: 93.79°)[*55a, 55b*].

References p. 621

Suitable as acceptors in such silylamine–acceptor adducts are the compounds usually described as acceptors with unoccupied non-bonding p (or s) orbitals, for example BF_3, $(CH_3)_3B$, or $H^{\oplus}$ (as an "onium compound"). Yet other compounds suitable as acceptors are compounds which favour an electrophilic attack by the incorporation of unoccupied d-orbitals or quite generally by a potential penta-covalent transition state. For a review, see [*25b*].

9.3.2 Reaction of the Si–N, Si–P, and Si–As bonds with acceptors

The reaction of the Si–N bond with acceptors takes place by an electrophilic attack of the acceptor on the nitrogen with subsequent cleavage of the Si–N bond weakened by the withdrawal of electrons:

$$R_{3-n}(R'_{3-m}SiH_m)_nN \xrightarrow{+BZ_3} R_{3-n}(R'_{3-m}SiH_m)_nN \cdot BZ_3 \xrightarrow{-R_{3-m}SiH_mZ}$$

$$R_{3-n}(R'_{3-m}SiH_m)_{n-1}NBZ_2 \xrightarrow{-R_{3-m}SiH_mZ} \frac{1}{x}[R_{3-n}(R'_{3-m}SiH_m)_{n-2}NBZ]_x$$

$$(n = 0,1,2,3;\ m = 0,2,3;\ R = \text{alkyl};\ Z = CH_3,F,Cl,Br)$$

Examples of the reaction with nitrogen compounds [*81, 142, 485a, 623*]:

$$(MeSiH_2)_3N + 4\,HCl \xrightarrow{-78^\circ} 3\,MeSiH_2Cl + NH_3 \cdot HCl$$
$$Me(MeSiH_2)_2N + 3\,HCl \xrightarrow{-78^\circ} 2\,MeSiH_2Cl + MeNH_2 \cdot HCl$$
$$Me_2(MeSiH_2)N + 2\,HCl \xrightarrow{-78^\circ} MeSiH_2Cl + Me_2NH \cdot HCl$$

$$(MeSiH_2)_3N + BMe_3 \rightarrow \text{no adduct}$$
$$Me(MeSiH_2)_2N + BMe_3 \rightarrow \text{no adduct}$$
$$Me_2(MeSiH_2)N + BMe_3 \rightarrow Me_2(MeSiH_2)N \cdot BMe_3$$
$$Me_2(Me_3Si)N + BMe_3 \rightarrow Me_2(Me_3Si)N \cdot BMe_3$$

$$(SiH_3)_3N + BF_3 \xrightarrow{-80^\circ} [(SiH_3)_3N \cdot BF_3]? \xrightarrow[-SiH_3F]{25^\circ} (SiH_3)_2NBF_2$$
$$\text{or}$$
$$\xrightarrow[-2SiH_3F]{25^\circ} \tfrac{1}{3}(SiH_3NBF)_3$$

$$Me(SiH_3)_2N + BF_3 \xrightarrow{-80^\circ} Me(SiH_3)_2N \cdot BF_3 \xrightarrow[-SiH_3F]{} Me(SiH_3)NBF_2$$
$$\xrightarrow[-SiH_3F]{0^\circ} \tfrac{1}{3}(MeNBF)_3$$

$$Me_2(SiH_3)N + BF_3 \xrightarrow{-80^\circ} Me_2(SiH_3)NBF_3 \xrightarrow[-SiH_3F]{} Me_2NBF_2$$
$$Me_3N + BF_3 \xrightarrow{-80^\circ} Me_3N \cdot BF_3$$

$$(SiH_3)_3N + BCl_3 \xrightarrow{-78^\circ} [(SiH_3)_3N \cdot BCl_3] \xrightarrow{-SiH_3Cl} (SiH_3)_2NBCl_2$$
$$Me(SiH_3)_2N + BCl_3 \xrightarrow{-78^\circ} [Me(SiH_3)_2N \cdot BCl_3] \xrightarrow[-SiH_3Cl]{} Me(SiH_3)NBCl_2$$
$$\xrightarrow[-SiH_3Cl]{20^\circ} \tfrac{1}{3}(MeNBCl)_3$$
$$(SiH_3)_3N + BMe_2Br \xrightarrow{\Delta} (SiH_2Br)_2NBMe_2,\ SiH_3Br,\ SiH_4,\ BMe_3$$

$$(SiH_3)_3N + B_2H_5Br \xrightarrow[-SiH_3Br,\,-BH_3]{-78^\circ} (SiH_3)_2NBH_2 \begin{cases} \xrightarrow{-40^\circ} \text{dimeric form} \xrightarrow{\Delta} \text{Si–B–H polymers} \\ \quad \downarrow +BH_3 \\ \quad (SiH_3)_2NB_2H_5\ (10\%) \rightarrow \text{Si–B–H polymers} \\ \xrightarrow{+BH_3} (SiH_3)_2NB_2H_5\ (90\%) \rightarrow \text{Si–B–H polymers} \end{cases}$$

$$Me(SiH_3)_2N + B_2H_5Br \xrightarrow[-SiH_3Br,\,-BH_3]{-78^\circ} [Me(SiH_3)NBH_2] \xrightarrow[\text{subsequent reactions}]{\Delta} Me(SiH_3)NB_2H_5$$

for comparison: $Me_2NBH_2 \xrightarrow{+BH_3} Me_2NB_2H_5$

Examples of the reaction with phosphorus and arsenic compounds [*129a*, *129b*, *131a*, *131b*, *131d*]:

$$SiH_3PH_2 + BCl_3 \xrightarrow{-78^\circ} SiH_3PH_2BCl_3 \xrightarrow{-23^\circ} SiH_3Cl + \frac{1}{n}(BCl_2PH_2)_n$$

$$SiH_3PH_2 + \frac{1}{2}B_2H_6 \xrightarrow[\text{pressure}]{-78^\circ} SiH_3PH_2BH_3 \xrightarrow[\text{pressure}]{-25^\circ} SiH_4 + \frac{1}{n}(BH_2PH_2)_n$$

$$(SiH_3)_2PH + \tfrac{1}{2}B_2H_6 \xrightarrow{-40^\circ} (SiH_3)_2PHBH_3$$

$$Si_2H_5PH_2 + \tfrac{1}{2}B_2H_6 \xrightarrow{-40^\circ} Si_2H_5PH_2BH_3 \xrightarrow{>-40^\circ} Si_2H_6 + \text{polymers}^4$$

$$SiH_3PH_2 + B_2H_5Br \xrightarrow{-78^\circ} SiH_3PH_2B_2H_5Br \xrightarrow{-45^\circ} SiH_3Br + \frac{1}{2}B_2H_6 + \frac{1}{n}(BH_2PH_2)_n$$

$$SiH_3PH_2 + H_2GeCl_2 \rightarrow GeH_2ClPH_2 + SiH_3Cl$$

$$SiH_3AsH_2 + \tfrac{1}{2}B_2H_6 \rightarrow SiH_3AsH_2BH_3$$

$$\begin{matrix} SiH_3AsH_2 \\ \downarrow \\ SiH_3AsH_2BH_3 \end{matrix} \rightarrow (H_3Si)_2AsH + AsH_3BH_3\ (\rightleftharpoons AsH_3 + \tfrac{1}{2}B_2H_6)$$

$$\begin{matrix} (SiH_3)_2AsH \\ \downarrow \\ SiH_3AsH_2BH_3 \end{matrix} \rightarrow (SiH_3)_3As + AsH_3BH_3\ (\rightleftharpoons AsH_3 + \tfrac{1}{2}B_2H_6)$$

9.3.3 *Reaction of the Si–N bond with potential acceptors*

The reaction of the Si–N bond with potential acceptors (Si or C) takes place through an intermolecular nucleophilic attack of the nitrogen atom on a silicon atom or – less pronounced – on a carbon atom, the formation of a new Si–N or C–N bond, and cleavage of the old Si–N bond weakened by the withdrawal of electrons (outflow of the π-electrons). Repetition of the process gives rise to Si–N polymers. Some examples will illustrate this [*27*].

In the pyrolysis at 350 to 450° or in the photolysis of bis(silyl)-methylamine

[4] With an excess of $Si_2H_5PH_2$ $(Si_2H_5)_3P$ and PH_3 are formed.

$(SiH_3)_2(CH_3)N$ or $(SiD_3)_2(CH_3)N$, silane and silyl–nitrogen polymers with N–Si linkage are formed. The compounds with $n = 1$ and 2 have been isolated [*27, 32f*]:

$$2n\ (SiH_3)_2(CH_3)N \xrightarrow{\text{further attack of amine nitrogen}} \left[\begin{array}{cc} H & SiH_3 \\ SiH_2 & CH_3N \\ CH_3N & SiH_2 \\ SiH_3 & H \end{array}\right]_n \longrightarrow H_3Si\left[\begin{array}{c} H \\ | \\ -N-Si- \\ | \quad | \\ CH_3 \ H \end{array}\right]_{2n}H + n\,SiH_4$$

The nucleophilic attack of the amine nitrogen takes place not only on the silicon but also on the carbon (formation of silyl–nitrogen polymers with N–C linkage) [*27*]:

$$2n\ (SiD_3)_2(CH_3)N \xrightarrow{\text{further attack of amine nitrogen}} \left[\begin{array}{cc} H & SiD_3 \\ CH_2 & SiD_3N \\ SiD_3N & CH_2 \\ SiD_3 & H \end{array}\right]_n \longrightarrow SiD_3\left[\begin{array}{c} H \\ | \\ -N-C- \\ | \quad | \\ SiD_3 \ H \end{array}\right]_{2n}H + n\ SiD_3H$$

Donors such as NH_3, $NH(CH_3)_2$, or $N(CH_3)_3$ (5% each) also catalyse the polymerisation [*27*]:

$$3\ (SiH_3)_3N \xrightarrow[\text{only in liq. THF}]{(NH_3)} 3\ SiH_4 + (H_2SiNSiH_3)_3 \text{ (cyclic) (and higher polymers)}$$

$$x\ (SiH_3)_2MeN \xrightarrow{[NMe_3 \text{ or } NHMe_2]} x\ SiH_4 + (SiH_2NMe_3)_x$$

The equilibrium

$$x\ SiH_2(NMe_2)_2 + x\ NH_2Me \rightleftharpoons (SiH_2NMe)_x + 2x\ NHMe_2$$

is displaced to the right when polymerisation is favoured [27]. This is done by the complete removal of dimethylamine at 20° or by heating to 100° for 4 hours. The molecular weight of $(SiH_3NMe)_x$ is between 550 and 2000. On being heated, bis(dimethylamino)silane dismutes into more and less highly aminated silanes [*27*]:

$$2\ SiH_2(NMe_2)_2 \longrightarrow [Me_2N \cdots Si(H)(NMe_2)(NMe_2) \cdots H_2Si(NMe_2) \cdots H] \longrightarrow SiH(NMe_2)_3 + SiH_3(NMe_2) \longrightarrow [Me_2N \cdots Si(NMe_2)_3 \cdots H_3Si \cdots H] \longrightarrow SiH_4 + Si(NMe_2)_4$$

In the gas phase at 20° aminosilanes (Me_2NSiH_3, Et_2NSiH_3, $C_4H_8NSiH_3$, $C_5H_{10}NSiH_3$) add iodosilane to give (yields %): $(Me_2NSiH_3) \cdot SiH_3I$ (95–97),

$(Et_2NSiH_3) \cdot SiH_3I$ (95–97), $(C_4H_8NSiH_3) \cdot SiH_3I$ (45) and $(C_5H_{10}NSiH_3) \cdot SiH_3I$ (60) [*174a*].

9.3.4 *Reaction of the Si–N and Si–P bonds with donors*

The donors CO_2, CS_2, and COS react with $SiH_3N(CH_3)_2$ [but not with $(SiH_3)_3N$ and $(SiH_3)_2NCH_3$] with polymerisation (formation of viscous oils) and the formation of "1 : 1 adducts" more or less unstable at room temperature [*155*]:

$$H_3SiO{-}C({=}O){-}N(CH_3)_2 \qquad H_3SiS{-}C({=}S){-}N(CH_3)_2 \qquad H_3SiO{-}C({=}S){-}N(CH_3)_2$$

Trisilylamine reacts with donors such as NH_3, ND_3, LiH, LiD, CH_3NH_2, and $(CH_3)_3N$ in the liquid phase. SiH_4 is always formed and, under certain conditions, N,N′,N″-trisilylcyclotrisilazane [*665a*]:

$$3\,(SiH_3)_3N + x\,ND_3 \rightarrow 3\,SiH_4 + (SiH_3NSiH_2)_3 + x\,ND_3$$

Silyldiphenylamine reacts with ammonia at −46° to give diphenylamine and bis(silyl)amine; no monosilylamine is formed [*32d, 34*]:

$$2(C_6H_5)_2NSiH_3 + NH_3 \rightarrow 2(C_6H_5)_2NH + (SiH_3)_2NH$$

Bis(dimethylamino)silane, $[(CH_3)_2N]_2SiH_2$, reacts with water to form oxygen-containing polymers, but the $(SiHO_{3/2})_x$ to be expected theoretically

$$2SiH_2(NMe_2)_2 + 3H_2O \rightarrow \frac{2}{x}(SiHO_{3/2})_x + 4NHMe_2 + 2H_2$$

is not obtained [found: $SiH_{1.2}(NMe_2)_{0.4}O_{1.2}$] [*35*]. Tris(dimethylamino)silane, $[(CH_3)_2N]_3SiH$, also hydrolyses incompletely [*35*]:

$$SiH(NMe_2)_3 + (m+1)\,H_2O \rightarrow \frac{1}{x}[Si(OH)_{m+1}(NMe_2)_{3-m}]_x + m\,NHMe_2 + H_2$$

In the reaction of $[(CH_3)_2N]_3SiH$ with hydrogen chloride, NMe_2 is replaced by Cl [*35*]:

$$SiH(NMe_2)_3 + 6HCl \rightarrow SiHCl_3 + 3Me_2NH_2Cl$$

Lithium alanate hydrogenates $[(CH_3)_2N]_2SiH_2$ to silane [*35*]:

$$2SiH_2(NMe_2)_2 + LiALH_4 \rightarrow 2SiH_4 + LiAl(NMe_2)_4$$

The pyrolysis of $[(CH_3)_2N]_2SiH_2$ at 380° takes place similarly to that of $(SiH_3)_2(CH_3)N$ (see above) through a bimolecular reaction step. A mechanism via free radicals is unlikely because of the small number of pyrolysis products [*35*].

References p. 621

While peralkylated silylphosphanes (*e.g.*, Me_3SiPEt_2) do not react with Et_2PLi phosphorus-substituted silicon hydrides (*e.g.*, H_3SiPEt_2) do react [*198a*]:

$$Et_2PLi + H_3SiPEt_2 \rightarrow H_2Si(PEt_2)_2 + LiH \quad (1)$$
$$2Et_2PLi + H_3SiPEt_2 \rightarrow HSi(PEt_2)_3 + 2LiH \quad (2)$$
$$Et_2PLi + MeH_2SiPEt_2 \rightarrow Me_2Si(PEt_2)_2 + LiH \quad (3)$$
$$Et_2PLi + MeHSi(PEt_2)_2 \rightarrow MeSi(PEt_2)_3 + LiH. \quad (4)$$

From eqns. (1) and (2) it is to be understood that in the preparation of H_3SiPEt_2 by the reaction of SiH_3Br and $LiPEt_2$, $H_2Si(PEt_2)_2$ and $HSi(PEt_2)_3$ are always formed as by-products [*198a*].

No further Et_2P groups can be introduced into $HSi(PEt_2)_3$ by means of Et_2PLi but $LiSi(PEt_2)_3$ is formed [*198a*]:

$$Et_2PLi + HSi(PEt_2)_3 \rightarrow LiSi(PEt_2)_3 + Et_2PH$$

The $LiSi(PEt_2)_3$ formed hardly takes part in coupling reactions, *e.g.* [*198a*]:

$$H_3SiBr + LiSi(PEt_2)_3 \rightarrow H_3SiSi(PEt_2)_3 + LiBr$$
$$Me_2SiHCl + LiSi(PEt_2)_3 \rightarrow Me_2HSiSi(PEt_2)_3 + LiCl$$

$(SiH_3)_3P$ reacts slowly with a solution of sulphur in CS_2 and yields $(SiH_3)_2S$ and an unidentified yellow solid. With iodine $(SiH_3)_3P$ yields SiH_3I and red phosphorus [*146a*].

10. Mixed hydrides of Si with Ge, Sn, P, As, and Sb

The syntheses of mixed hydrides can be divided into two groups: the undirected and the directed syntheses.

10.1 Undirected synthesis

Undirected syntheses are overwhelmingly radical reactions. The radicals are obtained thermally, electrically (gas discharge) or photochemically]*2a, 217, 234, 355b, 582, 592, 633*]:

$$SiH_4 + GeH_4 \xrightarrow[\text{discharge}]{\Delta} H_2, Si_nH_{2n+2}, Ge_nH_{2n+2}, (Ge + Si)_nH_{2n+2}, SiH_{0...2}, GeH_{0...2}$$

The pyrolysis of mixtures of germane and silane appears to take place through germanium hydride radicals and not through silicon hydride radicals [*633*]. When Ge_3H_8 and i-Si_4H_{10} are heated, in fact, $(SiH_3)_2SiHSiH_2GeH_3$ is formed. If the reaction took place through silyl radicals, the most stable Si_4H_9 radical, $(SiH_3)_3Si\cdot$, should yield $(SiH_3)_3SiGeH_3$:

$$GeH_3GeH_2GeH_3 \rightarrow GeH_3GeH_2\cdot + GeH_3\cdot$$

$$(SiH_3)_2SiHSiH_3 + GeH_3\cdot \xrightarrow{-H} \begin{cases} \rightarrow (SiH_3)_2SiHSiH_2GeH_3 \\ \not\rightarrow (SiH_3)_3SiGeH_3 \end{cases}$$

Tin-containing mixed hydrides cannot be prepared by radical reactions since SnH_4 readily decomposes into Sn and H_2. Another undirected synthesis of mixed hydrides is the hydrolysis of compounds with Si–Ge bonds (reaction of Ge(II)–Si(II) mixed oxide[*633*] or Mg–Si–Ge[*633*] or Ca–Si–Ge[*504, 507*] mixed crystals with aqueous HF or HCl; see also section 3.3). All undirected syntheses give rise to a multiplicity of hydrides difficult to separate, *e.g.*, by gas chromatography [*19*]; to aggravate the situation, they are generally formed in only low yield.

The following hydrides and mixed hydrides prepared by the silent electric discharge have been identified gas-chromatographically on the basis of their retention times [*19*]:

n-Si_5H_{12}, i-Si_5H_{12}, neo-Si_5H_{12};
n-Si_4GeH_{12}, i-Si_4GeH_{12}, neo-Si_4GeH_{12};
n-$Si_3Ge_2H_{12}$, i-$Si_3Ge_2H_{12}$, neo-$Si_3Ge_2H_{12}$;
n-$Si_2Ge_3H_{12}$, iso-$Si_2Ge_3H_{12}$, neo-$Si_2Ge_3H_{12}$;
n-$SiGe_4H_{12}$, iso-$SiGe_4H_{12}$, neo-$SiGe_4H_{12}$;
n-Ge_5H_{12}, i-Ge_5H_{12}, neo-Ge_5H_{12}.

n-Si_6H_{14}, i-Si_6H_{14};
n-Si_5GeH_{14}, i-Si_5GeH_{14};
n-$Si_4Ge_2H_{14}$, i-$Si_4Ge_2H_{14}$;
n-$Si_3Ge_3H_{14}$, i-$Si_3Ge_3H_{14}$;
"branched" Si_6H_{14};
"branched" Si_7H_{16}.

From SiH_4 and PH_3, the silent electric discharge forms $Si_2H_5PH_2$ (which can be separated by gas chromatography) and $(SiH_3)_2PH$ (decomposes during gas chromatography[*131, 234, 235*]); $(SiH_3)_2PH$ free from $Si_2H_5PH_2$ is formed by the action of the silent electric discharge on SiH_3PH_2 or SiH_3PH_2 with SiH_4. $Si_2H_5PH_2$ free from $(SiH_3)_2PH$ is formed by the action of the silent electric discharge on mixtures of Si_2H_6 with PH_3[*235*].

Water and hydrogen chloride cleave the Si–P bond[*235*]:

$$(SiH_3)_2PH + H_2O \rightarrow (SiH_3)_2O + PH_3$$
$$(SiH_3)_2PH + 2\,HCl \rightarrow 2\,SiH_3Cl + PH_3$$
$$2\,Si_2H_5PH_2 + H_2O \rightarrow (Si_2H_5)_2O + 2\,PH_3$$
$$Si_2H_5PH_2 + HCl \rightarrow Si_2H_5Cl + PH_3$$

Fast neutron irradiation of PH_3–SiH_4 and PH_3–C_2H_4 mixtures yield $H_3{}^{31}Si$-SiH_3 and ${}^{31}SiH_3CH_2CH_2PH_2$ respectively [${}^{31}P(n, p){}^{31}Si$][*208b, 208c*].

10.2 Directed syntheses of mixed hydrides

Higher yields of mixed hydrides and a smaller number of undesired by-products are obtained by directed syntheses. There are two routes for these. Hydrogenatable compounds already containing the SiE^{IV} bond can be hydrogenated (a), or monohydrides can be coupled (b) [*11,85*]:

$$Z_3E^{IV}\text{–}E^{IV}Z_3 \xrightarrow{+6H^{\ominus}, -6Z^{\ominus}} H_3E^{IV}\text{–}E^{IV}H_3 \quad (a)$$

$$H_3E^{IV}K + XE^{IV}H_3 \rightarrow H_3E^{IV}\text{–}E^{IV}H_3 + KX \quad (b)$$

$$H_2E^{V}K + XE^{IV}H_3 \rightarrow H_2E^{V}\text{–}E^{IV}H_3 + KX \quad (b)$$

where E^{IV} = Si, Ge, Sn; E^{V} = P, As, Sb; Z = hydrogenatable ligand (Cl, CH_3COO); X = halogen.

10.2.1 Reaction (a)

Halides are usually hydrogenated to prepare hydrides. Since the halogen compounds corresponding to the mixed hydrides are unstable, the starting materials must be the non-hydrogenatable but stable phenyl compounds. $(C_6H_5)_3SiGe(C_6H_5)_3$ and $(C_6H_5)_3GeSn(C_6H_5)_3$ are formed in about 80% yield by the dropwise addition of $(C_6H_5)_3SiK$ or $(C_6H_5)_3GeK$ or $(C_6H_5)_3SnK$ to a solution of $(C_6H_5)_3SiCl$ or $(C_6H_5)_3GeBr$, as appropriate, in tetrahydrofuran (coupling reactions 1). Because of transmetallation, reversed addition gives only about a 10% yield of mixed hydride (reaction 2) [*85, 676*]:

$$\left.\begin{array}{l} Ph_3SiK + Ph_3GeBr \\ Ph_3GeK + Ph_3SiCl \end{array}\right\} \xrightarrow[-KCl(KBr)]{20^\circ} Ph_3SiGePh_3$$

$$Ph_3SiK + Ph_3SnCl \xrightarrow[-KCl(KBr)]{20^\circ} Ph_3SiSnPh_3 \quad (1)$$

$$Ph_3GeK + Ph_3SnCl \xrightarrow[-KCl(KBr)]{20^\circ} Ph_3GeSnPh_3$$

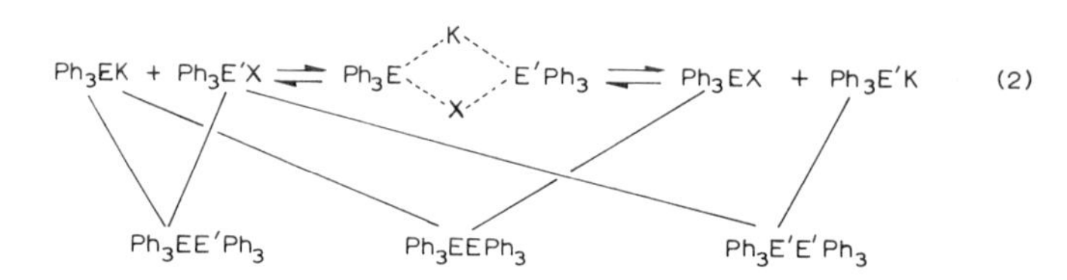

The perphenylated compounds containing tin atoms, but not those free from tin, form peracetates with anhydrous acetic acid (HOAc). The acetate ions attack the tin, so that semi-acetylated derivatives can be isolated (and be characterised by cleavage with I_2) [*85*]:

$$Ph_3GeSnPh_3 \xrightarrow[-3\,PhH]{+3HOAc} Ph_3GeSn(OAc)_3 \xrightarrow[-3\,PhH]{+3HOAc} (AcO)_3GeSn(OAc)_3$$

$$Ph_3GeSn(OAc)_3 \xrightarrow{+I_2} Ph_3GeI + SnI(OAc)_3$$

$$(AcO)_3GeSn(OAc)_3 \xrightarrow{+I_2} GeI(OAc)_3 + SnI(OAc)_3$$

The peracetates $(AcO)_3SiSn(OAc)_3$ and $(AcO)_3GeSn(OAc)_3$ can be hydrogenated either directly or, better, through the hexachloro compound with lithium alanate [*85*]:

$$(AcO)_3GeSn(OAc)_3 \xrightarrow[-105°]{+7\,HCl,\,-6\,HOAc} Cl_3GeSnCl_3 \cdot HCl \xrightarrow[-HCl]{-46°} Cl_3GeSnCl_3 \xrightarrow[-95°,\;-6\,AlH_3,\;-6\,LiCl]{+6\,LiAlH_4} H_3GeSnH_3$$

$$(AcO)_3GeSn(OAc)_3 \xrightarrow[-95°]{+6\,LiAlH_4,\,-6\,AlH_3,\,-6\,HOAc} H_3GeSnH_3$$

$$(AcO)_3SiSn(OAc)_3 \xrightarrow[-105 \cdots -46°]{+6\,HCl,\,-6\,HOAc} Cl_3SiSnCl_3 \xrightarrow[-95°,\;-6\,AlH_3,\;-6\,LiCl]{+6\,LiAlH_4} H_3SiSnH_3$$

10.2.2 *Reaction (b)*

Metallated hydrides couple with halogenated hydrides in a similar manner to the corresponding organic compounds. The couplings take place either by a nucleophilic substitution or by an α-elimination (with higher halogen substitution) [*11*]:

Nucleophilic substitution: $RNa + RI \rightarrow [\overset{\ominus}{R} \cdots\!\rightarrow R{-}I,\ Na^{\oplus}] \rightarrow R{-}R + NaI$

$$SiH_3K + Me_3SnCl \rightarrow [\overset{\ominus}{H_3Si} \cdots\!\rightarrow SnMe_3{-}Cl,\ K^{\oplus}] \rightarrow H_3SiSnMe_3 + KCl$$

α-Elimination: $C_4H_9Li + CH_2Cl_2 \rightarrow [\overset{\ominus}{C_4H_9} \cdots\!\rightarrow H{-}CHCl{-}Cl,\ Li^{\oplus}] \xrightarrow[-LiCl]{-C_4H_{10}} [|CHCl]$

$$\xrightarrow[-Cl^{\ominus}]{+C_4H_9^{\ominus}} CH_2{=}CHC_3H_7$$

$$GeH_3K + SiH_2Br_2 \rightarrow [\overset{\ominus}{H_3Ge} \cdots\!\rightarrow H{-}SiH(Br){-}Br,\ K^{\oplus}] \xrightarrow[-KBr]{-GeH_4} \frac{1}{x}[SiHBr]_x$$

A radical mechanism of inorganic couplings is unlikely because of the sensitivity to donor-active solvents and because of the insensitivity to light. In addition to the coupling reaction a transmetallation also takes place – as in the case of analogous organic reactions. It takes place by a polar 4-centre mechanism with exchange of hydrogen for metal without the occurrence of free ions.

$$R\text{-}H + R'\text{-}M \rightleftharpoons R\overset{H}{\underset{M}{\diamond}}R' \rightleftharpoons R\text{-}M + R'\text{-}H$$

$$Me_3SnSiH_3 + SiH_3K \rightleftharpoons Me_3SnSi\overset{H}{\underset{K}{\diamond}}SiH_3 \rightleftharpoons Me_3SnSiH_2K + SiH_4$$

The transmetallation is an equilibrium reaction. The driving force is the acidity gradient of the Si–H bond, here in the "acid" SiH_4 and Me_3SnSiH_3. The equilibrium is displaced to the left when a deficiency of SiH_3K is used (dropwise addition of SiH_3K solution to the $(CH_3)_3SnCl$ solution, but not conversely; see also above.) In this way, the primary coupling reaction becomes the main reaction. If the reactants are present in the ratio of 1:1 (heating together from −196° to 20°), and if the volatile reaction product – SiH_4, in our example – is removed, the equilibrium is displaced to the right. Then the primary coupling product – in our case, Me_3SnSiH_3 – can no longer be isolated but only subsequent products of the transmetallation reaction. For example, when SiH_3K and Me_3SnCl are heated together in monoglyme from −196° to 20° the products isolated are SiH_4, $(Me_3Sn)_3$-SiH, and $(Me_3Sn)_4Si$ according to the following system of equations of alternating nucleophilic substitution and transmetallation [*11*]:

$$SiH_3K + Me_3SnCl \rightarrow Me_3SnSiH_3 + KCl$$
$$SiH_3K + Me_3SnSiH_3 \rightarrow \mathit{SiH_4} + Me_3SnSiH_2K$$
$$Me_3SnSiH_2K + Me_3SnCl \rightarrow (Me_3Sn)_2SiH_2 + KCl$$
$$SiH_3K + (Me_3Sn)_2SiH_2 \rightarrow \mathit{SiH_4} + (Me_3Sn)_2SiHK$$
$$(Me_3Sn)_2SiHK + Me_3SnCl \rightarrow \mathit{(Me_3Sn)_3SiH} + KCl$$
$$SiH_3K + (Me_3Sn)_3SiH \rightarrow \mathit{SiH_4} + (Me_3Sn)_3SiK$$
$$(Me_3Sn)_3SiK + Me_3SnCl \rightarrow \mathit{(Me_3Sn)_4Si} + KCl$$

TABLE 7.44

MELTING POINTS AND VAPOUR PRESSURES AT 0° OF MIXED Si–Ge HYDRIDES IN COMPARISON WITH Si AND Ge HYDRIDES

[m.p.: upper numbers (°C); vapour pr. at 0°: lower numbers (torr)]

n	*Ge-hydrides* Ge_nH_{2n+2}	$SiGe_2H_8$	$SiGeH_6$	Si_2GeH_8	*n*-Si_3GeH_{10}	*n*-Si_4GeH_{12}	*Si-hydrides* Si_nH_{2n+2}
2	−109[*120*]		– [*582*]				−132.5[*355*]
	242[*120*]		– [*582*]				1593[*687*]
3	−105.6[*120*]	−108.5[*633*]		−113.4[*633*]			−117.4[*608,610*]
	13.9[*120*]	19.3[*633*]		39.6[*633*]			95[*608,610*]
4	—				−87.1[*633*]		−88.2[*633*]
	2.60[*7*]				4.7[*633*]		
5	—					−71.5[*633*]	−74.5[*633*]
	0.69[*7*]					—	~ 1.5(i-,n-)[*608*]
							7.8[i-,n-)[*608*]

Further examples of reactants: *SiH_3K*+$MeCl$[*498*], MeI[*11*], CH_3OCH_2Cl [*661a*], SiD_3Br[*498*], Me_3SiCl, Me_3GeBr, Me_3SnCl[*11*], $C_5H_5Fe(CO)_2Br$[*11b*]; *Ph_2SiHLi*+Ph_2SiCl_2, $Cl(SiPh_2)_3Cl$, $Ph_2ClSiSiClPh_2$[*686*]; *Ph_3SiLi*+Ph_2SiHCl [*686*]: *$Li(SiPh_2)_4Li$*+Ph_2SiHCl[*686*]; *GeH_3K*+Me_3SnCl[*11*]; *GeH_3Na*+ Me_3SiCl, Me_2SiCl_2, $MeSiCl_3$, $SiCl_4$[*133a, 438*]; *Ph_3GeNa*+$SiHCl_3$[*381*]; *PH_2K, $MePHK$, AsH_2K*+SiH_3Br[*8, 10, 68, 110a, 229b*]; *$LiAl(PH_2)_4$*+SiH_3Br [*429b*]; *$SbLi_3$*+SiH_3Br[*9, 68*].

The mixed hydrides decompose at an elevated temperature into lower simple hydrides; they are unstable in the air. Table 7.44 contains melting points and vapour pressures at 0° of some Si–Ge mixed hydrides. The vapour pressure at 0° of $(SiH_3)_2PH$ is 31 ± 1 torr and that of $Si_2H_5PH_2$: 28 ± 1 torr[*234, 235*].

11. Spectra

A literature review is given of silanes and silane derivatives with at least one Si–H bond. Where necessary, for comparison related silane derivatives are also given (denoted by ⋆).

Abbreviations: *IR* infrared spectra, *NMR* nuclear magnetic resonance spectra, *RA* Raman spectra, *UV* ultraviolet spectra, *MW* microwave spectra, *MS* mass spectra, *FL* fluorescence spectrum, *ESR* electron spin resonance, *STR* structure.

11.1 Silanes Si_nH_{2n+2} (⋆ = for comparison)

Silanes	*IR*	*NMR*	*MS*
SiH_4	12, 24a, 41, 258, 373, 372, 473, 531, 566a, 574	65a, 105a, 124a, 157, 158, 290a	516
SiH_2D_2	374	–	–
$SiHD_3$	374	–	–
SiD_4	374	–	–
Si_2H_6 (*RA* [*16, 607*]) (*STR* [*73*])	12, 131, 293	134a, 233	516
Si_3H_8	12, 293	157a, 233	–
n-Si_4H_{10}	12, 233	233	–
i-Si_4H_{10}	233	233	–
Si_5H_{12}	12	–	–
⋆C_nH_{n+2}	188, 331, 423, 562, 581	–	–

11.2 Organosilanes (⋆ = for comparison)

1. *Organosilanes with one silicon atom*	*IR*	*NMR*	*RA*
CH_3SiH_3(*MS* [*217*])	39a, 293, 580	65a, 86, 144, 145, 290a, 529b	–
CH_3SiHD_2	39a	–	–
CH_3SiD_3	39a	–	–
⋆$(CH_3)_4Si$	266, 564, 576, 578	132, 580a, 580b	–
$C_2H_5SiH_3$	335a, 355a	–	335a
$C_2H_5SiD_3$	335a, 355a	–	335a
n-$C_3H_7SiH_3$	439	–	–

References p. 621

1. *Organosilanes with one silicon atom*	*IR*	*NMR*	*RA*
n-$C_4H_9SiH_3$	439	–	–
tert-$C_4H_9SiH_3$	439	–	–
i-$C_5H_{11}SiH_3$	532a	–	–
n-$C_6H_{13}SiH_3$	–	532a	–
$\overline{CH_2(CH_2)_4CH}SiH_3$	439		
CH_2ClSiH_3 (*MW*[548])	–		
$CH_2{=}CHSiH_3$	192a	–	192a
$CH_2{=}CHSiD_3$	192a	–	192a
⋆$CH_2{=}CHSi(CH_3)_3$	159	–	–
$HC{\equiv}CSiH_3$(*UV*[146], *MW*[212a, 402a])	487b	146	–
$HC{\equiv}CSiD_3$	487b	–	–
⋆$HC{\equiv}CF$ (*MW*[643])			
⋆$HC{\equiv}CSi(CH_3)_3$ (*UV*[146])	–	146	702
⋆$HC{\equiv}CSi(OCH_3)_3$ (*UV*[146])	–	146	–
⋆$Si(C{\equiv}CH)_4$	521a	–	521a
$CH_3C{\equiv}CSiH_3$ (*MW*[314a, 487a])	314	–	
$CH_3C{\equiv}CSiD_3$	487a	–	
$C_6H_5SiH_3$ (*UV*[409b])	316–318, 580	65a, 523a, 290a	
$C_6H_5SiD_3$	316		
p-$CH_3OC_6H_4SiH_3$	439		
m-$ClC_6H_4SiH_3$	580		
1-$C_{10}H_7SiH_3$	439		
$(CH_3)_2SiH_2$ (*MW*[464])	580	65a, 290a, 529b	
$(n\text{-}C_5H_{11})_2SiH_2$	439	–	
$(CH_2)_5SiH_2$	442		
$(CH_3)(CH_2Cl)SiH_2$	580	–	
$(CH_2SiH_2)_4$	–	202a	–
$(CH_3)(C_6H_5)SiH_2$(*UV*[409b])	317, 318, 580	–	
$[\overline{CH_2(CH_2)_4CH}](C_6H_5)SiH_2$	580	–	
$(C_6H_5)_2SiH_2$(*UV*[409b])	316, 580	65a, 158a, 290a	
$(C_6H_5)_2SiD_2$	316		
$(CH_3)(p\text{-}ClC_6H_4)SiH_2$	580		
$(CH_3)(m\text{-}ClC_6H_4)SiH_2$	580		
$(CH_3)_3SiH$	279, 312, 312a, 636	65a, 158a, 158b, 290a, 529b	312, 312a
$(CH_3)_2(C_2H_5)SiH$	279	–	–
$(CH_3)(C_2H_5)_2SiH$	580, 636	97a, 152b	–
$(C_2H_5)_3SiH$	279, 312, 312a, 580, 598a, 636	97a, 158a, 158b	312, 312a
$(CH_3)(C_3H_7)_2SiH$	–	158d	–
$(CH_3)(n\text{-}C_3H_7)_2SiH$	–	158a, 158d	–
$(C_2H_5)(i\text{-}C_3H_7)_2SiH$	–	158a	–
$(n\text{-}C_3H_7)_3SiH$	312, 312a	–	312, 312a
$(sec\text{-}C_3H_7)_3SiH$	312a	–	312a
$(CH_3)(n\text{-}C_4H_9)_2SiH$	–	158c	–
$(CH_3)(n\text{-}C_5H_{11})_2SiH$	–	158c	–
$(n\text{-}C_4H_9)_3SiH$	312, 312a, 488b, 598a	97a	312, 312a
$(sec\text{-}C_4H_9)_3SiH$	312, 312a	–	312, 312a
$(i\text{-}C_4H_9)_3SiH$	312, 312a	–	312, 312a
$(CH_3)(n\text{-}C_6H_{13})_2SiH$	–	158c	–
$(n\text{-}C_5H_{11})_3SiH$	312, 312a	–	312, 312a

1. *Organosilanes with one silicon atom*	*IR*	*NMR*	*RA*
$(CH_3)(n-C_7H_{15})_2SiH$	–	158c	–
$(n-C_6H_{13})_3SiH$	598a	–	–
$(CH_3)_2(CH_2{=}CH)SiH$	279	–	
$(C_2H_5)_2(CH_2{=}CH)SiH$	580	–	
$(C_2H_5)(CH_2{=}CH)_2SiH$	–	486b	
$(CH_3)_2(CH_2{=}CHCH_2)SiH$	–	158f	
$(CH_3)(CH_2{=}CHCH_2)_2SiH$	488, 488b	–	
$(CH_2{=}CHCH_2)_3SiH$	488, 488b	–	
$(C_6H_5CH_2)_3SiH$	312, 312a	–	312, 312a
$(CH_3)_2(CF_3CH_2CH_2)SiH$	279	158d	
$(CH_3)_2(CF_3CH_2CH_2CH_2)SiH$	–	158d	
$(CH_3)(CF_3CH_2CH_2)_2SiH$	–	158b, 158d	
$(CH_3)(CF_3CH_2CH_2CH_2)_2SiH$	–	158d	
$(CF_3CH_2CH_2CH_2)_3SiH$	–	158d	
$(CH_3)_2(CH_2Cl)SiH$	279, 324b	–	
$(CH_3)_2(CHCl_2)SiH$	324b, 488b	–	
$(CH_3)(CHCl_2)_2SiH$	324b	–	
$(CH_3)_2(CH_2Br)SiH$	324b	–	
$(CH_3)_2(CH_2I)SiH$	324b	–	
$(C_4H_3O)_2(CH_3)SiH$	–	158g	
$(CH_3)_2(C_6H_5)SiH$ (*UV* [*409b*])	24a, 279, 317, 318, 488, 488a, 488b, 580, 630	–	488, 488a, 488b
$(CH_3)(CH_2{=}CH)(C_6H_5)SiH$	279, 580	–	–
$(CH_3)(C_6H_5)(CH_2CH{=}CH_2)SiH$	–	158f	–
$(C_6H_5)_2(CH_2CH{=}CH_2)SiH$	–	158f	–
$(CH_3)_2(p-CH_3C_6H_4)SiH$	488, 488a, 488b		488, 488a, 488b
$(CH_3)_2(m-CH_3C_6H_4)SiH$	488, 488a, 488b		488, 488a, 488b
$(CH_3)_2(p-FC_6H_4)SiH$	488, 488a, 488b		488, 488a, 488b
$(CH_3)_2(p-ClC_6H_4)SiH$	488, 488a, 488b		488, 488a, 488b
$(CH_3)_2(p-BrC_6H_4)SiH$	488, 488a, 488b		488, 488a, 488b
$(CH_3)_2(m-ClC_6H_4)SiH$	488, 488a, 488b		488, 488a, 488b
$(CH_3)_2(m-BrC_6H_4)SiH$	488, 488a, 488b		488, 488a, 488b
$(CH_3)_2(p-CH_3OC_6H_4)SiH$	488, 488a, 488b		488, 488a, 488b
$(CH_3)(C_6H_5)_2SiH$	24a, 580, 598a, 636		–
⋆$(CH_2{=}CBr)(C_6H_5)_3Si$	–		52a
$(CH_2{=}CH)(C_6H_5)_2SiH$	580		–
$(CH_3)(FC_6H_4)_2SiH$	–		158b
$(CH_3)(F_2C_6H_3)_2SiH$	–		158b
$(CH_3)(ClC_6H_3)_2SiH$	279		–
$(C_6H_5)_3SiH$ (*UV* [*409b*])	24a, 300, 312, 312a, 316, 318, 362a, 488b, 580, 598a, 636	65a, 97a, 158b, 290a, 523b	312, 312a
$(C_6H_5)_3SiD$	300, 316	–	–
⋆$(C_6H_5)_4Si$ (*UV* [*409b*])	429, 529a	–	–
$(p-ClC_6H_4)_3SiH$	312, 312a	–	312, 312a
$(p-CH_3C_6H_4)_3SiH$	312, 312a	–	312, 312a
$(p-CH_3OC_6H_4)_3SiH$	312, 312a	–	312, 312a

References p. 621

2. *Organosilanes with several silicon atoms*	*NMR*	*MS*
$(CH_3)_2HSiSiH_3$ (*IR*[*243a*])	243a	–
$(CH_3)H_2SiSiH_2(CH_3)$ (*IR*[*108*])		
$(CH_3)_3SiSiH_3$ (*IR*[*11a*])	11a	–
$(CH_3)_3SiSi(CH_3)H$ – radical (*ESR*[*63a*])		
⋆$(CH_3)_3SiSi(CH_3)_3$ (*IR*, *RA*[*78, 95, 409*])	701, 711	
$[(CH_3)_3Si]_2SiH_2$	11a	–
$[(CH_3)_3Si]_3SiH$(*IR*, *RA*[*79c*])	11a, 79c	–
$SiH_3CH_2SiH_3$	200	193
$SiH_3CClSiH_3$	200a	–
$CH_3SiH_2CH_2SiH_3$	–	193
$(CH_3)_3SiCH_2SiH_3$	–	193
$(CH_3)_2SiHCH_2SiH_2CH_3$	200	193
$(CH_3)_2SiHCH_2SiH(CH_3)_2$	200	193
$(CH_3)_3SiCH_2SiH(CH_3)_2$	200	193
$SiH_3CH_2SiH_2CH_2SiH_3$	200	193
$CH_3SiH_2CH_2SiH_2CH_2SiH_3$	200	193
$(CH_3)_2SiHCH_2SiH_2SH_2SiH_3$	–	193
$SiH_2CH_2SiH_2CH_2SiH_2CH_2$	200	193
$SiH_2CCl_2SiH_2CCl_2SiH_2CCl_2$	200a	–
$Si(CH_3)HCH_2SiH_2CH_2SiH_2CH_2$	200	193
$Si(CH_3)HCH_2Si(CH_3)HCH_2SiH_2CH_2$	200	193
$Si(CH_3)HCH_2Si(CH_3)HCH_2Si(CH_3)HCH_2$	200	193
$Si(CH_3)_2CH_2Si(CH_3)HCH_2SiH_2$	200	–
$Si(CH_3)HCH_2Si(CH_3)_2CH_2Si(CH_3)HCH_2$	200	193
$Si(CH_3)_2CH_2Si(CH_3)_2CH_2Si(CH_3)HCH_2$	–	193
$SiH_3C{\equiv}CSiH_3$ (*RA*[*348*])	–	–
⋆$(CH_3)_3SiC{\equiv}CSi(CH_3)_3$ (*IR*[*324*], *RA*[*324*])	–	–
⋆$(CH_3O)_3SiC{\equiv}CSi(OCH_3)_3$ [*IR*[*324*], *RA*[*324*])	–	–
H_2 Si CH_2 Si H_2	200	–

11.3 Halogenosilanes (⋆ = for comparison)

For halogenosilanes containing Si–O bonds, see section 11.4, last table.

1. *Halogenosilanes with one SiH_3 group*	*IR*	*NMR*
SiH_3F	293, 294a	65a, 157, 158, 279a, 290a, 354, 357a
⋆CH_3F	444	–
⋆$(CH_3)_3SiF$	187, 529a	79a, 532, 645a
$SiH_3F \cdot 2$pyridine	87	–
SiH_3Cl (*MW*[*364*])	293, 294a	65a, 157, 158, 279a, 290a, 354, 357a
⋆CH_3Cl	444	–
⋆$(CH_3)_3SiCl$	187, 529a, 578	79a, 532, 645a
SiH_3Br	293, 294a, 363, 408	157, 158, 279a, 290a
⋆CH_3Br	444	–

1. *Halogenosilanes with one SiH_3 group*	*IR*	*NMR*
★$(CH_3)_3SiBr$(*MW*[*436*])	529a	79a
$SiH_3Br \cdot N(CH_3)_3$	87	–
$SiH_3Br \cdot 2N(CH_3)_3$	87	–
SiH_3I	125, 293	279a, 290a
★CH_3I	444	–
★$(CH_3)_3SiI$ (*MW*[*494*])	529a	79a
$SiH_3I \cdot N(CH_3)_3$	87	–
$SiH_3I \cdot 2N(CH_3)_3$	87	–
$SiD_3I \cdot N(CH_3)_3$	87	–
$SiD_3I \cdot 2N(CH_3)_3$	87	–
$SiH_3I \cdot$ 2 pyridine	87	86
$SiD_3I \cdot$ 2 pyridine	87	–

2. *Halogenosilanes with one SiH_2 group*	*IR*	*NMR*
CH_3SiH_2F (*MW*[*325*])	154, 580	144, 145
CD_3SiH_2F(*MW*[*325*])		
SiH_2F_2 (*MW*[*333*])	580	65a, 157, 158, 279a, 290a, 296
★$(CH_3)_2SiF_2$	187	532, 645a
CH_3SiH_2Cl	154	144
★CH_3CH_2Cl (*MW*[*549*])		
SiH_2Cl_2	131e, 256, 257, 580	65a, 148a, 157, 158, 279a, 290a, 296
★$(CH_3)_2SiCl_2$	185, 568, 578	532, 645a
$SiH_2Cl_2 \cdot$ 2 pyridine	87	–
$SiD_2Cl_2 \cdot$ 2 pyridine	87	–
$C_6H_5SiH_2Cl$	317	–
SiH_2ClBr	–	148a
CH_3SiH_2Br	154	144
SiH_2Br_2 (*RA*[*408*])	256, 363, 408	148a, 157, 158, 279a, 290a, 296
SiH_2ClI	–	148a
SiH_2BrI	–	148a
CH_3SiH_2I	154, 169	86, 144
$CH_3SiH_2I \cdot$ 2 pyridine	–	86
$CH_3SiH_2I \cdot 2\,N(CH_3)_3$	87	86
$CH_3SiH_2I \cdot N(CH_3)_3$	87	86
CH_3SiHI_2	169	–
SiH_2I_2	–	148a, 157, 158, 279a, 290a, 296
$SiH_2I_2 \cdot$ 4 pyridine	87	–
$SiD_2I_2 \cdot$ 4 pyridine	87	–

3. *Halogenosilanes with one SiH group*	*IR*	*NMR*
$(CH_3)_2SiHF$	24a	–
$(CH_3)_2SiHCl$	24a, 580, 636	117a, 117c
$(CH_3)(C_2H_5)SiHCl$	–	158b, 158c
$(C_2H_5)_2SiHCl$	580, 636	–

References p. 621

3. *Halogenosilanes with one SiH group*	*IR*	*NMR*
$(CH_3)(C_6H_5)SiHCl$	580, 636	158b
$(CH_2{=}CH)(C_6H_5)SiHCl$	580	–
$(C_6H_5)_2SiHCl$	580	–
$(CH_3)_2SiHBr$	–	117c
$(CH_3)_2SiHI$	–	117c
CH_3SiHF_2	24a	–
CH_3SiHCl_2	24a, 580, 636	117c, 158b, 158c, 240, 368a
$C_2H_5SiHCl_2$	580, 636	158b
n-$C_3H_7SiHCl_2$	580, 636	–
i-$C_3H_7SiHCl_2$	580, 636	–
n-$C_5H_{11}SiHCl_2$	580	–
$CH_2{=}CHCH_2SiHCl_2$	580	–
$C_6H_5SiHCl_2$	317, 580, 636	–
CH_3SiHBr_2	–	117c
$C_6H_5SiHBr_2$	636	–
$CH_3SiHI_2 \cdot 2$ pyridine	87	–
$CH_3SiHI_2 \cdot 4$ pyridine	87	–
$SiHF_3$	580, 653	157, 158, 279a, 290a, 296
$SiDF_3$	653	
⋆CHF_3 (*RA* [*599*]		
⋆$(CH_3)SiF_3$ (*RA* [*101*]	101, 187	532, 645a
$SiHCl_3$ (*RA* [*312a*]) (*MW* [*399*])	24a, 256, 280a, 312a, 312b, 580, 636	97a, 157, 158, 158a, 290a, 296
⋆$(CH_3)SiCl_3$	578	532, 645a
⋆$C_6H_5SiCl_3$	317	–
$SiHCl_3 \cdot 2$ pyridine	87	–
$SiHBr_3$	408, 580, 636	290a
$SiHI_3$	–	290a
$SiHI_3 \cdot 4$ pyridine	87	–

4. *Halogenosilanes with several silicon atom*	*IR*	*NMR*
SiH_3SiH_2F	1, 2	134a
⋆$Si(CH_3)_3Si(CH_3)_2F$	645	–
SiH_3SiH_2Cl	1	134a
⋆$SiCl_3SiCl_3$	400	–
⋆C_2H_5Cl (*MW* [*539*])	–	–
SiH_3SiH_2Br	–	130, 134a
SiH_3SiH_2I	662, 663	134a
SiH_2BrSiH_2Br	–	130
SiH_3SiHBr_2	–	130
$SiHBrSiHBr_2$	–	130
$SiHBr_2SiHBr_2$	–	130
$SiCl_3CH_2SiCl_2H$	–	158e

11.4 Silanes with Si–O, Si–S, and Si–Se bonds (★ = for comparison)

1. *Silanes with one Si–O, Si–S, or Si–Se bond*	*IR*	*NMR*	*UV*
$HCOOSiH_3$	–	155, 156	156
$CH_3COOSiH_3$	–	155, 156	156
$CF_3COOSiH_3$	–	155, 156	156
CH_3OSiH_3	293, 294a, 603	134a, 155, 156, 290a, 354	–
★$HOSi(CH_3)_3$	344		
★$CH_3OSi(CH_3)_3$		645a	
★$ROSi(CH_3)_2CHBr$		575	
$CH_3OSiH_2CH_3$(*MS* [*661a*])	661a	661a	–
CH_3SSiH_3	294a	–	–
CF_3SSiH_3	129	129, 155, 157	–
CF_3SeSiH_3	143	143, 157	–
$C_3F_7SeSiH_3$	143	143, 157	–
$(CH_3)_2NC(O)OSiH_3$	155	–	155
$(CH_3)_2NC(O)OSiD_3$	155	–	–
$(CH_3)_2NC(S)OSiH_3$	155	155	155
$(CH\ )_2NC(S)OSiD_3$	155	–	–
$(CH_3)_2NC(S)SSiH_3$	155	–	155
$(CH_3)_2NC(S)SSiD_3$	155	–	–
$(CH_3)_2SiH(OCH_3)$(*MS* [*651a*])		651a	–
$(i\text{-}C_3H_7)_2SiHOH$	362a	–	–
$CH_3OSiH_2SiH_3$	–	134a	–

2. *Silanes and organosilanes with several Si–O, Si–S, and Si–Se bonds*	*IR*	*NMR*
$H_3SiOSiH_3$(*RA* [*371*])	111, 163, 293, 294a, 349, 369	155, 156, 157, 158, 290a, 354
★$(CH_3)_3SiOSi(CH_3)_3$(*RA* [*79b, 95*])	79b, 529a, 576	–
★$(C_2H_5)_3SiOSi(C_2H_5)_3$	476	–
★$(CH_3)_3SiOGe(CH_3)_3$	529a	–
$D_3SiOSiD_3$	349	–
$(SiH_3O)_2SiH_2$	–	290a
$SiH_3SiH_2OSiH_3$	–	134a
$(SiH_3SiH_2)_2O$	662, 663	134a
$(SiH_3SiH_2)_2S$	663	134a
$H_3SiSSiH_3$	163, 293	129, 157, 158
★$(CH_3)_3SiSSi(CH_3)_3$ (*RA* [*79b*])	79b	–
$H_3SiSeSiH_3$	163, 293	157
★$(CH_3)_3SiSeSi(CH_3)_3$ (*RA* [*79b*])	79b	–
★$(CH_3)_3SiTeSi(CH_3)_3$ (*RA* [*79b*])	79b	–
$(CH_3SiH_2)_2O$	154, 169	144, 145(?)
$(CH_3SiH_2)_2S$	154, 169	144
$[(CH_3)_2SiH]_2O$	580	–
$(CH_3)_3SiOSiH(CH_3)_2$	580	–
$[(CH_3)(i\text{-}C_3H_7)SiH]_2O$	362a	–
$[(i\text{-}C_3H_7)_2SiH]_2O$	362a	–
$\{(CH_3)[\overline{CH_2(CH_2)_4CH}]SiH\}_2O$	362a	–

References p. 621

2. *Silanes and organosilanes with several Si–O, Si–S, and Si–Se bonds*	*IR*	*NMR*
$\{[\overline{CH_2(CH_2)_4CH}]_2SiH\}_2O$	362a	–
$[(CH_3)(C_6H_5)SiH]_2O$	580	–
$[(C_2H_5)(C_6H_5)SiH]_2O$	580	–
$[(C_6H_5)_2SiH]_2O$	580	–
$[(C_6H_5)_3SiH]_2O$	362a	–
$(CH_3)(CH_3O)_2SiH$	24a	–
★$(CH_3O)_2Si(CH_3)_2$		645a
$(SiH_3O)_3SiH$	–	155, 290a
$[(CH_3)_2SiHO]_3SiH$	580	—
$[(CH_3)_3SiO]_3SiH$	580	—
$[Si_6O_3]H_6$ (siloxene) (*FL* [*264,265*])	—	—
$(SiH_3O)_2SiH_2$	—	155, 290a
$(CH_3)(C_2H_5O)_2SiH$	636, 580	—
$(CH_2{=}CHCH_2)(C_2H_5O)_2SiH$	—	158f
$[(CH_3)_3SiO](CH_3)(C_2H_5O)SiH$	580	—
$(CH_3O)_3SiH$	312b, 580, 636	290a
★$(CH_3O)_3SiCH_3$	522, 631	645a
$(C_2H_5O)_3SiH$	24a, 312b, 580, 636	97a
$(n\text{-}C_3H_7O)_3SiH$	312a	—
$(sec\text{-}C_3H_7O)_3SiH$	312a, 580	—
$(n\text{-}C_4H_9O)_3SiH$	312a	—
$(sec\text{-}C_4H_9O)_3SiH$	312a	—
$(i\text{-}C_4H_9O)_3SiH$	312a	—
$[(C_2H_5)_2CHCH_2O]_3SiH$	580	—
$(C_6H_5O)_3SiH$ (*RA* [*312a*])	312a	97a

3. *Silanes with Si–O and Si–Cl bonds*	*IR*
$CH_3OSiHCl_2$	312b
$C_2H_5OSiHCl_2$	312b
$(CH_3O)_2SiHCl$	312b
$(C_2H_5O)_2SiHCl$	312b
$(SiHCl_2)_2O$	580

11.5 Silanes with Si–N, Si–P, Si–As, and Si–Sb bonds (★ = for comparison)

1. *Silanes with Si–N bonds (ammonia and hydrazine derivatives)*	*IR*	*NMR*
$(i\text{-}C_3H_7)_2SiHNH_2$	362a	
$[\overline{CH_2(CH_2)_4CH}]_2SiHNH_2$	362a	
$(C_6H_5)_2SiHNH_2$	362a	
$[(CH_3)(i\text{-}C_3H_7)SiH]_2NH$	362a	
$[(i\text{-}C_3H_7)_2SiH]_2NH$	362a	
$\{(CH_3)[\overline{CH_2(CH_2)_4CH}]SiH\}_2NH$	362a	
$\{[\overline{CH_2(CH_2)_4CH}]SiH\}_2NH$	362a	
$[(C_6H_5)_2SiH]_2NH$	362a	

1. *Silanes with Si–N bonds (ammonia and hydrazine derivatives)*	*IR*	*NMR*
★$[(CH_3)_2SiNH]_3$	642	
$SiH_3N(CH_3)_2$	—	157, 290a
★$(CH_3)_3SiN(CH_3)_2$		645a
$SiH_3\overline{N(CH_2)_3C}H_2$	32b	—
$SiH_3\overline{N(CH_2)_4C}H_2$	32b	—
$SiH_3N(C_2H_5)_3$	32b	—
$SiH_3NH(C_6H_5)$	32e	
$SiH_3N(C_6H_5)$	32c	
$[(CH_3)_2N]_2SiH_2$	—	290a
$[(C_6H_5)_2N]_2SiH_2$	32c	—
$[(CH_3)_2N]_3SiH$	—	290a
$(SiH_3)_2NH$ (MS[32d]; *STR* [*485c*])	32d	—
$(SiH_3)_2NCH_3$	—	157
$(SiH_3)_2N(C_6H_5)$	32e	
$(CH_3SiH_2)_3N$	154	144, 145(?)
$[(CH_3)_2SiH]N[Si(CH_3)_3]_2$	83a	83a
$[(CH_3)_2SiH]_2N[Si(CH_3)_3]$	83a	83a
$[(CH_3)_2SiH]_3N$	83a	83a
$(SiH_3)_3N$ (*RA* [*117b, 323a*])	147, 293, 294a, 324a	157, 158
$(SiD_3)_3N$	147	—
★$[(CH_3)_3Si]_3N$	529a	—
$(CHF_2CF_2)SiH_3N(CH_3)_3$ (Si sp^3d)	104	104
$SiH_3SiH_2N(CH_3)_2$	2	134a
$(SiH_3SiH_2)_2NCH_3$	—	134a
★$[Si(CH_3)_3Si(CH_3)_2]_2NH$	645	—
$(SiH_3SiH_2)_3N$	663	134a
$SiH_3SiH[N(CH_3)_2]_2$	2	134a
$(SiH_3)_2N\text{-}N(SiH_3)_2$ (*RA* [*33*])	33	—
$(SiD_3)_2N\text{-}N(SiD_3)_2$ (*RA* [*33*])	33	—
★$(CH_3)_2N\text{-}N(CH_3)_2$	370	—
$(SiH_3NSiH_2)_3$ (*MS* [*665a*])	665a	665a

2. *Silanes with Si–N bonds (pseudohalides)*	*IR*	*NMR*	*LIV*
$(CH_3)_2SiHNC$	—	146	146
CH_3SiH_2NC	154, 169	146	146
SiH_3NC (*MW* [*405, 565*])	177, 293, 350	146	146
SiD_3NC (*MW* [*405*])			
★$(CH_3)_3SiNC$	66, 367	146	146
★$(C_2H_5)_3SiNC$	66	—	
$SiH_2Cl(NC)$	—	148a	
$SiH_2Br(NC)$	—	148a	
$SiH_2I(NC)$	—	148a	
$(SiH_3)_2CN_2$ (*RA* [*149a*])	149a	—	
$(SiD_3)_2CN_2$ (*RA* [*149a*])	149a	—	
★$[(CH_3)_3Si]_2CN_2$	528a	—	
SiH_3NCO	152, 294a	152, 157	
★Cl_3SiNCO	320b		
★H_2NCN	644		

References p. 621

2. *Silanes with Si–N bonds (pseudohalides)*	*IR*	*NMR*	*LIV*
⋆CH_3NCO (*MW* [*112*])	—	—	
⋆$(CH_3)_3CNCO$	320b		
⋆$ClSi(NCO)_3$	320b		
⋆$Si(NCO)_4$	94a, 378	—	
SiH_3NCS (*MS* [*283*])	294a, 524	—	
SiD_3NCS	524		
⋆Cl_3SiNCS	320b		
$(CH_3)_2SiHNCS$	326		
$SiH_2(NCS)_2 \cdot$ 2pyridine	87	—	
SiH_3NCSe	150	150, 157	
SiH_3N_3	149	—	
⋆$(CH_3)_3SiN_3$	103	—	
SiD_3N_3	149	—	

3. *Silanes with Si–P, Si–As, and Si–Sb bonds*	*IR*	*NMR*	*RA*
SiH_3PH_2 (*MS* [*516*])	131, 293, 347a	129a, 129b	—
$SiH_3PH_2BH_3$		129a	
$(SiH_3)_2PH$	—	31, 129a	—
$(SiH_3)_2PHBH_3$		129a	
$(SiH_3)_3P$ (*STR* [*55a*,*55c*, *117f*, *372a*])	10, 117b, 117f, 372a	129a	117b, 117f, 372a
$(SiD_3)_3P$	117f		117f
$Si_2H_5PH_2$	31, 235	31, 129a	—
$SiH_2ClSiH_2PH_2$		129b	
$Si_2H_5PH_2B(H,D)_3$		129a	
SiH_3PHMe		110a	
$(SiH_3)_2PMe$		110a	
SiH_3PMe_2		110a	
$SiH_3Si(PEt_2)_3$	—	190a	—
⋆$CH_3Si(PEt_2)_3$	—	198a	—
$(CH_3)_2SiH(PEt_2)$	—	198a, 198b	—
$CH_3SiH_2PEt_2$		198d	
$ClSiH_2PEt_2$		198c	
$Cl_2SiHPEt_2$		198c	
$SiH_2(PH_2)_2$	429b	429b	
$MeSiH(PEt_2)_2$		198d	
$ClSiH(PEt_2)_2$		198c	
$SiH(PH_2)_3$	429b	429b	
$SiH(PEt_2)_3$		198b	
⋆$(CH_3)_3Si(PEt_2)_3$	—	198a	—
$(SiH_3)AsH_2$	131, 131c, 293	—	131c
$(SiH_3)_3As$(*STR* [55b,55c,*372a*])	8, 117d, 372a	—	—
$(SiH_3)_3Sb$(*STR* [*485*])	9, 117d	—	—

11.6 Various silanes (⋆ = for comparison)

Silane	*IR*	*NMR*	*MS*	*MW*
GeH_3SiH_3	293, 592	—	217	106a
$(CH_3)_3GeSiH_3$	11a	11a		
$SiH_3SiH_2GeH_3$	355b	355b	355b	
$SiH_3GeH_2SiH_3$	355b	355b	355b	
$SiH_3GeH_2GeH_3$	355b	355b	355b	
$(CH_3)_3SnSiH_3$	—	11a		
$[(CH_3)_3Sn]_3SiH$	—	11a		
⋆$[(CH_3)_3Sn]_4Si$	—	11a		

Silane	*IR*	*NMR*	*MS*	*MW*
$SiH_3Co(CO)_4$	25e, 30, 31	25e		
$SiH_3Co(CO)_4 \cdot 2$ pyridine	30	—		
$SiH_3Mn(CO)_4$	25d, 30	25d		
$SiH_3Mn(CO)_4 \cdot 2$ pyridine	30	—		

REFERENCES

1. ABEDINI, M., C. H. VAN DYKE AND A. G. MACDIARMID, J. Inorg. Nucl. Chem. *25*, 307 (1963).
2. ABEDINI, M. AND A. G. MACDIARMID, Inorg. Chem. *2*, 608 (1963).
2a. ABEDINI, M. AND A. G. MACDIARMID, Inorg. Chem. **5**, 2040 (1966).
3. ABEL, E. W., G. NICKLESS AND F. H. POLLARD, Proc. Chem. Soc. *1960*, 288.
4. AGRE, C. L. AND W. HILLING, J. Am. Chem. Soc. *74*, 3895 (1952).
5. ALLEN, W. P. AND L. E. SUTTON, Acta Cryst. *3*, 46 (1950).
6. ALTSCHULLER, A. P. AND L. ROSENBAUM, J. Am. Chem. Soc. *77*, 272 (1955).
7. AMBERGER, E., Angew. Chem. *71* 327 (1959).
8. AMBERGER, E. AND H. BOETERS, Angew. Chem. *74*, 293 (1962).
9. AMBERGER, E. AND H. BOETERS, Z. Naturforsch. *18b*, 157 (1963).
10. AMBERGER, E. AND H. BOETERS, Angew. Chem. *74*, 32 (1962).
11. AMBERGER, E. AND R. RÖMER, Z. Anorg. Allgem. Chem. *345*, 1 (1966); and unpublished work.
11a. AMBERGER, E. AND E. MÜHLHOFER, J. Organometal. Chem. *12*, 55 (1968).
11b. AMBERGER, E., E. MÜHLHOFER AND H. STERN, J. Organomet. Chem. *17*, P. 5 (1969).
12. AMBERGER, E. *Hydrides of Ge, Sn, Pb and Bi*, Thesis, University of München, 1961.
13. ANDERSON, H. H., J. Am. Chem. Soc. *82*, 1323 (1960).
14. ANDERSON, H. H., J. Am. Chem. Soc. *80*, 5083 (1958).
15. ANDERSON, H. H., J. Am. Chem. Soc. *81*, 4785 (1959).
16. ANDERSON, T. F. AND A. B. BURG, J. Chem. Phys. *6*, 586 (1938).
17. ANDERSON, H. H. AND A. HENDIFAR, J. Am. Chem. Soc. *81*, 1027 (1959).
18. ANDERSON, H. H. AND H. FISCHER, J. Org. Chem. *19*, 1296 (1954).
19. ANDREWS, T. D. AND C. S. G. PHILLIPS, J. Chem. Soc. *A 1966*, 46.
20. ANDRIANOV, K. A., S. A. GOLUBTSOV, I. V. TROFIMOVA, A. S. DENISOVA AND R. A. TURETSKAYA, Dokl. Akad. Nauk SSSR, *108*, 465 (1956); C. A. *51*, 1026 (1957).
21. ANDRIANOV, K. A., S. A. GOLUBTSOV, T. V. TROFIMOVA AND A. S. DENISOVA, Zh. Prikl. Khim. *30*, 1277 (1957); C. A. *52*, 4472 (1958).
22. ANDRIANOV, K. A., V. I. PAKHOMOV AND N. E. LAPTEVA, Bull. Acad. Sci. USSR, Div. Chem. Sci. (Engl. transl.) *1962*, 1948.
23. ANDRIANOV, K. A., V. I. SIDOROV, L. M. KHANANASHVILI AND N. V. KUZNETSOVA, Bull. Acad. Sci. USSR, Div. Chem. Sci. (Engl. transl.) *1965*, 149.
24. ANDRIANOV, K. A., R. A. TURETSKAYA, S. A. GOLUBTSOV AND I. V. TROFIMOVA, Bull. Akad. Sci. USSR, Div. Chem. Sci. (Engl. transl.) *1962*, 1696.
24a. ATTRIDGE, C. F. Y. Organometal. Chem. *13*, 259 (1968).

25. Austin, J. D., C. Eaborn and J. D. Smith, J. Chem. Soc. *1963*, 4744.
25a. Aylett, B. J., Silicon hydrides and their derivatives, p. 249–307, in H. J. Emeleus and A. G. Sharpe, Editors, Advances in Inorg. Chem. and Radiochem., Vol. 11 (1968), Academic Press, New York/London.
25b. Aylett, B. J., Organometal. Chem. Rev. *A3*, 151 (1968).
25c. Aylett, B. J. and J. M. Campbell, J. Chem. Soc. *A1969*, 1920.
25d. Aylett, B. J. and J. M. Campbell, J. Chem. Soc. *A1969*, 1916.
25e. Aylett, B. J. and J. M. Campbell, J. Chem. Soc. *A1969*, 1910.
26. Aylett, B. J., J. Inorg. Nucl. Chem. *15*, 87 (1960).
27. Aylett, B. J., *Symposium on Hydrides of Groups 4, 5 and 6*, Berkeley, Calif., USA, 1963.
28. Aylett, B. J., J. Inorg. Nucl. Chem. *2*, 325 (1956).
29. Aylett, B. J., J. Inorg. Nucl. Chem. *5*, 292 (1958).
30. Aylett, B. J., Euchem Conference: *Chemistry of Metal Hydrides*, Bristol, 1966.
31. Aylett, B. J. and J. M. Campbell, Chem. Comm. *1965*, 217.
31a. Aylett, B. J., J. M. Campbell and A. Walton, Inorg. Nucl. Chem. Letters *4*, 79 (1968).
31b. Aylett, B. J. and I. A. Ellis, J. Chem. Soc. *1960*, 3415.
32. Aylett, B. J., H. J. Emeléus and A. G. Maddock, J. Inorg. Nucl. Chem. *1*, 187 (1955).
32a. Aylett, B. J. and J. Emsley, J. Chem. Soc. *A1967*, 652.
32b. Aylett, B. J. and J. Emsley, J. Chem. Soc. *A1967*, 1918.
32c. Aylett, B. J. and M. J. Hakim, J. Chem. Soc. *A1969*, 636.
32d. Aylett, B. J. and M. J. Hakim, J. Chem. Soc. *A1969*, 639.
32e. Aylett, B. J. and M. J. Hakim, J. Chem. Soc. *A1969*, 800.
32f. Aylett, B. J. and M. J. Hakim, J. Chem. Soc. *A1969*, 1788.
33. Aylett, B. J., J. R. Hall, D. C. McKean, R. Taylor and L. A. Woodward, Spectrochim. Acta *16*, 747 (1960).
34. Aylett, B. J. and M. J. Hakim, Inorg. Chem. *5*, 167 (1966).
35. Aylett, B. J. and L. K. Peterson, J. Chem. Soc. *1964*, 3429.
36. Bailey, D. L. and A. N. Pines, Ind. Eng. Chem. *46*, 2362 (1954).
37. Baines, J. E. and C. Eaborn, J. Chem. Soc. *1955*, 4023.
38. Baines, J. E. and C. Eaborn, J. Chem. Soc. *1956*, 1436.
39. Bak, B., J. Bruhn and J. Rastrut-Andersen, J. Chem. Phys. *21*, 752 (1953).
39a. Ball, D. F., T. Carter, D. C. McKean and L. A. Woodward, Spectrochim. Acta *20*, 1721 (1964).
40. Ball, D. F., P. L. Goggin, D. C. McKean and L. A. Woodward, Spectrochim. Acta *16*, 1358 (1960).
41. Ball, D. F. and D. C. McKean, Spectrochim. Acta *18*, 1019 (1962).
42. Barry, A. J., Brit. Pat. 671 710 (1952); C.A. *47*, 4909 (1953).
43. Barry, A. J., D. E. Hook and L. De Pree, U.S. Pat. 2 511 820 (1950); C.A. *44*, 8370 (1950).
44. Barry, A. J., D. E. Hook and L. De Pree, Brit. Pat. 653 300 (1951); C.A. *46*, 7113 (1952).
45. Barry, A. J., L. De Pree, J. W. Gilkey and D. E. Hook, J. Am. Chem. Soc. *69*, 2916 (1947).
46. Barry, A. J., L. De Pree and D. E. Hook, Brit. Pat. 632 824 (1949); C.A. *44*, 5378 (1950).
47. Barry, A. J. and L. De Pree, U.S. Pat. 2 488 487 (1949); C.A. *44*, 2547 (1950).
48. Barry, A. J. and L. De Pree, Brit. Pat. 609 507 (1948); C.A. *43*, 2221 (1949).
49. Barry, A. J., Brit. Pat. 622 970 (1949); C.A. *44*, 658 (1950).
50. Barrow, R. F. and J. L. Deutsch, Proc. Chem. Soc. *1960*, 122.
51. Bartell, L. S. and K. Kuchitsu, J. Chem. Phys. *37*, 691 (1962).
52. Bartlett, P. D. and L. H. Knox, J. Am. Chem. Soc. *61*, 3184 (1939).
52a. Bates, P., S. Cawley and S. S. Danyluk, J. Chem. Phys. *40*, 2415 (1964).
53. Batuev, M. I., V. A. Ponomarenko, A. D. Matveeva and A. D. Petrov, Dokl. Akad. Nauk USSR *95*, 805 (1954); C.A. *49*, 6089 (1955).
54. Bauer, R., Z. Naturforsch. *17b*, 201 (1962).
54a. Bauer, R., private communication, 1963.
55. Baukov, Yu. I. and I. F. Lutsenko, J. Gen. Chem. USSR (Eng. Transl.) *32*, 2705 (1962).
55a. Beagley, B., A. G. Robiette and G. M. Sheldrick, J. Chem. Soc. *A1968*, 3002.
55b. Beagley, B., A. G. Robiette and G. M. Sheldrick, J. Chem. Soc. *A1968*, 3006.
55c. Beagley, B., A. G. Robiette and G. M. Sheldrick, Chem. Commun *1967*, 601.
56. Becker, G. E. and G. W. Gobeli, J. Chem. Phys. *38*, 2942 (1963).
56a. Bedford, J. A., J. R. Bolton, A. Carrington and R. H. Prince, Trans. Faraday Soc. *59*, 53 (1963).
57. Bell, T. N., R. N. Haszeldine, M. J. Newlands and J. B. Plumb, J. Chem. Soc. *1965*, 2107.

57a. Bellama, J. M. and A. G. MacDiarmid, Inorg. Chem. 7, 2070 (1968).
58. Belyakova, Z. V., S. A. Golubtsov and T. M. Yakusheva, J. Gen. Chem. USSR (Eng. Transl.) *35*, 1187 (1965).
59. Belyakova, Z. V., M. G. Pomerantseva and S. A. Golubtsov, J. Gen. Chem. USSR (Eng. Transl.) *35*, 1053 (1965).
60. Benkeser, R. A., M. L. Burrous, L. E. Nelson and J. V. Swisher, J. Am. Chem. Soc. *83*, 4385 (1961).
61. Benkeser, R. A., S. Dunny and P. R. Jones, J. Organometal. Chem. *4*, 338 (1965).
62. Benkeser, R. A., R. A. Hickner, J. Am. Chem. Soc. *80*, 5298 (1958).
63. Benkeser, R. A., H. Landesman and D. J. Forster, J. Am. Chem. Soc. *74*, 648 (1952).
63a. Bennett, S. W., C. Eaborn, A. Hudson, H. A. Hussain and R. A. Jackson, J. Organometal. Chem. *16*, P36 (1969).
64. Bethke, G. W. and M. K. Wilson, J. Chem. Soc. *26*, 1107 (1957).
65. Birkhofer, L., A. Ritter and P. Richter, Angew. Chem. *74*, 293 (1962).
65a. Bishop, E. O. and M. A. Jensen, Chem. Comm. *1966*, 922.
66. Bither, T. A., W. H. Knoth, R. V. Lindsey Jr. and W. H. Sharkey, J. Am. Chem. Soc. *80*, 4151 (1958).
67. Bluestein, B. A., J. Am. Chem. Soc. *83*, 1000 (1961).
68. Boeters, H. D., Thesis, University of München, 1963.
69. Boyd, D. R., J. Chem. Phys. *23*, 922 (1955).
70. Borer, K. and C. S. G. Phillips, Proc. Chem. Soc. *1959*, 189.
70a. Bornhorst, W. R. and M. A. Ring, Inorg. Chem. *7*, 1009 (1968).
71. Brit. Pat. 646 629 (1950); C.A. *45*, 5184 (1951).
71a. Brit. Pat. 661 094 (1951); C.A. *46*, 5365 (1952).
71b. Brit. Pat. 771 587; C.A. *51*, 13904 (1957).
71c. Brit. Pat. 769 497 (1957); C.A. *52*, 430 (1958).
71d. Brit. Pat. 752 700 (1956); C.A. *51*, 7402 (1957).
71e. Brit. Pat. 737 963 (1955); C.A. *50*, 10760 (1956).
71f. Brit. Pat. 764 288 (1956); C.A. *51*, 14786 (1957).
71g. Brit. Pat. 746 510 (1956); C.A. *51*, 7402 (1957).
71h. Brit. Pat. 769 496 (1957); C.A. *52*, 429 (1958).
71i. Brit. Pat. 645 972 (1950); C.A. *45*, 5184 (1951).
72. *British Thomson-Houston Co. Ltd.*, Brit. Pat. 663 740; C.A. *46*, 11228 (1952).
73. Brockway, L. O. and J. Y. Beach, J. Am. Chem. Soc. *60*, 1836 (1938).
74. Brimm, E. O. and H. M. Humphreys, J. Phys. Chem. *61*, 829 (1957).
75. Brockway, L. O. and I. E. Coop, Trans. Faraday Soc. *34*, 1429 (1938).
76. Brook, A. G. and G. J. D. Peddle, Can. J. Chem. *41*, 2351 (1963).
77. Brook, A. G. and S. Wolfe, J. Am. Chem. Soc. *79*, 1431 (1957).
78. Brown, M. P., E. Cartmell and G. W. A. Fowles, J. Chem. Soc. *1960*, 506.
79. Brown, C. A. and R. C. Osthoff, J. Am. Chem. Soc. *74*, 2340 (1952).
79a. Bürger, H. Spectrochim. Acta *24A*, 2015 (1968).
79b. Bürger, H., V. Goetze and W. Sawodny, Spectrochim. Acta *24A*, 2003 (1968).
79c. Bürger, H. and W. Kilian, J. Organometal. Chem. *18*, 299 (1969).
80. Buff, H. and F. Wöhler, Ann. Chem. *104*, 94 (1857).
81. Burg, A. B. and E. S. Kuljian, J. Am. Chem. Soc. *72*, 3103 (1950).
81a. Burg, A. B. and J. S. Basi, J. Am. Chem. Soc. *90*, 3361 (1968).
82. Burkhard, C. A., J. Am. Chem. Soc. *72*, 1402 (1950).
83. Burkhard, C. A. and R. H. Krieble, J. Am. Chem. Soc. *69*, 2687 (1947).
83a. Bush, R. P., N. C. Lloyd and C. A. Pearce, J. Chem. Soc. *A1969*, 253.
84. Cade, J. A. Tetrahedron *2*, 322 (1958).
85. Cambensi, H., Thesis, University of München, 1964.
86. Campbell-Ferguson, H. J. and E. A. V. Ebsworth, *Intern. Symp. Organosilicon Chem., Prague*, *1965*, p. 259.
87. Campbell-Ferguson, H. J. and E. A. V. Ebsworth, J. Chem. Soc. *A1967*, 705.
88. Calas, R., N. Duffaut and J. Valade, Bull. Soc. Chim. France *1955*, 790.
89. Calas, R. and N. Duffaut, Rev. Franc. Corps Gras *3*, 5 (1956); C.A. *50*, 12890 (1956).
90. Calas, R. and N. Duffaut, Oleagineux *8*, 21 (1953); C. A. *47*, 12223 (1953).
91. Calas, R. and N. Duffaut, Bull. Mens. Inform. ITERG *7*, 438 (1953); C.A. *48*, 11303 (1954).
92. Calas, R. and N. Duffaut, Compt. Rend. *245*, 906 (1957).
93. Calas, R. and N. Duffaut, Compt. Rend. *243*, 595 (1956).

94. CALAS, R., E. FRAINNET AND J. BONASTRE, Compt. Rend. *251*, 2987 (1960).
94a. CARLSON, G. L. Spectrochim. Acta *18*, 1529 (1962).
95. CERATO, C. C., J. L. LAUER AND H. C. BEACHELL, J. Chem. Phys. *22*, 1 (1954).
95a. CETINI, G., O. GAMBINO, M. CASTIGLIONI AND P. VOLPE, J. Chem. Phys. *46*, 89 (1967).
96. CHALK, A. J. AND F. HARROD, J. Am. Chem. Soc. *87*, 16 (1965).
97. CHERNYSHEV, E. A. AND N. G. TOLSTIKOVA, Bull. Acad. Sci. USSR Div. Chem. Sci (Engl. transl.) *1964*, 1606.
97a. CHIDEKEL, M. L., A. N. EGOROCHKIN, V. A. PONOMARENKO, H. A. ZADOROZHNYI, G. A. RAZUVAEV AND A. D. PETROV, Izv. Akad. Nauk SSSR, Otd. Khim. Nauk *1963*, 1130 [Bull. Acad. Sci. USSR, Div. Chem. Sci. (Engl. transl.) *1963*, 1032].
98. CHERNYSHEV, E. A. AND A. D. PETROV, Izv. Akad. Nauk SSSR, Otd. Khim. Nauk *1956*, 630; C.A. *51*, 1064 (1957).
99. CLASEN, H., Angew. Chem. *70*, 179 (1958).
100. COTTERELL, T., *The Strength of the Chemical Bonds*, 2nd ed., Butterworth, London, 1958.
101. COLLINS, R. L. AND J. R. NIELSEN, J. Chem. Phys. *23*, 351 (1955).
102. CONNER, J. A., R. N. HASZELDINE AND G. J. LEIGH, *Intern. Symp. Organosilicon Chem., Prague, 1965*, p. 109.
103. CONNOLLY, J. W. AND G. URRY, Inorg. Chem. *1*, 718 (1962).
104. COOK. D. I., R. FIELDS, M. GREEN, R. N. HASZELDINE, B. R. ILES AND M. J. NEWLANDS, J. Chem. Soc. *A 1966*, 887.
105. COREY, J. Y. AND R. WEST, J. Am. Chem. Soc. *85*, 2430 (1963).
105a. COWLEY, A. H., W. D. WHITE AND S. L. MANATT, J. Am. Chem. Soc. *89*, 6433 (1967).
106. COWELL, R. D. AND G. URRY, J. Am. Chem. Soc. *85*, 822 (1963).
106a. COX, A. P. AND R. VARMA, J. Chem. Phys. *46*, 2007 (1967).
107. CRAIG, D. P., A. MACCOLL, R. S. NYHOLM, L. E. ORGEL AND L. E. SUTTON, J. Chem. Soc. *1954*, 339.
108. CRAIG, A. D. AND A. G. MACDIARMID, J. Inorg. Nucl. Chem. *24*, 161 (1962).
109. CRAIG, A. D., J. V. URENOVITCH AND A. G. MACDIARMID, J. Chem. Soc. *1962*, 548.
110. CRAWFORD, B. L. AND L. JOYCE, J. Chem. Phys. *7*, 307 (1939).
110a. CROSBIE, K. D., C. GLIDEWELL AND G. M. SHELDRICK, J. Chem. Soc. *A1969*, 1861.
111. CURL, R. F. AND K. S. PITZER, J. Am. Chem. Soc. *80*, 2371 (1958).
112. CURL JR., R. F., V. M. RAO, K. V. L. N. SASTRY AND J. A. HODGESON, J. Chem. Phys. *39*, 3335 (1963).
113. CURRAN, C., R. M. WITUCKI AND P. A. MCCUSKER, J. Am. Chem. Soc. *72*, 4471 (1950).
114. CURRY, J. W., J. Am. Chem. Soc. *78*, 1686 (1956).
115. CURRY, J. W. AND G. W. HARRISON, J. Org. Chem. *23*, 627 (1958).
116. CUSA, N. W. AND F. S. KIPPING, J. Chem. Soc. *1933*, 1040.
117. DANNELS, B. F. AND H. W. POST, J. Org. Chem. *22*, 748 (1957).
117a. DANYLUK, S. S., J. Am. Chem. Soc. *86*, 4504 (1964).
117b. DAVIDSON, G., E. A. V. EBSWORTH, G. M. SHELDRICK AND L. A. WOODWARD, Chem. Commun. *1965*, 122.
117c. DANYLUK, S. S., J. Am. Chem. Soc. *87*, 2300 (1965).
117d. DAVIDSON G., L. A. WOODWARD, E. A. V. EBSWORTH AND G. M. SHELDRICK, Spectrochim. Acta *23A*, 2609 (1967).
117e. DAVIDSON, I. M. T., C. EABORN AND C. J. WOOD, J. Organometal. Chem. *10*, 401 (1967).
117f. DAVIDSON, G., E. A. V. EBSWORTH, G. M. SHELDRICK AND L. WOODWARD, Spectrochim. Acta *22*, 67 (1966).
118. DEANS, D. R. AND C. EABORN, J. Chem. Soc. *1954*, 3169.
119. DELWAULLE, M. L., M. F. FRANÇOIS AND M. DELHAYE-BUISSET, J. Phys. Radium *15*, 206 (1954).
120. DENNIS, L. M., R. B. COREY AND R. W. MOORE, J. Am. Chem. Soc. *46*, 657 (1924).
121. DE PREE, L., A. J. BARRY AND D. E. HOOK, U.S. Pat. 2 469 355 (1949); C.A. *43*, 5791 (1949).
121a. DEVYATYKH, G. G. AND G. K. BORISOV, Zh. Fizičeskoj Khim. *37*, 1985 (1963).
121b. DEVYATYKH, G. G., K. G. BORISOV AND A. M. PAVLOV, Dokl. Akad. Nauk CCCP *138*, 402 (1961).
122. DEVYATIK, G. G., V. M. KEDYARKIN AND A. D. ZORIN, Russ. J. Inorg. Chem. (English Transl.) *10*, 833 (1965). J. Neorg. Khim. *10*, 1528 (1965).
122a. DEVYATYKH, G. G. AND S. M. VLASOV, Zh. Fizičeskoj Khim. *39*, 1171 (1965).
122b. DEVYATYKH, G. G. AND A. S. YUSHIN, Zh. Fizičeskoj Khim. *37*, 957 (1964).

123. Dewhurst, H. A. and G. D. Cooper, J. Am. Chem. Soc. *82*, 4220 (1960).
124. Dibeler, V. H., J. Res. Natl. Bur. Std. *49*, 235 (1952).
124a. Ditchfield, R., M. A. Jensen and J. N. Murell, J. Chem. Soc. *A 1967*, 1674.
125. Dixon, R. N. and N. Sheppard, J. Chem. Phys. *23*, 215 (1955).
126. Dolgov, B. N., S. N. Borisov, M. G. Voronkov, Zh. Obshch. Khim. *27*, 716 (1957); C.A. *51*, 16282 b (1957).
127. Dolgov, B. N., N. P. Kharitonov and M. G. Voronkov, Zh. Obshch. Khim. *24*, 1178 (1954); C.A. *49*, 12275 (1955).
128. Dow Corning Corp., Brit. Pat. 645 972 (1950); C.A. *45*, 5184 (1951).
129. Downs, A. J. and E. A. V. Ebsworth, J. Chem. Soc. *1960*, 3516.
129a. Drake, J. E. and N. Goddard, J. Chem. Soc. *A 1969*, 662.
129b. Drake, J. E., N. Goddard and C. Riddle, J. Chem. Soc. *A1969*, 2704.
130. Drake, J. E. and J. Simpson, Inorg. Nucl. Letters *2*, 219 (1966).
131. Drake, J. E. and W. L. Jolly, Chem. Ind. London *1962*, 1470.
131a. Drake, J. E. and J. Simpson, Chem. Commun. *1967*, 249.
131b. Drake, J. E. and J. Simpson, J. Chem. Soc. *A 1968*, 1039.
131c. Drake, J. E. and J. Simpson, Spectrochim. Acta *24A*, 981 (1968).
131d. Drake, J. E. and J. Simpson, Inorg. Chem. *6*, 1984 (1967).
131e. Drake, J. E., C. Riddle and D. E. Rogers, J. Chem. Soc. *A 1969*, 910.
132. Dreeskamp, H., Z. Physik. Chem. Frankfurt *38*, 121 (1963).
133. Dreeskamp, H., Z. Naturforsch. *19a*, 139 (1964).
133a. Dutton, W. A. and M. Onyszchuk, Inorg. Chem. *7*, 1735 (1968).
134. Dudani, P. G. and H. G. Post, Nature *194*, 85 (1962).
134a. Dyke, Ch. H. van and A. G. MacDiarmid, Inorg. Chem. *3*, 1071 (1964).
135. Eaborn, C., *Organosilicon Compounds*, Butterworth, London, 1960.
136. Eaborn, C., J. Chem. Soc. *1955*, 2047.
137. Eaborn, C., J. Chem. Soc. *1949*, 2755.
138. Eaborn, C., J. Chem. Soc. *1950*, 3077.
139. Eaborn, C., J. Chem. Soc. *1955*, 2517.
140. Eaborn, C., J. Chem. Soc. *1953*, 4154.
141. Ebsworth, E. A. V., Chem. Commun. *1960*, 530.
142. Ebsworth, E. A. V. and H. J. Emeléus, J. Chem. Soc. *1958*, 2150.
143. Ebsworth, E. A. V., H. J. Emeléus and N. Welcman, J. Chem. Soc. *1962*, 2290.
144. Ebsworth, E. A. V. and S. G. Frankiss, Trans. Faraday Soc. *59*, 1518 (1963).
145. Ebsworth, E. A. V. and S. G. Frankiss, J. Am. Chem. Soc. *85*, 3516 (1963).
146. Ebsworth, E. A. V. and S. G. Frankiss, J. Chem. Soc. *1963*, 661.
146a. Ebsworth, E. A. V., C. Glidewell and G. M. Sheldrick, J. Chem. Soc. *A 1969*, 353.
147. Ebsworth, E. A. V., J. R. Hall, M. J. MacKillop, D. C. McKean, N. Sheppard and L. A. Woodward, Spectrochim. Acta *13*, 202 (1958).
148. Ebsworth, E. A. V., D. R. Jenkins, M. J. Mays and T. M. Sugden, Proc. Chem. Soc. *1963*, 21.
148a. Ebsworth, E. A. V., A. G. Lee and G. M. Sheldrick, J. Chem. Soc. *A 1968*, 2294.
149. Ebsworth, E. A. V. and M. J. Mays, J. Chem. Soc. *1964*, 3450.
149a. Ebsworth, E. A. V. and M. J. Mays, Spectrochim. Acta *19*, 1127 (1963).
150. Ebsworth, E. A. V. and M. J. Mays, J. Chem. Soc. *1963*, 3893.
151. Ebsworth, E. A. V. and M. J. Mays, J. Chem. Soc. *1961*, 4879.
152. Ebsworth, E. A. V. and M. J. Mays, J. Chem. Soc. *1962*, 4844.
153. Ebsworth, E. A. V., R. Mould, R. Taylor, G. R. Wilkinson and L. A. Woodward, Trans. Faraday Soc. *58*, 1069 (1962).
154. Ebsworth, E. A. V., M. Onyszchuk and N. Sheppard, J. Chem. Soc. *1958*, 1453.
155. Ebsworth, E. A. V., G. Rocktäschl and J. C. Thompson, J. Chem. Soc. *A 1967*, 362.
156. Ebsworth, E. A. V. and J. C. Thompson, J. Chem. Soc. *A 1967*, 69.
157. Ebsworth, E. A. V. and J. C. Thompson, J. Phys. Chem. *67*, 805 (1963).
157a. Ebsworth, E. A. V. and J. J. Turner, Trans. Faraday Soc. *60*, 256 (1964).
158. Ebsworth, E. A. V. and J. J. Turner, J. Chem. Phys. *36*, 2628 (1962).
158a. Egorochkin, A. N., M. L. Chidekel', V. A. Ponomarenko, G. Ya. Zueva, S. S. Svirezheva and G. A. Razuvaev, Izv. Akad. Nauk SSSR *1963*, 1865.
158b. Egorochkin, A. N., M. L. Chidekel', V. A. Ponomarenko and N. A. Zadorozhnyi, Izv. Akad. Nauk SSSR *1963*, 1868.

158c. EGOROCHKIN, A. N., V. F. MIRONOV AND M. G. VORONKOV, Zh. Strukt. Khim. *7*, 450 (1966).
158d. EGOROCHKIN, A. N., M. L. CHIDEKEL' AND G. A. RAZUVAEV, Izv. Akad. Nauk SSSR *1966*, 437.
158e. EGOROCHKIN, A. N., M. L. CHIDEKEL', G. A. RAZUVAEV, V. F. MIRONOV AND A. L. KRAVCHENKO, Izv. Akad. Nauk SSSR *1966*, 1312.
158f. EGOROCHKIN, A. N., M. L. CHIDEKEL', G. A. RAZUVAEV, G. G. PETUCHOV AND V. F. MIRONOV, Izv. Akad. Nauk SSSR *1966*, 1522.
158g. EGOROCHKIN, A. N., E. YA. LUKEVITS AND M. G. VORONKOV, Khim. Geterotsikl. Soedin. *1965*, 499.
159. EISCH, J. J. AND J. T. TRAINOR, J. Org. Chem. *28*, 487 (1963).
160. EISINGER, J., J. Chem. Phys. *30*, 927 (1959).
161. EL-ABBADY, A. M. AND L. C. ANDERSON, J. Am. Chem. Soc. *80*, 1737 (1958); *78*, 2278 (1956).
162. EL-ABBADY, A. M. AND L. C. ANDERSON, J. Am. Chem. Soc. *78*, 2278 (1956); *80*, 1737 (1958).
163. EMELÉUS, H. J., A. G. MACDIARMID AND A. G. MADDOCK, J. Inorg. Nucl. Chem. *1*, 194 (1955).
164. EMELÉUS, H. J., K. M. MACKAY, J. Chem. Soc. *1961*, 2676.
165. EMELÉUS, H. J. AND A. G. MADDOCK, J. Chem. Soc. *1946*, 1131.
166. EMELÉUS, H. J. AND A. G. MADDOCK, J. Chem. Soc. *1944*, 293.
167. EMELÉUS, H. J., A. G. MADDOCK AND C. REID, J. Chem. Soc. *1941*, 353.
168. EMELÉUS, H. J. AND H. MILLER, J. Chem. Soc. *1939*, 819.
169. EMELÉUS, H. J., M. ONYSZCHUK AND W. KUCHEN, Z. Anorg. Allgem. Chem. *283*, 74 (1956).
170. EMELÉUS, H. J. AND D. S. PAYNE, J. Chem. Soc. *1947*, 1590.
171. EMELÉUS, H. J. AND C. REID, J. Chem. Soc. *1939*, 1021.
172. EMELÉUS, H. J. AND S. R. ROBINSON, J. Chem. Soc. *1947*, 1592.
173. EMELÉUS, H. J. AND K. STEWARD, Nature *135*, 397 (1935).
174. EMELÉUS, H. J. AND K. STEWARD, Trans. Faraday Soc. *32*, 1577 (1936).
174a. EMSLEY, J. Chem. Soc. *A1968*, 1009.
175. ENGLISH, W. D. AND R. V. V. NICHOLLS, J. Am. Chem. Soc. *72*, 2764 (1950).
176. EVERS, E. C., W. O. FREITAG, W. A. KRINER, A. G. MACDIARMID AND S. SUJISHI, J. Inorg. Nucl. Chem. *13*, 239 (1960).
177. EVERS, E. C., W. O. FREITAG, J. N. KEITH, W. A. KRINER, A. G. MACDIARMID AND S. SUJISHI, J. Am. Chem. Soc. *81*, 4493 (1959).
178. EYSTER, E. H., R. H. GILLETTE AND L. O. BROCKWAY, J. Am. Chem. Soc. *62*, 3236 (1940).
179. FEHÉR, R., *Proc. 17th Intern. Congr. Pure Appl. Chem., München, 1959*, Abstr. A 217.
179a. FEHÉR, F., G. JANSEN AND H. ROHMER, Z. Anorg. Allgem. Chem. *329*, 31 (1964).
180. FEHÉR, F. G., KUHLBÖRSCH AND H. LUHLEICH, Z. Naturforsch. *14b*, 466 (1959).
181. FEHÉR, F., G. KUHLBÖRSCH AND H. LUHLEICH, Z. Anorg. Allgem. Chem. *303*, 294 (1960).
182. FEHÉR, F., G. KUHLBÖRSCH AND H. LUHLEICH, Z. Anorg. Allgem. Chem. *303*, 283 (1960).
182a. FEHÉR, F. AND H. STRACK, Naturwiss. *50*, 570 (1963).
183. FEHÉR, F. AND W. TROMM, Z. Anorg. Allgem. Chem. *282*, 29 (1955).
184. FEINBERG, R. S. AND E. G. ROCHOW, J. Inorg. Nucl. Chem. *24*, 165 (1962).
185. FILIPOV, M. T., R. V. DZHAGATSPANYAN, G. V. MOTSAREV AND V. I. ZETKIN, Russ. J. Phys. Chem. (English Transl.) *36*, 942 (1962).
186. FINHOLT, A. E., A. C. BOND JR., K. E. WILZBACH AND H. I. SCHLESINGER, J. Am. Chem. Soc. *69*, 2692 (1947).
187. FÖRSTER, W. AND H. KRIEGSMANN, Z. Anorg. Allgem. Chem. *326*, 186 (1963).
188. FOX, K., K. T. HECHT, R. E. MEREDITH AND C. W. PETERS, J. Chem. Phys. *36*, 3135 (1962).
189. FRIEDEL, C. AND A. LADENBURG, Ann. Chem. *143*, 118 (1867).
190. FRIEDLINA, R. KH., Izv. Akad. Nauk SSSR *1957*, 1333; C.A. *52*, 7217 (1958).
191. FRAINNET, E., A. BASOUIN AND R. CALAS, Compt. Rend. *257*, 1304 (1963).
192. FRAINNET, E. AND R. CALAS, Compt. Rend. *240*, 203 (1956).
192a. FRANKISS, S. G., Spectrochim. Acta *22*, 295 (1966).
193. FRITZ, G., Fortschr. Chem. Forsch. *4*, 459 (1963).
194. FRITZ, G., Z. Naturforsch. *10b*, 423 (1955).
195. FRITZ, G., Z. Naturforsch. *7b*, 507 (1952).
196. FRITZ, G., Z. Anorg. Allgem. Chem. *273*, 275 (1953).
197. FRITZ, G., Z. Anorg. Allgem. Chem. *280*, 332 (1955).
198. FRITZ, G., Z. Naturforsch. *5b*, 444 (1950).
198a. FRITZ, G. AND G. BECKER, *Intern. Symp. Organometal. Chem., 3rd München, 1967.*
198b. FRITZ, G. AND G. BECKER, Z. Anorg. Allg. Chem. in press.

198c. Fritz, G. and G. Becker, Z. Anorg. Allg. Chem. in press.
198d. Fritz, G., G. Becker and D. Kummer, Z. Anorg. Allg. Chem. in press.
199. Fritz, G. and H. O. Berkenhoff, Z. Anorg. Allgem. Chem. *300*, 205 (1959).
200. Fritz, G., H. J. Buhl and D. Kummer, Z. Anorg. Allgem. Chem. *327*, 165 (1964).
200a. Fritz, G., H. Fröhlich and D. Kummer, Z. Anorg. Allgem. Chem. *353*, 34 (1967).
201. Fritz, G. and J. Grobe, Z. Anorg. Allgem. Chem. *311*, 325 (1961).
202. Fritz, G. and J. Grobe, Z. Anorg. Allgem. Chem. *315*, 157 (1962).
202a. Fritz, G., R. Haase and D. Kummer, Z. Anorg. Allgem. Chem. *365*, 1 (1969).
203. Fritz, G. and W. Kemmerling, Z. Anorg. Allgem. Chem. *322*, 34 (1963).
203a. Fritz, G. and D. Ksinsik, Z. Anorg. Allgem. Chem. *304*, 241 (1960).
204. Fritz, G. and D. Kummer, Chem. Ber. *94*, 1143 (1961).
205. Fritz, G. and D. Kummer, Z. Anorg. Allgem. Chem. *308*, 105 (1961).
206. Fritz, G. and D. Kummer, Z. Anorg. Allgem. Chem. *304*, 322 (1960).
207. Fritz, G. and D. Kummer, Z. Anorg. Allgem. Chem. *306*, 191 (1960).
208. Fritz, G. and D. Kummer, Z. Anorg. Allgem. Chem. *310*, 327 (1961).
208a. Fritz, G., R. Wiemers and U. Protzer, Z. Anorg. Allg. Chem. *363*, 225 (1968).
208b. Gaspar, P. P., S. A. Bock and W. C. Eckelman, J. Am. Chem. Soc. *90*, 6914 (1968).
208c. Gaspar, P. P., S. A. Bock and C. A. Levy, Chem. Commun. *1968*, 1317.
209. Gadsby, G. N., Research (London) *3*, 338 (1950).
210. George, M. V., D. J. Peterson and H. Gilman, J. Am. Chem. Soc. *82*, 403 (1960).
211. George, P. D., M. Prober and J. R. Elliott, Chem. Rev. *56*, 1065 (1956).
212. George, P. D., L. H. Sommer and F. C. Whithmore, J. Am. Chem. Soc. *77*, 1677 (1955).
212a. Gerry, M. C. L. and T. M. Sugden, Trans. Faraday Soc. *61*, 2091 (1965).
213. Geyer, A. M. and R. N. Haszeldine, J. Chem. Soc. *1957*, 3925.
214. Geyer, A. M. and R. N. Haszeldine, J. Chem. Soc. *1957*, 1038.
215. Geyer, A. M. and R. N. Haszeldine, Nature *178*, 808 (1956).
216. Geyer, A. M., R. N. Haszeldine, K. Leedham and R. J. Marklow, J. Chem. Soc. *1957*, 4472.
217. Gibbon, G. A., Y. Rousseau, C. H. van Dyke and G. J. Mains, Inorg. Chem. *5*, 114 (1966).
218. Gilman, H. and R. N. Clark, J. Am. Chem. Soc. *69*, 1499 (1947).
219. Gilman, H. and G. E. Dunn, J. Am. Chem. Soc. *73*, 3404 (1951).
220. Gilman, H. and G. E. Dunn, Chem. Rev. *52*, 77 (1953).
221. Gilman, H. and G. E. Dunn, J. Am. Chem. Soc. *73*, 4499 (1951).
222. Gilman, H. and G. D. Lichtenwalter, J. Am. Chem. Soc. *80*, 608 (1958).
223. Gilman, H. and H. W. Melvin, J. Am. Chem. Soc. *71*, 4050 (1949).
224. Gilman, H. and D. H. Miles, J. Org. Chem. *23*, 326 (1958).
225. Gilman, H., D. J. Peterson and D. Wittenberg, Chem. Ind. (London) *1958*, 1479.
226. Gilman, H. and G. N. R. Smart, J. Org. Chem. *15*, 721 (1950).
227. Gilman, H., T. C. Wu, H. A. Hartzfeld, G. A. Guter, A. G. Smith, J. J. Goodman and S. H. Eidt, J. Am. Chem. Soc. *74*, 561 (1952).
228. Gilman, H. and E. A. Zuech, J. Am. Chem. Soc. *79*, 4560 (1957).
229. Gilman, H. and E. A. Zuech, J. Am. Chem. Soc. *81*, 5925 (1959).
229a. Glemser, O. and W. Lohmann, Z. Anorg. Allgem. Chem. *275*, 260 (1954).
229b. Glidewell, C. and G. M. Sheldrick, J. Chem. Soc. *A1969* 350.
230. Gmelins Handbuch der anorganischen Chemie *15*/C (1958).
231. Goering, H. L., P. I. Abel and B. F. Aycock, J. Am. Chem. Soc. *74*, 3588 (1952).
232. Gokhale, S. D., J. E. Drake and W. L. Jolly, J. Inorg. Nucl. Chem. *27*, 1911 (1965).
233. Gokhale, S. D. and W. L. Jolly, Inorg. Chem. *3*, 949 (1964).
234. Gokhale, S. D. and W. L. Jolly, Inorg. Chem. *3*, 1141 (1964).
235. Gokhale, S. D. and W. L. Jolly, Inorg. Chem. *4*, 596 (1965).
236. Goldblatt, L. O. and D. M. Oldroyd, U.S. Pat. 2 533 240 (1950); C.A. *45*, 2262 (1951).
237. Golubtsov, S. A., K. A. Andrianov and G. S. Popoleva, *Intern. Symp. Organosilicon Chem., Prague, 1965*, p. 115.
238. Golubtsov, S. A., R. A. Turetskaya, K. A. Andrianov and Ya. I. Vabel', Bull. Acad. Sci. USSR, Div. Chem. Sci. (Engl. transl.) *1963*, 77.
239. Goodman, L., R. M. Silverstein and A. Benitez, J. Am. Chem. Soc. *79*, 3073 (1957).
240. Goodman, L., R. M. Silverstein and J. N. Shooley, J. Am. Chem. Soc. *78*, 4493 (1956).
241. Goodspeed, N. C. and R. T. Sanderson, J. Inorg. Nucl. Chem. *2*, 266 (1956).
242. Goubeau, J. and H. Siebert, Z. Anorg. Allgem. Chem. *261*, 63 (1950).
243. Goubeau, J. and H. Sommer, Z. Anorg. Allgem. Chem. *289*, 1 (1957).

243a. GROSCHWITZ, E. A., W. M. INGLE AND M. A. RING, J. Organometal. Chem. *9*, 421 (1967).
243b. GOWENLOCK, B. G. AND J. STEVENSON, J. Organometal. Chem. *13*, P 13 (1968).
244. GUNN, S. R., *Symposium on Hydrides of Groups 4, 5 and 6*, Berkeley, Calif., U.S.A., 1963.
245. GUNN, S. R. AND L. G. GREEN, J. Phys. Chem. *65*, 779 (1961).
246. GUTOWSKI, H. S., G. B. KISTIAKOWSKY, G. E. PAKE AND E. M. PURCELL, J. Chem. Phys. *17*, 972 (1949).
247. GUTOWSKI, H. S. AND O. STEJSKAL, J. Chem. Phys. *22*, 939 (1954).
248. HALUSKA, L. A., U.S. Pat. 2800494 (1957); C.A. *51*, 17982 (1957).
248a. *Handbook of Chemistry and Physics*, Chemical Rubber Publ. Co., Cleveland, Ohio, 48th edition (1967/68).
249. HANNAY, N. B. AND C. P. SMYTH, J. Am. Chem. Soc. *68*, 171 (1946).
249a. HARNISCH, H., A. MEHNE AND F. RODIS, Germ. Pat. 1 129 149 (1962); C.A. *57*, 3080 (1962).
250. HARPER, L. R., S. YOLLES AND H. C. MILLER, J. Inorg. Nucl. Chem. *21*, 294 (1961).
251. HARVEY, M. C., W. H. NEBERGALL AND J. S. PEAKE, J. Am. Chem. Soc. *76*, 4555 (1954).
252. HARVEY, M. C., W. H. NEBERGALL AND J. S. PEAKE, J. Am. Chem. Soc. *79*, 1437 (1957).
253. HARVEY, M. C., W. H. NEBERGALL AND J. S. PEAKE, J. Am. Chem. Soc. *79*, 2762 (1957).
254. HASZELDINE, R. N. AND R. J. MARKLOW, J. Chem. Soc. *1956*, 962.
255. HASZELDINE, R. N., M. J. NEWLANDS AND J. B. PLUMB, J. Chem. Soc. *1965*, 2101.
256. HAWKINS, J. A., S. R. POLO AND M. K. WILSON, J. Chem. Phys. *21*, 1122 (1953).
257. HAWKINS, J. A. AND M. K. WILSON, J. Chem. Phys. *21*, 360 (1953).
258. HEATH, D. F., J. W. LINNETT AND P. J. WHEATLEY, Trans. Faraday Soc. *46*, 137 (1950).
259. HEDBERG, K., J. Am. Chem. Soc. *77*, 6491 (1955).
260. HEDBERG, K. AND A. J. STOSICK, *Proc. Intern. Congr. Pure Appl. Chem., 12th, New York, 1951*, Abstr., p. 543.
261. HELFERICH, B. AND J. HAUSEN, Ber. *57*, 795 (1924).
262. HEMPTINNE, M. DE AND J. WOUTERS, Bull. Acad. Roy. Belg. [5] *20*, 1114 (1934).
263. HENGGE, E., Chem. Ber. *95*, 645 (1962).
264. HENGGE, E., Chem. Ber. *95*, 648 (1962).
265. HENGGE, E. AND K. PRETZER, Chem. Ber. *96*, 470 (1963).
266. HENRY, M. C. AND J. G. NOLTES, J. Am. Chem. Soc. *82*, 555 (1960).
267. HERRMANN, E., Wiss. Z. Univ. Rostock, Math.-Naturwiss. Reihe *9*, issue 3 (1959/60).
268. HERTWIG, K., DAS (West Germ. Provis. Pat. Spec.) 1040518 and 1040519.
269. HERZBERG, G., *Molecular Spectra and Molecular Structure*, I and II, Van Nostrand, New York, 1960 and 1961.
270. HESS, G. G., F. W. LAMPE AND L. H. SOMMER, J. Am. Chem. Soc. *87*, 5327 (1965).
271. HESS, G. G. AND F. W. LAMPE, J. Chem. Phys. *44*, 2257 (1966).
272. HETFLEJS, J., F. MARES AND V. CHVALOVSKY, *Intern. Symp. Organosilicon Chem., Prague, 1965*, p. 282.
273. HETFLEJS, J., F. MARES AND V. CHVALOVSKY, Collection Czech. Chem. Commun. *31*, 586 (1966).
274. HETFLEJS, J., F. MARES AND V. CHVALOVSKY, Collection Czech. Chem. Commun. *30*, 1643 (1965).
275. HÖPPNER, K. AND D. WALKIEWITZ, Z. Chem. *3*, 227 (1963).
276. HOGNESS, T. R., J. L. WILSON AND W. C. JOHNSON, J. Am. Chem. Soc. *58*, 108 (1936).
277. HURD, D. T., J. Am. Chem. Soc. *67*, 1545 (1945).
278. HURD, D. T., U.S. Pat. 2 537 763 (1951); C.A. *45*, 3409 (1951).
279. HYDE, J. F., P. L. BROWN AND A. L. SMITH, J. Am. Chem. Soc. *82*, 5854 (1960).
279a. INGLEFIELD, P. T. AND L. W. REEVES, J. Chem. Phys. *40*, 2424 (1964).
280. JACKSON, H. L., F. D. MARSH AND E. L. MUETTERTIES, Inorg. Chem. *2*, 43 (1963).
280a. JACOX, M. E. AND D. E. MILLIGAN, J. Chem. Phys. *49*, 3730 (1968).
281. JARVIE, A. W. AND D. LEWIS, J. Chem. Soc. *1963*, 1073.
282. JENKINS, J. W., N. L. LAVERY, P. R. GEUNTHER AND H. W. POST, J. Org. Chem. *13*, 862 (1948).
283. JENKINS, D. R., R. KEWLEY AND T. M. SUGDEN, Proc. Chem. Soc. *1960*, 220.
284. JENKINS, D. R., R. KEWLEY AND T. M. SUGDEN, Trans. Faraday Soc. *58*, 1284 (1962).
285. JENKINS, J. W. AND H. W. POST, J. Org. Chem. *15*, 556 (1950).
286. JENKNER, H. (Kali-Chemie AG, Hannover), W. Germ. Pat. 1 034 159 (1958).
287. JENKNER, H. (Kali-Chemie AG, Hannover), W. Germ. Pat. 1 033 660 (1958).
288. JENKNER, H. (Kali-Chemie AG, Hannover), W. Germ. Pat. 1 028 575 (1958).
289. JENKNER, H. (Kali-Chemie AG, Hannover), W. Germ. Pat. 1 009 631 (1957).

290. JENKNER, H. AND H.-W. SCHMIDT (Kali-Chemie AG, Hannover), DAS (W. Germ. Provis. Pat. Spec.) 1 038 553 (1958).
290a. JENSEN, M. A., J. Organometal. Chem. *11*, 423 (1968).
291. JOHNSON, W. C. AND T. R. HOGNESS, J. Am. Chem. Soc. *56*, 1252 (1934).
292. JOHNSON, W. C. AND S. ISENBERG, J. Am. Chem. Soc. *57*, 1349 (1935).
293. JOLLY, W. L., UCRL-10733, Contract No. W-7405-eng-48 (1963).
294. JOLLY, W. L., J. Am. Chem. Soc. *83*, 335 (1961) and private communication.
294a. JOLLY, W. L., J. Am. Chem. Soc. *85*, 3083 (1963).
295. JOKLÍK, J. AND V. BAZANT, Collection Czech. Chem. Commun. *29*, 603 (1964).
296. JUAN, C. AND H. S. GUTOWSKI, J. Chem. Phys. *37*, 2198 (1962).
297. KADONGA, M. AND K. IINO, Jap. Pat. 5645 (1954); C.A. *49*, 14377 (1955).
298. KAESZ, H. D. AND J. G. A. STONE, J. Chem. Soc. *1957*, 1433.
299. KANAZASHI, M., Bull. Chem. Soc. Japan *26*, 493 (1953); C.A. *49*, 11576 (1955).
300. KAPLAN, L., J. Am. Chem. Soc. *76*, 5880 (1954).
301. KAPLAN, L. AND K. E. WILZBACH, J. Am. Chem. Soc. *74*, 6152 (1952).
302. KAPLAN, L. AND K. E. WILZBACH, J. Am. Chem. Soc. *77*, 1297 (1955).
303. KASAI, P. H. AND R. J. MYERS, J. Chem. Phys. *30*, 1096 (1959).
304. KAUTSKY, H. AND L. HAASE, Chem. Ber. *86*, 1226 (1953).
305. KAUTSKY, H. AND G. HERZBERG, Z. Anorg. Allgem. Chem. *139*, 136 (1924).
306. KAUTSKY, H. AND A. HIRSCH, Z. Anorg. Allgem. Chem. *170*, 1 (1928).
307. KAUTSKY, H. AND P. SIEBEL, Z. Anorg. Allgem. Chem. *273*, 113 (1953).
308. KAUTSKY, H. AND H. TIELE, Z. Anorg. Allgem. Chem. *144*, 198 (1925).
309. KEIDEL, F. A. AND S. H. BAUER, J. Chem. Phys. *25*, 1218 (1956).
310. KERR, J. A., D. H. SLATER AND J. C. JOUNG, J. Chem. Soc. *A 1967*, 134.
311. KERR, J. A., D. H. SLATER AND J. C. JOUNG, J. Chem. Soc. *A 1966*, 104.
312. KESSLER, G., Thesis, Technical University, Dresden, 1963.
312a. KESSLER, G. AND H. KRIEGSMANN, Z. Anorg. Allgem. Chem. *342*, 53 (1966).
312b. KESSLER, G. AND H. KRIEGSMANN, Z. Anorg. Allgem. Chem. *342*, 63 (1966).
313. KILB, R. W. AND L. PIERCE, J. Chem. Phys. *27*, 108 (1957).
314. KING, W. T. AND I. W. LEVIN, *Symp. Mol. Struct. Spectry., Columbus, 1960*, Ohio State University.
314a. KIRCHHOFF, W. H. AND D. R. LIDE JR., J. Chem. Phys. *43*, 2203 (1965).
315. KLEJNOT, O. J., Inorg. Chem. *2*, 825 (1963).
316. KNISELEY, R. N., V. A. FASSEL AND E. E. CONRAD, Spectrochim. Acta *15*, 651 (1959).
317. KNÍZEK, J., V. CHVALOVSKY AND M. HORÁK, *Int. Symp. Organosilicon Chem., Prague, 1965*, p. 220.
318. KNÍZEK, J., V. CHVALOVSKY AND M. HORÁK, Collection Czech. Chem. Commun. *29*, 2935 (1964).
319. KOFEL, M. D. AND R. A. LAD, J. Chem. Phys. *16*, 420 (1948).
320. K[illegible], N. V. AND V. K. ROMAN, J. Gen. Chem. USSR (Eng. Transl.) *33*, 2008 (1963).
320a. KOVALEV, I. F., Opt. Spectry. USSR (English Transl.) *15*, 100 (1963).
320b. KOSTER, D. F., Spectrochim. Acta *24A*, 395 (1968).
321. KRAMER, K. A. W. AND A. N. WRIGHT, Chem. Ber. *96*, 1877 (1963).
322. KRAMER, K. A. W. AND A. N. WRIGHT, J. Chem. Soc. *1963*, 3604.
323. KRIEBLE, H. AND I. R. ELLIOT, J. Am. Chem. Soc. *67*, 1810 (1945).
323a. KRIEGSMANN, H., Z. Anorg. Allgem. Chem. *298*, 212 (1959).
324. KRIEGSMANN, H. AND H. BEYER, Z. Anorg. Allgem. Chem. *311*, 180 (1961).
324a. KRIEGSMANN, H. AND W. FÖRSTER, Z. Anorg. Allgem. Chem. *298*, 212 (1959).
324b. KRIEGSMANN, H., P. REICH, G. SCHOTT AND H. WERNER, Z. Anorg. Allgem. Chem. *343*, 101 (1966).
325. KRISHER, L. C. AND L. PIERCE, J. Chem. Phys. *32*, 1619 (1960).
326. KUCHEN, W., Z. Anorg. Allgem. Chem. *288*, 101 (1956).
327. KUMADA, M., J. Inst. Polytech., Osaka City Univ., Ser. C *2*, 131 (1952); C.A. *48*, 11303 (1954).
328. KUMADA, M. AND K. TAMARA, J. Chem. Soc. Japan, Ind. Chem. Sect. *54*, 769 (1951); C.A. *48*, 3889 (1954).
329. KUMMER, D. AND J. D. BALDESCHWIELER, J. Phys. Chem. *67*, 98 (1963).
330. KURSANOV, D. N., V. N. SETKINA AND YU. N. NOVIKOV, Bull. Acad. Sci. USSR, Div. Chem. Sci. (Engl. transl.) *1964*, 1821.
331. KWOK, J. AND W. ROBINSON, J. Chem. Phys. *36*, 3137 (1962).

332. LAMPE, F. W. AND J. H. FUTRELL, Trans. Faraday Soc. *59*, 1957 (1963).
333. LAURIE, V. W., J. Chem. Phys. *26*, 1359 (1957).
334. LAUTSCH, W. F., A. TRÖBER, W. ZIMMER, L. MEHNER, W. LINK, H.-M. LEHMANN, H. BRANDENBURGER, H. KÖRNER, H.-J. METZSCHKER, K. WAGNER AND R. KADEN, Z. Chem. *3*, 415 (1963).
335. LAW, J. T., J. Chem. Phys. *30*, 1568 (1959).
336. LAW, J. T., J. Chem. Phys. *28*, 511 (1958).
337. LAW, J. T., AND E. E. FRANÇOIS, J. Phys. Chem. *60*, 353 (1956).
338. LEFFLER, A. J. AND E. G. TEACH, J. Am. Chem. Soc. *82*, 2710 (1960).
339. LEITERS, L. A., V. G. DULOVA AND M. E. VOL'PIN, Izv. Akad. Nauk SSSR, Otd. Khim. Nauk *1963*, 731.
340. LENGYEL, B. AND T. SZÉKELY, Z. Anorg. Allgem. Chem. *287*, 273 (1956).
341. LENGYEL, B., T. SZÉKELY, S. JENEI AND G. GARZÓ, Z. Anorg. Allgem. Chem. *323*, 65 (1963).
342. LEVIN, I. W. AND W. T. KING, J. Chem. Phys. *37*, 1375 (1962).
343. LEVIS, R. N., J. Am. Chem. Soc. *69*, 717 (1947).
344. LICHT, K. AND K. KRIEGSMANN, Z. Anorg. Allgem. Chem. *323*, 190 (1963).
345. LIEBAU, F., Naturwiss, *49*, 481 (1962).
346. LINTON, H. R. AND E. R. NIXON, Spectrochim. Acta *12*, 41 (1958).
347. LINTON, H. R. AND E. R. NIXON, Spectrochim. Acta *10*, 299 (1958).
347a. LINTON, H. R. AND E. R. NIXON, Spectrochim. Acta *15*, 146 (1958).
348. LORD, R. C., D. W. MAYO, H. E. OPITZ AND J. S. PEAKE, Spectrochim. Acta *12*, 147 (1958).
349. LORD, R. C., D. W. ROBINSON AND W. C. SCHUMB, J. Am. Chem. Soc. *78*, 1327 (1956).
349a. LUKEVITS, E. AND M. G. VORONKOV, *Organic Insertion Reactions of Group IV Elements*, Consultant Bureau, New York, 1966; original: *Gidrosililirovanie, Gidrogermilirovanie i Gidrostannilirovanie*, Acad. Sci. Latvian SSR, Riga, USSR, 1964.
350. MACDIARMID, A. G., J. Inorg. Nucl. Chem. *2*, 88 (1956).
351. MACDIARMID, A. G., *Symposium on Hydrides of Groups 4, 5 and 6*, Berkeley, Calif., U.S.A., 1963.
352. MACDIARMID, A. G. *et al., Meeting Am. Chem. Soc., 142th, Sept. 1962*, Abstr., 15N-34.
352a. MACDIARMID, A. G. in *New Pathways in Inorg. Chem.* p. 149ff; editors: E. A. V. EBSWORTH, A. G. MADDOCK AND A. G. SHARPE, Cambridge University Press, 1968.
353. MACDIARMID, A. G. AND A. G. MADDOCK, J. Inorg. Nucl. Chem. *1*, 411 (1955).
354. MACDIARMID, A. G., J. J. MOSEONY, C. R. RUSS AND T. YOSHIOKA, *Intern. Symp. Organosilicon Chem., Prague, 1965*, p. 100.
355. MACDIARMID, A. G., Quart. Rev. *10*, 208 (1956).
355a. MACKAY, K. M. AND R. WATTS, Spectrochim. Acta *23A*, 2761 (1967).
355b. MACKAY, K. M., S. T. HOSFIELD AND S. R. STOBART, J. Chem. Soc. *A1969*, 2937.
356. MACKENZIE, C. A., L. SPIALTER AND M. SCHOFMAN, U.S. Pat. 2721873 (1955); C.A. *50*, 7844 (1956).
357. MACKENZIE, C. A., L. SPIALTER AND M. SCHOFMAN, Brit. Pat. 684597/684662; C.A. *48*, 2761 (1954).
357a. MALINOWSKI, E. AND T. VLADIMIROFF, J. Am. Chem. Soc. *86*, 3575 (1964).
358. MAL'NOVA, G. N., E. P. MIKHEEV, A. L. KLEBANSKII, S. A. GOLUBTSOV AND N. P. FILIMONOVA, Dokl. Akad. Nauk SSSR *117*, 623 (1957); C.A. *52*, 8996 (1958).
359. MAMEDOV, M. A., I. M. AKHMEDOV, M. M. GUSEINOV AND S. I. SADYKH-ZADE, J. Gen. Chem. USSR (Eng. Transl.) *35*, 458 (1965).
360. MANULKIN, Z. M., Zh. Obshch. Khim. *16*, 235 (1946); C.A. *41*, 90e (1947).
361. MANULKIN, Z. M., Zh. Obshch. Khim. *18*, 299 (1948); C.A. *42*, 6742f (1948).
362. MANULKIN, Z. M., Zh. Obshch. Khim. *20*, 2004 (1950); C.A. *45*, 5611i (1951).
362a. MARCHANT, A., M.-T. FOREL, F. METRAS AND J. VALADE, J. Chim. Phys. *1964*, 343.
363. MAYO, D. W., H. E. OPITZ AND J. S. PEAKE, J. Chem. Phys. *23*, 1344 (1955).
364. MAYS, J. M. AND B. P. DAILEY, J. Chem. Phys. *20*, 1695 (1952).
365. MCBEE, E. T., C. W. ROBERTS AND G. W. R. PUERCKHAUER, J. Am. Chem. Soc. *79*, 2329 (1957).
366. MCBEE, E. T., C. W. ROBERTS AND G. W. R. PUERCKHAUER, J. Am. Chem. Soc. *79*, 2326 (1957).
367. MCBRIDE JR., J. J. AND H. C. BEACHELL, J. Am. Chem. Soc. *74*, 5247 (1952).
368. MCCLUSKER, P. A. AND E. L. REILLY, J. Am. Chem. Soc. *75*, 1583 (1953).
368a. MCFARLANE, J. Chem. Soc. *A1967*, 1275.
369. MCKEAN, D. C., Spectrochim. Acta *13*, 38 (1958).
370. MCKEAN, D. C., Spectrochim. Acta *10*, 161 (1957).

371. McKean, D. C., Proc. Chem. Soc. *1959*, 321.
372. McKean, D. C., Chem. Commun. *1966*, 147.
372a. McKean, D. C. Spectrochim. Acta *24A*, 1253 (1968).
373. McKean, D. C. and A. A. Chalmers, Spectrochim. Acta *23A*, 777 (1967).
374. Meal, H. J. and M. K. Wilson, J. Chem. Phys. *24*, 385 (1956).
375. Métras, F. and J. Valade, Bull. Soc. Chim. France *1965*, 1423.
376. Midland Silicones Ltd., Brit. Pat. 751 370 (1956); C.A. *51*, 5828 (1957).
377. Miller, L. A. and G. L. Carlson, *Symp. Mol. Struct. Spectry., Columbus, 1960*, Ohio State University.
378. Miller, F. A. and G. L. Carlson, Spectrochim. Acta *17*, 977 (1961).
379. Miller, W. S., J. S. Peake and W. H. Nebergall, J. Am. Chem. Soc. *79*, 5604 (1957).
380. Miller, H. C. and R. S. Schreiber, U.S. Pat. 2 739 821 (1945); C.A. *39*, 4619 (1945).
381. Milligan, J. G. and C. A. Kraus, J. Am. Chem. Soc. *72*, 5297 (1950).
381a. Milligan, D. E. and M. E. Jacox, J. Chem. Phys. *49*, 1938 (1968).
382. Mironov, V. F., Izv. Akad. Nauk SSSR, Otd. Khim. Nauk *1962*, 1884.
383. Mironov, V. F., Collection Czech. Chem. Commun. *25*, 2167 (1960).
384. Mironov, V. F. and L. A. Leites, Izv. Akad. Nauk SSSR, Otd. Khim. Nauk *1959*, 2051.
385. Mironov, V. F. and N. G. Maksimova, Bull. Acad. Sci. USSR, Div. Chem. Sci. (Engl. transl.) *1962*, 1223.
386. Mironov, V. F. and N. G. Maksimova, Bull. Acad. Sci. USSR, Div. Chem. Sci. (Engl. transl.) *1963*, 351.
387. Mironov, V. F. and N. G. Maksimova, Izv. Akad. Nauk SSSR, Otd. Khim. Nauk *1963*, 387.
388. Mironov, V. F. and N. G. Maksimova, Izv. Akad. Nauk SSSR, Otd. Khim. Nauk *1962*, 1303.
389. Mironov, V. F. and N. G. Maksimova, Izv. Akad. Nauk SSSR, Otd. Khim. Nauk *1961*, 2059.
390. Mironov, V. F. and V. V. Nepomnina, Izv. Akad. Nauk SSSR, Otd. Khim. Nauk *1963*, 2142.
391. Mironov, V. F. and V. V. Nepomnina, Izv. Akad. Nauk SSSR, Otd. Khim. Nauk *1961*, 1886.
392. Mironov, V. F. and V. V. Nepomnina, Izv. Akad. Nauk SSSR, Otd. Khim. Nauk *1961*, 1795.
393. Mironov, V. F. and V. V. Nepomnina, Izv. Akad. Nauk SSSR, Otd. Khim. Nauk *1960*, 2140.
394. Mironov, V. F. and V. V. Nepomnina, Izv. Akad. Nauk SSSR, Otd. Khim. Nauk *1960*, 1419.
395. Mironov, V. F., V. V. Nepomnina and L. A. Leites, Izv. Akad. Nauk SSSR, Otd. Khim. Nauk *1963*, 756.
396. Mironov, V. F. and A. D. Petrov, Izv. Akad. Nauk SSSR, Otd. Khim. Nauk *1957*, 383.
397. Mironov, V. F. and A. D. Petrov, Izv. Akad. Nauk SSSR, Otd. Khim. Nauk *1958*, 787.
398. Mironov, V. F. and L. L. Shchukovskaya, Izv. Akad. Nauk SSSR, Otd. Khim. Nauk *1960* 760.
399. Mockler, R. C., J. H. Bailey and W. Gordy, J. Chem. Phys. *21*, 1710 (1953).
400. Morino, Y., J. Chem. Phys. *24*, 164 (1956).
401. Morrison, J. A. and M. A. Ring, Inorg. Chem. *6*, 100 (1967).
401a. Morterra, C. and M. F. D. Low, Chem. Commun. *1968*, 203.
402. Moissan, H. and S. Smiles, Compt. Rend. *134*, 569, 1549 (1902).
402a. Muenter, F. S. and V. M. Laurie, J. Chem. Phys. *39*, 1181 (1963).
403. Müller, R. and G. Seitz, Chem. Ber. *91*, 22 (1958).
404. Mozes, Gh. and F. L. Nicolau, Rev. Chim. (Bucharest) *10*, 709 (1959); C.A. *57*, 10641 (1962).
405. Müller, N. and R. C. Bracken, J. Chem. Phys. *32*, 1577 (1960).
406. Mulliken, R. S. J. Am. Chem. Soc. *72*, 4493 (1950).
407. Mulliken, R. S., J. Am. Chem. Soc. *77*, 884 (1955).
408. Murata, H. and K. Kawai, J. Chem. Phys. *23*, 2451 (1955).
409. Murata, H. and K. Shimizu, J. Chem. Phys. *23*, 1968 (1955).
409a. Nagy, F., F. Reffy, A. Kuszmann-Borbély and K. Pálossy-Becker, J. Organometal. Chem. *7*, 393 (1967).
409b. Nagy, J., J. Réffy, A. Borbély-Kuszman and K. Becker-Pálossy, J. Organomet. Chem. *17*, 17 (1969).
410. Nametkin, N. S., T. I. Chernysheva and L. V. Babaré, J. Gen. Chem. USSR (Eng. Transl.) *34*, 2270 (1964).
411. Nametkin, N. S., A. V. Topchiev and L. I. Kartasheva, Dokl. Akad. Nauk SSSR *101*, 885 (1956); C.A. *50*, 3216 (1956).
412. Nametkin, N. S., A. V. Topchiev and L. I. Kartasheva, Dokl. Akad. Nauk SSSR *93*, 667 (1953); C.A. *49*, 1541 (1955).
413. Nametkin, N. S., A. V. Topchiev and L. S. Povarov, Dokl. Akad. Nauk SSSR *99*, 403 (1954); C.A. *49*, 15727 (1955).

414. NEBERGALL, W. H., J. Am. Chem. Soc. *72*, 4702 (1950).
415. NEBERGALL, W. H. AND O. H. JOHNSON, J. Am. Chem. Soc. *71*, 4022 (1949).
416. NESMEYANOV, A. N., R. KH. FRIEDLINA AND E. TS. CHUKOVSKAYA, Dokl. Akad. Nauk SSSR *113*, 120 (1957); C.A. *51*, 14541 (1957).
417. NESMEYANOV, A. N., R. KH. FRIEDLINA AND E. TS. CHUKOSKAYA, Tetrahedron *1*, 248 (1957).
418. NESMEYANOV, A. N., R. KH. FRIEDLINA AND E. TS. CHUKOSKAYA, Dokl. Akad. Nauk SSSR *112*, 271 (1957); C.A. *51*, 11988 (1957).
419. NESMEYANOV, A. N., R. KH. FRIEDLINA AND E. TS. CHUKOSKAYA, Dokl. Akad. Nauk SSSR *115*, 734 (1957); C.A. *52*, 6165 (1958).
419a. NEUERT, H. AND H. CLASEN, Z. Naturforsch, *7a*, 410 (1952).
420. NEWMAN, C., J. K. O'LOANE, S. R. POLO AND M. K. WILSON, J. Chem. Phys. *25*, 855 (1956).
421. NEWMAN, C., S. R. POLO AND M. K. WILSON, Spectrochim. Acta *15*, 793 (1959).
422. NIEDERPRÜM, H. N. AND W. SIMMLER, *Intern. Symp. Organosilicon Chem., Prague, 1965*, p. 11.
423. NIELSEN, J. R. AND C. E. HATAWAY, J. Mol. Spectry. *10*, 366 (1963).
424. NISEL'SON, L. A., P. P. PUGACHEVICH, T. D. SOKOLOVA AND R. A. BEDERDINOV, Russ. J. Inorg. Chem. (English Transl.) *10*, 705 (1965).
425. NISHIKAWA, K., Jap. Pat. 2371 (1950); C.A. *47*, 2766 (1953).
426. NITZSCHE, S. AND M. WICK, Angew. Chem. *69*, 96 (1957).
427. NOLL, W., Naturwiss. *49*, 505 (1962).
428. NOLLER, D. C. AND H. W. POST, J. Am. Chem. Soc. *74*, 1361 (1952).
429. NOLTES, J. G., M. C. HENRY AND M. J. JANSSEN, Chem. Ind. London *1959*, 298.
429a. NORMAN, A. D., Chem. Commun. *1968*, 812.
429b. NORMAN, A. D. AND D. C. WINGELETH, Inorg. Chem. *9*, 98 (1970).
430. NOVIKOV, S. S. AND V. V. SEVOST'YANOVA, Bull. Acad. Sci. USSR, Div. Chem. Sci. (Engl. transl.) *1962*, 1400.
431. NOZAKURA, S. AND S. KONOTSUNE, Bull. Chem. Soc. Japan *29*, 322 (1956).
432. NOZAKURA, S. AND S. KONOTSUNE, Bull. Chem. Soc. Japan *29*, 326 (1956).
433. NOZAKURA, S., Bull. Chem. Soc. Japan *29*, 784 (1956).
433a. OBI, K., C. C. CLEMENT, H. E. GUNNING AND O. P. STRASZ, J. Am. Chem. Soc. *91*, 1622 (1969).
434. OIGENBLICK, A. A., Russ. J. Phys. Chem. (English Transl.) *39*, 935 (1965).
435. OKAWARA, R. AND M. KATAYAMA, Bull. Chem. Soc. Japan *33*, 659 (1960).
436. OLLOMA, J. F., A. A. SINISGALLI, H. N. REXROAD AND R. C. GUNTON, J. Chem. Phys. *24*, 487 (1956).
437. O'NEAL, H. E. AND M. A. RING, Inorg. Chem. *5*, 435 (1966).
437a. O'REILLY, J. M., L. PIERCE AND L. C. KRISHER, *Symp. Mol. Struct. Spectry., Columbus, 1960*, Ohio State University.
438. ONYSZCHUK, M., *Symposium on Hydrides of Groups 4, 5 and 6*, Berkeley, Calif., U.S.A., 1963.
439. OPITZ, H. E., J. S. PEAKE AND W. H. NEBERGALL, J. Am. Chem. Soc. *78*, 292 (1956).
440. ORLOV, N. F., R. A. BOGATKIN, Z. I. SERGEEVA AND M. G. VORONKOV, J. Gen. Chem. USSR (Eng. transl.) *32*, 2526 (1962).
441. O'REILLY, J. M. AND L. PIERCE, J. Chem. Phys. *34*, 1176 (1961).
442. OSHESKY, G. D. AND F. F. BENTLEY, J. Am. Chem. Soc. *79*, 2057 (1957).
443. PADDOCK, N. L., Nature *167*, 1070 (1951).
444. PALIK, E. D. AND K. N. RAO, J. Chem. Phys. *26*, 1401 (1957).
445. PAULING, L. AND L. O. BROCKWAY, J. Am. Chem. Soc. *59*, 13 (1937).
446. PATNODE, W. I. AND R. W. SCHEISLER, U.S. Pat. 2381000 (1945); C.A. *39*, 4889 (1945).
447. PEAKE, J. S., W. H. NEBERGALL AND YUN TI CHEN, J. Am. Chem. Soc. *74*, 1526 (1952).
447a. PETERSEN, D. H. AND L. PIERCE, *Symp. Mol. Struct. Spectry., Columbus, 1960*, Ohio State University.
448. PETROV, A. D. AND E. FISCHER, J. Gen. Chem. USSR (Eng. Transl.) *32*, 697 (1962).
449. PETROV, A. D., V. F. MIRONOV AND V. G. GLUKHOVTSEV, Izv. Akad. Nauk SSSR, Otd. Khim. Nauk *1954*, 1123; C.A. *49*, 7510 (1955).
450. PETROV, A. D., V. F. MIRONOV, V. M. VDOVIN AND S. I. SADYKH-ZADE, Izv. Akad. Nauk SSSR *1956*, 256; C.A. *50*, 13726 (1956).
451. PETROV, A. D. AND V. A. PONOMARENKO, Dokl. Akad. Nauk SSSR *90*, 387 (1953); C.A. *48*, 5080 (1954).
452. PETROV, A. D., V. A. PONOMARENKO AND V. I. BOIKOV, Izv. Akad. Nauk SSSR, Otd. Khim. Nauk *1954*, 504; C.A. *49*, 9494 (1955).

453. Petrov, A. D., V. A. Ponomarenko, L. H. Mkhitaryan and A. D. Snegova, Dokl. Akad. Nauk SSSR *100*, 1107 (1955); C.A. *49*, 10166 (1955).
454. Petrov, A. D. and S. I. Sadykh-Zade, Bull. Soc. Chim. France *1959*, 1932.
455. Petrov, A. D., S. I. Sadykh-Zade and I. L. Tsetlin, Dokl. Akad. Nauk SSSR *107*, 99 (1956); C.A. *50*, 13728 (1956).
456. Petrov, A. D., L. L. Shukovskaya and Yu. P. Egorov, Dokl. Akad. Nauk SSSR *93*, 293 (1953); C.A. *48*, 13616 (1954).
457. Petrov, A. D., L. L. Shukovskaya, S. I. Sadykh-Zade and Yu. P. Egorov, Dokl. Akad. Nauk SSSR *115*, 522 (1957); C.A. *52*, 5322 (1958).
458. Petrov, A. D., V. A. Ponomarenko and G. V. Odabashyan, Bull. Acad. Sci. USSR, Div. Chem. Sci. (Engl. transl.) *1962*, 159.
459. Petrov, A. D. and L. L. Shchukovskaya, Dokl. Akad. Nauk SSSR *86*, 551 (1952); C.A. *47*, 12225 (1953).
460. Petrov, A. D., N. P. Smetankina and G. I. Nikishin, Zh. Obshch. Khim. *25*, 2332 (1955); C.A. *50*, 9280 (1956).
461. Petrov, A. D. and V. M. Vdovin, Izv. Akad. Nauk SSSR, Otd. Khim. Nauk *1957*, 383; C.A. *52*, 7135 (1958).
462. Pierce, L., J. Chem. Phys. *34*, 498 (1961).
463. Pierce, L., J. Chem. Phys. *29*, 383 (1958).
464. Pierce, L., J. Chem. Phys. *31*, 547 (1959).
465. Pierce, L. and D. H. Peterson, J. Chem. Phys. *33*, 907 (1961).
466. Pietrusza, E. W., L. H. Sommer and F. C. Whitmore, J. Am. Chem. Soc. *70*, 484 (1948).
467. Pike, R. M. and P. M. McDonagh, J. Chem. Soc. *1963*, 2831.
468. Pike, R. M. and P. M. McDonagh, J. Chem. Soc. *1963*, 4058.
469. Pines, H., E. Aristoff and V. N. Ipatieff, J. Am. Chem. Soc. *71* 749 (1949).
470. Pines, H., E. Aristoff and V. N. Ipatieff, J. Am. Chem. Soc. *72*, 4022 (1950).
471. Pines, H. and V. Mark, J. Am. Chem. Soc. *78*, 4316 (1956).
472. Pines, H. and V. Mark, J. Am. Chem. Soc. *78*, 5946 (1956).
473. Pistorius, C. W. F. T., J. Chem. Phys. *27*, 965 (1957).
474. Pitzer, K. S., J. Am. Chem. Soc. *70*, 2140 (1948).
475. Plate, A. F., N. A. Momma and Yu. P. Egorov, Dokl. Akad. Nauk SSSR *97*, 847 (1954); C.A. *49*, 10169 (1955).
476. Polyakova, A. M., M. D. Suchkova, V. M. Vdovii, V. F. Mironov, V. V. Korshak and A. D. Petrov, Izv. Akad. Nauk SSSR, Otd. Khim. Nauk *1959*, 2257.
477. Dolo, S. R. and M. K. Wilson, J. Chem. Phys. *22*, 1559 (1954).
478. Ponomarenko, V. A., V. G. Cherkaev, A. D. Petrov and N. A. Zadorozhnyi, Izv. Akad. Nauk SSSR, Otd. Khim. Nauk *1958*, 247; C.A. *52*, 12751 (1958).
479. Ponomarenko, V. A. and V. F. Mironov, Izv. Akad. Nauk SSSR, Otd. Khim. Nauk *1954*, 497; C.A. *49*, 9495 (1955).
480. Ponomarenko, V. A., B. A. Sokolov and A. D. Petrov, Izv. Akad. Nauk SSSR, Otd. Khim. *1956*, 628; C. A. *51*, 1027 (1957).
481. Ponomarenko, V. A., B. A. Sokolov, Kh. M. Minachev and A. D. Petrov, Dokl. Akad. Nauk SSSR *106*, 76 (1956); C.A. *50*, 13726 (1956).
482. Post, H. W., *Organic Silicon Compounds*, Reinhold, New York, 1949.
483. Price, F. P., J. Am. Chem. Soc. *69*, 2600 (1947).
484. Prince, R. H., J. Chem. Soc. *1959*, 1783.
485. Pray, B. O., L. M. Sommer, G. M. Goldberg, G. T. Kerr, Ph. A. Digiorgio and F. C. Withmore, J. Am. Chem. Soc. *70*, 433 (1948).
485a. Rankin, D. W. H., J. Chem. Soc. *A1969*, 1926.
485b. Rankin, D. W. H., A. G. Robiette, G. M. Sheldrick, B. Beagley and T. G. Hewitt, J. Inorg. Nucl. Chem. *31*, 2351 (1969).
485c. Rankin, D. W. H., A. G. Robiette, G. M. Sheldrick, W. S. Sheldrick, B. J. Aylett, I. A. Ellis and J. J. Monaghan, J. Chem. Soc. *A1969*, 1224.
486. Pupezin, J. D. and K. F. Zmbov, Bull. Inst. Nucl. Sci. "Boris Kidrich" (Belgrade) *8*, 89 (1958); C.A. *52*, 17980 (1958).
486a. Randall, E. W. and J. J. Zuckerman, Chem. Commun. *1966*, 732.
486b. Razuvaev, G. A., A. N. Egorochkin, M. L. Chidekel' and V. F. Mironov, Izv. Akad. Nauk SSSR, Otd. Khim. Nauk *1966*, 928.
487. Razuvaev, G. A. and Ch. C. Vyazankin, *Intern. Symp. Organosilicon Chem., Prague, 1965*, p. 97.

487a. Reeves, R. B. and D. W. Robinson, J. Chem. Phys. *41*, 1699 (1964).
487b. Reeves, R. B., R. E. Wilde and D. W. Robinson, J. Chem. Phys. *40*, 125 (1964).
488. Reich, P., Thesis, Technical University, Dresden, 1963.
488a. Reich, P. and H. Kriegsmann, Z. Anorg. Allgem. Chem. *334*, 272 (1965).
488b. Reich, P. and H. Kriegsmann, Z. Anorg. Allgem. Chem. *334*, 283 (1965).
489. Reid, A. F. and C. J. Wilkins, J. Chem. Soc. *1955*, 4029.
490. Reilly, E. L., C. Curran and P. A. McCusker, J. Am. Chem. Soc. *76*, 3311 (1954).
491. Reilly, E. L. and H. W. Post, J. Org. Chem. *16*, 387 (1951).
492. Reikhsfel'd and V. A. Prokhorova, J. Gen. Chem. USSR (Eng. Transl.) *35*, 1826 (1965).
493. Reikhsfel'd and V. A. Prokhorova, J. Gen. Chem. USSR (Eng. Transl.) *35*, 694 (1965).
494. Rexroad, H. N., D. W. Hogate, R. C. Gunton and J. F. Ollom, J. Chem. Phys. *24*, 625 (1956).
495. Reynolds, H. H., L. A. Bigelow and C. A. Kraus, J. Am. Chem. Soc. *51*, 3067 (1929).
496. Ritter, M. A., *Symposium on Hydrides of Groups 4, 5 and 6*, Berkeley, Calif., U.S.A., 1963.
497. Ring, M. A., L. P. Freeman and A. P. Fox, Inorg. Chem. *3*, 1200 (1964).
498. Ring, M. A. and D. M. Ritter, J. Am. Chem. Soc. *83*, 802 (1961).
498a. Ring, M. A., G. D. Beverley, F. H. Koester and R. P. Hollandsworth, Inorg. Chem. *8*, 2033 (1969).
499. Ritter, A. and L. H. Sommer, *Intern. Symp. Organosilicon Chem., Prague, 1965*, p. 279.
500. Robinson, D. W. and R. B. Reeves, J. Chem. Phys. *37* 2625 (1962).
501. Rochow, E. G., *An Introduction to the Chemistry of the Silicones*, 2nd ed., Wiley, New York, 1951.
502. Rochow, E. G., J. Am. Chem. Soc. *67* 963 (1945).
503. Rochow, E. G., D. T. Hurd and R. N. Lewis, *The Chemistry of Organometallic Compounds*, Wiley, New York, 1954.
504. Royen, P. and C. Rocktäschel, Angew. Chem. *76*, 302 (1964).
505. Ryan, J. W. and J. L. Speier, J. Am. Chem. Soc. *86*, 895 (1964).
506. Royen, P. and C. Rocktäschel, Z. Anorg. Allgem. Chem. *346*, 279 (1966).
507. Royen, P. and C. Rocktäschel, Z. Anorg. Allgem. Chem. *346*, 290 (1966).
508. Russel, G. A., J. Am. Chem. Soc. *81*, 4815 (1959).
509. Russel, G. A., J. Am. Chem. Soc. *81*, 4825 (1959).
510. Russel, G. A., J. Am. Chem. Soc. *81*, 4831 (1959).
511. Russel, G. A., J. Am. Chem. Soc. *81*, 4834 (1959).
512. Rust, J. B. and C. A. MacKenzie, U.S. Pat. 2698334 (1954); C.A. *49*, 4328 (1955).
512a. Rustad, D. S. and W. L. Jolly, Inorg. Chem. *6*, 1986 (1967).
513. Ryan, J. W., G. K. Menzie and J. L. Speier, J. Am. Chem. Soc. *82*, 3601 (1960).
514. Ryan, J. W. and J. L. Speier, J. Org. Chem. *24*, 2052 (1959).
515. Saalfeld, F. E. and M. V. McDowell, Inorg. Chem. *6*, 96 (1967).
516. Saalfeld, F. E. and H. J. Svec, Inorg. Chem. *3*, 1442 (1964).
517. Saalfeld, F. E. and H. J. Svec, Inorg. Chem. *2*, 46 (1963).
518. Saalfeld, F. E. and H. J. Svec, Inorg. Chem. *2*, 50 (1963).
519. Saam, J. C. and J. L. Speier, J. Org. Chem. *24*, 427 (1959).
520. Saam, J. C. and J. L. Speier, J. Am. Chem. Soc. *80*, 4104 (1958).
521. Saam, J. C. and J. L. Speier, J. Am. Chem. Soc. *83*, 1351 (1961).
521a. Sacher, R. E., D. H. Lemmon and F. A. Miller, Spectrochim. Acta *23A*, 1169 (1967).
522. Sakiyama, M. and R. Okawara, Bull. Chem. Soc. Japan *33*, 1645 (1961).
522a. Sakurai, H., A. Hosomi and M. Kumada, Chem. Comm. *1969*, 4.
523. Sanin, P. S., Zh. Obshch. Khim. *27*, 1286 (1957); C.A. *52*, 3751 (1958).
523a. Saratov, I. E. and V. O. Reikhsfel'd, J. Gen. Chem. USSR (Eng. Transl.) *36* 1084 (1966).
523b. Saratov, I. E. and V. O. Reikhasfel'd, J. Gen. Chem. USSR (Engl. Transl.) *38*, 627 (1968).
524. Sathianandan, K. and J. L. Margrave, J. Mol. Spectry. *10*, 442 (1963).
525. Sauer, R. O. and W. Patnode, J. Am. Chem. Soc. *67*, 1548 (1945).
526. Schaarschmidt, K., Z. Anorg. Allgem. Chem. *310*, 78 (1961).
527. Schachtschneider, J. H. and R. G. Snyder, Spectrochim. Acta *19*, 117 (1963).
528. Schäfer, R. and W. Klemm, J. Prakt. Chem. *5*, 233 (1958).
528a. Scherer, O. F. and M. Schmidt, Z. Naturforsch. *20b*, 1009 (1965).
529. Schmeisser, M. and M. Schwarzmann, Z. Naturforsch. *11b* 278 (1956).
529a. Schmidbaur, H., J. Am. Chem. Soc. *85*, 2336 (1963).
529b. Schmidbaur, H., Chem. Ber. *97*, 1639 (1964).
530. Schmidt, H.-W. and H. Jenkner, DAS (West Germ. Provis. Pat. Spec.) 1046578 (1958).
531. Schneider, W. G., H. J. Bernstein and J. A. Pople, J. Chem. Phys. *28* 601 (1958).

532. Schnell, E. and E. G. Rochow, J. Inorg. Nucl. Chem. *6*, 303 (1958).
533. Schott, G., Z. Chem. *2*, 194 (1962).
534. Schott, G., Z. Chem. *3*, 41 (1963).
535. Schott, G. and P. Ebel, Z. Anorg. Allg. Chem. *325*, 169 (1963).
535a. Schott, G. and E. Fischer, Chem. Ber. *93*, 2525 (1960).
535b. Schott, G. and G. Gastmeier, Z. Anorg. Allgem. Chem. *312*, 322 (1961).
536. Schott, G. and D. Gutschick, Z. Anorg. Allgem. Chem. *325*, 175 (1963).
537. Schott, G. and C. Harzdorf, Z. Anorg. Allgem. Chem. *307*, 105 (1960).
538. Schott, G. and C. Harzdorf, Z. Anorg. Allgem. Chem. *306*, 108 (1960).
539. Schott, G. and E. Herrmann, Z. Anorg. Allgem. Chem. *307*, 97 (1960).
540. Schott, G. and W. Herrmann, Z. Anorg. Allgem. Chem. *288* 1 (1956).
541. Schott, G. and E. Hirschmann, Z. Anorg. Allgem. Chem. *288*, 9 (1956).
542. Schott, G. and J. Meyer, Z. Anorg. Allgem. Chem. *313*, 107 (1961).
543. Schott, G. and D. Naumann, Z. Anorg. Allgem. Chem. *291*, 103 (1957).
544. Schott, G. and D. Naumann, Z. Anorg. Allgem. Chem. *291*, 112 (1957).
545. Schumb, W. C. and R. C. Joung, J. Am. Chem. Soc. *52*, 1464 (1930).
546. Schumb, W. C. and D. W. Robinson, J. Am. Chem. Soc. *77*, 5294 (1955).
547. Schwarz, R. and F. Heinrich, Z. Anorg. Allgem. Chem. *221*, 277 (1935).
548. Schwendeman, R. H. and G. D. Jacobs, J. Chem. Phys. *36*, 1251 (1962).
549. Schwendeman, R. H. and G. D. Jacobs, J. Chem. Phys. *36*, 1245 (1962).
550. Scott, Th. R., G. King and J. M. Wilson, U.S. Pat. 2910394; C.A. *54*, 1054 (1960).
551. Seyferth, D., J. Inorg. Nucl. Chem. *7*, 152 (1958).
552. Seyferth, D. and D. L. Alleston, Inorg. Chem. *2*, 418 (1963).
552a. Seyferth, D., S. B. Andrews and H. D. Simmons, J. Organomet. Chem. *17*, 9 (1969).
553. Seyferth, D. and J. M. Burlitch, J. Am. Chem. Soc. *85*, 2667 (1963).
553a. Seyferth, D., J. M. Burlitch, H. Dertouzos and H. D. Simmons Jr., J. Organometal. Chem. *7*, 405 (1967).
554. Seyferth, D., H. Yamazaki and Y. Sato, Inorg. Chem. *2*, 734 (1963).
555. Seyferth, D. and E. G. Rochow, J. Am. Chem. Soc. *77*, 907 (1955).
556. Seyferth, D. and E. G. Rochow, J. Org. Chem. *20*, 250 (1955).
557. Seyferth, D. and M. Takamizawa, Inorg. Chem. *2*, 731 (1963).
558. Selin, T. G. and R. West, J. Am. Chem. Soc. *84*, 1860 (1962).
559. Selin, T. G. and R. West, J. Am. Chem. Soc. *84*, 1863 (1962).
559a. Semenova, E. A., D. Ya. Zhinkin and K. A. Andrianov, Bull. Akad. Sci. USSR, Div. Chem. Sci. (Engl. transl.) *1962*, 1945.
560. Shade, R. W. and G. D. Cooper, J. Phys. Chem. *62* 1467 (1958).
561. Shafer, P. W. and G. H. Wagner, Brit. Pat. 662916 (1951); C.A. *46*, 11229 (1952).
562. Sharikova, I. E., V. M. Al'bitskaya and A. A. Petrov, J. Gen. Chem. USSR (Eng. Transl.) *34*, 2275 (1964).
563. Shaw, R. A., J. Chem. Soc. *1957*, 2831.
564. Sheline, R. K., J. Chem. Phys. *18*, 602 (1950).
565. Sheridan, J. and A. C. Turner, Proc. Chem. Soc. *1960*, 21.
566. Shikiev, I. A., M. F. Shostakovskii, N. V. Komarov and L. A. Kayutenko, Izv. Akad. Nauk SSSR, Otd. Khim. Nauk *1957*, 1139; C.A. *52*, 4532 (1958).
566a. Shimanouchi, T., I. Nakagawa, J. Hiraishi and M. Ishii, J. Mol. Spectry. *19*, 78 (1966).
Shimizu, K. and H. Murata, J. Mol. Spectry, *4*, 201 (1960).
568. Shimizu, K. and H. Murata, J. Mol. Spectry. *4*, 214 (1960).
569. Shorr, L. M., J. Am. Chem. Soc. *76*, 1300 (1954).
570. Shostakovskii, M. F. and D. A. Kochkin, Izv. Akad. Nauk SSSR, Otd. Khim. Nauk *1956*, 1150; C.A. *51*, 4935 (1957).
571. Shostakovskii, M. F. and D. A. Kochkin, Dokl. Akad. Nauk SSSR *109*, 113 (1956); C.A. *51*, 1826 (1957).
572. Shostakovskii, M. F., D. A. Kochkin and V. L. Vinogradov, Izv. Akad. Nauk SSSR, Otd. Khim. Nauk *1957*, 1452; C.A. *52*, 7134 (1958).
573. Shostakovskii, M. F., B. A. Sokolov, A. D. Grishko, K. F. Lavrova and G. I. Kagan, J. Gen. Chem. USSR (Eng. Transl.) *32*, 3809 (1962).
574. Simanouti, T., J. Chem. Phys. *17*, 848 (1949).
575. Simmler, W., H. Walz and H. Niederprüm, Chem. Ber. *96*, 1495 (1963).
576. Simon, I. and H. O. McMahon J. Chem. Phys. *20*, 905 (1952).
577. Skinner, H. A., Trans. Faraday Soc. *41*, 645 (1945).
578. Smith, A. L., J. Chem. Phys. *21*, 1997 (1953).

579. Smith, A. L., Spectrochim. Acta *19*, 849 (1963).
580. Smith, A. L. and N. C. Angelotti, Spectrochim. Acta *15*, 412 (1959).
580a. Smith, G. W., *Proton Magnetic Resonance Studies of Compounds with the Structure* $(CH_3)_4X$, p. 219, in T. J. Hughel (ed.), Liquids: Structure, Properties, Solid Interactions, Elsevier Publ. Comp. Amsterdam, 1965.
580b. Smith, G. W., J. Chem. Phys. *42*, 4229 (1965).
581. Snyder, R. G. and J. H. Schachtschneider, Spectrochim. Acta *19*, 85 (1963).
582. Snobl, L., J. Cermak and M. Dvorak, *Khim. Prakt. Primenenie, Kremneorgan. Soedin., Tr. Konf., Leningrad, 1958*, 235; C. A. *57*, 7968 d (1962).
582a. Solan, D. and P. L. Timms, Inorg. Chem. *7*, 2157 (1968).
583. Sokolov, N. N., K. A. Andrianov and S. M. Akimova, J. Gen. Chem. USSR (Eng. Transl.) *25*, 675 (1955).
583a. Sommer, L. H., *Meeting Am. Chem. Soc., 135th, April 1959*, 22M/24M, No. 62/63; Angew. Chem. *71*, 532 (1959).
584. Sommer, L. H., O. F. Bennett, P. G. Campbell and D. R. Weynberg, J. Am. Chem. Soc. *79*, 3295 (1957).
585. Sommer, L. H. and O. F. Bennett, J. Am. Chem. Soc. *79* 1008 (1957).
586. Sommer, L. H., J. D. Citron and C. L. Frye, J. Am. Chem. Soc. *86*, 5684 (1964).
587. Sommer, L. H., F. P. MacKay, O. W. Steward and P. G. Campbell, J. Am. Chem. Soc. *79*, 2764 (1957).
588. Sommer, L. H. and J. E. Lyons, J. Am. Chem. Soc. *89*, 1521 (1967).
589. Sommer, L. H., K. W. Michael and H. Fujimoto, J. Am. Chem. Soc. *89*, 1519 (1967).
590. Sommer, L. H., E. W. Pietrusza and F. C. Whitmore, J. Am. Chem. Soc. *69*, 2108 (1947).
591. Sommer, L. H., E. W. Pietrusza and F. C. Whitmore, J. Am. Chem. Soc. *69*, 188 (1947).
592. Spanier, E. J. and A. G. MacDiarmid, Inorg. Chem. *2*, 215 (1963).
593. Spanier, E. J. and A. G. MacDiarmid, Inorg. Chem. *1*, 432 (1962).
594. Speier, J. L., U.S. Pat. 2 723 987; C.A. *50*, 10128 (1956).
595. Speier, J. L., U.S. Pat. 2 762 823 (1956); C.A. *51*, 7418 (1957).
596. Speier, J. L., J. A. Webster and G. H. Barnes, J. Am. Chem. Soc. *79*, 974 (1957).
597. Speier, J. L. and J. A. Webster, J. Org. Chem. *21*, 1044 (1956).
598. Speier, J. L., R. Zimmermann and J. A. Webster, J. Am. Chem. Soc. *78*, 2278 (1956).
598a. Spialter, L. and W. A. Swansiger, J. Am. Chem. Soc. *90*, 2187 (1968).
599. Stammreich, H. and R. Forneris, Spectrochim. Acta *8*, 52 (1956).
600. Steele, W. C. and F. G. A. Stone, J. Am. Chem. Soc. *84*, 3599 (1962).
601. Steele, W. C., L. D. Nicholls and F. G. A. Stone, J. Am. Chem. Soc. *84*, 4441 (1962).
602. Sternbach, B. and A. G. MacDiarmid, J. Inorg. Nucl. Chem. *23*, 225 (1961).
603. Sternbach, B. and A. G. MacDiarmid, J. Am. Chem. Soc. *83*, 3384 (1961).
604. Sternbach, B. and A. G. MacDiarmid J. Am. Chem. Soc. *81*, 5109 (1959).
605. Steudel, W. and H. Gilman, J. Am. Chem. Soc. *82*, 6129 (1960).
606. Stevenson, D. P., J. Chem. Phys. *8*, 285 (1940).
607. Stitt, F. and D. M. Yost, J. Chem. Phys. *5*, 90 (1937).
608. Stock, A. and C. Somieski, Ber. *49*, 111 (1916).
609. Stock, A. and C. Somieski, Ber. *56*, 247 (1923).
610. Stock, A., P. Stiebeler and F. Zeidler, Ber. *56*, 1695 (1923).
611. Stock, A. and C. Somieski, Ber. *54*, 524 (1921).
612. Stock, A. and C. Somieski, Ber. *50*, 1739 (1917).
613. Stock, A. and C. Somieski Ber. *51*, 989 (1918).
614. Stock, A. and C. Somieski, Ber. *52*, 695 (1919).
615. Stock, A. and C. Somieski, Ber. *56*, 1087 (1923).
616. Stock, A., C. Somieksi and R. Wintgen, Ber. *50*, 1754 (1917).
617. Stock, A. and P. Stiebler, Ber. *56*, 1087 (1923).
617a. Stokland, K., Kgl. Norske Videnskab. Selskabs, Forh. *12*, 122 (1939); *Gmelins Handbuch der anorganischen Chemie*, 15 Si, Part B, Verlag Chemie, Weinheim, 1959, p. 256.
618. Stolberg, U. G., Z. Naturforsch. *18b*, 765 (1963).
619. Storch, H. H. and L. S. Kassel, J. Am. Chem. Soc. *59*, 1240 (1937).
620. Sujishi, S., *Symposium on Hydrides of Groups 4, 5 and 6*, Berkeley, Calif., USA, 1963.
621. Sujishi, S., Office of Ordnance Research Project No. TB 2-0001 (817), Contract No. DA-11-022-ORD-1264, Control No. OOR-137-53; Final report, August 1957.
622. Sujishi, S., Air Force Office Scientific Research SRQA, Washington 25, D.C., Contract No. AF 49(638)276, August 1959.
623. Sujishi, S. and S. Witz, J. Am. Chem. Soc. *79*, 2447 (1957).

624. SUJISHI, S. AND S. WITZ, J. Am. Chem. Soc. *76*, 4631 (1954); *79*, 2447 (1957).
625. SUNDERMEYER, W., Thesis, Göttingen, 1956.
626. SUNDERMEYER, W., Angew. Chem. *74*, 717 (1962).
627. SUNDERMEYER, W., Angew. Chem. *74*, 875 (1962).
628. SUNDERMEYER, W., Chem. Ber. *96*, 1293 (1963).
629. SUNDERMEYER, W. AND O. GLEMSER, Angew. Chem. *70*, 625 (1958).
630. TAMBORSKI, C., F. E. FORD, W. L. LEHN, G. J. MOORE AND E. J. SOLOSKI, J. Org. Chem. *27*, 619 (1962).
631. TANAKA, T., Bull. Chem. Soc. Japan *33*, 446 (1960).
632. TANNENBAUM, S., S. KAYE AND G. LEWENZ, J. Am. Chem. Soc. *75*, 3753 (1953).
632a. TEBBE, E. M. AND M. A. RING, Inorg. Chem. *8*, 1787 (1969).
633. TIMMS, P. L., C. C. SIMPSON AND C. S. G. PHILLIPS, J. Chem. Soc. *1964*, 1467.
634. TINDAL, C. H., J. W. STRALEY AND H. H. NIELSEN, Phys. Rev. *62*, 151 (1942).
635. THOMAS, A. B. AND E. G. ROCHOW, J. Am. Chem. Soc. *79*, 1843 (1957).
636. THOMPSON, H. W., Spectrochim. Acta *16*, 238 (1960).
636a. THOMPSON, M. L. AND R. SCHAEFFER, Inorg. Chem. *7*, 1677 (1968).
637. TOPCHIEV, A. V., N. S. NAMETKIN AND T. I. CHERNYSHEVA, Dokl. Akad. Nauk SSSR *118*, 517 (1958); C.A. *52*, 10922 (1958).
638. TOPCHIEV, A. V., N. S. NAMETKIN AND T. I. CHERNYSHEVA, Dokl. Akad. Nauk SSSR *115*, 326 (1958); C.A. *52*, 4474 (1958).
639. TOPCHIEV, A. V., N. S. NAMETKIN, T. I. CHERNYSHEVA AND S. G. DUGAR-YAN, Dokl. Akad. Nauk SSSR *110*, 97 (1956); C.A. *51*, 4979 (1957).
640. TOPCHIEV, A. V., N. S. NAMETKIN AND O. P. SOKOLOVA, Dokl. Akad. Nauk SSSR *93*, 285 (1953); C.A. *48*, 12671 (1954).
641. TOPCHIEV, A. V., N. S. NAMETKIN AND V. I. ZETKIN, Dokl. Akad. Nauk SSSR *82*, 927 (1952); C.A. *47*, 4281 (1953).
642. TSITSISHVILI, G. V., G. D. BAGRATISHVILI, K. A. ANDRIANOV, L. M. KHANANASHVILI AND M. L. KANTARIYA, Bull. Acad. Sci. USSR, Div. Chem. Sci. (Engl. transl.) *1962*, 1123.
643. TYLER, J. K. AND J. SHERIDAN, Proc. Chem. Soc. *1960*, 119.
644. TYLER, J. K., L. F. THOMAS AND J. SHERIDAN, J. Opt. Soc. Am. *52*, 581 (1962).
644a. UHLE, K., Z. Chem. *7*, 236 (1967).
645. URENOVITCH, J. V. AND A. G. MACDIARMID, J. Chem. Soc. *1963*, 1091.
645a. VAHRENKAMP, H. AND H. NÖTH, J. Organometal. Chem. *12*, 281 (1968).
646. VALADE, J. AND R. CALAS, Compt. Rend. *243*, 386 (1956).
647. VALADE, J. AND F. METRAS, Compt. Rend. *253*, 1582 (1961).
647a. VARMA, R. AND A. P. COX, Angew. Chem. *76*, 649 (1964).
648. VARMA, R., A. G. MACDIARMID AND J. G. MILLER, J. Chem. Phys. *39*, 3157 (1963).
649. VARMA, R., A. G. MACDIARMID AND J. G. MILLER, Inorg. Chem. *3*, 1754 (1964).
650. VDOVIN, V. M. AND A. D. PETROV, Russ. Chem. Rev. (English Transl.) *31*, 393 (1962).
651. VISTE, A. AND H. TAUBE, J. Am. Chem. Soc. *86*, 1691 (1964).
651a. VISWANATHAN, N. AND C. H. VAN DYKE, J. Chem. Soc. *A1968*, 487.
652. VENKATESVARLU, P., R. C. MOCKLER AND W. GORDY, J. Chem. Phys. *21*, 1713 (1953).
653. VENKATESVARLU, P. AND K. SATHIANANDAN, Z. Physik, Chem. (Leipzig) *218*, 318 (1961).
654. VERMA, R. D. AND L. C. LEITCH, Can. J. Chem. *41*, 1652 (1963).
654a. VLASOV, S. M. AND G. G. DEVYATYKH, Zh. Neorg. Khim. *11*, 2681 (1966).
655. VOL'PIN, M. E., YU. D. KORESHKOV AND D. N. KURSANOV, Izv. Akad. Nauk SSSR, Otd. Khim. Nauk *1961*, 1355.
656. VOL'PIN, M. E., YU. D. KORESHKOV, V. G. DULOVA AND D. N. KURSANOV, Tetrahedron *18*, 107 (1962).
657. VOL'PIN, M. E. AND D. N. KURSANOV, J. Gen. Chem. USSR (Engl. transl.) *32*, 1113 (1962).
658. VORONKOV, M. G. AND YU. I. KHUDOBIN, Izv. Akad. Nauk SSSR, Otd. Khim. Nauk *1956*, 805; C.A. *51*, 3440 (1957).
658a. VYAZANKIN, N. S., G. A. RAZUVAEV, E. N. GLADYSHEV AND S. P. KORNEVA, J. Organometal. Chem. *7*, 353 (1967).
659. WAGNER, G. H., U.S. Pat. 2 632 013 (1953); C.A. *48*, 2760 (1954).
660. WAGNER, G. H., U.S. Pat. 2 637 738 (1953); C.A. *48*, 8254 (1954).
661. WAGNER, G. H., D. L. BAILEY, A. N. PINES, M. L. DUNHAM AND D. B. MCINTIRE, Ind. Eng. Chem. *45*, 367 (1953).
661a. WANG, J. T. AND C. H. VAN DYKE, Inorg. Chem. *6*, 1741 (1967).
662. WARD, L. G. L. AND A. G. MACDIARMID, J. Am. Chem. Soc. *82*, 2151 (1960).
663. WARD, L. G. L. AND A. G. MACDIARMID, J. Inorg. Nucl. Chem. *21*, 287 (1961).

664. WARTIK, T. AND R. K. PEARSON, J. Inorg. Nucl. Chem. *5*, 250 (1958).
665. WEISS, H. G. AND H. D. FISCHER, Inorg. Chem. *2*, 880 (1963).
665a. WELLS, R. L. AND R. SCHAEFER, J. Am. Chem. Soc. *88*, 37 (1966).
666. WEST, R., J. AM. CHEM. SOC. *76*, 6012 (1954).
667. WEST, R., J. Am. Chem. Soc. *76*, 6015 (1954).
668. WEST, R. AND R. E. BAILEY, J. Am. Chem. Soc. *85*, 2871 (1963).
668a. WEST, R. AND C. S. KRAIHANZEL, Inorg. Chem. *1*, 967 (1962).
669. WEST, R. AND E. G. ROCHOW, J. Org. Chem. *18*, 303 (1953).
670. WEST. R. AND J. S. THAYER, J. Am. Chem. Soc. *84*, 1763 (1962).
671. WESTERMARK, H., Acta Chem. Scand. *8*, 1830 (1954).
672. WESTERMARK, H., Acta Chem. Scand. *8*, 1086 (1954).
672a. WEYNBERG, D. R., A. E. BEY AND P. J. ELLISON, J. Organometal. Chem. *3*, 489 (1965).
673. WHITE, P. G. AND E. G. ROCHOW, J. Am. Chem. Soc. *76*, 3897 (1954).
674. WHITMORE, F. C., E. W. PIETRUSZA AND L. H. SOMMER, J. Am. Chem. Soc. *69*, 2108 (1947).
675. WHITMORE, F. C., L. H. SOMMER AND J. R. GOULD, J. Am. Chem. Soc. *69*, 1976 (1947).
676. WIBERG, E., E. AMBERGER AND H. CAMBENSI, Z. Anorg. Allgem. Chem. *351*, 164 (1967).
677. WIBERG, E. AND U. KRÜERKE, Z. Naturforsch. *8b*, 608 (1953).
678. WIBERG, E. AND U. KRÜERKE, Z. Naturforsch. *8b*, 609 (1953).
679. WIBERG, E. AND U. KRÜERKE, Z. Naturforsch. *8b*, 610 (1953).
680. WIBERG, E. AND H. MICHAUD, Z. Naturforsch. *9b*, 500 (1954).
680a. WIBERG, E. AND A. NEUMAIER, unpublished.
681. WIBERG, E. AND W. SIMMLER, Z. Anorg. Allgem. Chem. *282*, 330 (1955).
682. WIBERG, E. AND W. SIMMLER, Z. Anorg. Allgem. Chem. *283*, 401 (1956).
683. WIBERG, E., O. STECHER AND A. NEUMAIER, Inorg. Nucl. Chem. Letters *1*, 31 (1965).
683a. WIBERG, E., O. STECHER, CH. ANDRASCHEK AND L. KREUTZBICHLER, Uspechi Khim *36*, 703 (1965).
684. WIBERG, N., F. RASCHIG AND R. SUSTERMAN, Angew. Chem. *74*, 388 (1962).
685. WIBERG, N., F. RASCHIG AND R. SUSTERMAN, Angew. Chem. *74*, 71 (1962).
686. WINKLER, H. J. S. AND H. GILMAN, J. Org. Chem. *27*, 254 (1962).
687. WINTGEN, R., BER. *52*, 724 (1919).
688. WITTENBERG, D., H. A. MCNINCH AND H. GILMAN, J. Am. Chem. Soc. *80*, 5418 (1958).
689. WÖHLER, F. AND H. BUFF, Ann. Chem. *103*, 218 (1857).
690. WÖHLER, F., Ann. Chem. *107*, 112 (1858).
691. WOLF, E., Z. Anorg. Allgem. Chem. *313*, 228 (1961).
692. WOLF, E., W. STAHN AND M. SCHÖNHERR, Z. Anorg. Allgem. Chem. *319*, 168 (1962).
693. WOLF, E. AND R. TEICHMANN, Z. Chem. *2*, 343 (1962).
694. WOLFE, J. K. AND N. C. COOK, U.S. Pat. 2 786 862 (1957); C.A. *51*, 13904 (1957).
695. WORONKOW, M. G. AND A. J. DEITSCH, J. Prakt. Chem. *22*, 214 (1963).
696. WU, T. -Y., J. Chem. Phys. *9*, 195 (1941).
697. YAKUBOVICH, A. YA. AND V. A. GINSBURG, Zh. Obshch. Khim. *22*, 1783 (1952); C.A. *47*, 9256 (1953).
698. YAKUBOVICH, A. YA. AND G. V. MOTSAREV, Zh. Obshch. Khim. *23*, 1414 (1953); C.A. *47*, 12281 (1953).
699. YAKUBOVICH, A. YA AND G. V. MOTSAREV, Zh. Obshch. Khim. *23*, 1059 (1953); C.A. *48*, 8187 (1954).
700. YAKUBOVICH, A. YA., S. P. MAKAROV, V. A. GINSBURG, G. I. GAVRILOV AND E. N. MERKULOVA, Dokl. Akad. Nauk SSSR *72*, 69 (1950); C.A. *45*, 2856 (1951).
700a. YOSHIOKA, T. AND A. G. MACDIARMID, Inorg. Nucl. Chem. Letters *5*, 69 (1969).
701. YUKITOSHI, T., H. SUGA, S. SEKI AND J. ITOH, J. Phys. Soc. Japan *12*, 506 (1957); C.A. *51*, 13489 (1957).
702. ZEIL, W., H. PFÖRTNER, B. HAAS AND H. BUCHERT, Ber. Bunsenges. *67*, 476 (1963).
703. ZAKHARKIN, L. I., V. V. GAVRILENKO, I. M. KHORLINA AND G. G. ZHIGAREVA, Bull. Acad. Sci. USSR, Div. Chem. Sci (Engl. transl.) *1962*, 1784.
704. ZAKHARKIN, L. I. AND L. A. SAVINA, Bull. Acad Sci. USSR, Div. Chem. Sci. (Engl. transl.) *1962*, 231.
705. ZEMANY, P. D. AND F. P. PRICE, J. Am. Chem. Soc. *70*, 4222 (1948).
706. ZORIN, A. D., G. G. DEVYATYKH, V. YA. DUDOROV AND A. M. AMEL'CHENKO, Zh. Neorg. Khim. *9*, 2526 (1964); Russ. J. Inorg. Chem. (Engl. transl.) *9*, 1364 (1964).
707. ZORIN, A. D., G. G. DEVYATYKH, E. F. KRUPNOVA AND S. G. KRASNOVA, Zh. Neorg. Khim. *9*, 2280 (1964); Russ. J. Inorg. Chem. (Eng. transl.) *9*, 1235 (1964).

CHAPTER 8

GERMANIUM HYDRIDES

1. Historical

The first hydrogen compound of germanium was discovered late. In 1902, by adding sodium amalgam or, better, zinc to germanium-containing sulphuric acid Voegelen[*292*] produced a mixture of germanium hydride and hydrogen which decomposed in the same way as arsenic hydride in the Marsh test. Although Voegelen could prepare only a very small amount of hydride – germanium had been discovered only 16 years previously (1886) by Winkler[*298*], so that germanium compounds were still rare at that time – he nevertheless succeeded in determining the formula of the hydride: GeH_4. In this he relied mainly on the reaction product with silver nitrate, which possessed the approximate formula Ag_4Ge, and on the compound $(C_2H_5)_4Ge$, already known at that time[*299*], likewise with tetravalent germanium. An attempt to prepare tin hydride in the same way failed.

The date of the discovery of germanium hydride is in harmony with the discovery of the hydrides of the neighbouring elements. Quite generally, volatile hydrides of the heavy elements were discovered later than the (more stable) hydrogen compounds of the lighter elements of the 4th and 5th main groups. An exception is arsenic hydride, which was discovered very early (1775). The years of discovery for the first hydrides are, for the 4th main group: CH_4 1778[*293*], SiH_4 1857[*300*], GeH_4 1902[*292*], SnH_4 1919[*214*], and PbH_4 1920[*216*], and for the 5th main group: NH_3 1774[*224*], PH_3 1812[*49*], AsH_3 1775[*248*], SbH_3 1837[*220*], and BiH_3 1918[*218*].

2. General review

Germanium hydrides are prepared in a similar manner to silicon hydrides: by the protolysis of a germanium–metal bond or by the hydrogenation of a germanium–halogen bond.

The Ge–H bond reacts in a similar manner to the Si–H bond. Thus, H can easily be replaced: by halogen (for example with X_2, HX, RX, HgX_2), by oxygen (for example with O_2, $OH^{\ominus}$, RCOOH), by sulphur (for example with Ag_2S, RSH), by nitrogen (for example with NH_3), by carbon (for example with N_2CHR, LiR, by addition to $>C=C<$, $-C\equiv C-$), by alkali metals (with solutions of alkali metals in liquid ammonia or in polyethers), and by transition metal π-complexes (for example, by reaction with the acidic hydrogen atom in $HMn(CO)_5$).

The chemistry of the Ge–H bond therefore resembles that of the Si–H bond.

References p. 711

Differences are shown in the lower thermal stability of the germanium hydrides, in the lower polarity of the Ge–H bond, and in the change in polarity in the series $R_3\overset{\delta+}{Ge}-\overset{\delta-}{H}$, R_2XGe-H, RX_2Ge-H, $X_3\overset{\delta-}{Ge}-\overset{\delta+}{H}$ (R = organyl, X = halogen).

As in the case of silicon [*79*], in the case of germanium, also, $d_\pi-p_\pi$ overlappings with elements of the 5th and 6th main groups are possible. Like silicon, germanium is able to increase its covalency from four to six by using its vacant 4d-orbitals in bonding[*1*]. For reviews concerning organogermanium hydrides see [*31a*, *237*].

3. Unsubstituted germanium hydrides

3.1 Preparation of the germanes

3.1.1 Synthesis of volatile germanes by the protolysis of germanides

The action of acids (HCl or H_2SO_4 in H_2O; NH_4Cl or NH_4Br in liquid NH_3; N_2H_5Cl in N_2H_4) on germanium–magnesium alloys "Mg_2Ge" ($Mg_x \cdot Mg_{2-x}Ge$) and Ca_2Ge or CaGe (structure see [*81*, *82*]) leads to the formation of germanium hydrides of the homologous series Ge_nH_{2n+2} ($n = 1, 2, \ldots 5$) [*2*, *3*, *20*, *35*, *41*, *55*, *87*, *89*, *147*, *249*, *250*, *302*]. The action of acids on crystalline magnesium germanide Mg_2Ge (NH_4Br in liquid NH_3; N_2H_5Cl in anhydrous N_2H_4) gives monogermane almost exclusively [*3*, *89*].

$$n(Mg_x \cdot Mg_{2-x}Ge) \xrightarrow[-2n\,Mg^{2\oplus},\,-(n-1)H_2]{+4n\,H^{\oplus}} Ge_nH_{2n+2}$$

$$Mg_2Ge \xrightarrow[-2\,Mg^{2\oplus}]{+4\,H^{\oplus}} GeH_4$$

Examples of the reaction are given in Table 8.1. Germane prepared in the ammono system is freed from ammonia with acidified water and moist P_2O_5 and from water with dry P_2O_5.

3.1.2 Synthesis of volatile germanes from polymeric germanium dihydride

Polymeric solid germanium dihydride $(GeH_2)_x$ [*233*] or germanium monohydride $(GeH_{0.9\text{-}1.2})_x$ [*231*] (see section 3.1.3) decomposes on being heated under atmospheric pressure into germanium, monogermane, digermane, and hydrogen [*233*]. If a subsequent further cracking of the volatile temperature-sensitive (see section 3.2.1, Pyrolysis) germanium hydrides is prevented by rapid pumping off (the reaction vessel being kept at 10^{-5} torr), with a total yield of volatile hydrides of 40 mole-% (Ge content) higher germanes are also found (8% of GeH_4, 10% of Ge_2H_6, 9% of Ge_3H_8, 8% of Ge_4H_{10}, and 5% of Ge_5H_{12}) [*3*]:

$$(GeH_{1-2})_x \xrightarrow[\Delta]{140-220^\circ} Ge,\ Ge_nH_{2n+2},\ H_2$$

The action of acid (*e.g.*, HCl) on polymeric germanium hydride also cleaves Ge–Ge bonds with the formation of volatile germanium hydrides. The yield of higher hydrides ($n \geqslant 2$) is, however, lower in this process than in pyrolysis [*233*]:

$$(GeH_2)_x \xrightarrow[\Delta]{+HCl} GeCl_2,\ Ge_nH_{2n+2},\ H_2$$

TABLE 8.1

PREPARATION OF VOLATILE GERMANES BY THE REACTIONS OF "Mg_2Ge" (ALLOY) OR Mg_2Ge WITH ACIDS

Alloy or germanide		*Acid*			*Yield (mole-%) (% Ge of hydride × 100/% Ge of alloy)*											*Ref.*
Preparation temp. (°C)	*Time of tempering* (h)	*Compound*	*System*	*Temp. (°C)*	GeH_4	GeD_4	Ge_2H_6	Ge_2D_6	Ge_3H_8	Ge_3D_8	Ge_4H_{10}	*n-* Ge_4H_{10}	*i-* Ge_4H_{10}	Ge_5H_{12}	*higher hydr.*	
~600	0	HCl	H_2O	~20	18		2.5		0.36							55
~700	0?	HCl	H_2O	20	14		0.9								very small amount	41
800	0.25	10 wt. % HCl	H_2O	0	15		5		3		1.5			0.8	0.9	2, 3, 4
800	0.25	4 N DCl	D_2O	~20		+		+		+						302
		dil. H_2SO_4	H_2O		+										+	249
		20% H_3PO_4	H_2O		+		+		+			+	+			20
750–800	24	50% H_3PO_4	H_2O	50	12		11		4							87
800	0.25	NH_4Br	liq. NH_3	< -33	60–70											147
800	0.25	NH_4Br	liq. NH_3	−40	56.25		11.56		4.80							3, 4
1200	3	NH_4Br	liq. NH_3	−40	61.50		5.50									3, 4
1200	10[a]	NH_4Br	liq. NH_3	−40	66.73											3, 4
		NH_4Br	liq. NH_3	−40	31		11.2									35
800	0.25	NH_4Br	liq. NH_3	−40	64.27		3.54									3, 4
800		N_2H_5Cl	N_2H_4	10–100	70–80		2.3									89

[a]Selected crystals of Mg_2Ge

References p. 711

3.1.3 Synthesis of polygermanes $(GeH_{1.0-2.0})_x$

The decomposition of sodium germanide, NaGe[*124*, *304*], with aqueous acids[*65*, *124*) or with ammonium bromide in liquid ammonia[*144*, *147*] leads to the formation of dark brown polymeric germanium monohydride:

$$(NaGe)_x + xH^{\oplus} \xrightarrow{HOH} (GeH)_x + xNa^{\oplus}$$

$$(NaGe)_x + xNH_4Br \xrightarrow{liq.\ NH_3} (GeH)_x + xNaBr + xNH_3$$

If monochlorogermane, GeH_3Cl, is treated with ammonia, monomeric germanium dihydride is first formed. In contrast to a reaction of GeH_3CN, to be described below, however, GeH_2 disproportionates into germane and germanium monohydride[*66*]:

$$GeH_3Cl \xrightarrow[-NH_4Cl]{+NH_3} GeH_2 \xrightarrow[\Delta]{(NH_3)} \tfrac{1}{3}GeH_4 + \frac{2}{3x}(GeH)_x$$

A monomeric form of germanium dihydride is produced in the reaction of germylsodium with phenyl bromide in liquid ammonia; it subsequently disproportionates to monogermane and germanium monohydride[*104*, *144*]:

$$NaGeH_3 + PhBr \xrightarrow{liq.\ NH_3} GeH_2 + PhH + NaBr$$

$$3GeH_2 \xrightarrow{liq.\ NH_3} GeH_4 + 2GeH$$

Concerning the existence of a viscous liquid $(GeMe_2)_x$ and a white solid $(MeGeH)_x$, see section 6.3.7.

The action of (1) concentrated hydrochloric acid–water (1 : 1), (2) anhydrous acetic acid, (3) ammonium iodide in liquid ammonia, or (4) a mixture of acetic acid, ethanol, and water (1 : 1 : 1) on calcium germanide, CaGe or Ca_2Ge, forms solid germanium hydride insoluble in solvents: $(GeH_{0.9-1.2})_x$[*3*, *231*, *234*]. Mixed crystals of CaSi+CaGe hydrolyse with the formation of Si_nH_{2n+2}, Ge_nH_{2n+2}, $SiGeH_6$, Si_2GeH_8, and $SiGe_2H_8$[*232*].

3.1.4 Special methods for the preparation of monogermane

Monogermane can be prepared by the reaction of germanium tetrachloride or germanium tetra-alkoxides with metal hydrides (*e.g.*, LiH) or double hydrides [*e.g.*, $NaBH_4$, $LiAlH_4$, $LiAl(OBu^t)_3H$, $LiAlD_4$] in anhydrous organic solvents (diethyl ether, tetrahydrofuran, dioxan). Monogermane can be obtained in aqueous solution together with small amounts of di- and trigermanes by the reaction of a solution of germanium dioxide in hydrobromic acid, acetic acid or sulphuric acid with sodium or potassium hydride (for examples of the reaction, see Table 8.2).

Germanium tetrachloride is not hydrogenated by lithium alanate in accordance with the empirical equation:

$$GeCl_4 + LiAlH_4 \rightarrow GeH_4 + LiAlCl_4$$

TABLE 8.2

SYNTHESIS OF MONOGERMANE (GeH_4 OR GeD_4) BY HYDROGENATION

Halide	*Hydrogenating agent*	*Solvent*	*Reaction temp. (°C)*	*Yield (mole-%[a])*	*Ref.*
$GeCl_4$	$LiAlH_4$	diethyl ether	−196 to 20	10–35	90, 92, 277
$GeCl_4$	$LiAlH_4$	tetrahydrofuran	−65 to 70	4–40	179
$GeCl_4$	$LiAlH_4$	dioxan	30 to 96	26	179
$GeCl_4$	$LiAlH_4$	monoglyme	−196 to 20	?	286
$GeCl_4$	$LiAlH(OBu^t)_3$	tetrahydrofuran	20 to 30	81	275, 277
$GeCl_4$	$NaAlH_4$	diglyme	60 to 75	—	301
$Ge(OEt)_4$	$NaAlH_4$	diglyme	25	—	301
$GeCl_4$	$NaBH_4$	tetrahydrofuran	25	12	180
$GeCl_4$	$NaBH_4$	$THF + H_2O$	25	44	180
$GeCl_4$	$NaBH_4$	H_2O (+acid)	0 to 80	27–79	69, 133, 180, 221
$GeCl_4$	KBH_4	H_2O	20?	60–75	74, 235, 236
$GeCl_4$	LiH	tetrahydrofuran	—	—	179
$GeDCl_3$	$LiAlH_4$	diethyl ether	−196 to 20	30–40[b]	169, 170
$GeHCl_3$	$LiAlD_4$	diethyl ether	20	[c]	169, 170
GeO_2	$LiAlH_4$	no	148 to 178	4	17a

[a] Referred to germanium compound used
[b] Little GeH_3D + much GeH_4
[c] Little GeD_4 + much GeH_2D_2

In addition to monogermane (maximum yield 40 mole-%) a yellow to orange solid (probably GeH_2) and hydrogen are always formed. The ratio $(GeH_4 + H_2):GeCl_4 = 1:1$. In spite of very informative experiments [*179*, *180*, *277*], the mechanism of the $GeCl_4$–$LiAlH_4$ reaction has not yet been elucidated in all its details. Two independent competing reactions – the formation of germane (1) and the reduction of $GeCl_4$ to $GeCl_2$ (2) – must be excluded because of the molar ratio $(GeH_4 + H_2):GeCl_4 = 1:1$ that is always found:

$$GeCl_4 \xrightarrow{a} GeHCl_3 \xrightarrow{a} GeH_2Cl_2 \xrightarrow{a} GeH_3Cl \xrightarrow{a} GeH_4 \qquad (1)$$

$$GeCl_4 \xrightarrow{2a} GeCl_2 + H_2 \; (a = +\tfrac{1}{4}LiAlH_4, -\tfrac{1}{4}LiAlCl_4) \qquad (2)$$

It is likely that the hydrogenation series (1) can break off at GeH_2Cl_2, which readily decomposes into $GeCl_2$ and H_2 [*62*]. Instead of GeH_4, the equivalent amount of H_2 is formed together with polygermane (2):

$$GeCl_4 \xrightarrow{a} GeHCl_3 \xrightarrow{a} GeH_2Cl_2 \xrightarrow{a} GeH_3Cl \xrightarrow{a} GeH_4$$

$$GeH_2Cl_2 \xrightarrow{-H_2} GeCl_2 \xrightarrow{2a} GeH_2$$

The course of the reaction is certainly dependent on the solvent. In the relatively slightly nucleophilic diethyl ether or tetrahydrofuran, the total yield amounts to 20–40 mole-%, and in the highly nucleophilic water 60–80 mole-%. A scheme of reaction of $GeCl_4$ and KBH_4 or $NaBH_4$ with a GeH_3K-intermediate see [*13a*]. An apparatus for unusually pure GeH_4 is described in the literature [*69*]. For gas chromatographic analysis see [*304a*].

3.1.5 Preparation of higher germanes from monogermane

Higher germanes are formed by passing monogermane at 0.5 to 5 torr through a discharge tube (ozoniser type) at −78° with 4–15 kV alternating current (0.025 to 25 mmole/min). Under the best conditions – namely the highest voltage, high pressure, and low rate of flow – digermane is formed in 55% yield, trigermane in 10% yield and higher germanes in still lower yields [*74, 176a*].

If monogermane is allowed to flow through the discharge tube at a higher pressure (100 to 400 torr, −78° to 25°, 10–15 kV, reaction time 1 to 5.5 h) the following amounts of products are formed at, for example, 0°, 200 torr, 10 kV, and 1 h with an 8% decomposition of monogermane: 26 mole-% of Ge_2H_6, 19 mole-% of Ge_3H_8, and 11 mole-% of Ge_4H_{10}, together with small amounts of Ge_5H_{12}, Ge_6H_{14}, Ge_7H_{16}, Ge_8H_{18}, and Ge_9H_{20} [*74*]. If GeH_4 is circulated through the discharge tube at −78° and 380 torr until the monogermane has decomposed completely, 20 mole-% of Ge_2H_6, 30 mole-% of Ge_3H_8, 6 mole-% of n- and iso-Ge_4H_{10}, 0.4 mole-% of n-, iso-, and neo-Ge_5H_{12}, 0.12 mole-% of hexagermanes, 0.1 mole-% of heptagermanes, and 0.04 mole-% of octagermanes are formed [*73*].

The passage of an electric discharge through gaseous mixtures of GeH_4 and SiH_4 (or Si_2H_6) forms mixed higher Ge–Si hydrides [*13, 108*]. In the Hg-sensitised photolysis of gaseous CH_3I–SiH_4, GeH_4–SiH_4 or GeH_4–CH_3OH, CH_3SiH_3 and GeH_3SiH_3 or GeH_3OCH_3 are formed [*98*].

When monogermane is simply passed through a heated tube at 0.5 to 5 torr, higher germanes are formed in yields about half as great. At 300 torr, the yield of $Ge_2H_6 + Ge_3H_8$ is 3 mole-% [*74*]. Dry heating of GeO_2 with $LiAlH_4$ to 148–170° yields GeH_4, and (probably by thermolysis of the GeH_4 formed) also $Ge_2H_6$and Ge_3H_8[*17a*]. Irradiation of gaseous GeH_4 with a fast-neutron flux yields the nuclear transformation $^{76}Ge(n,2n)^{75}Ge$ (half-life of ^{75}Ge, 82 min) and higher germanes ($:^{75}GeH_2 + GeH_4 \rightarrow H_3{}^{75}GeGeH_3$); with a GeH_4–SiH_4 mixture, mixed hydrides are produced ($:^{75}GeH_2 + SiH_4 \rightarrow H_3{}^{75}GeSiH_3$) [*97b*].

Germane could possibly be synthesised by the hydrogenation of Ge_2Cl_6, which has become capable of preparation on a large scale [*137*].

3.1.6 Adsorption of hydrogen on germanium

A film of germanium deposited on glass wool by the pyrolysis of GeH_4 reversibly adsorbs one hydrogen atom per germanium atom of the surface at a coverage $\theta = 1$ (saturation point). The activation energy of this adsorption, which is to be regarded as chemisorption, is 14.6 kcal/mole. The heats of adsorption are: about 23.5 kcal/mole at $\theta = 0$; 23.5 kcal/mole at $\theta = 0.05$; and 23.3 kcal/mole at $\theta = 0.1$. The primary step of the adsorption is an immobile dissociative adsorption. This assumption is supported by the deuterium exchange $H_2 + D_2 \rightarrow 2HD$ on the surface of the germanium [*155, 285*]. Unfortunately, at the present time electrical measurements on Ge–H systems are lacking. Only these would give information on the nature of the Ge–H bond made probable by the adsorption experiments described above.

Hydrogen passed over liquid germanium (1000 to 1100°) transports germanium. Thermodynamic calculations and comparison with spectroscopic data support the assumption that here metastable GeH (equilibrium 1) and not GeH_2 (equilibrium 2) is produced [15, 247]:

$$Ge + 0.5H_2 \rightleftharpoons GeH \quad (1)$$

$$\underset{\text{(liquid)}}{Ge} + \underset{\text{(gaseous)}}{H_2} \rightleftharpoons \underset{\text{(gaseous)}}{GeH_2} \quad (2)$$

3.2 Physical properties of the germanes

3.2.1 Monogermane

Under normal conditions, GeH_4 is a colourless gas which is stable at room temperature even in the presence of stopcock grease [*3*]. On cooling, monogermane solidifies to foliate weakly birefringent crystals which have a high tendency to the formation of twinning lamellae. In the solid state there are at least three phases the ranges of existence of which are separated by transformations with hysteresis: phase I from 107.26°K (m.p.) to 76.5_5°K, phase II from 76.5_5 to 73.2_0°K, and phase III from 73.2_0°K downwards. Phase I possesses a lattice of higher symmetry (feeble birefringence) than phases II and III (pronounced birefringence) [*36, 37*].

Molar volume 56.65 at the m.p.; 45.32 at 0°K (from a comparison with other hydrides of the 4th main group of the periodic system) [*219*].

Surface tension 15.80 dyne/cm at the b.p. (from a comparison with other hydrides of the 4th main group of the periodic system) [*219*].

Critical constants: critical temperature 308°K, critical pressure 54.8 atm, critical volume 128 ml/mole (calculated on the basis of van Laar's theory) [*122*].

Density 3.420 g/1 (gaseous state); 1.523 g/ml (at −142°, liquid state) [*38*].

Vapour pressures: 53 torr (−127°); 190 torr (−111°). Vapour pressure equation: $\log p$ [torr] $= -789.26/T + 7.162$ (calculated from the mean values of the vapour pressures) [*38, 85, 217, 249*]; $\log p$ [torr] $= -722.225/T + 3.511025 \log T - 0.006316\,T$ [*67*].

Boiling point −88.5° (calculated from the vapour pressure figures) [*215, 217*]; −90° (calculated from the vapour pressure figures) [*38*]; −90 to −91° (calculated from the vapour pressure figures) [*249*]; 184.80°K at 760 torr [*36*]; −88.1° (extrapolated from the vapour pressure figures) [*85*]; 202.1°K (calculated from Egloff's boiling point equation in its expanded form according to English and Nicholls): $T^{760}_{\text{boil.}}$ [°K] $= 446.1 \ln(n + 3.0) - 416.3$ (n = number of central atoms in the germane) [*88*]; 185°K (from a comparison of the hydrides of the 4th main group) [*285*]; **184.93°K = −88.23°** (most probable value, mean of 184.80°K and −88.1°C).

Melting point −164° [*215*]; −165° [*38, 249*]; −164.8° [*217*]; **107.26°K = −165.90** (most probable value) [*36*]. Viscosity see [*291a*].

Enthalpy of evaporation 3.65 kcal/mole (from the vapour pressure equation [*217*]; **3.36 ± 0.01** kcal/mole at 760 torr (most probable value) [*36*]; 48.3 cal/g (from the vapour pressure equation) [*85*]; 3.59 kcal/mole (calculated on the basis of van Laar's theory) [*122*]; 3.70 kcal/mole (from a comparison with other hydrides of the 4th main group) [*282*].

Entropy of evaporation L_s/T_s 19.8 cal deg^{-1} mole^{-1}[*282*]; 19.6 cal deg^{-1} mole^{-1} [*122*].

Enthalpy of melting 199.7 cal/mole[*36*].

Molecular heat 12 to 21 cal in the range between 12.0 and 165.4°K, with striking anomalies at 62.9, 73.20, and 76.5$_0$°K[*36*].

Refractive index n 1.0009095 ± 2.3 × 10^{-6} (0°, 1 atm, 5484 Å = Hg$_I$)[*110*]; 1.0008940 ± 1.6 × 10^{-6} (0°, 1 atm, 5893 Å = Na$_D$)[*110*].

Molar refraction 13.35 ± 0.04 cm^3/mole[*110*].

Standard enthalpy of formation 20.8 kcal/mole (from the critical potential of ion formation in mass spectroscopy)[*235*]; 21.6 or 21.7 kcal/mole (from calorimetric data)[*118, 119*].

The mass spectrum of GeH_4 at an electron energy of 100 eV can be seen from Table 8.3. In it the strongest line has been given the arbitrary intensity value of 100[*208*]. Relative frequencies of the isotopes of Ge[*291*]: ^{70}Ge 20.55%, ^{72}Ge 27.37%, ^{73}Ge 7.67%, ^{74}Ge 36.74%, ^{76}Ge 7.67%[*291*].

Table 8.4 contains the critical potentials (in eV) of the ionisation and dissociation processes of GeH_4 in comparison with those of CH_4, SiH_4, SnH_4, and PbH_4 [*208, 235*]. The values of the critical potential obtained by Neuert and Clasen [*208*] by extrapolating[*142*] the plot of the ion current function against the electron energy to zero ion current are 1.5 to 2.2 eV larger than the corresponding figures of Saalfeld and Svec[*235*]. Saalfeld and Svec ascribe the higher potentials to an unknown and undetermined excess of kinetic energy of the ions. There are further investigations on the mass spectroscopy of GeH_4 in the literature[*139, 195*].

Ionisation potential 12.3 ± 0.3 eV[*208*]; 10.5 eV[*235*].

Pyrolysis. When monogermane is heated for a short time (passed through a heated tube), higher hydrides and hydrogen are formed (see also section 3.1.5) [*74*]. On more prolonged heating in a closed system, monogermane decomposes

TABLE 8.3
RELATIVE INTENSITIES OF THE MASS LINES IN THE MASS SPECTRUM OF GeH_4[*208*]
(for explanation, see text)

Mass	*Relative intensity at 100 eV*	*Possible ionic species*
80	15.0	$^{76}GeH_4^+$
79	16.9	$^{76}GeH_3^+$
78	75.7	$^{76}GeH_2^+$, $^{74}GeH_4^+$
77	95.0	$^{76}GeH^+$, $^{74}GeH_3^+$, $^{73}GeH_4^+$
76	90.0	$^{76}Ge^+$, $^{74}GeH_2^+$, $^{73}GeH_3^+$, $^{72}GeH_4^+$
75	100.0	$^{74}GeH^+$, $^{73}GeH_2^+$, $^{72}GeH_3^+$
74	65.8	$^{74}Ge^+$, $^{73}GeH^+$, $^{72}GeH_2^+$, $^{70}GeH_4^+$
73	74.7	$^{73}Ge^+$, $^{72}GeH^+$, $^{70}GeH_3^+$
72	16.0	$^{72}Ge^+$, $^{70}GeH_2^+$
71	23.0	$^{70}GeH^+$
70	weak	$^{70}Ge^+$
2	6.7	H_2^+
1	5.8	H^+

TABLE 8.4

CRITICAL POTENTIALS IN eV FOR THE IONISATION AND DISSOCIATION OF GeH_4 IN COMPARISON WITH CH_4, SiH_4, SnH_4, AND PbH_4

Ionic species	*Process*	*C*	*Si*[a]		*Ge*		*Sn*	*Pb*
		Ref. 208	*Ref. 208*	*Ref. 235*	*Ref. 208*	*Ref. 235*	*Ref. 235*	*Ref. 235*
MH_4^+	$MH_4 \rightarrow MH_4^+ + e^-$	13.1 ± 0.4	12.2 ± 0.3	11.4	12.3 ± 0.3	10.5	9.2	9.1
MH_3^+	$\rightarrow MH_3^+ + H + e^-$	14.4 ± 0.4	12.2 ± 0.2	11.8 ± 0.2	12.2 ± 0.3	10.8 ± 0.3	9.4 ± 0.3	9.6
MH_2^+	$\rightarrow MH_2^+ + H_2 + e^-$	15.7 ± 0.5	14.5 ± 0.5	12.1 ± 0.2	13.3 ± 0.6	11.8 ± 0.2	9.5 ± 0.3	—
	$\rightarrow MH_2^+ + 2H + e^-$	22.1 ± 0.8	18.6 ± 1.0	16.5 ± 0.3	19.0 (?)	15.4 ± 0.3	13.9 ± 0.4	10.1
MH^+	$\rightarrow MH^+ + H_2 + H + e^-$	23.3 ± 0.6	14.5 ± 0.6	16.1 ± 0.2	14.0 ± 0.8	11.3 ± 0.3	10.7 ± 0.3	—
	$\rightarrow MH^+ + 3H + e^-$	—	—	20.4 ± 0.5	—	16.8 ± 0.3	14.8 ± 0.5	11.1
M^+	$\rightarrow M^+ + 2H_2 + e^-$	26.7 ± 0.7	—	11.7 ± 0.2	—	10.7 ± 0.2	9.0 ± 0.3	—
	$\rightarrow M^+ + H_2 + 2H + e^-$	—	—	16.4 ± 0.2	—	14.1 ± 0.5	13.4 ± 0.4	11.2
	$\rightarrow M^+ + 4H + e^-$	—	—	20.8 ± 0.2	—	18.3 ± 0.3	18.3 ± 0.4	—
H_2^+	$\rightarrow MH_2 + H_2^+ + e^-$	—	16.0 ± 1.0	—	15.0 ± 0.6	—	—	—
H^+	—	22.7 ± 0.5	22.4 ± 1.0	—	22.5 ± 1.0	—	—	—

[a] For further data, see ref. 269.

slowly at 280°; at a somewhat higher temperature it decomposes at a measurable rate with the formation of a germanium mirror and hydrogen. The rate of the decomposition catalysed by a germanium surface between 283 and 374° is proportional at higher temperatures to $p^{1/3}$ (where p is the partial pressure of GeH_4), and at lower temperatures the reaction is inhibited by hydrogen. The activation energy is 39.7 kcal/mole of GeH [*68, 123, 284*]. Other pyrolysis experiments have indicated a homogeneous first-order decomposition reaction [*68, 86, 284*]. According to recent investigations of the kinetics of the decomposition of germane, the order of the reaction depends on the partial pressure of the germanium and is independent of the partial pressure of the hydrogen formed. At higher pressures (100 to 400 torr of GeH_4), the decomposition of GeH_4 is a homogeneous first-order gas-phase reaction. With falling pressure, the influence of the (Ge) surface becomes greater. At low pressures (50 to 100 torr of GeH_4) the decomposition of GeH_4 at the Ge surface is a zero-order heterogeneous reaction. The rate constants of the decomposition of GeH_4 are: $k_0 = 0.248$ (330°), 0.0864 (314°), 0.0296 (302°), 0.00908 (278°) cm/min; $k_1 = 0.0403$ (330°), 0.0111 (314°), 0.00510 (302°), 0.000658 (278°) min^{-1} [*286, 288*]. The activation energies are 51.4 kcal/mole [*286, 288*] or 53.4 kcal/mole [*68*] (first order) and 41.2 kcal/mole [*286, 288*] or 37.5 kcal/mole [*68*] (zero order). The incorporation of a small amount of oxygen into the germanium surface (by the introduction of a small amount of oxygen into the reaction space and subsequent renewal of the germane to be decomposed) enhances the heterogeneous reaction (zero order) substantially and lowers the activation energy of the decomposition of the germane to 38.2 kcal/mole. The incorporation of arsenic into the germanium surface (previous pyrolysis of GeH_4 containing a little AsH_3) leads to a similar but less pronounced effect [*286, 288*]. For the equilibrium constant of the thermal decomposition see [*69a*].

Neither in the H_2–D_2 system nor in the D_2–GeH_4 system is hydrogen exchanged for deuterium on a germanium surface at 320°. In the GeH_4–GeD_4 system, however, the formation of HD (but not of partially deuterated germane) is observed. These experimental results, together with the fact that exchange reaction $H + D_2 \rightarrow HD + D$ takes place readily (low activation energy) indicate a splitting off of molecular and not atomic hydrogen in the initial stage of the homogeneous reaction $GeH_4 \rightarrow GeH_2 + H_2$. The subsequent fate of GeH_2 is uncertain. All that is certain is that the back-reaction $GeH_2 + H_2 \rightarrow GeH_4$ is unlikely (since no GeH_2D_2 was formed).

The two initial steps of the heterogeneous reaction, independent of the concentration of GeH_4, are:

$$GeH_4 \rightarrow GeH_{2\,ads} + H_2$$

$$GeH_4 \rightarrow GeH_{3\,ads} + H_{ads}$$

These are followed by rate-determining steps such as dissociation of the $GeH_{2\,ads}$ or the $GeH_{3\,ads}$ or the desorption of hydrogen from the germanium surface [*90*].

Photolysis. Germanium is decomposed by irradiation with light of the Hg arc between 28° and 93° in a manner unaffected by the temperature or by foreign gases (N_2, H_2). The primary step is probably the splitting off of a hydrogen atom with subsequent fast intermediate steps. Irradiation with light from a hydrogen discharge causes 5% of the decomposition obtained with the light of the Hg arc [*178, 228*].

Dissolution of monogermane. Germane dissolves in liquid ammonia, raising its electrical conductivity considerably. Consequently, the formation of a salt-like compound such as $[NH_4]^{\oplus}[GeH_3]^{\ominus}$ is likely [*230*]. The following volumes of gaseous germane dissolve in 150 ml of tetrahydrofuran: 2334 ml (−73°), 351 ml (−40°), 169 ml (−30°), 74 ml (−10°), 33 ml (10°), 0 ml (65°). The plot of the logarithm of the volume of GeH_4 against $1/T$, °K, is linear [*179*].

Monogermane-d_4. Deuteromonogermane (GeD_4) is a colourless stable gas at room temperature. Density 1.684 g/ml (at −160.5°; liquid) [*302*]. Vapour pressure equation: $\log p$ [torr] $= -818.54/T + 7.327$ (valid between 10 and 800 torr; calculated from the literature) [*302*]. Enthalpy of evaporation 3744 cal/mole [*302*]. Boiling point −166.2° (calculated from the vapour pressure equation) [*302*]. Melting point −89.2° [*302*].

3.2.2 Digermane

At room temperature, Ge_2H_6 is a colourless mobile liquid. While monogermane is stable at room temperature even in the presence of stopcock grease, Ge_2H_6 decomposes above 0°, especially in glass vessels contaminated with grease. Ge_2H_6 can be stored without decomposition below 0° in sealed hard-glass tubes [*3*] as has been shown by a sample 22 years old [*70*].

Density 6.704 g/1 (in the gaseous state); 1.98 g/ml (in the liquid state, near the melting point −109°).

Critical constants: critical temperature 483°, critical pressure 45.7 atm, critical volume 229 cm³/mole (calculated on the basis of van Laar's theory) [*122*].

Vapour pressures [*55, 85, 86*]: 6.4 torr (−68.1°); 239.0 torr (0°). Vapour pressure equation: $\log p$ [torr] $= -1364.3/T + 7.394$ (calculated from the literature) [*55*].

Enthalpy of evaporation 6237.4 cal/mole (calculated from the literature) [*55*]; 39.6 cal/g [*85*]; 6.09 kcal/mole [*86*].

Enthalpy of evaporation L_s/T_s 20.2 cal deg^{-1} mole^{-1} (calculated on the basis of van Laar's theory) [*122*].

Boiling point 29° [*55*]; 31.5° (extrapolation of the vapour pressures from the literature [*55*]; 31° (extrapolated from the vapour pressures) [*85*]; 29° (extrapolated from the vapour pressures) [*86*]; 301.7°K = 28.5° (calculated from Egloff's boiling point equation, see under the boiling point of GeH_4, section 3.2.1) [*88*].

Melting point −109° [*55*]; −109° [*86*].

Standard enthalpy of formation 39.1 kcal/mole (from mass-spectroscopic data) [*236*]; 37.9 kcal/mole (from calorimetric data) [*119*]; 38.8 kcal/mole [*118*].

Mass spectrum. Table 8.5 contains the relative frequencies of positive ion fragments in the mass spectrum of $^{74}Ge_2H_6$ in comparison with the figures for C_2H_6, Si_2H_6, and $^{120}Sn_2H_6$, as well as $^{74}Ge_3H_8$. Table 8.6 gives the critical potentials (in eV) of the ionisation and dissociation processes of Ge_2H_6 in comparison with those of Si_2H_6, Sn_2H_6, and Ge_3H_8.

References p. 711

TABLE 8.5

RELATIVE ABUNDANCE OF POSITIVE ION FRAGMENTS IN THE MASS SPECTRUM OF $^{74}Ge_2H_6$ IN COMPARISON WITH THE FIGURES FOR C_2H_6, Si_2H_6, AND $^{120}Sn_2H_6$, AS WELL AS $^{74}Ge_3H_8$ [236, 305]

Ionizing current 0.4 mA, ion accelerating voltage 800 V, electron accelerating voltage 70 to 10 V. Calibration with Kr.

Ion	C_2H_6	Si_2H_6	$^{74}Ge_2H_6$[a]	$^{120}Sn_2H_6$[b]	$^{74}Ge_3H_8$[c]
$M_2H_6^+$	26.2	45.6	45.8	1.5	0.9
$M_2H_5^+$	21.5	18.0	29.1	3.0	8.3
$M_2H_4^+$	100.0	100.0	74.1	3.8	58.0
$M_2H_3^+$	33.3	18.5	31.6	20.1	21.2
$M_2H_2^+$	23.0	69.0	100.0	100.0	69.5
M_2H	4.15	36.0	61.7	29.9	28.2
M_2^+	0.74	24.6	65.8	28.6	30.1
MH_4^+	0.8	0.4	4.2	0.06	0.3
MH_3^+	4.57	0.6	27.9	0.3	16.2
MH_2^+	3.35	1.0	29.6	1.5	17.2
MH^+	1.16	0.6	16.6	2.0	4.9
M^+	0.54	1.8	36.6	1.0	8.2

[a]96.1% ^{74}Ge; [b]98.2% ^{120}Sn; [c]In addition, the following abundances occur: $M_3H_8^+$ 0.3, $M_3H_7^+$ 22.6, $M_3H_6^+$ 12.9, $M_3H_5^+$ 13.3, $M_3H_4^+$ 9.1, $M_3H_3^+$ 9.1, $M_3H_2^+$ 6.5, M_3H^+ 90.1, M_3^+ 100.0.

Ionisation potential. 12.5 eV for $Ge_2H_6 \xrightarrow{-e^-} Ge_2H_6^{\oplus}$ [236]. The figure is higher than those for the upper and lower neighbours in the periodic system (Si_2H_6 10.6. Sn_2H_6 9.0 eV), which is in harmony with other anomalies of elements of the 3rd period.

Dissociation energy of the $Ge–Ge^{\oplus}$ bond 0.4 eV ($Si–Si^{\oplus}$ 3.0, $Sn–Sn^{\oplus}$ 0.1, $P–P^{\oplus}$ 3.0, $As–As^{\oplus}$ 1.3, $Sb–Sb^{\oplus}$ 0.3 eV [236].

Pyrolysis. In the range from 195 to 222°, digermane decomposes by a similar kinetics to ethane and disilane. The reaction is apparently homogeneous and of the

TABLE 8.6

CRITICAL POTENTIALS IN eV FOR THE IONISATION AND DISSOCIATION OF Si_2H_6, Ge_2H_6, Sn_2H_6, AND Ge_3H_8 (Ip = ionisation potential)

Ion	Si_2H_6	$^{74}Ge_2H_6$	$^{120}Sn_2H_6$
$M_2H_6^+$	10.6 Ip	12.5 Ip	9.0 Ip
$M_2H_5^+$		12.6	10.0
$M_2H_4^+$		12.7	10.3
$M_2H_3^+$		12.8	10.4
$M_2H_2^+$		12.9	10.5
M_2H^+		13.0	10.6
M_2^+	12.2	13.1	10.7
M^+	15.2	13.3	10.8

Ion	$^{74}Ge_3H_8$
$Ge_3H_8^+$	9.6 Ip
$Ge_3H_7^+$	9.9
$Ge_3H_6^+$	10.0
$Ge_3H_5^+$	10.1
$Ge_3H_4^+$	10.4
$Ge_3H_3^+$	10.6
$Ge_3H_2^+$	10.7
Ge_3H^+	11.8
Ge_3^+	14.6
Ge_2^+	15.8
Ge^+	16.3

first order. Of the sequence of reactions, only the first step is fairly certain: the cleavage of the Ge–Ge bond (activation energy 33.7 kcal). All other steps are speculative. For example, the polymerisation of propene observed during the pyrolysis of Ge_2H_6 can be ascribed both to GeH_3 radicals and to H atoms [*85, 86*].

Solutions of digermane. Digermane dissolves in liquid ammonia, leading to a considerable increase in its electrical conductivity. Consequently, the formation of a salt-like compound, possibly $[NH_4]_2^{2+}$ $[Ge_2H_4]^{2-}$, is likely [*230*].

Digermane-d_6. Deuterodigermane (Ge_2D_6) is a colourless mobile liquid at room temperature. Density 2.184 g/ml (in the liquid state near the melting point, −106.4°) [*203*]. Vapour pressure equation $\log p$ [torr] $= -1417.4/T + 7.579$ (valid between 10 and 800 torr, calculated from literature figures) [*203*]. Enthalpy of evaporation 6483 cal/mole [*203*]. Boiling point 28.4° (calculated from the vapour pressure equation) [*203*]. Melting point −107.9° [*203*].

3.2.3 Trigermane

Ge_3H_8 is a colourless mobile liquid at room temperature. It can be stored without decomposition below 0° in sealed hard-glass tubes (like Ge_2H_6). Stopcock grease accelerates its decomposition [*3*].

Density 10.035 g/1 (in the gaseous state); 2.20 g/ml (liquid) [*55*].

Vapour pressures: 15.5 torr (2.4°); 238.5 torr (70.3°) [*55*]. Vapour pressure equation: $\log p$ [torr] $= -1789.2/T + 7.571$ (valid between 2 and 112°; calculated from literature figures) [*55*].

Enthalpy of evaporation 8180.2 cal/mole (calculated from the vapour pressure equation of the literature figures [*55*]); 8.00 kcal/mole (calculated on the basis of van Laar's theory [*122*]).

Entropy of evaporation L_s/T_s 20.8 cal deg^{-1} mole^{-1} (calculated on the basis of van Laar's theory) [*122*].

Critical constants: critical temperature 588°K, critical pressure 37.9 atm, critical volume 330 cm^3/mole (calculated on the basis of van Laar's theory [*122*].

Boiling point 110.5° [*55*]; 383.0°K (calculated from Egloff's boiling point equation) [*88*].

Melting point −105.6° [*55*].

Standard enthalpy of formation 48.4 kcal/mole (from mass-spectroscopic data) [*236*]; 54.2 kcal/mole (from calorimetric data [*118*]). Energy of formation 39.1 kcal [*118*].

Mass spectrum. See section 3.2.2 (*Digermane*) and Tables 8.5 and 8.6.

Ionisation potential 9.6 eV for $Ge_3H_8 \xrightarrow{-e^-} Ge_3H_8^{\oplus}$ [*236*].

Trigermane-d_8. Deuterotrigermane (Ge_3D_8) is a colourless liquid at room temperature. Density 2.618 g/ml (in the liquid state near the melting point, −99.9°) [*302*]. Vapour pressure equation $\log p$ [torr] $= -1721.9/T + 7.367$ (valid between 10 and 800 torr (?), calculated from literature figures [*302*]). Enthalpy of evaporation 7876 cal/mole [*302*]. Boiling point 110° (calculated from the vapour pressure equation) [*302*]. Melting point −100.3° [*302*].

References p. 711

3.2.4 *Tetragermane*

Ge_4H_{10} is a colourless mobile liquid at room temperature which is somewhat soluble in benzene. It can be stored without decomposition at −20° in sealed hard-glass tubes [*2*].

Vapour pressures: 1.05 torr (−15.0°); 21.6 torr (47.0°) [*3*]. Vapour pressure equation: $\log p$ [torr] $= -1714.6/T + 6.692$ (valid between −15 and +47° (calculated from the vapour pressure figures [*3*]) [*2*].

Enthalpy of evaporation 7842 cal/mole (from the vapour pressure equation) [*2*].

Boiling point 176.9° (calculated from the vapour pressure equation) [*2*]; 451.7°K = 178.6° (calculated from Egloff's boiling point equation, see section 3.2.1, *Monogermane*, Boiling point) [*2*].

Pyrolysis. Above 50°, Ge_4H_{10} decomposes at first slowly and then, after the formation of a sufficient amount of decomposition products, more rapidly. Three pyrolysis steps are recognisable: at about 100°, Ge_4H_{10} dismutes into monogermane and a yellow liquid higher germane. At somewhat higher temperatures or on faster heating, it disproportionates into monogermane and a polymeric solid yellow higher polygermane(2):

$$\cdots H_3Ge \,|\, \cdot GeH_2 \cdot GeH_2 \cdot GeH_2 \,|\, H + H_3Ge \,|\, \cdot GeH_2 \cdot GeH_2 \cdot GeH_2 \,|\, H + \cdots$$

The elements form only at a temperature at which monogermane decomposes rapidly (about 350°) [*2*].

Isomeric tetragermanes. Because of their different retention times, n-Ge_4H_{10} and i-Ge_4H_{10} can be separated by gas chromatography [*20, 73, 74, 176a*]. For the retention times, see the literature [*74, 176a*].

Mass spectrum. In the mass spectrum, the fraction taken from the gas chromatographs that can be assumed to be n-Ge_4H_{10} shows decreasing relative intensities of the ions in the sequence $Ge_3^{\oplus} > Ge_2^{\oplus} > Ge_4^{\oplus} > Ge_1^{\oplus}$ The other tetragermane fraction (i-Ge_4H_{10}) shows a different sequence of the intensities of the ions: $Ge_3^{\oplus} > Ge_2^{\oplus} > Ge_1^{\oplus} > Ge_4^{\oplus}$ (n- and i-butane behave analogously) [*74*].

3.2.5 *Pentagermane*

Ge_5H_{12} is a colourless oily liquid at room temperature. It can be stored without decomposition at −20° in sealed hard-glass tubes. It can be distilled in a high vacuum at a pressure of foreign gas of $< 10^{-5}$ torr [*2*]. Vapour pressure equation: $\log p$ [torr] $= -1805.6/T + 6.449$ (valid between 7 and 47°; calculated from the vapour pressure figures [*3*]) [*2*]. Enthalpy of evaporation 8260 cal/mole (from the vapour pressure equation) [*2*]. Boiling point 234° (calculated from the vapour pressure equation) [*2*]; 511.1°K = 238° (calculated from Egloff's boiling point equation, see section 3.2.1, *Monogermane)* [*2*].

Pyrolysis. On being heated to 100°, pentagermane dismutes into monogermane and solid yellow polygermane(2); above 350° it decomposes into the elements

[2]. For the separation of isomeric pentagermanes by gas chromatography, see the literature [*74, 176a*].

3.2.6 *Hexagermane to nonagermane*

The sparingly volatile higher germanium hydrides (Ge_6H_{14} to Ge_9H_{20}) can be separated by gas chromatography and detected by mass spectroscopy. In the case of Ge_8H_{18}, only three fractions have been found, and in the case of Ge_9H_{20} only one fraction [*74*]. More isomers certainly exist, but, probably because of their decomposition in the column, these cannot be separated.

4. Organogermanium hydrides

Only one year after the discovery of the element germanium, Winkler [*299*] prepared the first organogermanium compound, tetraethylgermanium (1887). The number of other tetraorganylgermanes rapidly increased to a very large figure. On the other hand, the number of organogermanium hydrides (with at least one Ge–H bond) is low (compendium see [*138, 227a*]) and the physical data relating to these few organogermanium hydrides are still sparse. Their properties depend on the number, size, and properties of the organic groups in the hydride. They come between those of the hydrides and those of the peralkylated germanes.

There are two methods for the preparation of organogermanium hydrides:

(a) Reaction of organogermyl cations with H anions (*e.g.*, $R_3GeCl + LiH$, *cf.* section 4.1.1b):

$$R_3Ge^{\oplus} + H^{\ominus} \rightarrow R_3GeH$$

(b) Reaction of germyl anions with protons (*e.g.*, $R_3GeNa + H_2O$, *cf.* section 4.1.2a) or carbonium ions (*e.g.*, $GeH_3Na + RX$, *cf.* section 4.1.2b):

$$R_3Ge^{\ominus} + H^{\oplus} \rightarrow R_3GeH$$
$$H_3Ge^{\ominus} + R^{\oplus} \rightarrow H_3GeR$$

4.1 *Preparation of the organogermanes*

4.1.1 *Synthesis of organogermanium hydrides by hydrogenation*

(a) *Preparation of organogermanium halides.* The Grignard reaction of germanium halides with a deficiency of Grignard reagent leads to readily hydrogenatable organogermanium halides:

$$GeX_4 + n\,RMgX \rightarrow GeR_nX_{4-n} + n\,MgX_2$$
$$(R = \text{alkyl, aryl};\ n = 1, 2, 3, 4)$$

With bulky substituents (*e.g.*, cyclohexyl or isopropyl [*127*]) it is impossible to prepare the tetraalkylgermanes even with an excess of Grignard reagent. After hydrolysis of the reaction solution, triorganylgermanes and triorganylgermanols

References p. 711

are found [*e.g.*, (i-$C_3H_7)_3GeH$ and (i-$C_3H_7)_3GeOH$], which makes the following reaction steps probable: metal transfer from the Grignard reagent to the germanium compound, formation of a Ge–Ge bond, and unsymmetrical cleavage of the Ge–Ge bond during hydrolysis [*10*]:

$$R_3GeX + 2R'MgX \rightarrow R_3GeMgX + R'_2 + MgX_2$$
$$R_3GeMgX + R_3GeX \rightarrow R_3GeGeR_3 + MgX_2$$
$$R_3GeGeR_3 + H_2O \rightarrow R_3GeH + R_3GeOH$$

Similar transfer reactions are known in organic chemistry, *e.g.* [*202*]:

$$C_6H_5CH_2X + 2\ RMgX \rightarrow C_6H_5CH_2MgX + R_2 + MgX_2$$

It is possible that R_3GeH is formed by an abnormal Grignard reaction even before hydrolysis (see Chapter 7, section 4.4).

Organogermanium halides are also formed in the partial bromination of tetraalkylgermaniums (for example, with $Br_2 + AlBr_3$) [*158, 161*] and by commutation (for example, $R_4Ge + GeCl_4 \rightarrow R_3GeCl + RGeCl_3$) [*29*].

(b) *Hydrogenation of organogermanium halides and hydroxides.* In the reaction of organohalogenogermanes, $R_{4-n}GeX_n$, with lithium alanate, which is readily soluble in numerous organic solvents, organogermanes, $R_{4-n}GeH_n$, are formed in satisfactory yield [*6, 92, 127–131, 223, 237, 244*]:

$$R_{4-n}GeX_n + \frac{n}{4}\ LiAlH_4 \xrightarrow{\text{ether}} R_{4-n}GeH_n + \frac{n}{4}\ LiAlX_4$$

Organogermyl hydroxides, R_3GeOH, and bis(triorganogermyl) oxides, $(R_3Ge)_2O$, can also be hydrogenated with lithium alanate [*131*]; for example:

$$4\ R_3GeOH + LiAlH_4 \xrightarrow{\text{ether}} 4R_3GeH + LiOH + Al(OH)_3$$

Lithium boranate, which is somewhat less readily soluble in organic solvents likewise hydrogenates organohalogenogermanes, $R_{4-n}GeX_n$, to the corresponding hydrides in satisfactory yields [*223*]:

$$R_{4-n}GeX_n + \frac{n}{4}\ LiBH_4 \xrightarrow{\text{water}} R_{4-n}GeH_n + \frac{n}{4}\ LiX + \frac{n}{4}\ BX_3$$

Here the hydrogenation takes place via the stage of a boranate, as has been shown in the reaction of trialkylhalogenogermanes with sodium boranate in tetrahydrofuran:

$$R_3GeX + NaBH_4 \rightarrow R_3Ge(BH_4) + NaX$$

Only when the complex is decomposed with water at room temperature the corresponding germanium hydride is formed [*237*]:

$$R_3Ge(BH_4) + 3\ H_2O \rightarrow R_3GeH + B(OH)_3 + 3\ H_2$$

The yields in these two-stage hydrogenations fall with increasing number of hydrogen atoms. They are 100 mole-% (R_3GeH), 70 mole-% (R_2GeH_2)[*237*], and 43.7 mole-% (GeH_4)[*180*].

The best yields in the hydrogenation of organohalogenogermanes are obtained in (highly polar) water with the relatively hydrolysis-resistant sodium or potassium boranate[*112*]:

$$R_{4-n}GeX_n + \frac{n}{4}NaBH_4 \xrightarrow{\text{water}} R_{4-n}GeH_n + \frac{n}{4}LiX + \frac{n}{4}BX_3$$

In order to achieve high yields of germanes, the careful condensation of the gases leaving the reaction flask appears to be important. The hydrides are condensed quantitatively only by means of five cold traps connected in series[*112*]. Table 8.7 gives examples of the reactions in the hydrogenation of organogermanium halides, oxides, and alkoxides.

In contrast to the silicon–halogen bond, the germanium–halogen bond in the organogermanium halides is hydrogenated by nascent hydrogen. Thus, for example, triphenylgermyl bromide is reduced to triphenylgermane in a two-phase system (12 *N* aqueous hydrochloric acid + ethanol–diethyl ether) with amalgamated zinc dust[*295*].

4.1.2 Synthesis of organogermanium hydrides from germylmetals

(a) *Replacement of metal by protons.* In the reaction of tetraphenylgermane with sodium in liquid ammonia, triphenylgermylsodium, $NaGePh_3$, is formed slowly, because of the low solubility of $GePh_4$ in liquid ammonia, but quantitatively[*149*]:

$$GePh_4 + Na + NH_3 \xrightarrow{\text{liq. NH}_3} NaGePh_3 + NaNH_2 + PhH$$

The colour of $NaGePh_3$ in liquid ammonia is yellow in dilute solutions and orange in concentrated, extremely viscous solutions. In concentrated solutions, a second phenyl group is also replaced by sodium (Na_2GePh_2), and the solution becomes red. Triphenylgermylsodium is soluble to a certain extent in benzene and ether [*149*]. Better yields of triorganylgermyl(alkali metal)s, $MGeR_3$, are obtained by the reaction of the metal with a hexaorganodigermane, R_3GeGeR_3, in ethylamine or monoglyme (1,2-dimethoxyethane)[*149, 290*]:

$$R_3GeGeR_3 + 2M \rightarrow 2R_3GeM$$

$$(M = Li, Na; R = C_6H_5)$$

When alkylgermanium hydrides, $R_{4-n}GeH_n$, are added to a solution of an alkali metal in ethylamine or to a suspension of the metal in monoglyme or diglyme (2,2′-dimethoxydiethyl ether), hydrogen and not alkyl is replaced[*149, 290*]:

$$RGeH_3 + M \rightarrow RGeH_2M + \tfrac{1}{2}H_2$$

$$(M = Li, Na; R = CH_3, C_2H_5, i\text{-}C_5H_{11})$$

TABLE 8.7

HYDROGENATION OF SOME ORGANOGERMANIUM HALIDES AND OXIDES $R_{4-n}GeX_n$ (X = Cl, Br, OH, O/2)

Halide (moles of)	*Hydride*	*Hydrogenating agent*	*Solvent*	*Yield (mole-%)*	*Ref.*
CD_3GeBr_3	CD_3GeH_3	$NaBH_4$	HBr aq	99	112
CH_3GeCl_3	CH_3GeH_3	LiH	—	—	112
CH_3GeCl_3	CH_3GeD_3	LiD	—	—	223
CH_3GeI_3	CH_3GeD_3	$LiAlH_4$	—	—	154
$C_2H_5GeCl_3$	$C_2H_5GeH_3$	LiH	—	—	223
$C_2H_5GeCl_3$	$C_2H_5GeD_3$	LiD	—	—	223
n-$C_3H_7GeCl_3$	n-$C_3H_7GeH_3$	$LiAlH_4$	isopropyl ether	85	130
n-$C_3H_7GeCl_3$	—	$LiAlH_4$	dioxan	0	130
i-$C_3H_7GeCl_3$	—	—	—	—	112, 244
n-$C_4H_9GeX_3$	n-$C_4H_9GeH_3$	$LiAlH_4$	diethyl ether	90–100	112, 244
n-$C_4H_9GeX_3$ (X = Cl, Br)	n-$C_4H_9GeH_3$	$LiAlH_4$	di-n-butyl ether	99	11, 161
n-$C_5H_{11}GeBr_3$	n-$C_5H_{11}GeH_3$	$LiAlH_4$	di-n-butyl ether	90–100	161, 237, 244
i-$C_5H_{11}GeBr_3$	i-$C_5H_{11}GeH_3$	$LiAlH_4$	di-n-butyl ether	90–100	161, 237, 244
n-$C_6H_{13}GeX_3$	n-$C_6H_{13}GeH_3$	$LiAlH_4$	di-n-butyl ether	90–100	237, 244
n-$C_7H_{15}GeX_3$	n-$C_7H_{15}GeH_3$	$LiAlH_4$	di-n-butyl ether	90–100	237, 244
n-$C_8H_{17}GeX_3$	n-$C_8H_{17}GeH_3$	$LiAlH_4$	di-n-butyl ether	90–100	237, 244
$(CH_3)_2GeBr_2$	$(CH_3)_2GeH_2$	$NaBH_4$	water	93.7	112
$(CD_3)_2GeBr_2$	$(CD_3)_2GeH_2$	$NaBH_4$	water	99.4	112
$(CH_3)_2GeX_2$ (X = Cl, Br)	$(CH_3)_2GeH_2$	LiH	—	—	223
$(CD_3)_2GeX_2$ (X = Cl, Br)	$(CD_3)_2GeH_2$	LiH	—	—	223
$(CH_3)_2GeX_2$ (X = Cl, Br)	$(CH_3)_2GeD_2$	LiD	—	—	223
$(C_2H_5)_2GeX_2$	$(C_2H_5)_2GeH_2$	$LiAlH_4$	diethyl ether	90–100	237
$(C_2H_5)_2GeX_2$	$(C_2H_5)_3GeH_2$	LiH	dibutyl ether	—	223
(n-$C_3H_7)_2GeX_2$	(n-$C_3H_7)GeH_2$	$LiAlH_4$	diethyl ether	90–100	237
(i-$C_3H_7)_2GeX_2$	(i-$C_3H_7)_2GeH_2$	LiH	dibutyl ether	—	6
(n-$C_4H_9)_2GeI_2$	(n-$C_4H_9)_2GeH_2$	$NaBH_4$	tetrahydrofuran	70	237
(n-$C_5H_{11})_2GeX_2$	(n-$C_5H_{11})_2GeH_2$	$LiAlH_4$	diethyl ether, dibutyl ether	90–100	237
(n-$C_6H_{13})_2GeX_2$	(n-$C_6H_{13})_2GeH$	$LiAlH_4$	diethyl ether, dibutyl ether	90–100	237
(n-$C_7H_{15})_2GeX_2$	(n-$C_7H_{15})_2GeH_2$	$LiAlH_4$	diethyl ether, dibutyl ether	90–100	237
(n-$C_8H_{17})_2GeX_2$	(n-$C_8H_{17})_2GeH_2$	$LiAlH_4$	diethyl ether, dibutyl ether	90–100	237
(C_2H_5)(i-$C_5H_{11})GeX_2$	(C_2H_5)(i-$C_5H_{11})GeH_2$	$LiAlH_4$	—	—	127
$(C_6H_5)_2GeBr_2$	$(C_6H_5)_2GeH_2$	$LiAlH_4$	diethyl ether (35°)	55	129

$(CH_3)_3GeBr$	$(CH_3)_3GeH$	$NaBH_4$	water	95.3	112
$(CH_3)_3GeX$ (X = Cl, Br)	$(CH_3)_3GeH$	LiH	—	—	223
$(CH_3)_3GeX$ (X = Cl, Br)	$(CH_3)_3GeD$	LiD	—	—	223
$(C_2H_5)_3GeX$	$(C_2H_5)_3GeH$	$LiAlH_4$	diethyl ether, dibutyl ether	90–100	237
$(n-C_3H_7)_3GeX$	$(n-C_3H_7)_3GeH$	$LiAlH_4$	diethyl ether, dibutyl ether	90–100	237
$(i-C_3H_7)_3GeX$	$(i-C_3H_7)_3GeH$	$LiAlH_4$	diethyl ether, dibutyl ether	90–100	237
$(n-C_4H_9)_3GeI$	$(n-C_4H_9)_3GeH$	$NaBH_4$	tetrahydrofuran	91	237
$(n-C_5H_{11})_3GeX$	$(n-C_5H_{11})_3GeH$	$LiAlH_4$	diethyl ether, dibutyl ether	90–100	237
$(i-C_5H_{11})_3GeX$	$(i-C_5H_{11})_3GeH$	$LiAlH_4$	diethyl ether, dibutyl ether	90–100	237
$(n-C_6H_{13})_3GeX$	$(n-C_6H_{13})_3GeH$	$LiAlH_4$	diethyl ether, dibutyl ether	90–100	237
$(n-C_7H_{15})_3GeX$	$(n-C_7H_{15})_3GeH$	$LiAlH_4$	diethyl ether, dibutyl ether	90–100	237
$(n-C_8H_{17})_3GeX$	$(n-C_8H_{17})_3GeH$	$LiAlH_4$	diethyl ether, dibutyl ether	90–100	237
$(CH_3)_2(C_2H_5)GeX$ (X = Cl, Br)	$(CH_3)_2(C_2H_5)GeH$	LiH	—	—	223
$(CH_3)_2(C_2H_5)GeX$ (X = Cl, Br)	$(CH_3)_2(C_2H_5)GeD$	LiD	—	—	223
$(C_4H_9)_2(C_8H_{17})GeX$ (X = Cl, Br, I)	$(C_4H_9)_2(C_8H_{17})GeH$	$LiAlH_4$	—	—	164
$(C_6H_{11})_3GeCl$	$(C_6H_{11})_3GeH$	$LiAlH_4$	diethyl ether (35°)	87	131
$(C_6H_5)_3GeBr$	$(C_6H_5)_3GeH$	$LiAlH_4$	diethyl ether (35°)	—	128, 131
$(C_6H_5)_3GeOH$	$(C_6H_5)_3GeH$	$LiAlH_4$	—	100	128
$[(C_6H_5)_3Ge]_2O$	$(C_6H_5)_3GeH$	$LiAlH_4$	—	?	128
$(n-C_4H_9)_3GeOC_6H_{11}$	$(n-C_4H_9)_3GeH$	$LiAlH_4$	—	100	163

The addition of acid (*e.g.*, NH_4Br in liquid NH_3[*149*] or H_2O in benzene[*149*] or in monoglyme[*149*]) to an organogermylmetal prepared in the manner described above leads to the formation of the corresponding organogermane (replacement of metal by proton[*149*]):

$$R_3GeM + H^{\oplus} \rightarrow R_3GeH + M^{\oplus}$$

In liquid ammonia, the hydride forms even without the addition of acids (but very slowly in agreement with the low concentration of protons):

$$Et_3GeLi + NH_3 \xrightarrow{\text{liq. } NH_3} Et_3GeH + LiNH_2$$

Addition of molecular hydrogen to the platinum complex $(Et_3P)_2Pt(GePh_3)_2$ leads to the formation of a Ge–H-bond[*42*]:

$$(Et_3P)_2Pt(GePh_3)_2 + H_2 \rightarrow (Et_3P)_2Pt(H)(GePh_3) + Ph_3GeH$$

(b) *Replacement of metal by organic groups.* When germane is passed into a solution of sodium in liquid ammonia, germylsodium is formed and the solution becomes decolorised[*87*] (see the exhaustive treatment in section 5.16). If alkyl halides are passed into these solutions, the corresponding organogermanium hydrides are formed (replacement of metal by alkyl)[*104, 290*]:

$$GeH_4 \xrightarrow[-H_2]{+2\,Na} Na_2GeH_2 \xrightarrow{+GeH_4} 2\,NaGeH_3$$

$$NaGeH_3 + RX \rightarrow GeH_3R + NaX$$

$$(R = CH_3, C_2H_5, C_3H_7, \text{i-}C_5H_{11}; X = Br)$$

Likewise, the addition of bromopentane to ethereal solutions of ethylgermyl-lithium or ethyl-i-amylgermyllithium leads to the formation of the corresponding hydrides $(C_2H_5)(C_5H_{11})GeH_2$ and $(C_2H_5)(\text{i-}C_5H_{11})(C_5H_{11})GeH$.

The Ge–H bond in organogermanium hydrides behaves with respect to alkyl (alkali metal)s in the same way as a C–H bond adjacent to an aryl or vinyl group. It differs from the Si–H bond in silanes (see section 5.17)[*100, 102, 193, 303*]:

$$(C_6H_5)_3CH + (CH_3)_2(C_6H_5)CK \rightarrow (C_6H_5)_3CK + (CH_3)_2(C_6H_5)CH$$

$$R_3SiH + LiR' \rightarrow R_3SiR + LiH \quad (R = C_6H_5, C_2H_5; R' = C_4H_9)$$

$$(C_6H_5)_3GeH + LiC_4H_9 \rightarrow (C_6H_5)_3GeLi + C_4H_{10}$$

If the metallation of germane is carried out with methyllithium instead of butyl lithium, the $(C_6H_5)_3GeLi$ that is slowly formed cannot be isolated, since it reacts with unchanged $(C_6H_5)_3GeH$ to give $(C_6H_5)_3GeGe(C_6H_5)_3$ (10% yield)[*100*]:

$$(C_6H_5)_3GeH + LiCH_3 \rightarrow (C_6H_5)_3GeLi + CH_4$$

$$(C_6H_5)_3GeH + LiGe(C_6H_5)_3 \rightarrow (C_6H_5)_3GeGe(C_6H_5)_3 + LiH$$

4.2 Physical properties of the organogermanes

The physical properties of organogermanium hydrides and the corresponding tetraorganogermanes (so far as they are known) are summarised in Table 8.8.

5. Reactions of the Ge–H bond

5.1 Reaction of the Ge–H bond with halogen

Germanium hydrides ($R_{4-n}GeH_n$, $n = 1$–4) react with free halogen (Cl_2, Br_2, I_2) with the replacement of hydrogen attached to germanium by halogen and the formation of the corresponding hydrogen halide (for examples of the reaction, see Table 8.14):

$$\geqslant Ge{-}H + X_2 \rightarrow \geqslant Ge{-}X + HX$$

In the absence of a catalyst, and in reactions in the gas phase and in solvents, any Ge–C bonds in the germane are not attacked.

In the gas phase, the substitution probably takes place, as in the case of the corresponding silanes, by means of a substantially non-polar four-centre reaction in which, in place of the weakly polarised Ge–H bond ($\overset{\delta+}{Ge}{-}\overset{\delta-}{H}$; electronegativities: Ge 2.02 and H 2.20 Pauling units[1] [*151*], a substantially more strongly polarised Ge–X bond is formed ($\overset{\delta+}{Ge}{-}\overset{\delta-}{X}$; electronegativities: F 4.10, Cl 2.83, Br 2.74, I 2.21):

$$GeH_4 + X_2 \longrightarrow \underset{X{-}X}{H_3Ge{-}H} \longrightarrow H_3GeX + HX$$

The attack of further halogen on the halogenogermane leads to the replacement of all the hydrogen atoms attached to the germanium. In the case of germanium, which has a substantially smaller tendency than silicon to the back-co-ordination of electrons, the polarity of the bond is retained (in contrast to this is the Si⇐X bond). This leads – again in contrast to the isologous halogenosilanes – to the ready splitting off of highly negative X and of H, more positive in comparison, as $\overset{\delta+}{H}\overset{\delta-}{X}$, with the formation of stable GeH_2:

$$\overset{\delta+}{H}{-}\underset{H}{\overset{H}{Ge}}{-}\overset{\delta-}{X} \rightarrow GeH_2 + HX$$

In solution, the replacement of the hydrogen atom by elementary halogen probably takes place, like the halogenation of the Si–H bond (see Chapter 7, section 5.1) by a polar reaction mechanism (electrophilic attack of the halogen

[1]In Cl_3GeH and $PhCl_2GeH$: positively polarized H [*183a*].

TABLE 8.8

PHYSICAL PROPERTIES OF ORGANOGERMANIUM HYDRIDES AND RELATED COMPOUNDS

Germane	*M.p.* *(°C)*	*B.p.* *[°C(torr)]*	*Density* d_4^{20} *[g/ml(°C)]*	*Refractive index* n_D^{20}	*Ref.*
GeH_4^a	−165.90	−89.23	1.523 (−142°)	—	36, 85
$CH_3GeH_3^b$	−154.5 ± 0.2 −158	−34.1/−35.1 (760)	—	—	112, 144, 290 5, 91, 127
$C_2H_5GeH_3$	—	~9.2 11.5 (743.5)	—	—	223, 290
$n-C_3H_7GeH_3$	—	30 (760)	—	—	223, 290
$n-C_4H_9GeH_3$	—	75.6 74 (760)	1.033 1.0220	1.4207 1.4200	11 237, 244
$n-C_5H_{11}GeH_3$	—	104–105 (760)	1.0138	1.4302	237, 244
$n-C_6H_{13}GeH_3$	—	128–129 (760)	0.9972	1.4350	237, 244
$n-C_7H_{15}GeH_3$	—	156 (750) 85 (74)	0.9819	1.4390	237, 244
$n-C_8H_{17}GeH_3$	—	80 (31)	0.9717	1.4422	237, 244
$CH_2{=}CHGeH_3^c$	—	−3.5 (760)	—	—	24
$(CH_3)_2GeH_2^d$	−144.3 ± 0.2 −149	−0.6; 3.0 6.5 (744)	—	—	5, 91, 112, 223, 295
$(C_2H_5)_2GeH_2$	—	74 (760)	1.0390	1.4219	237, 244
$(n-C_3H_7)_2GeH_2$	—	126–127 (760)	1.0030	1.4340	237, 244
$(i-C_3H_7)_2GeH_2$	—	110–111 (760)	0.982	1.432	6
$(n-C_4H_9)_2GeH_2$	—	75.2–76.7 (24) 85 (46) 173 (760)	0.977 0.9782	1.4423 1.4428	11 237
$(n-C_4H_9)_2GeH_2$	—	166 (18)	0.9614	1.4489	160
$(C_5H_{11})_2GeH_2$	—	102 (19) 92 (12)	0.9595	1.4478	237 237, 244

$(n\text{-}C_6H_{13})_2GeH_2$	—	245 (750) 113–114(8)	0.9484	1.4522	237, 244
$(n\text{-}C_7H_{15})_2GeH_2$	—	148 (10)	0.9348	1.4543	237, 244
$(n\text{-}C_8H_{17})_2GeH_2$	—	164–165 (9)	0.9274	1.4568	237, 244
$(C_6H_5)_2GeH_2$	—	93 (1)	—	1.5920 (n_D^{25})	129
$(C_6H_5)_2GeD_2$	—	—	—	—	106a
$(C_2H_5)(i\text{-}C_5H_{11})GeH_2$	—	—	20 (10°)	—	127
$(CH_3)_3GeH^e$	−123.1 ± 0.2	27 26 (755.5)	—	—	112, 223
$(C_2H_5)_3GeH$	—	124.4 (751) 124 (760)	—	—	7, 106a, 148, 219
$(n\text{-}C_3H_7)_3GeH$	—	65 (20); 183 (742)	—	—	130
$(i\text{-}C_3H_7)_3GeH$	—	174 (760)	0.9770	1.4505	237
$(n\text{-}C_4H_9)_3GeH$	—	232–233 (760) 123 (20)	0.9455	1.4508	237
$(n\text{-}C_5H_{11})_3GeH$	—	150 (11)	0.9310	1.4542	237
$(i\text{-}C_5H_{11})_3GeH$	—	140 (17)	0.9238	1.4517	237
$(n\text{-}C_6H_{13})_3GeH$	—	169–170 (9); 122–5 (0.5)	0.9228	1.4582	97, 237
$(n\text{-}C_7H_{15})_3GeH$	—	182 (1.7)	0.9108	1.4600	237
$(n\text{-}C_8H_{17})_3GeH$	—	179–180 (0.4)	0.9061	1.4610	237
$(CH_3)_2(C_2H_5)GeH$	—	62 (755.5)	1.0158	1.4090	223
$(CH_3)_2(C_2H_5)GeD$	—	60 (737)	—	—	223
$(C_2H_5)_2(i\text{-}C_5H_{11})GeH$	—	—	—	—	104
$(C_4H_9)_2(C_8H_{17})GeH$	—	153 (9)	0.9292	1.4558	164
$(n\text{-}C_4H_9)_2(CH_2CHCH_2)GeH$	—	96 (13)	0.9702	1.4608	237
$(C_6H_{11})_3GeH$	24–25	—	—	—	131
$(C_6H_5)_3GeH$	27 (β) 47.0–47.1 (α)	—	—	—	131, 294
$(C_{10}H_7)(C_6H_5)(CH_3)GeH$	74–75	—	—	—	26, 27
$(CH_3)_4Ge$	−88	43.4	1.0000 (0°)	1.3868 ($n_D^{23.5}$)	56, 156
$(C_2H_5)_4Ge$	−90	163.5	0.9932	1.4430 (n_D^{20})	141, 148, 200, 256, 267

References p. 711

TABLE 8.8 (cont'd)

$(n\text{-}C_3H_7)_4Ge$	−73	225 (746)	0.9539 (d^{20}_{20})	1.443 (n^{30}_D)	280
$(n\text{-}C_4H_9)_4Ge$	—	278	—	—	8
$(i\text{-}C_5H_{11})_4Ge$	—	163.4 (10)	0.9147 (d^{20}_{20})	1.451 (n^{30}_D)	280
$(C_6H_5)_4Ge$	225–228; 230–231 236–238	225–228 230–231	—	—	131, 149 105, 128, 129
$(CH_3)H_2GeGeH_3$[f]		54.7 (760)			175b
$(C_2H_5)H_2GeGeH_3$[g]		88.6 (760)			175b
$(CH_3)_3GeGeH_3$[h]	−89.6	74.4	—	—	78a
$(CH_3)_3GeGe(CH_3)_3$	−40	138 (750)	—	1.4564 (n^{25}_D)	28
$(C_2H_5)_3GeGe(C_2H_5)_2$	−60	265.0 (758)	—	—	148
$(C_6H_5)_3GeGe(C_6H_5)_3$	352–354; 340–341 330–331; 325–327	476–479	—	—	99, 105, 128, 129
$(CH_3)_3SiGeH_3$[i]	−77.4	73.6	—	—	78a
$(CH_3)_2Si(GeH_3)_2$[k]	−93.1	116.3	—	—	78a
$(CH_3)Si(GeH_3)_3$[l]	−101.0	154.6	—	—	78a
$Si(GeH_3)_4$	−53.3	—	—	—	78a

Vapour pressure equations $\log p$ [torr] $= A/T + B$

[a] A = 789.26, B = 7.162

[b] A = 1118.2, B = 7.578, or $\log p$ [torr] $= 3.9624 - 0.003034T + 1.75 \log T - 1080.3/T$

[c] A = 1263, B = 7.565

[d] A = 1340.4, B = 7.798, or $\log p$ [torr] $= 5.4643 - 0.005872T + 1.75 \log T - 1445.2/T$

[e] $\log p$ [torr] $= 5.4904 - 0.00544\, T + 1.75 \log T - 1594.4/T$

[f] A = +1637.7, B = 7.877

[g] A = +1609.7, B = 7.330

[h] A = +1889.3, B = 8.3172

[i] A = +1581.3, B = 7.4413

[k] A = +1797.3, B = 7.4955

[l] A = +2044.1, B = 7.6598

on the hydrogen atom with subsequent nucleophilic attack of more halogen on the germanium atom), since the reaction of GeH_4 with I_2 takes place in the polar liquid GeH_3I (even at low temperatures) substantially faster than in the gas phase [*50, 273*]:

$$\geq\overset{\delta+}{Ge}-\overset{\delta-}{H}+\overset{\delta+}{I}-\overset{\delta-}{I}_n \rightarrow \left[\geq Ge-H\ldots I-I_n\right]$$

$$I_2+\left[\geq Ge-H\ldots I-I_n\right] \rightarrow \left[I-I\ldots Ge-H\ldots I-I_n\right] \rightarrow \geq GeI+HI+I_{n+1}$$

(n = undetermined; in the reaction SiH_4–I_2, $n = 3$)

5.2 *Reaction of the Ge–H bond with hydrogen halide*

Germanium hydrides $R_{4-n}GeH_n$, $n = 1$–4) react with hydrogen halides (HCl, HBR, HI) with the replacement of hydrogen attached to germanium by halogen and the formation of molecular hydrogen:

$$\geq Ge-H+HX \rightarrow \geq Ge-X+H_2$$

For examples of the reaction, see Table 8.14. The presence of an aluminium halide increases the rate of the reaction. A polar reaction mechanism is fairly certain:

$$\geq GeH+AlX_3 \rightarrow \left[\geq\overset{\oplus}{Ge}-H\ldots\overset{\ominus}{Al}X_3\right]$$

$$H^{\oplus}+AlX_4^{\ominus}+\left[\geq\overset{\oplus}{Ge}-H\ldots\overset{\ominus}{Al}X_3\right] \rightarrow \left[X_3Al-X\ldots Ge-H\ldots AlX_3\right]^{\ominus}H^{\oplus}$$

$$\rightarrow AlX_3+\geq GeX+H_2+AlX_3$$

Whether in this case the electrophilic attack of AlX_3 on the hydrogen atom or the nucleophilic attack of $AlX_4^{\ominus}$ ions on the germanium atom is rate-determining cannot be decided at the present time because of the lack of kinetic measurements.

5.3 *Reaction of the Ge–H bond with metal and non-metal halides*

Germanium hydrides ($R_{4-n}GeH_n$, $n = 1$–4) react with certain metal halides [*7, 162, 237*] and non-metal halides [*237*] with the replacement of hydrogen by halogen and the formation of molecular hydrogen. A prerequisite for this is that the metal halide can pass into a lower stage of oxidation (metal or lower halide) with the liberation of halogen and that the values of the oxidation potentials of the

TABLE 8.9

REACTION OF GERMANIUM HYDRIDES WITH METAL AND NON-METAL HALIDES

Hydride	*Halide*	*Reaction products*	*Yield (mole-%)*	*Ref.*
GeH_4	AgCl	GeH_3Cl	71	177
GeH_4	AgBr	30 mole-% GeH_3Br + 2 mole-% GeH_2Br_2	32	177
Ge_2H_6	AgCl	Ge_2H_5Cl	25	177
Ge_2H_6	BCl_3	—	—	177
Ge_2H_6	AgBr	Ge_2H_5Br	40	177
$(C_2H_5)_3GeH$	$CuBr_2$	$(C_2H_5)_3GeBr + Cu_2Br_2 + H_2$	90	7
$(C_2H_5)_3GeH$	$HgCl_2$	$(C_2H_5)_3GeBr + Hg + HCl + H_2$	55	7
$n\text{-}C_4H_9GeH_3$	$HgCl_2$	$n\text{-}C_4H_9GeH_2Cl + n\text{-}C_4H_9GeCl_3 + Hg + H_2$	—	11
$(n\text{-}C_4H_9)_2GeH_2$	$HgCl_2$	$(n\text{-}C_4H_9)_2GeHCl + Hg + H_2$	—	11
$(C_2H_5)_3GeH$	Hg_2Cl_2	$(C_2H_5)_3GeCl + Hg + HCl + H_2$	90	7
$(C_2H_5)_3GeH$	$HgBr_2$	$(C_2H_5)_3GeBr + Hg + HBr + H_2$	98	7
$n\text{-}C_4H_9GeH_3$	$HgBr_2$	$n\text{-}C_4H_9GeH_2Br + n\text{-}C_4H_9GeBr_3 + Hg + H_2$	—	11
$(C_2H_5)_3GeH$	HgI_2	$(C_2H_5)_3GeI + Hg_2I_2 + H_2$	99	7
$n\text{-}C_4H_9GeH_3$	HgI_2	$n\text{-}C_4H_9GeH_2I + n\text{-}C_4H_9GeI_3 + Hg_2I_2 + H_2$	—	11
$(n\text{-}C_4H_9)_2GeH_2$	HgI_2	$n\text{-}C_4H_9GeHI + Hg_2I_2 + H_2$	—	11
$(C_2H_5)_3GeH$	$PdCl_2$	$(C_2H_5)_3GeCl + Pd + H_2$	88	7
$(C_2H_5)_3GeH$	$TiCl_4$	$(C_2H_5)_3GeCl + TiCl_3 + TiCl_2 + H_2$	70	7
$(C_2H_5)_3GeH$	$KAuCl_4$	$(C_2H_5)_3GeCl + KCl + Au + H_2$	95	7
$(C_2H_5)_3GeH$	K_2PtCl_6	$(C_2H_5)_3GeCl + Pt + H_2$	95	7
$(C_2H_5)_3GeH$	$VOCl_3$	$(C_2H_5)_3GeCl + VOCl_2 + VOCl + H_2$	95	7
$(C_2H_5)_3GeH$	CrO_2Cl_2	$(C_2H_5)_3GeCl + Cr_2O_3 + H_2O + H_2$	45	7
$(C_2H_5)_3GeH$	$CdCl_2$	—	0	7
$(C_4H_9)_3GeH$	$AlCl_3$	$(C_4H_9)GeCl + Al + H_2$	75	162, 237
$(C_4H_9)_3GeH$	CCl_4	$(C_4H_9)_3GeCl + CHCl_3$	90	237
$(C_2H_5)_3GeH$	$GeCl_4$	$(C_2H_5)_3GeCl + Ge + H_2$	87	162, 237
$(C_2H_5)_3GeH$	SO_2Cl_2	$(C_2H_5)_3GeCl + SO_2 + H_2$	78	162, 237

highest and second-highest oxidation stages (of the transition metal halides) are between approximately −2.0 and −0.06 [volt] [7]. Examples of the reaction are given in Table 8.9

$$\gt Ge\text{-}H + E'X_n \rightarrow \gt Ge\text{-}X + E'X_{n-1} + \tfrac{1}{2}H_2$$

5.4 *Reaction of the Ge–H bond with metal pseudohalides*

As in the case of the alkylsilicon hydrides, the reaction of alkylgermanium hydrides with metal pseudohalides leads to the replacement of hydrogen by pseudohalogen [9], for example:

$$\gt Ge\text{-}H + HgX_2 \rightarrow \gt Ge\text{-}X + Hg + HX$$

In the case of $(n\text{-}C_4H_9)_2GeH_2$ the substitutions take place rapidly with $Hg(CN)_2$ and $Hg(SCN)_2$ (yields 72 and 77 mole-%, respectively), and very slowly with AgNCO and AgNCS.

5.5 *Reaction of the Ge–H bond with organic halides*

Alkylgermanium hydrides react with alkyl bromides and iodides (quantitatively), alkyl chlorides (80% yield), halogenocarboxylic acids, α-halogenoethers, aryl chlorides, and benzoyl chloride with the replacement of the hydrogen atom by a halogen atom and the formation of the corresponding alkylhalogenogermanes [*7, 162, 237*]:

$$\diagdown\!\!-\!Ge\!-\!H + RX \rightarrow \diagdown\!\!-\!Ge\!-\!X + RH \quad (R = \text{alkyl, alkenyl, aryl})$$

$$\diagdown\!\!-\!Ge\!-\!H + RCClO \rightarrow \diagdown\!\!-\!Ge\!-\!X + RCHO \quad (R = \text{alkyl, aryl})$$

$$n\,\diagdown\!\!-\!Ge\!-\!H + CH_{3-n}X_nCOOH \rightarrow n\,\diagdown\!\!-\!Ge\!-\!X + CH_3\,COOH \quad (X = Cl, Br, I)$$

$$\diagdown\!\!-\!Ge\!-\!H + ROCH_2Cl \rightarrow \diagdown\!\!-\!Ge\!-\!X + ROCH_3 \quad (R = \text{alkyl})$$

The reactivity of the Ge–H bond increases the sequence $R_3GeH < R_2GeH < RGeH_3$. Thus, propyl iodide reacts sluggishly with tributylgermane but very rapidly and exothermically with monohexylgermane. The reactivity of the germanes with alkyl halides increases from the chlorides to the iodides (however, the very labile Cl in triphenylchloromethane also reacts rapidly, as can easily be understood). Allyl bromide and allyl chloride have only a halogenating action, even in the presence of Pt-asbestos or $H_2PtCl_6.6H_2O$, which catalyse the addition of hydrides to the C=C double bond [*237*]. Benzoyl chloride rapidly replaces hydrogen in alkylgermanes by chlorine, while it reacts with alkylsilanes only sluggishly ($HSi(CH_2C_6H_5)_3$ reacts without the addition of $AlCl_3$, and $HSi(C_2H_5)_3$ only in the presence of $AlCl_3$, to give the corresponding chlorosilanes [*126*]). Alkylgermanes are halogenated by α-halogenoethers without a catalyst (triethylsilane is chlorinated by a mixture of monochloromethyl ether and zinc chloride [*95*]). Examples of the reactions are given in Table 8.10

5.6 *Reaction of the Ge–H bond with oxygen*

Monogermane is slowly oxidised by oxygen in clean glass vessels at 230 to 330° [*85*]:

$$GeH_4 + 2O_2 \rightarrow GeO_2 + 2H_2O$$

In the presence of a slight brown deposit (germanium or polygermane) – probably formed by the pyrolysis of the GeH_4 – monogermane is oxidised even at 150 to 190°. Irradiation with ultraviolet light shortens the induction period of the oxidation by a factor of 500 (*e.g.*, with 20% of GeH_4 in the mixture of GeH_4 and O_2 at 230° the induction period is only 12 sec instead of 100 min) and lowers the in-

References p. 711

TABLE 8.10

REPLACEMENT OF HYDROGEN BY HALOGEN IN GERMANIUM HYDRIDES BY REACTION WITH ORGANIC HALOGEN COMPOUNDS

Hydride	*Halogenating agent*	*Reaction products*	*Yield (mole-%)*	*Ref.*
$(C_4H_9)_2GeH_2$	$(C_6H_5)_3CCl$	$(C_4H_9)_2GeCl_2$	100	237
$(C_4H_9)_3GeH$	n-$C_7H_{15}Cl$	$(C_4H_9)_3GeCl$	80	237
$(C_4H_9)_3GeH$	CCl_4	$(C_4H_9)_3GeCl$ (+ $CHCl_3$)	—	162
$C_6H_{13}GeH_3$	alkyl bromide	$C_6H_{13}GeBr_3$	100	237
$(C_2H_5)_2GeH_2$	C_4H_9Br	$(C_2H_5)_2GeHBr$	45	237
$(C_4H_9)_2GeH_2$	C_4H_9Br	$(C_4H_9)_2GeBr_2$	—	237
$(C_4H_9)_3GeH$	$C_7H_{15}Br$	$(C_4H_9)_3GeBr$	95	237
$C_6H_{13}GeH_3$	C_3H_7I	$C_6H_{13}GeI_3$ (+ C_3H_8)	96	237
$(C_2H_5)_2GeH_2$	C_3H_7I	$(C_2H_5)_2GeHI$	40	237
$(C_4H_9)_2GeH_2$	C_3H_7I	$(C_4H_9)_2GeHI$	40	237
(n-$C_5H_{11})_2GeH_2$	C_3H_7I	(n-$C_5H_{11})_2GeI_2$ (+ C_3H_8)	95	237
$(C_4H_9)_3GeH$	C_3H_7I	$(C_4H_9)_3GeI$ (+ C_3H_8)	95	237
(i-$C_5H_{11})_3GeH$	C_3H_7I	(i-$C_5H_{11})_3GeI$ (+ C_3H_8)	—	237
$(C_4H_9)_3GeH$	$CH_2{=}CHCH_2Br$	$(C_4H_9)_3GeBr$ (+ $CH_2{=}CHCH_3$)[a]	—	200, 237
(i-$C_5H_{11})_3GeH$	$CH_2{=}CHCH_2CN$	(i-$C_5H_{11})_3Ge[(CH_2)_3CN]$[b]	50	162
$(C_2H_5)_2GeH_2$	CH_3OCH_2Cl	$(C_2H_5)_2GeHCl$	80	162, 183, 245
$(C_4H_9)_2GeH_2$	CH_3OCH_2Cl	$(C_4H_9)_2GeHCl$	80	237
n-$C_6H_{13}GeH_3$	CCl_3COOH	n-$C_6H_{13}GeCl_3$[c]	80	162, 237
$(C_2H_5)_3GeH$	CCl_3COOH	$(C_2H_5)_3GeCl$[c]	80	7
$(C_2H_5)_3GeH$	CBr_3COOH	$(C_2H_5)_3GeBr$[c]	94	7
$(C_2H_5)_3GeH$	CI_3COOH	$(C_2H_5)_3GeI$[c]	94	7
$(C_4H_9)_3GeH$	C_6H_5COCl	$(C_4H_9)_3GeCl$ (+ C_6H_5CHO)	70	162
$(C_4H_9)_2GeH_2$	C_6H_5Cl	—	0	237
$(C_4H_9)_2GeH_2$	C_6H_5I	$(C_4H_9)_2GeI_2$	40	162, 237

[a]No addition even in the presence of H_2PtCl_6.
[b]Only addition in the presence of H_2PtCl_6.
[c]No formation of acetate.

flammation temperature of the GeH_4–O_2 mixture. In the photochemically induced oxidation of GeH_4 with O_2, hydrogen is also formed, in contrast to thermally induced oxidation [*85*].

Alkylgermanes are substantially more resistant to oxygen than the tin isologues. R_3GeH, R_2GeH_2, and $RGeH_3$ are oxidised in air to oxides with decreasing basicity: $(R_3Ge)_2O$, R_2GeO, and $(RGeO)_2O$. The last two oxides are in general cyclic, R_2GeO forming trimeric and $(RGeO)_2O$ tetrameric rings. Under the action of oxygen, $C_8H_{17}GeH_3$ is converted into a highly viscous liquid. From this (n-$C_8H_{17}GeO)_2O$ can be isolated, a white water-soluble powder which decomposes at 400° without melting. With HCl it forms n-$C_8H_{17}GeCl_3$ [*134*]. Potassium manganate(VII) oxidises triethylgermane in 28% yield to bis(triethylgermyl) oxide, $[(C_2H_5)_3Ge]_2O$ [*7*].

5.7 *Reaction of the Ge–H bond with water*

Monogermane is fairly stable to water and acidified water. Consequently, it can be prepared in this medium or be freed from ammonia by means of acidified water. In the presence of copper powder, however, $(C_4H_9)_2GeH_2$ and $(C_2H_5)_3GeH$ form the corresponding germyl oxides with water at 100° with the evolution of hydrogen[160]:

$$\rangle GeH_2 + H_2O \rightarrow \frac{1}{n}[-\overset{|}{\underset{|}{Ge}}O-]_n + 2H_2$$

$$2-\!\!\!\geq GeH + H_2O \rightarrow (-\!\!\!\geq Ge)_2O + 2H_2$$

5.8 *Reaction of the Ge–H bond with acids*

5.8.1 *Reaction of the Ge–H bond with mineral acids*

Dibutylgermane undergoes little change on being boiled for four hours with 60% aqueous hydrobromic acid[237]. On the other hand, triethylsilane forms the corresponding chloride with alcoholic hydrochloric acid and alkylstannanes the corresponding chlorides with ethereal hydrochloric acid[283].

Triethylgermane is converted into the sulphate in 70% yield by 100% H_2SO_4 and by boiling with $HgSO_4$ without a solvent[7]:

$$2(C_2H_5)_3GeH + H_2SO_4 \rightarrow [(C_2H_5)_3Ge]_2SO_4 + 2H_2$$

$$2(C_2H_5)_3GeH + HgSO_4 \rightarrow [(C_2H_5)_3Ge]_2SO_4 + Hg + H_2$$

In addition, a small amount of bis(triethylgermyl) oxide, $[(C_2H_5)_3Ge]_2O$, is formed. In contrast to this, under similar conditions the isologous silicon compound is converted into bis(triethylsilyl) oxide, $[(C_2H_5)_3Si]_2O$,[12]:

$$2(C_2H_5)_3SiH + 2H_2SO_4 \rightarrow [(C_2H_5)_3Si]_2O + 3H_2O + 2SO_2$$

$$2(C_2H_5)_3SiH + HgSO_4 \rightarrow [(C_2H_5)_3Si]_2O + H_2O + SO_2 + Hg$$

As was to be expected, fuming nitric acid and mixtures of nitric acid and sulphuric acid oxidise both the Ge–H and the Ge–C bonds in organogermanes. Germanes form GeO_2 quantitatively, often explosively[237].

Like the isologous triethylsilane[78], triethylgermane[237] gives a 70% yield of the sulphonate with benzenesulphonic acid in toluene:

$$(C_2H_5)_3GeH + C_6H_5SO_3H \rightarrow (C_2H_5)_3GeSO_3C_6H_5 + H_2$$

With triethylgermane in the presence of copper powder, boric acid forms tris(triethylgermyl) borate with the evolution of hydrogen[160]:

$$3(C_2H_5)_3GeH + B(OH)_3 \rightarrow B[OGe(C_2H_5)_3]_3 + 3H_2$$

5.8.2 *Reaction of the Ge–H bond with carboxylic acids*

The Ge–H bond in triethylgermane is converted by acetate ions into a Ge–OAc bond (Ac = CH_3CO). The rate of the reaction depends on the degree of dissociation of the acetic acid used. With weakly dissociated acetic acid, triethylgermyl

References p. 711

acetate, $(C_2H_5)_3GeOCOCH_3$, is formed only after prolonged boiling under reflux [*7*]. According to other workers, alkylgermanium hydrides are inert to glacial acetic acid both in the absence of a solvent and in diethyl and dibutyl ether [*237*].

With the more strongly dissociated fluoroacetic acids, the corresponding acetates are formed, the yield increasing with increasing fluorine substitution. With CH_2FCOOH, $(C_2H_5)_3GeOCOCH_2F$ is formed in 20% yield after 2 hours' boiling, with CHF_2COOH, $(C_2H_5)_3GeOCOCHF_2$ is formed in 60% yield, and with CF_3COOH, $(C_2H_5)_3GeOCOCF_3$ is formed in 95% yield after 3 hours' boiling. Similar high yields (97–99%) are obtained with perfluorinated propionic and butyric acids [7]. With $Hg(OCOCH_3)_2$, $(C_2H_5)_3GeH$ forms $(C_2H_5)_3$-$GeOCOCH_3$ after only 15 min [7]. In the presence of copper powder, $(C_2H_5)_3GeH$ gives a 60% yield of $(C_2H_5)_3GeOOC{\cdot}CH{=}CH_2$ with $CH_2{=}CHCOOH$, with the evolution of hydrogen, the C=C double bond remaining unchanged in this case in contrast to reactions without a catalyst (see section 5.13 and Table 8.11).

Halogenated (other than fluorinated) acetic acids do not form acetates with triethylgermane but replace hydrogen by halogen [*7, 237*]:

$$R_{4-n}GeH_n + \frac{n}{m} X_mCH_{3-m}COOH \rightarrow R_{4-n}GeX_n + \frac{n}{m} CH_3COOH$$

For example of these reactions see table 8.10 (section 5.5). The fact that the reactions of the fluoroacetic acids differ from those of the other halogenoacetic acids is explained as follows: the energies of the C–Cl, C–Br, and C–I bonds (67, 54, and 46 kcal/mole) are lower, while the energy of the C–F bond (107 kcal/mole) is higher than those of the C–H and Ge–H bonds (87 and 68.9 kcal/mole). Consequently, germanium hydrides are, in general, halogenated by alkyl iodides, bromides, and chlorides while the alkyl fluorides do not halogenate them.

5.9 Reaction of the Ge–H bond with bases

In contrast to the very sensitive Sn–H bond [*253*] and the only slightly more stable Si–H bond [*253*], the Ge–H bond is very resistant to alkaline hydrolysis [*237, 253*]. Thus, mono-, di- and trialkylgermanes hydrolyse only very slightly in 20% aqueous KOH solution [*237*]. On the other hand, monoalkylgermanes (*e.g.*, $n\text{-}C_6H_{13}GeH_3$) undergo 97.7% solvolysis in alcoholic potassium hydroxide solution after heating for 15 h to 80° yielding $(n\text{-}C_6H_{13}GeO)_2O$. The reaction of dialkylgermanes takes place somewhat more slowly.

5.10 Reaction of the Ge–H bond with alcohols and thiols

In the presence of copper powder, di- and trialkylgermanes [$(n\text{-}C_4H_9)_2GeH_2$, $(C_2H_5)_3GeH$, $(n\text{-}C_4H_9)_3GeH$] react with water, primary and secondary aliphatic alcohols, aromatic alcohols, and aliphatic or aromatic diols with the evolution of hydrogen to give the corresponding oxygen compounds; thioalcohols react similarly in the presence of platinum on starch [*160, 161, 238, 240*]:

$$>GeH_2 + 2ROH \rightarrow >Ge(OR)_2 + 2H_2$$

$$-GeH + ROH \rightarrow -Ge(OR) + H_2$$

$$>GeH_2 + 2R_2HCOH \rightarrow >Ge(OCHR_2)_2 + 2H_2$$

$$-GeH + R_2HCOH \rightarrow -Ge(OCHR_2) + H_2$$

$$>GeH_2 + HOROH \rightarrow >\overline{GeORO} + 2H_2 \text{ (+ viscous polymers)}$$

$$2-GeH + HOROH \rightarrow -GeOROGe- + 2H_2$$

$$-GeH + RSH \rightarrow -Ge(SR) + H_2$$

5.11 *Reaction of the Ge–H bond with aldehydes and ketones*

The Ge–H bond in alkylgermanium hydrides [$(C_2H_5)_3GeH$, $(C_4H_9)_3GeH$] generally reacts quantitatively with aldehydes and ketones in the presence of copper to give germyl alkoxides [*163*]:

$$R_3\overset{\delta+}{Ge}-\overset{\delta-}{H} + \overset{\delta-}{O}=\overset{\delta+}{C}(R')(R'') \xrightarrow[Cu]{150°} R_3Ge-O-C(R')(R'')-H$$

$(R', R'' = n-C_6H_{13}, H; C_6H_5, H; C_4H_9, C_2H_5; C_6H_{13}, CH_3; (CH_2)_5)$

5.12 *Reaction of the Ge–H bond with ammonia*

When germane is passed into liquid ammonia, ammonium trihydridogermanate is formed (judging from the increased conductivity of the solution):

$$GeH_4 + NH_3 \rightarrow [NH_4]^{\oplus}[GeH_3]^{\ominus}$$

The solution dissolves phosphorus, probably giving $[NH_4]^{\oplus}[GeH_3 \cdot P_x]^{\ominus}$. After the ammonia has been pumped off, the orange sorption compound of GeH_4 on amorphous phosphorus remains [*230*].

References p. 711

5.13 *Reaction of the Ge–H bond with alkenes*

Germanium hydrides add to the C=C double bond giving preferentially linear and, to a smaller extent, branched chains:

$$\rangle\overset{\delta+}{Ge}-\overset{\delta-}{H} + H_2\overset{\delta-}{C}=\overset{\delta+}{C}HR \rightarrow \rangle Ge-CH_2-CH_2R$$

(R = ligand with + I effect) (1)

$$\rangle\overset{\delta+}{Ge}-\overset{\delta-}{H} + H_2\overset{\delta+}{C}=\overset{\delta-}{C}HR' \rightarrow CH_3-CHR'-Ge\langle$$

(R′ = ligand with − I effect) (2)

Although the addition is not affected by benzoyl peroxide, a radical reaction mechanism cannot be excluded:

$$\text{Peroxide} \rightarrow R\cdot$$

$$\rangle Ge-H + R\cdot \rightarrow \rangle Ge\cdot + RH$$

$$\rangle Ge\cdot + CH_2=\dot{C}HR \rightarrow \rangle GeCH_2\dot{C}HR$$

$$\rangle GeCH_2CHR + \rangle Ge-H \rightarrow \rangle GeCH_2CH_2R + \rangle Ge\cdot$$

The basis for this is that germanium hydrides which, because of the accumulation of highly electronegative ligands probably possess protonic hydrogen ($\rangle\overset{\delta-}{Ge}-\overset{\delta+}{H}$) instead of hydridic hydrogen, add to alkyl halides or cyanides ($\overset{\delta+}{C}H_2=\overset{\delta-}{C}HX$) – if at all – linearly:

$$\rangle Ge-H + CH_2CHX \rightarrow \rangle Ge-CH_2CH_2X$$

which would be forbidden by the ionic mechanism (2). Furthermore, it has been shown that in the addition of Sn–H to alkenes [*209–211*] benzoyl peroxide, instead of catalysing the addition, causes chain termination by the formation of benzoates while other radical-forming agents favour the addition:

$$\rangle Sn-H + (C_6H_5COO)_2 \rightarrow \rangle SnOCOC_6H_5 + C_6H_5COOH$$

For a comprehensive account of the addition of $\rangle Ge-H$ to $\rangle C=C\langle$ see the literature [*182, 227a*].

5.13.1 *Reaction with unsubstituted alkenes*

Germanium hydrides add to alkenes in the presence of peroxides and under certain circumstances even without a catalyst to form the corresponding alkylgermanes (for examples of the reaction, see Table 8.11)[*164, 200, 227, 237*]:

$$R_3Ge{-}H + CH_2{=}CHR' \rightarrow R_3GeCH_2CH_2R'$$

(R = H, alkyl, halogen; R′ = alkyl, halogenoalkyl)

In the case of buta-1,3-diene, 1,4-addition takes place[*242, 243*].

5.13.2 Reaction with alkenyl halides

Germanium hydrides react with allyl halides either by adding to the C=C double bond with the formation of a germanium–alkyl bond or with the preservation of the double bond and the replacement of hydrogen attached to germanium by alkenyl[*164, 200, 227*]:

$$R_3Ge{-}H + CH_2{=}CH(CH_2)_nX \rightarrow R_3GeCH_2(CH_2)_nX$$

$$R_3Ge{-}H + CH_2{=}CH(CH_2)_nX \rightarrow R_3Ge(CH_2)_nCH{=}CH_2 + HX$$

The reactions still require clarifying experiments on the mechanism and on the direction in which they take place. Multiple-halogen-substituted alkenes do not add. Table 8.11 contains examples of the reactions.

5.13.3 Reaction with alkenyl cyanides

Di- and trialkylgermanes react with vinyl cyanide without a catalyst and with allyl cyanide with a catalyst to form the corresponding adducts, preferentially the linear adducts but also the branched compounds (*e.g.*, 1-alkylgermyl-2-cyanoethane and 1-alkylgermyl-1-cyanoethane)[*161, 164, 237*]:

$$R_2GeH_2 + CH_2{=}CHCN \rightarrow R_2HGeCH_2CH_2CN \text{ (or } R_2HGeCH(CH_3)CN)$$

$$R_2GeH_2 + 2CH_2{=}CHCN \rightarrow R_2Ge(CH_2CH_2CN)_2 \text{ (or } R_2Ge[CH(CH_3)CN]_2)$$

$$R_3GeH + CH_2{=}CHCN \rightarrow R_3GeCH_2CH_2CN \text{ (or } R_3GeCH(CH_3)CN)$$

$$R_2GeH_2 + CH_2{=}CHCH_2CN \xrightarrow{Pt} R_2HGeCH_2CH_2CH_2CN$$

$$R_2GeH_2 + 2CH_2{=}CHCH_2CN \xrightarrow{Pt} R_2Ge(CH_2CH_2CH_2CN)_2$$

$$R_3GeH + CH_2{=}CHCH_2CN \xrightarrow{Pt} R_3GeCH_2CH_2CH_2CN$$

The addition is little affected by radical initiators such as benzoyl peroxide or ultraviolet light, but radical initiators accelerate the polymerisation of the unsaturated reactant. Radical inhibitors such as hydroquinone decrease the yield of the addition reaction only very slightly. For examples of the reactions, see Table 8.11.

TABLE 8.11

ADDITION OF GERMANIUM HYDRIDES TO ALKENES

Germane	Alkene	Catalyst (inhibitor)	Reaction products	Yield (mole-%)	Ref.
	Alkenes				
$(C_2H_5)_3GeH$	$CH_2{=}CH(CH_2)_5CH_3$	—	$(C_2H_5)_3Ge(CH_2)_7CH_3$	—	237
$(C_4H_9)_2GeHCl$	$CH{=}CH(CH_2)_3CH_2$	—	$(C_4H_9)_2(Cl)GeCH(CH_2)_4CH_2$	70	164
$(C_4H_9)_2GeHCl$	$CH_2{=}CH(CH_2)_5CH_3$	—	$(C_4H_9)_2(Cl)Ge(CH_2)_7CH_3$	80	164
$(C_4H_9)_2GeHBr$	$CH_2{=}CH(CH_2)_5CH_3$	—	$(C_4H_9)_2(Br)Ge(CH_2)_7CH_3$	85	164
$(C_4H_9)_2GeHI$	$CH_2{=}CH(CH_2)_5CH_3$	—	$(C_4H_9)_2(I)Ge(CH_2)_7CH_3$	90	164
$GeHCl_3$	$CH_2{=}C(CH_3)_2$	—	$Cl_3GeCH_2CH(CH_3)_2$	78	200
$GeHCl_3$	$CH_2{=}CHCH_2CH_2CH_3$	0.5% peroxides	$C_5H_{11}GeCl_3$	16	227
$GeHCl_3$	$CH_2{=}CH(CH_2)_3CH_3$	0.5% peroxides	$C_6H_{13}GeCl$	24	227
$GeHCl_3$	$CH{=}CH(CH_2)_3CH_2$	—	$CH_2(CH_2)_4CHGeCl_3$	85	200
$GeHCl_3$	$CH_2{=}CH(CH_2)_4CH_3$	0.5% peroxides	$C_7H_{15}GeCl_3$	18	227
$GeHCl_3$	methylheptene-1	0.5% peroxides	$Me{\cdot}C_7H_{14}GeCl_3$	10	227
$GeHCl_3$	$CH_2{=}CH(CH_2)_5CH_3$	0.5% peroxides	$C_8H_{17}GeCl_3$	11	227
$GeHCl_3$	$C_{10}H_{20}$	0.5% peroxides	$C_{10}H_{21}GeCl_3$	9	227
	Alkenyl halides				
$(C_2H_5)_3GeH$	$CH_2{=}CHCH_2Br$	H_2PtCl_6	no addition!		200
			$CH_2{=}CHCH_2Ge(C_2H_5)_3$	37	
			$(C_2H_5)_3GeBr$	40	
$(C_2H_5)_2GeHCl$	$CH_2{=}CHCH_2Cl$	—	$(C_2H_5)_2(Cl)GeCH_2CH_2CH_2Cl$	90	200
$(C_4H_9)_2GeHCl$	$CH_2{=}CHCH_2Cl$	—	$(C_4H_9)_2(Cl)GeCH_2CH_2CH_2Cl$	90	200
$(C_2H_5)_2GeHBr$	$CH_2{=}CHCH_2Br$	—	no addition! $(C_2H_5)_2GeBr_2 + CH_2{=}CHCH_3$	—	200
$GeHCl_3$	$CHCl{=}CCl_2$	0.5% peroxides	no addition!	—	227

$GeHCl_3$	C_3H_5Cl	—	$ClC_3H_6GeCl_3$	30	200
$GeHCl_3$	$CH_2{=}C(Cl)CH_3$	—	$(CH_3)ClCHCH_2GeCl_3$	54	200
$GeHCl_3$	$(CH_2{=}CHCH_2Cl$	—	no addition! $CH_2{=}CHCH_2GeCl_3$	26	200
$GeHCl_3$	$CH_2{=}CHCH_2Br$	—	no addition! $CH_2{=}CHCH_2GeCl_3$	50	200
$GeHCl_3$	$CH_2{=}C(CH_3)CH_2Cl$	—	no addition! $CH_2{=}C(CH_3)CH_2GeCl_3$	60	200
$GeHCl_3$	$CCl_2{=}CClCCl{=}CCl_2$	—	no addition!	—	227
$GeHCl_3$	$CH_2{=}CHCH_2CHClCH_3$	—	$CH_3CHCl(CH_2)_3GeCl_3$	10	227
	Alkenyl cyanides				
$(n\text{-}C_3H_7)_2GeH_2$	$CH_2{=}CHCN$	without	$(n\text{-}C_3H_7)_2HGeCH_2CH_2CN$	21	237
		Pt-asbestos	$(n\text{-}C_3H_7)_2Ge(CH_2CH_2CN)_2$	31	
$(n\text{-}C_3H_7)_2GeH_2$	$CH_2{=}CHCH_2CN$	(hydroquinone)	$(n\text{-}C_3H_7)_2GeCH_2CH_2CH_2CN$	21	237
$(C_2H_5)_3GeH$	$CH_2{=}CHCN$	—	$(C_2H_5)_3GeCH_2CH_2CN$	—	161, 237
$(n\text{-}C_3H_7)_3GeH$	$CH_2{=}CHCN$	—	$(n\text{-}C_3H_7)_3GeCH_2CH_2CN$	—	237
$(n\text{-}C_4H_9)_3GeH$	$CH_2{=}CHCN$	—	$(n\text{-}C_4H_9)_3GeCH_2CH_2CN$	42	161, 237
$(n\text{-}C_5H_{11})_3GeH$	$CH_2{=}CHCN$	—	$(n\text{-}C_5H_{11})_3GeCH_2CH_2CN$	—	161
$(i\text{-}C_5H_{11})_3GeH$	CH_2CHCH_2CN	Pt-asbestos	$(i\text{-}C_5H_{11})_3GeCH_2CH_2CH_2CN$	50	237
$(C_4H_9)_2GeHCl$	$CH_2{=}CHCN$	—	$(C_4H_9)_2(Cl)GeCH_2CH_2CN$	70	164
	Unsaturated acids, esters, alkenecarboxylic esters				
$(n\text{-}C_3H_7)_2GeH_2$	$CH_2{=}CHCOOH$	—	$(n\text{-}C_3H_7)_2HGeCH_2CH_2COOH$	30	237
$(C_2H_5)_3GeH$	$CH_2{=}CHCOOH$	—	$(C_2H_5)_3GeCH_2CH_2COOH$	49	161, 237
$(n\text{-}C_4H_9)_2GeH_2$	$CH_2{=}CHCOOCH_3$	—	$(n\text{-}C_4H_9)_2HGe(CH_2)_2COOCH_3$	30	237
			$(n\text{-}C_4H_9)_2Ge[(CH_2)_2COOCH_3]_2$	20	
$(n\text{-}C_3H_7)_3GeH$	$CH_2{=}CHCOOCH_3$	—	$(n\text{-}C_3H_7)_3GeCH_2CH_2COOCH_3$	34	161, 237
$(n\text{-}C_4H_9)_3GeH$	$CH_2{=}CHCOOCH_3$	—	$(n\text{-}C_4H_9)_3GeCH_2CH_2COOCH_3$	40	237
$(n\text{-}C_4H_9)_3GeH$	$CH_2{=}CHCOOC_2H_5$	(hydroquinone)	$(n\text{-}C_4H_9)_3GeCH_2CH_2COOC_2H_5$	34	161, 237
$(C_2H_5)_3GeH$	$CH_2{=}CHCH_2OCOCH_3$	—	$(C_2H_5)_3Ge(CH_2)_3OCOCH_3$	50	200
$(C_4H_9)_2GeHCl$	$CH_2{=}CHCOOCH_3$	—	$(C_4H_9)_2(Cl)GeCH_2CH_2COOCH_3$	90	164a, 164b

TABLE 8.11 (cont'd)

Alkenols, alkenethiols, alkenals, alkenones, alkenylamines; alkenyl ethers, alkenyl silanes, and alkenyl germanes					
$(n\text{-}C_3H_7)_3GeH$	$CH_2{=}CHCH_2OH$	benzoyl peroxide	$(n\text{-}C_3H_7)_3GeCH_2CH_2CH_2OH$	14	161, 237
$(n\text{-}C_4H_9)_3GeH$	$CH_2{=}CHCH_2OH$	Pt-asbestos	$(n\text{-}C_4H_9)_3GeCH_2CH_2CH_2OH$	57	161, 237
$(C_4H_9)_2GeHCl$	$CH_2{=}CHCH_2OH$	—	$(C_4H_9)_2(Cl)GeCH_2CH_2CH_2OH$	65	164
$(n\text{-}C_4H_9)_3GeH$	$CH_2{=}CHCH_2SH$	H_2PtCl_6	$(n\text{-}C_4H_9)_3GeCH_2CH_2CH_2SH$	30	237
$(n\text{-}C_5H_{11})_2GeH_2$	$CH_3CH{=}CHCHO$	Pt (hydroquinone)	$(n\text{-}C_5H_{11})_2HGeCH_2CH_2CHO$	44	237
$(C_2H_5)_3GeH$	$CH_2{=}CHCHO$	—	$(C_2H_5)_3GeCH_2CH_2CHO$	—	161
$(n\text{-}C_4H_9)_3GeH$	$CH_2{=}CHCHO$	Pt (hydroquinone)	$(n\text{-}C_4H_9)_3GeCH_2CH_2CHO$	60	161, 237
$(C_2H_5)_3GeH$	$CH_3(CH_2)_5CHO$	Cu	$(C_2H_5)_3GeO(CH_2)_6CH_3$	—	163
$(n\text{-}C_4H_9)_2GeH_2$	$CH_2{=}CHCOCH_3$	(hydroquinone)	$R_2HGeCH_2CH_2COCH_3$	30	237
			$R_2Ge(CH_2CH_2COCH_3)_2$	27	
$(C_4H_9)_2GeHCl$	$CH_2{=}CHOC_4H_9$	—	$(C_4H_9)_2(Cl)GeCH_2CH_2OC_4H_9$	80	164
$(n\text{-}C_4H_9)_3GeH$	$CH_2{=}CHCH_2NH_2$	Pt-asbestos	$(n\text{-}C_4H_9)_3GeCH_2CH_2CH_2NH_2$	74	237
$n\text{-}C_7H_{15}GeH_3$	$CH_2{=}CHOC_4H_9$	—	$n\text{-}C_7H_{15}Ge(CH_2CH_2OC_4H_9)_3$	70	237
$(n\text{-}C_4H_9)_3GeH$	$CH_2{=}CHOC_4H_9$	—	$(n\text{-}C_4H_9)_3GeCH_2CH_2OC_4H_9$	72	237
$GeHCl_3$	$CH_2{=}CHSi(CH_3)Cl_2$	—	$Cl_2(CH_3)SiCH_2CH_2GeCl_3$	68	200
$GeHCl_3$	$CH_2{=}CHCH_2SiCl_3$	—	$Cl_3Si(CH_2)_3GeCl_3$	97	200
$(C_2H_5)_3GeH$	$CH_2{=}CHGe(C_2H_5)_3$	Pt-asbestos	$(C_2H_5)_3GeCH_2CH_2Ge(C_2H_5)_3$	100	237
$(C_2H_5)_3GeH$	$CH_2{=}CHCH_2Ge(C_2H_5)_3$	Pt-asbestos	$(C_2H_5)_3Ge(CH_2)_3Ge(C_2H_5)_3$	100	237
$GeHCl_3$	$CH_2{=}C(CH_3)CH_2GeCl_3$	—	$Cl_3GeCH_2CH(CH_3)CH_2GeCl_3$	57	200
$(C_2H_5)_3GeH$	$CH_2{=}CHCH_2B(OCH_3)_2$	—	$(C_2H_5)_3GeCH_2CH_2B(OCH_3)_2$	—	196

5.13.4 Reaction with acrylic acid

Di- and trialkylgermanes add acrylic acid without a catalyst to form the corresponding alkylgermylpropionic acids [*161, 237*]:

$$R_2GeH_2 + CH_2{=}CHCOOH \rightarrow R_2HGeCH_2CH_2COOH$$

$$R_2GeH_2 + 2CH_2{=}CHCOOH \rightarrow R_2Ge(CH_2CH_2COOH)_2$$

$$R_3GeH + CH_2{=}CHCOOH \rightarrow R_3GeCH_2CH_2COOH$$

For examples of the reactions, see Table 8.11. In contrast to this, the tin isologues react under similar experimental conditions to form trialkyltin acrylate and hydrogen (see section 4.4 of Chapter 9). In the presence of copper powder, trialkylgermanes react with acrylic acid with the retention of the C=C double bond to form the corresponding esters.

5.13.5 Reaction with acrylic esters

Di- and trialkylgermanes add acrylic esters without a catalyst to form the corresponding 3-trialkylgermylpropionic esters (60%) and 2-trialkylgermylpropionic esters (17%) [*161, 237*]:

$$R_2GeH_2 + CH_2{=}CHCOOR' \rightarrow R_2HGeCH_2CH_2COOR' \quad (\text{and } R_2HGeCH(CH_3)COOR')$$

$$R_2GeH_2 + 2CH_2{=}CHCOOR' \rightarrow R_2Ge(CH_2CH_2COOR')_2 \quad (\text{and } R_2Ge[CH(CH_3)COOR']_2$$

$$R_3GeH + CH_2{=}CHCOOR \rightarrow R_3GeCH_2CH_2COOR \quad (\text{and } R_3GeCH(CH_3)COOR')$$

For examples of the reactions, see Table 8.11.

5.13.6 Reaction with allyl alcohol

In contrast to the silane and stannane isologues and in harmony with the less hydridic hydrogen atom attached to germanium, trialkylgermanes add to allyl alcohol and also, in the presence of H_2PtCl_6, to alkenylthiols (for examples of the reactions see Table 8.11 [*161, 237*]:

$$R_3GeH + CH_2{=}CHCH_2OH \rightarrow R_3GeCH_2CH_2CH_2OH$$

$$R_3GeH + CH_2{=}CHCH_2SH \rightarrow R_3GeCH_2CH_2CH_2SH$$

5.13.7 Reaction with unsaturated aldehydes

Di- and trialkylgermanes add to acrolein and crotonaldehyde in the presence of a polymerisation inhibitor and a little platinised asbestos to form saturated aldehydes (for examples of the reactions see Table 8.11 [*161, 237*]:

$$R_2GeH_2 + R'CH{=}CHCHO \rightarrow R_2HGeCHR'CH_2CHO \; (R' = H, CH_3)$$

$$R_2GeH_2 + 2\,R'CH{=}CHCHO \rightarrow R_2Ge(CHR'CH_2CHO)_2 \; (R' = H, CH_3)$$

$$R_3GeH + CH_2{=}CHCHO \rightarrow R_3GeCH_2CH_2CHO$$

5.13.8 *Reaction with unsaturated ketones*

Dialkylgermanes add (like alkylsilanes [*32*]) to the C=C double bond of vinyl methyl ketone in the presence of a polymerisation inhibitor to form saturated germyl ketones (for examples of the reactions, see Table 8.11 [*237, 243*]:

$$R_2GeH_2 + CH_2{=}CHCOR' \rightarrow R_2HGeCH_2CH_2COR'$$
$$R_2GeH_2 + 2\ CH_2{=}CHCOR' \rightarrow R_2Ge(CH_2CH_2COR')_2$$

5.13.9 *Reaction with unsaturated ethers*

Mono- and trialkylgermanes add to vinyl ethers to form saturated germyl ethers in good yields (for examples of the reactions, see Table 8.11) [*237*]:

$$RGeH_3 + 3\ CH_2{=}CHOR' \rightarrow RGe(CH_2CH_2OR')_3$$
$$R_3GeH + CH_2{=}CHOR' \rightarrow R_3Ge(CH_2CHOR')$$

5.13.10 *Reaction with unsaturated amines*

Trialkylgermanes add to the double bond of allylamine in the presence of platinum (for an example of the reaction, see Table 8.11) [*237*]:

$$R_3GeH + CH_2{=}CHCH_2NH_2 \rightarrow R_3GeCH_2CH_2CH_2NH_2$$

while stannanes [*e.g.*, $(C_6H_5)_3SnH$] form amines R_3SnNH_2 with the evolution of propene:

$$(C_6H_5)_3SnH + CH_2{=}CHCH_2NH_2 \rightarrow (C_6H_5)_3SnNH_2 + CH_2{=}CHCH_3$$

5.13.11 *Reaction with alkenylgermanes*

Trialkylgermanes add to the C=C double bonds of vinylgermane and allylgermane (for examples of the reactions, see Table 8.11) [*237*]:

$$R_3GeH + CH_2{=}CHGeR'_3 \rightarrow R_3GeCH_2CH_2GeR'_3$$
$$R_3GeH + CH_2{=}CHCH_2GeR'_3 \rightarrow R_3GeCH_2CH_2CH_2GeR'_3$$

5.14 *Reaction of the Ge–H bond with alkynes*

In the absence of a catalyst, the Ge–H bond in alkylgermanes reacts with the ethyne (C≡C) triple bond substantially more sluggishly than the Sn–H bond. The reactivity of the Ge–H bond is roughly comparable with that of the Si–H bond. In the presence of $H_2PtCl_6 \cdot 6H_2O$, however, a fast and quantitative reaction takes place [*159*]. This reaction, a type of anti-Markovnikov reaction

$$R_3\overset{\delta+}{Ge}-\overset{\delta-}{H} + H-\overset{\delta-}{C}{\equiv}\overset{\delta+}{C}-R' \rightarrow R_3GeCH{=}CHR'$$
$$(R' = H, C_6H_5, C_4H_9, Ge(C_4H_9)_3, CH_2OH, CH_2C(CH_3)_2OH)$$

is generally applicable to the attachment of vinyl groups to germanium (for

TABLE 8.12

ADDITION OF GERMANIUM HYDRIDES TO ALKYNES

Germane	*Alkyne*	*Catalyst (inhibitor)*	*Reaction products*	*Yield (mole-%)*	*Ref.*
$C_7H_{15}GeH_3$	$CH{\equiv}C(CH_2)_4CH_3$	—	$C_7H_{15}Ge[CH{=}CH(CH_2)_4CH_3]_3$	—	159, 237
$(C_4H_9)_2GeH_2$	$CH{\equiv}CC_6H_5$	—	$(C_4H_9)_2HGeCH{=}CHC_6H_5$	—	237
			$[(C_4H_9)_2GeCH_2CH(C_6H_5)]_n$	—	
$(C_2H_5)_3GeH$	$CH{\equiv}CH$	H_2PtCl_6	$(C_2H_5)_3GeCH{=}CH_2$	—	159, 237
			$(C_2H_5)_3GeCH_2CH_2Ge(C_2H_5)_3$	75	
$(C_2H_5)_3GeH$	$CH{\equiv}C(CH_2)_3CH_3$	H_2PtCl_6	$(C_2H_5)_3GeCH{=}CH(CH_2)_3CH_3$	100	159, 237
$(C_2H_5)_3GeH$	$CH{\equiv}CCH_2Cl$	H_2PtCl_6	$(C_2H_5)_3GeCH{=}CHCH_2Cl$	65	184
$(C_4H_9)_3GeH$	$CH{\equiv}CC_6H_5$	H_2PtCl_6	$(C_4H_9)_3GeCH{=}CHC_6H_5$	35–100	159, 237
$(C_2H_5)_3GeH$	$[(CH_3)_2(OH)CC{\equiv}]_2$	—	$(CH_3)_2(OH)CC(R_3Ge){=}CHC(OH)(CH_3)_2$	40	159, 237
$(C_4H_9)_3GeH$	$CH{\equiv}CCH_2OH$	—	$(C_4H_9)_3GeCH{=}CHCH_2OH$	80	159, 237
$(C_4H_9)_3GeH$	$CH{\equiv}CC(OH)(CH_3)_2$	—	$(C_4H_9)_3GeCH{=}CHC(OH)(CH_3)_2$	90	159, 237
$(C_4H_9)_3GeH$	$CH{\equiv}CGe(C_4H_9)_3$	—	$(C_4H_9)_3GeCH{=}CHGe(C_4H_9)_3$	100	159, 237
$(C_2H_5)_2GeHCl$	$CH{\equiv}CH$	—	$(C_2H_5)_2(Cl)GeCH_2CH_2Ge(Cl)(C_2H_5)_2$	50	164
$(C_4H_9)_2GeHCl$	$CH{\equiv}CC_6H_5$	—	$(C_4H_9)_2(Cl)GeCH{=}CHC_6H_5$	75	164
$GeHCl_3$	$CH{\equiv}CH$	—	$CH_2{=}CHGeCl_3$	13	201
			$Cl_3GeCH_2CH_2GeCl_3$	17	

examples of the reaction, see Table 8.12) [*159*, *164*, *237*]. The reverse polarisation exists in $Cl_3Ge{-}H$ [*151*].

5.14.1 *Reaction with unsubstituted alkynes*

Trialkylgermanes react with ethyne in the presence of $H_2PtCl_6 \cdot 6H_2O$ in two steps: formation of trialkylgermylethene and the addition of further trialkylgermane to the latter with the formation of 1,2-bis(trialkylgermyl)ethane:

$$R_3GeH + HC{\equiv}CH \rightarrow R_3GeCH{=}CH_2$$

$$R_3GeH + CH_2{=}CHGeR_3 \rightarrow R_3GeCH_2CH_2GeR_3$$

In the presence of $H_2PtCl_6 \cdot 6H_2O$, trialkylgermanes react with hex-1-yne in a highly exothermic reaction to give 1-trialkylgermylhex-1-enes [*159*]:

$$R_3GeH + HC{\equiv}C(CH_2)_3CH_3 \rightarrow R_3GeCH{=}CH(CH_2)_3CH_3$$

Heptylgermane reacts with three molecules of hept-1-yne to form the thermally very stable heptyltriheptenylgermane in excellent yields [*159*, *237*]:

$$C_7H_{15}GeH_3 + 3HC{\equiv}C(CH_2)_4CH_3 \rightarrow C_7H_{15}Ge[CH{=}CH(CH_2)_4CH_3]_3$$

5.14.2 *Reaction with phenylalkynes*

Trialkylgermanes add to phenylethyne without a catalyst sluggishly and in low yield but in the presence of $H_2PtCl_6 \cdot 6H_2O$ rapidly and quantitatively (exothermically) to give 1-trialkylgermyl-2-phenylethenes [*159*, *237*]:

$$R_3GeH + HC{\equiv}CC_6H_5 \rightarrow R_3GeCH{=}CHC_6H_5$$

Bromine is not added to these styrenes but splits the Ge–C bonds, greatly weakened by the propinquity of the phenyl group, to form trialkylbromogermanes and bromostyrene [*159*]:

$$R_3GeCH{=}CHC_6H_5 + Br_2 \rightarrow R_3GeBr + BrCH{=}CHC_6H_5$$

The reaction confirms the linearity of the germyl compound.

Dibutylgermane likewise adds to phenylethyne:

$$R_2GeH_2 + HC{\equiv}CC_6H_5 \rightarrow R_2HGeCH{=}CHC_6H_5$$

On prolonged heating, the styrene formed polymerises (like its stannane isologue [*231*]) to a viscous oil [*237*]:

$$\left[-\overset{\displaystyle R}{\underset{\displaystyle R}{\overset{|}{\underset{|}{Ge}}}}-CH_2-\overset{\displaystyle H}{\underset{\displaystyle C_6H_5}{\overset{|}{\underset{|}{C}}}}- \right]_n$$

5.14.3 *Reaction with trialkylgermylakynes*

In the reaction of trialkylgermanes with tributylethynylgermane (for preparation, see [*188*]), 1,2-bis(trialkylgermyl)ethenes are formed in quantitative yield [*159*]:

$$R_3GeH + R'_3GeC{=}CH \rightarrow R_3GeCH{=}CHGeR'_3$$

5.14.4 *Reaction with alkynols*

Trialkylgermanes add to prop-1-yn-3-ol to form the corresponding propenols:

$$R_3GeH + HC{\equiv}CCH_2OH \rightarrow R_3GeCH{=}CHCH_2OH$$

Trialkylgermanes react with 3-methylbut-1-yn-3-ol exothermically to give 1-trialkylgermyl-3-methylbut-1-en-3-ols [*159*]:

$$R_3GeH + HC{\equiv}CC(OH)(CH_3)_2 \rightarrow R_3GeCH{=}CHC(OH)(CH_3)_2$$

With $POCl_3$, these split out water with the formation of 1-trialkylgermyl-3-methylbutadienes (60%) [*159*]:

$$R_3GeCH{=}CHC(OH)(CH_3)_2 \rightarrow R_3GeCH{=}CH(CH_3){=}CH_2 + H_2O$$

The reaction of trialkylgermanes with 2,5-dimethylhex-3-yne-2,5-diol in the presence of $H_2PtCl_6{\cdot}6H_2O$ gives 2,5-dimethyl-3-trialkylgermylhex-3-ene-2,5-diols:

$$R_3GeH + (CH_3)_2C(OH)C{\equiv}CC(OH)(CH_3)_2 \rightarrow (CH_3)_2C(OH)\underset{\displaystyle GeR_3}{\underset{|}{C}}{=}CHC(OH)(CH_3)_2$$

5.14.5 *Reaction with chloroalkynes*

Trialkylgermanes react with 3-chloroprop-1-yne in three types of reactions [*183a*]:

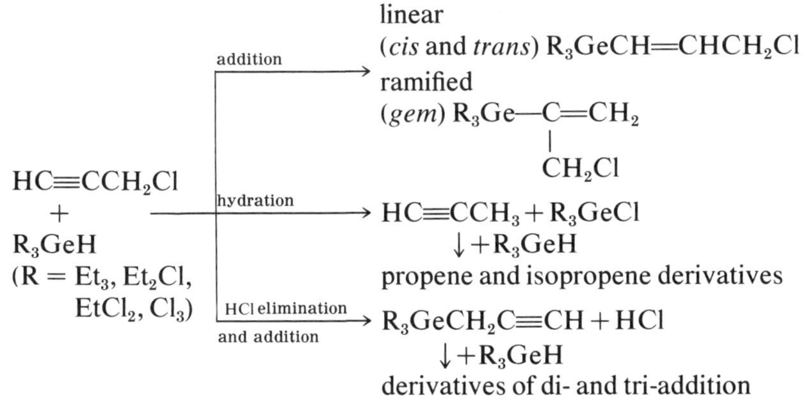

References p. 711

5.15 *Reaction of the Ge–H bond with diazo compounds*

Like the stannanes[*157*], di- and trialkylgermanes react with diazoalkanes, diazoacetic ester, diazoacetone, and diazoacetophenone in ethereal solution in the presence of copper (for examples of the reactions, see Table 8.13)[*161, 237*]:

$$N{\equiv}\overset{\oplus}{N}{-}\overset{\ominus}{C}HR + {>}Ge{-}H \rightarrow N{\equiv}\overset{\oplus}{N}{-}CHR \rightarrow \overset{\ominus}{Ge}{-}H \xrightarrow{-N_2} RH\overset{\oplus}{C}{-}\overset{\ominus}{Ge}{-}H \rightarrow RCH_2{-}Ge{<}$$

$$(R = C_2H_5COO, CH_3CO, C_6H_5CO)$$

5.16 *Reaction of the Ge–H bond with alkali metals or alkali metal hydrides*

When *monogermane* is passed into a solution of sodium or potassium in liquid ammonia at −63.5°, at a molar ratio GeH_4:Na(K) = 1, the colour of the solution changes from deep blue to pale yellow[*87*, *147*, *290*]. In this process, germylsodium, $NaGeH_3$, or germylpotassium, $KGeH_3$, is formed in accordance with the following equations[*234a*]:

$$e^{\ominus} + GeH_4 \rightarrow GeH_3^{\ominus} + H \qquad 2H \rightarrow H_2$$

$$H + e^{\ominus} \rightarrow H^{\ominus} \qquad GeH_4 + NH_2^{\ominus} \rightarrow GeH_3^{\ominus} + NH_3$$

$$H^{\ominus} + NH_3 \rightarrow NH_2^{\ominus} + H_2 \qquad GeH_4 + NH_2^{\ominus} \rightarrow GeH_3NH_2 + H^{\ominus}$$

CH_3GeH_3, $C_2H_5GeH_3$, and $C_5H_{11}GeH_3$ react with Li or Na in liquid ammonia to form CH_3GeH_2M, $C_2H_5GeH_2M$, and $C_5H_{11}GeH_2M$[*104*].

When *digermane* is passed into a solution of sodium in ammonia, the blue colour changes to pale green (sharp change at a molar ratio of Ge_2H_6 to Na of 0.5). In this process, the Ge–Ge bond splits and germylsodium is formed. At the same time, hydrogen is evolved in increasing amount with falling sodium concentration. Since the formation of hydrogen does not affect the conductivity (titration) curve, the formation of $Na_2Ge_2H_4$, which had been postulated previously, is likely.[2] $Na_2Ge_2H_4$ does not contribute to conductivity and is apparently responsible for the change in the colour from pale yellow ($NaGeH_3$) to pale green[*87*]:

$$Ge_2H_6 + 2Na \rightarrow 2NaGeH_3$$

$$Ge_2H_6 + 2Na \rightarrow Na_2Ge_2H_4 + H_2$$

Potassium hydride reacts with digermane to give germylpotassium and monogermane. The reaction may be a fairly simple cleavage reaction[*20a*]:

$$Ge_2H_6 + KH \xrightarrow{\text{monoglyme}} GeH_3K + GeH_4$$

[2]Note: In liquid ammonia solution however, digermane dismutes:

$$Ge_2H_6 \xrightarrow{GeH_2} GeH_2 \xrightarrow{-63°\Delta} GeH_4 + GeH_{<2}$$ [*77a*].

TABLE 8.13

ADDITION OF GERMANIUM HYDRIDES TO DIAZO COMPOUNDS

Catalyst (inhibitor) = Cu powder

Germane	*Diazo Compound*	*Reaction products*	*Yield (mole-%)*	*Ref.*
$(n-C_4H_9)_2GeH_2$	$N_2CHCOOC_2H_5$	$R_2HGeCH_2COOC_2H_5$	28	143, 237
		$R_2Ge(CH_2COOC_2H_5)_2$	—	
$(C_2H_5)_3GeH$	$N_2CHCOOC_2H_5$	$(C_2H_5)_3GeCH_2COOC_2H_5$	37[a]	237
$(n-C_4H_9)_3GeH$	$N_2CHCOOC_2H_5$	$(n-C_4H_9)_3GeCH_2COOC_2H_5$	40	161, 237
$(n-C_5H_{11})_3GeH$	$N_2CHCOOC_2H_5$	$(n-C_5H_{11})_3GeCH_2COOC_2H_5$	35	237
$(C_2H_5)_3GeH$	$N_2CHCOCH_3$	$(C_2H_5)_3GeCH_2COCH_3$	30	161, 237
$(C_2H_5)_3GeH$	$N_2CHCOC_6H_5$	$(C_2H_5)_3GeCH_2COC_6H_5$	31	161, 237

[a] Without catalyst, yield = 0%

On the other hand, the $KH-Si_2H_6$ reaction may involve numerous disproportionation and cleavage reactions in which Si_2H_6 is regenerated [*20a*]:

$$2Si_2H_6 \xrightarrow{(KH)} SiH_4 + Si_3H_8$$

$$Si_3H_8 + KH \longrightarrow SiH_3K + Si_2H_6$$

$$2Si_3H_8 \xrightarrow{(KH)} Si_2H_6 + Si_4H_{10} \text{ etc.}$$

Under the action of potassium in monoglyme, digermane dismutes with the formation of monogermane and polygermane(2). The GeH_4 produced subsequently reacts with K to give $KGeH_3$ [*97a*]:

$$Ge_2H_6 \xrightarrow{(K)} GeH_4 + \frac{1}{x}(GeH_2)_x$$

$$GeH_4 + 2K \rightarrow 2KGeH_3 + H_2$$

When *trigermane* is passed into a solution of sodium in ammonia, the colour changes to deep red at a molar ratio of Ge_3H_8 to Na of 0.25. The reaction can be represented by the equation [*87*]:

$$Ge_3H_8 + 4Na \rightarrow 2NaGeH_3 + Na_2GeH_2$$

but the evolution of (a small amount of) hydrogen and the red colour of the solution indicate a more complicated sequence of reactions.

Germylsodium prepared in liquid ammonia as solvent can be coupled only with halides (generally organyl halides) which do not undergo ammonolysis (*e.g.*, $RX + NaGeH_3 \rightarrow RGeH_3 + NaX$; R = Me, Et, Pr [*290*]), since NH_3 can be removed from the $NaGeH_3$ only incompletely [*5a*], or by 'prolonged' pumping at −23° [*78a*]. However, compounds synthesised in this way can be prepared more easily by other routes. For the deliberate synthesis of mixed hydrides (*e.g.*, by the reaction: $SiH_3Cl + KGeH_3 \rightarrow SiH_3GeH_3 + KCl$), it is necessary to work in the absence of ammonia. In very dry monoglyme or diglyme, and less well in

References p. 711

tetrahydrofuran or dioxan, potassium (or other alkali metals) or liquid alloys of sodium and another alkali metal react quantitatively at 20° within minutes or a few hours[3] [*5a, 5b*]:

$$2GeH_4 + 2K \rightarrow 2KGeH_3 + H_2$$

Potassium hydride and solid potassium hydroxide in monoglyme react with monogermane to give germylpotassium [*20a, 133a*]:

$$GeH_4 + KH \xrightarrow{\text{monoglyme}} GeH_3K + H_2$$

$$GeH_4 + KOH_{solid} \xrightarrow[\Delta]{\text{monoglyme}} GeH_3K + KOH \cdot H_2O_{solid}$$

No reaction occurs between GeH_4 and LiH or SiH_4 and KH [*20a*].

The donor anion germyl $GeH_3^{\ominus}$ cleaves the B–H–B double bridge of B_2H_6 forming a substituted boranate ion $H_3GeBH_3^{\ominus}$ [*234b*]:

$$GeH_3K + \tfrac{1}{2}(BH_3)_2 \xrightarrow[-78 \text{ to } 25°]{\text{monoglyme}} KH_3GeBH_3$$

In solutions of Na or K in the non-volatile and fairly chemically inert solvent $[(CH_3)_2N]_3PO$, GeH_4 (or SiH_4) forms GeH_3M (or SiH_3M) [*40a*].

GeH_3K reacts with CO_2 to give a salt $KGeH_3CO_2$, an analogue of potassium acetate [*150a*]. GeH_3K reacts with $(CH_3)_2GaCl$ yielding an ionic complex. In nonpolar solvents it decomposes to give $(CH_3)_2GaGeH_3$ and KCl [*5d*]:

$$GeH_3K + (CH_3)_2GaCl \xrightarrow{\text{THF}} K[Ga(GeH_3)Cl(CH_3)_2] \xrightarrow[(+\text{toluene})]{(-\text{THF})} (CH_3)_2GaGeH_3 + KCl$$

GeH_3K reacts with $ZnCl_2$, with the borazines $Cl_nMe_{3-n}B_3N_3Me_3$ ($n = 1, 2, 3$), and with $ClB(NMe_2)_2$ yielding the corresponding germyl compounds $(GeH_3)_2Zn$ [*5e*], $(GeH_3)_nMe_{3-n}B_3N_3Me_3$ [*5f*] and $(GeH_3)B(NMe_2)_2$ [*5g*]. Similar to the reaction of π-$C_5H_5Fe(CO)_2Br$ with SiH_3K [*5c*] giving the stable π-$C_5H_5Fe(CO)_2(SiH_3)$, GeH_3K yields the π-$C_5H_5Fe(CO)_2GeH_3$ of low stability [*4*].

For other reactions of germylmetals, see Chapter 7, section 10.2.

5.17 *Reaction of the Ge–H bond with organylmetals*

The Si–H, Ge–H, and Sn–H bonds in organosilanes, organogermanes, and organostannanes react with organylmetals. The Si–H bond is converted into an Si–C bond both with organyl(alkali metal)s and organyl(alkaline earth metal)s

---

[3] On the other hand, monosilane reacts in two ways, and potassium hydride is produced as well [*5b*]:

$$2SiH_4 + 2K \rightarrow 2KSiH_3 + H_2$$

$$SiH_4 + 2K \rightarrow KSiH_3 + KH$$

and with organomagnesium halides with the formation of a metal hydride or a metal halidohydride:

$$\geq Si\text{–}H + M^{\oplus}R^{\ominus} \rightarrow \geq Si\text{–}R + MH$$

(R = alkyl, aryl; M = an equivalent of a metal (E^{I}, $\frac{1}{2}E^{II}$, MgX))

The occurrence or non-occurrence of the reaction and the number of hydrogen atoms replaced by R are greatly affected by the solvent. However, the influence is not so strong that the silane–organylmetal reaction could take place in the reverse direction with the formation of silylmetal and organyl hydride ($\equiv SiH + RM \nrightarrow \equiv SiM + RH$). On the other hand, the reaction of the Ge–H bond in organogermanes with organylmetals can take place in both directions:

$$\geq Ge\text{–}H + M^{\oplus}R^{\ominus} \rightarrow \geq Ge\text{–}R + MH \qquad (1)$$

$$\geq Ge\text{–}H + M^{\oplus}R^{\ominus} \rightarrow \geq Ge\text{–}M + RH \qquad (2)$$

(M = one equivalent of a metal (E^{I}, $\frac{1}{2}E^{II}$, MgX))

To judge from the known reactions (although they are few in number), alkyllithiums react in accordance with eqn. (2)[*99*] and Grignard compounds in accordance with eqn. (1)[*237*]. The Grignard reactions are highly affected by the solvent. In diethyl ether and dibutyl ether, they hardly take place, while in tetrahydrofuran on long boiling under reflux alkylmagnesium bromides replace a hydrogen atom in the alkylgermane in low yield. (For example, with $C_5H_{11}GeH_3$ an excess of $C_5H_{11}MgBr$ gives a 12% yield of $(C_5H_{11})_2GeH$; C_4H_9MgBr reacts similarly[*237*]). More reactive Grignard compounds such as allylmagnesium and benzylmagnesium chlorides react with better yields:

$$(C_4H_9)_2GeH_2 + CH_2{=}CHCH_2MgCl \rightarrow (C_4H_9)_2HGeCH_2CH{=}CH_2 + MgClH$$

The yield amounts to 35%. Excessively prolonged heating must be avoided, since otherwise the polymerisation reaction competes successfully[*237*]:

$$n\,CH_2{=}CHCH_2\overset{C_4H_9}{\underset{C_4H_9}{Ge}}H \rightarrow \left[\text{–}CH_2CH_2CH_2\overset{C_4H_9}{\underset{C_4H_9}{Ge}}\text{–}\right]_n$$

Mono- and dialkylgermanes ($C_7H_{15}GeH_3$, $(C_4H_9)_2GeH_2$) react with benzylmagnesium chloride in tetrahydrofuran to give monoalkyl- and dialkylmonobenzylgermanes, respectively (yield about 20%[*237*]:

$$RGeH_3 + C_6H_5CH_2MgCl \rightarrow RH_2GeCH_2C_6H_5 + MgHCl$$
$$R_2GeH_2 + C_6H_5CH_2MgCl \rightarrow R_2HGeCH_2C_6H_5 + MgHCl$$

References p. 711

With dialkylgermanes $H_2Ge(C_4H_9)_2$, phenylmagnesium bromide in tetrahydrofuran gives a 20% yield of phenyldialkylgermanes [*237*]:

$$R_2GeH_2 + C_6H_5MgBr \rightarrow R_2HGeC_6H_5 + MgHBr$$

Because of the low yields in the replacement of hydrogen by organic substituents in the Grignard reactions, germanes with mixed organic substituents are better obtained from the corresponding halogenogermanes, which are relatively easy to prepare.

The reaction of germanium hydrides with $C_6H_5HgCX_2Br$ takes place with the insertion of the carbene $\cdot CX_2\cdot$ into the Ge–H bond [*254, 254a, 293a*]:

$$C_6H_5HgCX_2Br \xrightarrow{-C_6H_5HgBr} \cdot CX_2\cdot \xrightarrow{+\gt GeH} \gt GeCX_2H \quad (X = Cl, Br)$$

With $(C_2H_5)_3Bi$, $(C_2H_5)_2Cd$ and $(C_2H_5)_2Hg$, $(C_2H_5)_3GeH$ forms the corresponding bismuth, cadmium and mercury compounds [*150, 293a*]:

$$2(C_2H_5)_3GeH + (C_2H_5)_2M \not\rightarrow (C_2H_5)_3GeMGe(C_2H_5)_3 + C_2H_6$$

For the very interesting hydrogenolysis reactions of $R_3SnNR'_2$ with Ph_3GeH (yielding $R_3SnGePh_3$) see [*41a*]

5.18 Reaction of the Ge–H bond with transition-metal carbonyl hydrides

Monogermane and pentacarbonylmanganese hydride in a molar ratio of 1:0.7 at 15° give a 30% yield (calculated on the GeH_4) of bis(pentacarbonylmanganese) germane, $[Mn(CO)_5]_2GeH_2$, which is stable in the air for months. The reaction does not appear to take place in the following way:

$$GeH_4 + n\,HMn(CO)_5 \not\rightarrow GeH_{4-n}[Mn(CO)_5]_n + n\,H_2$$

since neither $[Mn(CO)_5]GeH_3$ nor $[Mn(CO)_5]_3GeH$ is formed [*181*].

With pentacarbonylmanganese chloride, trichlorogermane forms pentacarbonyltrichlorogermylmanganese [*206*]:

$$GeCl_3H + ClMn(CO)_5 \xrightarrow{THF} (GeCl_3)Mn(CO)_5 + HCl$$

trans-Bis(triphenylphosphane)carbonyliridium chloride and trialkylgermane react giving the corresponding germyl-iridium complex (C.N.; 4 → 6) [*106b*]:

$$trans\text{-}(Ph_3P)_2Ir(CO)Cl + 2R_3GeH \rightarrow (Ph_3P)_2Ir(CO)(H)_2GeR_3 + R_3GeCl$$

However *trans*-bis(triethylphosphane)palladium dichloride and triphenylgermane do not yield the corresponding germyl-palladium complex but the palladium hydride [*27a*]:

$$trans\text{-}(Et_3P)_2PdCl_2 + Ph_3GeH \rightarrow trans\text{-}(Et_3P)_2Pd(H)Cl + Ph_3GeCl$$

5.19 *Commutation reactions*

Aluminium halides catalyse the exchange of hydrogen, alkyl, and halogen in alkylhalogenogermanium hydrides. Mixtures of diethyldihalogenogermanes and triethylgermane commute to diethylmonohalogenogermanes and triethylhalogenogermanes if the chlorine or bromine compound but not the iodine compound is used[*162, 237*]. The reaction mechanism is probably similar to that in the commutation of the silanes (see Chapter 7, section 5.14):

$$R_2GeX_2 + R_3GeH \xrightarrow{+2\,AlX_3} \left| \begin{array}{c} R_2(X)\overset{\oplus}{Ge}-X\cdots\overset{\ominus}{Al}X_3 \\ X_3\underset{\ominus}{Al}\cdots H-\underset{\oplus}{Ge}R_3 \end{array} \right| \xrightarrow{-2\,AlX_3} R_2GeHX + R_3GeX$$

6. Halogenogermanes

6.1 *Preparation of the halogenogermanes*

Four methods are available for the preparation of halogenogermanes or organhalogenogermanes:

(1) partial halogenation of germanium hydrides;
(2) partial hydrogenation of germanium halides;
(3) action of hydrogen halides on germanium;
(4) transhalogenation of halogenogermanes.

6.1.1 *Partial halogenation of germanium hydrides*

To replace hydrogen in a germanium hydride, more or less strongly polarised halogen molecules or halogen anions are necessary:

(a) $\overset{\delta+}{X}-\overset{\delta-}{X}$ (in the reaction of a germanium hydride with a halogen in the gas phase);

(b) $\overset{\delta+}{H}-\overset{\delta-}{X}$ (in reactions of a germanium hydride with a hydrogen halide in the gas phase or in solution);
(c) $R^{\oplus}\,X^{\ominus}$ (in reactions of a germanium hydride with an alkyl halide in solution);
(d) $M^{\oplus}\,X^{\ominus}$ (in reactions of a germanium hydride with a metal halide).

(*a*) *Reaction* $>GeH + X_2$. In the reaction of germanium hydrides with free halogens (Br_2, I_2), without the addition of catalysts all possible halogenogermanes $GeH_{4-n}X_n$ ($n = 1, 2, 3, 4$) are formed[*3, 272, 273*]. For examples of the reactions see Table 8.14. For the reaction mechanism, see section 5.1. With long times for the action of the halogen on the germane (or on the partially converted germane) the reaction leads preferentially to the perhalogenated

TABLE 8.14

SYNTHESIS OF HALOGENATED GERMANIUM HYDRIDES (REPLACEMENT OF H BY X)

Germane	*Halogenating agent*	*Phase or solvent*	*Catalyst*	*Reaction temp. (cold part of vessel)(°C)*	*Yield*[a] *(mole-% referred to Ge)*				*Conversion of the hydride (mole-%)*	*Ref.*
					GeH_3X	GeH_2X_2	$GeHX_3$	GeX_4		
GeH_4	Br_2	gaseous	—	> -198	91	—	—	—	—	116
GeH_4	Br_2	liq. GeH_3Br	—	—	—	—	—	—	—	264
GeH_4	I_2	gaseous	—	20 (−94)	65	—	—	2	67	3
GeH_4	I_2	gaseous	AlI_3	20 (−94)	0.6	0.3	—	22	33	3
GeH_4	I_2	liq. GeH_3I	—	—	—	—	—	—	—	176, 177, 273, 276
$GeH_4(GeD_4)$	HCl	—	$AlCl_3$	—	48	14	—	—	—	62, 173, 187
GeH_4	HBr	gaseous	$AlBr_3$	—	< 10	—	—	—	—	187, 264
Ge_2H_6	HCl	liquid	$AlCl_3$	20	traces	—	—	—	—	177
Ge_2H_6	Br_2	liquid	—	−78	?	?	—	—	—	177
Ge_2H_6	I_2	liquid	—	−63 or 0	93	—	—	—	—	176, 177
n-$C_4H_9GeH_3$	I_2	liquid	—	0	70	—	—	—	—	11
$(n\text{-}C_4H_9)_2GeH_2$	I_2	liquid	—	> 20	57	—	—	—	—	11
CH_3GeH_3	HCl	gaseous	$AlCl_3$	20 (−78)	34–51	33–49	—	—	67	3,5
$(CH_3)_2GeH_2$	HCl	gaseous	$AlCl_3$	20 (−78)	17–29	42–71	—	—	58	3,5
n-$C_4H_9GeH_3$	$HgCl_2$	liquid	—	> 20	82	—	?	—	—	11, 237
$(n\text{-}C_4H_9)_2GeH_2$	$HgCl_2$	liquid	—	> 20	97	—	?	—	—	11,237
n-$C_4H_9GeH_3$	$HgBr_2$	liquid	—	31	78	—	?	—	84	11,237
$(n\text{-}C_4H_9)_2GeH_2$	$HgBr_2$	liquid	—	> 20	92	—	—	—	—	11, 237
n-$C_4H_9GeH_3$	HgI_2	liquid	—	—	?	—	?	—	—	237
$(n\text{-}C_4H_9)_2GeH_2$	HgI_2	liquid	—	—	?	—	?	—	—	237

[a] In the case of alkylated germanes: in part H = alkyl.

compound. Partially halogenated germanes are obtained in good yields only if in the sequence of reactions

$$GeH_4 \xrightarrow[-HX]{+X_2} GeH_3X \xrightarrow[-HX]{+X_2} GeH_2X_2 \xrightarrow[-HX]{+X_2} GeHX_3 \xrightarrow[-HX]{+X_2} GeX_4 \ (X = Br, I)$$

the partially halogenated germanes are removed from the further action of the halogen. This can be achieved conveniently by the appropriate removal by condensation of the halogenogermanes, which are less volatile than the unsubstituted germane. GeH_3I is formed in this way in the reaction of GeH_4 with I_2 at a reaction temperature of 20° and a condensation temperature of −78°. However, the reaction takes place very slowly (only 10% conversion of the GeH_4 in 20 to 25 h[*272, 273*]. Germyl iodide is formed substantially faster and more simply by allowing iodine to react with monogermane in a little liquid (polar) germyl iodide. This is obtained in a simple manner by cooling the lower part of the reaction vessel. Under these conditions, with a 50% conversion of GeH_4 in 2 h, 92–94% of GeH_3I and only 8–6% of higher halogenated germanes are formed[*272, 273*]. Monogermane is brominated with elementary bromine (60–90% yield of GeH_3Br),*264*], and digermane[*175b, 176*] and trigermane[*175*] are iodinated with iodine to Ge_2H_5I and Ge_3H_7I in a similar manner. In higher-boiling (less polar) organogermanium hydrides liquid at room temperature (*e.g.*, $C_4H_9GeH_3$), the reaction with iodine leads equally rapidly to (more highly polar) organoiodogermanium hydrides (*e.g.*, $C_4H_9GeH_2I$)[*11*]. For examples of the reactions see Table 8.14.

(*b*) *Reaction* $\rangle$*GeH*+*HX*. The reaction of germanium hydrides with hydrogen halides in the gas phase at room temperature in the presence of solid aluminium halide leads to the partial and complete replacement of hydrogen atoms by halogen[*62*] (concerning the reaction mechanism, see section 5.1; for examples of the reactions, see Table 8.14):

$$GeH_4 \xrightarrow[(AlX_3)]{+HX \atop -H_2} GeH_3X \xrightarrow[(AlX_3)]{+HX \atop -H_2} GeH_2X_2 \xrightarrow[(AlX_3)]{+HX \atop H_2} GeHX_3 \xrightarrow[(AlX_3)]{+HX \atop -H_2} GeX_4 \quad (X = Cl, Br, I)$$

However, the proportion of monohalogenogermane is low; for example, in the reaction of GeH_4 with HBr it is less than 10%[*264*]. Individual halogenogermanes can be isolated from the reaction product (a mixture of halogenogermanes) after (difficult) fractional high-vacuum distillation but generally in an impure form[*262*].

In the halogenation of alkylgermanium hydrides with hydrogen halides at room temperature using aluminium halide as catalyst, all the hydrogen atoms are preferentially replaced by halogen[*3*]. Partially chlorinated alkylgermanium chlorides can be recovered if the sequence of reactions in the CH_3GeH_3–HCl–$AlCl_3$ reaction, for example

$$RGeH_3 \xrightarrow[(AlCl_3)]{+HCl, -H_2} RGeH_2Cl \xrightarrow[(AlCl_3)]{+HCl, -H_2} RGeHCl_2 \xrightarrow[(AlCl_3)]{+HCl, -H_2} RGeCl_3$$

References p. 711

is broken off at, for instance, the CH_3GeH_2Cl stage. This is achieved by condensing out the CH_3GeH_2Cl, which boils at a higher temperature than the CH_3GeH_3. The Ge–C bonds and the C–H bonds are not attacked by hydrogen chloride at 20°[*3, 5*]. For examples of the reactions, see Table 8.14.

(c) Reaction $\geq$*GeH + RX.* Alkylgermanium hydrides [*e.g.*, $(C_2H_5)_2GeH_2$] are converted into alkylhalogenogermanium hydrides by being boiled with alkyl halides or α-halogenoethers under mild conditions[*237*] (see also section 5.5; for examples of the reactions see Table 8.10):

$$\geq Ge\text{–}H + RX \rightarrow \geq Ge\text{–}X + RH$$

$$\geq Ge\text{–}H + ROCH_2X \rightarrow \geq Ge\text{–}X + ROCH_3 \quad (R = \text{alkyl})$$

(d) Reaction $\geq$*GeH + MX.* Many metal and non-metal halides capable of giving up halogen replace hydrogen in germanium hydrides by halogen (see also section 5.3)[*7,237*]. However, only mercury(II) halides are suitable for partial substitution (for examples of the reactions see Table 8.9):

$$\geq Ge\text{–}H + HgX_2 \rightarrow \geq Ge\text{–}X + Hg + HX$$

$$2\geq Ge\text{–}H + HgX_2 \rightarrow \geq Ge\text{–}X + Hg + H_2 \quad (X = Cl, Br, I)$$

6.1.2 *Partial hydrogenation of germanium halides*

Germanium tetrahalides and organogermanium halides cannot deliberately be partially hydrogenated with the usual hydrogenating agents (*e.g.*, LiH, $LiBH_4$, $LiAlH_4$). Even with a deficiency of hydrogenating agent halogen-free germanes are formed together with unchanged halides. When $GeCl_4$ and H_2 are passed through a quartz tube heated to 900°, $GeHCl_3$ and a very small amount of $GeCl_2$ are formed[*226*]. However, this hydrogenation of $GeCl_4$ is not a replacement of chlorine by hydrogen but (more probably) an addition of HCl to $GeCl_2$ formed as an intermediate. $GeCl_2$ and chlorine (atoms) are formed in the hot reaction zone. Chlorine (atomic) reacts with hydrogen to give hydrogen chloride which adds to $GeCl_2$ in the cooler reaction zones.

6.1.3 *Addition of hydrogen halides to germanium*

The combined formation of a Ge–H and a Ge–X bond is achieved by the reaction of elementary germanium with hydrogen halides in accordance with the empirical equation:

$$Ge + 3HX \rightarrow GeHX_3 + H_2$$

When HCl or HBr is passed over heated germanium (in the case of HCl at about 500°), the trihalogenogermane is formed preferentially (*e.g.*, about 90%

of $GeHCl_3$)[*21, 51, 52*]. One suggests that the $GeHX_3$ is not formed in the hot reaction zone. Probably the germanium subhalide (e.g., GeX_2) formed in the hot zone adds hydrogen halide at a lower temperature (under certain circumstances, only when the hot reaction gases are quenched in the attached cold trap)

$$Ge + 2\,HX \xrightarrow{\text{high temperature}} GeX_2 + H_2$$

$$GeX_2 + HX \xrightarrow{\text{low temperature}} GeHX_3$$

since $GeHCl_3$, for example, begins to disproportionate appreciably into $GeCl_2$ and HCl even between −33 and −24°[*203a*].[4]

Likewise, the reaction of germanium(II) sulphide (or germanium(II) oxide) with hydrogen chloride, in which trichlorogermane is produced, must probably be regarded as an addition reaction[*32, 61, 203a*]. Possibly, here as well, HCl adds to the $GeCl_2$ first formed:

$$GeS + 2HCl \rightarrow GeCl_2 + H_2S$$

$$GeCl_2 + HCl \rightarrow GeHCl_3$$

6.1.4 *Transhalogenation of the Ge–X bond*

Germyl fluoride is produced by the reaction of germyl bromide with silver fluoride[*96, 116, 265, 266*]:

$$\rangle Ge\text{–}Br + AgF \rightarrow \rangle Ge\text{–}F + AgBr$$

For this purpose, gaseous germyl bromide is passed over solid silver fluoride at 25°. Transhalogenation avoids the use of F_2 or HF, which are difficult to handle. At the reaction temperature, the thermally unstable germyl fluoride formed dismutates partially into difluorogermane and germane[*266*]:

$$GeH_3Br + AgF \rightarrow GeH_3F + AgBr$$

$$2GeH_3F \rightarrow GeH_2F_2 + GeH_4$$

The reaction of GeH_2Br_2 with PbF_2 gives pure GeH_2F_2[*80*]. With PbF_2, Ge_2H_5I does not form Ge_2H_5F but Ge_2H_6, GeH_3F, and GeH_4[*177*]. Ge_2H_5I can be converted into Ge_2H_5Cl with AgCl (yield 18–65 mole-%) and into Ge_2H_5Br with AgBr (yield 13 mole-%)[*176, 177*].

The conversion of a pseudohalide into a halide is also to be regarded as a transhalogenation reaction. Germyl cyanide, for example, reacts with hydrogen bromide to form germyl bromide[*273*] (conversion of the Ge–CN bond[*107a*]

[4]Condensation of a low-pressure gaseous mixture containing SiF_2, SiF_4 and GeH_4 (mole ratio = 4:2:3) at −196° gives an orange-yellow solid, a mixture of GeH_3SiF_2H, $GeH_3SiF_2SiF_2H$ and GeH_3-SiF_3(?)[*262a*].

References p. 711

into the Ge–Br bond[*273*]). GeH_3Cl and GeH_2Cl_2 react with HI to form the corresponding iodine compounds[*40*]:

$$GeH_3CN + HBr \rightarrow GeH_3Br + HCN$$

$$GeH_3Cl + HI \rightarrow GeH_3I + HCl$$

$$GeH_2Cl_2 + 2HI \rightarrow GeH_2I_2 + 2HCl$$

6.2 *Physical properties of the halogenogermanes*

6.2.1 *Fluorogermanes*

Monofluorogermane, germyl fluoride (GeH_3F), is a liquid at room temperature which dismutates slowly into germane and difluorogermane (at 25°, it undergoes 15% dismutation in 16 h). Vapour pressure equation: $\log p$ [torr] $= -1662/T + 8.639$. Boiling point 15.6° (extrapolated from the vapour pressure equation). Melting point −22° (sharp). Enthalpy of evaporation 7605 cal/mole. Trouton's constant 26.4[*266*].

Difluorogermane (GeH_2F_2) has been little studied. The vapour pressure of the compound can be characterised as follows: when the mixture of mono- and difluorogermanes that is always present at room temperature is passed through a trap cooled to −95°, the difluorogermane, which is not volatile at this temperature, condenses out while monofluorogermane does not condense [*116, 266*].

Tetrafluorogermane, germanium tetrafluoride (GeF_4), is a colourless gas at room temperature which does not attack dry glass. The crystals that form on condensation at −186° evaporate on heating without melting. Under a pressure of 3032 torr, it melts at −15°. Density 6.650 g/l[*63*]. Vapor pressures[*63, 93*]: 75.9 torr (−62.7°); 452 torr (−41.4°). For thermodynamic data, see the literature [*19, 93, 94, 136*]. For comparison CF_3GeF_3: vapour pressure 760 torr (−1.7°), vapour pressure equation $\log p$ [torr] $= -2451/T + 11.94$, m.p. 3°[*34*].

6.2.2 *Chlorogermanes*

Monochlorogermane, germyl chloride (GeH_3Cl), is a liquid at room temperature. Vapour pressures[*62*]: 69.7 torr (−23.6°). Vapour pressure equation: $\log p$ [torr] $= -1527.4/T + 7.961$. Boiling point 28°. Melting point −52°. Density 1.75 g/ml[*62*]. Dipole moment $2.124 \pm 0.02 \times 10^{-18}$ electrostatic units[*187*]; 2.13×10^{-18} electrostatic units (from the Stark effect in the microwave spectrum)[*48*]; 2.03×10^{-18} electrostatic units (from measurements of the dielectric constant of GeH_3Cl vapour)[*262*].

Dichlorogermane (GeH_2Cl_2) is a liquid at room temperature. Vapour pressures: 1.3 torr (−54.7°); 110.0 torr (20.7°). Vapour pressure equation: $\log p$ [torr] $= -1742.7/T + 7.969$. Boiling point 69.5°. Melting point −68°. Density 1.90 g/ml (at −68°, liquid)[*62*]. Dipole moment 2.21×10^{-18} electrostatic units (from measurement of the dielectric constant in CCl_4 solution)[*261*].

Trichlorogermane ($GeHCl_3$) is a colourless liquid at room temperature.

Vapour pressures[*54, 64*]: 0.8 torr (–25.0°); 9.2 torr (0.0°). However, the vapour pressures for $GeHCl_3$ do not correspond to pure $GeHCl_3$ but to a mixture of $GeHCl_3$, $GeCl_2$, and HCl, since trichlorogermane dismutates even at −30° in accordance with the equation $GeHCl_3 \rightarrow GeCl_2 + HCl$[*201a*]. Both $GeCl_2$ and HCl are somewhat soluble in the $GeHCl_3$. On distillation, HCl is more volatile, so that $GeCl_2$ always remains in the distillation residue. However, HCl should affect the vapour pressures of $GeHCl_3$. Boiling point 75.2° (calculated from the vapour pressure figures[*54, 64*]. Melting point –71°[*64*]. Density 1.93 g/ml (at 0°, liquid); 7.980 g/l (gaseous)[*54, 64*].

Pyrolysis. Trichlorogermane begins to decompose appreciably into hydrogen chloride, germanium tetrachloride, and polychlorogermane $GeCl_n$ ($n = 2.0$–0.6) even at very low temperatures. Some of the (isolated) more stable compounds are given by the (non-stoichiometric) equation[*201a, 203a*]:

$$\underset{\text{colourless liquid}}{GeHCl_3} \xrightarrow[-HCl]{-33 \text{ to } -24°} \underset{\text{white solid}}{GeCl_2} \xrightarrow[-GeCl_4]{>-24°} \underset{\text{yellow solid at } n \sim 1}{GeCl_n} \xrightarrow{\sim +20°} \underset{\text{red-orange}}{GeCl_{0.6}}$$

At 0 to 60° and a pressure of N_2 of about 10 torr (flowing N_2), trichlorogermane etherate forms light yellow polymeric germanium dichloride (Ge:Cl = 1:1.5 to 1.6[*205*]:

$$2HGeCl_3 \cdot OEt_2 \rightarrow \frac{1}{x}(GeCl_2)_x + 2OEt_2 + HCl$$

Tetrachlorogermane, germanium tetrachloride ($GeCl_4$), is a colourless mobile stable liquid at room temperature. Density 1.8443–1.886 g/ml (30–19.5°)[*56, 91, 153, 229*]. Vapour pressures[*153*]: 1.1 torr (–40°); 24.3 torr (0.0°). Boiling point 83.0–86.5° (756–760 torr)[*16, 18, 19, 54, 56, 153, 203*]. Melting point –52.0° (metastable β-form); 49.5° (stable α-form)[*18, 19, 56, 153, 260*]. Enthalpy of evaporation 7.35 kcal/mole[*122*]. Refractive index n_D 1.45730[*197*].

Solubility. $GeCl_4$ dissolves in absolute alcohol, carbon disulphide, carbon tetrachloride, benzene, acetone[*21, 59, 279*], diethyl ether, tetrahydrofuran, tetrahydropyran[*259*], diphenyl ether, *m*-cresol methyl ether, *o*-cresol methyl ether, n-propyl phenyl ether, and methyl phenyl ether[*260*].

Monochlorodigermane, digermanyl chloride (Ge_2H_5Cl), is an unstable liquid. The half-life time for its thermolysis at 20° and 1 to 2 torr is 20 min[*117*]. Boiling point (extrapolated) 88°. Vapour pressures: 0.7 torr (–45°), 13.4 torr (5.3°). Vapour pressure equation: $\log p$ [torr] $= -2154/T + 8.850$. Enthalpy of evaporation 9.86 kcal/mole. Trouton's constant 27.3 cal deg^{-1} $mole^{-1}$[*117*].

6.2.3 Bromogermanes

Monobromogermane, germyl bromide (GeH_3Br), is a colourless mobile liquid at room temperature. Density 2.34 g/ml (29.5°, liquid). Vapour pressures[*62, 264*]: 8.1 torr (−44.6°); 25.7 torr (−23°). Boiling point 52°. Melting point −32° [*62*].

Dibromogermane (GeH_2Br_2) is a colourless mobile liquid at room temperature.

References p. 711

Density 2.80 g/ml (0°, liquid). Vapour pressures: 1.2 torr (–26.6°); 62.7 torr (35.2°). Vapour pressure equation: $\log p$ [torr] $= -2461.9/T + 9.798$. Boiling point 89°. Melting point –15°[*62*].

Tribromogermane ($GeHBr_3$) dissociates from 10° and quite considerably at room temperature into $GeBr_2$ and HBr[*23*]. Vapour pressure at 0° 1.9 torr[*21*]. Melting point –25 to –24°[*21*].

Tetrabromogermane, germanium tetrabromide ($GeBr_4$), is a white solid at room temperature. Density 3.1315 (29°, solid)[*56*]. Vapour pressures[*22*]: 1.6 torr (4.45°); 78.2 torr (104.31°). Boiling point: 185.9°[*56*]; 186°[*91*]; 183° [*16*]. Melting point: 26.1°[*56*]; 25.5°[*172*]. Refractive index 1.6268 (25.50°)[*56, 57, 136*].

Monobromodigermane, digermanylbromide (Ge_2H_5Br), is a liquid which decomposes rapidly at 0°[*177*].

6.2.4 *Iodogermanes*

Monoiodogermane, germyl iodide (GeH_3I), is a white crystalline substance below –16°. Liquid GeH_3I does not decompose appreciably in 24 h in the dark, but it does so to a large extent under laboratory lighting. At 300–400° it decomposes into Ge, GeI_2, and H_2. Vapour pressure at 0° 20.0 torr. Melting point –15.6 to –15.2°[*272, 274*].

Diiodogermane (GeH_2I_2) is a white crystalline substance which melts between 44.9 and 45.1° with decomposition. The vapour pressure at 20° is 0.1 torr[*272, 274*]. The white crystals slowly become yellow on the surface at 20°[*40*].

Tetraiodogermane, germanium tetraiodide (GeI_4), is a yellow to orange crystalline substance at room temperature which can be sublimed without decomposition. Density 4.3215 g/ml (25°, solid)[*58*]. Melting point: 146°[*152*]; 145–146° [*16*]; 140–141°[*204*]. Solubility: GeI_4 dissolves in carbon disulphide, monochlorobenzene, benzene, methanol, carbon tetrachloride, ethylene chloride, 1,2-ethanediol, and 2-chloroethanol. Decomposition sets in after months in solutions in hexane, trichloromethane, nitrobenzene, light hydrocarbons, butanol, and anhydrous acetic acid. Rapid decomposition sets in in absolute ethanol, isopropanol, pentanol, diethyl ether, acetone, and pyridine[*60*].

Monoiododigermane, digermanyl iodide (Ge_2H_5I), is very unstable. It can be distilled in high vacuum at 0° with a loss of 20%[*175, 176*].

Monoiodotrigermane, trigermanyl iodide (Ge_3H_7I), cannot be isolated in pure form, since it still cannot be transported at 0° (the highest possible temperature) under high vacuum[*175*].

6.2.5 *Organohalogenogermanes*

Monomethylmonochlorogermane (CH_3GeH_2Cl) is a stable colourless liquid at room temperature. Vapour pressure equation: $\log p$ [torr] $= -1346.0/T + 6.793$ (valid between –32 and +10°). Enthalpy of evaporation 6155 cal/mole. Melting point –101°. Boiling point 70.9° (extrapolated from the vapour pressures) [*5*].

Monomethyldichlorogermane (CH_3GeHCl_2) is a stable colourless liquid at room temperature. Vapour pressure at 20° 25 torr. Vapour pressure equation $\log p$ [torr] $= -1800/T + 7.551$ [*115*]; $\log p$ [torr] $= -1729.8/T + 7.357$ (valid between 0 and 17°) [5]. Enthalpy of evaporation 8250 [*115*]; 7910 [5] cal/mole. Melting point -63.1 ± 0.1 [*115*]; −62° [5]. Boiling point 112.2 ± 0.5 [*115*]; 113.2° (extrapolated from the vapour pressure) [5]. Trouton's constant 21.4 cal deg^{-1} $mole^{-1}$ [*115*].

Dimethylmonochlorogermane [$(CH_3)_2GeHCl$] is a stable colourless liquid at room temperature. Vapour pressure equation $\log p$ [torr] $= -1532.6/T + 7.111$ (valid between 0 and 15°). Enthalpy of evaporation 7013 cal/mole. Melting point −76°. Boiling point 89.4° (extrapolated from the vapour pressures) [*76*].

Monomethylmonobromogermane (CH_3GeH_2Br). Vapour pressure at 0° 26.8 torr. Vapour pressure equation $\log p$ [torr] $= -1740/T + 7.800$. Boiling point $80.5 \pm 0.5°$ (extrapolated from the vapour pressures). Enthalpy of evaporation 7960 cal/mole. Trouton's constant 22.5 cal deg^{-1} $mole^{-1}$. Melting point $-89.2 \pm 0.2°$ [*115*].

Diethylmonochlorogermane [$(C_2H_5)_2GeHCl$]. Boiling point 136° (760 torr). Density d_4^{20} 1.2409. Refractive index n_D^{20} 1.4572 [*237*].

Diethylmonobromogermane [$(C_2H_5)_2GeHBr$]. Boiling point 153° (760 torr). Density d_4^{20} 1.5340. Refractive index n_D^{20} 1.4888 [*237*].

Diethylmonoiodogermane [$(C_2H_5)_2GeHI$]. Boiling point 70° (23 torr). Density d_4^{20} 1.7717. Refractive index n_D^{20} 1.5382 [*237*].

Mono-n-butylmonochlorogermane ($n\text{-}C_4H_9GeH_2Cl$). Boiling point 140° (760 torr). Density d_4^{20} 1.246 g/ml. Refractive index n^{20} 1.459_8 [*11*].

Mono-n-butylmonobromogermane ($n\text{-}C_4H_9GeH_2Br$). Boiling point 159° (760 torr). Density d_4^{20} 1.536 g/ml. Refractive index n^{20} 1.491_0 [*11*].

Mono-n-butylmonoiodogermane ($n\text{-}C_4H_9GeH_2I$). Boiling point 181° (760 torr). Density d_4^{20} 1.776 g/ml. Refractive index n^{20} 1.541_2 [*11*].

Di-n-butylmonochlorogermane [$(n\text{-}C_4H_9)_2GeHCl$]. Boiling point 219° (760 torr) [*11*]; 115° (30 torr) [*237*]. Density d_4^{20} 1.107 [*11*]; 1.110 [*237*]. Refractive index n^{20} 1.461_8 [*11*]; n_D^{20} 1.4620 [*237*].

Di-n-butylmonobromogermane [$(n\text{-}C_4H_9)_2GeHBr$]. Boiling point 234° (760 torr). Density d_4^{20} 1.305. Refractive index n^{20} 1.483_2 [*11*].

Di-n-butylmonoiodogermane [$(n\text{-}C_4H_9)_2GeHI$]. Boiling point 249° (760 torr). Density d_4^{20} 1.470. Refractive index n^{20} 1.514_8 [*11*].

6.3 *Reactions of the Ge–X bond*[5]

6.3.1 *Reaction with water*

In the hydrolysis of a germanium–halogen bond, a germanium–hydroxide bond is first formed:

$$\rangle Ge{-}X + H_2O \rightarrow \rangle Ge{-}OH + HX$$

[5] For the halogen exchange in germyl and silyl halides, see [*39a*].

References p. 711

Tetrahalides hydrolyse to germanic acid; as was to be expected, the oxidation stage 4 is retained, *e.g.*, GeI_4 at 25°±1°[*136*]:

$$GeI_{4(solid)} + 3H_2O \rightarrow H_2GeO_{3(soln)} + 4H^{\oplus} + 4I^{\ominus} + 9.64 \pm 0.10 \text{ [kcal/mole]}$$

Halogenogermanium hydrides (apart from the fluorine compounds) also react with water, but they form orange oxides or oxide-hydrates of germanium having a lower oxidation number than 4, with the evolution of hydrogen[*289*]. However, the reaction of GeH_2Cl_2, for example, cannot be represented by the equation $GeH_2Cl_2 + H_2O \rightarrow GeO + 2HCl + H_2$, since less hydrogen is liberated than corresponds to the equation[*62*].

With water, trimethylhalogenogermanes form trimethylgermanol, which in an aqueous medium is in equilibrium with bis(trimethylgermyl) oxide:

$$(CH_3)_3GeCl + H_2O \rightarrow (CH_3)_3GeOH + HCl$$

$$2(CH_3)_3GeOH \rightleftharpoons [(CH_3)_3Ge]_2O + H_2O$$

If the germyl halides contain Ge–H bonds (GeH_3F[*266*], i-$C_4H_{11}GeH_2Br$[*11*]), no alcohols or ether analogues are formed.

6.3.2 *Mechanism of hydrolysis*

There are five possible mechanisms for the hydrolysis of a germanium–halogen bond[*132*].

(1) The first reaction step is the rate-determining dissociation of the Ge–X bond and the formation of germyl cations and halide ions with a subsequent fast addition of water:

$$R_3Ge\text{–}X \xrightarrow{\text{slow, }-X^{\ominus}} R_3Ge^{\oplus} \xrightarrow{\text{fast, }+H_2O} R_3Ge\text{–}OH_2^{\oplus} \qquad (S_N1)$$

(2) The first step is the fast dissociation of the Ge–X bond followed by a rate-determining addition of water to the germyl cation formed:

$$R_3Ge\text{–}X \underset{}{\overset{\text{fast, }-X^{\ominus}}{\rightleftharpoons}} R_3Ge^{\oplus} \xrightarrow{\text{slow, }+H_2O} R_3Ge\text{–}OH_2^{\oplus} \qquad (S_N2)$$

(3) Synchronous mechanism: one-stage splitting off of halide ion and simultaneous addition of water:

$$R_3Ge\text{–}X \xrightarrow{\text{slow, }+H_2O,\ -X^{\ominus}} R_3Ge\text{–}OH_2^{\oplus} \qquad (S_N2)$$

(4, 5) Before dissociation, water adds to R_3GeX with the formation of a sp^3d configuration on the Ge, the addition of H_2O being either rate-determining (4) or fast (5):

$$R_3Ge\text{–}X \xrightarrow{\text{slow, }+H_2O} H_2O\text{—}GeR_3\text{—}X \xrightarrow{\text{fast, }-X^{\ominus}} R_3Ge\text{–}OH_2^{\oplus} \qquad (S_N2)$$

$$R_3Ge{-}X \underset{}{\overset{\text{fast, }+H_2O}{\rightleftharpoons}} H_2O{-}\underset{R\quad R}{\overset{R}{Ge}}{-}X \xrightarrow{\text{slow, }-X^{\ominus}} R_3Ge{-}OH_2^{\oplus} \text{ (pseudo-}S_N1)$$

The most probable mechanism of the reaction follows from measurements of its kinetics. The rate of hydrolysis falls with increasing electronegativity of the halogen or increasing strength of the Ge–X bond $R_3GeBr > R_3GeCl > R_3GeF$. Consequently, the rate-determining step is the dissociation of the Ge–X bond and mechanisms 2 and 4 are therefore unlikely. Transition states with positive germanium are improbable (and therefore mechanisms 1, 2, and 3 are excluded). If halide ions $Y^{\ominus}$ are added to the hydrolysis medium, an equilibrium is set up:

$$R_3GeX \rightleftharpoons R_3Ge^{\oplus} + X^{\ominus}$$
$$R_3Ge^{\oplus} + Y^{\ominus} \rightleftharpoons R_3GeY$$

If the R_3GeY newly formed in part hydrolyses more slowly than the original R_3GeX, the rate of hydrolysis of R_3GeX falls (*e.g.*, by the addition of $Cl^{\ominus}$ to Ph_3GeBr). If the R_3GeY newly formed in part hydrolyses faster than R_3GeX, the rate of hydrolysis of R_3GeX (*e.g.*, by the addition of $Br^{\ominus}$ to Ph_3GeF) rises. The hydrolysis is substantially independent of the attack of the water (mechanisms 2, 3, and 4 are excluded). The hydrolysis of a germanium–halogen bond in triarylhalogenogermanes therefore takes place by mechanism 5 (pseudo-S_N1). The corresponding halogenogermanium hydrides are said to hydrolyse by a similar mechanism, since the replacement of phenyl by hydrogen in $(C_6H_5)_3GeX$ is said to make the germanium negative, so that a mechanism via germyl cations is still more unlikely.

6.3.3 *Reaction with alkoxides and mercaptides*

Alkylhalogenogermanes ($R_{4-n}GeX_n$) react with sodium methoxide in methanol (yield 66–77%), diglyme, or in the absence of a solvent (yield 98%) to give the corresponding alkylmethoxygermanes [*98a, 115, 296*]:

$$R_{4-n}GeX_n + n\,NaOCH_3 \rightarrow R_{4-n}Ge(OCH_3)_n + n\,NaX \quad (R = CH_3, n = 1\text{–}4)$$

If the halogenogermane contains Ge–H bonds like, for example, monomethylmonobromogermane, a volatile compound, probably $CH_3GeH_2OCH_3$, is formed with solvent-free dry sodium methoxide at –80°. In attempts to isolate it, it disproportionates slowly into methanol and white non-volatile solid, probably a derivative of germanium dihydride $(CH_3GeH)_x$ [*115*]:

$$CH_3GeH_2Br \xrightarrow[-NaBr]{+NaOCH_3} CH_3GeH_2OCH_3 \xrightarrow[-CH_3OH]{} \frac{1}{x}(CH_3GeH)_x$$

References p. 711

CH_3SNa reacts with GeH_3I (or GeH_3Cl) at −196° to 20° to give GeH_3SCH_3 [*294a*]:

$$GeH_3I + CH_3SNa \rightarrow GeH_3SCH_3 + NaI$$

6.3.4 *Reaction with ammonia and phosphane derivatives*

$GeH_3F + NH_3$. With ammonia at −78°, germyl fluoride forms a 1:2 adduct, $GeH_3F \cdot 2NH_3$. This is substantially more stable than the adducts of germyl chloride and changes at 25° with the liberation of one mole of ammonia into the 1:1 adduct ($GeH_3F \cdot 1NH_3$), which is non-volatile and decomposes at 180° without melting. Because of its non-volatility, and from the IR spectrum, it must be concluded that this compound is ionic $[GeH_3NH_3]^{\oplus} F^{\ominus}$[*266*].

$GeH_3Cl + NH_3$. With NR_3 (R = H, CH_3, C_2H_5, C_6H_5) below −78° germyl chloride forms 1:1 adducts ($GeH_3Cl \cdot NR_3$), which decompose into ammonium chlorides and germanium dihydride only slightly above −78°[*66, 213*] (in the case of the NH_3 adduct, between −78 and −50°[*66*]). The electrophilic attack of the strong Lewis base NH_3 on a proton and the splitting out of NR_3HCl first forms germanium dihydride. Subsequently, GeH_2 disproportionates under the influence of NH_3 into GeH and GeH_4:

$$3GeH_3Cl + 3NR_3 \longrightarrow 3GeH_2 + 3NR_3HCl$$

$$3GeH_2 \xrightarrow{(NH_3)} 2GeH + GeH_4$$

$$\overline{3GeH_3Cl + 3NR_3 \longrightarrow 2GeH + GeH_4 + 3NR_3HCl}$$

Unlike the reaction of SiH_3Cl with NH_3, GeH_3Cl forms no trigermylamine, $(GeH_3)_3N$, under similar conditions[*66, 213*].[6]

$GeH_2Cl_2 + NH_3$. With liquid ammonia at −78°, dichlorogermane forms germanium and ammonium chloride. With a deficiency of ammonia (passage of NH_3 into liquid GeH_2Cl_2), at −27° a white solid is first formed which soon becomes yellow and then black[*66*]:

$$GeH_2Cl_2 + 2NH_3 \xrightarrow{\Delta} \begin{bmatrix} \text{intermediate} \\ \text{stage,} \\ NH_3 \text{ adduct?} \end{bmatrix} \xrightarrow{\Delta} Ge + 2NH_4Cl$$

$GeH_3F + (Me_3Si)_2CN_2$. Bis(trimethylsilyl) carbodiimide reacts rapidly (5 min) with germyl fluoride at 20° to give $(GeH_3)_2CN_2$ (hield 95%)[*39c*]:

$$2GeH_3F + (Me_3Si)_2CN_2 \rightarrow (GeH_3)_2CN_2 + 2Me_3SiF$$

The vibrational spectra indicate that digermylcarbodiimide, $(GeH_3)_2CN_2$, is a carbodiimide:

$$H_3Ge{-}NCN{-}GeH_3$$

---

[6] But immediate removal of the reaction product of a gas-phase reaction of 4 moles of NH_3 and 3 moles of GeH_3Cl gives a 40% yield of $(GeH_3)_3N$ [*224b, 224c*].

rather than a cyanamide:

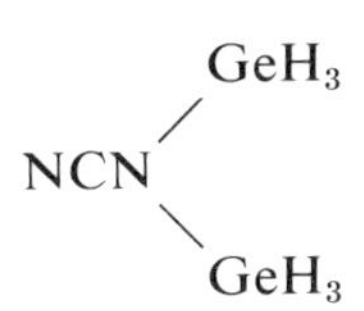

The selection rules are consistent with a linear heavy-atom skeleton, but the absence of resolved rotational detail proves that the skeleton is bent [*39c*].

$GeH_3Br + PH_3$. Germyl bromide reacts with phosphane in the presence of a base (Me_3N) but no $(GeH_3)_3P$ was detected [*39d*]:

$$GeH_3Br + PH_3 + Me_3N \xrightarrow[-78^\circ]{Me_2O} \not\rightarrow (GeH_3)_3P + 3Me_3NHBr$$

$GeH_3Br + KPH_2$. Attempts using GeH_3Br and KPH_2 in Me_2O at $-90°$, conditions analogous to those used in the preparation of $(SiH_3)_3P$ (see Chapter 7, section 10.2), gave no $(GeH_3)_3P$ [*39d*]:

$$3GeH_3Br + 3KPH_2 \xrightarrow[-90^\circ]{Me_2O} \not\rightarrow (GeH_3)_3P + 2PH_3 + 3KBr$$

$GeH_3Br + (SiH_3)_3P$, $(SiH_3)_3As$ or *$(SiH_3)_3Sb$*. Trisilylphosphane, -arsane and -stibane react with an excess of germyl bromide to give high yields of trigermyl compounds [*39, 39d, 79a*]:

$$3GeH_3Br + (SiH_3)_3E \xrightarrow[4h]{0^\circ} (GeH_3)_3E + 3SiH_3Br$$

$$E = P, As, Sb$$

GeH_3Cl or $GeD_3Cl + SiH_3PH_2$ or SiD_3PD_2. Monosilylphosphane (SiH_3PH_2 or SiD_3PD_2) undergoes an exchange reaction with monochlorogermane (GeD_3Cl or GeH_3Cl) to give monogermylphosphane (GeH_3PD_2 or GeD_3PH_2) [*76a*]:

$$SiD_3PD_2 + GeH_3Cl \xrightarrow[10\,h]{-78^\circ} GeH_3PD_2 + SiD_3Cl$$

$$SiH_3PH_2 + GeD_3Cl \xrightarrow[10h]{-78^\circ} GeD_3PH_2 + SiH_3Cl$$

For the preparation of GeD_3AsH_2 and GeH_3AsD_2 by the exchange reaction between SiH_3AsH_2 and GeD_3Cl or GeH_3Cl, see [*76b*].

GeH_3Br forms with $LiAl(PH_2)_4$ at $-45°$ GeH_3PH_2 (in accordance with the equation $LiAl(PH_2)_4 + 4\,GeH_3Br \rightarrow 4\,GeH_3PH_2 + LiBr + AlBr_3$) [*297a*].

6.3.5 Replacement of halogen by alkyl groups

A literature review on the replacement of halogen by organic substituents in halogenogermanes has been published [*227a, 252*]. Concerning the mechanism of reaction of $>$Ge–X with amines, see the literature [*239*].

References p. 711

>GeX + RMgX. Little has become known about the Grignard reactions of germanium halides ($GeH_aR_bX_c$, $a+b+c=4$), since the alkylgermanes that can be made by such reactions ($GeH_aR_bR'_c$) can be obtained more simply by the hydrogenation of the corresponding alkylhalogenogermanes ($GeX_aR_bR'_c$). The reaction of $GeH_aR_bX_c$ with alkylating agents (R'MgX, LiR') very probably does not lead to pure $GeH_aR_bR'_c$, since in the presence of highly electrophilic agents such as MgRX and MgX_2, exchange reactions (*e.g.*, $2\ GeH_aR_bR'_c \rightleftharpoons GeH_aR_{b-1}R'_{c+1} + GeH_aR_{b+1}R'_{c-1}$) must always be taken into account:

$$\begin{matrix}(R,R')_3Ge - R \cdots MgX_2 \\ X_2Mg \cdots R' - Ge(R,R')_3\end{matrix} \quad (R = H, alkyl)$$

Furthermore, in Grignard reactions (particularly when it is desired partially to replace halogen by alkyl groups), all conceivable alkylhalogenogermanes are formed. For example, in the reaction of $GeCl_4$ with a deficiency of CH_3MgI [with the aim of preparing $(CH_3)_3GeCl$], $GeCl_{4-n}(CH_3)_n$, $GeI_{4-n}(CH_3)_n$, and $GeCl_aI_b(CH_3)_c$ ($n = 0$–4; $a+b+c=4$) are formed [*252*].

The synthesis of Ge_2H_5R ($R = CH_3, C_2H_5$) is one of the few examples of reactions of higher germanes in which the Ge–Ge bonds are preserved [*175b*]:

$$Ge_2H_5I + RMgX \rightarrow Ge_2H_5R + MgIX$$

In the Grignard reactions of $GeCl_4$ with an excess of RMgX (R = bulky groups such as Pr^i or cyclo-C_6H_{11}) it is not R_4Ge that is formed but R_3GeH. A spectroscopic study of the course of the reaction shows that the Ge–H bond is not formed in the reaction of $GeCl_4$ and RMgX but only in the subsequent hydrolysis of the $R_3GeMgCl$ formed (in accordance with the reaction $(C_6H_{11})_3GeMgCl + D_2O \rightarrow (C_6H_{11})_3GeD + DOMgCl$) [*194*]. For chloride exchange reactions, see [*109a*].

>GeX + ZnR_2, HgR_2, LiR. As an example, with divinylmercury, germanium tetrachloride gives a 70% yield of vinyltrichlorogermane [*24*], and with dipropyl mercuridiacetate, tributylgermanium iodide gives a 90% yield of propyl (tributylgermyl)acetate [*17*]:

$$GeCl_4 + (CH_2CH)_2Hg \xrightarrow{80^\circ} CH_2CHGeCl_3 + CH_2CHHgCl$$

$$2\,(C_4H_9)_3GeI + Hg(CH_2COOC_3H_7)_2 \longrightarrow 2\,(C_4H_9)_3GeCH_2COOC_3H_7 + HgI_2$$

>GeX + N_2CHR. An α-halogenated alkyl group can be introduced by the reaction of halogenogermanes with diazo compounds:

$$N{\equiv}\overset{\oplus}{N}{-}\overset{\ominus}{C}HR + {>}Ge{-}X \rightarrow N{\equiv}\overset{\oplus}{N}{-}CHR \rightarrow \overset{\ominus}{Ge}{-}X \xrightarrow{-N_2} RH\overset{\oplus}{C}{-}\overset{\ominus}{Ge}{-}X \rightarrow RHXC{-}Ge{<}$$

(Examples of reaction products: $ClCH_2GeCl_3$, $(ClCH_2)_2GeCl_2$, $CH_3(CH_2Cl)GeCl_2$, $CH_3(CH_2Cl)_2GeCl$, $(CH_3)_3(CH_2Cl)Ge$[*255*], $(C_6H_5)(CH_2Cl)_2GeCl$, $(C_6H_5)(CH_2Cl)GeCl_2$, $(C_6H_5)(CH_2Cl)GeCl$[*287*].)

6.3.6 *Reaction with alkali metals*

The germanium–metal bond can also be obtained conveniently in ways other than by the reaction of the hydride with an alkali metal in liquid ammonia or in polyether (see section 5.16). It is also formed in reactions of a germanium–halogen bond with an alkali metal in tetrahydrofuran[*28, 103, 287*]. There are indications that when an alkylhalogenogermane is added to lithium in tetrahydrofuran or when tetrahydrofuran is added to a suspension of lithium in an alkylhalogenogermane, Ge–Ge compounds are first formed (analogously to the corresponding silicon compounds) and these are split in the subsequent stages of the reaction by lithium[*30, 31, 101*]:

$$\rangle Ge{-}X + 2Li \rightarrow \rangle Ge{-}Li + LiX$$

$$\rangle Ge{-}Li + X{-}Ge\langle \rightarrow \rangle Ge{-}Ge\langle + LiX$$

$$\rangle Ge{-}Ge\langle + 2Li \rightarrow 2\rangle Ge{-}Li$$

$$2\rangle Ge{-}X + 4Li \rightarrow 2\rangle Ge{-}Li + 2LiX$$

Since the germylmetals can be freed from ammonia only with difficulty, preparation reactions in tetrahydrofuran are advantageous when the germyls are subsequently to be treated with ammonolysing halogen compounds.

6.3.7 *Reaction with salts of transition metals*

Monohalogenogermanes, GeH_3X, react with silver salts in a similar manner to monohalogenosilanes. While, however, the silicon compounds formed are generally stable, the germanium isologues undergo subsequent reactions which lead to stable polymeric germanium hydrides. Table 8.15 (section 7.1) contains the physical properties of the compounds that can be isolated in these reactions.

$Ag_2CO_3 + (CH_3)_3GeCl$. When trimethylchlorogermane is shaken with silver carbonate[*115*] or when the two substances react in benzene in the presence of catalytic amounts of water[*252*], bis(trimethylgermyl) oxide is formed:

$$2(CH_3)_3GeCl + Ag_2CO_3 \rightarrow [(CH_3)_3Ge]_2O + 2AgCl + CO_2$$

$Ag_2CO_3 + (CH_3)_2GeHBr$. If the germyl halide still contains Ge–H bonds, the reaction with silver carbonate does not lead to ether analogues (such as $[(CH_3)_2HGe]_2O$) but to a clear viscous liquid[*115*], possibly the methyl derivative of germanium dihydride:

$$2(CH_3)_2GeHBr \xrightarrow[-2AgBr,\,-CO_2]{+Ag_2CO_3} [(CH_3)_2HGeOGeH(CH_3)_2] \xrightarrow{-H_2O} \frac{2}{x}[Ge(CH_3)_2]_x$$

This assumption is supported by the existence of the polymeric phenol analogue $Ge(C_6H_5)_2$ [*106, 146*] and the corresponding tin compounds $(R_2Sn)_n$ [*125*].

Ag_2CO_3 or Ag_2O+GeH_3Br. With silver carbonate or silver oxide, germyl bromide forms bis(germyl) oxide as an intermediate, but this decomposes into germanium dihydride and water [*264*]:

$$2GeH_3Br \xrightarrow[-AgBr,\ -CO_2]{+Ag_2CO_3} [(GeH_3)_2O] \xrightarrow{-H_2O} \frac{2}{x}(GeH_2)_x$$

$$2GeH_3Br \xrightarrow[-2AgBr]{+Ag_2O} [(GeH_3)_2O] \xrightarrow{-H_2O} \frac{2}{x}(GeH_2)_x$$

Ag_2CO_3 or Ag_2O+GeH_3Cl or GeH_3I. Germyl chloride does not react with either silver oxide or silver carbonate [*264*]; with these materials germyl iodide forms water and non-volatile products which have not been identified, but no bis(germyl) oxide [*272, 273, 276*].

$HgO+GeH_3Br$ or GeH_3I. Unlike germyl bromide, germyl iodide reacts with red and yellow mercury oxides to give no bis(germyl) oxide but (with red HgO) GeH_4, H_2O, Hg, and non-volatile products or (with yellow HgO) no GeH_4 but H_2O, Hg, and non-volatile products. The different reaction products are not the consequence of a different reaction mechanism. The absence of GeH_4 in the reaction with yellow HgO can be ascribed to the higher rate of formation of water (perhaps by the reaction: $GeH_4+2HgO \rightarrow Ge+2H_2O+2Hg$) [*272, 276*].

$HgS+GeH_3Br$ or GeH_3I. Neither silyl bromide nor germyl bromide reacts with mercury sulphide. However, germyl iodide (like the isologous silicon compound) in an exothermic reaction at 25° forms bis(germyl) sulphide in 55 to 75% yield within 15 to 30 min [*272, 273, 276*]:

$$2GeH_3I+HgS \rightarrow (GeH_3)_2S+HgI_2$$

$AgCN+GeH_3Cl$ or GeH_3Br. Germyl chloride reacts with silver cyanide to give germyl cyanide. In the presence of undefined impurities, however, this decomposes at 25° so that it cannot be isolated in the pure state. It can be obtained pure by the reaction of germyl bromide with silver cyanide [*273, 278*]:

$$GeH_3Br+AgCN \rightarrow GeH_3CN+AgBr$$

At room temperature, germyl cyanide forms colourless needle-shaped crystals. Surprisingly, the enthalpy of evaporation calculated from the vapour pressure equation is almost the same as that of silyl cyanide, although the enthalpies of evaporation of germyl chloride and bromide are about 1.7 kcal larger than those of the silicon isologues. This can be explained by the lower association in germyl cyanide than in silyl cyanide.

$AgNCO+GeH_3Cl$ or GeH_3Br. Gaseous germyl bromide (but not germyl chloride) reacts with silver cyanate to form germyl isocyanate (yield 93%) [*113, 264*]:

$$GeH_3Br+AgNCO \rightarrow GeH_3NCO+AgBr$$

The iso structure has been shown by the IR spectrum. This is not surprising, since both the free acid and its esters have the iso structure. Furthermore, the covalent normal cyanates are thermally less stable than the isocyanates. Germyl isocyanate is remarkably stable to heat. It does not decompose in 40 h at 110°; a 10% decomposition occurs only at 200 to 220° in 12 h according to the following equation [*264*]:

$$GeH_3NCO \rightarrow Ge + H_2 + HNCO$$

$AgSCN + GeH_3Br$. In an analogous manner to the preparation of silyl thiocyanate, the reaction of silver thiocyanate with germyl bromide gives a 96% yield of germyl isothiocyanate [*264*]:

$$GeH_3Br + AgSCN \rightarrow GeH_3NCS + AgBr$$

The isostructure has been proved by the IR spectrum. Germyl isothiocyanate decomposes completely at 55° in 20 h into germane and a yellow insoluble solid containing no free sulphur and probably polymeric thiocyanic acid [*264*]:

$$GeH_3NCS \rightarrow GeH_2 + HNCS$$

$$3GeH_2 \rightarrow GeH_4 + 2GeH$$

$$x\,HNCS \rightarrow (HNCS)_x$$

$CH_3COOAg + GeH_3Br$. Gaseous germyl bromide reacts with dry silver acetate to give a 96% yield of germyl acetate [*264*]:

$$GeH_3Br + AgOCOCH_3 \rightarrow GeH_3OCOCH_3 + AgBr$$

This decomposes at 100 to 110° with the evolution of a small amount of hydrogen first into a white solid (after 10 min) which gradually (over 16 h) becomes red [*264*]:

$$GeH_3OCOCH_3 \rightarrow GeH_2(\text{white, then yellow}) + CH_3COOH$$

$$2GeH_2 \rightarrow 2GeH(\text{red}) + H_2$$

$AgNO_2 + GeH_3Br$. With silver nitrite, germyl bromide forms solid reaction products containing germanium and monogermane, but no germyl nitrite [*264*].

$AgF + GeH_3Br$. For the reaction of silver fluoride with germyl bromide (formation of GeH_3F, and GeH_2F_2), see section 6.1.4.

7. Pseudohalogenogermanes

7.1 Synthesis of germyl pseudohalides

Germyl pseudohalides are formed in very good yields by reactions of germyl halides with silver pseudohalides (see section 6.3.7) or by an exchange reaction with other pseudohalides (*e.g.*, Me_3SiN_3 [*39b*], $(Me_3Si)_2SiCN_2$ [*39c*]):

$$Me_3SiN_3 + GeH_3F \xrightarrow[3\ min]{20°} GeH_3N_3 + Me_3SiF$$

$$(Me_3Si)_2CN_2 + 2GeH_3F \xrightarrow[5\ min]{20°} (GeH_3)_2CN_2 + 2Me_3SiF$$

The physical properties are given in Table 8.15. The compounds, which are surprisingly stable to heat, can be used in many syntheses, like the germyl halides. However, the reactions of the germyl halides generally take place differently from those of the germyl pseudohalides.

7.2 *Mechanism of the reaction of germyl cyanide with donors*

Germyl cyanide reacts with strong donors preferentially in a reaction similar to α-elimination in carbon chemistry (detachment of a proton and of the cyanide ion together with its electron pair) to form germanium dihydride. With weak donors it reacts in a manner similar to the S_N2 mechanism (attack on the central germanium and detachment of the cyanide ion) with formation of germyl compounds and hydrogen cyanide. [*274*]

```
            H  <------- Attack of strong donors
            |
          H-Ge-CN
           ↗|
Attack of weak donors-/  H
```

7.3 *Reactions of the germanium-pseudohalogen bond*

Table 8.16 summarises the reactions of germyl cyanide (and, for comparison, those of silyl cyanide and methyl cyanide) with diborane, trimethylboron, boron fluoride, trimethylgallium, ammonia, phosphane, water, methanol, hydrogen sulphide, and hydrogen bromide.

$(BH_3)_2 + GeH_3CN$. In the gas phase at 0°, diborane hydrogenates germyl cyanide to give an 80 to 100% yield (calculated on GeH_3CN) of monogermane. Probably BH_2CN is also formed (in analogy with the reaction of silyl cyanide). If less than 100% of GeH_4 is produced, the remainder of the Ge is found in Ge_2H_6 (up to 10% calculated on GeH_3CN); in that case, a boron hydride not further identified appears simultaneously [*273*]:

$$GeH_3CN + BH_3 \rightarrow GeH_4 + BH_2CN$$

or:

$$GeH_3CN + BH_3 \rightarrow 80\text{–}100\%\ GeH_4,\ Ge_2H_6,\ \text{solid},\ BH_2CN$$

In contrast monogermylphosphane, GeH_3PH_2, reacts with diborane at −20° in a sealed NMR tube, quantitatively forming the adduct $GeH_3PH_2 \cdot BH_3$ [*76a*].

TABLE 8.15

PHYSICAL PROPERTIES OF GERMYL COMPOUNDS

Germane	*M.p. (°C)*	*B.p.* [°C(*torr*)]	*log* p [*torr*] = −*A*/*T* + *B*			*Enthalpy of evaporation (kcal/mole)*	*Trouton's constant*	*Ref.*
			A	*B*	*valid between (°C)*			
GeH_3CN	45.0–46.8	—	2507	9.9858	—	11.5	—	264,275
GeH_3NCO	−44.0 ± 0.5	71.[illegible] 760)	1891	8.369	−23 and +23	8.651	25.1	264
GeH_3NCS	18.6 ± 0.3	150 [illegible]760)	2280	8.268	+19 and +50	10.4	24.6	264
$(GeH_3)_2(CN_2)$	10 ± 0.5	—	—	—	—	—	—	39c
GeH_3N_3	−31 ± 1	0([illegible])	—	—	—	—	—	39b
$(GeH_3)_3P$	−23.8/−84 ± 1	0([illegible])/20(2)	—	—	—	—	—	39/39d
GeH_3OOCCH_3	12.8 ± 0.1	82[illegible](760)	2256	9.227	−10 and +40	10.3	30.0	264
$GeH_3CH_2OCH_3$	−121.6 ± 0.3	44[illegible]	1678.6	8.1674	−64 and 0	7.679	24.2	98a
GeH_3OCH_3	−44.5 ± 0.5	24[illegible]3	1823.0	9.0090	−46 and −23	8.340	28.0	98a
GeH_3SCH_3	−97 ± 0.4	87[illegible]760)	1537.9	7.1555	—	7.040	19.6	294a
$(GeH_3)_2S$	−36· · ·−34	—	Vapour pressure at 0° 5.0 torr			—	—	273
GeH_3GaMe_2		2[illegible] (0.001)						5d
$(GeH_3)_2Zn{\cdot}2Et_2O$	−40/−35							5e
$GeH_3B(NMe_2)_2$		2[illegible] (3–4)						5f
$(GeH_3)Me_2B_3N_3Me_3$	−20/−15							5f
$(GeH_3)_2MeB_3N_3Me_3$	−20/−15							5f
$(GeH_3)_3B_3N_3Me_3$	−20/−15							5f

TABLE 8.16

PRODUCTS OF THE REACTIONS OF METHYL, SILYL AND GERMYL CYANIDE WITH DIBORANE, TRIMETHYLBORON, BORON FLUORIDE, TRIMETHYLGALLIUM, AMMONIA, PHOSPHANE, WATER, METHANOL, HYDROGEN SULPHIDE AND HYDROGEN BROMIDE[273]

	CH_3CN	SiH_3CN	GeH_3CN
$(BH_3)_2$	–	SiH_4+BH_2CN	GeH_4 + solid + Ge_2H_6
$B(CH_3)_3$	–	no reaction	no reaction
BF_3	$CH_3CN \cdot BF_3$[a]	SiH_3F+BF_2CN	$GeH_3CN \cdot BF_3$[b]
$Ga(CH_3)_3$	–	$CH_3SiH_3+(CH_3)_2GaCN$	$CH_3GeH_3+(CH_3)_2GaCN$
NH_3	–	–	GeH_2+NH_4CN
PH_3	–	–	compounds containing Ge and P
H_2O	–	$(SiH_3)_2O+HCN$	$(GeH_3)_2O+HCN$
CH_3OH	–	$(SiH_3)_2O+CH_3CN+HCN$	GeH_2+HCN
H_2S	–	–	$GeH_3SH+(GeH_3)_2S$
HBr	–	–	$GeH_3Br+HCN$

[a]Dissociation energy 26.5 kcal
[b]Dissociation energy 28.0 kcal

$B(CH_3)_3+GeH_3CN$. Up to its decomposition point, germyl cyanide does not react with trimethylboron[273].

BF_3+GeH_3CN. The reaction of boron trifluoride with silyl (iso)cyanide forms silyl fluoride and cyanoboron fluoride with no detectable intermediates. The reaction is explained by the entry of the free electron pair of the fluorine into the empty 3d-orbitals of the silicon:

$$SiH_3(CN) + BF_3 \longrightarrow \left[\begin{array}{c} F \\ | \\ F{-}B{-}F \\ \vdots \uparrow \\ H_3Si{-}N{\equiv}C \end{array} \right] \longrightarrow SiH_3F + BF_2CN$$

Because of the occupied 3d-orbitals in germanium, under the same reaction conditions germyl cyanide forms with boron fluoride a 1:1 adduct (analogy to the reaction of methyl cyanide) [273]:

$$GeH_3CN + BF_3 \rightarrow GeH_3CN{\cdot}BF_3$$

The germyl cyanide–boron fluoride adduct is very stable and is comparable with the methyl cyanide–boron fluoride adduct (for dissociation energies, see Table 8.16).

$Ga(CH_3)_3+GeH_3CN$. Germyl cyanide reacts with trimethylgallium in the same way as silyl cyanide. Methylgermane (methylsilane) and dimethylgallium cyanide are formed[273].

NH_3 or PH_3+GeH_3CN. Ammonia reacts with germyl cyanide to form germanium dihydride and ammonium bromide[273]:

$$GeH_3CN+NH_3 \rightarrow NH_4^{\oplus} + GeH_2CN^{\ominus} \rightarrow GeH_2+NH_4CN$$

However, phosphane appears to react quite differently, in a similar manner to water as described below. The reaction products include compounds containing phosphorus and germanium, among others. Further investigations are necessary to elucidate both reactions [*273*].

H_2O+GeH_3CN. Silyl cyanide and water react with one another to form bis(silyl)oxide and hydrogen cyanide, which are easily separated from one another. Germyl cyanide also reacts with water under similar conditions. However, it is substantially more difficult to separate the reaction products from one another [*273*].

$$H_2O+SiH_3CN \rightarrow \left[HO\cdots\overset{\displaystyle H}{\underset{\displaystyle H\ \ H}{Si}}-CN\right]H \xrightarrow{-HCN} HOSiH_3 \xrightarrow[-HCN]{+H_3SiCN} (SiH_3)_2O$$

$$H_2O+GeH_3CN \xrightarrow{-HCN} HOGeH_3 \xrightarrow[-HCN]{+H_3GeCN} (GeH_3)_2O$$

The more highly nucleophilic $(GeH_3)_2O$ forms adducts with HCN, while the more feebly nucleophilic $(SiH_3)_2O$ does not [*273*].

$H_2O+(GeH_3)_2CN_2$. Digermylcarbodiimide reacts readily with water to give bis(germyl)oxide, $(GeH_3)_2O$ [*38b, 39c*].

$CH_3OH+GeH_3CN$. Germyl cyanide does not react with methanol in the same way as with water. It is not bis(germyl)oxide, methyl cyanide, and hydrogen cyanide that are formed

$$CH_3OH+2GeH_3CN \nrightarrow (GeH_3)_2O+CH_3CN+HCN$$

but yellow polymeric germanium dihydride and hydrogen cyanide [*273*]:

$$GeH_3CN \xrightarrow{(CH_3OH)} GeH_2+HCN$$

CH_3OH or $CH_3OD+(GeH_3)_2CN_2$. Digermylcarbodiimide reacts readily with methanol to give methyl germyl ether $GeH_3{\cdot}O{\cdot}CH_3$ or $GeH_3{\cdot}O{\cdot}CD_3$ [*38b, 39c*].

H_2S+GeH_3CN. Germyl cyanide reacts with hydrogen sulphide to form germyl hydrogen sulphide, which can be isolated, and bis(germyl) sulphide [*273*]:

$$GeH_3CN \xrightarrow[-HCN]{+H_2S} GeH_3SH \xrightarrow[-HCN]{+GeH_3CN} (GeH_3)_2S$$

$HBr+GeH_3CN$. The reaction of germyl cyanide with hydrogen bromide forms germyl bromide and hydrogen cyanide (see section 6.1.4).

8. Chalcogenogermanes

8.1 Synthesis of the Ge–O and Ge–S bond

Germanium hydrides with Ge–O and Ge–S bonds can be prepared by the solvolysis of the corresponding halogenogermanes or, better pseudohalogenogermanes with H_2O[273] or H_2S[273] or by their reaction with a heavy-metal chalcogenide[272, 2 3] (see sections 6.3.7 and 7.3):

$$GeH_3X + H_2E^{VI} \xrightarrow[-HX]{} GeH_3E^{VI}H \xrightarrow[-HX]{+GeH_3X} (GeH_3)_2E^{VI}$$

$$2GeH_3X + M_2E^{VI} \xrightarrow[-2MX]{} (GeH_3)_2E^{VI}$$

(X = halogen, pseudohalogen; M = one equiv. heavy metal)

For the relative Lewis base strength of Me_2O, Me_2S, $MeOSiH_3$, $MeSSiH_3$, $MeSGeH_3$, $(SiH_3)_2O$, $(SiH_3)_2S$ and $(GeH_3)_2S$, see ref. [294b].

8.2 Reactions of the Ge–O bond

In contrast to the Ge–O bond in peralkylated bis(germyl) oxide, R_3Ge–O–GeR_3, which is not cleaved by acetic acid and only sluggishly by formic acid[8, 163], the Ge–O bond in germyl alkoxides, R_3GeOR', is reactive[163]. Trialkylgermyl alkoxides react readily and generally quantitatively with $LiAlH_4$, HI, $R''MgI$, $R''COOH$, $(R''CO)_2O$, C_6H_5COCl, and $C_6H_5SO_3H$ [163]:

$$R_3GeOR' + H^{\ominus} \rightarrow R_3GeH + OR'^{\ominus}$$

$$R_3GeOR' + 2HI \rightarrow R_3GeI + RI + H_2O$$

$$R_3GeOR' + R''^{\ominus} \rightarrow R_3GeR'' + R'O^{\ominus}$$

$$R_3GeOR' + 2R''COOH \rightarrow R_3GeOCOR'' + R''COOR' + H_2O$$

$$R_3GeOR' + (R''CO)_2O \rightarrow R_3GeOCOR'' + R''COOR'$$

$$R_3GeOR' + C_6H_5COCl \rightarrow R_3GeCl + C_6H_5COOR'$$

$$R_3GeOR' + 2C_6H_5SO_3H \rightarrow R_3GeSO_3C_6H_5 + R'SO_3C_6H_5 + H_2O$$

8.3 Reactions of the Ge–S bond

Bis(germyl) sulphide (for preparation, see section 6.3.7) reacts with HF, HCl, HBr, HI, and HCN with the formation of the corresponding germyl halides or pseudohalides[273]; with HgO, digermyl oxide $(GeH_3)_2O$, is formed[109]. Under the action of diborane, it disproportionates into GeH_4 and GeH_2S[273]:

$$GeH_4 + GeH_2S \xleftarrow{(B_2H_6)} (GeH_3)_2S \xrightarrow[-H_2S]{+2HX} 2GeH_3X$$

$$(GeH_3)_2S \xrightarrow[-HgS]{+HgO} (GeH_3)_2O$$

In the presence of B_2H_6, methyl germyl sulphide, GeH_3SCH_3, decomposes at −196° to −79° to form GeH_4 and an unidentified solid material[294a]; with BF_3, GeH_3SCH_3 decomposes to give GeH_4 and a solid[294a].

9. Spectra of germanes and germane derivatives

The figures given in the following review of the literature on germanes and germane derivatives with at least one Ge–H bond denote references. Where necessary for comparative purposes, related germanium derivatives have been included as well (indicated by ⋆).

Abbreviations: *IR*, infrared spectra; *NMR*, nuclear magnetic resonance spectra; *RA*, Raman spectra; *MS*, mass spectra; *MW*, microwave spectra; *UV*, ultraviolet spectra; *STR*, structure.

Germanes Ge_nH_{2n+2}	*IR*	*NMR*	*RA*	*MS*
GeH_4	3, 33a, 165, 191, 192	80	—	1a, 208, 235
GeH_3D	169, 170	—	—	
GeH_2D_2	169	—	—	
$GeHD_3$	169, 170	—	—	
GeD_4	165, 169, 192	—	—	
Ge_2H_6	3, 41, 70, 71, 117, 135	75, 80	117	
Ge_2D_6	41	—	—	
⋆$(CH_3)_6Ge_2$	176c	—	25	
⋆$(CH_2{=}CH)_6Ge_2$	—	33		
Ge_3H_8	3, 74	74		
Ge_3H_7D	175	—		
Ge_4H_{10}	3, 74, 176a	74, 176a	176a	176a
n-Ge_4H_{10}	176a	176a	176a	176a
i-Ge_4H_{10}	176a	176a	176a	176a
Ge_5H_{12}	3, 74, 176a	74, 176a	176a	176a
n-Ge_5H_{12}	176a	176a	176a	176a
i-Ge_5H_{12}	176a	176a	176a	176a
neo-Ge_5H_{12}	176a	176a	176a	176a

Organogermanes	*IR*	*NMR*	*Other*
CH_3GeH_3	3, 135	5a, 80, 83, 84, 251a	*MW* 154; *RA* 222a
CD_3GeH_3	111	—	—
CH_3GeD_3	111	—	—
$C_2H_5GeH_3$	175a	—	*RA* 222a
$C_2H_5GeD_3$	175a	—	
$C_3H_7GeH_3$	—	83	*RA* 222a
$CF_3CH_2CH_2GeH_3$	—	83	
n-$C_4H_9GeH_3$	185, 186		
n-$C_5H_{11}GeH_3$	185		
n-$C_7H_{15}GeH_3$	185		
n-$C_8H_{17}GeH_3$	185		
$CH_2{=}CHGeH_3$	24		
$CH_3OCH_2GeH_3$	98a	98a	

References p. 711

(Contd.)

Organogermanes	*IR*	*NMR*	*Other*
$(CH_3)_2GeH_2$	3, 83, 84	251a	*RA* 222a
$(C_2H_5)_2GeH_2$	185, 186	83	*RA* 222a
$(n\text{-}C_3H_7)_2GeH_2$	—	83	
$(i\text{-}C_3H_7)_2GeH_2$	—	83	
$(n\text{-}C_4H_9)_2GeH_2$	185, 186		
$\overline{CH_2(CH_2)_3}GeH_2$	189		
$(n\text{-}C_5H_{11})_2GeH_2$	185, 186		
$\overline{CH_2(CH_2)_4}GeH_2$	189		
$(n\text{-}C_6H_{13})_2GeH_2$	185		
$(n\text{-}C_7H_{15})_2GeH_2$	185		
$(n\text{-}C_8H_{17})_2GeH_2$	185		
$(C_6H_5)_2GeH_2$	295		
$(C_6H_5)_2GeD_2$	—	—	*MS* 106a
$(C_6H_5CH_2)_2GeH_2$	43	—	—
$(C_6H_5CH_2)_2GeD_2$	43	—	—
$(CH_3)_3GeH$	185	83, 251a	—
$(CH_3)_2(C_2H_5)GeH$	—	83	—
*$(CH_3)_4Ge$	171, 300a	72, 77, 251	*RA* 171, 257, 258
$(C_2H_5)_3GeH$	185, 186	83	*MS* 106a
*$(C_2H_5)_4Ge$	222		
$(n\text{-}C_3H_7)_3GeH$	185	—	
*$(n\text{-}C_3H_7)_4Ge$	222		
$(i\text{-}C_3H_7)_3GeH$	185, 186	83	
$(n\text{-}C_4H_9)_3GeH$	185, 186		
$(n\text{-}C_4H_9)_2(CH_2{=}CHCH_2)GeH$	185 186		
$(C_4H_9)_2(HOCH_2CH_2CH_2)GeH$	164b		
$(n\text{-}C_4H_9)_2(CH_3COCH_2CH_2)GeH$	185		
$(n\text{-}C_4H_9)_2(NCCH_2CH_2)GeH$	185		
$(n\text{-}C_4H_9)_2(C_6H_5CH_2)GeH$	186		
$(n\text{-}C_7H_{15})(C_6H_5CH_2)GeH$	185		
$(C_6H_{11})_3GeH$	—	83	
$(n\text{-}C_4H_9)_2(C_6H_5)GeH$	185		
$(CH_3)(C_6H_5)_2GeH$	43		
$(C_4H_9)(C_6H_5)_2GeH$	43		
$(C_6H_5)_3GeH$	43, 78b 295		*RA* 78b
$(C_6H_5)_3GeD$	43, 78b		*RA* 78b
*$(C_6H_5)_4Ge$	121		
$(C_6H_5CH_2)_3GeD$	—	—	*MS* 106a
$(CH_3)H_2GeGeH_3$	175b	175b	
$(C_2H_5)H_2GeGeH_3$	175b	175b	
$(CH_3)_2HGeGeH_3$	97c		*RA* 97c
$(CH_3)H_2GeGeH_2(CH_3)$	97c		*RA* 97c
$(CH_3)_2HGeGeH_2(CH_3)$	97c		*RA* 97c
$(CH_3)_3SiGeH_3$	5a	78a	—

*For comparison

Halogenogermanes	*IR*	*NMR*	*Other*
GeH_3F	96, 116, 134, 135, 225, 255, 266	80	*MW* 114, 116
GeH_3Cl	96, 134, 135, 173, 225, 255	80, 177	*MW* 48, 187; dipole mom. 300a
GeD_3Cl	173	—	—
GeH_3Br	96, 116, 134, 135, 225, 255	80	*MW* 187
GeH_3I	76b, 96, 134, 225, 255	80	*RA* 76b
GeH_2F_2	265	80	—
GeH_2Cl_2	76d, 265	80	*RA* 76d, dipole mom. 300a
GeD_2Cl_2	76d	—	*RA* 76d
CH_3GeH_2Cl	3	—	—
GeH_2Br_2	40, 265	80	—
GeH_2I_2	40	—	—
$GeHCl_3$	168	83	*RA* 33a, 51, 222a
$GeDCl_3$	168		*RA* 222a
CH_3GeHCl_2	3		
$(C_2H_5)_2GeHCl$	162, 185, 186		
$(n\text{-}C_4H_9)_2GeHCl$	185, 186		
$(C_2H_5)_2GeHBr$	185, 186		
$(n\text{-}C_4H_9)_2GeHBr$	162, 185, 186		
$(C_2H_5)_2GeHI$	162, 185, 186		
$(n\text{-}C_4H_9)_2GeHI$	185, 186		
⋆$GeCl_4$	168	—	*RA* 47, 53, 114, 120, 207
⋆CH_3GeCl_2	14	—	*RA* 14
Ge_2H_5Cl	—	177	—
Ge_2H_5Br	—	177	—
Ge_2H_5I	176	177	—

⋆For comparison

Germanes with Ge–O and Ge–S bonds	*IR*	*NMR*	*RA*	*MS*	*UV*
GeH_3OCH_3	38b, 98a	98a	38b		
GeH_3OCD_3	38b	—	38b		
⋆$GeMe_3OSiMe_3$	—	251	—		
$(GeH_3)_2O$(*STR* 104a)	38b, 109	38b	—		
⋆$(GeMe_3)_2O$	—	—	—	—	44, 45, 46
$(GeH_3)SCH_3$	38b, 294a	—	38b, 294a	294a	
$(GeD_3)_2$(*STR* 104a)	109	—	—	—	
⋆$(GeMe_3)_2S$	—	—	—	—	44, 45, 46

⋆For comparison

References p. 711

Germanes with Ge–N, Ge–P, Ge–As and Ge–Sb bonds	*IR*	*NMR*	*Other*
$(GeH_3)_3N$	224b, 224c	224b, 224c	
GeH_3PH_2	76, 135, 176b	75(*cf.* 174)	*RA* 176b
GeH_3PD_2	176b	76a	*RA* 176b
GeD_3PH_2	176b	76a	*RA* 176b
GeD_3PD_2	176b		*RA* 176b
$GeH_3PH_2{\cdot}BH_3$	—	76a	
$(GeH_3)_3P$	39	39	*MS* 39, *STR* 224d
GeH_3AsH_2	76, 76a, 76c, 135	75	*RA* 76c
GeD_3AsH_2	76a, 76c	76b	*RA* 76c
GeH_3AsD_2	76c	76b	*RA* 76c
GeD_3AsD_2	76c		*RA* 76c
$(GeH_3)_2AsH$	79a	79a	*RA* 79a
$(GeH_3)_3As$	79a	79a	*RA* 79a
$(GeH_3)_3Sb$	79a	79a	*RA* 79a
⋆$(R_3Ge)_3P$	4a	—	*RA* 4a
⋆$(R_3Ge)_3As$	4a	—	*RA* 4a

⋆For comparison

Pseudohalogeno-germanes	*IR*	*NMR*	*Other*
GeH_3CN	107, 107a, 134, 135	—	—
GeD_3CN	107	—	—
GeH_3NCO	113, 224a	—	—
⋆$Ge(NCO)_4$	198	—	*RA* 198, *STR* 198
$(GeH_3)_2CN_2$	39c	39c	*RA* 39c
GeH_3N_3	39b	39b	*UV* 39b
GeH_3COOK	150a	—	*UV* 150a

⋆For comparison

Germanes with Ge–Si and Ge–B bonds	IR	NMR	MW
GeH_3SiH_3	135, 263	—	38a
$GeH_3Si(CH_3)_3$	5a	5a, 78a	—
$(GeH_3)_2Si(CH_3)_2$	—	78a	—
$(GeH_3)_3Si(CH_3)$	—	78a	—
$(GeH_3)_4Si$	—	78a	—
GeH_3SiF_2H		262a	
GeH_3SiF_3		262a	
$GeH_3SiF_2SiF_2H$		262a	
$GeH_3B[N(CH_3)_2]_2$	5g		
$(GeH_3)(CH_3)_2B_3N_3(CH_3)_3$	5f		
$(GeH_3)_2(CH_3)B_3N_3(CH_3)_3$	5f		
$(GeH_3)_3B_3N_3(CH_3)_3$	5f		

REFERENCES

1. Aggarwal, R. C. and M. Onyszchuk, Proc. Chem. Soc. *1962*, 20.
1a. Agafonov, I. L., G. G. Devyatykh, I. A. Frolov and N. V. Larin, Russ. J. Phys. Chem (Engl. Transl.) *36*, 737 (1962).
2. Amberger, E., Angew. Chem. *71*, 372 (1959).
3. Amberger, E., *Hydrides of Ge, Sn, Pb und Bi*, Thesis, University of München, 1961.
4. Amberger, E., unpublished.
4a. Amberger, E., R. Salazar and J. Honigschmid, J. Organometal. Chem., in the press.
5. Amberger, E. and H. Boeters, Angew. Chem. *73*, 114 (1961).
5a. Amberger, E. and E. Mühlhofer, J. Organometal. Chem. *12*, 55 (1968).
5b. Amberger, E., R. Römer and A. Layer, J. Organometal. Chem., *12*, 417 (1968).
5c. Amberger, E., E. Mühlhofer and H. Stern, J. Organometal Chem. *17*, P5 (1969).
5d. Amberger, E., W. Stoeger and J. Hönigschmid, J. Organometal Chem. *18*, 77 (1969).
5e. Amberger, E. and W. Stoeger, J. Organometal. Chem. *18*, 83 (1969).
5f. Amberger, E. and W. Stoeger, J. Organometal Chem. *17*, 287 (1969).
5g. Amberger, E. and R. Römer, Z. Naturforsch. *23b*, 559 (1968).
6. Anderson, H. H., J. Am. Chem. Soc. *78*, 1692 (1956).
7. Anderson, H. H., J. Am. Chem. Soc. *79*, 326 (1957).
8. Anderson, H. H., J. Am. Chem. Soc. *73*, 5800 (1951).
9. Anderson, H. H., J. Am. Chem. Soc. *83*, 547 (1961).
10. Anderson, H. H., J. Am. Chem. Soc. *75*, 814 (1953).
11. Anderson, H. H., J. Am. Chem. Soc. *82*, 3016 (1960).
12. Anderson, H. H., J. Am. Chem. Soc. *80*, 5083 (1958).
13. Andrews, T. D. and C. S. G. Phillips, J. Chem. Soc. *1966*, 46.
13a. Antipin, L. M., E. S. Sobolev and V. F. Mironov, Russ. J. Inorg. Chem. (Engl. Transl.) *13*, 162 (1968).
14. Aronson, J. R. and J. R. Durig, Spectrochim. Acta *20*, 219 (1964).
15. Barrow, R. F. and J. L. Deutsch, Proc. Chem. Soc. *1960*, 122.
16. Bauer, H. and K. Burschkies, Ber. *66*, 277 (1933).
17. Baukov, Yu. I. and I. F. Lutsenko, J. Gen. Chem. USSR (Eng. Transl.) *32*, 2705 (1962).
17a. Bellama, J. M. and A. G. MacDiarmid, Inorg. Chem. *7*, 2070 (1968).
18. Bond, P. A. and E. B. Crone, J. Am. Chem. Soc. *56*, 2028 (1934).
19. Booth, H. S. and W. C. Morris, J. Am. Chem. Soc. *58*, 90 (1936).
20. Borer, K. and C. S. G. Phillips, Proc. Chem. Soc. *1959*, 189.
20a. Bornhorst, W. R. and M. A. Ring, Inorg. Chem. *7*, 1009 (1968).
21. Brewer, F. M. and L. M. Dennis, J. Phys. Chem. *31*, 1528, 1531 (1927).
22. Brewer, F. M. and L. M. Dennis, J. Phys. Chem. *31*, 1101 (1927).
23. Brewer, F. M., J. Phys. Chem. *31*, 1821 (1927).

24. BRINCKMAN, F. E. AND F. G. A. STONE, J. Inorg. Nucl. Chem. *11*, 24 (1959).
25. BROWN, M. P., E. CARTMELL AND G. W. A. FOWLES, J. Chem. Soc. *1960*, 506.
26. BROOK, A. G. AND G. J. D. PEDDLE, J. Am. Chem. Soc. *85*, 1869 (1963).
27. BROOK, A. G. AND G. J. D. PEDDLE, J. Am. Chem. Soc. *85*, 2338 (1963).
27a. BROOKS, E. H. AND F. GLOCKLING, J. Chem. Soc. *A1967*, 1030.
28. BROWN, M. P. AND G. W. A. FOWLES, J. Chem. Soc. *1958*, 2811.
29. BULTEN, E. J. AND J. G. NOLTES, Tetrahedron Letters *1966*, 3471.
30. BULTEN, E. J. AND J. G. NOLTES, Tetrahedron Letters *1966*, 4389.
31. BULTEN, E. J. AND J. G. NOLTES, Tetrahedron Letters *1967*, 1443.
31a. BULTEN, E. J., Thesis, Utrecht, 1969.
32. CALAS, R., E. FRAINNET AND J. BONASTRE, Compt. Rend. *251*, 2987 (1960).
33. CAWLEY, S. AND S. DANYLUK, Can. J. Chem. *41*, 1850 (1963).
33a. CHUMAEVSKII, N. A., Russ. Chem. Rev. (Engl. Transl.), *32*, 509 (1963).
34. CLARK, H. C. AND C. J. WILLIS, J. Am. Chem. Soc. *84*, 898 (1962).
35. CLUSIUS, K. AND C. FABER, Angew, Chem. *55*, 97 (1942).
36. CLUSIUS, K. AND C. FABER, Z. Phys. Chem. *B51*, 352 (1942).
37. CLUSIUS, K. AND C. FABER, Naturwiss. *29*, 468 (1941).
38. COREY, R. B., A. W. LAUBENGAYER AND L. M. DENNIS, J. Am. Chem. Soc. *47*, 112 (1925).
38a. COX, A. P. AND R. VARMA, J. Chem. Phys. *46*, 2007 (1967).
38b. CRADOCK, S., J. Chem. Soc. *A1968*, 1426.
39. CRADOCK, S., G. DAVIDSON, E. A. V. EBSWORTH AND L. A. WOODWARD, Chem. Commun. *1965*, 515.
39a. CRADOCK, S. AND E. A. V. EBSWORTH, J. Chem. Soc. *A1967*, 1226.
39b. CRADOCK, S. AND E. A. V. EBSWORTH, J. Chem. Soc. *A1968*, 1420.
39c. CRADOCK, S. AND E. A. V. EBSWORTH, J. Chem. Soc. *A1968*, 1423.
39d. CRADOCK, S., E. A. V. EBSWORTH, G. DAVIDSON AND L. A. WOODWARD, J. Chem. Soc. *A1967*, 1229.
40. CRADOCK, S. AND E. A. V. EBSWORTH, J. Chem. Soc. *A1967*, 12.
40a. CRADOCK, S., G. A. GIBBON AND C. H. VAN DYKE, Inorg. Chem. *6*, 1751 (1967).
41. CRAFORD, V. A., K. H. RHEE AND M. K. WILSON, J. Chem. Phys. *37*, 2377 (1962).
41a. CREEMERS, H. M. J. C. AND J. G. NOLTES, J. Organometal. Chem. *7*, 237 (1967).
42. CROSS, R. J. AND F. GLOCKLING, Proc. Chem. Soc. *1964*, 143.
43. CROSS, R. J. AND F. GLOCKLING, J. Organometal. Chem. *3*, 146 (1965).
44. CUMPER, C. W. N., A. MELNIKOFF AND A. I. VOGEL, J. Chem. Soc. *A1966*, 242.
45. CUMPER, C. W. N., A. MELNIKOFF AND A. I. VOGEL, J. Chem. Soc. *A1966*, 246.
46. CUMPER, C. W. N., A. MELNIKOFF AND A. I. VOGEL, J. Chem. Soc. *A1966*, 323.
47. CYVIN, S. J., J. Mol. Spectry. *6*, 338 (1961).
48. DAILEY, B. P., J. M. MAYS AND C. H. TOWNES, Phys. Rev. *76*, 136 (1949).
49. DAVY, H., Phil. Trans. Roy. Soc. London *1812*, 405.
50. DEANS, D. R. AND C. EABORN, J. Chem. Soc. *1954*, 3169.
51. DELWAULLE, M. -L. AND F. FRANCOIS, Compt. Rend. *228*, 1007 (1949).
52. DELWAULLE, M. -L. AND F. FRANCOIS, Compt. Rend. *230*, 743 (1950).
53. DELWAULLE, M. -L., F. FRANCOIS, M. B. DELHAYE-BUISSET AND M. DELHAYE, J. Phys. Radium *15*, 206 (1954).
54. DENNIS, L. M., Z. Anorg. Allgem. Chem. *174*, 97 (1928).
55. DENNIS, L. M., R. B. COREY AND R. W. MOORE, J. Am. Chem. Soc. *46*, 657 (1924).
56. DENNIS, L. M. AND F. E. HANCE, J. Am. Chem. Soc. *44*, 299 (1922).
57. DENNIS, L. M. AND F. E. HANCE, Z. Anorg. Allgem. Chem. *122*, 256 (1922).
58. DENNIS, L. M. AND F. E. HANCE, J. Am. Chem. Soc. *44*, 2854 (1922).
59. DENNIS, L. M. AND F. E. HANCE, Z. Anorg. Allgem. Chem. *122*, 276 (1922).
60. DENNIS, L. M. AND F. E. HANCE, J. Am. Chem. Soc. *44*, 2859 (1922).
61. DENNIS, L. M. AND R. E. HULSE, J. Am. Chem. Soc. *52*, 3553 (1930).
62. DENNIS, L. M. AND P. R. JUDY, J. Am. Chem. Soc. *51*, 2321 (1929).
63. DENNIS, L. M. AND A. W. LAUBENGAYER, Z. Physik. Chem. (Leipzig) *130*, 528 (1927).
64. DENNIS, L. M., W. R. ORNDORFF AND D. L. TABERN, J. Phys. Chem. *30*, 1052 (1926).
65. DENNIS, L. M. AND N. A. SKOW, J. Am. Chem. Soc. *52*, 2369 (1930).
66. DENNIS, L. M. AND R. W. WORK, J. Am. Chem. Soc. *55*, 4486 (1933).
67. DEVYATYKH, G. G. AND I. A. FROLOV, Russ. J. Inorg. Chem. (Engl. Transl.) *8*, 133 (1963).
68. DEVYATYKH, G. G. AND I. A. FROLOV, Russ. J. Inorg. Chem. (Engl. Transl.) *11*, 385 (1966).

69. Devyatykh, G. G., I. A. Frolov and N. Kh. Aglyulov, Russ. J. Inorg. Chem. (Engl. Transl.) *11*, 389 (1966).
69a. Devyatykh, G. G. and A. S. Yushin, Russ. J. Phys. Chem. (Engl. Transl.) *37*, 517 (1964).
70. Dows, D. A. and R. M. Hexter, J. Chem. Phys. *24*, 1029 (1956).
71. Dows, D. A. and R. M. Hexter, J. Chem. Phys. *24*, 1117 (1956).
72. Drago, R. S. and N. A. Matwiyoff, J. Organometal. Chem. *3*, 62 (1965).
73. Drake, J. E. and W. L. Jolly, Proc. Chem. Soc. *1961*, 379.
74. Drake, J. E. and W. L. Jolly, J. Chem. Soc. *1962*, 2807.
75. Drake, J. E. and W. L. Jolly, Chem. Ind. (London) *1962*, 1033.
76. Drake, J. E. and W. L. Jolly, Chem. Ind. (London) *1962*, 1470.
76a. Drake, J. E. and C. Riddle, J. Chem. Soc. *A 1968*, 1675.
76b. Drake, J. E. and C. Riddle, J. Chem. Soc. *A 1968*, 2452.
76c. Drake, J. E., C. Riddle, K. M. Mackay, S. R. Stobart and K. J. Sutton, Spectrochim. Acta *25A*, 941 (1969).
76d. Drake, J. E., C. Riddle and D. E. Rogers, J. Chem. Soc. *A1969*, 910.
77. Dreeskamp, H., Z. Naturforsch. *19a*, 139 (1964).
77a. Dreyfuss, R. M. and W. L. Jolly, Inorg. Chem. *7*, 2645 (1968).
78. Duffaut, N., R. Calas and B. Martel, Bull. Soc. Chim. France *1960*, 569.
78a. Dutton, W. A. and M. Onyszchuk, Inorg. Chem. *7*, 1735 (1968).
78b. Durig, J. R., C. W. Sink and J. B. Turner, Spectrochim. Acta *25A*, 629 (1969).
79. Ebsworth, E. A. V., Chem. Commun. *1966*, 530.
79a. Ebsworth, E. A. V., D. W. H. Rankin and G. M. Sheldrick, J. Chem. Soc. *A1968*, 2828.
80. Ebsworth, E. A. V., S. G. Frankiss and A. G. Robiette, J. Mol. Spectry. *12*, 299 (1964).
81. Eckerlin, P. and E. Wölfel, Z. Anorg. Allgem. Chem. *280*, 321 (1955).
82. Eckerlin, P., H. J. Meyer and E. Wölfel, Z. Anorg. Allgem. Chem. *281*, 322 (1955).
83. Egorochkin, A. N., M. L. Chidekel', V. A. Ponomarenko, G. Ya. Zueva, S. S. Svirezheva and G. A. Razuvaev, Izv. Akad. Nauk SSSR, Ser. Khim. *1963*, 1865.
84. Egorochkin, A. N., M. L. Chidekel', V. A. Ponomarenko, G. Ya. Zueva and G. A. Razuvaev, Izv. Akad. Nauk SSSR, Ser. Khim. *1964*, 373.
85. Emeléus, H. J. and E. R. Gardiner, J. Chem. Soc. *1938*, 1900.
86. Emeléus, H. J. and H. H. G. Jellinek, Trans. Faraday Soc. *40*, 93 (1944).
87. Emeléus, H. J. and K. M. MacKay, J. Chem. Soc. *1961*, 2676.
88. English, W. D. and R. V. V. Nicholls, J. Am. Chem. Soc. *72*, 2764 (1950).
89. Fehér, F. and J. Cremer, Z. Anorg. Allgem. Chem. *297*, 14 (1958).
90. Fensham, P. J., K. Tamaru, M. Boudart and H. Taylor, J. Phys. Chem. *59*, 806 (1955).
91. Finholt, A. E., Nucl. Sci. Abstr. *6*, 617 (1952).
92. Finholt, A. E., A. C. Bond Jr., K. E. Wilzbach and H. I. Schlesinger, J. Am. Chem. Soc. *69*, 2692 (1947).
93. Fischer, W. and W. Weidemann, Z. Anorg. Allgem. Chem. *213*, 106 (1933).
94. Fischer, W., Z. Anorg. Allgem. Chem. *211*, 321 (1933).
95. Frainnet, E. and C. Fritsch, Bull. Soc. Chim. *1960*, 596.
96. Freeman, D. E., K. H. Rhee and M. K. Wilson, J. Am. Chem. Soc. *39*, 2908 (1963).
97. Fuchs, R. and H. Gilman, J. Org. Chem. *23*, 911 (1958).
97a. Garrity, S. P. and M. A. Ring, Inorg. Nucl. Chem. Letters *4*, 77 (1968).
97b. Gaspar, P. P., C. A. Levy, J. J. Frost and S. A. Bock, J. Am. Chem. Soc. *91*, 1574 (1969).
97c. George, R. D. and K. M. MacKay, J. Chem. Soc. *1969*, 2122.
98. Gibbon, G. A., Y. Rousseau, C. H. van Dyke and G. J. Mains, Inorg. Chem. *5*, 114 (1966).
98a. Gibbon, G. A., J. I. Wang and C. H. van Dyke, Inorg. Chem. *6*, 1989 (1967).
99. Gilman, H. and W. Gerow, J. Am. Chem. Soc. *77*, 5509, 5740 (1955).
100. Gilman, H. and W. Gerow, J. Am. Chem. Soc. *78*, 5435 (1956).
101. Gilman, H. and W. Gerow, J. Am. Chem. Soc. *77*, 5740 (1955).
102. Gilman, H. and S. P. Massie, J. Am. Chem. Soc. *68*, 1128 (1946).
103. Gilman, H., D. J. Peterson and D. Wittenberg, Chem. Ind. (London) *1958*, 1479.
104. Glarum, N. S. and C. A. Kraus, J. Am. Chem. Soc. *72*, 5398 (1950).
104a. Glidewell, C., D. W. H. Rankin, A. G. Robiette, G. M. Sheldrick, S. Craddock, E. A. V. Ebsworth and B. Beagley, Inorg. Nucl. Chem. Letters *5*, 417 (1969).
105. Glockling, F. and K. A. Hooton, J. Chem. Soc. *1962*, 3509.
106. Glockling, F. and K. A. Hooton, J. Chem. Soc. *1963*, 1849.
106a. Glockling, F. and F. R. C. Light, J. Chem. Soc. *A1968*, 717.

106b. Glockling, F. and M. D. Wilbey, Chem. Commun. *1969*, 286.
107. Goldfarb, Th. D., J. Chem. Phys. *37*, 642 (1962).
107a. Goldfarb, Th. D. and B. P. Zafonte, J. Chem. Phys. *41*, 3653 (1965).
108. Gokhale, S. D., J. E. Drake and W. L. Jolly, J. Inorg. Nucl. Chem. *27*, 1911 (1965).
109. Goldfarb, Th. D. and S. Sujishi, J. Am. Chem. Soc. *86*, 1679 (1964).
109a. Grant, M. W. and R. H. Prince, J. Chem. Soc. *A1969*, 1139.
110. Green, M. and P. J. Robinson, J. Phys. Chem. *57*, 938 (1953).
111. Griffiths, J. E., J. Chem. Phys. *38*, 2879 (1963).
112. Griffiths, J. E., Inorg. Chem. *2*, 375 (1963).
113. Griffiths, J. E., and A. L. Beach, Chem. Commun. *1965*, 437.
114. Griffiths, J. E. and K. B. McAffee Jr., Proc. Chem. Soc. *1961*, 456.
115. Griffiths, J. E. and M. Onyszchuk, Can. J. Chem. *39*, 339 (1961).
116. Griffiths, J. E., T. N. Srivastava and M. Onyszchuk, Can. J. Chem. *40*, 579 (1962).
117. Griffiths, J. E. and G. E. Walrafen, J. Chem. Phys. *40*, 321 (1964).
118. Gunn, S. R., *Symposium on Hydrides of Groups 4, 5, and 6,* Berkeley, California, U.S.A., 1963.
119. Gunn, S. R. and L. G. Green, J. Phys. Chem. *65*, 779 (1961).
120. Haun, R. R. and W. D. Harkins, J. Am. Chem. Soc. *54*, 3917 (1932).
121. Henry, M. C. and J. G. Noltes, J. Am. Chem. Soc. *82*, 555 (1960).
122. Heukelom, W., Rec. Trav. Chim. *68*, 661 (1949).
123. Hogness, T. R. and W. C. Johnson, J. Am. Chem. Soc. *54*, 3583 (1932).
124. Hohmann, E., Z. Anorg. Allgem. Chem. *257*, 113 (1948).
125. Ingham, R. K., S. D. Rosenberg and H. Gilman, Chem. Rev. *60*, 459 (1960).
126. Jenkins, J. W. and H. W. Post, J. Org. Chem. *15*, 556 (1950).
127. Johnson, O. H., Chem. Rev. *48*, 259 (1951).
128. Johnson, O. H. and D. M. Harris, J. Am. Chem. Soc. *72*, 5566 (1950).
129. Johnson, O. H. and D. M. Harris, J. Am. Chem. Soc. *72*, 5564 (1950).
130. Johnson, O. H. and L. V. Jones, J. Org. Chem. *17*, 1172 (1952).
131. Johnson, O. H. and W. H. Nebergall, J. Am. Chem. Soc. *71*, 1720 (1949).
132. Johnson, O. H. and E. A. Schmal, J. Am. Chem. Soc. *80*, 2931 (1958).
133. Jolly, W. L., J. Am. Chem. Soc. *83*, 335 (1961).
133a. Jolly, W. L. Inorg. Chem. *6*, 1435 (1967).
134. Jolly, W. L., J. Am. Chem. Soc. *85*, 3083 (1963).
135. Jolly, W. L., UCRL-10733.
136. Jolly, W. L. and W. M. Latimer, J. Am. Chem. Soc. *74*, 5752, 5754 (1952).
137. Jolly, W. L., C. B. Lindahl and R. W. Kopp, Inorg. Chem. *1*, 958 (1962).
138. Jones, L. V., Dissertation Abstr. *13*, 308 (1953); Ann. Chim. (Paris) *6*, 519 (1961).
139. Kelen, G. P. van der, and D. F. van der Vondel, Bull. Soc. Chim. Belges, 504 (1960).
140. Kelen, G. P. van der, and J. G. A. Luijten, J. Appl. Chem. *6*, 93 (1956).
141. Kettering, C. F. and W. W. Sleator, Physica *4*, 39 (1933).
142. Kofel, M. B. and R. A. Lad, J. Chem.Phys. *16*, 420 (1948).
143. Kramer, K. A. W. and A. N. Wright, J. Chem. Soc. *1963*, 3604.
144. Kraus, C. A., J. Chem. Educ. *29*, 417 (1952).
145. Kruas, C. A., J. Chem. Educ. *29*, 488 (1952).
146. Kraus, C. A. and Brown, J. Am. Chem. Soc. *52*, 4031 (1930).
147. Kraus, C. A. and E. S. Carney, J. Am. Chem. Soc. *56*, 765 (1934).
148. Kraus, C. A. and E. A. Flood, J. Am. Chem. Soc. *54*, 1635 (1932).
149. Kraus, C. A. and L. S. Forster, J. Am. Chem. Soc. *49*, 457 (1927).
150. Kruglaya, O. A., N. S. Vyazankin and G. A. Razuvaev, J. Gen. Chem. USSR (Engl. Transl.) *35*, 392 (1965).
150a. Kuznesof, P. M. and W. L. Jolly, Inorg. Chem. *7*, 2574 (1968).
151. Labarre, J. -F., M. Massol and J. Satgé, Bull. Soc. Chim. France *1967*, 736.
152. Laubengayer, A. W. and P. L. Brandt, J. Am. Chem. Soc. *54*, 621 (1932).
153. Laubengayer, A. W. and D. L. Tabern, J. Phys. Chem. *30*, 1080 (1926).
154. Laurie, V. W., J. Chem. Phys. *30*, 1210 (1959).
155. Law, J. T., J. Phys. Chem. *59*, 543 (1955).
156. Lengel, J. H. and V. H. Dibeler, J. Am. Chem. Soc. *74*, 2683 (1952).
157. Lesbre, M. and R. Buisson, Bull. Soc. Chim. France *1957*, 1204.
158. Lesbre, M. and P. Mazerolles, Compt. Rend. *246*, 1708 (1958).
159. Lesbre, M. and J. Satgé, Compt. Rend. *250*, 2220 (1960).
160. Lesbre, M. and J. Satgé, Compt. Rend. *254*, 4021 (1962).

161. Lesbre, M. and J. Satgé, Compt. Rend. *247*, 471 (1958).
162. Lesbre, M. and J. Satgé, Compt. Rend. *252*, 1976 (1961).
163. Lesbre, M. and J. Satgé, Compt. Rend. *254*, 1453 (1962).
164. Lesbre, M., J. Satgé and M. Massol, Compt. Rend. *256*, 1548 (1963).
164a. Lesbre, M., J. Satgé and M. Massol, Compt. Rend. *257*, 2665 (1963).
164b. Lesbre, M., J. Satgé and M. Massol, Compt. Rend. *258*, 2842 (1964).
165. Levin, I. R., J. Chem. Phys. *42*, 1244 (1965).
166. Lide Jr., D. R., J. Chem. Phys. *19*, 1605 (1951).
167. Lide Jr., D. R. and D. K. Coles, Phys. Rev. *80*, 911 (1950).
168. Lindeman, L. P. and M. K. Wilson, Spectrochim. Acta *9*, 47 (1957).
169. Lindeman, L. P. and M. K. Wilson, Z. Physik. Chem. (Frankfurt) *9*, 29 (1956).
170. Lindeman, L. P. and M. K. Wilson, J. Chem. Phys. *22*, 1723 (1954).
171. Lippincott, E. R. and M. C. Tobin, J. Am. Chem. Soc. *75*, 4141 (1953).
172. Lister, M. W. and L. E. Sutton, Trans. Faraday Soc. *37*, 393 (1941).
173. Lord, R. C. and C. M. Steese, J. Chem. Phys. *22*, 542 (1954).
174. Lynden-Bell, R. M., Trans. Faraday Soc. *57*, 888 (1961).
175. MacKay, K. M. and P. Robinson, J. Chem. Soc. *1965*, 5121.
175a. MacKay, K. M. and R. Watts, Spectrochim. Acta *23A*, 2761 (1967).
175b. MacKay, K. M., R. D. George, P. Robinson and R. Watts, J. Chem. Soc. *A1968*, 1920.
176. MacKay, K. M. and P. J. Roebuck, J. Chem. Soc. *1964*, 1195.
176a. MacKay, K. M. and K. J. Sutton, J. Chem. Soc. *1968*, 2312.
176b. MacKay, K. M., K. J. Sutton, S. R. Stobart, J. E. Drake and C. R. Riddle, Spectrochim. Acta *25A*, 925 (1969).
176c. MacKay, K. M., D. B. Sowerby and W. C. Young, Spectrochim. Acta *24A*, 611 (1968).
177. MacKay, K. M., P. Robinson, E. J. Spanier and A. G. MacDiarmid, J. Inorg. Nucl. Chem. *28*, 1377 (1966).
178. Mahncke, H. E. and W. A. Noyes, J. Am. Chem. Soc. *57*, 456 (1935).
179. Macklen, E. D., J. Chem. Soc. *1959*, 1984.
180. Macklen, E. D., J. Chem. Soc. *1959*, 1989.
181. Massey A. G., A. J. Park and F. G. A. Stone, J. Am. Chem. Soc. *85*, 2021 (1963).
182. Massol, M., *Contribution à l'Etude d'Hydrures Organogermaniques: les Alcoylhalogenogermanes* [Contribution to the study of organogermanium hydrides: the alkylhalogenogermanes], Doctoral Thesis, University of Toulouse, 1967.
183. Massol, M. and J. Satgé, Bull. Soc. Chim. France *1966*, 2737.
183a. Massol, M., J. Satgé and M. Lesbre, J. Organometal. Chem. *17*, 25 (1969).
184. Massol, M., J. Satgé and M. Lesbre, Compt. Rend. *262*, 1806 (1966).
185. Mathis, R., J. Satgé and F. Mathis, Spectrochim. Acta *18*, 1463 (1962).
186. Mathis, R., M. Constant, J. Satgé and F. Mathis, Spectrochim. Acta *20*, 515 (1964).
187. Mays, J. M. and B. P. Dailey, J. Chem. Phys. *20*, 1695 (1952).
188. Mazerolles, P., Bull. Soc. Chim. France *1960*, 856.
189. Mazerolles, P., Bull. Soc. Chim. France *1962*, 1907.
190. McDowell, C. A. and J. W. Warren, Discussions Faraday Soc. *10*, 53 (1951).
191. McKean, D. C., Chem. Commun. *1966*, 147.
192. McKean, D. C. and A. A. Chalmers, Spectrochim. Acta *23A*, 777 (1967).
193. Meals, R. N., J. Am. Chem. Soc. *68*, 1880 (1946).
194. Mendelson, J.-C., F. Métras and J. Valade, Compt. Rend. *261*, 756 (1965).
195. Mévegnies, M. N., J. M. Delfosse, Ann. Soc. Sci. Bruxelles, Ser. I. *64*, 188 (1950).
196. Mikhailov, B. M., Yu. N. Bubnov and V. G. Kiselev, Bull. Acad. Sci. USSR, Div. Chem. Sci. (Engl. Transl.) *1965*, 58.
197. Miller, J. G., J. Am. Chem. Soc. *56*, 2360 (1934).
198. Miller, F. A. and G. L. Carlson, Spectrochim. Acta *17*, 977 (1961).
199. Mironov, V. F. and N. G. Dzhurinskaya, Bull. Acad. Sci. USSR, Div. Chem. Sci. (Engl. Transl.) *1963*, 66.
200. Mironov, V. F., N. G. Dzhurinskaya, T. K. Gar and A. D. Petrov, Izv. Akad. Nauk SSSR, Otd. Tekhn. Nauk *1962*, 460; C. A. *57*, 15138b (1962).
201. Mironov, V. F. and T. K. Gar, Bull. Acad. Sci. USSR, Div. Chem. Sci. (Engl. Transl.) *1964*, 1420.
201a. Mironov, V. F. and T. K. Gar, Organometal. Chem. Rev. *A3*, 311 (1968).
202. Morrison, R. T., Abstr. 137th Meeting Am. Chem. Soc., Cleveland, 1960.
203. Morrison, G. H., E. G. Dorfman and J. F. Cosgrove, J. Am. Chem. Soc. *76*, 4236 (1954).

203a. MOULTON, C. W. AND J. G. MILLER, J. Am. Chem. Soc. *78*, 2702 (1956).
204. NEBERGALL, W. H. AND R. H. WALSH, J. Am. Chem. Soc. *73*, 4043 (1951).
205. NEFEDOV, O. M. AND S. P. KOLESNIKOV, Bull. Acad. Sci. USSR, Div. Chem. Sci. (Engl. Transl.) *1966*, 187.
206. NESMEYANOV, A. N., K. N. ANISIMOV, N. E. KOLOBOVA AND A. B. ANTONOVA, Bull. Acad. Sci. Div. Chem. Sci. (Engl. Transl.) *1965*, 1284.
207. NEU, J. T. AND W. D. GWINN, J. Am. Chem. Soc. *70*, 3463 (1948).
208. NEUERT, H. AND H. CLASEN, Z. Naturforsch. *7a*, 410 (1952).
209. NEUMANN, W. P., H. NIERMANN AND R. SOMMER, Angew. Chem. *73*, 768 (1961).
210. NEUMANN, W. P., H. NIERMANN AND R. SOMMER, Ann. Chem. *659*, 27 (1962).
211. NEUMANN, W. P., Angew. Chem. *75*, 230 (1963).
212. NOLTES, J. G., Thesis, Utrecht, 1958; Ann. Chim. (Paris) *6*, 519 (1961).
213. ONYSZCHUK, M., *Symposium on Hydrides of Groups 4, 5 and 6*, Berkeley, California, USA, (1963).
214. PANETH, F. AND K. FÜRTH, Ber. *52*, 2020 (1919).
215. PANETH, F., W. HAKEN AND E. RABINOWITSCH, Ber. *57*, 1898 (1924).
216. PANETH, F. AND O. NÖRRING, Ber. *53*, 1693 (1920).
217. PANETH, F., E. RABINOWITSCH AND W. HAKEN, Ber. *58*, 1143 (1925).
218. PANETH, F. AND F. WINTERNITZ, Ber. *51*, 1728 (1918).
219. PEARSON, T. G. AND P. L. ROBINSON, J. Chem. Soc. *1934*, 736.
220. PFAFF, C. H., Poggendorff's Ann. *42*, 339 (1837).
221. PIPER, T. S. AND M. K. WILSON, J. Inorg. Nucl. Chem. *4*, 22 (1957).
222. PONOMARENKO, V. A., L. A. LEITES, YU. P. EGOROV, G. YA. ZUEVA, Izv. Akad. Nauk SSSR, Otd. Khim. Nauk *1961* , 2123; Bull. Acad. Sci. USSR, Div. Chem. Sci. (Engl. Transl.) *1961* , 1993.
222a. PONOMARENKO, V. A., G. YA. ZUEVA AND N. S. ANDREEV, Bull. Akad. Sci. USSR, Div. Chem. Sci. (Engl. Transl.) *1961*, 1939.
223. PONOMARENKO, V. A., G. YA. VZENKOVA AND YU. P. EGOROV, Dokl. Akad. Nauk SSSR *122*, 405 (1958); C. A. *53*, 112 d (1959).
224. PRIESTLEY, J., cited in *A History of Chemistry* by J. R. PARTINGTON, Vol. 3, Macmillan, London, 1962, p. 265.
224a. RAMAPRASAD, K. P., R. VARMA AND R. NELSON, F. S. C., J. Am. Chem. Soc. *90*, 62 47(1968).
224b. RANKIN, D. W. H., Chem. Commun. *1969*, 194.
224c. RANKIN, D. W. H., J. Chem. Soc. *A1969*, 1926.
224d. RANKIN, D. W. H., A. G. ROBIETTE, G. M. SHELDRICK, B. BEAGLEY AND T. G. HEWITT, J. Inorg. Nucl. Chem. *31*, 2351 (1969).
225. RHEE, K. H. AND M. K. WILSON, J. Chem. Phys. *43*, 333 (1965).
226. RICK, C. E., T. D. MCKINLEY AND J. N. TULLY, cit. in Gmelin, Handbuch der anorganischen Chemie. Ergänzungsband Ge, p. 521.
227. RIEMENSCHNEIDER, R., K. MENGE AND P. KLANG, Z. Naturforsch. *11b*, 115 (1956).
227a. RIJKENS, F. AND G. J. M. VAN DER KERK, *Investigations in the Field of Organogermanium Chemistry*, Germanium Research Committee, Schotanus and Jens, Utrecht N. V., Utrecht, 1964.
228. ROMEYN, H. AND W. A. NOYES, J. Am. Chem. Soc. *54*, 4143 (1932).
229. ROTH, W. A. AND O. SCHWARTZ, Z. Physik. Chem. (Leipzig) *134*, 466 (1928).
230. ROYEN, P., Z. Anorg. Allgem. Chem. *235*, 324 (1938).
231. ROYEN, P. AND C. ROCKTÄSCHEL, Z. Anorg. Allgem. Chem. *346*, 279 (1966).
232. ROYEN, P. AND C. ROCKTÄSCHEL Z. Anorg. Allgem. Chem. *346*, 290 (1966).
233. ROYEN, P. AND R. SCHWARZ, Z. Anorg. Allgem. Chem. *215*, 295 (1933).
234. ROYEN, P. AND R. SCHWARZ, Z. Anorg. Allgem. Chem. *211*, 412 (1933).
234a. RUSTAD, D. S. AND W. L. JOLLY, Inorg. Chem. *6*, 1986 (1967).
234b. RUSTAD, D. S. AND W. L. JOLLY, Inorg. Chem. *7*, 213 (1968).
235. SAALFELD, F. E. AND H. J. SVEC, Inorg. Chem. *2*, 46 (1963).
236. SAALFELD, F. E. AND H. J. SVEC, Inorg. Chem. *2*, 50 (1963).
237. SATGÉ, J., Ann. Chim. (Paris) *6*, 519 (1961).
238. SATGÉ, J., Bull. Soc. Chim. France *1964*, 630.
239. SATGÉ, J., AND M. BAUDET, Compt. Rend. *263*, 435 (1966).
240. SATGÉ, J. AND M. LESBRE, Bull. Soc. Chim. France *1965*, 2578.
241. SATGÉ, J., M. LESBRE AND M. BAUDET, Compt. Rend. *259*, 4733 (1964).
242. SATGÉ, J. AND M. MASSOL, Compt. Rend. *261*, 170 (1965).
243. SATGÉ, J., M. MASSOL AND M. LESBRE, J. Organometal. Chem. *5*, 241 (1966).

244. Satgé, J., R. Mathis-Noel and M. Lesbre, Compt. Rend. *249*, 131 (1959).
245. Satgé, J. and P. R. Vière, Bull. Soc. Chim. France *1966*, 1773.
246. Schäfer, K. and J. M. Gonzales Barredo, Z. Physik. Chem. (Leipzig) *193*, 334 (1944).
247. Schäfer, R. and W. Klemm, J. Prakt. Chem. *5*, 233 (1958).
248. Scheele, C. W., Svenska Akad. Handl. *36*, 263 (1775).
249. Schenck, R. and A. Imker, Ber. *58*, 271 (1925).
250. Schenck, R. and A. Imker, Rec. Trav. Chim. *41*, 569 (1922).
251. Schmidbaur, H., J. Am. Chem. Soc. *85*, 2336 (1963).
251a. Schmidbaur, H., Chem. Ber. *97*, 1639 (1964).
252. Schmidt, M. and I. Ruidisch, Z. Anorg. Allgem. Chem. *311*, 331 (1961).
253. Schott, V. G. and C. Harzdorf, Z. Anorg. Allgem. Chem. *307*, 105 (1960).
254. Seyferth, D. and J. M. Burlitch, J. Am. Chem. Soc. *85*, 2667 (1963).
254a. Seyferth, D., J. M. Burlitch, H. Dertouzos and H. D. Simmons Jr., J. Organometal. Chem. *7*, 405 (1967).
255. Seyferth, D. and E. G. Rochow, J. Am. Chem. Soc. *77*, 907 (1955).
256. Sidgwick, N. V. and A. W. Laubengayer, J. Am. Chem. Soc. *54*, 948 (1932).
257. Siebert, H., Z. Anorg. Allgem. Chem. *263*, 82 (1950).
258. Siebert, H., Z. Anorg. Allgem. Chem. *268*, 177 (1952).
259. Sisler, H. H., H. H. Batey, B. Pfahler and R. Mattair, J. Am. Chem. Soc. *70*, 3821 (1948).
260. Sisler, H. H., W. Wilson, B. J. Gibbon, H. H. Batey, B. Pfahler and R. Mattair, J. Am. Chem. Soc. *70*, 381 (1948).
261. Smyth, C. P., J. Am. Chem. Soc. *63*, 57 (1941).
262. Smyth, C. P., A. J. Grosman and S. R. Ginsburg, J. Am. Chem. Soc. *62*, 192 (1940).
262a. Solan, D. and P. L. Timms, Inorg. Chem. *7*, 2157 (1968).
263. Spanier, E. J. and A. G. MacDiarmid, Inorg. Chem. *2*, 215 (1963).
264. Srivastava, T. N., J. E. Griffiths and M. Onyszchuk, Can. J. Chem. *40*, 739 (1962).
265. Srivastava, T. N., J. E. Griffiths and M. Onyszchuk, Can. J. Chem. *41*, 2101 (1963).
266. Srivastava, T. N. and M. Onyszchuk, Proc. Chem. Soc. *1961*, 205.
267. Staveley, L. A. K., H. Paget, B. B. Goalby and J. B. Warren, Nature *164*, 787 (1949).
268. Staveley, L. A. K., H. Paget, B. B. Goalby and J. B. Warren, J. Chem. Soc. *1950*, 2290.
269. Steele, W. C., L. D. Nichols and F. G. A. Stone, J. Am. Chem. Soc. *84*, 4441 (1962).
270. Straley, J. W., C. H. Tindal and H. H. Nielson, Phys. Rev. *62*, 161 (1942).
271. Straumanis, H., Z. Phys. Chem. *B30*, 138 (1935).
272. Sujishi, S., Intern. Congr. Pure Appl. Chem., München, 1959, Abstract A 225.
273. Sujishi, S., *Symposium on Hydrides of Groups 4, 5 and 6*, Berkeley, California, USA, 1963.
274. Sujishi, S., private communication, 1963.
275. Sujishi, S., Final Report, Contract No. DA-11-022-ORD-1264, Project No. TB-2-0001 (817), Control No. OOR-137-53 (1957).
276. Sujishi, S., Final Report, Contract No. AF 49 (638) 376 (1959).
277. Sujishi, S. and J. N. Keith, J. Am. Chem. Soc. *80*, 4138 (1958).
278. Sujishi, S. and J. N. Keith, Abstr. 134th Meeting Am. Chem. Soc., Div. Inorg. Chem., 1958, p. 44N.
279. Tabern, D. L., W. R. Orndorff and L. M. Dennis, J. Am. Chem. Soc. *47*, 2043 (1925).
280. Tabern, D. L., W. R. Orndorff and L. M. Dennis, J. Am. Chem. Soc. *47*, 2039 (1925).
281. Taft, R., cited in M. S. Newman, *Steric Effects in Organic Chemistry*, Wiley, New York, USA, p. 587 ff.
282. Taft, R. W. and H. H. Sisler, J. Chem. Educ. *24*, 175 (1947).
283. Taketa, A., M. Kumada and K. Tamara, Bull. Inst. Chem. Res., Kyoto Univ. *31*, 260 (1953).
284. Tamaru, K., Bull. Chem. Soc. Japan *31*, 647 (1958); C. A. *53*, 5845c (1959).
285. Tamaru, K., J. Phys. Chem. *61*, 647 (1957).
286. Tamaru, K., M. Boudart and H. Taylor, J. Phys. Chem. *59*, 801 (1955).
287. Tamborski, C., F. E. Ford, W. L. Lehn, G. J. Moore and E. J. Soloski, J. Org. Chem. *27*, 619 (1962).
288. Taylor, H., Can. J. Chem. *33*, 833 (1955).
289. Tchakirian, A., Bull. Soc. Chem. France [4] *51*, 846 (1932).
290. Teal, G. K. and C. A. Kraus, J. Am. Chem. Soc. *72*, 4706 (1950).
291. Townes, C. H. and A. L. Schawlow, *Microwave Spectroscopy*, McGraw-Hill, New York, USA, 1955, p. 645.
292. Voegelen, E., Z. Anorg. Allgem. Chem. *30*, 324 (1902)

293. VOLTA, A., cited in Beilsteins Handbuch der Organischen Chemie, 4th Edn., Vol. 1, Springer-Verlag, Berlin 1918, p. 56.
293a. VYAZANKIN, N. S., G. A. RAZUVAEV AND O. A. KRUGLAYA, Organometal. Chem. Rev. *A3*, 323 (1968).
294. WALLBAUM, A. J., NATURWISS. *32*, 76 (1944).
294a. WANG, F. T. AND C. H. VAN DYKE, Inorg. Chem. *7*, 1319 (1968).
294b. WANG, F. T. AND C. H. VAN DYKE, Chem. Commun. *1967*, 928.
295. WEST, R., J. Am. Chem. Soc. *75*, 6080 (1953).
296. WEST, R., H. R. HUNT AND R. O. WIPPLE, J. Am. Chem. Soc. *76*, 310 (1954).
297. WHITMORE, F. C., E. W. PIETRUSZA AND L. W. SOMMER, J. Am. Chem. Soc. *69*, 2108 (1947).
297a. WINGLETH, D. C. AND A. D. NORMAN, Chem. Commun. *1967*, 1218.
298. WINKLER, C., Ber. *19*, 210 (1886).
299. WINKLER, C., J. Prakt. Chem. [2] *36*, 204 (1887).
300. WÖHLER, F. AND H. BUFF, Ann. Chem. *103*, 218 (1857).
300a. YOUNG, C. W., J. S. KOEHLER AND D. S. MCKINNEY, J. Am. Chem. Soc. *69*, 1410 (1947).
301. ZAKHARKIN, L. I., V. V. GAVRILENKO, I. M. KHORLINA AND G. G. ZHIGAREVA, Bull. Acad. Sci. USSR, Div. Chem. Sci. (English Transl.) *1962*, 1784.
302. ZELTMANN, A. H. AND G. C. FITZGIBBON, J. Am. Chem. Soc. *76*, 2021 (1954).
303. ZIEGLER, K., Angew. Chem. *49*, 459 (1936).
304. ZINTL, E. AND H. KAISER, Z. Anorg. Allgem. Chem. *216*, 282 (1934).
304a. ZORIN, A. D., G. G. DEVYATYKH, V. YA. DUDOROV AND A. M. AMEL'CHENKO, Russ. J. Inorg. Chem. (Engl. Transl.) *9*, 1364 (1964).
305. ZWOLENSKI, B. J., A. DANTI, J. T. KERRY AND W. T. BERRY, *Mass Spectral Data*, American Petroleum Institute, Pittsburgh, Pennsylvania, USA, p. 2.

Chapter 9

TIN HYDRIDES

1. Historical

As early as 1830, K. W. G. Kastner [cit. *90*] stated that when tin dissolved in dilute mineral acids a "Zinnwasserstoffgas" ("tin–hydrogen gas") was produced. 72 years later (1902) Voegelen[*118a*] made an unsuccessful attempt to prepare tin hydride analogously to germane ($Zn + H_2SO_4 + H_2O + Ge^{4\oplus}$). Only in 1919 did Paneth and Fürth[*90*] succeed in demonstrating with certainty the existence of gaseous tin hydride ("stannane"), SnH_4, although only in traces. In the succeeding years, the yields of tin hydride were somewhat improved, but they were still too small for work on a larger scale (apart from the determination of a few, mostly physical properties).

Another 45 years passed before, in 1947, Finholt, Bond, jr., Wilzbach and Schlesinger[*24*] succeeded in producing stannane, SnH_4, in preparative amounts – now by a totally different route, namely by the hydrogenation of tin tetrachloride with lithium alanate, which had been discovered shortly before (1947). In 1951, Wiberg and Bauer[*121*] elucidated the mechanism of this hydrogenation.

2. General review

The reactions of the stannanes Sn_nH_{2n+2} and their derivatives can be compared with those of the silanes Si_nH_{2n+2} and the germanes Ge_nH_{2n+2}, if both their lower thermal stability and their greater ease of reduction to $Sn^{2\oplus}$ and to $Sn^{(0)}$ are borne in mind. Like the silanes, the stannanes contain negative hydrogen.

2.1 *The stability of SnH_4 and Sn_2H_6*

Stannane, SnH_4 decomposes substantially more readily than germane GeH_4 or even silane SiH_4. The decomposition point of SiH_4 is approximately 450°, and elementary Si has little effect on the kinetics of the decomposition. GeH_4 decomposes at about 300°, and elementary Ge, particularly Ge doped with 0, accelerates the decomposition. On the other hand, SnH_4 decomposes even at –50°, and elementary Sn markedly catalyses the decomposition. Traces of O_2 also catalyse the decomposition, while higher concentrations of O_2 inhibit it.

In view of the pronounced catalytic action of metallic Sn on the decomposition of SnH_4, it is not surprising that – unlike the situation with the silanes and germanes – preparative amounts of volatile tin hydrides cannot be obtained by the protolysis of Mg–Sn alloys.

The stability of the higher hydrides E_2H_6 decreases rapidly in the sequence disilane > digermane ≫ distannane. The chain lengths of the known still higher hydrides E_nH_{2n+2} fall in the same sequence. The maximum value of n in the higher hydrides that have been *isolated* (and not merely detected) is $n = 8$ for the silanes, $n = 5$ for the germanes, and only $n = 2$ for the stannanes.

2.2 *Formation and stability of* SnH_3Cl (*cf.* section 4.2)

The temperatures of formation at which the monochloro derivatives EH_3Cl can be obtained preparatively by the treatment of EH_4 with HCl fall rapidly from the Si to the Sn derivative (SiH_3Cl at 200°, GeH_3Cl at 20°, and SnH_3Cl at −70°), and the thermal stability of the monochloro derivatives falls in the same direction: $SiH_3Cl > GeH_3Cl \gg SnH_3Cl$. The reaction of the hydrides SiH_4, GeH_4, and SnH_4 with HCl has several branches:

$$EH_4 \xrightarrow[-H_2]{+HCl} EH_3Cl \xrightarrow[-H_2]{+HCl} EH_2Cl_2 \xrightarrow[-H_2]{+HCl} EHCl_3 \xrightarrow[-H_2]{+HCl} ECl_4 \quad (a)$$

$$EH_2Cl_2 \xrightarrow[(e)]{-H_2} ECl_2$$

$$EH_3Cl \xrightarrow[(b)]{-HCl} EH_2$$

$$\frac{1}{x}(EH_2)_x \xleftarrow[(c)]{\text{polymerisation}} EH_2 \xrightarrow[(d)]{\text{disproportionation}} \frac{1}{2}EH_4 + \frac{1}{2}E$$

Reaction sequence (a) takes place with E = Si, Ge, and Sn. The degree of elimination of HCl (b) increases in the order Si < Ge ≪ Sn. In the case of Si and Ge, polymerization (c) takes place almost quantitatively. The disproportionation (d) takes place quantitatively with Sn [no polymerization by route (c)]. The elimination of hydrogen (e) increases in the sequence Si < Ge ≪ Sn. In contrast to SiH_2Cl_2 and GeH_2Cl_2, it is impossible to isolate SnH_2Cl_2.

2.3 *Hydrolysis*

In neutral and alkaline media, the Sn–H bond hydrolyses more rapidly than the Si–H bond, while the Ge–H bond, lying between them, reacts extraordinarily sluggishly with water, and only slightly faster in the presence of metallic copper. While the mechanism of the Cu-catalysed hydrolysis of Ge–H is not known even approximately, the mechanism of the hydrolysis of the Si–H and Sn–H bonds have been established, at least in their main features:

$$\geq Si{-}H \xrightarrow[\text{fast},\ k_1^{Si}]{+H_2O} \geq \overset{\delta-}{Si}{-}\overset{\delta+}{H}\cdots HOH \xrightarrow[\text{slow},\ k_2^{Si}]{+HO^{\ominus}} [HO\cdots Si\cdots H\cdots H\cdots OH]^{\ominus} \xrightarrow[\text{fast},\ k_3^{Si}]{} \geq SiOH + H_2 + HO^{\ominus}$$

$$\geq GeH \xrightarrow[\text{very slow},\ k_I^{Ge}]{+H_2O} \geq GeOH + H_2$$

$$\geq GeH \xrightarrow{+H_2O,\ +Cu} [\geq GeH\cdot Cu\cdot H_2O] \xrightarrow[\text{relatively fast}]{k_{II}^{Ge}} \geq GeOH + H_2 + Cu$$

$$\geq SnH \xrightarrow[\text{fast}\ k_1^{Sn}]{+H_2O} \geq \overset{\delta-}{Sn}-\overset{\delta+}{H}\cdots HOH \xrightarrow[\text{slow}\ k_2^{Sn}]{+HO^{\ominus}} [HO\cdots Sn\cdots H\cdots H\cdots OH]^{\ominus} \xrightarrow[\text{fast}\ k_3^{Sn}]{} \geq SnOH + H_2 + HO^{\ominus}$$

$$(k_I^{Ge} \ll k_{II}^{Ge}) < (k_2^{Si} < k_1^{Si} \ll k_3^{Si}) < (k_2^{Sn} < k_1^{Sn} \sim k_3^{Sn})$$

The hydrolysis of both the Si–H and the Sn–H bond takes place in an alkaline medium through a rapid electrophilic addition of protons (of the undissociated water) to the hydride hydrogen. The nucleophilic attack of hydroxide ions determines the rate of the subsequent process. In both cases, a pentacovalent intermediate HO···E···H···H···OH is formed. In the case of the silicon hydride, the HO···Si bond is weak and the Si–H bond strong (HO···Si–H···HOH), so that the H_2O molecules facilitating the detachment of the hydride hydrogen are important for the kinetics of the hydrolysis. In the case of the tin hydride, the HO···Sn bond is relatively strong (HO–Sn···H···HOH), and the Sn–H bond is therefore weaker (comparable with the HO···Si bond). Thus, the detachment (repulsion) of hydride hydrogen is easier than in the case of the silicon hydride. Consequently, the concentration of the water does not affect the kinetics.

In pronounced contrast to alkaline hydrolysis characterized by the nucleophilic attack on the central atom [Si, (Ge?), or Sn], protolysis occurs by electrophilic attack on the hydride hydrogen. Since the stability differences of the sp^3 configurations ⟩Si–H, (⟩Ge–H?), and ⟩Sn–H, on the one hand, and the sp^3d configurations –Si–H, (–Ge–H?), and –Sn–H, on the other, are very slight, it is possible to predict that acid hydrolysis will take place more uniformly: all three hydrides protolyse with comparable and not excessively high velocities. The mechanism is known approximately only in the case of the silanes (see section 5.7 of Chapter 7).

2.4 *Addition of Sn–H to C=C* (*cf.* section 4.4)

While Si hydrides react with alkenes predominantly by addition and, more rarely by substitution (only at high temperatures), Ge and Sn hydrides take part only in addition reactions with the alkenes. The absence of substitution reactions with Ge and Sn hydrides is due to the ease with which they undergo thermolysis. Silicon hydrides add by either a radical or by a polar mechanism, germanium hydrides almost exclusively by a polar mechanism, and tin hydrides almost exclusively by a radical mechanism. Additions of the hydrides to C=C are compared below with one another.

References p. 752

(a) The radical stereospecific *trans* addition of Si–H leads to *cis* adducts (with respect to Si and R-groups) of the alkene:

$$\geq SiH \xrightarrow[-H\cdot]{\text{thermal energy}, h\nu, Hg(^3P_1), (RCOO)_2} \geq Si\cdot \xrightarrow{+>C=C<R} \text{(Si-bridged radical)} \xrightarrow[-\geq Si\cdot]{+\geq SiH} \geq Si-C-C(H)(R)$$

Radical telomerization of the alkenes competes with the radical addition. In the following equation, the telomerization is formulated without respect to the stability of the radical, *i.e.* without respect to the direction of the $\geq$Si group:

$$\geq Si\cdot \xrightarrow{+>C=C<} \text{(Si-bridged radical)} \xrightarrow{+>C=C<} \geq Si-\overset{|}{\underset{|}{C}}-\overset{|}{\underset{|}{C}}-\overset{|}{\underset{|}{C}}-\overset{|}{\underset{|}{C}}\cdot \longrightarrow$$

$$\xrightarrow{\text{etc.}} \geq Si(C-C)_n\cdot \begin{cases} \xrightarrow{+H\cdot} \geq Si(C-C)_nH \\ \xrightarrow{+\geq Si\cdot} \geq Si(C-C)_nSi\leq \end{cases}$$

As compared with these two addition reactions, the tendency of formation of higher silanes is low:

$$\geq Si\cdot + \geq SiH \rightarrow \geq Si-Si\leq + H\cdot$$

The added $\geq$Si orients itself (as in the radical addition of HBr) according to the stability of the radical; it enters in the position vicinal to the radical C-atom.

(b) Germanium hydrides do not add by a radical mechanism.

(c) Tin hydrides add predominantly by a radical mechanism, yet none of the usual radical initiators initiate the reaction, nor does hydroquinone inhibit it. The mechanism, the sterospecificity and the direction of the addition of Sn–H are the same as those in the addition of Si–H.

(d) In accordance with the negative charge of the hydride hydrogen, the polar addition of Si–H and Ge–H to C=C takes place contrary to Markovnikov's rule:

$$\overset{\delta-}{H}-\overset{\delta+}{Si}\leq + \overset{\delta+}{R}\rightarrow\overset{\delta+}{C}H=CH_2 \rightarrow R\rightarrow CH_2-CH_2Si\leq$$

$$(H-Ge\leq) \qquad (R\rightarrow CH_2-CH_2Ge\leq)$$

(R is an electron-repelling + I substituent)

(e) The only addition of Sn–H to C=C taking place by a polar mechanism appears to be addition catalysed by AlR_3:

$$\text{H–Sn}\!\!<\!\!\!-\; + \text{R}'\text{CH=CH}_2 \xrightarrow{(\text{AlR}_3)} \text{R}'\text{CH}_2\text{–CH}_2\text{Sn}\!\!<\!\!\!-$$

However, this is probably not an addition of $>$Sn–H to C=C but the addition of $>$Al–H[1] to C=C with subsequent alkyl exchange:

$$R_2AlH \xrightarrow{+\,R'CH=CH_2} R'CH_2\text{—}CH_2AlR_2$$

$$R'CH_2CH_2AlR_2 \xrightarrow{+\,>SnH} \left[>Sn\overset{H}{\cdots}AlR_2\cdots\underset{R'CH_2}{CH_2}\right] \longrightarrow R_2AlH + {>}SnCH_2CH_2R'$$

The following sections deal with the synthesis of substituted and unsubstituted tin hydrides, including their physical properties (section 3) and the reactions of the tin hydrides (section 4). The latter part is divided into subsections: the decomposition and dismutation of tin hydrides (section 4.1), the reaction of the Sn–H bond with halogens and hydrogen halides (section 4.2), the reaction of the Sn–H bond with alkyl halides (section 4.3), the reaction of the Sn–H bond with alkenes and alkynes (section 4.4), the reaction of the Sn–H bond with the C=O, the C=S, and the C=N double bonds (section 4.5), and the reaction of the Sn–H bond with alkali metals (section 4.6). The conclusion consists of a review of the spectroscopic investigations on tin hydrides (section 5).

The reaction of the Sn–H bond with O and N compounds is treated with the other matters mentioned in section 4.5.

3. Synthesis of substituted and unsubstituted tin hydrides $R_{4-n}SnH_n$ and $R_{6-n}Sn_2H_n$

In addition to an overwhelming amount of hydrogen, stannane, SnH_4, is formed in small amount in the decomposition of a Mg–Sn alloy ('Mg_2Sn') with dilute hydrochloric acid or sulphuric acid [*90*]:

$$Mg_2Sn + 4H^{\oplus} \rightarrow SnH_4 + 2Mg^{2\oplus}$$

In contrast to the situation in the protolysis of Mg_2Si or Mg_2Ge, in the protolysis of Mg_2Sn the yield of tin hydride is not increased in the NH_3 system [*2*]. Similarly low yields of SnH_4 are obtained by the electrical discharge between Sn electrodes in a H_2–CH_4 atmosphere [*92*] and by electrolysis of a sulphuric acid solution of tin sulphate for some days [*88, 89, 91, 93*].

[1]Formed by the reaction: $>$Sn–H + $AlR_3 \rightleftharpoons\ >$SnR + $AlHR_2$

References p. 752

Preparative amounts of tin hydride are obtained by the hydrogenation of $SnCl_4$ with $LiAlH_4$[*24*, *121*]:

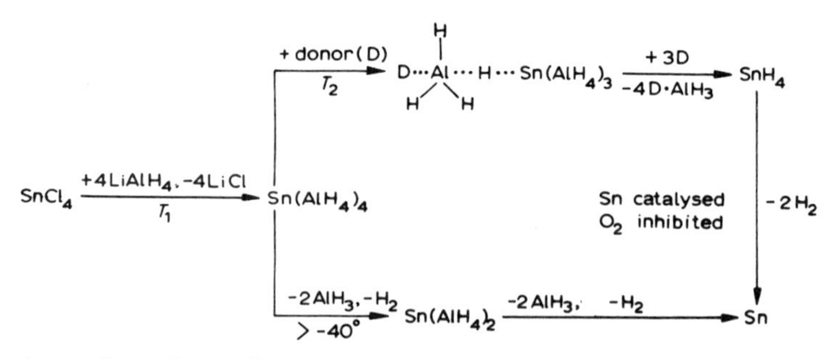

The reaction takes place via a nucleophilic attack of alanate ions on tin atoms ($-196° < T_1 < T_2$) with the intermediate formation of the white $Sn(AlH_4)_4$, which is stable below about $-40°$[*121*]. Donors facilitate the formation of stannane by detaching AlH_3 as a donor-alane adduct. The higher the donor activity the greater is the yield of SnH_4 and the lower is the temperature at which the detachment of AlH_3 begins. Thus, 18% of SnH_4 is formed[*2*], in the relatively weakly nucleophilic solvent di-n-butyl ether at $T_2 \geqslant -50°$, 20% of SnH_4[*2*, *24*] in the more highly nucleophilic diethyl ether at $T_2 \geqslant -59°$, and 31% of SnH_4[*2*] in the still more highly nucleophilic tetrahydrofuran at $T_2 \geqslant -80°$. A similar yield-enhancing effect is obtained by the dropwise addition of (comparatively highly nucleophilic) water to the ethereal reaction solution[*2*]. Oxygen inhibits the decomposition of SnH_4, catalysed by metallic tin. The yield rises to 80–90%[*23*, *25*] on working in a stream of N_2–O_2. Above $-40°$, $Sn(AlH_4)_4$ decomposes via unstable $Sn(AlH_4)_2$ into AlH_3, Sn, and H_2[*121*].

$Li[AlH(OBu^t)_3]$ in diethyl ether hydrogenates $SnCl_4$ to give a 34% yield of SnH_4[*2*]:

$$SnCl_4 + 4Li[Al(OBu^t)_3H] \xrightarrow{-80°} SnH_4 + 4Al(OBu^t)_3 + 4LiCl$$

SnH_4 is formed in 25% yield from $Sn(OC_2H_5)_4$ and $(C_2H_5)_2AlH$ according to the following equation[*74*]:

$$Sn(OEt)_4 + 4Et_2AlH \xrightarrow[\text{xylene}]{-10/+20°} SnH_4 + 4Et_2Al(OEt)$$

In an (acidic) aqueous medium, the (slowly hydrolysing) tetrahydroborates $NaBH_4$ and KBH_4 have a hydrogenating action:

$$SnCl_4 \xrightarrow[\Delta]{+H^{\ominus}, -Cl^{\ominus}} SnH_4, Sn_2H_6$$

The yields of SnH_4 and Sn_2H_6 depend markedly on the reaction conditions. However, the consumption of the hydrogenating agent is high in this method of synthesis, which requires only a low expenditure on apparatus (yields: 84% of SnH_4 referred to $SnCl_4$, 1.9% of SnH_4 referred to $BH_4^{\ominus}$; 0.095% of Sn_2H_6 referred to $BH_4^{\ominus}$)[*37*, *38*, *98*, *121*].

The silyl stannane, $H_3Sn{-}SiH_3$, and germyl stannane, $H_3Sn{-}GeH_3$, corresponding to distannane, $H_3Sn{-}SnH_3$, can be synthesized by the following route [*12*, *120*] (E = Si, Ge):

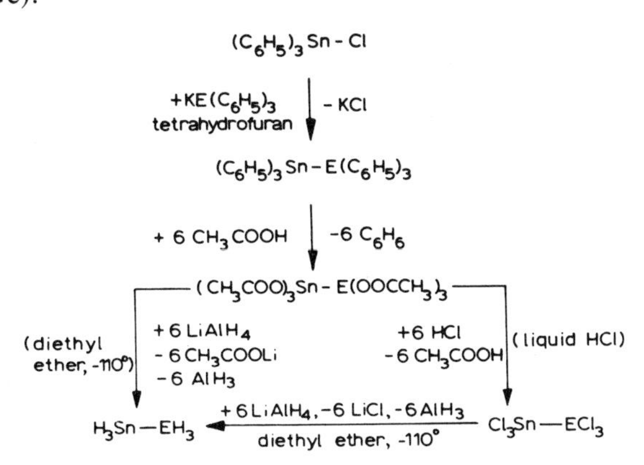

Both mixed hydrides are stable in ethereal solution at 20° provided that no traces of metals are present; these catalyse the decomposition $H_3Sn{-}EH_3 \rightarrow H_2Sn + EH_4$. On this, compare the dismutation of $Cl_3Sn{-}SnCl_3$ above −65° in accordance with the equation $Cl_3Sn{-}SnCl_3 \rightarrow Cl_2Sn + SnCl_4$ [*122*].

In the protolysis of calcium stannide CaSn with aqueous acids, anhydrous acetic acid, absolute alcohol, or hydrogen chloride in tetrahydrofuran, tin, tin salts, and calcium salts are formed, but no $(SnH_2)_x$[2]. NH_4I in liquid ammonia gives a 10% yield of solid polystannane $(SnH_2)_x$, contaminated with $(CaSn)_x$ [*59*]:

$$(CaSn)_x + 2x\,NH_4I \xrightarrow[-78°]{\text{liq. NH}_3} (SnH_2)_x + x\,CaI_2 + 2x\,NH_3$$

Alkyltin hydrides and aryltin hydrides, $R_{4-n}SnH_n$, are obtained in high yields by hydrogenating the corresponding halides, methoxides, or ethoxides, which are readily obtained by Neumann's method [*70*], with $LiAlH_4$, $NaBH_4$, or R_2AlH (Table 9.1 contains the references and examples of the reaction):

$$R_{4-n}SnX_n + \frac{n}{4}LiAlH_4 \rightarrow R_{4-n}SnH_n + \frac{n}{4}LiX + \frac{n}{4}AlX_3$$

$$R_{4-n}SnX_n + n\,R'_2AlH \rightarrow R_{4-n}SnH_n + n\,R'_2AlX$$

(X generally represents Cl; it may also represent Br, I, OCH_3 or OC_2H_5)

Diborane hydrogenates alkyltin or aryltin methoxides almost quantitatively [*5*]:

$$4R_3Sn(OMe) + (BH_3)_2 \xrightarrow[-78°]{\text{pentane}} 4R_3SnH + 2BH(OMe)_2$$

$$2R_2Sn(OMe)_2 + (BH_3)_2 \xrightarrow[-78°]{\text{pentane}} 2R_2SnH_2 + 2BH(OMe)_2$$

$(R = C_2H_5, n\text{-}C_3H_7, C_6H_5)$

---

[2]In contrast, CaSi and CaGe form the corresponding hydrides $(EH_2)_x$.

 [*text continued p. 730*]

TABLE 9.1

SYNTHESIS OF ORGANOTIN HYDRIDES $R_{4-n}SnH_n$ BY THE HYDROGENATION OF THE CORRESPONDING HALIDES OR ALKOXIDES $R_{4-n}SnX_n$ WITH $(BH_3)_2$, R_2AlH, $LiAlH_4$, $(MeHSiO)_n$ OR $NaBH_4$

Compound to be hydrogenated	*Hydrogenating agent*	*Solvent*	*Yield of hydride in mole%, referred to the initial tin compound*	*Ref.*
$(CH_3)SnCl_3$	$LiAlH_4$	ether?	a very small amount of $(CH_3)SnH_3$	25
$K[(CH_3)SnO_2]$	$NaBH_4$	water	high yield of $(CH_3)SnH_3$	25
$(C_2H_5)SnCl_3$	$(C_2H_5)_2AlH$	di-n-butyl ether	97% $(C_2H_5)SnH_3$	69, 74
$(n\text{-}C_4H_9)SnCl_3$	$NaBH_4$	monoglyme	16% $(n\text{-}C_4H_9)SnH_3$	9b
$(n\text{-}C_4H_9)SnCl_3$	$LiAlH_4$	diethyl ether	37% $(n\text{-}C_4H_9)SnH_3$	41
$(n\text{-}C_4H_9)SnCl_3$	$(C_2H_5)_2AlH$	diethyl ether	62% $(n\text{-}C_4H_9)SnH_3$	69, 74
$(i\text{-}C_4H_9)SnCl_3$	$LiAlH_4$	diethyl ether	71% $(i\text{-}C_4H_9)SnH_3$	77
$(C_6H_5)SnCl_3$	$(i\text{-}C_4H_9)_2AlH$	di-n-butyl ether	72% $(C_6H_5)SnH_3$	69, 74
$(CH_3)_2SnCl_2$	$NaBH_4$	monoglyme	96% $(CH_3)_2SnH_2$	9b
$(CH_3)_2SnCl_2$	$LiAlH_4$	dioxan	72% $(CH_3)_2SnH_2$	24, 25
$(C_2H_5)_2SnCl_2$	$LiAlH_4$	diethyl ether	90% $(C_2H_5)_2SnH_2$	22, 74
$(C_2H_5)_2SnCl_2$	$(C_2H_5)_2AlH$	none; di-n-butyl ether; decalin	84–97% $(C_2H_5)_2SnH_2$	69, 74
$(C_2H_5)_2SnCl_2$	$(C_2H_5)_3Al+(C_2H_5)_2AlH$	none; 0°	89% $(C_2H_5)_2SnH_2$	112
$(C_2H_5)_2Sn(OCH_3)_2$	$(BH_3)_2$	pentane	91% $(C_2H_5)_2SnH_2$	5
$(n\text{-}C_3H_7)_2SnCl_2$	$LiAlH_4$	diethyl ether	72% $(n\text{-}C_3H_7)_2SnH_2$	41
$(n\text{-}C_4H_9)_2SnCl_2$	$NaBH_4$	monoglyme	56% $(n\text{-}C_4H_9)_2SnH^2$	9b
$(n\text{-}C_4H_9)_2SnCl_2$	$LiAlH_4$	diethyl ether	66(86)% $(n\text{-}C_4H_9)_2SnH_2$	41(74)
$(n\text{-}C_4H_9)_2SnCl_2$	$(C_2H_5)_2AlH$	diethyl ether	85% $(n\text{-}C_4H_9)_2SnH_2$	69, 74
$(C_4H_9)_2SnCl_2$	$(C_2H_5)_3Al+(C_2H_5)_2AlH$	none; 0°	85% $(C_4H_9)_2SnH_2$	112
$(i\text{-}C_4H_9)_2SnCl_2$	$LiAlH_4$	diethyl ether	83% $(i\text{-}C_4H_9)_2SnH_2$	77
$(i\text{-}C_4H_9)_2SnCl_2$	$(i\text{-}C_4H_9)_2AlH$	none	72% $(i\text{-}C_4H_9)_2SnH_2$	77
$(C_6H_5)_2SnCl_2$	$LiAlH_4$	diethyl ether	72% $(C_6H_5)_2SnH_2$	74
$(C_6H_5)_2SnCl_2$	$(C_2H_5)_2AlH$	diethyl ether	81% $(C_6H_5)_2SnH_2$	69, 74
$(C_6H_5)_2SnCl_2$	$(C_2H_5)_3Al+(C_2H_5)_2AlH$	none; 0°	81% $(C_6H_5)_2SnH_2$	112
$(CH_3)_3SnBr$	$LiAlH_4$	dioxan	$(CH_3)_3SnH$	50
$(CH_3)_3SnCl$	$NaBH_4$	monoglyme	92% $(CH_3)_3SnH$	9b

$(C_2H_5)_3SnF$	$(C_2H_5)_2AlH$	diethyl ether	97% $(C_2H_5)_3SnH$	69, 74
$(C_2H_5)_3SnCl$	$LiAlH_4$	diethyl ether	66(80–90)% $(C_2H_5)_3SnH$	41, 52(74)
$(C_2H_5)_3SnCl$	$(C_2H_5)_2AlH$	diethyl ether	89% $(C_2H_5)_3SnH$	69, 74
$(C_2H_5)_3SnCl$	$(C_2H_5)_3Al + (C_2H_5)_2AlH$	none; 0°	83% $(C_2H_5)_3SnH$	112
$(C_2H_5)_3SnCl$	Al/Hg, H_2O	water	30% $(C_2H_5)_3SnH$	44
$(C_2H_5)_3SnI$	$LiAlH_4$	diethyl ether	48% $(C_2H_5)_3SnH$	8
$(C_2H_5)_3Sn(OCH_3)$	$(BH_3)_2$	pentane	97% $(C_2H_5)_3SnH$	5
$(C_2H_5)_3Sn(OC_2H_5)$	$(C_2H_5)_2AlH$	none	96% $(C_2H_5)_3SnH$	69, 74
$(n\text{-}C_3H_7)_3SnCl$	$LiAlH_4$	diethyl ether	75% $(n\text{-}C_3H_7)_3SnH$	41
$(n\text{-}C_3H_7)_3SnCl$	Al/Hg, H_2O	water	65% $(n\text{-}C_3H_7)_3SnH$	44
$(n\text{-}C_4H_9)_3SnCl$	$NaBH_4$	monoglyme	96% $(n\text{-}C_4H_9)_3SnH$	9b
$(n\text{-}C_4H_9)_3SnCl$	$LiAlH_4$	diethyl ether	74% $(n\text{-}C_4H_9)_3SnH$	41, 52
$(n\text{-}C_4H_9)_3SnCl$	Al/Hg, H_2O	water	60% $(n\text{-}C_4H_9)_3SnH$	44
$(n\text{-}C_4H_9)_3Sn(OCH)_3$	$(BH_3)_2$	pentane	100% $(n\text{-}C_4H_9)_3SnH$	5
$(i\text{-}C_4H_9)_3SnCl$	$LiAlH_4$	diethyl ether	89% $(i\text{-}C_4H_9)_3SnH$	77
$(C_6H_5)_3SnCl$	$NaBH_4$	monoglyme	82% $(C_6H_5)_3SnH$	9b
$(C_6H_5)_3SnCl$	$LiAlH_4$	diethyl ether; dioxan	40–70% $(C_6H_5)_3SnH$	28, 41, 52
$(C_6H_5)_3SnCl$	Al/Hg, H_2O	water	65% $(C_6H_5)_3SnH$	44
$(C_6H_5)_3SnOCH_3$	$(BH_3)_2$	pentane	99% $(C_6H_5)_3SnH$	5
$[(n\text{-}C_4H_9)_2SnCl]_2$	$LiAlH_4$	diethyl ether	76% $[(n\text{-}C_4H_9)_2SnH]_2$	100

TABLE 9.2 PHYSICAL PROPERTIES OF TIN HYDRIDES AND DERIVATIVES

Hydride	*M.p.*	B.p. [°C (torr)]	*Vapour pressure equation log* p [*torr*] = − A/T + B; *enthalpy of evaporation* Q[*cal/mole*]; *Trouton's constant* T_i; *density* d [*g/ml*]; *refractive index* n	*Ref.*
SnH_2			Calculated enthalpy of formation $\Delta H^{\circ}_{25^{\circ}} - 17 \pm 15$ kcal/mole	123
SnH_4	−150	−52(760)	Vapour pressures 39(−105°); 198(−78°); 760(−51.8°) torr; *Q* 4500; *Q* 4430; solubility in many inorganic and organic solvents; enthalpy of formation $\Delta H^{\circ}_{25^{\circ}}$ 38.9 ± 0.5 kcal/mole; solubility [g/100 ml (solvent)] at −78°: 1.4 g (Et_2O); 1.8 g (hexane); 2.4 g (THF); 7.7 g (CS_2); viscosity see	94, 96a, 102, 20, 30a; 18b
$(CH_3)SnH_3$		0; 1.4(760)	*A* 1255; *B* 7.475; *Q* 5750; *T* 21.0	23, 24
$(C_2H_5)SnH_3$		22–23(745)	n_D^{20} 1.4491	71, 74
		25(760)		23
$(n\text{-}C_3H_7)SnH_3$		—	decomposes rapidly	23
$(n\text{-}C_4H_9)SnH_3$		98–101(760)	n_D^{20} 1.4609; *d* 1.14; decomposes on storage	41, 67, 71, 74
$(i\text{-}C_4H_9)SnH_3$		87–89(760?)	n_D^{20} 1.4562; d_4^{20} 1.32	77
$(C_6H_5)SnH_3$		57–64(105–108)	n_D^{20} 1.4370	71, 74
$KSnH_3$			very soluble in monoglyme and diglyme; decomposes on storage	7
$(CH_3)_2SnH_2$		35(760)	*A* 1482; *B* 7.697; *Q* 6790; *T* 22.1; n_D^{20} 1.4480; d_4^{20} 1.4766	24
		36(760)	*A* 1473; *B* 6.64; *T* 21.6	46
$(C_2H_5)_2SnH_2$		96–98(760)	n_D^{20} 1.4679; d_4^{20} 1.27	22, 71, 74
$(n\text{-}C_3H_7)_2SnH_2$		39.0–40.5(12)	—	41, 67, 111
$(n\text{-}C_4H_9)_2SnH_2$		70; 75–76(12)	n_D^{20} 1.4703; d_4^{20} 1.19	41, 67, 111
		85(19)		74, 112
$(i\text{-}C_4H_9)_2SnH_2$		57–59(12)	n_D^{20} 1.4657; d_4^{20} 1.17	77
$(C_6H_5)_2SnH_2$		89–93(0.3); 90(0.2)	n_D^{20} 1.5950; 1.5951; d_4^{20} 1.19; 1.39	69, 71, 72, 112

$(CH_3)_3SnH$		59; 60(760)	*A* 1581; *B* 7.641; *Q* 7240; *T* 21.8; n_D^{18} 1.4484; Vapour pressure 268(−23°)	24, 25, 48a, 50, 117
$(C_2H_5)_3SnH$		148–150(760)	n_D^{20} 1.4709; d_4^{20} 1.250	71, 74
		37–38(12); 48(22)	n_D^{20} 1.4700; d_4^{20} 1.258	22, 103
		47(12)		
		79–81(92)	water-clear liquid, stable when dry	41
$(C_2H_5)_3SnD$		35–37(11)	n_D^{20} 1.4702	78
$(n\text{-}C_3H_7)_3SnH$		76–82(12); 101(28)	water-clear liquid, stable when dry; enthalpy of formation, $\Delta H^\circ_{25^\circ}$ −32.1 kcal/mole	41, 67, 103, 110a
$(n\text{-}C_4H_7)_3SnH$		73(0.5)	n_D^{18} 1.4738; enthalpy of formation, $\Delta H^\circ_{25^\circ}$ −48.6 kcal/mole	52, 110a
		76–81(0.7–0.9)	water-clear liquid, stable when dry	41, 53, 67
		65–67(0.6)	n_D^{20} 1.4726; d_4^{20} 1.104	71
		63–64(0.41–0.48)	n_D^{22} 1.4721–1.4720	115
		46–49(0.18)		115
$(i\text{-}C_4H_9)_3SnH$		101–104(12)	n_D^{20} 1.4697; d_4^{20} 1.09	77
$(i\text{-}C_4H_9)_3SnD$		104–106(11)	n_D^{20} 1.4697	78
$(C_6H_5)_3SnH$		174(1); 142–149(0.1)	n_D^{25} 1.6342–1.6345	52, 114
	26–28	168–172(0.8)		41, 53
		131–134(0.004)	n_D^{28} 1.6322; d_4^{20} 1.378	71
$(C_6H_5)_3SnD$	< 20	156–160(0.002)	n_D^{28} 1.6318	78
SnH_3Cl			decomposition at 20° into Sn, $SnCl_2$ and H_2	1
$(C_2H_5)SnH_2Br$			decomposition at −65° into $(RSnBr)_x$ and H_2	26
$(n\text{-}C_4H_9)_2SnHCl$			readily decomposes into R_2SnH_2 and R_2SnCl_2	99
$(SnH_3)_2$			very unstable at 20°	37, 38
$[(n\text{-}C_4H_9)_2SnH]_2$			colorless liquid; decomposes at 100° into $(n\text{-}C_4H_9)_2Sn$ and H_2	100
$H[(C_2H_5)_2Sn]_{n\leq 8}H$			chain structure; at $n > 8$: ring structure without Sn–H	69
$(CH_3)_2(C_2F_4H)SnH$	77.9		*Q* 9200	13

For hydride formation [$(Bu_2SnO)_n \rightarrow Bu_2SnH_2$] with polymeric silicon hydrides $(MeHSiO)_n$ see [*30b*].

In the reaction of tetramethoxytin with diborane, $Sn(BH_4)_2$ is formed in addition to the expected SnH_4 [*4, 5*]:

$$Sn(OMe)_4 + (BH_3)_2 \xrightarrow[-78^\circ]{\text{pentane}} SnH_4 + 2BH(OMe)_2$$

$$Sn(OMe)_4 + 2(BH_3)_2 \xrightarrow[-78^\circ]{\text{pentane}} Sn(BH_4)_2 + H_2 + 2BH(OMe)_2$$

When the solid, yellow, involatile tin(II) boranate, $Sn(BH_4)_2$, which is stable to −65°, is heated, monomeric tin dihydride cannot be obtained because of its instability (transformation into metallic tin) (*cf.* the more stable polymers $(SnH_2)_x$ above) [*4, 6*]:

$$Sn(BH_4)_2 \xrightarrow[20^\circ]{-(BH_3)_2} [SnH_2] \xrightarrow{-H_2} Sn$$

Nascant hydrogen (Al/Hg in water, but not Zn/Hg in water) reduces triorganotin halides. A prerequisite for this is that the Al must be sufficiently pure; Si and Mg as impurities, in particular, inhibit the reaction, while small amounts of Fe do not interfere [*44*].

When trialkyltin formate, $R_3SnOOCH$, is heated, carbon dioxide is eliminated with the formation of the corresponding hydride [*86a*]:

$$R_3Sn\text{-}O\text{-}CH{=}O \xrightarrow{160-180^\circ} R_3SnH + CO_2 \quad (R = n\text{-}C_3H_7, n\text{-}C_4H_9)$$

As might be expected from the high stability of the tetraalkyltins and the tetraaryltins, R_4Sn, and from the instability of stannane, SnH_4, in organotin hydrides $R_{4-n}SnH_n$ ($n = 1, 2, 3$) the stability increases with decreasing number of Sn–H bonds. Catalysts accelerate the decomposition of stannane itself, in particular, but also that of organotin hydrides. Thus, SnH_4 prepared from $AlHEt_2$ and $Sn(OEt)_4$ in xylene and distilled rapidly decomposes on rough, metallic, and greasy surfaces or on rubber [*74*]. SnH_4 prepared from $SnCl_2 \cdot aq$ and $NaBH_4$ in water and distilled in a high vacuum can be kept for only 20 days, even in clean glass vessels [*98*]. In the presence of 10% of oxygen, gaseous SnH_4 can be stored at 20° [*23*]. Table 9.2 gives the physical properties of stannanes with Sn–H bonds.

4. Reactions of the tin hydrides

4.1 Decomposition and dismutation of tin hydrides

Monostannane, SnH_4, decomposes at −50° to +100° into tin and hydrogen [*47, 113*]:

$$SnH_4 \xrightarrow[\text{catalysts}]{-50^\circ/+100^\circ} Sn + 2H_2$$

This reaction is catalysed by tin (even minute traces), cock grease, and oxygen with a partial pressure of about 0.0005 torr. At $T \geqslant 10°$, the decomposition is of first order with respect to SnH_4. Anomalies appear at $T \leqslant 10°$, because of the transformation of tetragonal (β-)tin into cubic (α-)tin. A higher content of oxygen (partial pressure > 1 torr) completely inhibits the decomposition of SnH_4[*47*].

The decomposition of organotin hydrides $R_{4-n}SnH_n$ depends on the organic substituent R, on the temperature and time of pyrolysis, and on the catalyst [*10, 13, 71, 72, 74*]. After brief heating at 120°, dimethylstannane gives trimethylstannane, among other products; on longer heating (40 h) to 130° with UV irradiation, hexamethyldistannane and tetramethylstannane are formed instead:

$$3Me_2SnH_2 \xrightarrow{120°} 2Me_3SnH + Sn + 2H_2 \text{ (+ traces of } Me_3SnSnMe_3)$$

$$5Me_2SnH_2 \xrightarrow[UV]{130°} Me_3SnSnMe_3 + SnMe_4 + 2Sn + 5H_2$$

With higher alkyl groups, or with aryl groups, disubstituted stannanes form chains saturated with H, containing up to eight members, or rings free from H:

$$n\,R_2SnH_2 \xrightarrow{(Sn)} H(R_2Sn)_nH + (n-1)H_2 \qquad \text{(a)}$$

$$n\,R_2SnH_2 \xrightarrow[\text{(Sn or cat.)}]{} (R_2Sn)_n + nH_2 \qquad \text{(b)}$$

(a) $R = C_2H_5, n \leqslant 8$;

(b) $R = C_2H_5, C_6H_5, i\text{-}C_4H_9, \alpha\text{-}C_{10}H_7, \beta\text{-}C_{10}H_7, p\text{-}C_6H_4C_6H_5$;
cat. for $R = C_6H_5$: $C_5H_5N, n = 6$; $CHON(CH_3)_2, n = 5$; (rings);
$R = C_2H_5$ or $i\text{-}C_4H_9$: $(C_2H_5)_2SnCl_2{\cdot}2NC_5H_5, n = 9$;
$R = \beta\text{-}C_{10}H_7$: $(C_2H_5)_2SnCl_2{\cdot}2NC_5H_5, n = 6$

Triethylstannane decomposes under the catalytic action of tin or diborane into hexaethyldistannane and hydrogen:

$$2Et_3SnH \xrightarrow{(Sn\,or\,B_2H_6)} Et_3SnSnEt_3 + H_2$$

Analogous Si and Ge hydrides do not decompose. In the case of $(C_6H_5)_2GeH_2$, the polymerization can be initiated by diethylmercury[*72*]:

$$4Ph_2GeH_2 \xrightarrow[-8EtH]{+4HgEt_2} \frac{4}{n}[\text{-}GePh_2\text{-}Hg\text{-}]_n \xrightarrow[-4Hg]{heat,\,UV} \text{cyclo-}[Ph_2Ge]_4$$

Organotin hydrides readily exchange hydrogen for deuterium[*78*]:

$$Bu_3^iSnH + Et_3SnD \rightleftharpoons Bu_3^iSnD + Et_3SnH$$

Thus, the equilibrium is reached without a catalyst after only 25 min at 40°. With dialkylaluminium hydrides, it is established even faster (after 3 min at 50°)[*78*]:

$$Et_3SnH + Bu_2^iAlD \rightleftharpoons Et_3SnD + Bu_2^iAlH$$

References p. 752

Both exchanges probably take place through four-center transition states [*78*]:

$$Bu_3^iSn \begin{smallmatrix} \cdot H \cdot \\ \cdot D \cdot \end{smallmatrix} SnEt_3 \quad \text{and} \quad Et_3Sn \begin{smallmatrix} \cdot H \cdot \\ \cdot D \cdot \end{smallmatrix} AlBu_2^i$$

In this connection see the very interesting reaction of Ph_3SnH with $Et_2Zn \cdot Donor_2$ (Donor = Et_2O, THF) which yields the co-ordination complex $(Ph_3Sn)_2Zn \cdot Donor_2$ and EtH [*115a*].

4.2 *Reaction of the Sn–H bond with halogens and hydrogen halides*

Gaseous stannane, SnH_4, reacts with iodine or hydrogen chloride with the replacement of hydrogen by halogen (*cf.* section 2) [*1, 2*]:

$$SnH_4 \xrightarrow[-HI]{+I_2} SnH_3I \xrightarrow[-HI]{+I_2} SnH_2I_2 \xrightarrow[-HI]{+I_2} SnHI_3 \xrightarrow[-HI]{+I_2} SnI_4 \qquad (a)$$

$$SnH_4 \xrightarrow[-H_2]{+HCl} SnH_3Cl \xrightarrow[-H_2]{+HCl} SnH_2Cl_2 \xrightarrow[-H_2]{+HCl} SnHCl_3 \xrightarrow[-H_2]{+HCl} SnCl_4 \qquad (b)$$

The sequence of reactions (a) takes place without a catalyst at 0°. SnH_3I can be isolated by removing it from further iodination by cooling a part of the reaction vessel not coated with solid iodine to −100° [*2*]. Without cooling, only SnI_4 and HI can be isolated. Reactions (b) take place without a catalyst at 20°, or as low as −70° on a wall coated with $AlCl_3$. Here again the SnH_3Cl can be protected against further halogenation only by continuously condensing it out at −115° (42% SnH_3Cl at +20°/−115°, without a catalyst; compare the temperatures of formation: SiH_3Cl at +200°, GeH_3Cl at +20°, and SnH_3Cl at −70°) [*1, 2*]. Both monohalides are unstable. SnH_3I decomposes even at −30° into SnI_2, SnH_4, and H_2; SnH_3Cl decomposes slowly at 20° as follows [*1*]:

$$2SnH_3Cl \xrightarrow{-2HCl} 2SnH_2 \rightarrow SnH_4 + Sn$$

$$SnH_4 + 2HCl \xrightarrow{-2H_2} SnH_2Cl_2 \rightarrow SnCl_2 + H_2$$

With hydrogen bromide in the absence of a solvent, or in cyclohexane at −78°, alkylstannanes, $RSnH_3$, form monoalkylmonobromostannanes. The latter decompose above −65° with the liberation of hydrogen to form yellow insoluble polymeric monoalkylmonobromotin(II) $(RSnBr)_n$ [*26*]:

$$EtSnH_3 \xrightarrow[-78°]{+HBr, -H_2} EtSnH_2Br \xrightarrow[-65°/20°]{-H_2} \frac{1}{n}(EtSnBr)_n$$

In reactions of phenylstannanes, $Ph_{4-n}SnH_n$, with hydrogen bromide—as in the case of the corresponding phenylsilanes, see section 5.1.2 of chapter 7—the cleavage of Sn–Ph bonds [reaction (a) below] predominates, while the reaction of the Sn–H bonds [reaction (b) below] decreases, *e.g.* [*26, 26a*]:

$$Ph_3SnH \xrightarrow[-2\,PhH]{+\,2\,HBr} PhSnHBr_2 \qquad (a)$$

$$Ph_3SnH \xrightarrow[-H_2]{+\,HBr} Ph_3SnBr \qquad (b)$$

In addition, a solid containing Sn–Sn bonds is formed in small amount. The last phenyl group cannot be split off even with an excess of HBr. An accumulation of negative substituents on the Sn-atom obviously hinders cleavage in a similar manner to the case of the phenylsilanes (see section 5.1.2 of Chapter 7) [*26*].

Triethylstannane, Et_3SnH, reacts similarly to triethylgermane, Et_3GeH (see section 5.3 of chapter 8), with a deficiency of transition-metal halides or oxides, as well with predominantly covalent main-group metal halides, to give the corresponding tin halides Et_3SnX or oxides $(Et_3Sn)_2O$. When halides (*e.g.* $PdCl_2$) are gradually added [*8*] to an excess of triethylstannane, only hydrogen is replaced [eqn. (a)]; in other cases ethyl groups are replaced as well [eqn. (b)] [*8, 80*]:

$$2Et_3SnH + PdCl_2 \rightarrow 2Et_3SnCl + Pd + H_2 \quad (a)$$

$$2Et_3SnH + 2PdCl_2 \rightarrow 2Et_2SnCl_2 + 2Pd + Et_2 + H_2 \quad (b)$$

$$3Pr_3{}^nSnH + AlCl_3 \rightarrow 3Pr_3{}^nSnCl + Al + \tfrac{3}{2}H_2$$

Table 9.3 contains examples of these reactions. With $HgCl_2$ or $SbCl_3$ Ph_3SnH gives $Ph_3SnSnPh_3$ [*9a*].

TABLE 9.3

REACTION OF $(C_2H_5)_3SnH$ WITH METAL HALIDES AND OXIDES (extract from [*8*])

Halide or oxide	*Reaction products*
$GeCl_4$	97% Et_3SnCl, Ge, GeH_4, H_2
$SnCl_4$	94% Et_3SnCl, Sn, H_2
$SnCl_2$	96% Et_3SnCl, Sn, H_2
S_2Cl_2	84% Et_3SnCl, $(Et_3Sn)_2S$, H_2S, H_2
$CuBr_2$	91% Et_3SnBr, a little Et_2SnBr_2, CuBr, Cu, H_2
AgBr	82% Et_3SnBr, Ag, H_2
$KAuCl_4$	94% Et_3SnCl, Au, KCl, H_2
ZnO	40% $(Et_3Sn)_2O$, Zn, H_2
$TiCl_4$	55% Et_3SnCl, $TiCl_3$, H_2
$VOCl_3$	88% Et_3SnCl, VOCl, H_2
$PdCl_2$	97% Et_3SnCl, Pd, H_2

Triethylstannane, Et_3SnH, reacts with dichlorodiethylstannane with evolution of H_2 to give quantitative yields of triethylchlorostannane, Et_3SnCl, and diethyltin(II) [*118*]:

$$2Et_3SnH + Et_2SnCl_2 \rightarrow 2Et_3SnCl + Et_2Sn + H_2$$

On the other hand, di-n-butylstannane, Bu_2SnH_2, and di-n-butyldihalostannanes, Bu_2SnX_2, commute to give dibutylmonohalostannanes [*98a, 99*]:

$$Bu_2SnH_2 + Bu_2SnX_2 \rightarrow 2Bu_2SnHX \ (X = F, Cl, Br, I)$$

Similarly to the other known stannanes simultaneously containing H and halogen,

References p. 752

Bu_2SnHCl is unstable. However, it does not – in an analogous manner to SnH_3Cl or $EtSnH_2Br$ – decompose to Bu_2Sn and HCl but dismutes to dibutylstannane and dibutyldichlorostannane [*99*].

4.3 Reaction of the Sn–H bond with alkyl halides

Organotin hydrides $R_{4-n}SnH_n$ react with saturated alkyl halides and in certain cases with alkenyl and alkynyl halides with replacement of the halogen of the halide by hydrogen (with retention of the multiple bond of the alkenyl or alkynyl group) in accordance with the general equation:

$$R_{4-n}SnH_n + nR'X \rightarrow R_{4-n}SnX_n + nR'H$$

Table 9.4 contains examples of the reaction and relevant references. In the overwhelming majority of cases, the reaction probably takes place by means of a radical chain (a–c), whose rate-determining step is the abstraction of the halogen (b) [*57, 66*]:

$$\gt SnH + A\cdot \rightarrow \gt Sn\cdot + AH \text{ (A}\cdot = \text{free radical)} \quad \text{(a)}$$

$$\gt Sn\cdot + RX \rightarrow \gt SnX + R\cdot \quad \text{(b)}$$

$$R\cdot + \gt SnH \rightarrow \gt Sn\cdot + RH \quad \text{(c)}$$

Dehalogenations of vicinal dihalides naturally take place by a different mechanism [*57*]:

$$2 \gt Sn{-}H + {-}\underset{Br}{\overset{|}{C}}{-}\underset{Br}{\overset{|}{C}}{-} \rightarrow {-}\overset{|}{C}{=}\overset{|}{C}{-} + 2 \gt SnBr + H_2$$

In the radical chains given above, polar factors play only a slight part. Only with an accumulation of halogen on one C-atom can the following transition states be imagined:

$$[Cl_3C{:}X \quad Sn\cdot] \qquad [Cl_3C\cdot \quad Sn{:}X] \qquad [Cl_3C{:}^{\ominus}Sn^{\oplus} \; X\cdot]$$

A polar mechanism of the replacement of halogen in alkenyl halides via a primary addition in the β-position to the halogen and subsequent elimination of a substituted tin halide:

$$CH_2{=}CH{-}CH_2Br \xrightarrow{+\text{—SnH}} CH_3{-}\underset{\text{—Sn}}{\underset{|}{CH}}{-}CH_2Br \xrightarrow{-\gt SnBr} CH_3{-}CH{=}CH_2$$

is unlikely, since terminal *anti*-Markovnikov addition of the substituted tin group is to be expected ($\gt \overset{\delta+}{Sn}{-}\overset{\delta-}{H}$; $\overset{\delta-}{C}H_2{=}CH{-}CH_2{\rightarrow}Br$) and since saturated alkyl halides

TABLE 9.4

HALOGENATION OF TRI-n-BUTYLSTANNANE BY ORGANIC HALIDES

Organic halide	*Conversion (mole-%)*[a]	*Ref.*	*Organic halide*	*Conversion (mole-%)*[a]	*Ref.*
$CFBr_3$	69	106			
$CHCl_3$	92	106			
CCl_4[b]	85	106			
$CHBr_3$	62	106	Cl, Cl	83	106
$(C_6H_5)CH_2Cl$	78	53			
$(C_6H_5)CHCl_2$	70	53			
$(C_6H_5)CCl_3$	95	53			
$(C_6H_5)CHClCH_3$	77	57	Br, Cl	97	106
$(C_6H_5)_3CCl$	70	117b			
$\overline{CH_2(CH_2)_4C}HCl$	36	57			
$\overline{CH_2(CH_2)_4C}HBr$	71	57			
$(C_6H_5)CH_2Br$	68	57	Br, Br	82	106
$(C_6H_{13})CHBr(CH_3)$	78	57			
$(C_6H_5)CH_2CH_2Br$	85	53			
$n\text{-}C_8H_{17}Br$	80	57			
$n\text{-}C_7H_{15}I$	96	53	Br, Br	84	106
$(CH_3)_2\overline{CCH_2C}Br_2$	82	106			
$(CH_3)_2\overline{CCH(CH_3)C}Br_2$	79	106			
$(CH_3)_2\overline{CC(CH_3)_2C}Br_2$	78[c]	106			
$(C_6H_5)HCCH_2CBr_2$	71	106			
$CH_2C(CH_3)CH_2Cl$ (with Ph_3SnH)		83	$CH{\equiv}CCH_2Br$		110
			$CH_2{=}CHCH_2Br$		43
RC_6H_4Br (with Ph_3SnH) ($R = CH_3, OCH_3, H, F, Cl, Ph, CF_3$)		83, 97, 101	$CH_3CH{=}CHCH_2Cl$		57

[a]Where there are several halogen atoms, only one is replaced by H; Br, preferentially before Cl or F.
[b]Me_3SnH and Et_3SnH also react with CCl_4; Ph_3SnH does not react [*51*].
[c]With an excess of tin hydride: 69% of $(CH_3)_2\overline{CC(CH_3)_2C}H_2$.

are also hydrogenated under comparable conditions (*i.e.* without the addition elimination mechanism given above) (formation of alkane and substituted tin halide).

If isomerizations in the alkyl or alkenyl radicals R· are possible, which is generally the case, the radical halogen substitutions with tin hydrides generally give mixtures of isomers RH [*17, 57*]:

Br —(−Br·)→ (· ⇌ ·) —(+Br·, + Bu^n_3SnH / − Bu^n_3SnBr)→ $\frac{1}{2}$ (+)

$CH_3CH{=}CHCH_2Cl$ —(−Cl·)→ $CH_3CH{=}CHCH_2\cdot$ ⇅ $CH_2{=}CH\dot{C}HCH_3$ ←(−Cl)— $CH_2{=}CHCH(CH_3)Cl$

—(+Cl·, + Ph_3SnH / − Ph_3SnCl)→ $CH_3CH{=}CHCH_3$; $CH_2{=}CHCH_2CH_3$

$$CH{\equiv}CCH_2Br \xrightarrow{-Br\cdot} \left\{ \begin{matrix} CH{\equiv}CCH_2\cdot \\ \updownarrow \\ \dot{C}H{=}C{=}CH_2 \end{matrix} \right\} \xrightarrow[-Bu^n_3SnBr]{+Br\cdot, + Bu^n_3SnH} \left\{ \begin{matrix} CH{\equiv}CCH_3 \\ + \\ CH_2{=}C{=}CH_2 \end{matrix} \right.$$

$$\underset{\text{(epoxide)}}{CH_2{-}CHCH_2X} \xrightarrow{-X\cdot} \left\{ \begin{matrix} CH_2{-}CHCH_2\cdot \text{ (epoxide)} \\ \updownarrow \\ \cdot OCH_2CH{=}CH_2 \end{matrix} \right\} \xrightarrow[-\geqslant SnX]{+X\cdot, + \geqslant SnH} HOCH_2CH{=}CH_2$$

$$(+)\ C_6H_5CHClCH_3 \xrightarrow{-Cl\cdot} C_6H_5\dot{C}HCH_3 \xrightarrow[-Ph_3SnCl]{+Cl\cdot, + Ph_3SnD} (\pm)\ C_6H_5CHDCH_3$$

The last reaction, the conversion of the optically active (+) α-phenylethyl chloride into racemic ($\pm$) α-deuterophenylethane in particular, supports a radical explanation of the process of halogen substitution with tin hydrides [57]. Radicals initiate the hydrogenation reactions. Thus, azobis(isobutyronitrile) raises the yield in the replacement of the halogen in C_6H_5Cl from 0 to 64%, in C_6H_5Br from 4.7 to 41%, in cyclo-$C_6H_{11}Cl$ from 1 to 70%, and in $C_6H_5CH_2Cl$ from 26 to 100% [57].

The fact that the replacement of halogen in organic halides by hydrogen takes place by a mechanism different from solvolysis (S_N1 reactions) is shown by a comparison of the reaction rates. In halogen substitution by $(n\text{-}C_4H_9)_3SnH$, benzyl bromide reacts faster than *tert*-butyl bromide, while in the solvolysis of the halides benzyl chloride reacts slower than *tert*-butyl chloride [57]. The relative reaction rates in the abstraction of halogen by the following mechanisms

$$CH_3\cdot + R{-}Br \rightarrow CH_3Br + R\cdot$$

and

$$Bu^n{}_3SnH + R{-}Br \rightarrow Bu_3{}^nSnBr + RH$$

rise in the sequence of compounds with $R = ClCH_2\text{-} < C_6H_5CH_2\text{-} < Cl_2CH\text{-}$. It is therefore evident that substitution in trialkyltin hydrides also takes place via radicals [57].

The rapid radical halogen substitution with tin hydrides can be used to accelerate the slow (polar) hydrogenation of organic halides by lithium alanate, $LiAlH_4$. The addition of a few per cent of an organotin halide to the $LiAlH_4$–RX solution is sufficient:

$$\geqslant SnX \xrightarrow[-\frac{1}{4}LiAlX_4]{+\frac{1}{4}LiAlH_4} \geqslant SnH \xrightarrow[-\geqslant SnX]{+RX} RH$$

The catalytic action of the organotin halides (or the hydrides arising from them) increases in the sequence: $(n\text{-}C_4H_9)_3SnH < (n\text{-}C_4H_9)_2SnH_2 \sim (C_6H_5)_3SnH < C_4H_9SnH_3 \sim (C_6H_5)_2SnH_2$. $C_6H_5SnH_3$ and SnH_4 decompose too fast to be effective [58].

4.4 *Reaction of the Sn–H bond with alkenes and alkynes*

Tin hydrides add to alkenes and alkynes (see the comprehensive study [*61e*])

$$>Sn{-}H + >C{=}C< \rightarrow >Sn{-}C{-}C{-}H$$

$$>Sn{-}H + {-}C{\equiv}C{-} \rightarrow >Sn{-}C{=}C{-}H$$

(*cf.* section 2) or react differently with alkenes to retain the double bond (*e.g.* evolution of H_2 with acid hydrogen, reduction of reducible groups [40b, 42a]. Tin hydrides do not react, or react with difficulty in very rare cases, with benzenoid systems.

Addition with the formation of monomers, oligomers, and polymers takes place in accordance with the following equations:

$$R^1_3Sn{-}H + R^2HC{=}CR^3R^4 \rightarrow (R^1_3Sn)R^2HC{-}CHR^3R^4$$

$$R^1_3Sn{-}H + CH_2{=}CH(CHR^2)_nCH{=}CH_2 \rightarrow R^1_3SnCH_2{-}CH_2(CHR^2)_nCH{=}CH_2$$

$$R^1_3SnCH_2{-}CH_2(CHR^2)_nCH_2{-}CH_2SnR^1_3 \xleftarrow{+R^1_3Sn{-}H}$$

(R^1 = alkyl, phenyl; R^2, R^3 = H, CH_3; R^4 = H, alkyl, phenyl, OH, CN, COOR, OCOR, $CONH_2$, OR, N, R^1_3Si, R^1_3Ge, R^1_3Sn, R^1_3Pb)

$$Pr^n_2SnH_2 \xrightarrow[\text{slow}]{+CH_2=CHCOOCH_3} [Pr^n_2Sn(H)CH_2{-}CH_2COOCH_3] \xrightarrow[\text{fast}]{+CH_2=CHCOOCH_3} Pr^n_2Sn(CH_2{-}CH_2COOCH_3)_2$$

(non-isolable)

$$CH{\equiv}CR' \xrightarrow[\text{fast}]{+R_3Sn-H} R_3SnCH{=}CHR' \xrightarrow[\text{slow}]{+R_3Sn-H} R_3SnCH_2{-}CHR'SnR_3$$

(R = alkyl, phenyl; R′ = H, alkyl, phenyl, CH_2OH, $COOCH_3$)

$$nR_2SnH_2 + nHC{\equiv}CPh \rightarrow nR_2Sn(H)CH{=}CHPh \rightarrow [-R_2SnCH_2CHPh-]_n$$

(R = CH_3, C_2H_5, n-C_3H_7, n-C_4H_9, C_6H_5)

$$n\,Pr^n_2SnH_2 + n\,CH_2{=}CRQCR{=}CH_2 \rightarrow [-Pr^n_2SnCH_2CRHQCRHCH_2-]_n$$

(dienes $CH_2{=}CRQCR{=}CH_2$:
$CH_2{=}CHCOOCH{=}CH_2$, $(CH_2{=}CHCOOCH_2)_2$, $[CH_2{=}C(CH_3)COOCH_2]_2$, $[CH_2{=}C(CH_3)CO]_2O$, $(CH_2{=}CHCH_2OCOCH_2)_2O$, $(p\text{-}CH_2{=}CHC_6H_4)_2E^{IV}(C_6H_5)_2$

(E^{IV} = Ge, Sn, Pb), $CH_2{=}CHCH<(OCH_2)(OCH_2)>C<(CH_2O)(CH_2O)>CHCH{=}CH_2$)

Tin hydrides add to conjugated dienes in accordance with the radical stability: with butadiene the *trans*-1,4-adducts (migration of the unattacked double bond) are formed preferentially, together with the 1,2-adducts [*79*]:

$$R_3Sn{-}H + CH_2{=}CH{-}CH{=}CH_2 \xrightarrow{\text{radical initiator}} \begin{cases} R_3SnCH_2{-}CH{=}CH{-}CH_3\ (80\%) \\ R_3SnCH_2{-}CH_2{-}CH{=}CH_2\ (20\%) \end{cases}$$

With 2,3-dimethylbutadiene, on the other hand, a higher proportion of the 1,2-adducts is formed:

$$R_3SnH + CH_2{=}C(CH_3){-}C(CH_3){=}CH_2 \xrightarrow{\text{radical initiator}} \begin{cases} R_3SnCH_2{-}C(CH_3){=}C(CH_3){-}CH_3\ \ (68\%) \\ R_3SnCH_2{-}CH(CH_3){-}C(CH_3){=}CH_2\ \ (32\%) \end{cases}$$

Note. The isomers mentioned above have nothing to do with the stereospecificity of the addition to the C=C double bond itself. Tables 9.5–9.7 contain examples of the reactions and relevant references.

No addition of tin hydrides takes place, for example:

(a) in the case of alkenyl halides (see section 4.3):

$$(C_6H_5)_3SnH + CH_2{=}CHCH_2Br \rightarrow (C_6H_5)_3SnBr + CH_2{=}CHCH_3$$

(b) in the case of alkenylamines R–NH_2 [not in the case of R–N(alkyl)$_2$ or in the case of amides R–$CONH_2$] [*42*, *43*, *85*, *111*, see, however, *45* and *115*]:

$$2(C_6H_5)_3SnH + CH_2{=}CHCH_2NH_2 \rightarrow (C_6H_5)_3SnSn(C_6H_5)_3 + CH_2{=}CHCH_3 + NH_3$$

(c) generally in the case of unsaturated acids [*43*, *71*, *85*, *111*]:

$$(n{-}C_3H_7)_3SnH + HOOCCH{=}CH_2 \rightarrow (n{-}C_3H_7)_3SnOOCCH{=}CH_2 + H_2$$

$$(C_6H_5)_3SnH + CH_2{=}CHCOOH \rightarrow (C_6H_5)_3SnCH_2CH_2COOH$$

$$\downarrow$$

$$(C_6H_5)_2Sn(\text{–O–C(=O)–}CH_2\text{–}CH_2\text{–, ring, O}\rightarrow Sn) + C_6H_6$$

(d) in the case of vinyl ketones (apart from hex-1-en-5-one) [*42*]:

$$2(C_6H_5)_3SnH + CH_2{=}CHCOR' \rightarrow (C_6H_5)_3SnSn(C_6H_5)_3 + CH_2{=}CHCH(OH)R' \quad \text{(a)}$$

$$(C_6H_5)_3SnH + CH_2{=}CH(CH_2)_2COCH_3 \rightarrow (C_6H_5)_3Sn(CH_2)_4COCH_3 \quad \text{(b)}$$

(The two reactions take place in a manner opposite to that expected, since the conjugation of double bonds (a) should promote the addition, while their isolation (b) should hinder it; see below).

(e) in the case of nitrobenzene (reduction) [*80*]:

$$6(C_6H_5)_3SnH + C_6H_5NO_2 \rightarrow 3(C_6H_5)_3SnSn(C_6H_5)_3 + C_6H_5NH_2 + 2H_2O$$

The experimental results now available indicate a radical addition of the tin hydrides to alkenes [*53, 61e, 71, 76, 77*]:

$$\text{initiator} \rightarrow 2\ \text{radical}\cdot \qquad \text{(a)}$$

$$\text{radical}\cdot + \rangle\text{Sn–H} \rightarrow \text{radical H} + \rangle\text{Sn}\cdot \qquad \text{(b)}$$

$$\rangle\text{Sn}\cdot + \rangle\text{C=C}\langle \rightarrow \rangle\text{Sn–}\overset{|}{\underset{|}{\text{C}}}\text{–}\overset{|}{\underset{|}{\text{C}}}\cdot \qquad \text{(c)}$$

$$\rangle\text{Sn–}\overset{|}{\underset{|}{\text{C}}}\text{–}\overset{|}{\underset{|}{\text{C}}}\cdot + \rangle\text{Sn–H} \rightarrow \rangle\text{Sn–}\overset{|}{\underset{|}{\text{C}}}\text{–}\overset{|}{\underset{|}{\text{C}}}\text{–H} + \rangle\text{Sn}\cdot \qquad \text{(d)}$$

Neither hexachloroplatinic(IV) acid (polar catalysis) nor benzoyl peroxide or irradiation (radical catalysis) affect the addition [*27, 28*]. Benzoyl peroxide has no catalytic effect, since its reaction with the organotin hydride given below competes with reaction (b), and the benzoic acid produced forms an organotin benzoate with the evolution of H_2 [*71, 76, 77*]:

$$\rangle\text{SnH} + (C_6H_5COO)_2 \rightarrow C_6H_5COOSn\langle + C_6H_5COOH$$

$$C_6H_5COOH + \rangle\text{SnH} \rightarrow C_6H_5COOSn\langle + H_2$$

The addition is catalysed satisfactorily by other (radical) catalysts: *tert*-butyl peroxide, azoisobutyric ester, azobis(isobutyronitrile), phenyl-azoisobutyronitrile, and especially by benzyl hyponitrite [*71, 73b, 77, 79a*]. The radical-acceptor hydroquinone does not inhibit the reaction, since the organotin hydride itself is a still better "radical trap" [eqs. (b) and (d)]. In the absence of donors such as ether or a tertiary amine, the addition of $\rangle$Sn–H to alkenes is catalysed by AlR_3, $AlHR_2$, AlH_3, $LiAlR_4$, and $LiAlH_4$ [*75*]:

$$R_3SnH + R_3Al \rightarrow R_3Sn\langle{}^{H}_{R}\rangle AlR_2 \rightarrow R_4Sn + R_2AlH$$

$$R_2AlH + CH{=}CH_2R' \rightarrow R_2Al(CH_2CH_2R')$$

$$R_3SnH + CH_2{=}CHR' \xrightarrow{(R_3Al)} R_3Sn(CH_2CH_2R') \quad (R = R'CH_2CH_2\text{–})$$

The reaction mechanism given points to a polar (?) four-centre reaction, which, from the analogous additions of the Si–H and Ge–H bonds, is quite possible. The addition to α-alkenes or dienes takes place particularly smoothly with AlR_3. Table 9.5 and 9.6 contain examples of these reactions.

References p. 752

Phenylstannanes are reactive and add even without a catalyst. On the other hand, alkylstannanes are unreactive; they require catalysts for the addition.

Alkenes with terminal C=C double bonds in conjugation with other double bonds, (*e.g.* $\rangle C{=}\overset{|}{C}{-}\overset{|}{C}{=}O$) react more easily than alkenes with isolated non-terminal C=C double bonds[*16, 33, 40, 42, 43, 83*]. The former add $-\!\!\!\!\rangle Sn{-}H$ without catalysts. However, catalysts shorten the time of reaction and permit lower reaction temperatures. Consequently, they prevent a decomposition of the hydrides completely or partially, in accordance with the following equations

$$2R_3SnH \rightarrow R_3SnSnR_3 + H_2$$

or

$$nR_2SnH_2 \rightarrow (R_2Sn)_n + nH_2$$

and of the primary additions product, for example[*71, 77*]:

$$R_3SnH + CH_2{=}CHCOOCH_3 \rightarrow R_3SnCH_2CH_2OCOCH_3$$

(at the high temperature necessary in the absence of a catalyst)

$$\downarrow$$

$$R_3SnOCOCH_3 + CH_2{=}CH_2$$

The conjugation of C=C double bonds does not accelerate the reaction, as shown by additions to penta-1,3-diene[16].

In the addition of $-\!\!\!\!\rangle Si{-}H$ to alkenes, the latter frequently undergo polymerization or telomerization (see section 5.11.2 of Chapter 7). This does not occur to an appreciable extent in the addition of $-\!\!\!\!\rangle Sn{-}H$ to alkenes, probably because in radical polymerization $R\cdot + \rangle C{=}C\langle \rightarrow R{-}\overset{|}{\underset{|}{C}}{-}\overset{|}{\underset{|}{C}}\cdot$, ~ 60 kcal/mole must be supplied for the breaking of the C=C double bond, while in the formation of the C–C single bond 80 kcal/mole is obtained (a gain of 20 kcal/mole). ~ 70 kcal/mole must be supplied to open the Sn–H bond. Here 98 kcal/mole is obtained by the formation of the C–H bond (gain 28 kcal/mole). On the other hand, the opening of an Si–H bond requires 76 kcal/mole; again, 98 kcal/mole is obtained in the formation of the C–H bond. The gain in energy amounts to 22 kcal/mole, and is therefore almost identical with the gain on polymerization. In other words, in the reaction of alkenes with silicon hydrides addition and polymerization are equally probable, so that telomerization frequently occurs. In the reaction with tin hydrides, the favored reaction is the addition, so that telomerization takes place rarely.

Alkyltin hydrides do in fact catalyse polymerization reactions in the presence of cocatalysts. Thus, with the mixed catalyst (R_2SnH_2 or $RSnH_3$)[3] + (Co and/or Ni "derivatives") + (Al halide) conjugated dienes give a yield of at least 90%[5] of *cis*-1,4-polymers [108]. Mixed catalysts from $ZrCl_4$ (or other metal halides of the fourth or fifth subgroups) and Bu_2SnH_2 bring about the polymerization of acrylonitrile to polyacrylonitrile within 6 hours at 70° [*36*].

In general, alkyl *halides* do not add tin hydrides but react with them with replacement of the halogen. The addition $>Sn\cdot + >C{=}C< \rightarrow >Sn{-}\overset{|}{\underset{|}{C}}{-}\overset{|}{\underset{|}{C}}\cdot$ therefore probably takes place more slowly than the substitution $>Sn\cdot + RX \rightarrow$

While it is unaffected by polar influences of the alkene substituents R′ and R″ in R′CH=CHR″, the triorganotin group R_3Sn(R = alkyl or aryl) is directed both by the stability of the intermediate (alkyl or aryl) radicals (primary < secondary < tertiary) and by steric effects. Usually both effects lead to terminal addition of the triorganotin group.

Alkynes add $>SnH$ substantially more readily than alkenes (radical and polar mechanism [*16c, 61c-e, 62b, 62e*]). The solvent effect indicates an appreciable charge separation in the transition state of the rate determining step of the polar mechanism. From the substituent effects it can be concluded that in this transition state the tin atom has a positive and the triple bond a negative charge. Consequently, the mechanism could involve a slow hydride transfer, followed by a fast step [*61d*].

$$R_3SnH + R'{-}C{\equiv}C{-}R'' \xrightarrow{\text{slow}} \left[\begin{array}{c} R' \quad\quad\quad \\ C{\equiv}C \\ H \quad\quad\quad R'' \\ SnR_3 \end{array} \right] \longrightarrow$$

$$\rightarrow R_3Sn^{\oplus} + \begin{array}{c} R' \quad\quad\quad \\ C{=}C^{\ominus} \\ H \quad\quad\quad R'' \end{array} \xrightarrow{\text{fast}} \begin{array}{c} R' \quad\quad SnR_3 \\ C{=}C \\ H \quad\quad\quad R'' \end{array}$$

All known reactions of tin hydrides with alkynes take place rapidly and exothermically. The resulting alkenyl-substituted tin hydrides, which can be isolated by breaking off the reaction, add further tin hydride only slowly ($alkyl_3Sn$– more slowly than $phenyl_3Sn$–), with the formation of ditin compounds.

---

[3] R = Ph, for example.

References p. 752

[*text continued p. 746*]

TABLE 9.5

ADDITION OF TRIORGANOTIN HYDRIDES, R_3SnH, TO ALKENES AND DIENES

[Catalyst AIBN = 2,2′-azobis (isobutyronitrile)]

Tin hydride	*Unsaturated compounds*	*Catalyst*	*Reaction product (yields in mole-%)*	*Ref.*
$(CH_3)_3SnH$	CH_2=C=CH_2	AIBN	67% [45% $(CH_3)_3SnC(CH_3)$=CH_2 + 55% $(CH_3)_3SnCH_2CH$=CH_2]	58a
	CH_3CH=C=CH_2	AIBN	72% [31% $(CH_3)_3SnC(CH_2CH_3)$=CH_2 + 39% $(CH_3)_3SnC(CH_3)$=$C(CH_3)H$]	58a
	$(CH_3)_2C$=C=CH_2	AIBN	72% [27% $(CH_3)_3SnC(CMe_2H)$=CH_2 + 73% $(CH_3)_3SnC(CH_3)$=$C(CH_3)_2$]	58a
	CH_3CH=C=$CHCH_3$	AIBN	65% $(CH_3)_3SnC(CH_2CH_3)$=$CHCH_3$	58a
	$(CH_3)_2C$=C=$CHCH_3$	AIBN	82% [55% $(CH_3)_3SnC(CHMe_2)$=$CHCH_3$ + 45% $(CH_3)_3SnC(CH_2CH_3)$=$C(CH_3)_2$]	58a
$(C_2H_5)_3SnH$	CH_2=$CH(CH_2)_5CH_3$	AIBN or R_3Al	79% $(C_2H_5)_3Sn(CH_2)_7CH_3$	
	CH_2=$CH(CH_2)_7CH_3$	AIBN	68% $(C_2H_5)_3Sn(CH_2)_9CH_3$	77
	CH_2=$CH(C_6H_5)$	AIBN	48% $(C_2H_5)_3Sn(CH_2)_2(C_6H_5)$	77
	CH_2=$C(CH_3)(C_6H_5)$	AIBN	79% $(C_2H_5)_3SnCH_2CH(CH_3)(C_6H_5)$	77
	CH_2=CHCN	AIBN	86% $(C_2H_5)_3SnCH_2CH_2CN$	77
	CH_2=$CHCH_2OH$	AIBN	89% $(C_2H_5)_3Sn(CH_2)_3OH$	77
	CH_2=$CH(CH_2)_3OH$	AIBN	59% $(C_2H_5)_3Sn(CH_2)_5OH$	77
	CH_2=$CHCOOCH_3$	AIBN	93% $(C_2H_5)_3Sn(CH_2)_2COOCH_3$	77
	CH_2=$CH(CH_2)_7COOC_2H_5$	AIBN	78% $(C_2H_5)_3Sn(CH_2)_9COOC_2H_5$	77
	CH_2=$CH(CH_2)_8COOC_2H_5$	AIBN	78% $(C_2H_5)_3Sn(CH_2)_{10}COOC_2H_5$	71
	CH_2=$CHOCOCH_3$	AIBN	65% $(C_2H_5)_3Sn(CH_2)_2OCOCH_3$	77
	CH_2=$CHCH_2OCOCH_3$	AIBN	83% $(C_2H_5)_3Sn(CH_2)_3OCOCH_3$	77
	CH_2=$C(CH_3)COOCH_3$	AIBN	80% $(C_2H_5)_3SnCH_2CH(CH_3)COOCH_3$	77
	CH_2=$CHCONH_2$	AIBN	75% $(C_2H_5)_3Sn(CH_2)_2CONH_2$	77
	CH_2=$C(CH_3)CONH_2$	AIBN	67% $(C_2H_5)_3SnCH_2CH(CH_3)CONH_2$	77
	CH_2=$CHOC_2H_5$	AIBN	65% $(C_2H_5)_3Sn(CH_2)_2OC_2H_5$	77
	CH_2=$CHO(CH_2)_3CH_3$	AIBN	86% $(C_2H_5)_3Sn(CH_2)_2O(CH_2)_3CH_3$	77
	CH_2=$CHOCH_2CH(CH_3)_2$	AIBN	87% $(C_2H_5)_3Sn(CH_2)_2OCH_2CH(CH_3)_2$	71
	CH_2=$CHCH_2OCH_2(OH)CH_2NH_2$	AIBN	78% $(C_2H_5)_3Sn(CH_2)_3OCH_2CH(OH)CH_2NH_2$	77
	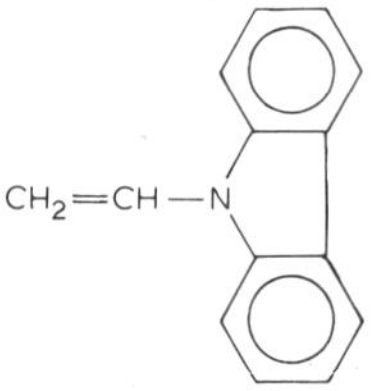 CH_2=CH—N (carbazole ring structure)	AIBN	82% 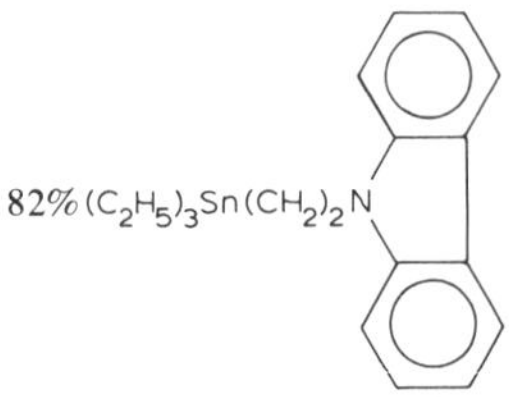 $(C_2H_5)_3Sn(CH_2)_2N$ (carbazole ring structure)	77
	CH_2=$CHCH_2Sn(C_2H_5)_3$	R_3Al	68% $(C_2H_5)_3Sn(CH_2)_3Sn(C_2H_5)_3$	75
	CH_2=$CHCH$=CH_2		80% $(C_2H_5)_3SnCH_2CH$=$CHCH_3$, 20% $(C_2H_5)_3Sn(CH_2)_2CH$=CH_2	79
	CH_2=$CHC(CH_3)$=CH_2		81% $(C_2H_5)_3SnCH_2CH$=$C(CH_3)_2$, 19% $(C_2H_5)_3Sn(CH_2)_2C(CH_3)$=$CH_2$	79
	CH_2=$C(CH_3)C(CH_3)$=CH_2		68% $(C_2H_5)_3SnCH_2C(CH_3)$=$C(CH_3)_2$, 32% $(C_2H_9)_3SnCH_2CH(CH_3)C(CH_3)$=$CH_2$	79
	CH_2=$CH(CH_2)_2CH$=CH_2	AIBN	72% $(C_2H_5)_3Sn(CH_2)_4CH$=CH_2	77
	CH_2=$CH(CH_2)_4CH$=CH_2	AIBN	65% $(C_2H_5)_3Sn(CH_2)_6CH$=CH_2, 20% $(C_2H_5)_3Sn(CH_2)_6CH(CH_3)Sn(C_2H_5)_3$	71, 77
	CH_2=CHC_6H_4CH=CH_2	AIBN	37% $(C_2H_5)_3Sn(CH_2)_2C_6H_4CH$=$CH_2$ (hydride : diene = 1 : 1)	77

TABLE 9.5 (cont'd)

Tin hydride	*Unsaturated compounds*	*Catalyst*	*Reaction product (yields in mole-%)*	*Ref.*
	$\overline{CH{=}CHCH_2CH_2CH}$		87% $(C_2H_5)_3Sn\overline{CHCH_2CH_2CH_2CH_2}$	45
	$CH_2{=}\overline{CHCHCH_2CH{=}CHCH_2CH_2}$	AIBN	53% $(C_2H_5)_3Sn(CH_2)_2\overline{CHCH_2CH{=}CHCH_2CH_2}$ (hydride : diene = 1 : 1)	77
		R_3Al	80%	75
	$(C_6H_5)_3N_3B_3(CH{=}CH_2)_3$	AIBN	84% $(C_6H_5)_3N_3B_3[(CH_2)_2Sn(C_2H_5)_3]_3$	105
$(n\text{-}C_3H_7)_3SnH$	$CH_2{=}CHC_6H_5$		86% $(n\text{-}C_3H_7)_3Sn(CH_2)_2C_6H_5$	42
	$CH_2{=}CHCN$		70% $(n\text{-}C_3H_7)_3Sn(CH_2)_2Cn$	43
	$CH_2{=}CHCN_2CN$		53% $(n\text{-}C_3H_7)_3Sn(CH_2)_3CN$	42
	$C(CN)H{=}CH(C_6H_5)$		83% $(n\text{-}C_3H_7)_3SnC(CN)HCH_2(C_6H_5)$?	42
	$CH_2{=}CHCOOH$		no addition! $CH_2{=}CHCOOSn(C_3H_7)_3 + H_2$	42
	$CH_2{=}CHCOOCH_3$		63% $(n\text{-}C_3H_7)_3Sn(CH_2)_2COOCH_3$	43
	$CH_2{=}CHCH_2COOC_2H_5$		30% $(n\text{-}C_3H_7)_3Sn(CH_2)_3COOC_2H_5$	42
	$CH_2{=}CHCONH_2$		77% $(n\text{-}C_3H_7)_3Sn(CH_2)_2CONH_2$	43
	$C(CH_3)H{=}CHCOOC_2H_5$		40% $(n\text{-}C_3H_7)_3SnC(CH_3)HCH_2COOC_2H_5$?	42
	$CH_2{=}CH{-}C_6H_5$		64% $(n\text{-}C_3H_7)_3Sn(CH_2)_2{-}C_6H_5$	42
$(n\text{-}C_4H_9)_3SnH$	$CH_2{=}CHCN$		70% $(n\text{-}C_4H_9)_3Sn(CH_2)_2CN$	43
	$CH_2{=}CHCH_2OH$		89% $(n\text{-}C_4H_9)_3Sn(CH_2)_3OH$	77
	$CH_2{=}CHCOOCH_3$		84% $(n\text{-}C_4H_9)_3Sn(CH_2)_2COOCH_3$	43
	$CH_2{=}CHOCH_2CH(CH_3)_2$	AIBN	86% $(n\text{-}C_4H_9)_3Sn(CH_2)_2OCH_2CH(CH_3)_2$	77
	$CH_2{=}CH(CH_2)_2CH{=}CH_2$	AIBN	68% $(n\text{-}C_4H_9)_3Sn(CH_2)_4CH{=}CH_2$, 20% $[(n\text{-}C_4H_9)_3Sn(CH_2)_3]_2$	77
	$CH_2{=}CH(CH_2)_4CH{=}CH_2$	AIBN	63% $(n\text{-}C_4H_9)_3Sn(CH_2)_6CH{=}CH_2$ 31% $(n\text{-}C_4H_9)_3Sn(CH_2)_6CH(CH_3Sn(n\text{-}C_4H_9)_3$	77
	$CH_2{=}C(CH_3)C(CH_3){=}CH_2$		72% $(n\text{-}C_4H_9)_3SnCH_2C(CH_3){=}C(CH_3)_2$, 28% $(n\text{-}C_4H_9)_3SnCH_2CH(CH_3)C(CH_3){=}CH_2$	79
$(i\text{-}C_4H_9)_3SnH$	$CH_2{=}CH(CH_2)_5CH_3$	AIBN	81% $(i\text{-}C_4H_9)_3Sn(CH_2)_7CH_3$	77
	$CH_2{=}C(CH_3)COOCH_3$	AIBN	97% $(i\text{-}C_4H_9)_3SnCH_2CH(CH_3)COOCH_3$	77
	$CH_2{=}CHCH_2OCOCH_3$	AIBN	91% $(i\text{-}C_4H_9)_3Sn(CH_2)_3OCOCH_3$	77
	$CH_2{=}CHO(CH_2)_3CH_3$	AIBN	80% $(i\text{-}C_4H_9)_3Sn(CH_2)_2O(CH_2)_3CH_3$	77
$(C_6H_5)_3SnH$	$CH_2{=}CH(CH_2)_5CH_3$		72% $(C_6H_5)_3Sn(CH_2)_7CH_3$	43
	$CH_2{=}CHC_6H_5$		82% $(C_6H_5)_3Sn(CH_2)_2C_6H_5$	40a, 43
	$CH_2{=}CHCN$		85% $(C_6H_5)_3Sn(CH_2)_2CN$	43
	$CH_2{=}CHCH_2CN$		73% $(C_6H_5)_3Sn(CH_2)_3CN$	43
	$CH(CH_3){=}CHCN$		89% $(C_6H_5)_3SnCH(CH_3)CH_2CN$	43
	$CH_2{=}CHCH_2OH$		93% $(C_6H_5)_3Sn(CH_2)_3OH$	43
	$CH_2{=}C(CH_3)CH_2OH$		41% $(C_6H_5)_3SnCH_2CH(CH_3)CH_2OH$	42
	$CH_2{=}CHCOOH$		94% $(C_6H_5)_3Sn(CH_2)_2COOH$ [migration of protons $\rightarrow C_6H_6 + (C_6H_5)_2Sn(CH_2)_2COO$?]	43, 71, 85, 111
	$CH_2{=}CHCOOCH_3$		85% $(C_6H_5)_3Sn(CH_2)_2COOCH_3$	43
	$CH_2{=}C(CH_3)COOCH_3$		56% $(C_6H_5)_3SnCH_2CH(CH_3)COOCH_3$	42
	$CH_2{=}CHOCOCH_3$		100% $(C_6H_5)_3Sn(CH_2)_2CH_2OCOCH_3$	43
	$CH_2{=}CHCH_2OCOCH_3$		42% $(C_6H_5)_3Sn(CH_2)_3OCOCH_3$	42
	$CH_2{=}CHCONH_2$		90% $(C_6H_5)_3Sn(CH_2)_2CONH_2$	43
	$CH_2{=}CHCH(OC_2H_5)_2$		90% $(C_6H_5)_3Sn(CH_2)_2CH(OC_2H_5)_2$	43
	$CH_2{=}CH(CH_2)_2COCH_3$		39% $(C_6H_5)_3Sn(CH_2)_4COCH_3$	42
	$CH_2{=}CHOC_6H_5$		89% $(C_6H_5)_3Sn(CH_2)_2OC_6H_5$	43
	$CH_2{=}CHCH_2O(CH_2)_2CN$		50% $(C_6H_5)_3Sn(CH_2)_3O(CH_2)_2CN$	42

References p. 752

TABLE 9.5 (cont'd)

Tin hydride	*Unsaturated compounds*	*Catalyst*	*Reaction product (yields in mole-%)*	*Ref.*
	$CH_2{=}CHCH_2NHCOCH_3$		35% $(C_6H_5)_3Sn(CH_2)_3NHCOCH_3$	42
	p-$CH_2{=}CHC_6H_4NHCOCH_3$		98% p-$(C_6H_5)_3Sn(CH_2)_2C_6H_4NHCOCH_3$	42
	4-$CH_2{=}CHC_5H_4N$		83% 4-$(C_6H_5)_3Sn(CH_2)_2C_5H_4N$	42
	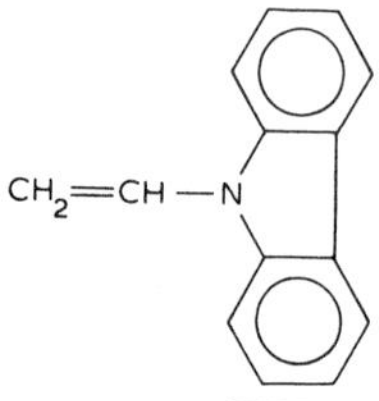 $CH_2{=}CH-N$ (carbazole)		93% $(C_6H_5)_3Sn(CH_2)_2-N$ (carbazole) 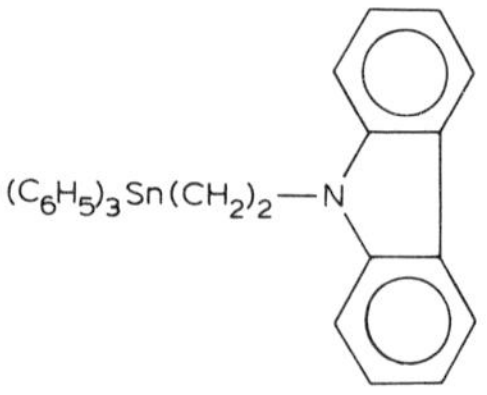	42
	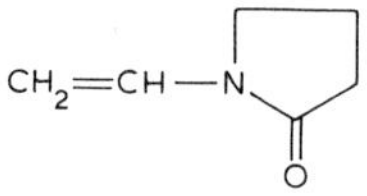 $CH_2{=}CH-N$ (pyrrolidone, C=O)		78% $(C_6H_5)_3Sn(CH_2)_2-N$ (pyrrolidone, C=O) 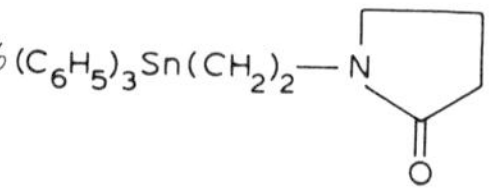	43
	$CH_2{=}CHSi(C_6H_5)_3$		54% $(C_6H_5)_3Sn(CH_2)_2Si(C_6H_5)_3$	33
	$CH_2{=}CHGe(C_6H_5)_3$		32% $(C_6H_5)_3Sn(CH_2)_2Ge(C_6H_5)_3$	33
	$CH_2{=}CHSn(C_6H_5)_3$		50% $(C_6H_5)_3Sn(CH_2)_2Sn(C_6H_5)_3$	33
	$CH_2{=}CHCH_2Sn(C_6H_5)_3$		18% $(C_6H_5)_3Sn(CH_2)_3Sn(C_6H_5)_3$	42
	p-$CH_2{=}CHC_6H_4Ge(C_6H_5)_3$		80% p-$(C_6H_5)_3Sn(CH_2)_2C_6H_4Ge(C_6H_5)_3$	84
	p-$CH_2{=}CHC_6H_4Sn(C_6H_5)_3$		81% p-$(C_6H_5)_3Sn(CH_2)_2C_6H_4Sn(C_6H_5)_3$	84
	p-$CH_2{=}CHC_6H_4Pb(C_6H_5)_3$		85% p-$(C_6H_5)_3Sn(CH_2)_2C_6H_4Pb(C_6H_5)_3$	84
	$[CH_2{=}C(CH_3)CH_2]_2$		30% $[(C_6H_5)_3SnCH_2CH(CH_3)CH_2]_2$	42
	p-$CH_2{=}CHC_6H_4CH{=}CH_2$		40% p-$(C_6H_5)_3Sn(CH_2)_2C_6H_4(CH_2)_2Sn(C_6H_5)_3$	42
	$[CH_2{=}CHCO]_2O$		60% $[(C_6H_5)_3Sn(CH_2)_2CO]_2O$	42
	$[CH_2{=}CHCOOCH_2]_2$		91% $[(C_6H_5)_3Sn(CH_2)_2COOCH_2]_2$	42
	$[CH_2{=}CHCH_2]_2O$		33% $[(C_6H_5)_3Sn(CH_2)_3]_2O$	42
	$[p$-$CH_2{=}CHC_6H_4]_2Ge(C_6H_5)_2$		90% $[p$-$(C_6H_5)_3Sn(CH_2)_2C_6H_4]_2Ge(C_6H_5)_2$	84
	$[p$-$CH_2{=}CHC_6H_4]_2Sn(C_6H_5)_2$		77% $[p$-$(C_6H_5)_3Sn(CH_2)_2C_6H_4]_2Sn(C_6H_5)_2$	84
	$[p$-$CH_2{=}CHC_6H_4]_2Pb(C_6H_5)_2$		89% $[p$-$(C_6H_5)_3Sn(CH_2)_2C_6H_4]_2Pb(C_6H_5)_2$	84
	$(C_6H_5)_3N_3B_3(CH{=}CH_2)_3$		73.5% $(C_6H_5)_3N_3B_3[(CH_2)_2Sn(C_6H_5)_3]_3$	105
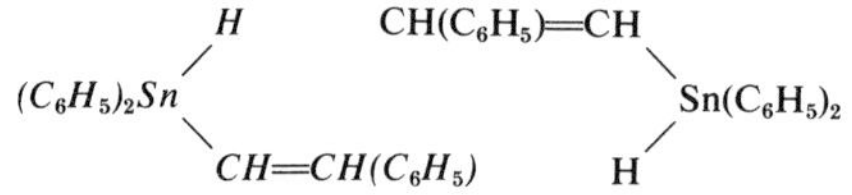 $H-(C_6H_5)_2Sn-CH{=}CH(C_6H_5)$	$CH(C_6H_5){=}CH-Sn(C_6H_5)_2-H$		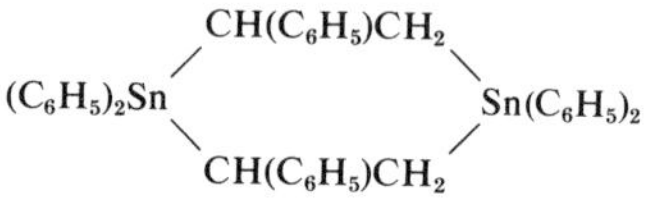 $(C_6H_5)_2Sn[CH(C_6H_5)CH_2]_2Sn(C_6H_5)_2$ (cyclic)	31

TABLE 9.6

ADDITION OF DIORGANOTIN DIHYDRIDES, R_2SnH_2, OR ORGANOTIN TRIHYDRIDES, $RSnH_3$, TO ALKENES OR DIENES

[Catalyst AIBN = 2,2′-azobis(isobutyronitrile)]

Tin hydride	*Unsaturated compound*	*Catalyst*	*Reaction product (yields in mole-%)*	*Ref.*
$(C_2H_5)_2SnH_2$	$CH_2{=}CHCH_2O(CH_2)_2OH$	AIBN	55% $(C_2H_5)_2Sn[(CH_2)_3O(CH_2)_2OH]_2$	71
$(C_2H_5)_2SnH_2$	$CH_2{=}CHC_6H_5$	AIBN	85% $(C_2H_5)_2Sn[(CH_2)_2C_6H_5]_2$	71
$(n\text{-}C_3H_7)_2SnH_2$	$CH_2{=}CHCOOCH_3$		60% $(n\text{-}C_3H_7)_2Sn[(CH_2)_2COOCH_3]_2$	42
$(n\text{-}C_3H_7)_2SnH_2$	$CH_2{=}CHCN$		43% $(n\text{-}C_3H_7)_2Sn[(CH_2)_2CN]_2$	42
$(n\text{-}C_3H_7)_2SnH_2$	$CH_2{=}CHC_6H_5$		82% $(n\text{-}C_3H_7)_2Sn[(CH_2)_2C_6H_5]_2$	42

TABLE 9.5 (cont'd)

Tin hydride	*Unsaturated compounds*	*Catalyst*	*Reaction product (yields in mole-%)*	*Ref.*
$(n\text{-}C_4H_9)_2SnH_2$	$CF_2{=}CF_2$		$(n\text{-}C_4H_9)_2Sn[CF_2CF_2H]_2$	49
$(n\text{-}C_4H_9)_2SnH_2$	$CH_2{=}CH(CH_2)_4CH{=}CH_2$	AlR_3	73% $(n\text{-}C_4H_9)_3Sn[(CH_2)_6CH{=}CH_2]_2$	75, 83a
$(n\text{-}C_4H_9)_2SnH_2$	$CH_2{=}CHCH_2OH$		$(n\text{-}C_4H_9)_2Sn(CH_2CH_2CH_2OH)_2$	61a
$(C_6H_5)_2SnH_2$	$CH_2{=}CHCOOCH_3$		34% $(C_6H_5)_2Sn[(CH_2)_2COOCH_3]_2$	42
$(C_6H_5)_2SnH_2$	$(CH_2{=}CH)_2E^{IV}(C_6H_5)_2$		$(C_6H_5)_2Sn(CH_2CH_2)_2E^{IV}(C_6H_5)_2$, E^{IV} = Si, Ge, not Sn	31
$(C_6H_5)_2SnH_2$	$CH_2{=}CH$, $CH_2{=}CH$ (o-C_6H_4)		$(C_6H_5)_2Sn(CH_2CH_2)_2C_6H_4$	86
$(C_6H_5)_2SnH_2$	$CH_2{=}CHC_6H_4CH{=}CH_2$		$[-Sn(C_6H_5)_2(CH_2)_2C_6H_4(CH_2)_2-]_x$ (rubber)	86
$(n\text{-}C_3H_7)SnH_3$	$CH_2{=}CHCOOCH_3$		11% $(n\text{-}C_3H_7)Sn[(CH_2)_2COOCH_3]_3$	42
$(n\text{-}C_4H_9)SnH_3$	$CH_2{=}CHCOOCH_3$		82% $(n\text{-}C_4H_9)Sn[(CH_2)_2COOCH_3]_3$	42
$(C_4H_9)SnH_3$	$CH_2{=}C(CH_3)COOCH_3$	AIBN	75% $(C_4H_9)Sn[CH_2CH(CH_3)COOCH_3]_3$	71
$(i\text{-}C_4H_9)SnH_3$	$CH_2{=}CH(CH_2)_5CH_3$	AlR_3	59% $(i\text{-}C_4H_9)Sn(CH_2)_7(CH_3]_3$	75

TABLE 9.7

ADDITION OF ORGANOTIN HYDRIDES, R_3SnH AND R_3SnH_2, TO ALKYNES OR DIYNES

Tin hydride	*Unsaturated compound*	*Reaction product (yields in mole-%)*	*Ref.*
$(n\text{-}C_3H_7)_3SnH$	$CH{\equiv}C(CH_2)_3CH_3$	82% $(n\text{-}C_3H_7)_3SnCH{=}CH(CH_2)_3CH_3$	42
$(n\text{-}C_3H_7)_3SnH$	$CH{\equiv}C(C_6H_5)$	75% $(n\text{-}C_3H_7)_3SnCH{=}CH(C_6H_5)$	42
$(n\text{-}C_3H_7)_3SnH$	$CH{\equiv}CCH_2OH$	34% $(n\text{-}C_3H_7)_3SnCH{=}CHCH_2OH$	42
$(n\text{-}C_3H_7)_3SnH$	$CH{\equiv}C(COOCH_3)$	69% $(n\text{-}C_3H_7)_3SnCH_2CH(COOCH_3)Sn(n\text{-}C_3H_7)_3$	42
$(n\text{-}C_4H_9)_3SnH$	$CH{\equiv}CH$	81% $(n\text{-}C_4H_9)_3SnCH{=}CH_2$	110
$(C_6H_5)_3SnH$	$CH{\equiv}CH$	42% $(C_6H_5)_3SnCH_2CH_2Sn(C_6H_5)_3$	42
$(C_6H_5)_3SnH$	$CH{\equiv}C(C_6H_5)$	94% $(C_6H_5)_3SnCH{=}CH(C_6H_5)$	42
$(C_6H_5)_3SnH$	$CH{\equiv}C(C_6H_5)$	49% $(C_6H_5)_3SnCH_2CH(C_6H_5)Sn(C_6H_5)_3$	42
$(C_6H_5)_3SnH$	$CH{\equiv}CCH_2OH$	51% $(C_6H_5)_3SnCH{=}CHCH_2OH$	42
$(C_6H_5)_3SnH$	$CH{\equiv}C(COOCH_3)$	52% $(C_6H_5)_3SnCH_2CH(COOCH_3)Sn(C_6H_5)_3$	42, 62c
R_2SnH_2	$CH{\equiv}CCH_2CH_2C{\equiv}CH$	$[-SnR_2CH{=}CHCH_2CH_2CH{=}CH-]_x$; 12–18% (ring: Sn, R, R); R = CH_3, C_2H_5, $n\text{-}C_3H_7$, C_6H_5	54, 86
$(CH_3)_2SnH_2$	C≡CH, C≡CH (o-C_6H_4)	Sn, CH_3, CH_3 (benzo ring)	62h

References p. 752

TABLE 9.7 (cont'd)

Tin hydride	*Unsaturated compound*	*Reaction product (yields in mole-%)*	*Ref.*
$(n\text{-}C_3H_7)_2SnH_2$	$CH\equiv CC_6H_5$	$[-Sn(n\text{-}C_3H_7)_2CH_2CH(C_6H_5)-]_{\sim 11}$ (viscous oil)	85, 86
$(n\text{-}C_4H_9)_2SnH_2$	$CH\equiv CC_6H_5$	$[-(n\text{-}C_4H_9)_2CH_2CH(C_6H_5)-]_{\sim 11}$ (viscous oil)	85, 86
$(C_6H_5)_2SnH_2$	$CH\equiv CC_6H_5$	$[-Sn(C_6H_5)_2CH_2CH(C_6H_5)-]_x$ (hard solid)	85, 86
$C_6H_4[Sn(CH_3)_2H]_2$	$CH\equiv CC_6H_5$	$[-Sn(CH_3)_2C_6H_4Sn(CH_3)_2CH_2CH(C_6H_5)-]_n$	62g

4.5 *Reaction of the Sn–H bond with the C=O, C=S, and C=N double bonds*

Organotin hydrides react with (bond attacked given in parenthesis) aldehydes (C=O), ketones (C=O), azomethines (C=N), diazo compounds (C=N), isocyanates (C=N), and isothiocyanates (C=S).

Carbonyl double bonds, $>C=O$, add organotin hydrides $R_{4-n}SnH_n$ ($n = 1, 2, 3$) with the formation of alkoxytin compounds [eq. (a)]. These are attacked by additional organotin hydride with the formation of alcohol and linkage of the Sn atoms [eq. (b)][4]:

$$\geq Sn\text{-}H + O=C< \rightarrow \geq Sn\text{-}O\text{-}\overset{|}{\underset{|}{C}}\text{-}H \qquad (a)$$

$$\geq Sn\text{-}H + \geq Sn\text{-}O\text{-}\overset{|}{\underset{|}{C}}\text{-}H \rightarrow \geq Sn\text{-}Sn\leq + HO\text{-}\overset{|}{\underset{|}{C}}\text{-}H \qquad (b)$$

A radical mechanism has been discussed [*53*] for these reactions – a mechanism which by no means explains only, for example, the differences from a carbonyl–$LiAlH_4$ reaction [*11, 55, 56, 61e, 73, 116, 117, 119*]. An increase in temperature, irradiation, azobis(isobutyronitrile), and zinc chloride, accelerate the addition reaction [eq. (a)] [*11, 73, 117*]. The reduction of a ketone or aldehyde with tin hydrides is particularly advantageous because of the ease of separation of the tin compound formed, as can be seen from the following overall equations [*55, 71, 83, 116*]:

$$2R_3SnH + >C=O \rightarrow R_3Sn\text{-}SnR_3 + >CH\text{-}OH$$

---

[4] Sn=O reacts similarly to $>C=O$ in eqns. (a) and (b) [*98b*]:

$$2Bu_3SnH + Bu_2Sn=O \xrightarrow{100°} Bu_3Sn\text{-}\underset{Bu}{\overset{Bu}{Sn}}\text{-}SnBu_3 + H_2O$$

$$R_2SnH_2 + \;\rangle C{=}O \rightarrow \frac{1}{n}(R_2Sn)_n + \;\rangle CH{-}OH$$

$$RSnH_3 + \;\rangle C{=}O \rightarrow \frac{1}{n}(RSn)_n + \;\rangle CH{-}OH + \frac{1}{2}H_2$$

The reducing power of tin hydrides with respect to the carbonyl compounds increases in the sequence $(n\text{-}C_4H_9)_3SnH < (C_6H_5)_3SnH < n\text{-}C_4H_9SnH_3 < (n\text{-}C_4H_9)_2SnH_2 < (C_6H_5)_2SnH_2$ [*53*].

Carboxylic acids with diphenylstannane, Ph_2SnH_2, do not, as in the reaction with inorganic protonic acids (*e.g.* $Ph_2SnH_2 + 2HCl \rightarrow Ph_2SnCl_2 + 2H_2$) give the corresponding diphenyldiacyloxytins:

$$Ph_2SnH_2 + RCOOH \not\twoheadrightarrow Ph_2Sn(OOCR)_2 + 2H_2 \quad (R = Ph) \qquad (a)$$

but tetraphenyldiacyloxyditins [*101*]:

$$2Ph_2SnH_2 + 2RCOOH \rightarrow Ph_2(RCOO)Sn{-}Sn(OOCR)Ph_2 + 3H_2 \quad (R = Ph) \qquad (b)$$

With di-n-butylstannane, $Bu^n{}_2SnH_2$, both reactions (a) and (b) take place, in relative proportions which depend on the carboxylic acid and the acid–hydride ratio; acids of different strengths show no systematic preference for (a) or (b) ($R = CH_3$, $ClCH_2$, Cl_2CH, Cl_3C, F_3C, C_6H_5, $o\text{-}ClC_6H_4$, $p\text{-}ClC_6H_4$).

The anhydrous *protolysis* of tributylstannane, Bu_3SnH, with acetic acid catalysed by halide ions $X^{\ominus}$ is of first order for Bu_3SnH, first order for CH_3COOH, and first order and catalytic for $X^{\ominus}$. A mechanism involving coordination of $X^{\ominus}$ on the tin, which activates the hydride hydrogen bound to the tin and thereby facilitates the attack of CH_3COOH [*19*], has been proposed for the protolysis. The protolysis of Bu_3SnH in acidic and basic methanol, see [*56a*].

Aryl isocyanates and *alkyl isocyanates* add organotin hydrides to the C=N double bond with the formation of a Sn–N or Sn–C linkage. *Aryl isothiocyanates* add to the C=S double bond with the formation of an Sn–S bond[5] [*61e, 62a, 62d, 65, 80a, 81, 82*]:

$R_3SnH + R'N{=}C{=}O \rightarrow [R'N{=}O{=}N \cdots R_3SnH]$ (R = C_2H_5)

→ ($R' = C_6H_5$) $R'N(SnR_3){-}C(H){=}O$, i.e. R'N—C=O with R_3Sn on N and H on C

→ ($R' = C_6H_{13}$) $R'N(H){-}C(SnR_3){=}O$, i.e. R'N—C=O with H on N and SnR_3 on C

$R_3SnH + R''N{=}C{=}S \rightarrow [R''N{=}C{=}S \cdots HSnR_3] \rightarrow R''N{=}C(H){-}S{-}SnR_3$

(R = C_2H_5; R'' = C_6H_5, α-naphthyl)

[5] However, according to Lorenz and Becker [*65*], since the C=S double bond is weaker than the C=O double bond, it forms N-methylaniline, probably by the reaction:

$$4Ph_3SnH + PhN{=}C{=}S \rightarrow (Ph_3Sn)_2S + Ph_3SnSnPh_3 + PhNH(CH_3)$$

References p. 752

Azomethines also add tin hydrides to the C=N double bond, with the formation of Sn–N bonds [*73*] as in the case of the addition to aryl isocyanate:

$$\left.\begin{array}{c} p\text{-}CH_3C_6H_4N{=}CHC_6H_5 \\ + \\ R_3SnH \end{array}\right\} \longrightarrow \left[\begin{array}{c} p\text{-}CH_3C_6H_4N{=}CHC_6H_5 \\ \downarrow \\ R_3SnH \end{array}\right] \longrightarrow \begin{array}{c} p\text{-}CH_3C_6H_4N{-}CHC_6H_5 \\ \quad\quad\quad | \quad\;\; | \\ \quad\quad\quad R_3Sn \;\; H \end{array}$$

Diazo compounds replace hydrogen in tin hydrides by functional groups (the substitutions take place more smoothly [*104*] than analogous reactions with tin halides [*104*, *48*, *82*]:

$$R_3SnH + N_2CHR' \rightarrow R_3SnCH_2R' + N_2 \begin{cases} R = C_3H_7, C_4H_9 \\ R' = H, COOC_2H_5, CN, COCH_3, \\ \qquad COC_6H_5 \end{cases}$$

The yields of methyl derivatives (with N_2CH_2) depend on the steric hindrance of the alkyl substituents R [*48*]. Table 9.8 contains examples of such reactions and relevant references.

TABLE 9.8

ADDITION OF TRIORGANOTIN HYDRIDES, R_3SnH, TO ALDEHYDES, KETONES, ISOCYANATES, AZIDES, AZOMETHINES, AND DIAZO COMPOUNDS

[Catalyst AIBN = 2,2′-azobis(isobutyronitrile)]

Tin hydride	*Unsaturated compound*	*Catalyst initiator*	*Reaction product (yields in mole-%)*	*Ref.*
	Aldehydes			
$(C_2H_5)_3SnH$	$(CH_3)_2CHCHO$	AIBN	80% $(C_2H_5)_3SnOCH_2CH(CH_3)_2$	73
$(C_2H_5)_3SnH$	C_6H_5CHO	AIBN	$(C_2H_5)_3SnOCH_2C_6H_5$	73
$(C_2H_5)_3SnH$	$CH_3OC_6H_4CHO$	AIBN	75% $(C_2H_5)_3SnOCH_2C_6H_4OCH_3$	73
$(C_2H_5)_3SnH$	furan-2-CHO	AIBN	$(C_2H_5)_3SnOCH_2$-furyl	73
	Ketones			
$(C_2H_5)_3SnH$	$C_2H_5COCH_3$	$ZnCl_2$	95% $(C_2H_5)_3SnOCH(CH_3)C_2H_5$	73
$(n\text{-}C_4H_9)_3SnH$	$C_2H_5COC_2H_5$		30% $(n\text{-}C_4H_9)_3SnOCH(C_2H_5)_2$, with the formation of alcohol	11
$(n\text{-}C_4H_9)_3SnH$	$CH_2{=}CH(CH_2)_2C(O)CH_3$	UV-Light	61% C=C-addition, 8% C=O-addition	94a, 94b
$(n\text{-}C_4H_9)_3SnH$	$C_6H_5COCH_3$		30% $(n\text{-}C_4H_9)_3SnOCH(C_6H_5)CH_3$, with the formation of alcohol	11, 73a
$(C_2H_5)_3SnH$	$p\text{-}CH_3OC_6H_4COCH_3$		70% $(C_2H_5)_3SnOCH(CH_3)\text{-}C_6H_4OCH_3$	73
$(C_2H_5)_3SnH$	cyclohexanone (H)=O	$ZnCl_2$	61% $(C_6H_5)_3SnO$–(H)	73
$(n\text{-}C_4H_9)_3SnH$	cyclohexanone (H)=O		73% $(n\text{-}C_4H_9)_3SnO$–(H)	11

TABLE 9.8 (cont'd)

Tin hydride	*Unsaturated compound*	*Catalyst initiator*	*Reaction product (yields in mole-%)*	*Ref.*
$(C_6H_5)_3SnH$	$CH_2{=}CHC(O)CH_3$		$(C_6H_5)_3SnCH_2CH_2C(O)CH_3$	62f
$(C_6H_5)_3SnH$	H =O		55% H —OH + $(C_6H_5)_3SnSn(C_6H_5)_3$	116
$(C_6H_5)_3SnH$	CH_3 H =O		56% H —OH CH_3 + $(C_6H_5)_3SnSn(C_6H_5)_3$	95a, 116
$(C_6H_5)_3SnH$	CH_3 H =O		43% H —OH CH_3 + $(C_6H_5)_3SnSn(C_6H_5)_3$	95a, 116
$(C_6H_5)_3SnH$	CH_3— H =O		52% (at 150°), 70% (at 100°) CH_3— H —OH + $(C_6H_5)_3SnSn(C_6H_5)_3$	95a, 116
	Isocyanates			
$(C_2H_5)_3SnH$	$C_6H_{13}N{=}C{=}O$		$[(C_2H_5)_3Sn](C_6H_{13})NCHO$	62a
$(C_2H_5)_3SnH$	$C_6H_5N{=}C{=}O$		$[(C_2H_5)_3Sn](C_6H_5)NCHO$	62d
	Azides			
$(n\text{-}C_4H_9)_3SnH$	$C_6H_5CON_3$	AlBN	$C_6H_5CONHSn(n\text{-}C_4H_9)_3 + N_2$	25a
	Azomethines			
$(C_2H_5)_3SnH$	$p\text{-}CH_3C_6H_4N{=}CHC_6H_5$		63% $(C_2H_5)_3SnN(C_6H_4CH_3)(CH_2C_6H_5)$	73
	Diazo compounds			
$(C_4H_9)_3SnH$	N_2CH_2	Cu	$(C_4H_9)_3SnCH_3 + N_2$	62
$(C_3H_7)_3SnH$	$N_2CHCOOC_2H_5$		$(C_3H_7)_3SnCH_2COOC_2H_5 + N_2$	62
$(C_4H_9)_3SnH$	$N_2CHCOOC_2H_5$		$(C_4H_9)_3SnCH_2COOC_2H_5 + N_2$	62
$(C_4H_9)_3SnH$	$N_2CHCOCH_3$		$(C_4H_9)_3SnCH_2COCH_3 + N_2$	62
$(C_3H_7)_3SnH$	$N_2CHCOC_6H_5$		$(C_3H_7)_3SnCH_2COC_6H_5 + N_2$	62
$(C_4H_9)_3SnH$	$N_2CHCOC_6H_5$		$(C_4H_9)_3SnCH_2COC_6H_5 + N_2$	62
$(C_4H_9)_3SnH$	N_2CHCN		$(C_4H_9)_3SnCH_2CN + N_2$	62

4.6 *Some further reactions of the Sn–H bond*

(1) Preparations of polytin monohydrides by reaction of trialkyl(N-phenylformamido)tin with dialkyl- or diaryltin dihydride [*16b, 16f*]:

$$R_2'SnH_2 + R_3SnH(Ph)CH{=}O \rightarrow R_3SnSnR_2'H + PhNHCH{=}O$$

References p. 752

but:

$$Ph_2SnH_2 + 2R_3SnN(Ph)CH{=}O \rightarrow R_3Sn\text{-}SnPh_2\text{-}SnR_3 + 2PhNHCH{=}O$$

or:

$$R_2SnH_2 + R_3Sn\text{-}SnR_2\text{-}N(Ph)CH{=}O \rightarrow R_3Sn\text{-}SnR_2\text{-}SnR_2H + PhNHCH{=}O$$

$$2R_3Sn\text{-}SnR_2\text{-}SnR_2H \rightarrow R_3Sn(SnR_2)_4SnR_3 + H_2$$

(2) Fission of the Sn–N and Sn–O bond [*16a, 16d, 16e, 16g, 16h*] (for the mechanism see [*16g, 16h*]):

$$Ph_3SnH + Bu^n{}_3GeNEt_2 \rightarrow Bu^n{}_3GeSnPh_3 + Et_2NH$$

$Ph_3SnH + Bu^n_3PbN$(imidazolyl) $\rightarrow$ (not formed) $Bu^n_3PbSnPh_3$ + HN(imidazole)

$Ph_3SnH + Bu^n_3PbN$(imidazolyl) $\rightarrow$ Bu^n_3PbH (decomposes) + HN(imidazole)

or:

$$Ph_3SnH + Bu^n{}_3GeNEt_2 \rightarrow Bu^n{}_3GeSnPh_3 + Et_2NH$$

4.7 Reaction of the Sn–H bond with alkali metals and alkali metal hydrides

With LiH or NaH, SnH_4 does not form the $SnH_6{}^{2\ominus}$ ion [*96a*].

When dimethylstannane, Me_2SnH_2, is passed into the blue solution of sodium in liquid ammonia at $-63.5°$, the solution is decolorized and metallated derivatives of monostannane and distannane are formed [46]:

$$Me_2SnH_2 \xrightarrow[-H_2]{+2\,Na(NH_3)} Me_2SnNa_2 \xrightarrow{+Me_2SnH_2} \begin{cases} 2\,Me_2HSnNa \\ NaMe_2SnSnMe_2Na + H_2 \end{cases}$$

When Me_2SnH_2 is passed at $-45°$ into an excess of $Na\text{-}NH_3$ solution, an ammonolysis product which gives a red solution is also formed [*46*]:

$$Me_2SnH_2 + Na + NH_3 \rightarrow Me_2(NH_2)SnNa + \tfrac{3}{2}H_2$$

A conception of the reaction mechanism, see [*97a*].

Stannane SnH_4 reacts in monoglyme or diglyme with liquid K–Na-alloy (80:20 per cent of weight) and yields quantitatively the very soluble stannyl potassium SnH_3K [7]:

$$SnH_4 + K \xrightarrow{-70°} SnH_3K + 0.5\,H_2$$

Like the other alkalimetalhydryles SiH_3Na, SiH_3K, SiH_3Rb, SiH_3Cs and GeH_3K, the SnH_3K couples in organic solvents with organic and inorganic halides giving the stannyl derivatives and potassium halides [7].

5. Spectra of tin hydrides and derivatives

The figures given in the following review of the literature denote references. Abbreviations: *IR*, infrared spectrum and microwave spectrum; *NMR*, nuclear magnetic resonance; *RA*, Raman spectrum; *FC*, force constants.

	IR	*NMR*	*RA*	*FC*
SnH_4*	18, 62a, 62i, 62j, 68, 96, 124	25, 34, 66a, 96	–	–
SnD_4*	18, 62a, 62i, 62j, 124		–	–
SnF_4	–		–	29
$SnCl_4$	–		21	29
$SnBr_4$	–		21	29
SnI_4	–		–	29
$(CH_3)SnH_3$	22a, 63	22b, 25, 47a, 66a	–	–
CH_3SnD_3	–	47a	–	–
CD_3SnH_3	–	47a	–	–
$(CH_3)_2SnH_2$	22a	22b, 25, 66a	22a	–
$(CH_3)_2SnD_2$	22a			
$(CH_3)_3SnH$	22a, 25, 52	22b, 25, 61, 66a	50	50
$(CH_3)_3SnD$	22a			
$(CH_3)_2(CF_2HCF_2)SnH$	–	14	–	–
$(CH_3)_4Sn$	64, 107, 109	25, 39	64, 109	109
$(CH_3)_3SnF$	50	–	50	–
$(CH_3)_3SnCl$	50	39	50	50
$(CH_3)_3SnBr$	50	39	50	–
$(CH_3)_3Sn(OOCCH_3)$	9, 35			
$(CH_3)_3Sn(ClO_4)$	15, 87			
$(CH_3)_3Sn(NO_3)$	15, 87			
$[(CH_3)_3Sn]_2CO_3$	87			
$(CH_3)_2SnCl_2$	–	39		
$(CH_3)SnCl_3$	–	39		
$(C_2H_5)SnH_3$	74	22b, 66a		
$(C_2H_5)SnH_2$	74	22b, 66a		
$(C_2H_5)_3SnH$	52, 74	22b, 61, 66a	52	
$(C_2H_5)_3SnD$	78			
$(n\text{-}C_3H_7)SnH_3$		22b, 66a		
$(n\text{-}C_3H_7)_2SnH_2$		22b, 66a		
$(n\text{-}C_3H_7)_3SnH$	61, 96	22b, 61, 66a, 96		
$(i\text{-}C_3H_7)SnH_3$		22b, 66a		
$(i\text{-}C_3H_7)_2SnH_2$	61	22b, 61, 66a		
$(i\text{-}C_3H_7)_3SnH$		22b, 66a		
$(n\text{-}C_4H_9)SnH_3$	74, 96, 98c	22b, 66a, 96, 98c		
$(n\text{-}C_4H_9)_2SnH_2$	74, 98a, 98c	22b, 66a, 98a, 98c		
$(n\text{-}C_4H_9)_3SnH$	52, 61, 98c	22b, 66a, 61, 98c	52	
$(i\text{-}C_4H_9)_2SnH_2$	–	22b		
$(i\text{-}C_4H_9)_3SnH$	–	22b		
$(i\text{-}C_4H_9)_3SnD$	78			
$(t\text{-}C_4H_9)_2SnH_2$	–	22b, 66b		
$(n\text{-}C_8H_{17})SnH_3$		22b		
$(n\text{-}C_8H_{17})_2SnH_2$	–	22b	–	
$(n\text{-}C_8H_{17})_3SnH$	61	22b, 61		

References p. 752

(Cont'd)

$(C_6H_5)SnH_3$	74	3, 22b, 66a	
$(C_6H_5)_2SnH_2$	74	3, 22b, 66a	
$(C_6H_5)_3SnH$	52	3, 22b, 61, 66a	52
$(p\text{-}CH_3C_6H_4)_3SnH$	61	22b, 61	
$(C_6H_5)_3SnD$	78	–	
$(p\text{-}CH_3C_6H_4)_4Sn$	30	–	
$(C_6H_5CH_2)_4Sn$		117a	
$R_nSn(CH_3)_{4-n}$** $[R = C_6H_5, p\text{-}CH_3C_6H_4,$ $2,4,6\text{-}(CH_3)_3C_6H_2,$ $3,4,5\text{-}(CH_3)_3C_6H_2]$	–	60	
$(n\text{-}C_4H_9)SnH_2Cl$	98c	98c	
$(n\text{-}C_4H_9)_2SnHF$	98a	98a	
$(n\text{-}C_4H_9)_2SnHCl$	98a, 98c	98a, 98c	
$(n\text{-}C_4H_9)_2SnHBr$	98a	98a	
$(n\text{-}C_4H_9)_2SnHI$	98a	98a	
$(C_6H_5)SnCl_3$	95		
$(C_6H_5)SnI_3$	95		
$(C_6H_5)_2SnCl_2$	95		
$(C_6H_5)_2SnI_2$	95		
$(C_6H_5)_3SnF$	49a	–	49a
$(C_6H_5)_3SnCl$	32, 49a, 95	–	49a
$(C_6H_5)_3SnBr$	95		
$(C_6H_5)_3SnI$	95		
$(C_6H_5)_2SnO$	95		
$(C_6H_5)_2SnS$	95		
$(SnH_3)_2$	37		
$[(n\text{-}C_4H_9)_2SnH]_2$	100		

*Mass spectrum, see [*61b, 98*]; Mössbauer, see [*33a*].
**LCAO–MO calculation of $(C_6H_5CH_2)_4Sn$, see [*68a*].

REFERENCES

1. AMBERGER, E., Angew. Chem. *72*, 78 (1960).
2. AMBERGER, E., Thesis, University of München, 1961.
3. AMBERGER, E., H. P. FRITZ, C. G. KREITER AND M.-R. KULA, Chem. Ber. *96*, 3270 (1963).
4. AMBERGER, E. AND M.-R. KULA, Chem. Ber. *96*, 2556 (1963).
5. AMBERGER, E. AND M.-R. KULA, Chem. Ber. *96*, 2560 (1963).
6. AMBERGER, E. AND M.-R. KULA, Angew. Chem. *75*, 476 (1963).
7. AMBERGER, E. AND R. RÖMER AND A. LAYER, J. Organometal. Chem. *12*, 417 (1968).
8. ANDERSON, H. H., J. Am. Chem. Soc. *79*, 4913 (1957).
9. BEATTIE, I. R. AND T. GILSON, J. Chem. Soc. *1961*, 2585.
9a. BORISOV, A. E. AND A. N. ABRAMOVA, Bull. Acad. Sci. USSR (engl. edit.) *1964*, 791.
9b. BIRNBAUM, E. R. AND P. H. JAVORA, J. Organometal, Chem. *9*, 379 (1967).
10. BURG, A. B. AND J. R. SPIELMAN, J. Am. Chem. Soc. *83*, 2667 (1961).
11. CALAS, R., J. VALADE AND J. POMMER, Compt. Rend. *255*, 1450 (1962); C.A. *58*, 5710 (1963).
12. CAMBENSI, H., Thesis, University of München, 1964.
13. CLARK, H. C., S. G. FURNIVAL AND J. T. KWON, Can. J. Chem. *41*, 2889 (1963).
14. CLARK, H. C., J. T. KWON, L. W. REEVES AND E. J. WELLS, Can. J. Chem. *41*, 3005 (1963).

15. Clark, H. C. and R. J. O'Brien, Inorg. Chem. *2*, 740 (1963).
16. Cooke, D. J., G. Nickless and F. H. Pollard, Chem. Ind. (London) *1963*, 1493.
16a. Creemers, H. M. J. C., A. J. Leusink, J. G. Noltes and G. J. M. van der Kerk, Tetrahedron Letters *27*, 3167 (1966).
16b. Creemers, H. M. J. C. and J. G. Noltes, Rec. Trav. Chim. *84*, 382 (1965).
16c. Creemers, H. M. J. C. and J. G. Noltes, Rec. Trav. Chim. *84*, 590 (1965).
16d. Creemers, H. M. J. C. and J. G. Noltes, Rec. Trav. Chim. *84*, 1589 (1965).
16e. Creemers, H. M. J. C. and J. G. Noltes, J. Organometal. Chem. *7*, 237 (1967).
16f. Creemers, H. M. J. C., J. G. Noltes and G. J. M. van der Kerk, Rec. Trav. Chim. *83*, 1284 (1964).
16g. Cremers, H. M. J. C., F. Verbeek and J. G. Noltes, J. Organometal. Chem. *8*, 469 (1967).
16h. Cremers, H. M. J. C., *"A General Method for Establishing Tin-Metal Bonds"*, Thesis, Utrecht, 1967.
17. Cristol, S. J., G. D. Brindell and J. A. Reeder, J. Am. Chem. Soc. *80*, 635 (1958).
18. Cyvin, S. J., J. Brunvoll, B. N. Cyvin, L. A. Kristiansen and E. Meisingseth, J. Chem. Phys. *40*, 96 (1964).
19. Dessy, R. E., Th. Hieber and F. Paulik, J. Am. Chem. Soc. *86*, 28 (1964).
20. Devyatykh, G. G., A. E. Ezheleva, A. D. Ziron and M. V. Zueva, Zh. Neorgan, Khim. *8*, 1307 (1963); C. A. *59*, 8182d (1963); Russ. J. Inorg. Chem. (Engl. Transl.) *8*, 678 (1963).
21. Delwaulle, M., M. F. François and M. Delhaye-Boisset, J. Phys. Radium *15*, 206 (1954).
22. Dillard, C. R., E. H. McNeill, D. E. Simmons and J. B. Yeldell, J. Am. Chem. Soc. *80*, 3607 (1958).
22a. Dillard, C. R. and L. May, J. Molec. Spectr. *14*, 250 (1964).
22b. Duffermont, J. and J. C. Maire, J. Organometal. Chem. *7*, 415 (1967).
23. Emeléus, H. J. and S. F. A. Kettle, J. Chem. Soc. *1958*, 2444.
24. Finholt, A. E., A. C. Bond, Jr., K. E. Wilzbach and H. I. Schlesinger, J. Am. Chem. Soc. *69*, 2692 (1947).
25. Flitcroft, N. and H. D. Kaesz, J. Am. Chem. Soc. *85*, 1377 (1963).
25a. Frankel, M., D. Wagner, D. Gertner and A. Zilkha, J. Organometal. Chem. *7*, 518 (1967).
26. Fritz, G. and H. Scheer, Z. Naturforsch. *19b*, 537 (1964).
26a. Fritz, G. and H. Scheer, Z. Anorg. Allg. Chem. *338*, 1 (1964).
27. Fuchs, R. and H. Gilman, J. Org. Chem. *22*, 1009 (1957).
28. Gilman, H. and J. Eisch, J. Org. Chem. *20*, 763 (1955).
29. Godnev, I. N., A. M. Alexandrovskaya and A. S. Sverdlin, Russ. J. Phys. Chem. (Engl. Transl.) *36*, 1420 (1962).
30. Griffiths, V. S. and G. A. Derwish, J. Mol. Spectry. *11*, 81 (1963).
30a. Gunn, S. R. and L. G. Green, J. Phys. Chem. *65*, 779 (1961).
30b. Hayashi, K., J. Iyoda and I. [illegible], J. Organometal. Chem. *10*, 81 (1967)
31. Henry, M. C. and J. G. Noltes, J. Am. Chem. Soc. *82*, 561 (1960).
32. Henry, M. C. and J. G. Noltes, J. Am. Chem. Soc. *82*, 555 (1960).
33. Henry, M. C. and J. G. Noltes, J. Am. Chem. Soc. *82*, 558 (1960).
33a. Herber, R. H. and G. I. Parisi, Inorg. Chem. *5*, 769 (1966).
34. Jameson, C. J. and H. S. Gutowsky, J. Chem. Phys. *40*, 1714 (1964).
35. Janssen, M. J., J. G. A. Luijten and G. J. M. van der Kerk, Rec. Trav. Chim. *82*, 90 (1963).
36. Jenkins, L. T. (Monsanto Chem. Co.), U.S. Pat. 3,088,940 (1963); C. A. *59*, 4061d (1963).
37. Jolly, W. L., Angew. Chem. *72*, 268 (1960).
38. Jolly, W. L., J. Am. Chem. Soc. *83*, 336 (1961).
39. Kelen, G. P. van der, Nature *193*, 1069 (1962).
40. Kelen, G. P. van der, J. G. A. Luijten and J. G. Noltes, Chem. Ind. (London), *1956*, 352.
40a. Kerk, G. J. M. van der, J. G. A. Luijten and J. G. Noltes, Chem. Ind. (London) *1956*, 352.
40b. Kerk, G. M. J. van der, J. G. A. Luijten and J. G. Noltes, Angew. Chem. *70*, 298 (1958).
41. Kerk, G. J. M. van der, J. G. Noltes and J. G. A. Luijten, J. Appl. Chem. *7*, 366 (1957).
42. Kerk, G. J. M. van der, and J. G. Noltes, J. Appl. Chem. *9*, 106 (1959).
42a. Kerk, G. J. M. van der, and J. G. Noltes, Annals New York Acad. Sci. *125*, 25 (1965).
43. Kerk, G. J. M. van der, J. G. Noltes and J. G. A. Luijten, J. Appl. Chem. *7*, 356 (1957).
44. Kerk, G. J. M. van der, J. G. Noltes and J. G. A. Luijten, Chem. Ind. (London), *1958*, 1290.
45. Kerk, G. J. M. van der, J. G. Noltes and J. G. A. Luijten, Rec. Trav. Chim. *81*, 853 (1962).
46. Kettle, S. F. A., J. Chem. Soc. *1959*, 2936.
47. Kettle, S. F. A., J. Chem. Soc. *1961*, 2569.

47a. KIMMEL, H. AND C. R. DILLARD, Spectrochim. Acta, *24A*, 919 (1968).
48. KRAMER, K. A. W. AND A. N. WRIGHT, J. Chem. Soc. *1963*, 3604.
48a. KRAUS, C. A. AND W. N. GREER, J. Am. Chem. Soc. *44*, 2629 (1922).
49. KRESPAN, C. G. AND V. A. ENGELHARDT, J. Org. Chem. *23*, 1565 (1958).
49a. KRIEGSMANN, H. AND H. GEISSLER, Z. Anorg. Allgem. Chem. *323*, 170 (1963).
50. KRIEGSMANN, H. AND S. PISCHTSCHAN, Z. Anorg. Allgem. Chem. *308*, 212 (1961).
51. KRIEGSMANN, H. AND K. ULBRICHT, Z. Chem. *3*, 67 (1963).
52. KRIEGSMANN, H. AND K. ULBRICHT, Z. Anorg. Allgem. Chem. *328*, 90 (1964).
53. KUIVILA, H. G., *Organometallic Chemistry*, Vol. 1, Academic Press, New York, 1964, p. 47.
54. KUIVILA, H. G. AND O. F. BEUMEL, JR., J. Am. Chem. Soc. *80*, 3250 (1958).
55. KUIVILA, H. G. AND O. F. BEUMEL, JR., J. Am. Chem. Soc. *80*, 3798 (1958).
56. KUIVILA, H. G. AND O. F. BEUMEL, JR., J. Am. Chem. Soc. *83*, 1246 (1961).
56a. KUIVILA, H. G. AND P. L. LEVINS, J. Am. Chem. Soc. *86*, 23 (1964).
57. KUIVILA, H. G., L. W. MENAPACE AND C. R. WARNER, J. Am. Chem. Soc. *84*, 3584 (1962).
58. KUIVILA, H. G. AND L. W. MENAPACE, J. Org. Chem. *28*, 2165 (1963).
58a. KUIVILA, H. G., W. RAHMAN AND R. H. FISCH, J. Am. Chem. Soc. *87*, 2835 (1965).
59. KULA, M.-R., Thesis, University of München, 1962.
60. KULA, M.-R., E. AMBERGER AND K.-K. MAYER, Chem. Ber. *63*, 634 (1965).
61. KULA, M.-R., E. AMBERGER AND H. RUPPRECHT, Chem. Ber. *98*, 629 (1965).
61a. LALIBERTE, B. R., W. DAVIDSON AND M. C. HENRY, J. Organomet. Chem. *5*, 526 (1966).
61b. LARIN, N. V., I. L. AGAFONOV AND S. M. VLASOV, Russ. J. Inorg. Chem. *13*, 1 (1968).
61c. LEUSINK. A. J., H. A. BUDDING AND J. W. MARSMAN, J. Organometal. Chem. *9*, 285 (1967).
61d. LEUSINK, A. J., H. A. BUDDING AND W. DRENTH, J. Organometal. Chem. *9*, 295 (1967).
61e. LEUSINK, A. J., *Hydrostannation*, Thesis, Utrecht, 1966.
61f. LEUSINK, A. J., H. A. BUDDING and W. DRENTH, J. Organometal. Chem. *11*, 541 (1968).
62. LESBRE, M. AND R. BUISSON, Bull. Soc. Chim. France *1957*, 1204.
62a. LEUSINK, A. J., H. A. BUDDING AND J. G. NOLTES, Rec. Trav. Chim. *85*, 151 (1966).
62b. LEUSINK, A. J. AND J. W. MARSMAN, Rec. Trav. Chim. *84*, 1123 (1965).
62c. LEUSINK, A. J., J. W. MARSMAN, H. A. BUDDING, J. G. NOLTES AND G. J. M. VAN DER KERK, Rec. Trav. Chim. *84*, 567 (1965).
62d. LEUSINK, A. J. AND J. G. NOLTES, Rec. Trav. Chim. *84*, 585 (1965).
62e. LEUSINK, A. J. AND J. G. NOLTES, Tetrahedron Letters *1966*, 335.
62f. LEUSINK, A. J. AND J. G. NOLTES, Tetrahedron Letters *1966*, 2221.
62g. LEUSINK, A. J., J. G. NOLTES, H. A. BUDDING AND G. J. M. VAN DER KERK, Rec. Trav. Chim. *83*, 609 (1964).
62h. LEUSINK, A. J., J. G. NOLTES, H. A. BUDDING AND G. J. M. VAN DER KERK, Rec. Trav. Chim. *83*, 1036 (1964).
62i. LEVIN, I. W., J. Chem. Phys. *46*, 1176 (1967).
62j. LEVIN, I. W., AND H. ZIFFER, J. Chem. Phys. *43*, 4023 (1965).
63. LIDE, D. R., J. Chem. Phys. *19*, 1605 (1951).
64. LIPPINCOTT, E. R. AND M. C. TOBIN, J. Am. Chem. Soc. *75*, 4141 (1953).
65. LORENZ, D. H. AND E. I. BECKER, J. Org. Chem. *28*, 1707 (1963).
66. LORENZ, D. H., P. SHAPIRO, A. STERN AND E. I. BECKER, J. Org. Chem. *28*, 2332 (1963).
66a. MADDOX, M. L., N. FLITCROFT AND H. D. KAESZ, J. Organomet. Chem. *4*, 50 (1965).
66b. MAIRE, J.-C. AND J. DUFERMONT, J. Organometal. Chem. *10*, 369 (1967).
67. MATHIS-NOEL, R., M. LESBRE AND I. SEREE DE ROCH, Compt. Rend. *243*, 257 (1956).
68. MAY, L. AND C. R. DILLARD, J. Chem. Phys. *34*, 694 (1961).
68a. J. NAGY, J. REFFY, A. KUSZMANN-BORBELY AND K. PALOSSY-BECKER, J. Organomet. Chem. *7*, 393 (1967).
69. NEUMANN, W. P., Angew. Chem. *73*, 542 (1961).
70. NEUMANN, W. P., Ann. *653*, 157 (1962).
71. NEUMANN, W. P., Angew. Chem. *75*, 225 (1963).
72. NEUMANN, W. P., Angew. Chem. *75*, 679 (1963).
73. NEUMANN, W. P. AND E. HEYMANN, Angew. Chem. *75*, 166 (1963).
73a. NEUMANN, W. P. AND E. HEYMANN, Ann. *683*, 11 (1965).
73b. NEUMANN, W. P. AND E. HEYMANN, Ann. *683*, 24 (1965).
74. NEUMANN, W. P. AND H. NIERMANN, Ann. *653*, 164 (1962).
75. NEUMANN, W. P., H. NIERMANN AND B. SCHNEIDER, Angew. Chem. *75*, 790 (1963).
76. NEUMANN, W. P., H. NIERMANN AND R. SOMMER, Angew. Chem. *73*, 768 (1961).
77. NEUMANN, W. P., H. NIERMANN AND R. SOMMER, Ann. *659*, 27 (1962).

78. NEUMANN, W. P. AND R. SOMMER, Angew. Chem. *75*, 788 (1963).
79. NEUMANN, W. P. AND R. SOMMER, Angew. Chem. *76*, 52 (1964).
79a. NEUMANN, W. P., R. SOMMER AND H. LIND, Ann. *688*, 14 (1965).
80. NOLTES, J. G., Thesis, University of Utrecht, 1958.
80a. NOLTES, J. G., Rec. Trav. Chim. *83*, 515 (1964).
81. NOLTES, J. G. AND M. J. JANSSEN, Rec. Trav. Chim. *82*, 1055 (1963).
82. NOLTES, J. G. AND M. J. JANSSEN, J. Organometal. Chem. *1*, 346 (1964).
78. NEUMANN, W. P. AND R. SOMMER, Angew. Chem. *75*, 788 (1963).
83. NOLTES, J. G. AND G. J. M. VAN DER KERK, Chem. Ind. (London) *1959*, 294.
83a. NOLTES, J. G. AND G. J. M. VAN DER KERK, Rec. Trav. Chim. *80*, 623 (1961).
84. NOLTES, J. G. AND G. J. M. VAN DER KERK, Rec. Trav. Chim. *80*, 623 (1961).
85. NOLTES, J. G. AND G. J. M. VAN DER KERK, Rec. Trav. Chim. *81*, 41 (1962).
86. NOLTES, J. G. AND G. J. M. VAN DER KERK, Chimia (Aarau) *16*, 122 (1962).
86a. OHARA, M. AND R. OKAWARA, J. Organometal. Chem. *3*, 484 (1965).
87. OKAWARA, R., B. J. HATHAWAY AND D. E. WEBSTER, Proc. Chem. Soc. *1963*, 13.
88. PANETH, F., Z. Elektrochem. *26*, 452 (1920).
89. PANETH, F., Z. Elektrochem. *29*, 97 (1923).
90. PANETH, F. AND K. FÜRTH, Ber. *52*, 2020 (1919).
91. PANETH, F., W. HAKEN AND E. RABINOWITSCH, Ber. *57*, 1891 (1924).
92. PANETH, F., M. MATTHIES AND E. SCHMIDT-HEBBEL, Ber. *55*, 775 (1922).
93. PANETH, F. AND E. RABINOWITSCH, Ber. *57*, 1877 (1924).
94. PANETH, F. AND E. RABINOWITSCH, Ber. *58*, 1138 (1925).
94a. PEREYRE, M. AND J. VALADE, Compt. Rend. *258*, 4785 (1964).
94b. PEREYRE, M. AND J. VALADE, Compt. Rend. *260*, 581 (1965).
95. POLLER, R. C., J. Inorg. Nucl. Chem. *24*, 593 (1962).
95a POMMIER, J.-C. AND J. VALADE, Bull. Soc. Chim. France *1965*, 975.
96. POTTER, P. E., L. PRATT AND G. WILKINSON, J. Chem. Soc. *1964*, 524.
96a. REIFENBERG, G. H. AND W. L. CONSIDINE, J. Am. Chem. Soc. *91*, 2402 (1969).
97. ROTHMAN L. A. AND E. I. BECKER, J. Org. Chem. *25*, 2203 (1960).
97a. RUSTAD, D. S. AND W. L. JOLLY, Inorg. Chem. *6*, 1986 (1967).
98. SAALFELD AND H. J. SVEC, J. Inorg. Nucl. Chem. *18*, 98 (1961).
98a. SAYWER, A. K., J. E. BROWN AND E. L. HANSON, J. Organometal. Chem. *3*, 464 (1965).
98b. SAYWER, A. K. J. Am. Chem. Soc. *87*, 537 (1965).
98c. SAYWER, A. K. AND J. E. BROWN, J. Organomet. Chem. *5*, 438 (1966).
99. SAYWER, A. K. AND H. G. KUIVILA, Chem. Ind. (London) *1961*, 260.
100. SAYWER, A. K. AND H. G. KUIVILA, J. Am. Chem. Soc. *85*, 1010 (1963).
101. SAYWER, A. K. AND H. G. KUIVILA, J. Org. Chem. *27*, 610 (1962).
102. SCHAEFFER, G. W. AND M. EMILIUS, O.S.F., J. Am. Chem. Soc. *76*, 1203 (1954).
103. SCHOTT, G. AND C. HARZDORF, Z. Anorg. Allgem. Chem. *307*, 105 (1960).
104. SEYFERTH, D. AND E. G. ROCHOW, J. Am. Chem. Soc. *77*, 1302 (1955).
105. SEYFERTH, D. AND M. TAKAMIZAWA, Inorg. Chem. *2*, 731 (1963).
106. SEYFERTH, D., H. YAMAZAKI AND D. L. ALLESTON, J. Org. Chem. *28*, 703 (1963).
107. SHELINE, R. K. AND K. S. PITZER, J. Chem. Phys. *18*, 595 (1950).
108. SHELL INTERNATIONAL RESEARCH MAATSCHAPPIJ N.V., Belg. Pat. 620, 290 (1963); C.A. *59*, 4061h (1963).
109. SIEBERT, H., Z. Anorg. Allgem. Chem. *268*, 177 (1952).
110. SMOLIN, E. M., Tetrahedron Letters *1961*, 143.
110a. STACK, W. F., G. A. NASH AND H. A. SKINNER, Trans. Faraday Soc. *61*, 2122 (1965).
111. STERN, A. AND E. I. BECKER, J. Org. Chem. *27*, 4052 (1962).
112. STUDIENGESELLSCHAFT KOHLE M.B.H., Brit. Pat. 951 150 (1964); C.A. *60*, 13271b (1964).
113. TAMARU, K., J. Phys. Chem. *60*, 610 (1958).
114. TAMBORSKI, C., F. E. FORD AND E. J. SOLOSKI, J. Org. Chem. *28*, 181 (1963).
115. TAMBORSKI, C., F. E. FORD AND E. J. SOLOSKI, J. Org. Chem. *28*, 237 (1963).
115a. TOMBE, F. J. A. DES AND G. J. M. VAN DER KERK, Chem. Comm. *1966*, 914.
116. VALADE, J., M. PEREYRE AND R. CALAS, Compt. Rend. *253*, 1216 (1961).
117. VALADE, J., J. C. POMMIER, Bull. Soc. Chim. France *1963*, 199.
117a. VERDONCK, L. AND G. P. VAN DER KELEN, J. Organomet. Chem. *5*, 532 (1966).
117b. VYAZANKIN, N. S. AND V. T. BYCHKOV, J. Gen. Chem. USSR (Engl. Transl.) *35*, 685 (1965).
118. VYAZANKIN, N. S., G. A. RAZUVAEV AND S. P. KORNEVA, J. Gen. Chem. USSR (Engl. Transl.) *33*, 1029 (1963).

118a. VOEGELEN, E., Z. Anorg. Chem. *30*, 323 (1902).
118b. VLASOV, S. M. AND G. G. DEVYATYKH, J. Neorgan. Chim. *11*, 2681 (1966); Russ. J. Inorg. Chem. (Engl. Transl.) *11*, 1439 (1966).
119. WEBER, S. AND E. I. BECKER, J. Org. Chem. *27*, 1259 (1962).
120. WIBERG, E., E. AMBERGER AND H. CAMBENSI, Z. Anorg. Allgem. Chem., *351*, 164 (1967).
121. WIBERG, E. AND R. BAUER, Z. Naturforsch. *6b*, 392 (1951).
122. WIBERG, E. AND H. BEHRINGER, Z. Anorg. Allgem. Chem. *329*, 290 (1964).
123. WILCOX, D. E. AND L. A. BROMLEY, Ind. Eng. Chem. *55*(*7*), 32 (1963).
124. WILKINSON, G. R. AND M. K. WILSON, J. Chem. Phys. *25*, 784 (1956).

Chapter 10

LEAD HYDRIDES

1. Historical and general review

Excellent careful experimental investigations by Paneth *et al.* about 1920 first indicated the existence of the two extraordinarily unstable hydrides PbH_4 [*19*, *21*] and BiH_3[*22*]. In about 1960, Amberger[*1*, *2*] was able to confirm the existence of the obviously somewhat more stable bismuthane, BiH_3, by its production in preparative amounts. On the other hand, similar attempts to obtain plumbane, PbH_4, at about the same time were unsuccessful[*1*, *4*, *14*, *16*, *24*]. In contrast, a solid lower lead hydride $PbH_{0.19}$ appears to be remarkably stable [*29*].

Alkylplumbanes, $R_{4-n}PbH_n$(n = 1,2), can be obtained relatively simply today by the hydrogenation of the corresponding halides $R_{4-n}PbX_n$ with soluble double hydrides (moderate to good yields) or (almost quantitatively) by the reaction of the corresponding methoxides with diborane, for example:

$$2R_2PbX_2 + LiAlH_4 \xrightarrow{\text{ether}} 2R_2PbH_2 + LiAlX_4$$

$$Me_3PbCl + KBH_4 \xrightarrow{\text{liquid } NH_3} Me_3PbH + KCl + BH_3\ (\rightarrow BH_3{\cdot}NH_3)$$

$$R'_3PbOMe \xrightarrow[+\frac{3}{2}(BH_3)_2,\ -BH(OMe)_2]{\text{diethyl ether or pentane}} R'_3PbBH_4 \xrightarrow[-B(OMe)_3,\ -3H_2]{+3MeOH} R'_3PbH$$

$$(R = CH_3, C_2H_5;\ R' = CH_3, C_2H_5, n\ C_3H_7, n\ C_4H_9;\ X = Cl, Br)$$

The thermal stability of the alkyl-lead hydrides, $R_{4-n}PbH_n$, is between that of the corresponding tetraalkylplumbanes, R_4Pb, and plumbane, PbH_4, itself. The decomposition points of the methyl compounds, for example, are +110° (Me_4Pb), −30° (Me_3PbH), −50° (Me_2PbH_2), and −196°? (PbH_4). Only a few reactions have been carried out with alkyl-lead hydrides. They react in the same manner as the hydrides of the other elements of the 4th main group. However, lead hydrides are distinguished by their polarisation:

$$\diagdown\!\!\!-\!\overset{\delta-}{Pb}\!-\!\overset{\delta+}{H}\ \ (\diagup)$$

This knowledge now allows us to understand better why, for example, attempts to hydrogenate $PbCl_4$ with $LiAlH_4$ are unsuccessful and, instead, give rise to the formation of hydrogen:

$$Cl_3Pb{-}H^{\delta+} + H^{\delta-}{-}H{-}AlH_3^{\ominus} \rightarrow Cl_3Pb^{\ominus}(\rightarrow Cl_2Pb + Cl^{\ominus}) + H_2 + AlH_3$$

References p. 764

2. Plumbane, PbH_4

In 1920, Paneth and Nörring[21] indicated the possible existence of a volatile lead hydride for the first time: in the acid treatment of a sheet of magnesium coated with radioactive thorium B ($^{212}_{82}Pb$) and thorium C ($^{212}_{83}Bi$), a radioactive gas (mainly hydrogen together with lead hydride and/or bismuth hydride) was formed which, on heating, gave an invisible and unweighable, but radioactive, deposit[19, 21, 23].

The decomposition of Mg–Pb alloys in acid forms mass-spectroscopic amounts of PbH_4[25]. The electrolysis of sulphuric acid with lead anodes gives a measurable amount of a material that can be condensed in liquid air. On being passed through a heated tube, the re-evaporable compound deposits a visible lead mirror. However, the amounts of volatile lead hydride that can be prepared in this way are insufficient for analysis or for physical measurements. Both methods show that a volatile lead hydride can be prepared, but—particularly in view of the extraordinarily accurate method of working in Paneth's group—they are obviously not suitable for the preparation of a larger amount of hydride. This is not surprising, since Paneth and Fürth[18], by the analogous decomposition of Mg–Sn alloys in acid were likewise able to prepare stannane, which—as we know today—is substantially more stable, only in traces (~ 0.001%).

The electric discharge between lead electrodes in an atmosphere of H_2 and CH_4 forms no volatile lead hydride, while analogous reactions give BiH_3 and SnH_4, at least in traces[20]. Atomic hydrogen removes a lead mirror prepared by the pyrolysis of Me_4Pb. The gas so formed redeposits a lead mirror on heating; another proof for the existence of a lead hydride[23, 27].

Hydrogenations of lead(II) compounds with double hydrides have been unsuccessful: plumbane, PbH_4, is not formed by reactions in aqueous media of lead(II) bromide, $PbBr_2$, or of sodium plumbate(II), $NaPb(OH)_3$, with $NaBH_4$ in aqueous HBr solution[16, 24] (a method which gives an 84% yield of SnH_4 in the case of tin[26]) or by reactions in various ethers at temperatures of −140 to −78° of $PbCl_4$ with $LiAlH_4$[1, 12, 14], with $LiAlH(OBu^n)_3$[1], or with $LiBH_4$[1, 14], or of $Pb(OOCCH_3)_4$ with $LiAlH_4$[14] and of $(C_5H_5N)_2PbCl_6$ with $LiBH_4$[14]. The hydrogenation of $Sn(OMe)_4$ with $(BH_3)_2$ that is successful in the case of tin[6, 7] fails in the case of lead, since $Pb(OMe)_4$ cannot be prepared from lead(IV) compounds either by alcoholysis or by reaction with sodium methoxide[14].

The dismutation of methylbismuthanes, $Me_{3-n}BiH_n$, used for the preparation of bismuthane[1, 2] cannot be transferred to methylplumbanes $Me_{4-n}PbH_n$[14]:

$$3Me_{3-n}BiH_n \xrightarrow{\sim -45^\circ} nBiH_3 + (3-n)\,Me_3Bi$$

$$4Me_4 - nPbH_n \overset{\sim -50^\circ}{\not\longrightarrow} nPbH_4 + (4-n)\,Me_4Pb$$

In the reduction of alkali-metal plumbates, $MPb(OH)_3$, with aluminium foil in aqueous solution, grey solid dilead dihydride, Pb_2H_2, which decomposes in vacuum into Pb and H_2, is said to be formed[28]. Another solid lead hydride is more certain[29]: pure coatings of lead prepared by evaporation at 10^{-6} torr

(4–40 mg, ~ 100 cm^3) absorb large amounts of atomic hydrogen very rapidly at 0°. After a critical H content has been passed, the obviously supersaturated metastable Pb–H solution gives off hydrogen (at 0°) to form a lead hydride (not a surface adsorbate) of the formula $PbH_{0.19}$. This only very slowly absorbs atomic hydrogen passed over it again. On heating, it gives off hydrogen in steps, which indicates discrete absorption centres of different energies. For example, at 140° it loses only 25% of its H content:

$$\text{Pb} \xrightarrow[\text{vaporisation}]{10^{-6}\text{ torr}} \text{Pb coating} \xrightarrow[0^\circ]{+\text{H}} \text{PbH}_{>0.19} \xrightarrow[0^\circ]{-(\text{H}_2, \text{H?})} \text{PbH}_{0.19} \xrightarrow{< 160^\circ} \text{lower Pb hydrides}$$

The activation energy is said to be 16–20 kcal/mole; probably, however, it is considerably higher [*29*].

3. Alkylplumbanes, R_3PbH and R_2PbH_2

The reaction of tetramethyllead, Me_4Pb, with aluminium boranate (but not with diborane) at low temperatures forms lead, methylboranes, $MeAl(BH_4)_2$, and hydrogen – probably via methyl-lead boranates [*e.g.*, $Me_2Pb(BH_4)_2$ or Me_2-$Pb(BH_3Me)_2$][*15, 15a*]. Preparative amounts of alkylplumbanes, $R_{4-n}PbH_n$, can be made by the reaction of suitable alkyl-lead halides, $R_{4-n}PbX_n$, with hydrogenating agents at low temperatures:

$$2R_2PbX_2 + LiAlH_4 \xrightarrow{\text{ether}} 2R_2PbH_2 + LiAlX_4$$

$$4R_3PbX + LiAlH_4 \xrightarrow{\text{ether}} 4R_3PbH + LiAlX_4$$

$$Me_3PbCl + KBH_4 + (x+1)\,NH_3 \xrightarrow[\text{KCl, }-BH_3\cdot NH_3]{\text{liquid }NH_3,\ -33^\circ} Me_3PbH \cdot x\,NH_3 \xrightarrow[-x\,NH_3]{-5^\circ} Me_3PbH$$

Table 10.1 contains examples of the reactions and the relevant references. Monoalkylplumbanes, $RPbH_3$, have not been prepared, probably because the corresponding halides can be prepared only with extraordinary difficulty (if at all) [*1*]. Dialkylaluminium hydrides (Et_2AlH, $Bu_2{}^nAlH$) are very useful in the hydrogenation of organotin halides (see section 3 of Chapter 9). A similar advantage for synthesis of lead hydrides fails [*17a*].

Organotin hydrides and organolead compounds ($Bu_3{}^nPbOCOCH_3$, $(Bu_3{}^nPb)_2O$ and $Bu_3{}^nPbCH{=}CH_2$) react giving the corresponding organolead hydrides [*8a, 17a*].

The trialkyl-lead methoxides, $R_3Pb(OMe)$, that can easily be prepared today when certain experimental precautions are taken [*5, 14*] react quantitatively with diborane in diethyl ether or n-pentane to give the corresponding boranates R_3Pb-(BH_4), which are stable up to about −35°. With methanol at −78° they form the corresponding hydrides without appreciable decomposition, these separating as

References p. 764

heavy oils insoluble in methanol (yields of 80–90%) [4, 14]:

$$2R_3Pb(OMe)+\tfrac{3}{2}(BH_3)_2 \rightarrow 2R_3Pb(BH_4)+BH(OMe)_2$$

$$R_3Pb(BH_4)+3MeOH \rightarrow R_3PbH+B(OMe)_3+3H_2$$

$$(R = CH_3, C_2H_5, n\text{-}C_3H_7, n\text{-}C_4H_9)$$

At −78°, ether-insoluble lead dimethoxide (preparation [5, 14]) forms with diborane in diethyl ether the likewise ether-insoluble (and therefore impure) white lead diboranate [4, 14]:

$$Pb(OMe)_2+\tfrac{3}{2}(BH_3)_2 \rightarrow Pb(BH_4)_2+BH(OMe)_2$$

TABLE 10.1

SYNTHESIS OF ALKYL-LEAD HYDRIDES, $R_{4-n}PbH_n$ BY HYDROGENATION OF THE CORRESPONDING HALIDES AND METHOXIDES $R_{4-n}PbX_n$(R = Cl, Br, OCH_3; n = 1, 2)

Alkyl-lead halide	*Hydrogenating agent*	*Solvent*	*Reaction temperature (°C)*	*Alkyl-lead hydride*	*Ref.*
$(CH_3)_2PbCl_2$	$LiAlH_4$	dimethyl ether,	−78	$(CH_3)_2PbH_2$	3, 8
		diethyl ether	−90/−110		
$(CH_3)_2PbBr_2$	$LiAlH_4$	diethyl ether	−90/−110	$(CH_3)_2PbH_2$	3, 14
$(n\text{-}C_4H_9)_2PbCl_2$	$LiAlH_4$	diglyme	−60	$(n\text{-}C_4H_9)_2PbH_2$	17
$(CH_3)_3PbCl$	$LiAlH_4$	dimethyl ether,	−78	$(CH_3)_3PbH$	3, 8
		diethyl ether	−90/−110		
$(CH_3)_3PbBr$	$LiAlH_4$	diethyl ether	−90/−110	$(CH_3)_3PbH$	3
$(C_2H_5)_2PbCl_2$	$LiAlH_4$	dimethyl ether	−78	$(C_2H_5)_2PbH_2$	8
$(C_2H_5)_3PbCl$	$LiAlH_4$	dimethyl ether	−78	$(C_2H_5)_3PbH$	8
$(CH_3)_3PbCl$	KBH_4	liquid NH_3	−33/−78	$(CH_3)_3PbH$	10, 11
$(n\text{-}C_3H_7)_3PbCl$	$LiAlH_4$	diglyme	−60	$(n\text{-}C_3H_7)_3PbH$	17
$(n\text{-}C_4H_9)_3PbCl$	$LiAlH_4$	diglyme	−60	$(n\text{-}C_4H_9)_3PbH$	17
$(i\text{-}C_4H_9)_3PbCl$	$LiAlH_4$	diglyme	−60	$(i\text{-}C_4H_9)_3PbH$	17
$(cyclo\text{-}C_6H_{11})_3PbCl$	$LiAlH_4$	diglyme	−60	$(cyclo\text{-}C_6H_{11})_3PbH$	17
$(CH_3)PbOCH_3$	B_2H_6/MeOH	diethyl ether	−78	$(CH_3)_3PbH$	4, 14
$(C_2H_5)_3PbOCH_3$	B_2H_6/MeOH	diethyl ether	−78	$(C_2H_5)_3PbH$	4, 14
$(n\text{-}C_3H_7)_3PbOCH_3$	B_2H_6/MeOH	diethyl ether	−78	$(n\text{-}C_3H_7)_3PbH$	4, 14
$(n\text{-}C_4H_9)_3PbOCH_3$	B_2H_6/MeOH	diethyl ether	−78	$(n\text{-}C_4H_9)_3PbH$	4, 14

4. Reactions of the Pb–H bond

Trimethylplumbane, Me_3PbH, and triethylplumbane, Et_3PbH, decompose according to kinetics that are first-order for $[R_3PbH]$. Since the decomposition of Me_3PbD at 20° gives only D_2 and MeH, apart from lead-containing materials, and no HD and MeD (by reaction 4, for example), the first decomposition step for Me_3PbH is probably the rate-determining cleavage of the Pb–H bond (1) [8, 9, 14]:

$$Me_3PbH \rightarrow Me_3Pb\cdot + H\cdot \quad (1)$$
$$H\cdot + H\cdot \rightarrow H_2 \quad (2)$$
$$Me_3PbH + H\cdot \rightarrow Me_3Pb\cdot + H_2 \quad (3)$$
$$\geq PbMe + H\cdot \rightarrow \geq Pb\cdot + MeH \quad (4)$$
$$2Me_3Pb\cdot \rightarrow Me_3PbPbMe_3 \quad (5)$$
$$2Me_3Pb\cdot \rightarrow Me_4Pb + Me_2Pb \quad (6)$$
$$3Me_2Pb\cdot \rightarrow Pb + Me_3PbPbMe_3 \quad (7)$$
$$2Me_3PbPbMe_3 \rightarrow Pb + 3Me_4Pb \quad (8)$$

Most of the hydrogen gives H_2 by reaction (2) (93%), and a small amount gives CH_4 by reaction (4) (no formation of C_2H_6) or by reaction (3).

Trimethylplumbanes and triethylplumbanes, R_3PbH, react with diazoalkanes with the replacement of hydrogen by alkyl [*8*]. On the assumption of a radical mechanism of reaction (9) or on the assumption of a polar mechanism with the more probable polarisation $^{\delta-}Pb–H^{\delta+}$ (10) or the less probable polarisation $^{\delta+}Pb–H^{\delta-}$ (11), only tetraalkylplumbanes of the form $R_3Pb(R'H)$ should be formed:

$$\left[\overset{\ominus}{N}=\overset{\oplus}{N}-\dot{C}H_2 \longleftrightarrow N=N-\dot{C}H_2\right],\quad \geq PbH \xrightarrow{-H\cdot} \geq Pb\cdot \;\xrightarrow{-N_2} \geq PbCH_2\cdot \xrightarrow{+H\cdot} \geq PbCH_3 \quad (9)$$

$$\overset{\ominus}{N}=N-\overset{\oplus}{C}H_2 + \geq\overset{\delta-}{Pb}-\overset{\delta+}{H} \rightarrow \overset{\ominus}{N}=NCH_2 \leftarrow \overset{\oplus}{Pb}H \xrightarrow{-N_2} [CH_2PbH] \rightarrow CH_3Pb\leq \quad (10)$$

carbenium azeniate

$$N\equiv\overset{\oplus}{N}-\overset{\ominus}{C}H_2 + \geq\overset{\delta+}{Pb}-\overset{\delta-}{H} \rightarrow N\equiv NCH_2 \rightarrow PbH \xrightarrow{-N_2} [CH_2PbH] \rightarrow CH_3Pb\leq \quad (11)$$

diazonium carbeniate

However, Me_3PbH and $N_2C_2H_4$ do not give Me_3EtPb exclusively but mainly Me_4Pb, and likewise Et_3PbH and N_2CH_2 give only very little Et_3MePb and mainly Et_4Pb (see Table 10.2) [*8*]. Consequently, in this case, unlike that of alkylations of stannanes, germanes, or silanes with diazoalkanes, thermolysis reactions (1–8) predominate.

Under pressure, trimethylplumbane adds to ethene in 92% yield [*8*]:

$$(CH_3)_3PbH + CH_2{=}CH_2 \xrightarrow[0°,\ 35\ atm]{diglyme} (CH_3)_3(C_2H_5)Pb$$

while it does not add at 1 atm and −78° to 0° [*9*].

On the other hand, C=C double bonds activated by neighbouring groups and the C=N double bond react without a catalyst and without pressure even at 0°

References p. 764

TABLE 10.2

REACTION PRODUCTS IN WT.% IN THE REACTION OF ALKYL-PLUMBANES, $R_{4-n}PbH_n$ (R = Me, Et; $n = 1, 2$), WITH DIAZOALKANES $N_2C_2H_4$ AND N_2CH_2[8]

Reaction product	Weight % formed			
	$N_2C_2H_4$		N_2CH_2	
	$(CH_3)_3PbH$	$(CH_3)_2PbH_2$	$(C_2H_5)_3PbH$	$(C_2H_5)_2PbH_2$
$(CH_3)_4Pb$	89	83		
$(CH_3)_3(C_2H_5)Pb$	11	12		
$(CH_3)_2(C_2H_5)_2Pb$		5		2
$(CH_3)(C_2H_5)_3Pb$			31	1
$(C_2H_5)_4Pb$			69	97

and below [*17*]:

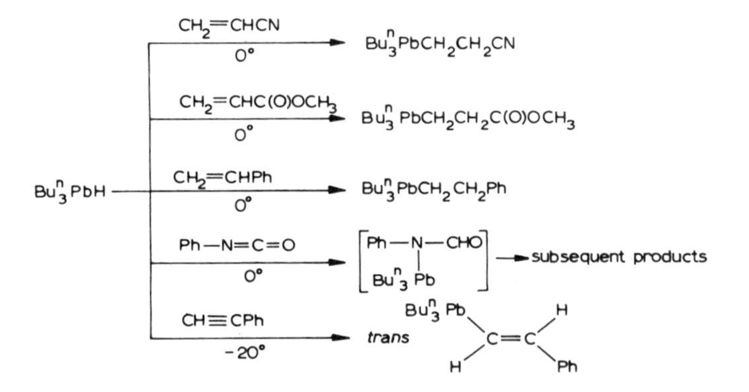

Like the hydrostannation reactions, the hydroplumbation reactions of the alkynes proceed via both (1) a polar mechanism and (2) a free radical mechanism [*16a*]. For an extensive study on hydroplumbation reactions, see [*17a*].

On the basis of the positive polarisation of the hydrogen on the lead, Me_3PbH reacts with ammonia at low temperatures with the formation of an ammonium salt [*9*]. The PMR spectra of the reaction mixture indicate the equilibria [*9*, *13*]

$$Me_3PbH + HPbMe_3 \rightleftharpoons Me_3PbH_2^{\oplus}\ PbMe_3^{\ominus}$$

and

$$NH_3 + HPbMe_3 \rightleftharpoons NH_4^{\oplus}\ PbMe_3^{\ominus}$$

On heating, $NH_4^{\oplus}PbMe_3^{\ominus}$ decomposes into Pb, $Me_3PbPbMe_3$, $PbMe_4$, NH_3, MeH, and H_2[*11*]:

$$NH_3 + HPbMe_3 \xrightarrow{-78^\circ} \underset{\text{green}}{NH_4^{\oplus}\ PbMe_3^{\ominus}} \xrightarrow{-MeH,\,-NH_3} PbMe_2$$

$$NH_4^{\oplus}\,PbMe_3^{\oplus} + PbMe_2 \rightarrow \underset{\text{red}}{NH_4[Pb_2Me_5]} \begin{cases} \rightarrow \tfrac{1}{2}Pb + \tfrac{1}{2}PbMe_4 + \tfrac{1}{2}Pb_2Me_6 + NH_3 + \tfrac{1}{2}H_2 \\ \rightarrow Pb + PbMe_4 + MeH + NH_3 \end{cases}$$

Because of the positive nature of the hydrogen in Me_3PbH, Me_3PbNa (unlike Me_3SnNa) does not form a lead–hydrogen compound Me_3PbH with acids but subsequent products [*8*]:

$$Me_3SnNa + NH_4Cl \xrightarrow{\text{liquid } NH_3} Me_3SnH + NaCl + NH_3$$

$$Me_3PbNa + NH_4X \xrightarrow[\Delta]{\text{liquid } NH_3} \text{alkanes, lead halides, hexamethyldilead}$$

With hydrogen chloride, trialkylplumbanes, R_3PbH, form the corresponding chlorides R_3PbCl with the evolution of H_2 [*9*, *10*]:

$$R_3PbH + HCl \rightarrow R_3PbCl + H_2$$

The fairly stable triethylplumbane, Et_3PbH, gives 100% of Et_3PbCl at $-112°$. The analogous reaction of the less stable trimethylplumbane, Me_3PbH, is disturbed by thermolysis reactions (96% of Me_3PbCl at $-112°$ in a slow reaction, and 78% of Me_3PbCl at $-78°$ in a fast reaction) [*9*].

The lead hydride can be determined quantitatively by the reaction of R_3PbH with EtI and measurement of the gaseous EtH produced [*17*]:

$$R_3PbH + EtI \rightarrow R_3PbI + EtH$$

5. Physical properties of the lead hydrides

PbH_4. Condensable thermally very unstable gas. Has been prepared only in traces [*19*, *21*, *25*].

$PbH_{0.19}$. Solid stable up to 160° [*29*].

$R_3Pb(BH_4)$ ($R = CH_3$, C_2H_5, n-C_3H_7, n-C_4H_9). White ether-insoluble solid stable up to $-35°$. Decomposes more readily in the presence of $(BH_3)_2$ [*4*, *14*].

$Pb(BH_4)_2$. Ether-insoluble solid contaminated with the starting material $Pb(OCH_3)_2$ [*4*, *14*].

$(CH_3)_2PbH_2$. Colourless liquid decomposing above $-50°$. Vapour pressure equation: $\log p$ [torr] $= -1332.6/T + 7.2502$ (-100 to $-50°$). Enthalpy of evaporation 6095.4 cal/mole. Pb–H stretching vibration at 1709 cm^{-1} [*3*]. PMR spectrum [*13*].

$(n\text{-}C_4H_9)_2PbH_2$. Unstable liquid. Spectrum [*17*].

$(CH_3)_3PbH$. Colourless liquid decomposing above $-30°$ or $-37°$ [*9*]. Melting point about $-106°$ [*8*]; $-104°$ [*9*]. Vapour pressure equation: $\log p$ [torr] $= -1623.3/T + 7.7300$ (-80 to $-30°$). Enthalpy of evaporation 7425.0 cal/mole. Pb–H

References p. 764

stretching vibration at 1709 cm^{-1} [*3*]. PMR spectrum [*9*, *13*]. Detonates on access of air [*1*, *8*].

$(C_2H_5)_2PbH_2$. Decomposes at room temperature [*8*].

$(C_2H_5)_3PbH$. Colourless liquid decomposing above −20°. Melting point −145°. Does not explode on access of air, but caution is called for [*8*].

$(n\text{-}C_3H_7)_3PbH$. Colourless oil stable below −35°. Decomposes at −20° within days or weeks [*4*, *14*]. Spectrum [*17*].

$(n\text{-}C_4H_9)_3PbH$. Colourless oil stable below −9°. Decomposes at 20° within 1 to 2 days [*4*, *14*]. Spectrum [*17*].

$(i\text{-}C_4H_9)_3PbH$. Spectrum [*17*].

$(cyclo\text{-}C_5H_{11})_3PbH$. Spectrum [*17*].

REFERENCES

1. Amberger, E., Thesis, University of München, 1961.
2. Amberger, E., Chem. Ber. *94*, 1447 (1961).
3. Amberger, E., Angew. Chem. *72*, 494 (1960).
4. Amberger, E. and R. Hönigschmid-Grossich, Chem. Ber. *99*, 1673 (1966).
5. Amberger, E. and R. Hönigschmid-Grossich, Chem. Ber. *98*, 3795 (1965).
6. Amberger, E. and M. -R. Kula, Chem. Ber. *96*, 2556 (1963).
7. Amberger, E. and M. -R. Kula, Chem. Ber. *96*, 2560 (1963).
8. Becker, W. E. and S. E. Cook, J. Am. Chem. Soc. *82*, 6264 (1960).

8a. Cremers, H. M. J. C., A. J. Leusink, J. G. Noltes and G. J. M. van der Kerk, Tetrahedron Letters *1966*, 3167.

9. Duffy, R., J. Feeney and A. K. Holliday, J. Chem. Soc. *1962*, 1144.
10. Duffy, R. and A. K. Holliday, Proc. Chem. Soc. *1959*, 124.
11. Duffy, R. and A. K. Holliday, J. Chem. Soc. *1961*, 1679.
12. Emeléus, H. J. and S. F. A. Kettle, J. Chem. Soc. *1958*, 2444.
13. Flitcroft, N. and H. D. Kaesz, J. Am. Chem. Soc. *85*, 1377 (1963).
14. Hönigschmid-Grossich, R., Thesis, University of München, 1964.
15. Holliday, A. K. and W. Jeffers, J. Inorg. Nucl. Chem. *6*, 134 (1958).

15a. Holliday, A. K. and G. N. Jessop, J. Organometal. Chem. *10*, 291 (1967).

16. Jolly, W. L., J. Am. Chem. Soc. *83*, 335 (1961).

16a. Leusink, A. J. and G. J. M. van der Kerk, Rec. Trav. Chim. *84*, 1617 (1965).

17. Neumann, W. P. and K. Kühlein, Angew. Chem. *77*, 808 (1965).

17a. Neumann, W. P. and K. Kühnlein, *Advances in Organometallic Chemistry* (Editors: F. G. A. Stone and R. West), Vol. 7, pp. 266ff. (1968).

18. Paneth, F. and K. Fürth, Ber. *52*, 2020 (1919).
19. Paneth, F. and A. Johannsen, Ber. *55*, 2622 (1922).
20. Paneth, F., M. Matthies and E. Schmidt, Ber. *55*, 775 (1922).
21. Paneth, F. and O. Nörring, Ber. *53*, 1693 (1920).
22. Paneth, F. and E. Winternitz, Ber. *51*, 1728 (1918).
23. Pearson, T. G., P. L. Robinson and E. M. Stoddart, Proc. Roy. Soc. (London), Ser. A, *152*, 275 (1933).
24. Piper, T. S. and M. K. Wilson, J. Inorg. Nucl. Chem. *4*, 22 (1957).
25. Saalfeld, F. E. and H. J. Svec, Inorg. Chem. *2*, 46 (1963).
26. Schaeffer, G. W. and M. Emilius, J. Am. Chem. Soc. *76*, 1203 (1954).
27. Schultze, G. and E. Müller, Z. Physik. Chem. *B 6*, 267 (1929).
28. Weeks, E. J., J. Chem. Soc. *1925*, 2845.
29. Wells, B. R. and M. W. Roberts, Proc. Chem. Soc. *1964*, 173.

APPENDIX

Literature in 1970

Alkali and Alkaline-Earth Metal Hydrides

LiH or NaH react with ZnR_2 (R = Me, Et, Ph) forming $MH(ZnR_2)$ or $MH(ZnR_2)_2$ respectively [101]. LiF and LiH form a solid solution [89]. Now again the most recent models of BeB_2H_8 must be questioned [124]. About the pyrolysis of $Na(R_2BeH)$ see [42]. There are some doubts about the possibility for the preparation of HMgX by hydrogenolysis or pyrolysis of Grignard compounds [15]. The hydrogenolysis of $K(MgBu_2H)$ in benzene yields $KMgH_3$ [14].

Boron Hydrides

In a fast flow reactor $BH_3 \cdot CO$ gives the coordination-free BH_3 in a high absolute yield [107]. The reinvestigation of the pyrolysis of B_2H_6 leads to the following first and second step of the reaction chain: The initial act of decomposition is neither symmetric nor asymmetric fission, but $2B_2H_6 \rightarrow BH_3 + B_3H_9$ followed so rapidly by $BH_3 + B_2H_6 \rightarrow B_3H_9$; that a measurable concentration of BH_3 never builds up [105]. At low pressures (5–120 mTorr) B_2H_6 thermally decomposes obeying the very simple set of reaction: $B_2H_6 \rightarrow 2BH_3$; $BH_3 \rightarrow B(s) + 1.5H_2$ [65]. μ-(HS)-B_2H_5 have been made by treatment of $Et_4N[HS(BH_3)_2]$ with HCl at $-78°$ [96]. For the preparation of μ-(Me_2N)-B_2H_4Me see [54]. About the rate of bridge-terminal position exchange in μ-(Me_2N)-B_2H_5 see [143].

Dimethylchloramine reacts with amine-boranes to give amine-chloroboranes as the first product (*e.g.*, $Me_3N \cdot BH_3 \rightarrow Me_3N \cdot BH_2Cl$) [118]. About the mechanism of hydrolysis of amine-monohaloboranes see [106]. The reaction of $NaBH_4$, I_2 and a donor gives in high yields amine-boranes and phosphane-boranes [122].

The gas phase photolysis of a mixture of $H_3B_3N_3H_3$ and MeBr yields $BrH_2B_3N_3H_3$ [130]. A mixture of $H_3B_3N_3H_3$ and MeCl, $CHCl_3$, CCl_4 or HSO_3Cl gives $ClH_2B_3N_3H_3$ [130]. The partially B-chlorinated borazine $ClH_2B_3N_3H_3$ can also be made by the reaction of $Cl_3B_3N_3H_3 \cdot$ pyridine with $LiBH_4$ in Et_2O [17].

$Zr(BH_4)_4$ undergoes exchange reactions with $LiBH_4$ ($LiBD_4$) in Et_2O which probably involve an intermediate of the type $Li[Zr(BH_4)_5]$. In the gas phase exchange occurs with a different mechanism [48]. About the preparation of $Sc(BH_4)_3 \cdot THF$ from $ScCl_3$ and $LiBH_4$ see [121], about the preparation of $NaBH_3NC$, $NaBD_3NC$ and $NaBH_3CNBH_3$ see [151].

Ph_3CBF_4 abstracts a hydride ion from BH containing compounds to form Ph_3CH and boron cations, *e.g.*, $Me_3N \cdot BH_3$ + Ph_3CBF_4 + MeCN → $(Me_3N)(MeCNBH_2^{\oplus}BF_4^{\ominus}$ + Ph_3CH [19]. For another synthesis of $(Me_3N)(MeCN)BH_2^{\oplus}$see [142]. About the formation of five-membered ring chelates of $BH_2^{\oplus}$ and macrocyclic ions with the general formula $[(diamine)BH_2]_n^{n\oplus}$ see [141]. About the equilibrium of $[(Me_2NH)(C_5H_5N)\text{-}BH_2]^{\oplus}PF_4^{\ominus}$ and Et_3N see [117].

The reaction of B_2H_6, $BH_{4-n}D_n^{\ominus}$ and specific donor solvents in an autoclave, yielding $B_3H_8^{\ominus}$, strongly suggests the following mechanism: B_2H_6 + 2donor → $B_2H_4 \cdot$ 2donor + H_2; $B_2H_4 \cdot$ 2donor + $BH_{4-n}D_n$ → $B_3H_{8-n}D_n^{\ominus}$ + 2donor [3]. The structure of $Me_4N[(CO)_4CrB_3H_8]$ resembles the structure of a B_4H_{10} derivative by substitution of a BH_2 group by $Cr(CO)_4$ [88]. About the structure of $(Ph_4P)_2Cu(B_3H_8)$ see [104].

The reaction of $B_3H_7 \cdot OMe_2$ and 1,2-$Me_2B_2H_4$ yields 2-MeB_4H_9, 1,2-, 2,2- and 2,4-$Me_2B_4H_8$ [50]. It is possible to extend the three-membered framework to a four-membered framework: NaB_3H_8 and Me_2BCl gives 2,2-$Me_2B_4H_8$ [74]. NaB_3H_8 or TlB_3H_8 react with R_2GaCl, R_2InCl and R_2TlBr giving $MB_3H_8R_2$. Spectroscopic investigation indicates tetraborane(10) derivatives ($GaB_3H_8R_2$. $InB_3H_8R_2$) and an octahydrotriborate(1−) salt ($R_2Tl[B_3H_8]$) [4]. Me_2Hg alkylates B_4H_{10} to give 2-MeB_4H_9 [116]. Me_2Hg and B_5H_{11} produces B_5H_9, B_4H_{10} and other products [116]. B_4H_8CO and F_2PH forms $F_2HP \cdot B_4H_8$ by ligand-displacement reaction [36]. About the formation of a B_4H_8 species see [92].

About the preparation of MB_5H_8 and MB_6H_9 (M = Li, Na, K) by metallation of B_5H_9 and B_6H_{10} with LiMe, NaH or KH see [94]. In LiB_5H_8 there is a covalent bond between Li and the boron framework rather than a salt-like structure [6]. About electron elimination and coupling reaction on 1-BrB_5H_8 see [5]. LiB_5H_8 reacts with Me_2BCl at low temperatures to produce μ-$(Me_2B)B_5H_8$. The latter isomerizes in the presence of Et_2O to 4,5-$Me_2B_6H_8$ [75].

Similar to the reaction of NH_3 with B_2H_6, B_4H_{10} or B_5H_{11}, which gives the diammoniates: $(NH_3)_2BH_2^{\oplus}BH_4^{\ominus}$, $(NH_3)_2BH_2^{\oplus}B_3H_8^{\ominus}$ and $(NH_3)_2BH_2^{\oplus}B_4H_9^{\ominus}$ at low temperatures, B_5H_9 and NH_3 yields $(NH_3)_2BH_2^{\oplus}B_4H_7^{\ominus}$ [98]. The gas phase reaction at 100° of B_5H_9 with NH_3 or $MeNH_2$ yields borazine, methylborazines and methylaminodiborane [28]. Me_2S and B_5H_9 yields $Me_2S \cdot BH_3$ and $(Me_2SB_2H_2)_n$ [119]. The silyl group of μ-(SiH_3)-B_5H_8 and 2-(SiH_3)-B_5H_8 can be partially halogenated by BCl_3 or BBr_3 [78]. Gaseous B_5H_9 and B_2H_6 at elevated temperatures in a flow-quench system yields B_8H_{16}, $B_{10}H_{18}$ and probably $B_{10}H_{16}$, a new B_{6-7} hydride and a second new B_8 hydride [53]. D_2O degrades B_8H_{12} giving B_6H_{10} [129]. About the B_6H_{10}–D_2O exchange reaction see [129] and about exchange reactions of B_6H_{12} with B_2D_6 forming B_6H_8-1,1,4,4-D_4 or B_6D_6 see [43].

Irradiation of a THF solution of $NaB_{10}H_{13}$ and hexacarbonyls gives $Na[(B_{10}H_{10}COH)\text{-}M(CO)_4]$. Treatment of the latter with NaH yields the quasiicosahedral $Na_2[B_{10}H_{10}\text{-}COMCO(CO)_3]$ [155].

Thermal decomposition at 230–240° of $Na_2B_{10}H_{12}$ gives $Na_2B_9H_9$ and $Na_2B_{10}H_{10}$ among other products. The decomposition of $Rb_2B_{10}H_{14} \cdot nH_2O$ yields a mixture of $Rb_2B_9H_9$ and $Rb_2B_{10}H_{10}$ [35].

The one-electron electrochemical oxidation of $B_{12}H_{12}^{2\ominus}$ yields $B_{24}H_{23}^{3\ominus}$. By halogenating $B_{24}I_2H_{21}^{3\ominus}$, $B_{24}Br_7H_{16}^{3\ominus}$, $B_{24}Br_{10}H_{13}^{3\ominus}$, $B_{24}Br_{11}H_{12}^{3\ominus}$ are formed and $B_{24}H_{22}^{4\ominus}$ derivatives. $Na(NH_3)$ cleaves $B_{24}H_{23}^{3\ominus}$ and regenerates $B_{12}H_{12}^{2\ominus}$ [157].

B_5H_9 and C_2H_2 reacts in a continuous-flow system (500–600°, 0.5 sec) to produce directly the smaller closo-carboranes 1,5-$C_2B_5H_5$, 1,6-$C_2B_4H_8$ and 2,4-$C_2B_5H_7$ in combined yields approaching 70% [52]. 1,2-Tetramethylenediborane(6) in a high vacuum pyrolysis system yields 2,3,4,5-$C_4B_2H_6$ [132]. About the new carborane system with a pentaborane(9) cage 1,2-$C_2B_3H_7$ see [66]. About the reaction of 2,5-$C_2B_6H_8$ and its C-methyl or C,C′-dimethyl derivatives with Me_4NBH_4 see [60]. For the synthesis of CB_5H_9 and its alkyl substituted derivatives see [60]. $MeC_3B_6H_6$ reacts with $Mn_2(CO)_{10}$ yielding $(MeC_3B_3H_5)Mn(CO)_3$ [93].

2,4-$C_2B_5H_7$ reacts with $LiBu^n$ yielding $LiCB_5H_5CLi$ which undergoes coupling reactions with MeI (forming $LiCB_5H_5CLi$) and $SiMe_3Cl$ (forming $LiCB_5H_5CSiMe_3$) [131]. These metallated carboranes react with HCl substituting Li by H [131]. 2,4-$C_2B_5H_7$ reacts rapidly with Cl_2 in the presence of $AlCl_3$ to form HCl and 5-Cl-2,4-$C_2B_5H_6$. Without $AlCl_3$, the reaction is accelerated by light forming also the isomers 3-Cl-2,4-$C_2B_5H_6$ and 1-Cl-2,4-$C_2B_5H_6$ [154].

A gallium containing 7-membered carborane $MeGaC_2B_4H_6$ occurs among other products by treatment of 2,3-$C_2B_4H_8$ with Me_3Ga at 215°. The proposed structure is a pentagonal bipyramid with an apiceal Ga atom [87].

Electrophilic substitution reaction of 1,6-$C_2B_7H_9$ with MeCl, C_2H_4 and Br_2 in presence of AlX_3 occurs preferentially at the boron atom in the position 8, *e.g.*, reaction with MeCl yields 8-Me-1,6-$C_2B_7H_8$ and HCl [59]. 1,3-$C_2B_7H_{13}$ reacts with NaH to give 1,3-$C_2B_7H_{11}^{2\ominus}$. The latter with $Co^{2\ominus}$ gives $Co(C_2B_7H_9)_2^{\ominus}$, H and Co [79].

$S^{2\ominus}$ degrades the decaborane(14) cage yielding the thiaborane $B_9H_{12}S^{\ominus}$ [90]. The latter with PCl_5 gives $Ph_2PH \cdot B_9H_{12}S$ which reacts with $Et_3N \cdot BH_3$ yielding $Et_3NH^{\oplus}B_{10}H_{11}S^{\ominus}$, H_2 and Ph_2PH. Treatment of $B_{10}H_{11}S^{\ominus}$ or $B_{10}H_{12}S$ with LiBu gives $B_{10}H_{10}S^{2\ominus}$ [90]. The latter ion with various transition metal halides gives icosahedral metalothiaboranes, *e.g.* $(B_{10}H_{10}S)_2Co^{\ominus}$ [90]; $CsB_9H_{12}S$ and *trans*-$(Et_3P)_2PtCl$ yields $(Et_3P)_2Pt(H)(B_9H_{10}S)$ [95].

The copyrolysis of 1,3-$C_2B_7H_{13}$ (or its C-substituted derivatives) and B_2H_6 yields 1,6-$C_2B_8H_{10}$, 1,6-$C_2B_8H_9Me$, 1,6-$C_2B_8H_8Me_2$ and 1,6-$C_2B_8H_9Ph$ [76].

Treatment of 1,8-$C_2B_9H_{11}$ with $NaBH_4$ yields Na[(3)-1,7-$C_2B_9H_{12}$]. The anion can be converted into a two cage carborane [133]. About the preparation of mixed Ni–Pd carboranes, *e.g.* $[Ni(1,2\text{-}C_2B_9H_{11})_2]^{2\ominus}[Pd(1,2\text{-}C_2B_9H_{11})_2]^{2\ominus}$ see [153], and about $[C_2B_9H_{11} \cdot Co \cdot C_2B_8H_{10} \cdot Co \cdot C_2B_8H_{10} \cdot Co \cdot C_2B_9H_{11}]^{3\ominus}$ see [39]. Treatment of (3)-1,2-$C_2B_9H_{11}^{2\ominus}$ with carbenoid germanium, tin and lead reagents (NaH + GeI_2, $SnCl_2$, $Pb(MeCOO)_2$) leads to the formation of icosahedral tricarbaborane analogs $MC_2B_9H_{11}$ [139].

$Na_3(B_{10}H_{10}CH) \cdot (THF)_2$ and $AsCl_3$ or SbI_3 react to give 1,2-$B_{10}H_{10}CHAs$ and

$1,2-B_{10}H_{10}CHSb$ respectively. Heating of these compounds yields the appropriate 1,7- and 1,12-compounds [149].

$B_{10}H_{14}$ and C_2H_2 react in a continuous-flow system (500–600°, 0.5 sec) to give predominately $1,7-C_2B_{10}H_{12}$ in yields of the order of 70% [52]. About the synthesis of $Ph_3PAu(C_2B_{10}H_{10}Ph)$ see [120] and about $M(C_2B_{10}H_{10})_2$ (M = Cu, Ni, Co) see [134]. About direct chlorination and direct fluorination of 1,2-, 1,7- and $1,12-C_2B_{10}H_{12}$ see [145].

Aluminium Hydrides

The reaction of $LiAlH_4$ with Et_2Mg in Et_2O proceeds according to the equation $LiAlH_4 + \frac{n}{2}Et_2Mg \rightarrow \frac{n}{2}MgH_2 + LiAlEt_nH_{4-n}$ [12]. About the direct synthesis of M_3AlH_6 (M = Li, Na, K, not Mg) see [13]. $NaAlH_4$ reacts with $MgCl_2$ or $MgBr_2$ yielding $Mg(AlH_4)_2$ [16]. Et_2AlBH_4 can be made: i) from the redistribution reaction between Al_2Et_6 and $Al(BH_4)_3$ and ii) from the reaction of $Al(BH_4)_3$ with C_2H_4 at 80° [49]. AlH_3 and Me_3SiOH react forming the O double bridged aluminium dihydride $H_2Al(OSiMe_3)_2{}^{\mu}AlH_2$ [137]. For $LiAl(AsH_2)_4$ see [9], for $LiAl(PH_2)_4$ see [126].

Silicon Hydrides

The SiH_4-SiD_4 pyrolysis in the presence of H_2 strongly suggests that silane decomposes by the following route: $SiH_4 \rightarrow SiH_3 + H$; $H + SiH_4 \rightarrow H_2 + SiH_3$; $SiH_3 + SiH_4 \rightarrow Si_2H_6 + H$; $2SiH_3 \rightarrow Si_2H_6$ [136]. The study of the kinetics of the disilane pyrolysis allows to establish the following mechanism: $Si_2H_6 \rightarrow SiH_4 + SiH_2$; $SiH_2 + Si_2H_6 \rightarrow Si_3H_8$ (initial reaction); $SiH_2 + Si_3H_8 \rightarrow n\text{-}Si_4H_{10}$ or $i\text{-}Si_4H_{10}$ (early secondary process) [27]. About the factors influencing the formation of $\cdot SiH_3$ or $\cdot SiH_2 \cdot$ in the pyrolysis of silicon compounds see [45]. Si_3Cl_3 and $LiAlH_4$ in Et_2O yields Si_3H_8 (60%) [77].

SiH_4 and small carbon molecules generated in a carbon arc produce methylsilane, disilane, acetylene, diacetylene and benzene [22]. About the pyrolysis of $MeSiH_3$ and $MeGeH_3$ see [100]. About new linear and cyclic carbosilanes see [70, 71]. By action of a silent electric discharge on equimolar mixtures of SiH_4 or GeH_4 and H_2S, H_2Se, MeSH the mixed hydrides are produced. Among others the new compounds SiH_3SGeH_3 and SiH_3SeGeH_3 are obtained [56]. Si_2H_6, Si_3H_8 and $n\text{-}Si_4H_{10}$ react with I_2 in $CH_3IC_5H_{12}$ solution giving the partially iodinated silanes [64]. Si_2H_6 and BX_3 (X = F, Cl, Br, I) reacts yielding partially halogenated disilanes $SiH_{3-n}X_n \cdot SiH_{3-m}X_m$ (n = 0–2; m = 0–3) [55]. About the chlorination of silanes with PCl_5 see [112].

The "scarce water" hydrolysis of $HSiCl_3$ yields higher homologs of $(HSiO_{3/2})_8$: the $(HSiO_{3/2})_{10-16}$ [72]. The tetrameric prosiloxan $(H_2SiO)_4$ has a puckered SiO ring [84].

SiH_3Br and amines (*e.g.* C_3H_6NH) reacts to give the corresponding aminosilanes (*e.g.* $C_3H_6NSiH_3$) [80]. About the reaction mechanism of the reaction of $HSiCl_3$ and NR_3 see [21]. The following compounds have been synthesized: $(Cl_3Si)_2N(SiH_3)$ [152], $H_nSi(PEt_2)_{4-n}$, H_nSiX_{4-n}, Me_nSiH_{4-n} [69] and $Me_2SiH(PH_2)$, $Me_2GeH(PH_2)$ [127].

About reactions of H_3SiPEt_2 and $HSi(PEt)_3$ with $LiPEt_2$ and LiMe see [67], about rearrangement of $HClSi(PEt_2)_2$ and $H_2ClSiPEt_2$ see [68]. $LiAl(AsH_2)_4$ reacts with halogeno-silanes and -germanes to produce the corresponding primary silyl- and germyl-arsanes, *e.g.* $MeSiH_2(AsH_2)$, $MeGeH_2(AsH_2)$ and $Me_3Ge(AsH_2)$ [8].

Germanium Hydrides

In D_2O–KOD mixtures GeH_4 undergoes H–D exchange [2]. $(GeH_3)_2Se$ and $(GeH_3)_2Te$ have been prepared by exchange with the analogous silyl compounds [44]. GeH_3PH_2 and GeH_3AsH_2 react with H_2Se forming GeH_3SeH. There exists the equilibrium $2GeH_3SeH \rightleftharpoons (GeH_3)_2Se + H_2Se$. $(GeH_3)_2Se$ reacts with other $(GeH_3)_2Se$ molecules forming $(GeH_3Se)_{4-n}GeH_n$ (n = 0–2) and GeH_4 [58]. Ge_2H_5I reacts with $NaMn(CO)_5$ yielding the remarkably stable $Ge_2H_5Mn(CO)_5$ [146]. About $(GeH_3)Re(CO)_5$ see [113].

Spectra

Abbreviations: *IR* infrared spectrum, *RA* raman spectrum, *NMR* nuclear magnetic resonance spectrum, *MS* mass spectrum, *ESR* electron spin resonance spectrum, *MO* molecule orbital calculation, *STR* structure, ★ for comparison. The figure given is the number of the reference.

Group I and Group II metal hydrides

BeB_2H_8 *STR* 124. $Mg(AlH_4)_2$ *IR,STR* 16.

Boron hydrides

BH_3, BH_2F, BHF_2 *STR* 10,144. MeB_2H_5, 1,1-$Me_2B_2H_4$, $Me_3B_2H_3$, $Me_4B_2H_2$ *RA* 33. μ-$(HS)B_2H_5$ *NMR* 96.

$BH_3 \cdot$ donor *IR* 29. $BH_3 \cdot NMe_3$ *NMR* 140. $BH_3 \cdot PH_3$ *MO* 51. $BH_3 \cdot PD_3$, $BD_3 \cdot PH_3$, $BD_3 \cdot PD_3$ *IR,NMR* 46. $BH_3 \cdot PH_2Me$, $BH_3 \cdot PH_2SiH_3$, $B(H,D)_3 \cdot PH_2Si(H,D)_3$ *NMR* 47. $BH_2X \cdot NMe_3$ *IR* 29.

$(BH_2NH_2)_2$ *STR* 11. BH_2NMe_2, BH_2NEt_2, $BH_2NC_5H_{10}$, BH_2NBu_2 *NMR* 125.

$Ph_2Zr(BH_4)_2$, Ph_3UBH_4, Ph_3ThBH_3 *NMR* 7. $NaBH_3NC$, $NaBD_3NC$, $NaBH_3CNBH_3$ *IR,NMR* 151.

$H_3B_3N_3H_3$ *IR,RA* 24. $H_2XB_3N_3H_3$ *IR,NMR,MS* 62,130.

$B_3H_8^{\ominus}$ *NMR* 109. $(Ph_3P)_2Cu(B_3H_8)$ *STR* 104. $(Me_4N)(CO)_4Cr(B_3H_8)$ *STR* 88.

B_4H_8 *MS* 92. B_4H_{10} *NMR* 103; *STR* 147,148. $Me_2B_4H_8$ *IR,NMR* 50. B_4H_8CO *MS* 92.

B_5H_9 *MS* 92; *STR* 147,148. 2-$(SiH_2Br)B_5H_8$, 2-$(SiH_2Cl)B_5H_8$ *IR,NMR* 78. 1-Cl-2-Me-B_5H_7, 2-Cl-1-Me-B_5H_7, 2-Cl-3-Me-B_5H_7, 2-Cl-4-Me-B_5H_7 *IR,NMR* 150. MB_5H_8 *IR,NMR* 6,94. B_5H_{11} *NMR* 103,135; *STR* 147,148; *MS* 92.

B_6H_{10} *NMR* 34,103. $B_6H_5D_5$ *NMR* 34. B_6H_{12} *NMR* 103. $B_6H_8D_8$, B_6H_{12} *NMR* 43.

B_8H_{16} *NMR,MS* 53. n-B_9H_{15} *NMR* 53,97. CsB_9H_{14} *NMR* 86. $B_{10}H_{14}$, $B_{10}H_{10}D_4$, 2-Br-$B_{10}H_{13}$ *NMR* 25. (EtO)$B_{10}H_{13}$ *NMR* 128. *i*-$B_{10}H_{16}$, $B_{10}H_{18}$ *NMR,MS* 53.

Carboranes *STR* 37. 1,2-$C_2B_3H_7$ *NMR* 66. CB_5H_9 *NMR* 60. $MeGaC_2B_4H_6$ *IR,NMR,MS* 87. 1-Cl-2,4-$C_2B_5H_6$, 3-Cl-2,4-$C_2B_5H_6$, 5-Cl-2,4-$C_2B_5H_6$ *NMR* 154. 1,6-$C_2B_8H_8Me_2$ *STR* 99. $(Et_3P)_2Pt(H)(B_9H_{10}S)$ *STR* 95. $Ni(C_2B_9H_{11})_2$ *STR* 40. (3,4′)-$Me_2B_9C_2H_9)_2Ni$ *STR* 38. $Cs[\pi$-(3)-1,2-$B_9C_2H_{11}]_2Fe_2(CO)_4$ *STR* 85. [MeN($C_2H_4)_3$NMe] [Ni(neo-$C_2B_9H_{11})_2$] *STR* 158. $Cs_2[(B_9C_2H_{11})Co(B_8C_2H_{10})Co(B_9C_2H_{11})] \cdot H_2O$ *STR* 41. $[(B_9C_2H_{11})Co(B_8C_2H_{10})Co(B_8C_2H_{10})Co(B_9C_2H_{11})]^{3\ominus}$ *STR* 39. [Ni(1,2-$C_2B_9H_{11})_2$] [Pd(1,2-$C_2B_9H_{11})_2$] *NMR* 153. (C_5H_5)Co(1,2-B_9H_9CHAs) *NMR* 149.

$B_9H_{12}S^{\ominus}$ *STR,NMR* 90. (FC_6H_4)-$C_2B_{10}H_{11}$ *NMR* 1. (o-$C_2B_{10}H_{10} \cdot CO)_2$ *STR* 138. $GeB_9C_2H_{11}$, $SnB_9C_2H_{11}$, $PbB_9C_2H_{11}$ *IR,NMR* 139.

Aluminium hydrides

$H_2Al(OSiMe_3)_2{}^{\mu}AlH_2$ *IR,RA,NMR* 137. Al_2Cl_6 *RA* 108.

Silicon hydrides

SiH_3 *ESR* 20. $PhSiH_3$ *NMR* 23,73. $PhSiD_3$ *NMR* 73. SiH_3CH_2X, $MeSiH_2X$ *electric dipole moment* 18. $(SiH_3)_2CH_2$, $(SiD_3)_2\dot{C}H_2$ *IR,RA* 115. ★ Si_2X_2 *IR,RA* 91. ★ $(Me_3Si)_4E$ (E = C, Si, Ge, Sn) *IR,RA* 31. $H_{4-n}SiX_n$, $H_{4-n}SiMe_n$ *NMR* 69. $HSiCl_3$, $DSiCl_3$ *IR,RA* 32. Me_3SiCN, Me_3SiNC *IR,RA,NMR* 26. $H_{3-n}X_nSiSiX_mH_{3-m}$ (n = 0–2; m = 0–3) *NMR* 55. SiH_3GeH_3, SiH_3GeD_3, SiD_3GeH_3 *IR* 102.

$HSi(OR)_3$ (R = Me,Et,Pr^n,Pr^i,Bu^n,Bu^i,Bu^s,Bu^t) *IR,RA* 123. $(SiH_3)_2{}^{16}O$, $(SiH_3)_2{}^{18}O$, $(SiD_3)_2{}^{16}O$ *IR* 114.

H_3SiNEt_2 *IR,NMR* 80. $(H_3Si)_2NN(SiH_3)_2$ *STR* 81.

SiH_3PH_2, SiD_3PD_2 *IR* 57. ★ Me_3SiPH_2, Me_3SiPD_2 *IR,NMR,MS* 30,127.

SiH_3AsH_2, SiD_3AsD_2 *IR* 57. $MeSiH_2AsH_2$ *IR,NMR* 8. $MSiH_3$ (M = K,Rb,Cs) *STR* 156. $Ph_2SiH_2Re(CO)_8$ *STR* 63.

Germanium hydrides

$RGeH_3$, R_2GeH_2, R_3GeH *IR,RA* 23,111. ★ Ph_4Ge, $(C_6D_6)_4Ge$ *IR,RA,STR* 61. $Na(GeH_2Ph)$ *NMR* 23. $(GeH_3)_2S$ *STR* 82. $(GeH_3)_2Se$, $(GeH_3)_2Te$ *IR,RA,NMR,MS* 44,58. $(GeH_3Se)_{4-n}GeH_n$ *NMR* 58. $(GeH_3)_3N$ *STR* 83. $Me_2GeH(PH_2)$ *IR,NMR,MS* 127. $MeGeH_2(AsH_2)$, ★ $Me_3Ge(AsH_2)$ *IR,NMR* 8. $(GeH_3)Re(CO)_5$ *IR* 113.

REFERENCES

1. ADLER, R.G. AND M.F. HAWTHORNE, J. Am. Chem. Soc. *92*, 6174 (1970).
2. ALLRED, A.L. AND R.L. DEMING, Inorg. Nucl. Chem. Letters, *6*, 39 (1970).
3. AMBERGER, E. AND E. GUT, Chem. Ber., in press.
4. AMBERGER, E., E. GUT AND D. OPP, Chem. Ber., in press.
5. AMBERGER, E., J. HÖNIGSCHMID AND B. REISINGER, Chem. Ber., in press.
6. AMBERGER, E. AND B. REISINGER, Chem. Ber., in press.
7. AMMON, R.V., B. KANELLAKOPULOS, G. SCHMID AND R.D. FISCHER, J. Organomet. Chem. *25*, C1 (1970).
8. ANDERSON, J.W. AND J.E. DRAKE, J. Chem. Soc. *A1970*, 3131.
9. ANDERSON, J.W. AND J.E. DRAKE, Inorg. Nucl. Chem. Letters, *5*, 887 (1969).
10. ARMSTRONG, D.R., Inorg. Chem. *9*, 874 (1970).
11. ARMSTRONG, D.R. AND P.G. PERKINS, J. Chem. Soc. *A1970*, 2748.
12. ASHBY, E.C. AND R.G. BEACH, Inorg. Chem. *9*, 2300 (1970).
13. ASHBY, E.C. AND B.D. JAMES, Inorg. Chem. *8*, 2468 (1969).
14. ASHBY, E.C., R. KOVAR AND R. ARNOTT, J. Am. Chem. Soc. *92*, 2182 (1970).
15. ASHBY, E.C., R.A. KOVAR AND K. KAWAKAMI, Inorg. Chem. *9*, 317 (1970).
16. ASHBY, E.C., R.D. SCHWARTZ AND B.D. JAMES, Inorg. Chem. *9*, 325 (1970).
17. BEACHLEY, Jr., O.T., Inorg. Chem. *8*, 2665 (1969).
18. BELLAMA, J.M. AND A.G. MACDIARMID, J. Organometal. Chem. *24*, 91 (1970).
19. BENJAMIN, L.E., D.A. CARVALHO, S.F. STAFIEJ AND E.A. TAKACS, Inorg. Chem. *9*, 1844 (1970).
20. BENNETT, S.W., C. EABORNE, A. HUDSON, R.A. JACKSON AND K.D.J. ROOT, J. Chem. Soc. *A1970*, 348.
21. BERNSTEIN, S.C., J. Am. Chem. Soc. *92*, 699 (1970).
22. BINENBOYM, J. AND R. SCHAEFFER, Inorg. Chem. *9*, 1578 (1970).
23. BIRCHALL, T. AND I. DRAMMOND, J. Chem. Soc. *A1970*, 1401.
24. BLICK, K.E., J.W. DAWSON AND K. NIEDENZU, Inorg. Chem. *9*, 1416 (1970).
25. BODNER, G.M. AND L.G. SNEDDON, Inorg. Chem. *9*, 1421 (1970).
26. BOOTH, M.R. AND S.G. FRANKISS, Spectrochim. Acta *26A*, 859 (1970).
27. BOWERY, M. AND J.H. PURNELL, J. Am. Chem. Soc. *92*, 2594 (1970).
28. BRAMLETT, C.L. AND A.T TABEREAUX, Jr., Inorg. Chem. *9*, 978 (1970).
29. BROWN, M.P., R.W. HESELTINE, P.A. SMITH AND P.J. WALKER, J. Chem. Soc. *A1970*, 410.
30. BÜRGER, H., U. GOETZE AND W. SAWODNY, Spectrochim. Acta *26A*, 671 (1970).
31. BÜRGER, H., U. GOETZE AND W. SAWODNY, Spectrochim. Acta *26A*, 685 (1970).
32. BÜRGER, H. AND A. RUOFF, Spectrochim. Acta *26A*, 1449 (1970).
33. CARPENTER, J.H., W.J. JONES, R.W. JOTHAM AND L.H. LONG, Chem. Commun. *1968*, 881.
34. CARTER, J.C. AND N.L.H. MOCK, J. Am. Chem. Soc. *91*, 5891 (1969).
35. CARTER, J.C. AND P.H. WILKS, Inorg. Chem. *9*, 1777 (1970).
36. CENTOFANTI, L.F., G. KODAMA AND R.W. PARRY, Inorg. Chem. *8*, 2072 (1969).
37. CHEUNG, C.-C.S., R.A. BEAUDET AND G.A. SEGAL, J. Am. Chem. Soc. *92*, 4158 (1970).
38. CHURCHILL, M.R. AND K. GOLD, J. Am. Chem. Soc. *92*, 1180 (1970).
39. CHURCHILL, M.R., A.H. REIS, Jr., J.N. FRANCIS AND M.F. HAWTHORNE, J. Am. Chem. Soc. *92*, 4993 (1970).
40. CLAIR, D.St., A. ZALKIN AND D.H. TEMPLETON, J. Am. Chem. Soc. *92*, 1173 (1970).
41. CLAIR, D.St., A. ZALKIN AND D.H. TEMPLETON, Inorg. Chem. *8*, 2080 (1969).
42. COATES, G.E. AND R.E. PENDLEBURY, J. Chem. Soc. *A1970*, 156.
43. COLLINS, A.L. AND R. SCHAEFFER, Inorg. Chem. *9*, 2153 (1970).
44. CRADDOCK, S., E.A.V. EBSWORTH AND D.W.H. RANKIN, J. Chem. Soc. *A1969*, 1628.
45. DAVIDSON, I.M.T., J. Organometal. Chem. *24*, 97 (1970).
46. DAVIS, J. AND J.E. DRAKE, J. Chem. Soc. *A1970*, 2959.
47. DAVIS, J., J.E. DRAKE AND N. GODDARD, J. Chem. Soc. *A1970*, 2962.

48. DAVIS, J., D. SAUNDERS AND M.G.A. WALLBRIDGE, J. Chem. Soc. *A1970*, 2915.
49. DAVIES, N., C.A. SMITH AND M.G.A. WALLBRIDGE, J. Chem. Soc. *A1970*, 342.
50. DEEVER, W.R. AND D.M. RITTER, Inorg. Chem. *8*, 2461 (1969).
51. DEMUYNCK, J. AND A. VEILLARD, Chem. Commun. *1970*, 873.
52. DITTER, J.F., E.B. KLUSMANN, J.D. OAKES AND R.E. WILLIAMS, Inorg. Chem. *9*, 889 (1970).
53. DOBSON, J., R. MARUCA AND R. SCHAEFFER, Inorg. Chem. *9*, 2161 (1970).
54. DOBSON, J. AND R. SCHAEFFER, Inorg. Chem. *9*, 2183 (1970).
55. DRAKE, J.E. AND N. GODDARD, J. Chem. Soc. *A1970*, 2587.
56. DRAKE, J.E. AND C. RIDDLE, J. Chem. Soc. *A1970*, 3134.
57. DRAKE, J.E. AND C. RIDDLE, Spectrochim. Acta *26A*, 1697 (1970).
58. DRAKE, J.E. AND C. RIDDLE, J. Chem. Soc. *A1969*, 1573.
59. DUNKS, G.B. AND M.F. HAWTHORNE, Inorg. Chem. *9*, 893 (1970).
60. DUNKS, G.B. AND M.F. HAWTHORNE, Inorg. Chem. *8*, 2667 (1969).
61. DURIG, J.R., C.W. SINK AND J.B. TURNER, Spectrochim. Acta *26A*, 557 (1970).
62. EACHLEY, Jr., O.T., J. Am. Chem. Soc. *92*, 5372 (1970).
63. ELDER, M., Inorg. Chem. *9*, 762 (1970).
64. FEHER, F., P. PLICHTA AND R. GUILLERY, Chem. Ber. *103*, 3028 (1970).
65. FEHLNER, T.P. AND S.A. FRIDMANN, Inorg. Chem. *9*, 2288 (1970).
66. FRANZ, D.A. AND R.N. GRIMES, J. Am. Chem. Soc. *92*, 1438 (1970).
67. FRITZ, G. AND G. BECKER, Z. anorg. allg. Chem. *372*, 180 (1970).
68. FRITZ, G. AND G. BECKER, Z. anorg. allg. Chem. *372*, 196 (1970).
69. FRITZ, G., G. BECKER AND D. KUMMER, Z. anorg. allg. Chem. *372*, 171 (1970).
70. FRITZ, G. AND P. SCHOBER, Z. anorg. allg. Chem. *372*, 21 (1970).
71. FRITZ, G. AND P. SCHOBER, Z. anorg. allg. Chem. *372*, 59 (1970).
72. FRYE, C.L. AND W.T. COLLINS, J. Am. Chem. Soc. *92*, 5586 (1970).
73. FUNG, B.M. AND I.Y. WEI, J. Am. Chem. Soc. *92*, 1497 (1970).
74. GAINES, D., J. Am. Chem. Soc. *91*, 6503 (1969).
75. GAINES, D. AND T.V. IORNS, J. Am. Chem. Soc. *92*, 4571 (1970).
76. GARRETT, P.M., G.S. DITTA AND M.F. HAWTHORNE, Inorg. Chem. *9*, 1947 (1970).
77. GASPAR, P.P., C.A. LEVY AND G.M. ADAIR, Inorg. Chem. *9*, 1272 (1970).
78. GEISLER, T.C. AND A.D. NORMAN, Inorg. Chem. *9*, 2167 (1970).
79. GEORGE, T.A. AND M.F. HAWTHORNE, J. Am. Chem. Soc. *91*, 5475 (1969).
80. GLIDEWELL, C. AND D.W.H. RANKIN, J. Chem. Soc. *A1970*, 279.
81. GLIDEWELL, C., D.W.H. RANKIN, A.G. ROBIETTE AND G.M. SHELDRICK, J. Chem. Soc. *A1970*, 318.
82. GLIDEWELL, C., D.W.H. RANKIN, A.G. ROBIETTE, G.M. SHELDRICK, B. BEAGLEY AND S. CRADDOCK, J. Chem. Soc. *A1970*, 315.
83. GLIDEWELL, C., D.W.H. RANKIN AND A.G. ROBIETTE, J. Chem. Soc. *A1970*, 2935.
84. GLIDEWELL, C., A.G. ROBIETTE AND G.M. SHELDRICK, Chem. Commun. *1970*, 931.
85. GREENE, P.T. AND R.F. BRYAN, Inorg. Chem. *9*, 1464 (1970).
86. GREENWOOD, N.N., H.J. GYSLING, J.A. MCGINNETY AND J.D. OWEN, Chem. Commun. *1970*, 505.
87. GRIMES, R.N. AND W.J. RADEMAKER, J. Am. Chem. Soc. *91*, 6498 (1970).
88. GUGGENBERGER, L.J., Inorg. Chem. *9*, 367 (1970).
89. HAHN, H. AND G. STRICK, Z. anorg. allg. Chem. *372*, 248 (1970).
90. HERTLER, W.R., F. KLANBERG AND E.L. MUETTERTIES, Inorg. Chem. *6*, 1696 (1967).
91. HÖFLER, F., W. SAWODNY AND E. HENGGE, Spectrochim. Acta *26A*, 819 (1970).
92. HOLLINS, R.E. AND F.E. STAFFORD, Inorg. Chem. *9*, 877 (1970).
93. HOWARD, J.W. AND R.N. GRIMES, J. Am. Chem. Soc. *91*, 6499 (1969).
94. JOHNSON, II, H.D., R.A. GEANANGEL AND S.G. SHORE, Inorg. Chem. *9*, 908 (1970).
95. KANE, A.R., L.J. GUGGENBERGER AND E.L. MUETTERTIES, J. Am. Chem. Soc. *92*, 2571 (1970).
96. KELLER, P.C., Inorg. Chem. *8*, 2457 (1969).
97. KELLER, P.C. AND R. SCAEFFER, Inorg. Chem. *9*, 390 (1970).
98. KODAMA, G., J. Am. Chem. Soc. *92*, 3482 (1970).
99. KOETZLE, T.F. AND W.N. LIPSCOMB, Inorg. Chem. *9*, 2279 (1970).
100. KOHANEK, J.J., P. ESTACIO AND M.A. RING, Inorg. Chem. *8*, 2516 (1969).
101. KUBAS, G.J. AND D.F. SHRIVER, J. Am. Chem. Soc. *92*, 1949 (1970).

102. LANNON, J.A., G.S. WEISS AND E.R. NIXON, Spectrochim. Acta *26A*, 221 (1970).
103. LEACH, J.B., T. ONAK, J. SPIELMAN, R.R. RIETZ, R. SCHAEFFER AND L.G. SNEDDON, Inorg. Chem. *9*, 2170 (1970).
104. LIPPARD, S.J. AND K.M. MELMED, Inorg. Chem. *8*, 2755 (1969).
105. LONG, L.H., J. Inorg. Nucl. Chem. *32*, 1097 (1970).
106. LOWE, J.R., S. SUPPAL, C. WEIDIG AND H.C. KELLY, Inorg. Chem. *9*, 1423 (1970).
107. MAPPES, G.W. AND T.P. FEHLNER, J. Am. Chem. Soc. *92*, 1562 (1970).
108. MARONI, V.A., D.M. GRUEN, R.L. MCBETH AND E.J. CAIRNS, Spectrochim. Acta *26A*, 418 (1970).
109. MARYNICK, D. AND T. ONAK, J. Chem. Soc. *A1970*, 1160.
110. MATHIS, R., M. BARTHELAT AND F. MATHIS, Spectrochim. Acta *26A*, 1993 (1970).
111. MATHIS, R., M. BARTHELAT AND F. MATHIS, Spectrochim. Acta *26A*, 2001 (1970).
112. MAWAZINY, S., J. Chem. Soc. *A1970*, 1641.
113. MCKAY, K.M. AND S.R. STOBART, Inorg. Nucl. Chem. Letters *6*, 687 (1970).
114. MCKEAN, D.C., Spectrochim. Acta, *26A*, 1833 (1970).
115. MCKEAN, D.C., G. DAVIDSON AND L.A. WOODWARD, Spectrochim. Acta *26A*, 1815 (1970).
116. MILLER, F.M. AND D.M. RITTER, Inorg. Chem. *9*, 1284 (1970).
117. MILLER, F.M. AND G.E. RYSCHKEWITSCH, J. Am. Chem. Soc. *92*, 1558 (1970).
118. MILLER, V.R., G.E. RYSCHKEWITSCH AND S. CHANDRA, Inorg. Chem. *9*, 1427 (1970).
119. MISHRA, I.B. AND A.B. BURG, Inorg. Chem. *9*, 2188 (1970).
120. MITCHELL, C.M. AND F.G.A. STONE, Chem. Commun. *1970*, 1263.
121. MORRIS, J.H. AND W.E. SMITH, Chem. Commun. *1970*, 245.
122. NAINAN, K.G. AND G.E. RYSCHKEWITSCH, Inorg. Chem. *8*, 2671 (1969).
123. NEWTON, W.E. AND E.G. ROCHOW, J. Chem. Soc. *A1970*, 2664.
124. NIBLER, J.W. AND T. DYKE, J. Am. Chem. Soc. *92*, 2920 (1970).
125. NÖTH, H. AND H. VAHRENKAMP, Chem. Ber. *100*, 3353 (1967).
126. NORMAN, A.D., Chem. Commun. *1968*, 812.
127. NORMAN, A.D., Inorg. Chem. *9*, 870 (1970).
128. NORMAN, A.D. AND S.L. ROSELL, Inorg. Chem. *8*, 2818 (1969).
129. ODON, J.D. AND R. SCHAEFFER, Inorg. Chem. *9*, 2157 (1970).
130. OERTL, M. AND R.F. PORTER, Inorg. Chem. *9*, 904 (1970).
131. OLSEN, R.R. AND R.N. GRIMES, J. Am. Chem. Soc. *92*, 5072 (1970).
132. ONAK, T.P. AND G.T.F. WONG, J. Am. Chem. Soc. *92*, 5226 (1970).
133. OWEN, D.A. AND M.F. HAWTHORNE, J. Am. Chem. Soc. *91*, 6002 (1969).
134. OWEN, D.A. AND M.F. HAWTHORNE, J. Am. Chem. Soc. *92*, 3194 (1970).
135. RIETZ, R.R., R. SCHAEFFER AND L.G. SNEDDON, J. Am. Chem. Soc. *92*, 3514 (1970).
136. RING, M.A., M.J. PUENTES AND H.E. O'NEAL, J. Am. Chem. Soc. *92*, 4845 (1970).
137. ROBERTS, C.B. AND D.D. TONER, Inorg. Chem. *9*, 2361 (1970).
138. RUDOLPH, R.W., J.L. PFLUG, C.M. BOCK AND M. HODGSON, Inorg. Chem. *9*, 2274 (1970).
139. RUDOLPH, R.W., R.L. VOORHEES AND R.E. COCHOY, J. Am. Chem. Soc. *92*, 3351 (1970).
140. RYSCHKEWITSCH, G.E. AND A.H. COWLEY, J. Am. Chem. Soc. *92*, 745 (1970).
141. RYSCHKEWITSCH, G.E. AND T.E. SULLIVAN, Inorg. Chem. *9*, 899 (1970).
142. RYSCHKEWITSCH, G.E. AND K. ZUTSHI, Inorg. Chem. *9*, 411 (1970).
143. SCHIRMER, R.E., J.H. NOGGLE AND D.F. GAINES, J. Am. Chem. Soc. *91*, 6240 (1969).
144. SCHWARTZ, M.E. AND L.C. ALLEN, J. Am. Chem. Soc. *92*, 1466 (1970).
145. SEMENUK, N.S., S. PAPETTI AND H. SCHROEDER, Inorg. Chem. *8*, 2441 (1969); *8*, 2449 (1969).
146. STOBART, S.R., Chem. Commun. *1970*, 999.
147. SWITKES, E., I.R. EPSTEIN, J.A. TOSSELL, R.M. STEVENS AND W.N. LIPSCOMB, J. Am. Chem. Soc. *92*, 3837 (1970).
148. SWITKES, E., W.N. LIPSCOMB AND M.D. NEWTON, J. Am. Chem. Soc. *92*, 3847 (1970).
149. TODD, L.J., A.R. BURKE, A.R. GARBER, H.T. SILVERSTEIN AND B.N. STORHOFF, Inorg. Chem. *9*, 2175 (1970).
150. TUCKER, P.M., T. ONAK AND J.B. LEACH, Inorg. Chem. *9*, 1430 (1970).
151. WADE, R.C., E.A. SULLIVAN, J.R. BERSCHIED, Jr. AND K.F. PURCELL, Inorg. Chem. *9*, 2146 (1970).
152. WANNAGAT, U., M. SCHULZE AND H. BÜRGER, Z. anorg. allg. Chem. *375*, 157 (1970).

153. Warren, L.F. and M.F. Hawthorne, J. Am. Chem. Soc. *92*, 1157 (1970).
154. Warren, R., D. Paquin, T. Onak, G. Dunks and J.R. Spielman, Inorg. Chem. *9*, 2285 (1970).
155. Wegener, P.A., L.J. Guggenberger and E.L. Muetterties, J. Am. Chem. Soc. *92*, 3473 (1970).
156. Weiss, E., G. Hencken and H. Kühr, Chem. Ber. *103*, 2868 (1970).
157. Wiersema, R.J. and R.L. Middaugh, Inorg. Chem. *8*, 2074 (1969).
158. Wing, R.M., J. Am. Chem. Soc. *92*, 1187 (1970).

SUBJECT INDEX

161066

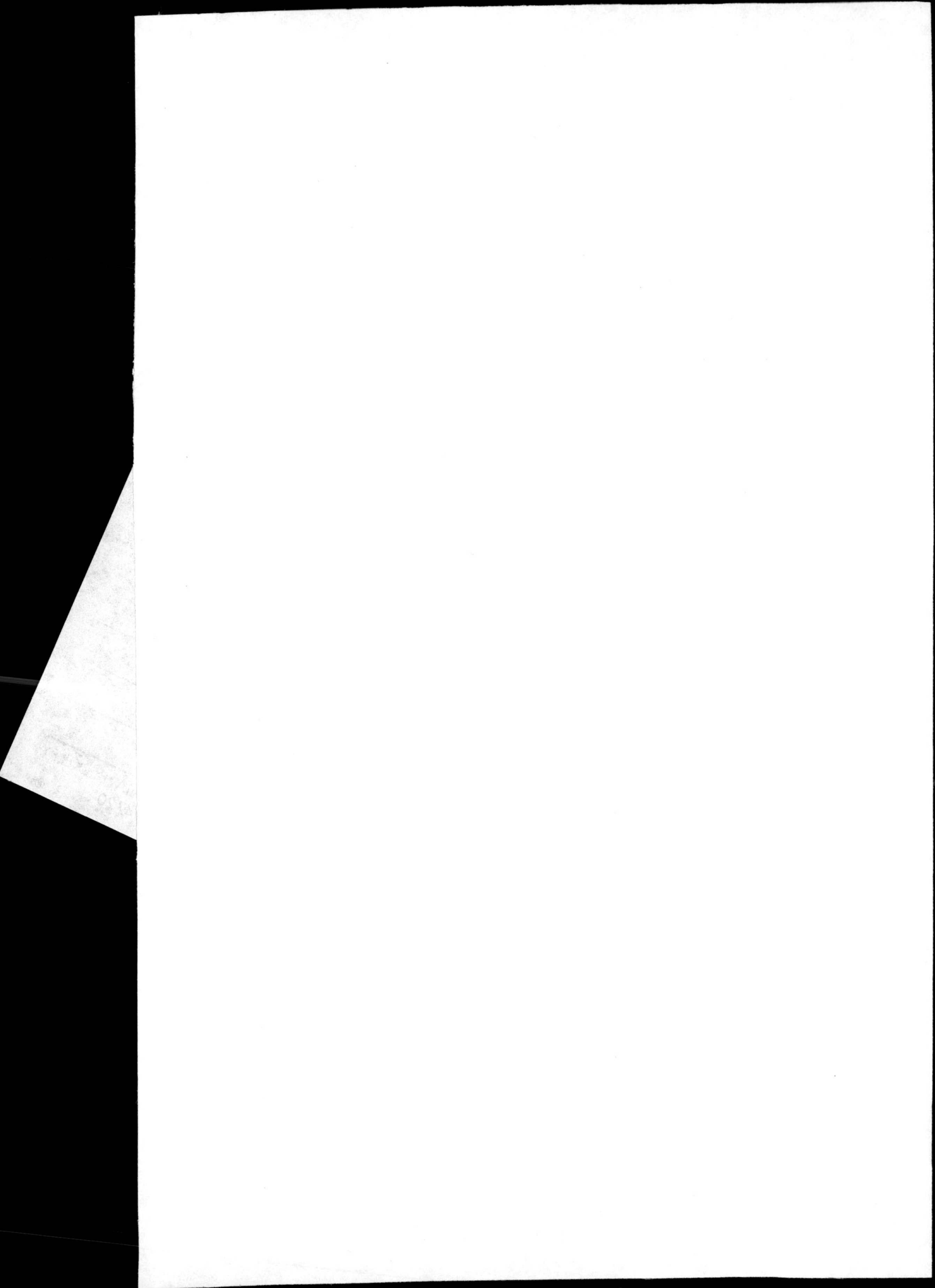

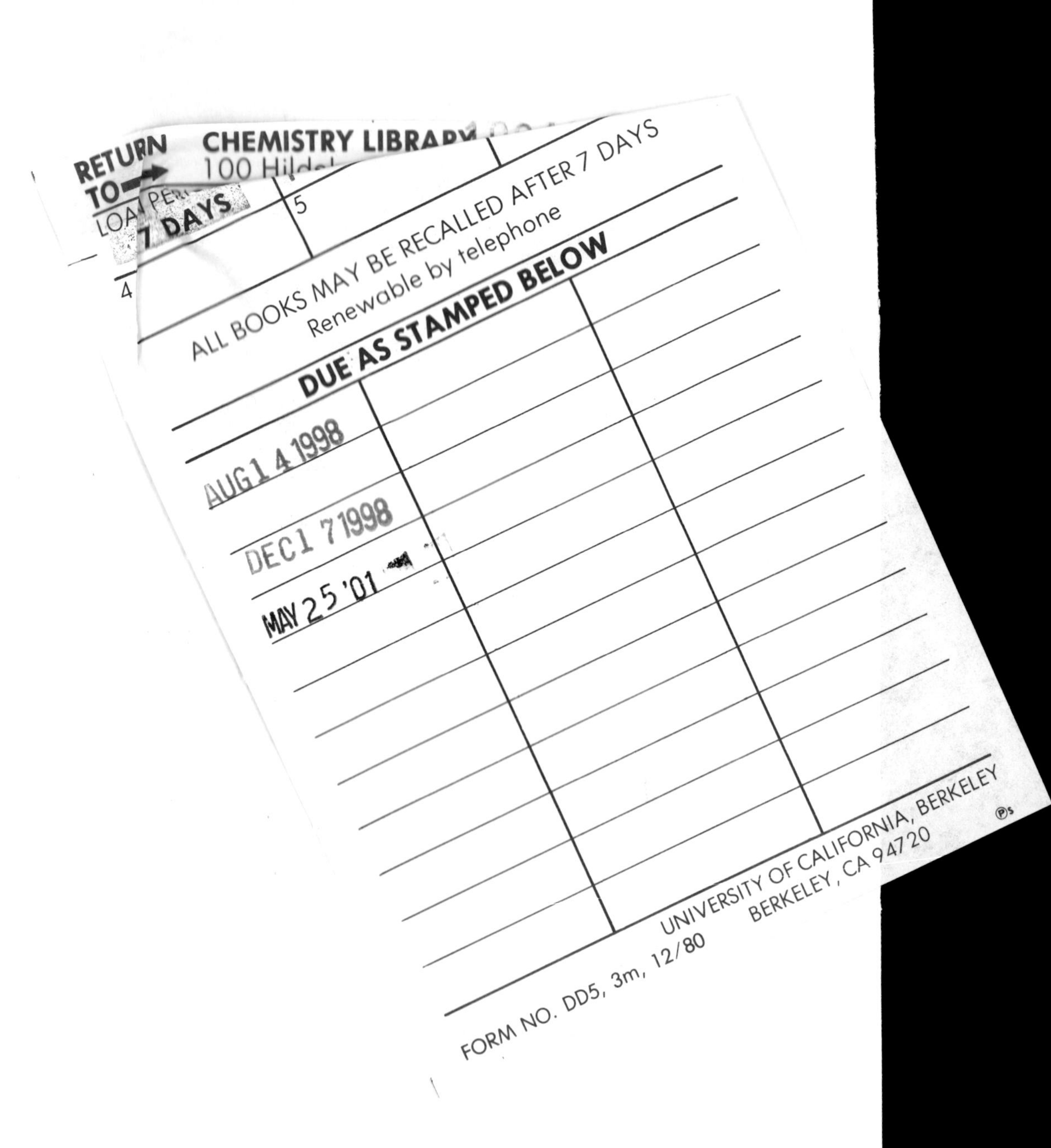

RETURN TO → CHEMISTRY LIBRARY
100 Hilde...
LOAN PERIOD
7 DAYS
4
5
ALL BOOKS MAY BE RECALLED AFTER 7 DAYS
Renewable by telephone
DUE AS STAMPED BELOW
AUG 1 4 1998
DEC 1 7 1998
MAY 2 5 '01
FORM NO. DD5, 3m, 12/80
UNIVERSITY OF CALIFORNIA, BERKELEY
BERKELEY, CA 94720